Progress in the Chemistry
of Organic Natural Products

Founded by L. Zechmeister

Editors:
A.D. Kinghorn, Columbus, OH
H. Falk, Linz
J. Kobayashi, Sapporo

Honorary Editor:
W. Herz, Tallahassee, FL

Editorial Board:
V.M. Dirsch, Vienna
S. Gibbons, London
N.H. Oberlies, Greensboro, NC
Y. Ye, Shanghai

98

Progress in the Chemistry
of Organic Natural Products

Bilirubin: *Jekyll* and *Hyde* Pigment of Life;
Pursuit of Its Structure Through Two World Wars
to the New Millenium

Author:
David A. Lightner

Prof. A. Douglas Kinghorn, College of Pharmacy,
Ohio State University, Columbus, OH, USA

em. Univ.-Prof. Dr. H. Falk, Institut für Organische Chemie,
Johannes-Kepler-Universität, Linz, Austria

Prof. Dr. J. Kobayashi, Graduate School of Pharmaceutical Sciences,
Hokkaido University, Sapporo, Japan

ISSN 2191-7043   ISSN 2192-4309 (electronic)
ISBN 978-3-7091-1636-4   ISBN 978-3-7091-1637-1 (eBook)
DOI 10.1007/978-3-7091-1637-1
Springer Wien Heidelberg New York Dordrecht London

Library of Congress Control Number: 2013956743

© Springer-Verlag Wien 2013

This work is subject to copyright. All rights are reserved by the Publisher, whether the whole or part of the material is concerned, specifically the rights of translation, reprinting, reuse of illustrations, recitation, broadcasting, reproduction on microfilms or in any other physical way, and transmission or information storage and retrieval, electronic adaptation, computer software, or by similar or dissimilar methodology now known or hereafter developed. Exempted from this legal reservation are brief excerpts in connection with reviews or scholarly analysis or material supplied specifically for the purpose of being entered and executed on a computer system, for exclusive use by the purchaser of the work. Duplication of this publication or parts thereof is permitted only under the provisions of the Copyright Law of the Publisher's location, in its current version, and permission for use must always be obtained from Springer. Permissions for use may be obtained through RightsLink at the Copyright Clearance Center. Violations are liable to prosecution under the respective Copyright Law.

The use of general descriptive names, registered names, trademarks, service marks, etc. in this publication does not imply, even in the absence of a specific statement, that such names are exempt from the relevant protective laws and regulations and therefore free for general use.

While the advice and information in this book are believed to be true and accurate at the date of publication, neither the authors nor the editors nor the publisher can accept any legal responsibility for any errors or omissions that may be made. The publisher makes no warranty, express or implied, with respect to the material contained herein.

Cover picture: Crystals (*left*) of bilirubin-IXα (from $CHCl_3$-CH_3OH) photographed through polarized light. (*Center*) Crystalline bilirubin in bulk. (*Right*) Dilute solution of bilirubin in dimethyl sulfoxide. Image obtained by scanning on a flatbed scanner (Graphics courtesy of Prof. *A.F. McDonagh*, UCSF)

Printed on acid-free paper

Springer is part of Springer Science+Business Media (www.springer.com)

*This volume is dedicated to those
who introduced me to bilirubin,
Profs. Albert Moscowitz and Cecil J. Watson
of the University of Minnesota, and to
those who helped me become familiar
with it, Prof. Med. Antony F. McDonagh
of the University of California,
San Francisco, and all of my students
and postdoctoral fellows.*

Acknowledgments

In the preparation of this work, special acknowledgment is due to the University of Nevada Library specialists – Mr. *Mark Lucas*, who provided invaluable assistance in document retrieval, and Ms. *Maggie Ressel*, who made electronic search possible; to my Vienna- and Berlin-educated son, Dr. *Derek Lightner*, for helpful corrections and comments on reading the manuscript; likewise to em. Univ. Prof. Dr. *Heinz Falk* (Johannes Kepler Universität Linz, Austria) and Prof. Med. *Antony F. McDonagh* (University of California, San Francisco) for invaluable suggestions and careful editing; to o. Prof. Dr. *Peter Laur* (Rheinisch-Westfälische Technische Hochschule Aachen) and Prof. *Falk* for assistance in retrieving some biographical information; and to *Stefano Abbate*, a budding Latin scholar. Immense gratitude is due to Ms. *Susan Grobman* for technical assistance par excellence, for patiently preparing and expertly editing "*das Buch.*"

Contents

1	**Introduction**	1
	1.1 Early Reviews, Current Objectives	1
	1.2 Background	2
2	**Early Scientific Investigations**	9
	2.1 Advances in the 18th Century	9
	2.2 Color Diagnostics	15
	2.3 Emergence of a New Analytical Methodology: Quantitative Combustion Analysis	17
	2.4 Early 19th Century Pigment Separation from Bile and Gallstones	19
	2.5 Bilirubin and Biliverdin Separation from Bile by the Middle 19th Century	28
	2.6 Bile Pigments from Gallstones and Urine, and Their Combustion Analyses during the late 18th to mid-19th Centuries	50
	2.7 Bile Pigments from Gallstones in the Middle of the 19th Century	57
	2.8 Hämatoidin, Bilifulvin, and the Origin of Bile Pigments	62
	2.9 Bile Pigment Isolation, Purification, and Combustion Analysis in the 1860s and 1880s	89
	2.9.1 *Georg Andreas Karl Städeler* Gives the Name Bilirubin as a First Step Is Taken Toward Structure Identification	91
	2.9.2 *Johann Ludwig Wilhelm (aka John Lewis William) Thudichum* and Bilirubin	113
	2.9.3 *Richard L. Maly* and Bilirubin	133
	2.10 The Emergence of Bile Pigment Spectroscopy: Colorimetry and Its Applications	149
	2.11 Bilirubin Polemics of the 1870s	157

2.12 Conjectural Chemistry and Bilirubin Polemics
at the Close of the 19th Century 161
2.13 Knowledge of Bilirubin Near the End of the 19th Century 175

3 Advent of the Bilirubin Structure Proof 181
3.1 Experimentally Derived Molecular Weight
of Bilirubin at the End of the 19th Century 183
3.2 Bilirubin Elemental Combustion Analyses
and Molecular Weight Determinations at the
Beginning of the 20th Century 187
3.3 Molecular Fragmentation: Initial Approach
to Bilirubin Structure 193
3.4 *Küster's* Maleimides and *Nencki's* Hämopyrrol 217

4 A Modern Proof of Bilirubin Structure Emerges 235
4.1 *Oskar Piloty* and the Structure of Bilirubin 238
4.2 *Hans Fischer* and the Early Structures of Bilirubin 263
4.2.1 *Fischer* and *Piloty* Cleave Bilirubin in Half, 1912-1913 264
4.2.2 The *Piloty-Fischer* Polemics 271
4.2.3 *Fischer's* Bilirubinsäure, Xanthobilirubinsäure,
and Bilirubin Structures ca. 1914 274
4.2.4 The Alkylated Monopyrrole Components
of Bilirubinsäure and Xanthobilirubinsäure 276
4.2.5 The Monopyrrole Propionic Acid Components
of Bilirubinsäure and Xanthobilirubinsäure 283
4.3 Status and Conjectures Regarding Bilirubin
and Hemin Structures circa 1916 291
4.4 Raison d'être Behind *Fischer's* Failed Tetrapyrrole
Structures Prior to 1926 298

5 Preparing the Way to the Constitutional Structure of Bilirubin 307
5.1 Evolution of the Bilirubin Structure in the 1920s 309
5.2 Tetrapyrrylethylene-Based Hemin and Bilirubin Kaput 317
5.3 *Küster-Fischer* Polemics 329
5.4 *Schumm-Fischer* Polemics 338
5.5 Vindication of the Macrocyclic Porphyrin Structure 343
5.6 The Structure of Bilirubin Reformulated 354

6 The Status of Bilirubin in 1930 357
6.1 The *Fischer* Laboratory in the 1930s
and the Structure of Bilirubin 363
6.2 Serendipitous Discovery of New Dipyrrole
Fragments of Mesobilirubin: *Neobilirubinsäure*
and *Neoxanthobilirubinsäure* 365
6.2.1 Synthetic Dipyrroles, Tetrapyrroles, Inhomogeneity,
and Nefarious Mixture Melting Points 371

Contents

	6.2.2	A Major Breakthrough in Understanding the Structure of Bilirubin. Resolution of Discordant Melting Points and Synthesis of Mesobilirubin	379
	6.2.3	The Structure of Natural-Analytical Mesobilirubin Confirmed. Synthesis of Mesobilirubin-IXα	392
6.3	*Fischer* Solves the Structure of Bilirubin	398
	6.3.1	Der Nitritkörper	400
	6.3.2	The Exo-vinyl Group of Bilirubin	408
	6.3.3	Bilirubin Structure Proof and Total Synthesis *QED*	412
6.4	*Hans Fischer, Finis Vitae*	427

7 Evolution of the Constitutional and Stereochemical Structure of Bilirubin 433
- 7.1 Lactim or Lactam Tautomeric Structure? 435
- 7.2 Configuration at the Exocyclic Double Bonds 448
- 7.3 Proposing and Detecting Hydrogen Bonds 455

8 Analysis of Bilirubin in Three Dimensions 465
- 8.1 Chirality Considerations 467
- 8.2 Bilirubin as a Chiral Molecule 468
- 8.3 Bilirubin Conformational Dynamics by NMR Spectroscopy 475
- 8.4 Conformational Dynamics by Molecular Mechanics Computation 477
- 8.5 Bilirubin Chirality Revisited. Chiral Discrimination and Optical Activity 488
  - 8.5.1 Bilirubin Circular Dichroism from a First-Order Asymmetric Transformation 490
  - 8.5.2 Bilirubin as a Molecular Exciton. Absolute Configuration from Exciton Chirality 497
  - 8.5.3 Mapping Bilirubin Conformation to Exciton Chirality CD 506
  - 8.5.4 Stereocontrol of Bilirubin Conformational Enantiomerism by Intramolecular Buttressing 514

9 Understanding and Translating Bilirubin Structure 525
- 9.1 Bilirubin Chemistry Reawakened Thanks to Jaundice Phototherapy 527
- 9.2 How Phototherapy Exposed Bilirubin (*E*)-Configuration Isomers 545
  - 9.2.1 (4*Z*,15*E*)- and (4*E*,15*Z*)-Bilirubin Diastereomers (Photobilirubin) 559
  - 9.2.2 Lumirubin and Isobilirubin Diastereomers 568
  - 9.2.3. (*E*)-Configuration Bilirubins from X-Ray Crystallography 575

	9.3	The Glue that Binds. Persistence, Recognition, and Role of Hydrogen Bonds	578
	9.3.1	Understanding Dipyrrinone-Carboxylic Acid Hydrogen Bonding	580
	9.3.2	Misconceptions About Hydrogen Bonding	587
	9.3.3	Intramolecular Hydrogen Bonds and pK_a. Un Cordon Sanitaire?	594
	9.4	Extrapolating from Bile Pigment Stereochemistry and Hydrogen Bonding to Synthetic Bilirubinoids	603

10 Bilirubin. *Quod Erat Faciendum* 617

11	**Translations**	623
	11.1 Translations to Section 2	623
	11.2 Translations to Section 3	661
	11.3 Translations to Section 4	674
	11.4 Translations to Section 5	693
	11.5 Translations to Section 6	714
	11.6 Translations to Section 7	732
	11.7 Translations to Section 9	733

References 735

Author Index 777

Subject Index 791

Listed in PubMed

Contributor

David A. Lightner, Department of Chemistry, University of Nevada, Reno, NV, USA
lightner@unr.edu

About the Author

At Rainy Lake, Canada, in 1967, nearby the *Watson* family islands. From *left* to *right*: *David Lightner*, *Albert Moscowitz*, and *Cecil Watson*.

David A. Lightner was born in Los Angeles, California, in 1939. In 1956, he matriculated at the University of California at Berkeley where he majored in biochemistry, mathematics, and chemistry, graduating with the *Artium Baccalaureus* (AB) degree in chemistry (*Summa cum Laude*) in 1960. As an undergraduate student, he was introduced to laboratory research in natural products chemistry, organic synthesis, and photochemistry under the guidance of *William G. Dauben*. In 1960, he commenced doctoral studies in natural products, synthesis and organic stereochemistry, circular dichroism (CD), and optical rotatory dispersion (ORD) spectroscopy at Stanford University under the guidance of *Carl Djerassi*. He completed his Ph.D. studies in October 1963, and in 1963–1964 he was a US National Science Foundation (NSF) Postdoctoral Fellow in mass spectrometry in the *Djerassi* lab. From 1964 to 1965, he was an NSF Postdoctoral Fellow at the University of Minnesota in theoretical chemistry and optical activity with *Albert Moscowitz*, during which time he was engaged in research collaborations on bile pigments with *Cecil J. Watson* of the University of Minnesota Medical School.

Research collaborations with *Moscowitz* and *Watson* continued for another two to three decades, until *Watson's* death in 1983 and *Moscowitz's* death in 1997. In 1964, *Lightner* accepted an assistant professorship in chemistry at the University of California, Los Angeles (UCLA), and in 1965 set up his lab to study organic stereochemistry and synthesis, optical activity and chiroptical spectroscopy, mass spectrometry, and bilirubin photochemistry. While at UCLA, and from 1972 to 1974, when he was Associate Professor at Texas Tech University (TTU), he continued his interests and collaborative research in the theory of optical activity with *Moscowitz* and with *Thomas D. Bouman* (Southern Illinois University) and *Aage E. Hansen* (H.C. Ørsted Institute, Copenhagen), studies that continue today with *Sergio Abbate* (University of Brescia). In 1968, he initiated a collaboration with *Rudi Schmid* (University of California, San Francisco (UCSF)) to explain how neonatal phototherapy works. This study was joined by *Antony F. McDonagh* (UCSF) in 1970 – a collaboration that continued until *McDonagh's* death in 2012. From 1976 to 2011, *Lightner* was Professor of Chemistry at the University of Nevada, Reno (UNR); in 1984, adjunct Professor of Biochemistry and Cell and Molecular Biology, and the first *R.C. Fuson* Professor of Chemistry at UNR. In 1992, he became Regents' Research Professor, and in 2011, he was appointed *R.C. Fuson* Professor Emeritus. His research is represented by more than 375 research publications from 1962 to 2012, five authored or co-authored book chapters, and eight authored or co-authored books, including: Lightner DA, Gurst JE (2000) Conformational Analysis and Organic Stereochemistry from Circular Dichroism Measurements. John Wiley & Sons, New York; Lambert JB, Gronert S, Shurvell HF, Lightner DA (2011) Organic Structural Spectroscopy. Prentice-Hall, Inc, New York; Huggins MH, Gurst JE, Lightner DA (2011) Organic Spectroscopy Problems. Prentice Hall, New York; Lambert JB, Gronert S, Shurvell HF, Lightner DA (2012) Spektroskopie. Strukturaufklärung in der Organische Chemie. Translated into German by Biele C, Marsmann HC, Kuck D Pearson Education Deutschland GmbH, Munich; Lightner DA (2013) Bilirubin: *Jekyll and Hyde* Pigment of Life; Pursuit of Its Structure Through Two World Wars to the New Millenium. In: Kinghorn AD, Falk H, Kobayashi J (eds) Progress in the Chemistry of Organic Natural Products (*Fortschritte der Chemie organischer Naturstoffe*), Springer-Verlag, Vienna (this volume).

1 Introduction

1.1 Early Reviews, Current Objectives

Only two reviews treating bilirubin and other bile pigments have appeared in *Zechmeister's Fortschritte der Chemie organischer Naturstoffe*. The first, in volume III, 1939, under the title "Gallenfarbstoffe" (*1*) by *Walter Siedel*, a student of *Hans Fischer*, appeared three years prior to the published total synthesis of bilirubin. Two years earlier, *Hans Fischer* and *Hans Orth* had reviewed succinctly what was known of the constitutional structure of bilirubin in volume II of the monumental three-part series, *Die Chemie des Pyrrols* (*2–4*). The second review in *Zechmeister*, by *Wolfhart Rüdiger* (*5*), appeared some 32 years after *Siedel's*, in volume XXIX, 1971, under a similar title, "Gallenfarbstoffe und Biliproteide". Its focus was on the more reduced bile pigments and plant linear tetrapyrroles. It did not review the structure or synthesis of bilirubin and appeared before the stereochemistry of bilirubin was understood. The solution to the structure of bilirubin was considered so important that *Siedel* followed his 1939 publication in *Zechmeister* by three shorter summary publications in 1940, 1943, and 1944 (*6–8*), the last being the summary of a speech delivered on October 10, 1943 to a special session of the Deutschen Chemischen Gesellschaft. Bilirubin research from the *Fischer* lab during the period 1939–1946 was later summarized briefly in a technical report (*9*). Although much of the pigment's constitutional structure was known at that time, and complete elucidation had come in 1942 from chemical synthesis (*10*), its stereochemistry was revealed some 35 years later. Indeed, since 1947 very few reviews concerning the structure of bilirubin are to be found, *e.g.* (*12–16*). Like those of *Siedel* (*1, 6–8*) and *Fischer* (*2–4*) they share the characteristic of not being directed solely toward bilirubin structure, nor are they comprehensive, being oriented toward synthesis (*1*) or biliproteides (*5*) – or outdated. The present work will attempt to break with these precedents and focus entirely on bilirubin structure, from the early attempts to isolate and purify the pigment, to the dawn of organic chemistry that introduced structure proof by degradation and ultimately, total synthesis; and to the discovery of its stereochemistry.

D.A. Lightner, *Bilirubin: Jekyll and Hyde Pigment of Life*, Progress in the Chemistry of Organic Natural Products, Vol. 98, DOI 10.1007/978-3-7091-1637-1_1,
© Springer-Verlag Wien 2013

The reader will note the inclusion of selected brief biographical information, in lettered footnotes, as well as numbered quotations of original material in the primary authors' languages. The quoted passages, sometimes difficult to retrieve, indicate the accomplishments and beliefs of the authors and appear as intra-text inserts. Nearly exact replicas of text and structures are included for historical and practical emphasis, and for easy reference to relevant parts of the original literature. It is important in this book to learn and understand as much as possible what the original workers knew and did not know, and how they interacted with the subject and each other as they expressed it, the scientific *Zeitgeist* of the 18th–20th centuries, the progress toward organic structure elucidation over the past two centuries. Translations of the texts are numbered, section by section, and collected near the end of the book in Section 11.

Yet no attempt to render an exhaustive account of bilirubin will be made here. The focus is on structure. Thus, the inclusion or exclusion of a particular related topic is more a reflection of the author's tastes and inclinations than it is of the importance of the subject matter to the field; the presence or absence of a particular reference is less indicative of its inherent worth than it is of the author's narrowness or ignorance.

1.2 Background

For as long as colors have been seen and distinguished by eye, pigments have been objects of fascination, use, and study. The subject of this work is the yellow-orange pigment called bilirubin, a "linear" tetrapyrrole natural product and the major pigment of mammalian bile. It is the colored substance of jaundice (icterus), observed as the yellowish discoloration of the sclera (whites of the eye), skin, mucous membranes, bile, and urine. As such it provides an early medical diagnosis by linking color to a pathological or physiologic condition, for anatomy and medicine have been studied for millenia and recorded at least back to the age of the pharaohs in Egypt. For the last, consult the papyrus *Ebers* of 1550 B.C. (*17–19*) and the surgical papyrus *Smith* of ~1600 B.C. (*20*) believed to be copies of even more ancient books of *Thoth*, ~3000 B.C. In such sources, one recognizes, *inter alia*, the liver, gall bladder, bile and urine. Not limited to ancient Egypt, references to such tissues may also be found in ancient India and China (*21*).

More than just a superficial observation, in ancient times the yellow coloration was found in biological fluids and linked to human health. The visual observation of yellow and green pigmentation in bile reflux or in vomit and in the even more easily accessible fluids, the red of blood and the yellow to yellow-brown coloration of urine, played important roles in early studies of anatomy, physiology, and medicine. (Of course the relationship between a color and molecule would have been inconceivable then.) In ancient Egypt, Babylon, and India (and probably earlier and elsewhere) changes in that plentiful and readily accessible fluid, urine, were examined in relation to disease. Medical diagnoses from urine (uroscopy) based on

1.2 Background

its color, odor, taste, and volume, as well as analyses based on the presence of particulate matter, turbidity sediments, blood, or pus were practiced in Greek and Roman medicine of ancient antiquity, as instructed in the writings of the *Hippocratic Corpus* (*Hippocrates* of Cos, 460–*ca.* 370 B.C.), the medical encyclopedia, *De Medicina* of *Aulus Cornelius Celsus* (25 B.C.–*ca.* 50 A.D.) and the medical works *Galenic Corporus* of *Galen* of Pergamum (129–210/217 A.D.), the Roman physician whose theories dominated western medicine for over a millennium. Thus, a brownish tint to urine would be taken as an indicator of jaundice. Such diagnoses along with those based on *Hippocrates'* four ancient humors of human health (yellow bile, black bile, blood, and phlegm) were the root of *Galen's* temperament theory. (The reader is encouraged to visit *Heinz Falk's* (*15*) excellent, succinct historical summary of the bile pigments and bilirubin from the early Egyptian dynasties to the scientific investigations of bilirubin in the early 19th century through the 1980s.)

Medicine, such as it was, overlapped with a technique (distillation) in classical antiquity when alchemy, representing "an amalgamation of certain chemical technologies and philosophical speculations began to be a formal discipline toward the end of the first century of the Christian era" (*22*). Distillates of plant and animal materials served as pharmaceuticals; alcohol and herbal distillates assumed medicinal value, thus associating mixtures of chemical substances with medicine. Although visual diagnoses based on urine and blood have not lost their importance in modern medicine, in the 16th century the world of alchemy intruded into the stagnating medical theory and practice of antiquity and stimulated investigations and explanations of bodily functions and diseases in chemical terms (*23*). Thus, the famous Swiss physician/alchemist *Paracelsus* (*Philippus Aureolus Theophrastus Bombastus von Hohenheim*, 1493–1541), who believed that the real purpose of alchemy was to produce medicines, introduced *iatrochemistry*, or chemistry in the service of medicine. Primitive medicinals and mineral drugs were introduced, and bodily fluids and their contents began to be probed and analyzed by the use of the developing chemical methodology and then-available chemicals.

As a "modern" chemist, imagine trying to investigate chemical structure absent a knowledge of *any* organic chemistry, chromatography or spectroscopy, acquainted only with the most primitive separation methods and possessing no means of assuring a high level of sample homogeneity, yet embarking on an investigation of the pigments of bile and blood to learn of their characteristics. By the end of the 18th century, systematic procedures for separating (distillation; crystallization) and identifying physical substances by their physical characteristics and chemical reactions were being discovered and systematized. Thus, the fluids of the body could be probed by these newly available means, including color changes that might occur upon the addition of a chemical, such as a mineral acid or alcohol. Biological fluids such as blood and urine were distilled of their volatile substances, and the distillates and the residues left behind were independently probed. Yet, even by the middle of the 18th century there were only two recognized states of matter (liquid and solid), and there appeared to be no concept of matter in a gaseous state. Nor were there criteria for purity and methodologies for ascertaining it. The situation was to change by the end of the 18th century: the gaseous state of matter had been recognized, new elements and substances had been discovered, and certain quantitative analytical methods were coming more commonly into use and refinement. Such methodology, together with an understanding of the gaseous state had a revolutionary effect on the science of medicine, marking the inauguration of physiological and clinical chemistry and scientific studies of animal metabolism.

One technique in particular had a profound and lasting influence: quantitative combustion analysis, from which one might calculate an empirical formula. Prior to the 19th century, "organic" matter had since the time of *Paracelsus* been analyzed by dry distillation, a method the results of which gave weighed fractions of distilled parts: gases, phlegma, oil, and carbon residue, but which evolved into assignments of chemical formulas based on combinations of more chemically sophisticated renderings of the distillates. Dry distillation analyses began to fall into disuse in the late 1780s after *Antoine Laurent de Lavoisier* (1743–1794) introduced an apparatus in which an organic substance was combusted to yield CO_2. The CO_2 and the oxygen consumed were measured gasometrically (*24*). Major advances in this quantitative analysis apparatus and method ensued at the Sorbonne in France in 1810 by *Joseph Louis Gay-Lussac* (1778–1850), together with *Louis Jacques Thenard* (1777–1857), and shortly thereafter by *Justus von Liebig* (1803–1873), who introduced the final major modification while working in 1822–1824 with *Gay-Lussac*. The need for a satisfactory analysis of nitrogen content, which might greatly expand the usefulness to a wide range of biological substances was evident, and in 1833 *Jean-Baptiste André Dumas* (1800–1884) invented a gasometric method for determining the amount of nitrogen following combustion, a method still in use today. Subsequently, two of *Liebig's* former students, *Heinrich Will* (1812–1890) and *Franz Varentrapp* (1815–1877) introduced a variation by converting the nitrogen evolved into ammonia. The *Will-Varrentrap* variation was improved over the years by others, culminating in 1883 with the analytical modifications of *Johan Gustav Christoffer Thorsager Kjeldahl* (1849–1900), followed in 1913 by those of *Lajos Winkler* (1863–1939). A major advance in analysis and structure determination ensued with the invention of the combustion train for *microanalysis* by *Friderik (Fritz) Pregl* (1869–1930), who was awarded the *Nobel* Prize in Chemistry in 1923 for making important contributions to organic microanalysis, *i.e.* having lowered the amount of substance necessary for analysis by a factor of 50, *inter alia*. All this became essential to the development of organic chemistry and thus to our understanding of clinical chemistry and animal chemistry. And so, by the late 1830s, scientists could with reasonable accuracy routinely determine the elemental composition (% C, H, N) and even derive empirical formulas of combustible substances isolated from animal and plant fluids and tissues. However, whether such substances qualified as homogeneous or "pure" remained a troublesome and discomfiting issue for over 100 years.

The concept of atomistic theory for the cosmos goes back more than 2,000 years (*22*) to the pre-Socratic Greek philosophers *Leucippus* and *Democritus*, who speculated that everything is composed of extremely small particles, which they called atoms, of which there are many different kinds, distinct in shape, size,

1.2 Background

and weight, physically but not geometrically indestructable. Atoms were said to move at random, able to collide and interlock to form clusters that are the basis for the objects of our perception. Such clusters could disintegrate into atoms and again reform differently. The basis of chemical atomism and the theory of atomic weights of elements go back only 200 years to *John Dalton* (1766–1844). The concept of a chemical structure is relatively more recent, and what constitutes a chemical structure emerged through many stages of evolution. Thus, late in the 19th century, until early in the 20th, influential mainstream scientists such as *Wilhelm Ostwald* (1853–1932, 1909 *Nobel* Prize in chemistry) did not believe in atoms. He spoke ferociously against them; yet in 1906 he was forced to capitulate on the basis of *Perrin's* data (*Jean Baptiste Perrin*, 1870–1942, 1926 *Nobel* Prize in Physics) that confirmed *Einstein's* 1905 theory of Brownian motion (*Albert Einstein*, 1879–1955, 1921 *Nobel* Prize in Physics) by empirically calculating *Avogadro's* Constant, conclusively proving *Dalton's* atomic theory. The emergence of organic chemistry in the 19th century focused on scientific investigations of substances found in nature that could be detected by the senses, especially by their odors and colors – especially colors that were distinct and highly visible. The early "organic" chemists were typically also physicians and thus unsurprisingly their curiosity was piqued by the "pigments of life": the red pigment of blood and the yellow pigment of bile, and that most noticeable to almost all, the green pigment of plants. These we now known as heme, bilirubin and chlorophyll, respectively (Fig. 1.1). Nor is it surprising, given their structural complexity, that elucidation of structure consumed the early decades of the past century in which the notion of a molecule and structural organic chemistry was in its infancy. The importance of such pigments was recognized by *Nobel* Prize awards in 1915 (*Richard Martin Willstätter*, 1872–1942, for plant pigments, especially chlorophyll), 1930 (*Hans Fischer*, 1881–1945, for hemin and chlorophyll), 1965 (*Robert Burns Woodward*, 1917–1979, for organic synthesis, including chlorophyll), 1988 (*Johann Deisenhofer*, *Robert Huber*, and *Michel Hartmut*, for the determination of the three-dimensional structure of a photosynthetic reaction center, bacteriochlorophyll and cytochrome).

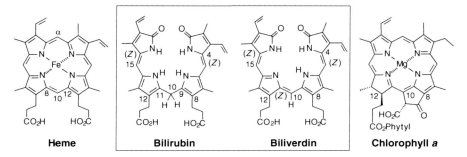

Fig. 1.1 The pigments of blood (*left*), bile (*middle*), and green leaves (*right*)

The focus of this book is the chemical structure of bilirubin, how it was defined, and those who solved it in scientific efforts that spanned more than 200 years, from the late 1700s to the late 1900s. Any proof of structure that consumed so many years provides a window into the development of organic chemistry: the approaches and constraints of the emerging chemical logic and intuition related to molecular structure, and the absence and appearance of applicable methodology, including synthesis. Because we now know most of the structural details of this natural product, it would be a straightforward task simply to draw the pigment's structure and chronicle the steps that led most directly to it – the typical approach taken when the structure of the pigment was reviewed at various points in the last century (*1–9, 12–16*). However, such reviews usually fail to reveal the enormous scientific struggles that accompanied the early proofs of structure, the roadblocks and sudden advances, the ambivalent and even misleading data that were available to chemists and were generated along the way, the emerging chemical logic and intuition, both the faulty and perspicacious insight of those who attempted to see clearly, their polemics and personality clashes. The actual constitutional structure work provides a glimpse into the dominance of pre-World War II organic chemistry in Germany and the dominance of its practitioners in the field of pyrrole chemistry. To omit such aspects would create a sterile history and fail to reveal that chemistry is a human endeavor, with human foibles. Unfortunately, the principal investigators of bilirubin's structure, unlike *Willstätter*, who made early important discoveries on the structure of chlorophyll (*4, 15, 25*), were not able to write a personal history of their work, leaving it instead to their students to write from their memories and notes (*1–8, 12*). What follows is a modest attempt to examine the how and why of the successes and failures leading to the currently understood structure of bilirubin, the yellow pigment of jaundice. It is written without the benefit of insight from those researchers who helped to create the science, as they are no longer with us, having for the most part left the stage after the constitutional structure of bilirubin was proven – and leaving behind few personal accounts beyond their scientific publications. Recorded oral histories were not then in vogue; few spared the time to write memoirs; others vanished from the scene as histories and lives were ended prematurely by the two great European wars of the 20th century.

The chemistry of natural products, especially structure determination, was highly prominent in the field of organic chemistry in the first half of the 20th century; it promoted the development and application of spectroscopy and organic synthesis in the 20th century. Proof of structure was both necessary and a force that drove and defined the development of modern organic chemistry. Some 50% of *Nobel* Prizes have been awarded for research in organic and biochemistry, and of these two-thirds were in the broad sense awarded for investigations of natural products. Nowadays, when one thinks of natural products chemistry, viewed by some as belonging to a bygone era, it may surprise the reader to find it clearly still prospering (with >250,000 citations in the American Chemical Society's SciFinder database between 1987 and 2012). In these, the close relationship between biochemistry and natural products chemistry, terpenes, steroids, antibiotics, and alkaloids may

1.2 Background

come to mind. Natural products based on pyrroles (*1–16, 25, 26*) are historically among the most ancient and most prevalent, and they are still today among the most important. The major sources of bilirubin in Nature were eventually shown to be bile and gallstones. The details of bilirubin's discovery, isolation, and (especially) the establishment of its molecular structure follow.

2 Early Scientific Investigations

2.1 Advances in the 18th Century

Following *Paracelsus*, the early investigators of medicinal chemistry were often largely trained as physicians who had gained a knowledge of the emerging field of chemistry, a field yet in its infancy. By the middle of the last century of the Enlightenment (*Zeitalter der Aufklärung*), an area of critical questioning and an age of clearing up, chemistry was a field in revolution, poised for rapid development from its 16th and 17th century alchemistry antecedents and stimulated by a cascade of discoveries from many notables, including the seminal contributions of the French nobleman scientist *Lavoisier*[1] and his findings, *inter alia*, on stoichiometry, the law of conservation of mass, respiration and on gases (*24, 27*), and the discovery of oxygen (*27*) by *Scheele* (*Carl Wilhelm Scheele*, 1742–1786) in Sweden and the English natural philosopher *Priestly* (*Joseph Priestley*, 1733–1804) in England. Although it could not have been recognized at the time, *Lavoisier*'s studies of combustion became the basis for what was to become perhaps the most important quantitative method for analyzing carbon compounds in the 19th and early 20th centuries and the first quantitative analytic technique applied to bile pigments: named elemental combustion analysis. *Lavoisier* was the first to explain combustion as a process of combination with oxygen, that led to the abandonment of the long-held phlogiston theory, which like many fiercely-held beliefs died only slowly. Successfully applying his knowledge of the combustion process, *Lavoisier* devised an apparatus in which *weighed quantities* of natural products such as spirit of wine, oils, fats, sugars, *etc.* were combusted in air and the CO_2 and H_2O products formed were *weighed*. From the *weights* involved, the %C and %H of the original substance could be calculated, and from the atomic weights of carbon, hydrogen, and oxygen, an empirical formula for the substance could be derived. The basic

---

[1] *Antoine-Laurent de Lavoisier*, the founder of scientific chemistry, was born on August 26, 1743 in Paris and died on May 8, 1794 in Paris. He studied chemistry, botany, mathematics, and astronomy at the Collège de Mazarin from 1754 to 1761, was elected to l'Académie Française des Sciences at age 25 and commenced his famous investigations on combustion at age 30. He was a strong advocate of quantitative (weighing and measuring) methods and of experimental work.

D.A. Lightner, *Bilirubin: Jekyll and Hyde Pigment of Life*, Progress in the Chemistry of Organic Natural Products, Vol. 98, DOI 10.1007/978-3-7091-1637-1_2, © Springer-Verlag Wien 2013

principle and seminal experiment became the basis for the modern process of combustion analysis introduced some 50 years later by *Liebig*.[2]

Lavoisier's genius and adherence to experimentation were revealed in his influential *Traité Élémentaire de Chimie* in 1789 *(24)*, the first modern chemistry textbook that so clearly and logically set forth principles which were fully confirmed in later times. "Il n'est jamais permis, en physique et en chimie de *supposer* ce qu'on plut *déterminer* par des experiences directes." [It is never allowed in physics and in chemistry, to *suppose* what can be *determined* by direct experiment.] France lost its most prominent scientist during its revolution. Denounced by his colleagues during the "Reign of Terror" (including *Antoine François le Comte de Fourcroy*, who was among the first to separate the components of bile and gallstones, *see below*), *Lavoisier* at age 50 was tried, convicted, and sent to the guillotine on May 8, 1794, some 16 years to the month after the death of *Voltaire* (*François-Marie Arouet*, 1694–1778), one of the most prominent individuals of the French Enlightenment and a forerunner of the French Revolution of 1789–1799. The famous Italian-French mathematician *Lagrange*[3] wrote: "Cela leur à pris seulement un instant pour lui couper la tête, mai la France pourrait pas en produire une autre pareille en un siècle." [It took them only an instant to cut off his head, but France will probably not produce another like it in a century].

Such was the importance of *Lavoisier*'s contributions that *Kekulé*[4] described him in his famous textbook *(28)* ". . . *Lavoisier*, der eigentliche Begründer wissen-

---

[2] *Justus von Liebig* was born on May 12, 1803 in Darmstadt and died on April 18, 1873 in Munich. He studied under *Kastner* from 1819 to 1822 at the universities in Bonn and in Erlangen, where he received the doctorate, engaged in advanced chemistry studies in *Gay-Lussac*'s laboratory in 1822, became a.o. Professor at Giessen (since the end of World War II renamed Justus-Liebig Universität Giessen) in 1824 and o. Professor in 1826. There he established the first major (teaching and research) school of chemistry and the journal *Annalen der Chemie und Pharmacie* (subsequently *Justus Liebig's Annalen der Chemie*) before moving to the University of Munich in 1852. He discovered N_2, improved elemental combustion analysis and trained certain scientists named in this work: *C. von Voit, H. von Fehling, H.F.M. Kopp, A. Kekulé, A. von Hofmann, E. Erlenmeyer*, and *A. Strecker*.

[3] *Joseph-Louis Comte de Lagrange*, born *Guiseppe Lodovico Lagrangia*, was born on January 25, 1736 in Turin and died on April 10, 1813 in Paris. A mathematician and astronomer, he succeeded *Euler* in 1766 as director of mathematics at the Prussian Academy of Sciences in Berlin, where he remained for 20 years. In 1787, he moved from Berlin to France and became a member of the French Academy.

[4] *Friedrich August Kekulé von Stradonitz*, was born on September 7, 1829 in Darmstadt and died on July 13, 1896 in Bonn. He matriculated at the University of Giessen and, swayed by *Liebig*, studied chemistry. After postdoctoral studies in Paris (1851–1852), Chur (1852–1853), and London (1853–1855) where he came in contact with *Alexander Williamson*, Kekulé became a *Privatdozent* at the University of Heidelberg in 1856 and o. Professor at the University of Ghent in 1858. In 1867 he was called to a chair at the University of Bonn. Among other attributes, *Kekulé* is known for his theory of chemical structure, tetravalent carbon, and the structure of benzene – the last a subject of controversy. (See, for example, *Bader A* (1998) The *Wiswesser-Loschmidt* Connection. Bull Hist Chem 22:21, and references therein.) Footnote (2) of *Kekulé*'s 1865 paper [*Kekulé A* (1865) Sur la Constitution des Substances Aromatique. Bull Soc Chim 3:98] shows *Kekulé's* preference for his own structure of benzene over those of *Loschmidt* and

2.1 Advances in the 18th Century

schaftlicher Chemie . . ." [*Lavoisier*, the true founder of scientific chemistry]. By 1859, *Kekulé* had defined organic chemistry as the chemistry of carbon compounds (*28*), but *Lavoisier*, too, may have been a founder of modern organic chemistry by his (*27*): "(i) recognition of the qualitative composition of vegetable and animal substances, (ii) recognition that these contain compound radicals which can combine with oxygen to form oxides such as sugar or alcohol and acids such as oxalic and acetic acids, and (iii) introduction of a method of combustion analysis". Animal anatomy and physiology were becoming linked to physiological chemistry and pharmacology, or animal chemistry, which continued to attract "chemical" probing of tissues and fluids. These domains were not in the least exempt from the scientific and chemical revolution taking place in the late 18th century.

The middle of the 18th century ushered in a period of intense scientific investigation, which for chemistry involved building upon a rapidly expanding and often confusing world of empirical knowledge of chemical substances and their manipulations, and the development of laboratory apparatus. Modern chemistry was thus emerging from its roots in alchemy and medicine, from a medicinal chemistry that predated *Paracelsus*, and over time led to the preparation of drugs by distillation of all types of plant and animal sources. Nonetheless, one should not think that scientific investigations of the 1800s necessarily involved pure chemicals, especially those of biological origin.

The historic philosophical-medical interest in animal fluids: blood, phlegm, bile, and urine, led to probings beyond dry (or destructive) distillation to the mixing of chemicals with such fluids, either before or after water had been gently removed, in order to effect separation into the component parts that could be probed independently. Like many biological fluids, bile turned out to be a complex mixture, and the early investigations were understandably constrained by a lack or absence of chemical knowledge. Modestly successful investigations of the 18th century typically involved manipulations using additives such as mineral acids, acetic acid, alcohol, ether, lead and barium salts, to effect separations by combinations of sequences involving precipitations, washings, and extractions. While such efforts were of limited success for isolating the pigments of bile, they were useful in isolating the fatty substances that are typically the major components of bile and gallstones, which led investigators of bile of the 18th century to focus less often on the pigments and more often on substances we now know as cholesterol, bile salts, and fatty acids.

Crum Brown – an apparent recognition of earlier published conceptual structures of benzene. Most notably, a cyclic structure for benzene from 1861 by *Johann Joseph Loschmidt*, professor of physical chemistry at the University of Vienna, who was born on March 15, 1821 in Karlsbad (now Karlovy Vary) and died on July 8, 1895 in Vienna. [Loschmidt J (1861) Chemische Studien I. Carl Gerold's Sohn, Vienna]. And in an unusually clear and modern molecular representation of phenol in the 1861 M.D. thesis of *Alexander Crum Brown*, professor of chemistry at the University of Edinburgh from 1869-1908, who was born on March 20, 1838 in Edinburgh, where he died on October 28, 1922 [*Crum Brown* (1866) On the Classification of Chemical Substances. Trans Roy Soc Chem 24:331]. *Kekulé*'s famous students include *van 't Hoff*, *Emil Fischer*, *Adolf von Baeyer*, and *Richard Anschütz*.

To the anatomists and physiologists of the first half of the 18th century and earlier, bile was seen as a yellow or yellowish-green, slightly alkaline, and slimy fluid possessing a peculiar sickening odor, with a taste at first sweet then bitter and exceedingly nauseating. Of variable consistency, commonly ropy and viscid, but at times limpid, it was found to be of greater density than water and miscible in all proportions with it (*29–31*). The fact that bile was used at times as a soap suggested a composition of animal fat and alkali (*31*). The opinion of the physiologists of the era might be best summarized as expressed by the physician *Thomas Coe* in 1757 (*29*) in which the yellow color of bile is mentioned:

> That bile is of a saponaceous nature appears by a plain experiment known to the vulgar, that is the use of the gall of oxen in washing linen, scouring wool, & where, like soap, being mixed with water, it helps to wash out grease and other stains, which the water alone could have little or no effect upon. ... The bile is of two kinds, namely that of the gall-bladder, called *bilis cystica*, and that which comes directly from the liver to the gut, called *bilis hepatica*. The cystic bile is thicker, of a deep yellow color and very intensely bitter.

And in 1767, *Cadet*[5] wrote of bile (*32, 33*):

> Je puis donc conclure que la bile est un véritable savon composé d'une graisse animale et de la base alkaline du sel marin, et du sel marin lui-même, d'un sel essentiel de la nature du sucre de lait et d'une terre calcaire qui participe un peu du fer. 1*

The color and bitterness of bile were attributed to the last two principles together with the nature of the oily principle (*32, 33*). Such was the status of animal chemistry of the times and such was its colorful terminology: sugar of milk (= lactose), calcareous earth (= CaO), and ferruginous (= rust colored).

Investigations were not limited to bile alone but were quite naturally drawn to concretions that appeared in bile, or were more generally found in the gallbladder. Again, in 1757, *Coe* described bile stones in the English scientific language of the times (*29*):

> And when by any means the bile is stopped or retarded so as to stagnate long either in the gall-bladder or ducts, especially if before the stoppage it was unusually thick or viscid, or abounded more than ordinarily with earthy particles, it is readily formed into biliary concretions, or gall-stones, of various kinds. ... It has been observed that some of these calculi seem to be made almost solely of earthy particles, cemented together, perhaps by a kind of mucus, without any appearance of bile; and that others seem to consist of mere inspissated or thickened bile without any mixture of earth, which will be different from one another, according to the bile from which they are formed, whether it was black or yellow, or green, or of some other color; but that the greater part of them are an undoubted mixture of earthy particles and bile, as both are plainly seen in the composition. ... [S]ome are compact and hard, and rather heavy, others are soft, or friable, and light.

For some early attempts at "chemistry":

> Biliary concretions do not dissolve in water, even with boiling, though the bile itself readily dissolves and mixes with it. Nor are they soluble in spirituous menstruum, as neither indeed does the bile dissolve well in rectified spirit, though it does in a weak spirit.

---

[5] *Louis-Claude Cadet de Grassicourt*, 1731–1799, studied at the Collège de Quatre-Nations and was a pharmacist at the Hôtel Royale des Invalides in Paris.

*Please note that translations numbered in this manner are provided in Section 11.

2.1 Advances in the 18th Century

However,

> ... some large and soft ones dissolved to about half their bulk in hot water. ... [T]hey will not dissolve in lime-water ... but some of them will dissolve in lixivium of salt of tartar. Most of the gall-stones will burn and flame more or less when they are dry ...

where the almost alchemical terms, spirituous menstruum = aqueous alcohol; rectified spirit (of wine) = repeatedly distilled alcohol, to concentrate; weak spirit (of wine) = dilute alcohol; lime-water = a clear solution of $Ca(OH)_2$; lixivium of salt of tartar = aqueous alkaline extract of wood ashes, or a solution of K_2CO_3, give evidence to the richness of abandoned chemical terminology.

Yet gallstones eventually proved to offer easier access than bile itself to the fatty materials contained therein, and, as shall be seen, also to the yellow and green pigments of bile. In the first half of the 18th century *Vallisneri*[6] noticed shortly before his death that gallstones dissolved in a mixture of spirit of wine (alcohol) and turpentine (*34*). And by the middle of the 18th century physicians considered gallstones to consist of the same oily, flammable material as in bile. In 1764, in his *Elements of Physiology* (*35*), the famous anatomist of his time, *Haller*[7] summarized the knowledge of gallstones at that time:

> The bile concretions contain a lot of air, up to four times their volume. Some are almost tasteless, except for the nucleus, which is bitter. They dissolve best in alkali, but fail in oil of tartar, and dissolve in potassium carbonate, for example, and very completely in oil of turpentine, sometimes in spirit of wine, sometimes it dissolves in none of these materials. They soften and dissolve in dilute nitric acid while sulfuric acid is without effect on them. Subjected to distillation, they soften and flow like sealing wax and then produce a little phlegm, a yellow oil, then a red oil and finally a black and empyreumatic oil.

Such were the early chemical investigations.

At about the same time as *Haller*'s 1764 treatise, *Poulletier de la Salle*[8], a contemporary of *Lavoisier*, practiced experiments on bile during 1745–1755, confirming that it had a soapy nature and contained an alkali salt – an observation by then well-known that was reconfirmed by *Cadet* (*32*). A more important observation was *Poulletier*'s apparent isolation of what we now know as cholesterol from gallstones. Stimulated perhaps by *Vallisneri*'s work on the solubility of gallstones, and assisted by a young *Fourcroy* in 1786–1787, from a fairly large collection of human gallbladder gallstones, *Poulletier* powdered some and dissolved them with warming in alcohol (and thereby confirmed *Vallisneri*'s experiment). Small blades of a glistening white crystalline substance appeared upon cooling, doubtless what we now know as cholesterol and the primary constituent of most human gallstones.

---

[6] *Antonio Vallisneri*, 1661–1730, a professor of practical medicine in Padua.

[7] *Albertus (Albrecht) von Haller* was born on October 16, 1708 in Bern and died on December 12, 1777. He studied anatomy at Tübingen (1723–1725) and at Leiden, where he graduated in 1727, studied mathematics at Basel (1728), practiced medicine in Bern (1729) until answering a call to the University of Göttingen in 1736 and resigned his chair in 1753 to return to Switzerland.

[8] *François Paul Lyon Poulletier, Sieur de la Salle*, was born on September 30, 1719 in Lyon and died on March 20, 1788 in Paris. He was the first to isolate crystals of cholesterol in ~1758.

Poulletier apparently communicated his results to *Macquer*[9], who reported them in his *Dictionnaire de Chymie* in 1788 (*36, 37*). If he had an interest in the pigment of gallstones, it was not evident.

Between 1775 and 1789 the shiny crystalline material from gallstones was also isolated by others (*37*): the theses of *Conradi* (*38*), *Delius* (*39*), *Dietrich* (*40*), and the compilation of *Gren* (*41*). In 1775, *Conradi* repeated *Poulletier*'s preparation, and similar experiments were carried out by *Delius* in 1782, and *Dietrich* in 1788, working with Prof. *Gren* – with the same result: isolation of a fatty substance that *Gren* called gallstone fat. *Delius, Dietrich*, and *Gren* gave an early analysis of gallstones: 85% waxy material; 15% resinous material (*37*). Any colored material present was not investigated. Together with *Vauquelin*[10], *Fourcroy*[11] (*42, 43*) continued the investigations of *Poulletier* on gallstones, perhaps from the same collection. Bile stones were divided according to their external color (brown or black, yellowish or greenish, white and ovoid) and distinguished by "chemical analysis": specific gravity compared to water, exposure to a flame, dry distillation, alcohol treatment. Some were noted to have a green-brown internal color. The pulverized stones dissolved with warming in alcohol, except for the hard and brown parts, and after filtration the liquid exhibited a yellow to light green color. (This was apparently the first chemical separation of pigments from gallstones.) Cooling the alcohol extract yielded brilliant white crystals having a waxy, scaly appearance of what *Fourcroy* thought (incorrectly) to be adipocire (now known as a mixture of calcium and potassium palmitates) (*27*) and spermaciti (*37, 42, 43*). The name for the white crystals would later (1815) be given as cholestérine (Greek: $\chi o \lambda \eta$ for bile; $\sigma \tau \varepsilon \rho \varepsilon \acute{\alpha}$ for solid) by *Chevreul*[12] who showed that it was unsaponifiable (*44–46*). Investigations of the colored material of bile and gallstones would have to wait until the 19th century.

---

[9] *Pierre-Joseph Macquer* was born on October 9, 1718 in Paris and died on February 15, 1784. He was one of the most famous chemists of his era and was known in particular for his *Dictionnaire de Chymie* first published in 1766, with subsequent expanded editions that followed.

[10] *Louis Nicholas Vauquelin*, was born on May 10, 1763 in Normandy and died on November 14, 1829. He was an assistant in *Fourcroy*'s laboratory from 1783 to 1791, and from 1809 Professor at the University of Paris.

[11] *Antoine François le Comte de Fourcroy*, was born on June 15, 1755 in Paris and died on December 16, 1809 in Paris. He was a physician turned chemist with help from the famous French anatomist *Felix Vicq D'Azur* (1748–1794), studied at the Faculté de Mèdicine in Paris, was promoted to chair of chemistry at the Jardin du Roi, Musée d'Histoire Naturelle upon the death in 1784 of *Macquer*, Professor of Chemistry at the Collège de France.

[12] *Michel Eugène Chevreul* was born on August 31, 1786 in Angers and died on April 9, 1889 in Paris. He lived through the "Reign of Terror" as a youth in France, applied to *Fourcroy* and worked in the laboratory under *Vauquelin*. His first teaching appointment was at the Lycée Charlemagne; in 1810 he became Assistant Naturalist at the Museum; in 1821 Examiner in Chemistry at the École Polytechnique. In 1826 he was elected to the Acadèmie des Sciences and in 1830 elected successor to *Vauquelin* as the Administrative Professor of the Musée d'Histoire Naturelle. Over such an incredibly long life he saw and accomplished much while knowing personally most of the famous chemists in Europe.

2.2 Color Diagnostics

In 1753, *Georg Heuermann* (1723–1768) wrote that yellow bile turned green in the presence of air and under the influence of acid (*47*):

> Das merckwürdigste hiebey ist, daß selbige, wie der Herr Seger schon augemercket ('De orfu et progressu bilis cysticae, § 13') durch beymischung des Spiritus nitri, salis und Olei Vitrioli, so besonders ihre Farbe verwandelt, denn mit dem ersten wird es *fast augenblicklich grün...* 2

The color change in yellow bile resulting from addition of nitric acid had also been observed, as recorded in *von Haller*'s 1764 treatise on physiology (*35*) in his chapter on the action of acids on bile (*ut se habeat ad acida*):

> Spiritus nitri bilem efficacius cogit, ut virides et duri grumi in aero subsideant. Viridem fecit, quae flava fuerit . . . Cum aqua forti alias arbusculae virides natae sunt; et grumus in fundo subsedit. In aliis puto meracioris acidi exemplis, bilis in coagulum amarum, viridis resinae similis, abiit... 3

A series of color changes were reported, in 1794, as having been seen by *Marabelli* when nitric acid was added to bile (*48*). Such color changes continued to be observed into the early 19th century when *Tiedemann*[13] and *Gmelin*[14] reported a detailed, systematic investigation in their famous treatise on digestion, *Die Verdauung nach Versuchen*, describing and analyzing the reaction (*48*). *Tiedemann* and *Gmelin* noted that when yellow-brown bile from a dog was treated with hydrochloric acid that had been de-aerated (freed from oxygen) no color change occurred during several days, but when oxygen was introduced the solution turned green near the oxygen inlet. They had thereby established a link between color change and oxidation/oxygenation. The color change from yellow to green was not restricted to hydrochloric acid treatment but was also observed following addition of sulfuric acid or acetic acid – and nitric acid. In the last, the color change to green was more rapid and was followed by further changes in color (in succession: green, blue, violet, red and finally yellow) in bile from mammals, birds, amphibians and fish (*48*):

> Dieselbe Wirkung, jedoch augenblicklich und weiter schreitend, zeigt die Salpetersäure, ohne Zweifel weil sie selbst den zur Farben veränderung nöthigen Sauerstoff abgiebt. Alle Arten von Galle, sowohl von Säugthieren, als Vögeln, als Amphibien und Fischen, die wir in dieser Beziehung untersuchten, färbten sich bei allmähligen Zufügen von Salpetersäure

---

[13] *Friedrich Tiedemann* was born on August 23, 1781 in Kassel and died on January 22, 1861 in Munich. He studied medicine and science in Bamberg and Würzburg, earning the Dr. med. in 1804 in Marburg, while continuing studies in Paris and Würzburg. In 1805 he became professor at Landshut and in 1815 accepted a position as Professor and Director of the Institute of Anatomy at Heidelberg, for 33 years.

[14] *Leopold Gmelin* was born on August 2, 1788 in Göttingen and died on April 13, 1853. He studied medicine and chemistry at Göttingen, Tübingen, and Vienna and in 1814 was appointed a. o. Professor and in 1817 o. Professor of chemistry and medicine at Heidelberg until 1852. *Gmelin*'s *Handbuch der Chemie* was first published in 1817–1819, and many successive editions appeared as the *Handbuch der Anorganischen Chemie*.

erst grün, dann blau, dann violett, dann roth, und zwar alles dieses bei hinreichen der Säuremenge innerhalb weniger Secunden. Hierauf tritt in einigen Stunden oder, bei grösserem Säureuberschüss, in einigen Minuten, Zerstörung der rothen Farbe ein, worauf die Flüssigkeit gelb erscheint.   4

Again, using nitric acid, the same authors detected the same progression of color changes in pathologic blood serum, chylus serum and urine, thereby indicating the presence of the pigment of bile (*48*):

Mittelst dieses Verhältnisses haben wir den Farbstoff der Galle in krankhaftem Blut-Serum; Chylus-Serum und Urin entdeckt, und es möchte hierdurch auch eine medicinische Wichtigkeit erhalten, da es zur Auffindung der Galle das sicherste Mittel ist...   5

and citing potential medical importance to this diagnostic color test for detecting the presence of bile in other tissues.

Tiedemann and *Gmelin* reported further on color reactions of bile following the addition of chlorine and from attempts using base to probe the colors obtained during the various stages of the nitric acid reaction. They learned that although oxygen is necessary to turn yellow into green in acids such as HCl, H_2SO_4, and acetic acid, only nitric acid (and the traces of NO_2 present) is required for the spectrum of colors. The work established what became famous as the *Gmelin* reaction (or *Gmelin* test) for bilirubin (*48*):

Man versetze z.B. Hundegalle mit so viel Salpetersäure, dass die blaue Färbung eintritt, übersättige sie dann mit Kali und giesse dann Vitriolöl in hinreichender Menge hinzu, so hat man ein Stück von *Regenbogen*; nämlich über dem farblosen Vitriolöl befindet sich eine rosenrothe Schicht, darüber eine blaue, dann eine grüne, und zu oberst eine gelbgrüne...   6

– the tints of the rainbow.

The *Gmelin* reaction was used for many decades following 1826 as a medical test to detect and characterize bilirubin in urine or other body tissues and over time was elaborated by the German physician *Ottomar Rosenbach*[15] (1851–1907), when it became known as the *Gmelin-Rosenbach* or *Rosenbach-Gmelin* color test (*49*). In one variation of the test, suspected urine or aqueous pigment is layered onto concentrated nitric acid (containing nitrous acid) or fuming nitric acid contained in a small tube so that it forms a layer on top. From the liquid-liquid junction outward disc-like rings are formed from the interface upward of colors yellow, red, violet, blue and green. In another variation, urine is passed through the same filter paper several times, the filter paper is dried and spotted with a drop of (slightly fuming)

---

[15] *Ottomar Ernst Felix Rosenbach* was born on January 4, 1851 in Krappitz, Silesia and died on March 20, 1907 in Berlin. He was educated at the universities in Berlin and Breslau (Dr. med., 1874). From 1874 to 1877 he was *Assistenzart* to *Leube* and *Nothnagel* at the medical hospital at the University of Jena, and in 1878 was *Oberassistent* at the Allerheiligen Hospital in Breslau, and became *Privatdozent* at the University. Rising to chief of the department of medicine of the hospital, he was appointed Assistant Professor in 1888, and resigned his position in 1896 to return to Berlin.

2.3 Emergence of a New Analytical Methodology: Quantitative Combustion Analysis

nitric acid to form a yellow spot with characteristic concentric rings of red, violet, blue and green. The colors are also reproduced in organic solvents: a yellow solution of the pigment in $CHCl_3$, treated with one drop of fuming HNO_3, becomes green and then in rapid succession blue, violet, reddish-orange and finally pale yellow or colorless.

Yet neither at the time (1826) of *Tiedemann* and *Gmelin*'s *Die Verdauung nach Versuchen*, nor until the 1840s, were the color changes shown to depend on a *specific* pigment in bile. That, of course, required some form of isolation.

2.3 Emergence of a New Analytical Methodology: Quantitative Combustion Analysis

At the end of the 18th century, *Fourcroy* summarized (*50*) the typical methods employed for analysis of organics from plant or animal products (*22*):

1. Natural mechanical analysis (separation by nature).
   Exudates of plants – saps, gums, manna, resins, rubber.
2. Artificial mechanical analysis (separation by presses, mortars).
   Juices and oils. The product is unaltered.
3. Distillation.
   Forms products which may not have been present as such in the plants.
4. Combustion analysis.
   Produces quantity of carbon and ash.
5. Analysis by water.
   a. Soaking after maceration.
   b. Soaking with agitation.
   c. Infusion (boiling water poured over macerated tissues).
   d. Digestion (tissues in cold water are heated slowly until boiling point is reached).
   e. Decoction (tissues are boiled with water for several hours).
   The various forms of analysis by water result in progressively greater alteration of the tissue components.
6. Analysis by acids and alkalies.
   Treatment may be similar to analysis by water, but alteration of principles is generally greater.
7. Analysis by alcohol, ether or oils.
   Results in a selective dissolving action; i.e., alcohol dissolves essential oils but no fixed (fatty) oils.
8. Analysis by fermentation.

As *Ihde* wrote (*22*):

As is clearly evident, the above analytical procedures are, at best, capable only of separating mixtures of related substances (proximate principles). Frequently the separation is achieved only after significant chemical alteration. The analyses could have only superficial value in leading to an understanding of organic materials; in many instances they were downright misleading. The time was becoming ripe for a more sophisticated approach, one which demanded pure, unaltered compounds which could be analyzed for their component elements, and studied for their characteristic properties.

To these "analytical procedures", now deservingly absent from organic chemistry, one might add dry distillation (heating a solid, often absent air, to produce and remove gaseous products) and, similarly, calcination, a method in which a substance is heated in air to a high temperature in a crucible to drive off water, carbon dioxide, and other volatiles until it is reduced to ash, which is then analyzed (see #4, above). Such methods of analysis date back to alchemy; yet, as will become clear in the attempt to analyze bile and gallstones, they had not been abandoned entirely in the 1800s.

With its roots in the very late 18th century, a new and revolutionary technique for organic analysis had, within a few decades of its discovery, reached a useful level of reliability and offered something no other previous method of analysis could: an empirical formula for the (presumed) pure substance. Thus, *Lavoisier*'s novel 1794 method (*24*) for analyzing the composition of alcohol, fats and waxes by combusting them and measuring the oxygen consumed and CO_2 produced, although yielding results that were usually inaccurate, opened the door to improvements in combustion techniques and gasometric measurements. As elaborated by *Holmes* and *Levere* (*51*), the rather large and cumbersome *Lavoisier* device was followed fairly rapidly by improvements: (i) in 1810 by *Gay-Lussac*,[16] working with *Thenard*, who upgraded the combustion process by admixing $KClO_3$ with the sample, then later abandoned it in favor of admixing CuO, for reasons of safety; and (ii) between 1811 and 1815 from *Berzelius* working with *Gay-Lussac* who in 1815 reintroduced the use of $KClO_3$, but admixed with $NaCl$ to temper the combustion; and (iii) *Liebig*, working between 1822 and 1824 with *Gay-Lussac*, who replaced bell jar gasometry with the combustion train (the *Kaliapparat*) wherein water vapor formed by the combustion of a weighed sample was absorbed by $CaCl_2$ and CO_2 was absorbed by KOH (*Kali*), both being weighed. Thus, in the early 19th century a major contribution to scientific methodology had come about in the technique called combustion analysis that enabled one to determine the %C and %H (and ash) in organic or biological samples from a quantitative measure of the CO_2 and H_2O produced (*22, 49, 50*). This methodology was followed shortly by one developed to determine the %N (*22, 49, 50*), a particularly major advance, as one could begin to group biological substances according to whether they contained nitrogen – and how much. Consequently, in the 19th century scientists were able to perform certain analyses involving partial separations of components of a mixture, probe the mixture and its components by treatment with chemicals such as acids and bases and heavy metals, alcohol and various organic solvents, and combust the components in order to obtain a quantitative measure of their %C, H, and N, with an eye toward calculating an empirical formula.

---

[16] *Joseph Louis Gay-Lussac* was born on December 6, 1778 in Saint Léonard de Noblat and died on May 9, 1850 in Paris. He was assistant to *Berthollet* and demonstrator to *Fourcroy* at the École Polytechnique in Paris, and became professor of chemistry in 1809. From 1808 to 1832 he was professor of physics at the Sorbonne, in 1832 chair of chemistry at the Jardin des Plantes. He is best known for his two gas laws and recognition of iodine as an element.

Improvements to the determination of %N were advanced by *Dumas* and *Kjeldahl* during the 19th century, and the development of modern-day organic microanalysis was advanced by *Pregl*[17] in Graz, who demonstrated early in the 20th century that quantitative analysis for C, H, N, S, and halogens could be accomplished with 7–13 mg of a sample, then down to 3–5 mg, with weighings ±0.001 of a milligram and the accuracy of macroanalysis. Until the advent of modern spectroscopic methods, elemental combustion analysis and microanalysis became fundamentally important to understanding organic structure. *Pregl*'s contributions, honored with a *Nobel* Prize in 1923, were probably the most important advance in organic analysis following the time of *Liebig*.

2.4 Early 19th Century Pigment Separation from Bile and Gallstones

As the 18th century drew to a close, and *Napoléon Bonaparte* (1769–1821) of France emerged to dominate the European continent in war and in law during the first decade and a half of the 19th century, organic chemistry was very much still the chemistry of animal and vegetable components, largely a descriptive science oriented toward the isolation and identification of the products of nature. Destructive distillation (calcination), which had served for centuries, was being abandoned as an analytical method and replaced by new approaches aimed at isolation of components in a state unchanged by the process of separation.

The revolution in scientific thought and experimentation of the 18th century thus brought into the turbulent 19th a new perspective in organic chemistry, which was still steeped in "Vitalism" but poised to broaden into the realm of interconversion and synthesis in a laboratory environment. For the animal and plant chemistry precursors to organic chemistry were, fewer than 200 years prior to this writing, clearly natural products, and vitalism was a widespread belief that organic compounds were to be found only in animal or plant sources, produced there by a "vital force" until *Wöhler*[18] overthrew ancient dogma (Vitalism) by creating an organic substance (urea) from its elements by heating an inorganic source (NH_4CNO, ammonium

---

[17] *Fritz Pregl* was born on September 3, 1869 in Ljubljana and died on December 13, 1936 in Graz. An Austrian chemist and physician, he received the Dr. med. degree in 1894 at the University of Graz, studied in Germany in 1904 with *Gustav v. Hüfner*, in Tübingen, *W. Ostwald* in Leipzig, and *E. Fischer* in Berlin before returning to work at the Medico-Chemical Institute under *K.B. Hofmann* at the University of Graz. He was appointed forensic chemist at Graz and professor at the University of Innsbruck from 1910 to 1913 before being recalled to Graz in 1913.

[18] *Friedrich Wöhler* was born on July 31, 1800 in Eschersheim and died on September 25, 1882 in Göttingen. He studied under *Gmelin* in Heidelberg and *Berzelius* in Stockholm, taught chemistry at the Gewerbeschule in Berlin from 1825, in Kassel from 1831, and in 1836 he became o. Professor of Chemistry in the medical faculty at Göttingen.

cyanate) (52). That event contradicted the firm beliefs of respected scientific authorities, such as *Gmelin*,[19] who said in 1817 that a characteristic of organic compounds was that they could not be produced from their elements; and *Berzelius*, who in 1827 believed that the elements present in living bodies obeyed laws totally different from those that rule inanimate nature.

The early 1800s were clearly a lively period for chemical science, with new discoveries occurring at a rapid rate. In the spirit of the Enlightenment, it was also a contentious period where firmly held beliefs were being challenged and reinterpreted or discarded, often reluctantly. Nonetheless, it ushered in a quantitative analytical method (combustion analysis) important during the following two centuries for determining not only the elemental composition but also the empirical formula of a sample, no less for bile pigments. And though the characteristic yellowish and greenish colors associated with bile had been recognized for millenia, it was not until the first half of the 19th century that modestly successful separations of the pigments from their biological sources were achieved. The typical source targets were well known from their yellow color: urine, bile, gallbladder and gallstones. Bile and gallstones were the most intensively investigated; the first turned out later to be the poorest source, the latter the best. Thus, early in the 19th century, three famous scientists, *Thenard* in France, *Berzelius* in Sweden, and *Gmelin* in Germany, commenced their chemical analyses of bile and gallstones – although not specifically to isolate the coloring matter.

Late in the first decade of the 19th century, when during the Napoleonic wars the Holy Roman Empire of 234 states was dissolved in 1806 and replaced in the Congress of Vienna in 1815 by the German Confederation of 39 states, the English Romantic poets *George Gordon Lord Byron* (1788–1824), *Percy Bysshe Shelley* (1792–1822), and *John Keats* (1795–1821) began to produce their famous literary contributions. And *Thenard* and *Berzelius* began to report the first chemical studies of bile and its concretions. In 1807 *Thenard*[20] reported his results from undertaking an analysis of the bile of several animals (53–56). In his bile analysis *Thenard* used reagents not previously employed, including acetic acid and lead oxide to effect precipitation and thereby initiated separation. When treating yellow-green bile from an ox gallbladder with H_2SO_4, HNO_3, or HCl, in all cases a yellow material was formed, along with little resin. Using, variously, alcohol, ether, $BaCl_2$, lead acetate, from the precipitated barium or lead salts and the supernatant he obtained

---

[19] *Jöns Jakob Berzelius* was born on August 20, 1779 in Väversunde and died on August 7, 1848 in Stockholm. He was perhaps the most influential scientist of the first half of the 19th century, and one of the fathers of modern chemistry, graduated as a Dr. med. in 1802 in Uppsala, became assistant professor of botany and pharmacy at Stockholm and full professor in 1807. From 1815 to 1832 he was professor of chemistry at the Karolinska Institute in Stockholm. Early on, his interests were physiological chemistry but expanded rapidly to include the law of definite proportions, and he compiled tables of relative atomic weights, or atomic equivalents, *etc.*

[20] *Louis Jacques Thenard* was born on May 4, 1777 in La Louptière and died on June 21, 1857 in Paris. He was the son of a peasant, became *Vauquelin's* laboratory boy in Paris at age 17 and was helped by *Fourcroy* to succeed *Vauquelin* in the Collège Polytechnique (1804–1837) as professor, where he worked with *Gay-Lussac* (also a professor from 1809) and became famous for his discovery of hydrogen peroxide in 1818.

2.4 Early 19th Century Pigment Separation from Bile and Gallstones

yellow material and, respectively, three essential principals: soda, a resin, and a substance that he named *picromel* (a colorless, viscous substance having a bitter-sweet taste). As *Thomson* wrote (*30*): "The name picromel is, I presume, from πικροζ: bitter, and μελι: honey." The substances were later shown to be mixtures: with picromel containing principally salts of bile acids that later became known as glycocholic acid and taurocholic acid.

The biles of many different animals were analyzed by *Thenard*, including that of the ox and humans. After evaporation of 800 parts of ox bile to dryness, and calcining; or 1,100 parts of human bile, quantitation showed:

Composition	Ox	Human
Water	700	1000
Albumin		42
Picromel	60.3	
Resin	24	41
Yellow matter	4	2-10
Soda	4	5.6
Phosphate of soda	2	
Muriate of soda	3.2	
Sulphate of soda	0.8	4.5
Phosphate of lime or perhaps magnesia	1.2	
Oxide of iron	trace	

Thenard found that human and quadruped bile contained similar substances, that the resin was sometimes green and sometimes yellow, that while human bile is sometimes green it is nearly always yellow-brown – and at times colorless. He noted that the yellow material from human bile and ox bile was insoluble in water and in kerosene but soluble in alkali – and that the alkaline solution upon acidification with HCl formed a flocculent green-brown precipitate (*54*). *Thenard's* research into bile extended to include that from fish, birds, *etc.*

Thenard also investigated gallstones as well as bile, from humans and from cattle. He found the concretions or calculi "absolument sans saveur et sans odeur", without taste or odor, and that the color was always yellow. When the yellow stones were exposed to air they gradually went green. When the stone was dissolved in caustic alkali, a yellow solution was obtained that gave a green precipitate upon addition of acid. He cites *Poulletier's* crystals obtained from human gallstones by partially dissolving in alcohol and concludes, with *Fourcroy*, that the stones have yellow lamina with a yellow interior and that they contain 88–94% *adipocire*. Ox gallstones exhibited brown-black coloration and contained variable yellow material. It might thus be said that *Thenard* had performed the first crude partial separation of the components of bile and gallstones, into yellow and green components, *inter alia*, and that he had noticed the yellow coloring undergoing a change to green from exposure to air.

Simultaneously, *Berzelius*, who later became virtually the supreme authority in Europe on matters of chemistry, had initiated investigations of bile, which he reported to the Swedish Academy in 1806–1808 (57). This study, previously reported in Swedish and perhaps not read widely, was communicated to the Royal Society of Medicine in England, by invitation. Some of his principal results were thus (58):

1. *Of Bile.*

It is well known that the elder chemists considered the bile as an animal soap composed of soda and a resin. The accuracy of this opinion had often been questioned, owing to the very small proportion of soda; and lately our skilful contemporary Thenard, has published an analysis of bile, in which he gives as its component parts, soda, a peculiar matter name by him *Picromel* and a resin, which united, produce a fluid that has the taste and other distinguishing properties of this secretion. Nevertheless I am convinced that there is no such resin as Thenard and his predecessors have described. I shall not here relate my experiments on this supposed resin in particular, but shall give the results of my enquiries on the bile itself, which will enable the reader to confirm or reject my opinions according as he finds them founded on accurate experiment.

The substance which is peculiar to bile has an excessively bitter taste followed by some sweetness; the smell is also peculiar, and the colour in most animals varies from green to greenish yellow. It is soluble in water, and its solubility is not in the least promoted by the alkali of bile, since, when this is neutralized by an acid, the peculiar matter does not separate: it also dissolves in alcohol in all proportions. Like the albuminous materials of the blood of which this peculiar matter is composed, it will unite with acids, producing compounds of two degrees of saturation, and hence, of solubility. The acetous acid, which gives soluble compounds with the albumen of the blood, does the same with the peculiar matter of the bile; and hence this matter is not precipitated on adding this acid to bile, though it falls down on the addition of the sulphuric, nitric, or muriatic acids. It is this sparingly soluble compound of biliary matter with a mineral acid which has been mistaken by chemists for a resin; since it possesses the external characters of a resin, melts when heated, dissolves in spirit of wine, and is again precipitated (in part at least) by the addition of water. The alkalies, alkaline earths, and alkaline acetates decompose and dissolve it: the former by depriving it of its combined acid; the latter, by furnishing it with acetous acid which renders it soluble in water. . . .

The biliary matter may be obtained pure in the following way: mix fresh bile with sulphuric acid diluted with 3 or 4 times its weight of water; a yellow precipitate of a peculiar nature first appears, which must be allowed to subside and be removed; then continue to add fresh acid as long as any precipitate is formed; heat the mixture gently for some hours, and afterwards decant the fluid part, and thoroughly edulcorate the green resin which is left. This resin reddens litmus, and is partially and sparingly soluble in water. It may be deprived of its acid in two ways: one of them is by digesting it with carbonate of baryates and water, whereby the carbonate is decomposed, and the water forms a green solution possessing all the peculiar characters of bile: the other way is by dissolving it in alcohol and digesting the solution, either with carbonate of potass or carbonate of lime till it no longer reddens litmus, and then evaporating it to dryness. Either of these methods will give the pure biliary matter, and there are also other ways of obtaining it, which I have described in my work on Animal Chemistry, Vol. II. p. 47. [57]

This peculiar biliary matter when pure, resembles exactly entire desiccated bile. Being soluble in alcohol it might be supposed that it would dissolve in ether, but this is not the case, for ether only changes it to a very fetid adipocirous substance, exactly as it acts upon the albuminous matter of the blood. One circumstance relating to the biliary matter has much surprised me, which is, that it gives no ammonia by destructive distillation. Therefore it contains no azote; but what can have become of the albuminous matter of the blood? for, no vestige of azote is found in any other of the constituent parts of the bile, nor does bile contain any ammonia.

2.4 Early 19th Century Pigment Separationfrom Bile and Gallstones

The following is the result of my analysis of bile.

Water	907.4
Biliary matter	80.0
Mucus of the gall-bladder, dissolved in the bile	} 3.0
Alkalies and salts (common to all secreted fluids)	} 9.6
	1000.00

Clearly, *Thenard*'s investigations of bile had not gone unnoticed. Though *Berzelius*' studies were contemporaneous, he employed a somewhat different separation method, relying on H_2SO_4 (not at all on acetic acid) and barium salts, especially $BaCO_3$ and heat. In this report, he strongly disputed resin matter and believed that it and the yellow matter and picromel were one and the same, merely modifications of the same substance, to which he later gave the descriptive name *Gallenstoff* (constituent of bile). He also disputed the presence of human albumin in bile, as reported by *Thenard*, for albumin is not precipitated upon addition of acetic acid, or alcohol. It is interesting to note that the 1812 synopsis (*58*) and the English or German translations of the original Swedish (*59, 60*) differ somewhat, suggesting that *Berzelius* had not ceased work on bile since his Swedish reports in 1806–1808.

Nor had *Thenard* ceased investigations. He indicated that according to his scientific investigations, specific yellow pigments were characteristic of bile, and that pigments also occurred in large quantities in gallstones (*53–56*), thereby linking the yellow pigment to bile and concretions found in the biliary tract or gallbladder. As reported in 1827, his subsequent examination of the biliary tract of an elephant (elephants do not have gallbladders) that had died in the Paris zoo revealed dilated bile ducts that were packed with yellow "magma", yielding 500 g of a powdery yellow, water insoluble material after drying. Treatment with hydrochloric acid immediately gave a strong green color (*61, 62*). As written from the perspective of 1977 by *Watson*,[21] an esteemed physician and porphyrinologist who studied under *Hans Fischer* in 1931–1932 (*62*):

---

[21] *Cecil James Watson* was born on May 31, 1901 in Minneapolis, Minnesota and died there on April 14, 1983. In 1919, he began his undergraduate studies at the University of Minnesota and entered the University of Michigan Medical School in 1921. Returning to the University of Minnesota in 1922, he completed the Dr. med. degree in 1926, after which he began a fellowship in pathology, and earned a Ph.D. in pathology in 1928. He then spent two years as the resident pathologist and director of laboratories in a private clinic in Minot, North Dakota. His interest in bile pigments came apparently from his suffering a bout of catarrhal jaundice (epidemic viral hepatitis) while in medical school, during which he made detailed observations on the course of his disease and found that urobilinogen (from intestinal reduction of bilirubin) disappeared from his excreta at the height of the jaundice – but reappeared in urine as the condition improved. Apparently, this personal experience led to his research interests in bile pigments, and it lured him back into an academic career. Returning to Minneapolis in 1930, while taking an advanced course in organic chemistry he was awarded a fellowship to study in *Hans Fischer*'s laboratory at the Technical University of Munich, where he succeeded in crystallizing stercobilin from human feces, proved the structure of stercobilinogen and showed it was not identical to urobilinogen or

24   2 Early Scientific Investigations

Thenard drew the curious conclusion that his green was due to impurities derived from the mucus of bile, apparently quite unaware that the HCl had converted yellow to green. From his descriptions, it seems likely that the elephantine orange pigment was a relatively pure unconjugated bilirubin, quite analogous to the pigment calculi of cattle and those so relatively common in the human bile ducts in India and the Orient. What a gold mine this elephant, at least for its time, and what a golden opportunity for Thenard!

Although *Thenard* appeared to be unaware that it was the yellow substance which had been converted to green by the action of acid, others had conducted scientific probings much earlier that consisted of observing and recording color changes brought about by adding reagents such as mineral acids to bile and urine. These early experiments were followed much later by attempts to isolate the colored matter – and purify it. For a description of efforts to separate the components of bile, from an early 19th century perspective, see *Thomson*, 1817 (*30*).

At nearly the same time, two well-respected Heidelberg professors of anatomy and physiology (*Tiedemann*) and of medicine and chemistry (*Gmelin*) jointly published their results on bile and, significantly, the cascade of colors following addition of HNO_3, in their famous treatise on digestion *Die Verdauung nach Versuchen* (*48*). Here they noted that a characteristic, very distinctly colored material is present in all bile, as *Fourcroy, Berzelius, Thenard, etc.* had seen earlier, and they reported achieving a partial separation (*48*):

> Schon Fourcroy nahm einen färbenden Bestandtheil der Galle an, und wiewohl es durch einige spätere Versuche zweifelhaft gemacht schien, ob eine eigenthümliche Materie der Art existire, sofern die Färbung der Galle zum Theil dem Gallenstoff zugeschrieben wurde, so hat doch Thenard *) [*) Traité de chimie edit. 4. Tom. 4. p. 580.] angenommen, dass in der Galle fast aller Thiere eine eigenthümliche gelbe Materie existirt. Er nimmt an, dass dieser Farbstoff die Gallensteine der Ochsen gänzlich constituirt und in fast allen der Menschen enthalten ist. Dieser Ansicht müssen wir uns, nach unsern Versuchen, vollständig anschliessen. Dass wirklich ein eigenthümlicher sehr ausgezeichneter färbender Körper in der Galle aller Thiere vorkomme, beweist Folgendes:
>
> 1) Wäre der Schleim das färbender Princip, so müsste, wenn man die zur Trockne abgedampfte Galle mit Weingeist auszieht, alle Farbe im unauflöslichen Schleimrückstand bleiben, wovon aber gerade das Gegentheil erfolgt. Schlägt man den Schleim durch Säure nieder, so reisst dieser zwar eine etwas grössere Menge des Farbstoffs mit sich nieder, die grösste Menge desselben bleibt jedoch gelöst.
> 2) Alle übrige Bestandtheile der Galle besitzen noch weniger Farbe, und können deshalb noch weniger als das färbende Princip derselben betrachtet werden.
> 3) Der Gallenfarbstoff zeigt, z. B. wie er in der Galle vorkommt, höchst auffallende, bis jetzt noch nicht hinlänglich bekannte Reactionen, die ihn überhaupt von allen bekannten Materien unterscheiden. Zu den wichtigsten gehören folgende.

---

to mesobilirubinogen, and identified mesobiliviolin. The studies in Munich with *Fischer*, and with *Friedrich von Müller* at the medical clinic at the University of Munich, made him well situated to assume a brilliant academic career at the University of Minnesota Medical School, to which he returned in 1932. In 1934 he was appointed assistant professor of medicine, associate professor in 1936, and as professor and head of the department of medicine in 1942. After 24 years as head, he resigned in 1966 to assume research full time. *Watson* was elected to the U.S. National Academy of Sciences in 1959, authored more than 350 research publications, and during his tenure at Minnesota introduced *Fischer*-based science to the fields of bile pigments and porphyrins in U.S. medicine.

2.4 Early 19th Century Pigment Separation from Bile and Gallstones

> Versetzt man die gelbbraune Galle des Hundes mit Salzsäure. . . .
>
> Die Galle der Thiere besitzt je nach ihrer Art und Individualität eine verschiedene Farbe; sie ist bei den Hunden gewöhnlich gelbbraun, nur wenig grün; bei den Ochsen braungrün; bei den Vögeln meistens lebhaft smaragd- oder grasgrün. Es lassen sich hieraus wahrscheinlich Schlüsse machen auf den mehr alkalischen oder mehr sauren Zustand der Galle, und auf den mehr desoxydirten oder mehr oxydirten Zustand des Farbestoffs derselben. 7

Gmelin also noted that gallstones contain the pigment of bile and give the same color test (*Gmelin* reaction) with HNO_3. He achieved a separation of pigmented material from pulverized ox gallstone by a series of chemical manipulations: (i) first heat in alcohol; (ii) then heat the residue in ammonium hydroxide (which gives a strong yellow color to the ammonia solution which then goes green in air); (iii) dissolve most of the undissolved residue from (ii) in aq. potash to give a yellow-brown solution that goes green-brown overnight and gives a positive *Gmelin* reaction; (iv) addition of HCl precipitated green flakes copiously from the solution in (iii). *Gmelin* carried out some further experimentation with the green flakes and concluded that the (yellow) pigment of bile and gallstones is converted to green by oxidation (*48*):

> 4) Wir untersuchten auch den Gallenstein eines Ochsen. Er liess sich leicht zu einem lebhaft braunrothen Pulver zerreiben. Kochender absoluter Weingeist färbte sich damit sehr blassgelb, nahm jedoch nur etwas festes Fett auf, welches sich nicht krystallinisch erhalten liess; als man auf den Rückstand Ammoniak einwirken liess, so nahm dieses eine etwas stäkrere Färbung an, und gab eine Flüssigkeit, die anfangs gelb war, sich jedoch an der Luft grasgrün und mit Salpetersäure blassroth färbte und durch Chlor entfärbt wurde.
>
> Der grösste Theil des Pulvers war ungelöst geblieben und dieser löste sich bei fortgesetzter Digestion mit Kali, mit Ausnahme einiger Flocken von phosphorsaurem Kalk, völlig darin auf. Diese Auflösung war anfangs gelbbraun und wurde über Nacht ebenfalls grünbraun. Sie gab mit Salpetersäure die oben bemerkten Farben-Veränderungen; sie gab mit Salzsäure einen reichlichen Niederschlag in dunkelgrünen Flocken, und nachdem sich diese völlig gesetzt hatten, so zeigte sich die überstehende anfangs noch grüne Flüssigkeit sehr blassgelb gefärbt. Die hiebei niedergefallenen grünen Flocken gaben nach dem Trocknen mit Salpetersäure eine blassrothe, bald gelb werdende Auflösung. Sie lösten sich in concentrirter Salzsäure vollständig mit smaragdgrüner Farbe; diese salzsaure Lösung trübte sich nicht mit Wasser, färbte sich mit Ammoniak gelb, mit Salpetersäure roth und wurde durch Chlor entfärbt. Auch in Ammoniak lösten sich die durch Salzsäure aus der Kali-Lösung gefällten grünen Flocken sehr leicht mit grasgrüner Farbe auf. Der Farbestoff der Galle scheint daher durch die Oxydation an der Luft, die er in der kalischen Lösung erleidet, in Salzsäure und Ammoniak löslich gemacht zu werden.
>
> Diesen Versuchen zufolge möchten wir den von uns untersuchten Gallenstein als fast reinen Farbestoff der Galle ansehen, dem nur eine kleine Menge von Fett und Kalksalzen, vielleicht auch etwas Schleim, beigemengt war, und wir möchten diesen Farbestoff wegen seines Stickstoffgehalts zunächst dem Indig setzen. 8

Although never described as pure compounds, the yellow and green material behaved like what we now know as bilirubin and biliverdin.

The pigments of bile notwithstanding, it was evident that bile was not composed of pigmented material alone but also contained fatty or soap-like components. Indeed, bile had long been considered to be a "soap", and the soapy or lipid-like substances were attracting the attention of *Gmelin*, *Berzelius*, and other investigators, including *Demarçay* (*63*) and *Loir* (*64, 65*).

In the period following *Thenard*'s and *Berzelius*' early investigations on the composition of bile, especially by the 1840s, a number of new investigators had independently joined the quest. While *Thenard* probably published his last findings in 1827 (*61*), *Berzelius*' investigations of bile continued well past his 1806/1808 initial studies (*57–60*), probably until he reached infirmity. Though *Berzelius*' work on bile does not qualify as his most important, it was highly advanced for its time, and thorough. *Berzelius* died in 1848, some nine years before *Thenard* and nearly 40 years past his first published studies of bile. He was apparently in declining health, as noted in his August 1839 correspondence with *Liebig*, in which he wrote of health problems (gout or arthritis) and having to take "the waters" at Marienbad as a palliative (*66*). (He apparently took seriously ill during 1818–1820 when he also had periodic head pains, then rebounded from such but again began to suffer in 1834 from the earlier nervous disorder. He suffered variable health onward, especially the two to three years prior to his death.) Despite his problems, *Berzelius* did not fail to describe his analysis of bile as not yet completed and commented on *Demarçay*'s[22] "transformation" of bile into taurine and *Gallenharz* (bile resin) as completely correct. *Berzelius* then went on to indicate that he, too, had obtained cholic acid, by his own method, and that *Demarçay*'s *acide choleïque* was an artifact that could never be obtained as the same material twice and contained at least four different organic substances. He commented that *Demarçay*'s *acide choleïque* contained two resinous acids that are also contained in his *acide choleïque*, as well as a neutral resin. *Berzelius* wrote that the major component of bile is his old *Gallenstoff*, which he now called *Bilin*, and that *Thenard*'s *picromel* and *Gmelin*'s *Gallenzucker* are not only non-acidic but are so extraordinarily sensitive as to defy purification. Writing in August 1839 to *Liebig*, *Berzelius* noted that bile also contained an acidic component, which he complained had cost him much pain to separate, and, although he had found at least four different compounds, he could not yet say which are transformation products and which are not (*66*):

> Meine Analyse der Galle betreffend, so bin ich auf lange nicht damit fertig. Am 22 Juni wurde ich von einem drohenden Gicht-Anfall auf dem Kopfe heimgesucht, der doch leicht und ohne Folgen gehoben wurde. Ich musste aber auf das Land gehen und Marienbader-Wasser trinken. Wegen meinen Amtsgeschäften musste ich 2mal in der Woche in der Stadt seyn, und diese Tage wollte ich einige Zeit die Versuche fortsetzen. Aber die warme Jahreszeit, die verdammten Fliegen die sich in meinen Auflösungen, aller Sorge sie abzuhalten unerachtet, immer ertränkten, und ein mit den Jahren schlechter werdendes Gedächtniss haben mich veranlasst alles liegen zu lassen, bis ich im October in meiner Wohnung in der Stadt wieder einziehe und mich wieder täglich damit beschäftigen kann. Demarçay's Metamorphose der Galle in Taurin und Gallenharz ist vollkommen richtig, Cholsäure habe ich auch nach seiner Methode bekommen. Es geht sogar besser mit kohlensaurem als mit caustischem Kali. —— Seine acide choleïque ist ein Kunstprodukt, das man nie zwey mal gleich erhalten kann. Es enthält wenigstens 4 verschiedene org. Substanzen, seitdem man die unorganische Säure, womit es niedergeschlagen ist, abgeschieden hat. Die acide choloïdique enthält zwey harzartige Säuren, welche auch in der ac. choleïque

---

[22] *Marc-Horace Demarçay*, 1813–1866, worked in *Liebig*'s lab, independently on bile for six months in the period 1836–1837.

2.4 Early 19th Century Pigment Separation from Bile and Gallstones

enthalten sind und ein indifferentes Harz. Der Hauptbestandtheil der Galle ist mein alter Gallenstoff, den ich Bilin nennen will. (Thénard's Pikromel, Gmelins Gallenzucker), der nicht sauer ist und der sich mit der grössten Leichtigkeit metamorphosirt, so dass es äusserst schwierig ist ihn rein zu bekommen. Er ist in Wasser und Alkohol in allen Verhältnissen auflöslich. Aber die Galle enthält auch sauer Bestandtheile. Es sind eigentlich diese die mir so viele Mühe gekostet haben auszuscheiden, und obgleich ich wenigstens 4 verschiedene bekommen habe, so kann ich doch in diesem Augenblick nicht sagen welche Metamorphos-Produkte sind und welche nicht. Aus einer alten *Bilis bubula spissata*, wo das Bilin grösstentheils metamorphosirt ist, kann man sie leicht ausscheiden, und auf diese Weise habe ich sie kennen lernen, aber aus der frischen Galle hält es schwierig sie hervorzuziehen, weil sie immer mit Bilin verbunden masquirt sind. Keinem von diesen gleicht die kristallinische Substanz, die Du so gut warst mir zu schicken; es kommt darauf an wie sich diese zu den Basen verhält, was ich versuchen werde. — 9

Just a few months earlier, on May 10, 1839, *Berzelius* had written to *Liebig* that *Demarçay*'s results had caused him to revise his analysis, that they reminded him of an idea he had from his own work showing that the *Gmelin* components of bile were all due to metamorphosis of the native components; that *Demarçay*'s new acid, *acide choleïque* was also a metamorphosis product, transformed by the isolation procedure (*66*):

Demarçay's Versuche über die Galle veranlassten mich zu einer Revision der Analyse der Galle; Sie werden sich erinnern dass ich in meiner Chemie die Idé ausgesprochen habe dass die Gmelin'schen Bestandtheile der Galle alle Metamorphosen sind, was nun Demarçay bewiesen hat, aber auch Demarçay's neue Säure, Acide choleïque, ist ein Metamorphos-Produkt. Es war aus seiner Abhandlung klar, weil die neue Säure durch Essigsäure sich aus der Galle nicht ausscheiden lässt, wohl aber aus ihren Verbindungen mit den Alkalien. Es ist mir geglückt den Farbstoff der Galle auszufällen und rein zu bekommen, die Basen der Galle auszuscheiden und eine in Wasser und Alkohol in allen Verhältnissen lösliche, intensiv bittere Säure zu bekommen, die in Wasser aufgelöst von Schwefelsäure oder Salzsäure sogleich in Demarcay's Acide choleïque verwandelt wird. Sie hat die Eigenschaft Fett aufzulösen in noch höherem Grade als Seife, und wäre sie in Ether löslich, würde man sie nie vom Fett scheiden können. Sie gibt lösliche Verbindungen mit allen bis jetzt versuchten Basen, sogar mit Silberoxyd. Sobald ich mit meinen Versuchen fertig werde, werde ich für Ihre Annalen meine Arbeit mittheilen. Ich nenne die neue Säure Gallensäure, Acidum bilicum. — In einer inspissirten Galle von einer Apotheke fand ich eine ganz neue, kristallisirende Säure als Hauptbestandtheil. Ich weiss aber noch nicht ob sie in allem alten Gallen-Extract enthalten ist. Sie scheint in der Stelle der frischen Gallensäure aufgetreten zu seyn. 10

It would seem that *Berzelius* could not have commented in as much depth on the non-pigmented components of bile in 1839 unless he had continued working on them before and during the 1830s. And although he was clearly responding to *Liebig*'s comments regarding the work of *Demarçay*, he was apparently not impressed or startled by *Liebig*'s comment to him in a letter of February 1837 regarding his student, *Demarçay*, to the effect that *Demarçay*'s then recent findings on bile seemed to overturn all previously accepted results. Yet, he could not have responded as he did unless he had known some of the details of *Demarçay*'s work, published in 1838 (*63*).

Ein fünfter (Demarçay) hat eine grosse Arbeit über die Galle vor, seine Resultate sind noch zu unbestimmt, als dass sich etwas davon mittheilen liesse, obwohl er seit 6 Monaten damit arbeitet, allein wie es scheint, so wird alles seither Angenommene umgeworfen werden. 11

Although *Demarçay*'s investigation of bile focused on the separation of what we now know as bile acids, he was not fully aware that he was working with conjugates and bile salts – and he was clearly not alone in his probing the fatty substances of bile. The advantage he had while working for six months in *Liebig*'s lab in Giessen was that he had access to *Liebig*'s state of the art combustion analysis apparatus, an advantage not available to workers in other labs where the methodology was more primitive, even if one had access to it. He published his work in 1838 (*63*), writing that bile was essentially 90% a soap, with sodium, and indicating the *Gmelin* had obtained 22 different substances from bile, almost all neutral and unknown. In Giessen, while noting the green colors of bile along the way, *Demarçay* probed ox bile variously with hydrochloric acid, ammonia, alcohol and lead salts, *etc.* so as to separate out acidic substances that the called *Choleinsäure*, *Choloidinsäure*, *Cholsäure, etc.* – all bile acids – as well as *Gmelin*'s *Taurin* (*63*):

> Ich gehe nun zur Beschreibung der eigenthümlichen Säure in der Galle über, welche ich *Choleinsäure* (von χολη, Galle) nenne und zu der ihrer drei Zersetzungsprodukte: der festen, stickstofffreien Substanz, welche ich *Choloidinsäure* (von χολειδης, gallenähnlich) nennen will, des *Taurins*, und der krystallisibaren, in Aether löslichen Säure, für welche ich den Namen *Cholsäure* beibehalten habe, denn es ist, wie ich glaube, der nemliche (?) Körper, welchen Gmelin unter diesem Namen beschrieben hat.   12

Demarçay's published work may have created more excitement than its intrinsic worth warranted. Indeed, he did publish combustion analysis results (%C, H, N) for his bile acid isolates, which were doubtless inhomogeneous. His work yielded empirical formulas such as $C_{21}H_{33.5}NO_6$ for *Choleinsäure* and even what he referred to as "atomic weights" of ~5,000, determined by burning the sodium salt of the substance and titrating the equivalents of base with acid.

Taurine, isolated in 1826 from bile first by *Gmelin* (*48*), was crystallized, apparently to purity, by *Demarçay* in beautiful needles, and its combustion analysis yielded $C_4H_{14}N_2O_{10}$, thought then to be a di-salt with ammonia. Taurine is still known today as a major component of bile, but with the formula $C_2H_7NO_3S$ ($^-O_3SCH_2CH_2NH_4^+$). Considering that in the *Demarçay* combustion analysis the percent oxygen was doubtless calculated by difference, after determining the %C, H, and N, one can imagine a *Demarçay* empirical formula $C_2H_7NO_5$, or $C_2H_5NO_3S$.

Keeping in mind that these studies represented "state of the art" organic chemistry of the late 1830s, the advances in analysis employed represented a move toward modern technology, including the use of crystallization as a means of purification to homogeneity and the emergence of combustion analysis as a powerful analytical tool.

2.5 Bilirubin and Biliverdin Separation from Bile by the Middle 19th Century

Despite the rather halting and somewhat controversial progress in isolating the yellow and green pigments identified with bile and gallstones in the early 1800s, significant headway had been made by the middle part of the 19th century. Yet at

2.5 Bilirubin and Biliverdin Separation from Bile by the Middle 19th Century

a time following *Wöhler*'s disproof of Vitalism in 1828 and coincident with a decade when new and radical thoughts were being formulated by *Friedrich Engels* (1820–1895) and *Karl Marx* (1818–1883), who met up in Paris in 1844 and criss-crossed Europe while chasing the revolutions of the time, progress in the analysis of bile was, however, perhaps not universally accepted. Some offered a less than sanguine perspective on bile, such as that of *J. Oliver Curran*, secretary to the council of the Dublin Pathological Society in *The Dublin Quarterly Journal of Medical Science* in 1846 (*67*):

> We have made but little allusion to the chemical history of the bile, for the very simple reason, that we believe analysis of the biliary fluid has as yet thrown no light on the subject. Although bile has been carefully examined by Berzelius, Bracconot, Bizio, Bostock, Chevreul, Chevallier, Demarçay, Fourcroy, Frommhertz, Gmelin, Gugert, Henry, Kuhn, Kemp, Lychnell, Lassaigne, Liebig, Pleischl, Prout, Thenard, Theyer, Schlosser, and many others, and each has added his mite in the form of a proximate principle, or something of the kind, to increase the complexity of this puzzling fluid, none of them found in it any sulphur; yet it was recently shown by Redtenbacher, that taurine (a proximate principle obtained by Gmelin from bile, by boiling it in hydrochloric acid) contains no less than *thirty per cent. of sulphur*. This discovery completely overthrows most of the beautiful and ingenious formulæ which we find in Liebig's book and proves how much has yet to be done before analytic chemistry can pretend to form any exclusive theory of the vital processes.

But this is getting ahead of the history of bile analysis in the first half of the 19th century, and, as shall be seen for the pigments of bile, analysis was found lacking.

Berzelius, a prodigious scientist and writer, left scientific accomplishments recorded in the 27 volumes of his *Årsberätelse* (yearly reports to the Swedish Academy on the progress of chemistry and physics between 1821 and 1848, the year of his death), and in five editions comprising many volumes of his *Lehrbuch der Chemie*, the all-encompassing summaries on the same subjects published as a first edition in 1803–1818 and culminating in the fifth edition in 1843–1848. The *Årsberätelse* were translated from the original Swedish into German as *Jahres-Berichte* by *Friedrich Wöhler*. They were also translated into other languages, *e.g.* French, and *Berzelius' Lehrbuch* became the most comprehensive reference on chemistry in the 19th century.

Berzelius' analyses of bile covered more than 40 years from 1807. Clearly between 1812 and 1842 *Berzelius* continued his studies on bile, presenting his findings annually to the Swedish Academy of Science and leaving them to be translated into German and published variously in his *Jahres-Berichte*, which in 1828 (*68*) included a report on bile, and in his *Lehrbuch*, which in 1831 (*69*) included an update of his analysis of bile and was translated from German into French in 1833 (*70*), in 1840 (*71*), and in 1842 in *Wagner's Handwörterbuch der Physiologie* (*72*). His subsequent long reports in 1840 and 1842 in the early research journals, *e.g. Annalen der Chemie und Pharmacie* (*73, 74*) (*Liebig's Annalen*), and more concise versions in 1840 and 1842 in the *Journal für praktische Chemie* (*75, 76*) focus exclusively on bile. These works and others, albeit repetitious, summarize *Berzelius'* nearly four decades of research on bile, which was, of course, only a small fragment of his much vaster and doubtless more earth-shaking contributions to science (*22, 77*).

30 2 Early Scientific Investigations

Shortly after the 1826 report on bile and gallstones by *Tiedemann* and *Gmelin* (*48*), in his *Jahres-Bericht* of 1827 ("Ueber die Fortschritte der physischen Wissenschaft"), *Berzelius* summarized his studies of ox, dog, and human bile, from which he separated a number of components (*68*): (1) a musk-like odorous/malodorous material from ox bile that co-distilled with water; and (2) from ox bile dried by gentle warming: *Gallenfett* (*Cholesterin* – cholesterol), *Oelsäure* (oleic acid) and *Margarinsäure* (margaric or heptadecanoic acid) which he obtained collectively by extraction into alcohol, then separated by various manipulations; (3) *Gallenharz*, softer at room temperature than wax but firmer than turpentine and of a dark green-brown color, obtained from the lead sulfate precipitates obtained during the isolation of *Margarinsäure*; (4) *Gallensäure* (*Acidum cholicum, Cholsäure*) or cholic acid, named to avoid confusion with *Gallasäure*, which was discovered in bile by one of the authors ("…ist eine von den Verfassern in der Galle entdeckte, vorher unbekannt gewesene, Säure"), a previously unknown acid and which contains nitrogen, released NH_3 in dry distillation – doubtless it was not pure cholic acid, and clearly both carbon-hydrogen and nitrogen combustion analysis were being used along with the obsolescent dry distillation; (5) *Gallenspäragin* ("wie die Verfasser auch selbst zugeben"), also claimed to be discovered by *Berzelius*, obtained as a consequence of the various manipulations above that yielded the *Gallenharz*, mixed with asparagine and *Gallenzucker* and separated; (6) *Gallenzucker* (*Thenard*'s *picromel*) obtained from *Gallenharz* by manipulations and precipitation involving treatment with basified lead oxide, eventually yielding a bright yellowish mass of irregular granular crystals; (7) *Farbstoff*, the coloring matter of bile; (8) *Gliadin*, obtained during the separation of *Gallenharz*; (9) *Schleim* (mucus) from the gallbladder, obtained by heating in water the substance remaining after dried bile is treated with alcohol; (10) *Käsestoff* (cheesy material) mixed with *Speichelstoff* (ptyalin), the water-insoluble substance obtained after drying the decoction above and heating the mass in alcohol; (11) a unique nitrogen-containing, yellow-colored substance that is soluble in water and insoluble in alcohol; (12) *Fleischextract* (meat extract) *Osmazom*, which remained behind with the *Gallenzucker* precipitated by vinegar of lead (a solution of basic lead acetate); (13) a substance with a urine-like odor obtained after calcining or heating red hot; (14) Na_2CO_3 and $(NH_4)_2CO_3$; (15) sodium acetate; (16) sodium and potassium oleate, margarate, bile acid salts, sulfates and phosphates, and NaCl, calcium phosphate, and 91.51% water.

The last was especially telling and reinforced what others concluded: bile is >90% water. In 1827, *Berzelius* doubted, however, that all of the materials that he isolated are actually present intact in bile, and he strongly suspected, having investigated the chemical composition of bile 20 years earlier, that many of the separated components were in fact artifacts of the isolation processes, *i.e.* the original components of bile had suffered transformations – a concept disputed at the time by *Chevreul* as well as *Gmelin*. In fact, *Berzelius* believed even 20 years earlier that bile actually had a simpler composition than that summarized above (*68*):

> Es entsteht hierbei nun die Frage: Finden sich alle diese Stoffe in der Galle, oder sind sie durch die Einwirkung der Reagentien auf einen oder einige Bestandtheile der Galle, deren

2.5 Bilirubin and Biliverdin Separation from Bile by the Middle 19th Century

Zusammensetzung leicht verändert wird, erzeugt worden? Als ich vor 20 Jahren die chemischen Verhältnisse einiger thierischen Stoffe untersuchte, glaubte ich zu finden, dass sie durch gewisse Reagentien Veränderungen erlitten und neue Producte entstünden, und ich hielt insbesondere Kochen mit Wasser, Aether oder Alkohol für weniger anwendbar, da die beiden letzteren aus Eiweiss, Faserstoff, Leim u. a. ein Fett von einem eigenen widrigen Geruch hervorbrachten (vergl. Jahresb. 1826, p. 277.). Diese Ideen sind von Chevreul bestritten worden, und Leopold Gmelin hält hierbei Chevreul's Ansicht für die richtigere...

Bekanntlich wird die von dem Gallenblasenschleim befreite Galle durch Säuren, und vorzüglich durch Schwefelsäure, auf die Art zersetzt, dass die Säure, bei einer gewissen Concentration, eine harzartige Substanz ausfällt, die etwas in Wasser und vollkommen in Alkohol auflöslich ist. Dabei bleiben in der sauren Flüssigkeit nur Fleischextract und Salze zurück. Bei einer Analyse, die ich vor mehr als 20 Jahren mit der Galle auf diese Weise anstellte, glaubte ich zu finden, die Galle habe eine ganz einfache Zusammensetzung, es seien nämlich die eiweissartigen Bestandtheile des Blutes in eine eigene Substanz verwandelt worden, die, wie jene, die Eigenschaft hätte, von Mineralsäuren, nicht aber von Essigsäure, gefällt zu werden; wobei die Flüssigkeit, worin diese Substanz aufgelöst war, fast von gleicher Natur mit der wäre, worin das Eiweiss und der Faserstoff im Blute aufgelöst sind. Aus der Verbindung mit Schwefelsäure konnte diese Substanz durch Digestion mit kohlensaurem Baryt wieder erhalten werden, wobei sie mit ihren vorigen Eigenschaften wieder im Wasser auflöslich worde. Ich nannte sie Gallenstoff. – Diese Versuche sind von Gmelin und Tiedemann wiederholt worden; sie fanden dabei, dass die Schwefelsäure den Gallenstoff ausfällte, dass aber die, durch Digestion mit kohlensaurem Baryt erhaltene Auflösung davon barythaltig war, und dass in dem im Ueberschuss angewandten kohlensauren Baryt Gallenharz unaufgelöst zuruckblieb. Sie schlossen daraus, dass Schwefelsäure Essigsäure mit dem Harze gefällt habe (ein gewiss ganz ungegründeter Schluss), dass diese Essigsäure Baryt aufgelöst habe, und dass die von mir Gallenstoff genannte Substanz eine Zusammensetzung aus Gallenharz, Farbstoff, Gallenzucker, Asparagin, Gallenfett, Margarinsäure, Oelsäure etc. und Esigsaurem Baryt sei. Dieser Schluss kann nicht richtig sein, denn wenn auch die Zusammensetzung der Galle nicht so einfach ist, wie aus meinen Versuchen hervorgehen würde, so lässt sich doch mit Gewissheit sagen, dass sich nicht 7 verschiedene organische Stoffe mit einander vereinigen, um einen einzigen Stoff von so bestimmten Characteren hervorzubringen, wie der ist, mit Schwefelsäure und anderen Mineralsäuren Harz zu bilden und von Essigsäure nicht gefällt zu werden; und wie sollten Oelsäure und Margarinsäure in eine solche Verbindung mitfolgen, da doch ihre Verbindung mit Baryterde unauflöslich ist. Was den Barytgehalt betrifft, so ist diese Beobachtung richtig; nicht allein Baryterde, sondern auch Kalkerde und Bleioxyd * [*] Lychnell hat einige Versuche angestellt, um die durch ungleiche Behandlung der Galle entstehende Verschiedenheit im analytischen Resultate auszumitteln. Bei einem dieser Versuche wurde schwefelsaurer Gallenstoff in Alkohol aufgelöst und mit kohlensaurem Baryt digerirt, bis die Flüssigkeit neutral wurde. Beim Abdampfen hinterliess die Auflösung eine in Wasser vollkommen auflösliche, der Galle ähnliche Substanz, die aber beim Verbrennen kohlensauren Baryt hinterliess. Dasselbe fand mit kohlensaurem Blei statt, aber die Auflösung wurde nicht neutral. Beim Verdünnen mit Wasser fiel schwefelsaurer Gallenstoff nieder, und nach dem Filtriren und Abdampfen blieb dieselbe Substanz, wie vorher, zurück, entheilt aber nun Bleioxyd. Als zu der Auflösung der sauren Verbindung in Alkohol kohlensaures Kali gesetzt wurde, entstand schwefelsaures Kali und eine regenerirte Galle. Ich hoffe, künftig die Resultate von Lychnell's Versuchen ausführlicher mittheilen zu können.] womit man die Schwefelsäure wegnimmt, verbinden sich mit der Substanz, die jene verlässt, und löst sich damit in Wasser auf, wenn nicht die Digestion mit einem Ueberschuss der Base zu lange fortgesetzt wird, wodurch sich eine unauflösliche Verbindung bildet, und es ist hier keine Säure, sondern der thierische Stoff, der die Base auflöst. Er hat in diesem Fälle mit mehreren anderen organischen Stoffen Aehnlichkeit, vor allen aber besonders mit

32   2 Early Scientific Investigations

dem Süssholzzucker, der mit der Schwefelsäure und den Säuren im Allgemeinen harzähnliche Verbindungen bildet, und der bei ihrer Zersetzung mit einer kohlensauren Basis, z. B.
kohlensaurem Baryt, Baryterde aufnimmt und damit in Wasser auflöslich wird.legt man
noch die zwischen Gallenstoff und Süssholzzucker bestehende Aehnlichkeit im Geschmack
zusammen, so wird die Uebereinstimmung noch auffallender.

Wäre Asparagin in der Galle aufgelöst enthalten, so würde diese Substanz mit dem
Schleim unaufgelöst zurückbleiben, wenn eingetrocknete Galle in Alkohol aufgelöst wird;
diess geschieht gleichwohl nicht, und Gmelin und Tiedemann bemerken, dass es nicht
einmal der Fall sei, wenn die mit Essigsäure versetzte und zur Trockne abgedampfte Galle
mit Alkohol behandelt wird, wobei doch die Affinitäten der Säure das Band aufgelöst
haben müssten, wovon man glauben könnte, dass es diese Substanzen in Verbindung halte.
Es geht hierans ziemlich gewiss hervor, dass sich das Asparagin nicht in der Galle vor der
Einwirkung gewisser Reagentien befindet; aber zu gleicher Zeit, wenn Asparagin aus irgen
einem Bestandtheil der Galle entsteht, müssen auch andere Stoffe gebildet werden, und
könnten in Folge hiervon nicht zuvor in der Galle enthalten gewesen sein.

Hierbei ist indessen zu bemerken, dass wenn auch die Zusammensetzung der Galle
einfacher wäre, als es aus den vorhergehenden Versuchen scheinen würde, es doch nicht zu
bestreiten ist, dass das Interessanteste unserer Kenntniss von der Galle die Bekanntschaft
mit den vorzüglichsten Veränderungen ist, die sie durch Reagentien ausserhalb des Körpers
erleidet, wodurch wir einen Theil der Veränderungen voraussehen können, die wie in dem
lebenden Körper beim Digestionsprozess erleidet.   13

Despite the curious and interesting results in *Berzelius'* 1827 report (*68*), for
our purposes item (7) above, *Farbstoff*, is the most relevant. *Berzelius* indicated
that, "as everyone knows" (*bekanntlich*), bile from the human gallbladder was yellow, which he said was the result of a characteristic pigment component of bile.
He noted then that no method for its extraction from bile had as yet been discovered but reassured that its existence had nevertheless been proven. He cited
Thenard's work in which he found that the yellow pigment was the main component of ox gallstones, which formed a brownish-yellow, easily-pulverized mass,
from which when a little crystalline fat was removed by heating in water, caustic
ammonia (NH_4OH) dissolved the yellow pigment. The latter changed color to
grass-green in air and became pale red with HNO_3 and lost its color with Cl_2. It was
found to be dissolved best in aq. potash to form a yellow-brown solution that
gradually turned green. Addition of hydrochloric acid to the green solution yielded
an emerald green precipitate that dissolved in caustic ammonia, and also in HNO_3
with a rose-red color that gradually went over to yellow. *Berzelius* noted that bile
behaved in the same way and indicated that dog bile when protected from air and
mixed with hydrochloric acid did not go green. The very same result was found by
Tiedemann and *Gmelin* (*48*). How much of *Berzelius'* report recapitulated the latter's work is unclear, but what he made clear was that the thusly protected bile
went green when air was absorbed, and that all acid-treated bile went green upon
evaporation in air. Moreover, that every sort of bile, mixed in small portions with
HNO_3 undersent the color changes of the *Gmelin* reaction: the yellow bile changing first to green, then blue, violet and next to red – all in the course of a few
seconds before finally turning to yellow. *Berzelius* noted further that when the
green stage of the *Gmelin* reaction was quenched with excess KOH, the liquid
became brownish-yellow, and at the blue or violet stage it became yellow-green.
Addition of H_2SO_4 restored the first color, *etc.* (*68*):

2.5 Bilirubin and Biliverdin Separation from Bile by the Middle 19th Century

VII. Farbstoff. Bekanntlich färbt die Galle alle die Gallenblase umgebenden Theile gelb, Leberkranke bekommen von der absorbirten Galle eine gelbe Farbe etc., und diess rührt von einem in der Galle entkräfteten eigenen Farbstoff her, zu dessen Ausziehung sie gleichwohl keine Methode ausfindig machen konnten, dessen Existenz aber doch dargethan werden kann. Thénard glaubte gefunden zu haben dass dieser Farbstoff die Hauptmasse der bei den Ochsen so gewöhnlichen Gallensteine ausmache. So wie er darin vorkommt, bildet er eine braungelbe, leicht pulverisirbare Masse. Kochendes Wasser sich daraus ein wenig, nicht krystallinisches Fett aus und färbt sich blassgelb. Kaustisches Ammoniak nimmt mehr davon auf die Flüssigkeit ist gelb, färbt sich an der Luft grasgrün wird von Salpetersäure blassroth und verliert durch Chlor die Farbe. Am besten löst er sich in Kali auf; die Auflösung ist gelbbraun und wird allmählig grünlich. Salzsäure fällt dann diese Auflösung mit grüner Farbe. Der Niederschlag wird von Salzsäure mit smaragdgrüner, von Salpetersäure mit rosenrother Farbe aufgelöst, die allmählig in eine gelbe übergeht. Der grüne Niederschlag mit Salzsäure wird leicht von kautischem Ammoniak aufgelöst. * [*Dieselben Verhältnisse sind von Lassaigne und Leuret hei dem gelben Farbstoff in der Haut und den Flüssigkeiten von Kindern bemerkt worden, die mit Gelbsuch geboren waren. Journ. de Ch. med. II, p. 264.] – Diese Verhältnisse zeigt auch die Galle. Vermischt man Hundegalle in einer umgestülpten Glasröhre über Quecksilber mit Salzsäure, so venändert sich die Farbe nicht, lässt man aber Sauerstoffgas zu, so wird eine Portion davon absorbirt und die Flüssigkeit färbt sich grün. Auf gleiche Weise wird alle mit Säure versetzte Galle beim Abdampfen in der Luft grün. Jede Art von Galle, in kleinen Antheilen mit Salpetersäure vermischt, färbt sich zuerst grün, dann blau, violett, und darauf roth; und zwar nach einigen Secunden; nach längerer Zeit oder durch mehr Säure wird sie zuletzt gelb. Durch diese Reaction kann die Gegenwart von Galle bei Krankheiten im Serum und im Urin entdeckt werden. Wird eine mit Salpetersäure grün gefärbt Hundegalle mit Kali gesättigt, so wird sie braungelb, in's Grünliche; war sie blau oder violett, so wird die alkalische Flüssigkeit gelbgrün. Zugesetzte Schwefelsäure bringt wieder die erste Farbe hervor. Sättigt man eine mit Salpetersäure blaugefärbte Galle mit Kalk und setzt hierzu, ohne umzurühren, concentrirte Schwefelsäure, so hat man über der Säure, die zu Boden gesunken ist, Schichten von verschiedenen Farben, nämlich der Säure zunächst roth, dann blau, dann grün und zuletzt gelbgrün. 14

Soon thereafter, in 1831, *Berzelius* published an update on his isolations from bile in the second edition of his *Lehrbuch*, volume 4 (*69, 70*). He described the color of bile as green, from yellowish-green to emerald green, and the substance as bitter tasting and of a peculiar nauseating odor; then he recounted his 1827 description of the pigment of bile, with some modification. He associated the yellow color of bile with the yellow color of jaundice that is seen in the skin and the eyes, with its source being the gallbladder. He indicated further that *Thenard* found it precipitated in human bile as a yellow powder, which he named *la matière jaune de la bile* and showed that it is the same as the yellow substance found in ox gallstones and is also the same as the yellow pigment found in the bile duct of a dead elephant of *le Jardin du Roi*, Paris – with an accumulated weight/mass amounting to 1.5 pounds (*61*). *Berzelius* wrote further that, led by *Gmelin's* investigation of ox gallstones and assertion that the yellow pigment comprises their main component, he then ground gallstones into a red-brown powder, which he heated in alcohol (to remove only a little fat) and found that caustic ammonia (NH_4OH) dissolved only a little of it but that caustic potash (KOH) was more effective and became brightly yellow colored but turned green-brown due to absorption of O_2 from air. When strongly saturated with HNO_3, within in a few seconds it displayed the color changes of the *Gmelin* reaction, as is characteristic of bile. He further indicated that the *Gmelin* reaction was the most certain means for detecting the presence of bile or its pigment. With added hydrochloric acid, the KOH solution formed a precipitate of dark green flakes, leaving a solution with a tinge of green. After washing and drying, the green precipitate was soluble in HNO_3, in which it raised a red color without blue or violet in between that quickly turned yellow. Characteristically, as indicated in 1826 by *Gmelin* (*48*), the yellow color of bile underwent the *Gmelin* reaction with HNO_3, and the yellow color change from yellow to green in bile occurred by oxidation from oxygen in the air – but remained yellow in the absence of air. *Berzelius* thus concluded that the green color encountered in bile originates from the yellow pigment – by oxidation, and that the green pigment was more soluble in alkali, which rendered difficult the separation of the two commingled pigments.

In 1831, *Berzelius* wrote (*69*):

> 9) *Farbstoff.* Die grünliche Farbe der Ochsengalle gehört, aller Wahrscheinlichkeit nach, einer eigenen Substanz an, die mit den übrigen Stoffen in der Galle aufgelöst ist, und die sich zwar auf analytischem Wege bisher noch nicht mit Zuverlässigkeit abscheiden liess, sich aber bei krankhaftem Zustande in der Galle bisweilen in so grosser Menge absetzt, dass sie eine eigene Art Gallensteine bildet, durch die man sie eben in isolirter Gestalt, hinsichtlich ihrer charakteristischen Reactionen, kennen lernnen konnte. Es ist derselbe Stoff, welcher in der Gelbsucht einen grossen Theil des Körpers, wie namentlich die Haut, das Weisse im Auge u. a., gelb färbt und die Ursache der gelben Farbe ist, welche man der Gallenblase und den sie umgebenden Theilen nach dem Tode gefunden hat. Thénard machte zuerst aufmerksam darauf; er fand ihn in der Menschengalle in Gestalt eines gelben Pulvers augeschlämmt, welches er *matière jaune de la bile* nannte, und von dem er zeigte, dass sie dieselbe Substanz sei, welche man in den Gallensteinen von Ochsen finde, und auch bei einem im *le Jardin du Roi* zu Paris verstorbenen Elephanten gefunden habe, bei dem sie eine in dem Lebergallengang angesammelte Masse von 1½ Pfund Gewicht ausmachte.
>
> Zur Darlegung der Beschaffenheit dieser Substanz werde ich Gmelin's Untersuchung Ochsen-Gallensteins anführen, wovon sie den Hauptbestandtheil ausmachte, Er liess sich leicht zu einem hell rothbraunen Pulver reiben. Kochender Alkohol zog daraus nur wenig Fett aus, und färbte sich gelb. Kaustisches Ammoniak löste eine geringe Menge davon auf; das beste Lösungsmittel dafür war aber Kalihydrat. Die durch Digestion erhaltene Auflösung war hellgelb, und wurde durch Sauerstoff-Absorption aus der Luft grünlich-braun. Mit Salpetersäure stark übersättigt, zeigt diese Auflösung eine Reaction, die für den Farbstoff der Galle characteristisch ist; setzt man nicht zu viel Säure auf einmal hinzu, indem man wohl ummischt, so wird die Flüssigkeit zuerst grün, darauf blau, violett und zuletzt roth, und diese Farbenveränderung geht innerhalb weniger Secunden vor sich. Nach einer Weile verschwindet auch die rothe Farbe, die Flüssigkeit wird gelb, und die Eigenschaften des Farbstoffs haben sich nun gänzlich verändert. Es bedarf nur einer sehr geringen Menge Farbstoff, um diese Reaction deutlich merkbar zu machen, und sie findet nicht allein mit Galle, sondern auch mit Blutwasser, Chylus-Serum, Urin und anderen Flüssigkeiten statt, wenn sie bei der Gelbsucht eine gelbe Farbe angenommen haben, und ist daher das sicherste Entdeckungsmittel für die Gegenwart von Galle oder ihres Farbstoffs. Die Auflösung des Farbstoffs in Kali wird von Chlorwasserstoffsäure in dicken dunkelgrünen Flocken gefällt, und nachher zeigt die Flüssigkeit nur einen schwachen Stich in's Grüne. Der niedergeschlagene Farbstoff löst sich, nach dem Auswaschen und Trocknen, in Salpetersäure mit rother Farbe, ohne blau oder violett dazwischen, auf, und die rothe Farbe geht bald in die gelbe über. Der durch Salzsäure bewirkte dunkelgrüne Niederschlag löst sich sehr leicht und mit grasgrüner Farbe, sowohl in Ammoniak als Kali auf.

2.5 Bilirubin and Biliverdin Separation from Bile by the Middle 19th Century

> Die Ursache der in der Galle oft vorgehenden Farbenveränderungen von Gelb in Braun und Grün, scheinen auf der Oxydation des Farbstoffs zu behruhen, wobei er von Gelb in Grün übergeht, und dadurch in Alkali leichter löslich wird. Galle, mit einer Säure versetzt und Berührung mit der Luft gelassen, wird nach einigen Tagen völlig grün. Gmelin vermischte Hundegalle, die gelbbraun ist, mit Salzsäuer in einer, an einem Ende zugeschmolzenen und über Quecksilber umgestürzten, Glasröhre. Auf diese Weise vor dem Luftzutritte geschützt, blieb die Farbe des Gemisches unverändert; sowie aber Sauerstoffgas hinzugelassen wurde, färbte es sich grün, zuerst an der Berührungsfläche mit dem Gase und nachher durch und durch, indem die Galle dabei ihr halbes Volum Sauerstoffgas absorbirte. Chlor bringt dasselbe Farbenspiel wie Salpetersäure hervor, jedoch weniger lebhaft; das Blau ist kaum merklich, sondern die Farbe geht gleich von Grün in Roth über, und ein Ueberschuss von Chlor zerstört die Farbe der Galle gänzlich und bleicht dieselbe unter Bildung einer weissen Trübung. 15

All of the above was also summarized in 1834 by *Dulk* in his *Lehrbuch* (*78*), for use in his lectures and for self study, which paints a somewhat different picture from that of *Curran* (*67*) but perhaps did little to improve the latter's view. Nonetheless, bile clearly became known and described as a complicated mixture, a nearly intractable biological fluid from the viewpoint of some. Undeterred, *Berzelius* persisted in his analyses of bile into the early 1840s, motivated by scientific curiosity, the new discoveries of *Demarçay* (*63*), *Chevreul* (*44, 45*), and others, the need to update his *Lehrbuch* with new editions (4th ed., 1835–1842; 5th ed., 1843–1848), and perhaps to re-establish his own studies and perspectives. His collection of writings published between 1840 and 1842 may well have expressed his then most recent and final thoughts on the components of bile. These works, while oriented toward the newly-discovered major (lipid) components of bile, especially the bile acids, fatty acids and their salts did address the yellow and green pigments of bile.

In the 3rd edition, 9th volume of his *Lehrbuch*, published in 1840 (*71*), *Berzelius* reviewed the early work and progress since 1807, his own, *Fourcroy*'s, *Thenard*'s, *Gmelin*'s, as well as the then more recent work of *Demarçay*, *Frommherz*, and *Gugert*. In this comprehensive volume, which as in all the *Lehrbuch* was far broader than the subject of bile alone, *Berzelius* critiqued the work of other investigators while reconciling or repudiating it relative to his latest studies, which were presented in considerable detail. He wrote that the prevailing early view was that bile consisted mainly of *Gallenharz* and picromel (*71*):

> Diese Ansicht wurde hierauf die herrschende, und alle später angestellten Analysen gingen von der Idee aus, dass die Galle hauptsächlich aus Picromel und Gallenharz bestehe. 16

Berzelius' approach in 1840 (*71*) to initiating the separation of bile into its components seemed twofold: (1) first adding H_2SO_4, followed by manipulations involving barium salts, or (2) adding lead salts (*71*):

> Wie erwähnt wurde, kann die Analyse der Galle auf zweierlei Art geschehen, nämlich durch Schwefelsäure oder durch Bleisalze; allein sie muss, damit so viel wie möglich Metamorphosen vermieden werden, mit andern, als den bis jetzt angewandten Vorsichtsmaasregeln angestellt werden. 17

(1) Analysis of bile using H_2SO_4. First ox-bile was evaporated over H_2SO_4 between 100° and 110°, taken to dryness in order to be pulverized. The powder

was digested 2–3 times with dry ether in order to remove fats, and the digested powder was taken up in anhydrous alchohol to leave behind mucus (*Schleim*), NaCl, and other alcohol-insoluble salts and animal substance but dissolving a compound of the bitter component of bile with alkali, alkali oleate, and margarinate, the pigments of bile in a similar compound, *etc.* The solution obtained was filtered and the residue was washed with anhydrous alcohol. The residue was washed with 85% alcohol, which dissolved certain substances, and then retained. The anhydrous alcohol solution above was then mixed in small portions, with shaking, with a solution of $BaCl_2$ in H_2O until a dark green precipitate had formed. The green precipitate was filtered and washed with alcohol, which however was not required to be anhydrous. Baryta water (aq. BaO) was added dropwise to the filtered solution. The precipitate thus formed was first dark gray colored but became green after a few moments. Baryta water was added as long as the solution was still cloudy. The precipitate was soon no longer green, but only yellow-brown, and finally only yellow, whereupon the solution had for the most part lost its color, and showed only yellow in it. The precipitate was filtered and washed with 84% alcohol (*71*):

> 1. Analyse der Galle durch Schwefelsäure – Die Ochsengalle wird im Wasserbade oder im leeren Raum über Schwefelsäure verdunstet, indem zuletzt die Temperatur in dem leeren Raum auf + 100° bis +110° steigen muss, damit die Masse so trocken wird, dass sie zu Pulver gerieben werden kann. Dann wird sie mit wasserfreiem Aether übergossen. Ist der Aether wasserhaltig, so nimmt die Galle das Wasser auf und fliesst zusammen. Der Aether zieht alles Fett aus, welches nicht mit Alkali zu Seife verbunden ist. Das mit Aether zwei bis drei Mal digerirte Pulver wird darauf in wasserfreiem Alkohol aufgelöst, welcher Schleim, Kochsalz und andere in Alkohol unlösliche Salze und Thierstoffe zurücklässt, dagegen eine Verbindung des bittern Bestandtheils der Galle mit Alkali, ölsaures und margarinsaures Alkali, den Farbstoff der Galle in einer ähnlichen Verbindung, u. s. w., auflöst. Die erhaltene Lösung wird filtrirt und das Ungelöste zuerst mit wasserfreiem Alkohol gewaschen, der dann der filtrirten Lösung zugefügt wird, und darauf mit Alkohol von 0,85, welcher gewisse Stoffe daraus auflöst, und der für sich genommen wird. Die Lösung in wasserfreiem Alkohol wird nun in kleinen Portionen und unter Umschütteln mit einer Lösung von Chlorbarium in Wasser vermischt, so lange noch ein dunkelgrüner Niederschlag gebildet wird, den man abfiltrirt und mit Alkohol, der jedoch nicht wasserfrei zu sein braucht, abwäscht. Zu die filtrirten Lösung tropft man dann Barytwasser. Der Niederschlag, welcher dadurch gebildet wird, ist anfänglich dunkelgrau, färbt sich aber nach einigen Augenblicken grün. Das Barytwasser wird so lange zugesetzt, als die Lösung noch dadurch getrübt wird. Der Niederschlag wird bald nicht mehr grün, sondern erst braungelb, und zuletzt nur gelblich, worauf die Lösung ihre Farbe grösstentheils verloren hat, und sich nur noch in's Gelbe zieht. Der Niederschlag wird abfiltrirt, und mit Alkohol von 0,84 ausgewaschen. 18

The residual ethanolic solutions from approach (1) that contained free BaO/$Ba(OH)_2$ was precipitated as $BaCO_3$ with CO_2 gas, filtered, and evaporated to dryness before processing further with PbO, *etc.* to yield *Bilin* (named by *Berzelius* from *Bilis*, bile) that is identical to *Gmelin*'s *Gallenzucker*, a procedure similar to that which led *Thenard* to isolate a component that he called *Picromel* ($\pi\iota\kappa\rho o\varsigma$, bitter, and $\mu\epsilon\lambda\iota$, honey). Isolated as the metamorphosis products of *Bilin* were (as named by *Berzelius*): *Fellinsäure* (from *Fel fellis*, bile), *Acidum fellicum*, *Cholinsäure* (from $\chi o\lambda\eta$, bile), *Acidum cholonicium*, and *Dyslysin* (from $\delta v\varsigma$, difficult, and $\lambda v\delta\iota\varsigma$, solution).

2.5 Bilirubin and Biliverdin Separation from Bile by the Middle 19th Century

(2) Analysis of bile from lead salts. In this approach, dilute acetic acid is added to fresh gallbladder bile to separate mucus (*Schleim*), mixed with twice the volume of alcohol then processed further using PbO to yield bile acids and their salts, *inter alia*, bile acids akin to those isolated by *Demarçay*. Thus, as described in his 1840 *Lehrbuch* (volume 9) (*71*), *Berzelius* found that approaches (1) and (2) could be used to precipitate biliverdin (as its barium salt) and *Bilifulvin* (bilifulvin), also as its barium salt, but mainly (1) and (2) served as the entry point to separate out the many other more major components of bile. For *Gmelin* in ~1826 (*48*) it led to previously unknown components such as taurine and cholic acid as well as a substance he named (bittersweet) *Gallenzucker*, and much more (*71*):

> Gmelin fand ausser diesen Bestandtheilen, nämlich dem Gallenharz und Gallenzucker, noch Taurin, Cholsäure, Cholesterin, Oelsäure, Margarinsäure, Farbstoff, Fleischextract, eine extractähnliche unrinöse Substanz, eine dem Pflanzenleim analoge Materie, Käsestoff, Speichelstoff, Albumin, Schleim, kohlensaures Natron, kohlensaures Ammoniak, essigsaures (milchsaures) Natron, ölsaures, margarinsaures, cholsaures, schwefelsaures und phosphorsaures Natron und Kali, Kochsalz und phosphorsauren Kalk.   19

And the discussion led to *Berzelius'* new term, *Bilin*, which he said was identical to *Gallenzucker* and numerous other lipid and inorganic products. For *Demarçay*, it opened the door to bile acids and their salts and created quite a stir with *Berzelius*, who was motivated to devote numerous pages of his *Lehrbuch* to further explanations of the lipid components (*71*):

> Demarçay leugnet gänzlich die Existenz eines Gallenzuckers und hält Gmelin's Gallenzucker und Thénard's Picromel für identisch mit *Acide choleique*.
>
> Nachdem wir auf diese Weise innerhalb eines Zeitraums von mehr als 30 Jahren hinsichtlich des Hauptbegriffs von der Natur der Galle in einem Zirkel gegangen sind, freilich nicht ohne bedeutende Vermehrung unserer Kenntnisse, stehen wir wieder auf demselben Punct, und ungeachtet aller der Erfahrunge, die wir durch die angeführten Arbeiten gewonnen haben, wäre es doch nicht möglich ohne neue Untersuchungen einen nur einigermaassen richtigen Begriff von der Zusammensetzung der Galle zu geben. Ich werde sie nun abhandeln nach den Untersuchungen, die ich neuerlich in dieser Absicht mit der Ochsengalle angestellt habe.   20

Though *Berzelius'* discussion of the pigments of bile and their separation was much less extensive than that of other components of bile, he described the isolation of the two pigments of bile, initiated by approach (1) and followed by several manipulations before precipitation with $BaCl_2$. Two different barium precipitates were obtained, the first from addition of $BaCl_2$, followed by a second addition of BaO or $Ba(OH)_2$, and each yielded a different pigment. The first, which gave bile its green color was bound to baryta, *Berzelius* named *Biliverdin* (coming from *bilis*, bile, and *verdire*, green). The second, brownish precipitate formed from baryta water and contained, beside biliverdin, a reddish-yellow (orange) pigment that *Berzelius* named *Bilifulvin* (from *Bilis*, bile, and *fulvus*, reddish-yellow), an extractable substance and characteristic nitrogen-containing animal substance which *Berzelius* would return to later.

These pigments were doubtless mixtures of barium salts (*71*):

> Der erste Niederschlag mit Chlorbarium enthält den Stoff, welcher der Galle ihre grüne Farbe gibt, verbunden mit Baryterde. Ich nenne ihn *Biliverdin* (von Bilis, Galle, und verdire,

grün werden). Der andere, oder der Niederschlag mit Barytwasser, enthält neben dem Biliverdin einen rothgelben Farbstoff, welchen ich *Bilifulvin* (von Bilis, Galle, und fulvus, rothgelb) nenne, einen extractähnlichen Stoff und einen eigenthümlichen, stickstoffhaltigen Thierstoff, auf welche ich weiter unte zurückkommen werde.   21

Possibly a better way to access the green pigment was found in yet a third approach, where dried bile, dissolved in alcohol, was treated with $BaCl_2$ – a method similar to (1) but still employing $BaCl_2$ to precipitate the green pigment. The precipitate was digested with hydrochloric acid to remove BaO, washed with ether to remove fat, and processed with cold anhydrous alcohol to yield a green-brown solution and a green insoluble residue. The alcohol solution, allowed to evaporate on its own, yielded biliverdin in the form of a nearly black-brown, earthy compound. When evaporated with heating, it formed a shiny, translucent dark-green film (*71*):

> 6. *Biliverdin*, Gallengrün. Der Niederschlag wird noch feucht mit verdünnter Salzsäure übergossen, welche die Baryterde auszieht und Biliverdin zurücklässt. Er ist nur mit wenig Fett vermischt, welches man mit Aether auszieht, in dem sich jedoch auch ein kleiner Theil von dem Biliverdin gleichzeitig auflöst. Das Zurückbleibende wird mit kaltem wasserfreien Alkohol behandelt, welcher sich davon grünbraun färbt, der aber einen grünen, in kaltem Alkohol, unlöslichen Rückstand zurücklässt. Die Lösung in Alkohol, der freiwilligen Verdauung überlassen, lässt das Biliverdin in Gestalt eines fast schwarzbraunen, erdigen Körpers zurück. In der Wärme verdunstet, bildet es einen glänzenden, durchscheinenden, dunkelgrünen Ueberzug.   22

After providing a long list of the properties of biliverdin, *Berzelius* wrote that those properties corresponded altogether with all "three modifications" (*79*) of chlorophyll. He indicated that his assessment was valid not only for biliverdin from ox bile but might extend to bile from other herbivores. He said further that biliverdin from the bile of carnivores possessed quite different properties (or was tied up with a pigment not yet separated), on which he himself had not yet been able to carry out a few experiments (*71*):

> Diese Eigenschaften des Biliverdins stimmen in Allem mit denen des Chlorophylls überein, so dass ich entschieden bin, dasselbe als damit identisch zu betrachten, und ich habe es aus verschiedenen Gallen in allen drei Modificationen des Chlorophylls erhalten. Das jetzt Angeführte gilt natürlicherweise nur für das Biliverdin aus Ochsengalle, vielleicht auch für das aus der Galle anderer grasfressender Thiere. Aber in der Galle fleischfressender Thiere besitzt es ganz andere Eigenschaften, oder es ist darin mit noch einem anderen Farbstoff verknüpft, von dem man es bis jetzt noch nicht geschieden hat. Da ich noch nicht Gelegenheit hatte, darüber selbst einige Versuche anzustellen, so muss ich nach Angaben Anderer berichten.   23

Berzelius' correlation of biliverdin to chlorophyll (and to the bile of "grass eaters") is rather startling and apparently did not come from a lack of experience with chlorophyll because simultaneous with his work on the pigments of bile, *Berzelius*, whose experimental work ranged far and wide in chemistry, had also been working on the isolation and properties of the green pigment of leaves in the 1830s (*79*). Though chemical studies of the green pigment of leaves dates back to the 1780s (*80*), the name *chlorophyll* was coined for green colorant of plants (after the Greek words for "leaf" and "green") by two apothecaries in Paris, *Pierre-Joseph Pelletier* (1788–1842) and *Jean Bienaimé Caventou* (1795–1877), who taught at the École de Pharmacie in 1817 (*81*).

2.5 Bilirubin and Biliverdin Separation from Bile by the Middle 19th Century

Reddish-yellow (orange) bilifulvin on the other hand was not present in sufficient amounts in 1840 for *Berzelius* to study and exhibited color changes during its separation from an alcohol solution: first brown, then green before turning brown again and precipitating as a brownish-yellow barium salt. There was no evidence that air or light were excluded in the preparation. The separated solution was treated with sugar of lead [lead(II) acetate] solution to give a dark gray-green precipitate and became orange. Then it was precipitated with vinegar of lead (aqueous solution of basic lead acetate), but it could not be precipitated so that the solution lost its color entirely. When the precipitate had sunk to the bottom, it showed a mixture of two (compounds), of which one was reddish-yellow and heavy and lay below (the other). The upper precipitated layer, a yellowish and lighter precipitate, could not be completely separated mechanically with certainty. When it was filtered, washed, and then decomposed with H_2S, a yellow solution was obtained that left a reddish-brown extract upon evaporation. It was dissolved in alcohol and the solution was left to evaporate on its own, which led to the formation initially of reddish-yellow crystals, and then, with further evaporation, a brownish-red extract formed. The crystals were the substance that *Berzelius* named *Bilifulvin* (*71*):

> 7. *Bilifulvin* habe ich eine noch problematische, aus Bilis bubula spissata erhaltene, krystallisirte, rothgelbe Substanz genannt, die ich noch nicht gehörig zu studieren Gelegenheit hatte. Nachdem die Alkohollösung der Galle mit Chlorbarium ausgefällt worden, gibt eingetropftes Barytwasser einen neuen Niederschlag, der im ersten Augenblick braun ist, aber seine Farbe verändert und grün wird, worauf er braun und am Ende braungelb niederfällt. Wird er nun auf ein Filtrum genommen und gewaschen, zuerst mit Alkohol und darauf mit Wasser, so löst sich in diesem ein grosser Theil, und auf dem Filtrum bleibt Biliverdin-Baryt zurück.
>
> Die durchgegangene Lösung, mit Bleizuckerlösung versetzt, gibt einen dunklen graugrünen Niederschlag und wird rothgelb. Nun wird sie mit Bleiessig gefällt, aber sie kann nicht so ausgefällt werden; dass sie ganz ihre Farbe verliert. Wenn der Niederschlag zu Boden gesunken ist, zeigt er sich aus zweien gemischt, von welchen der eine rothgelb und schwer ist, und zu unterst liegt. Oben darauf liegt ein nur gelblicher und leichterer Niederschlag, der jedoch nicht mit Sicherheit mechanisch abzuscheiden ist. Wenn sie abfiltrirt, gewaschen und darauf mit Schwefelwasserstoff zersetzt werden, so bekommt man eine gelbe Lösung, die verdunstet ein rothbraunes Extract zurücklässt. Wird dieses in Alkohol aufgelöst und die Lösung der freiwilligen Verdunstung überlassen, so schiessen daraus zuerst kleine rothgelbe Kyrstalle an, um welche sich dann bei fortgesetzter Verdunstung ein braunrothes Extract bildet. Diese Krystalle sind es, die ich Bilifulvin gennant habe. 24

Aside from the biliverdin-chlorophyll correlation drawn by *Berzelius*, he noted rather importantly that occasionally a yellow substance was found precipitated in bile and which he believed was responsible for producing a specific class of gallstones. *Thenard* first called attention to it much earlier and named it, descriptively, *la matière jaune de la bile*.

Much of what *Berzelius* wrote on bile, in the 1840 *Lehrbuch* (*71*), was an update of his chapters on bile expressed earlier in his 1831 *Lehrbuch*, and his 1828 *Jahres-Bericht*, and his chapter on bile in *Wagner*'s 1842 *Handwörterbuch*. But sections from these sources were also published or republished in the emerging new journals of the times (*73–76*) in part nearly verbatim, in part including updates or further

40   2 Early Scientific Investigations

comments. In his 1842 publication in the *Annalen der Chemie und Pharmacie* (*74*)
he restated his isolation of biliverdin from ox bile that was presented in detail in his
1840 *Lehrbuch* section on bile. However, to this he added his new experiments,
including those with human gallstones, which were pulverized to a reddish-yellow
powder – as *Gmelin* had reported. Processing the powder led to conversion to
green material, which he isolated as a leaf-green pigment as well as a yellow one.
The green pigment was separated by various manipulations into three green modifications and each (called *Blattgrün* = leaf green, green of leaves, or chlorophyll)
showed a different behavior toward HNO_3, becoming red, which did not occur with
the biliverdin from ox bile. It is not entirely clear why *Berzelius* used the word
Blattgrün when addressing the green pigment(s) of gallstones, and *Biliverdin* when
addressing that from bile, except that they came from different sources. In the 1842
Annalen, he used the word *Chlorophyll* and not *Blattgrün* when describing identity
with biliverdin (*74*):

> Das mit Wasser aus der Salzsäure gefällte Grüne verhielt sich zu Alkohol, Aether,
> Salzsäure, Schwefelsäure, Essigsäure und Alkalien ganz so, wie Blattgrün der ersten
> Modification, und das, was nicht durch Wasser gefällt worden war, aber durch kohlensau-
> ren Kalk niederfiel, ganz so, wie das Blattgrün der zweiten Modification. Und wenn der
> Niederschlag mit Salzsäure aus der Lösung in Kali mit Alkohol ausgekocht wurde, so blieb
> eine dunkelgrünes Pulver zurück, welches Alkohol, worin es schwerlöslich war, noch grün
> und Salzsäure gelb färbte, und welches sich also wie Blattgrün der dritten Modification
> verhielt. Bei dem aus dem Gallenstein erhaltenen Grün zeigte sich jedoch der Unterschied,
> dass es mit Salpetersäure eine rothe Flüssigkeit bildete, was mit dem aus Ochsengalle
> abgeschiedenen Biliverdin nicht stattfindet. Alle drei aus dem Gallenstein erhaltenen
> Modificationen gaben mit Salpetersäure eine rothe Flüssigkeit. Von meinen alteren
> Versuchen mit dem Blattgrün hatte ich noch eine kleine Portion von dem Blattgrün der
> zweiten Modification übrig behalten. Ich übergoss diese mit reiner Salpetersäure, und sie
> bildete damit eine tiefrothe Flüssigkeit, aber nur für einen Augenblick, worauf sie
> Stickoxydgas entwickelte und gelb wurde. Bei dem Grün aus dem Gallenstein blieb das
> Rothe viel länger.   25

More likely, *Berzelius* was being careful not to equate the green pigments (*74*):

> Ich habe im Uebrigen das nicht untersucht, was Salzsäure nicht ausfällt, da ich hier nur
> beabsichtigte, den grünen Stoff mit Biliverdin zu vergleichen, welcher aus diesem brand-
> gelben Krankheitsproducte aus der Galle verschiedener Thiere durch Alkali unter dem
> Einfluss der Luft hervorgebracht wird.   26

For he concluded that the green pigment of bile and that from (processed) gallstones was the same. Very significantly, he also concluded that they came about by
alteration (air oxidation) of the yellow pigment of bile, which he named
Cholepyrrhin (*74*) and which we now know as *Bilirubin*:

> Diese Versuche zeigen, dass der gelbe Körper, aus welchem diese Art von Gallensteinen
> besteht, in Berührung mit Luft und unter dem Einflusse von Alkalien sowohl als auch von
> Säuren metamorphosirt, und dass durch diese Metamorphose Blattgrün, so zu sagen auf
> künstlichem Wege gebildet wird; ein neues Beispiel, welches wir mehreren bereits
> bekannten hinzuzufügen haben, dass solche Stoffe, welche die lebende Natur hervorbringt,
> sich künstlich durch die Metamorphose anderer Stoffe hervorbringen lassen. Hierdurch
> betrachte ich es also als dargelegt, dass Biliverdin und Blattgrün wirklich identische Körper
> sind, und ein Product der Metamorphose des eigentlichen Farbstoffs der Galle, der

2.5 Bilirubin and Biliverdin Separation from Bile by the Middle 19th Century

während der Analyse metamorphosirt wird, ist. Dieser Farbstoff verdient einen eigenthümlichen Namen, man kann ihn *Cholepyrrhin* (von χολη, Galle, und πυρροζ, brandgelb) nennen. 27

Cholepyrrhin was thus found in fresh ox bile as well as in gallstones, which incorporated the pigment from the bile, and it was the source of the biliverdin obtained by working up bile. Likewise, *Berzelius* maintained that the taurine, cholic acid, *etc.* were also transformation products of bile, formed during the manipulations of their isolation (*74*):

> Es bleibt noch übrig, die Natur des Cholepyrrhins in unverändertem Zustande zu bestimmen, so wie auch die Producte zu untersuchen, welche sich bei seiner Metamorphose ausser dem Biliverdin bilden. Wiewohl ich bei den oben angeführten Versuchen aus Ochsengalle nicht habe denselben gelben Körper ausziehen können, welcher in den Gallensteinen enthalten ist, so ist es doch klar, dass er ursprünglich darin enthalten ist. Denn die vom Schleim abfiltrirte frische Galle zieht sich kaum merklich ins Grüne, sondern ist gelb oder bräunlich gelb. Erst während der Verdunstung wird sie dunkler und grün, indem sich dabei Biliverdin bildet und das Cholepyrrhin metamorphosirt wird. Das Biliverdin ist also, gleichwie Taurin, Cholsäure u. s. w., ein Product der Metamorphose in der Galle. 28

Although it is not entirely clear from the publication dates whether *Berzelius'* 1842 publication (*74*) in *Annalen der Chemie und Pharmacie* or whether his 1842 publication in the *Journal für praktische Chemie* (*76*) is the more recent (submission dates are not announced in either), he referred in the latter to having published an extension of his studies on bile in 1840 (*52c*), studies that he had undertaken in order to update his older *Lehrbuch, i.e.* to complete the 1840 volume 9 of the *Lehrbuch*. In 1842 (*76*), he explained that those studies could not be completed until the publication of the *Lehrbuch*, that he later continued the research and thus obtained various (or different) new views on the subject and corrected others. Those results, he said, were published in 1841 in Sweden and in 1842 (German translation) in the 1842 *Annalen der Chemie und Pharmacie* (*74*) cited above. *Berzelius* wrote that the important new results could be found in his article "Galle", printed in 1842 in *Wagner's Handwörterbuch der Physiologie* (*72*). And the main results merited publication in the *Journal für praktische Chemie* (*76*). Nowadays, these repetitive efforts in publishing might be seen as an unwarranted duplication. Nonetheless, each publication is somewhat different and, in the last cited (*76*), *Berzelius* expressed his final experiments and views on the subject of bile pigments.

Thus, the first chemical separations of the components of bile, carried out by *Berzelius* (*57–60, 68–76*), *Thenard* (*53–56, 61*), *Tiedemann* and *Gmelin* (*48*), *Demarçay* (*63*), *Chevreul* (*44, 45*) during 1806–1842, were summarized in 1842–1843 by *Liebig* (*82*) and *Thomson* (*83*). These summaries show the advances in knowledge of the composition of bile since 1835 (*84*). Whether little or much, one can judge for oneself, as did *Curran* in 1846 (*67*), who saw little progress. Nonetheless, the work served as the basis to follow in 1843–1850 by *Simon, Platner, Scherer, Heintz*, and *Virchow* who wrote on bile, either summarizing the work of their predecessor(s), while adding a few of their own studies, or describing a relationship between the pigment of urine to that of bile.

42   2 Early Scientific Investigations

It becomes apparent that from the 1830s onward various investigations of the components of bile were typically directed far more toward the isolation and composition of its non-pigment components from a wide variety of vertebrate species, as *Strecker's*[23] summaries and studies (*85, 86*) at the time indicate. In addition to fatty acids, investigations were clearly dominated by the newly discovered bile acids, their composition, properties, and transformations. Though bile acids and their conjugates were clearly a major focus, the particular individuals cited above also pursued the separation and analysis of pigments found in bile, especially from humans and cattle.

E.A. Platner (*Privatdozent* in Heidelberg in 1844) made some new observations on bile and introduced the use of stannous oxide to precipitate a bright green solid, which he freed (into *Weingeist*, ethanol) from (colorless) salts using a few drops of sulfuric acid (*87, 88*). Filtration of the green liquid and addition of water precipitated the pigment. Repeated processing freed it from fat and perhaps other material, and *Platner* ended up with the green pigment, which he described as difficultly soluble in ether, more difficult even if the ether contains alcohol, odorless, of a somewhat bitter taste, insoluble in hydrochloric acid and sulfuric acid but easily soluble in aq. potash and NH_4OH (in which the green color went over into yellow). With heating, the green faded into yellow (in unrevealed solvent). Ammonia was released upon heating it in aq. potash which, if not derived from a different compound than the bile pigment, would not be identical with chlorophyll, as *Berzelius* thought (*68–76*). According to *Platner* (*87*):

> . . . Er stellt dann eine grüne, leicht zu pulvernde, harzartige Masse dar, ist unlöslich in Wasser, aber leicht löslich in Weingeist. In Aether löst er sich schwer, und um so schwieriger, je weniger der Aether Weingeist enthält. Ist ohne Geruch und von etwas bitterlichem Geschmack. Unlöslich in Salzsäure und Schwefelsäure, aber leicht löslich in Kali und Ammoniak, wobei die grüne Farbe sich in eine gelbe umwandelt. Auch beim Erhitzen verblasst das Grüne und geht in's Gelbe über. Mit Kali erwärmt, entwickelt er Ammoniak. Wenn dieses nicht von einem anderen Körper herkömmt, so kann demnach der Gallenfarbstoff nicht identisch seyn mit dem Blattgrün, wie Berzelius meint.   29

Platner then went on to describe the isolation of bile acids and other materials in bile and ended with a few interesting remarks on the bile pigments. According to his experiments, bile in air gradually underwent the same sequence of color changes as in the *Gmelin* reaction. An ethanolic solution of bile left in air for a long time gradually went over completely from green to red, a color change that he attributed to a progressive oxidation of the pigment. Interestingly, *Platner* finished with the comment that the pigment obtained by *Berzelius* from bile using BaO contained no nitrogen, that the pigment he obtained by the method described did, and that a further examination was recommended (*87, 88*):

> Schliesslich will ich noch einige Bemerkungen über den Farbstoff der Galle machen. Die Galle wird bekanntlich durch Säuren nach und nach grün, wenn zugleich die Luft

---

[23] *Adolph Strecker* was born on October 21, 1822 in Darmstadt and died on November 7, 1871 in Würzburg. He received the Dr. phil. in 1842 at Giessen, habilitated with *Liebig* at Giessen and became lecturer, then Professor at the University of Christiana in Norway in 1851, and Professor at Tübingen following *Gmelin*'s death. He moved to Würzburg in 1870.

2.5 Bilirubin and Biliverdin Separation from Bile by the Middle 19th Century

> beitreten kann. Augenblicklich ist aber die Farbenveränderung der Galle durch Salpetersäure. Sie wird zuerst grün, dann blau, violett, und zuletzt gelb. Nachher zerstört die Salpetersäure den Farbstoff. Nach meinen Erfahrungen bewirkt jedoch der blosse Zutritt der Luft nach und nach ganz dieselben Farbenveränderungen. Setzt man nämlich eine weingeistige Auflösung der Galle längere Zeit der Luft aus, so wird sie zuerst grün, geht aber nach und nach in eine vollkommen rothe Färbung über. Diese Farbenveränderungen entstehen demnach ohne Zweifel durch eine fortschreitende Oxydation des Farbstoffs. Der von Berzelius aus der Galle mit Hülfe von Baryt dargestellte Farbstoff enthielt keinen Stickstoff. Der von mir auf die oben angegebene Weise dargestellte Gallenfarbstoff ist aber stickstoffhaltig. Möge daher auch dieser Gegenstand einer weiteren Prüfung empfohlen seyn.   30

Simon[24] wrote on the constituents of bile (*89*), following his report in 1840 (*90*) on urea from urine and components of the meconium (which is green) from children and feces from 6-day-old children nursed with mother's milk. He separated various components from the meconium (cholesterol, picromel, *etc.*) and *Gallengrün* (4%), and he also found *Gallenfarbstoff* (bile pigment) in the feces. But these were not further identified. In his section on the coloring matter of bile (*89*), for reasons unclear and unstated, *Simon* introduced a new name for the brownish-yellow pigment: *Biliphäin*, which *Berzelius* had named *Cholepyrrhin*. Despite the intrusion of a new name, *Simon*'s report on bile serves as a useful summary of what was known (*ca.* 1845) in its English translation of the original German by *George E. Day* of the Royal College of Physicians. Whether *Simon* was simply summarizing the results of previous workers (*Berzelius* in particular) or whether he repeated the experiment of *Berzelius* is not entirely clear; however, the latter seems probable (*91*):

II. THE BILE

> *a.* The most important colouring matter of the bile is that to which it owes its characteristic brownish yellow tint. It is termed *cholepyrrhin* by Berzelius, and *biliphæin* by Simon. We shall adopt the latter term. On the gradual addition of nitric acid to a fluid that contains this substance in solution, a very characteristic series of tints are evolved. The fluid becomes first blue, then green, afterwards violet, and red, and ultimately assumes a yellow or yellowish brown colour.
>
> All attempts to isolate this substance from the bile, by chemical means, have failed; it is apparently decomposed by the processes that are adopted in the analysis of this complicated fluid. We sometimes, however, find it deposited in the form of a yellow powder, in the gall-bladder, or concreted, with a little mucus, constituting a biliary calculus.
>
> In this manner we have an opportunity of examining its chemical reactions. *Biliphæin* [italics added] is of a bright reddish-yellow colour, and is only slightly soluble in most fluids; it is devoid of taste and odour, and yields ammonia on dry distillation. Water takes up an extremely minute trace of biliphæin, just sufficient to communicate a faint yellow tinge. Alcohol dissolves more than water, but only a very inconsiderable quantity. Its best solvent is a solution of caustic potash or soda, both of which are more efficient than ammonia. On exposing this solution to the atmosphere, oxygen is absorbed, and the yellow colour becomes gradually green. On the addition of an acid to this yellow or green solution, there is a precipitation of green flocculi which possess all the properties of chlorophyll, or the green colouring matter of leaves. In this state it is termed *biliverdin* by Berzelius. It is no

---

[24] *Johann Franz Simon* (1807–1848) received the Dr. phil. in 1838 in Berlin, habilitated as *Privatdozent* at the Charité, Berlin in 1842.

longer *biliphæin* [italics added] (or *cholepyrrin* [sic] [italics added]), but a product of its metamorphosis.

The colouring matter of the bile may be separated from a composite animal fluid, by evaporation to dryness; by successive extractions with alcohol of ·845, ether, and water; by dissolving the colouring matter in a solution of potash, and then precipitating it, as biliverdin, by hydrochloric acid.

Diagnosis. The action of nitric acid affords a certain test of the presence of biliphæin.

b. After the separation of the *biliphæin* [italics added], by conversion into biliverdin, another colouring matter remains, to which Berzelius has given the name of bilifulvin. It is a double salt of lime and soda, combined with an organic nitrogenous acid, to which the term bilifulvic acid has been applied. When isolated, this acid is insoluble in water and in alcohol, and separates in pale yellow flocculi when it is precipitated from an aqueous solution of its salts by a stronger acid. Whether bilifulvin is an actual constituent of the bile, or whether it is a mere product of metamorphosis, is unknown.

Simon then went on to describe *Bilin*, which *Berzelius* considered to be "the principal and most important constituent of the bile," processing it, challenging it with chemicals and eventually digesting it with dilute hydrochloric acid to separate at least five components, including what turned out (later) to be such transformation products as taurine, and what *Berzelius* called fellinic and cholaric acids, both bile acids, as it turned out, *etc.* (*89*). And in his final summary on bile, *Simon* (*91*) described what to him was then the latest writings of *Berzelius* on the subject. In his own experience, though he was able to detect urea in blood, he was never able to detect the least trace of bile pigment (or *Bilin*) in the blood of a healthy calf. From which he concluded that *Bilin* was produced and secreted only by the liver. He restated *Berzelius'* findings on bile, as complicated and containing *Bilin*, *Cholepyrrhin* (or *Biliphäin*), biliverdin, cholesterol, sodium oleate, stearate and margarate, sodium chloride, sulfate, phosphate, sodium lactate, calcium phosphate, and, of course, mucus (*Schleim*). Other investigators (*48*) would add casein, ptyalin, carbonates, *etc.* to the list. *Simon* then went on to propose a separation scheme to permit quantitation of the components, especially the bile and salts, but *Berzelius* noted earlier that many of the isolated components may have arisen due to the methods and chemicals used in the separation.

Simon's own studies advanced in 1846 (*91*) included an analysis of morbid bile from the gallbladder of a man who died in a jaundiced condition. However, the analysis did not cite bile pigments, only red and black particles in suspension. The presence of bile pigment was indicated and summarized by *Simon*, however, due to the analyses of others: bile from a man with scirrhous pancreas (by *Chevallier*), bile from death due to cholera (by *Phoebus*), and bile from a man who died in a state of icterus (by *Scherer*). He then summarized the analyses of bile of animals, from *Berzelius*, *Gmelin*, *Thenard*, himself, and others.

Scherer[25] worked on the pigments of urine and separated a green pigment from the yellow-to-brown fresh urine of jaundiced patients and identified as the green pigment from bile on the basis of its color, solubility, properties, and reaction with HNO_3 (*92*):

---

[25] *Johann Joseph Scherer* was born on March 18, 1814 in Aschaffenburg and died on February 17, 1869 in Würzburg. He was a pioneer in clinical chemistry, graduated from the University of Würzburg, and after practicing medicine from 1831-1838, he studied chemistry in Munich, then in 1840 at Giessen with *Liebig* before returning to Würzburg in 1842 as professor.

2.5 Bilirubin and Biliverdin Separation from Bile by the Middle 19th Century

> Der frische *Harn* ward zur Entfernung von Schleim und allenfalls schon ausgeschiedener Harnsäure filtrirt und hierauf mit Chlorbarium versetzt. Der erhaltene hellgrüne Niederschlag wurde sodann mit Wasser ausgewaschen, filtrirt und darauf aus demselben der Gallenfarbstoff nach zwei verschiedenen Methoden abgeschieden. 31

As further proof of the identity of his green pigment obtained from urine and the green pigment of bile, *Scherer* reported some of the early elemental combustion analyses of the two. After various manipulations of the green pigment from urine, he carried out elemental combustion analyses using lead chromate as the oxidant for producing CO_2 and H_2O, and the soda lime and hexachloroplatinate method for nitrogen analysis. He found no weighable ash, and reported the %C, H, N, O for two analyses (A and B). The data were compared with those (C) of a previously obtained sample of bile pigment (*Gallenfarbstoff*), but whether the difference in %C and %N was important or due to insufficient material could not be decided.

	A	B	C
%C	67.409	67.761	68.192
%H	7.692	7.598	7.473
%N	6.704	6.704	7.074
%O	18.195	17.937	17.261
	100.000	100.000	100.000

Scherer's early investigations of pigments from urine were later expanded, and a detailed separation method was outlined (*93*). *Scherer* assumed the right to call the pigment *Harnfarbstoff* (urinary pigment): ". . . so glaube ich mit Recht dieselben den namen *Harnfarbstoff* aufstellen zu dürfen, und werde sie dem nach der Kürze halber so benennen." He then presented the results (%C, H, N, O) of numerous elemental combustion analyses of the pigment isolated from the urine of individuals of varying degrees of health. While it is possible that *Scherer*'s *Harnfarbstoff* and biliverdin from *Gallenfarbstoff* were one and the same, the evidence at the time was only suggestive.

The French Revolution of 1789 subsequently inspired rebellion throughout the European continent, including the 39 cities of the German Confederation and led to the famous National Assembly of 1848 in Frankfurt. Though it passed a Basic Rights Law, the *Märzrevolution* failed in its purpose to meld German-speaking states, including those of the Austro-Hungarian Empire, into a German-speaking *Großdeutschland* confederation. It left behind *Kleindeutschland* under Prussian Hohenzollern leadership. The age coincided with the arrival of *Virchow*[26] who wrote a famous series of monographs (*Archiv für pathologische Anatomie und*

---

[26] *Rudolf Ludwig Karl Virchow* was born on October 13, 1821 in Schivelbein and died on September 5, 1902 in Berlin. He was known as the "Father of Pathology" and the founder of the social medicine field. He received the Dr. med. in 1843 in Berlin, was Professor at the University of Berlin until 1849, when he accepted the chair of pathological anatomy at Würzburg. In 1856 he returned to Berlin as Professor and established the *Virchow* Klinikum in eastern Berlin.

Physiologie) wherein, in 1847, he wrote an article on pathologic pigments (94). These he classified as three types: colored fats, altered or unaltered bile pigment (Cholepyrrhin), and altered or unaltered blood pigment Hämatin (hematin). Regarding Cholepyrrhin, Virchow wrote that it showed all the blendings from saffron yellow to dark brown to dark green, clearly not recognizing, as Berzelius did, that the colors represented distinct albeit related entities. From the physiological perspective, whether it was present in almost every tissue, it was found principally in the constituents of the biliary pathway. The pigment was associated with icterus and liver cells, during which condition the Cholepyrrhin collected in small, insoluble brownish or greenish grains that grouped into a nucleus (94):

> ... Cholepyrrhin zeigt alle Uebergänge von Safrangelb durch das Dunkelbraun bis zum Schwarzgrünen, und obwohl es in fast allen Geweben vorkommen kann, so findet es sich doch am häufigsten in den die Gallenwege constituirenden Elementen. Jede Stauung der Galle in ihren Ausführungswegen bedingt zunächst eine Infiltration der um die Gallengänge gelegenen Leberzellen, einen partiellen Icterus (Hft. 1. pag. 159), so dass in allen Fällen, wo der allgemeine Icterus durch Gallenstauung bedingt ist, dem Icterus des Körpers ein Icterus der Leber voraufgeht. Die Infiltration der Leberzellen mit Cholepyrrhin ist zuerst eine gleichmässige, diffuse; sehr bald sammelt sich aber der Farbstoff in kleine, unlöslich, bräuniche oder grünliche Körper, die sehr häufig gruppenweise neben dem Kern liegen.   32

Virchow's article is written mainly from the perspective of liver pathologies and less from the chemical perspective, unlike the 1847 short paper (95–97) by Heintz.[27] Heintz addressed (95–97) what became known as the Gmelin reaction, in which bilirubin underwent a characteristic progression of color changes. He qualified it by indicating that HNO_3 did not produce the color change in every case in the presence of the components of bile. What had to be taken into account was that the reaction occurred not with the characteristic principal component of bile but with Gallenbraun, which Simon named Biliphäin in comparison to Berzelius' Cholepyrrhin. If HNO_3 brought about no color change, it was strongly assumed thereby that only the absence of Gallenbraun was proved, but not the absence of any other components of bile. However, the color change that HNO_3 caused in fluids that contained Gallenbraun was therefore at least a firm characteristic sign for the presence of this substance (96):

> Es ist bekannt, dass die Salpetersäure viel gebrauchtes Reagens ist, um die Gegenwart der Galle, in irgend einer Flüssigkeit nachzuweisen. Man giebt an, dass solche Flüssigkeiten dadurch zunächst grün, dann blau, violett, roth und endlich gelb gefärbt werden, und es ist diess in den meisten Fällen ganz richtig. Allein nicht in allen Fällen bewirkt die

---

[27] Wilhelm Heinrich Heintz was born on November 4, 1817 in Berlin and died on December 1, 1880 in Halle. He studied pharmacy in Berlin in 1840, was promoted to the doctorate in 1844 working under Heinrich Röse, habilitated in 1846 as Privatdozent at the Charité in Berlin and became a. o. Professor at the University of Halle-Wittenberg, then in 1856 o. Professor für Chemie und Pharmazie in 1855, and finally director of the pharmaceutical institute there. At Halle he supervised the doctoral work of Johannes Wislicenus, whose pro-forma advisor in Zurich was Georg Karl Andreas Städeler (Sections 2.8 and 2.9.1). He was the only chemist among the six founding members of the Deutsche Physkalische Gesellschaft.

2.5 Bilirubin and Biliverdin Separation from Bile by the Middle 19th Century

Salpetersäure bei Gegenwart von Gallenbestandtheilen jene Farbenveränderung. Zunächst muss berücksichtigt werden, dass, jene Reaction nicht durch die eigentlich wesentlichen Bestandtheile der Galle veranlasst wird, sondern durch das Gallenbraun, welches Simon Biliphäin, Berzelius dagegen Cholepyrrhin nennt. Wenn man also durch Salpetersäure keine Farbenveränderung hervorbringen kann, so ist, streng genommen, dadurch nur die Abwesenheit des Gallenbrauns, aber nicht die jener wesentlichen Gallenbestandtheile erwiesen.

Allein jene Farbenveränderung welche Salpetersäure in Flüssigkeiten hervorbringt, die Gallenbraun enthalten, bliebe doch wenigstens ein sicheres Kennzeichen für die Gegenwart dieses Stoffs, wenn sie wirklich in jedem Fall einträte. 33

There is a certain logic to *Heintz*'s statements, if it actually occurred in every case. Yet it is unclear what problems this short paper clarified. In the various occasions when the *Gmelin* reaction had been employed from the time of its postulate by *Tiedemann* and *Gmelin* (*48*) in 1826 to 1847, it has been generally conceded that the color change reaction is diagnostic of bilirubin.

By the exact middle of the 19th century, the chemistry of bile pigments could be summarized rather simply. Bile of the mammals studied (man, ox, dog, *etc.*) was typically yellow, and the yellow pigment *Gallenfarbstoff* (a bile pigment) was difficult to separate by the available manipulations of the time, which were mainly extractions, precipitations as lead, barium, or tin salts, and washings. The liquids involved were typically water, ethanol, and ether. The pigment morphed easily into green (*Gallengrün*) during the separation, a color change that required (oxygen of) air. While this color change seemed to establish a 1:1 relationship between the yellow and green pigments, not all investigators agreed. A relationship between the pigments of bile and urine was investigated, with uncertain conclusions, while the relationship with the pigment of blood was an open conjecture. The *Gmelin* color reaction stood out as a sensitive analytical diagnostic for the yellow pigment. Combustion analyses were beginning to be employed, from which the %C, H, and N were determined for the bile pigment samples. However, the measurements were compromised by the formation of ash, which indicated impure samples. Even when the combustion left no ash, the samples were found to give different results after standing in air for a few days. Thus, *Lehmann* summarized the status of bile pigments in 1850 (*98*):

Gallenfarbstoff.

Chemisches Verhalten.

Eigenschaften. Dieser Stoff gehört, wie so viele Farbstoffe, zu den chemisch noch sehr wenig erforschten Gegenständen; diess liegt theils daran, dass man sich denselben nur in sehr geringer Menge verschaffen kann, theils an seiner grossen Wandelbarkeit, indem er nicht nur im thierischen Organismus bereits unter verschiedenen Modificationen vorkommt, sondern auch bei der einfachsten chemischen Behandlung sich bereits umändert. Die gewöhnlichste Modification, welche auch die Ursubstanz der Gallenpigmente in den höhern Thieren zu sein scheint, ist das sog. *Gallenbraun, Cholepyrrhin (Berzelius) Biliphäin (Fz. Simon)*. Dasselbe bildet ein rothbraunes, nicht krystallinisches Pulver, ohne Geschmack und Geruch, löst sich nicht in Wasser und sehr wenig in Aether, besser in Alkohol, der sich dadurch gelb färbt, in Aetzkali aber leichter noch als in Aetzammoniak; die hellgelben alkalischen Lösungen werden an der Luft allmälig grünlichbraun. Diese Modification des Gallenpigmentes ist es, von der die bekannten Farbenveränderung mancher gefärbter, thierischer Flüssigkeiten abhängen. Die gelbe Lösung dieses Pigments wird bei allmäligen Zusatz von *Salpetersäure* (besonders wenn diese etwas salpetrige Säure enthält, *Heintz*) anfangs grün, dann blau (welches jedoch kaum bemerkbar ist, seines schnellen Uebergangs wegen in Violett) und roth; nach längerer Zeit geht die rothe Farbe wieder in eine gelbe über; dabei ist jedoch der Gallenfarbstoff völlig verändert. Durch *Salzsäure* wird derselbe aus der Kalilösung grün gefällt; dieser Niederschlag löst sich Salpetersäure mit rother, in Alkalien mit grüner Farbe auf, und scheint dadurch volkommen in die grüne Modification des Gallenpigments überzugehen. Der in frischer Galle enthaltene Farbstoff wird durch Säuren grün gefärbt; *Gmelin* fand, dass diese Färbung ohne Sauerstoffzutritt nicht statt finde; es ist daher höchst wahrscheinlich, dass die meisten jener Farbenveränderungen auf einer allmäligen Oxydation beruhen. Chlorgas wirkt auf dieses Pigment gleich der Salpetersäure, nur etwas schneller; grössre Mengen von Chlor bleichen den Farbstoff vollkommen und schlagen weisse Flocken nieder.

Dieses braune Pigment ist sehr geneigt, sich mit Basen zu verbinden, und zwar nicht blos mit Alkalien, sondern auch mit Metalloxyden und alkalischen Erden; auch mit letztern bildet es unlösliche Verbindungen, weshalb man den Stoff selbst oft für unlöslich gehalten hat.

Das *Gallengrün, Biliverdin (Berzelius)* is eine dunkelgrüne, amorphe Substanz, ohne Geruch und Geschmack, unlöslich in Wasser, in Alkohol wenig, in Aether mit rother Farbe löslich; Fette, Salzsäure und Schwefelsäure lösen es mit grüner, Essigsäure und Alkalien mit gelbrother Farbe auf. Beim Erhitzen wird dieser Körper ohne zu schmelzen und ohne merklich Ammoniak zu entwickeln unter Zurücklassung weniger Kohle zersetzt. *Berzelius* hält diesen Stoff für völlig identisch mit dem Chlorophyll der Blätter und glaubt auch alle 3 Modificationen des Chlorophylls in verschiedenen Gallen gefunden zu haben. Dieses grüne Pigment hat nicht mehr die Eigenschaft durch Salpetersäure Farbveränderungen zu erleiden; indessen findet man auch zuweilen grünliche Gallenpigmente, welche noch jene Eigenschaft besitzen. Meist schon nach der Behandlung mit Alkalien oder Säuren zeigt das Pigment der Galle andre Eigenschaften als der ursprüngliche Körper, theils wohl weil er mit diesen Stoffen selbst verschiedene Verbindungen eingeht, theils aber auch, weil er sich so leicht modificirt.

Aus diesem Grunde sind die Angaben über die Eigenschaften dieser Stoffe so verschieden; man vergleiche *Berzelius* [1]), *Scherer* [2]), *Hein* [3]), *Platner* [4]) und Andre.

Berzelius fand in der Galle auch einen Alkohol lösliche, in kleinen rothgelben Krystallen auschiessenden Stoff, den er *Bilifulvin* nennt. Ich habe denselben nur in Lösung, aber nicht in fester Gestalt erhalten können; auffallender Weise fand ich ihn oft in der mit neutralem und basisch essigsaurem Bleioxyd ausgefällten Galle, so dass er also durch diese Metallsalze nicht gefällt oder vielmehr im Ueberschuss des basischen Salzes wieder aufgelöst zu werden scheint.

Zusammensetzung. Bei unsrer Unbekanntschaft mit dem reinen unveränderten Gallenfarbstoffe ist es nicht zu verwundern, dass seine elementare Zusammensetzung noch nicht bekannt ist. *Scherer* und *Hein* haben Gallenpigmente untersucht; allein es geht aus ihren Analysen hervor, dass sie sehr verschiedene Substanzen unter den Händen gehabt haben, und *Scherer* hat insbesondre gezeigt, dass das Gallenpigment durch Einwirkung von Luft, Alkalien und Säuren viel Kohlenstoff und Wasserstoff verliert. Man hat übrigens 7 bis 9% Stickstoff in dem Gallenpigmente gefunden.

Darstellung. Früher empfahl man gewöhnlich zur Darstellung des Gallenfarbstoffs, die aus solchem vorzugsweise bestehenden Gallenconcremente mit Wasser und Aether zu extrahiren; der Rückstand hat aber in der Regel nicht die oben angegebene Eigenschaft, sich in Alkohol zu lösen, da er mit Kalk in unlöslicher Verbindung ist (wie *Bramson* [1] [*Bramson*, Zeitschr. f. rat. Med. Bd. 4, S. 193–208] ganz richtig angegeben hat und jeder vorurtheilsfreie Beobachter sich leicht überzeugen kann), selbst in solchen Concrementen, die grösstentheils aus Cholesterin bestehen.

Die *Bramson*'sche Untersuchungsweise, die ich oft widerholt habe, scheint mir gar keinen Zweifel an der Richtigkeit seiner Ansichten übrig zu lassen; übrigens stimmen damit auch die Gallensteinanalysen von *Schmid* [2] [*Schmid*, Arch. der Pharm. Bd. 42, S. 291–293]) und *Wackenroder* [3] [*Wackenroder*, ebendas. S. 294–296]) überein.

2.5 Bilirubin and Biliverdin Separation from Bile by the Middle 19th Century

Berzelius stellt das Biliverdin aus der Rindsgalle dar, indem er den alkoholischen Auszug derselben mit Chlorbaryum fällt; der Niederschlag wird erst mit Alkohol, dann mit Wasser ausgewaschen und durch Salzsäure zerlegt, welche den Baryt auszieht; der Rückstand wird durch Aether von Fett befreit und dann in Alkohol gelöst.

Platner fällt den Gallenfarbstoff aus der Galle durch Digestion derselben mit Zinnoxydulhydrat; dieses bildet damit einen hellgrünen Niederschlag, der nach gehörigem Aussüssen mit Wasser mit schwefelsäurehaltigem Weingeist geschüttelt wird; aus der filtrirten grünen Lösung wird durch Wasser der Farbstoff in grünen Flocken gefällt.

Scherer schied aus gallenfarbstoffhaltigem Harn den Farbstoff durch Chlorbaryum aus, stellte ihn aber daraus auf 2 Wegen dar: entweder zerlegte er die Barytverbindung mit kohlensaurem Natron, und schlug aus der Natronlösung das Pigment durch Salzsäure nieder, wo es dann durch Auflösen mit ätherhaltigem Alkohol, Auswaschen mit Wasser u. s. w. gereinigt wurde, oder die Barytverbindung ward mit salzsäurehaltigem Alkohol extrahirt, die Lösung verdunstet, mit Wasser extrahirt und dann wie oben behandelt.

Prüfung. Ist die Gegenwart von Gallenfarbstoff in einer Flüssigkeit nicht zu gering, so giebt Salpetersäure, namentlich wenn sie etwas salpetrige Säure enthält, das oben erwähnte sehr characteristische Farbenspiel. Bei kleinen Mengen von Farbstoff giebt jedoch die Salpetersäure oft keine recht deutliche Reaction, so wie auch dann, wenn das Pigment schon zum Theil modificirt ist. *Schwertfeger* [1] [*Schwertfeger*, Jahrb. f. prakt. Pharm. Bd. 9, S. 375] empfiehlt in solchen Fällen die Flüssigkeit mit basisch essigsaurem Bleioxyd zu fällen, und den Niederschlag mit schwefelsäurehaltigem Alkohol zu extrahiren; dieser färbt sich bei Gegenwart des Pigments grün. *Heller* [2] [*Heller*, Arch. f. phys. u. pathol. Ch. Bd. 2, S. 95] räth der zu untersuchenden Flüssigkeit lösliches Eiweiss zuzusetzen, sobald sie nicht schon solches enthält, und dann durch überschüssige Salpetersäure zu präcipitiren; das coagulirte Eiweiss ist dann durch das Pigment bläulich oder grünlich blau gefärbt. Nach *Heller* bildet sich auf vorsichtigen Zusatz von Ammoniak zu Harn, der bereits umgewandeltes Gallenpigment enthält, wenn man nicht umschüttelt, auf der Oberfläche der Flüssigkeit eine rothe Scheibe.

Physiologisches Verhalten.

Vorkommen. Der Gallenfarbstoff findet sich in frischer Galle gewöhnlich aufgelöst vor, doch oft auch nur aufgeschlemmt; fast immer bildet er die Kerne zu Gallensteinen; zuweilen findet man auch ästige, knotige Concremente in der Gallenblase und den Gallengängen, die fast nur aus Gallenfarbstoff bestehen. Diesen Gallenfarbstoff hat man nicht blos in der Galle des Menschen und der Rinder gefunden, sondern auch in der andrer fleisch- und pflanzenfressender Thiere, jedoch in den verschiedensten Modificationen, wie schon die verschiedene Färbung der Galle nicht nur verschiedener Genera, sondern selbst verschiedener Individuen derselben Species lehrt; so ist die Hundergalle gelbbraun, die Rindsgalle bräunlich grün, die Galle der Vögel, Fische und Amphibien meist smaragdgrün. 34

In an age when no organic chemical structures could be proposed – or even imagined – there was a flowering of names for the pigments: *Gallenbraun, Cholepyrrhin* (by *Berzelius*) and *Biliphäin* (by *Simon*) for the yellow pigment; *Gallengrün* and *Biliverdin* (by *Berzelius*) for the green (*98*); and *Bilifulvin*, which *Berzelius* isolated from ox-bile as reddish-yellow (or orange) crystals and which were subsequently shown to give a positive *Gmelin* test. Soon to follow were two additional names for the yellow, and a scathing rebuke of the name *Biliphäin* by *Legg* (*99*):

The name cholepyrrhin is commonly said to have been given by Berzelius to the orange red pigment of the bile. F. Simon invented a barbarous word biliphæin, compounded of Greek and Latin, the use of which has been unfortunately endorsed by Heintz. Dr. Thudichum uses the word cholophæin, to avoid this bastard word. Städeler called this pigment bilirubin,

50 2 Early Scientific Investigations

forming the word with a cognomen like the other names which Berzelius used; biliverdin, bilifulvin, and the like. Maly has continued the use of the name cholepyrrhin.

Had *Legge* issued his rebuke some 20 years earlier, he might have persuaded investigators of the middle 1800s who followed *Simon* to drop the name. Yet *Biliphäin* persisted and, as will be noted in the following section, it was applied by *Heintz* to what was then the most highly purified bilirubin, apparently free from salts. Yet it eventually fell by the wayside, as did *Cholephäin*, in favor of the name *Bilirubin*. By the middle of the 19th century, coinciding nearly with the onset of the longest reigning (1848–1916) monarchy of Austria and its penultimate emperor *Franz Joseph I* (1830–1916) of the House of Habsburg-Lothringen, investigations of bile as a source of bile pigments began to wane, although not vanishing entirely (*100*). Rather, bile became the focus of studies directed toward its component bile acids and fatty acids (*85, 86, 101*). A more tractable source of the pigments, especially the yellow pigment, was proving to be gallstones, concretions found in the gallbladder or bile duct, as mentioned above in Section 2.1 and at the end of Section 2.4. For we now know that fresh bile contains little, if any, bilirubin, but numerous conjugates, such as bilirubin glucuronides, that are labile, sensitive to acid and base hydrolysis – and are poorly soluble in $CHCl_3$.

2.6 Bile Pigments from Gallstones and Urine, and Their Combustion Analyses during the late 18th to mid-19th Centuries

Gallstones, or concretions found in the gallbladder and in the bile duct, were recognized as such centuries ago (*102, 103*). *Alexander Trallianus* (*Alexander of Tralles*, 525–605), the famous Greek physician mentioned concretions in the liver in his *Twelve Medical Books*, which were lost for a thousand years, then rediscovered and published in Paris in the year 1548 together with a Latin translation by *Stephanus*. Shortly thereafter, in 1549, *J. G. Andrenacus* dedicated a translation to *Thomas Cranmer* (1489–1556), Archbishop of Canterbury, 1532–1534, during the reigns of *Henry VIII* and *Edward VI*. In his translation of the second chapter of *Alexander*'s eighth book, which treats obstructions of the liver, it is found:

Nam humores nimium exiccati assatique, lapidum instar concreverunt, adeo ut non amplius discuti potuerint. 35

referring to dried up humours, concreted like little stones and the cause of obstructions. However, even with gallstones having been found in humans and animals, and used as pharmaceuticals during the years following publication of the translation dedicated to *Cranmer*, little was known of their composition. They were described in detail regarding origin, size, macroscopic geometric structure, whether they floated on water, combustibility, texture, and color, which varied from white to livid yellow to green to reddish to blackish, in the case of humans, depending on the health of the individual. Gallstones were described as being friable and having

2.6 Bile Pigments from Gallstones and Urine...

concentric (colored) layers. Yet from the middle of the 16th century and not until the 18th century does it appear that investigations of gallstones, though numerous, told much about the pigments contained therein. The investigations were largely of a medical and morphological nature for gallstones obtained from humans and wide variety of animals (*101–103*).

Coe (*29*), in 1757, was probably the first to publish a comprehensive monograph on the anatomical, clinical, and physical aspects of gallstones. He assumed that the concretions were formed in the gallbladder or bile ducts due to stagnation and inspissation from stopping or retarding the flow of bile. His comments on them did not go much beyond attempts to dissolve them in water and alcohol (see Section 2.1). Earlier in the 18th century, chemical investigations of gallstones were reported by *Vallisneri* (*34*) who described dissolving what must have been cholesterol gallstones in an alcohol-turpentine mixture, and other investigators [*Galletti* in 1748, *Haller* in 1764 (*35*)] reported on their dissolution, distillation and flammability. So that by the late 1700s, at the onset of the French Revolution (1789–1799), *Fourcroy* (*42*) was able to publish what may have been the first separation of pigments from colored gallstones, which had been pulverized and warmed in alcohol to dissolve steroids and other substances but which also leached out yellow-green pigments. And at the beginning of the 19th century *Thenard*, working with both bile and gallstones, found that yellow stones became green when exposed to air and that caustic alkali extracted a yellow pigment that gave a green precipitate upon acidification – thus achieving a partial separation of the bile pigment from gallstones (*53–56*) – and thereby providing an experimental link between the pigment of bile and that of gallstones.

Recall that one of *Thenard*'s most remarkable discoveries was nearly half a kilogram of a water-insoluble yellow powder in the bile duct of an elephant that died in the Paris zoo (*61, 62*) in the 1820s. In the same decade when *Heinrich Heine* (1797–1856, one of Germany's most famous romantic poets) wrote the play *Almansor* in 1821 [with its famous line referring to the burning of the Qur'an during the Spanish Inquisition: "Dort, wo man Bücher verbrennt, verbrennt man am Ende auch Menschen" (portending the sequelae to the book burnings of 1933 Nazi Germany)], *Gmelin* in Heidelberg also achieved a separation of the yellow pigment of gallstones. This he accomplished by taking the pigment up into ammonium hydroxide, and found that the solution became green in air and precipitated in green flakes upon acidification (*48*). *Gmelin* not only demonstrated that the color change from yellow to green was due to oxidation, he also wrote of the series of color changes that accompanied the yellow pigment upon treatment with nitric acid – which became a standard analysis for the presence of the yellow pigment.

Fourcroy, *Thenard*, and *Gmelin* were not the only investigators of the pigment components of gallstones and bile during 1785–1826. In the mid 1830s, a *Monsieur Dr. Loir* in Paris also analyzed gallstones and classified them as: (A) those of pure *cholésterine* (cholesterol), (B) stones of cholésterine mixed with colored material, and (C) stones formed of colored material alone (nonflammable) (*64, 65*). He indicated that *Thenard*'s picromel could be obtained from dried bile and describes brown-black, dark-green, sea-green, yellow and brick-red as colors of and within

52   2 Early Scientific Investigations

concretions. And he concluded that the colored substance of bile appeared to be the only component, or the major principle of gallstones.

In the 1840s, *Berzelius (74)* described a method for isolating the pigments of gallstones, along with cholesterol and other substituents; however, the preparations were crude, hardly pure, and the isolation procedure tended to convert the yellow pigment at least partially to green. However, he was able to assert that the green pigment obtained from gallstones and bile was the same – and that both arose from oxidation of the yellow pigment (see Section 2.5).

In the same decade, *Scherer (104)*, *Hein (105)*, and *Bley (106)* published their findings on gallstones. *Scherer* reported finding a black-brown pigment in black gallstones that he purified by heating in ether, alcohol, and water (presumably separately). The final product gave ash upon incineration (41.79% C as $CaCO_3$ ash) and the following elemental analysis from combustion:

%	I	II
C	73.237	73.212
H	6.306	6.313
N	14.434	14.434
O	6.023	6.041

The %N was very high compared to later analyses of the pigments of gallstones.

In 1847 *Bley (74)* isolated what he believed to be *Biliphäin* from the gallstones of a deceased 60-year-old woman. (*Biliphäin*, the word coined by *Simon (89)* and rejected by *Legg* as bastardized (*99*) was rapidly picked up by *Heintz (95–97)* and used by him (*97*) for the purified *Gallenfarbstoff/Gallenbraun*.) Analysis showed the gallstone to be 96% cholesterol, with 4% a mixture of *Biliphäin, Gallenfarbstoff*, and gallbladder mucus. *Bley* noted that the stone contained both brown and reddish spots and a pale yellow core. After being pulverized it was yellowish in color and was extracted with 84% ethanol to leave a small brown residue. The last, upon treatment with aqueous KOH to effect dissolution, gave a positive *Gmelin* reaction. *Bley*'s was clearly a very qualitative experiment.

In contrast to *Bley*, *Hein (105)* gave a much more detailed analysis of gallstones, having at his disposal a large number of them, which he divided according to their color and (quantity) from ocher-brown (5), to black-brown (1), to brown (11), to yellow-gray (2), to yellow (15), to brown green (21). Samples were air-dried to remove water, burned to determine the amount of ash, and extracted with boiling ethanol to remove cholesterol and saponifiable fat. The last ranged from 8% to 85% by weight, with the black stone having the least. Following removal of the hot alcohol, a solid brown residue was obtained from every sample. It resembled the brown powder residue seen by *Berzelius* in a similar processing of gallstones following successive treatment with water, ethanol, and ether. *Hein* chose not to dissolve the brown residue in aqueous potash, as *Berzelius* reported (*72*), but to effect a partial dissolution using the weaker base ammonium hydroxide, to which he added hydrochloric acid to throw down a "sehr schön grünen Niederschlag" [very beautiful green precipitate], which fit the properties of biliverdin. The undissolved

2.6 Bile Pigments from Gallstones and Urine...

brown residue (*Gallenbraun*) from ammonium hydroxide extraction corresponded to *Cholepyrrhin*. Thus *Hein* had apparently achieved at least a partial separation of biliverdin from bilirubin, but the samples were of questionable purity and the former may well have arisen from the latter during all of the extractions carried out in air. Indeed, the final *Gallenbraun* corresponding to *Cholepyrrhin*, when dissolved in aqueous potash, soon turned a lovely green color and, with added hydrochloric acid, yielded a dark green precipitate. The filtrate, still so deeply colored that light transmitted through it only at the edges, iridesced red-green. Redissolving the green precipitate in ammonia and reprecipitating with hydrochloric acid restored more brown than green. The brown material, heated in hydrochloric acid, dissolved as bright yellow, and the solution saturated with NH_3 was colored violet. Added hydrochloric acid threw down a very similar green precipitate. The brown material gave the characteristic *Gmelin* test and was submitted to combustion analysis, whose results are shown in Table 2.6.1. Thus a comparison of *Hein*'s first analysis (*Hein*-1) to *Scherer*'s brown pigment from gallstones (*Scherer*-1A, 1B) differ hugely in the %C, N, and O and especially in the amount of ash, with the latter being much richer. The percentages were determined after subtracting the ash content.

Table 2.6.1 Elemental combustion analysis data of *Hein*'s (*105*) *Cholepyrrhin* compared with *Scherer*'s (*92, 93*) *Gallenfarbstoffe*

%	Hein-1	Hein-2A*	Hein-2B**	Scherer-1A	Scherer-1B	Scherer-2[†]	Hein-3A	Hein-3B
C	69.68	67.96	68.13	73.237	73.212	62.491	58.26	58.5
H	7.60	6.21	6.44	6.306	6.313	6.148	6.30	6.29
N	8.84	9.94	9.94	14.434	14.434	8.169		
O	13.88	15.89	15.49	6.023	6.041	23.122		
Ash	9.326	4.301	4.33	41.79	41.79	[†]	9.326	

*   *bei der Verbrennung im Platinschiffchen* (from combustion in a Pt boat)

**   *bei der Verbrennung zwischen Kupferoxyd* (from combustion mixed with CuO)

[†]   *ohne Zahl für die Asche* (without accounting for ash)

In an attempt to reduce the amount of ash in the brown pigment, *Hein* processed his material by repetitive partial dissolution of *Gallenbraun* in ammonia or aqueous Na_2CO_3 followed by precipitation with HCl, with or without heating, treatment with alcohol, precipitation with basified acetic acid-lead oxide, then $BaCl_2$ induced precipitation followed by treatment with H_2SO_4, *etc.* – a longish, circuitous but historic processing of doubtful effect. Though the process reduced the amount of ash by roughly one-half and altered the %C, H, N, O somewhat (*Hein*-2A, 2B), the combustion analyses were still found lacking. Careful processing of yet another gallstone, after careful drying under vacuum with heating gave a much lower %C, H for ash-free material (*Hein*-3A, 3B). It thus seemed to *Hein* that the brown substance was not pure because it exhibited a variable composition depending on the method of isolation.

Hein also subjected the green material to combustion analysis (*Hein*-4, Table 2.6.2). This material, too, contained ash, but there was too little material for a nitrogen analysis. These results may be compared with those from *Scherer*'s (*92*) green pigment that was isolated from fresh urine (*Scherer*-3A, 3B) and gave no weighable ash. Pigment from the same source, but which *Scherer* isolated using a different method gave slightly different results (*Scherer*-3C). As *Scherer* noted earlier, after the green pigment had undergone changes during long exposure to air, acid (*Scherer*-4A), and alkali (*Scherer*-4B), the %C and %H dropped considerably. The considerable variability in the analytical results is further illustrated in *Scherer*'s work (*92*) in which he stated in 1843 that the green pigment isolated from gallstones exhibited a high %C (*Scherer*-5) due apparently to a significant non-combustible component (as evidenced by the amount of ash). *Scherer* then reprocessed a small number of the same gallstones (apparently) to obtain an ash-free pigment, and he performed a new combustion analysis to give new results (*Scherer*-6) that were similar in %C and %H to those obtained from urine in *Scherer*-4A, 4B.

Table 2.6.2 Elemental combustion analysis data of *Hein*'s (*105*) biliverdin compared with *Scherer*'s (*92, 93*) *Gallengrün*

%	*Hein*-4	*Scherer*-3A	*Scherer*-3B	*Scherer*-3C	*Scherer*-4A	*Scherer*-4B	*Scherer*-5	*Scherer*-6
C	65.5	67.409	67.761	68.192	61.837	62.086	74	62.491
H	6.62	7.692	7.598	7.473	6.464	6.567	6.3	6.148
N		6.704	6.704	7.074	9.080	7.101	14.4	8.169
O		18.195	17.987	17.261	22.619	24.246		23.192
Ash	7.833							

It is clear from the studies of *Scherer* and *Hein* that just as elemental combustion analyses began to be performed somewhat routinely, so too were the pigments of gallstones, bile, and urine subjected to such analyses in the 1840s following isolation. Evidently, the isolation procedures did not always produce organic compounds entirely free of their salts or inorganic impurities, which led to the findings of ash left after incineration or combustion. Just how one factored out the ash may have led to the variable results seen in Tables 2.6.1 and 2.6.2. Such results did not serve to confirm that the green pigment isolated from gallstones, urine, or bile was in fact the same green pigment, or that the brown pigment isolated from the same sources was in fact the same compound. But clearly, by the middle of the 19th century combustion analysis had been introduced to bile pigments, and as such analyses continued during the following 50–75 years, the need for better methods of isolation and purification of the pigments became evident.

Shortly after the isolation and elemental combustion analyses of bile pigments from gallstones published by *Hein* (*105*) and *Scherer* (*104*), and from urine by *Scherer* (*92*), *Heintz* (*97*) studied the ash content of gallstones and bile pigments (*Gallenbraun*) isolated therefrom, noting the work of others (*Bramson, Schmid, Wachenroder*, and *Bley, Bolle, Scherer*, and *Hein*) and their analyses of the residue (ash) after combustion or burning. The ash was composed largely of calcium salts, especially CaO and $CaCO_3$, along with some $MgCO_3$. Isolating *Gallenbraun* from gallstones, *Heintz*, too, found considerable ash (9.41–9.91%) and considered that it

2.6 Bile Pigments from Gallstones and Urine...

arose from impure pigment in which the calcium was bound not to carbonate but to the pigment itself. Repeating the experiment carefully with a different source of gallstones, *Heintz* concluded (97): "... dass in der That in dem rohen Gallenbraun wenigstens ein Theil des Farbstoffs mit Kalkerde verbunden ist" [that in crude *Gallenbraun* at least part of the pigment is bound to CaO].

He then sought to prepare the unbound pigment with an altered isolation procedure while taking special precautions to protect from air (oxidation) during each step of the manipulations (97):

> Nach den Versuchen von Gmelin geschieht diese Umänderung auf Kosten des Sauerstoffs der Luft. Um daher das Biliphäin, im reinen Zustande zu erhalten, muss jene Auflösung und Fällung in einem Raume geschehen, der keinen Sauerstoff enthält.   36

First, in an H_2-blanketed flask, crude *Gallenbraun*, previously washed (agitation) by hydrochloric acid and water, was dissolved in aqueous Na_2CO_3 as completely as possible by heating for a long time; then, it was rapidly filtered before dilute HCl was introduced into the dark brown-black filtrate, with CO_2 evolution, to separate dark brown flakes of *Gallenbraun* precipitate. The clear acidic supernatant was removed and allowed to stand, which precipitated pure *Gallenbraun* after washing with hot water and drying in air. The last, which *Heintz* called *Biliphäin*, possessed a dark brown color, tending toward olive green (97):

> Nachdem dieser Apparat zusammengestellt war, wurde mit der Wasserstoffgasentwickelung begonnen, und nachdem so viel dieses Gases entwickelt worden war, dass man annehmen konnte, auch in der im Kolben befindlichen kohlensauren Natronlösung befinde sich kein Sauerstoff mehr, wurde der Kolben geöffnet und schnell das mit Salzsäure und Wasser ausgewaschene rohe Gallenbraun hineingeschüttet. Nachdem sofort der Pfropfen wieder auf den Kolben gesetzt worden war, liess man mehrer Stunden Wasserstoffgas durch den Apparat strömen, bis auch in der Glocke sich kein Sauerstoff mehr befinden konnte. Darauf wurde die Natronlösung längere Zeit erhitzt, während der Gasstrom noch immer fortdauerte, und nachdem die Auflösung des Gallenbrauns möglichst vollkommen erreicht worden war, wurde das Rohr, welches die Gase aus dem zur Auflösung dienenden Kolben ableitete, so tief in diesen gesenkt, dass der Gasstrom die Gallenbraunlösung in die Glocke übertreiben musste. Hier wurde sie von dem darunter befindlichen Filtrum aufgenommen, und die klar davon abfliessende dunkelbraunschwarze Flüssigkeit filtrirte unmittelbar in die verdünnte Salzsäure hinein. Unter Kohlensäureentwicklung geschah die Zersetzung. Das Gallenbraun fiel in dunkelbraunen Flocken nieder. Nachdem die ganze Menge der Lösung auf diese Weise in die verdünnte Salzsäure klar abgeflossen war, wurde die Flasche herausgenommen, schnell umgeschüttelt, und mit einem Glaspfropf verstopft einige Zeit stehen gelassen, bis sie sich geklärt hatte. Der so erhaltene Niederschlag zieht nun nicht mehr so leicht Sauerstoff aus der Luft an, als seine Lösung. Er kann an der Luft mit heissem Wasser ausgesüsst werden.
>
> Im getrockneten Zustande bilden das reine Gallenbraun, welches ich Biliphäin nennen will, eine dunkelbraune, etwas ins Olivengrün ziehende Farbe.   37

Heintz described the solubility properties of his *Biliphäin* and indicated that in ammonia it forms brown precipitates, a calcium salt from $CaCl_2$ and a barium salt with $BaCl_2$. Dissolved in dilute alcoholic KOH, the solution rapidly turned green upon acidification by hydrochloric acid; then it changed to a beautiful blue color upon dropwise addition of HNO_3. Treatment of a dilute alkaline solution of *Biliphäin* with excess HNO_3 containing some aqueous HNO_2 gave the *Gmelin* color reaction characteristic of *Cholepyrrhin* (and bilirubin). Elemental combustion

56 2 Early Scientific Investigations

analysis of *Biliphäin* revealed essentially no ash and the %C, H, N shown in Table 2.6.3. According to *Heintz*, the $C_{31}H_{18}N_2O_9$ (formula wt 562) molecular formula was a better fit than $C_{32}H_{18}N_2O_9$ (formula wt 574). However, the molecular weight of *Biliphäin* could not be determined experimentally at the time.

Table 2.6.3 Elemental combustion analyses of *Heintz*'s purified *Biliphäin* (I, II, III, IV) and the corresponding biliverdin (V) and calculated values for suggested molecular formulas (97)

%	I	II	III*	IV*	$C_{32}H_{18}N_2O_9$	$C_{31}H_{18}N_2O_9$	V	$C_{16}H_9NO_5$
C	60.70	60.71	61.06	61.03	61.94	61.18	60.04	60.38
H	6.05	6.02	6.09	6.06	5.80	5.92	5.84	5.66
N			9.12	9.12	9.03	9.21	8.53	8.80
O			23.73	23.79	23.23	23.69	25.59	25.16
Ash	0.29	0.37	0.045	0.079			0.11	

* *Biliphäin* from *Gallenbraun* separated from gallstones provided by Hrn. Dr. *R. Virchow*

The green product (*Gallengrün*/biliverdin) from air oxidation of *Biliphäin* was isolated and purified, and its elemental combustion analysis was obtained for the essentially ash-free pigment. *Heintz* compared the data (Table 2.6.3) to those corresponding to a molecular formula $C_{16}H_9NO_5$ (formula wt 295), although $C_{32}H_{18}N_2O_{10}$ (formula wt 590) would give the same %C, H, N, O. He wrote that his analyses for *Biliphäin* and biliverdin differed significantly from those obtained by *Scherer* (92) (Table 2.6.1) and *Hein* (105) (Table 2.6.1), noting that the first could not be correct due to the amount of ash and that anyway the method that *Scherer* used to isolate the pigment from icteric urine would not produce pure bile pigment. The pigment that *Hein* called *Gallenbraun* was separated from an insoluble residue from gallstones by heating crude *Gallenbraun* in ammonia. However, the differing results from *Hein*'s green material could also be attributed to an admixture with some fat or cholesterol, which explains both his 140–145°C melting point and his analysis. *Heintz* expressed hope that another chemist might repeat his experiments with suitable material to confirm or modify them. This wish was to be realized repeatedly during the following decades, but with bile pigments obtained from gallstones by somewhat different isolation methods (97):

> Die von mir für die Zusammensetzung des Biliphäins und Biliverdins gefundenen Zahlen weichen wesentlich von denen ab, welche früher von Scherer und Hein angegeben worden sind. Die des ersteren konnten aber kein richtiges Resultat geben, weil aus dem aus Gallensteinen dargestellten Farbstoff weder die Asche, die ja die kohlensaure oder kaustische Kalkerde enthalten konnte, noch das Epithelium entfernt worden war, und dass nach der Methode, welche er zu seiner Darstellung aus icterischem Harn anwendete, kein reiner Gallenfarbstoff erhalten werden könne, war *a priori* zu vermuthen. Die Versuche von Hein mit dem Körper, den er Gallenbraun nennt und den er durch Auskochen des rohen Gallenbrauns mit Ammoniak als unlöslichen Rückstand erhielt, trifft dasselbe, was gegen die zuerst erwähnten Versuche von Scherer gesagt worden ist. Der grüne Stoff aber, den Hein untersucht hat, muss eine zufällige Beimengung gehabt haben, da er bei 140°-145°C. schmolz. Wahrscheinlich enthielt er noch etwas Fett oder Cholesterin. Die Abweichung der Resultate seiner Analyse (er hat zudem nur eine Kohlenstoff- und Wasserstoffbestimmung ausgeführt) is demnach erklärlich. Es wäre zu wünschen, dass andere Chemiker, denen

2.7 Bile Pigments from Gallstones in the Middle of the 19th Century

> passendes Material, welches mir jetzt fehlt, zu Gebote steht, meine Versuche wiederholten um meine Schlüsse zu bestätigen oder zu modificiren. 38

It seems clear that despite the investigations of bile pigments up to 1850, there were very few advances toward what we understand as chemical structures. In 1850, chemical structure was still a remote or at least an evolving concept. The focus in the first half of the 19th century was on isolating the pigments from the natural sources, mainly from bile but also from gallstones, attempting to purify them (which was largely unsuccessful), making salts, and running combustion analyses. Perhaps all those efforts were a necessary precursor to the advances that were to come during the next 50 years. The latter was an era that also produced new and bizarre ideas (*e.g.* that bilirubin arose from the action of acids upon bile acids and amino acids), incredibly detailed isolation schemes, and a large number of pigment reactions, especially salt formation with ions such as Ag^+, Ba^{+2}, Ca^{+2}, *etc.* that revealed mainly that bilirubin is a diacid. The details that follow in Sections 2.8 and 2.9 reveal both a high level of experimental activity of a repetitious nature and a low level of actual breakthrough. Yet, possibly, it was a necessary step in the evolution of the structure of bilirubin. Perhaps the era represented "marking time" until organic chemistry had evolved to such a point where it might have a positive impact on "animal chemistry", bringing with it the concept of molecular structure and a level of synthesis "know how".

2.7 Bile Pigments from Gallstones in the Middle of the 19th Century

The late 1850s brought forth papers by *Charles Darwin* (1809–1892) and *Alfred Russell Wallace* (1823–1913) announcing a theory of evolution by natural selection in papers read at London's Linnean Society, and in 1859 *Darwin* published *On The Origin of Species*. Also brought forth was a breakthrough in bile pigment isolation and separation that was to have a lasting impact in the quest to determine the structure of bilirubin.

Soon after the work of *Heintz* (*95–97*), toward the end of the 6th decade of the 19th century, a new and improved method of pigment isolation from gallstones was introduced: extraction using chloroform. This rather remarkably simple alteration of the earlier established procedures (that involved washings with ethanol and ether, in which bile pigments are insoluble, and dissolving pigments in base followed by precipitation with acid or as barium or lead salt) also introduced a convenient way to separate the yellow pigment from the green, which was previously very difficult. In 1858, *Valentiner*[28] (*107*), while working in *von Frerichs'* laboratory in Breslau

---

[28] *Gabriel Gustav Valentin* (*Valentiner* = *Valentin* ?), 1810–1883, received the Dr. med. from the University of Breslau in 1832, studied with *Jan E. Puikyně* at the University from 1833, and from 1836 was Professor of Physiology at the University of Bern for 45 years.

58   2 Early Scientific Investigations

and in Berlin, discovered that $CHCl_3$ digestion of pulverized gallstones, which had been exhaustively washed with ethanol and with ether to remove cholesterol, fats, and other solubles, produced a yellow $CHCl_3$ solution. Upon evaporation of the $CHCl_3$, while protecting the solution from air, red and red-brown crystals separated out, the majority (as he reported) with characteristics of *Hämatoidin*. As reported in *Virchow's Archiv* (*108*):

> Herr Valentiner hat in dem Chloroform ein neues Lösungsmittel für thierische Farbstoffe gefunden. Zunächst gelang es ihm, aus gepulverten Gallensteinen nachdem er dieselben erschöpfend mit Alkohol und Aether ausgezogen hatte, durch Digestion mit Chloroform eine gelbe Lösung zu erhalten, aus der sich beim Verdampfen (unter Vermeidung zu starken Luftzutrittes) rothe und braunrothe Krystalle, der Mehrzahl nach mit den Eigenschaften des Hämatoidins, ausschieden. Es waren lancettförmige und rhomboidale Plättchen und prismatische Krystalle in drusiger Gruppirung. Um grössere Krystalle rein zu erhalten, war es vortheilhaft, der Chloroformlösung vor dem Verdunsten etwas thierisches Fett zuzusetzen und dies aus dem Rückstande rasch durch Aether auszuwaschen. Mehrmals wurde auch durch Aether-Auszug ein krystallinischer Farbstoff (Frerichs, Atlas zur Klinik der Leberkrankheiten Taf. I. Fig. 7) erhalten, sowie in vielen Fällen direct aus der Chloroform-Lösung Krystalle, die nach Farbe und Form von Hämatoidin verschieden zu sein schienen.   39

Although *Berzelius* had isolated red crystals (*Bilifulvin*) in 1840 from bile (*73–76*), apparently, red crystals from gallstones had not been isolated previously, and their color reminded one of *Hämatoidin* (hematoidin) (*Virchow's Hämatoidin* (*94*), a crystalline or amorphous iron-free red pigment formed from *Hämatin* (hematin) in old hemorrhages).

Valentiner also described the extraction of hematoidin crystals from icteric liver or fatty liver and further described them in terms of shape and solubility. Prophetically, with respect to much later photochemical experiments, he reported that the crystals decompose over long exposure to diffuse daylight to give a porous, amorphous green powder. The hematoidin obtained by *Valentiner* was treated with conc. H_2SO_4 which decomposed it to green; concentrated aqueous KOH gave it a dirty red to green color; other reagents gave these and other colors. Significantly the hematoidin gave a positive *Gmelin* reaction. It would thus appear, on the basis of *Valentiner*'s extensive "tests" and manipulations that hematoidin and *Cholepyrrhin* or *Biliphäin* were the same substance (*108*):

> Ikterische, fettreiche Lebern, am besten die ikterische Fettleber höchsten Grades, bilden, bei Wasserbadhitze ausgeschmolzen, unter der sich abscheidenden Fettschicht und in den noch fettig durchtränkten Parenchymstückchen sehr zahlreiche Hämatoidin Krystalle. Es sind, auch wiederholter Reinigung und Umkrystallisrung gestreckte, fast rechtwinklige Täfelchen, denen bei beträchtlicher Dicke ganz flache Pyramiden, fast nur durch diagonal sich kreuzende Linien angedeutet, aufgesetzt sind. Bei Verunreinigung sind es gestreckte, rhomboidale Plättchen, zuweilen mit abgerundeten Winkeln, bisweilen dumbell-artig aneinander gesetzt, oder man sieht die bekannten schiefen Prismen mit rhombischen Endflächen, oder bei schneller Verdunstung feine rhombische Nadeln und kurze, fast rechtwinklige Tafeln. Die reine Substanz ist in Wasser, Alkohol und Aether unlöslich; in letzterem zerfallen die Krystalle, längere Zeit dem zerstreuten Tageslicht ausgesetzt, zu einen lockern, amorphen, grünen Pulver. Aetherische und fette Oele sind

2.7 Bile Pigments from Gallstones in the Middle of the 19th Century

wirkungslos. Reine concentrirte Schwefelsäure löst unter raschem Farbenwechsel und Zersetzung zu körnigen flockigen Massen mit vorwiegend bräunlicher Färbung. Unterbricht man die Zersetzung durch Wasserzusatz während einer gleichmässig grünen Färbung, so erhalt man einen amorphen grünen Farbstoff, der durch Lösung in Ammoniak und widerverdunsten der Lösung in compacte grüne Körnchen und zarte formlose Häutchen geschieden kann. Salpetersäure (unreine) zersetzt rasch, unter anfänglich grüner, dann blaugrüner, blauer, endlich rothgelber und blassgelber Färbung bis zur Vernichtung jeder Farbe. Auch ein Gemisch von Salpeter- Schwefelsäure ruft den lebhaftesten Farbenwechsel der Gallenpigmentreaction hervor. Salzsäure giebt unter langsamer Zerstörung Dunkelgrün, schliesslich Blaugrün, jedoch lässt noch lange ein Zusatz von Salpetersäure die Masse chromatisiren. 40

In yet another experiment using $CHCl_3$ extraction, human and animal bile shaken with $CHCl_3$ always yielded hematoidin according to *Valentiner*. In addition to human bile, *Valentiner* investigated bile from the dog, cat, pig, cattle, sheep, chicken, goose, frog, and sturgeon and found that $CHCl_3$ extracted hematoidin and left behind dark green bile. In his experiments *Valentiner* thus discovered that the yellow-red pigment was soluble in $CHCl_3$ but the green was not, a distinction in relative solubility shared by bilirubin and biliverdin.

Very soon after *Valentiner*'s report, in 1859, *Brücke*[29] attempted to answer a question that he posed as to whether small amounts of bile pigment could still be detected after successful extraction using $CHCl_3$, *i.e.* how completely does $CHCl_3$ remove the pigment, with detection to be monitored using the very sensitive *Gmelin* color change reaction from HNO_3 (*109*):

Im December vorigen Jahres machte Dr. Valentiner in Günzburg's Zeitschrift bekannt, dass sich aus Gallensteinen, aus der Galle, ferner aus den Lebern der Icterischen, oft auch aus anderen Geweben derselben mittelst Chloroform eine krystallinische Substanz erhalten lasse, welche verschieden von den bisher bekannten Gallenfarbstoffen sei und in allen ihren Eigenschaften mit dem Hämatoidin übereinstimme. Die chloroformige Lösung gab mit Salpetersäure in besonders schöner Weise die bekannte Farbenfolge der Gmelin'schen Gallenprobe; dagegen „enthielt nach Entfernung der in Chloroform löslichen Farbstoffe die immer noch stark dunkelgrün pigmentirte Galle kein Substrat der Gallenpigmentreaktion mehr". Dr. Valentiner schlägt desshalb vor, da, wo es sich darum handelt kleine Mengen von Gallenfarbstoff in einer Flüssigkeit nachzuweisen, diese mit Chloroform anhaltend zu schütteln und letzteres nach wieder erfolgter Trennung direct mit Salpetersäure zu prüfen. 41

In contrast to *Valentiner*, *Brücke* focused on extracting bile rather than gallstones. He found that $CHCl_3$ extracted a yellow pigment from human gallbladders. After separation by decantation, the $CHCl_3$ extract was placed in a retort and evaporated (without boiling) by heating on a water bath. The residue was then covered with spirits of wine containing 94% alcohol to produce crystals that partially adhered to the inside wall of the retort and partly sank as a brick red powder after shaking with the alcohol. After decanting the alcohol, the crystals were removed

---

[29] *Ernst Wilhelm Ritter von Brücke* (1819–1892) was a medical student in Berlin in 1838 and in Heidelberg in 1840. He was promoted to Dr. med. in Berlin and Professor in Vienna from 1848 to 1890.

from the retort by washing them out with alcohol and ether. *Brücke*'s examination (under a microscope) revealed that they were no longer co-mixed with extraneous material. He then proceeded to examine the residual, chloroform-extracted bile in order to determine whether the extraction had removed the bile pigment completely.

To settle this question, *Brücke* took a portion of the bile decanted after the initial $CHCl_3$ extraction and evaporated it to dryness. At this point one might wonder how complete the separation of bile from $CHCl_3$ was and whether $CHCl_3$ (with some solubilized bile pigment) was not dissolved in the bile. Of course that would perhaps have been difficult to determine in 1859. Though "separating" funnels had been known since the time of *Berzelius* (*110*), they would hardly be recognized as such today, and in the 1850s they were crude devices at best and not widely known. It was not until around 1854 that any device similar to a modern glass separatory funnel was employed. Irrespective, *Brücke* took the dried sample of $CHCl_3$-extracted bile, pulverized it, and digested the powder with $CHCl_3$, decanted the $CHCl_3$ and then added fresh $CHCl_3$ to the residue along with as much water as needed to dissolve the bile. The aqueous bile was then repeatedly extracted with fresh $CHCl_3$ and the color of the $CHCl_3$ extracts became weaker and weaker to imperceptible with each successive extraction. At this point it was presumed that the bile was completely depleted of the *Gmelin* test reactive bile pigment. However, the *Gmelin* reaction was still positive. Repeating the experiment, *Brücke* produced the same results (*109*):

> Abgesehen von einigen von Dr. Valentiner angegebenen Versuchen, welche ich mit den Krystallen anstellte, richtete ich meine Aufmerksamkeit zunächst darauf, ob in der That die durch Chloroform erschöpfte Galle die Farbenveränderungen mit Salpetersäure nicht mehr zeige. Ich dampfte einen Theil der von Chloroform abgegossenen Galle im Wasserbade zur Trockne ab, pulverte sie, extrahirte sie mit Chloroform, filtrirte dasselbe ab, leerte den Filterrückstand wieder in eine Flasche, übergoss ihn mit neuem Chloroform fügte dann wieder so viel Wasser hinzu, dass sich die trockene Galle darin löste. Nun extrahirte ich durch Schütteln weiter, indem ich das Chloroform von Zeit zu Zeit erneuerte; es nahm immer weniger Farbstoff auf, die Farbenveränderungen, welche es mit Salpetersäure zeigte, wurden immer schwächer und zuletzt unmerklich. Von der nun abgegossenen Galle wurde eine kleine Quantität mit vielem Wasser verdünnt, der Gmelin'schen Probe unterworfen und *zeigte den Farbenwechsel sehr schön*. Ich habe den Versuch mehrmals wiederholt und ihn theils in der ursprünglichen von Gmelin angegebenen Form angestellt, theils mit der Modification, welche ich vor zehn Jahren an dieser Probe angebracht habe und welche darin besteht, dass nur verdünnte Salpetersäure hinzugesetzt wird und dann concentrirte Schwefelsäure, welche sich zu Boden senkt und von unten her den Zersetzungsprocess einleitet, so dass man sämmtliche Farben gleichzeitig in über einander liegenden Schichten beobachten kann. Stets erhielt ich dasselbe positive Resultat.   42

Though *Brücke* found his results at odds with *Valentiner*'s description, a finding which begged explanation, he determined that the $CHCl_3$-soluble yellow pigment had all the characteristics of *Heintz*'s *Biliphäin* (or *Berzelius' Cholepyrrhin*) and was apparently analogous to *Virchow*'s hematoidin, as *Brücke* expressed (*109*):

> Diese Thatsache war in offenem Widerspruche mit Dr. Valentiner's Angabe, und es fragte sich, wie ich sie erklären sollte. Die durch Chloroform erschöpfte Galle bildete mit Wasser grüne Lösungen, dieselben wurden auch durch Zusatz von Kali nicht gelb, sondern

2.7 Bile Pigments from Gallstones in the Middle of the 19th Century   61

> nur ein wenig mehr gelbgrün, durch Salzsäure mehr blaugrün. Ich vermuthete desshalb, dass vielleicht von den beiden als Biliphäin und Biliverdin bekannten Farbstoffen, welche Object der Gmelin'schen Probe sind, der eine, das Biliphäin, in Chloroform löslich sei, der andere nicht, und es lag desshalb nahe, zu untersuchen, ob nicht die aus dem Chloroform erhaltenen Krystalle krystallisirtes Biliphäin oder doch eine krystallisirte Verbindung des Biliphains seien. Es würde diese ihre von Dr. Valentiner vertheidigte Identität mit dem Hämatoidin keineswegs ausschliessen. Virchow hat schon vor eilf Jahren auf die Analogien mit dem Biliphäin (Cholepyrrhin) aufmerksam gemacht, welche ihm sein Hämatoidin bei Einwirkung gewisser Reagentien darbot.   43

In order to resolve the apparent discrepancy with *Valentiner* regarding the persistent positive *Gmelin* reaction from *Biliphäin*-depleted bile, *Brücke* studied bile depleted of yellow pigment by $CHCl_3$ and found it to form a green solution with water that turned yellow-green upon addition of KOH, then blue-green with added hydrochloric acid. These colors were typical of biliverdin, which could be expected to remain in the bile after removal of *Biliphäin* by $CHCl_3$ extraction. And because biliverdin also exhibited a positive *Gmelin* reaction, the post-extraction bile would be expected to give a positive *Gmelin* reaction.

Probably even more important, *Brücke* found that while $CHCl_3$ extracted *Biliphäin* from bile, it did not extract biliverdin. That is, *Biliphäin* was soluble in $CHCl_3$ but biliverdin was not. *Brücke* also learned or knew that while *Biliphäin* was insoluble in alcohol, biliverdin was soluble. Thus, not only did the $CHCl_3$ extraction provide a route to pure *Biliphäin* from bile, or from a mixture of *Biliphäin* and biliverdin, but digestion of the latter with alcohol provided a route to remove biliverdin from *Biliphäin*. As a consequence pure biliverdin could be isolated following air oxidation of *Biliphäin*.

The salient points of *Brücke*'s investigations were succinctly summarized by *Virchow* in his *Archiv* in 1859 (*108, 111*), thus bringing to a close the important discoveries from investigations of gallstones and bile at the end of the 6th decade of the 19th century (*111*):

> Herr Brücke wiederholte einen Theil der vorstehenden Versuche des Herrn Valentiner, zunächst um zu sehen, ob die durch Chloroform erschöpfte Galle keine Reaction mehr darbiete. Allein er fand, dass auch diese Galle bei der Gmelin'schen Probe den Farbewechsel schön zeigt, und es fragt sich nun, ob die erhaltenen Krystalle nicht Biliphäin oder eine Verbindung desselben seien. In der That erhielt er aus der ammoniakalischen Lösung der Krystalle durch Salzsäure gelbbräunliche Flocken, welche alle Eigenschaften des Biliphäins (Heintz) darboten, und aus denen sich durch Chloroform wieder eine gelbe Lösung und nach dem Abdestilliren des Chloroforms wieder Krystalle gewinnen liessen. Brücke schliesst daher, dass die neue Methode ein vortreffliches Mittel zur Scheidung von Biliphäin und Biliverdin sei. Letzteres lässt sich auch rein aus den rothen Krystallen gewinnen, indem man sie in wässerigem kohlensauren Natron löst und die Lösung an der Luft Sauerstoff absorbiren lässt, mit Salzsäure fällt, das Filtrat auswäscht und etwaige Reste von Biliphäin durch Chloroform auszieht.   44

Brücke repeated part of the preceding experiments of *Valentiner* [pp. 201–202 of the same *Virchow*'s *Archiv* (*111*)] to see initially whether bile that had been exhausted by $CHCl_3$ extraction can give any further reaction (specifically, the *Gmelin* test). He found that this bile, too, showed the beautiful change of colors of the *Gmelin* test. The question then arising was whether the (reddish) crystals obtained (following evaporation of the $CHCl_3$ extracts) would not be *Biliphäin* or a compound of the same. In fact, on hydrochloric acid addition to an ammonia solution of the crystals, he obtained yellow-brownish flakes that presented the characteristics of *Biliphäin* (*Heintz*), and which dissolved in $CHCl_3$ to afford again a yellow solution from which crystals were able to be obtained upon distilling off the $CHCl_3$. *Brücke* concluded therefore that the new method was a superior means for separating *Biliphäin* and biliverdin. The latter could also be obtained pure from the red crystals by a process in which the crystals were dissolved in aqueous Na_2CO_3 and the solution allowed to absorb oxygen from air before HCl is added. The resulting (green biliverdin) precipitate was washed and any possible residue extracted with $CHCl_3$. Thus, the conclusions to be drawn from the work of *Valentiner* and *Brücke* were: (i) $CHCl_3$ is a superior solvent for removing *Biliphäin* or *Cholepyrrhin* from gallstones or bile following removal of cholesterol, fats, mucus, and other ethanol or ether-soluble components; (ii) $CHCl_3$ extraction leaves biliverdin behind; (iii) *Biliphäin* is soluble in $CHCl_3$; biliverdin is insoluble; (iv) biliverdin is soluble in ethanol; *Biliphäin* is insoluble; (v) pure *Biliphäin* can be separated from biliverdin by a $CHCl_3$ wash; and (vi) pure biliverdin can be isolated by extraction into ethanol following air oxidation of pure *Biliphäin*.

2.8 *Hämatoidin*, Bilifulvin, and the Origin of Bile Pigments

The biological origin of bile pigments remained a mystery for millenia until it began to unravel in the middle of the 19th century and in the absence of any knowledge of chemical structure. At that time it was known, of course, that yellow and green pigments could be isolated from bile and gallstones, whose colors ranged from light yellow to brown and blackish. *William Saunders* (1743–1817) speculated in 1809 that a relationship might exist between the pigments of bile and blood (*112*):

> Green and bitter bile being in common to all animals with red blood, and found only in such, makes it probable that there is some relative connexion between this third and the colouring matter of blood, by the red particles contributing especially to its formation.

Reddish crystals in old blood extravasations seem to have been noticed first by *Home* (*113*) in 1830 and subsequently in 1842 by *Rokitansky*, in 1843 by *Scherer*, in 1846 by *Zwicky* [as reported by *Wedl* (*114*) and *Robin* (*115, 116*)], and in 1847 by *Virchow* (*94*). Yet despite the increasing investigations of bile pigments in the decades subsequent to *Saunders*, no further connection had been drawn between bile pigments and the red pigment of blood, which had been a separate focus of attention. That is until 1847, when *Virchow* reported extensively on the reddish crystals that he observed in extravasated or hemorrhaged (stagnant) blood from a very large number of diverse cases involving humans – and named *Hämatoidin* (hematoidin) (*94*):

> In Beziehung auf die Gefässe will ich die jenigen Fälle angeben, wo man am sichersten auf die Anwesenheit von Hämatoidin-Krystallen rechnen darf.   45

2.8 Hämatoidin, Bilifulvin, and the Origin of Bile Pigments

Hematoidin was described morphologically as to color (red), crystal shape, and dimensions. The red pigment of blood, named *Hämatin* (hematin), derived from the hemoglobin of red blood cells, is a protein-free reddish-brown crystalline solid obtained from dried blood. Of course, nothing was then known of its structure, but hematoidin was known to be different from hematin and suspected to be derived from it. *Virchow* credited an earlier investigator, Sir *Everard Home* (1756–1832) as having published (*113*) beautiful illustrations of clots from aneurysmatic sacs in 1830. Which left no doubt that he had seen genuine crystals of changed hematin (*94*):

> Die erste Beobachtung derselben finde ich bei Everard Home. In seinem letzten Werke (*A short tract on the formation of tumours. Lond.* 1830) sieht man auf der ersten Tafel 3 sehr schöne Abbildungen von Gerinnseln aus aneurysmatischen Säcken, welche keinen Zweifel übrig lassen, dass er wirklich Krystalle von verändertem Hämatin gesehen hat. Leider giebt er keine weitere Beschreibung davon, sondern bezieht sie nur auf Krystallisation von Blutsalzen (p. 22). In der Erklärung der Tafel heisst es: *The figure shows the different shades of colours of the layers, according to the length of time they had been deposited, and the crystallised salts as they appear in different parts of the coagulum.* 46

Virchow cited several other investigators, including *Scherer* (*104*), who had only a few years earlier reported finding red crystals in extravasated blood, and who linked that substance to his urinary pigment and the one from bile. A rather startling connection between blood, bile, and urine. *Virchow* described the red pigment as "Das pathol. Pigment, dass aus Hämatin stammt, kann also diffus, körnig und krystallinisch sein.... Es kann gelb, roth oder schwarz sein oder irgend eine Uebergangstufen. Zwischen diesen Farben ausdrücken." [The pathologic pigment that is derived from Hämatin can thus be diffuse, granular, and crystalline. . . . It can be yellow, red or black or possibly express a transitional state in between.] (*94*). But could a pigment found in blood (extravasations) be identical to a pigment found in bile? After many chemical probings of numerous and varied samples, including finding a positive *Gmelin* test from the action of hematoidin with $H_2SO_4 + HNO_3$, *Virchow* then concluded that "Eine Vergleichung unserer Pigmente mit den Gallenfarbstoff ist daher unabweisbar." [A difference between our pigments and the bile pigments is therefore irrefutable.] Yet, he repeatedly expressed an uncertainty as to whether his hematoidin was strictly identical to the brown bile pigment *Gallenbraun*, *Simon*'s *Biliphäin*, or *Berzelius*' *Cholepyrrhin*, both of which were separated from bile but had a rather different physical appearance and color and were of uncertain purity. Of major importance to physiology, might the origin of the bile pigments be interpreted as products of red cell consumption, as precipitated and altered hematin? If true, it would run counter to the conclusion of one of the most famous physician chemists of the times, for *Berzelius* had thought that green biliverdin (*Gallengrün*), which he knew was derived from yellow *Cholepyrrhin* by oxidation, was the same green pigment as that (chlorophyll) from green plants. And up to that time (and later), conversion of chlorophyll into the blood mass had never been confirmed. So *Virchow* rationalized the seeming contradiction by noting that, by *Berzelius*' accounts, biliverdin and not *Cholepyrrhin* was found only in the bile of herbivores (*94*):

> Kehren wir damit wiederum zu der Vergleichung unserer Pigmente mit dem Gallenfarbstoff zurück, so können wir die Bemerkung nicht unterdrücken, dass jeder Beobachter sich an

den einzelnen Fällen, wo ihm die Pigmente, namentlich das krystallinische, vorkommen, gewiss besser überzeugen wird, dass eine Ableitung derselben aus präformirtem Gallenfarbstoff nicht statuirt werden kann, als wir es hier durch lange Deductionen zu thun vermöchten. Die Unterschiede, welche wir zwischen beiden Arten von Farbstoffen aufgeführt haben, genügen nach dem jetzigen Stande der Chemie schon zu einer Unterscheidung, allein wenn man sie näher betrachtet, so wird man leicht einsehen, dass sie nicht bloss keine absoluten sind, sondern, genau genommen, mehr auf Verschiedenheiten der Cohäsion zurückführen, ja dass sogar eine ausserordentlich grosse Aehnlichkeit zwischen beiden Farbstoffen nicht weggeläugnet werden kann. Wir kommen damit auf eine andere Frage, die für die Physiologie des gesunden und kranken Körpers von der grössten Bedeutung ist, ob nämlich der Gallenfarbstoff als ein Produkt des Blutkörperchen-Verbrauchs, als ausgeschiedenes und verändertes Hämatin aufgefasst werden dürfe. Gegen diese Ansicht, welche von den verschiedensten Seiten seit langer Zeit, aber immer vollkommen hypothetisch, aufgestellt worden ist, schien namentlich die von Berzelius zu streiten, welcher die Aehnlichkeit desjenigen Gallenfarbstoffs, den er als Biliverdin bezeichnet, mit dem grünen Pflanzenpigmente, dem Chlorophyll hervorhob. . . .

. . . Die eigenthümlichen Farbenveränderungen, welche das bei Contusionen in die Hautgebilde extravasirte Blut eingeht, haben schon lange als Argument für die Umwandlung von Hämatin in eine dem Gallenfarbstoff ähnliche Substanz dienen müssen, allein man muss zugestehen, dass eine solche Art von Beweisen, wenn sie nicht einmal von einer wirklichen Untersuchung des Extravasates begleitet sind, gar nichts gilt. Die Frage wird aber von dem Augenblick an vollkommen erledigt sein, wo wir den Beweis exakt durch das chemische Experiment führen können, dass aus Hämatin nicht eine gelbliche oder grünliche Substanz, sondern eine dem Gallenfarbstoff identische entsteht. Ich schmeichele mir, dass die bisher mitgetheilten Untersuchungen den Weg zu einer endlichen Entscheidung der Frage angebahnt haben. Ich hätte gern diese Entscheidung selbst versucht, wenn meine zahlreichen Beschäftigungen mich nicht nöthigten, zu viel Gegenstände gleichzeitig zu verfolgen; mögen daher die vorstehenden Thatsachen anderen Beobachtern übergeben sein, um weiter verwerthet zu werden. Von einem besonderen Interesse erscheint mir dabei die Untersuchung der Bilifulvin-Krystalle. Könnte man aus der Galle Krystalle gewinnen, welche den im alten Blut entstehenden identisch sind, so bliebe nichts zu wünschen übrig. Die pathologische Anatomie scheint einen solchen Nachweis nicht möglich zu machen... 47

Virchow concluded, on the basis of a large number of observations that: ". . . ich die Wahrscheinlichkeit einer Umwandlung des Blutstoffes in Gallenfarbstoff bis zu einem möglichst hohen Grade gebracht habe." [. . . I have brought the likelihood of a conversion of the pigment of blood into the pigment of bile to the highest extent possible.] Also that the origin of jaundice (and its yellow color) is associated with changes in blood, especially the destruction of red blood cells: ". . . so scheint es vollkommen gerechtfertigt, die Quelle der Gelbsucht in Veränderungen des Blutes und zwar speciell in einer ausgedehnten Zerstörung von Blutkörpchen zu suchen." [. . . thus it appears completely valid to seek the origin of jaundice from changes in blood, especially from the destruction of red cells]. In July 1853, the origin, status, and knowledge of hematoidin was summarized by the Austrian pathologist *Carl Wedl* (1815–1891) in Vienna (translated into English in 1855 by *George Busk* for the Sydenham Society) (*114*):

> The *hematoiden* crystals of Virchow. Brilliant, transparent crystals, having the form of regular oblique rhombic prisms, and of a red colour, varying in tint and depth, according to the state of aggregation of the crystals. They are of a comparatively stable nature, and

2.8 Hämatoidin, Bilifulvin, and the Origin of Bile Pigments

are insoluble in water, alcohol, ether and acetic acid. And they occur either free, or enclosed in flaky particles, or in cells, exclusively in extravasated blood, which has been retained for a longer or shorter time in the organism. . . . *Hematoidin* also occurs in the *amorphous* condition aggregated into reddish-brown granules or amorphous masses, mixed with crystals . . . Chemists have hitherto been unable to establish a theory of the formation of *hematoidin*, since the chemical composition of *hematin* itself is not as yet accurately determined, and that of *hematoidin* is still unknown.

In the mid-1850s, *Robin*,[30] while registering objection to the name *Hämatin* and preferring instead the name *Hämatosin* given first to it in 1827 by *Chevreul*, reviewed the history of hematoidin in 1856 and added his own extensive observations on its properties (*115, 116*):

> Das in Prismen so wie das in Nadeln krystallisirte Hämatoïdin ist ziemlich hart, brüchig und bricht das Licht stark unter dem Mikroskop. Die Krystalle sind im Innern von lebhaft orangerother oder ponceaurother Farbe, an den Kanten und Ecken von dunkel carminrother Farbe. Im auffallenden Lichte haben die von allen Unreinigkeiten befreiten Krystalle eine dem Quecksilberjodid oder dem Alizarin ähnliche Farbe. Sie besitzen ein starkes Färbungsvermögen, sind etwas schwerer als Wasser und bilden voluminöse Massen. Die Winkel der Prismen sind 118° und 62°.
>
> An der Luft erhitzt entwickeln sie anfangs einen theerähnlichen Geruch, wie stickstoffhaltige Körper und brennendes Horn, entzünden sich alsdann und brennen mit leuchtender Flamme unter Zurücklassung einer aufgeblähten voluminösen Kohle, welche endlich vollkommen verschwindet. Es is deshalb die Verbindung schwierig im Verbrennungsapparat zu analysiren. Bei abgehaltener Luft entwickeln sich beim Erhitzen der Substanz übelriechende Gase, es destillirt eine theerartige Substanz und zurück bleibt ebenfalls eine voluminöse Kohle.
>
> Die Krystalle sind unlöslich in Wasser, Alkohol, Aether, Glycerin, ätherischen Oelen und Essigsäure, aber leicht löslich in Ammoniak. Die concentrirte ammoniakalische Lösung ist amaranthroth und nimmt bald einer safrangelbe und bräunliche Farbe an. In Berührung mit Kali und Natron zerfallen die Krystalle des Hämatoïdins und lösen sich allmählich auf, aber in geringerer Menge als in Ammoniak; die Lösung ist röthlich. Salpetersäure löst dieselben ziemlich schnell mit dunkelrother Farbe auf, unter Entwicklung von Gasblasen, wenn dieselbe concentrirt ist. Auch von Chlorwasserstoffsäure werden sie gelöst, aber in geringer Menge. Die Lösung ist goldgelb oder röthlichgelb; die ungelöst bleibenden Krystalle haben im auffallenden Lichte eine ockerbraune, unter dem Mikroskop eine röthlichgelbe Farbe. Von Schwefelsäure werden sie nicht gelöst; sie macht die Krystalle blos dunkler und nimmt eine grüne Farbe an, wenn die Krystalle noch Spuren von alkali- oder eisenhaltigen Verbindungen enthalten. 48

Though *Robin* looked at hematoidin from many angles, it is interesting to note that he reported no *Gmelin* bile pigment test on this pigment. He did, however, conduct an elemental combustion analysis on the crystals after attempting to remove all impurities by treating with water, alcohol, and ether, and (for comparison) he also analyzed *Hämatosin* (= *Hämatin*), for which he determined the formula $C_{44}H_{22}N_3O_6Fe$ on the basis of five analyses (*115, 116*):

---

[30] *Charles-Philippe Robin* was born on June 4, 1821 in Josseron and died on October 6, 1885 in Josseron. He was a biologist-physician and one of the founders of modern histology and member of l'Academie des Sciences de France.

Zur Analyse verwendete ich durch Wasser, Alkohol und Aether gereinigte Krystalle, nachdem ich mich zuvor unter dem Mikroskop überzeugt hatte, dass auf diese Weise alle Unreinigkeiten entfernt werden können und erhielt folgende Resultate:

	I	II	III
Kohlenstof	65,0460	65,8510	–
Wasserstoff	6,3700	6,4650	–
Stickstoff	–	–	10,5050
Sauerstoff	18,0888	17,1788	–
Asche	0,0002	0,00002	–

Hämatosin besteht im Mittel von 5 Analysen aus:

$$C_{44}H_{22}N_3O_6Fe$$

oder in 100 Theilen aus:

Kohlenstoff	65,84
Wasserstoff	5,37
Stickstoff	10,40
Sauerstoff	11,75
Eisen	6,64

49

The two hematoidin samples analyzed, with unusual accuracy, gave a bit of ash, determined to contain iron and traces of alkali salts, but no calcium, sulfur, or phosphorous. *Robin* compared the results of his analyses to those of *Mulder* (*117*) who reported removing iron from non-crystallizable hematin in 1839, and analyzing the product as 70.49% C, 5.76% H, 11.16% N, and 12.59% O, which gives the formula $C_{14}H_8NO_2$ – or as, *Mulder* wrote: $C_{44}H_{22}N_3O_6$. The first has the same composition as *Robin* found for hematoidin ($C_{14}H_9NO_3$, or $C_{14}H_8NO_2+HO$). *Robin* concluded that it was thus easy to recognize that hematoidin was nothing other than the pigment of blood, or a hematin, in which one equiv. of iron was replaced by one equiv. of H_2O (*115, 116*):

> Es ist deshalb leicht einzusehen, dass das Hämatoidin nichts anderes ist als der Farbstoff des Blutes, oder ein Hämatin, in welchem 1 Aeq. Eisen durch 1 Aeq. Wasser ersetzt ist. 50

By the late 1850s, *Valentiner* (*107*), the first to discover that $CHCl_3$ extracted nearly pure pigment from gallstones and bile, would find the brown-red crystalline residue from evaporation of the $CHCl_3$ extract to be suspiciously like hematoidin in its crystalline and chemical properties. At nearly the same time, in 1859, *Brücke* (*109*) came to a similar thought in connection with his extraction and purification of *Biliphäin* from bile. He noted, importantly, that the *Biliphäin* could be red and crystalline or amorphous and yellow depending on whether it had been obtained directly from $CHCl_3$ evaporation or whether it had been precipitated by acidification of an aqueous Na_2CO_3 solution using HCl. He also determined that the *Gmelin* test was positive for both *Biliphäin* (*Cholepyrrhin*) and the biliverdin derived from it by air oxidation. Such studies, even absent a knowledge of chemical structures, would suggest that *Biliphäin* and hematoidin were one and the same, or that

2.8 Hämatoidin, Bilifulvin, and the Origin of Bile Pigments

Hämatoidin contained *Biliphäin* (or *vice versa*) – and thus that bile pigments originated from the pigment of red cells, hematin.

Apparently, *Frerichs*[31] did not agree and was less certain – at least regarding the origin of the pigment of icteric urine. It was in Breslau that *Frerichs* published (in 1858) the first edition of his famous *Klinik der Leberkrankheiten*, volume I (*118*). In the preface (March 1858) he acknowledged the aid of two individuals, whose importance to bile pigments would become apparent subsequently: "Prof. G. Städeler of Zürich ('my friend Städeler'), who on many occasions aided Frerichs with chemical advice and carried out elemental combustion analyses of the abnormal transformation products found in the liver and urine, and Dr. Valentin(er) who performed a large part of the chemical work in Frerichs' lab." While in Berlin, *Frerichs* finished the revised second edition of 1861 (*119*) as well as volume II (*120*). In the former, he acknowledged the general acceptance that the hematin of blood was the origin of all pigments, and that it underwent metamorphosis into a yellow pigment (of jaundice) that was similar to or identical with a bile pigment (*119*):

> In neuerer Zeit, wo die Lehre von den Pigmenten eine sorgfältigere Bearbeitung fand und man sich mehr und mehr dahin einigte, dass das Hämatin des Blutes die Grundlage aller Pigmente ausmache, konnte es nicht an Beobachtern fehlen, welche nach der Idee von Senac icterische Färbungen der Haut, die, wie bei der Pyämie, bei putrider Infection und verwandten Processen, ohne Betheiligung der Leber sich enwickelten, auf eine directe Metamorphose des Hämatins zu einem gelben, dem Gallenpigmente ähnlichen oder mit demselben identischen Farbstoff zurückführten. 51

More controversial and seemingly contradictorily, *Frerichs* reported experiments from which he proposed that bile pigments originated from bile acids. This conclusion, especially when empirical formulas are taken into account, might seem far-fetched today. Nonetheless, given the state of chemical knowledge of the era, it was not irrational, though it was based almost entirely on the generation of colors and dubious results of the *Gmelin* reaction. His view in 1858 (*118*) and in 1861 thus rested on the work with *Städeler* and was expressed in the following (*119*):

> Diese Ansicht stützt sich auf folgende Thatsachen: Reine farblose Gallensäuren lassen sich in Gallenpigment umwandeln mit allen Eigenschaften, welche diesen Farbstoff auszeichnen. Eine solche Umwandlung erfolgt nicht bloss unter Einwirkung von Reagentien, sondern auch im Blute lebender Thiere, sie geschieht unter Aufnahme von Sauerstoff und

---

[31] *Friedrich Theodor von Frerichs* was born on March 24, 1819 in Aurich and died on March 14, 1885 in Berlin. He was professor of clinical medicine at the University of Berlin (Humboldt University of Berlin) and the founder of modern pathology. He studied medicine and science in Göttingen, learned chemistry from *Wöhler*, and departed in 1842 with a Dr. med. to establish himself as a surgeon of high repute and an ophthalmologist. In 1846, he returned to Göttingen, where he habilitated as *Privatdozent*, and in 1848, he was appointed a. o. Professor, working in association with *Wöhler* and *Rudolf Wagner*. He contributed to *Wagner's Handwörterbuch der Physiologie*, and established himself as an excellent leader and researcher who expanded his expertise into clinical autopsies. He accepted a call as head of the academic medical institution in Kiel in 1850, then in Breslau (today's Wroclaw) as Ordinarius of Pathology and director of the medical clinic. In 1859, he succeeded *Schönlein* as director at the Medical Clinic at the Charité (Berlin).

ist zum Theil abhängig von dieser[1]). Durch Einwirkung von concentrirter Schwefelsäure bilden sich aus farbloser Galle Chromogene, welche an der Luft, und noch rascher unter Einwirkung von Salpetersäure, einen Farbenwechsel zeigen, volkommen übereinstimmend mit Gallenpigment. Dieselben Chromogene und Farbstoffe, welche sich ganz wie Cholepyrrhin verhalten, entstehen, wenn farblose Galle in reichlicher Menge ins Gefässsystem lebender Thiere injicirt. Die Gallensäuren werden in diesem Falle im Blute unter dem Einflusse der Respiration zu Gallenpigment umgewandelt. Dass solche Umwandlung auch die im Normalzustande aus dem Darm resorbirte und von der Leber direct ins Blut übertretende Galle erleidet, dafür scheint zunächst das reichliche Vorkommen von Taurin in der normalen Lunge, welches Staedeler und Cloëtta nachwiesen, zu sprechen. Die Pigmente, welche hierbei entstehen, treten indess erst dann mit dem Harn zu Tage, wenn der stetig weiterschreitende Umsetzungsprocess, welchem der Farbstoff unterworfen ist, schon eine Stufe erreicht hat, auf welcher er die Eigenschaften des Gallenpigments nich mehr besitzt.

---

[1]) Wenn man vollständig entfärbtes reines glycocholsaures Natron mit concentrirter Schwefelsäure übergiesst, so bildet sich eine farblose harzähnliche Masse, welche in der Kälte mit safrangelber, beim Erwärmen mit rother Farbe sich auflöst. Aus dieser Lösung fällt Wasser farblose, grünliche oder bräunliche Flocken, je nach der Temperatur, bei welcher die Lösung erfolgte. Die durch Schwefelsäure veränderte Glycocholsäure hat die Eigenschaft, an der Luft rasch Sauerstoff aufzunehmen und damit in prachtvoll gefärbte Verbindungen überzugehen. Bringt man die durch Schwefelsäure entstandene amorphe, farblose Masse, nachdem sie möglichst von anhängender Säure befreit worden ist, auf ein Stück Filtrirpapier, so zerfliesst sie und es entsteht ein rubinrother Fleck, welcher bald blaue Ränder zeigt und nach kurzer Zeit indigblau wird. Nach einigen Tagen verschwindet auch diese Farbe und der Fleck wird braun.

Durch anhaltendere Einwirkung von Schwefelsäure auf Glycocholsäure wird eine Substanz gebildet, welche in Wasser mit tief grüner, in verdünnter Kalilösung mit brauner Farbe sich löst und auf Zusatz von Salpetersäure zuerst eine grüne, dann röthliche und zuletzt gelbe Färbung annimmt. Das Verhalten dieser Zersetzungsproducte gegen Salpetersäure erinnert an das der natürlichen Gallenpigmente, indess ist der Farbenwechsel weniger lebhaft. Ein mit dem Cholepyrrhin in jeder Beziehung sich gleich verhaltendes Product erhält man dagegen, wenn taurochalsaures Natron auf obige Weise behandelt wird. Mit wenig Wasser gelöst und mit concentrirter Schwefelsäure versetzt, färbt sich dasselbe prachtvoll roth und wird an der Luft allmälig blau. Vermischt man die roth gefärbte Lösung mit mehr Schwefelsäure, so geht die Farbe in braun über. Auf Zusatz von Wasser entsteht ein zarter, nach und nach blassgrün werdender Niederschlag; giesst man davon die säure Flüssigkeit ab und erwärmt den Rückstand, so treten intensiv grüne, blaue und violette Farben auf. Die gefärbten Producte lösen sich mit gallenbrauner Farbe in Kali, und die Lösung verhält sich gegen Salpetersäure vollkommen gleich einer alkalischen Cholepyrrhinlösung.

Dass dieselbe Metamorphose im Blute eines lebenden Individuums vor sich gehe, beweisen Injectionen von Auflösung entfärbter Galle in die Venen von Hunden. Der nach einem solchen Versuche gelassene Harn lässt beim Stehen gewöhnlich grüne Flocken fallen, welche auf Zusatz von Salpetersäure den für Gallenfarbstoff charakteristischen Farbenwechsel von Grün, Blau, Violett und Roth in schönster Form erkennen lassen. Unveränderte Gallensäure wird durch die Pettenkofer'sche Probe vergebens gesucht. Nur in einem Falle, wo eine ungewöhnlich grosse Menge, gegen zwei Drachmen, trockener Galle zur Injection verwandt wurde, liess sich eine Spur davon nachweisen. Bemerkenswerth ist, dass die Quantität des in den Harn übergehenden Farbstoffes am grössten erschien, wenn das betreffende Thier an Respirationsnoth litt, so namentlich bei einem Hunde, welcher in Folge des Versuches an Lungenödem zu Grunde ging. In einem Falle, wo eine geringe Quantität Galle injicirt war, das Thier auch frei von Athmungsbeschwerden blieb, wurde gar kein Pigment gefunden.

Or, as *Murchison* translated (*121*):

The bile-pigment is so intimately related on the one hand to the red matter of the blood, and on the other, to the colorless biliary acids, as to justify us in referring its origin to one or the other of these sources.

2.8 Hämatoidin, Bilifulvin, and the Origin of Bile Pigments

The intimate relation subsisting between the bile-pigment and the coloring-matter of the blood is indicated by facts which have been already mentioned, but more particularly by observations which have been recently made in my laboratory by Dr. Valentin (*Günsburg's Zeitschrift*, Dec., 1858), according to whom a portion of the coloring-matter of the bile dissolves in chloroform, and from this solution a crystalline substance may be obtained presenting all the characters of hæmatoidine. From this it appears possible, nay probable, that, as in extravasations, hæmatoidine may be developed from blood-pigment, so in like manner, in the vascular system and in the liver, the coloring-matter of bile may originate from the same source. Hitherto, however, no one has succeeded in obtaining bile-pigment directly from the red matter of the blood.

The second view rests upon the following facts:– The pure colorless acids of the bile may be transformed into bile-pigment with all the properties characterizing this substance. Such a transformation takes place not only under the influence of reagents, but it also follows the absorption of the acid substance (into the blood of living animals), and is in a measure dependent upon this.[1] By the action of concentrated sulphuric acid upon colorless

---

[1] If concentrated sulphuric acid is poured upon pure, perfectly colorless, glycocholate of soda, there is formed a resinous mass, devoid of color, which dissolves in the cold with a saffron yellow color, and with a reddish color upon the application of heat. This solution separates into a colorless water, and flakes of a greenish or brownish color, according to the temperature at which the solution has been made. Glycocholic acid, when changed by sulphuric acid, has the property, upon exposure to the atmosphere, of rapidly taking up acid substances, and of passing into gorgeously-colored combination. If the amorphous, colorless mass resulting from the action of sulphuric acid, after it has been deprived, as far as possible, of the adherent acid, is placed upon a piece of filtering paper, it dissolves, and there is produced a ruby-red spot, which soon presents a blue margin, and after a short time assumes an indigo-blue color. After some days, this color also disappears, and the spot becomes brown.

By the continued action of sulphuric acid upon glycocholic acid, a substance is produced, which dissolves in water with a deep green color, and in a weak solution of soda with a brown color, and which, upon the addition of nitric acid, assumes first a green, then a reddish, and lastly, a yellow tint. The behavior of this substance with nitric acid reminds us of that which characterizes the natural bile-pigment, although the change of color is less rapid. When taurocholate of soda is treated in the above manner, there is obtained in its place a product behaving in every respect the same as cholepyrrhin. When dissolved in a little water, and mixed with concentrated sulphuric acid, this assumes a brilliant red color, and gradually, upon exposure to the air, becomes blue. When the red solution is mixed with more sulphuric acid, the color passes into brown. Upon the addition of water, there is produced a delicate precipitate, gradually becoming pale green; if the acid fluid is pour off from this, and what remains is warmed, intense green, blue, and violet colors are produced. The colored products dissolve in potash, with a bilious brown color, and the solution behaves, with nitric acid, in precisely the same manner as a basic solution of cholepyrrhin.

That the same metamorphoses may take place in the blood of a living individual is proved by injections of colorless solutions of bile into the veins of dogs. The urine passed after such an experiment usually deposits, upon standing, green flakes, which, upon the addition of nitric acid, exhibit in a beautiful manner the alternation of green, blue, violet, and red colors, characteristic of bile-pigment. The unchanged acids of the bile may then be sought for in vain by means of Pettenkofer's test. In one case only, where an unusually large quantity (about two drachms of dry bile), was injected, a trace of it could be detected. It is worthy of notice, that the quantity of coloring-matter voided in the urine appears greatest, when the animal experimented upon has suffered from dyspnœa, as, for instance, in one dog, which died from œdema of the lung, consequent upon the experiment. In one case, where the quantity of bile injected was small, and the animal remained free from respiratory ailments, no pigment was found at all. The statements which have been made by Dr. Kühne (*Virchow's Archiv*, xiv., p. 810) in opposition to the correctness of this view, have been completely refuted by Dr. Neukomm (*Archiv für Anatomie und Physiologie*. Leipzig, 1820).

bile, there are formed color-producing substances (*Chromogene*),[2] which, upon exposure to the atmosphere, and still more rapidly on the addition of nitric acid, exhibit alternations of tints, corresponding in every respect with bile-pigment. The same pigments and color-producing substances (*Chromogene*), which in their properties precisely resemble cholepyrrhin, are produced by the injection of large quantities of colorless bile into the vascular system of living animals. In this case the acids of the bile are transformed in the blood into pigment under the influence of respiration. That the bile which has been re-absorbed from the intestine, or which has passed directly from the liver into the blood, may, under normal circumstances, experience a similar transformation, is an opinion which is favored in the first place by the presence of large quantities of taurine in the healthy lung, as shown by Staedeler and Cloëtta. The pigments, however, which are produced in this way, are not voided with the urine, until the constantly advancing process of transformation to which the coloring-matter is subjected, has gone so far, that the substance is no longer endowed with the properties of bile-pigment.

---

[2]Chromogen is a term applied by Frerichs to a colorless material which, when subjected to the action of certain agencies above mentioned, is transformed into the coloring-matter of bile. The relations of the two substances are somewhat analogous to those of colorless and blue indigo.—TRANSL.

Frerichs had thus conducted two experiments. One convinced him that colorless bile acids convert into bile pigments by submitting the former to cold conc. H_2SO_4 and observing a color change to saffron-yellow, and a reddish color upon heating. When diluted with water, greenish or brownish flakes appeared, depending upon whether the H_2SO_4 solution was kept cold or heated. When sodium glycocholate was treated with conc. H_2SO_4 variously colored combinations were observed. Prolonged treatment produced a substance that imparts a deep green color to water and a brown color in aqueous Na_2CO_3. Upon treatment with HNO_3, the color changes observed (green→reddish→yellow) were reminiscent of a slower reacting positive *Gmelin* test for bilirubin or biliverdin. From colorless sodium taurocholate treated with conc. H_2SO_4 the same result was observed, except that the product obtained behaved "in every respect like Cholepyrrhin".

Frerichs suggested that the same metamorphosis may take place in the blood of a living individual. The jump to this conclusion might seem far-fetched today. There was no evidence, other than color, that bile acids convert to bile pigments in conc. H_2SO_4, and there is no reason to believe that the pigmented material obtained from chemical transformation of a bile acid in H_2SO_4 might also be obtained when dissolved in blood. Nonetheless, the second experiment convinced *Frerichs*. He injected colorless bile acid salts into the veins of dogs and found that the voided urine usually deposited green flakes upon standing – and the green flakes gave a positive *Gmelin* reaction. However, he could find no unchanged bile acids in the

2.8 Hämatoidin, Bilifulvin, and the Origin of Bile Pigments

voided urine by examination using the *Pettenkofer*[32] test (*122*). Today, one might suspect that administration of bile salts intravenously would lead to lysis of red cells, leading to exposure of hemoglobin to be acted upon by heme oxygenase to yield biliverdin, and possibly biliverdin reductase to yield bilirubin – as happens in a hematoma.

Frerichs' two telling experiments received comments in the translator's preface in *Murchison*'s translation from German into English (*121*) of *Frerichs'* updated and revised volume I before publication of the latter (*119*). (In 1860 only the 1st edition of *Frerichs'* volume I had been published, but most of the additions and corrections for the 2nd edition were provided by *Frerichs* for the English translation.) *Frerichs'* conclusions regarding the origin of bile pigments from bile acids were both contested and supported, as in *Murchison*'s preface (*120*):

> This view as to the origin of Jaundice is supported by two experiments, tending to show that the colorless biliary acids may become converted into bile-pigment. 1. the coloring-matter of bile may be formed artificially out of compounds of the biliary acids with soda. If the glycocholate or tauro-cholate of soda be digested for a long time, at an ordinary temperature, with concentrated sulphuric acid, the solution gradually assumes several different colors, and after a certain time, on the addition of water, a flaky precipitate, resembling the coloring matter of bile, is produced. 2. Frerichs found that, on injecting ox-bile, entirely freed from its coloring-matter and mucus, into the veins of dogs, the urine afterwards secreted became deeply colored with a substance, which was ascertained on chemical analysis to be bile-pigment. None of the biliary acids injected were found in the urine, and, indeed, Frerichs denies that these acids are ever found in the urine along with bile-pigment, although they are sometimes present in urine having no jaundiced hue. From these experiments, which were repeatedly confirmed, it has been concluded, that there is an intimate relation between the biliary acids and the bile-pigments, and that in fact the former become converted into the latter when subjected to the influence of certain agencies; and it has been thought, that, under certain pathological conditions, the biliary acids normally present in the blood are transformed into bile-pigment.

In his preface written in 1860 (*121*), *Murchison* cited detractors to this theory, especially *Kühne*.[33] *Murchison* thus wrote, referring to *Kühne*'s work (*123*) that " ... *Kühne* maintains that biliary acids do constitute an integral part of jaundiced urine, and he attributes the circumstance of their not having been hitherto demonstrated, to the insufficiency of the tests employed for the purpose." (*121*). More to the apparent controversy it generated (*121*):

---

[32] *Max Joseph von Pettenkofer*, was born on December 31, 1818 in Lichtenstein and died (suicide) on February 10, 1901 in Munich. He was Professor of Medical Chemistry in Munich, who studied medical chemistry under *Liebig* in Giessen, and devised a test for bile acids involving heating in cane sugar and conc. H_2SO_4 to produce a purple coloration.

[33] *Wilhelm Friedrich Kühne* was born on March 28, 1837 in Hamburg and died on June 10, 1900 in Heidelberg. He was a respected German physiologist who studied under *Wöhler* and *Wagner* at the Universität Göttingen in the 1850s, following which he studied physiology in Berlin, Paris, and Vienna (with *K.F.W. Ludwig* and *E.W. von Brücke*) before taking charge of the chemical department of the pathological laboratory under *Virchow* in 1863. Some five years later, he was appointed Professor in Amsterdam, and in 1871 answered a call to succeed *H. von Helmholtz* at Heidelberg.

Quite apart from the correctness of Frerichs' theory of icterus, which by the way, is only advanced as one that is highly probable, it is obvious that we have here to do with a question of facts, and the Kühne's facts are diametrically opposed to those brought forward by Frerichs. It is due, however, to Frerichs to state, that the results arrived at by him have been confirmed by several subsequent observers. Dr. Folwarczny, of Vienna (*Zeitschrift der kaiserl. u. königl. Gesellschaft der Aerzte zu Wien*. 1859. No. 15, p. 225), examined the urine in three cases of jaundice in Prof. Oppolzer's Clinique, but in all he failed to detect any trace of the biliary acids, although the examination was performed repeated, and Hoppe's process adopted in each case.

Professor Staedeler of Zurich, and Dr. Neukomm, have likewise arrived at results similar to those of Frerichs, and have in Frerichs' opinion, completely refuted the statements made by Kühne. . . .

. . . As to Kühne's opinion, that the coloring-matter which appears in the urine, after the injection into the veins of the colorless biliary acids, is derived from the hæmatine of the blood, it may be observed that, although it is possible that the coloring-matter of the blood may become transformed into bile-pigment, positive proofs are still wanting to show, that such a transformation really takes place. No one has yet succeeded in obtaining bile-pigment from the coloring matter of the blood. At all events, Kühne's experiments fail in proving that the coloring-matter in the urine originates from this source, and not from a transformation of the biliary acids; and they likewise fail in accounting for the disappearance of the biliary acids injected into the blood, in any other manner than that suggested by Frerichs.

Further observations and experiments on the whole subject are still required; but in the meantime it should be understood, that the main facts adduced by Frerichs in support of his theory of Icterus have received confirmation at the hands of most subsequent observers.

Working in collaboration with, and thus supporting *Frerichs*, were *Städeler*[34] in Zürich, and *Valentiner* (in *Frerichs'* lab) and perhaps others. But what were *Kühne's* facts that stood so clearly in opposition to *Frerichs'* theory? He reminded that bile acids hemolyzed red cells and proposed that the hemoglobin released is what leads to bilirubinuria, and that *Frerichs'* inability to detect intravenously administered bile pigments in urine was due to an insufficiency in the *Pettenkofer* test (*123–125*). In 1858, *Kühne* was, however, unable to show that intravenous injection of hemoglobin led to bilirubinuria.

Yet *Frerichs'* theory, his conclusions on the origin of urinary bile pigments, with its support from other scientists and relatively little dissent, would appear to have held sway in 1861. And these beliefs remained unaltered, even as studies by *Valentiner* in *Frerichs'* lab linked hematin to hematoidin, and hematoidin to bile pigments.

Despite his unaltered belief that bile pigments can arise from bile acids, *Frerichs* introduced new, potentially contradictory information in the updated and revised second edition of volume I (*119*). The information had emerged from studies in his own lab by *Valentiner*, who introduced $CHCl_3$ as an extraction solvent to remove

---

[34] *Georg Andreas Karl Städeler* was born on March 25, 1821 in Hannover and died on January 11, 1871 in Hannover. He received his doctoral degree in 1849 with *Wöhler* at Göttingen, was *Habilitant* and then a. o. Professor in 1851, before accepting a call as ordinarius Professor at the Polytechnic Institute in Zurich in 1853. (See Section 2.9.1).

2.8 Hämatoidin, Bilifulvin, and the Origin of Bile Pigments

Biliphäin (or *Cholepyrrhin*), which seemed from its color, crystal morphology, and *Gmelin* reaction to be the same as hematoidin or *Bilifulvin* (*71, 109*). *Frerichs* considered it probable that the hematoidin found in blood extravasations was derived from the pigment of blood, and the pigment of bile might originate in the vasculature and liver from the same source (*119*):

> Der Gallenfarbstoff steht in so enger Beziehung auf der einen Seite zum Blutroth, auf der anderen zu den farblosen Gallensäuren, dass es gerechtfertigt erscheinen kann, seinen Ursprung von der einen wie von der anderen Quelle herzuleiten. Die nahe Verwandtschaft zwischen Gallenpigment und Blutroth ergiebt sich aus den bereits oben erwähnten Thatsachen, besonders aber aus den Beobachtungen, welche in neuerer Zeit von Dr. Valentiner in meinem Laboratorio gemacht wurden (Günsburg's Zeitschrift, Dec. 1858.), denen zufolge ein Theil des Gallenfarbstoffs sich in Chloroform löst und aus dieser Lösung beim Verdunsten in krystallinischer Form mit den Eigenschaften des Hämatoidins gewonnen werden kann. Hierdurch wird die Möglichkeit nahe gelegt, dass, wie in Extravasaten aus Blutroth Hämatoidin sich bildet, in ähnlicher Weise auch im Gefässsystem und in der Leber Gallenfarbstoff aus dieser Quelle hervorgehe. Bis jetzt ist es indess Niemandem gelungen, Gallenpigment direct aus Blutroth darzustellen.

Or as *Murchison* wrote in his preface to the English translation of *Frerichs*' second edition of volume I (*121*):

> Since the publication of the German edition of the first volume, certain experiments have been performed in Frerichs' laboratory by his assistant, Dr. Valentin [*sic*], which tend to show that one of the coloring matters of bile consists of hæmatine, the substance which is known to be derived from blood-pigment. Valentin has succeeded in detecting crystals of hæmatine in gall-stones, in the bile of men and animals, and in the tissues and secretions of jaundiced patients. The addition of chloroform is found to dissolve the hæmatine with a yellow color, and from this solution red and brownish-red, lancet-shaped, and rhomboidal prismatic crystals separate, which correspond in every respect with those of hæmatine (*Günsburg's Zeitschrift, Dec.,* 1858). From these experiments, Frerichs admits there is an intimate relation between bile-pigment and the coloring matter of the blood, and even thinks it probable, that the former substance may be developed from the latter. Still he urges, that no one has succeeded in obtaining bile-pigment from the red matter of the blood, and that Valentin's results are not at all opposed to his theory of the convertibility of the colorless biliary acids into bile-pigment.

Shortly before the publication of *Frerichs*' 1861 updated 2nd edition of volume I of his *Klinik der Lebenkrankheiten* (*119*) and *Murchison*'s 1860 English translation of the greatly updated first edition (*119*), in 1860 (*126*), *Funke*[35] summarized work conducted with *Rudolf Zenker* on the origin of hematoidin from hematin derived from red cells. He indicated the identity of hematoidin with the *Biliphäin* obtained by *Valentiner* (*107*) and *Brücke* (*109*), and the *Bilifulvin* isolated first by *Berzelius* (*71*), then by *Virchow* (*94*). He concluded that *Bilifulvin* and *Biliphäin* were one and the same, and that the green pigment obtained by oxidation of *Biliphäin* is identical to the green pigment formed in stagnant blood. He cited *Kühne* as having carried out investigations showing that the bile pigment of icterus and icteric urine originated from the pigment of blood – contrary to *Frerichs*, who

---

[35] *Otto Funke*, born on October 27, 1828, died on August 17, 1879, was a German physiologist who was the first to crystallize hemoglobin.

74 2 Early Scientific Investigations

believed the bile pigment of icteric urine came from metamorphosis of bile acids. *Funke* thought *Kühne*'s explanation superior: that bile acids injected into blood cause red cell lysis, and the red pigment so released is transformed into the bile pigment found in icteric urine (*126*):

> Der Ursprung des Gallenfarbstoffs lässt sich mit voller Bestimmtheit, wie der Ursprung aller im thierischen Organismus unter normalen oder pathologischen Verhältnissen sich bildenden Pigmente, auf den Inhalt der farbigen Blutzellen zurückführen, ohne dass wir nöthig haben, den vermeintlichen Farbstoff des Blutzelleninhaltes, das sogenannte Hämatin, als darin präformirt anzusehen. Wir haben bereits oben erwähnt, dass das krystallinisch aus der Galle dargestellte Biliphäin nach VALENTINER und BRUECKE sich mit dem Hämatoidin, jenem krystallinischen Umwandlungsproduct des Blutfarbstoffes, identisch verhält, während VIRCHOW früher schon beide Stoffe zwar noch als verschieden, aber doch als einander höchst ähnlich und in genetischer Beziehung zu einander stehend bezeichnet hatte. Ebenso evident geht diese Identität oder nächste Verwandtschaft aus einer früher gleichzeitig von mir und ZENKER gemachten Beobachtung hervor. VIRCHOW hatte unter pathologischen Verhältnissen in stagnirender Galle einen in geknickten Nadeln und Nadelgruppen krystallisirten rothgelben Farbstoff (FUNKE, *Atlas*, 2. Aufl. *Taf.* IX, *Fig.* 3) gefunden, welchen er „Bilifulvin" nannte, weil er ihn für identisch mit einem von BERZELIUS so benannten (dritten) Gallenfarbstoff hielt. Schon VIRCHOW machte auf die Aehnlichkeit dieses Bilifulvins mit Hämatoidin in seinem Verhalten gegen Reagentien aufmerksam; ZENKER und ich wiesen nach, dass Bilifulvin entweder von selbst oder durch Behandlung mit Aether sich in schöne grosse Krystalle verwandelt, welche alle Eigenschaften des Hämatoidins haben, also mit diesem identisch sind. Nehmen wir die oben genannten Beobachtungen von BRUECKE hinzu, so ist wohl nicht zu bezweifeln, dass dieses VIRCHOW'sche Bilifulvin nichts Anderes als krystallinisches Biliphäin ist. Kurz der normale braune Farbstoff der Galle, welcher durch Oxydation in den grünen übergeht, ist identisch mit einem in stagnirendem Blut sich bildenden Umwandlungsproduct des farbigen Blutzelleninhaltes. Einen weiteren trefflichen Beweis für den Ursprung des Gallenfarbstoffs aus Blutfarbstoff hat KUEHNE durch sein Untersuchungen über Icterus geliefert. FRERICHS hatte beobachtet, dass nach Injection von farbstofffreier Galle oder reinen gallensauren Salzen in's Blut im Harn Gallenfarbstoff erscheint, und hatte daraus geschlossen, das die Gallensäure sich im Blute in Gallenfarbstoff unwandelte, eine Metamorphose, welche er auch künstlich durch Digestion von gallensauren Salzen mit Schwefelsäure erzielt haben wollte. KUEHNE hat der genannten Thatsache eine andere besser gestützte Deutung gegeben. Die gallensauren Salz haben das schon oben erwähnte eigenthümliche Vermögen, die Blutkörperchen vollständig aufzulösen, daher auch nach Injection derselben in's Blut häufig zuerst blutig gefärbter Harn secernirt wird. Dieser aus den gelösten Blutkörperchen befreite Blutfarbstoff, nicht die Gallensäure selbst, welche neben dem Farbstoff im Harn erscheint, ist es, welcher sich in Gallenfarbstoff umwandelt, und so in den Harn übergeht. 52

Thus, in 1860–1861, a picture emerged that *Virchow*'s hematoidin was probably the same as *Berzelius*' and *Virchow*'s *Bilifulvin*, which appeared to be the same as the *Biliphäin* from *Valentiner* and *Brücke*. But the controversy over the origin of bile pigment in icteric urine was unresolved.

Why *Frerichs* conducted experiments to show that the origin of the bile pigment of icteric urine has its roots in bile acids is not entirely clear. Though it might seem an odd tangent, it should be recalled that *Frerichs*' earlier research on liver diseases and renal dysfunction involved the presence of leucine and tyrosine in urinary sediment in acute yellow atrophy of the liver – studies that may have had their origin in a youthful collaboration with *Städeler* involving studies of leucine and tyrosine

2.8 Hämatoidin, Bilifulvin, and the Origin of Bile Pigments

produced in humans and animals (*127, 128*). The two probably met through *Wöhler* at Göttingen, where *Frerichs* was located from 1842 to 1850 and where *Städeler* received his doctoral degree in 1849. *Frerichs* and *Städeler* had clearly worked together in some fashion, starting in the early 1850s on leucine and tyrosine in biological tissues. By 1855, they had reported their findings on the presence of leucine and tyrosine in the human liver (*127*), a study possibly emanating from *Frerichs'* autopsies conducted in Kiel in 1851, wherein he found needle-like crystals in degraded liver cells from death due to liver atrophy and blood intoxication. Later, in 1853, while in Breslau, during an autopsy *Frerichs* found crystals in the hepatic vein of a liver with bile duct blockage, crystals that were separated and identified as leucine and tyrosine. (The bile was dark brown, with brown corn-like solids present). Treatment of tyrosine with conc. H_2SO_4 produced a red-colored solution from the dissolved solid. And it was stated (*127*) that a year earlier one of the authors (*Städeler*) had communicated to the *Köngl. Gesellsch. d. Wissensch. zu Göttingen* that tyrosine mixed with hydrochloric acid and sodium chlorate produced a red solution that turned yellow, with evolution of a gas. A connection between animal pigments and tyrosine should not be drawn from these experiments. But the authors noted that after injection of tyrosine into the blood system of an animal, it was not found in its urine, in contrast to leucine, thereby suggesting that perhaps the tyrosine was decomposed in the liver.

In 1856, *Frerichs* and *Städeler* published a follow-up paper (*128*) on the presence of leucine and tyrosine in animal organisms. Citing the earlier work (*127*) carried out prior to 1855, they indicated that proteins probably cleave in human organs as they do in the presence of acid or base to yield crystalline leucine and tyrosine that may accumulate in the liver in certain liver pathologies and which also (in the case of tyrosine) is used in the biosynthesis of bile acids. They found crystals of tyrosine in the urine of a woman with acute liver atrophy.

Given their interest in tyrosine, and the known fact that it could be produced from bile acids (taurocholic acid), it was not entirely illogical that *Frerichs* and *Städeler* might take an interest in bile acids and their metabolism. This led to their 1856 publication on the transformation of bile acids into pigments (*129*). It was known that while icteric urine is rich in pigment, it is devoid of bile acids, and these observations, reconfirmed by *Frerichs* and *Städeler*, led to the belief that there might be a close relationship between the bile acids and the bile pigments of urine. That is, with impeded bile flow the bile acids arrive either unaltered in urine or transformed into a bile pigment (*129*):

> Es kann als feststehend angenommen werden, dass in dem Harn Ikterischer, wenn derselbe reich an Pigment ist, keine Gallensäuren oder doch nur Spuren derselben vorkommen. Wir selbst konnten bei frühern wiederholten Versuchen keine Gallensäuren darin auffinden, gelangten also zu demselben Resultat wie Griffith, Pickford, Gorup-Resanez und Scherer. – Lehmann hat dagegen beobachtet, dass bei entschiedenem Ikterus in schwach pigmentirtem Harn die Gallensäuren oft in grosser Menge vorkommen.
>
> Diese Beobachtung, an deren Richtigkeit wohl nicht gezweifelt werden kann, schien uns entschieden darauf hinzudeuten, dass ein naher Zusammenhang zwischen den Säuren und den Farbstoffen der Galle vorhanden sei, und dass bei verhindertem Abfluss der Galle, die Säuren entweder unzersetzt in den Harn gelangen, oder zuvor im Blut oder irgend welchen Orangen eine Umwandlung in Farbstoff erleiden. 53

In order to examine this, it first had to be determined whether bile acids would convert outside the organism into pigments. And indeed, it was found that conc. H_2SO_4 dissolves sodium glycocholate (glycocholic acid is the amide of glycine with cholic acid) to afford a saffron yellow color that turns to a bright, fire-red-to-brown color upon warming. The glycocholic acid-turned-pigment took up O_2 from air to produce various colors. Precipitation by added H_2O produced flakes, which when gently heated turned violet, then blue after a few seconds. Similarly, filter paper coated with the aqueous H_2SO_4 solution and dried produced a green color (*129*):

> Wird reines glycocholsaures Natron mit concentrirter Schwefelsäure übergossen, so klebt es zu einer farblosen, harzähnlichen Masse zusammen, die sich in der Kälte mit saf-rangelber, beim Erwärmen lebhaft feuerrother bis bräunlichrother Farbe auflöst. Aus der Lösung fällt Wasser farblose, grünliche oder bräunliche Flocken, je nach der Temperatur bei welcher die Lösung erfolgte.
>
> Die durch concentrirte Schwefelsäure veränderte Glycocholsäure hat die Eigenschaft, an der Luft rasch Sauerstoff aufzunehmen, und damit in prachtvoll gefärbte Verbindungen überzugehen. Bringt man die durch Schwefelsäure entstandene farblose amorphe Masse, nachdem sie möglichst von anhängender Säure befreit worden ist, auf ein Stück Filtrirpapier, so zerfliesst sie, und es entsteht ein rubinrother Fleck, der bald blaue Ränder zeigt, und nach kurzer Zeit rein indigblau wird. Nach einigen Tagen verschwindet auch diese Farbe und der Fleck wird hellbraun. – Die Papiersubstanz scheint bei dieser Reaction ohne Einfluss zu sein, denn man beobachtet eine ganz ähnlichen Farbenwechsel beim Zerfliessen der amorphen Masse auf Glas oder Porzellan, nur tritt er in diesem Falle etwas weniger rasch ein.
>
> Die Lösung der Glycocholsäure in concentrirter Schwefelsäure enthält dasselbe Chromogen aufgelöst, die überschüssige Säure verzögert aber die Oxydation und die damit verbundene Färbung. Fällt man die Lösung mit Wasser, und erwärmt die von der sauren Flüssigkeit getrennten Flocken gelinde im Wasserbade, so färben sie sich nach wenigen Secunden violett und blau. Sehr schön beobachtet man auch den Farbenwechsel, wenn man ein Stück Filtrirpapier mit Wasser befeuchtet, dann mit der sauren Lösung bestreicht, und über der Lampe trocknet. Hat die Schwefelsäure, längere Zeit bei der Temperatur des Wasserbades auf Gallensäure eingewirkt, so wird der auf gleiche Weise auf Papier erzeugte Fleck grün. 54

To determine whether bile acids might become bile pigments, *Frerichs* and *Städeler* treated glycocholic acid with H_2SO_4 and were satisfied that colors were produced. The follow-up experiments were then directed to treating bile itself, decolorized and shown to precipitate substantial sodium taurocholate with added ethyl alcohol. This bile, which also contained other colorless components, when mixed with conc. H_2SO_4 turned red-brown with warming (likely due to heat of mixing) and reflected light with a vivid grass-green color. Exposure to oxygen of air turned the red-brown bile mixture to an indigo-blue color. The blue pigment separated as a solid mass upon addition of H_2O. The blue pigment partially dissolved to form a grass-green solution in alcohol and a green-blue residue, which turned greenish-brown upon dissolving in aq. potash. Treatment with acetic acid regenerated the original color.

Heating the original bile and H_2SO_4 solution for six hours produced substantially the same results, except the blue residue produced by addition of H_2O became yellow-green instead of green-brown upon partially dissolving in aq. potash. Addition of acetic acid gave the same green-brown color seen previously. In hot

2.8 Hämatoidin, Bilifulvin, and the Origin of Bile Pigments

acetic acid a bile-brown (*Gallenbraun*) color was seen, and this solution upon treatment with HNO_3, turned deep blue-green, then violet, then dirty yellow. This was vaguely reminiscent of the color display seen in the *Gmelin* reaction for bile pigments – a promising but misleading sign to the investigators. Treatment of the brown acetic acid solution (above) with $Pb(OAc)_2$ yielded a little colored precipitate that showed the (bile pigment-like?) display of colors upon treatment with HNO_3. At this point, these pigments (of uncertain purity and composition) and some of their solutions were beginning to behave like bile pigments toward the *Gmelin* reaction (*129*):

> Die syrupförmige Galle wurde mit dem 3-4 fachen Volumen concentrirter Schwefelsäure vermischt, wobei sie sich unter freiwilliger Erwärmung bräunlichroth färbte. Nach halbstündigem Erhitzen im Wasserbade war die Mass tiefer rothbraun und reflectirte das Licht mit lebhaft grasgrüner Farbe. Wasser fällte braune Flocken, die bei Luftzutritt erwärmt indigblau wurden. Die blaue Masse war in Wasser unlöslich, bei Siedhitze entstand eine braune Lösung, aus der sich beim Verdampfen ein Zersetzungsproduct als dunkelbraune Membran abschied. Die grasgrüne weingeistige Lösung des blauen Farbstoffs hinterliess beim Verdunsten einen grünlichblauen Rückstand, der beim Uebergiessen mit Kali gelbbraun wurde, ohne sich in wesentlicher Menge zu lösen. Säuren, selbst verdünnte Essigsäure, stellten die ursprüngliche Farbe wieder her.
>
> Nach sechsstündigem Erhitzen der Mischung von Galle und Schwefelsäure wurde im Wesentlichen dasselbe Resultat erhalten. Auch jetzt färbte sich die blaue Masse auf Zusatz von Kali gelbbraun, löste sich kaum im Ueberschuss, und ward auf Zusatz von Essigsäure wieder grünlichblau. Mit heisser Essigsäure entstand eine gallenbraune Lösung, die auf Zusatz von Salpetersäure sogleich tief blaugrün, dann violett und zuletzt schmutzig gelb wurde. – Essigsaures Bleioxyd erzeugt in der braunen essigsauren Lösung einen wenig gefärbten Niederschlag, der beim Uebergiessen mit Salpetersäure ebenfalls Farbenwechsel zeigte.
>
> Nachdem die Mischung von Galle und Schwefelsäure acht Tage lang auf einem mässig geheizten Wasserbade erhitzt worden war, hatte sich eine dunkelgrüne, aus kleinen mikroskopischen Kugeln bestehende Masse abgeschieden, die in saurem Wasser unlöslich, in reinem Wasser mit tief grüner Farbe löslich war. In verdünntem Kali löste sie sich vollständig mit rein gallenbrauner Farbe, und auf Zusatz von Salpetersäure trat zuerst grüne, dann röthliche und zuletzt gelbe Färbung ein.
>
> Das mitgetheilte Verhalten dieser Zersetzungsproducte gegen Salpetersäure erinnert an das der natürlichen Gallenpigmente, indess war der Farbenwechsel immer weniger lebhaft, wie man ihn beim Vermischen von stark pigmentirtem ikterischen Harn mit Salpetersäure beobachtet. Günstigere Resultate erhielten wir aber, als wir den amorphen, vorzugsweise aus taurocholsaurem Natron bestehenden Niederschlag, den wir mit Aether aus der weingeistigen Lösung der entfärbten Ochsengalle gefällt hatten, mit Schwefelsäure behandelten. 55

The authors were clearly impressed with the similarity in behavior (in colors, to some extent, and with apparently positive, or at least similar, *Gmelin* reactions) between natural bile pigments and the decomposition products which they obtained from bile acids. Though they were to be led astray by the correlation, they suspected that the pigments might arise as by-products of the biosynthesis of glycocholic acid – in which taurine ($^+H_3NCH_2CH_2SO_3^-$) is decomposed into glycine ($^+H_3NCH_2CO_2^-$) and *Saligenin* in the liver (*129*):

> Für jetzt beschränken wir uns darauf, auf die Aehnlichkeit der natürlichen Gallenpigmente mit den von uns erhaltenen Zersetzungsproducten der Gallensäuren aufmerksam zu

78   2   Early Scientific Investigations

> machen; das aber glauben wir schon jetzt bestimmt aussprechen zu dürfen, dass das Chromogen, aus welchem durch Oxydation der blaue Farbstoff entsteht, mitunter in der Leber, und wie es scheint auch im Pancreas . . . vorkommt. Wir haben schon bei früherer Gelegenheit auf diesen Farbstoff aufmerksam gemacht, . . . damals war es uns aber noch unbekannt, dass derselbe in so einfacher Relation zu den Gallensäuren stehe. Auch der blaue Farbstoff, der sich mitunter aus Menschenharn auf Zusatz von Säuren abscheidet, und sich nach v. Sicherer's Versuchen in einen Körper umwandeln lässt, der dem Indigo vollkommen ähnlich ist, ist vielleicht ein Zersetzungsproduct der Gallensäuren. Wir sprachen schon früher. . . die Ansicht aus, dass dieser Farbstoff als Nebenproduct bei der Bildung der Glycocholsäure entstehen könne, indem sich das Tyrosin in der Leber in Glycin und Saligenen zerlege . . .   56

Which of course is a little far-fetched.

In a footnote to the 1856 paper (*129*) by *Frerichs* and *Städeler*, an experiment was cited in which a measure ("eine Drachme") of pure (?) colorless ox-bile dissolved in H_2O, was injected intravenously into a dog. Six hours later, 3 ounces of dark-brown urine were collected. It was strongly alkaline (the pH of urine is usually ~6.0). Upon standing, a thick sediment of green flakes precipitated which looked like brownish-green granules under a microscope. Addition of HNO_3 produced the most beautiful display of color changes characteristic of bile pigments. And the *Pettenkofer* reaction was negative (*129*):

> Neuere Erfahrungen haben aus diese allerdings bestätigt. Wir injicirten einem Hunde etwa eine Drachme reiner farbloser Ochesengalle, die in destillirten Wasser gelöst war. Sechs Stunden nachter liess das Thier gegen 3 Unzen dunkelbraunen Harns von 1,015 spec. Gew. und sehr schwach alkalischer Reaction. Beim Stehen liess derselbe eine ziemlich dicke schicht grüner Flocken fallen, welche unter dem Mikroskop als braungrüne Körnchen erschienen. Auf Zusatz von Salpetersäure zeigten sie auf das Schönste den für Gallenpigment characteristichen Farbenwechsel. Die Pettenkofer'sche Probe ergab ein negatives Resultat.   57

From the latter (hound) experiment, *Frerichs* and *Städeler* concluded that the origin of bile pigments (at least those excreted into urine) had their origin in bile acids. This apparently straightforward conclusion neglected the possibility that bile, or the bile acids therein, might have induced the release of a different pigment precursor from a component of blood, *e.g.* the "heme" of hemoglobin) from red cells (by cell lysis), which was converted to the bile pigment by way of hematin. The footnote supporting this contention is from work published in 1856 (*129*, pp. 105–106). *Frerichs* referred to the study in his 1861 second edition (*119*), and it was cited in *Murcheson*'s translation (*121*) published in 1860. In both publications, the work of *J. Neukomm* (thesis in Zürich, 1859) was indicated as support. *Neukomm* apparently worked in *Städeler*'s lab in Zurich and perhaps from this he also became associated with *Frerichs*. But what had *Neukomm* accomplished in Zürich that was so supportive of *Frerichs* and *Städeler*'s belief that bile acids were the source of urinary bile pigments in icterus? In March 1860, he published on the detection of bile acids in urine and their transformation in the blood stream (*130*). He took issue with *Kühne*'s experiments that showed bile acids injected into the bloodstream underwent no change and were expelled again in urine (*130*):

> W. Kühne hat gestuetzt auf eine Reihe von Versuchen, die Behauptung ausgesprochen, dass Gallensäuren, welche in die Blutbahn gelangen, keine Veränderung erleiden und durch den Urin wieder aus dem Körper entfernt werden.   58

2.8 Hämatoidin, Bilifulvin, and the Origin of Bile Pigments

It was the method of analysis and conflicting results associated therewith, *inter alia*, that was bothersome to *Neukomm*, and so he set about calibrating the *Pettenkofer* test with samples of ammonium cholate and sodium glycocholate, made up in urine, in order to compare the accuracy of the modification of *Hoppe* (*Hoppe-Seyler*, 1825–1895, see Section 2.10.2) used by *Kühne* to the usual method. In the usual method, according to *Pettenkofer*, a bile acid solution was mixed with 2/3 its volume of conc. H_2SO_4, after which a 10% solution of sugar was added with care and allowed to warm up to 70–75°C. Depending on the type and initial concentration of bile acid; for cholic acid at 0.4% a purple-violet coloration was observed, at 0.1% purple-red, at 0.04% weakly wine-red, at 0.01% weakly yellow. With glycocholic acid at the same concentration a noticeably weaker coloration was observed. A quantitative colorimetric experiment seemed to be required (*130*):

> Es sind hier indess nur die am besten gelungenen Färbungen angeführt, da auf dieselben raschere oder langsamere Mischung mit Schwefelsäure und die dabei unvermeidlichen Temperaturschwankungen von grossem Einfluss sind. Eine quantitative colorimetrische Bestimmung der Gallensäuren ist daher mit Hülfe der Pettenkofer'schen Reaction nicht zu erzielen. 59

From a series of careful quantitative measurements, *Neukomm* learned that the colors in the *Pettenkofer* reaction depended on the initial bile concentration and reaction temperature, irrespective of whether the H_2SO_4 was mixed rapidly or slowly. He concluded that quantitative bile pigment determination could not be attempted colorimetrically – a conclusion important to the level of bile acid detectability in urine (*130*):

> Die Grenzen der Reaction werden bedeutend erweitert, wenn man jenes Verfahren etwas abändert. Ich beobachtete, dass ein einziger Tropfen einer 1/20 procentigen Cholsäure oder Glycocholsäurelösung noch ein prachtvolles Purpurviolett liefert wenn man denselben in einer Porcellanschale mit einem Tropfen verdünnter Schwefelsäure (4 Theile HO + 1 Theil $HOSO_3$)[36] und einer *Spur* Zuckerlösung vermischt und unter Umschwenken über einer kleinen Spirituslampe vorsichtig und bei gelinder Wärme verdampft. Bei einigem Stehen der Probe nimmt die Farbe an Intensität ansehnlich zu. – Da 1 CC. nahezu acht Tropfen ausmacht, so gelingt es also auf diese Weise, noch 6/100 Milligrm. Gallensäure mit voller Schärfe nachzuweisen. Eine grössere Concentration der Lösung ist natürlich nicht störend; bei stärker Verdünnung hat man die zu prüfende Flüssigkeit zuvor auf einen oder zwei Tropfen zu verdampfen. – 1 CC. einer 1/100 procentigen Lösung beider Säuren gab auf die angegebene Weise noch die herrlichste purpurviolette Färbung, während bei gleicher Verdünnung und bei Anwendung von 3 CC. Lösung das Pettenkofer'sche Verfahren ohne Resultat blieb. 60

In the process of this experimentation, *Neukomm* devised a very sensitive modification to the original *Pettenkofer* test that effectively lowered the detection limit of bile acids in urine. Adding dilute H_2SO_4 to a small sample of urine, plus a trace of sugar, and warming to evaporation in a porcelain dish to display the colors (*130*):

> Gelang es nur auf die letzte Weise, das Vorhandensein von Gallensäuren zu constatiren, so wird diess in dem Folgenden der Kürze wegen durch „*Prüfung in der Porcellanschale*" angedeutet werden. 61

---

[36] The formulas, HO for H_2O and $HOSO_3$ for H_2SO_4, were based on *Gmelin's* atomic masses for H (1), O (8), and S (16).

To compare *Hoppe*'s variation of the *Pettenkofer* test, a clear solution of 0.1 g sodium glycocholate in 500 cc urine was mixed with milk of calcium, $Ca(OH)_2$, and heated to reduce the volume to ~2/3, then filtered, and the filtrate was reduced to a volume of ~50 cc. At which point excess HCl was added and the liquid was heated for ½ hour. It was strongly red-brown; to it was added 6–8 times its volume of H_2O to precipitate brown flakes. The precipitate was isolated and dissolved in alcohol; further processing afforded a yellow residue that was dissolved in a little aq. NaOH. This residue was submitted to the *Pettenkofer* reaction (adding H_2SO_4) to produce a reddish brown coloration that intensified upon addition of sugar – however without the characteristic color tone for bile acids. In contrast, when a portion of the solution was treated in *Neukomm*'s modified *Pettenkofer* test, a purple-violet color ensued. The same results were obtained from a 50% lower initial concentration of bile acid. On the basis of the unusual coloration in the *Hoppe* modification it was thought that that modification would lead to uncertainties in the case of ambiguous amounts of bile acids (*130*):

> Aus diesen Versuchen geht hervor, dass die Hoppe'sche Methode auch bei Anwendung nicht unbedeutender Mengen von Gallensäuren nur ein zweideutiges Resultat liefert und dass sie zur Nachweisung von kleinen Mengen ganz unbrauchbar ist. 62

In order to semi-quantitate his modified *Pettenkofer* test, *Neukomm* used lead (II) acetate to precipitate cholic acid and (separately) glycocholic acid from their aqueous solutions of ammonium cholate and sodium glycocholate. The precipitated lead salts were then converted back to small aq. volumes (3 cc) of sodium salts and treated with H_2SO_4 (2 cc) and some sugar to produce a purple-red color. The initial concentrations of bile salts ranged from 0.03 g/1,000 cc, to 0.005 g/1,000 cc, and in all cases the characteristic coloration was observed. Even at 100,000–200,000 times more dilute bile salt, the test was positive following isolation and processing of the lead salt precipitate.

These experiments established the great sensitivity of *Neukomm*'s modification. In the same way he proceeded to analyze the same bile salts made up in urine to similar concentrations to yield the same colors and concluded that he could detect 0.001% glycocholic acid with his modification and only 0.02% using *Hoppe*'s – his being a factor of 20 more sensitive (*130*):

> Nach dieser Methode gelang es, 1/1000 pC. Glycocholsäure im Urin nachzuweisen, während dieses bei den nach Hoppe's Verfahren angestellten Versuchen bei 1/50 pC. kaum möglich war. Es ist daher jene Methode allein brauchbar, wenn es sich um die Nachweisung kleiner Gallensäuremengen handelt. Ja ich muss hinzufügen, dass die Hoppe'sche Methode in *allen* Fällen unsicher und daher untauglich zu sein scheint. 63

In order to sort out possible interference in the *Pettenkofer* test due to the presence of bile pigments, *Neukomm* investigated icteric urine. Here he showed the expected positive *Gmelin* reaction (for bile pigments), the display of colors from added HNO_3, and also a positive (modified) *Pettenkofer* test for strongly brown-colored icteric urine using the lead (II) acetate precipitation method, but a negative or highly uncertain test resulted from the standard *Pettenkofer* test (*130*):

> Da nun in der anfangs erwähnten Abhandlung W. Kühne eine Umwandlung der Gallensäuren im Blute ganz in Abrede stellt und behauptet, dass die demselben zugeführten

2.8 Hämatoidin, Bilifulvin, and the Origin of Bile Pigments 81

Säuren durch den Harn wieder aus dem Körper entfernt werden, so schien es für die Physiologie sowohl wie für die Pathologie von Interesse zu sein, theils durch Untersuchung von icterischem Harn, theils durch Injectionsversuche an Thieren die Angaben Kühne's einer weiteren Prüfung zu unterwerfen.

In dem Folgenden theile ich die Resultate der angestellten Untersuchungen mit....

Auch bei diesem Harn wurde also durch die gewöhnliche Pettenkofer'sche Probe ein negatives oder doch höchstens sehr zweifelhaftes Resultat erhalten, während nach unserem modificirtem Verfahren wenigstens Spuren von Gallensäuren unzweideutig nachweisbar waren. 64

With the preceding calibrations and controls accomplished, *Neukomm* advanced his studies to examining the urine of dogs injected intravenously with bile acids (aqueous sodium glycocholate). Following injection of the solution into a leg vein, urine was collected 12–15 hours (bright yellow) and again 36 hours (yellow) post injection. The first, weakly alkaline, became wine red upon the addition of H_2SO_4 and did not change upon addition of sugar solution. Crude concentrated HNO_3 showed at the contact point with urine a faint rose red ring without a tint of green. The second, acidic, behaved the same way with H_2SO_4 and HNO_3. Both urines were evaporated, taken up in alcohol and treated further as described earlier via precipitation by lead(II) acetate to produce the smallest hint of violet, which thus excluded the presence of bile acids in any considerable amount.

Four weeks later, the same dog was injected into the jugular vein with sodium glycocholate and urine was collected 15 hours, 26 hours, and 64 hours post injection. The first urine sample was dark brown and acidic (15 hours); the second was yellow and acidic with a tinge of dirty brownish green (26 hours); and the third (64 hours) was yellow and neutral. After standing several hours, the first urine yielded a greenish sediment with a positive *Gmelin* reaction. The greenish, yellow-brown filtrate, upon heating, gave red-brown flakes that did not dissolve upon addition of a little acetic acid. Filtration yielded a yellow filtrate and a greenish sediment. Added crude HNO_3 produced a barely perceptible *Gmelin* reaction. Addition of conc. H_2SO_4 showed a violet-red ring at the site of contact, and with complete mixing a wine-red color. Addition of sugar gave no further change. The second (26 hours) urine was yellow and acidic with a dirty brown-green sediment. Heating induced only slight turbidity that persisted upon increased acidification with acetic acid. Added HNO_3 produced a distinct *Gmelin* reaction; added conc. H_2SO_4 produced at the contact point a brown-red color, which in contrast to the urine lying on top, changed over to violet and blue.

The third urine (64 hours) was yellow and neutral, gave a barely detectable *Gmelin* reaction, and with conc. H_2SO_4 behaved as above.

The first and second urines were combined and divided into two equal parts. To one part an ethanol extract was prepared according to the earlier procedure. The other half was subjected to *Hoppe*'s method for detecting bile acids. From the ethanol extract, following processing, the lead salt was subjected to *Neukomm's* revised *Pettenkofer* test to give first a bluish color at the contact point with the lower H_2SO_4 layer, then violet and brownish. With complete mixing and addition of sugar it turned brown-yellow with a reddish tinge. In comparison, sodium glycocholate easily produced a purple-violet color. The *Hoppe* variation produced a

82 2 Early Scientific Investigations

weak reddish-brown; added sugar turned it yellow-brown. Thus, no bile acid was detected in either version of the *Pettenkofer* test (*130*):

> In diesem Falle liessen sich also bei vorsichtiger Anwendung der üblichen Methoden keine Gallensäure im Harn nachweisen. 65

Fourteen days later the same, poor, overworked hound was again injected (jugular vein) with aq. sodium glycocholate, and yellow, weakly alkaline urine was collected 15 hours and 24 hours post injection. The first urine gave a negative *Gmelin* reaction with added HNO_3. Added conc. H_2SO_4 produced a weak violet-reddish to brownish color that did not change upon adding sugar. The second urine behaved in the same ways. The combined urines were split in half. One half was treated according to the modified *Pettenkofer* reaction, the other according to the *Hoppe* method, as above. Both methods gave the same results as above, a reddish-brown coloration with H_2SO_4 with no characteristic purple-violet color being produced even upon adding sugar.

The dog injection experiments were repeated with several different dogs, giving both similar and mixed results: sometimes an uncertain or negative test result for bile pigments; sometimes unequivocally positive. Usually, a negative result in the more sensitive modified *Pettenkofer* test seldom indicated the presence of bile acids.

Having gathered as much experimental evidence as he was able, *Neukomm* explained that: (1) while bile acids might have been found in icteric urine, the levels were too low to be detected in the usual *Pettenkofer* test; (2) in the animal experiments transfer of intravenously injected bile acids into urine was disproved by the usual *Pettenkofer* test and the absence of a bitter taste, as the facts proved that only traces of the injected bile acids passed into urine and that *Kühne's Pettenkofer* tests showing otherwise were deceptive (*130*):

> . . . In keinem Falle wurde aber ein bitterer Geschmack der schliesslich erhaltenen Natronverbindungen wahrgenommen; in keinem Falle liess sich darin mit Hülfe des gewöhnlichen Pettenkofer'schen Verfahrens Gallensäure mit einiger Sicherheit nachweisen, und nur in zwei Fällen wurde bei der Prüfung in der Porcellanschale eine charakteristische Färbung wahrgenommen.
>
> Diese Thatsachen beweisen, dass die ins Blut getretenen Gallensäure nur spurweise in den Harn übergehen können, und es wird damit der Ausspruch von Kühne: „*Die Natronverbindungen der Glycochol-, der Chol- und Choloïdinsäure verlassen in die Venen injicirt durch die Nieren den Körper des Thiers*" genügend widerlegt. Kühne hat sich mehrfach damit begnügt, direct mit den nöthigenfalls nur von Eiweiss befreiten Harn die Pettenkofer'scher Probe anzustellen; offenbar hat in solchen Fällen eine Täuschung durch die vorhandenen Farb- und Extractivstoffe stattgefunden, die, wie angeführt wurde, bei alleinigem Zusatz von Schwefelsäure zum Harn von Menschen und Hunden nicht selten zu rothen und selbst violetten Färbungen Veranlassung geben. 66

Neukomm noted the variability in the excretion of bile pigments or even traces of bile acids in the urine of dogs, post-injection, and that bile pigment was always seen by *Kühne*, along with "supposed" bile acid (*130*):

> Zuweilen enthält der Harn von Hunden, denen glycocholsaures Natron ins Blut injicirt worden ist, bald grössere, bald kleinere Mengen von Gallenfarbstoff. Frerichs . . . stellte 29 Versuche an, unter denen 19 ein positives Resultat gaben. Gewöhnlich enthielt dann der Harn gleichzeitig etwas Eiweiss und aufgelöstes Blutroth. Bei den von mir angestellten 7

2.8 Hämatoidin, Bilifulvin, and the Origin of Bile Pigments

> Injectionsversuch trat einmal der Farbstoff in solcher Menge auf, dass er sich zum Theil in Flocken ausschied, in zwei anderen Fällen war nur gelöstes Pigment vorhanden, die übrigen vier Versuche führten zu einem negativen Resultat. In den von Kühne mitgetheilten Versuchen war neben der vermeintlichen Gallensäure stets Gallenfarbstoff vorhanden. 67

The conclusion drawn was that injected bile acids probably led to the presence of bile pigment in urine, though injection might also have caused no production of bile pigment. Although *Kühne* completely denied such a transformation, he communicated that pigment regularly arose in post-injection urine, and he maintained that any bile pigment seen in such urine originated from the pigment of blood, the hematin released from red cells into blood. *Neukomm* did not agree with the last because when *Kühne* injected hematin no bile pigments appeared in urine, while when hematin and bile acids were injected together, bile pigment was detected in urine (*130*):

> Aus diesen von ganz verschiedenen Seiten gemachten Beobachtungen über Pigmentbildung bei Einführung von Gallensäuren ins Blut dürfte man schliessen, dass sich die Gallensäuren ebenso wie auf künstlichem Wege so auch in der Blutbahn in Chromogene und schliesslich in Farbstoffe verwandeln. Indess sind die beobachteten Ausnahmen nicht zu gering anzuschlagen; eine Umwandlung der Gallensäuren in Gallenpigment kann jedenfalls nur unter Zusammentreffen besonderer günstiger Umstände stattfinden. Mir wollte es scheinen, als ob dazu ein gewisser Grad von Irritation nothwendig sei, denn in drei von meinen Versuchen trat das erste Mal bei zufälliger, das andere Mal bei absichtlicher stossweiser Injection das Gallenpigment im Harn auf. Es fehlte an Hunden, um diese Versuche zu vervielfältigen.
>
> Kühne leugnet die Umwandlung der Gallensäure in Gallenfarbstoff gänzlich, obgleich er eine grosse Zahl von Versuchen mittheilt, bei denen regelmässig nach Galleninjection Pigment im Harn auftrat. Er vertheidigt die Ansicht, dass aller Gallenfarbstoff vom Blutfarbstoff abstamme, und zwar soll das beim Zerfallen der Blutkörperchen frei in Lösung gehende Hämatin ein Umwandlung in Gallenfarbstoff erleiden. Diese Ansicht erhielt aber durch das Experiment keine Stütze, denn als Kühne gelöstes Hämatin in die Venen injicirte, trat kein Gallenfarbstoff im Urin auf, während wenn er zur Injection gleichzeitig Hämatin und Gallensäure anwandte, die Bildung von Pigment beobachtet wurde. Kühne sieht sich daher auch gezwungen, der Gallensäure einen besonderen, noch räthselhaften Einfluss auf das gelöste Blutroth zuzuschreiben. 68

While *Neukomm* was forced to assume that blood was not the origin of the observed urinary bile pigment, he indicated that *Kühne*'s experiments did not disprove that bile acids injected into the bloodstream did not convert into bile pigments in the urine under certain circumstances (*130*):

> Ich bin weit davon entfernt anzunehmen, dass das im Körper zu Grunde gehende Blutroth *nicht* zur Bildung von Gallenfarbstoff Veranlassung geben könne, obwohl dieses durch das Experiment noch nicht nachgewiesen ist. Auf der andern Seite ist aber durch Kühne's Versuche nicht widerlegt worden, dass auch die in die Blutbahn gelangenden Gallensäure unter Umständen in Gallenpigment übergehen können. – Dass hier noch Lücken auszufüllen sind, ehe man diese Umwandlung als fest begründet betrachten darf, hat schon Frerichs ausgesprochen; häufigere Wiederholung der Versuche und vorurtheilsfreie Interpretation der erlangten Resultate wird uns allmälig zur Wahrheit führen. 69

One may summarize from the collection of experiments that bile pigments may sometimes be found in urine post-intravenous injection and that bile acids are usually not found. The simplistic conclusion would have been that injected bile acids are converted into bile pigments. The less direct conclusion would be that injected bile acids induced the formation of bile pigments in urine. Had the chemical structures been known, one would have been forced to assume the latter.

Some ten years later, in 1871, *Edward R. Taylor (131)*, a medical doctor, would write his prize essay, awarded by the American Medical Association, about the source of *Cholepyrrhin*: "Its *origin* has been pretty well made out to be from the hematin of the blood cell. ... Virchow, in his Cellular Pathology remarks that hematoidine is the only substance in the body with which we are acquainted, that is allied to the bile pigment." Not a subject without lingering controversy, however, for *Taylor* noted that: "Frerichs contests the above views, and maintains that no one has succeeded in manufacturing bile pigment from the red coloring matter of blood. ... On the contrary, he holds that biliary acids are the source of the bile pigments." By 1869–1871, however, experiments had been conducted which better explained *Frerichs'* experiments and reinforced *Kühne's* thesis and thus laid to rest any doubt that *Cholepyrrhin* did not originate from hematin (*131*):

> ... Niemeyer, however, holds the views of Kühne to be well established, for he says in the seventh edition (1869) of his practical medicine (I quote from Humphrey's and Hackley's translation), that the biliary acids "possess to a peculiar degree the property of dissolving the red blood-corpuscles. By injecting weak solutions of them into the blood of animals, we may artificially induce the so-called hæmatogenous icterus (jaundice without reabsorption), as the liberated coloring matter of the blood is transformed into biliary coloring mater. * * * * * The views regarding the occurrence of jaundice without retention and reabsorption of bile have totally changed since the observations of Virchow, Kühne, and Hoppe-Leyler [*sic*] have shown that bile coloring matter may be formed from the free coloring matter of the blood without the action of the liver; and we may induce artificial jaundice in animals by injecting substances that dissolve the blood-corpuscles. There is now no doubt that some of the formerly enigmatical forms of icterus are due to the disintegration of the freed coloring matter circulating in the blood, into bile coloring matter." Besides, the iron that both contain would point directly to a close kinship between them. It would seem, therefore, that we may finally rest upon the belief that the source of the cholepyrrhine [*sic*] of the bile is the hæmatin of the blood.

Though the relationship between bile acids and bile pigment seems now to be explained, the same could not be said with respect to hematoidin and its relationship to the bile pigments, for in the decade of the 1860s this issue was still one of considerable controversy. As noted earlier, *Zenker* and *Funke* reported (*126*) in 1860 that the red *Bilifulvin* pigment isolated from bile by *Berzelius (71)* changed into fine, large crystals, spontaneously or by treatment with ether, that were identical with hematoidin. As reported in 1862, *Jaffe*[37] (*132, 133*), too, thought that hematoidin and *Bilifulvin* were identical, following his experiments on old brain hemorrhages, which he dried and extracted with $CHCl_3$ and evaporated to yield golden yellow crystals that showed a positive *Gmelin* test. As with *Jaffe's* work,

---

[37] *Max Jaffe* was born on July 25, 1841 in Grünberg and died on October 26, 1911 in Berlin. After his early education in Grünberg and Breslau, he received the Dr. med. at the University of Berlin in 1865 and became *Assistent* to *Leyden* at Königsberg, where he later became Professor of Pharmacology.

2.8 Hämatoidin, Bilifulvin, and the Origin of Bile Pigments

which to him proved the existence of a bile pigment not originating from bile, *Hoppe-Seyler* (*134*) isolated the same pigment, with the same properties from a cyst in the breast. *Städeler* (*135*) criticized the crystal morphology cited by *Valentiner* (*107, 108*), finding the crystal angles of the orange-colored elliptic, *i.e.* crystals (of bilirubin) isolated following evaporation of $CHCl_3$, as being very different from those of hematoidin, which never had convex surfaces. *Holm*, working in Zürich with *Städeler* (*136*) noted that the yellow $CHCl_3$ extract of hematoidin from old hemorrhages of the brain turned green upon exposure to light (a characteristic of bilirubin). His hematoidin from corpora lutea of cows as well as brain hemorrhages differed from bilirubin in crystal form, color, and solubility properties in ether, CS_2, and alkali. In comparing hematoidin from a cyst to bilirubin, *Salkowski* (*137*) found somewhat different results from *Holm*: similar crystal forms, solubility, and positive *Gmelin* test. In contrast, *Preyer* (*138*), in 1871, strongly expressed his opinion (on the basis of his spectral measurements) that hematoidin and bilirubin are not identical. (Spectra at the time were measured by the absence or diminution of certain regions of the visible spectrum when visible light was passed through a dissolved sample of pigment.) To complicate matters further, *Thudichum* (1829–1901, see below, Section 2.9.2) claimed that the others had examined not hematoidin but *Lutein* (lutein, also xanthophyll) – and that lutein differed altogether from bilirubin. At essentially the same time, *Kühne* (*123–125*) and *Hoppe-Seyler* (*134*) began to use hematoidin as a synonym for bilirubin.

The dispute over the origin of bile pigments, whether from bile acids or otherwise, was nowhere as long lasting as the dispute about the identity of hematoidin with bilirubin. The controversy over the identity of hematoidin did not slip easily away. In the sixth edition of his *Manual of General Pathology* (English translation published in 1876) *Ernst Wagner* wrote that blood extravasations consisted of not just hematoidin but also contain bilirubin (and probably other compounds). Following *Thudichum*, *Wagner* preferred to call the pigment lutein rather than hematoidin. He cited the studies of *Holm* and *Städeler* (*136*), who believed to have shown that hematoidin was not identical to bilirubin (*139*):

> Hæmatoidin was for a long time considered identical with bilirubin, and because bilirubin (from bile) like hæmatoidin (from extravasations of blood) separated from its solutions in chloroform always in crystal of the same form and color. But it was afterward demonstrated that hæmatoidin (the coloring matter of the yolk of the egg and corpora lutea) is throughout different from bilirubin of the bile. The orange-red coloring matter of blood-extravasations consists not merely of "hæmatoidin," but also of bilirubin, so that it does not appear improper to call the pigment identical with the coloring matter of the corpora lutea not hæmatoidin, but lutein.
>
> HOLM and STÄEDELER (*Journ. f. pract. Chemie*, 1867, C., p. 142) demonstrated that hæmatoidin and bilirubin are entirely distinct. Well-formed crystals of hæmatoidin by reflected light appear beautifully green (like cantharides), those of bilirubin, orange-red. Hæmatoidin dissolves with bisulphide of carbon with a flame-red or, in dilute solutions, with an orange-red color, bilirubin with a golden yellow. The latter enters into combinations in fixed proportions with alkalies, and is soluble in alkalies, the former not; therefore from a solution of the latter in chloroform it may be separated by agitation with caustic alkalies,

which is not true of the former. Bilirubin furnishes with nitric, containing nitrous acid in alcoholic solution, a beautiful play of colors; green, blue, violet, red, yellow (reaction of biliary matters); hæmatoidin, on the other hand, by nitrous acid is colored light-blue, and then becomes either yellow or colorless. Lutein, according to THUDICHUM (*Med. Ctrlbl.*, 1869, No. 1), is also identical with the yellow coloring matter of butter, fat, blood-serum, and many plants (flowers, stamens, seeds).

SALKOWSKY (HOPPE-SEYLER, *Med.-chem. Unters.*, 3 H., p. 436) found, on the other hand, that hæmatoidin from a strumous cyst had all the peculiarities of bilirubin. He concludes that the hæmatoidin (from corpora lutea) investigated by HOLM was not pure, or that there are different kinds of hæmatoidin.

Following the decade of the 1860s, in 1878, *Charles Thomas Kingzett* of the Council of the Institute of Chemistry of Great Britain and Ireland would state authoritatively (*140*):

> There is no established connection whatever between bilirubin or other biliary pigments and the colouring matter of blood; it is necessary to state this emphatically on account of the existence of erroneous statements and impressions to the contrary.

And in 1880, *Legg* (*99*) concluded, on the basis of the existing evidence that hematoidin and bilirubin were not identical, but in 1883, *Hermann* (*141*) identified hematoidin with bilirubin. Toward the end of the 19th century, in 1891, *Ewald* summarized the advances in knowledge of the origin of bile pigments (*142*), leaving little doubt of a consensus that they arise from blood corpuscles and that the hematoidin derived from them is the same as bilirubin (*142*):

> The following is a summary of our present tolerably satisfactory knowledge concerning the *bile pigments*.
>
> If we shake bile that has been exposed to the air with chloroform, this takes up a green colouring matter, biliverdin. Fresh bile, however, owes it golden yellow colour to bilirubin, which when pure is an amorphous orange yellow powder, forming, by oxidation in the air or other oxidising means, the green biliverdin (formerly called cholepyrrhin or cholephäin). Chemists have produced a series of intermediate states, especially biliprasin and bilifuscin, and studied their spectroscopic relations and their connections with the blood and urine, which we referred to in the first lecture. Two points especially interest us: the derivation of and tests for the bile-colouring matter. At first sight there seems no doubt that bile pigment is derived from the pigment of the blood corpuscles, hæmochromogen. By injection into the circulation of a whole series of substances which dissolve the blood corpuscles and set free the pigment from them, we succeed in producing bile-coloured urine. Among these solvents are salts of the biliary acids, solutions of hæmoglobin, large quantities of water, chloroform, and ether, common salt solution, glycerine, toluylendiamine, arseniuretted hydrogen; and in the same way jaundice occurs after burning and scalding, after poisoning with oxalic acid, pyrogallic acid, naphthol, phosphorus, &c.; finally, *icterus neonatorum* and the jaundice that occurs in paroxysmal hæmoglobinuria are both due to destruction of blood corpuscles. The same solution of blood pigment and formation of bile pigment may occur naturally in old blood extravasations, where, as you know, peculiar crystals (Virchow's hæmatoidin crystals) have been found, first by Virchow, later by Hoppe-Seyler, also in the margin of the placenta and in the fluids of cysts, while their identity with bilirubin has been ascertained by Jaffé. Moreover, this formation of bile pigment or bilirubin crystals has been observed in artificial extravasations of blood (Langhans, Quincke), in blood injected into the abdominal cavity (Cordua), in frog's blood kept free from putrefaction (v. Recklinghausen). On the other hand, Funke and Zenker found the same crystals in old bile residue. Valentiner prepared hæmatoidin crystals from pulverised gallstones, and

2.8 Hämatoidin, Bilifulvin, and the Origin of Bile Pigments

Schwanda succeeded in extracting characteristic crystals from the urine of a case of jaundice. Neumann found bilirubin crystals in the blood of a three-days' old and probably suffocated child.

Thus, in the last decade of the 19th century it could be said regarding bilirubin (*143*): "*Bilirubin* . . . It is identical with Virchow's hæmatoidin." Yet, as if the subject of the identity of hematoidin would never be put to rest, as more chemical knowledge became available in the early 20th century, a new (and in retrospect radical) theory would emerge in which hematoidin was said to be identical to mesoporphyrin, which was thought to be identical to bilirubin (*144*) – a conclusion that would, rather incredibly, leave both hematoidin and bilirubin as reduced forms of a porphyrin (*144*):

... Haematoporphyrin is found occasionally in the urine (especially after sulphonal poisoning, which produces considerable blood destruction) and is no doubt derived from haematin, set free from hæmoglobin.
... Haematoporphyrin, on partial reduction with hydriodic acid, yields *mesoporphyrin*:

$$C_{33}H_{38}O_6N_4 + 2H_2 = C_{33}H_{38}O_4N_4 + 2H_2O$$
Hæmatoporphyrin.   Mesoporphyrin.

... Mesoporphyrin is probably identical with a substance described under the name of *haematoidin*, which was discovered by Virchow in 1847 in blood extravasations, and also with *bilirubin*, which is one of the best-known bile-pigments.

Of course at the time the chemical structures of porphyrins and bile pigments were unknown; so, chemical imagination was not constrained to sensible structures.

In 1923, *Fischer* and *Reindel* (*145*) indicated a probable identity of hematoidin with bilirubin based on the likeness of their crystal forms and their similar behavior on coupling with benzenediazonium chloride. Later, *Rich*,[38] who investigated the origin of bile pigments in the 1920s, provided a useful summary on the status of the subject in 1925 (*146*):

... Virchow could not prove conclusively that the pigment formed under such circumstances is identical with bilirubin, and he therefore gave it the name of "hematoidin." Indeed, even at the present time there are writers who maintain that the "hematoidin" found in hemorrhages is quite different from true bile pigment (100) or at least that there is no proof that the two substances are identical (37). Of course, since we do not yet know the details of the chemical structure of bilirubin itself, we are unable to say with absolute certainty that bilirubin and "hematoidin" are identical; but they have, apparently, the same percentage composition (although, unfortunately, analyses have been made only upon material obtained from echinococcus cysts of the *liver* (22, (84)) and they are so much alike physically and chemically that most workers who have studied them have felt safe in the

---

[38] *Arnold Rice Rich* was born on March 28, 1893 in Birmingham, Alabama and died on April 17, 1968 in Baltimore, Maryland. He received the Bachelor's degree in biology at the University of Virginia in 1915 and the Dr. med. in 1919 at Johns Hopkins University. He was a pathologist at Johns Hopkins Medical School, and Professor and Chairman of Pathology in 1944. In 1947, he was appointed the third Baxley Professor of Pathology and remained director of the department until his retirement in 1958.

belief that bile pigment itself can be formed from hemoglobin locally in blood extravasations (Jaffe (39), Quincke (74), Stadelman (94), Hooper and Whipple (35), Van den Bergh and Snapper (103), Leschke (51), McNee (64)). The statements which are scattered throughout the literature concerning the failure of "hematoidin" to give a typical Gmelin test, or to form crystals typical of bilirubin, or to possess the solubility characteristics of bilirubin, – statements which have so often in the past disturbed the acceptance of the identity of these pigments (48), can very probably be referred either to the presence of loosely bound impurities in the "hematoidin" examined, or to some change of such a little-understood nature as that which is known to alter the properties of even gall-bladder bilirubin on standing (1). Rich and Bumstead (80) have subjected "hematoidin," obtained from old hemorrhages, to the long series of physical and chemical tests and reactions which are well established as characteristic of bilirubin, and in every instance the "hematoidin" behaved precisely as did a control of pure bilirubin. In this study it was found that "hematoidin" yields oxidation and reduction products (bilicyanin and urobilin (hydrobilirubin)) which have the same properties and are identical spectroscopically with the substances obtained by the same methods from pure bilirubin. It seems clear that in "hematoidin" we have to deal with a substance which is so much like bilirubin that it cannot be distinguished from the latter pigment by any of our present physical or chemical tests. The burden of proof must, therefore, rest upon those who may deny that true bile pigment can be formed at the site of blood extravasations.[3]

Experimental evidence of the origin of bile pigment from hemoglobin began to appear about 10 years after Virchow's discovery. Interest in the matter was precipitated by the experiments of Frerichs and Städeler (23) who found that a pigment resembling bilirubin could be produced *in vitro* by the action of sulphuric acid upon bile acids,[4] and, more important, that the injection of bile acids into the blood stream of an animal would be followed by the appearance of undoubted bilirubin in the urine. Their conclusion was that the body could transform bile acids into bile pigment. Kühne (44), shortly after, repeated and confirmed a forgotten or unnoticed observation of von Dusch (107) that bile acids are powerful hemolytic agents; and he insisted that experiments of Frerichs and Städeler did not prove the origin of bilirubin from bile acids, for those investigators had not taken into account the fact that a large amount of hemoglobin is set free in the plasma by the injection of bile acids. Kühne was unable to satisfy himself that the injection of hemoglobin alone, in the absence of bile acids, would be followed by bilirubinuria, and he was forced to hold to the idea that the bile acids were necessary in some way for the formation of bile pigment. Herrmann (31), however, in 1859, was able to produce bilirubinuria at will by inducing intravascular hemolysis with injections of distilled water. This was the first clear demonstration that the simple liberation of hemoglobin into the blood stream may be followed by an increased output of bile pigment in the urine. Neither Naunyn (70) nor Steiner (96) could confirm Herrmann's results, and they opposed the conclusion that hemoglobin can be changed by the body into bile pigment. Their failure, as well as that of Kühne, is less difficult to understand now, for we have learned that the appearance of bilirubin in the urine after intravascular hemolysis depends upon a number of factors, and that the absence of bilirubinuria as determined by the Gmelin test, is by no means a proof

---

[3] In birds in which biliverdin is the predominant pigment of the gall-bladder bile, biliverdin (*i.e.* a bright green pigment which gives a positive Gmelin test) is formed in blood extravasations as well as "hematoidin." This is a further proof of the local formation of true bile pigment in hemorrhages.

[4] Hoppe-Seyler (36) was able to show that this pigment did not really have the properties of bilirubin, and later Städeler (93) himself denied the identity of the two pigments.

that there has been no increased formation of the pigment. Tarchanoff (101), on the other hand, not only confirmed Herrmann's work but, with the use of bile-fistula animals, carried the proof of the relation of hemoglobin to bile pigment still further by demonstrating, for the first time, that the introduction of pure hemoglobin into the circulation is followed by a marked increase in the amount of bile pigment excreted by the liver. Stadelmann (95) confirmed this observation of Tarchanoff in a more carefully controlled series of experiments, and it has since been established beyond question by numerous other investigators, using a variety of experimental animals and procedures, that the liberation of hemoglobin into the blood stream of an intact animal is regularly followed by an increased production of bile pigment which, according to conditions, may be eliminated by the liver or the kidneys, or partially retained in the plasma and tissues producing jaundice (Minowski and Naunyn (66), Gilbert, Chabrol and Bernard (25), Brugsch and Yoshimoto (14), McNee (63), Whipple and Hooper (115), van den Bergh and Snapper (103), and Rich (76)).

The *clinical* evidence of the relation of hemoglobin to bile pigment is to be found in the many different pathological states in which the condition of the liberation of an excessive amount of hemoglobin into the circulation is reproduced. In all of these maladies it is the rule that the formation of bile pigment is increased above the normal level and the pigment content of the feces, the urine and even of the plasma and tissues may be very high.

This was a time, as we shall learn, that the chemical structures of hematoidin, hematin, and bilirubin were still unknown, though it was believed that bilirubin and hemoglobin were closely related chemically, and *Rich*, being unable to find any differences in the physical and chemical properties of hematoidin and bilirubin, thus found them indistinguishable based on the state of knowledge of the times. Yet he was still reluctant to conclude with certainty that they are identical. That would come later, as their chemical structures were revealed.

2.9 Bile Pigment Isolation, Purification, and Combustion Analysis in the 1860s and 1880s

The seventh decade of the 19th century brought about a serious attempt to resolve chaotic differences among chemists regarding atomic weights (relative atomic masses) and equivalents, radicals and molecules, and nomenclature, which resulted in the formula of as simple a molecule as water to be expressed variously as HO, H_2O, and HΘ. Key to a firmer understanding of organic and natural products chemistry was knowing atomic and equivalent weights of which, confusingly, there were, for example three in common use, reflecting disagreements regarding differing atomic weights (relative atomic masses) of carbon and oxygen: those of *Berzelius* (H = 1, C = 12, O = 16), *Liebig* (H = 1, C = 6, O = 8), and *Dumas* (H = 1, C = 6, O = 16), as well as *Gmelin's* system of "equivalents" (H = 1, C = 6, O = 8, N = 14) – each of which had its adherents and practitioners. Thus a formula as simple as ethyl alcohol could be expressed as C_2H_6O (*Berzelius*), $C_4H_{10}O$, H_2O (*Liebig*), or C_8H_8, H_4O_2 (*Dumas*), rendering molecular weight calculations uncertain and limited to guesswork. Yet, by the middle of the decade, the discrepancies had apparently been resolved by non-unanimous agreement among the 140 participants of the famous Karlsruhe Congress of 1864 (*22, 147–149*). However, not all of the participating scientists agreed to follow the resolution and adopt the currently accepted atomic weights, which were thus only slowly put into practice, leaving a somewhat confusing array of combustion analyses-derived formulas that were often at variance with each other for the same compound.

At the time, bile pigment isolation had evolved from tedious, imperfect separations of the coloring matter of bile, gallstones and icteric urine. Thus, from the methodology involving repeated precipitations and washings pioneered by *Berzelius* (*68–76*) and his contemporaries in the first half of the century, a new and more efficient method involving $CHCl_3$ extraction was introduced by *Valentiner* (*107, 108*) and *Brücke* (*109*). The work of the latter two investigators, published in 1858–1859, opened the door to isolation of purified samples of bilirubin and biliverdin, a necessary first step to eventual full characterization of their chemical structure by what was probably the most important analytical method available: combustion analysis. Elemental combustion analysis was developed and applied to organic structure early in the 19th century and by 1860 had evolved into a reliable and effective method of characterization by providing an empirical formula. However, as in most analytical methods, the efficacy of a combustion analysis depends especially, aside from proper analytical technique, on the purity of the sample being analyzed. Up to 1859 and the time of *Brücke*'s work (*109*), the bile pigment samples that had been investigated, whether by combustion analysis or the sensitive, qualitative *Gmelin* test, were of uncertain purity – or of certain impurity. They often contained non-combustible inorganic material, which showed up as residual ash following combustion. And considering the method of isolation, and lacking anything approaching chromatographic methods, it could not be certain that the bile pigment had been freed from other combustible material. As a consequence, though the measured %C, H, and N more often than not had to be adjusted (imperfectly) for ash left behind in the combustion, there was no way to adjust the measurement for organic impurities, rendering any so-derived empirical formulas tenuous at best. A major challenge to the utility of this technique in the second half of the 19th century, and into the 20th, was thus to achieve sample purity, as it often is today.

By 1860, most of the principal investigators of bile pigment "chemistry" had passed on, either permanently or to new endeavors, leaving the two decades between 1860 and 1880 to mainly three individuals, whose names came into prominence in connection with bile pigment analysis: *Städeler*, who published in 1856 with *Frerichs* on the origin of hematoidin from bile acids (*129*); *Thudichum*, who wrote extensively from 1862 to 1881 on gallstones (*101–103*); and *Maly*, who initiated his work with the thesis that *Cholepyrrhin* is the amide of biliverdin (*150*) but soon after adopted great care in experimentation. These

2.9.1 Georg Andreas Karl Städeler Gives the Name Bilirubin as a First Step Is Taken Toward Structure Identification

Städeler's[39] interests and expertise tended more toward chemistry, and early on he became engaged in performing elemental combustion analyses to help develop his theories on the conversion of tyrosine, *inter alia*, to pigments. He also achieved one of the earlier combustion analyses of *Cholepyrrhin*, which he had isolated from bile and purified by the *Valentiner-Brücke* $CHCl_3$ extraction technique (*107–109*). Concerning the first, in 1860, while writing on tyrosine and its reactions (*151*), he noted the formation of a lemon-yellow color following treatment of tyrosine with nitric acid. The color was due to the presence of a red-orange pigment, and what *Städeler* called *dinitrotyrosine* crystallized in golden-yellow blades whose lead salt was colored a chromic-acid red. The red pigment, which had been so easily obtained by oxidation of tyrosine with an excess of HNO_3, he tentatively reserved the name *Erythrosin*. The reddish color apparently reminded him of *Hämatoidin* (hematoidin). *Erythrosin* turned greenish in light and underwent various other color changes upon manipulation with acids and bases. *Städeler* thus noted many similarities between *Erythrosin* and hematoidin and wondered whether a relationship existed between them. Though *Robin's* analysis (*115, 116*) of hematoidin (64.12% C, 6.87% H, 10.69% N, 18.32% O), from which he gave the formula $C_{14}H_9NO_3$, did not correspond in any way, a re-analysis of hematoidin yielded a satisfactory correspondence to $C_{30}H_{18}N_2O_6$, as comes from the following composition (*129*):

---

[39] *Georg Andreas Karl Städeler* was born in 1821 in Hannover on 25 March (an historically significant date for the author and for Greek independence), received the Dr. phil. in 1849 and became *Habilitand* at the Universität Göttingen as *Privatdozent* and the first director of *Rudolph Wagner's* newly established Laboratory for Physiological Chemistry. He was appointed a. o. Professor in Göttingen and, failing to achieve an academic chair (o. Professor) in Breslau, he (and not *Kekulé*, who was also interested) was appointed to o. Professor at the University of Zürich, where he more fully engaged in his academic career until his death in Hannover on 11 January 1871. Among his colleagues in Göttingen, he found *Friedrich Frerichs*, who was nearly the same age and with whom he struck up a close friendship. *Frerichs* was *Assistent* in *Rudolf Wagner's* Laboratorium für physiologische und pathologische Chemie from 1843 to 1850 before he moved to Kiel as a.o. Professor für Pathologie und Vorstand der Poliklinik. *Städeler's* name was linked to *Frerichs'* through their jointly published work on the conversion of certain amino acids to pigments (*127, 128*) and, in animal metabolism, the conversion of intravenously-injected bile acids into bile pigments found soon afterward in urine (*129*).

				Berechnet[1]		Gefunden	
30	Aeq.	Kohlenstoff	[C]	180	65,69	65,85	65,05
18	„	Wasserstoff	[H]	18	6,57	6,47	6,37
2	„	Stickstoff	[N]	28	10,22	10,50	10,50
6	„	Sauerstoff	[O]	48	17,52	17,18	18,08
				274	100,00	100,00	100,00

Using his new formula for hematoidin, *Städeler* then wrote a chemical equation showing how oxidation of tyrosine, coming from decomposition of proteins in organisms, might produce hematoidin (*129*):

> Nimmt man diese Formel für das Hämatoïdin an, so würde sich dieselbe vom Tyrosin ableiten lassen:

$$2\ C_{18}H_{11}NO_6 \quad + \quad 2\ O \quad = \quad C_2O_4 \quad + \quad C_4H_4O_4 \quad + \quad C_{30}H_{18}N_2O_6$$

| Tyrosin | Kohlensäure | Essigsäure | Hämatoïdin |

> Der Grundfarbstoff des Bluts, das Hämatoïdin, könnte also durch einen Oxydationsprocess aus dem Tyrosin, das beim Zerfall der Proteïnstoffe im Organismus ensteht, hervorgehen. 70

Of course, the formulas above are based on the atomic weights C = 6, H = 1, N = 14, O = 8, which were revised in the Karlsruhe Conference of 1864 (*22, 147–149*), but not widely or immediately accepted. As in the calculation of an empirical formula, this point too illustrates the state of chemistry and how easy it is fall into a trap in the absence of more correct information. *Städeler* did not assert, however, that hematoidin and *Erythrosin* were identical. He only proposed the possibility and indicated that he would pursue the question as soon as his time permitted:

> Damit soll übrigens keineswegs behauptet werden, dass Hämatoïdin und Erythrosin identisch seien; für möglich halte ich diess allerdings und ich werde daher, sobald es meine Zeit erlaubt, die Frage weiter verfolgen. 71

Städeler's analysis of *Cholepyrrhin* was reported in a footnote in volume II of *Frerichs' Klinik der Leberkrankheiten* published in 1861 and translated by *Murchison* in the same year (*120*):

> The elementary analysis performed by my friend Professor Städeler, of Zurich of cholepyrrhin purified by repeated crystallization from boiling, and washing with cold chloroform, yielded results from which the chemical formula $C_{18}H_9NO_4$ is calculated.

---

[1] N.B. *Berechnet* = Calculated; *Gefunden* = Found. It may be noted that the calculations above are based on *Gmelin*'s system of "equivalents", where atomic mass carbon = 6, hydrogen = 1, nitrogen = 14, and oxygen = 8. Had the current atomic weights been used, the empirical formula would be $C_{15}H_{17}NO_3$.

					Calculated formula.		Actual Result of Analysis.
18	equivalents	of	carbon	=	108	66·26	66·52
9	„	„	hydrogen	=	9	5·52	6·00
1	„	„	nitrogen	=	14	8·59	8·70
4	„	„	oxygen	=	32	19·63	18·78
					163	100·00	100·00

Hence cholepyrrhin only differs from isatine, the product of the oxydation of indigo, by the elements of one equivalent of hyduret of methyle.

$$\left.\begin{array}{r} C_{16}H_5NO_4 + C_2H_3 \\ H \end{array}\right\} = C_{18}H_9NO_4$$

Moreover, cholepyrrhin contains 2 equivalents of water less than tyrosine, and 2 equivalents of oxygen less than hippuric acid. According to this, the occurrence of indigo in human urine, which has been repeatedly observed, is a less remarkable circumstance than might at first be thought. It will be interesting to study more closely the relations between cholepyrrhin and isatine.

From these data, with no reported residue of ash, *Städeler* derived the chemical formula $C_{18}H_9NO_4$, which is somewhat different from his analytical data that predicted $C_{30}H_{18}N_2O_6$ for hematoidin (*151*), and also that from *Robin* (*115, 116*), thus *Städeler* wrote (*151*): $C_{14}H_9NO_3$.[40]

With an apparently more direct focus on *Cholepyrrhin*, in 1864, *Städeler* published what might be considered a landmark paper (*135*), this nearly four years subsequent to his earlier reported combustion analysis. Also, in 1864, *Maly* published his preliminary studies (*150*), and only a year earlier *Thudichum* had published his important work on the same pigment (*102, 103*). *Städeler* comprehensive work (*135*) briefly reviewed previous studies of others on the pigments of bile, and then related his own studies on the pigments from bile and gallstones. From the perspective of the mid-19th century, this publication is an impressive scientific endeavor and a tribute to *Städeler*'s clarity of thought and attention to detail. The work presented represents an elevation in knowledge and thought regarding bile pigments.

Yet perhaps *Städeler*'s longest lasting contribution to bile pigments was the name *Bilirubin* (*Gallenroth*, red-bile = red pigment of bile; Latin: *bilis*, bile; *rubris*, reddish) that he gave to the purified reddish pigment of bile and gallstones. The name apparently caught on, was adopted increasingly widely, and has been accepted for more than 100 years as the standard name of the pigment. In fact, the name was apparently so appealing and logical that even *Thudichum*, who coined his own

---

[40] N.B. *Städeler* was consistent in his use of *Gmelin*'s system of atomic equivalents rather the actual atomic values agreed upon at the 1860 Karlsruhe Conference. A recalculation of the data using the atomic weights of today would give the empirical formula $C_{18}H_{19}N_2O_4$ for *Cholepyrrhin* and $C_{15}H_{17}NO_3$ for hematoidin, but even these are not the correct values. Those were learned only decades later.

94 2 Early Scientific Investigations

words for the bile pigments, adopted it before 1868 (*152*), and it began to appear in medical and physiology textbooks within a decade or so of *Städeler*'s introducing it. Thus, one may find the name *bilirubin* in various subsequent authoritative sources, such as: *Pflüger*'s 1871 *Archiv für die gesammte Physiologie des Menschen und der Thiere* (*153*), *Wood*'s *Report on Medical Chemistry* in 1873 in *The Boston Medical and Surgical Journal* (*154*), *Wagner*'s 1876 *A Manual of General Pathology* (*99*), *Kingzett*'s 1878 *Animal Chemistry* (*140*), *Legg*'s 1880 *On the Bile, Jaundice and Bilious Diseases* (*99*), and in later publications.

Städeler reviewed what were the then most recent combustion analyses from *Scherer*, *Hein*, and *Heintz*, especially those obtained by *Heintz* some 13 years earlier (*97*) for *Biliphäin* and biliverdin. The *Biliphäin* analyzed was suspected to be a mixture of pigments, and *Städeler* believed that *Valentiner*'s (*107, 108*) successful $CHCl_3$ extraction of *Gallenroth* from bile and gallstones proved it. He also thought that *Brücke* (*109*) proved that *Biliphäin* is converted to biliverdin by absorption of oxygen. Since *Valentiner*, too, took *Gallenroth* to be identical to hematoidin, *Städeler* was puzzled by *Robin*'s formula for hematoidin (based on its combustion analysis) because he believed that if it were correct then biliverdin could not arise from oxidation of hematoidin (*135*):

> Eine Analyse des Gallenrothes ist nicht gemacht worden, und vergleicht man die Formel, welche sich aus Robin's Analysen für das *Hämatoïdin* ... berechnet: $C_{30}H_{18}N_2O_6$ mit der Formel des *Biliverdins*: $C_{16}H_9NO_5$ oder $C_{32}H_{18}N_2O_{10}$, so ergiebt sich, dass das letztere im Verhältniss zum Stickstoff mehr Kohlenstoff enthält, als das Hämatoïdin, dass also, wenn Robin's Analysen richtig sind, das Biliverdin nicht durch Oxydation aus dem Hämatoïdin entstehen kann. 72

His interest in bile pigments was also driven by his work with *Frerichs* (*129*), and that of *Neukomm* (*130*) in his lab in Zürich, on the production of bile pigments from bile acids, for which he proposed two explanations: (i) intravenously injected bile acids are convered directly into bile pigments in the bloodstream, or (ii) bile acids influence bile pigment production from hemoglobin or hematin. In order to pursue a comparative chemical investigation of synthetic and naturally occurring bile pigments, he took up an investigation of the latter (*135*):

> Meinungsverschiedenheiten herrschen nur darüber, ob die Gallensäuren in der Blutbahn direct in Pigmente verwandelt werden, oder ob die Pigmentbildung der auflösenden Wirkung dieser Säuren auf das Blutroth zugeschrieben werden müsse. Durch blosse Injectionsversuche, wie es bisher geschehen ist, liess sich die Frage offenbar nicht genügend beantworten, während von einer vergleichenden chemischen Untersuchung der künstlichen und der natürlich vorkommenden Gallenpigmente bestimmte Aufschlüsse zu erwarten standen.
>
> Um diese Vergleichung vornehmen zu können, habe ich mich zunächst mit einer Untersuchung der natürlichen Gallenpigmente beschäftigt. – Indem ich die erhaltenen Resultate mittheile, benutze ich zugleich die Gelegenheit, allen Freunden und Collegen, die mich durch Zusendung von Material bei dieser Untersuchung unterstützt haben, meinen Dank hiermit auszusprechen. 73

In order to obtain the natural pigment(s), he turned to pigmented gallstones as a source of bilirubin by processing according to *Valentiner*'s method. After removing fats and cholesterol from the pulverized stones, followed by a hot water wash to

2.9 Bile Pigment Isolation, Purification, and Combustion Analysis...

remove traces of bile, he extracted with $CHCl_3$ to obtain a small amount of sticky, greenish-brown residue (after evaporation) that contained *Gallenroth* crystals, as seen under a microscope. The powdered gallstone residue, after the $CHCl_3$ extraction, was treated with dilute HCl to dissolve a large quantity of calcium and magnesium salts and evolve CO_2. The resulting dark brown residue, after washing and drying, yielded a large amount of pigment into boiling $CHCl_3$, thus suggesting to *Städeler* that the majority of the pigment had been originally bound up as salts. Evaporation of this $CHCl_3$ extract gave a dark solid-crystalline residue, from which a "brown pigment (among other material)" was extracted into hot alcohol. *Städeler* named it *Bilifuscin* (Latin: *bilis*, bile; *fuscus*, dark). The gallstone residue, after the boiling $CHCl_3$ extraction above, contained a considerable amount of *Gallenroth* (bilirubin), albeit in impure condition. After as much "brown pigment" as possible had been extracted with $CHCl_3$, the solid residue was colored bright olive and still contained considerable *Gallenroth* as well as a green pigment that *Städeler* called *Biliprasin* (from Latin: *bilis*, bile; *prasinus*, green), which was washed exhaustively with alcohol to give a beautiful green colored solution. Then the remaining *Gallenroth* was extracted into boiling $CHCl_3$. The residue, after all of the washings/extractions, was insoluble in H_2O, alcohol, ether, $CHCl_3$, and dilute acids. It reminded *Städeler* of *humin* (soil), and thus he found the name *Bilihumin* (Latin: *bilis*, bile; *humis*, soil) appropriate.

Essentially following *Brücke*'s method (*109*), *Städeler* further purified the $CHCl_3$-extracted *Gallenroth*, taking it through several cycles of dissolving it in $CHCl_3$, filtering, and evaporating, washing the residue each time with ether and alcohol. The alcohol washings were always more or less green to greenish-brown, while the bilirubin remained as a vivid red to orange-red granular-crystalline powder. With this purified bilirubin, *Städeler* proceeded to its combustion analysis, from which he discovered that the data corresponded to no acceptable formula. Just what constituted an acceptable formula is unclear. In an analysis mentioned in 1861 by *Frerichs* (*120*), *Städeler* had found $C_{18}H_9NO_4$ from 66.52% C, 6.00% H and 8.70% N, which was later found to be unacceptable. In any event, *Städeler* repurified his bilirubin by precipitating it with alcohol from a $CHCl_3$ solution (*135*):

> 1) *Bilirubin*. – Um diesen Farbstoff, der in vorwiegender Menge in den menschlichen Gallensteinen vorkommt, zu reinigen, wurde er einige Male in Chloroform gelöst, die filtrirte Lösung verdunstet und der Rückstand mit Aether und Weingeist gewaschen. Der abfliessende Weingeist zeigt sich immer mehr oder minder grün bis grünlichbraun gefärbt, während das Bilirubin als ein lebhaft rothes bis orangerothes, körnig-krystallinisches Pulver zurückblieb.
>
> Bei der Analyse des so gereinigten Farbstoffes wurden Zahlen erhalten, die mit keiner annehmbaren Formel genügend übereinstimmen, woraus auf eine Verunreinigung geschlossen werden musste. Diese zu beseitigen gelang mir dadurch, dass ich die Chloroformlösung nur bis zur beginnenden Abscheidung von Bilirubin verdunsten liess und sie dann durch Zusatz von Weingeist fällte. Auf diese Weise wurde das Bilirubin als amorphes orangefarbenes Pulver erhalten; ein ziemlich bedeutender Verlust war dabei nicht zu vermeiden. 74

A similar procedure is still used today to purify the commercially available pigment (*13*).

Significantly, the purified pigment left no ash upon combustion and was dried at 100°C over conc. H_2SO_4 to lose 1% of its weight. Further heating between 120°C and 130°C produced no further reduction in weight. The material was thus deemed suitable for combustion analysis, but before it was accomplished, however, *Städeler* observed that by heating it in a glass (melting point) tube, the solid became swollen and evolved a yellow, foul-smelling vapor that blackened lead paper – a test typically used to detect H_2S. (It is unclear how bilirubin, which contains no sulfur, might evolve H_2S upon heating.) Nonetheless, the trace of sulfur was thusly shown to be present in all the pigments of *Städeler*'s study. It did not come from sulfate, as combustion of the bilirubin with lime (CaO) and salt-peter (niter or any nitrate, usually KNO_3) followed by acidification of the residue with HCl produced no turbidity upon addition of $BaCl_2$ (*135*):

> Der erhaltene Farbstoff verbrannte auf Platinblech, ohne einen Rückstand zu hinterlassen. Nach mehrtägigem Stehen über Schwefelsäure verlor er bei 100° nahezu 1 pC. an Gewicht. Bei weiterem Erhitzen auf 120 bis 130° blieb das Gewicht constant. Beim Erhitzen im Glasrohr schmolz das Bilirubin, es blähte sich auf und entwickelte gelbe übelriechende Dämpf, welche Bleipapier schwärzten. Dagegen wurde beim Verbrennen von 0,176 Grm. Substanz mit Kalk und Salpeter, Auflösen der geglühten Masse in verdünnter Salzsäure und Zusatz von Chlorbaryum keine Trübung wahrgenommen. – Die durch Bleipapier angezeigte Spur von Schwefel war auch in allen übrigen Pigmenten der Gallensteine nachzuweisen.
>
> Das zu den folgenden Analysen benutzte Bilirubin war bei zwei Darstellungen erhalten worden.
> I. 0,3765 Grm., bei 120° getrocknet, gaben 0,927 Grm. Kohlensäure und 0,2125 Grm. Wasser.
> 0,2563 Grm, bei derselben Temperatur getrocknet, lieferten bei der Verbrennung mit Natronkalk eine Quantität Salmiak, aus welcher mit salpetersaurem Silber 0,252 Grm. Chlorsilber gefällt wurden.
> II. 0,3105 Grm., bei 130° getrocknet, gaben 0,764 Grm. Kohlensäure und 0,171 Grm. Wasser.
> Aus diesen Daten berechnet sich für das Bilirubin die Formel $C_{32}H_{18}N_2O_6$.[41]

			berechnet		I.	II.
32	Aeq	Kohlenstoff	192	67,13	67,15	67,11
18	„	Wasserstoff	18	6,29	6,27	6,12
2	„	Stickstoff	28	9,79	9,59	–
6	„	Sauerstoff	48	16,79	16,99	–
			286	100,00	100,00	

The properties of this highly purified bilirubin are described by *Städeler* in some detail, properties that anyone today might recognize as characteristic of this pigment: The pigment was orange-colored in the amorphous state (colored somewhat like Sb_2S_3); in the crystalline state, the crystals were well-formed and measurable, with the vivid dark color of chromic acid. It was insoluble in ether, soluble in traces

---

[41] N.B. The recalculated formula using conventional atomic weights would be $C_{16}H_{19}N_2O_3$.

2.9 Bile Pigment Isolation, Purification, and Combustion Analysis...

in ethanol but dissolved in cold $CHCl_3$ to give a yellow to yellow-orange solution. The more crystalline it was, the more difficult it was to effect dissolution in $CHCl_3$; continuous heating was required. And it was noted that pure $CHCl_3$ became rapidly acidic and generated phosgene. In such $CHCl_3$ the erstwhile yellow color turned to green. But when the $CHCl_3$ contained a bit of ethanol, there was no color change. (And so commercial $CHCl_3$ typically contains a little ethanol as stabilizer). Benzene and CS_2 were good solvents; turpentine and fatty oils (almond oil) dissolved the pigment upon warming and produced a yellow color. It dissolved in alkali giving deep orange solutions that became yellow at high dilution. The dilution experiments are interesting. With a 15 mm thick layer of the alkaline (NH_4OH) solution as the standard reference, a dilution factor of 15,000 still left an orange color; a 20,000 dilution factor left a deep golden yellow; from a factor 25,000–100,000 it was pure yellow, as in solutions of neutral K_2CrO_4. At a factor of 5×10^5, and at 10^6 at twice the sample thickness the yellow color was still noticeable. Dilutions of 3–4 $\times 10^4$ imparted a distinctly yellow coloration to the skin. Thus such extraordinary tinctural power easily explained the yellow coloration of skin and eyes at the occasional rapid onset of jaundice. From the coloration of the eyes due to intense icterus, one might conclude an approximate $2–2.5 \times 10^4$ dilution of the pigment. From this, *Städeler* appeared to imply a visual method for diagnosing the severity of jaundice (*135*):

> 30- bis 40000 fach verdünnte Lösungen färben die Haut noch deutlich gelb. – Bei so ausserordentlichem Farbvermögen ist das mitunter so rasche Eintreten von Gelbsucht, die gelbe Färbung des Auges und der Haut, leicht erklärlich. Aus der Farbe des Auges bei intensivem Icterus darf man auf etwa 20– bis 25000 fache Verdünnung des Pigmentes schliessen. 76

The alkaline ammonia solutions above were found to bleach, even if not completely, moderately rapidly in direct sunlight, while in diffuse light bleaching occurred only slowly. The solutions gradually became light brownish-yellow and lost the ability to be precipitated upon addition of hydrochloric acid, while from the undecomposed solution, even at great dilution, bilirubin precipitated at once in orange colored flakes upon addition of the HCl. Apparently *Städeler* was the first to record a photooxidation or photooxygenation reaction of bilirubin, one assumes in the presence of air – thereby anticipating the early photochemical investigations (see Chapter 9) of the molecular mechanisms of phototherapy for neonatal jaundice (*155–157*). Thus, from *Städeler* (*135*):

> Die mitgetheilten Bestimmungen der Farbenintensität wurden mit ammoniakalischen Bilirubinlösungen gemacht; solche Lösungen bleichen, wenn auch nicht vollständig, ziemlich rasch im directen Sonnenlicht, während sie sich im zerstreuten Licht nur langsam zersetzen. Sie werden allmälig hellbräunlich gelb und verlieren die Eigenschaft durch Salzsäure gefällt zu werden, während sich aus der unzersetzten Lösung, auch bei grosser Verdünnung, auf Zusatz von Salzsäure sogleich Bilirubin in orangefarbigen Flocken abscheidet. 77

Städeler noted some differences between solutions of bilirubin in aqueous ammonia, NaOH, and Na_2CO_3, and that these aqueous basic solutions extract all of the pigment from its solution in $CHCl_3$. He indicated that compounds of bilirubin with "earths" and heavy metal oxides were insoluble or barely soluble in H_2O. A voluminous rust-colored calcium compound was precipitated by addition of $CaCl_2$ to an aqueous ammonia solution of the pigment. The dried compound was a splendid dark green, with a metallic reflection. Pulverizing it yielded a dark brown powder of the color of pigment-rich human gallstones that to the greatest part also consisted of this compound. The calcium compound was as good as insoluble in ether, alcohol, and $CHCl_3$, and when heated in the last two solvents gave only a weakly yellow color. In a similar way, the salts with $BaCl_2$, sugar of lead, $Pb(OAc)_2$, and $AgNO_3$ produced barium, lead, and silver compounds. The last precipitated in brownish-violet flakes that could be heated without reduction of the silver. The calcium compound analyzed for $C_{32}H_{17}N_2O_6Ca$ (*135*):

> 0,2549 Grm. hinterliessen beim Verbrennen, Anfeuchten der Asche mit kohlensaurem Ammoniak und Trocknen bei 130° 0,0414 Grm. kohlensauren Kalk, übereinstimmend mit der Formel: $C_{32}H_{17}CaN_2O_6$. Die Rechnung verlangt 9,18 pC. Kalk; gefunden wurden 9,10 pC.　　78

Städeler treated bilirubin systematically with HNO_3, producing new results and a calibration of the *Gmelin* reaction. Warming bilirubin with dilute HNO_3 (20% H_2O) produced dark violet resinous flakes that became light brownish with further heating and dissolved to form a yellow solution. In the cold there was essentially no change, but with more dilute HNO_3 (30% H_2O) bilirubin formed resinous flakes in the cold and became reddish colored; upon heating the mixture ended up as a yellow solution, as above. If pure HNO_3 hydrate was used, bilirubin dissolved immediately in the cold with a dark red color, and after a little while, or by heating, the solution lightened but retained a bright cherry-red color upon standing after several days (*135*):

> Uebergiesst man Bilirubin mit einer verdünnten Salpetersäure, welche 20 pC. Hydrat enthält, so bemerket man in der Kälte keine wesentliche Einwirkung; beim Erwärmen damit verwandelt es sich dagegen in dunkelviolette Harzflocken, die bei weiterer Einwirkung hellbräunlich werden und sich beim Aufkochen mit gelber Farbe lösen. Eine Säure mit 30 pC. Hydrat bildet die Harzflocken schon in der Kälte und färbt sich röthlich; beim Erwärmen verschwinden die Flocken und die Lösung wird gelb. Wendet man reines Salpetersäurehydrat an, so löst sich das Bilirubin schon in der Kälte mit tief rother Farbe, und nach einiger Zeit oder beim Erhitzen wird die Lösung heller, behält aber selbst bei mehrtägigem Stehen eine lebhaft kirschrothe Farbe.　　79

If bilirubin was dissolved in commercial conc. HNO_3, to which one added a little fuming red acid, the well-known bile pigment reaction (*Gmelin* reaction) was thus seen outstandingly. It was best to use alkaline solutions before the addition of HNO_3 and mix them with an approximately equal volume of alcohol. Upon addition of HNO_3 a magnificent reaction was seen even when the added acid contains no nitrous acid, and the sample was not turbid with precipitated flakes of pigment. As *Gmelin* observed decades earlier, the yellow color goes green first, then blue, violet, ruby red, and finally dirty yellow. By not stirring, all of the colors could be seen at the same time, as layer upon layer. The limits of detection were excellent: 0.25 mg bilirubin in a 4 cm^3 solution still produced a splendid display of colors. The entire reaction occurred best at a dilution factor of $7–8 \times 10^5$ (*135*):

> Vermischt man Lösungen des Bilirubins mit käuflicher concentrirter Salpetersäure, der man zweckmässig etwas rothe rauchende Säuer zusetzt, so erhält man die bekannte

2.9 Bile Pigment Isolation, Purification, and Combustion Analysis...

> Gallenpigmentreaction in ausgezeichnetem Grade. Am besten wendet man alkalische Lösungen an und vermischt dieselber vor dem Säurezusatz mit ungefähr dem gleichen Volumen Weingeist. Bei Weingeistzusatz erhält man eine prachtvolle Reaction auch dann, wenn die anzuwendende Säure keine Untersalpetersäure enthält, und die Probe wird durch ausgeschiedene Pigmentflocken nicht getrübt. Die gelbe Farbe geht zuerst in grün über, wird dann blau, violett, rubinroth und endlich schmutzig gelb. Wird nicht geschüttelt, so zeigen sich alle diese Farben gleichzeitig schichtenweise über einander. ¼ Milligr. Bilirubin in 4 CC. Lösung bringt noch ein prächtiges Farbenspiel hervor. Die Grenze der Reaction tritt erst bein 70– bis 80000 facher Verdünnung ein.   80

The blue pigment formed fleetingly in the *Gmelin* reaction was of interest to *Städeler* in connection with suspected indigo in urine. Why indigo might be present is anyone's guess, but *Städeler* isolated the blue pigment from bilirubin without difficulty. He did this essentially by dropwise addition of the acid mixture (above) to a "not too dilute" solution of bilirubin in aqueous ammonia, and eliminated too great an excess of HNO_3 by neutralizing with ammonia. All this produced at first a green flocculent precipitate that gradually became blue. After washing the precipitate with H_2O, the co-mixed green pigment was removed with alcohol to leave behind a dark blackish-blue powder. The likely view is that this blue pigment was related to the indigo content of urine. *Städeler* expressed the misfortune of not having sufficient material to be able to undertake further experiments on it.

The blue pigment could also be obtained from a yellow $CHCl_3$ solution of bilirubin by mixing in 1–2 drops of HNO_3 and shaking. This resulted in a very dark liquid that soon went violet, then ruby red. If alcohol were quickly added and mixed in as soon as the violet color appears, the solution became dark blue and changed color only very slowly. Using this approach, a splendid green or red was produced, colors that depend on an earlier or later addition of alcohol (*135*):

> Das bei der angegebenen Reaction entstehende blaue Pigment lässt sich ohne Schwierigkeit isoliren. Vermischt man eine nicht zu verdünnte ammoniakalische Bilirubinlösung tropfenweise mit der oben angegebenen Säuremischung, und beseitigt von Zeit zu Zeit einen zu grossen Ueberschuss von Salpetersäure durch annähernde Neutralisation mit Ammoniak, so erhält man zuerst einen grünen flockigen Niederschlag, der allmälig blau wird. Nach dem Auswaschen mit Wasser kann ihm beigemengtes grünes Pigment durch Weingeist entzogen werden und es bleibt dann ein tief-schwarzblaues Pulver zurück. Die Ansicht liegt nahe, dass dieses blaue Pigment in Beziehung steht zu dem Indiggehalt des Harns. Leider besass ich nich genug Material, um Versuche in dieser Richtung anstellen zu können.
>
> Ein prachtvolles Blau kann man auch bei Anwendung von Chloroform erhalten. Wird eine gelbe Chloroformlösung des Bilirubins mit einem oder zwei Tropfen Salpetersäure vermischt und geschüttelt, so wird die Flüssigkeit sehr dunkel, bald in's Violette übergehend und dann rubinroth werdend. – Setzt man, sobald der violette Farbenten eingetreten ist, rasch viel Weingeist hinzu, so erfolgt Mischung, die Lösung wird tief blau und verändert nur langsam ihre Farbe. – Auf gleiche Weise kann man auch ein prachtvolles Grün oder Roth erzeugen; die Farbe hängt ab von dem früheren oder späteren Weingeistzusatz.   81

Städeler noted that bilirubin dissolved in cold, conc. H_2SO_4 to produce a brownish liquid that gradually turned violet-green. Addition of H_2O separated dark green, nearly black flakes that dissolved in alcohol with a marvellous violet color. Addition of HNO_3 gave a beautiful display of colors, with the red being especially vivid and beautiful. On the other hand, by heating bilirubin in fuming HCl, the solution became dark brown (*Städeler* thought possibly due to *Bilifuscin* formation). Decomposition appeared to proceed to *Humin* formation, and by heating longer a brown compound resulted that was insoluble in dilute ammonia.

He carried out what might be the first experiments involving reduction of bilirubin by treating a dark red-brown alkaline solution of the pigment with Na(Hg) – a method often used subsequently, even some hundred years later in the *C.J. Watson* lab at the University of Minnesota. The color rapidly decreased, and the solution became pale yellow, a coloration which did not vanish upon warming. *Städeler* was not able to investigate the resulting compound further, which he believed to remain probably in a similar relationship to blirubin as is indigo white to indigo blue. Assuming this is correct, then the (new) yellow pigment would have the composition formula $C_{32}H_{20}N_2O_6$. (Or two more hydrogens than in the blirubin formula given by *Städeler* above) (*135*):

> Reducirende Materien wirken sehr energisch auf das Bilirubin ein. Vermischt man die tief-rothbraune alkalische Lösung des Farbstoffs mit Natriumamalgam, so nimmt die Farbe rasch ab und die Lösung wird blassgelb; auch beim Erwärmen verschwindet dieser Farbenton nicht. Ich habe den hierbei entstehenden Körper, der wahrscheinlich in demselbe Verhältniss zum Bilirubin steht, wie das Indigweiss zum Indigblau, nicht näher untersuchen können. Ist das angedeutet Verhältniss richtig, so würde dieser gelbe Körper der Formel $C_{32}H_{20}N_2O_6$ entsprechend zusammengesetzt sein. 82

Before turning from the pigments of gallstones to the pigments of human bile, *Städeler* addressed biliverdin in his by now evidently comprehensive fashion. To make biliverdin, as was done by others in the past, he oxidized a solution of bilirubin in aq. NaOH using air; after rapid uptake of oxygen the solution turned green. When at its greatest intensity, hydrochloric acid was added to produce a strongly green precipitate that was insoluble in ether and in $CHCl_3$. As *Brücke* noted earlier (*109*), it dissolved in alcohol leaving unreacted bilirubin behind as orange flakes, and the green solution gave a positive *Gmelin* reaction; turning blue, then violet, red, and finally a dirty yellow. *Städeler* was convinced that the green pigment was the same as that which *Heintz* had analyzed by combustion and for which he established the formulas $C_{16}H_9NO_5$, or $C_{32}H_{18}N_2O_{10}$ (*95*). These formulas would be produced from *Städeler*'s bilirubin formula by the addition of four oxygen atoms. Yet from his own analyses, *Städeler* remained doubtful (*135*):

> 2) *Biliverdin*. – Wird eine Lösung von Bilirubin in überschüssiger Natronlauge auf flachen Tellern der Einwirkung der Luft ausgestezt oder anhaltend mit Luft geschüttelt, so nimmt sie ziemlich rasch Sauerstoff auf und die Lösung wird grün. Hat diese Farbe ihre grösste Intensität erreicht, so entsteht auf Zusatz von Salzsäure ein lebhaft grüner Niederschlag, der in Aether und in Chloroform unlöslich ist, während er sich in Weingeist sehr leicht mit prachtvoll grüner Farbe auflöst. Etwa beigemengtes unzersetztes Bilirubin bleibt dabei in orangefarbenen Flocken zurück. Salpetersäure färbt die grüne Lösung zuerst blau, dann violett, roth und schliesslich schmutzig gelb.
>
> Dieses grüne Pigment ist ohne allen Zweifel das von Heintz . . . analysirte Biliverdin, wofür er die Formel $C_{16}H_9NO_5$ oder $C_{32}H_{18}N_2O_{10}$ aufgestellt hat.

2.9 Bile Pigment Isolation, Purification, and Combustion Analysis... 101

> Nimmt man diese Formel als richtig an, so würde die Bildung des Biliverdins aus dem Bilirubin auf einfacher Oxydation beruhen:

$$C_{32}H_{16}N_2O_6 \quad + \quad 4\,O \quad = \quad C_{32}H_{18}N_2O_{10}$$

$$\underbrace{\hspace{3cm}}_{\text{Bilirubin}} \qquad\qquad \underbrace{\hspace{3cm}}_{\text{Biliverdin}}$$

83

> Aber ich habe einige Beobachtungen gemacht, welche die Richtigkeit dieser Formel bezweifeln lassen.

Air oxidation of bilirubin interested *Städeler*, and following his keen investigative instincts, he found that the pigment dissolved in cold aq. NaOH without change and precipitated in orange-colored flakes with excess added (supersaturated with) HCl. A solution of bilirubin in ammonia behaved likewise and it made no difference therefore whether the solution was prepared cold or had been heated previously. In contrast if an NaOH solution were heated, even with complete absence of air, a remarkable color change was observed. The red solution became dark brown to green-brown and, when supersaturated with hydrochloric acid, a dark green, and not an orange, precipitate was obtained. Treatment of the same with alcohol left a dirty yellow matter on the filter paper, while the pigment, which was found in the splendid green filtrate, possessed all the properties of biliverdin. Its solution in alkalis, especially, was green, by which biliverdin was most easily distinguished from *Biliprasin*, which dissolved in alkalis with a brown color.

Formation of biliverdin simply by heating an aq. NaOH solution of bilirubin seemed to *Städeler* to stand in the way of acceptance of the formula proposed by *Heintz* (97). If one compares *Heintz*'s analytical results with his formula a satisfactory correspondence is in no way shown that might compel one to regard the formula as definitely established. The carbon and nitrogen content of the analysis fit better to the formula $C_{32}H_{20}N_2O_{10}$ than to *Heintz*'s $C_{32}H_{18}N_2O_{10}$ formula (97), while the hydrogen content found lay in the middle between the two formulas (135):

> Der gefundene Kohlenstoff- und Stickstoffgehalt stimmt besser mit der Formel $C_{32}H_{20}N_2O_{10}$ überein, während der gefundene Wasserstoff in der Mitte zwischen beiden Formeln liegt:

	$C_{32}H_{20}N_2O_{10}$	gefunden	$C_{32}H_{18}N_2O_{10}$
Kohlenstoff	60,00	60,04	60,38
Wasserstoff	6,25	5,84	5,66
Stickstoff	8,75	8,53	8,80
Sauerstoff	25,00	25,59	25,16
	100,00	100,00	100,00.

> Wahrscheinlich was das von Heintz analysirte Biliverdin nicht vollkommen rein, da es aus einem Farbstoffgemenge, aus dem s. g. Biliphäin, durch Auflösen in kohlensaurem Natron und freiwillige Oxydation erhalten wurde. Ich bedaure daher um so mehr, gegenwärtig nicht im Besitze einer genügenden Menge von reinem Bilirubin zu sein, um das Biliverdin einer neuen Analyse unterwerfen zu können.   84

Städeler believed that *Heintz* had analyzed impure biliverdin since it had been obtained from a pigment mixture precursor, the so-called *Biliphäin*, by dissolving in aq. Na_2CO_3 and allowing it to oxidize spontaneously. *Städeler* considered himself

102   2 Early Scientific Investigations

unfortunate not to have had in his possession a sufficient amount of bilirubin to produce biliverdin and undertake a new combustion analysis of it. Given the formula $C_{32}H_{20}N_2O_{10}$ for biliverdin, *Städeler* felt that the pigment stood in relationship to bilirubin as did *Biliprasin* to *Bilifuscin*, and its formation by oxidation of bilirubin would yield the equation (*135*):

$$C_{32}H_{18}N_2O_6 \quad + \quad 2\,HO \quad + \quad 2\,O \quad = \quad C_{32}H_{20}N_2O_{10}{}^{[1]}$$

$$\underbrace{\hphantom{C_{32}H_{18}N_2O_6}}_{\text{Bilirubin}} \qquad\qquad\qquad \underbrace{\hphantom{C_{32}H_{20}N_2O_{10}}}_{\text{Biliverdin}}$$

---

[1] N.B. These formulas are based on *Gmelin's* system of atomic "equivalents", which for H = 1, C = 6, and O = 8, water becomes HO.

In order to explain his own conversion of bilirubin to biliverdin simply by heating in aq. NaOH, *Städeler* theorized that two equivalents of bilirubin were involved to give one equiv. of biliverdin and one of the same compound that was formed when bilirubin was treated with Na(Hg) (*135*):

$$2\,C_{32}H_{18}N_2O_6 \quad + \quad 4\,HO \quad = \quad C_{22}H_{20}N_2O_6 \quad + \quad C_{32}H_{20}N_2O_{10}$$

$$\underbrace{\hphantom{2\,C_{32}H_{18}N_2O_6}}_{\text{Bilirubin}} \qquad\qquad\qquad\qquad \underbrace{\hphantom{C_{32}H_{20}N_2O_{10}}}_{\text{Biliverdin}}$$

Städeler continued to rationalize the formation of the other compounds found in gallstones: (brown) *Bilifuscin*, (green) *Biliprasin*, and *Bilihumin*. He had not found more than traces of biliverdin in gallstones and theorized that it had been converted earlier in bile to *Biliprasin* (*135*):

$$C_{32}H_{20}N_2O_{10} \quad + \quad 2\,HO \quad = \quad C_{32}H_{22}N_2O_{12}$$

$$\underbrace{\hphantom{C_{32}H_{20}N_2O_{10}}}_{\text{Bilirubin}} \qquad\qquad \underbrace{\hphantom{C_{32}H_{22}N_2O_{12}}}_{\text{Biliprasin}}$$

> Ich bemerke noch, dass ich das Biliverdin nicht fertig gebildet in den Gallensteinen angetroffen habe. Kommt es überhaupt darin vor, so kann es nur spurweise darin vorhanden sein. Wahrscheinlich verwandelt es sich in der alkalischen Galle durch Wasseraufnahme in Biliprasin.   85

Städeler purified and analyzed *Bilifuscin*. He washed out occluded fatty acids with ether and found the pigment was no longer soluble in $CHCl_3$, which allowed traces of $CHCl_3$-soluble bilirubin to be removed. After dissolution in alcohol and filtration, evaporation gave the pigment as an almost black lustrous brittle mass. Pulverizing afforded a dark brown powder with a somewhat olive color in it. It proved to be free of ash content, behaved like bilirubin upon heating, and gave a beautiful *Gmelin* reaction with HNO_3. From its combustion analysis (note the missing nitrogen analysis), *Städeler* determined the formula as $C_{32}H_{20}N_2O_8$, which suggested (to him) a close relationship to bilirubin, *i.e.* the two pigments differed by only two equivalents of water (*135*):[42]

---

[42] N.B. To *Städeler*, the formula for water in 1864 was HO, not H_2O.

2.9 Bile Pigment Isolation, Purification, and Combustion Analysis...

So dargestellt bildet das Bilifuscin eine fast schwarze glänzende spröde Masse, die beim Zerreiben ein dunkelbraunes, etwas in's Olivenfarbene ziehendes Pulver giebt. Es ist frei von Aschenbestandtheilen, verhält sich beim Erhitzen eben so wie das Bilirubin und giebt mit Salpetersäure eine eben so schöne Pigmentreaction.

0,2655 Grm. der bei 120° getrockneten Substanz gaben bei der Verbrennung 0,614 Kohlensäure und 0,1575 Wasser; überesinstimmend mit der Formel $C_{32}H_{20}N_2O_8$:

			berechnet		gefunden
32	Aeq.	Kohlenstoff	192	63,16	63,07
20	,,	Wasserstoff	20	6,58	6,59
2	,,	Stickstoff	28	9,21	–
8	,,	Sauerstoff	64	21,05	–
			304	100,00.	

Der Analyse zufolge steht das Bilifuscin in sehr einfacher Beziehung zum Bilirubin; es unterscheidet sich davon in der Zusammensetzung nur durch die Elemente von 2 Aeq. Wasser, welche es mehr enthält:

$$\underbrace{C_{32}H_{18}N_2O_6} \qquad \underbrace{C_{32}H_{20}N_2O_6}$$

$$\text{Bilirubin} \qquad\qquad \text{Bilifuscin.}$$

Indicating that *Bilifuscin* was, of all the pigments in gallstones, present in the smallest quantity, which perhaps placed a constraint on obtaining a %N in the combustion analysis, *Städeler* found just enough to learn a few of its properties: *Bilifuscin* was insoluble in H_2O, ether, and $CHCl_3$ (or only soluble in trace amounts); it was soluble in alcohol (giving a dark-brown color) which in high dilution had the color of strongly pigmented icteric urine, did not change color upon addition of HCl but became strongly reddish-brown upon addition of alkali. It dissolved easily in aqueous NH_3 or NaOH, producing a dark-brown solution from which brown flakes precipitated upon addition of HCl. Mixing an aq. NH_3 solution with $CaCl_2$ precipitated dark-brown flakes, much less voluminously than with bilirubin. Aerating an aq. NaOH solution of *Bilifuscin* caused decomposition, with color changes indicating the formation of *Biliprasin* and then probably *Bilihumin*.

The *Biliprasin* isolated from gallstones was also purified and analyzed by *Städeler*. It was pulverized, washed with ether and with $CHCl_3$, and then dissolved in cold alcohol and filtered. After evaporation of the dark green solution the "pure" *Biliprasin* was obtained as a lustrous, nearly black, brittle crust that looked quite similar to *Gallenbraun*. When pulverized it had a greenish-brown color. It yielded 0.6% ash upon combustion, which gave a strongly alkaline reaction and no effervescence with acids. The combustion analysis, correcting for ash, gave the formula $C_{32}H_{22}N_2O_{12}$, and the deviation from the calculated %N was not viewed as unusual, given the small amount of material available (*135*):

0,301 Grm. des bei 100° getrockneten Farbstoffes gaben bei der Verbrennung 0,627 Kohlensäure und 0,1765 Wasser.

Der Stickstoff wurde auf gleiche Weise bestimmt, wie beim Bilirubin. 0,096 Gm, gaben 0,073 Chlorsilber.

Diese Verhältnisse führen zu Formel $C_{32}H_{22}N_2O_{12}$:

			berechnet		gefunden
32	Aeq.	Kohlenstoff	192	56,81	56,81
22	„	Wasserstoff	22	6,51	6,52
2	„	Stickstoff	28	8,28	7,42
12	„	Sauerstoff	96	28,40	29,25
			338	100,00	100,00

Die Abweichung im Stickstoffgehalt ist nicht auffallend, wenn man berücksichtigt, dass zu dem Versuch nur eine sehr kleine Menge des Farbstoffes zu Gebote stand. [87]

Städeler summarized the properties of *Biliprasin*: insoluble in H_2O, ether, and $CHCl_3$; soluble in alcohol to give a pure green coloration different from that of biliverdin, which had more of a blue-green color. These two pigments could thus be differentiated on the basis of the color of their solutions in alcohol (presumably at the same concentration) and the color change that ensued upon addition of ammonia: The *Biliprasin* solution turned brown; whereas, that of biliverdin did not. *Biliprasin* exposed to air absorbed some ammonia and dissolved in alcohol with a brown color, which could be confused with a *Bilifuscin* solution. For differentiation, the latter did not change color upon addition of HCl; whereas, the former became a beautiful green. As with bilirubin, biliverdin, and *Bilifuscin*, a positive *Gmelin* reaction was seen after mixing an alcohol solution of *Biliprasin* with HNO_3, except the blue color was recessive or indistinct. Although *Biliprasin* was easily soluble in alcohol, it was much less soluble in aq. Na_2CO_3. Highly dilute solutions had the same color as intensely brown pigmented icteric urine. If the solution were mixed with acid, the green color reappeared by removal of the alkali. Since brown, icteric urine showed the same color change upon acidification, one might conclude that *Biliprasin* was present in predominant amounts. Introducing air to a solution of *Biliprasin* in aq. NaOH caused it to go over gradually to *Bilihumin*.

Bilihumin was found in a considerable quantity in gallstones and was not extracted into $CHCl_3$, ether, alcohol, H_2O, or dilute acid. It was freed completely of the various pigments already discussed by extraction a few times with aq. NH_3 to leave behind a black-brown pulverizable substance, which of course was not sufficiently pure for analysis. Purification was undertaken by repeated digestion in conc. ammonia at $50°-60°$ to extract a dark brown color and leave behind a dark brown solid that when dried and pulverized was black. The ammonia extracts were tediously processed by the usual methods: precipitation, washing, *etc.* to free the *Bilihumin* of inorganics. Yet despite multiple processing steps, *Städeler* did not consider the *Bilihumin* sufficiently pure for combustion analysis (*135*):

Eine Elementaranalyse habe ich nicht gemacht, da ich nicht die Ueberzeugung gewinnen konnte, dass der Körper rein sei, und da zu weiteren Reinigungsversuchen das vorhandene Material nicht ausreichend war. Ich bemerke nur, dass das gereinigte Bilihumin in Ammoniak nicht vollständig oder doch sehr langsam löslich ist, dass es sich dagegen in verdünnter Natronlauge beim Erwärmen ziemlich leicht löst, und dass die tiefbraune

2.9 Bile Pigment Isolation, Purification, and Combustion Analysis... 105

Lösung, wenn sie mit Weingeist und dann mit NO_4[43] haltiger Salpetersäure vermischt wird, einen ganz hübschen Farbenwechsel zeigt. Namentlich ist das Roth sehr rein und intensiv, während die vorher auftretenden Farben in der tiefbraunen Lösung nicht deutlich zu erkennen sind. 88

The *Bilihumin* so obtained was found to be soluble in dilute aq. NaOH (but not in ammonia) to give a dark brown solution, which, upon addition of alcohol and HNO_3 containing NO_4 [= NO_2], gave a nice but different color change. Though the red color was very pure and intense, the preceding colors were not distinctly recognized in the dark brown solution.

Städeler said that *Bilihumin* captured his interest chiefly because it occurred as the final decomposition product of all the rest of the bile pigments when, in aq. NaOH solution, they were exposed to air. He then proposed a simple relationship between *Bilihumin* and the others (*135*):

$$C_{32}H_{16}N_2O_6 \quad + \quad 2\,HO \quad = \quad C_{32}H_{20}N_2O_2$$

Bilirubin Bilifuscin

$$(+\ 2\,HO + 2\,O =) \qquad\qquad (+\ 2\,HO + 2\,O =)$$

$$C_{32}H_{20}N_2O_{10} \quad + \quad 2\,HO \quad = \quad C_{32}H_{22}N_2O_{12}$$

Biliverdin Biliprasin

Bilihumin

Ohne Zweifel steht die Formel des Bilihumins in einem ähnlichen Verhältniss zu der des Biliprasins, wie die Formeln der analysirten Körper unter einander. Für sehr wahrscheinlich halte ich es auch, dass die im lebenden Organismus vorkommenden dunkelen unlöslichen Pigmentsubstanzen, das s.g. *Melanin*, sich dem Bilihumin anschliessen und vielleicht gleichen Ursprungs sind. 89

And he suggested the likelihood that dark, insoluble pigments, the so-called *Melanin*, reminded one of *Bilihumin* and perhaps originated from the same source.

Although the focus of *Städeler*'s comprehensive work was the pigments of gallstones, he also looked into human bile in order to reinvestigate similarities between his bilirubin and *Valentiner*'s hematoidin, and to address further *Frerichs'*, *Neukomm*'s, and his own notion of bile acids as a source of bile pigment. *Städeler* knew that there was little doubt (and no further proof was needed) that the bilirubin of human gallstones came from inspissation of the same pigment in human bile. It was the apparent differences in crystal form between bilirubin and hematoidin that interested him, and he suspected that the crystal form was promoted by impurities carried along in the pigment isolation. He reasoned that if bilirubin and hematoidin were in fact identical, a more careful extraction and purification would confirm it (*135*):

---

[43] N.B. $NO_4 = NO_2$ when the atomic weight of oxygen is changed to 16 from *Gmelin*'s system of "equivalents", where O = 8.

Die menschliche Galle.

Es bedarf keiner chemischen Beweisführung, um die Annahme zu rechtfertigen, dass in der menschlichen Galle dieselben Farbstoffe vorkommen, wie in den Concrementen, welche sich darin bilden. Die Versuche, welche ich mit menschlicher Galle angestellt habe, hatten daher einen anderen Zweck. Wie bereits erwähnt, ist die krystallinische Form des Bilirubins um so mangelhafter, je reiner die Lösungen sind, aus welchen es anschiesst, während unreine Chloroformlösungen ganz gewöhnlich krystallinisches Bilirubin liefern. Die krystallinische Ausscheidung scheint bedingt zu sein oder doch sehr befördet zu werden durch die Gegenwart gewisser fremder Stoffe, ebenso wie zur krystallinischen Ausscheidung des Teichmann'schen Hämins aus essigsauerer Lösung die Gegenwart irgend welcher Chlormetalle erforderlich ist. Ich wählte daher die Galle, um das Bilirubin in messbarer Form darzustellen. War der darin vorkommende rothe Farbstoff wirklich identisch mit dem Hämatoïdin, wie Valentiner annimmt, so musste er sich bei richtig gewählter Behandlung auch in der so regelmässig auftretenden Hämatoïdinform gewinnen lassen. 90

Städeler had learned from his own experiments that in addition to its solubility in $CHCl_3$, bilirubin was sufficiently soluble in CS_2 and in benzene to be extracted from gallstones. Thus, he extracted human bile with these three solvents. From repeated experiments using $CHCl_3$, he obtained crystals of bilirubin that usually did not match up exactly with the crystal form of hematoidin, although in one experiment they came rather close (*135*):

Schüttelt man Galle mit Chloroform, so beobachtet man, wie schon Valentiner gefunden hat, beim langsamen Verdunsten der Lösung die Bildung von orangefarbigen elliptischen Blättchen oder sehr kleiner, fast rechtwinkeliger Tafeln, deren Winkelverhältnisse sehr wesentlich verschieden sind von denen des Hämatoïdins. Bei wiederholten Versuchen war das Resultat immer nahezu dasselbe; immer wurden jene rhomboïdischen Gestalten mit geringem Unterschiede der Seiten und Winkel wahrgenommen, bei denen die Diagonalen des Rhomboïdes durch abweichende Färbung markirt waren. Nur ausnahmsweise wurde mitunter einmal eine vereinzelte Form beobachtet, die sich der gewöhnlichen Hämatoïdinform näherte. 91

Using CS_2 and commercial benzene, which he purified by distillation, taking no fraction with a boiling point greater than $100°C$ and making sure as best he could that it contained no sulfur, he extracted bile. Actually the bile from two humans was dried, pulverized, and partitioned three ways into three flasks. One part was extracted or digested using $CHCl_3$, one by CS_2, and the third by the purified benzene, and in all of these washings a yellow coloration was produced. To each was added 20 drops of 25% aq. HCl, with continuous shaking, and after 12 hours each was filtered through filter paper moistened with the corresponding solvent. (It is not clear whether air was excluded in this procedure).

The $CHCl_3$ solution was colored an intense green and left a resinous violet residue upon passive evaporation. The residue was washed successively with ether (to remove cholesterol and fats) and alcohol (to remove a green pigment and other possible substances). The bilirubin so obtained consisted of orange-colored crystalline granules and flakes mixed with rhomboids described earlier (*135*):

Die *Chloroformlösung* hatte eine intensiv grüne Farbe und hinterliess beim freiwilligen Verdunsten einen mehr violetten klebenden Rückstand. Bei der Behandlung mit Aether wurden Cholesterin und Fett ausgezogen, Weingeist nahm neben anderen Substanzen den grünen Farbstoff auf, der nach seinem Verhalten gegen Alklien Biliverdin . . . zu sein schien, und als Rückstand wurde Bilirubin erhalten, aber nicht in guten Krystallen, sondern in orangefarbigen krystallinischen Körnern und Flocken, die mit den beschriebenen rhomboïdischen Formen gemengt waren. 92

The CS_2 extract had a pure golden yellow color, and after passive evaporation left behind a reddish crystalline mass. From the last, cholesterol, fats, and some bile acids were removed in the usual way to leave behind bilirubin as dark red microscopic crystals, which *Städeler* described in considerable detail. He found them different from the crystals obtained following the $CHCl_3$ extraction and while similar to those of hematoidin, he was unable to confirm an exact match due to the small size of his hematoidin crystals (*135*):

> Die *Schwefelkohlenstofflösung* hatte eine rein goldgelbe Farbe. Beim freiwilligen Verdunsten hinterliess sie eine röthliche krystallinische Masse, aus der Aether und Weingeist Cholesterin, Fett und vielleicht auch etwas Gallensäure aufnahmen, während das Bilirubin in tiefrothen mikroscopischen Krystallen zurückblieb. Die Krystalle erschienen als klinorhombische Prismen mit der Basisfläche, woran der vordere Winkel sehr scharf und die Prismenflächen convex gebogen waren, so dass die Ansicht auf die Basisfläche Ellipsen zeigte. Auf den convexen Flächen aufliegende Krystalle zeigten rhomboïdische Gestalten mit bedeutend grösserem Unterschiede der Seiten und Winkel, als bei den aus Chloroform angeschossenen Krystallen. Häufig findet man die prismatischen Krystalle in der Mitte eingeschnürt, was auf Zwillingsbildung hinzudeuten scheint. Die Diagonalen waren auf gleiche Weise markirt wie bei den aus Chloroform angeschossenen Krystallen. ... – Die Winkelverhältnisse dieser Krystalle zeigten Aehnlichkeit mit denen des Hämatoïdins; genaue Messungen und Vergleichungen waren aber wegen der Convexität der Flächen und wegen der Kleinheit der mir zu Gebote stehenden Hämatoïdinkrystalle leider nicht möglich. 93

The benzene extract had the same color as the CS_2 extract and yielded a quite similar residue of crystals upon evaporation in a mildly heated water bath, but these bilirubin crystals were larger and more irregular. Yet even if the crystals from benzene and from CS_2 were similar to hematoidin, *Städeler* concluded that was not a sufficient basis to conclude that bilirubin and hematoidin are identical. He indicated that a sufficient basis had to come from their combustion analyses; which showed large differences – differences that he concluded could not possibly be due to a small impurity or an unavoidable analytical error (*135*):

> Zunächst sind beim Hämatoïdin noch niemals convexe Flächen beobachtet worden, während dieselben beim Bilirubin so hervortretend sind, dass man dasselbe bei flüchtiger Betrachtung leicht für Harnsäure halten könnte. Das Hauptgewicht muss aber auf das Resultat der Analyse gelegt werden, und da ergiebt sich, wie die folgende Zusammenstellung zeigt, eine so grosse Abweichung in der Zusammensetzung, dass man die Differenz unmöglich auf Rechnung geringer Verunreinigungen* oder der unvermeidlichen Analysenfehler setzen kann.

	Bilirubin		Hämatoïdin	
Kohlenstoff	67,15	67,11	65,85	65,05
Wasserstoff	6,27	6,12	6,47	6,37
Stickstoff	9,59		10,51	
Sauerstoff	16,99		17,17	
	100,00		100,00	

---

*Bei einem nicht genügend gereinigten Bilirubin fand ich folgende procentische Zusammensetzung: 66,52 Kohlenstoff, 6 Wasserstoff, 8,7 Stickstoff und 18,78 Sauerstoff. 94

A few years earlier, *Städeler* had called attention (cited in *120*) to the fact that the formula from *Robin*'s combustion data (*115, 116*) did not agree with the formula $C_{14}H_9NO_3$ but that the formula $C_{30}H_{18}N_2O_6$ did, although it was out of correspondence by 0.1% and 0.2% smaller in hydrogen. *Städeler* concluded that a close relationship existed between bilirubin and hematoidin based on the great similarity of their formulas: If hematoidin contained two fewer hydrogens, its formula would thus be $C_{30}H_{16}N_2O_6$; so, it and bilirubin ($C_{32}H_{18}N_2O_6$) would belong to a homologous series, which would clarify their manifold similarities in characteristics. *Städeler* believed that a decision could be reached only from new combustion analyses (*135*):

> Robin . . . hat aus jenen Analysen die Formel $C_{14}H_9NO_3$ für das Hämatoïdin berechnet, doch habe ich schon vor Jahren darauf aufmerksam gemacht . . . , dass diese Formel nicht mit Robin's Analysen übereinstimmt, und dass man bei richtiger Berechnung zu der Formel $C_{30}H_{18}N_2O_6$ gelangt; nur der Wasserstoff ist in diesem Falle um 1/10 und 2/10 pC. geringer gefunden, als der Formel entspricht. – Dass Bilirubin und Hämatoïdin nahe verwandte Körper sind, ergiebt sich schon aus der grossen Aehnlichkeit der Formeln. Enthielte das Hämatoïdin 2 Aeq. Wasserstoff weniger, hätte es also die Formel $C_{30}H_{16}N_2O_6$, so würde es mit dem Bilirubin, $C_{32}H_{18}N_2O_6$, in eine homologe Reihe gehören, und damit wären die mehrfachen Aehnlichkeiten in den Eigenschaften genügend erklärt. Doch darüber kann nur durch neue Analysen entschieden werden.   95

Städeler's comprehensive publication on bilirubin from gallstones and bile would not have been complete without his concluding comments directed toward other pigments that gave a positive *Gmelin* reaction. Such included the green pigments that he isolated from gallstones as indicated above: biliverdin and *Biliprasin*, another green pigment isolated by *Scherer* from icteric urine (*92, 93*), which *Städeler* thought was a decomposition product formed in the isolation procedure (*135*):

> Wahrscheinlich war dieser Farbstoff ebenfalls nur ein Zersetzungsproduct, entstanden durch Einwirkung der Salzsäure auf den ursprünglichen Farbstoff; jedenfalls war er nicht rein, wie aus dem hohen Kohlenstoff- und Wasserstoffgehalt neben dem geringen Stickstoffgehalt hervorgeht.   96

All these and a third green compound isolated several years earlier by *Städeler* (when the $CHCl_3$ extraction method was not known) from a brown-colored ox gallstone the size of a walnut that had been given to him by his friend Prof. *Merklein* in Schaffhausen, Switzerland. This green material gave the formula $C_{32}H_{18.5}N_{2.5}O_{10}$ from combustion analysis – an odd analysis that *Städeler* attributed to insufficient care that a pure sample was used. He believed that gallstones from animals appeared to be richer in nitrogen (10.5% N) than those from humans.

Städeler could not end the discussion of the pigments of gallstones and bile without a commentary on "synthetic" bile pigments that also show a beautiful color change, the *Gmelin* reaction. He had obtained a brownish-red pigment by warming a bile salt in conc. H_2SO_4, a chromogen that precipitated in resin-like flakes upon addition of H_2O. If the H_2SO_4 solution were warmed briefly in the absence of air, the precipitated flakes were colorless or greenish, but after standing 24 hours in conc. H_2SO_4 the solution showed a beautiful dichroism that was orange-colored or brownish with a striking pure green transmitted light. Addition of H_2O precipitated green-blue flakes. Further processing, isolation, and purification produced a pigment that imparted a bile-green color in alcohol, became yellow or orange upon basification, and returned to green upon addition of HCl. With "NO_4"-containing

HNO₃ (presumably it was NO_2; NO_4 assumes the atomic weight for oxygen is 8), the pigment gave, even at great dilution, a vivid color change: at first green, then green-blue or greenish-brown, next red, and finally dirty yellow. These pigment color reactions appeared (to *Städeler*) to signify a relationship between the synthetic and natural pigment. With this perspective, *Städeler* thought it not inappropriate to think that the bile pigments found in the urine of dogs after intravenous injection of bile acids came from their transformation in the blood stream. This "completely proven and irrefutably established fact" was not brought to the fore as such, however, because bile pigments were not detected in bile in some experiments after intravenous administration of bile acids. Yet there was also an unresolved question as to how a nitrogen-free bile acid might be converted to a nitrogen-containing bile pigment (*135*):

> Da durch diese Pigmentreaction ein Zusammenhang der künstlichen Pigmente mit den natürlichen Gallenpigmenten angedeutet schien, und da wir, wie schon oben (S. 324 f.) angegeben wurde, ausserdem noch beobachteten, dass nach der Injection von gallensauren Salzen in eine Vene fast regelmässig Gallenpigment im Harn auftritt, so war es gewiss nicht übereilt, wenn wir schlossen, dass die Gallensäuren auch in der Blutbahn eine Umwandlung in Pigment erleiden könnten. Als völlig erwiesene und unumstösslich feststehende Thatsache ist diese Umwandlung übrigens niemals hingestellt worden, da uns einige, wenn auch nur wenige Fälle vorkamen, wo nach Galleninjection kein Pigment im Urin nachgewiesen werden konnte. Es ist mir jetzt gelungen, auch die stickstofffreie Cholsäure auf gleiche Weise wie die Glycocholsäure und Taurocholsäure in Farbstoffe zu verwandeln, und da sich ungezwungen nicht annehmen lässt, dass die stickstoffhaltigen Gallenpigmente ihr Entstehen einem stickstofffreien Körper verdanken, so kann von einer Umwandlung der Gallensäuren in die wirklichen Gallenfarbstoffe nicht wohl ferner mehr die Rede sein.   97

And there was an equally fundamental question related to whether the bile pigments found in urine under the circumstances described come about by transformation of the intravenously injected bile acids or whether the red cells of blood were lysed by the bile acid and their extruded pigment was the source of the urinary bile pigments. Arguing against the latter is that injection of H_2O did not lead to bile pigments and that, in the case of a rabbit, injection of water produced urine that was rich in blood pigment but contained no bile pigment (*135*):

> Es bleibt nun noch immer die Frage unerledigt, welche Rolle die in das Blut getretene Galle bei der Erzeugung der Gallenpigmente spielt; denn die Annahme, dass die Gallensäure *nur* die Blutkörperchen auflöst, und dass das gelöste Blutroth dann in Gallenfarbstoff übergehe, scheint mir doch nicht gerechtfertigt zu sein. Einmal müsste dann nach Galleninjectionen regelmässig Gallenpigment im Urin auftreten was bekanntlich nicht der Fall ist, und ausserdem müssten Wasserinjectionen dieselbe Wirkung hervorbringen wie die Injection von Gallensäuren. Auch dieses ist nicht der Fall. Röhrig . . . spritzte einem Kaninchen, dessen Blutgehalt sich zu 130 Grm. berechnete, 100 CC. Wasser in die Vena jugularis und beobachtete, dass der darauf gelassene Harn reich an Blutpigment war, aber keinen Gallenfarbstoff enthielt.   98

A new idea apparently struck *Städeler* when he realized that during icterus the heartbeat was known to be reduced, usually by 20–30 contractions. His colleague *Frerichs* mentioned this and cited two cases where the heartbeat dropped 28 and 21 beats. He ascribed the perturbations to the presence of bile acids and suggested that small amounts of sodium salts of glycocholic, taurocholic, and cholic acids act likewise, proportionately depressing the pulse. The presence of larger amounts of bile acid salts led to sudden death by paralysis of the heart.

110 2 Early Scientific Investigations

Yet on the basis of all the various observations made toward understanding the induction of bile pigments in urine, which *Städeler* knew from his and *Neukomm*'s studies that the bile pigments were not always found in urine post intravenous injection, doubts persisted. Other factors may have been the cause: differences in age, size, and constitution of the dogs used were uncontrolled potential variables, as was the heartbeat (*135*):

> Nach diesen Beobachtungen halte ich es für wahrscheinlich, dass wir in diesen enormen Kreislaufstörungen, mit denen natürlich auch grosse Störungen in der chemischen Stoffmetamorphose verbunden sein müssen, hauptsächlich den Grund der Pigmentbildung nach Einführung von Gallensäuren in das Blut zu suchen haben. Es würde sich damit auch erklären, dass die Pigmentbildung nicht constant eintritt, denn Thiere von verschiedenem Alter und Grösse, von schwacher und kräftiger Constitution, können nicht auf gleiche Weise von derselben Menge Gallensäure afficirt werden. – Demnach wäre also die Pigmentbildung nach Galleninjection nur eine secundäre Wirkung der in's Blut gebrachten Gallensäure, und ist dieses der Fall, so steht zu erwarten, dass andere Substanzen, welche ähnlich Störungen der Herzthätigkeit hervorbringen, ebenfalls zur Bildung von Gallenpigment Veranlassung geben müssen. Eine solche Substanz besitzen wir in der Digitalis, mit der ich einige Versuche angestellt habe. 99

For the last, *Städeler* reasoned that if a bile acid-perturbed heartbeat were the root cause of the presence of urinary bile pigments, he might conduct control experiments using digitalis as a heartbeat perturber. (Of course, as with bile acids, the chemical structure of the steroid digitalis was also not known.) So he brought two dogs up to a modicum of good health, and after their urine proved to be free of bile pigment, he then infused the animals with 2 g of herbal digitalis – which induced vomiting and diarrhea. Some 48 hours later, the urine of one dog showed a distinct and intense pigment reaction with HNO_3. Using the $Pb(OAc)_2$ precipitation method to sequester the pigment, a positive *Gmelin* reaction was confirmed for eight days following the initial dose. The second dog gave no detectable bile pigment in urine following the same procedure as in the first dog. The poor dogs expired eight days following the initial dosing.

Städeler admitted that these contradictory results did little to settle the issue, which was apparently still unresolved, at least from his perspective, and he was resigned to the belief that a large series of experiments would be required. Then he essentially bowed out of bile pigment research by indicating that other work prevented his giving the question the attention it deserved (*135*):

> Diese beiden Versuche widersprechen einander. Die angeregte Frage ist also noch nicht erledigt; sie lässt sich aber nur durch eine grössere Versuchsreihe beantworten, und ich bedauere, dass andere Arbeiten mich verhindern, diesem Gegenstande ferner die Aufmerksamkeit zu widmen, die er zu verdienen scheint. 100

To summarize *Städeler*'s achievements briefly, he introduced new names for bile pigments isolated from gallstones: (i) *Bilirubin*, for the reddish pigment of gallstones and bile (*Gallenroth*), which soon thereafter replaced the older names *Cholepyrrhin* [*Berzelius*' yellow pigment from bile (*73–76*)], and *Biliphäin* [*Simon*'s name for *Cholepyrrhin* (*89–91*)], and the contemporary name *cholophain* or *Cholophäin* [*Thudichum*'s name for *Cholepyrrhin* or *Biliphäin* (*103*)]; and (ii) *Bilifuscin* and *Biliprasin* (brownish pigments), *Bilihumin* (brownish-green). The only original name that has persisted, *Biliverdin*, is that given by *Berzelius* to the green pigment of bile (*Gallengrün*), and which *Städeler* also isolated from gallstones.

2.9 Bile Pigment Isolation, Purification, and Combustion Analysis...

In 1864, *Städeler* undoubtedly had prepared the purest bilirubin up to that time, taking care in the $CHCl_3$ extractions that the solvent was freed of HCl (from its decomposition to $COCl_2$ and HCl by light). He also learned by so doing that the pigments of gallstones clung to certain metals, as salts, mainly to calcium. This fact may have been suspected by the investigators immediately preceding him such as *Brücke (109)*, *Heintz (97)*, and *Hein (105)*. They had to exert considerable effort to prepare pigment samples from bile and gallstones that were ash-free by combustion – a difficulty that plagued elemental combustion analyses prior to *Städeler* – causing him difficulty in his *Biliprasin* analysis and thwarting an analysis of *Bilihumin*.[44]

The combustion analysis data (see Table 2.9.1) obtained by *Städeler* differed from the data of his earlier (*120*) analyses of bilirubin, and as indicated earlier, he had insufficient biliverdin for analysis. As did his predecessors, from the %C, H, N data *Städeler* calculated formulas for the pigments – and he made many attempts to provide correlations between the pigments based on these formulas. Although well-intentioned in this, *Städeler* and others preceding him struggled with sample purity, which is always a consideration, and were dependent on, unknowingly (and hamstrung by), the prevailing assignments of the atomic weights of C, H, N, and O.

Table 2.9.1 *Städeler*'s elemental combustion analysis data of bilirubin, *bilifuscin*, and *biliprasin* compared with hematoidin. (The formulas are based upon the *Gmelin* system of atomic equivalents, C = 6, H = 1, N = 14, and O = 8)

	Bilirubin					Hematoidin			
	Experimental			Calculated for		Experimental[c]		Calculated for	
%	A[a]	B[a]	C[b]	$C_{32}H_{18}N_2O_6$[a]	$C_{18}H_9NO_4$[b]	D	E	$C_{30}H_{18}N_2O_6$[d]	$C_{14}H_9NO_3$[e]
C	67.15	67.11	66.52	67.13	66.26	65.85	65.05	65.69	64.12
H	6.27	6.12	6.00	6.29	8.52	6.47	6.37	6.57	6.87
N	9.59	–	8.70	9.79	8.59	10.50	10.50	10.22	10.69
O	16.99	–	18.78	16.79	19.63	17.18	18.08	17.52	18.32

	Bilifuscin[a]		*Biliprasin*[a]		*Biliverdin*[a]	
%	Experimental	Calculated for $C_{32}H_{20}N_2O_8$	Experimental	Calculated for $C_{32}H_{22}N_2O_{12}$	Experimental	Calculated for $C_{32}H_{20}N_2O_{10}$
C	63.07	63.16	56.81	56.81	60.00	60.04
H	6.59	6.58	6.52	6.51	6.25	5.84
N	–	9.21	7.42	8.28	8.75	8.53
O	–	21.05	29.25	28.40	25.00	25.59

[a] *Städeler*'s data from reference (*135*)

[b] *Heintz*'s data from reference (*97*)

[c] Hematoidin experimental data from *Robin* (*115, 116*)

[d] *Städeler*'s formula (*135*) from *Robin*'s experimental data

[e] *Robin*'s formula from his experimental data (*115, 116*)

---

[44] N.B. Combustion analyses may at times be more important in revealing the presence of impurities than in characterizing the intended compound.

The issues surrounding atomic weights were addressed on September 5, 1860, at a major European and first *international* scientific congress which opened in Karlsruhe, capital city of the Grand Duchy of Baden. (Karlsruhe entered the German empire in 1871 and is now part of the German Federal Republic State of Baden-Württenberg.) The congress was organized by *Kekulé, Wurtz, Weltzien, Baeyer, Roscoe*, and *Williamson* to discuss the major issues in science (*147-149*). Among the topics were the highly disputed atomic weights, especially those of C and O, which are of great importance to organic chemistry and the then undeveloped Periodic Table of Elements. According to *Kauffman* and *Adloff*, writing on the history of the Karlsruhe Congress, as an alternative to *Dalton*'s "incorrect and inadequate" atomic weights (relative atomic masses), the year 1814 brought forth (from *W.H. Wollaston*) (*149*):

> a new, more pragmatic term, "atomic equivalent". . . . In dealing with proportional relationships between chemical compounds many chemists such as Leopold Gmelin used equivalent weights (He called them "*Mischungsgewichte*") . . . rather than atomic weights. The resulting debate between the so-called "atomists" and "equivalentists" raged for another half century.
>
> Until 1849, when English chemist Edward Frankland (1825-1899) recognized the concept of valence, . . . it was impossible to know whether the assigned atomic weights were correct or should be multiples of the values. Thus, for example, some chemists used atomic weights of 6 and 8 for carbon and oxygen, respectively, while others preferred atomic weights of 12 and 16. Therefore different formulas were often assigned to the same substance. . . . As an extreme example of the problem of inconsistent formulas we may cite the 19 formulas for acetic acid in August Kekulé's organic chemistry textbook of 1861 [28]...

In the third session of the Congress the conflicting theories and concepts were addressed in a lecture by *Cannizzaro*, who reminded the attendees of *Avogadro*'s hypothesis (equal volumes of gases at the same temperature and pressure have the same number of molecules) and on that basis, as well as the Law of *Dulong* and *Petit* (relating the specific heat of a solid to its atomic weight), reassigned the atomic weight of C from 6 to 12, O from 8 to 16, S from 16 to 32, *etc.* while convincing the majority of the attendees (*149*):

> The Karlsruhe Congress dramatized the importance in the minds of the younger attendees of Avogadro's hypothesis, . . . which had been largely overlooked for half a century, thus making possible the impressive strides in chemistry that took place during the next four decades of the nineteenth century. Removing the uncertainty about atomic weights established the certainty of molecular weights and made it possible to distinguish between empirical and molecular formulas and to formulate correctly hydrocarbons, alcohols, organic acids, aromatic compounds, and almost all the simpler organic molecules, leading to the tremendous progress in organic chemistry. . . .

Though the new (correct) set of atomic weights was well-accepted in Germany, it lagged in some other countries. Yet its influence was considerable (*149*):

> The congress established a paradigm shift for the understanding of chemistry and led to the periodic tables of Mendeleev . . . and Lothar Meyer. . . .
>
> In addition to its impact on the development of chemical theory and practice discussed above, the Karlsruhe Conference was the prototype for future international chemical meetings.

2.9 Bile Pigment Isolation, Purification, and Combustion Analysis...

Although the change was adopted only slowly, *Städeler*'s formulas were based on the relative atomic weights H = 1, C = 6, N = 14, and O = 8. Which explains his use of the formulas HO (or OH) for H_2O and NO_4 for NO_2 that should look odd to us today but illustrate how much the structure of the chemical sciences depends on exact fundamental constants.

Finally, *Städeler* continued to address the apparent (to him) transformation of bile acids into "synthetic" bile pigments that give an apparent positive *Gmelin* color reaction. He demonstrated this by transformation in two ways: in conc. H_2SO_4 from which he isolated the pigment, and detection in urine following intravenous injection of a bile acid. However, the latter experiment sometimes produced an apparent bile pigment and other times it did not, which created uncertainty. *Städeler* was clearly a careful scientist in analyzing the experiment and hesitated to commit firmly to the thesis, while also questioning how the transformation of a bile acid that contained no nitrogen might be transformed to a bile pigment that does. The 1864 publication was apparently his last on the subject of bile pigments, and he was to die some seven years later.

2.9.2 *Johann Ludwig Wilhelm (aka John Lewis William) Thudichum and Bilirubin*

Thudichum[45] cast a broad shadow across the entire gallstone literature in the last half of the 19th century, including the chemistry of gallstones. Unlike *Städeler*, *Thudichum* lived a long life. In 1863, he wrote a long treatise on gallstones (*103*), citing the history of the early chemical analyses, the older analytical proceedings of

---

[45] *Johann Ludwig Wilhelm (aka John Lewis William) Thudichum* was born eight years after *Städeler*, on August 27, 1829 in Büdingen, in Hessen, Germany, and died on September 7, 1901 in Kensington, in his adopted England. Though he is most noted for his studies on the chemical constitution of the brain (identifying sphingomyelin, sulfatides, cerebrosides, *etc.* therein) in the late 1800s, his fame came mainly posthumously. His greatest work, A *Treatise on the Chemical Constitution of the Brain*, stirred controversy and provoked criticism for his rejection of the then firm belief that the brain is composed of a single giant molecule (Protagon) and his insistence that it consisted of elaborate chemical structures (in the scientific press he was called by some a liar and falsifier). At age 18, he began medical studies in 1847 at the University of Giessen, working after hours in *Justus Liebig*'s lab, where he developed his interest in physiological chemistry. He studied in Heidelberg, volunteered as a surgeon in 1850 during the Prussian-Danish War, then obtained the Dr. med. degree in 1851 at Giessen, where he began his medical practice. Drawn to chemistry from his studies under *Liebig*, and at odds politically over the war, he emigrated to London in 1853, where he obtained the diploma M.R.C.S. Eng. in 1854 and where he practiced medicine as an otologist and rhinologist first at St. Pancras Dispensary and elsewhere. After accepting several subsequent appointments, in 1860 he became M.R.C.P. and in 1865 was appointed Lecturer at St. Thomas's Hospital and director of its newly founded chemical and pathological laboratory. While continuing his medical practice, from 1871 he conducted experimental physiological chemistry in his home laboratory. In addition to his aforementioned work on the brain, he wrote authoritative treatises on urine and on gallstones.

Berzelius and *Heintz*, and a method of his own for analyzing human gallstones. For reasons not entirely clear, he coined new names for pigments: The "colouring matter of bile and all its varieties" he called "cholochrome"; the brown coloring matter he retained the name "cholophæine" (synonymous with *Cholepyrrhin, Biliphäin,* and *Bilifulvin*); for the green he adopted the name *cholochloine* (synonymous with *Biliverdin* and *Cholechlorin*). Just why a new set of names was required is unknown (except possibly to put his stamp on the bile pigment field?), but they fortunately began to melt away some five years later when he began to use the new term *Bilirubin* in a major publication on bile pigments (*152*). Previously he wrote briefly on the composition of gallstones (*102*), with separations based on modifications of *Berzelius'* approach some two decades earlier (*71, 72–76*) by treating pulverized gallstones with H_2SO_4, followed by precipitating with or without $Ba(OH)_2$, $(NH_4)_2S$, acidifying with HCl, basifying with NH_4OH and decanting as needed along the separation route, *etc.* to provide cholochrome and inorganic salts, *inter alia.*

In his comprehensive treatise on gallstones (*103*), *Thudichum* reviewed the early experiments of his predecessors who attempted to separate the components of bile, from work preceding that of *Haller* (*35*) in 1764 to the more recent studies of *Fourcroy* (*42*) and *Thenard* (*53–56*), *Berzelius* (*68–76*), and *Heintz* (*95–97*), who was concerned about the amount of ash remaining in his combustion analyses of bile pigments. *Thudichum*, too, was especially concerned with combustion analyses and composition; he repeated the analysis of cholophæine (= *Cholepyrrhin* = *Biliphäin*) of material isolated from gallstones according to *Heintz* only to obtain different results for the %C, and even larger differences in %C from material isolated from ox bile (*103*):

> . . . Since the first attempts of Berzelius, about 1812, to determine the properties of the colouring matter of bile, several analyses have been instituted with the particular object of ascertaining its chemical or elementary composition. Those of Scherer (1843), Hein (1847), Heintz (1854), and Städeler (1861), were the most methodical, although none of them have led to final results. The elementary analyses of Scherer and Hein were performed upon specimens of cholochrome which, to conclude from the process adopted for their preparation, must have contained impurities and inorganic matter. The analyses of Heintz, on the contrary, were executed upon materials apparently homogeneous, and certainly free from inorganic substances. But the analyses of cholophæine, the brown modification of cholochrome, lead to a formula which is very ill-supported by the formula of the only metamorphosis to which, at that period, cholophæine could be subjected. Four elementary analyses, agreeing with each other, led to the empirical formula $C_{31}H_{18}N_2O_9$ for cholophæine; but one analysis of cholochloine, the green colouring matter hitherto termed biliverdine, obtained from the brown by oxidation, led Heintz to the formula $C_{16}H_9NO_5$. The improbability of the suggestion that cholophaeine, in order to pass into cholochloine, should take up only half an equivalent of oxygen, Heintz met by assuming the formula of cholophæine to be $C_{32}H_{18}N_2O_9$, and by further assuming that this body took up one equivalent of oxygen, and then split up into two equivalents of cholochloine.
>
> I have repeated the analysis of cholophæine upon materials prepared in accordance with the precedent of Heintz. In some of them I have obtained figures which are very near to those of Heintz, the hydrogen in most cases keeping steadily near 6 per cent.; but the carbon varied between 60 and 62 per cent., or to the same extent to which the first analyses of Heintz differed from his check calculation. But when I came to analyse cholophæine obtained from ox bile directly (the former specimens having been prepared from gall-stones), I obtained totally different results, the carbon rising to 66, the hydrogen to 10 and 11 per cent.

The only combustion analyses data directly cited were those of *Städeler* for purified *Cholepyrrhin*, obtained from gallstones by $CHCl_3$ extractions, and for which the formula $C_{18}H_9NO_4$ was calculated (*103*):

> An elementary analysis by Städeler of cholepyrrhine, purified by repeated crystalliza-
> tion from boiling and washing with cold chloroform, yielded results from which the for-
> mula $C_{18}H_9NO_4$ was calculated.

				Calculated.		Found.
18	equivalents of	carbon	=	108	66·26	66·52
9	"	hydrogen	=	9	5·52	6·00
1	"	nitrogen	=	14	8·59	8·70
4	"	oxygen	=	32	19·63	18·78
				163	100·00	100·00

Confirming the absence of iron as a component of his cholochrome, *Thudichum* reached a rash but perhaps a logical conclusion then, but unjustified now, that there is, on the basis of the absence of iron in the bile pigment, no apparent connection between it and the pigment of blood (which was known to contain iron) (*103*):

> In none of the specimens analysed by me was there any trace of iron; I can, therefore,
> fully confirm the statement of Heintz, that iron is not an elementary ingredient of cholo-
> chrome. Hence it follows that cholochrome has no immediately apparent connection with
> the colouring matter of blood.

Thudichum modified *Berzelius'* method for the separation of bile pigments from bile, in which bile was left standing 1–2 days before a tedious work-up that led to what he believed to be "somewhat impure cholechloine" (= biliverdin), a "beauti-fully green substance." From human and ox gallstones, he applied *Heintz*'s method to separate olive-green tinted and brown cholophæine that, as with *Heintz*'s, was "perfectly free from ash on combustion" (*103*). He then subjected a narrow, 15-inch-long tube of cholochrome (from ox gallstones) to extraction into $CHCl_3$ over a 1-week-period to yield a reddish-brown solution, from which he obtained red crystals, which he supposed to be the "original form of cholochrome or *cholery-thrine*" – and which one might now assume was cholophæine = *Cholepyrrhin* = *Biliphäin* = hematoidin = bilirubin (*103*):

> . . . the chloroform of the extract was distilled off, and the concentrated solution left to
> spontaneous evaporation. A granular substance was deposited, which yielded to boiling
> absolute alcohol a greenish-brown matter, and became a most beautiful red colour, resem-
> bling cinnabar or red oxide of mercury. When dry, it had the sweet, musk-like odour of a
> healthy cow. Viewed under the microscope, it appeared mostly amorphous; but when a
> concentrated solution in chloroform was allowed to evaporate slowly under a little glass
> cover, crystals were formed in great numbers, being needles and rhombic plates. The pow-
> der was insoluble in water, little soluble in boiling absolute alcohol, sparingly soluble in
> ether, easily soluble in chloroform, a little more soluble in boiling than in cold chloroform.
> It was soluble in dilute solutions of caustic and carbonated alkalies and in an alcoholic
> solution of caustic potassa. When treated with concentrated sulphuric acid, it dissolved
> with a yellow colour, and green flakes separated on the addition of water. Nitric acid
> imparted a deep-crimson colour to the powder, dissolving a part, which changed from red

to blue, violet, and lastly crimson. This change of colour was particularly beautiful on a thin layer of colouring matter, produced by allowing a very dilute chloroform solution to evaporate in a china dish. Such a layer, like a stain of the same solution on the skin, was of a bright-yellow colour.

This red substance is, evidently, the original form of biliary colouring matter, and a chemically pure body. I shall hereafter speak of it as the red or original form of cholochrome or *cholerythrine*.

Consistent with earlier observations that the reddish pigment was easily oxidized to the green, *Thudichum* found that cholophæine was easily oxidized to a green pigment that he called *cholochloine*, then later called it *Biliverdin*. In a different oxidation, one initiated by nitrous oxide gas (N_2O) followed by HNO_3, there was isolated a new crystalline, water-insoluble substance, and an "uncrystallizable acid, which gave a crystallized salt with ammonia." The intermediate at the first step by treatment with nitrous acid, called *cholochromic acid* by *Thudichum*, was isolated apparently in two crystal modifications, if not two chemically-different reddish compounds. One showed the crystal form and color attributed to hematoidin, but neither type of crystal could be isolated from the surrounding syrup. Various manipulations of cholochromic acid were engaged: nearly insoluble in H_2O; soluble in spirit of wine, to give a port wine colored strongly acidic solution that precipitated a red solid with aq. $Pb(OAc)_2$; a pink solid with $AgNO_3$, turned deep red upon addition of ammonia. *Thudichum* concluded that cholochromic acid is not hematoidin (*103*):

Cholochromic acid differs from hæmatodine by its solubility in alcohol and by crystallizing in (clino ?) rhombic octahedra, not rhombic plates. Rotten bile, and bile treated by the proceeding of Berzelius for obtaining cholochrome have both a dark-pink colour, and chloroform extracts from the former some coloured acid.

Thudichum classified gallstones into seven series and provided examples from the literature in each series (*103*):

Classification of Gall-stones.

First Series.–Pellucid or pure cholesterine calculi.
Second Series.–Mixed calculi, with prevalence of cholesterine.
Third Series.–Calculi with prevalence of cholochrome.
Fourth Series.–Calculi with prevalence of modified cholochrome.
Fifth Series.–Gall-stones with prevalence of bile acids.
Sixth Series.–Gall-stones with prevalence of fatty acids.
Seventh Series.–Gall-stones with prevalence of carbonate of lime.

At this point he had carried out only a limited investigation into the pigments (*cholochrome*, as he named them collectively) of bile and gallstones, but that was due to change with his 1868 publication (*152*) on the isolation of a red pigment (*Cholephäin*, or bilirubin) from ox gallstones, its conversion into what he called *cholechlorin* (or biliverdin), and his combustion analyses thereof. It was work which followed that of *Städeler* (*135*) by four years.

In 1868, in a paper on bile pigments written in his native German, *Thudichum* published his experimental results on the red pigment of ox gallstones, its isolation

2.9 Bile Pigment Isolation, Purification, and Combustion Analysis...

and purification, physical and chemical properties, combustion analysis and its transformation into salts (ammonium, sodium, potassium, silver, barium, calcium, zinc, and lead). He also described the conversion of bilirubin into biliverdin and discussed its chemical and physical properties, combustion analysis, calcium and barium salts.

The isolation of bilirubin [*Thudichum* used *Städeler*'s name for the pigment interchangeably with his own, *Cholephäin*] from gallstones was pursued by an elaborate series of washings, $CHCl_3$ and alcohol digestions, precipitations, *etc.* excluding exposure to air as much as possible in order to remove traces of bile and bile acid components, and to break apart bile pigment salts. The detailed care taken exceeded even *Städeler*'s. Thus, ox gallstones were pulverized (during which one's air passages were protected from the powder by a kerchief), stirred with a bit of hot H_2O (the same way a cook mixes flour for dough) then bathed in hot H_2O with vigorous stirring before being allowed to stand for two days. The water was drained and the solid left behind was thoroughly washed with H_2O before washing and filtering and washing again until the filtrate was clear. The remaining slurry was transferred to a flask where it was digested with a large quantity of alcohol while being heated to remove bile acids and their calcium as well as some fatty acid salts (but rarely cholesterol). The washed powder was then treated with cold dilute HCl, which evolved CO_2 and H_2S. *Thudichum* found it better to let the HCl do its work on the solid without heating. The solid was washed free of HCl with H_2O by decanting, and then it was treated again twice with alcohol to remove traces of any bile acids. After complete exhaustion, the solid was treated with ether and then dried. At this point the powder had a beautiful reddish-yellow color ("Nach dem Trocknen ist das Gallensteinpulver schön rothgelb."), and it was heated in water and acid-free $CHCl_3$. (It may be important to note that *Thudichum*, like *Städeler* before him, took precaution with the $CHCl_3$, which then doubtless lacked the ethanol stabilizer found nowadays in commercial $CHCl_3$, because it commonly became acidic by reaction with air and light while it partially decomposed into HCl and phosgene.) The $CHCl_3$ solution was filtered from the solid, and the residue was again heated in fresh $CHCl_3$ while certain measures were used to avoid losses of $CHCl_3$. Then with evidence of great care, *Thudichum* removed the red $CHCl_3$ solution from the solid by siphoning, presumably to minimize exposure to air, and distilled off most of the $CHCl_3$. This left a red residue with some green spots admixed, which was washed on a filtration funnel with $CHCl_3$ until it was red and no longer co-mixed with green, while the $CHCl_3$ was all the more yellow-red. A little alcohol was added to the dark, nearly black-green colored mother liquor and red, very finely dispersed bilirubin was removed by filtration and washed with alcohol, and crystals easily formed in the alcohol mother liquor. The pigment obtained was a splendid red, of a color similar to that of the HgO obtained by heating its nitrate. Neither absolute alcohol nor ether extracted any impurities, only traces of pigment. Further purification could be approached by careful, repetitious dissolving in $CHCl_3$ and precipitating the concentrated solution by adding absolute alcohol.

118 2 Early Scientific Investigations

Repeated $CHCl_3$ extractions of powdered gallstones, as above, pretty much lost effectiveness in terms of bile pigment extraction. Yet the remaining powder was treated with alcoholic KOH to dissolve pigment and produce a dark red-brown color, and the solution could be filtered away from a voluminous residue of impurities. Acidification of the solution with HCl precipitated voluminous flakes of red pigment, which was filtered as rapidly as possible from considerable green pigment and then taken up into either alcohol or $CHCl_3$ and further processed to purification.

Prior to the use of $CHCl_3$ to extract bilirubin, only a brown modification (*Gallenbraun, Biliphäin, Cholephäin*) was typically obtained from icteric urine, bile, or gallstones and was, by the time of *Städeler* and *Thudichum*, recognized as *impure* bilirubin, typically containing (calcium) salts. The then purest form of the pigment was red, hence bilirubin or *Cholerythrin*, became available by extractions involving $CHCl_3$, and *Thudichum* examined the crystals in considerable detail (*107*):

> Vor der Entdeckung des Gebrauchs des Chloroform als ein Lösungmittel für diesen Farbstoff hatte man nur braune Modificationen desselben erhalten und Biliphäin oder Cholephäin benannt. Nachdem indessen der rothe Farbstoff vermittelst Chloroform erhalten worden war, nahm man allgemein an, dass die braune Farbe früherer Präparate ein Zeichen ihrer Unreinheit gewesen sei. Der rothe Farbstoff, unter dem Namen Bilirubin oder Cholerythrin, wurde für die einzige Form von reinem Gallenfarbstoff gehalten.   101

Thudichum stated that from his many isolations and purifications he always found two modifications of the pigment that were chemically identical. One was red-brown, the other was red like the color of HgO. Under a microscope the former exhibited numerous microcrystals among many completely formed crystals. Yet, in contrast, bilirubin consisted almost entirely of small amorphous granules, and only when it was precipitated by alcohol did it yield small yellow rhombic prisms. A mixture was precipitated from a saturated $CHCl_3$ solution of *Cholephäin* by added alcohol. The first precipitate (bilirubin) was captured by filtration; with gradual addition of more alcohol a second crop was obtained – half red and amorphous, half little brown crystals. The crystals seldom remained gathered together in husks but could be separated quickly from the remaining suspended bilirubin by washing with alcohol. Thus, the isolated crystals had a dark red-brown color and their surfaces reflected light with a purple, steel-blue luster (*152*):

> . . . In zahlreichen Operationen, welche ich behufs der Isolirung des reinen Gallenfarbstoffs unternahm, erhielt ich stets zwei Modificationen, die sich zwar chemisch gleich verhielten, wovon jedoch die eine rothbraun, die andere rein roth wie Quecksilberoxyd war. Die mikroskopische Untersuchung ergab, dass der dunkelbraunrothe Farbstoff aus zahllosen krystallinischen Partikelchen mit vielen vollständigen Krystallen bestand. Das Bilirubin dagegen bestand beinahe ganz aus kleinen amorphen Körnchen; nur wenn es mit Alkohol gefällt worden war, enthielt es kleine gelbe rhombische Prismen. Liess ich eine Mischung aus einer gesättigten Chloroformlösung des Cholephäins und absoluten Alkohol bestehend, aus der der erste Niederschlag vom Bilirubin durch das Filter entfernt

2.9 Bile Pigment Isolation, Purification, and Combustion Analysis...

> worden war, stehen, und setzte ich allmählich mehr Alkohol zu, so erhielt ich allmählich einen zweiten halb rothen und amorphen, halb krystallinischen braunen Niederschlag. Die Krystalle sassen nicht selten in Drusen zusammen. Sie konnten durch Schlämmen mit Alkohol, in dem sie sich schnell absetzten, von dem länger suspendirt bleibenden Bilirubin getrennt werden. 102

By examination under a microscope, the crystals were seen as opaque, thin reddish or red blades that transmitted light, but the most opaque, of which there were few present, sent yellow light to the eye. Their dimensions and shapes were specified.

These *Thudichum* referred to as *Cholephäin*. The smallest bilirubin crystals showed the same shape and yellow color. By careful recrystallization the red modifications could be changed partially into the brown. From which he concluded that crystallized or microcrystalline purple-brown *Cholephäin* or bilirubin is only a different state of aggregation of amorphous red bilirubin or *Cholerythrin*. As a consequence, he regarded *Cholephäin* and bilirubin as chemically identical and cautioned that when one names the pigment, the process for obtaining it should also be specified (*152*):

> Die kleinsten Bilirubin-Krystalle zeigten dieselbe Gestalt und gelbe Farbe. Durch vorsichtiges Umkrystallisiren konnte die rothe Modification stets theilweise in die braune verwandelt werden. Es ist daher klar, dass das krystallisirte oder krystallinsiche purpurbraune Cholephäin oder Biliphäin nur ein anderer Aggregatzustand des amorphen rothen Bilirubins oder Cholerythrins ist. Ich werde daher in der Folge Cholephäin und Bilirubin als chemisch identisch betrachten, füge aber hinzu, dass wenn in der Beschreibung eines Processes der eine oder andere Name gebraucht wird, die dadurch bezeichnete Modification für den Process benutzt worden ist. 103

Thudichum then described a color change (to brown) when the solid orange pigment was exposed to light [apparently another early example of bilirubin photochemistry] in the absence of moisture (but not apparently in the absence of oxygen). The same brown color was obtained by briefly heating the solid pigment in water. The change in color occurred only on the surface of the pigment, but when heated for a longer time, it became thoroughly brown. Thus it would appear that bilirubin had been converted to the *Gallenbraun* from whence it came, though *Thudichum* did not say so.

The solubility of the (orange) pigment was determined, with results more or less coincident with *Städeler*'s: insoluble in H_2O and slightly soluble in boiling absolute alcohol (with yellow coloration). The latter coloration was apparently due to a dispersion of solid because filtration yielded a colorless filtrate and a colored filter paper. It was slightly soluble in ether, somewhat soluble in CS_2 and in benzene, and had its best solubility in $CHCl_3$: 1.7 parts per 1,000 parts $CHCl_3$ to form a beautiful dark red solution, or about the same solubility as seen today with bilirubin. Sunlight (presumably on the $CHCl_3$ solution) produced a brown to black coloration, which *Thudichum* presumed was caused by the [photochemical] formation of HCl gas. Saturation of the solution with HCl gas followed by complete removal of the $CHCl_3$ and acid by distillation left a mixture of two beautiful green compounds that could not be separated by differential solubility in alcohol (in which both were soluble) but by ether (in which only one was soluble). The bilirubin had been converted entirely into the two new compounds, which were apparently not investigated further[46] (*152*):

> In Wasser ist der Stoff ganz unlöslich, wenig löslich in kochendem absoluten Alkohol, mit gelber Farbe; filtrirt man diese Lösung durch Papier, so bleibt der Farbstoff der ersten Portionen der Lösung an den Papierfasern haften und der Alkohol fliesst beinahe farblos ab. In Aether ist er wenig löslich, etwas löslicher in Schwefelkohlenstoff und in Benzol. Das beste Lösungmittel ist Chloroform, wovon 1000 Theile, 1,7 Theil, 586 Theile daher einen Theil Bilirubin lösen. Die Lösung ist prächtig dunkelroth gefärbt. Die Sonnenstrahlen verfärben diese Lösung zu Braun und Schwarz, wahrscheinlich durch Bildung von Salzsäure. Der Zusatz von wässeriger Salzsäure bringt einen Niederschlag in der Lösung hervor. Leitet man indessen trockenes Salzsäuregas in die Lösung bis zur Sättigung, und destillirt alsdann Chloroform und Säure vollständig ab, so bleibt eine Mischung von zwei prächtig grünen Körpern übrig, die sich nicht durch Alkohol, worin beide löslich sind, wohl aber durch Aether, worin nur einer löslich ist, trennen lassen. Das Bilirubin geht ganz in diese neuen Verbindungen über.   104

Like *Städeler*, *Thudichum* conducted elemental combustion analyses on his purified bilirubin, or *Cholephäin*. The pigment was dried under vacuum at 100°C, then between 120°C and 130°C to constant weight, which made it a little darker. Six combustion analyses were performed, three for carbon and hydrogen (I–III); three for nitrogen (IV–VI) below. *Thudichum* included all of the relevant weighings to four significant figures and qualified the results of III as having come from too small a sample. The nitrogen analyses were conducted in different ways; that of VI came from again repurified pigment. From the combustion data, *Thudichum* calculated an empirical formula (C_9H_6NO) for *Cholephäin*, a name he used interchangeably with bilirubin (*152*):

Vergleich der Empirie und Theorie der Elementar-Zusammensetzung des Cholephäins.

	I.	II.	III.	IV.	V.	VI.	Mittel
C	66,02	66,41	65,61	–	–	–	66,01
H	5,97	6,13	5,95	–	–	–	6,01
N	–	–	–	9,05	9,49	8,56	9,03
O	–	–	–	–	–	–	18,95
							100,00

Diese Zahlen führen zur Formel $C_9H_6NO_2$,[47] deren Theorie mit obigen Thatsachen folgendermassen sich vergleicht:

---

[46] One might guess that at least one was biliverdin-IXα, possibly contaminated with XIIIα.

[47] N.B. *Thudichum* probably meant "Formel $C_9H_9NO_2$".

2.9 Bile Pigment Isolation, Purification, and Combustion Analysis...

Atom	At.-Gew.	Theorie in 100	Mittel der Analysen
$\mathrm{\overline{C}}_9$	108	66,26	66,01
H_9	9	5,52	6,01
N	14	8,59	9,03
$\mathrm{\overline{O}}_2$	32	19,63	18,95
	163	100,00	100,00

105

Writing an element's symbol with a bar through it, such as $\mathrm{\overline{C}}$ and $\mathrm{\overline{O}}$, was a convention introduced by *Alexander Williamson* and *August Kekulé* well before the Karlsruhe Congress (*147–149*). It signified that the correct atomic mass of the element (12 and 16, respectively) was to be used in the formula and not the equivalent mass (6 and 8) introduced by *Berzelius* and *Gmelin* and widely used in the 1800s to give what today are odd-looking formulas, such as HO for water and $HOSO_3$ for sulfuric acid. As the correct atomic masses gained acceptance, "barred" elements disappeared. Here, *Thudichum* was expressing adherence to the decision at Karlsruhe. It is noteworthy that the (correct) atomic weights agreed upon in the famous 1860 chemical congress in Karlsruhe were used, and the formula weight (= 163) corresponding to the empirical formula $C_9H_9NO_2$ was then called an "atomic weight" (*das Atomgewicht*) rather than molecular weight (*Moleculargewicht*) – in recognition that the formula was not necessarily that of the molecule. *Thudichum* believed his analyses gave the correct formula and actual *Atomgewicht* of *Cholephäin* or bilirubin from a long series of noteworthy compounds as well as from several interesting transformations by acids and bases (*152*):

> Dass obige Formel die richtige, und dass 163 das wirkliche Atomgewicht des Cholephäins oder Bilirubins ist, werde ich in dem Folgenden durch eine lange Reihe merkwürdiger Verbindungen, sowie durch mehrer interessante Umwandelungen dieses Stoffes unter dem Einfluss verschiedener Säuren und Alkalien näher beweisen. 106

For the last, *Thudichum* converted bilirubin into its ammonium, sodium, and potassium salts. To obtain the first he treated the pigment with saturated aq. ammonia to form a dark red voluminous mass. A stream of air was passed through, first cold, then heated to 100°C, to drive off the NH_3 and leave behind a greenish-brown lustrous, brittle mass. In order to learn how much NH_3 was combined with the bilirubin, a carefully weighed and dried sample of the pigment (1.8483 g) was saturated in liq. NH_3 to yield a brown-red solid (1.8589 g) after blowing off the NH_3 and drying in a stream of air at 100°C. The difference in weights of the initial and final pigments indicated how much (0.0106 g) NH_3 or oxygen had been absorbed – to yield the hypothetical formula $[C_9H_8(NH_4)NO_2 + H_2O]$ which predicted an increase in weight of 0.41 g. The much smaller experimental difference in weight undoubtedly confirmed the formulas as only empirical but it was too small or too suspect from which to predict a molecular formula based on a different stoichiometry between NH_3 and the pigment. The ammonia adduct of bilirubin was readily soluble in strong alcohol (95%) and insoluble in ether. As was observed previously by others, bilirubin dissolved in aq. or ethanolic KOH or NaOH and could be precipitated by acid. Or (in alkaline solution) converted to a green pigment, biliverdin, by warming.

More important possibly were *Thudichum*'s preparations of the silver, barium, calcium, zinc, and lead salts of bilirubin – and their combustion analyses. The silver salt was prepared from a neutral ammoniacal solution of *Cholephäin* (prepared by digestion of an excess of *Cholephäin* in aqueous ammonia and precipitated by the addition of $AgNO_3$). The reddish-brown precipitate thus obtained was dried under vacuum over H_2SO_4 in the dark. The solid was analyzed for silver content/residue after combustion, and the data from three analyses, guided by *Thudichum*'s empirical formula for bilirubin ($C_9H_9NO_2$), were found to be consistent with the neutral hydrated formula, $C_9H_{10}AgNO_3$ (*152*):

Dass Mittel dieser Bestimmungen ist 37,39 p.C. Ag.

Wenn man nun in Betracht nimmt, dass die Analysen des Cholephäins oder Bilirubins zur empirischen Formel $C_9H_9NO_2$ führren, so kann es keinem Zweifel unterliegen, dass die in dem oben beschriebenen Silbersalze enthaltene Menge Silber genau derjenigen entspricht, welche eine neutrale, einfach gewässerte Verbindung von der Formel $C_9H_{10}AgNO_3$ erfordert. Wie anomal ... auch immer ein Silbersalz mit einem Atom Wasser sein möge, es ist jetzt gewiss, dass die Elementarzusammensetzung und das Molekül des Cholephäins durch die Formel $C_9H_9NO_2$ ausgedrückt wird.

Vergleich der Theorie und Empirie des im Vacuo getrockneten Silber-Cholephäinats.

Symbole	At.-Gew.	In 100 Th.	Gef.			Mittel
			a.	b.	c.	
C_9	108	37,30	–	–	–	–
H_{10}	10	3,47	–	–	–	–
~~Ag~~	108	37,50	37,63	37,52	37,03	37,39
N	14	–	–	–	–	–
O_3	48	–	–	–	–	–
	288					

107

What *Thudichum* called the basic silver salt was prepared from *Cholephäin* dissolved in aq. NH_3 and precipitated by addition of $AgNO_3$ and HNO_3 (*152*):

Eine kleine Menge Cholephäin, welche wiederholt durch Lösen in Chloroform und in alkoholischer Kalilösung gereinigt worden, war, wurde in Ammoniak gelöst und mit Silbersalpeter gemischt. Da kein Niederschlag erschien, so wurde mehr Silberlösung zugesetzt und das ganze dann mit Salpetersäure bis beinahe zur Neutralität abgestumpft. Der jetzt erscheinende Niederschlag liess die Flüssigkeit farblos; er wurde mit Wasser gewaschen und in der Leere getrocknet.

108

Combustion analysis for silver predicted the formula $C_9H_7Ag_2NO_2$, with two silver atoms replacing two hydrogen atoms in *Thudichum*'s empirical formula $C_9H_9NO_2$ (*152*):

Es ist auf diese Weise ermittelt, dass das Cholephäinsilber in freiem Ammoniak löslich ist, und dass wenn diese Lösung bei Gegenwart von überschüssigem Silbersalpeter auf einen gewissen an Neutralität gränzenden Alkalinitätsgrad herabgestimmt wird, das basische Salz niederfällt. Seine Theorie leitet sich aus den über das freie und mit einfach Silber verbundene Cholephäin bekannten Thatsachen her und wird durch die Analysen

2.9 Bile Pigment Isolation, Purification, and Combustion Analysis...

bestätigt; seine Formel ist $C_9H_7Ag_2NO_2$. In dieser Verbindung sind daher zwei Wasserstoffatome durch zwei Atome Silber ersetzt. Ich werde später eine analoge Bleiverbindung beschreiben, in welcher zwei Atome Wasserstoff durch ein didynamisches Atom Blei ersetzt sind. Ihre Formel ist $C_9H_7PbNO_2$, und sie ist eine wesentliche theoretische Stütze für die Annahme, dass das oben beschriebene basische Silbersalz eine wirkliche feste Verbindung und nicht nur eine zufällige Mischung sei.

Vergleich der Theorie und Empirie des basischen Cholephäinsilbers.

Symbole	At.-Gew.	In 100 Th.	a.	b.	c.	Mittel
C_9	108	28,69	–	–	–	–
H_7	7	1,85	–	–	–	–
Ag_2	216	57,29	56,81	56,41	55,86	56,27
N	14	–	–	–	–	–
O_2	32	–	–	–	–	–
	377					

The barium salt was also prepared from an aqueous solution of *Cholephäin* in excess NH_3 by precipitating with added $BaCl_2$. The resultant green precipitate, which *Thudichum* designated a neutral barium *Cholephäinate*, was, after drying, combusted. The analysis data, for C, H, and Ba, were determined to be consistent with the formula $C_{18}H_{20}BaN_2O_6$, or $(C_9H_{10}NO_3)_2Ba$ (*152*):

Diese Thatsachen entsprechen den Anforderungen der Theorie einer dem neutralen Silbersalz genau analogen Baryumverbindung, in welcher ein zweidynamisches Atom Baryum, zwei Moleküle Cholephäin durch Ersatz eines Atoms Wasserstoff in jedem derselben zusammenschweisst; ausserdem treten zwei Moleküle Wasser in die Verbindung ein.

Vergleich der Theorie und Empirie des Baryumcholephäinats,
$$C_{18}H_{20}BaN_2O_6.$$

Symbole	At.-Gew.	In 100 Th.	a.	b.	c.
C_{18}	216	43,46	–	–	44,58
H_{20}	20	4,02	–	–	3,98
Ba	137	27,56	27,56	27,55	–
N_2	28	–	–	–	–
O_6	96	–	–	–	–
	497				

A somewhat more complicated formula was derived for the half-acid barium *Cholephäinat* (or *Sesquicholiphäinat*) that arose by precipitation from digesting a completely neutralized aqueous solution of $BaCl_2$ with excess *Cholephäin*. The precipitate was washed with H_2O, then digested in alcohol, heated, and washed until the alcohol was colorless. The brown-red product, after powdering and drying, had a dark brown surface. Combustion analysis indicated three molecules of *Cholephäin* to one of barium, corresponding to the formula $C_{27}H_{29}BaN_3O_8$ (*152*):

In dieser Analyse zersprang die Röhre am Ende der Operation, als das Kali in die Sicherheitsblase des Apparats zurückstieg, so dass das Residuum an Kohlensäure und

124　　2　Early Scientific Investigations

Wasser nicht ausgesogen werden konnte. Uebrigens zeigen diese Analysen ganz klar, dass in diesem Cholephäinate ein Atom Baryum mit drei Molekülen Cholephäin verbunden ist.

Wenn wir zu dem oben beschriebenen zwiefach gewässerten neutralen Baryumcholephäinat ein Molekül Cholephäin hinzufügen, wie hier

1	Baryum-Cholephäinat,	$C_{18}H_{29}BaN_2O_6$	= 497	At.	Gew.
1	Cholephäin,	$C_9H_9\quad NO_2$	= 163	At.	Gew.
	so erhalten wir	$C_{27}H_{29}BaN_3O_8$	= 660	At.	Gew.

Die Analysen der oben beschriebenen Verbindung entsprechen nun dieser Theorie ganz vollständig.

Symbole	At.-Gew.	In 100 Th.	a.	b.	c.	d.	Mittel
			\multicolumn{4}{c}{Gef.}	
C_{27}	324	49,09	–	–	51,50	49,76	50,63
H_{29}	29	4,39	–	–	4,65	4,09	4,37
Ba	137	20,75	20,66	20,60	–	–	20,66
N_3	42	–	–	–	–	–	–
O_8	128	–	–	–	–	–	–
	660						

111

In like manner, the neutral and half-acid calcium *Cholephäinats* were prepared and analyzed to indicate probable formulas $C_{18}H_{20}CaN_2O_6$ for the former and $C_{27}H_{29}CaN_3O_8$ for the latter (*152*):

Die von diesen Daten abgeleitete Formel führt zu einem neutralen Calciumcholephäinat $C_{18}H_{20}CaN_2O_6$, welches in jeder Beziehung der oben beschriebenen Baryumverbindung analog ist. Auch in ihm müssen wir die Existenz von 2 Mol. Wasser annehmen, welche durch eine Temperatur von 100° nicht ausgetrieben werden.

Folgender Vergleich der Theorie dieser Verbindung mit den analytischen Daten wird die Richtigkeit dieses Schlusses leicht anschaulich machen.

Symbole	At.-Gew.	In 100 Th.	a.	b.	c.	d.
	\multicolumn{2}{c}{Theorie}		\multicolumn{4}{c}{Gef.}
C_{18}	216	54,00	–	–	–	52,35
H_{20}	20	5	–	–	–	5,04
Ca	40	10	9,63	9,88	9,92	–
N_2	28	–	–	–	–	–
O_6	96	–	–	–	–	–
	400					

	e.	f.	g.	h.	Mittel
	\multicolumn{4}{c}{Gef.}	
C	–	–	54,96	54,26	53,86
H	–	–	5,03	4,65	4,9
Ca	11,02	10,44	–	–	10,17

... Die Verbindung ist demnach dem bereits beschriebenen halbsauren Baryumcholephäinat analog und hat die Formel $Ꞓ_{27}H_{29}ꞒaN_3Θ_8$. Mit dieser Ansicht stimmen die Resultate der Analysen wie folgt.

		Theorie		Experimente.		
		der Atome	p.C.	a.	b.	c.
$Ꞓ_{27}$		324	57,54	–	–	60,37
H_{29}		29	5,15	–	–	5,74
$Ꞓa$		40	7,1	7,03	6,79	–
N_3		42	–	–	–	–
$Θ_8$		128	–	–	–	–
		563				

Durch die nachfolgende Zusammenstellung werden die Unterschiede in der Zusammensetzung des neutralen Calciumcholephäinats auf der einen und des halbsauren auf der anderen Seite sehr deutlich.

Neutrales Calcium-Cholephäinat, $Ꞓ_{18}H_{20}ꞒaN_2Θ_6$. At.-Gew. = 400.				Halbsaures Calcium-Cholephäinat, $Ꞓ_{27}H_{29}ꞒaN_3Θ_8$. At.-Gew. = 563			
	Theorie	Gef.			Theorie	Gef.	
$Ꞓ$	54	53,86		$Ꞓ$	57,54	60,37	
H	5	4,9		H	5,15	5,74	
$Ꞓa$	10	10,17		$Ꞓa$	7,1	6,91	112

The analytical data for these calcium salts were compared with *Städeler*'s data for his calcium salt of bilirubin from human gallstones. At issue for *Thudichum* was the interpretation of *Städeler*'s 9.1% CaO datum and his assumption that he was analyzing the neutral calcium salt. *Thudichum* reminds us of *Städeler*'s empirical formula ($C_{18}H_9NO_4$) published in 1861 in *Frerichs'* 2nd edition (*120*) of his famous *Klinik der Leberkrankheiten* (which did not involve analysis data for a calcium salt) and his subsequent formulas, $C_{32}H_{18}N_2O_6$ for free bilirubin and $C_{32}H_{17}CaN_2O_6$ for its calcium salt, published in 1864 (*135*). The formulas, irrespective of their being derived using the old notation [of atomic weights for O (=8) and C (=6)], were said to be incorrect by *Thudichum* because they had been derived for the neutral calcium salt of bilirubin rather than the half acid salt, which *Thudichum* showed fit the %Ca datum better. Which thus meant that *Städeler*'s formulas for biliverdin, *Biliprasin, Bilifuscin* (bilifuscin), and *Bilihumin* (bilihumin) would by necessity be incorrect (*152*):

In seinen Untersuchungen über den Farbstoff menschlicher Gallensteine stellte Städeler eine Calciumverbindung des Bilirubins dar, welche ihm bei der Analyse 9,1 p.C. Calciumoxyd ergab. Von der Annahme ausgehend, dass diese Verbindung ein normales Neutralsalz sei, bestimmte er nach ihr das Atomgewicht des Bilirubins. Er verwarf demnach seine früheren Analysen des krystallisirten Cholephäins, wie sie in Frerich's Klinik der Leberkrankheiten mitgetheilt waren, sowie auch die empirische Formel $C_{18}H_9NO_4$ und substituirte $C_{32}H_{18}N_2O_6$ als die Formel des freien Bilirubins, und $C_{32}H_{17}CaN_2O_6$ als die des Calciumbilirubats (die vorstehenden drei Formeln sind in der alten Notationsweise gegeben). Da nun diese Formeln nur durch eine, hierfür ungenügende, Atomgewichtsbestimmung

durch Kalkgewicht unterstützt sind, auf der anderen Seite aber alle Analysen Städeler's über das Bilirubin and Cholephäin mit meinen Resultaten in vollständigem Einklang gebracht werden können, so kann ich nicht zögern, die Formeln, welche dieser Forscher für Bilirubin und Calciumbilirubat gegeben hat, für irrthümlich zu erklären.
Das von Städeler analysirte Bilirubat war offenbar das halbsaure Salz.

Theorie von $C_{27}H_{29}CaN_2O_6$ Städeler

	erfordert	fand
CaO	9,94 p.C.	9,1 p.C.
Ca	7,1 „	6,5 „

Mit der Formel Städeler's für das Bilirubin fallen die Formeln aller anderen von ihm beschriebenen Derivate des Gallenfarbstoffs, namentlich des Biliverdins, Biliprasins, Bilifusins und Bilihumins.   113

Thudichum also prepared zinc- and lead-*Cholephäinat* and carried out combustion analyses of each. The red-brown zinc salt, from $ZnSO_4$, was reddish-brown; the lead salt was from $Pb(OAc)_2$. Analyses for zinc, determined as ZnO, predicted a half-acid salt, $C_{27}H_{29}ZnN_3O_8$, composed of one molecule of neutral zinc *Cholephäinat* ($C_{18}H_{18}ZnN_2O_4$), one molecule of *Cholephäin* ($C_9H_9NO_2$), and two of H_2O ($2 \times H_2O$). In comparison a neutral salt of formula $C_{18}H_{10}ZnN_2O_6$ (M.W. 389) would give 16.70% Zn (*152*):

Nach diesen Daten ist es klar, dass das Zinksalz den halbsauren Salzen des Baryums und Calciums analog zusammengesetzt ist, wie folgt.

1	neutrales Zinkcholephäinat...........	$C_{18}H_{16}ZnN_2$	O_4	
2	Wasser...	H_4	O_2	
1	Cholephäin...................................	C_9 H_9	N	O_2
1	Mol. halbsaures Zn Choleph.........	$C_{27}H_{29}ZnN_3$	O_8	

Mit dieser Auffassung harmoniren die Resultate der Analysen wie folgt.

		Theorie		Gef.	
	der Atome	p.C.		a.	b.
C_{27}	324	–		–	–
H_{29}	29	–		–	–
Zn	65	11,05		12,03	11,30
N_3	42	–		–	–
O_8	128	–		–	–
	588				

Ein neutrales Cholephäinat von der Formel $C_{18}H_{16}ZnN_2O_4$ Mol.-Gew. = 389 hätte 16,70 p.C. Zn erfordert.   114

And from the lead salt analyses, with the lead being analyzed as lead oxide, as a basic *Cholephäinat* or as *Cholephäin*, the formula $C_9H_7PbNO_2$ was correlated, in which two hydrogens are replaced by one divalent lead. A formula corresponding to the di-silver salt of *Cholephäin* ($C_9H_7Ag_2NO_2$) seen above (*152*):

Diese Verbindung kann als ein basisches Cholephäinat oder als Cholephäin aufgefasst werden, in welchem zwei Atome Wasserstoff durch ein zweidynamisches Atom Blei ersetzt sind.

	Theorie		Gef.	
	der Atome	p.C.	a.	b.
C_9	108	29,34	–	–
H_7	7	1,9	–	–
Pb	207	56,25	58,38	57,91
N	14	–	–	–
O_2	32	–	–	–
	368			

Diese Verbindung entspricht dem basischen Silbercholephäinat oder zweifach Silbercholephäin $C_9H_7Ag_2NO_2$, welches oben näher beschrieben worden ist. 115

Thudichum also prepared biliverdin, which he had previously named *cholechloine*, studied its chemical and physical properties, conducted combustion analysis (from which he derived an empirical formula), and prepared calcium and barium salts for combustion analyses. Biliverdin was thus dissolved in aq. KOH and exposed to air until it was completely green in thin films. Since the reaction could take two or three weeks, in order to accelerate it the solution was heated while air was introduced. Addition of HCl precipitated large green flakes of biliverdin. The precipitate was purified by washing through and through with water on a filter. In yet another synthesis to convert bilirubin to biliverdin, the first was heated with an alkaline solution of copper and potassium acetate. Cuprous oxide precipitated and after removal by filtration, the filtrate was acidified with HCl to precipitate biliverdin. A part of the copper was bound in a characteristic/peculiar way, however, to part of the biliverdin and the free pigment could not be easily released from it.

Under moist conditions, biliverdin was a voluminous mass of a magnificent dark green color. After drying it shrank to a lustrous brittle mass with a completely black color. Its powder was very dark green, and it had not yet been possible at the time to obtain it in a crystalline state. It was completely insoluble in H_2O, ether, and $CHCl_3$. In a moistened condition it was easily soluble in alcohol, but when dry it was much less soluble. It was more soluble in hot alcohol than in cold. By heating its conc. alcohol solution for a long while, it appeared to be transformed and became much less soluble. It dissolved in HCl with a green color, and the solution gave an amorphous green precipitate with $PtCl_4$ and with corrosive sublimate of mercury ($HgCl_2$). After dissolving biliverdin in aq. KOH, when H_2SO_4 was added the color gradually changed to brown-green from the original green.

Thudichum described (*152*) a wide array of reactions but, absent a knowledge of the pigments' structures, they offered no information other than possibly being characteristic of that given pigment. Nor did the reactions reveal much in the way of structural information, for in 1868 the concept of organic structure was in its infancy. *Thudichum*'s reactions consisted of the following: (i) Zn metal added to a solution of biliverdin in hydrochloric acid changed the color from green to brown-red; (ii) Na(Hg) added to an alkaline solution changed the color from green to reddish-brown, which then changed to green upon introduction of air and also precipitated a brown flocculent mass upon addition of HCl – experiments that showed that reduction of biliverdin by the usual methods did not convert it back to bilirubin. (iii) Reaction with I_2 led to a greenish-black resin, while reaction with Cl_2 in H_2O converted biliverdin to dirty yellow-colored flakes that were insoluble in H_2O and ether but readily soluble in alcohol. But when a saturated alcohol solution of biliverdin was subjected to a small blast of Cl_2 gas, the solution went colorless immediately and yielded chlorine-containing whitish-yellow flakes that were insoluble in H_2O and melted to a reddish-yellow mass upon gentle heating. A yellowish white resin containing several chlorine-containing compounds was obtained by treating biliverdin with hydrochloric acid, followed by gradual addition of $KClO_3$ during warming. Included were one soluble in $CHCl_3$, two in alcohol, but none in ether. (iv) When an alcoholic solution of biliverdin was heated with pure, moist Ag_2O, the pigment was converted to a purple-colored compound, *Bilipurpurin* (bilipurpurin). For the most part bilipurpurin remained insoluble and bound to Ag_2O, but it dissolved in ammonia to yield a green color and an excess of Ag_2O. Hydrochloric acid or any one of several other acids freed up the bilipurpurin, which remained as water-insoluble but easily alcohol-soluble, brownish-red flakes and masses following evaporation of the alcohol and HCl and extraction of NH_4Cl with H_2O.

The properties of the pigment facilitated detection of even small amounts. Thus, to bilirubin or biliverdin dissolved in aq. ammonia was carefully added a little $AgNO_3$ such that all of the excess silver was dissolved. The solution became or remained green after heating and was filtered to remove reduced silver; then alcohol was added followed by an acid, such as HCl, whence the green solution assumed a purple color.

If Ag_2O were left for a longer time in contact with an alcoholic solution of biliverdin, the reaction went over to the formation of bilipurpurin, and the greenish-black solution after treating with ammonia and precipitating the silver by H_2S, gave a clear yellow filtrate. After the alcohol had been removed, a yellow precipitate remained from which, after washing with H_2O and recrystallization from alcohol, left crystals of sulfur to separate upon longer standing. The mother liquor, freed from sulfur, remained yellow and was somewhat soluble in H_2O. Though the entire operation resulted in a significant loss of the original material, the ultimate result was always a yellow-brown compound that appeared as spherical crystalline granules that were easily soluble in alcohol, poorly soluble in H_2O, insoluble in ether, but dissolved in aq. ammonia or KOH, and were precipitated by HNO_3 or HCl. *Thudichum* designated the product *biliflavin*. When an alcoholic solution of biliverdin was heated with HgO alone, or after addition of ammonia, no transformation was noticed. If PbO_2 were used instead of HgO, the biliverdin was transformed into a brown material, or was insoluble, partly perhaps as a lead salt. When heated with aqueous peroxide and ammonia, the biliverdin solution assumed a brownish-red color. When an alcoholic solution was used, it became light yellow and developed the odor of aldehyde or ethyl acetate. Thus, biliverdin was apparently transformed into biliflavin by Ag_2O.

2.9 Bile Pigment Isolation, Purification, and Combustion Analysis…

Thudichum also noted (*152*) that biliverdin underwent the *Gmelin* color change reaction following addition of conc. HNO_3 to an alcohol solution of the pigment: first blue, then violet, next red, finally yellow after standing for a long time or heating. When no alcohol was present, the pigment precipitated, and its blue and red colors appeared much less intense to the eye. Before the pigment underwent conversion to the yellow substance, it dissolved in HNO_3 and the transformation proceeded to a maximum. If one accelerated the reaction by heating, then considerable HNO_2 was formed, and yellow flakes of a nitro compound separated from the solution upon addition of H_2O. The aqueous acidic liquor contained a fixed acid that formed a crystalline salt with Ag_2O.

He discussed the solubility of biliverdin in alkali and various salts that he prepared, including those of Ca^{+2}, Ba^{+2}, Pb^{+2}, Cu^{+2}, and Hg^{+2}, finding the pigment to be soluble in aq. potash, NaOH, and NH_4OH. On standing or heating, it became brownish and the precipitate that formed had lost much of its solubility in alcohol. Calcium and barium salts caused no precipitation from aqueous ammonia solutions of biliverdin, but addition of $Ca(OH)_2$ or $Ba(OH)_2$ to an alcohol solution of biliverdin produced green, water-soluble precipitates that were subjected to elemental combustion analysis, as was purified biliverdin itself.

Thus an alkaline solution of the bilirubin analyzed previously as numbers III and IV (*152*) was oxidized by air, over time and without heating to produce the biliverdin used for analyses (a) and (b), below. For analysis (c), the biliverdin was purified by dissolving in hot alcohol then cooling to precipitate, followed by drying the precipitate under vacuum. For analyses (d) and (e), the pigment remaining from analysis (c) was dissolved in hot alcohol and filtered before cooling. The results of *Thudichum*'s combustion analyses of biliverdin and a comparison with those from his bilirubin analyses are shown in the following (*152*):

Zusammenstellung und Mittel der Analysen.

	a.	b.	c.	d.	e.	Mittel
C	–	63,08	62,09	–	62,14	62,43
H	–	6,25	6,12	–	6,00	6,13
N	9,32	–	–	9,36	–	9,34
O	–	–	–	–	–	22,10
						100,00

Vergleicht man diese Befunde mit den das Bilirubin betreffenden Thatsachen,

	Bilirubin		Biliverdin	
	Theorie in 100	Mittel der Analyse		Theorie
C	66,26	66,01	62,43	63,57
H	5,52	6,01	6,13	5,96
N	8,59	9,03	9,34	9,27
O	19,63	18,95	22,10	21,20

In comparing the bilirubin and biliverdin data it may be noted that the %C has dropped from the former to the latter, with a small increase in %H and larger increases in the %N and %O. How was this explained? *Thudichum* cited *Heintz*'s theory (*97*) that biliverdin is an oxide of *Cholephäin* (though he did not use the latter word), and that *Städeler* proposed (*135*) that biliverdin was a hydrated oxide. He calculated, on the basis of his formulas, that if biliverdin were a simple oxide, then one would have $C_9H_9NO_2 + O \rightarrow C_9H_9NO_3$, which could compute as 60.33% C, 5.02% H, and 7.82% N – or rather far from the percentages shown in the small tables above and thus completely contradicting the "oxide" theory. *Städeler's* hypothesis, too, which might be formulated $C_9H_9NO_2 + O + H_2O \rightarrow C_9H_{11}NO_3$, falls even shorter with its computed 54% C and a corresponding diminution in %H and %N. *Thudichum* calculated 62.81% H, and 8.13% N for a biliverdin composed of two molecules of *Cholephäin* and one of H_2O $(2 \times C_9H_9NO_2 \rightarrow C_{18}H_{20}N_2O_5)$. While the %C is an acceptable match, the computed percentage for N falls rather short, and the %H somewhat less short. Thus biliverdin could not be an oxide or hydrate or both (*152*):

> ... so findet man, dass das Bilirubin, um in Biliverdin überzugehen, viel Kohle verloren, ein wenig Wasserstoff gewonnen, seinen Gehalt an Stickstoff etwas vergrössert und den Sauerstoffgehalt beinahe so viel vermehrt hat, als den Kohlenstoff vermindert. Nach der Theorie von Heintz war das Biliverdin ein Oxyd des Cholephäins gewesen, nach der Hypothese von Städeler ein gewässertes Oxyd. Wäre des Biliverdin ein einfaches Oxyd des Bilirubins ($C_9H_9NO_2 + O =$)$C_9H_9NO_3$, so würde es 60,33 p.C. C, 5,02 p.C. H und 7,82 p.C. N erfordern. Ein Vergleich dieser Zahlen mit denen, welche die Analysen des Biliverdins ergaben, verneinen indessen die Ansicht, dass das Biliverdin ein Oxyd des Bilirubins sei, vollständig. Die Hypothese von Städeler ist noch viel weniger anwendbar, da die Formel $C_9H_9NO_2 + O + H_2O$, 54 p.C. C und eine entsprechend verminderte Menge von H und N verlangt. Bestände das Biliverdin aus zwei Molekülen Cholephäin, verbunden mit einem Atom Wasser oder $2(C_9H_9NO_2) + H_2O = C_{18}H_{20}N_2O_5$, so wären 62,81 p.C., 5,81 p.C. H und 8,13 p.C. N erforderlich. Selbst wenn die gefundene Kohlenstoffmenge diese Annahme erlaubte, so würde doch der H- und N-Beträgt dieselbe vollständig verneinen. Das Biliverdin ist weder eine Oxyd, noch ein Oxydhydrat, noch ein Hydrat des Bilirubins. 117

Reassessing his experimental combustion analysis data for biliverdin, *Thudichum* computed a C:H:N ratio equal to 7.8:9.2:1, to give the formula $C_8H_9NO_2$, or one carbon atom fewer than in his bilirubin formula (*152*):

> Bei der Berechnung der Formel führen die Durchschnitt der Elementaranalysen zu den Verhältnissen
>
> $$N_1 : H_{9,2} : C_{7,8}.$$
>
> Diess giebt die Formel $C_8H_9NO_2$ als die des Biliverdins. Die Analysen stimmen mit dieser Theorie wie folgt.

Atome	At.-Gew.	In 100 Th	Mittel der Empirie
C_8	96	63,57	62,43
H_9	9	5,96	6,13
N	14	9,27	9,34
O_2	32	21,20	22,10
	151	100,00	100,00

118

How can bilirubin be converted to biliverdin? This is much easier using oxygen from the air than it is to understand based on *Thudichum's* formulas. Yet *Thudichum* proposed that a molecule of bilirubin combined with a molecule of oxygen to form a

2.9 Bile Pigment Isolation, Purification, and Combustion Analysis...

molecule of biliverdin and expel a molecule of CO_2: $C_9H_9NO_2 + O_2 \rightarrow C_8H_9NO_2 + CO_2$. Then he concluded his long and important 1868 publication with combustion analyses of the calcium and barium salts of biliverdin. The analysis data for the calcium salt are shown below (*152*):

Zusammenstellung der Analysen und Vergleich mit der Theorie des Zweifach-Calcium-Neunfach-Biliverdin.

	Theorie		Analysen				
	der Atome	p.C.	a.	b.	c.	d.	Mittel
C_{72}	864	60,20	–	–	61,33	63,06	62,19
H_{77}	77	5,36	–	–	5,68	5,8	5,74
Ca_2	80	5,56	5,52	5,77	–	–	5,64
N_9	126	–	–	–	–	–	–
O_{18}	288	–	–	–	–	–	–
	1435						

119

Thudichum found that the %Ca fit best for a formula with a ratio nine biliverdins to two of Ca: $C_{72}H_{77}Ca_2N_9O_{18}$, which was clearly an attempt to fit the combustion data to some sort of formula. But it was apparently one compound and not a mixture with free biliverdin because no biliverdin could be washed out into alcohol (*152*):

Aus der Menge des gefundenen Calciums berechnet sich das Atomgewicht 709, welches aber offenbar verdoppelt werden muss, damit das Atomgewicht des Biliverdins im Residuum mit einfachen Quotienten aufgehe. $\frac{1418 - 80 + 4}{9} = 149$, welches von dem direct gefundenen Atomgewicht des Biliverdins 151, so gut wie nicht verschieden ist. Die Verbindung besteht daher aus 9 At. Biliverdin und 2 At. Calcium. Wären Gründe vorhanden, den Austritt von 1 At. Wasser aus der Verbindung anzunehmen, so erhielte man eine absolute Uebereinstimmung der Theorie mit den Analysen. . . .

Dass dieser Körper eine Verbindung und nicht etwa eine Mischung von einer Kalkverbindung mit freiem Biliverdin ist, geht unter anderen aus dem Umstande hervor, dass er in Alkohol unlöslich ist. Enthielte er freies Biliverdin, so müsste Alkohol dasselbe leicht auszuziehen.

120

Analysis of the barium salt and correlation with a formula proved to be somewhat less complicated. The data fit the formula of a half-acid barium *Biliverdat* salt, $C_{24}H_{27}BaN_3O_7$, according to the following reckoning (*152*):

Das Mittel des gefundenen Baryums, 22,41 p.C., führt zum Atomgewicht 611, welches durch die Operation $\frac{611 - 137 + 2}{151} = 3$ und ein Residuum von 23 führt, das man vielleicht als ein Atom Wasser unterbringen darf. Eine kleine Stütze für diese Annahme erhält man aus der Zusammensetzung der Barytsalze des Cholephäins, die alle Wasser, aber auf ein Atom Baryum zwei Moleküle desselben enthalten. Nach dieser Annahme ist das Biliverdat des Baryums ein halbsaures, einfach gewässertes, bestehend aus

1	Mol.	neutrales Biliverdat	$C_{16}H_{16}BaN_2O_4$	
1	Mol.	Biliverdin	C_8H_9	$N\,O_2$
1	Mol.	Wasser	H_2	O
1	Mol.	halbsaures Ba-Biliverdat	$C_{24}H_{27}BaN_3O_7$	

121

	Theorie		Experimente			
	der Atome	p.C.	a.	b.	c.	d.
C_{24}	288	47,52	–	–	49,39	48,20
H_{27}	27	4,45	–	–	4,43	4,34
Ba	137	22,60	22,15	22,67	–	–
N_3	42	6,93	–	–	–	–
O_7	112	18,50	–	–	–	–
	606	100,00				

However, a formula $C_{24}H_{25}BaN_3O_6$ would also fit the combustion data, according to *Thudichum*, if a molecule of H_2O were left out of the calculation. This gives a better fit to the %C but a slightly less good fit for the %Ba (*152*):

> Lässt man das eine Atom Wasser aus der Berechnung weg, so stimmt die Theorie des Kohlenstoffs besser mit der Erfahrung, aber die des Baryums weniger gut.

			Mittel
C_{24}	288	48,97	48,79
H_{25}	25	4,25	4,38
Ba	137	23,29	22,41
N_3	42	7,14	–
O_6	96	–	–
	588		

Although *Thudichum* concluded his major work on bile pigments in 1868, he continued investigating reactions of bilirubin and biliverdin into the mid-1870s, especially reactions involving halogens, reduction to hydrobilirubin and the coloring matter of urine (*158–160*). In his erstwhile last work on bile pigments, in 1876, he published on some reactions of biliverdin (*159, 160*), and in the same year he published (*161, 162*) a critique and rebuke of *Maly*'s published work of 1868–1876. As we shall see in the latter, he took issue with *Maly*'s combustion analysis of the product of bilirubin bromination, other reactions of bilirubin (including reductions by Na(Hg) and Zn and the relationship of the product(s) to urobilin from urine), the formulas for bilirubin and biliverdin, *etc.* in 41 itemized points. Though he published on bile in 1881 (*101*), in a chapter in his edited book addressing mainly bile acids a cursory examination of the pigments from pig gallstones was provided (*101*). Seemingly from the late 1870s and forward *Thudichum* had reoriented and refocused his typical scientific rigor and his efforts to the chemical constituents of the brain (*163*), seminal work of considerable importance and for which he holds a well-deserved reputation. Never one to pass up an opportunity to bring correction to perceived errant science in the bile pigment field, near the end of his life his zest for polemics had not waned, as we shall see at the end of Section 2.10.

2.9.3 Richard L. Maly and Bilirubin

Maly[48] accomplished his major work on bile pigments in Austria between the early 1860s and mid-1870s. His first reading on the subject of the chemical nature of bile pigments was very brief and appeared in part in early 1864 (*150*) as a preliminary communication (*Vorläufige Mittheilungen über die chemische Natur der Gallenfarbstoffe*). It was submitted in longer form from Graz in April 1864 for publication (*164*) in the 1864 *Annalen der Chemie und Physiologie* (now *Liebig's Annalen der Chemie*) and read before the Academy (*165*) at its May 12, 1864 meeting (*vorgelegt in derselben Sitzung, nr. XIII*) but not published in those proceedings. At the time, *Maly* held Dr. med. and Dr. phil. degrees and was stationed at the Universität Graz, where he was *Assistent der Physiologie*. The early work of *Maly* is interesting in that he postulated a surprising new relationship between bilirubin and biliverdin, which, *inter alia*, subsequently became a contentious issue between him and *Thudichum*.

Maly wrote that crystallized *Cholepyrrhin* (*Biliphäin*) – in 1864, while in Graz, he either did not know or did not subscribe to *Städeler's* new term "bilirubin" – behaved like an amide toward alkali because it released NH_3 and yielded a yellow or green pigment (*150, 164*):

> . . . Dieses verhält sich zu Alkalien wie ein Amid, d. h. entwickelt damit Ammoniak, während der Rest sich mit den Basen zu gelben oder grünen salzartigen Körpern vereint.
>
> Alles Cholepyrrhin war zu den angestellten Versuchen zweimal umkrystallisirt; von ihnen theile ich vorderhand mit Ausschluss von Analysen Folgendes mit:
>
> Alkoholische oder wässerige Kalilösung entwickelt aus Cholepyrrhin schon bei gewöhnlicher Temperatur Ammoniak; die Flüssigkeit färbe sich für kurze Zeit roth und wird dann grüngelb.
>
> Eben so wirkt Natronlauge. 123

That is, heating *Cholepyrrhin* in alcoholic or aq. KOH or NaOH at the usual temperature released NH_3 and left briefly a red solution that turned green-yellow. Heating with $Ba(OH)_2$ or $Ca(OH)_2$ produced NH_3 and yielded Ba^{+2} or Ca^{+2} salts.

The *Cholepyrrhin* used had been isolated as a red-yellow pigment from human bile using the $CHCl_3$ extraction method of *Valentiner* (*107, 108*) and *Brücke* (*109*) and purified according to the latter by crystallization. *Maly* was apparently unaware that *Städeler* (*135*) and *Thudichum* (*102*) had to use more heroic methods to free

---

[48] *Richard L. Maly* was born on June 28, 1839 in Graz and died on March 23, 1891 in Prague. He studied pharmacy and medicine at the University of Vienna and in 1864 was awarded the Dr. med. degree. In the same year he habilitated at the University of Graz for surgical science preparation. In 1866 he was promoted to Professor of Medicine-Surgery at the Lehranstalt Olmütz, then in 1869 to Professor of Physiological Chemistry at the University of Innsbruck and in 1875 to Professor of General Chemistry at the Technische Hochschule in Graz. Not one to remain long in one location, in 1886. *Maly* accepted his final professorship (in general chemistry) at the Deutsche Universität in Prague (currently Charles University). His interest in natural products appears to have migrated from work on abietic acid in the mid-1860s to bilirubin, and much of the latter work was presented at the *Sitzungberichte der kaiserlichen Akademie der Wissenschaften in Wien*.

134 2 Early Scientific Investigations

the gallstone-derived purer bilirubin from its calcium and other occluded salts. From his experiments, he concluded that biliverdin is an acid and *Cholepyrrhin* is its amide, which he named *Biliverdinamid* (biliverdin amide) (*150*), or an ammonium salt (*164*):

> Das Biliverdin ist eine Säure, das Cholepyrrhin ihr Amid (Biliverdinamid), ersteres gehört dem Wasser – letzteres dem Ammoniaktyp an; oder Biliverdin und Cholepyrrhin verhalten sich wie Kohlensäure und Harnstoff. 124

This contention was reinforced by an experiment in which *Cholepyrrhin* was heated in a mixture of $CHCl_3$-acetic acid and thereby converted to a green color. After washing with H_2O (to extract acetic acid) and evaporating the aqueous layer, a white substance containing what was said by *Maly* to be ammonium acetate was left behind; whereas, evaporation of the $CHCl_3$ layer, washed free of acetic acid, left a black-green residue of what was said by *Maly* to be pure biliverdin (*164*):

> Der Inhalt eines solchen Rohrs wurde in Wasser gegossen; unten sammelte sich die dunkelgrüne Chloroformschichte, während das Wasser den Eisessig aufnahm. Erstere Schichte wurde so lange mit Wasser gewaschen, als dieses sauer abfloss. Dann vereinigte man die wässerigen Flüssigkeiten und brachte sie im Wasserbade zur Trockne. Der Rückstand in concentrischen weissen Ringen enthielt essigsaures Ammonium; es war also ein Theil des Stickstoffs im Cholepyrrhin durch die Einwirkung des Eisessigs in Form von Ammoniak abgespalten. Die mit Wasser gewaschene und von der Essigsäure befreite Chloroformschichte gab, nachdem das Lösungsmittel abgedunstet war, einen dunkelfast schwarzgrünen Rückstand von reinem Biliverdin. 125

Maly found that HCl and tartaric acid gave essentially the same reaction with *Cholepyrrhin*, and from the collective data, he became convinced that *Cholepyrrhin* was an amide (but was not an ammonium salt) (*164*):

> Diese und die vorigen Reactionen lassen unverkennbar das Cholepyrrhin als ein Amid erscheinen (ein Ammoniumsalz hätte zur Spaltung wohl keiner so lange dauernden Einwirkung gebraucht), das sowohl, wie der Character der Amide mit sich bringt, durch Alkalien, als durch Säuren gespalten wird, in die entsprechende Säure – hier Biliverdin – und in den Rest NH_3, der im ersten Falle entweicht, im zweiten als einfaches Ammonium salz sich vorfindet.
>
> Das Biliverdin ist eine Säure, des Cholepyrrhin ihr Amid (Biliverdinamid). Ersteres gehört dem Wasser-, letzteres dem Ammoniaktypus an, oder sie verhalten sich wie Kohlensäure und Harnstoff. 126

The state of knowledge of organic chemistry in 1864 was apparently insufficient to cause one to puzzle that conversion of an amide to its acid might cause a color change from red-yellow to green. Knowledge of organic structure was then only primitive and thus a correlation between a chromophore and its "color" was absent. "Spectroscopy" in the visible region of the spectrum was yet a few years distant in organic or physiological chemistry. And *Maly*'s experimental "conversion" of biliverdin back to *Cholepyrrhin* through the action of NH_3 involved heating what *Maly* called the ammonium salt of the former, which could well have been the ammonium salt. But the *Cholepyrrhin* allegedly formed was at the time identified only by its color and solubility in $CHCl_3$. Again, at the time of this work there was only a sketchy knowledge of organic structure and its relationship to reactivity.

2.9 Bile Pigment Isolation, Purification, and Combustion Analysis...

The work, novel though it was, was retracted by *Maly* some four years later (*166, 167*), with the adventitious NH_3 being attributed to an impure sample of *Cholepyrrhin*; thus, in February 1868, *Maly* would write (*166*):

> Ammoniak und die ätzenden Alkalien lösen das Cholepyrrhin mit braunrother Farbe. Bei Anwendung der letzteren schien mir früher . . . eine Entwicklung von Ammoniak statt zu finden. Dieser Irrthum wurde aber durch eine nicht ganz reine aus Menschengalle erhaltene Substanz hervorgerufen. Den damals daraus gezogenen Schluss nehme ich daher zurück. Gegenwärtig nach viel weitläufigeren Beobachtungen bin ich vielmehr zu der weiter unten durch Belege begründeten Ueberzeugung gelangt, dass bei dem Uebergange von Cholepyrrhin in Biliverdin kein Ammoniak sich abspaltet, und dass in letzterem Körper gleichwie in ersterem noch dieselbe atomistische Menge Stickstoff enthalten ist.   127

Maly's sources of *Cholepyrrhin* were human and ox gallstones, the former of which he recognized were often rich in the calcium salt of the pigment. The latter had little or no cholesterol, and served as a convenient source of purified *Cholepyrrhin*. (In 1868, *Maly* was able to acknowledge that the orange pigment of bile had accumulated three names, from *Berzelius' Cholepyrrhin* to *Simon*'s *Biliphäin* to the most recent name: *Städeler*'s *Bilirubin* (bilirubin), given in 1864.) *Maly* conducted C, H elemental combustion analyses of his *Cholepyrrhin*, isolated from both human and ox gallstones, and obtained data very closely coincident with *Städeler*'s (*135*). *Maly* calculated the formula $C_{16}H_{18}N_2O_3$ for it (*166*):

Analyse.

I. 0,2770 Grm. aus Menschengallensteinen gaben bei der Verbrennung 0,681 Grm. CO_2 und 0,1545 Grm. H_2O.

II. 0,2734 Grm. Cholepyrrhin aus Ochesengallensteinen gaben 0,1532 Grm. Wasser.

Diese giebt in 100 Theilen Substanz:

	I.	II.
Kohlenstoff	67,16	–
Wasserstoff	6,18	6,22

Diese Zahlen zeigen mit der Berechnung für $C_{16}H_{18}N_2O_3$ und mit den analytischen Mittelzahlen von Städeler:

	Bez. für $C_{16}H_{18}N_2O_3$	Mittel von Städeler (i.e.)
Kohlenstoff	67,13	67,13
Wasserstoff	6,29	6,19

eine so grosse Uebereinstimmung, dass ich dadurch über die Zusammensetzung dieses Körpers völlig versichert, nicht weiter Material zu Analysen opfern wollte.   128

Maly reconfirmed some of the then recent earlier observations of the pigment's solubility properties: a little soluble in benzene, insignificantly soluble in petroleum ether, somewhat more soluble in hot amyl alcohol, in fatty oil, and glycerin. He noted that it dissolved in conc. H_2SO_4 with the same red-brown color as in lye but after a short time it became a dirty, dark brown-green. If the original red-brown solution were poured into H_2O, dark brown flakes precipitated, which left behind a colorless solution after removal by filtration. The precipitate no longer behaved like *Cholepyrrhin*: it was easily soluble in alcohol, turning it green-brown, and it transmitted garnet red light. Added NH_3 and potash (K_2CO_3) did not change the color of the solution essentially, and the *Gmelin* color reaction failed. Only a pale red residue was seen at the layer bordering the HNO_3, which was yellow beneath, but yielded no green, blue, or violet coloration. Heated with a little soda lime (which is a mixture of 75% $Ca(OH)_2$, 20% H_2O, 3% NaOH, 1% KOH), *Cholepyrrhin* gave, besides NH_3, a tarry compound with a decidedly aniline-like odor; however, the presence of the latter could not be confirmed.

Maly reinvestigated *Städeler*'s *Biliprasin* (*135*), which, as may be recalled, appeared along with biliverdin during the various manipulations that he used to purify bilirubin from gallstones. *Städeler* found that biliverdin and *Biliprasin* were differentiated only by a color difference in alkaline solution, with the former being green and the latter being brown. However, *Maly* called into question the existence of *Biliprasin* since it was based only on an easily changeable and nuanced color, and because he never found a pure green alkaline solution of the biliverdin from isolated bile, rather, only when it was prepared from the purest *Cholepyrrhin*. Otherwise it was always brown-green. *Maly* then went on to describe three conditions under which *Cholepyrrhin* is converted to biliverdin: acids, alkalis, Br_2 and I_2. The last (halogens) represented what he believed to be novel reactions of *Cholepyrrhin*. The first two he explored again.

Some three years earlier, in 1865, *Maly* had reported that he was able to convert *Cholepyrrhin* completely to biliverdin by heating it in a mixture of $CHCl_3$ and acetic acid in a sealed tube in a water bath (presumably at 100°C). Since the reaction tube was only half full – the rest being air – he concluded that oxygen from the air was responsible, extolled the virtues of this simple transformation to very pure biliverdin, and concluded that biliverdin is an oxidation product of *Cholepyrrhin* – which others had concluded previously. He noted that other acids, such as HCl, will function in place of acetic acid to afford biliverdin in a less-clean transformation, and he speculated on whether the HCl in biliary vomitus might function likewise. Whether this "greening" due to an oxidation by means of oxygen was determined by or based on the influence of the acid he thought to contest by examining the influence of sulfurous acid (H_2SO_3) – because in the presence of this acid, a second compound (biliverdin) could in no way be formed by an oxidation. The experiment showed that biliverdin formation failed completely in the presence of H_2SO_3. Heating *Cholepyrrhin* in an aqueous or alcoholic solution of SO_2 in a water bath, either open to the atmosphere or in a sealed tube, gave no trace of "greening". (In the reaction open to the atmosphere, Professors *A.F. McDonagh* and *Jin-Shi Ma* produced yellow ranarubin, see Section 6 and references *168–170*). What dissolved in alcohol from these reactions with *Cholepyrrhin* was nothing other than golden yellow. *Maly* concluded from all the above that biliverdin formation still involved an oxidation process that requires oxygen in an amount sufficient for the relatively small amounts of *Cholepyrrhin* used.

More positive results came from using alkali. Thus, in an experiment reminiscent of that by *Tiedemann* and *Gmelin* (*48*), *Cholepyrrhin* in a dilute solution of

2.9 Bile Pigment Isolation, Purification, and Combustion Analysis... 137

NaOH was divided into two parts, one was placed in a tube with air excluded by Hg, and the other was placed in a covered porous dish. After a few days, or even a month, the former still had a reddish brown color, while the latter had turned brownish green after a few days and precipitated green flakes of biliverdin upon addition of HCl. When a flask of oxygen was introduced into the glass bulb part of a glass cylinder, the gas was slowly but completely absorbed, leading exactly to a second and third "greening" of the solution. Other similar experiments were employed, *e.g.* using dilute soda lye (NaOH) in a U-tube, where only the end exposed to air turned green and the color change proceeded very slowly along the tube – an illustration of the diffusion of air through the liquid. These and other experiments convinced *Maly* that the question of oxygen uptake had been settled, and that the peculiar ability of oxygen from air to be absorbed and chemically bonded was nothing strange, because alkaline solutions of indigo white, gallic acid, and pyrogallic acid behaved just like *Cholepyrrhin* (in taking up oxygen) (*166*):

> Demnach betrachte ich die Frage von der Sauerstoffaufnahme als erledigt; die Eigenthümlichkeit den Luftsauerstoff zu absorbiren und chemisch zu binden hat, wie wir wissen, gar nichts seltsames; Indigweiss, Gallussäure und Pyrogallussäure in alkalischer Lösung verhalten sich eben so wie Cholepyrrhin. 129

The slow oxidation by atmospheric oxygen was compared to a more rapid oxidation by nascent oxygen. In an interesting experiment, an alkaline solution of (red-brown) *Cholepyrrhin* was stirred cautiously with PbO_2, and in two minutes the solution went over to green-brown; whereas, when the original solution was allowed to stand in air without PbO_2, the "greening" took place only after three to four or five days. At which point the addition of a little HCl and a lot of alcohol led to a biliverdin solution. Apparently the added PbO_2 considerably shortened the time-consuming reaction taking place in air alone. In the presence of stronger oxidizing agents, biliverdin itself suffered further oxidation. Thus, $KMnO_4$ gave further oxidation products. Though he did not realize it, *Maly* may have been the first to report a chemical degradation of the pigment to small fragment molecules (*166, 167*): "Uebermangansaures Kali giebt sogleich weitergehende Oxydationsproducte" ($KMnO_4$ also gives further oxidation products).

Maly was thus able to obtain the green pigment that he called *Biliverdin* from *Cholepyrrhin* by: (i) heating in $CHCl_3$-glacial acetic acid solution in a sealed tube containing air; (ii) allowing an alkaline solution to stand in air a few days; and (iii) using PbO_2 as well as Br_2 as an oxidizing accelerant (*166, 167*):

> Die *Darstellung* des Biliverdins kann nach dem Vorhergehenden verschiedene Wege einschlagen. 1) Entweder man erhitzt die chloroformige Cholepyrrhinlösung mit Eisessig in zugeschmolzenen Röhren, und wäscht, mit Wasser die Essigsäure weg; oder 2) man lässt die alkalischen Lösungen einige Tage an der Luft stehen, fällt mit Salzsäure und wäscht mit Wasser aus. Immer wurde zur weiteren Reinigung das Biliverdin in wenig starkem oder absolutem kalten Alkohol gelöst, von dabei etwa bleibenden braunen Flocken filtrirt, und mit Wasser vollständig ausgefällt. Der nun erhaltene flockige schwarzgrüne Niederschlag wurde noch mit Wasser, zuletzt mit Aether gewaschen.
>
> 3) Die oben erwähnte Einwirkung des Bleisuperoxyd's so wie die des Broms lassen sich noch zweckmässiger zur Darstellung des Biliverdins ausbeuten. Man rührt in die kalische Lösung des Cholepyrrhins langsam Bleisuperoxyd ein, bis eine Probe mit Säuren

138   2   Early Scientific Investigations

eine rein grüne Fällung giebt, übersättigt dann das Ganze *schwach* mit Essigsäure, wobei
unter vollständiger Entfärbung der Flüssigkeit Biliverdinblei niederfällt, das man abfiltrirt.
Es wird dann gewaschen bis das Filtrat bleifrei ist, mit schwefelsäurenhaltigem Alkohol
zerlegt, filtrirt und durch Wasser ausgefällt.   130

Maly described the characteristics of his pure biliverdin and provided its elemental combustion analysis (%C, H, N). As a powder it was dark green, odorless and tasteless, and somewhat hygroscopic. The purest biliverdin dissolved easily in alcohol (as well as in methanol), not with a brilliant green color but with more of a sap-green. But with a trace of added acid (HCl, H_2SO_4, glacial acetic acid) the color turned a beautiful clear green. Inorganic salts of calcium, lead, and silver could be prepared from the pigment in aq. ammonia. The pigment was soluble in alkali carbonates and hydroxides, giving a sap-green to brown-green color. When solid biliverdin was ground up with conc. H_2SO_4, the pigment dissolved to give a green color and was unchanged upon addition of H_2O, which precipitated flakes that produced a green color in alcohol. (One is led to believe that the biliverdin had undergone no chemical changes during the process.) It was soluble in ether to only an insignificant degree, insoluble in $CHCl_3$ but soluble in $CHCl_3$ containing a few drops of alcohol, soluble in glacial acetic acid-$CHCl_3$, and also in glacial acetic acid with an especially beautiful color. It was not soluble in benzene or CS_2, poorly soluble in amyl alcohol and in CH_3CH_2I – but easily soluble in the latter two if a little ethyl alcohol is added (*166, 167*):

Das reine Biliverdin ist ein schwarzer glänzender, gepulvert schwarzgrüner Körper. Es
ist geschmack- und geruchlos, und benetzt sich schwer mit Wasser. Bei 100° getrocknet
giebt es etwas hygroskopische Feuchtigkeit ab, bleibt bei dieser Temperatur dann unverändert an Gewicht, ist aber so getrocknet sehr hygroskopisch.

Das reinste getrocknete Biliverdin löst sich in Alkohol nicht mit feurig grüner, sondern
mit mehr saftgrüner Farbe. So wie aber dieser Lösung nur eine Spur einer Säure (Salz-,
Schwefel-, Essigsäure) zugefügt wird, so wird sie prächtig rein grün.

Die alkoholische Biliverdinlösung giebt nach Zusatz von ein wenig Ammoniak mit
Chlorcalcium einen dunkelgrünen in Wasser nicht löslichen Niederschlag; mit Silbernitrat
eine flockige dunkelbraune Fällung unter vollständiger Entfärbung der Flüssigkeit. Dieses
Biliverdinsilber löst sich nicht in Wasser, aber leicht in Ammoniak mit dunkelkastanienbrauner Farbe. Das auf ähnliche Weise mittelst Bleizucker dargestellte *Biliverdinblei* ist
braungrün flockig.

Mit concentrirter Schwefelsäure verrieben löst sich das Biliverdin mit grüner Farbe, und
wird von Wasser unverändert daraus in grünen in Alkohol löslichen Flocken ausgefällt.

In kohlensauren und ätzenden Alkalien löst es sich mit saftgrüner oder braungrüner
Farbe. Es wird nur in unbedeutender Menge von Aether aufgenommen, und nicht von
Chloroform, löst sich aber sehr leicht, sobald dem Chloroform nur einige Tropfen alkohol
zugesetzt werden. Es löst sich ferner in Eisessig, in einem Gemenge desselben mit
Chloroform und auch in gewöhnlicher starker Essigsäure, in diesen Flüssigkeiten mit
besonders schöner Farbe.

Das Biliverdin ist nicht löslich in Benzol, Schwefelkohlenstoff, sehr wenig in
Amylakohol und Jodäthyl, wohl aber leicht in beiden letzteren, wenn diesen ein wenig
Aethylalkohol zugefügt wurde.

Methylalkohol löst das Biliverdin so leicht wie der gewöhnliche Alkohol.   131

The elemental combustion analysis showed the presence of 2% ash where only the %N is reported below (III and IV) but no ash in I and II. And using the atomic mass convention where O = 16, *Maly* wrote that *Cholepyrrhin* added an oxygen atom to give biliverdin as $C_{16}H_{18}N_2O_4$. However, the actual %N of this formula is higher (9.26%) in N than that (8.74–8.77%) determined by experiment. Though this did not bother *Maly*, he took issue with *Städeler*, who had proposed the biliverdin formula $C_{16}H_{20}N_2O_5$, which would give a value of 60% for C (*166, 167*):

Analyse.

I. 0,2400 Grm. Biliverdin gaben 0,561 Grm. Kohlensäure und 0,129 Grm. Wasser.

II. 0,2905 Grm. Substanz einer anderen Darstellung gaben 0,1585 Grm. Wasser.

III. 0,3356 Grm. Substanz einer dritten Darstellung gaben mit Natronkalk geglüht etc. 0,204 Grm. Platin.

IV. 0,3465 Grm. einer vierten Darstellung gaben eine 0,210 Grm. Platin hinterlassende Menge Platinsalmiak.

Diesen Resultaten entsprechen nach Abzug von circa 2 p.C. Asche bei III und IV (die Substanz von I und II war aschefrei) folgende Procentzahlen:

	I.	II.	III.	IV.
Kohlenstoff...............	63,74	–	–	–
Wasserstoff...............	5,97	6,05	–	–
Stickstoff...................	–	–	8,77	8,74

Würde das Cholepyrrhin wenn es in Biliverdin übergeht, ein Atom Sauerstoff (16. Gewth.) aufnehmen:

$$C_{16}H_{18}N_2O_3 + O = C_{16}H_{18}N_2O_4,$$

so wäre die Formel des Biliverdins $C_{16}H_{18}N_2O_4$ und dieser entspricht die Berechnung:

Kohlenstoff........................	63,58
Wasserstoff........................	5,96
Stickstoff........................	9,26
Sauerstoff........................	21,19

welche mit den gefundenen Zahlen nur ein wenig im Stickstoffgehalt abweicht.

Nähme das Cholepyrrhin, wie Städeler angiebt, auch noch ein Molekül Wasser auf:

$$C_{16}H_{18}N_2O_5 + O + H_2O = C_{16}H_{20}N_2O_5,$$

so würde der Kohlenstoffgehalt im Biliverdin bis auf 60,00 p.C. sinken. Ich glaube daher die erstere Formel für die richtige halten zu müssen. Die vollständige Erschöpfung meines Materiales, durch welche der Abschluss dieser ersten Abhandlung veranlasst ist, hindert mich vorläufig an einer letzten, noch nothwendigen Controlanalyse des Bilverdins.   132

Some six years later, when in Innsbruck, *Maly* read a paper before the Austrian Academy of Science in 1874, which appeared in a *Sitzungsbericht* (*171*) and was published a year later in *Annalen der Chemie und Pharmacie* (*172*) – a publication that summarized his careful studies on the conversion of bilirubin to biliverdin. These reports (*171, 172*) were destined to be among his final three papers on bile pigments. Therein he described in great detail the isolation of bilirubin (by the year 1874, he had eschewed *Berzelius'* name for it, *Cholepyrrhin*) from ox gallstones. He noted that bilirubin was the major pigment, comprising some 28–48% by weight after having been freed from its salts, mainly the calcium salt, and it appeared along with the typical other components, which he cited. The separation followed the methods of other workers, especially *Städeler* (*135*) and *Thudichum* (*152*). *Maly* added an improvement and was apparently the first to remove bilirubin by an extraction process using a continuous extraction apparatus (*171–173*) that operated on the same principle and design as the *Soxhlet* extractor.

Maly revisited the elemental combustion analyses of biliverdin that he reported in 1868 (*166, 167*) and compared them to those published by *Thudichum* in the same year (*152*). *Städeler* (*135*) had reported no combustion analyses of biliverdin, but he nonetheless had proposed a formula ($C_{16}H_{20}N_2O_5$) for the pigment on the basis of *Heintz*'s earlier analysis in 1851 (*109*), and from this a %C, H, N, O could be calculated – for comparison purposes. *Maly*'s formula ($C_{16}H_{18}N_2O_4$) corresponded to a %N that was ~0.5% higher than the experimental value, and whereas the latter matched the value calculated from *Städeler*'s formula, *Thudichum*'s experimental %N matched that predicted by *Maly*'s formula. *Thudichum*'s formula ($C_8H_9NO_2$) for biliverdin, however, corresponded to one-half the *Maly* formula (*171*):

$\mathfrak{C}_{16}H_{18}N_2O_4 =$ Bilirubin + Θ verlangt	Meine früheren Analysen gaben		Thudichum fand l.c.			Städeler's Formel $\mathfrak{C}_{16}H_{20}N_2O_5$ = Bilirubin + H_2O + Θ will	
	I	II	I	II	III		
C 63·58	63·74		63·08	62·09	62·14	C	60·00
H 5·96	5·97	6·05	6·25	6·12	6·00	H	6·25
N 9·26	N {8·77 u. 8·74		N {9·32 u. 9·36]			N	8·75
O 21·19						O	25·00.

133

In order to obtain a better combustion analysis of biliverdin, *Maly* prepared the latter from bilirubin in dilute aqueous Na_2CO_3 solution left in the presence of oxygen over a few days. He precipitated the green pigment by adding HCl and purifying (using its alcohol solubility) until it was ash-free upon combustion. Believing that his earlier nitrogen analysis fell short of the mark due to a failure of the method (incineration with copper oxide), he used the *Dumas* method and obtained a %N that nicely matched his formula. And on this basis, with the earlier %N deficiency having been explained and rectified, *Maly* then believed that the composition of his biliverdin should be considered as firmly established and definitive (*171, 172*):

> Da sich bei meinen früheren Analysen eine Differenz nur im N gezeigt hat, der als NH_3 bestimmt worden war, mittlerweile aber von mehreren Seiten, so von Ritthausen und Kreusler . . . und namentlich von Nowak . . . constatirt wurde, dass gewisse Körper nur durch Glühen mit Kupferoxyd. ihren ganzen Stickstoff ausgeben, so wurde diesmal der N nach Dumas' Methode bestimmt.
>
> 1. 0·2785 Grm. Biliverdin, bei 100° getrocknet, gaben 0·6516 Grm. $C\Theta_2$ und 0·1452 Grm. $H_2\Theta$.
> 2. 0·3693 Grm. eines anderen Präparates gaben 31·5 CC. feuchten N bei 15° C. und 27·35 Par. Zoll.

	Gefunden	Berechnet $C_{16}H_{18}N_2O_4$
C...........	63·82	63.58
H...........	5·80	5·96
N...........	9·35	9·26.

> Die geänderte N-Bestimmung hat also auch beim Biliverdin den kleinen Ausfall an N verschwinden machen, und da nun die Übereinstimmung in Bezug auf die verschiedenen Präparate, die Thudichum's und meinen Analysen zu Grunde liegen, eine ganz vollständige ist, so darf die Zusammensetzung dieses Körpers als definitiv festgesetzt betrachtet werden. [134]

With this problem ostensibly behind him – although the apparent discrepancy between his formula and *Thudichum*'s (empirical) formula was unresolved and could not be resolved in the absence of knowing the molecular weight of the pigment – *Maly* turned to: (i) investigating a new method for converting bilirubin to biliverdin in the presence of oxygen and (ii) determining the material balance in the conversion. (i) Thus, heating pulverized bilirubin in molten $ClCH_2CO_2H$ (62°C) in the presence of air for a few days turned the melt green; addition of H_2O led to a green precipitate that was easily separated to leave behind an aqueous solution that contained only traces of pigment. In contrast, when the reaction was blanketed by CO_2, the color changed to brown, with no evidence of green. In two experiments, 0.7566 g bilirubin gave 0.7528 g biliverdin, and 0.4863 g bilirubin gave 0.4767 g biliverdin. The recoveries of pigment were 99.5% and 98.0%, which meant that very little was lost to the aqueous filtrate.

Then, yet another quantitative measure of the bilirubin to biliverdin conversion was determined – using bilirubin in dilute aq. Na_2CO_3. The green pigment, isolated by precipitation when HCl was added to the reaction, was dried, weighed, and compared to the weight of the dried bilirubin starting material. Traces of green pigment in the aqueous filtrate were estimated colorimetrically (*171, 172*):

> Bilirubin wurde in sehr verdünnter Sodalösung gelöst, unter gelegentlichem Einleiten von Sauerstoff einige Tage stehen gelassen, mit HCl das Biliverdin gefällt, am getrockneten und gewogenen Filter gesammelt und bis zum Verschwinden der Chlorreaction gewaschen. Es wurde dann bein 110° getrocknet und gewogen. Das grüngelbe Filtrat dampfte man ein und bestimmte darin den Gehalt an organischer Substanz durch schwaches Glühen des bei 125° getrockneten Rückstandes. Die Waschwasser, welche in dickerer Schichte auch eine Spur grüngelber Färbung zeigten, wurden colorimetrisch nach dem ersten Filtrate geschätzt. Dabei erhielt man:

Angewandtes Bilirubin (110° getrocknet)...............	0·4558	Grm.
Abfiltrirtes Biliverdin (110° getrocknet	0·4458	,,
Organische Substanz im Filtrate	0·0223	,,
Gesammtes Biliverdin	0·4681	Grm. [135]

An estimated increase in weight of 2% – based entirely on the quantitative colorimetric analysis – was attempted to be correlated with the proposed stoichiometry for converting bilirubin to biliverdin: $C_{16}H_{18}N_2O_3 + O \rightarrow C_{16}H_{18}N_2O_4$, or an increase of 5.3%. From the current perspective, the quantitative determination of the

142 2 Early Scientific Investigations

purported biliverdin left dissolved in the aqueous filtrate is clearly suspect and the experiment compromised. Nonetheless, *Maly* held to the belief that biliverdin contained one more oxygen than bilirubin (*174*):

> Jedenfalls stimmen also Analyse und Gewichtszunahme zusammen, und beide führen zu der Biliverdinformel $C_{16}H_{18}N_2O_4$, welche von der des Bilirubins durch einen Mehrgehalt von O sich unterscheidet. 136

In 1868, *Maly* had also explored the further oxidation of *Cholepyrrhin*, noting that (in his formula for biliverdin) only one atom of oxygen was added to yield the color change to green, *i.e.* the first stage of the *Gmelin* color change reaction. He thus contemplated that the subsequent color changes of the reaction were due to further oxidation, which to him meant the addition of more oxygen. Not an illogical extrapolation but one clearly based on the belief that: (i) the green color at the first stage of the *Gmelin* reaction was due to biliverdin, and (ii) the addition or incorporation of one atom of oxygen was responsible for the conversion of *Cholepyrrhin* to biliverdin. It was only later that (ii) was shown to be incorrect, that oxygen was in fact not incorporated.

In order to explore the colors of the *Gmelin* reaction, to attempt to stop the reaction at the various color stages, it was carried out using arsenic acid anhydride (As_2O_5) and HNO_2 to produce the usual color changes, which however concluded at the red coloration stage (and not the pale yellow), at a non-changing bright wine-red tone. Upon addition of H_2O at this stage, a bright, iron-oxide-colored flocculent precipitate ensued, but it could not be crystallized and thus remained of questionable purity. Nonetheless, it underwent an elemental combustion analysis which showed the new compound to be comparatively richer in oxygen than either *Cholepyrrhin* or biliverdin (*166*, *167*):

> ... Ohne jetzt näher auf ihn einzugehen, will ich nur erwähnen, dass er in der That sehr viel sauerstoffreicher ist, als Cholepyrrhin oder Biliverdin, während Kohlenstoff und Wasserstoff zurücktreten. Folgende Zahlen zeigen dieses:

	Sauerstoff		Kohlenstoff	
Cholepyrrhin enthält }	16,79	p.C.	67,13	p.C.
Biliverdin „ }	21,19	„	63,58	„
Neuer Körper „	30,39	„	55,23	„

> Mag nun dieser neue Körper nicht völlig rein erhalten worden sein, so viel zeigt seine Analyse sicher, dass die Oxydation noch weit über die Bildung des Biliverdins hinaus fortschreitet. 137

Difficulties were acknowledged in stopping the reaction at a given stage of color because the HNO_2 continued to effect oxidation. Later, *Maly* found a way to arrest all of the individual stages. This was accomplished using Br_2, which as described earlier, could oxidize *Cholepyrrhin* to biliverdin and beyond. Thus, addition of an alcohol solution of Br_2 led to a beautiful dark blue colored solution that remained unchanged for weeks. Addition of more of the alcoholic Br_2 produced a dirty violet color through clear dark red and finally a light wine-red. The series of color changes was much the same as that described previously from HNO_3 and from HNO_2. Although *Maly* indicated an ability to stop the color changes at individual

2.9 Bile Pigment Isolation, Purification, and Combustion Analysis...

colored stages, he did not apparently isolate the corresponding pigments. Instead, he found that when the dark blue-colored $CHCl_3$ solution formed above was mixed with a $CHCl_3$ solution of *Cholepyrrhin* it simulated the clear green color of biliverdin but contained none of it. Evaporation in a dish separated blue and yellow rings, of which alcohol extracted only the blue, leaving behind the (yellow) *Cholepyrrhin*.

Maly concluded that there could be no doubt that the *Gmelin* reaction formed a series of compounds from *Cholepyrrhin* that contain increasingly more oxygen, from the single oxygen incorporated by biliverdin to blue and then red, and finally to the 30% O contained in the wine-red pigment. The violet he attributed to a mixture of red and blue. As described in his presentation to the Austrian Academy of Science in 1869, addition of Br_2 could be used to stop the oxidation at the blue stage (*174*). He believed he had found the means, using Br_2, to stop the progression and thus isolate pure pigments and said he would try to extend his research in that direction (*166, 167*):

> Es kann sonach kein Zweifel sein, dass die bei der Gallenfarbprobe sich bildenden Körper weitere Oxyde des Cholepyrrhins darstellen, die zwischen Biliverdin und dem Körper der weinrothen Lösung mit 30 p.C. Sauerstoff stehend, mit diesen eine *mehrgliedrige an Sauerstoff zunehmende Reihe* bilden. Jedenfalls existiren noch ein blauer und rother Körper und das hellbraune Endproduct, während der violette wahrseheinlich ein Gemenge des rothen und blauen ist.
>
> Nachdem im Brom ein Mittel zu ihrer Fixirung und Reindarstellung gefunden ist, werde ich in dieser Richtung meine Versuche zu erweitern suchen. 138

Though *Maly* believed that he had achieved oxidation of *Cholepyrrhin* on the basis of a change in color, his thinking of oxidation was conditioned by processes involving the incorporation of oxygen. *Städeler*, too, and his predecessors were similarly inclined, as was *Thudichum*. But by the time that *Maly* began his work on bilirubin, *Städeler* was absenting himself from bile pigment research. Not so with *Thudichum*, who began to follow *Maly*'s published work, and as will be seen, became exasperated by it and the multitude of errors he believed to have found in it. Eventually, neither *Maly* nor *Thudichum* were proved correct in the concept of oxidation as applied to bile pigments.

Maly also had an interest in reduction, though he was not alone in this. His reduction of biliverdin appears to be novel for its period in time.

In 1868, he explored the reduction of biliverdin using spongy platinum, apparently freshly precipitated and activated. Thus, treatment of the pigment with the Pt over a period of a few days to a few hours gave a red-brown solution, seen after screening out the spongy Pt. (Whether the product was bilirubin or one of further reduction was not further stated.) (*166, 167*):

> Platinschwamm reducirt die Biliverdinbildung von einigen Tagen auf einige Stunden; hat man die rothbraune Lösung in einer flachen Schale, so siebt man vom hineingeworfenen Platinschwamm aus die Farbenumwandlungen vor sich gehen. 139

Bilirubin, too, was shown by *Maly* to suffer reduction, albeit much less readily than biliverdin. Thus in his studies from Innsbruck, where he was *Professor der Physiologischen Chemie* at the university from 1869 to 1875, he published his preliminary work on the synthetic transformation of bilirubin into the pigment of urine in an article submitted on February 26, 1872 (*175*). He then followed it with a longer article, also submitted in February 1872 to the same journal (*176*). Thus, as *Maly* wrote while transitioning the pigment's names, *Cholepyrrhin* (bilirubin that had been isolated from ox gallstones and purified, as described earlier) was dissolved, or in later experiments suspended, in dilute aq. KOH or NaOH, protected from air, and allowed to react with the nascent hydrogen evolved upon addition of Na(Hg). (The author carried out essentially the same reaction of bilirubin in the *C.J. Watson* lab at the University of Minnesota Medical School in 1964–1965.) At first the procedure revealed no H_2 evolution (because it was being taken up by bilirubin). Later, as the reaction progressed, and the original opaque, dark solution had cleared and become a light brown color, it could be shown that the reaction vessel contained evolved hydrogen. After 2–4 days, during which excess Na(Hg) had been added, with frequent shaking at room temperature, and with subsequent gentle warming until no further lightening of the color could be observed, the Hg was removed and excess HCl (or acetic acid) was added. This produced a garnet red color, showing that bilirubin had undergone a change, and dark red-brown flakes separated, leaving a red-colored solution. The precipitate was filtered and washed to remove NaCl entirely. The collected precipitate had the characteristics of a weak acid, as it dissolved easily in ammonia or alkali with a yellow-brown color. Unlike bilirubin, however, it was readily soluble in alcohol with a reddish color, in $CHCl_3$ with a yellow-red color, and in alkaline solution with a brown color. This new pigment was submitted to combustion analysis from which a formula ($C_{32}H_{40}N_4O_7$) was calculated from the assumption that two bilirubin molecules had together absorbed one molecule each of H_2 and H_2O. *Maly* named the new pigment *Hydrobilirubin* (hydrobilirubin) (*176*):

Zur *Analyse* wurden Proben von Substanz genommen, die von dreierlei Darstellungen herrührten, und bei welchen das Auflösen in Alkali und Ausfällen mit Säuren bald einmal, bald zwei- und dreimal vorgenommen war.

1. 0,2193 Grm. Substanz gaben 0,523 CO_2 und 0,1404 H_2O.
2. 0,2652 Grm. Substanz gaben 0,1646 H_2O.
3. 0,2262 Grm. Substanz gaben 0,1474 Platin.
4. 0,2483 Grm. Substanz gaben 0,5886 CO_2 und 0,1559 H_2O.
5. 0,2174 Grm. Substanz gaben 0,5142 CO_2 und 0,1347 H_2O.

Auf Procente bezogen:

	1.	2.	3.	4.	5.	Mittel
C	64,89	–	–	64,65	64,50	64,68
H	7,09	6,80	–	6,98	6,87	6,93
N	–	–	9,22	–	–	9,22.

Diese Zahlen stimmen so gut untereinander überein, dass die Reaction als eine sehr glatte bezeichnet werden muss. Die Substanz is kohlenstoffärmer und wasserstoffreicher als Bilirubin, entsprechend ihrer Bildung und kann also nur durch Bindung von Wasserstoff entstanden sein. Nimmt man, an dass auch noch Wasser eingetreten ist, und zwar H_2O auf 2 (Mol. ?) Bilirubin neben H_2, so würde der Körper $C_{32}H_{40}N_4O_7$ resultiren, nach der Gleichung:

$$2 \ C_{16}H_{18}N_2O_3 + H_2 + H_2O = C_{32}H_{40}N_4O_7$$

2.9 Bile Pigment Isolation, Purification, and Combustion Analysis...

und dieser verlangt:

		Gefunden (Mittel)
C_{32}	64,86	64,68
H_{40}	6,75	6,93
N_4	9,45	9,22
O_7	–	–

was mit den erhaltenen Resultaten recht gut übereinstimmt. Der neue Körper ist also durch Aufnahme von Wasserstoff und Wasser unter Verdoppelung des Moleculs von Bilirubin (falls nicht, wie vielleicht wahrscheinlicher, die Bilirubinformel doppelt so gross als gewöhnlich zu schreiben ist) entstanden und soll fortan als *Hydrobilirubin* bezeichnet werden.   140

Various other properties of hydrobilirubin were investigated, including those customary for the times: salts with a wide range of various metal ions, from alkali to heavy metals. One property not shared with its precursor, bilirubin, was the strong fluorescence exhibited by certain of the salts, especially those formed from $ZnCl_2$ or $ZnSO_4$ in aq. NH_3. The pigment did not exhibit the *Gmelin* reaction, but elementary colorimetry was investigated (and will be described later). Of considerable importance to *Maly* was a probable relationship between his hydrobilirubin and the pigment of urine, *i.e.* what *Jaffe* termed *Urobilin* (urobilin). In comparing the various characteristics of hydrobilirubin and urobilin, *Maly* found them to be identical, though he preferred the former name (his) to the latter because it expressed something more of its constitution (*176*):

> Durch die Wiederholung dieser Stellen aus Jaffe's Abhandlung habe ich am Besten gezeigt, dass die Eigenschaften meines Hydrobilirubins und Jaffe's Urobilins, also die Substanzen selbst identisch sind. Dass ich für meine Substanz (richtiger für beide) den Namen Hydrobilirubin einführe, begründet sich durch die künstliche und natürliche Bildung, und zweitens dadurch, dass wenigstens etwas von der Constitution durch den Namen ausgedrückt ist.   141

Maly also noted that *Thudichum* had isolated a compound from urine that he named *Urochrom* (urochrome) but had little discussed it, aside from noting its yellow-red color. More important to him, however, was *Scherer*'s work (*93*) on the urinary pigment. Apparently repeating *Scherer*'s isolation procedure for fresh urine from feverish patients, *Maly* found *Scherer*'s pigment and hydrobilirubin to have identical characteristic properties and rather similar %C and H in combustion analysis: %C 65.25; %H 6.59; and %C 64.99; %H, 7.00 for *Scherer*'s urinary pigment, or not far removed from the %C 64.68; %H 6.93 for hydrobilirubin. Yet other analyses gave results in poorer agreement (*176*):

> . . . was nicht weit entfernt von der Zusammensetzung des Hydrobilirubins ist, und darauf deuten würde, dass das Scherer'sche Präparat wenigstens keine grossen Mengen verunreinigender Substanz enthielt. Anderer Analysen freilich gaben weiter abstehende Resultate *).

---

*) Damit ist auch wenigstens für den wichtigsten Harnfarbstoff der angebliche Eisengehalt widerlegt. Auch hat Dr. Schlemmer neuerdings wieder in meinem Laboratorium in grösseren Mengen Harns vergeblich nach Eisen gesucht.   142

146   2 Early Scientific Investigations

It may be interesting to note that *Maly* assumed, given the identity of hydrobilirubin with the urinary pigment, that the circulation of bilirubin took it to the gut, where it was reduced (hydrogenated) by the hydrogen produced there and added H_2O to form the hydrobilirubin that later appeared in urine. (He also found that treatment of biliverdin with Na(Hg) led to entirely similar results as with bilirubin: both formed a brown solution.) *Maly* thus explained that the hydrobilirubin was absorbed from the gut and went finally into the urine, thereby to end its cycle in the organism. He assumed that hydrobilirubin formed in the gut played no important role there and was only a means to bring the compound to excretion from the organism; *i.e.* the hydrobilirubin was absorbed from the gut, where it apparently played no role, and finally went into the urine. Bile pigments could thus be viewed as useless by-products of liver metabolism (*176*):

> Indem wir so gesehen haben, in welch naher chemischer Beziehung der Orange-Gallenfarbstoff und der (hauptsächlichste) Harnfarbstoff wenigstens beim Menschen zu einander stehen, ergiebt sich der Kreislauf dieser Pigmente von selbst, und manche zusammenhanglose Thatsache reiht sich schön ein. Das mit der Galle in den Darm ergossene Bilirubin erleidet während seiner Wanderung herab bis zum Colon und in diesem selbst seine Wasserstoff- und Wasseraufnahme unter dem Einflusse von Wasserstoff entbindenden Processen. Ganz gleich verhält sich Biliverdin: ich habe eine alkoholische Biliverdinlösung mit Natriumamalgam behandelt, und bald eine braune Lösung erhalt, identisch mit der aus Bilirubin. . . .
>
> Vom Darm aus wird das Hydrobilirubin aufgesaugt und geht schliesslich in den Harn, um dort seinen Cyclus im Organismus zu beenden. Da das Hydrobilirubin im Darm keine ersichtliche Rolle spielt, und die Aufsaugung nur ein Mittel ist den Körper aus dem Organismus hinaus zu bringen, so ist nicht einzusehen, dass die Gallenfarbstoffe überhaupt einem Zwecke dienlich sein sollten, und man wird dermalen sie nicht anders denn als nutzlose Nebenproducte des Leberchemismus anzusehen haben.   143

Maly's experiments involving bilirubin and biliverdin, especially the oxidation reactions that produced the latter from the former, and their formulas derived from the elemental combustion analyses, drew sharp criticism from *Thudichum*. So did hydrobilirubin. But what piqued *Thudichum*'s interest and ire most were *Maly*'s reactions of bilirubin with Br_2. In his work published in 1868 (*166, 167*), *Maly* mentioned a third route for converting *Cholepyrrhin* to biliverdin, a route destined to provoke controversy and a polemic from *Thudichum*: oxidation using Br_2 or I_2. To accomplish such a transformation, described by *Maly* as surprisingly nice, *Cholepyrrhin* was allowed to stand in Br_2 vapor mixed with moist air. This resulted in rapid darkening and yielded a compound no longer soluble in $CHCl_3$ but one that dissolved in alcohol with a clear green color. (No mention was made as to whether the *Cholepyrrhin* used was as a solid or in solution.) The reaction described was allowed to continue, but *Maly* found it advantageous to carry out the transformation of a yellow solution of *Cholepyrrhin* in $CHCl_3$ using a decently dilute solution of Br_2 in alcohol. Dropwise addition of the latter into the former caused immediate darkening of the $CHCl_3$ solution to a sap-green color. Careful addition led to a point where the $CHCl_3$ solution was a clear, beautiful bright green, with the biliverdin formed remaining in the $CHCl_3$-alcohol mixture. At this point *Maly* claimed that all of the *Cholepyrrhin* had been converted to biliverdin as the solution was stable for weeks (*166, 167*):

> Ich habe erwähnt, dass es ausser Säuren und Basen noch eine dritte Reihe von Körpern giebt, welche Biliverdin aus Cholepyrrhin erzeugen; es sind diess die Haloide Brom und Jod. Namentlich überraschend schön ist die Umwandlung mittelst Brom. Bringt man Cholepyrrhin unter eine Glasglocke, in der sich mit feuchter Luft gemischter Bromdampf befindet, so färbt es sich bald dunkel, und wird nicht mehr von Chloroform, aber von Weingeist mit rein grüner Farbe gelöst. Da aber dabei die Bromwirkung leicht etwas zu weit geht, so kann man den Versuch viel vortheilhafter in folgender Weise anstellen. Man versetzt eine gelbe chloroformige Cholepyrrhinlösung mit einer recht verdünnten alkoholischen Lösung von Brom. Schon die ersten Tropfen machen die Flüssigkeit dunkel saftgrün, und es lässt sich sehr leicht bei weiterem vorsichtigen Bromzusatz der Punkt treffen, bei dem die ganze Flüssigkeit ein reines prachtvoll feuriges Grün zeigt *). In diesem Momente ist alles Cholepyrrhin in Biliverdin übergegangen, und die Flüssigkeit kann wochenlang stehen, ohne sich zu verändern. 144

---

*) In diesem Gemenge von Chloroform mit nur wenig Alkohol bleibt das Biliverdin gelöst.

Thus, reaction of bilirubin with Br_2 was viewed by *Maly* to be an oxidation because it produced a green pigment thought to be biliverdin, a known oxidation product of bilirubin, for he knew that halogens in the presence of moisture cause oxidation of oxidizable compounds. And what atmospheric oxygen brought about so slowly, the conversion with Br_2 took a few seconds.

Some 15 months after his July 9, 1874 presentation to the Austrian Academy, *Maly*, then *Professor der allgemeinen Chemie* at the TH-Graz, made his final presentation on bile pigments on October 17, 1875 (*177*), which was published in 1876 in the *Annalen der Chemie und Pharmacie* (*178*). The subject was the treatment of bilirubin with halogens, especially Br_2, and it drew a lambasting from *Thudichum*, as will be noted later. Although *Maly* had first believed that bilirubin was oxidized to biliverdin by reacting with Br_2, from which a blue coloration gradually became evident, by reinvestigating this erstwhile "oxidation" reaction, he became convinced that the green color of reaction was actually due to a mixture of a blue compound and unreacted bilirubin. Thus, with careful control of the ratio of added Br_2 as a solution in $CHCl_3$ to a solution of bilirubin in $CHCl_3$, with an added few drops of alcohol, he conducted a series of reactions from which each step in the sequence of *Gmelin* color changes was observed (*178*): "Es zeigten sich brillante farbige Lösungen von grosser Haltbarkeit und in der Reihenfolge, wie sie bei der Gmelin'schen Salpetersäurereaction auftreten" (It exhibits brilliantly colored solutions of great stability and in the sequence that appears in the *Gmelin* HNO_3 reaction). From a stable "blue step" produced by reaction with an appropriate amount of Br_2, the solution was observed to run through the remainder of the color steps of the *Gmelin* reaction, from red to yellow-brown. Based on his bromination experiments, *Maly* concluded that the blue pigment did not arise by oxidation but by bromination – and that it was a very bromine-rich new compound (*178*): "So begreift sich, dass es für den sich damit Beschäftigen denn viel Ueberraschendes hatte, zu finden, dass die Bromwirkung dabei keine oxydirende ist und der blaue dabei entstehende Körper eine an Brom sehr reiche Verbindung ist" (So it is understandable that it was very surprising to find that the action of the bromine used is not oxidizing and the blue compound arising is a substance very rich in bromine).

Maly set about to prepare, isolate, and characterize the blue pigment from bromination of bilirubin and achieved (i) good success with a procedure that involved addition of a few drops of Br_2 to bilirubin suspended in ether, and (ii) even better success with bilirubin suspended in alcohol-free $CHCl_3$ and addition of Br_2 in the same solvent. The elemental combustion analysis revealed that although the %C varied considerably (35.51–47.83% among the six C,H analyses performed), the %H (4.14–4.7%) did not; moderate consistency was found among the seven Br analyses (27.70–29.60%); and the two N analyses gave 7.4% and 7.8%. From those data, *Maly* concluded that the blue compound was a tribromo derivative of bilirubin, wherein three hydrogens had been lost and replaced with three bromine atoms to give a formula $C_{32}H_{33}Br_3N_4O_6$ (*178*):

$$2\ C_{16}H_{18}N_2O_3 = C_{32}H_{36}N_4O_6$$
$$C_{32}H_{36}N_4O_6 + 3\ Br_2 = C_{32}H_{33}Br_3N_4O_6 + 3\ HBr$$

C_{32}	47,46
H_{33}	4,08
Br_3	29,66
N_4	6,92
O_6	11,88.

In order to reach the formula, *Maly* had to assume a doubling of his bilirubin "basic" formula ($C_{16}H_{18}N_2O_3$) to $C_{32}H_{36}N_4O_6$. This, and the hydrobilirubin formula ($C_{32}H_{40}N_4O_7$) derived from twice the original formula plus $H_2 + H_2O$, induced him to rethink his formula for bilirubin. He concluded that the original "basic" formula could not be maintained and settled on $C_{32}H_{36}N_4O_6$ as the most appropriate (*178*):

> Bei solcher Zusammenstimmung bin ich wohl berechtigt zu behaupten, dass die bisher übliche Formel des Bilirubins, welche durch $C_{16}H_{18}N_2O_3$ ausgedrückt wurde, nicht aufrecht gehalten werden kann, sondern dass dieselbe verdoppelt werden müsse. Dieselbe wird dann $C_{32}H_{36}N_4O_6$, und man hat:
> Bilirubin $C_{32}H_{36}N_4O_6$,
> Tribrombilirubin $C_{32}H_{33}Br_3N_4O_6$
> Hydrobilirubin $C_{32}H_{40}N_4O_7$. 145

Maly's formula is about as close to the correct formula ($C_{33}H_{36}N_4O_6$) for bilirubin as anyone had reached by 1875. But there were others, including *Thudicum*, who held to a formula different from *Maly*'s. Yet before he bowed out of bile

2.10   The Emergence of Bile Pigment Spectroscopy: Colorimetry and Its Applications   149

pigment research, *Maly* initiated some of the earliest spectroscopic investigations of bilirubin and its derivatives.

2.10   The Emergence of Bile Pigment Spectroscopy: Colorimetry and Its Applications

Städeler was the last of the well-known investigators of bilirubin to employ no spectroscopic measurements, though he was well acquainted with and used the other available analytical technique: elemental combustion analysis. *Maly* and *Thudichum*, whose investigations of bilirubin followed very closely to those of *Städeler*, were apparently the first to employ the new analytical technique, *spectrum analysis*, in their studies. Spectrum analysis, the precursor to ultraviolet-visible spectroscopy, was based on absorption of light in the visible region of the electromagnetic spectrum – specifically the colors seen when sunlight is dispersed through a 60° prism. When a solution of a colored substance is positioned in the light beam and before the prism, certain of these colors are reduced in intensity or extinguished by the substance absorbing the complementary color of light. Hence absorption colorimetry.

Absorption colorimetric measurements of the day made use of early instrumentation due to *Bunsen*,[49] *Kirchhoff*,[50] and *von Steinheil*[51] that followed a report on October 20, 1859 to the Royal Prussian Academy of Sciences (*Königliche Preussische Akademie der Wissenschaften*) by *Bunsen* and *Kirchhoff*. *Bunsen*, who with his laboratory assistant *Peter Desaga*, designed the *Bunsen* burner which gave a hot, clean flame, and *Kirchhoff*, who studied thermal radiation and coined the phrase "black body radiation," jointly studied the emission spectrum of heated elements and laid the basis for the emergent field of "spectrum analysis" to become used as a new analytical technique in biological chemistry for characterizing substances, together with elemental combustion analysis. A month later, on 19 November 1859, *von Steinheil* was asked by *Bunsen* and *Kirchhoff* to fabricate an

---

[49] *Robert Wilhelm Eberhard Bunsen* was born on March 30, 1811 in Göttingen and died on August 16, 1899 in Heidelberg. In 1836 he succeeded *Friedrich Wöhler* at Kassel and in 1852 he succeeded *Leopold Gmelin* at the University of Heidelberg.
[50] *Gustav Robert Kirchhoff* was born on March 12, 1824 in Königsberg and died on October 17, 1877. He was a physicist and professor at Breslau. In 1854 he was called to the University of Heidelberg where he collaborated with *Bunsen*, and in 1875 he accepted the first chair in theoretical physics at Berlin.
[51] *Carl August von Steinheil* (1801–1870), was a physicist, Professor of Mathematics in Munich from 1832, and scientific instrument builder.

instrument to examine the "fixed lines" of the solar spectrum. The primitive spectroscope using a prism to disperse the incident light (179) was thus built to serve the scientific investigations of *Bunsen* and *Kirchhoff*. It allowed *Hoppe-Seyler*[52] at the University of Tübingen to study the absorption of solutions of colored substances held in a rectangular cuvette and positioned between the (sunlight) light source and collimating telescope (180). The apparatus used, which arose from studies of the visible part of the electromagnetic spectrum, was thus limited mainly to colored substances – of which the yellow, green, and other colors of the bile pigments were ideal candidates for analysis.

The spectrum analysis scale for the visible region was adjusted to the *Fraunhofer*[53] emission lines from certain elements, *e.g.* Kα (7685 Å), Liα (6705 Å), Na (5892 Å), Sr (4607 Å), Ca (4226 Å), *etc.* Thus, the *Bunsen-Kirchhoff* scale, which ranged from 17.5 to 166.0, could be calibrated, *e.g.* the sodium D-line above corresponded to 50 on the *Bunsen* scale. The colors of the scale ranged of course from one extreme end of the spectrum, the red, identified by the *Fraunhofer* line "A" and corresponding to the potassium Kα line (seen in a flame test), or to 17.5 on the *Bunsen-Kirchhoff* scale, and ended at the other, the violet, identified as *Fraunhofer* line "H$_2$" at its extreme, or 166.0 on the *Bunsen-Kirchhoff* scale (181):

In sunlight, which is thrown horizontally upon the slit by a heliostat, Fraunhofer's lines may be employed, the most characteristic of which are shown in the

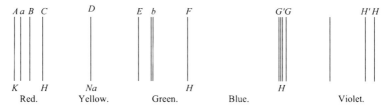

Fig. 8.

---

[52] *Ernst Felix Immanuel Hoppe-Seyler* was born on December 26, 1825 in Freyburg an der Unstrut, Saxony and died on August 10, 1895 in Wasserburg am Bodensee, Bavaria. He was perhaps the pre-eminent physiological chemist of the 19th century. Trained as a physician, he received the Dr. med. in 1850 in Berlin after studies at the Universities in Halle, Leipzig, Berlin, Prague, and Vienna. He practiced medicine, habilitated at Greifswald in 1855, and in 1856 was *Assistent* to *Rudolf Virchow* at the Pathological Institute in Berlin, then in 1857 director of the chemical laboratories at *Virchow*'s newly established Pathological Institute of the Berlin Charité, where he was appointed a. o. Professor in 1860. He was appointed a. o. Professor of Applied Chemistry at Tübingen in 1861, then o. Professor until he accepted a call in 1872 as o. Professor of Physiological Chemistry at the newly-established University of Strassburg, where he remained until his death due to a stroke at his house in Wasserburg. In 1877, he founded the respected journal *Zeitschrift für Physiologische Chemie*, that became know after his death as *Hoppe-Seyler's Zeitschrift für Physiologische Chemie*. Born *Ernst Hoppe*, his mother died when he was a child, and he added *Seyler* to his name after he was adopted by his brother-in-law.

[53] *Joseph von Fraunhofer* (1787–1826), the Bavarian optician, invented the spectroscope and discovered 574 dark lines (above absorption) appearing in the solar spectrum that are still called *Fraunhofer* lines.

2.10 The Emergence of Bile Pigment Spectroscopy: Colorimetry and Its Applications 151

accompanying figure in their relative positions, as seen through a flint-glass prism. In order to see *A* and *a* the slit must not be too narrow, and a red glass should be held before it. With a narrow slit and greater magnifying power, *D* is seen to be a very close double line.

Where sunlight cannot be used, the line *A* may be obtained by means of the potassium flame, *D* by the sodium flame, *C*, *F*, and *G'* by the light of the electric spark in a narrow Geissler's tube filled with rarefied hydrogen.

These thus correspond to the red and near ultraviolet ends of the visible spectrum, where the wavelengths correlate approximately to the *Bunsen-Kirchhoff* scale as explained long ago by *Kohlrausch* (*181*):

TABLE 19.
LINES OF THE FLAME-SPECTRA OF THE MOST IMPORTANT
LIGHT METALS,

according to Bunsen and Kirchhoff's scale; the sodium-line being taken as 50, and the slit having a breadth of 1 division.

The first number denotes the position of the middle of the line upon the scale, the Roman figure indicates the brightness, I being the brightest, and the third number gives the breadth of the band when it exceeds 1 scale-division, the breadth of the slit.

S signifies that the line is quite sharp and clearly defined, *s* that it is tolerably so; the remaining lines being nebulous and ill defined.

The lines most characteristic of each body are printed in thick type.

The brightness of the lines of *Ca*, *Sr*, and *Ba* is that of a constant spectrum. If the chlorides be employed, the spectra are at first much brighter. In many cases the flame-spectra are really those of compounds, the spectra of the metals themselves obtained by the electric spark being frequently entirely different, and consisting of much finer lines.

The colours of the spectrum are approximately—red to 48, yellow to 52, green to 80, blue to 120, and violet beyond.

K.	*Na.*	*Li.*	*Ca.*	*Sr.*	*Ba.*
17·5 II. *s*		**32·0 I.** *S*	33·1 IV. 2	29·8 III.	
			36·7 III.	32·1 II.	
				33·8 II.	
Faint continuous spectrum	**50·0 I.** *S*	45·2 IV. *s*	**41·7 I. 1·5**	36·3 II.	35·2 IV. 2
from 55 to120			46·8 III. 2	38·6 III.	41·5 III. 3
				41·5 III.	45·6 III.*s* 1·5
			49·0 III.	**45·8 I.**	
					52·1 IV.
			52·8 IV.		56·0 III. 2
			54·9 IV.		**60·8 II s**
			60·8 I. 1·5		66·5 III. 3
			68·0 IV. 2		**71·4 III. 3**
				105·0 III. *s*	**76·8 III. 2**
153·0 IV.			**135·0 IV.** *S*		82·7 IV. 4
					89·3 III. 2

TABLE 19a.
Wave-Lengths of the Principal Lines of the Solar Spectrum in Tenth-Metres in Air at 760 mm. Pressure and 16° Temperature (Angström)

In order to obtain the wave-lengths in vacuo the numbers must be multiplied by the respective refractive indices of the rays for air at 16° C. (Watts).

			Approximate Positions on Bunsen and Kirchhoff's Scale.
A	7604	1^{-10} metre	17·5
B	6867	,,	27·6
C	6562	,,	34·0
D_1	5895	,,	50·0
D_2	5889	,,	
E	5269	,,	71·0
b_1	5183	,,	75·7
F	4861	,,	90·0
G	4307	,,	127·5
H_1	3968	,,	162·0
H_2	3933	,,	166·0

TABLE 19b.
Wave-Lengths of some of the Principal Bright Lines in the Spectra of the Elements, and their Approximate Positions on Bunsen and Kirchhoff's Scale.

Element.	Wave-Length.		Scale Number.	
$K\alpha$	7685	1^{-10} metre	17·5	
$Li\alpha$	6705	,,	32·0	
$H\alpha$	6562	,,	24·0	
$Li\beta$	6102	,,	45·2	
Na	5892	,,	50·0	
C	5662	,,	58·	Edge of band seen in blue of candle flame.
Tl	5348	,,	67·	
C	5170	,,	75·	Edge of band in candle flame.
$H\beta$	4861	,,	90·	
Sr	4607	,,	105·	
Ca	4226	,,	135·	Approximate in flame spectrum.
$H\gamma$	4101	,,	151·	
$K\beta$	4080	,,	153·	Flame spectrum.

With *Beer*'s earlier report on the transmission of light through colored solutions (*182*), the stage had been set for the evolution of spectrum analysis to a quantitative level, as promoted by *Bunsen*. *Bunsen* found it possible to gain quantitative information with use of a standard reference and sample dilutions, which led him to the notion of a molar absorptivity extinction coefficient (*183*). Subsequently, almost

2.10 The Emergence of Bile Pigment Spectroscopy: Colorimetry and Its Applications 153

inevitably as spectral analysis became widely used, the instrumentation evolved in stages to more modern types by *Vierodt, d'Arsonval, Duboscq, Zeiss, etc. (184).*

In 1868, in his first long paper (*166, 167*), *Maly* reported on what might be the earliest (visible) absorption spectra of *Cholepyrrhin* (bilirubin) and biliverdin. He indicated that a $CHCl_3$ solution of the former extinguished (light absorbed by the sample solution, not transmitted through it) the entire blue and violet regions up to approximately line 70 on the *Bunsen* scale (or from ~3900 to 5300 Å); whereas, more dilute solutions removed only the violet. A similar behavior was seen in aq. NH_3. Though it was possible in the 1860s to prepare accurately weighed solutions, there was no indication of measured concentrations of the solutions, possibly because there was no concept of an exact quantitative relationship between sample concentration, the incident and exit light intensities, and the thickness (pathlength) of the sample solution (*Beer-Lambert* law). In any event, at best only a qualitative or only vague quantitative reference was typically expressed in terms of regions of the visible spectrum having been extinguished, which meant that all of the available light had been absorbed. At times a reference standard was invoked for comparative purposes. The solutions of *Cholepyrrhin* were said to have a color approximating a concentrated solution of acidic K_2CrO_7, and at the corresponding concentration the field of vision of the spectroscope was completely extinguished from the violet end to the sodium D-line (50 on the *Bunsen* scale, or ~ 5889 Å) and fairly sharply defined. If the solution were dilute it generally appeared yellow or green, but somewhat blurred. When the solutions that were very dilute such that the "coloring power" of *Cholepyrrhin* in NH_3 solution contained barely measurable traces, the lamp light appeared nearly colorless, but a good part of the violet was still extinguished (*166, 167*):

Absorptionsspectra der Gallenfarbstoffe.

>   Eine Chloroformige Cholepyrrhinlösung vor den Spalt eines Spectralapparates gebracht, löscht das ganze Blau und Violett aus, bis etwa zur Linie 70 nach der Bunsen'schen Scala. Sehr verdünnte eben noch gelbe Lösungen nehmen noch das Violett hinweg.
>   Lösungen von Cholepyrrhin in wässerigem Ammoniak verhalten sich ähnlich. Sind sie so gefärbt wie etwa eine concentrirte Lösung von saurem chromosauren Kalium, so erscheint das Sehfeld von violetten Ende bis nahe an die Natriumlinie (50) vollständig schwarz, und ziemlich scharf abgegrenzt; wird die Lösung verdünnt, so erscheint allmählich gelb und grün, aber etwas verwischt. Selbst Lösungen, die so verdünnt sind, dass sie bei Lampenlicht fast farblos erscheinen, also bei der färbenden Kraft des Cholepyrrhins in ammoniakalischen Lösungen . . . kaum mehr wägbare Spuren enthalten, löschen noch einen guten Theil von Violett aus.   146

From these data it seems clear that *Maly*'s reddish solutions of *Cholepyrrhin* were very concentrated, and they blanked out or extinguished the blue-violet region of the spectrum. *Maly* also performed a spectrum analysis of biliverdin. Thus an alcoholic solution of biliverdin was found to exhibit absorption at both ends of the spectrum. Through strongly colored layers or films, only green light was transmitted. In somewhat more dilute solutions, first yellow, orange, and a part of the red, later blue and violet. The outermost red was still removed by very dilute solutions, but no specific references to the *Bunsen-Kirchhoff* scale were cited (*166, 167*):

>   Biliverdin in alkoholischer Lösung zeigt Absorptionen nach beiden Enden des Spectrums. In stark gefärbten Schichten geht nur grünes Licht hindurch, in etwas verdünnteren erscheint

154 2 Early Scientific Investigations

zunächst gelb, orange und ein Theil des Roth, später blau und violett; das alleräusserste Roth wird noch von sehr verdünnten Lösungen hinweg genommen.   147

All this was from 1868, within a decade of *von Steinheil's* building a spectroscope and not long after *Hoppe-Seyler's* report on his spectral studies of blood and hematin (*180*). Shortly after his report on the spectrum analysis of *Cholepyrrhin* and biliverdin, *Maly* applied the emerging spectroscopy to hydrobilirubin, as reported in 1872, for purposes of comparing it to *Jaffe's* urinary pigment, urobilin (*185*). The former, dissolved in alcohol or dilute aqueous ammonia or sodium phosphate to give a yellow, or a red-yellow or rose color, was placed in the spectroscope in 0.5–2.0 cm (pathlength) cuvettes. The solution showed a vivid and marked spectral absorption between green and blue, between the *Fraunhofer* lines b and F, or between 5183 and 4861 Å (*175*):

> Löst man etwas Hydrobilirubin in verdünnten Alkohol, oder setzt man zu einer so verdünnten alkalischen Lösung (in Ammoniak oder phosphorsaurem Natron u. s. w.) deselben, dass Säuren nichts mehr ausfällen, etwas Salz- oder Essigsäure bis zur sauren Reaction, d. h. so weit, dass die Flüssigkeit die gelbe Farbe verliert und rothgelb oder rosenfarbig wird, so zeigt sie in dünner Schicht (½ bis 2 CM.) vor den Spectralspalt gestellt eine sehr lebhafte und markirte Absorption des Spectrums zwischen grün und blau, und zwar bei meinem grössen Apparat (wenn Li bei 102,5; Na auf 120 und K-β auf 219,5 steht) innerhalb der Theilstriche 146 bis 160, oder allgemeiner ausgedrückt genau zwischen den Fraunhofer'schen Linien b und F. Eben so bleibt es wenn die Lösung stärker sauer wird; Ammoniak hingegen macht das Band verschwinden und lässt nur eine schwache diffuse Absorption zwischen Grün und Blau, aber auf Zusatz von Säuren kehrt mit der röthlichen Farbe das schwarze Band zurück.   148

With the preparation of the zinc salt of hydrobilirubin from aqueous NH_3, spectrum analysis of the rose-red solution showed extinction from *Fraunhofer* line b (5183 Å) to the middle of the spectral range between b and F (4861 Å) – signifying a rather sharp band or narrow absorption region (*175*):

> [148] Hingegen geben die ammoniakalischen Lösung des Farbstoffs, wenn sie etwas eines Zinksalzes (auch Cadmium) gelöst enthalten, besonders schöne Bänder. Es genügt, der stark ammoniakalisch gemachten Hydrobilirubinlösung ein paar Tropfen von Zinkchlorür oder –Sulfat hinzuzusetzen (wobei sich der entstandene Niederschlag leicht wieder löst) und diese Flüssigkeit vor den Apparat zu bringen. Oder man löst ausgefälltes Hydrobilirubinzink in Ammoniak und verdünnt. Beide Flüssigkeiten sind rosenroth und geben ein durch Schärfe und Dunkelheit ausgezeichnetes Band, das gegenüber den sauren Lösungen etwas nach links gerückt erscheint, daselbst bei 142 meiner Skale, also etwas vor b, scharf abgegrenzt, nach rechts hin verschieden breit ist, je nach der Concentration der Lösung, das aber immer am Dunkelsten von 142 bis 155 erscheint, d. i. von b an bis zur Mitte des Spectralabschnittes b bis F. Die ganze Erscheinung ist mindestens eben so empfindlich als die der sauren Pigmentlösung.   149

These data were to be compared to *Jaffe's* urobilin pigment found in stronglycolored urine of feverish individuals said, as a dilute solution, to give a "dark shadow" from *Fraunhofer* lines b to F in the spectrum analysis (*185*):

> Bringt man eine concentrirte Lösung vor den Spalt der Spectralapparate, so erscheint das Spectrum vom violetten Ende her bis etwas zur Linie b völlig dunkel; beim Verdünnen hellt sich der verdunkelte Theil allmählich auf und es bleibt schliesslich ein Absorptionsstreif

2.10 The Emergence of Bile Pigment Spectroscopy: Colorimetry and Its Applications

> (γ) mit etwas verschwommenen Rändern an der oft genannten Stelle zwischen den Fraunhofer'schen Linien b und F....
>
> Die Verdünnung, bei der die Fluorescenz in Urobilinlösungen erscheint, ist enorm. Lösungen, die im durchfallenden Lichte fast farblos sind, zeigen im auffallenden noch deutlich grünen Schimmer, namentlich wenn sie den directen Sonnenstrahlen ausgesetzt werden. 150

Like hydrobilirubin, urobilin also fluoresced intensely as its zinc salt, and its absorption band, apparently much sharper than urobilin itself, lay between b and F, but closer to b than F (*185*):

> Dieses Absorptionsband liegt, wie bereits angegeben . . . , zwischen den Linien b und F, aber der Linie b näher, als der Streifen der sauren Lösung (γ). – Es ist weit dunkler, schärfer begrenzt, als letzteres und bleibt noch bei den grössten Verdünnungsgraden sichtbar. 151

From the spectrum analyses of both hydrobilirubin and urobilin, *Maly* concluded in 1872 that they were identical. He was not again to publish results involving spectrum analysis until 1876, when he reinvestigated and clarified the reaction of bilirubin with Br_2 to give biliverdin, as he thought earlier, and a new blue pigment that he analyzed as the tribromo derivative (*177, 178*).

Maly's colorimetric spectral analysis of his bromobilirubin came to him courtesy of *von Vierodt*,[54] who during 1870–1881 modified and improved the *Bunsen-Kirchhoff-von Steinheil* spectroscope to incorporate a double collimator with adjustable slits used to calibrate the absorption of a sample to that of a reference, and used it to perform qualitative and quantitative studies of pigments in blood, bile, and urine. His instrument became a "standard" for nearly two decades (*184*). A dilute alcohol solution of bromobilirubin, which was blue, was found to transmit only green and blue light. With added NH_3 and a little $ZnCl_2$, the solution became grass-green, and its absorption spectrum showed two narrow, well-separated lines between 105 and 111 on a scale where Na is 120 (the Na *Fraunhofer* line is at 5892 Å) and Li is 102.5 (the Liβ line is at 6102 Å) (not the *Bunsen-Kirchhoff* scale), or exactly to the right of C, which corresponds to 6562 Å.

Of course, absent a quantitative characteristic such as the molar absorptivity (molar extinction) constant (ε) of the *Beer-Lambert* law and the wavelength at maximum absorption and bandwidth, the data from spectrum analysis were only marginally useful. Yet they furnished a potentially useful new characteristic for classifying or comparing bile pigments – and this spectroscopic method, like all others, would become better developed instrumentally, more exact in defining absorption characteristics, and more widely used. *Thudichum* also used the technique at about the same time as *Maly*.

In 1872 *Thudichum* published a *Manual of Chemical Physiology* in which he described experimental procedures for separating the components of bile and gallstones, including bilifuscin (probably $C_9H_{11}NO_3$) and bilirubin (or *Cholephäin,*

---

[54] *Karl von Vierodt* was born July 1, 1818 in Lahr, Baden and died on November 22, 1884 in Tübingen. He became Dr. med., a. o. Professor of theoretical medicine at the University of Tübingen in 1849, and in 1855 o. Professor and chair of physiology.

$C_9H_9NO_2$), from human gallstones. From the latter "in caustic or carbonated alkali exposed to the air for some days" (*186*), he prepared biliverdin ($C_8H_9NO_2$). And to an aq. NH_3 solution of bilirubin, he prepared blue *Cholecyanin* by adding conc. HNO_3 dropwise; whereas treatment of the solid directly with fuming or conc. H_2SO_4 led to the formation of green *Cholethalline*, *inter alia*. *Thudichum* provided spectra for each. It is unclear whether the spectra contain any information of diagnostic use, except that both samples are seen to absorb visible light strongly in the violet-blue region, and *Cholethalline* also absorbs strongly in the violet-green region.

Three years later, *Thudichum* published 13 spectra on the first page of a paper read before *The Chemical Society* and published in the May 1875 issue of the *Journal of the Chemical Society* (*158*):

Spectra referred to in this paper as diagnostic of certain Educts and Products.

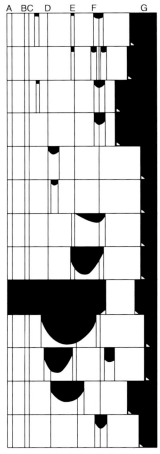

1. Zn precipitate (Jaffé's), from rheumatic fever urine. Dissolved in alcohol and H_2SO_4. Colour, yellowish-red.

2. Urerythrine, from brick-red deposit in rheumatic fever urine. Dissolved in absolute alcohol. Colour, fiery-red.

3. Product by H_2SO_4 from No. 4. Dissolved in ether. Colour of solution, red.

4. Urochrome, normal; by H_2SO_4 from lead precipitate. Dissolved in water and acid. Colour, yellow.

5. Omicholine by H_2SO_4 from extract of urine. Dissolved in ether. Colour, red. Fluoresces green.

6. Omicholic acid; accompanies 5; soluble in NH_3. Dissolved in ether. Behaves like 5.

7. Uropittin, from extract or urine and urochrome by H_2SO_4. Dissolved in alcohol. Colour of solution, red.

8. Hydrobilirubin (Maly's), from bilirubin by Na amalgam. Dissolved in alcohol. Colour of solution, red.

9. Bromo-bilirubin in alcohol and HBr. Colour of solution, deep blue.

10. Same as 9, more dilute. Colour, fine blue.

11. Bromo-bilirubin changed by BaH_2O_2 and HCl. Colour, rose-red.

12. Sulphate of bromo-bilirubin in alcohol. Colour, violet-blue.

13. The same as 12, changed by hyposulphite and HCl.

The various intensities of absorption observed are expressed by shadows between perpendiculars, in tenths of the entire height of each spectrum. The rationality of the distances of the spectral lines is the empirical one of the author's spectrometer, described on p. 192 *et seq.* of the 10*th Report of the Medical Office of the Privy Council*. 1867. [Redrawn from ref (*158*)].

The work illustrates that by 1875 spectrum analysis was becoming widely adopted and in *Thudichum*'s lab served to distinguish: (i) a variety of urinary pigments from each other and from *Maly*'s hydrobilirubin, and (ii) his blue, brominated bilirubin from these pigments and its reaction products. *Thudichum*'s blue pigment, a dibromo-bilirubin which he felt certain had the formula $C_9H_7Br_2NO_2$, was formed by exposing a weighed quantity of dry bilirubin in a watch glass to Br_2 vapor. When Br_2 uptake had ceased (the weight of the bilirubin had tripled and no longer changed), *Thudichum* determined the ratio of the increase in weight due to bromine to the original weight of the bilirubin and took it to be the same as the ratio of the atomic masses of $2 \times Br - 2 \times H_2$ (or 158) to the atomic mass of bilirubin, which was thereby determined as 162.4. Thus, *Thudichum* felt he had accomplished an experimental determination of the molecular weight of bilirubin as 162.4, or very close to the 163 deduced by all other of his experiments. Hence the $C_9H_7Br_2NO_2$ formula. This seems rather like an attempt to fit an experimental result to a previously determined molecular weight value. With the assumption of an uptake of 2 Br_2, the equation used would have predicted 1.1942 (increase due to bromine) : 1.2280 (weight of dried bilirubin) = 316 (4 Br – 4 H) : 325, or twice the molecular weight assigned by *Thudichum* and thus a formula $C_{18}H_{18}N_2O_4$, or close to *Maly*'s first proposed $C_{16}H_{18}N_2O_3$ (mol. wt. 286), but not the doubled formula. *Thudichum*'s dibromo-bilirubin was violet with a golden luster.

2.11 Bilirubin Polemics of the 1870s

Thudichum took issue with *Maly*'s research, chiding him for the early (incorrect) belief that biliverdin was the amide of bilirubin (*150, 164, 165*), which *Maly* had recanted some seven years prior (*166, 167*). More to the point of bromination, *Thudichum* took issue with *Maly*'s earlier belief that bilirubin was oxidized to biliverdin by Br_2 (*166, 167*). This too *Maly* had corrected in a subsequent paper (*171, 172*). Yet, perhaps unaware of the correction, *Thudichum* wrote a pointed yet valid criticism (*158*):

The change is explained as oxidation, and in absence of any proof whatever, a somewhat analogous reaction is adduced to make the assumption probable; a reaction, however, the nature of which is as unknown as that which it has been called to illustrate.

In contact with bromine vapour and moist air, bilirubin perhaps turns green for an instant, namely as long as the orange powder is able to send yellow rays through the blue compound, which quickly covers its surface. But often as I have repeated the experiment, it has had the same result in moist as well as dry air; never has there been formed a matter or colour similar to biliverdin, but always the brominated products above described.

Further, if the green colour produced in the chloroform solution of biliverdin by bromine had been due to biliverdin, the latter must have been precipitated, as it is insoluble in chloroform. The green colour, according to my explanation, was simply a mixture of the yellow of the original solution, with the blue of the brominated product. The dark blue when once obtained remains unaltered for weeks, a good proof of the difference of this reaction from that of Gmelin, in which the blue produced by nitrous acid is of the most transient nature. The spectroscopy easily shows that the two blues are due to entirely different chemical entities. Even the blues produced by nitrous nitric acid in different bile-colouring matters are different. Their different spectra were originally observed and described by me in 1866 and 1867, in the 9th and 10th *Report of the Medical Officer of the*

Privy Council. See the latter volume, p. 251 to 260. Cholocyanine; its sulphate; sulphate of sulpho-cholocyanine; and hyocoerulin. Therefore in reactions with bile-colouring matters a blue colour is no more a proof of identity than a green.

Note *Thudichum*'s reference to spectroscopy as a means of distinguishing the blue bromination product from the blue pigment of the *Gmelin* reaction.

Maly's work was not alone in *Thudichum's* gun sight, and he did not spare any criticism of the *Bilicyanin* that *Heynsius* and *Campbell* obtained by what they called an oxidation of bilirubin by bromine water (*153, 154*) or the greenish *Choleverdin* of *Stokvis* (who later declared it identical to *Bilicyanin*). To these gentlemen he issued a stern rebuke (*158*):

> A most elaborate account of the alleged oxidation-products of bile-pigments and their absorption-bands was published by A. Heynsius and J.F.F. Campbell, in *Pflüger's Arch. f. Physiol.*, iv, 497-547, extending over fifty pages. A blue substance, *bilicyanin*, was obtained by what is termed the oxidation of bilirubin by bromine-water. The spectra obtained varied, as also did the solubilities of the products. Not a single product was isolated, and none was analysed. It is easy to see that these products were principally mixtures of the mono- and dibrominated bilirubin. Of oxidation there is no evidence whatever. The same remarks apply to a greenish product, obtained formerly by Stokvis, and termed choleverdin, which, after perusal of the paper just quoted, he declared to be identical with and thenceforth termed bilicyanin (*Neues Report. f. d. Pharm.*, 21, 732-737). Without entering into any detailed discussion of these discursive papers, which relate merely to experiments made with dilute impure solutions in test tubes, and do not start with any pure substance, nor arrive at any stoïchiometrical conclusion, I hope that the following conclusions will be acceptable to the reader.

And he clarified all of the alleged bromine-induced oxidations of bilirubin as nothing more than brominations (*158*):

> The allegation made by Maly, that bilirubin under the influence of bromine was converted into biliverdin is unfounded.
>
> The allegation made by Maly, Heynsius and Campbell, and Stockvis, that bilirubin under the influence of bromine yielded products of oxidation, is unfounded.
>
> The products obtained by his halogen are not products of oxidation but of substitution.

It was not just misinterpreted bromination reactions of bilirubin that attracted *Thudichum*'s attention and drew his response. For he also keyed in on *Maly*'s hydrobilirubin and *Maly*'s belief that he had transformed bilirubin into the coloring matter of urine, that there was a probable relationship to *Jaffe*'s urobilin (*158*):

> Maly (*Ann. Chem. Pharm.*, 1872, No. 7, p. 77) claims to have transformed bilirubin into the colouring matter of urine, "at least," he says, qualifying considerably his general title, "that kind of urinary colouring matter which according to Jaffé, is the best defined." Now although Jaffé has extracted from urine, by means of zinc oxide, a mixture of at least two of the decomposition products of urochrome, and has described their spectral phenomena, long since and originally published by me, as if they belonged to a single body, and as if they were new discoveries, yet he has not isolated a single pure substance and has not instituted a single elementary analysis.
>
> At first sight, therefore, the metamorphosis announced by Maly was extremely improbable to any one acquainted with the chemical bearing, composition, and physical qualities of the bodies in question. But the spectroscopic identity of the products of Maly and Jaffe [sic] was announced with such assurance, that I felt it my duty to repeat some of the relative experiments of these authors.

2.11 Bilirubin Polemics of the 1870s

Repeating *Maly*'s preparation of (reddish) hydrobilirubin by reduction of bilirubin using Na(Hg), *Thudichum* found the same product as *Maly*, unchanged from the first half of the reaction to one lasting two days. He disputed the identity of hydrobilirubin as neither urobilin nor urochrome based on the results of his spectrum analysis and, even better, by the fact that hydrobilirubin is insoluble in water whereas urochrome is soluble. (Urobilin was an educt, found only in the urine of feverish patients after standing. *Urochrome* was the term coined by *Thudichum* for the matter to which urine owed its yellow color; it was not the chromogen of urobilin.) Further investigation simply reaffirmed his conclusions. Not one to eschew analysis and criticism of hydrobilirubin, which *Maly* believed to be a tribasic acid, *Thudichum* also objected to *Maly*'s formula ($C_{32}H_{40}N_4O_7$) for it and the sparse supporting evidence: it formed only one silver compound that analyzed for 35.75% Ag, and one zinc compound that analyzed for 14.2% Zn; whereas other analyses of the former yielded 37.1% Ag and up to 37% Zn for the latter. He was very plainly unconvinced of its formula and remained emphatic that hydrobilirubin and *Jaffe*'s urobilin were not at all the same (*158*):

> These data therefore do not afford the means for determining either the atomic weight or the basicity of the new product, but seem to show that the sodium-reaction produces a variety of new products, which remain partly mixed in the precipitate, partly in the mother-liquor from which it falls. For this liquid remains red, and retains a considerable quantity of a by-product.

Thudichum further cast doubt on *Jaffe*'s urobilin, which was obtained by *Jaffe* only from pathologic urine, noting that the pigment had been diagnosed by *Jaffe* only by a spectroscope and was never isolated. In his attempt to isolate urobilin, *Thudichum* found that it separated into a mixture of urochrome and "urerythrin", the latter also found dissolved in fresh urine to which it imparted a reddish-yellow color. He refuted any similarity between these pigments and hyrobilirubin (*158*):

> The following are the *irreconcilable differences* between hydrobilirubin on the one side, and urochrome and all its products and urerythrin on the other side.
> Urochrom is yellow, soluble in water, shows narrow faint band in acid mixture, none in neutral or alkaline solution.
> Hydrobilirubin is brownish red, insoluble in water, easily soluble in watery acid, and in alcohol with deep red colour, and spectrum differing entirely from urochrom.
> Urochrom, when concentrated enough to show any band, is by boiling with acids immediately split up into omicholin, uropittin, and uromelanin, each of which products can be separated out and recognized with the greatest ease either spectroscopically or by chemical tests.
> Hydrobilirubin is not altered by boiling with acids in any characteristic manner; it is certainly not split up, and yields not one of the products of urochrome.
> Hydrobilirubin is, therefore, not identical with, or even similar to, the urinary colouring matters, including urerythrin. Its only similarity spectrally is to uropittin, but the general differences between the two bodies are striking.

And then he moved on to stating his objections to *Städeler*'s newly expressed hypothesis on the theory of bilirubin and its metal salts – apparently the last commentaries by *Städeler* before his death. It seems that *Städeler* had read of the many studies of bilirubin and its salts reported in 1868 (*152*), and he attempted to bring them into harmony with his own concepts of bilirubin, as revised to accommodate the new data. It did not sit well with *Thudichum*. Although *Städeler*'s very early elemental combustion analyses of bilirubin found favor with *Thudichum*, who found it coincident with his own but at odds with *Städeler*'s new hypotheses – hypotheses that included the doubling of the bilirubin formula to $C_{32}H_{30}N_4O_6$ and assertion that it had six replaceable hydrogens to accommodate metal salt formation drew special ire (*158*):

> I do not believe that this hypothesis has any foundation in fact. Not a single formula of Städeler's, and not a single element of any formula, can be derived from my analyses. . . .
> ... I hold the hypothesis of Städeler to be merely not proved by facts, but to be directly disproved by all my analyses, with exception . . .

Yet *Maly* was a special early whipping boy. An apparently exasperated *Thudichum* felt compelled in 1876 (*161, 162*) to write "An open letter to the Imperial Academy of Sciences at Vienna, containing an examination of the researches on the colouring matter of bile, by *Richard Maly*, of Graz." This polemic was in response to what apparently became (or caused) *Maly*'s final publication on bile pigments in 1875 and is illustrated in the following excerpts from 14 of the 40 very detailed points of *Thudichum's Offenes Sendschreiben* (*162*):

> 8. That Prof. Maly had received the letter containing the foregoing passages is proved by the reply which he addressed to me, dated from Innsbruck, June 24 (1874), now before me. Prof. Maly, therefore, before he began the experiments which are so exhaustively described in the fifth paper, was not only informed of his error, but actually in possession of the key to his alleged discovery, and it was therefore impossible that he should have been led to this discovery by his experiments.
>
> 9. ... That these researches and publications should have remained unknown to the editor of an annual report on the progress of animal chemistry is not impossible, but that he excluded the contents of my letter from the circumference of the "usual duty" admits of only *one* explanation, but not of justification.
>
> 10. ... This description leaves the main points which have been established by my researches entirely out of consideration, and in all particular statements it is completely incorrect. Indeed I can hardly believe that Prof. Maly has read my paper; it is certain he has not understood it.
>
> 16. ... All these necessary precautions Prof. Maly has neglected, and in consequence has arrived at conclusions which have no foundation.
>
> 17. ... Prof. Maly further endeavors to influence the judgment of the Academy by raising doubts in general regarding my experiments; first, on the ground that I had performed each experiment only once; secondly, because I had not analysed the final product. Against these objections I maintain that the above experiment, considered by the light of my former researches in the *Journ. d. Pract. Chem.* (civ., 193), requires no further analysis. I thought and think every analysis of the product to be a mere waste of time, – every repetition on my part a waste of labour and material. However, in order to meet the object, and from a high regard for the Academy, I have repeated the experiment described under 15, yet two several times, and have analysed the products by determining quantitatively the amounts of carbon, hydrogen, nitrogen, and bromine. . .
>
> 21. ... In making this statement Prof. Maly loses sight of "the usual duty of characterising previous knowledge," or other knowledge. . . .
>
> 22. ... In the letter alluded to, Städeler, in view of my researches, abandons all his former formulæ, and coerces my results by an utterly unjustifiable process of re-calculation, in which no single analytical result harmonises with the new hypothesis into some sort of support for his doubled formula and hexa-basic acid hypothesis, without having produced a single compound or made a single new analysis.

2.12 Conjectural Chemistry and Bilirubin Polemics at the Close of the 19th Century 161

23. Prof. Maly causes to himself many difficulties by his preconceived opinions and uncontrolled imagination, as I am obliged to prove now more in particular. . . .

25. ... How can an author who works with such preparations call others to account for the alleged impurity of their preparations!

27. The Academy may justly demand of me to prove these statements. I am ready, on receiving a request to that effect, to communicate to the Academy details, the extent of which are excluded from the present letter on account of their length. . . .

30. ... On the contrary, it must be maintained that such results and corollaries are directly opposed to the principles of chemical science, and slap the endeavor for final accuracy rudely upon the face.

32. The observation of the influence of sodium amalgam upon bilirubin, which led Prof. Maly to the discovery of the so-called hydro-bilirubin, would have been an interesting progress in our knowledge concerning bilirubin. But as the author starts from erroneous views regarding the composition and molecular weight of bilirubin, his conclusions regarding his product and its composition, and regarding the formula of the change, are necessarily erroneous. . . .

39. It is impossible here to point out all the irrelevant and erroneous detail with which Prof. Maly surrounds his faulty observations. . . .

40. I conclude my letter to the Imperial Academy with the expression of the deepest regret concerning the circumstances which have compelled me to write it. I should not be able nor dare to molest the Academy a second time with this matter, and I therefore pray the Academy to excuse the length and serious tone of this letter, with the importance which the matter has for me, for science, and for the maintenance of the ethical rules which govern the intercourse or cultivators of science. I hope that the Academy will give to my letter no less publicity than it has given to the papers which have called it forth.

2.12 Conjectural Chemistry and Bilirubin Polemics at the Close of the 19th Century

The last quarter of the 19th century brought new investigators into the bile pigment field, most with medical-physiological interests and sophistication but with an incomplete understanding or knowledge of the earlier chemical studies and errors therein. Even as he neared the sunset of his long life, *Thudichum* had not abandoned his penchant for "setting the record straight", while apparently retaining his intellectual vigor and keen memory. Three-quarters of the way into the 19th century he believed he had settled some of the important problems associated with bilirubin, its purification, combustion analysis, controversial formula, "spectrum analysis" characteristics, the controversy with *Maly* over the reaction products with Br_2 and even the non-equivalence of its Na(Hg) reduction product (*Maly*'s hydrobilirubin) with urobilin, *Jaffe*'s purported urinary pigment – that *Thudichum* had discounted as such, *etc.* So 20 years later, after having turned his interests over to research on brain chemistry during the previous two-plus decades, it must have come as something of a shock to him to discover that the error-laden publications of others in the 1870s were being cited to support work in the 1890s.

Thudichum responded forcefully in print, from 1896 to 1899 (*187–190*), by pointedly citing where and how the authors had been led astray by earlier errors (especially *Maly*'s) – that the new authors were basing their work on the *conjecture* of

162   2 Early Scientific Investigations

others, unsupported by high quality experiments. In apparent exasperation with the extent to which errors had permeated the literature and were being promulgated uncritically, in 1900 he coined the word *Conjecturalchemie* (conjectural chemistry) (*191*) – a term he used to chastise researchers for the propagation of their (defective) conclusions or statements based on *supposition*, not fact, and previously found (by *Thudichum*) to be deficient of firm experimental verification. Thus *Thudichum* responded forcefully to a long, comprehensive article in 1893 by *F. Grimm*, a physician in Berlin, on the urobilin of normal patients and those with a wide variety of pathologies (*192*). *Grimm*'s article, which contained references to the studies of *Jaffe, Maly, Hoppe-Seyler*, and others, but no reference to *Thudichum*'s earlier work on the pigments of urine summarized in his treatise on the same (*193*) clearly provoked *Thudichum*. In 1897, he indicated that when he referred to urobilin it was the pigment isolated from urine by *Jaffe*'s process; whereas, most of the later reports on urobilin related to the product obtained by a different process. On this basis, he stated that the pigment isolated from human feces is not urobilin but the intestinal *Lutein* that he had reported earlier and was in no way identical to the compound obtained from urine – and all reports on its identity (with urobilin) were in error (*188*):

> Also ist dieser Körper in den Fäces, der übrigens nie isolirt worden ist, von dem aus Harn gänzlich verschieden, und alle Angaben über Identität u. s. w. sind irrthümlich.   152

Thudichum again did not hesitate to reprimand (the deceased, 1891) *Maly* for having indicated that *Jaffe*'s urobilin was identical to his hydrobilirubin obtained from Na(Hg) reduction of bilirubin. He declared it absolutely erroneous: "Auch dies ist ein absoluter Irrthum" (*188*). Then he proceeded to critique *Grimm*'s work, which used mainly spectrum analysis to correlate urobilin with the various pigments that he had isolated from urine, for assuming that hydrobilirubin and urobilin were identical, and for naming the product from urine hydrobilirubin. But 14–20 years earlier *Thudichum* had proven them not to be identical, saying that the proof is entirely indisputable, and thus cannot even be contested: "Der Beweis ist ganz unanfechtbar, und daher auch nicht angefochten worden" (*188*). He declared that using the name *hydrobilirubin* for any product of urine, that hypotheses from that related to the transformation of bilirubin, and that physiological and pathological speculation based on it were absolutely in error. And he decried the use of spectrum analysis as the only means of identification, *etc.* However, shortly thereafter, in 1898 (*189*) *Thudichum* had to be pleased with the proof, based on the combustion analysis by *Hopkins*[55] and *Garrod*[56] (*189*).

---

[55] *Sir Frederick Gowland Hopkins* was born on June 20, 1861 in Eastbourne, Sussex, and died on May 16, 1947 in Cambridge, UK. He taught physiology and toxicology at Guy's Hospital, London, from 1894-1898, became Reader in chemical physiology at Cambridge University from 1902-1914, then professor from 1914, and in 1929 was awarded the *Nobel* Prize in Physiology or Medicine (with *Christian Eijkman*) for the discovery of vitamins.

[56] *Sir Archibald Edward Garrod* was born on November 25, 1857 in London, and died on March 28, 1936 in Cambridge, UK. He was a physician who saw dynamic biochemistry in metabolic pathways, and recognized Mendelian heredity as an explanation for inborn errors of metabolism (albinism, alkaptonuria, cystinurea, and pentosurea).

2.12 Conjectural Chemistry and Bilirubin Polemics at the Close of the 19th Century

Hopkins and *Garrod*'s analyses (*194*) based on *Maly*'s hydrobilirubin and urobilin isolated from various human sources: normal and pathological urine, feces, and bile from the post-mortem gallbladder, although resembling each other in certain properties, were very clearly different in the %N (*194*):

SUMMARY OF RESULTS.

	Urinary Products				Fecal Products	
	No. 1	No. 2	No. 3	No. 4	No. 1	No. 2
C	63·69	–	–	63·24	–	63·81
H	7·73	–	–	7·60	–	8·20
N	4·02	4·22	4·05	4·09	4·17	–

	Urobilin	Hydrobilirubin		
	Mean of above results	Theory	Mean of Maly's results	Our estimation of nitrogen in hydrobilirubin
C	63·58	64·86	64·68	–
H	7·84	6·75	6·93	–
N	4·11	9·45	9·22	9·57
O	24·47	18·94	19·17	–

[Where the urinary products are from: No. 1, a patient with hepatic cirrhosis; No. 2, a patient with pernicious anemia; No. 3, a patient with intestinal obstruction; No. 4, mixed urines of hospital patients in surgical wards. And the fecal products are from: No. 1, stools of a case of typhoid fever in the early convalescent stage; No. 2, normal feces].

Despite an expressed "uncertainty with regard to the question of ash", the clear difference between *Maly*'s hydrobilirubin and natural urobilin, *Hopkins* and *Garrod* still held the belief that they shared a relationship (*194*):

> We may be permitted to say that we entered upon the analysis of urobilin obtained from natural sources in the hope that our results might help to place upon a firmer foundation the belief, which has prevailed since the publication of Maly's results, that there exists a *simple* relationship between that pigment and bilirubin. This hope has not been justified by the results, and we are convinced that the relationship is by no means so simple as has been supposed. The change from bilirubin to urobilin cannot be a mere question of reduction and hydrolysis, but must necessarily be attended by a removal of nitrogen; of this our analyses leave no doubt whatever.
>
> On the other hand we cannot doubt that the one pigment is actually derived from the other, a conclusion which evidence of other kinds appears to us to render unavoidable.

The data also pointed to *Hopkins* and *Garrod*'s conclusion that the urinary and fecal urobilins are identical. *Thudichum* had earlier (*193*) objected to this, a then unproven prospect, and broadened his 1898 report on urobilin to include comments on his urinary urochrome, *Omnicholin*, *Urorhodin*, and *Uropittin*. He commented that *Hopkins* and *Garrod* had not come up with a formula for their analysis of urobilin (they said they did not feel themselves in a position to attempt to assign an empirical formula (*194, 195*)) – but *Thudichum* was less inhibited and gave $C_{18}H_{25}NO_5$ (*189*). In fact, *Hopkins* and *Garrod* indicated that (*194*):

> The figures obtained do not appear to lend themselves to a formula showing any simple relationship to that accepted for bilirubin, and until experiment has shown by what chemical steps a product strictly agreeing in its general characters with natural urobilin can be prepared from bile pigment it is undesirable to pursue the question of its constitution.

Hopkins and *Garrod* brought forth (*194*) several interesting points related to bilirubin metabolism. By allowing Na(Hg) to act upon bilirubin beyond the stage specified by *Maly*, the product resembled natural urobilin more closely (*194*):

> Passing on to the consideration of the further question we may say at once that the results which we have obtained by allowing the action of sodium amalgam to proceed further agree closely with those of Disqué and Eichholz. As the action proceeds the liquid assumes a pale yellow colour, the extra alkaline bands disappear and the precipitability of the urobilin-like product by hydrochloric acid is conspicuously diminished. When acidified, filtered and exposed to the air the liquid darkens and the absorption band gains in intensity. The product so obtained bears a far closer resemblance to the natural pigment than Maly's hydrobilirubin does.

And of seemingly greater importance, because it almost certainly showed a relationship between bilirubin as a metabolic precursor to urobilin, as is understood today (*194*):

> It is a well-known fact that in health the bile pigment which enters the duodenum disappears, as such, before the intestinal contents are expelled, and in its place we find in the fæces urobilin and its chromogen.
>
> When, as in certain cases of typhoid fever, the bile pigment is found in abundance in the fæces, the urobilin is greatly diminished in quantity or altogether wanting. When the flow of bile into the intestine is arrested urobilin and its chromogen disappear from the fæces, to reappear when the patency of the bile ducts is re-established.
>
> Friedrich Müller . . . has further shown that when bile is introduced into the stomach of a patient with complete biliary obstruction and whose fæces are urobilin-free, urobilin appears in the stools.

Thudichum had less patience with the 1894 publication of *Jolles*[57] on the oxidation of bilirubin to biliverdin using I_2 (*196*). Apparently the lessons associated with the *Thudichum-Maly* (one-sided) polemics of the 1870s had become unlearned by the last decade of the 19th century, a time when there appeared a renewed interest in bilirubin and related pigments. In 1894, *Jolles* published a very long paper on a quantitative method for determining bilirubin in bile using I_2 as an oxidant (*187*). In this work, he described, *inter alia*, his study of the oxidation of bilirubin to biliverdin using a dilute alcoholic solution of I_2, for which he

---

[57]*Adolf Jolles* was born on November 9, 1862 in Warsaw and died on November 13, 1942 in Theresienstadt. He was an Austrian chemist and in 1894 a young docent at the k.k. technologischen Gewerbsmuseum in Vienna. *Jolles* had a long and productive career as an analytical/medicinal chemist in Vienna with numerous publications that brought recognition for his urinary tests (determining bile pigments and albumin in urine), detecting "hematoporphyrin" in the urine of patients with drug-exacerbated porphyria, his studies of fats, and a test for pigments (*Jolles'* test), *etc.*

2.12 Conjectural Chemistry and Bilirubin Polemicsat the Close of the 19th Century 165

found precedent in *Maly*'s published oxidation by halogens. *Jolles* wrote a chemical equation describing the interconversion of *Maly*'s and *Städeler*'s formulas for bilirubin and biliverdin (*196*):

$$C_{32}H_{36}N_4O_6 + 4\ J + 2\ H_2O = C_{32}H_{36}N_4O_8 + 4\ HJ$$

The work caught the eye of *Thudichum*, who despite his advanced age and having earlier redirected his research to a study of the chemical constituents of the brain, must have been surprised or perhaps even shocked to realize that his important studies on the halogenation of bilirubin had gone unread or unappreciated and that his rejection of the *Maly* and *Städeler* formulas had gone unrecognized. After repeating *Jolles*' I_2 reaction in $CHCl_3$ and with I_2 vapor and finding no reaction or no biliverdin among the reaction products, but also providing scant experimental details, he issued a stern rebuke in 1896 (*187*). The complaint was that *Jolles* based his work on the disproven results of *Maly*, published in 1868 (*166, 167*), that bilirubin is oxidized to biliverdin by Br_2, which was proven by *Thudichum* to be a substitution reaction (hydrogen for bromine) some 20 years earlier. He chastised *Jolles* for overlooking *Maly*'s correction in 1875 (*171, 172*), where, prompted by *Thudichum*, *Maly* had conceded that the reaction of bilirubin with Br_2 was not an oxidation but a substitution. He objected to *Jolles*' not having isolated or analyzed the reaction products (*187*):

> Der Aufsatz des Hrn. Jolles beginnt mit der ganz unbedingten Angabe, dass das Bilirubin durch (eine alkoholische Lösung von) Jod in Biliverdin verwandelt werde. Diese Behauptung ist indessen nur eine aus der angeblichen und widerlegten Reaction des Broms hergeleiteten Analogie und fällt deshalb mit ihrem Muster. Es ist auch gar kein Versuch gemacht, die angebliche Reaction zu begründen, es sind keinerlei Produkte isolirt oder analysirt worden. Ich könnte mich daher mit dem Resultat begnügen, dass, da die Prämissen des Hrn. Jolles gar nicht existiren, seine Thesen nothwendiger Weise dasselbe Schicksal haben. 153

Thudichum went further, with point by point admonishments directed toward *Jolles*' published work. He objected to *Jolles*' spectroscopic characterization of biliverdin as the product of I_2-promoted oxidation of bilirubin, finding a mismatch between *Jolles*' product and authentic biliverdin (*192*):

> Spectroskopische Angaben. Auf S. 3 seiner Abhandlung sagt Hr. Jolles, er habe die Identität seines grünen Produkts, aus Galle oder Bilirubin durch Jod erhalten, mit reinem Biliverdin spectroskopisch bewiesen. Nun hat aber das aus Bilirubin durch Einfluss von Soda als Lösungsmittel und Luft als Oxydationsmittel dargestellte Biliverdin keine specifischen Absorptionsschatten in seinem Spectrum. Daraus allein folgt, dass, da Jolles' grüne Produkte, durch Jod aus Galle oder Bilirubin erhalten, solche Absorptionen zeigen, diese Produkte nicht mit Biliverdin identisch sind. 154

He objected to *Jolles*' use of unproven formulas and especially for an unproven reaction, bilirubin + I_2 " biliverdin (*187*):

> Unbegründete Formeln. Damit verschwindet die in der Abhandlung verschiedentlich wiederholte Formel, wonach eine Molekül sogenannten Bilirubins 4 Atome Jod und zwei Molekül Wasser zur Oxydation zu Biliverdin erfordern und aufnehmen solle. Da die Reaction überhaupt nicht existirt, so müssen die den Pigmenten zugeschriebenen Formeln ungültig sein. 155

166   2 Early Scientific Investigations

Thudichum was especially angered that *Jolles* would use a bilirubin formula ($C_{32}H_{36}N_4O_6$) ascribed to *Maly* and *Städeler* because, as he stated, *Maly* had determined no formula for the pigment, had not once completely analyzed (by combustion) the pigment – and especially the %N had not been determined or weighed. *Städeler* had determined a different formula ($C_{16}H_{18}N_2O_3$) to bilirubin then doubled it in 1870 in order to establish it as a hexabasic acid, which it is not. *Thudichum* had no kind words on this sore point (*187*):

> Herr Jolles wiederholt die irrige Angabe, Maly und Städeler hätten die Formel des Bilirubins als $C_{32}H_{36}N_4O_6$ „bestimmt". Allein Maly hat überhaupt keine Formel für Bilirubin bestimmt; er hat es nicht einmal vollständig analysirt, und insbesondere den Stickstoff seines Präparats weder gemessen noch gewogen. Er war daher gar nicht in der Lage, eine Formel zu berechnen. Nur Städeler hatte dem Bilirubin die Formel $C_{16}H_{18}N_2O_3$ beigelegt, dieselbe aber um 1870 verdoppelt, um dasselbe als eine sechsbasische Säure darstellen zu können. Dieser ganz ungerechtfertigte Versuch ist vollständig misslungen.   156

Thudichum continued, unrelentingly, taking *Jolles* to task on (i) the latter's assertion that cattle bile contains no bilirubin; (ii) the latter's spectra of bilirubin and biliverdin and failure to recognize that he (*Thudichum*) had studied the reaction of bilirubin with I_2 and Br_2 much earlier; (iii) *Jolles*' belief that *Choletelin* is the endproduct of the reaction of bilirubin with Br_2, which *Thudichum* claimed to have shown to be invalid and that choletelin came from reaction with HNO_2 and not from Br_2; (iv) *Jolles*' stating incorrectly that bile contains lecithin and making incorrect comments on urobilin, which *Thudichum* said was long ago disproved – ox bile contains no lecithin but a phosphatide with four nitrogen atoms, and the urobilin statements (that swine bile contains relatively high amounts) were completely refuted in 1875.

Using his 1896 publication (*187*) as a vehicle for more corrections, *Thudichum* did not fail to remind that the Italian Professor *Capranica* incorrectly stated that a $CHCl_3$ solution of bilirubin gave biliverdin upon exposure to sunlight, that it gave only chlorinated products. (This, however, was later shown to be untrue; the reaction does in fact yield some biliverdin.) Not one to let "sleeping dogs lie", *Thudichum* resurrected the ancient history of how *Städeler* reached his formulas for bilirubin and explains for *Jolles*' benefit why they are incorrect and indirectly admonished him for not having recognized it. *Thudichum* accepted *Städeler*'s earliest bilirubin formula ($C_9H_9NO_2$) as the only correct version. It matched his own. He disavowed the later *Städeler* formula ($C_{16}H_{18}N_2O_3$), which was based on the neutral calcium salt of bilirubin ($C_{32}H_{34}N_4O_6$, from $2(C_{16}H_{17}N_2O_3) + Ca$) and what *Thudichum* described as its questionable calcium determination of 9.1% Ca. From his own studies of bilirubin calcium salts, in which he found both a neutral salt and a half acid salt, *Thudichum* cited that the latter ($C_{27}H_{29}N_3O_8Ca$, based on *Städeler*'s $C_9H_9NO_2$ for 3 × bilirubin + Ca(OH)$_2$) theoretically has 7.1% Ca (*Städeler* found 6.5%); in contrast the neutral salt ($C_{18}H_{20}N_2O_6Ca$) yields 10% Ca – and thus cannot be identical to *Städeler*'s. *Thudichum* claimed that after his own investigations were published *Städeler* gave up on the second bilirubin formula in favor of its doubled formula ($C_{32}H_{36}N_2O_6$) without carrying out a single experiment or analysis. He noted that there is no correlation between the last *Städeler* formula and his

2.12 Conjectural Chemistry and Bilirubin Polemics at the Close of the 19th Century

preparation and analysis and attributed *Städeler*'s changing formulas to desperation and the result of following an incorrect calcium analysis (*187*):

> Nach der Veröffentlichung meiner Untersuchungen nun gab Städeler auch diese zweite Formulirung des Bilirubins auf, und damit natürlich alle anderen Formeln seines Biliverdins, Biliprasins und Bilfuscins, verdoppelte seine contrahirte Formel für Bilirubin zum zweiten Mal, auf $C_{32}H_{36}N_4O_6$, und erklärte es für eine sechsbasische Säure; dazu hatte ihn vielleicht die damals eben gemachte Entdeckung der Honigsteinsäure verleitet. In diese Hypothese suchte er nun meine Resultate einzuzwängen, ohne eine einzige Darstellung oder Analyse auszuführen. Keine einzige der Formeln Städeler's und kein einziges Element irgend einer seiner Formeln kann aus meinen Präparaten und Analysen abgeleitet werden. Die von ihm berechneten Metallmengen der Verbindungen betragen alle von 1% bis zu 6% weniger als die von mir bestimmten Mengen. Ich kann dieses Verfahren Städeler's nur als ein Resultat der Verzweiflung an allen seinen Arbeiten bezeichnen. Nach meiner Ueberzeugung waren seine ersten Präparate rein und sein Analysen richtig, obwohl er sie selbst, ohne Erklärung, irriger Weise verleitet durch eine trügerische Kalkbestimmung, aufgegeben hat. 157

Unlike *Maly*, who seems to have been chased out of the bilirubin arena by *Thudichum*'s forceful dismantling of his work, *Jolles* did not back down and in 1899 published a polite but assertive rejoinder (*197*). In this he focused almost exclusively on *Thudichum*'s main point: that treatment of bilirubin with iodine could not lead to an oxidation of the pigment (to biliverdin) but caused only a substitution reaction. He expressed astonishment that *Thudichum* had not carefully read his earlier paper of 1894 (*196*). For *Jolles* insisted that he had not (as *Thudichum* wrote) cited *Maly*'s work on the oxidation of bilirubin to biliverdin, published in 1868, in support of his own work, and that he also had not failed to recognized *Maly*'s revocation, in 1872, of the 1868 work which suggested that altered reaction conditions could lead to different results/products (*197*):

> Bevor ich auf den sachlichen Inhalt der Einwendungen des Herrn Thudichum näher eingehe, muss ich zunächst meinem Erstaunen darüber Ausdruck geben, dass Thudichum meine Abhandlung anscheinend nur ganz flüchtig durchgelesen hat, indem er ganz willkürlich und unberechtigt von meiner Arbeit angiebt, dass sie auf die angebliche Entdeckung des Prof. Maly aus dem Jahre 1868, wonach das Bilirubin durch Brom vermittelst eines Oxydationsprocesses in „Biliverdin" verwandelt werden sollte, basire. Diese Unterschiebung muss ich aber mit Entschiedenheit zurückweisen, denn ich habe die Untersuchungsergebnisse Maly's nicht als Beweismittel für die Richtigkeit meiner Resultate herangezogen, sondern ich habe nur – wie üblich – in der Literatur, soweit dieselbe meine Arbeit zu tangiren schien, die von Maly im Jahre 1868 publicirten Ergebnisse anzuführen mich für verpflichtet gehalten. Herr Thudichem irrt sehr, wenn er glaubt, dass mir die zweite Arbeit Maly's vom Jahre 1872, in welcher er die erste Arbeit widerruft, nicht bekannt war. Aber der Umstand, dass Maly bei seiner zweiten Arbeit ganz andere Versuchsbedingungen eingehalten hat, die für die Entstehung von bromirten Substitutionsprodukten günstiger gewählt waren, und es als selbstverständlich vorausgesetzt werden kann, dass die Aenderung der Versuchsbedingungen gerade bei der Einwirkung von Halogenen zu ganz anderen Ergebnissen führen könne, sowie andererseits der Umstand ... 158

To *Jolles*, the issue was not *Städeler*'s formulas or *Maly*'s or *Thudichum*'s, and it was certainly not the bromination reaction investigated by the latter two; it was *Thudichum*'s insistence that *Jolles*' oxidation of bilirubin to biliverdin by I_2 could

168   2 Early Scientific Investigations

not happen. In rebuttal, *Jolles* indicated that the results from treating bilirubin with Br_2 was a poor model for reaction with I_2, that he had shown that, in dilute solutions, a molecule of bilirubin reacted with four atoms of iodine to produce a green pigment that he characterized as biliverdin – and that process could be used to detect bilirubin in animal bile and (quantitatively) in urine, which was the reason for his original study (*197*):

> ... dass ich in meiner Arbeit ja keinen anderen Zweck verfolgt habe, als eine Methode bekanntzugeben, welche gestattet, das in den thierischen Gallen enthaltene Bilirubin quantitativ zu bestimmen, hat es mir als überflüssig erscheinen lassen, auf die zweite Maly'sche Arbeit in dem Litteraturverzeichnisse hinzuweisen. Die Quintessenz meiner Arbeit war ja doch nur die, zu zeigen, dass bei Einhaltung bestimmter Versuchsbedingungen Bilirubin quantitativ in einen grünen Farbstoff übergeführt wird, wobei auf 1 Mol. Bilirubin 4 Atome Jod verbraucht werden, so dass dieser Process ein Mittel an die Hand giebt, den Bilirubingehalt in den thierischen Gallen und in Harnen quantitativ zu bestimmen. Dass diese meine Methode den Zweck erfüllt, beweisen meine zahlreichen Beleg-Analysen, die bisher noch Niemand widerlegt hat, und thatsächlich haben bereits anerkannte Handbücher, wie Huppert in der 10. Auflage seiner Anleitung zur qualitativen und quantitativen Analyse des Harns (S. 865) die Methode zur annähernden quantitativen Bestimmung des Bilirubins im Harne empfohlen.   159

Thudichum viewed the green pigment as simply an iodinated substitution product of bilirubin, much as the reaction with Br_2 gave bromine substitution; *Jolles* disputed the first statement, not the last, and stood firm on the reaction with I_2 being an oxidation (to biliverdin). He acknowledged not having proved that the green product is actually biliverdin, but in the 1899 publication he gave full details (*197*):

> Thudichum den grünen Farbstoff als ein jodirtes Substitutionsprodukt, ich jedoch als ein Oxydations-produkt, und zwar als Biliverdin, ansehen. Ich gebe gern zu, dass Thudichum insofern Recht hat, als ich für die Identität des grünen Farbstoffes mit Biliverdin die analytischen Belege nicht geliefert habe. Jedoch bemerke ich, dass mir schon damals ein genügendes noch nicht zur Publikation gebrachtes Material vorlag, auf Grund dessen ich mich berechtigt hielt, den bei der Einwirkung einer verdünnten alkoholischen Jodlösung auf Bilirubin vor sich gehenden Process als eine Oxydation anzusehen und den hierbei zunächst entstehenden grünen Farbstoff als Biliverdin anzusprechen.   160

Jolles thus focused on *Thudichum*'s dogma that I_2 cannot cause oxidation of bilirubin, only substitution (*197*):

> Was nun Herr Thudichum in erster Linie bestreitet, ist die Thatsache, dass bei der Einwirkung der Jodlösung auf gelöstes Bilirubin eine Oxydation vor sich gehe, es könne sich nach ihm einzig und allein nur um einen Substitutionsprocess handeln. . . . Herr Thudichum stellt sich in seiner Erwiderung auf den eigentümlichen Standpunkt, dass alle seine Behauptungen bezüglich der Gallenfarbstoffe förmlich als unumstössliche Dogmen zu betrachten wären.   161

Jolles thus asked a fundamental question: why could *Thudichum* not see the possibility of several competing reactions taking place by reaction of bilirubin with I_2 (or Br_2 for that matter): oxidation, addition, and substitution? He queried correctly whether the reaction with I_2 might be more selective than reaction with Br_2, whether an oxidation might occur first and be followed by a substitution or an

2.12 Conjectural Chemistry and Bilirubin Polemics at the Close of the 19th Century

addition, and whether the final result was actually a mixture of oxidation and substitution products, from which only (brominated) products were isolated by *Thudichum* and *Maly* (*197*):

> Ueberdies sind uns beide Forscher den einwandsfreien Beweis schuldig geblieben, ob nicht bei der Einwirkung von Brom auf Bilirubin unter den angegebenen Bedingungen neben dem Substitutionsprodukt auch ein Oxydationsprodukt parallel verläuft und ferner, ob nicht zuerst eine Oxydation erfolgt und erst bei weiterer Einwirkung eine Substitution stattfindet, so dass die von den Verfassern erhaltenen Körper einestheils Mischungen von Oxydations- und Substitutionsprodukten waren, andererseits als bromirte Derivate von Oxydationsproduckten des Bilirubins angesehen werden könnten. Man muss sich wundern, dass auf Grund solcher noch ziemlich lückenhafter Arbeiten die Einwirkung der Halogene auf Bilirubin als ein abgeschlossenes Gebiet angesehen wird ... 162

He took offense at what he described as an unfounded and prejudiced criticism regarding his work, with no attempt having been made to repeat the reputed experiments, asserted that a comparison between the reactions of Br_2 and I_2 is not generally permissible, especially when I_2 is used at high dilution, and cited *Kekulé* as having already shown that (*197*):

> Die vornehmliche Stütze für seine Behauptungen, dass bei Einwirkung von Jod auf Bilirubin eine Jodsubstitutionsprodukt entstehe, bildet der Analogieschluss, dass, weil Brom auf Bilirubin substituirend wirkt, dies auch zweifellos beim Jod der Fall sein müsse. Ist einsolcher Schluss gerade bei Jod und Brom schon im Allgemeinen nicht statthaft, so ist hier noch zu berücksichtigen, dass Jod in gelöster Form bei seiner Einwirkung auf gelöste organische Substanzen überhaupt nicht substituirend wirkt, zumal in einer so ausserordentlichen Verdünnung, worauf ja schon Kekulé zuerst ausführlich hingewiesen hat[1]... 163

---

[1] Ann. Chem. *131*, 122.

Jolles completed his work (*197*) with the publication of an experimental procedure for the oxidation of bilirubin to biliverdin by I_2, which he characterized by the equation $C_{16}H_{18}N_2O_3 + 2\,I + H_2O = C_{16}H_{18}N_2O_4 + 2\,HI$, and he provided an elemental combustion analysis of the latter as well as a list of the usual characteristic properties: solubility, fluorescence with added $ZnCl_2$, and various color changes upon treatment with acids, including a positive *Gmelin* test (*197*).

It may be noted in the combustion analysis that the %C, H, and N found do not match up well with the theoretical values for the $C_{16}H_{18}N_2O_4$ formula (*197*):

> 1. 0,1846 Grm. Substanz, bei 100° getrocknet, lieferten 0,3928 Grm. CO_2 und 0,0963 Grm. H_2O.
> 2. 0,1708 Grm. Substanz, bei 100° getrocknet, lieferten 14,4 Ccm. Stickstoff bei 728 Mm. und 20° .

Berechnet für Biliverdin $C_{16}H_{18}N_2O_4$:		Gefunden:
C	63,58 %	62,76 %
H	5,96 ,,	6,27 ,,
N	9,26 ,,	8,44 ,, 164

170 2 Early Scientific Investigations

Yet the total collection of data was apparently sufficient to have convinced *Jolles* that he had in fact prepared biliverdin from bilirubin by the action of I_2 (*197*):

> Aus den vorstehend angeführten Resultaten geht somit mit Sicherheit die Thatsache hervor, dass das durch Einwirkung der alkoholischen Jodlösung auf Bilirubin unter den angegebenen Versuchsbedingungen entstehende Produkt weder ein Jodsubstitutions-, noch ein Jodadditionsprodukt, sondern nur ein Oxydationsproduct darstellt, und zwar ist dasselbe mit Rücksicht auf die Ergebnisse der Elementaranalyse, sowie der charakteristischen Eigenschaften des Körpers als Biliverdin anzusprechen. 165

Jolles' reply was apparently not the *mea culpa* that *Thudichum* had sought. In 1900, a year before his death and beset by *Jolles'* studies as well as others on bile pigment issues that he believed to have put to rest decades earlier, including whether urobilin is present in normal urine, investigators who appeared to repeat or rely upon the errors of previously published work (while neglecting his corrections of such errors), an exasperated *Thudichum* coined the word *Conjecturalchemie* (*191*). He assailed the current crop of bile pigment researchers for having read the published literature only selectively, for not being able to distinguish between conjecture and fact when citing it, and for being out of touch with chemistry. Waxing philosophical *Thudichum* attributed such errors and deficiencies to a continuous deterioration (of scientific knowledge and understanding) ever since the so-called physiological chemistry had suffered separation from overall chemistry in every civilized country as promoted by academic chairs and literary organs (*191*): "Seitdem die sog. physiologische Chemie von der allgemeinen durch Professuren und litterarische Organe abgetrennt worden ist, hat sie in allen Culturländern eine unblässige Verschlechterung erlitten."

Unwilling to let *Jolles* have the last word on the subject of halogen-promoted oxidation of bilirubin to biliverdin, in 1900 *Thudichum* (*191*) wrote (on the subject of the treatment of bilirubin with I_2) that *Jolles'* 57-page article in 1894 (*196*) was completely refuted in his paper published in 1896 (*187*). Yet, in 1899 *Jolles* persisted (*197*), according to *Thudichum*, in attempting to revive a few of his earlier assertions by changing his position, necessitating changes that might befuddle a reader who does not follow the subject (*191*):

> Dadurch wurde eine 57 Seiten lange Abhandlung . . . von Dr. Adolf Jolles in Wien vollständig widerlegt. Nichtsdestoweniger hat derselbe in diesem Journal . . . einige seiner früheren Behauptungen aufzufrischen versucht, zu diesem Zwecke aber seinen Standpunkt so zu verändern sich genöthigt gesehen, dass die Leser, welche dem Gegenstand nicht folgen, darüber orientirt werden sollten. 166

Thudichum continued to object to *Jolles'* published work, accusing him of first using the false oxidation of bilirubin by Br_2 as an analog of the purported oxidation by I_2 – then quietly abandoning it, along with all other false statements based on formulas, results, and processes reported by *Maly* and *Rödeler*. Of course, he strongly objected first to *Jolles'* use of *Maly*'s doubled formula ($C_{32}H_{36}N_4O_6$) for bilirubin – which he said had been determined by no one except to be written on paper – and then to *Jolles'* switching to the *Städeler-Maly* formula ($C_{16}H_{18}N_2O_3$) that, *Thudichum* said, was not defined by any analysis. He complained that *Jolles* had not analyzed his bilirubin, nor had he investigated the biliverdin prepared from

2.12 Conjectural Chemistry and Bilirubin Polemics at the Close of the 19th Century

it by *Zuntze*'s method (but not by I_2) but referred only to a *Maly* preparation (which *Thudichum* said did not exist). One gets a better sense of *Thudichum*'s dismay and strong feelings here and later in the original German (*191*):

> Die Basis, auf welche er seine durch gar nicht vorhandene Analogie ihm eingegebene Arbeit zu gründen glaubte, nämlich die schon lange als nicht existirend nachgewiesene Oxydation des Bilirubins durch Brom, ist ihm jetzt unter den Händen entschlüpft. Alle die falschen Angaben, welche er über angebliche Formeln, Resultate und Processe von Rödeler und Maly gemacht hatte, sind ebenfalls aus dem neuen Text weggelassen. Also z. b. anstatt $C_{32}H_{36}N_4O_6$, der von Niemand ermittelten, sondern nur auf dem Papier gemachten, an sich in jeder Beziehung falschen Formel für Bilirubin, giebt er jetzt die ebenfalls ganz falsche, durch keine Analyse oder Verbindung gestützte Formel $C_{16}H_{18}N_2O_3$. Er hat nun nicht etwa Bilirubin analysirt oder durch Verbindungen definirt, oder das daraus durch Zuntz's Methode (aber nicht durch Jod) darstellbare Biliverdin untersucht, sondern spricht von einer angeblichen Methode Maly's, Biliverdin herzustellen, die gar nicht existirt.   167

Unrelentingly, *Thudichum* scoffed at *Jolles*' spectrum analyses of his pigments, complained that the bilirubin spectrum was due as much to impurities in the commercial (allegedly pure) bilirubin as to the pigment itself – saying that all that the spectroscopy proved was the impurity of all of *Jolles*' preparations, without his realizing it. And he broadened his assault to say that it then followed entirely irrefutably that *Jolles*' errors in print (should anyone believe them) would bring forth only confusion and that his quantitative estimates (of bilirubin, using I_2) were falsely called "determinations" and possessed no value whatsoever (*191*):

> Ich hatte nachgewiesen, dass die von Dr. Jolles dem Bilirubin und Biliverdin zugesprochene und von ihm auf einer Tafel mit anderen ähnlichen Neuigkeiten abgebildeten Spectra diesen Körpern nicht zukommen. Er giebt nun jetzt an, dass seine Spectra mit „Substanz" erhalten worden seien, welche er als rein von einem Fabrikanten gekauft habe, während sie doch in der That unrein gewesen sei. Die von ihm skizzirten Absorptionsspectra waren daher Produkte der Unreinigkeiten in seinen Präparaten, und keineswegs der zu erforschenden Substanz selbst. Auf Seite 3 seiner Schrift in Pflüger's Archiv behauptet er die Identität seines aus Galle oder vermeintlichem Bilirubin durch Jod erhaltenen grünen Produkts mit „reinem Biliverdin spectroskopisch" bewiesen zu haben. Alles was er, ohne es zu wissen, bewiesen hat, war die Unreinheit aller seiner Präparate. Daraus folgt nun ganz unwiderleglich, dass er mit seinen Irrthümern, wenn ihnen Jemand traute, nur Wirrwarr hervorbringen würde, jedenfalls aber, dass seine angeblichen Mengeschätzungen, fälschlich Bestimmungen genannt, keinerlei Werth besitzen.   168

And in a final admonishment to *Jolles*, *Thudichum* objected to the former's having written or copied from a statement by *Maly* that no analyses were run on the brominated product from reaction of bilirubin with Br_2. *Thudichum* was emphatic in stating that he had in fact concluded complete elemental combustion analyses on two preparations of dibromo-bilirubin and had long ago refuted *Maly*'s false statement on the subject. His parting words on the polemic with *Jolles*: I herewith protest against the carelessness with which *Jolles* treats the literature (*191*):

> In seinem Aufsatz über diesen Gegenstand in den Wiener Monatsheften, der im Wesentlichen eine Wiederholung des Aufsatzes in diesem Journal ist, sagt Hr. Jolles, ich hätte mein Bromsubstitutionsprodukt des Bilirubins nicht analysirt. Diese Angabe ist nicht etwa eine Ermittlung des Hrn. Jolles selbst, sondern sie ist aus einem Aufsatz des weil. Prof. Maly abgeschrieben. Sie is völlig unbegründet. Die Theorie der Bildung des

> Dibrombilirubins ist nicht nur durch die Zunahme des Bilirubins an Brom und das Weggehen des Bromwasserstoffs, sondern auch durch vollständige Elementaranalyse von zwei Präparaten, von denen eines über 20 Grm. wog, bewiesen worden. Die falsche Angabe von Maly habe ich schon lange widerlegt, und ich erhebe hiermit nochmals Protest gegen die Nachlässigkeit, mit welcher Hr. Jolles die Litteratur behandelt.   169

Leaving for the moment his polemic with *Jolles*, *Thudichum* then moved on to address recent work of others on the urinary pigments by firmly reminding us of his own, also in the context of conjectural chemistry, of which he gave many examples from the chemistry of urine, bile, brain, and other organs and essential parts of the body, chiefly in articles in this journal (*Journal für praktische Chemie*) and in more than 30 articles published in English medical journals. Referring to his three most recent publications on the subject (*188–190*), two of them (*188, 189*) addressed mainly the errors that he had refuted some 25 years earlier, including the purported identity of urobilin (isolated from urine but previously not analyzed) with *Maly*'s hydrobilirubin [one of a mixture of products obtained by treatment of bilirubin with Na(Hg)], he turned his ire again toward *Maly* for having published falsely on the subject and thus having provided the means for physiological chemists who later picked up on the work to incorporate and propagate errors. Though he expressed hope that *Hopkins* and *Garrod* in London (*194, 195*) would provide a further final rejection of *Maly*'s work, he also chided them for not having recognized that their elemental analysis of urobilin proved it to be nothing more than that discovered by him in 1864, where he had described it as *Omicholin* (*191*), analyzed from eight preparations. Satisfyingly, their urobilin analyzed for 4.11% N and *Thudichum*'s *Omnicholin* for 4.18% N, in contrast to *Maly*'s hydrobilirubin, 9.75% N. Accordingly, no further explanation was required (*191*):

> Ich habe dies an vielen Beispielen aus der Chemie des Harns, der Galle, des Gehirns und anderer Organe und Bestandtheile des Körpers bewiesen, hauptsächlich in Artikeln in diesem Journal, und in mehr als dreissig Artikeln in englischen medicinischen Zeitschriften. Um nicht mit Wiederholungen zu belästigen, weise ich an dieser Stelle auf drei von mir gemachte neueste Mittheilungen hin, welche in Virchow's Archiv für pathol. Anat. un Physiol. und für klin. Med. **150** (1897) 586, daselbst **153** (1898) 154 und **156** (1899) 284 erschienen sind. Zwei derselben betreffen hauptsächlich den schon vor 25 Jahren von mir widerlegten Irrthum, dass das sogenannte Urobilin, eine aus dem Harn isolirte, bisher nicht analysirte Substanz, mit dem Hydrobilirubin, einer aus Bilirubin durch Natriumamalgam erhaltenen Mischung von Pigmenten identisch sei. Diese von Maly in die Welt gesetzte falsche Angabe ist später von den physiologischen Chemikern weiter geschleppt worden, bis sie durch die Untersuchung der HHrn. F. G. Hopkins und A. E. Garrod in London eine weitere, und wie zu hoffen steht, endliche Abweisung erhielt. Letztere haben durch Elementaranalyse bewiesen, dass das Urobilin weiter nichts ist als das von mir im Jahre 1864 entdeckte und genau beschriebene, an acht Präparaten analysirte Omicholin.... Ihr Urobilin enthält 4,11% Stickstoff; mein Omicholin 4,18% Stickstoff; Hydrobilirubin dagegen 9,75% Stickstoff. Darnach bedürfen die übrigen Unterschiede keiner weiteren Darlegung.   170

Seemingly unrelentingly, *Thudichum* again took issue with *Jolles*, who nevertheless, by conjectural chemistry and from five centigrams of impure material and mathematical equations added a new pigment (*Bilixanthin*) to the scene. Which *Thudichum* claimed was identical to the *Uroxanthin* obtained from urine that *Jolles* passed off as a new discovery. *Uroxanthin*, the particularly colored material that is differentiated from the characteristic yellow urinary pigment (urochrome) (*191*)

2.12 Conjectural Chemistry and Bilirubin Polemicsat the Close of the 19th Century 173

was of course discovered by an earlier professor of physiological chemistry, *Heller*.[58] *Jolles* apparently indicated that unlike *Uromelanin* (*196*) *Uroxanthin* contains *Indigoblau* (indigo blue) as the diagnostic radical. *Thudichum* expressed that the radical was falsely identified with an indigo plant extract and named *Indican*, that the indigo-containing substance of urine, *Heller's Indigogen* or *Uroxanthin*, yielded no sugar and no glucoside and therefore was not identical with indican, the glucoside of the indigo plant and thus there was no justification for use of the word *Indican* in urology. He admonished *Jolles* for having usurped *Heller's* name, *Uroxanthin*, to apply to a different product, and thereby for violating ethics (*191*):

> Eben weil nun Heller den Namen Uroxanthin für ein jedenfalls genügend gekennzeichnetes Educt gewählt, und sich dieser Name in der Litteratur eingebürgert hat, halte ich seine Anwendung auf ein anderes Produkt nach den Gesetzen der litterarischen Ethik für unerlaubt. 171

Thudichum's final article on bile pigments continued in the same vein, objecting to physiological chemists' penchant for misrepresenting compounds previously discovered and cited *Heller's Urohidin* as an example. He subsequently identified *Heller's Urohidin* as indigo-red or *Indirubin*, because it was colored red and obtained in addition to indigo-blue. Apparently, they had not read *Thudichum's* 1877 article on *Urohidin*, which could not be an indigo-blue isomer because it analyzed for no nitrogen and contained 80% carbon (*191*):

> Als weiterer Beweis für die nachlässige Art, mit welcher manche physiologische Artikelschreiber mit den besten Entdeckungen der Vorgänger umgehen, erwähne ich das Schicksal von Heller's Urohodin, ein von ihm zuerst hervorgebrachtes Produkt. Dasselbe wurde von Conjecturchemikern, weil es neben dem Indigoblau erhalten wurde, und roth von Farbe war, für Indigoroth oder Indirubin erklärt. Allein die Elementaranalyse machte auch dieser Vermuthung ein Ende. Mein „Experiment über das Urohodin" in Pflüger's Archiv **15** (1877) 346 bewies, dass dasselbe, von einem ungefärbten Urohodinogen durch starke Salzsäure erhalten, keineswegs dem Indigoblau isomer sein kann, da es keinen Stickstoff, wohl aber 80% Kohlenstoff enthält. Seine Aetherlösung zeigte ein specifisches Absorptionsspectrum, in welchem allen Grün durch ein dunkles Band ausgelöscht, ist, wenn Roth und Blau durchscheinen. 172

Thudichum did not spare the new, young investigator *William Küster* from criticism on the crystal morphology and purity of his crystallized bilirubin (*198, 199*) and the 0.7–1.3% and 1.53–2.89% higher than expected %C and %N values, respectively, found by *Küster's* elemental combustion analysis – values similar to those from his own macroscopic crystalline bilirubin (*191*):

> Er hat dann ein ganzes Capitel der Beschreibung der Darstellung von angeblich krystallisirtem Bilirubin verfasst, ohne dass dabei auch nur ein einziger Krystall zum Vorschein

---

[58] *Johann Florian Heller* was born on May 4, 1813 in Iglau, Austria and died on November 21, 1871 in Vienna. He was one of the founders of clinical chemistry and a distinguished pathological chemist who established a laboratory of pathological chemistry at the Wiener Allgemeines Krankenhaus ("AKH" = Vienna's General Hospital). He had studied chemistry in Prague and in Giessen (with *Liebig* and *Wöhler*), researched the chemistry of urine in Vienna, and developed the (well-known) *Heller's* ring test for albumin in urine. A prize in his name is awarded by the Austrian Association for Clinical Chemistry (ÖGKC).

gekommen wäre. ... Ich lehne daher die von dem Aussehen der Produkte des Hrn. Küster abgeleiteten Schlüsse für Reinheit seiner Produkte ab, und behaupte, dass alle Produkte von Bilirubin aus seinen Processen unrein waren. Den Beweis hat er selbst geführt durch seine Elementaranalysen von Proben, deren Kohlenstoff von 0,7% bis 1,3% zu hoch ist; aber namentlich der Stickstoff ist von 1,53%-2,89% zu hoch, wenn controllirt durch die Verbindungen des nach meiner Methode dargestellten und makroskopisch krystallisirten Bilirubins.   173

He scolded *Küster* for having used *N,N*-dimethylaniline as a bilirubin crystallization solvent. He deemed it entirely unsuitable because it is a base, can react easily with other compounds at its high boiling temperature, and remains attached to the precipitate. *Thudichum* stated that *Küster*'s crystals so obtained contained some of the base. He was apparently dismayed that *Küster*'s yield of crystallized product was only one-third of the starting bilirubin, took issue with *Kuster*'s attempted oxidation to biliverdin using PbO_2 (which *Thudichum* claimed had been shown long ago to fail). And after progressing to what he called "erroneous reports on biological-chemical matter in periodical journals of chemistry and medicine", for which he provided examples, he could not end the discourse without providing a reason for having chastised *Maly*, who apparently had the temerity to attack *Thudichum*'s research on urine. *Thudichum* was then immediately forced to convince *Maly* in public that his relevant reports were incorrect from beginning to end and would consequently draw no belief whatsoever from informed readers (*191*):

> Zuletzt muss ich die Leser warnen vor einigen Ausfällen, welche Professor Maly in seinem Jahresbericht als Zeugniss der Fortsetzung seiner Behandlung der Wahrheit gegen meine Forschungen über das Hirn gemacht hat, die er vorher durch ein Plagiat bewiesen hatte, so dass ich ihn öffentlich zu überführen geradezu gezwungen war. Die betreffenden „Berichte" des Hrn. Maly sind unrichtig von Anfang bis zu Ende und haben daher bei unterrichteten Lesern keinerlei Glauben gefunden.   174

Thudichum was the consummate scientist of his era, an extraordinarily talented and meticulous researcher. An apparent deep thinker and broadly interested in medicine, disease, and chemical physiology. Like the author's former colleague at UCLA, the renowned *Nobel* Prize candidate and physical organic chemist *Saul Winstein* (1912–1969), he showed no mercy when confronted with what he considered to be suspect or shoddy work and especially its continued promulgation. Yet *Winstein* was more often correct in polemic discourse than was *Thudichum*. *Thudichum* clearly and forcefully expressed his beliefs, backed by experiment and the scientific logic of the age, when assailing contemporary and (especially) new investigators of bile pigments, as well as those investigators and their theories of the research area for which he is most famous: the chemistry of the brain (*163, 200*):

> It is surprising to find how little the chemical relations of the brain are understood by physiologists, and chemists of profession. They ignore the broadest facts, and maintain the most absurd fallacy which has ever disfigured animal chemistry, namely, the so-called doctrine of protagon. They thereby impede the progress of science, and confuse the minds of those who are desirous to learn and to work. ...

2.13 Knowledge of Bilirubin Near the End of the 19th Century

By the close of the 19th century, the then greatest names associated with bilirubin had passed from the scene: *Thenard* (1777–1857), who carried out early isolations of bilirubin and biliverdin from bile and discovered a goldmine of the yellow pigment in the bile duct of a deceased elephant; *Berzelius* (1777–1848), who labored for nearly 40 years to isolate and purify bilirubin (*Cholepyrrhin, Gallenbraun*), biliverdin (*Gallengrün*), and bilifulvin from bile – and for very different reasons dominated the field of chemistry; *Tiedemann* (1781–1861) and *Gmelin* (1788–1853), who showed that air (oxygen) was required to convert the yellow pigment to the green and discovered the characteristic display of colors from treatment with HNO_3 that became the enduring and famous *Gmelin* reaction (or diagnostic color test) for bilirubin in bile, urine, *etc.*; *Scherer* (1814–1869), who isolated a green pigment from bile and jaundiced urine and carried out one of the earliest elemental combustion analyses; *Heintz* (1817–1880), who created an improved separation method for isolating bilirubin, carried out elemental combustion analyses of it and wrote formulas for bilirubin as $C_{31}H_{18}N_2O_9$ as better than $C_{32}H_{18}N_2O_9$, and for biliverdin as $C_{16}H_9NO_5$ or its double formula, $C_{32}H_{18}N_2O_{10}$ to fit the data; *Valentiner*, who while working in *Friedrich Theodor von Frerichs'* lab in Göttingen in the 1840s introduced $CHCl_3$ extraction to isolate bilirubin and showed that it was probably identical to *Virchow*'s hematoidin cited in 1847; *Brücke* (1819–1892), who improved *Valentiner*'s isolation method to obtain the purest bilirubin to date, as well as a collection of related pigments and analyzed "ash-free" samples by combustion; *Städeler* (1821–1871), who isolated "purified" bilirubin and biliverdin and conducted C, H, N elemental combustion analyses that corresponded first to the formula $C_{18}H_9NO_4$, then later to $C_{32}H_{18}N_2O_6$ for bilirubin, and $C_{32}H_{20}N_2O_{10}$ for biliverdin; *Maly* (1839–1891), who also conducted C, H elemental combustion analyses of isolated bilirubin and suggested the formula $C_{16}H_{18}N_2O_3$ for it, while the %C, H, N of his biliverdin was fit to $C_{16}H_{20}N_2O_5$; and *Thudichum* (1829–1901), who carried out detailed isolations and combustion analyses to show that bilirubin had the (empirical) formula $C_9H_9NO_2$, biliverdin had the formula $C_8H_9NO_2$, and who became a dominating voice on bile pigments in the last half of the 19th century.

With the completion of the important new bile pigment research of the mid-late 1800s by *Städeler*, *Maly*, and *Thudichum*, understanding bilirubin had reached its final stage before the advent of the era of chemical degradation and synthesis. Certainly, by the late 1870s "animal" or organic chemistry had progressed to the use of a wide range of chemicals, reagents, and solvents that illustrated a rapidly maturing chemical science. New and improved methods for isolating bilirubin from gallstones and bile had been developed. Many elemental combustion analyses had been run, albeit few of them on apparently homogeneous samples, from which conflicting molecular or empirical formulas were extracted. An elementary form of absorption spectroscopy in the visible region had been introduced and was used for comparing pigments. Yet despite the many advances in knowledge of bilirubin and biliverdin, and the discovery of a probable relationship between the pigments of blood and bile, a correct molecular formula was still debatable, a molecular weight had been determined only from the material balance in a chemical reaction devoid of knowledge of almost any aspect of chemical structure, and the purity or homogeneity of bilirubin, while vastly improved over that in the early part of the 19th century, was still suspect.

Polemics, however satisfying, disillusioning, or disenabling, failed to produce new knowledge on the structure of bilirubin itself. Thus, near the close of the 19th century, the status of the knowledge of bilirubin could be summarized briefly by *Arthur Gamgee* in 1893 (*201*):

Bilirubin $C_{32}H_{36}N_4O_8$

(Synonyms: Cholepyrrhin, Biliphœïn, Bilifulvin, Hæmatoidin[6]).

Occurrence. Bilirubin occurs in the yellow or reddish-yellow bile of man and carnivorous animals, in the bile of the pig and occasionally in the bile of the herbivora which have been long without food. It also occurs in the contents of the small intestine and is a normal constituent of the blood serum of the horse ... It is further a common constituent of gall-stones; it occurs in the urine, and stains the conjuctivae and skin, in cases of jaundice. In old blood extravasations it occurs in microscopic crystals which were first discovered by Virchow and by him called hæmatoidin. ...

Physical and Chemical Characters.

Colour and crystalline form. Bilirubin occurs in an amorphous and in a crystalline condition. In the former it presents the appearance of an orange-coloured powder resembling sulphide of antimony; in the latter it has the colour of crystallized chromic acid. Examined under the microscope, crystalline bilirubin exhibits orange-coloured rhombic tables, in which the obtuse angles are often rounded off. When crystallising from solutions which are not quite pure (containing cholesterin, &c.) better formed crystals are obtained than is the case when the solutions contain no such impurities (Hoppe-Seyler ...).

Solubility. Bilirubin is insoluble in water, almost insoluble in ether and very sparingly soluble in alcohol. It is readily soluble in chloroform especially with heat; it is likewise soluble (though to a much less extent than in chloroform) in benzol, carbon disulphide, amyl alcohol, and glycerin. These fluids dissolve enough however to acquire a yellow or a brown red colour. Solutions of bilirubin which contain 1 part in 500000 exhibit a perceptible yellow colour when a layer 1·5 cm. thick is observed (Hoppe-Seyler).

Bilirubin is readily soluble in dilute solutions of sodium and potassium hydrate and ammonia, and if the solutions be kept from contact with air or with oxygen, it can be reprecipitated from them by addition of hydrochloric acid.

It is important to notice that solutions of bilirubin in alkalies do not yield the colouring matter to chloroform. A chloroformic solution of the colouring matter shaken with dilute sodium or potassium hydrate is at once decolourised; on the other hand a similar alkaline solution

---

[6] The name hæmatoidin is only applied to bilirubin when occurring in old extravasations of blood.

2.13 Knowledge of Bilirubin Near the End of the 19th Century

of bilirubin if acidulated and shaken with chloroform at once gives up its colouring matter, which is dissolved by the chloroform and imparts to it a much less brownish-yellow colour.

Bilirubin forms compounds with bases of which several have been studied. The Na-compound is obtained by precipitating a dark orange solution of bilirubin in sodium hydrate by means of a concentrated solution of caustic soda.

The Ca-compound is obtained by precipitating an ammoniacal solution of bilirubin with calcium chloride. The precipitate is rust-coloured, flocculent, and insoluble in water, alcohol, ether and chloroform. It has the composition indicated by the formula $C_{32}H_{34}N_4O_6$. Ca. When this compound is dried *in vacuo* over sulphuric acid it is of a dark-green colour with a metalic lustre, but when powdered it has a dark-brown colour.

By the action of barium chloride, lead acetate, and nitrate of silver on ammoniacal solutions of bilirubin, compounds similar to the calcium compound can be obtained. The silver compound occurs in violet-coloured flakes and is not reduced even when the liquid in which it is suspended is boiled. Bilirubin, as Maly observes, shews by the compounds which it forms, that it has the characters of a weak acid.

Composition and formula. Heintz[1] was the first chemist to make an ultimate analysis of bilirubin, and assigned to it the formula $C_{16}H_{18}N_2O_5$. The method which he followed in the preparation of the substance, which was not until later obtained crystallised, renders it certain that it was not free from impurities, and the results of his analysis may therefore be left out of consideration. The same objection does not apply to Städeler's methods. The results of his work have been absolutely confirmed by the more recent and exhaustive researches of Maly, as well as by Hoppe-Seyler[2].

Both Städeler and Maly from their analyses deduced for bilirubin the formula $C_{16}H_{18}N_2O_3$. Thudichum[3], on the other hand, has assigned to bilirubin the formula $C_9H_9NO_2$, which neither agrees with the concordant analytical results of Städeler and Maly, nor fits in with many facts with which we are acquainted. The reader will see at a glance how considerable are the differences in the percentage of the various elements calculated from Städeler and Maly's formula on the one hand, and from that of Thudichum on the other.

	(Städeler and Maly.) $C_{16}H_{18}N_2O_3$ or $C_{32}H_{36}N_4O_6$	(Thudichum.) $C_9H_9NO_2$
Carbon	67·13	66·25
Hydrogen	9·79	5·52
Nitrogen	9·79	8·59
Oxygen	16·79	19·64
	100·00	100·00

---

[1] Heintz, Poggendorff's *Annalen*, Vol. LXXXIV, p. 106.

[2] "Ausser diesen Ergebnissen der Untersuchungen von Städeler sind noch von Maly und von Thudichum solche veröffentlicht, von denen die Resultate Maly's Bestätigung der Untersuchungen Städeler's geben. Die Analysen von Hoppe-Seyler lassen gleichfalls keinen Zweifel an der Richtigkeit der Formel von Städeler und von Maly." Hoppe-Seyler, *Handbuch d. Phys. u. Path. Chem. Analys.*, 6th ed. (1893), p. 226.

[3] Thudichum, *Journ. f. prakt. Chem.*, Vol. CIV. (1868), p. 193.

Quite apart from the remarkable concordance of the results of Städeler and of Maly, an examination of all facts bearing on the question[1] has led chemists to the opinion that the formula of Städeler and Maly, or probably a multiple of it, is correct. The various reactions are best explained by doubling Städeler's formula.

Action of nitric acid on bilirubin 'Gmelin's reaction'. When bilirubin is treated with pure dilute nitric acid (containing 20 per cent. of HNO_3) no change occurs at ordinary temperatures. When the solution is heated, however, dark-violet resinous flakes are formed which as the temperature rises assume a light-brown colour and ultimately dissolve, yielding a yellow-coloured liquid.

Pure concentrated nitric acid acts in the cold and a cherry-red liquid is obtained which retains its colour for many days. Nitric acid which has a slightly yellow colour and which contains nitrous acid[2] (as the nitric acid of commerce does) gives rise in solutions which contain bilirubin, to a remarkable play of colours already referred to as 'Gmelin's reaction.' The reaction may be tried with a dilute alkaline solution of bilirubin, with diluted bile, or with any liquid, such as the urine of jaundice, which contains bilirubin.

Various methods of exhibiting Gmelin's reaction may be adopted. The most common is to pour some of the solution to be tested into a test tube containing nitric acid, so that the two liquids are not mixed. Near the line of junction the colour-reaction at once commences to develope [sic], and a succession of zones of colour appear, the tints being, from above downwards, as follows: –green, blue, violet, red and reddish-yellow. These tints represent the successive stages of the reaction, the first being green and the last the reddish-yellow, which is observed in the region where the oxidising action is most intense, viz. in close proximity to the nitric acid.

Instead of employing a test tube, a few drops of diluted bile, or bilious urine may be poured upon a flat plate, so that a thin layer of liquid is obtained. On now adding a drop or two of coloured nitric acid, wherever the acid falls a series of concentric coloured rings of beautiful is developed, the succession of tints being the same as in the experiment previously described.

The delicacy of 'Gmelin's reaction' is such that it permits of the detection of bilirubin in solutions which contain only 1 part of the colouring matter in from seventy- to eighty-thousand parts of water. It must be remembered that in order to be sure of the presence of bilirubin the whole series of tints must be observed, as *lutein* (yellow crystalline matter obtained from corpora lutea, from the yolk of egg, and which is also present in the liquor sanguinis of some animals), when treated with nitric acid, exhibits a green and also a blue tint very similar to those developed in Gmelin's reaction. The spectroscopic characters of lutein are, however, sufficiently distinctive to enable the observer to ascertain whether this substance is present in a solution or not.

Each tint in Gmelin's reaction corresponds apparently to a definite chemical change, probably to a definite oxidation product. The green tint is due to the production of biliverdin, which as will be afterwards shewn is the first stage in the oxidation of bilirubin. The blue tint is due to an imperfectly studied body termed bilicyanin; the final reddish-orange colour is due to choletelin.

---

[1] Such as the results of the analysis of the calcium compound of bilirubin, of Maly's tribro-mobilirubin, no less than the relation of bilirubin to biliverdin; to the latter point reference will again be made.

[2] If the acid is too highly coloured (i.e. if the amount of nitrous acid and of nitrogen peroxide be large) it exerts so energetic an action on the bilirubin that the successive stages of Gmelin's reaction cannot be properly observed.

2.13 Knowledge of Bilirubin Near the End of the 19th Century

Though *Thudichum* might have become upset with *Gamgee*'s assertion that the *Städeler-Maly* formula for bilirubin was superior to his own and that *Städeler*'s doubled formula best explains the various reactions of the pigment, his major research projects at that time were focused on the chemistry of the brain. However, if he could not be bothered with *Gamgee*, he found the energy to rebut the new researchers of the 1890s who had probably innocently stepped too hard on the wrong set of toes. Yet, while the tirades continued, newer important studies on bilirubin, regarding its molecular weight, its elemental combustion analysis, and its degradation into identifiable small fragments were underway by a new set of investigators: *Nencki*, *Teeple*, and *Küster*.

3 Advent of the Bilirubin Structure Proof

In 1837, as *Berzelius* was completing the majority of his bilirubin studies and the fundamentals of chemistry were being developed, *Victoria Regina* began her reign as British monarch. At her death in 1901, as the longest reigning female monarch in history, all of the bilirubin investigators of the first half of the 19th century, whose contributions centered mainly on isolations by precipitation methods and early attempts at combustion analysis, had died. Even their successors, the generation of bile pigment investigators represented mainly by *Städeler*, *Maly*, and *Thudichum*, who left such an indelible mark related to isolation by extraction methods, elemental combustion analysis, and controversial formulas, had also passed away. During their lifetimes, and near the middle of the reign of *Queen Victoria* (*Alexandrina Victoria*, the last monarch of the House of Hanover), almost coincident with the early passing (1861) of her Consort, *Prince Albert* of Saxe-Coburg and Gotha, a revolution in chemical thinking and perspective was underway that was destined to provide a new and fundamentally important advance in organic chemistry. Such breakthrough was to have a profound, positive influence on the next generation of investigators of organic structure and synthesis, as well as on all successive generations of chemists, and no less to investigators of the pigments of blood and bile A revolution in chemical theory was advanced in mid-19th century by *August Kekulé*, possibly as the result of a paper on "The Theory of Aetherification" read by 26-year-old *A. W. Williamson*[1] in Edinburgh on August 3, 1850 before the British Association for the Advancement of Science. This paper may have led *Kekulé* to propose a way to visualize atoms and their connectivity in molecules (*202*). Namely, to draw images (pictograms) on paper that macroscopically

---

[1] *Alexander William Williamson* was born on May 1, 1824 in London and died on May 6, 1904 in Surrey, England. He studied under *Gmelin* in Heidelberg and *Liebig* in Giessen and spent three years studying mathematics in Paris with *Comte*. In 1849, he was appointed Professor of Practical Chemistry at University College, London, and from 1844 until his retirement in 1887 he was Professor of Chemistry there.

D.A. Lightner, *Bilirubin: Jekyll and Hyde Pigment of Life*, Progress in the Chemistry of Organic Natural Products, Vol. 98, DOI 10.1007/978-3-7091-1637-1_3,
© Springer-Verlag Wien 2013

represented the microscopic world, as elaborated in detail in the book *Image and Reality*, by *Alan J. Rocke* (*147*) and described succinctly by *Peter J. Ramberg* in his book review (*203*):

> Modern chemists routinely shuttle between the immediate empirical, sensual characteristics of substances in the laboratory and the invisibly small world of atoms and molecules. Like mathematicians, chemists do much of their thinking on paper, doodling chemical formulas or tinkering with hand-held molecular models. They think in a highly visual, non-verbal way. The origin of this way of thinking lies in the middle of the 19th century, when an extraordinary group of chemists, most of them in Great Britain and Germany, developed productive techniques for revealing the invisible world of the molecule. But by 1890, nearly all chemists could "see" what molecules looked like at the microscopic level, meaning they had access to the explicit connections between atoms in the molecule and how those atoms are arranged in space. The ability to elucidate the structure of molecules down to the atomic level by purely macrosopic manipulation of chemicals is arguably one of the greatest intellectual accomplishments of 19th-century science.

Yet, between 1860 and 1890, very few new discoveries or insights related to the *chemical structure* of bilirubin had arisen, and important new advances in understanding the structure of bilirubin were not to happen until shortly before 1900, when a new cadre of pigment investigators initiated research that was illustrated by structural drawings. *Nencki* applied a new method to determine the bilirubin molecular weight, *Teeple* had begun his doctoral studies on its combustion analysis and formula, and *Küster* had initiated actual structural investigations involving degradation. These advances were followed soon thereafter by the structural studies of *Piloty* and *H. Fischer*. While organic chemistry was rapidly evolving in the last half of the 19th century, the end of the 19th century was also notable for scientific investigations of the pigments of blood, *Hämin* (hemin) and *Hämatin* (hematin), and of green leaves, chlorophyll. A relationship of sorts had been established between a pigment derived from blood, *Hämatoidin* (hematoidin), and bilirubin in the middle of the 19th century. That connection, albeit contested well into the 20th century, would imply an actual structural relationship existed when eventually the concept of organic chemical structures became a reality. But that concept was just emerging in the mid-1800s, and so at the end of the 19th century one was left with a knowledge of bilirubin based mainly on its solubility properties, its color, crystal shape or lack thereof, its elemental combustion analysis and spectrum analysis (colorimetry), and its chemical reactions, including salt formation, bromination, and oxidation (especially the *Gmelin* reaction). Simultaneous with *Küster*'s early structural probings, two important studies appeared: (i) *Nencki* and *Rotschky* and *Abel*'s direct determination of bilirubin's molecular weight using *Rotschky*'s method, and (ii) *Teeple* and *Orndorff*'s comprehensive study of all previous elemental analyses of bilirubin and dedicated, critical reanalyses. Yet, except for the conclusion that bilirubin is a diacid, nothing was known of its *chemical* structure until *Küster* initiated degradation studies.

3.1 Experimentally Derived Molecular Weight of Bilirubin at the End of the 19th Century

Knowing the molecular weight of a compound of unknown structure seems so essential and obvious nowadays that one can perhaps scarcely believe that it escaped investigation for so long in studies of bilirubin. Except perhaps for two reasons: the concept of a molecule and its structure was of relatively recent origin, and there were no physical measurements able to provide a molecular weight. In 1875, *Thudichum* reported what might be the first experimentally determined molecular weight of bilirubin, 162.5, which was close to the formula weight (163) of $C_9H_9NO_2$, a formula he had assigned to the pigment based on its elemental combustion analyses. The experimental number was obtained by carefully comparing the weight of dried pigment before bromination to the weight of its dibromo product (*158*). *Thudichum*'s molecular weight determination was followed within 15 years by *Nencki*'s[2] more direct determination using a new technical achievement.

Nencki studied the chemical structures of hemin, hematin, and *Hämatoporphyrin* (hematoporphyrin), which were obtained from hemoglobin by degradation methods. While at the Universität Bern, with *Sieber* (*204, 205*), he published an experimental determination of the molecular weight of hematoporphyrin, the most soluble of the three pigments above, using the now classical method that was based on a relationship discovered by *Raoult*,[3] who in 1884 expressed the "General Law of the Freezing of Solvents" as follows: "If 1 molecule of any substance is dissolved in 100 molecules of any liquid of a different nature, the lowering of the freezing point of this liquid is always nearly the same, approximately 0.63°. Consequently, the lowering of the freezing point of a dilute solution of any strength whatever is, obviously, equal to the product obtained by multiplying 63 times the ratio between the number of molecules of the dissolved substance and that of the solvent" (*206*). *Raoult*'s important comprehensive work on solutions also led him

---

[2] *Wilhelm Marceli Nencki* was born on January 15, 1847 in Boćki, Poland and died on October 14, 1901 in St. Petersburg, Russia. One of Poland's must distinguished biochemists, he studied at Krakau and Jena and obtained the Dr. med. in 1870 in Berlin. In 1872, he accepted a research assistant position in pathological anatomy at the University of Bern; then, in 1876 he was promoted to a. o. Professor. In 1877, he accepted a position in Bern as o. Professor of Physiological Chemistry and Director of the Medizinisch-Chemisches Institut. Later, in 1891, he accepted an invitation to co-organize (together with *Ivan P. Pavlov*) the Institute of Experimental Medicine in St. Petersburg, where he was head of the chemical department of the Institute until his untimely death of stomach cancer in 1901. The Nencki Institute in Warsaw was founded in his memory in 1918.

[3] *François-Marie Raoult* was born on May 10, 1830 in Fournes, France and died on April 1, 1901 in Grenoble. He obtained the doctoral degree in Paris in 1863 and in 1867 assumed charge of Université de Grenoble's chemistry classes, becoming chair and remaining there until his death.

184   3 Advent of the Bilirubin Structure Proof

to formulate a simple equation relating the molecular weight of a solute to the freezing point depression of its solution (*144a*):

$$\Delta T_f = T_m - T_m^\circ = -k_f M$$

where ΔT_f is the freezing or melting point difference, T_m° and T_m are the melting or freezing points of pure solvent and the solution, respectively; M is the molality of the solute and k_f is the freezing point depression constant. In practice, the method relied on accurate weighings, which were quite feasible in the late 1800s; knowledge of the values of k_f; and accurate temperature readings, which were only as good as the thermometers then available. And at the time of *Nencki's* work, the best thermometers were accurate only to $\pm0.01°C$, which could at best yield only a crude determination of molecular weight, considering that the best data would come from dilute solutions. (With *Beckmann's*[4] invention in 1905 of a thermometer accurate to $\pm0,001°C$ (*207*) – the *Beckmann* thermometer – *Raoult's* method became a standard technique for determining the molecular weight of an organic substance). Thus in the late 1800s, absent a *Beckmann* thermometer, application of *Raoult's* method could not have been expected to show great accuracy; yet the technique opened the door to new investigations of bilirubin's molecular weight. For *Nencki* it held out the promise of determining whether *Städeler's* formula for bilirubin ($C_{16}H_{18}N_2O_3$) or the double formula ($C_{32}H_{36}N_4O_6$) was correct, or whether any of the other formulas had validity, *e.g. Thudichum's* $C_9H_9NO_2$.

Consequently, *Nencki* and *Rotschky* (*208*) measured the freezing point depression of bilirubin samples, supplied by *Maly* (3 g amorphous and 1 g rhombic crystals), that had been dried at 110°C. Using the formula $M = T \cdot P$ (100)/ED, where M = molecular weight; T = molecular depression constant for the solvent used; P = grams of solute; E = grams of solvent; D = freezing point depression, *Nencki* created a table of experimentally determined molecular weights for bilirubin and hematoporphyrin. *Nencki's* equation reminds one of that used in the classical 20th century molecular weight (MW) determinations: MW = $K(1,000/W)w/\Delta$, where K is the constant for camphor (39.7) when 1 mol of solute is dissolved in 1,000 g camphor; w is the g-weight of solute, W is the g-weight of the solvent; and Δ is the

---

[4] *Ernst Otto Beckmann* was born on July 4, 1853 in Solingen and died on July 12, 1923 in Berlin Dahlem. He studied pharmacy in Elberfeld and in 1874 studied under *Fresenius* in Wiesbaden, then at the Universität Leipzig. After passing the pharmacy exam in 1877, he studied under *Kolbe* and *von Meyer* for the Dr. phil. in chemistry, awarded in 1878. In 1882, after studies in toxicology at the TH-Braunschweig, he habilitated there. However, the *Habilitation* from a technical university did not qualify him for a lectureship in Leipzig. To achieve that he completed the *Abitur* in 1883, passing exams in Latin, Greek, and history, and returned to Leipzig in 1887 to serve as *Assistent* to *Ostwald*. After a year at the Universität Giessen and a professorship at the University of Erlangen, in 1897 he returned to Leipzig as Director of the Laboratory of Applied Chemistry. In 1912, he accepted an offer to head a division of the Kaiser Wilhelm Institute in Berlin, where he remained until his retirement in October 1921.

3.1 Experimentally Derived Molecular Weight of Bilirubin...

freezing point depression – the micro-molecular weight determination invented by *Karl Rast* in 1922 for a melting-point apparatus (*209*). However, long before *Rast*, *Nencki* and *Rotschky*'s six independent determinations of the molecular weight of bilirubin in phenol solvent produced values: 127, 235 (should have been 382?), 61.4, 230 (should have been 291?), 286, and 136 for freezing point depressions ranging from 0.03°C to 0.095°C. Two determinations in ethylene dibromide gave molecular weights 145.5 and 150 for freezing point depressions 0.03°C and 0.07°C. The method employed was only crude, yet it provided the first direct molecular weight determination of bilirubin based on a physical measurement. *Nencki* believed the data were a reasonably good fit to *Städeler*'s bilirubin formula: $C_{16}H_{18}N_2O_3$, molecular weight = 286.[5] Yet some of the experimental values seem also to fit *Thudichum*'s formula: $C_9H_9NO_2$, molecular weight 163 (*208*):

> Jedenfalls sprechen die mit dem Bilirubin erhaltenen Resultate zu Gunsten der einfachen Formel von Städeler $C_{16}H_{18}N_2O_3$. Wie man sieht, sind die nicht ganz scharfen Zahlen unserer Versuche durch die Schwerlöslichkeit und leichte Zersetzbarkeit der beiden Farbstoffe bedingt; sie genügen aber, um zu entscheiden, ob die aus den Analysen abgeleitete einfachste Formel die richtige ist.   1

In the same work, *Nencki* and *Rotschky* determined three values for the molecular weight of hematoporphyrin in phenol, values that ranged from 226 to 325. Just as *Maly* had found that bilirubin was reduced to a different urobilin by Na(Hg) that he named *Hydrobilirubin* (*175, 176*), so *Nencki* suggested that hematoporphyrin was reduced to urobilin using Sn and HCl, or Fe and acetic acid (*208*):

> Bekanntlich fand Maly, dass Bilirubin mit Natriumamalgam in Urobilin übergeführt wird. Ein ganz ähnlicher Farbstoff wird aus Hämatoporphyrin durch Einwirkung von Zinn und Salzsäure oder Eisen und Essigsäure erhalten. Wie jodoch die späteren Untersuchungen ergaben, ist das Urobilin aus Hämatoporphyrin mit dem Urobilin aus Bilirubin nicht identisch.   2

These molecular weight determinations were followed very soon afterward by a similar study of *Maly*'s hydrobilirubin by *John J. Abel*, who worked in *Nencki*'s laboratory in Bern (*210*). *Abel*, of course, knew of *Nencki*'s molecular weight determination of bilirubin, with the focus being on whether *Städeler*'s simple formula ($C_{16}H_{18}N_2O_3$) or the double formula obtained. The solubility of bilirubin and its tendency toward oxidation proved to be a limiting factor. Phenol was deemed to be the best solvent for the measurement, which was found to favor the simple formula (*210*):

> Vor Kurzem haben Nencki und Rotschky ... die interessante Streitfrage, ob die Zusammensetzung des Bilirubins der einfachen Formel = $C_{16}H_{18}N_2O_3$ oder der verdoppelten = $C_{32}H_{36}N_4O_6$ entspricht, mittelst der Raoult'schen Methode zu entscheiden gesucht. Ihr Vorhaben kann kaum als gelungen angesehen worden, indem das Bilirubin in sämmtlichen hier in Betracht kommenden Lösungsmitteln zu wenig löslich ist, respective dadurch,

---

[5] N.B. *Städeler* had also proposed $C_{32}H_{18}N_2O_6$ for bilirubin, based on atomic masses: C = 6, H = 1, N = 14, and O = 8, which also gives the same molecular weight.

wie z. B. durch Eisessig, verändert wird. Als das beste Lösungsmittel des Bilirubins erwies sich das Phenol. Eine gesättigte Lösung des Bilirubins in Phenol enthält etwa 0·4% des Farbstoffes und erst bei einem Gehalte von 0·3–0·4%, entsprechend einem Molekül Bilirubin in 1000 Molekulen Phenol gelöst, ergaben die Bestimmungen der Formel $C_{16}H_{18}N_2O_3$ entsprechende Werthe. ... Welche Depressionen, respective das darans erhaltene Moleculargewicht erhalten worden wäre, wenn nicht 1 Molekül, aber 2, 3 bis 10, z. B. Phenol gelöst wären, das zeigen die Versuche von Nencki und Rotschky eben wegen der Unlöslichkeit dieses Farbstoffes nicht. 3

Given the more favorable solubility of the product obtained from treating bilirubin by Na(Hg), hydrobilirubin, in phenol and in acetic acid, it appeared to *Abel* to be a more favorable target than bilirubin itself because it might provide the more dilute solutions essential to the *Raoult* method. Could a hydrobilirubin molecular weight determination help decide between simple and double formulas for bilirubin? *Maly* provided a 1.5 g sample of the former, which was purified in Bern and dried to constant weight. *Maly* had proposed a formula for hydrobilirubin based upon his elemental combustion analysis data and fit to twice *Städeler*'s bilirubin formula: $2 \times (C_{16}H_{18}N_2O_3) + H_2 + H_2O = C_{32}H_{40}N_4O_7$ (hydrobilirubin), which would lead to a molecular weight 592 for hydrobilirubin. In three independent determinations of the molecular weight in phenol solvent, with accurate weighing to four significant figures, but melting point depressions only to two significant figures, *Abel* found 480, 410, and 550 for the molecular weight of hydrobilirubin – or reasonably close to *Maly*'s formula (*210*):

Die Bestimmung des Moleculargewichtes des Bilirubins durch Nencki und *Rotschky* kann, wie schon eingangs hervorgehoben, wegen der Zersetzlichkeit und Schwerlöslichkeit desselben nicht als entscheidend für die eine oder die andere Formel angesehen werden. Bekanntlich erhielt Maly durch Reduction des Bilirubins mittelst Naturiumamalgam das Hydrobilirubin, identisch mit den von Jaffé aus Harn erhaltenen Urobilin. Die gut stimmenden Analysen Maly's entsprechen der Formel $C_{32}H_{40}N_4O_7$. Und gerade die Zusammensetzung des Hydrobilirubins war für Maly eine Veranlassung, die Städeler'sche Formel des Bilirubins zu verdoppeln, da dann die Bildung des Reductionsproductes aus dem Bilirubin sich auf die einfachste Weise ergibt:

$$C_{32}H_{36}N_4O_6 + H_2 + H_2O = C_{32}H_{40}N_4O_7.$$

Nun hat Hydrobilirubin die gute Eigenschaft, sich sowohl in Eisessig, wie in Phenol leicht zu lösen, und es war von hohem Interesse, zu sehen, ob dieser Farbstoff, dessen aus der Elementaranalyse abgeleitete Formel nicht theilbar ist, dem Raoult'schen Gesetze folgen würde. Auf die Bitte von Prof. Nencki hatte Herr Prof. Maly die grosse Freundlichkeit, das Hydrobilirubinrein darzustellen und mir etwas über 1·5 g zu übersenden. Das Präparat war zuletzt in Ammoniak gelöst und mit HCl gefällt. Ich habe es vor dem Gebrauch über H_2SO_4 im Vacuum bis zur Gewichtsconstanz getrocknet. 4

For comparison and calibration, eight molecular weight determinations of cholesterol ($C_{27}H_{46}O$, formula weight = 386) in phenol yielded 343, 395.7, 370, 423, 412, 394, 431, and 433. With such validation having been assured, *Abel* thus concluded in favor of the double formula for bilirubin: $C_{32}H_{36}N_4O_6$ (*210*).

3.2 Bilirubin Elemental Combustion Analyses and Molecular Weight Determinations at the Beginning of the 20th Century

By the end of the 19th century, many combustion analyses of bilirubin had been run, with varying results being due to sample impurity or salts present. Several competing empirical formulas had been proposed. It was time for a thorough re-analysis of the data and subject. To this end, in 1903, *Teeple*[6] conducted a series of carefully designed combustion analyses of bilirubin, which he isolated from pigmented ox gallstones, purified meticulously, and crystallized (*211*):

> The procedure first adopted was to dry the stones at 100°, powder and sift through a 30 mesh sieve, extract with ether in a Soxhlet apparatus until exhausted (12 to 20 hours), dry again and exhaust with chloroform in the same way (30 to 40 hours); dry again, place in a large evaporator and extract with successive portions of boiling distilled water until exhausted. This method of procedure removed fat, lecithin, cholesterin, uncombined bilirubin and other bile pigments soluble in chloroform, and bile. The red material remaining was then digested for several hours with an excess of very dilute hydrochloric acid, washed by decantation until the filtrate was free from calcium and hydrochloric acid, dried thoroughly and extracted in a Soxhlet apparatus, first with ether (2 hours), then with absolute alcohol until the alcoholic filtrate had only a light yellow color (6 hours), finally with pure chloroform until exhausted (one hundred twenty-five hours). Fresh chloroform was used every eight hours and all extractions and other work with this solvent and bilirubin were done in a photographic dark room. The chloroform extracts containing the bilirubin were concentrated, the bilirubin precipitated by absolute alcohol, and the mixture heated on the water bath until most of the chloroform had been removed. The alcohol was then decanted and the precipitated bilirubin was boiled repeatedly with absolute alcohol until exhausted, to remove the biliverdin, biliprasin, bilifuscin, etc. The product so obtained was not completely soluble in chloroform, but it left no ash when burned. (In nearly all prolonged extractions with the Soxhlet apparatus we find the extracted mass not entirely soluble in the solvent used). [*sic*] It was therefore heated to boiling with a large quantity of chloroform in successive portions, the solution filtered twice, concentrated, precipitated by absolute alcohol, washed with alcohol, filtered, dried, and if not completely soluble in chloroform the process was repeated. Some of the product completely soluble in chloroform was then recrystallized from dimethyl aniline (B.P. 192.2 deg. to 192.8 deg.) [*sic*] This product was beautifully crystallized in reddish yellow microscopic needles and seemed perfectly homogeneous. A mixture of two parts chloroform and one part dimethyl aniline gave even better crystals. A chloroform solution of quinine was also found to dissolve large quantities and [from] this solution the bilirubin separated in crystals nearly as good as the preceding ones.

---

[6] *John Edgar Teeple* was born in 1874 and died in 1931. He completed his Ph.D. thesis (*148a*) at Cornell University (Ithaca, New York): *On Bilirubin, the Red Coloring-Matter of the Bile*, in which he reviewed briefly a history of the pigment and its known reactions, its relationship to the blood pigments, and especially its elemental combustion analyses. *Teeple*'s own experimental work for his thesis was on combustion analyses (some 17 of them) and molecular weight determinations. Unlike previous investigators of bilirubin, *Teeple* did not pursue a medical or academic career, except for a brief stint at Columbia University in New York as an interim professor of chemical engineering in 1917–1918. Nor did he pursue work on bilirubin past his Ph.D. Rather he briefly worked in industry, then spent the rest of his life as a consulting chemist. Ironically, his death followed a long and painful illness apparently induced by gallstones.

Combustion analyses were then conducted on 17 samples prepared as follows (*211*):

[1]. Bilirubin crystallized from dimethyl aniline dried at 120°, completely soluble in chloroform, leaves no ash.

0.2690 gram gave 17.7 c.c. $\frac{n}{10}NH_3$ or 0.02485 gram of N.

Kjeldahl-Gunning method modified to include nitrogen of nitrates, as adopted by the Association of Official Agricultural Chemists.

2. Same product and method as number 1. Digested three quarters of an hour longer.

0.3444 gram gave 22.7 c.c. $\frac{n}{10}NH_3$ or 0.3187 g. of N.

3. Product like No. 1 but different preparation. First reduced by zinc and sulphuric acid, then used Kjeldahl method.

0.2977 g. gave 19.07 c.c. $\frac{n}{10}NH_3$ or 0.02679 g. of N.

4. Same as No. 3. First reduced with sodium amalgam then used Kjeldahl method.

0.2145 g. gave 12.75 c.c. $\frac{n}{10}NH_3$ or 0.01790 g. of N.

5. Product like No. 1 but different prepraration.[2] Reduced by the action of P_3I_4 and H_2O, then used Kjeldahl method.

0.2158 g. gave 14.12 c.c. $\frac{n}{10}NH_3$ or 0.01982 g. of N.

6. Product like preceding ones but a different preparation. Dumas' method. CO_2 generated in a Kipp apparatus from marble that had been thoroughly boiled in water.

0.2208 g. gave 19.2 c.c. N at 732 mm and 24.5°.

7. Same product and method as No. 6.

0.1728 g. gave 15.25 c.c. N at 729.8 mm and 26°.

8. Remains of products used in analyses 3-7. Dumas method, generated CO_2 from boiled H_2SO_4.

0.1659 g. gave 14.2 c.c. N at 733.95 mm and 25.8°.

9. Product crystallized from dimethylaniline, not completely soluble in $CHCl_3$, but leaves a negligible residue. Dumas determination using magnesite to obtain CO_2 and first heating the CuO and the copper spiral in a current of CO_2 and allowing them to cool in the same gas to expel the air.[3]

0.1659 g. gave 14.2 c.c. N at 748.8 mm and 27°.

10. Product precipitated from $CHCl_3$ by alcohol. Same method as No. 9.

0.2084 g. gave 17.45 c.c. N at 742 m.m. and 24°.

11. Same product and method as No. 10; completely soluble in $CHCl_3$. 0.2050 g. gave 0.5048 g. CO_2 and 0.1133 g. H_2O.

12. Same product as No. 9. Same method as No. 11.

0.2836 g. gave 0.5761 g. CO_2 and 0.1248 g. H_2O.

13. [4]Same product as Nos. 10 and 11. Method of Benedict using a weighed amount of rock candy to furnish reduced copper.

0.2594 g. bilirubin, 0.1270 g. sugar gave 0.8300 g. CO_2 and 0.2078 g. H_2O. The sugar gave 0.1960 g. CO_2 and 0.7355 g. H_2O leaving 0.6340 g. CO_2 and 0.13425 g. H_2O.

14. Same as No. 13.

0.3025 g. bilirubin and 0.1348 g. sugar, gave 0.9504 g. CO_2 and 0.23995 g. H_2O. The sugar gave 0.2080 g. CO_2 and 0.07805 g. H_2O leaving 0.7424 g. CO_2 and 0.1619 g. H_2O.

15. [5]Product same as Nos. 10, 11, 13, 14.

0.1502 g. bilirubin and 0.1126 g. sugar gave 0.5411 g. CO_2 and 0.1543 g. H_2O. The sugar gave 0.1737 g. CO_2 and 0.0652 g. H_2O leaving 0.3674 g. CO_2 and 0.0891 g. H_2O.

---

[1]   Analyses 1 and 2 were made by Prof. G. W. Cavanaugh of the Agricultural Experiment Station at Cornell University.

[2]   Chenel, Bull. soc. chim. (Paris) 1892, [1], 321

[3]   Following a suggestion of Prof. Morse of the Johns Hopkins University.

[4]   Benedict, Am. Chem. Jour 23, 343: According to Mr. Benedict the reduced Cu. spiral as ordinarily used might contain enough CO_2 or other carbon compound to vitiate the result.

[5]   Analyses 18, 19, 20 were made by Prof. F. G. Benedict of Wesleyan University, Middletown, Conn., according to his own method.

3.2 Bilirubin Elemental Combustion Analyses and Molecular Weight...

16. Same as No. 15.

0.1512 g. bilirubin and 0.1017 g. sugar gave 0.5253 g. CO_2 and 0.1409 g. H_2O. The sugar[1] gave 0.1569 g. CO_2 and 0.0589 g, H_2O, leaving 0.3684 g. CO_2 and 0.0820 g. H_2O.

17. Same product as preceding. (Determination made by method used in cases in which the nitrogen is not in the oxidized condition.)

0.1245 g. gave 0.2990 g[2] CO_2 and 0.0675 g. H_2O.

---

[1] The sugar used in experiments 18 and 19 gave the following results:

	Calculated.	Found.
C.	42.07	42.04
H.	6.49	6.50

[2] Tube broke before completion.

Teeple indicated that the 17 samples came from seven different preparations of bilirubin that were analyzed by three different men "using 7 variations of methods for determining nitrogen and 3 modifications for the carbon and hydrogen." As it turned out *Teeple* believed that "the more refined and accurate the methods of analysis used the greater seemed to be the divergence from the percentages required for $C_{16}H_{18}N_2O_3$", which is the simple formula assigned by *Städeler* (*135*) – and that which best fits *Nencki*'s and *Rotschky*'s molecular weight[7] for bilirubin (*208*). The summarized results (below) to which the above refers were given by *Teeple* as (*211*):

SUMMARY

NO.	C.	H.	N.
1			9.24
2			9.25
3			9.10
5			9.19
6			9.36
7			9.39
8			9.17
9			9.31
10			9.16
11	67.15	6.18	
12	67.25	5.99	
13	66.66	5.80	
14	66.93	5.99	
15	66.62	6.63	
16	66.45	6.07	
17	65.50	6.07	
Averages	66.84	6.02	9.22
Computed for ($C_{16}H_{18}N_2O_3$)	67.08	6.34	9.81

---

[7] N.B. If the analytical results fit the *Städeler* formula, they also fit the doubled formula.

190   3 Advent of the Bilirubin Structure Proof

Note the spread in the %C values, over a range from 65.50 to 67.25; whereas, the %N variation is much smaller. *Teeple* provided three possible rationales for the divergence of the average %C, H, N from the calculated %C, H, N of the *Städeler* formula and concluded that the first possibility "ultimately proved to be the correct one" (*211*):

> 1. The bilirubin might not be a pure chemical individual.
> 2. The substance might be so difficult to analyze that the ordinary analytical methods and precautions were inadequate.
> 3. The commonly accepted formula for bilirubin $C_{16}H_{18}N_2O_3$ might be wrong.

> Regarding the first possibility which ultimately proved to be the correct one, we had used every precaution that had ever been suggested to insure a pure product, besides some additional ones. Most of the determinations were made on crystallized products that appeared perfectly homogeneous when examined microscopically. The bilirubin precipitated from $CHCl_3$ by alcohol (such as Städeler and Maly analyzed) gave practically the same analytical results as the crystallized products (such as Küster and von Zumbusch analyzed). All were perfectly soluble in $CHCl_3$ and left not a trace of ash when burned. Analysis No. 1 was made on the first crystals and No. 6 on the last crystals from the same dimethyl aniline with successive portions of bilirubin dissolved in it. There was no evidence of a fractional precipitation from this solvent. We sometimes noticed when dissolving a product in $CHCl_3$ that one portion seemed to go into solution more readily than another but attributed this merely to differences in the mechanical state of division. As far as we could judge from the descriptions given our bilirubin compared very favorably in purity with the best that any previous investigators had obtained.

Teeple was convinced that his bilirubin samples were at least as pure as any obtained earlier by others. Bilirubin precipitated from $CHCl_3$ by alcohol in the same procedure as used by *Städeler* and by *Maly* yielded essentially the same results upon analysis as did the crystalline bilirubin of *Küster* (*198, 199*) and *von Zumbusch* (*212*). Both sample types were soluble in $CHCl_3$ and ash-free upon combustion. A differential solubility in $CHCl_3$ was noticed at times by *Teeple*, with one part dissolving more readily than another. Yet fractional crystallization in dimethylaniline produced practically no difference in %N analyzed from No. 1, the first crop of crystals, and No. 6, the last.

At first *Teeple* thought that the low experimental %N relative to that predicted by *Städeler*'s formula was due to an insufficiency in the *Kjeldahl* analytical method, *i.e.* incomplete heating at white hot. However prolonged heating produced the same results as obtained in Nos. 1–10 of the table. Though the average %C, H was also low relative to *Städeler's* formula, *Teeple* felt most confident in the values in Nos. 13, 18, 19, and 20, which are close to the values predicted by $C_{16}H_{18}N_2O_3$.

The average values of the %C, H, and N from *Teeple's* 17 combustion analyses could be compared to those obtained by the earlier investigators: *Städeler*, *Thudichum*, *Maly*, *Küster*, and *von Zumbusch* on $CHCl_3$-soluble, ash-free bilirubin. The average value of 12 C, H analyses yielded %C, 66.79; %H, 6.19. The average value of 11 N analyses gave %N, 9.21. It should be noted, however, that the spread in values was rather large: from 65.61 to 67.45 %C, from 5.97 to 7.03 %H, and from

3.2 Bilirubin Elemental Combustion Analyses and Molecular Weight...

8.3 to 11.48 %N. Yet the values averaged from the five previous investigators were acceptably close to the average of the 17 combustion analyses from *Teeple (211)*:

NAME		C.	H.	N.	REMARKS
Städeler	1864	67.15	6.27	9.59	N. by soda lime method
		67.11	6.12		Bilirubin precipitated from chloroform by alcohol.
Thudichum	1868	66.02	5.97	9.05	N. by absolute method.
		66.41	6.13	9.49	N. by soda lime method.
		65.61	5.95	8.82	From these analyses he derived
				8.3	the formula $C_9H_9NO_2$.
				9.03	
Maly	1868	67.16	6.18		Bilirubin precipitated from
	1874	67.52	6.22		chloroform by alcohol.
		66.95	6.29		
			6.29		
Küster	1898	66.94	7.03	11.48	Bilirubin crystallized from
		66.99	6.88	11.35	dimethylaniline.
		67.45	6.77	10.21	Only the last of the four
		67.19	6.46	10.12	products was perfectly soluble in $CHCl_3$ and free from ash.
von Zumbusch	1901			9.30	Crystallized. Kjeldahl method.
Average of all analyses on products soluble in $CHCl_3$ and free from ash		66.79	6.19	9.21	
Average found by us		66.66	5.98	9.22	
Theory for $C_{16}H_{18}N_2O_3$		67.08	6.35	9.81	
Theory for $C_{34}H_{36}N_4O_7$		66.61	5.94	9.17	

Believing that the discrepancy between the experimental %N average value and that predicted by *Städeler*'s formula was due to an incorrect formula, because *Städeler* had based it on but one determination – and that by the obsolete soda lime method, *Teeple* found a better fit between the analytical data and published a revised formula for bilirubin: $C_{34}H_{36}N_4O_7$ (formula weight 612) in 1901 (*211, 213, 214*). Shortly thereafter, however, *Küster* proposed (*215*) that bilirubin was a mixture of two substances of differing solubility in $CHCl_3$, and the less soluble material (for which he retained the name *Bilirubin*) gave a combustion analysis compatible with *Städeler*'s formula, $C_{16}H_{18}N_2O_3$. The more soluble material generally gave a lower %N and %C. Following up on *Küster*'s discovery, *Teeple* repeated the separation, and his analysis of the less soluble material from $CHCl_3$ also reaffirmed the *Städeler* formula. *Teeple* noted that all these results came from the bilirubin isolated from gallstones, whether they also applied to the (bilirubin) pigment from bile was unknown.

Teeple determined the molecular weight of bilirubin, albeit not directly (*211*, *213, 214*). Following *Ehrlich*'s example of reacting bilirubin with diazotized benzenesulfonic acid, the pigment was reacted with diazotized tribromoaniline to precipitate the azo derivative, which was purified and analyzed for %Br and %N (*148a*):

I.	.1348 g. gave	.1202	g.	Ag. Br.	– Carius method.
II.	.1070 g. gave	.0954	g.	Ag. Br.	– Lime method.
	.1640 g. gave	10.31		c.c. $\frac{n}{10}NH_3$, reduced by Chenel method,	
				then Kjeldahl-Gunning method.	

	I.	II.	Computed for $C_{32}H_{34}N_4O_6(C_6H_2Br_3N_2)$
Br.	37.95	37.94	38.24
N.		8.83	8.95

The best fit to the data was the formula $C_{32}H_{34}N_4O_6(C_6H_2Br_3N_2)_2$, and so *Teeple* considered two possible formulas for bilirubin: (1) $C_{16}H_{18}N_2O_3$, the formula then in vogue, with a MW = 627 for the azo compound, and (2) $C_{32}H_{36}N_4O_6$ (azo compound MW = 1,254.5). From the boiling point elevation of the azo derivative, the experimentally determined MWs ranged from 1014 to 1467 (nine determinations) in $CHCl_3$ and 1042 in ethyl acetate (one determination) (*211*):

If the formula of bilirubin is written $C_{16}H_{18}N_2O_3$, then the molecular weight of this azo compound would be 627, if it is written $C_{38}H_{36}N_4O_6$, the molecular weight would be 1254.5. Notwithstanding the large molecular weight of the substance and the small amount at our disposal we made some attempts to determine it with the following results, using the Landsberger-McCoy apparatus:–

grams of material	SOLVENT $CHCL_3$ [sic] c.c. solution	rise	molecular weight
1.15	11.2	.207	1289
	13.9	.147	1463
	17.8	.130	1292
	19.3	.135	1148
.9763	8.6	.207	1422
	12.1	.153	1371
	16.3	.122	1014
	23.8	.086	1240
.526	11.8	.079	1467
	SOLVENT ACETIC ETHER.		
.2675	8.4	.096	1042

Teeple therefore concluded that the azo compound had the formula $C_{34}H_{34}N_4O_6(C_6H_2Br_3N_2)_2$, which meant the molecular formula for bilirubin is $C_{32}H_{36}N_4O_6$ (MW 572) – or rather close to that known today, $C_{33}H_{36}N_4O_6$ (MW 584).

3.3 Molecular Fragmentation: Initial Approach to Bilirubin Structure

By the beginning of the 20th century, virtually all of the important investigators of bile pigments had expired, passed from the scene, and a new generation of scientists had begun to pursue studies directed toward clarifying elemental combustion analyses and chemical formulas and, especially, learning the *chemical structures* of the pigments of bile and gallstones and blood. The first among the new investigators were, most prominently, *Nencki* and *Küster*.[8] Although their focus was primarily the pigment of blood, their discoveries carried over easily to the pigments of bile because a probable link between the two pigments had been drawn earlier. The new investigators were, of course, the beneficiaries of the discoveries, advances, and failures of their predecessors' isolation and purification methods, reports on the properties of the pigments, including spectrum analyses and the more advanced but often contested elemental combustion analyses and formulas derived therefrom. Most important to their success was access to and a working familiarity with the emerging but rapidly expanding field of organic chemistry, where by the late 1800s chemists had developed a good understanding of the *structures* of small molecules and understood how to *synthesize* them. Knowledge and ability that were absent until late in the 19th century. This knowledge of organic chemistry enabled *Küster* to isolate, identify, and characterize structural fragments of the pigments of blood and bile, which prompted him to propose sketchy chemical structures of the intact pigments. With orientation toward, and training in the organic chemistry of the 1890s, *Küster* was well-prepared to isolate and characterize small molecule products coming from degradation of bilirubin, biliverdin, hemin, *etc*. From such a mind-set, by following this path he broke with all previous investigators and was thus uniquely qualified in the late 1800s and early 1900s to begin to provide important structural pieces to the unsolved puzzle of pigment structure. But first some background on the chemist who successfully introduced oxidative cleavage to bilirubin.

In 1891, *Hüfner*, who had established a reputation for research on the binding capacity of hemoglobin for O_2, suggested that *Küster* study the prosthetic group (hematin) of the blood pigment (*216*):

> Auf Veranlassung von Hrn. Prof. Hüfner habe ich daher das Studium des Hämins aufgenommen und will im Folgenden über die seit dem Sommersemester 1891 gewonnenen Resultate berichten.   5

---

[8]*William Küster* was born on September 22, 1863 in Leipzig and died on March 25, 1929 in Stuttgart. He received his early education in Berlin, studied at the Universities in Berlin and Leipzig (with *Wislicenus*) and at the Physiological Chemistry Institute in Tübingen under *Gustav von Hüfner*. *Küster* habilitated in 1896, became *Privatdozent* and *Assistent*, and in 1900 a. o. Professor at Tübingen. From 1903 to 1913 he was o. Professor of Chemistry at the Thierärztlichen Hochschule zu Stuttgart, and from 1913 until his death as o. Professor of Organic Chemistry at the TH-Stuttgart.

194   3 Advent of the Bilirubin Structure Proof

And this became a subject of *Küster*'s focus in his 1896 *Habilitationsschrift* (*Beiträge zur Kenntniss des Hämatins* – Contributions to the Knowledge of Hämatin), from which he began a quest to establish the molecular formulas, molecular weights, and chemical structures of the pigments of blood and bile as *Privatdozent, Assistent*, and in 1900 as a. o. Professor. For ten years, from 1894 (*216*) to 1904 (*217, 218*), *Küster* published from the Physiologisch-Chemischen Institut zu Tübingen, from 1903 to 1913 he published as o. Professor of Chemistry at the Chemischen Institut der Königl. Thierärtzlichen Hochschule zu Stuttgart, and from 1913 as o. Professor of Organic Chemistry and Pharmacy at the Technischen Hochschule Stuttgart (now the Universität Stuttgart) until his untimely death of cardiac arrest on March 5, 1929, only shortly before the nominations deadline for the 1930 *Nobel* Prize in Chemistry, which was thus awarded to only *Hans Fischer* for his monumental total synthesis of the red pigment prosthetic group (hemin) of hemoglobin. A pigment whose then breathtakingly novel structure had already been proposed by *Küster* in 1912. Although promptly rejected by *Richard Willstätter* and *Hans Fischer*, it was proved correct in nearly very detail.

Nencki, some 17 years senior to *Küster*, died in 1901 when *Küster* was hitting his stride in investigations of the blood and bile pigments. In addition to settling the elemental composition and molecular formulas for the prosthetic pigment of hemoglobin, these two scientists discovered two very different chemical reactions that pried out small molecular components from biliverdin, bilirubin, and the blood pigments: hemin, hematin, and hematoporphyrin. *Nencki* investigated the blood pigment almost exclusively, except for urobilin (*Maly*'s *Hydrobilirubin*); *Küster*'s early work was focused on the blood pigment. Yet, drawing from the probable biological origin of bile pigments from the blood pigment, *Küster* also investigated the structures of bilirubin and biliverdin by fragmentation degradation.

The work of these two investigators marked the intersection of organic structural chemistry and synthesis applied to the pigments of blood and bile. Earlier investigations had provided probable cause to believe that a strong relationship existed between *Hämatoidin* (hematoidin) from blood and bilirubin; hence the apparent identification of a chemically derived blood degradation product (hematoporphyrin) with bilirubin – achieved on the basis of elemental combustion analyses and molecular formulas derived therefrom; and studies that appeared to link the bile pigments to the urinary pigment (urobilin) by reduction. Yet, all these lacked true chemical validation on the basis of structure.

In order to understand how the structure of bilirubin was established, it is important to understand when and how investigations of the pigments of bile and blood intersected – and to what benefit.

Many years before the time of *Nencki* and *Küster*, in 1840, near the end of *Berzelius*' long reign as perhaps the most influential scientist in Europe, one of his students, *Hünefeld*[9] reported the discovery of hemoglobin (*219*), though that name

---

[9] *Friedrich Ludwig Hünefeld* was born on February 16 1784 in Boltenhage and died on March 24, 1882 in Greifswald. He was Professor of Chemistry, Pharmacy, and Mineralogy, and *Weigel*'s successor at the University of Greifswald from 1831.

3.3 Molecular Fragmentation: Initial Approach to Bilirubin Structure

was not applied to the pigment of blood until some 25 years later by *Hoppe-Seyler* (*220*). *Hünefeld*'s discovery set off a series of investigations into the pigment of blood, as described by *E.T. Reichert* and *A.P. Brown* in 1909 (*221*). Thus, the accidental discovery by *K.E. Reichert* of tetrahedral crystals of *Hämoglobin* (hemoglobin) in fetal membranes and mucous membranes of the uterus of a suddenly deceased guinea pig (*222*) was followed by other chance discoveries, until in 1851 *Funke* prepared crystals of hemoglobin directly from red cells (*223*), and *Teichmann*[10] discovered that red-brown prismatic microscopic crystals of hemin were seen after mixing dried blood with acetic acid and NaCl (*224*) – a forensic test for blood still in use.

His contemporary, *Hoppe-Seyler* showed that the prosthetic group of hemoglobin was invariant over a wide range of animal bloods. He cleaved the crystallizable pigment prosthetic group from hemoglobin by the action of a large excess of glacial acetic acid and identified it as hemin and converted the latter hematin by dissolving in dilute aqueous KOH or in NH_4OH, followed by neutralized by added hydrochloric acid (*220*):

> Durch die Aetzlaugen fixer Alkalien schnell, durch Aetzammoniak langsamer wird das Hämoglobin in Hämatin und Globulin gespalten, ebenso durch Säuren z. B. Essigsäure, Weinsäure. Bei Gegenwart von ClHverbindungen[a] wird Hämoglobin durch grossen Ueberschuss chlorwasserstoffsaures Hämatin; durch Lösen in Aetzalkalien schon allmälig durch das Ammoniak der Luft bei hinlänglicher Feuchtigkeit derselben wird das Hämin in Chlormetall und Verbindung des Alkali mit Hämatin zerlegt. Man erhält reines Hämatin aus den Häminkrystallen durch Auflösen in Ammoniak, Abdampfen zur völligen Trockne, Extrahiren des Rückstandes mit Wasser und Trocknen. Das Hämatin besitzt dieselbe dunkel graublaue Farbe wie das Hämin.   6

---

[a] As it appears in the original; probably a misprint.

Hoppe-Seyler concluded that hemin is the hydrochloride of hematin since the former was converted to the latter by dissolving in aqueous NH_3. Combustion analyses led him to calculate early formulas for hemin ($C_{48}H_{51}N_6O_9Fe_3 \cdot HCl$) and hematin ($C_{48}H_{51}N_6O_9Fe_3$) and to write a reaction for the formation of bilirubin from hematin (*225*):

> Vergleicht man obige Formel des Hämatin mit der, welche Städeler*) kürzlich in seiner schönen Untersuchung über die Gallenfarbstoffe für das Bilirubin angibt, so ergibt sich die einfache Beziehung:

$$2\ (C_{48}H_{51}N_6Fe_3O_9) + 3\ H_2O = 6\ (C_{16}H_{18}N_2O_3) + 3\ Fe_2O$$
$$\text{Hämatin} \qquad\qquad\qquad \text{Bilirubin}$$

> oder durch Substitution von Wasserstoff an die Stelle des Eisens würden aus 1 Molekül Hämatin 3 Moleküle Bilirubin entstehen. Auf die physiologischen Gründe, die für die

---

[10] *Ludwig Karl Teichmann* was born in 1823 and died in 1895. He received the Dr. med. from Göttingen and became Professor of Anatomy in Krakau. The confirmatory *Teichmann* test for hemoglobin, based on the formation of distinctive hematin micro-crystals after treatment of blood stains with concentrated acetic acid/NaCl, is named after him.

196 3 Advent of the Bilirubin Structure Proof

Annahme der Bildung Gallenfarbstoffes aus dem Hämatin sprechen, habe ich bereits in früheren Arbeiten über die Gallensubstanzen so ausführlich hingewiesen, dass ich hier nicht nochmals darauf zurückkommen will.

---

*) Mittheilungen aus dem analytischen Laboratorium in Zürich 1863. I. Ueber die Farbstoffe der Galle. 7

Some 20 years earlier, crude hematin had been prepared and isolated from blood by *Mulder* and *van Goudoever*, who in 1844 (*226*) reported its elemental combustion analysis and suggested a formula ($C_{44}H_{44}N_6O_6Fe$) for the substance, and for its iron-free derivative, $C_{44}H_{44}N_6O_6$. Of course useful/truthful combustion analysis data invariably depend on sample purity, which was questionable for most pigment studies. And for iron-containing samples, combustion will leave an ash, whether exclusively from Fe or from inorganic contaminants. Newer sample preparations and/or purifications led to new formulas. Thus, as reported in 1871 (*227*), extensive analytical preparations of hemin and hematin from blood led *Hoppe-Seyler* to report their formulas as $C_{68}H_{72}N_8O_{10}Fe_2Cl_2$ for the former and $C_{68}H_{70}N_4O_{10}$ for the latter – and two iron-free analogs, $C_{68}H_{74}N_8O_{12}$ and $C_{68}H_{78}N_8O_7$ (which he named *Hämatoporphyrin* [hematoporphyrin] and *Hämatolin*, respectively), obtained from hematin by dissolving in conc. H_2SO_4. Yet none of these combustion-based formulas would prove to be correct.

Hoppe-Seyler also studied bile pigments and in 1874 reported work from his laboratory in Tübingen that the reduction product, easily obtained from hemoglobin with Sn and HCl in alcohol, is identical with *Jaffe*'s *Urobilin*, or *Maly*'s *Hydrobilirubin* prepared by reduction of bilirubin with Na(Hg). (A conclusion that did not sit well with *Thudichum*.) He inferred that bilirubin and biliverdin are intermediates between the pigment of blood and the pigments of urine and feces (*228*):

> ... mit den von Hrn. Dr. Kistiakowsky in meinem Laboratorium aus Bilirubin durch Einwirkung von Natriumamalgam, oder von Zinn und Salzsäure gewonnenen Präparate, die Ueberzeugung gewonnen, dass das Urobilin von Jaffé, oder Hydrobilirubin von Maly mit meinem Reductionsprodukte des Hämatin identisch sind. Da man nun bei Behandlung von unzersetztem Hämoglobin mit Zinn und Salzsäure in alkoholischer Lösung denselben Farbstoff mit Leichtigkeit erhält, so ergiebt sich, dass der Farbstoff normaler Fäcalstoffe und des Harnes als ein durch Reduction verändertes Spaltungsprodukt des Blutfarbstoffs aufgefasst werden darf, dass die Gallenfarbstoffe Bilirubin und Biliveridin Zwischenstufen dieser Umwandlung darstellen, oder wenigstens zum Blutfarbstoff in so naher Beziehung stehen ... 8

– an inference that was later substantiated. Within 10 years of *Hoppe-Seyler*'s postulation, *Nencki* had begun studies (*229, 230*) designed to clarify some issues related to the earlier formulas for hemin, hematin, and the iron-free derivative, hematoporphyrin. *Nencki* and *Sieber* summarized the work of previous investigators, citing *Hoppe-Seyler*'s formulas for hemin ($C_{68}H_{70}N_8O_{10}Fe_2HCl$) and hematin ($C_{68}H_{70}N_8O_{10}Fe_2$), and analyzed the pigments obtained from their own preparations. Thus, dried blood from cattle (I), horse (II), pig (III), and humans (IV) after heating in amyl alcohol up to 100°C yielded hemin crystals, which, after drying over conc. H_2SO_4 and taken through a combustion analysis, gave the

3.3 Molecular Fragmentation: Initial Approach to Bilirubin Structure

formula: $(C_{32}H_{30}N_4O_3FeHCl)_4C_5H_{12}O$, containing ¼ molecule amyl alcohol of crystallization per molecule of hemin (*229*):

			C	H	Cl	Fe	N	
I.	Rinderblut	1)	62.73	5.69	5.29	8.95	8.99	pCt.
		2)	62.75	5.71	5.28	8.72	–	»
II.	Pferdeblut	3)	62.81	5.86	5.38	8.61	9.13	»
		4)	62.90	5.98	5.30	8.96	–	»
III.	Schweineblut	5)	62.72	5.72	–	–	–	»
IV.	Menschenblut	6)	–	–	5.22	8.96	–	»

Mit Rücksicht darauf, dass die Krystalle in ihrem Molekül Amylalkohol enthalten, entspricht die procentische Zusammensetzung der Formel: $(C_{32}H_{30}N_4FeO_3HCl)_4C_5H_{12}O$; welche verlangt: 63.09 pCt. C, 5.69 pCt. H, 5.59 pCt. Cl, 8.86 pCt. Fe und 8.86 pCt. N. 9

Similarly, crystals of hematin, prepared from hemin by dissolving in alkali, gave the formula: $C_{33}H_{32}N_4O_4Fe$ (*229*):

Die Elementaranalysen des aus den Häminkrystallen dargestellten Hämatins ergaben ferner folgende Zahlen:

			C	H	Fe	N	
I.	Rinderblut	1)	64.98	5.61	9.35	9.49	pCt.
II.	Pferdeblut	2)	64.99	5.62	9.29	9.34	»
		3)	64.68	5.37	–	–	»
III.	Schweineblut	4)	65.04	5.53	9.29	–	»
		5)	65.13	5.55	9.31	–	»

Die für das aus den Häminkrystallen dargestellte Hämatin erhaltenen Zahlen entsprechen der Formel: $C_{32}H_{32}N_4FeO_4$, welche verlangt: C 64.68 pCt., H 5.40 pCt., N und Fe 9.47 pCt. 10

Thus, *Nencki* and *Sieber* wrote that the transformation from hemin, which they considered to be a hydrochloride salt, to hematin occurred by replacement of HCl in the former by H_2O in the latter (*229*):

Beim Auflösen der Häminkrystalle in Alkalien wird daher nicht allein Salzsäure und Amylalkohol abgespalten, sondern auch Wasser in das Molekül aufgenommen, entsprechend der Gleichung:

$$(C_{32}H_{30}N_4FeO_3HCl)_4C_5H_{12}O + (NaOH)_4 = (C_{32}H_{32}N_4FeO_4)_4 + C_5H_{12}O + (NaCl)_4.$$ 11

Accordingly, they assigned the formula $C_{32}H_{30}N_4O_3Fe$ to hemin, which was obtained originally by *Teichmann* as the hydrochloride salt, and to hematin was assigned $C_{32}H_{32}N_4O_4Fe$ (*229*):

Wir werden daher den Körper: $C_{32}H_{30}N_4FeO_3$, mit dem Namen Hämin bezeichnen. Die Teichmann'schen Krystalle sind die salzsaure Verbindung desselben. Durch Auflösen des Hämins in Alkalien wird das letztere in das Hämatin verwandelt, dessen Zusammensetzung $= C_{32}H_{32}N_4FeO_4$ ist. 12

And the *Hoppe-Seyler* treatment of hematin with H_2SO_4 to form hematoporphyrin was carried out and explained in terms of an oxygenation of the revised formula above for hematin, though its yield was not great (*229*):

> ... Analysen ist das Hämatoporphyrin nach der Formel: $C_{32}H_{32}N_4O_5$ zusammengesetzt und seine Bildung geschieht nach folgender Gleichung:
>
> $$C_{32}H_{32}N_4O_4Fe + SO_4H_2 + O_2 = C_{32}H_{32}N_4O_5 + SO_4Fe + H_2O.$$
>
> Aus Hämatin ist die Ausbeute an Hämatoporphyrin nicht gross.   13

Hoppe-Seyler's Hämatolin, the alkali-insoluble undissolved solid remaining after treating hematin with conc. H_2SO_4, was also found by *Nencki* and *Sieber*, who avoided *Hämatolin* formation by dissolving hemin completely in conc. H_2SO_4. They investigated the reduction of hemin in alcohol by Sn + hydrochloric acid to yield hexahydrohematoporphyrin ($C_{32}H_{38}N_4O_5$) (*160a*):

> Die Bildung des Hexahydrohämatoporphyrins aus den Häminkrystallen erfolgt unter gleichzeitiger Aufnahme von Wasser und Wasserstoff in das Molekül, gemäss der Gleichung:
>
> $$C_{32}H_{30}N_4O_3FeHCl + 2\ H_2O + HCl + H_2 = C_{32}H_{38}N_4O_5 + FeCl_2.$$ 14

which when heated with alcoholic KOH produced an easily alkali-soluble product that had a great similarity to urobilin (*229*):

> Mit alkoholischer Kalilauge gekocht wird das Hexahydrohämatoporphyrin in ein in wässerigen Alkalien leicht lösliches Produkt verwandelt, das mit dem Urobilin grosse Aehnlichkeit hat.   15

Hoppe-Seyler had previously found a urobilin-like product from treatment of hematin with Sn and HCl but did not analyze the product (*228*). *Nencki* and *Sieber* also found a urobilin-like product in the preparation of hexahydrohematoporphyrin formed according to the following equation (*229*):

> Es entspricht dies auch der theoretischen Voraussetzung; dem die Umwandlung des Hämins zu Urobilin beruht vorwiegend auf einer Hydratation:
>
> $$C_{32}H_{30}N_4FeO_3 + 4\ H_2O + 2\ HCl = C_{32}H_{40}N_4O_7 + FeCl_2.$$ 16

More interesting from a structural viewpoint, hematin in molten KOH evolved considerable amounts of a *Pyrrol*, detected only by odor and a positive pine splint test. The *Pyrrol* obtained was probably a mixture of substituted pyrrole derivatives and certainly not pyrrole (C_4H_5N) itself. At the time of this work, *Pyrrol* (pyrrole) was known, having been isolated from the volatile distillates of coal tar as a red oil (*Rotöl*) by *Runge* in 1834 (*231, 232*), who gave it the name *Pyrrol* for the brilliant red color that develops when its vapor comes in contact with a pine splint(er) wetted with hydrochloric acid, and from the Greek for flame-colored, *purrhos*, πυρρός. It is also a component of oils arising from the destructive distillation of horns and bones (bone oil) practiced several centuries earlier (*233*). Subsequently, purified, colorless pyrrole (bp 133°C) was isolated by *Anderson* in 1858, and by combustion analysis it was given the formula C_8H_5N (*234, 235*), based on the then acceptable atomic weight of 6 for carbon, but actually C_4H_5N based on the current atomic

3.3 Molecular Fragmentation: Initial Approach to Bilirubin Structure 199

weight, $C = 12$. Though the purified liquid was investigated further in its reactions with $HgCl_2$, with which it formed a solid mercury "salt" that analyzed for $C_4H_5NHg_4Cl_4$. (This solid salt formation with Hg^{2+} was destined to become useful later to *Nencki*.) At the time, even with such a simple formula for pyrrole, a structure for the compound was not conceivable. Until a few years later (in 1860), *Schwanert* synthesized pyrrole (*236*) by reaction of galactaric acid or mucic acid ($C_6H_{10}O_8$) (*Schleimsäure*), obtained by nitric acid oxidation of the lactose (*Milchzucker*) of milk, with NH_3 to form the bis-ammonium salt, followed by pyrolysis of the salt. However, like many organic chemicals in 1860, though the formula was known, the structure was unknown. Some ten years later the situation regarding organic structure had improved considerably in *Baeyer*'s[11] writings.

Baeyer and *Fittig* (*237, 238*) recognized sugars, including galactose, as polymers of formaldehyde $(CH_2O)_n$ and described a nonstereochemical semi-structural formula for the hexose hydrate. Thus, in 1870 *Baeyer* wrote structure I in the manner of *Fittig* for the two then known aldohexoses, glucose and galactose (*237*):

I

$CH_2 . OH$

$CH . OH$

$CH . OH$

$CH . OH$

$CH . OH$

$CH . (OH)_2$

Extrapolating from his proposed structure of *Indol* (*239*) and recognizing that pyrrole was synthesized from C_6 mucic acid in a pyrolytic reaction involving loss of CO_2 and H_2O, he wrote the first, correct structure for pyrrole in 1870 (*239*):

... so bekommt man die Formel des Pyrrols C_4H_5N, so dass das Indol danach aus Benzol und Pyrrol zusammengesetzt wäre, gerade wie das Naphtalin aus zwei Benzolringen.

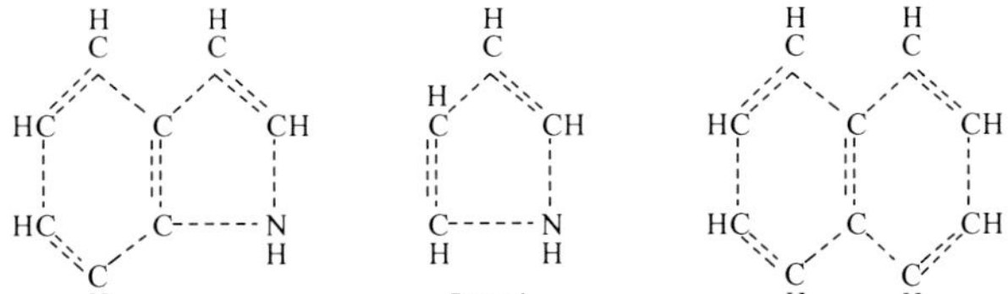

Indol.

Pyrrol.

Naphtalin.

17

---

[11] *Johann Friedrich Wilhelm Adolf von Baeyer* was born on October 31, 1835 in Berlin, Prussia and died on August 20, 1917 in Starnberg, Bavaria. He studied mathematics and physics at Berlin University from 1853-1855 and in 1856 transferred to Heidelberg to study chemistry first with *Bunsen*, then in *Kekulé*'s laboratory, received the Dr. phil. in 1858 from Berlin University and spent several years working with *Kekulé* who had become Professor at Ghent, took a position as Lecturer at the Gewerbeakademie Berlin, moved to the University of Strassburg as Professor in 1871 before being called to succeed *Justus von Liebig* in 1873 at the Universität München, where he was when awarded the *Nobel* Prize in Chemistry in 1905.

The mucic acid route to pyrrole persisted through the 1930s as the best route for its synthesis and was incorporated into undergraduate organic laboratory teaching manuals (*240*) of the 1950s. In fact the synthesis of *N*-methylpyrrole was carried out in the author's first organic laboratory course at the University of California at Berkeley in December 1958, his first (and foul-smelling) synthesis of a pyrrole.

Though reduction of hemin and hematin with Sn + hydrochloric acid led to new products, oxidation of hematin with either HNO_3, or $KMnO_4$ in alkaline solution went much too far – a reaction that would be revisited by *Küster*.

What was the then known relationship between the pigments of blood and bile? *Nencki* and *Sieber* (*229, 230*) accepted *Maly*'s results: (i) his formula for bilirubin ($C_{32}H_{36}N_4O_6$), which represents a doubling of the older formula ($C_{16}H_{18}N_2O_3$) of *Städeler*; (ii) his conversion of bilirubin by Na(Hg) treatment into hydrobilirubin of formula $C_{32}H_{40}N_4O_7$, by the uptake of H_2 and H_2O; and (iii) his experiment in which bilirubin was converted to biliverdin [$C_{32}H_{33}(OH)_3N_4O_6$] by treating its tribromo derivative ($C_{32}H_{33}Br_3N_4O_6$, from reaction with Br_2) with KOH. All of which meant to *Nencki* and *Sieber* that if hematin were converted to bilirubin, it would lose Fe and take up H_2O (*229*):

III.

Die Beziehungen des Blutfarbstoffes zu Gallenfarbstoff.

Das Bilirubin ist nach der Formel $C_{32}H_{36}N_4O_6$ zusammengesetzt. Aus den Untersuchungen Maly's geht hervor, dass diese Formel der älteren von Städeler: $C_{16}H_{18}N_2O_3$, vorzuziehen ist. Das Bilirubin geht unter Aufnahme von H_2O und H_2 in das Urobilin (Hydrobilirubin = $C_{32}H_{40}N_4O_7$) über. Diese Umwandlung ist nur durch Verdoppelung der Städeler'schen Formel verständlich. Ferner giebt Bilirubin mit Brom das Tribrombilirubin = $C_{32}H_{33}Br_3N_4O_6$, das durch Alkalien zu Biliverdin umgewandelt wird.

$$C_{32}H_{33}Br_3N_4O_6 + 3 \ KHO = C_{32}H_{33}(HO)_3N_4O_6 + 3 \ KBr$$
Tribromobilirubin. Biliverdin.

Wenn Blutfarbstoff zu Gallenfarbstoff wird, so verliert er Eisen und nimmt Wasser in das Molekül auf.

$$C_{32}H_{32}N_4O_4Fe + 2 \ H_2O - Fe = C_{32}H_{36}N_4O_6$$
Hämatin. Bilirubin. 18

In this simple equation, chemistry fulfilled an old claim of physiology, that a structural and even a genetic relationship must exist between the pigment of blood and the pigment of bile. Bilirubin (hematoidin) from extravasated blood is deposited in tissues or eliminated as an increase in urobilin. This allowed *Nencki* and *Sieber* to state that the formation of bile pigments from blood under physiological conditions is understandable, that the reverse may be possible, that bilirubin in the liver cell is unfinished hemin (*229*):

Mit dieser einfachen Gleichung erfüllt die Chemie eine alte Forderung der Pathologie, dass zwischen dem Blutfarbstoff und Gallenfarbstoff ein genetischer Zusammenhang bestehen müsse. So oft Blut aus der lebendigen Gefässwand in das umliegende Gewebe austritt, wird entweder in den Geweben das Bilirubin (Hämatoïdin) abgelagert, oder es findet vermehrte Ausscheidung des Urobilins statt. Auch die Bildung des Gallenfarbstoffes

3.3 Molecular Fragmentation: Initial Approach to Bilirubin Structure

> aus Blutfarbstoff unter physiologischen Verhältnissen ist jetzt verständlich. Es ist aber auch möglicherweise das Umgekehrte der Fall, dass nämlich das Bilirubin in seinem Aufbau in der Leberzelle unvollendetes Hämin ist. 19

Whereas the first is understood as correct, the latter is unsupported, however.

Subsequent *Nencki* and *Sieber* publications of the late 1880s, while also repeating what was expressed in the 1884 publications (*229, 230*), indicated that hemin can be extracted from red corpuscles into amyl alcohol ($C_5H_{12}O$) containing HCl and crystallizes with inclusion of amyl alcohol to give $(C_{32}H_{30}N_4O_3FeHCl)_4C_5H_{12}O$. The alcohol is lost by heating at 130–135° to yield hemin as $C_{32}H_{31}N_4O_6FeCl$ (*229*). Hematin was said to result from having removed HCl from hemin by hydroxide, forming $C_{32}H_{32}N_4O_4Fe$. Hematoporphyrin, obtained from the last, was reported and reanalyzed to give a molecular formula ($C_{32}H_{34}N_4O_5$) with two more hydrogens than found previously (*229*):

> Wir haben die von uns für das Hämatoporphyrin aufgestellte Formel aus den Zahlen der Elementaranlysen berechnet. In zwei von verschiedener Darstellung herrührenden Präparaten haben wir gefunden:

> | | | | | | |
> |---|---|---|---|---|---|
> | C | 69,57 | und | 69,44 | Proc. | |
> | H | 6,20 | „ | 6,13 | „ | |
> | N | 9,67, | 9,83 | und | 10,17 | Proc. |

> Die von uns aufgestellte Formel $C_{32}H_{32}N_4O_5$ verlangt: C 69,55 Proc., H 5,80 Proc., N 10,14 Proc., O 14,51 Proc.
>
> Noch besser stimmen die erhaltenen Zahlen zu der Formel: $C_{32}H_{34}N_4O_5$, welche verlangt: C 69,31 Proc., H 6,13 Proc. und N 10,10 Proc.
>
> Sie unterscheidet sich von der obigen nur durch ein Plus von H_2. Die Entstehung des Hämatoporphyrin aus Hämatin wäre danach sehr einfach:

> $$C_{32}H_{32}FeN_4O_4 + H_2O - Fe = C_{32}H_{34}N_4O_5.$$ 20

Within a few years' time, and with improvements in the preparation of hematoporphyrin from hematin using HBr instead of H_2SO_4, purified, iron-free hematoporphyrin (its hydrochloride) salt was found by *Nencki* and *Sieber* to give a combustion analyses with a better fit to revised formulas: $C_{16}H_{18}N_2O_3$ for metal-free hematoporphyrin and $C_{16}H_{18}N_2O_3HCl$ for its hydrochloride (*241*):

> Die aus den erhaltenen Zahlen berechnete Formel des salzsauren Salzes = $C_{16}H_{18}N_2O_3HCl$ verlangt in Procenten

> | | Theorie | | Versuch | | |
> |---|---|---|---|---|---|
> | | | | I. | II. | III. |
> | C_{16}····· | 59·53% | C······ | 59·80% | 59·79% | 59·57% |
> | H_{19}····· | 5·89 | H······ | 6·16 | 5·89 | 6·29 |
> | N_2 ····· | 8·68 | N······ | 8·5 | – | 8·41 |
> | Cl ····· | 11·00 | Cl····· | 10·77 | 10·73 | 10·9 |

Die Hämatoporphyrinformel = $C_{16}H_{18}N_2O_3$ verglangt in Procenten

	Theorie		I.		II.	III.	
					Versuch		
C⸱⸱⸱⸱⸱	67·13%	C⸱⸱⸱⸱⸱	66·84%	und	67·16%	66·98%	66·85%
H⸱⸱⸱⸱⸱	6·29	H⸱⸱⸱⸱⸱	6·32	″	6.56	6.21	6.53
N⸱⸱⸱⸱⸱	9·79	N⸱⸱⸱⸱⸱	9·77	–	–	9·51	21

Which meant to the authors that hematoporphyrin had the same formula as bilirubin (*Städeler's* formula = $C_{16}H_{18}N_2O_3$) and that the two compounds were therefore isomers (*241*):

> Die Elementaranalysen des Salzsauren, sowie des daraus dargestellten freien Hämatoporphyrins ergaben uns die überraschende Thatsache, dass dasselbe nach der Formel: $C_{16}H_{18}N_2O_3$ zusammengesetzt, das heisst dem Gallenfarbstoff – dem Bilirubin – isomer ist. 22

Yet their colors were clearly different, as were their spectra.

That was of lesser concern to *Nencki* and *Sieber* who recognized that the hematoporphyrin derived from treatment of hematin with HBr had a different formula from that obtained using H_2SO_4 (*241*), which they rationalized by indicating that the latter product was most likely the anhydride of the former. That suggested a possible anhydride relationship between hemin crystals and hematoporphyrin by the uptake of H_2O (*241*):

$$(C_{16}H_{18}N_2O_3)_2 = C_{32}H_{34}N_4O_5 + H_2O.$$

> Das Mittelst concentrirter Schwefelsäure erhaltene Hämatoporphyrin wäre also ein Anhydrid des neu erhaltenen und diese Annahme erscheint uns auch als die wahrscheinlichste. Die Häminkrystalle sind nach der Formel $C_{32}H_{31}ClN_4O_3Fe$ zusammengesetzt. Bei der Bildung des Hämatoporphyrins daraus müsste freier Wasserstoff auftreten, entsprechend der Gleichung:

$$C_{32}H_{31}ClN_4O_3Fe + (BrH)_2 + 3\ H_2O = (C_{16}H_{18}N_2O_3)_2 + FeBr_2 + HCl + H_2 \qquad 23$$

A possible "isomeric" relationship between hematoporphyrin and bilirubin was somewhat clouded, however, by the expectation that hematin was the precursor to both, suggesting that the metal (Fe) was possibly responsible for holding together two units of $C_{16}H_{18}N_2O_3$, *Städeler's* original formula for bilirubin, which *Maly* had doubled in order to account for the conversion of bilirubin to biliverdin *via* tribromobilirubin (*241*):

> Als Ergebniss unserer ersten Untersuchung über das Hämatin haben wir den Satz aufgestellt, dass, wenn Blutfarbstoff in Gallenfarbstoff übergehe, dies unter Abspaltung von Eisen und Aufnahme von zwei Molekülen Wasser geschehen müsse.

$$C_{32}H_{32}N_4O_4Fe + 2\ H_2O - Fe = C_{32}H_{36}N_4O_6$$

Hämatin.   Bilirubin.

3.3 Molecular Fragmentation: Initial Approach to Bilirubin Structure

> Diese Gleichung ist durch die Darstellung des Hämatoporphyrins mittelst Bromwasserstoff realisirt. Gleichzeitig erscheint dadurch die ursprügliche einfache Formel des Bilirubins von Städeler als die richtigere. Aus der Zusammensetzung des salzsauren Hämatoporphyrins, sowie der Metallverbindungen desselben, geht mit Sicherheit hervor, dass ihm die Formel $C_{16}H_{18}N_2O_3$ zukommt. Maly hat die Städeler'sche Formel des Bilirubins verdoppelt. Veranlassung dazu waren ihm die Zusammensetzung des Tribrombilirubins, des Biliverdins und des Urobilins.
>
> Sicher ist es, dass aus der Bilirubinformel $C_{32}H_{36}N_4O_6$ die Bildung der genannten Producte sich viel einfacher erklärt. 24

Nencki and *Sieber* thus felt that on the basis of their studies they had achieved insights into the earlier work on bile pigments from: (i) *Städeler*, who proposed $C_{16}H_{18}N_2O_3$ as the bilirubin formula; (ii) *Maly*, who proposed a doubled formula, $C_{32}H_{36}N_4O_6$; and (iii) *Thudichum*, who insisted on $C_9H_9NO_2$. They concluded, rather audaciously, that *Thudichum*'s reports did not merit serious consideration and that every experienced chemist who read through his publications would certainly share this view (*241*):

> Für uns sind die Angaben Thudichum's einer ernsten Beachtung nicht werth, und jeder sachverständige Chemiker, der sich die Mühe gegeben hat, seine Publicationen durchzulesen, wird gewiss unserer Ansicht sein. 25

Such a statement seems like an invitation to trouble, and *Nencki* was not to resume publishing on the pigments of blood and bile for over a decade, until well after *Küster* had begun to publish his own studies of these pigments.

Küster initiated investigations into the pigments of blood in 1891 and continued and broadened the work into investigations of bilirubin and biliverdin in Tübingen as an *Habilitant* and beyond. First he reinvestigated hemin, its isolation, crystallization, and combustion analysis, which confirmed *Nencki*'s formula of the pigment crystallized from amyl alcohol as $(C_{32}H_{31}N_4O_3FeCl)_2C_5H_2O$ and a better fit to the data than other options (*216*):

> Meine Präparate z. B. ergaben bei der Analyse Werthe, welche am besten zu der Formel $(C_{32}H_{31}ClN_4FeO_3)_2C_5H_{12}O$ passen würden.
>
> Es verlangt z. B.:

		C		H		Cl		N		Fe		
$C_{32}H_{31}ClN_4FeO_3)_{16}C_5H_{12}O$:	C	62.96,	H	5.155,	Cl	5.74	N	u.		Fe	9.09	pCt
$(C_{32}H_{31}ClN_4FeO_3)_8C_5H_{12}O$:	»	63.006,	»	5.23,	»	5.69,	»	»		»	9.01	»
$(C_{32}H_{31}ClN_4FeO_3)_4C_5H_{12}O$:	»	63.08	»	5.376[1],	»	5.59,	»	»		»	8.85	»
$(C_{32}H_{31}ClN_4FeO_3)_2C_5H_{12}O$:	»	63.27	»	5.66,	»	5.45,	»	»		»	8.55	»
Nencki u. Sieber erhalten												
durchschnittlich [2]:········	C	62.78,	H	5.79	Cl	5.294	N	9.06	Fe	8.84	pCt	
Meine Analysen ergeben:												
Präparat I	»	62.95,	»	6.04,	»	5.04,	»	8.91,	»	8.28	»	
Präparat II	»	62.94,	»	6.07,	»	5.08,			»	8.57	»	
									»	8.48	»	

---

[1] 5.376 pCt, Wasserstoff, nicht 5.69 wie Nencki.

[2] Berechnet aus der Gesamtheit ihrer Analysenresultate. 26

204   3 Advent of the Bilirubin Structure Proof

Although by 1896, *Küster* believed that the empirical composition of hemin was inconclusive.

Nencki's hematin formula ($C_{32}H_{32}N_4O_4Fe$) for material obtained from hemin in alkaline solution exposed to air had the appearance of being the product of oxygenation, an oxidation as the term was understood in the 1800s. In 1896, apparently knowing that *Nencki* had failed in attempts to oxidize hemin to hematin using HNO_3 or $KMnO_4$ (the reaction went too far) and that both *Nencki* and *Hoppe-Seyler* had obtained no useful results from oxidation of hematin, *Küster* reported (*242*) an oxidation of hematin using chromic acid, which he believed to be a weaker oxidizing agent than those previously used. It is not entirely clear what the objective was in applying chromic acid oxidations to hematin, except possibly to gently pry apart its structure, to carve off a structural fragment using a more moderate oxidant in order to achieve results not previously seen with more powerful oxidizing agents. *Küster*'s choice proved to be an important key toward realizing the structures of hematin and hemin – and as we shall see, both bilirubin and biliverdin.

Confronted with what must have been a nasty looking final reaction mixture (from the addition of aqueous $Na_2Cr_2O_7$ to hematin dissolved in acetic acid), he extracted it with ether. In the ether extracts were found two separable colorless products, named by *Küster* as: a dibasic *Hämatinsäure* (hematinic acid), mp 112–113°C, $C_8H_{10}O_5$, and the anhydride ($C_8H_8O_5$) of tribasic *Hämatinsäure* (hematinic acid) ($C_8H_{10}O_6$). Here, "basicity" refers in the classical sense to the number of mole equivalents of inorganic base required to neutralize the acids, or the number of mole equivalents of Ag in the acid precipitated as its silver salt – once their formulas had been determined by elemental combustion analyses and molecular weight determinations (*243*):

Ich betrachte daher die Untersuchung betreffs der Darstellung und empirischen Zusammensetzung der »Hämine« noch nicht als abgeschlossen. Ebenso bedarf der Process noch der weiteren Aufklärung, welcher sich beim Uebergang »Hämine« in das Hämatin abspielt. Unter jeder Bedingung scheint er doch nicht so glatt zu verlaufen, wie Nencki und Sieber annehmen. Meine zahlreichen Analysen ergaben für den Kohlenstoff stets, manchmal auch für den Wasserstoff, zu niedrige Werthe in Bezug auf die von Nencki aufgestellte Formel $C_{32}H_{32}N_4FeO_4$, sodass ich eine namentlich in alkalischer Lösung eintretende Oxydation des Hämatins durch den Sauerstoff der Luft nicht für ausgeschlossen halte. Diese Meinung führte mich auch dazu, Oxydationsversuche des Hämatins wieder aufzunehmen, obgleich Hoppe-Seyler sowohl als Nencki keine Resultate erhalten hatten, aus denen sich für die Constitution des Hämatins Bemerkenswerthes ableiten liess.

Ich verwendete ein auf viele Körper schwächer wirkendes Oxydationsmittel, die Chromsäure, in der Weise, dass ich eine wässrige Lösung des dichromsauren Natriums bei Wasserbadtemperatur[1] auf das in Eisessig gelöste Hämatin einwirken liess. Durch dieses Verfahren erhielt ich ein ätherlösliches Säuregemisch, aus dem bisher zwei chemische Individuen isolirt werden konnten, die ich zweibasische Hämatinsäure und Anhydrid der dreibasischen Hämatinsäure genannt habe. Estere besitzt den Schmelzpunkt 112–113° und ist nach der Formel $C_8H_{10}O_5$ zusammengesetzt.

Analyse: Ber. für $C_8H_{10}O_5$.

	Procente:	C	51.61,	H	5.37.
Gef.	»	»	51.68,	»	5.29.

3.3 Molecular Fragmentation: Initial Approach to Bilirubin Structure

Diese Säure ist zweibasisch, denn sie giebt ein Silbersalz, $C_8H_8Ag_2O_5$, wofür sich berechnet:

	Procente	Ag	53.84.
Gef.	»	»	53.76.

Auch entspricht die Formel dem Molekulargewicht.

Ber. für $C_8H_{10}O_5$:	186.	
Gefunden:	194,	200.

Diese dreibasische Hämatinsäure entsteht aus der ersteren jedenfalls durch weitere Oxydation; sie hat die Zusammensetzung $C_8H_{10}O_6$ und bildet leicht eine Anhydridsäure $C_8H_8O_5$, welche den Schmelzpunkt 94.5° zeigt und mit grossem Krystallisationsvermögen begabt ist, scheinbar aber immer noch etwas Wasser einschliesst.

Analyse:		Ber. für $C_8H_{10}O_6$		$C_8H_8O_6 + 1/8\ H_2O$		$C_8H_8O_5$
	Procente:	C	47.52,	51.54,		52.17.
		H	4.95,	4.43,		4.35.
Gef.	»	C	51.24,	51.35,	H 4.44, 4.50.	

Auch hier entspricht die angenommene Formel dem Molekulargewichte, denn es berechnet sich für $C_8H_8O_5$

Molekulargewicht:	184.
Gefunden:	199, 192, 202.

Die Salze leiten sich von der dreibasischen Säure ab.

---

[1] Auch bei Zimmertemperatur wirkt die Chromsäure ein; die hierbei entstehenden Producte bin ich im Begriffe zu untersuchen. 27

The tribasic acid was said to arise by further oxidation of the dibasic acid, then form an anhydride, $C_8H_8O_5$, mp 94.5°C. No structures for these low molecular weight acids were suggested at the time; those would come within five years.

Küster reported on the presence of a further oxidation product, an iron-containing, aqueous Na_2CO_3-soluble compound, which he indicated might contain a pyrrole ring (*243*):

Neben den Säuren wurde als weiteres Oxydationsproduct ein eisenhaltiger Körper erhalten, der in kohlensaurem Natron löslich ist. In dieser Lösung ist das Eisen durch die gewöhnlichen Mittel der Analyse nicht nachweisbar. Mit der Untersuchung dieser Körpers, der vielleicht in Beziehung zum Pyrrol steht, bin ich ebenfalls beschäftigt. 28

A year later, he had prepared pure hematoporphyrin by treatment of hemin (obtained from cattle blood) with HBr-saturated acetic acid, and by subjecting it to chromic acid oxidation, he obtained the same two acids as from hematin. The dibasic acid was separated as its calcium salt (*243*). Yet, still concerned about *Nencki*'s formula for hemin, *Küster* again prepared hemin, purified a large quantity, and found that its combustion analysis data were a better fit (in %H) to $C_{32}H_{33}N_4O_3FeCl$ than to *Nencki*'s $C_{32}H_{31}N_4O_3FeCl$ (*243*).

In the same year, *Küster* reported the results of chromic acid oxidation of biliverdin (*244*). He was apparently motivated to do this experiment because he felt that a stronger proof was needed to establish the relationship between hemoglobin and bile pigments, which everyone assumed (at least by 1897) had been fully established. The stronger proof that he had in mind was a chemical proof rather than one derived from considerations of animal metabolism (*244*):

> Es wird als nahezu sichergestellt angenommen, dass die Gallenfarbstoffe aus dem Hämoglobin und speciell aus eisenhaltigen Bestandtheile desselben, dem Hämatin, entstehen, da sich u. a. die ersteren nur bei Thieren vorfinden, welche rothes Blut besitzen. ...
> Ein strenger Beweis für diese Auffassung fehlt indessen, insofern der chemische Zusammenhang der beiden Farbstoffe noch nicht aufgeklärt ist. Nun konnte es aber gerade im gegebenen Falle möglich sein, die chemischen Beziehungen experimentell klar zu legen ... 29

One relationship had already been claimed by *Nencki*: that hematoporphyrin (formed from hematin by the action of HBr in acetic acid) and bilirubin had identical formulas. After all, hematoporphyrin gave a *Gmelin*-like color reaction when treated with HNO_3, and upon reduction it gave a compound resembling *Maly's* hydrobilirubin. From these data, *Küster*, like *Nencki*, was led to believe that hematoporphyrin and bilirubin were doubtless very similar compounds constructed of similar atom complexes (*244*):

> Bekanntlich existiren verschiedene Thatsachen, aus denen geschlossen werden kann, dass Blut- und Gallen-Farbstoff chemisch verwandte Körper sind. Namentlich hat Nencki gezeigt, dass das Hämatoporphyrin, welches durch Einwirkung von Bromwasserstoff und Eisessig aus dem Hämatin hervorgeht, isomer mit dem Bilirubin ist: beiden Körpern schreibt man die empirische Zusammensetzung $C_{16}H_{18}N_2O_3$ zu. Ferner giebt es bei der Einwirkung von Salpetersäure gewisse Farbenveränderungen, welche an die Gmelin'sche Gallenfarbstoffreaction erinnern, und bei der Reduction liefert es einen Körper, welcher einem Reductionsproduct des Bilirubins ähnelt [1].
> Diese Farbstoffe sind also zweifellos chemisch verwandte Körper und müssen sich aus ähnlichen Atomcomplexen aufbauen.

---

[1] Arch. f. exp. Pathol. Pharmakol. 24, 441 sagt Nencki: so viel ist sicher, ein mit dem Gallenfarbstoffurobilin identisches Product entsteht hierbei nicht. 30

Although obtaining enough cattle or sheep gallstones as a source of bilirubin was not as easy as obtaining blood for hematoporphyrin, *Küster* secured 36 g of gallstones and converted them to biliverdin in the classical way by dissolving in alkali and allowing air oxidation. Though the conversion was incomplete, a portion of the mixed bile pigments was oxidized with $Na_2Cr_2O_7$ in acetic acid, using the same procedure as in the preparation of hematinic acid. From the reaction mixture was isolated a compound, as pale yellow needles, mp 100–101°C. It analyzed for $C_8H_9NO_4$, titrated as a monobasic acid with 0.2 N NH_4OH and yielded a di-silver salt with $AgNO_3$ (*244*):

> ... und erhält so das Oxydationsproduct des Gallenfarbstoffs in Form schwach gelblich gefärbter Nadeln, welche den Schmelzpunkt 100–101° zeigen. Die Analyse führte zur Formel: $C_8H_9NO_4$

3.3 Molecular Fragmentation: Initial Approach to Bilirubin Structure

Analyse: Ber. Procente: C 52.45, H 4.92, N 7.65.
Gef. » » 52.12, » 5.52, » 7.81.

Nach der Titration mit 1/5 n-Ammoniak zu urtheilen, verhält sich die Substanz, deren wässrige Lösung übrigens stark sauer reagirt, in der Kälte wie eine einbasische Säure. Es erforderten 0.083 g zur Neutralisation 1.72 ccm, berechnet 1.8 ccm.

Das Silbersalz der Säure, dargestellt durch Fällen der mit Ammoniak neutralisirten Lösung mit einer weingeistigen Silbernitrat lösung, enthält aber zwei Atome Metall.

Analyse: Ber. für $C_8H_7Ag_2O_4N$.
Procente: Ag 54.33.
Gef. » » 53.96. 31

From the preceding, *Küster* concluded that the isomeric hematoporphyrin and bilirubin do not give the same products under identical oxidation conditions, although the hematinic acid from the former and the *Biliverdinsäure* (a name given by *Küster*) from biliverdin both decolorized $KMnO_4$. The unexpected difference between the two cited oxidation products, *i.e.* that *Biliverdinsäure* contained a nitrogen and hematinic acid did not, ran counter to *Küster*'s expectation, based upon the conversion of *Hämatin* to both pigments. He was left with the prediction that the nitrogen of *Biliverdinsäure* was present as a cyano group, which was also present in its precursor, biliverdin. The relationship between *Biliverdinsäure* and biliverdin was seen simply as (*244*):

In diesem Falle wäre auch ein direkter Zusammenhang mit der Hämatinsäure möglich, wie sich das schon aus den empirischen Formeln $C_8H_8O_5$ und $C_8H_9NO_4$ ergiebt. Die letztere weist endlich eine sehr einfache Beziehung zum Biliverdin auf und, wenn die erhaltene Säure das einzige product der Oxydation wäre, würde sich ihr Entstehen durch die Gleichung wiedergeben lassen:

$$C_{16}H_{18}N_2O_4 + 4\ O = 2\ C_8H_9NO_4.$$ 32

Two years later, the apparent difference between *Biliverdinsäure* and hematinic acid due to the presence of a nitrogen atom in the former was on its way to being resolved. In 1899 *Küster* reported new chromic acid oxidations of hematin, hematoporphyrin, and biliverdin (*198*). From the first two were isolated two compounds, mp 109–111.5° and mp 94.5°C. The first isolate analyzed for $C_8H_{10}O_5$, *i.e.* the dibasic hematinic acid; the second for $C_8H_8O_5$, *i.e.* the anhydride of the tribasic hematinic acid, $C_8H_{10}O_6$. The dibasic hematinic acid was also analyzed by its water-soluble calcium salt, solubility that was important in separating it from the tribasic hematinic acid anhydride. In addition to repeating the oxidation reaction of hematin and hematoporphyrin, *Küster* also repeated his oxidation of biliverdin, having just obtained 130 g of the pigment from a large supply of gallstones. A larger scale chromic acid oxidation led again to a product of mp = 110°, but the mp was not sharp; softening occurred at 108°C. Yet, it still released a mole equivalent of NH_3 upon reaction with hydroxide – and then produced the tribasic hematinic acid anhydride, $C_8H_8O_5$, obtained from hematin. This new observation suggested a direct relationship between the anhydride of the tribasic hematinic acid and *Biliverdinsäure*: that perhaps the latter was a precursor to the former.

208 3 Advent of the Bilirubin Structure Proof

Reinvestigating the crude mixture of acids isolated from oxidation of hematin, *Küster* recalled that the crude product isolated by ether extraction hinted at the presence of nitrogen, but not conclusively. Now, upon closer inspection, he learned that the primary oxidation product from hematin was isolated as $C_8H_8O_5$, that three-fourths of the acids present in the crude extract precipitated as a calcium salt of $C_8H_8O_5$, and that NH_3 was expelled in the conversion to the salt. Significantly, his re-analysis of the dibasic hematinic acid showed the presence of nitrogen, and the analytical results led to the formula $C_8H_9NO_4$ (*198*):

> Dies Ergebniss bestätigte also lediglich die früheren. Andere Portionen von Hämatin und Hämatoporphyrin dagegen lieferten vorwiegend nur Stickstoff haltende Säure, und diese gab ein leicht lösliches, beim Erhitzen nicht ausfallendes Calciumsalz, welches durch Umkrystallisiren gereinigt werden konnte. Die nunmehr regenerirte Säure zeigt den Schmelzpunkt 111–113° [2].
> Das sind aber Eigenschaften der »zweibasischen Hämatinsäure«.
> Die Analyse führte nun zur formel $C_8H_9NO_4$.

$C_8H_9NO_4$. Ber.	C	52.45,			H	4.92,			N	7.65.
	Ht. (Pferd)	Hp. (Rind)								Hp.[3]
Gef.	C	52.49,	52.52,	52.12,	»	5.16,	5.30,	5.06,	»	7.59.

---

[2] Bei anderen Präparaten wurde beobachtet: 110–113°; 111.5–112.5, Erweichen von 103° an.

[3] Ht. = Hämatin; Hp. = Hämatoporphyrin. 33

Küster admitted that what led him astray was the one-time failure of his qualitative test for nitrogen, coupled with an accidentally good fit of the observed %C and %H to that calculated for $C_8H_{10}O_5$, the previously given formula of the dibasic hematinic acid (*198*):

> Ich bin also durch das einstmalige Versagen der qualitativen Probe auf Stickstoff und die zufällig gut auf $C_8H_{10}O_5$ stimmenden Werthe für den Kohlenstoff zur Aufstellung der falschen Formel $C_8H_{10}O_5$ geführt worden, während alle früheren gefundenen Werthe für den Wasserstoff besser zur richtigen Formel passen. Eine nachträgliche Prüfung der von früher noch vorhandenen als »zweibasische Hämatinsäure« bezeichneten Reste bestätigte den Stickstoffgehalt. 34

The classical *qualitative* test for nitrogen is the sodium fusion test devised originally by *Lassaigne*.[12] In 1843, he devised a test to detect the presence of nitrogen (and halogens) that involved dropping small samples of a compound directly into Na vapor (sodium fusion) (*245–247*). After the reaction mixture was cooled, excess Na was destroyed by addition of CH_3OH or CH_3CH_2OH, then H_2O was added and the resulting solution, after filtration, was analyzed for the presence of nitrogen, as NaCN – and/or it was analyzed for the presence of halogens. Cyanide was detected as a blue color or precipitate that attended formation of ferrocyanide

---

[12] *Jean Louis Lassaigne* was born on September 22, 1800 in Paris and died there on March 18, 1859. He was a French chemist who worked initially in *Vauquelin*'s lab, and in 1828 he succeeded *Pierre Louis Dulong* as Professor of Chemistry and Physics at the École Royale Vétérinaire d'Alfort in Maison-Alfort until 1854.

3.3 Molecular Fragmentation: Initial Approach to Bilirubin Structure 209

ion upon addition of $FeSO_4$ and H_2SO_4. The development of the blue coloration depended upon adjusting the pH correctly.

Küster showed that the new dibasic hematinic acid ($C_8H_9NO_4$) was converted to $C_8H_8O_5$ by warming in aqueous NaOH, in which the imide group (NH) was replaced by oxygen (O). And from numerous chromic acid oxidations and work-ups on a wide variety of hemins, as well as hematoporphyrins, obtained from the blood of various animals, the result was always the same: the acid $C_8H_9NO_4$ appeared as the first product of oxidation, *in a yield of ~40% of the hematin used.* Finally, he concluded that the acids, $C_8H_9NO_4$, from hematin and from bilirubin were probably identical. However he could not exclude the possibility that the dibasic hematinic acid and the *Biliverdinsäure* of formula $C_8H_9NO_4$ represented not chemically distinct compounds but mixtures of isomers that were transformed into one and the same compound of formula $C_8H_8O_5$ (*198*):

> Ich halte darum die Möglichkeit nicht für ausgeschlossen, dass die Körper $C_8H_9NO_4$ (zweibasische Hämatinsäure – Biliverdinsäure) keine chemischen Individuen vorstellen, sondern Gemische von isomeren Körpern sind, welche aber in ein und denselben Körper $C_3H_8O_5$ übergehen. 35

With this self-made correction to his earlier work, *Küster* wrote the three steps in the conversion of hematin (I) to the anhydride of the tribasic acid (IV) via hemato-porphyrin (II) and the dibasic hematinic acid (III), which is the same as *Biliverdinsäure*. The $C_8H_8O_5$ compound was found to be reduced by HI into a $KMnO_4$-resistant substance, $C_8H_{12}O_6$ that analyzed correctly (combustion analysis) and proved to be a tribasic acid. However, various preparations of it showed variable melting points, and *Küster* thus considered it to be a mixture of isomers (*198*):

> Durch diese Berichtigung meiner früheren Angaben ist nun auch der Zusammenhang der bisher erhaltenen Spaltungsproducte des Hämatins mit letzterem übersichtlicher geworden.
>
> Unter Abspaltung von Eisen und Aufnahme von Wasser zerfällt, wie Nencki … zeigte, das Hämatin (I) in Hämatoporphyrin (II), bei der Oxydation giebt das letztere, wie das isomere Bilirubin, die zweibasische Hämatinsäure = Biliverdinsäure (III), aus diesen endlich entsteht durch Abspaltung von Ammoniak das Anhydrid der dreibasischen Hämatinsäure (IV).

$$\begin{array}{cccccccc} \text{I} & & \text{II} & & \text{III} & & \text{IV} \\ C_{32}H_{32}N_4FeO_4 & \rightarrow & C_{16}H_{18}N_2O_3 & \rightarrow & C_8H_9NO_4 & \rightarrow & C_8H_8O_5 \end{array}$$

> … Die ungesättigte Verbindung $C_8H_8O_5$ geht endlich bei der Reduction mit Jodwasserstoff in eine gegen Kaliumpermanganat beständige Substanz $C_8H_{12}O_6$ über, welch zwar scharfe Werthe bei der Analyse gab und sich als dreibasische Säure erwies, aber bei verschiedenen Präparaten nicht den gleichen und keinen scharfen Schmelzpunkt zeigt und daher wohl auch als Gemisch isomerer Körper zu betrachten ist. 36

Two new types of reactions were mentioned: (i) resistance to $KMnO_4$, from which *Küster* assumed that the $C_8H_{12}O_6$ compound had no carbon-carbon double bonds; and (ii) reduction of its precursor by HI, which suggested the presence of a carbon-carbon double bond in $C_8H_8O_5$ – and thus in $C_8H_9NO_4$. Before catalytic

210   3 Advent of the Bilirubin Structure Proof

hydrogenation of double bonds had been invented in 1867, *Berthelot*[13] showed that
a concentrated solution of hydriodic acid acted as a universal reducing agent at high
temperatures.

Given that in 1899 organic chemists were very much concerned with chemical
structure and isomerism, now that a chemical relationship had been established
between hematin and bilirubin, it remained to *Küster* to determine the chemical
structure(s) of the dibasic hematinic acid/*Biliverdinsäure* and the anhydride of the
tribasic hematinic acid. But first he summarized his work to date (*248–250*) in
detail, his isolation and combustion analyses of hemin and hematin from blood,
providing experimental details of various means of oxidation: $Na_2Cr_2O_7$, $CaCr_2O_7$,
$K_3Fe(CN)_6$, $(NH_4)_2S_2O_8$ on hematin and hematoporphyrin, and analyses of the oxidation products extracted into ether. Clearly, thinking about how the oxidation
fragmentation products might fit together in the hematin ($C_{32}H_{32}N_4O_4Fe$) and hematoporphyrin ($C_{16}H_{18}N_2O_3$) structures, he proposed the following diagrammatic
relationships, where R represents the monopyrrole precursor to the dibasic hematinic acid, $C_8H_9NO_4$ (*248*):

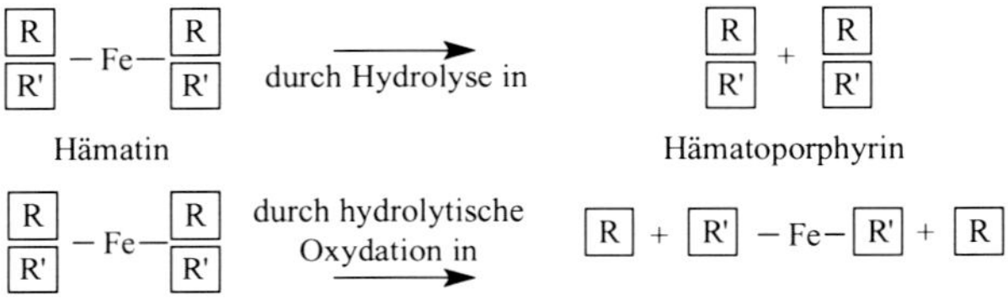

The age of organic structural chemistry arrived in porphyrin and bile pigment
investigations when, in 1900, *Küster* considered possible chemical structures of (i)
the primary oxidative cleavage product ($C_8H_9NO_4$) of hematin and hematoporphyrin, and (ii) the anhydride of the nitrogen-free tribasic hematinic acid obtained from
the primary product by treatment with alkali. Two possible crude structures were
considered: an anhydride (**I**) and a lactone (**II**), but only the first is capable of giving
a trimethyl ester and a tri-silver salt – both of which were isolated (*251*):

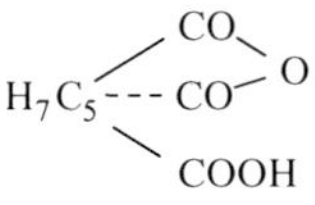

In addition, only structure **I** would be expected to give the saturated triacid
($C_8H_{12}O_6$) obtained following reaction with hot hydriodic acid. Consequently, the

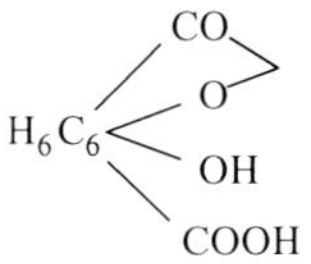

---

[13] *Marcellin Berthelot* was born on October 25, 1827 in Paris and died there on March 18, 1907.
He received the doctorate in 1854, became Professor of organic chemistry in 1859 at the École
Superieur de Pharmacie, and chair of organic chemistry at the Collège de France in 1865.

3.3 Molecular Fragmentation: Initial Approach to Bilirubin Structure

imide would have the following structure, one that could be obtained by heating the anhydride ($C_8H_8O_5$) in alcoholic ammonia at 110°C in a sealed tube (*251*):

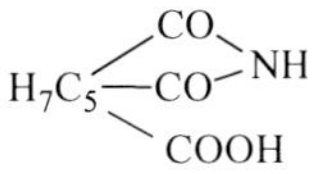

$$H_7C_5-CO \underset{COOH}{\overset{CO}{<}} NH$$

Significantly, *Küster* also observed a new product, mp = 72–73°C, in addition to the compound $C_8H_9NO_4$ when the reaction was heated above 120°C, accompanied by the loss of CO_2. This new product was reminiscent of iodoform in odor, liquidity, and sublimability – all properties that reminded *Küster* of an imide of the maleic acid series. It analyzed correctly for $C_7H_7NO_2$ and gave the correct corresponding molecular weight (*252*):

> Bei wenig höherer Temperatur, etwa von 120° ab, fängt hierbei die Abspaltung von Kohlendioxyd an; bei 130° geben sowohl $C_8H_9NO_4$ als $C_8H_8O_5$ einen neuen Körper $C_7H_9NO_2$, der den Schmp. 72-73° besitzt und alle die Eigenschaften aufweist (Geruch nach Jodoform, Flüchtigkeit, Sublimirbarkeit), wie sie für ein Imid aus der Reihe der Maleïnsäure angegeben werden.

$C_7H_9NO_2$.	Ber.	C	60.43,	H	6.48,	N	10.07,	Mol.-Gew.	139.	
	Gef.	»	60.20,	»	6.52,	»	10.36,	»	136.	37

Saponification of the new product using $Ba(OH)_2$ gave two different Ba^{+2} salts, and after acidification a peppermint-like oil (bp = 228–229°C) was obtained that analyzed for $C_7H_8O_3$, the anhydride of a dibasic acid, $C_7H_{10}O_4$, mp = 175°C (*252*):

> Der Körper $C_7H_8O_3$ ist also als Anhydrid einer zweibasischen Säure $C_7H_{10}O_4$ anzusehen. Das bei der Verseifung des Imids entstehende zweite Baryumsalz gehört zu einer in kaltem Wasser schwer löslichen, aus absolutem Alkohol in breiten Nadeln krystallisirenden, ungesättigten Säure $C_7H_{10}O_4$, vom Schmp. 175° (unter theilweiser Zersetzung).

$C_7H_{10}O_4$.	Ber.	C	53.17,	H	6.32.	
	Gef.	»	53.18,	»	6.29.	38

Küster's isolation of chromic acid degradation products containing seven to eight carbon atoms marked the start of the structural chemistry era of bile pigments. All previous investigations had focused on isolation, purification, combustion analysis, and molecular weight – but not structure, possibly because most of the various formulas for the pigments of bile (except *Thudichum*'s) and blood contained too large a number of carbons. However, the state of knowledge of organic chemistry in the year 1900 was in principle and practice readily amenable to solving the structures of the C_7 and C_8 degradation products. Which *Küster* accomplished, because the last half of the 19th century had seen considerable interest among organic chemists in carboxylic acids and their reactions, and much new knowledge on structure had been forthcoming. Although the stereochemistry of butenedioic acids and their alkylated analogs was well-studied by the time of *Küster*, those studies could not have been accomplished, nor could *Küster*'s structure assignments

212   3 Advent of the Bilirubin Structure Proof

have been reached, without certain advances in the knowledge and practice of organic synthesis.

Searching the available literature as well as his own knowledge of organic chemistry, *Küster* was drawn to dibasic acids that formed anhydrides, with a bias toward maleic acid ($C_4H_4O_4$) and its anhydride ($C_4H_2O_3$). Scanning the then known literature on C_7 homologs, he identified *Fittig*[14] and *Glaser*'s (*253*) *Aethylmesaconsäure* (ethylmesaconic acid) [= *Propylfumarsäure* (propylfumaric acid)] in the maleic form and *Bischoff*'s (*254, 255*) *Methyläthylmaleinsäureanhydrid* (methyl ethyl maleic acid anhydride). In fact, *Bischoff*[15] had prepared the imide of the latter by heating the bis-ammonium salt of *Methyläthylmaleinsäure* (methyl ethyl maleic acid), and it analyzed for $C_7H_8NO_2$, with mp 62°C, which was 10° lower than *Küster*'s $C_7H_8NO_2$.

The structures of these acids were known to *Küster*, courtesy of earlier investigators, especially *Fittig* and *Glaser*, who in 1899 reported three diacid isomers of $C_7H_{10}O_4$ (*253*):

Aethylmesaconsäure,

$C_7H_{10}O_4 = CH_3-CH_2-CH_2-C=CH-CO-OH$
$\phantom{C_7H_{10}O_4 = CH_3-CH_2-CH_2-C}|$
$\phantom{C_7H_{10}O_4 = CH_3-CH_2-CH_2-}COOH$

Ethylmesaconic Acid (**3.3.1**)
mp 174-175°C

Aethylcitraconsäure,

$C_7H_{10}O_4 = CH_3-CH_2-CH_2-C=CH-CO-OH$
$\phantom{C_7H_{10}O_4 = CH_3-CH_2-CH_2-C}|$
$\phantom{C_7H_{10}O_4 = CH_3-CH_2-CH_2-}CO-OH$

Ethylcitraconic Acid (**3.3.2**)
mp 93-95°C

Aethylitaconsäure,

$C_7H_{10}O_4 = CH_3-CH_2-CH=C-CH_2-CO-OH$
$\phantom{C_7H_{10}O_4 = CH_3-CH_2-CH=C}|$
$\phantom{C_7H_{10}O_4 = CH_3-CH_2-CH=}CO-OH$

Ethylitaconic Acid (**3.3.3**)
mp 162-167°C

Acid **3.3.2** upon treatment with HNO_3 gave **3.3.1**. Acid **3.3.3** was converted into **3.3.2** upon dry distillation, which distilled the anhydride of **3.3.2**. Acid **3.3.2** was in turn isolated by heating the acid anhydride in hot water; molten **3.3.2** could be converted back to some **3.3.3** by heating at 150°C. Upon treatment with Na(Hg), all three acids gave the same product, $C_7H_{12}O_4$, mp = 91–92°C, which proved to be identical to *Bischoff*'s *Propylbernsteinsäure* (propylsuccinic acid) prepared from *Propyläthenyltri-carbonsäure* (*255*). [Succinic acid = *Bernsteinsäure* = spirit of

---

[14] *Wilhelm Rudolf Fittig* was born on December 6, 1835 in Hamburg and died on November 19, 1910 in Strassburg. He received the doctorate in chemistry at Göttingen in 1858 under *Wöhler* and *Limpricht*, was *Privatdozent* at Göttingen in 1869, and a. o. Professor in 1870. In 1870 he accepted the chair of organic chemistry at Tübingen before moving as chair to the new University of Strassburg in 1876. He was a major contributor to the development and knowledge of organic reactions.

[15] *Carl Adam Bischoff* was born in 1855 in Würzburg and died in 1908 in Monaco. He was Professor of Organic Chemistry at the Polytechnical Institute in Riga as successor to *Ostwald*. He published the monumental *Handbuch der Stereochemie* with *Paul Walden* in 1894 and in 1890 raised the question of the existence of C–C rotational isomers.

3.3 Molecular Fragmentation: Initial Approach to Bilirubin Structure

amber, mp 185°C, bp 235°C, had been known for centuries as a distillate of pulverized amber.] Reduction to (a substituted) succinic acid had its analogy in *Kekulé*'s reports of nearly 40 years earlier, in (28, 202) that the isomeric unsaturated *Ita-*, *Citra-*, and *Mesa-consäure* (all prepared from *Citronensäure* – citric acid) were each reduced by Na(Hg) to the same *Methylbernsteinsäure* (methylsuccinic acid).

The organic structures of the relevant acids come to us from *Fittig*, who was much interested in the isomers of butenedioic acid and its alkylated analogs (256). Following his discovery that condensation of benzaldehyde with the sodium salt of succinic acid (*Bernsteinsäure*) heated in the presence of glacial acetic anhydride (Scheme 3.3.1) in a sealed tube gave phenylparaconic acid (*Phenylparaconsäure*) (257), he generalized the condensation for the use of aliphatic aldehydes (258) to include ethylparaconic acid (*Aethylparaconsäure* = **3.3.4**, R = CH_2CH_3), *inter alia*. He "proved" the structures of the substituted paraconic acids by dry distillation to yield g-substituted homocrotonic acids (crotonic acid is $CH_3CH=CH-CO_2H$; homocrotonic acid is $CH_2=CH-CH_2CO_2H$):

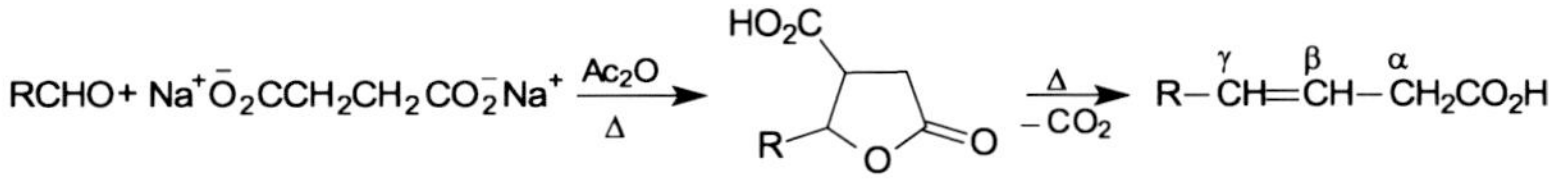

$$RCHO + Na^+\bar{O}_2CCH_2CH_2CO_2^-Na^+ \xrightarrow[\Delta]{Ac_2O} \quad \xrightarrow[-CO_2]{\Delta} R-\overset{\gamma}{CH}=\overset{\beta}{CH}-\overset{\alpha}{CH}_2CO_2H$$

(3.3.4)

Scheme 3.3.1

Fittig's proof of structure came from a reaction sequence (Scheme 3.3.2) in which benzaldehyde is condensed with the sodium salt of methylsuccinic acid (*Brenzweinsäure*):

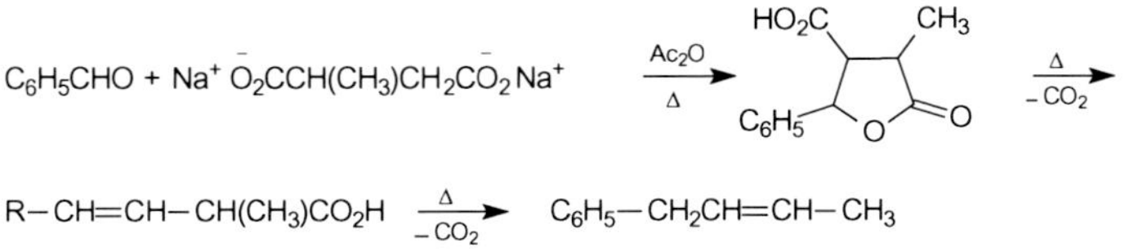

$$C_6H_5CHO + Na^+\bar{O}_2CCH(CH_3)CH_2CO_2^-Na^+ \xrightarrow[\Delta]{Ac_2O} \quad \xrightarrow[-CO_2]{\Delta}$$

$$R-CH=CH-CH(CH_3)CO_2H \xrightarrow[-CO_2]{\Delta} C_6H_5-CH_2CH=CH-CH_3$$

Scheme 3.3.2

The (2-butenyl)benzene produced was verified by its identity with the same compound produced independently by *Aronheim* in 1874 (259) from condensation of allyl iodide with benzyl chloride using Na metal (the *Wurtz* reaction; subsequently *Wurtz-Fittig*). Both *Aronheim* and *Fittig* (260) preferred the rearranged 2-butenyl isomer to the 3-butenyl, the product predicted today according to the mechanism of decarboxylation of β,γ-unsaturated acids.

Condensation of propionaldehyde with sodium succinate (Scheme 3.3.3.) by *Fittig* and *Delisle* (261) thus led to ethylparaconic acid (*Aethylparaconsäure*). Some ten years later, *Fittig* and *Glaser* (253) converted the ethyl ester of ethylparaconic acid to *Aethylitaconsäure* [ethylitaconic acid = propylidene-succinic acid] from which *Aethylitaconsäureanhydrid* (propylmaleic anhydride) was prepared, as noted above. Its molecular formula corresponds to that of *Küster's* anhydride

$(C_7H_8O_3)$ obtained from the product $(C_7H_9NO_3)$ derived by decarboxylation of the dibasic hematinic acid $(C_8H_9NO_4)$:

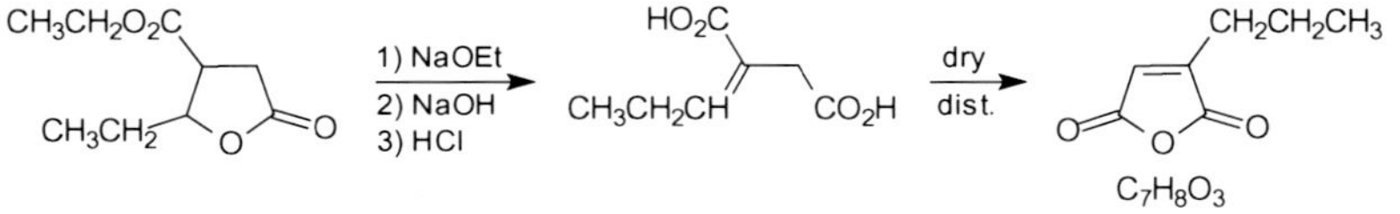

Scheme 3.3.3

The boiling point of *Küster*'s anhydride, an oil, was 228–229°C. *Fittig* did not give the boiling point of his anhydride, just the melting point of the corresponding diacid, 162–167°C.

In addition to *Fittig*'s propylmaleic anhydride, known only since 1899 (*261*), *Küster* was acquainted with an isomeric anhydride, the methylethylmaleic anhydride of *Bischoff* (*254, 255*), *Fittig* (*253, 256, 262*), and *Michael*[16] and *Tissot* (*263, 264*) – a compound prepared by three different routes at nearly the same time, but not at the time related in any way to a natural product. In 1890, *Bischoff* reported having converted *Aethylmethylbernsteinsäureanhydrid* (methylethylsuccinic anhydride) into methylethylmaleic anhydride, bp 237°C (*176a*) by reaction of the diacid of the former with Br_2, followed by elimination of HBr – a route by which the synthesis of *Pyrocinchonsäureanhydrid* (dimethylmaleic anhydride) from *s*-dimethylsuccinic acid was successfully accomplished (*265*). The required methylethylsuccinic acid came from *Bischoff*'s reaction of sodio diethyl ethylmalonate with ethyl α-bromopropionate (*266*); however, an alternative route from *Fittig* was also acknowledged by *Bischoff* (*255*). *Fittig*, too, had synthesized methylethylmaleic anhydride by a more direct route, which though not published until 1892 (*262*) was apparently known earlier to *Bischoff*. It involved condensation of *Brenztraubensäure* (pyruvic acid) with *Brenzweinsäure* (methylsuccinic acid) to produce the lactone *Dimethylbutyro-lactondicarbonsäure*, followed by pyrolysis to eliminate CO_2 (*262*):

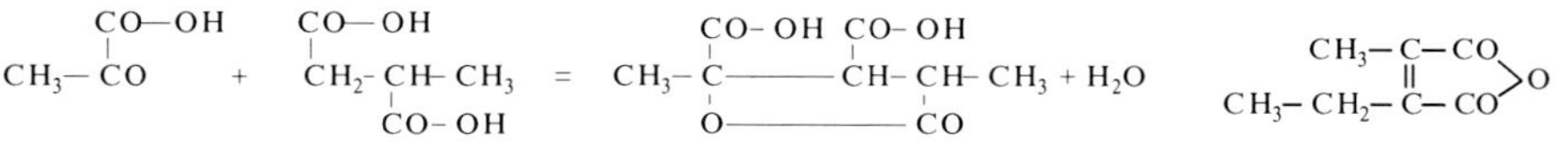

| Brenztraubens. | Brenzweins. | Dimethylbutyrolactondicarbonsäure. | Aethylmethylmaleïnsäure Anhydrid |

---

[16] *Arthur Michael* was born on August 7, 1853 in Buffalo, New York and died on February 8, 1942 in Orlando, Florida. Possessing little or no knowledge of chemistry, he nonetheless began his studies of chemistry under *A.W. Hofmann* at the Uni-Berlin from 1871 to 1872, then spent two years in basic experimental training under *R.W. Bunsen* at the Uni-Heidelberg, before returning to Berlin to work with *Hofmann* 1875–1878, all the while establishing a reputation for impetuosity. *Michael* concluded his studies in Europe in 1880, after having spent the previous two years with *A. Wurtz* in Paris and *D.I. Mendeleev* in St. Petersburg, to return to the U.S. as an assistant in the chemical laboratory at Tufts College, Boston. After a short time, he was promoted to Professor and in 1887 reported the reaction that bears his name. When my former colleague at UCLA, *Saul Winstein*, was a National Research Council Fellow at Harvard University in 1939–1940, he was invited by *Michael* (then 86–87 years old) to Tufts for a chat about chemistry. His mental alertness and huge library left solid impressions on the then 27–28 year-old *Winstein*.

3.3 Molecular Fragmentation: Initial Approach to Bilirubin Structure

Thus, the dried sodium salt of methylsuccinic acid was mixed with freshly distilled pyruvic acid, and the mixture was heated (in acetic anhydride) at length to form first the lactone, from which CO_2 and H_2O were evolved in producing the anhydride. After a careful work-up involving steam distillation and extraction into ether, a colorless oil (crude anhydride) was obtained. Following this, treatment with $Ba(OH)_2$ gave a solid barium salt, $C_7H_8O_4Ba$, which upon addition of aqueous HCl released the maleic anhydride. The anhydride defied crystallization down to $-18°C$, had a bp of $232°–233°$, and gave the correct combustion analysis for $C_7H_8O_3$ (262):

Die Analyse ergab die Formel $C_7H_8O_3$.

0,1755 g gaben 0,3847 CO_2 und 0,0969 H_2O.

	Berechnet für $C_7H_8O_3$	Gefunden
C	60,00	59,64
H	5,71	6,11.

Die Verbindung ist demnach nicht die Säure des Baryumsalzes, sondern deren Anhydrid. 39

Bischoff and *Fittig* were not alone in the early 1890s in having synthesized methylethylmaleic anhydride. *Michael* and *Tissot* prepared it from a derivative of *Aepfelsäure* [malic acid: $HO_2C–CH_2–CH(OH)–CO_2H$], whose dehydration to the unsaturated (maleic) acid anhydride they had been investigating. Thus, they prepared *symmetrische Aethylmethyläpfelsäure* from ethylated ethyl acetoacetate (*Aethylacetessigäther*, prepared by reaction of ethyl acetoacetate with sodium ethoxide, then with ethyl iodide) by reaction with KCN to form cyanohydrin. The last was hydrolyzed to the diacid with aq. HCl to yield the crystalline product a,b-*Hydroxymethyläthylbernsteinsäure*, mp 131.5–132°C (264):

0,1738 Grm., im Vacuum getrocknete Substanz, gab 0,8044 Grm. CO_2 und 0,1080 Grm. H_2O.

Berechnet für	$\begin{array}{c} CH_3-COH-COOH \\ | \\ C_2H_5-CH-COOH \end{array}$	:	Gefunden:
C	47,78		47,89%
H	6,81		6,92 " 40

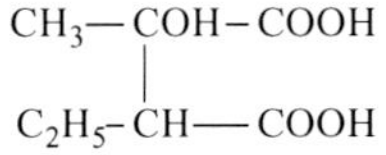

When the acid was heated, CO_2 was liberated and at 234°C a colorless oil distilled. Redistillation of the oil gave a bp = 236–237° at atmospheric pressure, or 122°C at 30 mm Hg (264):

0,8068 Grm. Oel gaben 0,0750 Grm. CO_2 und 0,1603 Grm. H_2O.

Berechnet für	$\begin{array}{c} CH_3-\overset{.}{C}-CO \\ | \quad\quad O \\ C_2H_5-C-CO \end{array}$	Gefunden:
C	60,00	59,97%
H	5,71	5,80 " 41

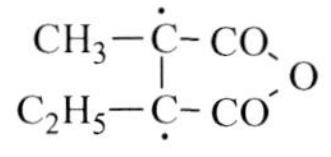

216   3 Advent of the Bilirubin Structure Proof

The authors described its taste (at first sweet, then glycerin-like, then later, bitter) and indicated that their anhydride was identical with *Bischoff*'s preparation from methylethylsuccinic acid (*254*), and with *Fittig*'s and *Parker*'s, prepared from pyruvic acid and sodium methylsuccinate (*262*):

> Dieses Anhydrid ist identisch mit der Verbindung, welche Bischoff ... aus Methyläthylbernsteinsäure, und Fittig und Parker ... Brenztraubensäure und brenzweinsaurem Natrium dargestellt haben ...   42

Apparently many of the smaller molecule components that made up the three synthetic routes to methylethylmaleic anhydride were well-known in the early 1890s, having been isolated from natural sources or synthesized centuries or decades earlier. *Fittig* drew from among: (i) *Bernsteinsäure* [succinic acid, $HO_2C–CH_2–CH_2–CO_2H$] distilled from amber in 1550 by *Agricola* and prepared synthetically by *Simpson* in 1861 (*267*); (ii) *Äpfelsäure* [= malic acid, $HO_2C–CH(OH)–CH_2–CO_2H$] isolated from apple juice by *Scheele* in 1785 and *Berzelius* in 1808 (*267, 268*); (iii) *Milchsäure* [= lactic acid, $CH_3CH(OH)–CO_2H$] isolated from milk (*267, 268*); (iv) *Brenztraubensäure* [= pyruvic acid, $CH_3–C(O)–CO_2H$], from dry distillation of *Traubensäure* (racemic tartaric acid, prepared in 1824 by *Kestner*) or in 1834 from heating tartaric acid (*d*-tartaric acid isolated by *Scheele* in 1769) with $KHSO_4$ (*268*); (v) *Brenzweinsäure* [= methylsuccinic acid = pyrotartaric acid, $HO_2C–CH(CH_3)–CH_2–CO_2H$], from dry distillation of *Weinsäure* (tartaric acid) in 1807 or *Traubensäure* (racemic tartaric acid) (*268*). Acids (iv) and (v) are described in detail in 1859 in *Liebig's Handwörterbuch* (*269*); the others are found in earlier volumes. In their synthesis of methylethylmaleimide, *Michael* and *Tissot* used ethyl acetoacetate [$CH_3–C(O)–CH_2–CO_2CH_2CH_3$], a β-keto-ester that had been prepared in 1866 by *Frankland* and *Duppa* by reaction of malic acid ethyl ester [$CH_3–CH(OH)CO_2CH_2CH_3$] with Na (*270*) and two years later by self-condensation (*271*) of ethyl acetate, an ester prepared from long-available components in 1759 by *Lauroguais* (*267*). In the latter publication, the authors also reported alkylation of ethyl acetoacetate with ethyl iodide to produce the ethyl acetoacetate used by *Michael* and *Tissot* some 25 years later. In 1866 and 1868, the structures were written in a different style than currently (*264*):

[ethyl acetate]	[sodium acetoacetate]	[ethyl acetoacetate]

$$4 \begin{cases} CH_3 \\ COAeo \end{cases} + Na_2 = 2 \begin{cases} COMe \\ CNaH \\ COAeo \end{cases} + 2\,AeHo + H_2; \qquad \begin{cases} COMe \\ CAaH \\ COAeo \end{cases}$$

Essigs. Aethyl	Natracetonkohlens. Aethyl	Alkohol	Aethylacetonkohlens. Aethyl

Acetoacetic ester was studied subsequently by *Geuther* in 1883–1885 (*272, 273*), *Baeyer* in 1885 (*274*), *Nef* in 1891 (*275*), and by *Claisen* in 1887-1905 (*276–278*), after whom the Na or $NaOCH_2CH_3$ assisted self-condensation of esters to β-keto-esters is named as the *Claisen* condensation.

3.4   Küster's Maleimides and Nencki's Hämopyrrol   217

In his synthesis of methylethylsuccinimide, and eventually the corresponding maleimide, *Bischoff* employed a malonic ester condensation to couple diethyl ethylmalonate to ethyl α-bromopropionate (*266*):

$$CH_3CH_2O_2C-\overset{\overset{\displaystyle CH_3CH_2}{|}}{\underset{\underset{\displaystyle CO_2CH_2CH_3}{|}}{C}}Na^+ \quad + \quad CH_3\overset{}{\underset{\underset{\displaystyle Br}{|}}{CH}}-CO_2CH_2CH_3 \quad \longrightarrow$$

$$CH_3CH_2O_2C-\overset{\overset{\displaystyle CH_3CH_2}{|}}{\underset{\underset{\displaystyle CO_2CH_2CH_3}{|}}{C}}-\overset{\overset{\displaystyle CH_3}{|}}{CH}-CO_2CH_2CH_3 \quad \xrightarrow[\text{2) } H_3O^+/\Delta]{\text{1) aq NaOH}}$$

Malonic acid was discovered in 1858 by *Dessaignes* by oxidizing malic acid with $K_2Cr_2O_7$ (*267, 268, 269*); diethyl malonate alkylations were studied by *Conrad* in 1879 (*268, 280*). Propionic acid was discovered by *Gottlieb* when he fused cane sugar with caustic potash (KOH) (*268*); ethyl α-bromopropionate had been prepared by *Zelinsky* of the *Hell-Volhard-Zelinsky* reaction in 1887 (*281*).

3.4   *Küster*'s Maleimides and *Nencki*'s Hämopyrrol

Clearly, at the end of the 19th century the development of organic chemistry had accelerated far faster than physiological or medicinal (or animal) chemistry. In the latter, bilirubin advances were largely limited to efforts to obtain homogeneity in order to get meaningful, ash-free combustion analyses. The concept of a chemical structure did not seem to be in the forefront; colorimetry was of little use. Meanwhile, huge advances were made in understanding how atoms were connected, first in simple, then in increasingly more complex organic compounds. The low molecular weight organic compounds provided by Nature were being fully characterized and used as building blocks in deliberate and organized patterns of synthesis. New reactions were being discovered each decade and put to use. Elementary organic stereochemistry had begun to be formulated about carbon-carbon double bonds and the tetrahedral carbon. One gets the impression of a dynamic new area of chemistry, of an explosive growth of knowledge, understanding, and capability that was largely lacking in or not applied to investigations of the pigments of bile and blood at the time. Yet, at the turn of the century, inevitable changes in thought and capability were becoming evident as organic chemistry and natural products structure proof intersected. *Küster* was able to apply his knowledge of organic chemistry, and by the year 1900, he had shown that the only degradation products then isolated from hematin, hematoporphyrin, biliverdin, and bilirubin were identified as the "dibasic *Hämatinsäure*" (dibasic hematinic acid) and *Biliverdinsäure*, which were identical compounds of formula $C_8H_9NO_4$. Although their structure had not yet been resolved, *Küster* knew that it was almost certainly a maleimide analog. What had been resolved, however, was the structure of their decarboxylation product ($C_7H_7NO_2$), formed by heating hematinic acid. It was determined by synthesis to be methylethylmaleimide. The only inconsistency in the path to complete identity was a difference in melting points between the synthetic imide, mp 62°C (*255*) and that derived from the natural product, mp 72–73°C (*251, 252*). A not insignificant difference, for at the time a melting point was one of the most important experimentally determined characteristics of an organic compound, and a sharp mp was taken as diagnostic of homogeneity. Then, there were few other immutable characteristics available for comparison and identification of organic compounds. In a long article summarizing the state of knowledge of the pigments in 1901, *Küster* wrote (*252*) that: (i) the Cl of hemin ($C_{32}H_{31}N_4O_3FeCl$) was replaced by an OH group to form hematin ($C_{32}H_{32}N_4O_4Fe$), which loses iron to form hematoporphyrin ($C_{16}H_{18}N_2O_3$), according to *Nencki*; (ii) the primary oxidative scission product of hematin, hematoporphyrin, and biliverdin was believed to be an imide ($C_8H_9NO_4$), the earlier-named dibasic hematinic acid (that was then said to have the character of a mono-basic acid); and (iii) the latter hydrolyzed in base to lose nitrogen and yield $C_8H_8O_5$, which titrated as a tribasic acid $C_8H_{10}O_6$. The $C_8H_8O_5$ compound was thought to be the anhydride of $C_8H_{10}O_6$. The $C_8H_9NO_4$ compound was proved to be an imide: it saponified easily in NH_4OH or $Mg(OH)_2$, a behavior, however, thought to be contrary to that of an amide; it formed a di-silver salt by making use of the imide NH in addition to the carboxylic acid OH; it re-formed when the anhydride, $C_8H_8O_5$, was treated with alcoholic NH_3 under pressure; and it underwent decarboxylation to $C_7H_7NO_2$ when the last reaction was heated above 120°C, again consistent with the presence of a carboxylic acid group in addition to the imide. The new imide, $C_7H_7NO_2$, was hydrolyzed to afford an anhydride, $C_7H_8O_3$, which *Küster* believed to be methylethylmaleic anhydride.

The lack of agreement with melting point of the totally synthetic material was resolved rather simply (and instructively): when the mp was taken by heating slowly, the lower (62°C) mp was observed; however, when heating was rapid to 65°C, then slow, a value of 72.5°C was found for both the "natural" and synthetic material. There was no mention of a mixture mp; apparently that type of experiment, with its information to confirm the identity of two samples, was to appear later. The structural information may be summarized as follows, with an uncertain location of the CO_2H group on the ethyl or methyl groups:

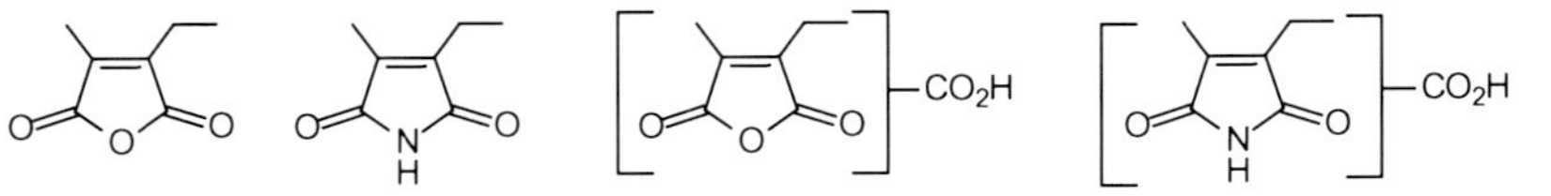

C_7H_8O $C_7H_9NO_2$ $C_8H_8O_4$ $C_8H_9NO_4$

Fig. 3.4.1. Structures of anhydrides and imides known to *Küster* in 1900

Küster did not believe that imide units were present in the pigments. Rather he favored the presence of pyrrole units from which the imides might be formed and released as imides (*252*):

3.4 Küster's Maleimides and Nencki's Hämopyrrol

… dass das Hämatin als aus zwei symmetrisch gebauten Theilen bestehend betrachtet werden kann, welche durch das Eisen zusammengehalten werden:

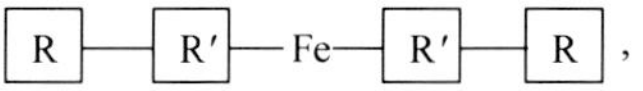

$$\boxed{R} - \boxed{R'} - Fe - \boxed{R'} - \boxed{R} \,,$$

so sprechen alle Beobachtungen dafür, dass es der mit R bezeichnete Complex ist, welcher die Hämatinsäuren liefert …

Wenn nun, wie es ja bereits als höchst wahrscheinlich dargethan worden ist, die Hämatinsäure ein Derivat der Maleïnsäure ist, so muss der Complex R des Hämatins die folgende Gruppe enthalten:

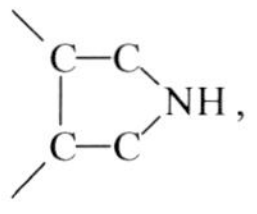

$$\begin{array}{c} C-C \\ | \quad \diagdown NH \,, \\ C-C \diagup \end{array}$$

d. h. also den Pyrrolring. Das ist aber eine Annahme, welche auf Grund zahlreicher Beobachtungen an Reactionen, bei denen das Hämatin allerdings eingreifendste Zersetzung erleiden musste, bereits des öfteren gemacht worden ist. So berichtet Hoppe-Seyler …: die trockne Destillation des Hämatins liefert reichlich Pyrrol. 43

He viewed hematin as being formed from four pyrrole units, arranged as two pairs connected to Fe. At least two of the pyrrole units were precursors to the imide, hematinic acid. In fact, it was recalled that *Hoppe-Seyler* claimed that pyrroles were produced by dry distillation of hematin; that *Nencki* and *Sieber* had produced a volatile, water-soluble compound that imparted a red color to a pine splint moistened with hydrochloric acid (a color test characteristic of pyrrole) by reducing hemin with Sn in alcohol; and that they had also noted that considerable *Pyrrol* is produced from hematin in molten KOH (*252*):

Nencki und Sieber … reduciren Hämin mit Zinn und Alkohol und erhalten u. a. *eine flüchtige, in Wasser lösliche Verbindung, die einen mit Salzsäure befeuchteten Fichtenspahn intensive roth färbte* … [ital in orig.] Dieselben Forscher erhitzten 20 g Hämatin mit dem fünffachen Gewichte schmelzenden Kalis bis zur vollständigen Zersetzung, wobei ziemlich viel Pyrrol entwich. … Sie sprechen deshalb das Hämatin und ihr Hämatoporphyrin als Abkömmlinge des Pyrrols an. … 44

Küster concluded with a statement confirming his belief that the pigment of blood is composed of four substituted pyrrole units, of which one goes over to the dibasic hematinic acid upon oxidative cleavage, and (providing an insight into structure) that the pigments can be obtained by a combination of pyrroles and ketones (*252*):

Ich glaube daher, dass die Annahme, der Complex R des Hämatins – um bei meinem Schema zu bleiben – enthalte eine substituirte Pyrrolgruppe, welche bei der Oxydation in das Imid der dreibasischen Hämatinsäure übergeht, die grösste Wahrscheinlichkeit für sich hat. Sie giebt uns endlich einen Einblick in die Farbstoffnatur des Hämatins, ist es doch bekannt, dass durch Combination von Pyrrol mit Ketonen Farbstoffe erhalten werden können. … 45

After a hiatus of nearly a dozen years and near the end of his life, *Nencki* resumed publishing on the pigments of blood. In 1900, with *Zaleski* he reviewed and updated his earlier work (*282*), while taking note of *Küster*'s contributions from oxidative cleavage: a 50% yield of an acid of $C_8H_9NO_4$ from hematoporphyrin (and biliverdin), that when heated in alkali lost NH_3 and produced the acid $C_8H_8O_5$, and which when treated with HI produced the tribasic acid $C_8H_{12}O_6$ that exhibited a composition similar in every respect to the synthetically prepared *Aethyltricarballylsäure* (*282*):

> Durch die schönen Untersuchungen von W. Küster . . . und seiner Mitarbeiter wissen wir, dass das Hämatoporphyrin, mit einer Ausbeute von gegen 50%, zu einer Säure von der Formel: $C_8H_9NO_4$ oxydirt werden kann, welche durch Kochen mit Alkalien unter Aufnahme von H_2O und Abspaltung von NH_3 in die Säure: $C_8H_8O_5$ übergeht. Durch Reduction mit Jodwasserstoff geht die letztere in die von M. Kölle ... dargestellte dreibasische Hämotricarbonsäure $= C_8H_{12}O_6$ über, welche einer vergleichenden Zusammenstellung dieses Autors zufolge in allen Stücken der von Auwers, Köbner und Meyenburg ... synthetisch dargestellten Aethyltricarballylsäure gleicht.   46

In 1891, *Auwers et al.* had reported their syntheses and characterizations of a series of alkylated tricarboxylic acids by conjugate addition of (a substituted) sodio malonate to diethyl fumarate (*283*):

> Eine grössere Anzahl von Versuchen wurde mit Fumarsäureester ausgeführt. Durch Condensation dieses Esters mit Natriummalonsäureester bezw. den Natriumverbindungen der Alkylmalonsäureester gelangt man leicht zur Tricarballylsäure und den bislang noch nicht dargestellten Monoalkyltricarballylsäuren:

    47

where R = H (mp 162–164°C) corresponds to the parent triacid, and the diastereomeric α-alkylated triacids are: R = CH_3 (mp 180°C and 134°C), CH_3CH_2 (mp 147–148°C) [*Aethyltricarballylsäure*], $CH_3CH_2CH_2$ (mp 151–152°C), and $(CH_3)_2CH$ (mp 161–162°C). When condensed with diethyl itaconate, sodio-malonate yielded *Butantricarbonsäure* (*284*):

3.4 Küster's Maleimides and Nencki's Hämopyrrol

In his final paper on hemin, reported in 1901, the year of his death, *Nencki*, with *Zaleski* (*285, 286*), reported a discovery nearly as earthshaking as *Küster*'s oxidative scission of hematin and isolation of dibasic hematinic acid: reduction-fragmentation. Thus, crude *Acethämin*, $C_{34}H_{33}N_4O_4FeCl$, the product obtained by treating blood, hemin, or hematin with glacial acetic acid (*192*) produced two products of special interest when dissolved in glacial acetic acid and heated with conc. HI and PH_4I. Following vigorous heating, a colorless, water-insoluble oil was distilled that changed rapidly in air, gave an intense red color in the pine splint test for pyrroles, and formed a double salt with $HgCl_2$ that analyzed for $(C_8H_{12}N)_2Hg(HgCl_2)$. *Nencki* named the air-sensitive pyrrole *Hämopyrrol*. It formed a picrate of mp 108°C, that analyzed for $C_8H_{13}NC_6H_2(NO_2)_3OH$, which meant that the formula for *Hämopyrrol* would be $C_8H_{13}N$. Rather startlingly, *Hämopyrrol* was also obtained by the same reductive cleavage of a chlorophyll, which apparently established a structural relationship between the pigment of blood and the pigment of plants (*286*). Two likely pyrrole structures were considered: butyl- or methylpropylpyrrole (*285*):

> Alle unsere Versuche sprechen mehr zu Gunsten der Annahme, dass das Hämopyrrol entweder ein Butyl- oder ein Methylpropyl-Pyrrol ist. 48

To *Nencki* and *Zaleski*, having just discovered *Hämopyrrol*, it became clear that it must bear a close relationship to *Küster*'s hematinic acid; thus, they came to *Küster*'s view that the latter arises by oxidation of the pyrrole nucleus. Based on an earlier observation from *Kölle* in *Küster*'s group that the (C_9) hematinic acid could be converted to the C_9 *Aethyltricarballylsäure* (**3**), prepared earlier by chemical synthesis, *Nencki* and *Zaleski* proposed a structure for *Hämopyrrol* that was oxidized to "hematinic acid" (**1**), converted to the corresponding maleic anhydride (**2**) and reduced to (**3**) (*285*):

> Nachdem wir das Hämopyrrol aufgefunden hatten, wurde es unsgleich klar, dass es in näher Beziehung zu den Küster'schen Hämatinsäuren stehen muss. Von ganz anderen Gesichtspunkten ausgehend, sind wir zu der gleichen Ansicht wie auch Küster … gekommen, dass die Hämatinsäuren durch die Oxydation des Pyrrolkerns entstehen. Maassgebend dafür, welches Kohlenstoffatom ausser den beiden Pyrrolkohlenstoffen zu Carboxyl oxydirt wird ist die Beobachtung von Kölle …, wonach das Anhydrid der Küster'schen dreibasischen Säure $C_8H_8O_5$ durch JH zu der Säure $C_8H_{12}O_6$ reducirt wird, welche den gleichen Schmelzpunkt und sonstige Eigenschaften wie die von Auwers, Köbner und Meyerburg … synthetisch dargestellte Aethyltricarballylsäure hat. Danach müsste das Hämopyrrol folgende Structur

> haben, und bei der Oxydation zu der Küster'schen Säure von der Formel $C_8H_9O_4N$ das Methyl der Seitenkette zu Carboxyl oxydirt werden.

222 3 Advent of the Bilirubin Structure Proof

= Partielles Imid der Aethylaconitsäure = Küster'schen Säure $C_8H_9O_4N$.

$$(2)\ CH \underset{CO}{\overset{C-CH\cdot C_2H_5}{\boxed{}}} CO\ CO_2H$$
$$O$$

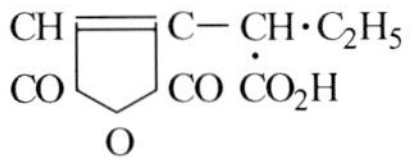

= Partielles Anhydrid der Aethylaconitsäure = Küster'schen Säure $C_8H_8O_5$; aus der Letzteren durch Reduction:

$$(3)\ H_2C-CH-CH-C_2H_5\ ,$$
$$CO_2H\ CO_2H\ CO_2H$$

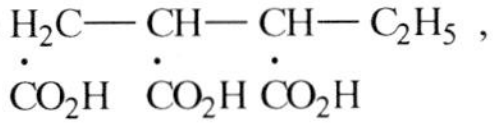

Aethyltricarballylsäure von Auwers.   49

Loss of CO_2 from (1) would lead to propylmaleimide. However, because *Küster* defined the imide as methylethylmaleimide, if *Küster* were correct, hemopyrrol would be methylpropylpyrrole and not *sec*-butylpyrrole – a determination left to the future (*285*):

Durch Abspaltung von Kohlensäure geht das Imid $C_8H_9O_4N$ in das Imid der Propylmaleïnsäure über.

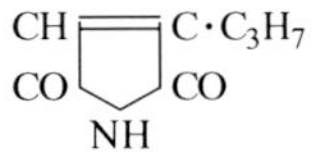

$$CH \underset{CO}{\overset{C\cdot C_3H_7}{\boxed{}}} CO$$
$$NH$$

Nach Küster ... ist es aber wahrscheinlicher, dass dieses Imid identisch mit dem Imid der Methyläthylmaleïnsäure von Fittig ist. Wenn die Ansicht von Küster die richtige ist, und für molekulare Umlagerungen ist auch kein Grund vorhanden, so wäre das Hämopyrrol ein Methylpropylpyrrol,

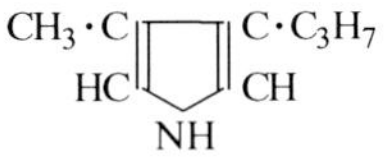

$$CH_3\cdot C \underset{HC}{\overset{C\cdot C_3H_7}{\boxed{}}} CH$$
$$NH$$

Welche von den beiden Formeln dem Hämopyrrol zukommt, wird wohl in der nächsten Zukunft entschieden werden; davon wird auch die Vorstellung abhängig sein, die wir über den molekularen Bau des Bilirubins und der drei bis jetzt bekannten Porphyrine haben werden. 50

Speculating as to how hematoporphyrin might be constituted, *Nencki* and *Zaleski* were the first to write a conjectured structure (*285*):

Ist das Hämopyrrol ein Isobutylpyrrol, so ist die Configuration der Porphyrine durch Verkettung zweier Hämopyrrolmoleküle z. B. nach folgendem Schema sehr einfach zu veranschaulichen:

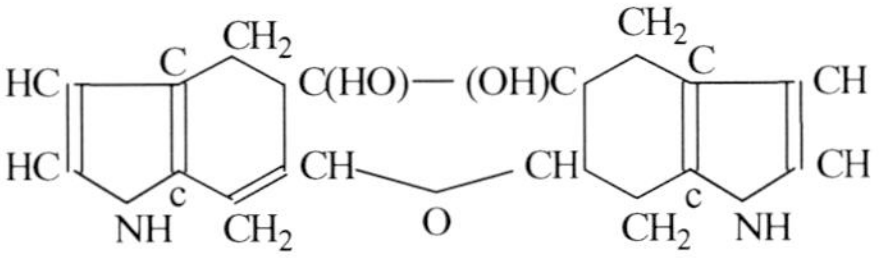

$$HC\quad C\quad CH_2\quad C(HO)-(OH)C\quad CH_2\quad C\quad CH$$
$$HC\quad CH\quad CH\quad CH$$
$$NH\quad CH_2\quad O\quad CH_2\quad NH$$

Hämatoporphyrin, $C_{16}H_{18}O_3N_2$ 51

3.4 Küster's Maleimides and Nencki's Hämopyrrol

And thus hemin would be composed of two hematoporphyrins held together by an Fe in such a way as to be cleaved into the two halves by HBr in acetic acid (285):

Je nachdem wir das Hämopyrrol als Butyl- oder Methylpropyl-Pyrrol betrachten, könnte des Hämin folgende Structur haben:

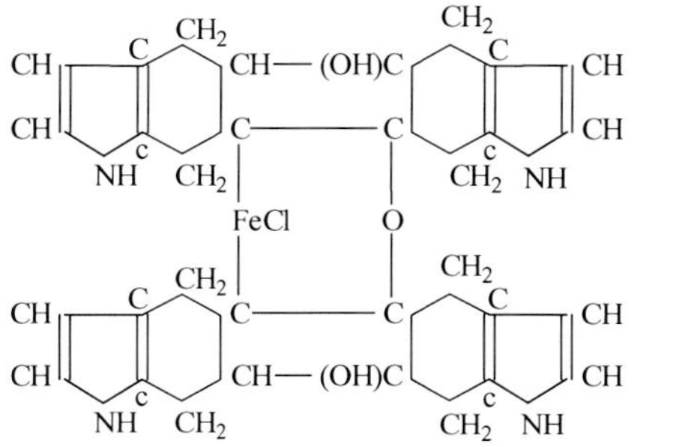

oder

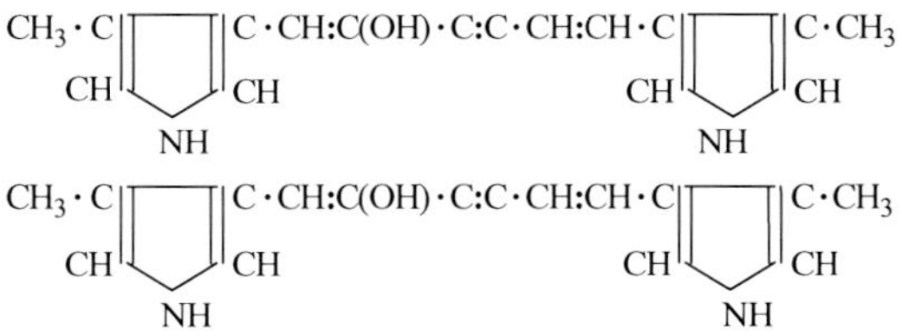

52

Since it was believed then that biliverdin and hematoporphyrin were identical, *Nencki* and *Zaleski* had also proposed a structure for the green bile pigment.

Nencki and *Zaleski* had discovered that treating crude *Acethämin* in acetic acid with conc. HI under somewhat milder conditions produced a new iron-free compound, $C_{16}H_{18}N_2O_2$, that resembled hematoporphyrin ($C_{16}H_{18}N_2O_3$) and was named *Mesoporphyrin* by *Nencki* (285). *Mesoporphyrin* (mesoporphyrin) and hematoporphyrin were said to have very similar properties. In 1902, *Zaleski* (287) carefully resynthesized mesoporphyrin, in ~40% yield from hemin, obtaining some *Hämopyrrol* in addition. Crystalline mesoporphyrin was re-analyzed by combustion analysis to give the formula $C_{17}H_{19}N_2O_2$, and the formula $C_{17}H_{18}N_2O_2(CH_3)$ for what was believed to be the methyl ether, prepared similarly to the derivative of hematoporphyrin. Reducing hematoporphyrin with HI-PH$_4$I under the same gentler conditions as for hemin also gave mesoporphyrin, which was converted to numerous different salts that analyzed correctly for the C_{17} formula. When "free" mesoporphyrin was prepared pure, its combustion analysis data correlated best with the formula $C_{17}H_{19}N_2O_2$ – not exactly the earlier formula, $C_{16}H_{18}N_2O_2$. The question of the molecular formula of mesoporphyrin was apparently decided based on *Zaleski*'s molecular weight determination from the melting point depression (*Raoult*'s method) of phenol by added pigment. For mesoporphyrin the experiment gave 486.4 [*vs.* a formula weight = FW = 283 for $C_{17}H_{19}N_2O_2$]. The experimental values measured for the mono- and di-ethyl ethers were likewise high: 572.2 for the former [*vs.* FW = 311 for $C_{17}H_{18}N_2O_2(C_2H_5)$], and, in two measurements 529.4 and 527.4 for the latter [*vs.* FW = 339 for $C_{17}H_{17}N_2O_2(C_2H_5)_2$]. The formulas derived from combustion analysis data were thus called into question. The analytical MW data indicated that all the old formulas had to be doubled, and so *Zaleski* proposed a table of molecular formulas for (free) mesoporphyrin itself, along with various of its derivatives (*287*):

> Die Ergebnisse sämmtlicher Beobachtungen veranlassen uns also, alle oben angeführten Formeln zu verdoppeln und die chemische Zusammensetzung dieser Körper, wie folgt, zu bezeichnen:

$C_{34}H_{38}O_4N_4 \cdot 2\ HCl$	salzsaures Mesoporphyrin.
$C_{34}H_{38}O_6N_4 \cdot 2\ HCl$	salzsaures Hämatoporphyrin.
$C_{34}H_{30}O_4N_4(C_2H_5)_2$	Aethyläther des Mesoporphyrins.
$C_{34}H_{36}O_4N_4Zn$	Zink- oder Kupfersalz des Mesoporphyrins.
$C_{34}H_{34}O_4N_4(C_2H_5)_2Cu$	Kupfersalz des Mesoporphyrinäthyläthers.
$C_{34}H_{38}O_4N_4$	(freies) Mesoporphyrin.

> Nehmen wir nun für salzsaures Hämin die Formel $C_{34}H_{33}O_4N_4ClFe$ ...   53

Significantly, the data led to a revised formula for hemin as $C_{34}H_{33}N_4O_4FeCl$, which has only one hydrogen more than the formula accepted today: $C_{34}H_{32}N_4O_4FeCl$, FW = 651.94. They also suggested a revision of the hematin formula, and rather importantly, a doubling of the then accepted formula for hematoporphyrin to $C_{34}H_{38}N_4O_6$, the same as that accepted today. The formation of hematoporphyrin from hemin by the action of HBr would thus have to be written (*287*):

> Dann können wir uns die Bildung des Hämatoporphyrins aus Hämin mittelst Bromwasserstoff in folgender Gleichung vergegenwärtigen:
>
> $$C_{34}H_{33}O_4N_4ClFe + 2\ HBr + 2\ H_2O = C_{34}H_{38}O_6N_4 + FeBr_2 + HCl.$$

> Einen Beweis findet diese Gleichung noch darin, dass bei dreimaliger Untersuchung der Gase, welche bei dieser Reaction ausgeschieden werden, kein Wasserstoff nachgewiesen werden konnte.   54

In support of the new equation over the old, *Zaleski* could find no release of H_2 in the treatment of hemin with HBr, as suggested in the old equation. The revised, doubled formulas were destined to alter the previous belief that biliverdin had half the number of carbons as bilirubin, for if hematoporphyrin and biliverdin were simply isomeric, as the term was understood in 1902, they would have to possess the same molecular formula.

With *Nencki*'s death in 1901 and *Zaleski*'s last work on the subject in 1902, it remained to *Küster* to serve, at least in the short term, as torchbearer of further structural investigations into the structures of bile and blood pigments.

By early 1902, *Küster* had returned to his studies of bilirubin, unaware of *Zaleski*'s revised formulas (*287*) that were not be submitted for publication until October 1902. *Küster* was finally able to gather sufficient gallstones (1 kg, dried, from cattle) that allowed him to isolate sufficient bilirubin to conduct detailed analyses. (It was far easier to obtain sufficient blood for the hemin study). Powdered

3.4 Küster's Maleimides and Nencki's Hämopyrrol

gallstones were digested successively with ether, H_2O, aqueous HCl, H_2O, and ether, then extracted using $CHCl_3$ in a Soxhlet apparatus to yield a red-black $CHCl_3$ solution in the boiling flask and a thick crust. After filtration, standing to allow more precipitation in the filtrate, then refiltering and adding alcohol to the last filtrate, a large amount of brown-green pigment was obtained. The pigment was repeatedly taken up in $CHCl_3$ and precipitated until it was freed from *Bilifuscin, etc.* and could be recrystallized from boiling dimethyaniline, whereby somewhat more than one-half of the pigment used was obtained as beautiful crystals, whose elemental combustion analysis was consistent with the formula $C_{16}H_{18}N_2O_3$. Recrystallization again gave the same result, and no change resulted from subsequent recrystallization from $CHCl_3$. All these manipulations, in which traces of dimethylaniline had surely been removed proved to *Küster* that *Thudichum's* earlier objection to hot dimethylaniline as a bilirubin crystallization solvent was unfounded – except by this time *Thudichum* had expired (*215*):

> Durch diese Angaben dürften wohl die Einwendungen, welche Thudichum ... gegen meine Methode, Bilirubin rein und krystallisirt herzustellen, gemacht hat, als erledigt angesehen werden ...   55

Recounting that the suspected relationship between the pigments of blood and bile had been established in (chemical) fact: Both bilirubin and hematin yielded the same imide of tribasic hematinic acid, $C_8H_9NO_2$ (also called *Biliverdinsäure*). Both imides produced one and the same $C_8H_8O_5$ compound, the partial anhydride of tribasic hematinic acid, upon treatment with alkali. A new preparation of *Biliverdinsäure* convinced *Küster* that the originally established identity was correct.

An important remaining task was to determine exactly the chemical structure of *Biliverdinsäure*-hematinic acid imide. The corresponding anhydride, $C_8H_8O_5$, was believed to have one of three possible structures (*288*):

> Hiernach stellt sich also das partielle Anhydrid der dreibasischen Hämatinsäure, $C_8H_8O_5$, als eine carboxylirte Methyläthylmaleïnsäure dar; für dasselbe können somit nur folgende drei Formeln in Betracht kommen:

I.
$$\begin{array}{c} COOH \cdot CH_2 \cdot C \cdot CO \\ \| \qquad\qquad O \\ H_3C \cdot CH_2 \cdot C \cdot CO \end{array}$$

II.
$$\begin{array}{c} H_3C \cdot C \cdot CO \\ \| \qquad\quad O \\ H_3C \cdot CH \cdot C \cdot CO \\ \cdot \\ COOH \end{array}$$

III.
$$\begin{array}{c} H_3C \cdot C \cdot CO \\ \| \qquad\qquad O \\ COOH \cdot CH_2 \cdot CH_2 \cdot C \cdot CO \end{array}$$

> Die folgenden Versuche geben nun den Ausschlag zu Gunsten der Formel III.   56

Küster proved that **III** was correct, as follows. Oxidation of $C_8H_8O_5$ by $KMnO_4$ in H_2SO_4 solution at 0°C gave succinic acid ($HO_2C–CH_2–CH_2–CO_2H$), which can arise only from **III**. Reduction of **III** using **HI** in a sealed tube at 105°C yielded an optically inactive mixture possessing a non-sharp melting point. Formula **III** is capable of producing two optically inactive, racemic acids (actually diastereomeric acids) of *Hämotricarbonsäure* (so-named by *Küster*). They were separated by fractional crystallization, one with mp 175–176°C, the other with mp 140–141°C, each a β,γ,ε-tricarboxylic acid (or in the Geneva convention of the day, *2-Methylhexandisäure-3-methylsäure*) (*288*):

> ... welches vielleicht dasjenige der Brenztraubensäure ist, die sich ja bei der Oxydation eines nach Formel III constituirten Körpers ergeben muss. Das Entstehen von Bernsteinsäure würde sich aber bei Zugrundelegung von Formel I und II nur höchst gezwungen erklären.
>
> Die Reduction durch Jodwasserstoff im Rohr bei 150° führt, wie früher erwähnt, ... zu einem unscharf schmelzenden Gemisch optisch inactiver Säuren. In der That muss die Reduction einer Säure von der Formel III zwei inactive, racemische Säuren geben, da ja durch die Anlagerung zweier Wasserstoffatome zwei Kohlenstoffatome asymmetrisch werden.
>
> Es ist nun Hrn. Apotheker O. Mezger im hiesigen Institut gelungen, durch eine sehr mühsame, fractionirte Krystallisation aus Wasser das Reductionsgemisch in zwei inactive Säuren $C_8H_{12}O_6$ zu zerlegen, für welche ich den (von Kölle gewählten) Sammelnamen: »Hämotricarbonsäuren«, der Kürze halber und um die Herkunft anzuzeigen, beibehalten möchte. Ihrer Constitution nach sind es βγε-Tricarbonsäuren. ...   57

These studies seemed to confirm the structure of hematinic acid as a maleimide with a β-CH₃ and a β′-$CH_2CH_2CO_2H$.

In an attempt to determine the structure of *Nencki*'s *Hämopyrrol*, which was air-sensitive, *Küster* submitted it to chromic acid oxidation to yield, without doubt (as *Küster* said) an imide of a substituted maleic acid. But larger amounts of *Hämopyrrol* were needed to prove that the acid obtained, or that its anhydride was identical with the expected *Methyl-n-propylmaleinsäureanhydrid* (methyl-*n*-propylmaleic acid anhydride) (*288*):

> Danach besteht kein Zweifel, dass die Oxydation des Hämopyrrols über ein Imid zu einer substituirten Maleïnsäure geführt hat. Es wird nur grösserer Mengen bedürfen, um den Beweis zu liefern, dass die erhaltene Säure, resp. deren Anhydrid $C_8H_{10}O_3$, mit dem erwarteten Methyl-*n*-propylmaleïnsäureanhydrid identisch ist. ...   58

Two years later, in 1904, *Küster* isolated more of the *Nencki-Zaleski Hämopyrrol* ($C_8H_{13}N$), oxidized it to an imide (presumed to be $C_8H_{11}NO_2$), and worked to prove the structure by synthesis (*217, 218*). To this end, it was necessary to synthesize the suspected anhydride of methyl-*n*-propylmaleic acid. Following the method used by *Michael* and *Tissot* (*181*) to prepare dimethylmaleic anhydride (*Pyrocinchonsäuren anhydrid*), the sodium salt of ethyl acetoacetate was mono-alkylated with propyl iodide, and the product, *Propyl-acetessigester* (see below), was converted to the cyanohydrin using HCN. Saponification of the cyanohydrin followed by distillation eliminated H_2O and yielded the anhydride, bp 241–242°C (*217, 218*), as represented in Scheme 3.4.1:

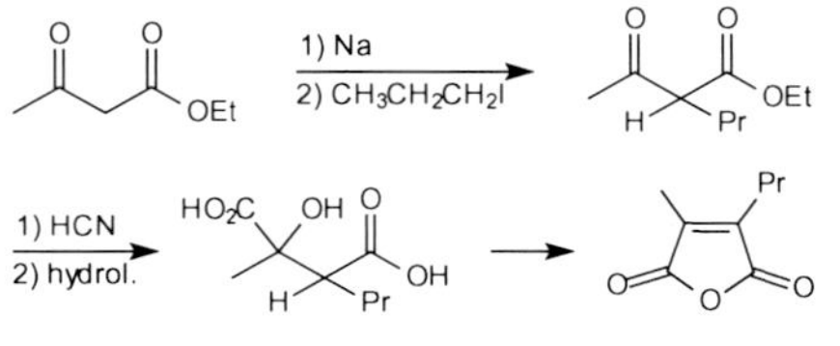

Scheme 3.4.1

3.4 Küster's Maleimides and Nencki's Hämopyrrol

The maleimide product analyzed correctly for the $C_8H_{10}O_3$ formula (FW = 154) and gave molecular weights (158.2, 159.4, 160.4) in three determinations by the boiling point elevation method.

The corresponding imide, mp 56–57°C, was prepared by heating the anhydride at 130°C in freshly-prepared ethanolic ammonia in a sealed tube for three hours. Its odor resembled iodoform and analyzed correctly for $C_8H_{11}NO_2$ (*217*):

> Es schmilzt bei 56–57° und ist unzersetzt sublimirbar; es löst sich leicht in Aether, Alkohol, Chloroform, Benzol und Essigesester, ist schwer in kaltem, leichter in heissem und in ammoniakhaltigem Wasser löslich.
> Der Geruch erinnert an Jodoform.
> 0.1117 g Sbst.: 0.2567 g CO_2, 0.0755 g H_2O. – 0.1222 g Sbst.: 10.1 ccm N (16°, 729 mm).

$C_8H_{11}O_2N$.	Ber.	C	62.7,	H	7.2,	N	9.2.	
	Gef.	»	62.7,	»	7.5,	»	9.2.	59

For comparison to *Hämopyrrol*, *Küster* repeated the *Nencki-Zaleski* cleavage reaction and isolation and oxidized the resultant oil with chromic acid to isolate a syrup that began to crystallize only after months, at which point it could be taken up in aq. HCl and extracted into ether to yield brown crystals. The last were dissolved in hot water and decolorized with charcoal to provide colorless needles of mp 63–64°C – or 7° higher than the synthetic methylpropylmaleimide. However, the amount of *Hämopyrrol*-derived imide was insufficient for an elemental combustion analysis – absence of which would have important negative implications (*217*):

> Es verblieben uns nunmehr noch 2 g eines immer noch gefärbten Syrups, der das gesuchte Imid dem Geruch nach enthalten musste; in der That begann nach Monaten eine Krystallisation, und nun konnte durch Aufnahme in Salzsäure und Ausschütteln mit Aether eine weitere Reinigung erzielt werden, wonach braune, büschelförmige Krystallnadeln isolirt wurden. Diese wurden schliesslich in heissem Wasser gelöst und die Lösung durch Thierkohle entfärbt; eine Ausschüttelung mit Aether lieferte jetzt eine geringe Menge farbloser Nadeln, welchen der charakteristische Geruch eigen war. Der Schmelzpunkt derselben wurde bei 63–64° gefunden, also um etwa 7° höher als der des synthetisch hergestellten Methyl-propyl-maleïnsäureimids. Leider reichte die erhaltene Menge an reinem Product zur Analyse nicht aus ...   60

From *Küster*'s perspective, his investigations had not produced the expected result, *i.e.* that the imide from *Hämopyrrol* would be methylpropylmaleimide, but if not the latter, then *Hämopyrrol* was not methyl-*n*-propylpyrrole. Convinced that the imide isolated by oxidation of *Hämopyrrol* was a disubstituted imide, two alternative possibilities apparently came to *Küster*'s mind: methylisopropylmaleimide and xeronimide (diethylmaleimide), neither of which were known compounds at the time. So *Küster* synthesized methylisopropylmaleimide using the same general procedure as in the methylpropylmaleimide synthesis, analyzed it correctly for $C_8H_{11}NO_2$ but found it to have a mp 44–45°C, or 19–20°C lower than the imide produced from *Hämopyrrol* (*217*):

> Somit hat unsere Untersuchung das erstrebte positive Resultat bisher nicht erbracht; allem Anschein nach ist das Imid der Methyl-propyl-maleïnsäure nicht identisch mit dem Imid, welches aus dem Hämopyrrol erhalten werden konnte. Da nun Letzteres seinen

228   3 Advent of the Bilirubin Structure Proof

Eigenschaften zufolge zur Klasse der bisubstituirten Maleïnsäureimide gehört und, auf Grund allerdings nur einer Analyse eines bei seiner Verseifung entstehenden Baryumsalzes, im Molekül acht Kohlenstoffatome enthält, ist an die Möglichkeit zu denken, dass es mit dem Methyl-isopropyl-maleïnsäureimid oder mit dem Xeronsäureimid identisch wäre. Jenes ist von uns hergestellt und analysirt worden.

$C_8H_{11}O_2N.$ Ber.   C   62.7,   H   7.02,   N   9.2.
                Gef.   »   62.65,   »   7.3,   »   9.24.

Es stellt eine strahlige Krystallmasse dar, welche in ihren Eigenschaften, auch im Geruch, dem Methyl-propyl-maleïnsäureimid ausserordentlich ähnlich ist. Der Schmelzpunkt des Präparats wurde bei 44-45° gefunden, es dürfte also mit dem Imid aus dem Hämopyrrol nicht identisch sein.   61

Though *Küster*'s synthesis of xeronimide was still in progress as of June 1903 (*217*), he reported, prophetically as will become evident later, that the difficulty in isolating pure imide products from *Hämopyrrol* might be due to the fact that the pyrrole could not be claimed to have been isolated in a pure state, and so it remained seemingly inconclusive in 1903 as to whether it might turn out to consist of a mixture of isomers (*217*):

Mit der Herstellung des Xeronsäureimids sind wir zur Zeit beschäftigt, auch beabsichtigen wir, noch einmal die Oxydation des Hämopyrrols zu versuchen, wenn auch die Schwierigkeit, aus den Oxydationsproducten des Hämopyrrols das reine Imid zu isoliren, nicht gering ist. Auch halten wir es für nicht ausgeschlossen, dass das Hämopyrrol, welches bisher nicht im reinen Zustand isolirt werden konnte, sich als aus einem Gemenge von Isomeren bestehend erweisen wird.   62

Later in 1903, *Küster* summarized at length his methods for isolating and purifying various hemins and hematin (*289*), and in a long paper, submitted in April 1905 (*290*) he recognized *Zaleski*'s new C_{34} formulas (*287*) for hemin, hematin, hematoporphyrin, and mesoporphyrin. Given the 50% yield of C_8 *Hämatinsäureimid* (hematinic acid imide), there seemed to be a good correlation in carbon count. *Küster* summarized in detail all of his chromic acid oxidation results from hematin and hematoporphyrin, provided new information on oxidation of the same using HNO_3, which gave oxalic and succinic acids and a trace of hematinic acid, as its Ca salt; using 30% H_2O_2 in acetic acid, which gave succinic acid and hematinic acid, $C_8H_8O_5$, isolated as its Ca salt; with NaOBr in alkaline solutions, which yielded a small amount of hematinic acid, $C_8H_8O_5$; and with $Ca(MnO_4)_2$, which yielded unchanged hematin, oxalic acid and traces of an H_2O-soluble nitrogen-free compound. The chromic acid oxidation seemed to be the most reliable and tractable.

The *Küster* summary (*290*), which provided new structural information, set the stage for an even longer review article and summary of experimental details that was submitted in November 1905 (*291*) on the constitution of hematinic acid, *Methyläthylmaleïnsäureanhydrid* (methylethylmaleic anhydride) (with *H. Galler* and *K. Haas*), *Methyläthylmaleïnsäureimid* (methylethylmaleimide) (with *K. Haas*), the dry distillation of $C_8H_9NO_4$ and $C_8H_8O_5$ hematinic acid (with *K. Haas* and *O. Mezger*), the oxidation and the reduction of $C_8H_8O_5$ hematinic acid, reduction of the latter and of $C_8H_9NO_4$ hematinic acid (with *O. Mezger*), and the reduction

3.4 Küster's Maleimides and Nencki's Hämopyrrol

of methylethylmaleic anhydride (with *K. Haas*). This important paper, published in 1906 (*291*), confirmed the structure of the tribasic hematinic acid, $C_8H_{10}O_6$, in exquisite detail (*291*):

> Sie erbringen den Beweis, dass in der dreibasischen Hämatinsäure eine γ-Penten-αγδ-tricarbonsäure folgender Constitution

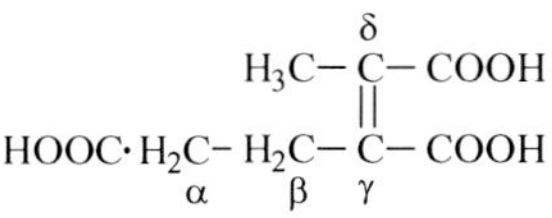

$$
\begin{array}{c}
\overset{\delta}{\text{H}_3\text{C}-}\underset{}{\overset{}{\text{C}}}-\text{COOH} \\
\| \\
\text{HOOC·H}_2\text{C}-\text{H}_2\text{C}-\text{C}-\text{COOH} \\
\quad\alpha\qquad\beta\quad\gamma
\end{array}
$$

> vorliegen muss, deren Anhydrid ... die Säure $C_8H_8O_5$, deren Imid die Säure $C_8H_9O_4N$ vorstellt, und zwar sind es die bebanachbarten Carboxyle, aus denen Wasser austritt und an denen sich die Imidbildung vollzieht.   63

Its structure was proved, as revealed earlier (*288*), because it formed succinic acid upon $KMnO_4$ or CrO_3 oxidation and because its imide gave methylethylmaleimide upon decarboxylation as proved by synthesis from the corresponding anhydride that had been prepared synthetically by two independent routes (*291*):

> ... wonach das gesuchte Product durch Condensation der Brenzweinsäure mit Brenztraubensäure in Gegenwart von Essigsäureanhydrid und entwässertem Natriumacetat entstehen soll:

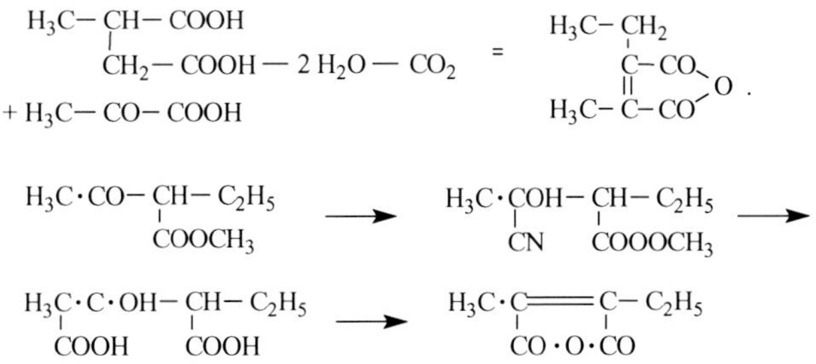

$$
\begin{array}{l}
\text{H}_3\text{C}-\text{CH}-\text{COOH} \\
\quad\quad|\\
\quad\quad\text{CH}_2-\text{COOH}-2\,\text{H}_2\text{O}-\text{CO}_2 \\
+\text{H}_3\text{C}-\text{CO}-\text{COOH}
\end{array}
=
\begin{array}{l}
\text{H}_3\text{C}-\text{CH}_2 \\
\quad\quad|\\
\quad\quad\text{C}-\text{CO}\diagdown \\
\quad\quad\|\qquad\quad\text{O} \\
\text{H}_3\text{C}-\text{C}-\text{CO}\diagup
\end{array}
\;.
$$

$$
\begin{array}{l}
\text{H}_3\text{C·CO}-\text{CH}-\text{C}_2\text{H}_5 \\
\quad\quad\quad|\\
\quad\quad\quad\text{COOCH}_3
\end{array}
\longrightarrow
\begin{array}{l}
\text{H}_3\text{C·COH}-\text{CH}-\text{C}_2\text{H}_5 \\
\quad\quad|\qquad\quad|\\
\quad\quad\text{CN}\qquad\text{COOOCH}_3
\end{array}
\longrightarrow
$$

$$
\begin{array}{l}
\text{H}_3\text{C·C·OH}-\text{CH}-\text{C}_2\text{H}_5 \\
\quad\quad|\qquad\quad\quad|\\
\quad\quad\text{COOH}\qquad\text{COOH}
\end{array}
\longrightarrow
\begin{array}{l}
\text{H}_3\text{C·C}\!=\!=\!=\!\text{C}-\text{C}_2\text{H}_5 \\
\quad\quad|\qquad\quad\quad|\\
\quad\quad\text{CO·O·CO}
\end{array}
\qquad 64
$$

The latter synthesis was repeated by *Küster*, and the product anhydride, $C_8H_8O_5$, was carefully characterized and converted to the corresponding imide. Though early on *Küster* considered (*288*) three possible structures of the anhydride (*291*), each of which might undergo decarboxylation to produce methylethylmaleic anhydride, only **III** is capable of giving succinic acid upon oxidation:

I.

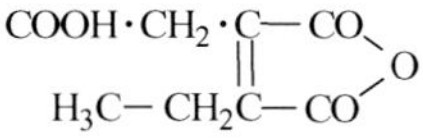

$$
\begin{array}{l}
\text{COOH·CH}_2\cdot\text{C}-\text{CO}\diagdown \\
\quad\quad\quad\quad\|\qquad\quad\text{O} \\
\text{H}_3\text{C}-\text{CH}_2\text{C}-\text{CO}\diagup
\end{array}
$$

II.

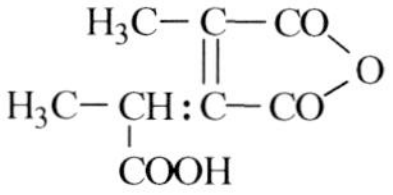

$$
\begin{array}{l}
\text{H}_3\text{C}-\text{C}-\text{CO}\diagdown \\
\quad\quad\|\qquad\quad\text{O} \\
\text{H}_3\text{C}-\text{CH}:\text{C}-\text{CO}\diagup \\
\quad\quad\quad|\\
\quad\quad\quad\text{COOH}
\end{array}
$$

III.

$$
\begin{array}{l}
\text{H}_3\text{C}-\text{C}-\text{CO}\diagdown \\
\quad\quad\|\qquad\quad\text{O} \\
\text{H}_2\text{C}-\text{H}_2\text{C}-\text{C}-\text{CO}\diagup \\
\quad\quad\quad|\\
\quad\quad\quad\text{COOH}
\end{array}
$$

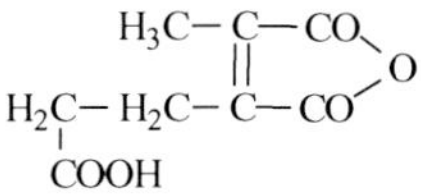

230   3   Advent of the Bilirubin Structure Proof

Küster's research focus was clearly aimed more at the pigments of blood than the pigments of bile, although he studied both. His major contributions to both by 1906 were two-fold. The first came from chromic acid oxidations of hemin, hematin, and hematoporphyrin, from which he obtained hematinic acid, and oxidations of bilirubin and biliverdin to yield an imide, also $C_8H_9NO_4$, which he named *Biliverdinsäure* – names that reflected of their pigment sources. These imides proved to be identical to and possessed the structure of the imide of anhydride **III** above. The second contribution came from the *Nencki-Zaleski* breakthrough discovery of *Hämopyrrol*, $C_8H_{11}N$, in 1901 (*285, 286*) from reductive scission of hemin, hematin, or hematoporphyrin by HI + PH_4I.

In order to prove the structure of *Hämopyrrol*, *Küster* oxidized it to an imide, mp 63–64°, using his chromic acid method. Convinced that *Hämopyrrol* was a β,β'-disubstituted pyrrole, the prime suspect, methylpropylmaleimide, mp 56–57°C, proved to have a melting point 7°C lower than the imide from *Hämopyrrol*. The secondary suspects were methylisopropylmaleimide and diethylmaleimide (xeronimide), of which *Küster* synthesized the first and found its mp 44–45°C again too low. Xeronimide, whose synthesis *Küster* said in 1904 (*217, 218*) was in progress, actually has mp 68–70°C, which would be too high. The possibilities are summarized in the structures of Scheme 3.4.2.

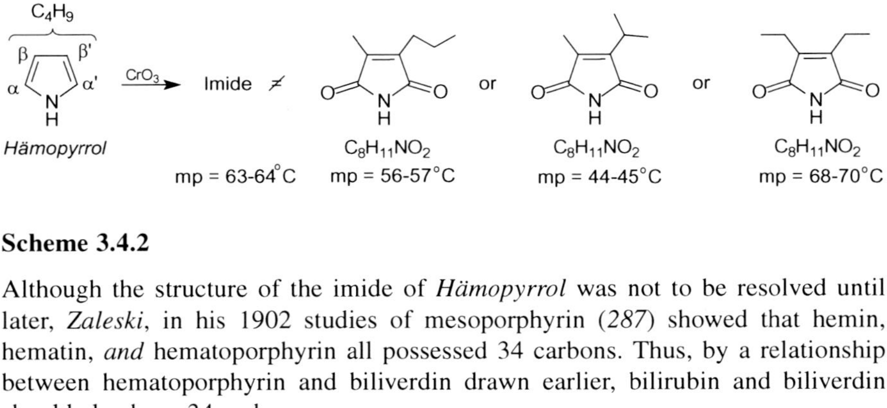

Scheme 3.4.2

Although the structure of the imide of *Hämopyrrol* was not to be resolved until later, *Zaleski*, in his 1902 studies of mesoporphyrin (*287*) showed that hemin, hematin, *and* hematoporphyrin all possessed 34 carbons. Thus, by a relationship between hematoporphyrin and biliverdin drawn earlier, bilirubin and biliverdin should also have 34 carbons.

Three years were to elapse before *Küster* was able to resolve the question of the structure of *Hämopyrrol*. By 1906, it had come to his attention that in 1903 *Plancher* and *Cattadori* (*293, 294, 295*) had oxidized α,β-dimethylpyrrole to citraconimide (*292*):

This meant that the formation of imides from pyrroles did not require both α-positions of the pyrrole to be unsubstituted (*292*). And it apparently deepened *Küster*'s suspicion as to whether *Hämopyrrol* was necessarily alkylated at only the

3.4 Küster's Maleimides and Nencki's Hämopyrrol

β-positions – as was suggested earlier. In a detailed experimental section (*292*) were given the *de novo* synthesis, properties, and reactions of methylpropylmaleimide, methylisopropylmaleimide, and diethylmaleic acid (the precursor to xeronimide). In addition, toward the end of this long, full paper, *Küster* and *Haas* prepared *Hämopyrrol* by the *Nencki-Zaleski* method of treating hemin in hot acetic acid with HI and PH_4I. Then they oxidized it in aqueous chromic acid to isolate the by then well-known usual mp 63–64°C imide product, but again did not apply a combustion analysis – until a year later. By then *Küster* had re-examined the *Hämopyrrol* imide more critically (*296*).

Saponification of the imide with $Ba(OH)_2$ yielded a Ba^{+2} salt whose solubility properties and crystal form resembled those of the methylethylmaleic acid brown salt prepared earlier. Based on the analysis of only very small amounts, it satisfied the formula $C_8H_{10}O_3$ and thus the formula $C_8H_{13}N$ given to its *Hämopyrrol* precursor by *Nencki* and *Zaleski*, corresponding to *β,β'-Methyl-propyl-pyrrol*. However, in a continuation of his research on the mp 63–64°C imide from *Hämopyrrol*, the great difficulty *Küster* experienced in purifying the crude material led him to suspect that it might not be a simple compound (*217*):

> Aus der Schwierigkeit, mit welcher die Reinigung des Rohimids verbunden war, folgerten wir, dass in ihm ein kompliziertes Gemisch vorlag, und dass demnach auch das Hämopyrrol kein einheitlicher Körper sein dürfte. 65

Recalling the earlier separation (*292, 297*), involving ether extraction of an acidic solution of the crude imide after neutralization with Na_2CO_3 to give the crystalline, mp 63–64°C imide, and the isolation of a mp 93–96°C imide from the alkaline syrup after extraction, he repeated the oxidation reaction 10 times to collect a much larger amount of crude imide than in any of his previous studies (2.85 g *vs.* 0.2–0.4 g).

In order to purify the crude imide, its solution in ether was extracted with conc. HCl, which reduced the amount (in ether) appreciably and thus left a preparation in the form of nearly colorless needles that had all the properties of a disubstituted imide of mp 67–68°C – which could only have been methylethylmaleimide. However, a mixture mp with authentic, synthetic methylethylmaleimide melted at 64°, and a combustion analysis did not match exactly to $C_7H_9NO_2$. Repurification gave an even less good result (*296*):

> Bei der Reinigung, die schliesslich dadurch erreicht wurde, dass die ätherische Lösung des Imids mit konzentrierter Salzsäure ausgeschüttelt wurde, reduzierte sich diese Menge erheblich; doch wurde jetzt ein Präparat in Form fast farbloser Nadeln erhalten, mit allen Eigenschaften der Imide bisubstituierter Maleinsäuren begabt, das den Schmp. 67-68° aufwies. Danach konnte nur Methyl-äthyl-maleinsäureimid vorliegen . . , ein Gemisch mit dem synthetisch erhaltenen schmolz allerdings schon bei 64°, auch lieferte die Analyse nur unscharfe Werte.
>
> 0.1334 g Sbst. (im Vakuum getr.): 0.2985 g CO_2, 0.083 g H_2O. – 0.1205 g Sbst.: 11.1 ccm N (20°, 746 mm).

$C_7H_9O_2N$.	Ber.	C	60.40,	H	6.50,	N	10.10
	Gef.	»	61.00,	»	6.91,	»	10.34.

232   3 Advent of the Bilirubin Structure Proof

Als dann die Reinigung mit der kleinen, noch verfügbaren Menge von 0.5 g wiederholt
wurde, ergab sich ein noch weniger gutes Resultat.

0.1447 g Sbst.: 0.3260 g CO_2, 0.1065 g H_2O = 62.44% C, 8.29% H.   66

Hämopyrrol was again prepared, from *Dehydrochloridhämin* instead of hemin
and the oxidized acidic *Säure Hämopyrrol* so obtained gave a mp 63–65°C. Further
struggles with purification, removal of stubborn carbon-rich impurities clinging to
the imide, led to a new sample being purified by conc. HCl that had mp 69–70° and
gave a better analysis for $C_7H_9NO_2$. From all this, *Küster* drew the conclusion that the
main product of oxidation of the acidic *Hämopyrrol* is methylethylmaleimide (*296*):

Jetzt erhielt ich aus sechs Versuchen 1.62 g Rohimid, das bereits schön krystallisierte,
durch die Reinigung mit konzentrierter Salzsäure dann 0.65 g eines Präparats, das bei
69-70° schmolz.
    0.1315 g Sbst. (bei 75° ½ Stunde getrocknet): 0.2907 g CO_2, 0.080 g H_2O.

$C_7H_9O_2N$.	Ber.	C	60.40,	H	6.50
	Gef.	»	60.29,	»	6.76

Danach kann kein Zweifel mehr bestehen, dass bei der Oxydation des »sauren
Hämopyrrols« als Hauptprodukt das Methyl-äthyl-maleinsäureimid entsteht.   67

Curiously, the basic *Hämopyrrol* was oxidized to a crystalline imide of mp
64–66°C, and a mixture of the imides from the acidic and basic *Hämopyrrol* had a
mp 63°C. Still, *Küster* was confident that both imides were identical with methylethylmaleimide (*296*):

… durch Wiederholung der Reinigung ergaben sich dann 0.7 g schwach gelb gefärbte, zu
Büscheln angeordnete Krystalle, deren Schmp. bei 64–66° lag. Ein Gemisch der Imide aus
»saurem« und aus »basichen« Hämopyrrol schmolz bei 63°.
    0.1333 g Sbst. (vakuumtrocken): 0.294 g CO_2, 0.083 g H_2O. – 0.1231 g Sbst.: 11.5 ccm
N (24°, 746 mm). – 0.0936 g Sbst: 0.2068 g CO_2, 0.0638 g H_2O. – 0.1178 g Sbst.: 10.9 ccm
N (18°, 728.6 mm).

$C_7H_9O_2N$.	Ber.	C	60.40,		H	6.50,		N	10.10.	
	Gef.	»	60.15,	60.26,	»	6.92,	7.46,	»	10.27,	10.34.

Trotz des um 2° zu niedrig befundenen Schmelzpunkts ist somit auch hier an der
Identität mit dem Methyl-äthyl-maleinimid nicht zu zweifeln.   68

Given loss of an α-methyl during oxidation to the imide, *Küster* concluded in
favor of either *β,β′-Methyl-äthyl-pyrrol*, *β,β′-Methyl-äthyl-pyrrolin*, or an *α,β′-
Dimethyl-β′-äthyl-pyrrol* or *Pyrrolin* as *Hämopyrrol* (*296*):

Aus den angeführten Beobachtungen ergibt sich, dass das Hämopyrrol ein Gemisch ist,
in dem sich zwei Pyrrolderivate befinden, welche beide basischen Charakter haben, dazu
besitzt das »säure Hamopyrrol« schwach säure Eigenschaften. Letzteres lässt sich glatter
oxydieren, und ich möchte annehmen, dass in ihm das β,β-Methyl-äthyl-pyrrol vorliegt. In
dem nur basische Eigenschaften aufweisenden zweiten Bestandteil haben wir es dagegen
entweder mit einem β,β′-Methyl-äthyl-pyrrolin oder einem α,β′-Dimethyl-β′-äthyl-pyrrol
oder -pyrrolin zu tun, das bei der Oxydation die α-ständige Methylgruppe verliert …   69

3.4 Küster's Maleimides and Nencki's Hämopyrrol

Thus, in the first decade of the 20th century the *molecular structures* of hemin, hematin, bilirubin, and biliverdin were on their way to being understood, by virtue of research that pointed to the presence of pyrrole or pyrrole-derived structures contained within them. Previously, little or nothing was known of their molecular structures, and it was from the work of *Küster* and *Nencki* that structural components were being pried away and fully characterized chemically.

At the close of 1907, the new discovery related to the structures of bilirubin and biliverdin was that they probably contained one or two β-methyl, β'-propionic acid pyrrole units that are oxidatively cleaved from the pigment structures to give *Biliverdinsäure = Hämatinsäureimid* (hematinic acid imide), thereby providing the first glimpse into their structures. Not unrelated to this discovery of *Küster's*, the pigments associated with blood (hemin and hematin) and their iron-free derivative, which was thought to be an isomer of bilirubin, also cleave oxidatively into a ~50% yield of hematinic acid. This, too, gave *Küster* an insight into the components of the red pigments. The structural information on hemin and hematin was further enhanced with the discovery by *Nencki* and *Zaleski*, late in *Nencki*'s life, that the same red pigments could be reductively cleaved into *Hämopyrrol*, an 8-carbon pyrrole compound, whose structure was almost completely solved by *Küster* when he oxidized it to the 7-carbon methylethylmaleimide.

Küster's work illustrates the changes taking place in "animal" or physiological chemistry. For natural products, the concept of a chemical structure was rapidly becoming a commonplace objective, the next logical step beyond elemental combustion analysis, the chemical formula, and the molecular weight – all great advances in scientific practice and thought in the 19th century. The last half of the 19th century gave witness to the explosion of practice and knowledge in organic chemistry: structure, synthesis methodology, the ability to make carbon-carbon bonds, and, with all that, the capability and necessity to prove the chemical structure of a newly-obtained natural product, or a fragment of it, by logical synthesis. Thus, at the turn of the century, with *Küster's* ability to meld the two disciplines, pigment chemistry and organic chemistry, the stage had been set for the entry of two new investigators, trained to think as organic chemists, who emerged to tackle the structure of bilirubin nearly simultaneously in Munich in the early 20th century: *Oskar Piloty* and *Hans Fischer*.

4 A Modern Proof of Bilirubin Structure Emerges

If it was not evident to many in the late 1800s that the pigments of bile, blood, and green leaves constituted attractive targets for establishing chemical structures, the perspective began to change in the early 1900s. The most attractive targets were the pigments of blood and leaves, which were far more available resources than the pigments of bile. The latter were shown to be difficult to collect and purify and came typically from less available gallstone resources that contained numerous, difficultly separable pigments which resisted release from their (calcium) salts. Bile itself had apparently ceased to be an attractive resource of bile pigments. Although gallstones were a better source of bilirubin than bile, isolation and purification were still not easy, and obtaining sufficiently large quantities of pigment gallstones was not assured. Given the availability of blood and the relatively easier procedures to remove its pigment, work on the structure of the blood pigment advanced more rapidly than work on the structure of bilirubin. Investigations of chlorophyll, also obtained from readily available sources, had led to *Willstätter*'s 1915 *Nobel* Prize in chemistry, but the unknown structure of the green pigment had not been forgotten to the chemists of the early 20th century. Its relationship to the pigment of blood was not obvious from a biological perspective. Yet, given *Küster*'s investigations, it had been shown that the closely related blood pigments hemin, hematin, and hematoporphyrin were in some way structurally related to bilirubin and biliverdin because oxidative cleavage (using chromic acid) yielded the same monopyrrole-derived compound, hematinic acid imide (= *Biliverdinsäure*). *Küster* had proved its structure, by synthesis, to be a maleimide with methyl and propionic acid β-substituents. At least two additional links between the classes of pigments had come earlier: (i) the apparent though controversial identity of bilirubin and hematoidin, with the latter arising from "old" blood, and (ii) the belief that bilirubin and hematoporphyrin had the same formula and were but isomers of one another. It thus followed logically (and hopefully) that any structural information derived for the blood pigments would translate into structural information on bilirubin. Consequently, although investigations of the early part of the 20th century focused on both pigments, a larger effort was directed toward the more available red pigment.

To this purpose, *Nencki* and *Zaleski* discovered a reductive cleavage reaction involving HI + PH_4I, which when applied to hemin, hematin, and hematoporphyrin produced *Hämopyrrol*, $C_8H_{13}N$. They considered *Hämopyrrol* to be a β,β'-dialkylated pyrrole but did not prove its structure. Nor did they apply this new reductive cleavage reaction to bilirubin or biliverdin. *Nencki* did, however, together with *Marchlewski*, apply the reaction to the green plant pigment, chlorophyll, and obtained *Hämopyrrol*. This proved at least a partial structural relationship between the red pigment of blood and the green pigment of plants. By 1906 the structure of *Hämopyrrol* was believed to have been almost completely determined by *Küster*, who oxidized it, using chromic acid, and obtained methylethylmaleimide, C_7H_9O, whose structure he proved by logical synthesis. The location of the "missing" carbon of *Hämopyrrol* became a source of considerable speculation; to *Küster* it was a pyrrole α-CH_3, but that was also a controversial point that begged a fuller clarification.

Nonetheless, given the success of chemical cleavage of the pigments into two recognizable components, one (hematinic acid) in a yield of ~50%, a major advance in knowledge of the pigments' structures had at long last been achieved. The solution to an historic and apparently intractable problem of pigment structure began to appear rather promising. Although *Küster* was not about to abandon his structural investigations, following *Nencki*'s death in 1901 two new investigators entered the scene, initiated research on bilirubin and the blood pigments, and in 1908–1909 began to publish results from Munich: *Oskar Piloty* (1866–1915), who was of *William Küster*'s generation (1863–1929), and the younger *Hans Fischer* (1881–1945).

It is unclear what drew *Piloty* to the unsolved structures of hemin, hematin, hematoporphyrin, and bilirubin; yet, he addressed the subject in nearly all of his ~25 publications between 1909 and 1915, the year of his apparently heroic death on the Somme battlefield in World War I (*298*). Imbued with a sense of patriotism and chauvinism, *Piloty* had responded to the death of one of his sons in battle on September 25, 1914 by signing up for the war to lead a machine gun company into battle and then expiring on October 6, 1915 from an enemy bullet to the forehead at Somme-Py in the Marne Valley. His death predated the bitterly contested Somme Offensive (Battle of the Somme, *la Bataille de la Somme, die Sommeschlacht*), July 1 - November 18, 1916, that ended with no victory for either side but left 624,000 Allied and 465,000 German casualties.

Both father and son were thus among the 70 million military personnel mobilized and counted among the nine million combatants killed in the Great War. Though it did wonders for technological advances in fire power, it settled no great issues: it dismantled the social system in Europe, reconfigured governments, redrew national boundaries, and set the stage for the apocalypse of World War II. It also wrought its toll on bilirubin chemistry by having removed one of the most capable and important bile pigment and porphyrin researchers. *Piloty* perished, with his son, in a cause that drove young men to despair and robbed them of their future, as *Erich Maria Remarque* described so remorsefully in his famous work, *Im Westen Nichts Neues* (*299*):

4 A Modern Proof of Bilirubin Structure Emerges

Albert spricht es aus. „Der Krieg hat uns für alles verdorben."

Er hat recht. Wir sind keine Jugen mehr. Wir wollen die Welt nicht mehr stürmen. Wir sind Flüchtende. Wir flüchten vor uns. Vor unserem Leben. Wir waren achtzehn Jahre und begannen die Welt und das Dasein zu lieben; wir mußten darauf schießen. Die erste Granate, die einschlug, traf in unser Herz. Wir sind abgeschlossen vom Tätigen, vom Streben, vom Fortschritt. Wir glauben nicht mehr daran; wir glauben an der Krieg. . . .

Ich bin jung, ich bin zwanzig Jahre alt; aber ich kenne vom Leben nichts anderes als die Verzweiflung, den Tod, die Angst und die Verkettung sinnlosester Oberflächlichkeit mit einem Abgrund des Leidens. Ich sehe, daß Völker gegeneinandergetrieben werden und sich schweigend, unwissend, töricht, gehorsam, unschuldig töten. Ich sehe, daß die klügsten Gehirne der Welt Waffen und Worte erfinden, um das alles noch raffinierter und längerdauernd zu machen. Und mit mir sehen das alle Menschen meines Alters hier und drüben, in der ganzen Welt, mit mir erlebt das meine Generation. Was werden unsere Väter tun, wenn wir einmal aufstehen und vor sie hintreten und Rechenschaft fordern? Was erwarten sie von uns, wenn eine Zeit kommt, wo kein Krieg ist? Jahre hindurch war unsere Beschäftigung töten – es war unser erster Beruf im Dasein. Unser Wissen von Leben beschränkt sich auf den Tod. Was soll danach noch geschehen? Und was soll aus uns werden?

Or, as translated into English by *A. W. Wheen* (*300*):

Albert expresses it: "The war has ruined us for everything."

He is right. we are not youth any longer. We don't want to take the world by storm. We are fleeing. We fly from ourselves. From our life. We were eighteen and had begun to love life and the world; and we had to shoot it to pieces. The first bomb, the first explosion, burst in our hearts. We are cut off from activity, from striving, from progress. We believe in such things no longer, we believe in the war. . . .

I am young, I am twenty years old; yet I know nothing of life but despair, death, fear, and fatuous superficiality cast over an abyss of sorrow. I see how peoples are set against one another, and in silence, unknowingly, foolishly, obediently, innocently slay one another. I see that the keenest brains of the world invent weapons and words to make it yet more refined and enduring. And all men of my age, here and over there, throughout the whole world see these things; all my generation is experiencing these things with me. What would our fathers do if we suddenly stood up and came before them and proffered our account? What do they expect of us if a time ever comes when the war is over? Through the years our business has been killing; – it was our first calling in life. Our knowledge of life is limited to death. What will happen afterwards? And what shall come out of us?

All repeated more gruesomely and more widely in World War II.

Before his death, *Piloty* reported his investigations into the structure of bilirubin in two important publications in 1912 (*301, 302*), in which with considerable novelty and expectation, he cleaved it in half by the use of HI. This was surely one of the most important advances toward unlocking the chemical structure of bile pigments since *Küster*'s isolation and characterization of the hematinic acid oxidative cleavage product. After 1912, for reasons unclear, *Piloty* largely abandoned bilirubin studies to return to his earlier studies of the blood pigment. Possibly the shift was due to *Hans Fischer*, who had also began to publish on bile pigments (in 1911) and had expanded into research on the blood pigment. Possibly *Piloty* wished to reclaim his stake in its research. He had a permanent faculty position at the University of Munich and was doubtless better situated than *Fischer* in terms of manpower and resources to solve either of the pigment structures. *Fischer*, on the other hand, had no faculty appointment and a lesser claim to resources between 1911 and 1914. After *Piloty* left his position and laboratory in 1914, the quest for

238   4   A Modern Proof of Bilirubin Structure Emerges

structures was left principally to *Küster* and *Fischer*. Inescapably, understanding the various steps, difficulties, and controversies in the structure proof of bilirubin (and biliverdin) could not be adequately accomplished without reference to some of the structural information derived from the structure proof of the pigment of blood; thus, the key pieces of information from the latter had also to be addressed. They weighed in importantly in influencing the advances in the bilirubin structure proof.

4.1   *Oskar Piloty* and the Structure of Bilirubin

It was in Munich, but not until after 1906, that *Piloty*'s[1] interests and major work turned toward the structure of the pigment of blood, a challenging, historic subject of great attraction to natural products chemists of the 19th and early 20th centuries (*298*). And one in which limited progress had made it an attractive and challenging research target. *Nencki* and *Sieber* had sliced out a component (*Hämopyrrol*) by HI-PH$_4$I reduction that, with *Nencki*'s death in 1901, was no longer being investigated – except by *Küster*. *Küster* had earlier used oxidation (most successfully with chromic acid) to carve out a monopyrrole component that he isolated as an imide, hematinic acid imide, and proved its structure. He knew that the imide was itself not a component but reflected the presence of a monopyrrole unit that was substituted with a β-CH$_3$ and a β-CH$_2$CH$_2$CO$_2$H. Though he was well on the way to solving the chemical structure of *Nencki*'s *Hämopyrrol*, by 1906 he had largely ceased those investigations.

Trained under two of Germany's most famous organic chemists in the art and practice of the most advanced organic chemistry of 1900, armed with broad knowledge and with little or no attachment to medicinal or animal chemistry, *Piloty* might be considered the first organic chemist to attempt the structure determination of bilirubin. At first, however, he began to publish on the cleavage products of hematoporphyrin degradation in 1909. He initiated the investigations by addressing the vexing problem of the chemical structure of *Hämopyrrol*[2] (*303*). (For the French version, see (*304*).] And he exceeded the expectations of *Otto Hahn* (1879–1968)

---

[1] *Oskar Piloty* was born in Munich on April 30, 1866 and died in World War I in France on October 6, 1915. Although strongly inclined toward humanistic studies, he gravitated into chemistry, possibly influenced by the chemist *Ludwig Knorr*, who had married his older sister. *Piloty* initiated his formal chemical education in 1888 in *von Baeyer*'s laboratory in Munich and completed his doctoral studies on sugars in 1890 in Würzburg under *Hermann Emil Fischer*, who had also studied under *Adolf von Baeyer* in Strassburg and received the Dr. phil. in 1874. In 1892, *Piloty* followed *Fischer* to Berlin as *Assistent* when *Fischer* accepted the call to succeed *A.W. von Hoffmann* as chair of chemistry at the University of Berlin. In the same year, 1892, *Piloty* married *Adolf von Baeyer*'s daughter, whom he met while studying briefly under him in Munich. *Piloty* completed his *Habilitation* with *Emil Fischer* in Berlin in 1898 and surprised his colleagues by returning to Munich in 1899, as a. o. Professor of Inorganic Chemistry.

[2] *Nencki*'s term for the liquid distillate will continue to be used for such (rather than its English cognate terms (hemopyrrole and haemopyrrole – words used to indicate 2,3-dimethyl-4-ethylpyrrole) until its purity and structure(s) become clear – toward the end of this section.

4.1 Oskar Piloty and the Structure of Bilirubin

who had the same doctoral advisor (*Theodor Zincke*) as *Hans Fischer* and who was awarded the *Nobel* Prize in Chemistry in 1944). In 1909, he was a guest of *von Baeyer* in Gries, near Bozen (Bolzano), where he met *Piloty*. *Hahn* wrote scathingly in his autobiography (*305*):

> The elder Baeyer's son-in-law was the chemist Piloty, head of a department in Baeyer's institute: he was a very musical and cultured man who did not count for much as a chemist.

After reviewing the *Nencki-Zaleski* studies (*285*) on the formation of *Hämopyrrol* by reductive cleavage of hematin and hematoporphyrin using HI-PH_4I, as published in 1901, and the oxidative cleavage work of *Küster* going back to 1897 (*243, 244*), wherein hematinic acid imide was isolated and its structure proved by synthesis in 1900–1902 (*215, 251, 252*), *Piloty* introduced a new and different reductive cleavage reaction. He treated hematoporphyrin with $SnCl_2$ and Sn foil in conc. (fuming) HCl and isolated three products. One product proved to be *Hämopyrrol*, a liquid with a skatole-like odor that analyzed for $C_8H_{13}N$, gave MW = 118 (FW = 123) by *Raoult*'s boiling point elevation method in benzene, and precipitated a crystalline picrate with a sharp mp at 108°C, by reaction with picric acid.

A second product was isolated and analyzed, a pyrrole carboxylic acid named *Hämopyrrolcarbonsäure* (hemopyrrole carboxylic acid) by *Piloty*. Its elemental combustion analysis and MW = 140 (obtained from its freezing point depression in phenol) found a good fit to the formula $C_9H_{13}NO_2$ (FW = 167). The acid dissolved in aq. $NaHCO_3$ with evolution of CO_2 and also formed a crystalline picrate as yellow, prismatic blades that sintered at 140° and melted at 148°C.

A third isolate, named *Hämatopyrrolidinsäure* (hematopyrrolidinic acid) by *Piloty*, proved to be difficult to separate from Sn salts and obtain pure – until the hematoporphyrin reductive cleavage conditions were changed from Sn to Zn dust in hydrochloric acid. This change produced the expected *Hämopyrrol*, hemopyrrole carboxylic acid, and hematopyrrolidinic acid. The last formed a picrate of mp 125°C, and from its combustion analysis was determined to have the formula $C_{17}H_{18}N_2O_3$, or $C_{14}H_{23}N_2O_3$ (*303*):

Für die Analysen wurde es bei 60° im Vacuum bis zur Gewichtsconstanz getrocknet. Für alle drei folgenden Analysen wurden Präparate verschiedener Darstellung verwandt.

I.	0,1491 g	gaben	0,2717 CO_2 und 0,0679 H_2O.
	0,1143 g	„	15,2 ccm Stickgas bei 17° und 712 mm Druck.
II.	0,1520 g	„	0,2764 CO_2 und 0,0704 H_2O.
	0,1384 g	„	18,4 ccm Stickgas bei 16° und 108 mm Druck.
III.	0,1394 g	„	0,2537 CO_2 und 0,0657 H_2O.

Berechnet für

	$(C_{17}H_{28}N_2O_2)_2(C_6H_3N_3O_7)_3$	$(C_{14}H_{22}N_2O_2)C_6H_3N_3O_7$
C	49,1	50,1
H	5,11	5,22
N	14,32	14,61

	Gefunden		
	I	II	III
C	49,70	49,59	49,63
H	5,09	5,18	5,27
N	14,78	14,42	–

Wie man sieht, können diese Analysen keine Entscheidung herbeiführen darüber, ob die Hämatopyrrolidinsäure die Zusammensetzung $C_{17}H_{28}N_2O_2$ oder $C_{14}H_{22}N_2O_2$ besitzt. 1

At the time, *Piloty* could not determine which formula was the more correct, but if either was in fact correct, he had sliced out a much larger part of the hematoporphyrin structure than had any of his predecessors. By August 11, 1909, he believed it to be a C_{17} compound, not C_{14} (*306*). Assigning structures to go with the names was often a difficult task at the time, and although *Piloty* did not assign a structure to hematopyrrolidinic acid until 1912 (*301*), in 1909 he was already prepared to assign structures to the lower molecular weight *Hämopyrrol* and hemopyrrole carboxylic acid in a flow chart summary of his degradation investigations of hematoporphyrin (*303*):

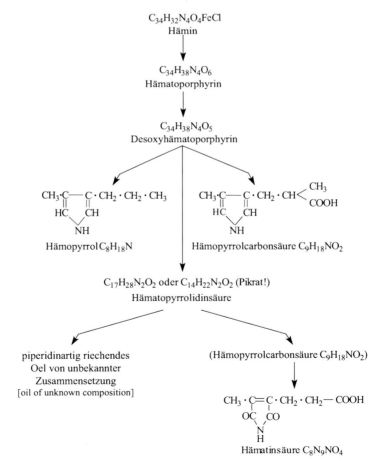

4.1 Oskar Piloty and the Structure of Bilirubin

It is interesting to note that in 1909, *Piloty* assumed *Küster*'s old structure for *Hämopyrrol* (β,β'-methylpropylpyrrole); yet, in 1906 *Küster* had already proved it to be incorrect by synthesis of its imide oxidation product (*292*), which was not the predicted methylpropylmaleimide, and he was also expressing doubt that *Hämopyrrol* was a single compound that was substituted at only the two β-positions (*296*). It is unclear why *Piloty* had missed it. In addition to proposing a structure for *Hämopyrrol*, he proposed a structure for his newly found hemopyrrole carboxylic acid. This too might have been rejected given that he knew *Küster* had proved the structure of hematinic acid by synthesis. For hematinic acid imide has a propionic acid β-substituent and was known to decarboxylate to methylethylmaleimide. Studying that relationship, in the absence of any other knowledge except for *Küster*'s expressed views on the possible structures of *Hämopyrrol*, might have guided *Piloty* into thinking in terms of an unsubstituted propionic acid group in hemopyrrole carboxylic acid, methyl and ethyl β-substituents in *Hämopyrrol*, with the residual methyl at a pyrrole α-position. Especially since his hematopyrrolidinic acid could be oxidized to hematinic acid. But that did not happen, not at least in 1909 (*303, 304, 306, 307*). Hedging a bit regarding the nature of the side chains in the structures shown, *Piloty* expressed his reservations in a footnote to the *Hämopyrrol* structure that it was only provisional, with the nature of the side chains not yet having been determined with certainty. Equivocating a bit, he viewed his structures as being presented only for the purpose of assisting the reader, and indicated in a footnote to his Hämopyrrol structure (see chart) that a more exact investigation with a student (*Quitmann*) had been undertaken (*303*):

> [10]) Die Natur der Seitenketten und die Stellung derselben am Pyrrolring ist noch nicht endgültig festgestellt. Ich gebe diese Formel sowie die übrigen im Folgenden gebrauchten Formeln mit allem Vorbehalt und nur zu dem Zwecke, um dem Leser eine vorläufige Orientirung zu erleichtern. Auch ist die Untersuchung dieser einfach en Spaltungsstücke des Hämins meinerseits noch keine abgeschlossene und auch in Bezug auf sie diese Mittheilung nur eine vorläufige. Ich habe die genauere Untersuchung dieser Körper gemeinsam mit Herrn. cand. chem. Quitmann in Angriff genommen und hoffe sehr bald mehr darüber mittheilen zu können.   2

What now becomes clear is that by 1909 the conventions associated with the long history of the pigments had changed; there was a willingness and ability to draw chemical structures. Pigment chemistry could now be identified with organic structural chemistry.

Although a structure could not be proposed for hematopyrrolidinic acid, it was investigated by oxidation using MnO_2, which yielded a compound that analyzed for $C_8H_9NO_4$ and had a melting point close to that of *Küster*'s hematinic acid imide. *Piloty* called the structure "probable", despite *Küster* having proved it by synthesis in 1906 (*291*), and he wrote that since hemopyrrole carboxylic acid is the parent substance of the hematinic acid obtained from hematoporphyrin, the appearance of hematinic acid from cleavage of the hematopyrrolidinic acid indicated that hemopyrrole carboxylic acid was a part of the latter's structure (*303*):

> Da aus den oben mitgetheilten Untersuchungen hervorgeht, dass die im Hämatoporphyrin enthaltene Muttersubstanz der Hämatinsäure die Hämopyrrol-carbonsäure

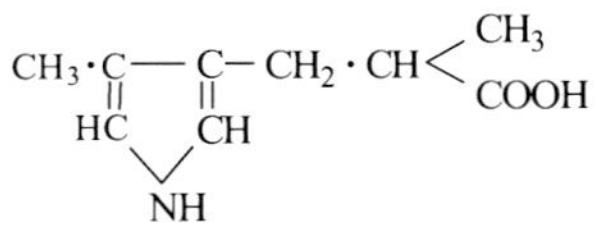

242   4   A Modern Proof of Bilirubin Structure Emerges

$$CH_3 \cdot \underset{\underset{HC}{\|}}{C} - \underset{\underset{CH}{\|}}{C} - CH_2 \cdot CH {\displaystyle <} {\overset{CH_3}{\underset{COOH}{}}}$$
$$NH$$

ist, so geht aus dem Auftreten der Hämatinsäure bei dieser Spaltung hervor, dass der eine Theil der Hämatopyrrolidinsäure Hämopyrrolcarbonsäure ist, die mit einem sauerstofffreien basischen Rest condensirt ist.   3

Piloty's paper (*303*) was received in the journal office of *Liebig*'s *Annalen* on March 2, 1909, and it drew a rapid response from *Küster* in Stuttgart on July 2, 1909 (*309*), wherein he explained his having suspended research on the hematin cleavage products three years earlier. More to the point, he admired and admonished *Piloty*. Admired *Piloty* for his new cleavage method (Zn or Sn + conc. HCl) and from it the isolation of new pyrrole cleavage products, in addition to *Nencki*'s *Hämopyrrol*. Admonished *Piloty* for not having read his (*Küster*'s) work carefully and his comprehensive paper of 1908 (*310*), and for not realizing that hemopyrrole carboxylic acid was more likely to have not structure **I**, but structure **II**, which would oxidize directly to give hematinic acid imide, **III** (*309*):

Die von Piloty auf Grund ihrer empirischen Zusammensetzung und der Konstitution der Hämatinsäure angenommene Formel I dürfte aber wohl besser durch die Formel II zu ersetzen sein:

I.   II.

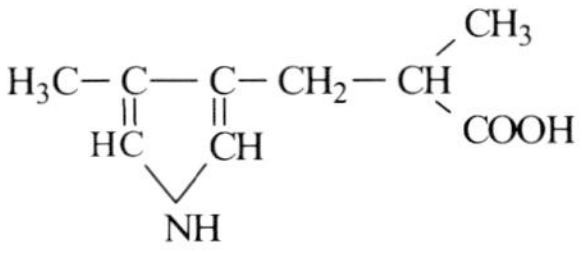

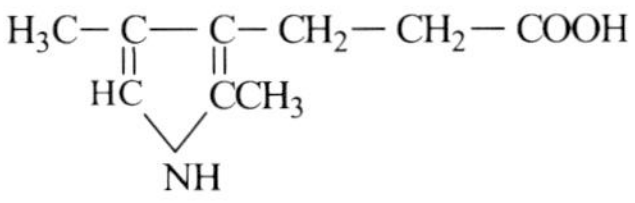

$$H_3C - \underset{\underset{HC}{\|}}{C} - \underset{\underset{CH}{\|}}{C} - CH_2 - CH {\overset{CH_3}{\underset{COOH}{}}}$$
$$NH$$

$$H_3C - \underset{\underset{HC}{\|}}{C} - \underset{\underset{CCH_3}{\|}}{C} - CH_2 - CH_2 - COOH$$
$$NH$$

Ich glaube wenigstens daß letztere das Entstehen der Hämatinsäure III

III.

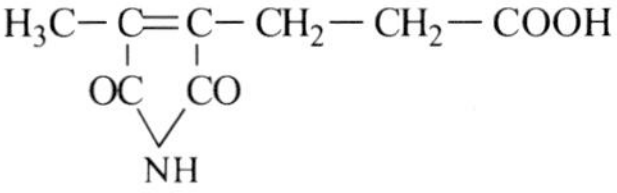

$$H_3C - \underset{\underset{OC}{|}}{C} = \underset{\underset{CO}{|}}{C} - CH_2 - CH_2 - COOH$$
$$NH$$

zwangloser erklärt, da ja nach Plancher die auch in α-Stellung substituierten Pyrrole den Substituenten bei der Oxydation verlieren, um in Imide von Maleinsäuren überzugehen. 4

A year earlier, on April 16, 1908 (*310*) – a year before *Piloty* submitted his first paper on the subject (*303*), *Küster* had provided the experimental details of his work on methylpropylmaleimide (**4.1.1**) as well as (i) methylethylmaleic anhydride and its imide (**4.1.2**), their synthesis and reactions; (ii) xeronimide (**4.1.3**) and the synthesis of its anhydride; and (iii) the syntheses of Δ^1-, Δ^2-, and *cis* and *trans* Δ^4-*Tetrahydrophthalsäureanhydrid* (tetrahydrophthalic anhydride) and their imides (**4.1.4**, **4.1.5**, and **4.1.6**, respectively). The last were prepared in order to settle questions as to whether *Hämopyrrol* and hematinic acid imide (**4.1.7**) had come from an isoindole component in hematin (or bilirubin), *e.g.* somehow **4.1.1** had come from the parent pyrrole **4.1.4** or **4.1.5**, whether **4.1.2** had come from the parent pyrrole

4.1 Oskar Piloty and the Structure of Bilirubin

of **4.1.7** (and **4.1.7** from **4.1.5**), and whether **4.1.3** had come from the parent pyrrole of **4.1.6** or **4.1.4**.

| 4.1.2 | 4.1.3 | 4.1.4 |

| 4.1.5 | 4.1.6 | 4.1.7 |

Küster did not accept the validity of isoindole components, nor β-methyl-β-propylpyrrole as answers to the unresolved structure of *Hämopyrrol*.

In the same 1908 publication (*310*), *Küster* reported obtaining what he had earlier termed the basic acid and acidic *Hämopyrrole* from the *Nencki-Zaleski* reductive cleavage of hematin and his own separation method. He acknowledged that *Marchelewski* (*311*) had found that *Hämopyrrol*, $C_8H_{13}N$, took up two equivalents of benzenediazonium chloride, which suggested that both pyrrole α-positions had been unsubstituted (an argument given for *Hämopyrrol* as β-methyl-β'-propylpyrrole).

Recall that in 1906 *Küster* had weighed the possibility that *Hämopyrrol* was β-methyl-β'-propylpyrrole, but he was more inclined toward α,β-dimethyl-β'-ethylpyrrole (*309*). *Piloty* too had considered that β-methyl-β'-propylpyrrole as well as hematinic acid imide might arise from the isoindole structural type represented by **I** (below), but *Küster* had discounted it because the oxidation of his *Hämopyrrol* gave β-methyl-β'-ethylmaleimide, not β-methyl-β'-propylmaleimide (*309*):

> Natürlich könnte ein Hämopyrrol, wie Piloty es schreibt, d. h. ein β-Methyl-β'-n-Propylpyrrol, auch entstehen aus einem Isoindolring (I), der dann bei der Oxydation die Hämatinsäure geben würde.

I.

> Aber diese Formel, die nach Piloty namentlich durch meine Arbeiten wahrscheinlich gemacht ist, habe ich gerade durch die Oxydation meines Hämopyrrolgemisches zum Imid der β-Methyl-β'-äthylmaleinsäure als unhaltbar erwiesen!   5

Küster also reasserted his experimental work which showed that hematoporphyrin oxidation yielded two mole equivalents of hematinic acid imide, suggesting that half of the pigment structure was represented by two such units.

244   4 A Modern Proof of Bilirubin Structure Emerges

In the small flood of papers from *Piloty* and his students in 1909–1910 (*303, 304, 306- 308, 312, 313*), the focus continued to be on isolating and identifying or characterizing the cleavage products of the pigments of blood.

In this era of organic chemistry, the ability and willingness to write structures (or was it now expected?) suggest some knowledge of how the structure was determined. For pyrroles, this usually meant a chemical synthesis, possibly coupled to expected chemical reactions and their products. As indicated above, and reported in his first paper in 1909 (*303*), cleavage of the latter with Sn + hydrochloric acid formed *Hämopyrrol* ($C_8H_{13}N$), hemopyrrole carboxylic acid ($C_9H_{13}NO_2$), and hematopyrrolidinic acid ($C_{17}H_{28}N_2O_2$), two of whose structures were given by *Piloty* in the chart shown on page 240, and all were undergoing continual re-evaluation. Still early in 1909, *Piloty* showed that *Hämopyrrol* reacted with $NaNO_2$ in the presence of H_2SO_4 to give a single, crystalline derivative, mp 206-207°, that hydrolyzed to the known methylethylmaleimide, mp 66° (*303*). Later in 1909 (*308*) with *Quitmann*, he interpreted the results in terms of structures, deducing two possible structures for *Hämopyrrol* and two possible oximes (**A** and **B**) (*308*):

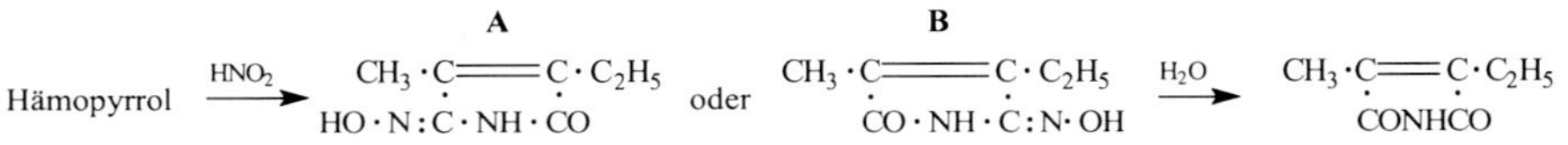

$$
\textbf{A} \qquad\qquad\qquad \textbf{B}
$$

Hämopyrrol $\xrightarrow{\text{HNO}_2}$ $\begin{array}{c} CH_3 \cdot C = C \cdot C_2H_5 \\ HO \cdot N : C \cdot NH \cdot CO \end{array}$ oder $\begin{array}{c} CH_3 \cdot C = C \cdot C_2H_5 \\ CO \cdot NH \cdot C : N \cdot OH \end{array}$ $\xrightarrow{\text{H}_2\text{O}}$ $\begin{array}{c} CH_3 \cdot C = C \cdot C_2H_5 \\ CONHCO \end{array}$

From these he deduced two possible structures for *Hämopyrrol*,

Hämopyrrol: $\begin{array}{c} CH_3 \cdot C \text{------} C \cdot C_2H_5 \\ CH_3 \cdot C \cdot NH \cdot CH \end{array}$ oder $\begin{array}{c} CH_3 \cdot C \text{------} C \cdot C_2H_5 \\ HC \cdot NH \cdot C \cdot CH_3 \end{array}$

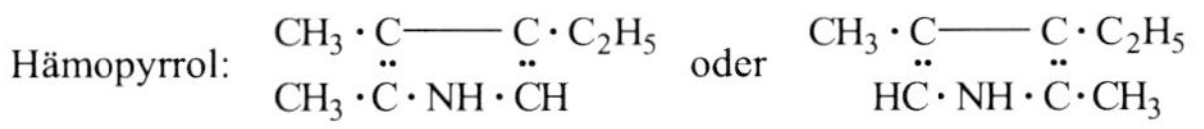

To remove any further doubt that *Hämopyrrol* was one of two possible α,β,β'-dimethylethylpyrroles, *Piloty* showed that it could be N-acetylated and that it formed a potassium salt (*308*). These results constituted an experimental refutation of the β,β'-methylpropylpyrrole structure for *Hämopyrrol* advocated earlier by *Nencki* and *Zaleski* (*285*), and *Marchelewski* (*311*), and earlier by *Küster* (*288*), who had subsequently (by 1907) shown that *Hämopyrrol* was oxidized by CrO_3 to methylethylmaleimide (*310*).

Pursuing the "pyrrole + HNO_2" reaction further, *Piloty* and *Quitmann* showed that the known α,β'-dimethylpyrrole, when treated with HNO_2, formed a single oxime (mp 223–224°C) of citraconimide, which hydrolyzed to the known citraconic acid, mp 90.5°C (*308*):

$\begin{array}{c} CH_3 \cdot C = CH \\ HC \cdot NH \cdot C \cdot CH_3 \end{array}$ $\longrightarrow$ $\begin{array}{c} CH_3 \cdot C = CH \\ CO \cdot NH \cdot C : N \cdot OH \end{array}$ $\xrightarrow{\text{H}_2\text{O}}$ $\begin{array}{c} CH_3 \cdot C = CH \\ HOOC \quad COOH \end{array}$

Piloty asserted that a pyrrole α-CH_3 was replaced by the oxime group (NOH) in the reactions shown above and extended the HNO_2 oxidation-oximation studies to his hemopyrrole carboxylic acid. Here he found a single mono-oxime of hematinic acid, from which he concluded that the former had one of two possible structures and that hemopyrrole carboxylic acid was thus an α,β-dimethyl-β-pyrryl-propionic

4.1 Oskar Piloty and the Structure of Bilirubin

acid (note that *Piloty* had altered his earlier structure (*303*), which had the pyrrole attached to the α-carbon of the propionic acid) (*308, 314*):

$$CH_3 \cdot C\text{---}C \cdot CH_2 \cdot CH_2 \cdot COOH$$
$$CH_3 \cdot C \cdot NH \cdot CH$$

oder

$$CH_3 \cdot C\text{---}C \cdot CH_2 \cdot CH_2 \cdot COOH \quad \xrightarrow{HNO_2}$$
$$HC \cdot NH \cdot C \cdot CH_3$$

$$CH_3 \cdot C\text{=}C \cdot CH_2 \cdot CH_2 \cdot COOH$$
$$CO \quad C:NOH$$
$$NH$$

oder

$$CH_3 \cdot C\text{=}C \cdot CH_2 \cdot CH_2 \cdot COOH$$
$$HON:C \quad CO$$
$$NH$$

However, which of the two structures of hemopyrrole carboxylic acid, or even *Hämopyrrol*, corresponded to the actual compounds could not be decided (*308*):

> Zwischen diesen beiden Formeln kann ebenfalls jetzt noch nicht entschieden werden.   6

Apparently a work in progress, in October 1910, *Piloty* and *Quitmann* decided on a name change for the hemopyrrole carboxylic acid isolated from hematoporphyrin following reductive cleavage (*313*):

> Der eine von uns hat durch Reduktion des Hämatoporphyrins mittelst Zinn und Salzsäure das Hämopyrrol in so großer Menge und Reinheit darstellen können, daß dadurch die sichere Untersuchung seiner Konstitution ermöglicht wurde. Neben dem Hämopyrrol wurde eine Carbonsäure bei dem gleichen Verfahren entdeckt, welche wegen ihres ähnlichen Verhaltens und ihrer gleichen Herkunft wie das Hämopyrrol Hämopyrrolcarbonsäure genannt wurde. Der Name erweckt die Vorstellung, daß diese Säure ein carboxyliertes Hämopyrrol sei. Die weitere Untersuchung hat aber ergeben, daß der Säure nicht Hämopyrrol sondern ein anderes Pyrrolderivat zugrunde liegt. Wir sehen uns daher gezwungen, den Namen der Säure zu ändern und schlagen den Namen *Phonopyrrolcarbonsäure* vor. (Der Name is mit φόνος = vergossenes Blut gebildet.)   7

This led to a new name, *Phonopyrrolcarbonsäure* (phonopyrrole carboxylic acid), though only ten months prior, *Piloty* had represented hemopyrrole carboxylic acid as (*312*):

$$CH_3 \cdot C\text{---}C \cdot CH_2 \cdot CH_2 \cdot COOH$$
$$HC \cdot NH \cdot C \cdot CH_3$$

Hämopyrrolcarbonsäure, $C_9H_{13}NO_2$

and yet the same problem persisted: was the α-methyl adjacent to the β-methyl or adjacent to the β-propionic acid, or to the ethyl group of *Hämopyrrol*? (*313*):

Phonopyrrolcarbonsäure $C_9H_{13}NO_2$.

In einer früheren Mitteilung[1]) [*308*] haben wir die Frage nach der Konstitution der Phonopyrrolcarbonsäure soweit gefördert, daß eine der beiden folgenden Formeln als gesichert bezeichnet werden konnte

$$CH_3 \cdot C\text{---}C\text{--}CH_2 \cdot CH_2 \cdot COOH$$
$$CH_3 \cdot C \quad CH$$
$$NH$$

oder

$$CH_3 \cdot C\text{---}C\text{--}CH_2 \cdot CH_2 \cdot COOH$$
$$H \cdot C \quad C \cdot CH_3$$
$$NH$$

und nur unentschieden blieb, ob das α-ständige Methyl in der α- oder α'-Stellung sich befindet. Und zwar konnten wir dies deshalb, weil die Säure beim Behandeln mit salpetriger Säure eine Methylgruppe verliert und in das Oxim der Hämatinsäure $C_8H_9NO_4$ und beim nachfolgenden Verseifen des Oxims in die Küstersche Hämatinsäure selbst

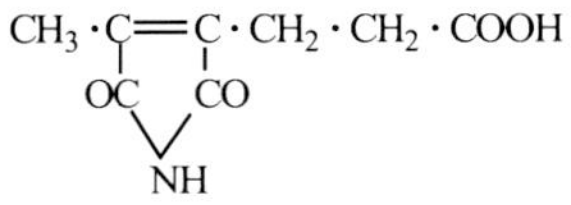

$$CH_3 \cdot C = C \cdot CH_2 \cdot CH_2 \cdot COOH$$
$$OC \quad CO$$
$$NH$$

übergeht und wir den Nachweis führen konnten, daß immer eine α-ständige Methylgruppe bei dieser Prozedur abgespalten wird.

Es liegt also der Carbonsäure ein carboxylfreies Pyrrol zugrunde, dem eine der beiden Formeln zukommen muß:

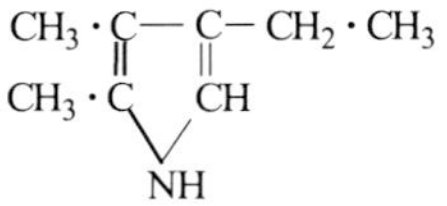

$$CH_3 \cdot C - C - CH_2 \cdot CH_3 \qquad\qquad CH_3 \cdot C - C - CH_2 \cdot CH_3$$
$$CH_3 \cdot C \quad CH \qquad\quad oder \qquad H \cdot C \quad C \cdot CH_3$$
$$NH \qquad\qquad\qquad\qquad\qquad NH$$

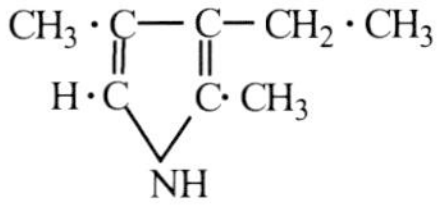

Aus der oben zitierten Mitteilung geht hervor, daß auch dem Hämopyrrol eine dieser beiden Formeln zukommen muß. 8

Piloty reasoned that when phonopyrrole carboxylic acid is decarboxylated to *Phonopyrrol* (phonopyrrole), if the latter is different from *Hämopyrrol*, then one can know the structure of both pyrroles by determining the structure of the acid. And phonopyrrole in fact proved to be different from *Hämopyrrol* (*313*):

Die endgültige Bestimmung der Konstitution der Phonopyrrolcarbonsäure und des Phonopyrrols schließt also gleichzeitig diejenige des Hämopyrrols mit ein, falls dieses und das Phonopyrrol nicht identisch sind, in der Art, daß, wenn dem Phonopyrrol die eine Formel zukommt, das Hämopyrrol dann nach der anderen konstituiert sein muß.

Durch Abspaltung von Kohlensäure ließ sich aus der Phonopyrrolcarbonsäure das *Phonopyrrol* gewinnen; dasselbe is *nicht* identisch mit dem Hämopyrrol. 9

How different were they? Their picrates had different melting points. *Hämopyrrol* (bp 86–87°C/23 mm Hg) formed a picrate which crystallized from alcohol, melted at 108.5°C, and it formed an oxime from reaction with HNO_2 that had mp 201°C (*313*):

Analyse des Hämopyrrols:
0,1253 g gaben 0,3614 CO_2 und 0,1246 H_2O

	Ber. für $C_8H_{13}N$	Gef.
C	78,05	78,35
H	10,47	11,01

Durch Vermischen einer ätherischen Lösung des Hämopyrrols mit einer feuchtätherischen Pikrinsäurelösung wurde das Pikrat in schief abgeschnittenen rhombischen Prismen erhalten, die nach dem Unkrystalisieren aus Alkohol bei 108,5° schmelzen.

0,1116 g gaben 0,1960 CO_2 und 0,0484 H_2O.

	Ber. für $C_{14}H_{16}N_4O_7$	Gef.
C	47,72	47,89
H	4,55	4,81

4.1 Oskar Piloty and the Structure of Bilirubin

> 1 g des Hämopyrrols wurde in verdünnter Schwefelsäure gelöst und die Lösung mit Natriumnitrit versetzt, wie dies früher beschrieben worden ist. Beim Abkühlen fiel das Oxim des Methyläthylmaleinimids aus, das nach dem Umkrystallisieren aus Wasser den richtigen Schmelzpunkt 201° zeigte. 10

Phonopyrrole, formed by heating phonopyrrole carboxylic acid in a vacuum, had a boiling point of 96–98°C/19 mm, did not form a crystalline derivative with picric acid. Attempts to prepare its oxime by treatment with HNO_2 gave only a syrupy derivative, and chromic acid oxidation did not produce a crystalline derivative (*313*):

> Das *Phonopyrrol* hat demnach dieselbe Zusammensetzung wie das Hämopyrrol. Est ist aber nicht mit dem letzteren identisch. Während das Hämopyrrol charakterisiert ist durch das leicht und schön krystallisierende Pikrat, konnten wir aus dem Phonopyrrol weder in ätherischer noch in wäßriger Lösung ein solches Derivat in krystallinischer Form erhalten. Die erstere färbte sich auf Zusatz von feucht-ätherischer Pikrinsäurelösung nur dunkel ohne Abscheidung. Die letztere ließ ein öliges Pikrat fallen, welches aber auch im Kältegemisch nicht erstarrte. Für das Hämopyrrol ist die überaus leichte Bildung des Oxims des Äthylmethylmaleinimids bei der Einwirkung von Natriumnitrit auf seine schwefelsaure Lösung sehr charakteristisch. Das Phonopyrrol liefert in sehr geringer Menge ein sirupöses Maleinsäureimidderivat, dessen Natur noch nicht erkannt werden konnte. Auch die Einwirkung von Chromsäure in schwefelsaurer Lösung, die nach Plancher und Cattadoris [*293, 294*] Vorschrift ausgeführt wurde, führte zu keinem krystallisierten Derivat.
>
> Das Phonopyrrol färbt sich an der Luft zunächst braun, später rot. 11

The two pyrroles behaved sufficiently differently that *Piloty* concluded they were different substances. To determine their structures, and that of phonopyrrole carboxylic acid, he subjected the latter to both acid and base in an unsuccessful attempt to effect ring closure to form skatole (*313*):

> Man wäre berechtigt zu erwarten, daß die Phonopyrrolcarbonsäure, wenn sie die α-Methylgruppe auf derselben Seite des Pyrrolkernes hätte wie den Propionsäurerest, wenn ihr die Formel

$$CH_3 \cdot C - C - CH_2 \cdot CH_2 \cdot COOH$$

> zukäme, daß dann Neigung vorhanden sein müßte, durch wasserentziehende Mittel ein Indolderivat, das Skatol, zu bilden.

> Skatolbildung konnte auf keine Weise hervorgerufen werden. Ist es auch daher wenig wahrscheinlich, daß in der Phonopyrrolcarbonsäure die α-Methylgruppe die dem Propionsäurerest benachbarte Stellung einnimmt, so hätten wir auf dieses negative Ergebnis allein doch nicht eine Meinung gründen wollen. Sie wird aber durch eine andere Betrachtung unterstützt. 12

248   4 A Modern Proof of Bilirubin Structure Emerges

And this failure led *Piloty* to favor phonopyrrole carboxylic acid and phonopyrrole structures with adjacent α and β methyls – leaving *Hämopyrrol* with an α-methyl adjacent to the ethyl (*313*):

Die Phonopyrrolcarbonsäure und das Phonopyrrol haben also folgende Formeln:

$$CH_3 \cdot C \overset{\|}{\longrightarrow} C \cdot CH_2 \cdot CH_2 \cdot COOH$$
$$CH_3 \cdot C \diagdown CH$$
$$NH$$

Phonopyrrolcarbonsäure
α,β-Dimethyl-β'-propionylpyrrol

$$CH_3 \cdot C \overset{\|}{\longrightarrow} C \cdot CH_2 \cdot CH_3$$
$$CH_3 \cdot C \diagdown CH$$
$$NH$$

Phonopyrrol
α,β-Dimethyl-β'-äthylpyrrol

Daraus geht nach dem oben Gesagten ohne weiteres hervor, daß das Hämopyrrol die durch folgende Formel auszudrückende Konstitution besitzt

$$CH_3 \cdot C \overset{\|}{\longrightarrow} C \cdot CH_2 \cdot CH_3$$
$$H \cdot C \diagdown C \cdot CH_3$$
$$NH$$

Hämopyrrol
α,β'-Dimethyl-β-äthylpyrrol   13

In 1912, *Piloty* began to publish results from cleaving bilirubin, variously with molten KOH, Sn + HCl, or HI + PH$_4$I (*301*) – and then ran up against *Hans Fischer* and his investigations of bilirubin. *Piloty*'s two publications (*301, 302*) on bilirubin produced striking, new information on the structure, some of it of seminal value, some apparently leading into a blind alley. Whether *Piloty* might have overcome the latter predicament is unclear, and that aspect of his work still remains unresolved and unlikely to be reinvestigated. On May 25, 1912, *Piloty* submitted his first paper on bilirubin (*301*) and its relationship to hematin, a compound whose structure he had been actively investigating during the previous several years (*303, 304, 306–308, 313*). In his first bilirubin paper (*301*), he described the results of two different cleavage reactions applied to the pigment: (i) molten KOH and (ii) HI + PH$_4$I. Thus, from 240 g of bilirubin in molten KOH, between 290°C and 370°C a colorless or pale yellow oil (16.3 g) distilled that was believed to be a mixture of pyrrole compounds. Fractional distillation yielded 8.3 g (bp 59–64°C, 11 mm Hg) and 6.7 g (bp 64–80°C, 11 mm Hg). A picrate precipitated upon addition of picric acid in dry ether to each fraction and was crystallized from absolute alcohol. The picrate from the lower boiling point fraction melted at 147–148°C after a single recrystallization and yielded 1.9 g of a pyrrole of bp 53–63°C, 11 mm Hg. The picrate of the higher boiling point fraction melted at 140°C after recrystallization and yielded 2 g of a pyrrole of bp 64–80°C, 11 mm Hg. The first picrate, mp 148°C, analyzed for the ratio 2 pyrroles:1 picric acid, and was thought to be the picrate of a *bis*-α,β-dimethylpyrrole. The second picrate, mp 140°C, analyzed for 1:1 pyrrole:picric acid and gave no depression in a mixture mp with authentic, synthetic α,β,β'-trimethylpyrrole,

4.1 Oskar Piloty and the Structure of Bilirubin 249

but it is unclear how the pyrrole was synthesized and who synthesized it before May 1912 (*301*):

α,β,β'-Trimethylpyrrol,

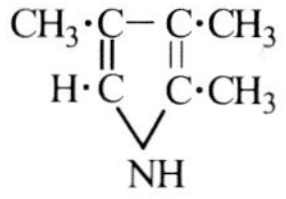

$$CH_3 \cdot \underset{H \cdot \underset{NH}{C}}{C} \overset{C \cdot CH_3}{\underset{C \cdot CH_3}{C}}$$

Das Pikrat aus Teil II vom Schmelzp. 140° wurde wegen Mangels an Material nicht in die freie Base verwandelt, sondern nur analysiert und als Salz mit dem Pikrat des auf synthetischem Wege erhaltenen α,β,β'-Trimethylpyrrols, dessen Beschreibung ebenfalls später erfolgen wird, verglichen. Mischschmelzpunkt der beiden Präparate 140°.

| 0,1196 g | gaben | 0,2030 CO_2 und 0,0461 H_2O. |
| 0,1171 g | ,, | 17,8 ccm Stickgas bei 20° und 705 mm Druck. |

	Ber. für $C_{13}H_{14}N_4O_7$	Gef.
C	46,16	46,28
H	4,15	4,31
N	16,57	16,66

Das Pikrat löst sich leicht in heißem Alkohol und kann daraus umkrystallisiert werden. 14

The free base isolated from the mp 148°C picrate, had mp 84–85°C, gave a positive pine splint test and exhibited a MW 190. Its elemental combustion analysis fit the formula $C_{12}H_{18}N_2$, which was thought to correspond to some sort of covalent dimer of α,β-dimethylpyrrole (*301*):

Nach dem Umkrystallisieren aus Petroläther schmilzt die Substanz bei 84-85°. Sie ist leicht löslich in Äther, Alkohol, Benzol, Eisessig, heißem Petroläther und Wasser.

0,1254 g	gaben	0,3476 CO_2 und 0,1079 H_2O.
0,1134 g	,,	15,20 ccm Stickgas bei 20° u. 721 mm Druck.
0,2382 g	,,	in 14,565 Benzol 0,3495 Schmelzp.-Depress.

	Ber. für $C_{12}H_{18}N_2$	Gef.
C	75,79	75,61
H	9,47	9,62
N	14,74	14,81
M	190	234

$\alpha,\beta,$-Dimethylpyrrol,

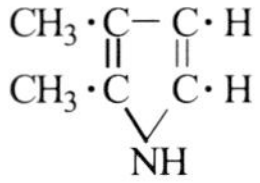

$$CH_3 \cdot \underset{CH_3 \cdot \underset{NH}{C}}{C} \overset{C \cdot H}{\underset{C \cdot H}{C}}$$

Das Pikrat vom Schmelzp. 148° ist kein normales Pyrrolsalz, sondern enthält nur 1 Mol. Säure auf 2 Mol. Base. Die dem Gemisch zugrunde liegende Base ist ein fester Körper und, wie die Molekulargewichtsbestimmung und Analyse ergab, ein Polymeres des α,β-Dimethylpyrrols, ein bis-α,β-Dimethylpyrrol. Die Pikrinsäure hat demnach dem einfachen Pyrrolderivat gegenüber polymerisierende Wirkung. Genau so verhält sich das auf synthetischem Wege gewonnene α,β-Dimethylpyrrol, über dessen Darstellung und Eigenschaften in einer späteren Mitteilung berichtet werden wird. Auch der Mischschmelzpunkt der beiden Präparate ergibt die Identität derselben. 15

The dimer was suspected to have been formed from the monopyrrole by the action of picric acid during picrate formation. Its identity was proved following (i) rational synthesis of α,β-dimethylpyrrole from the *Knorr* condensation of 3-aminobutanone with oxalacetic ester, then (ii) treatment of the pyrrole with picric acid in ethyl acetate and is represented as (*315*):

$$\textit{bis-}\alpha,\beta\textit{-Dimethyl-pyrrol,} \quad \begin{array}{c} CH_3\cdot C \!-\! CH \!-\! CH \!-\! C\cdot CH_3 \\ CH_3\cdot C \quad CH \!-\! CH \quad C\cdot CH_3 \\ NH \qquad\qquad NH \end{array}$$

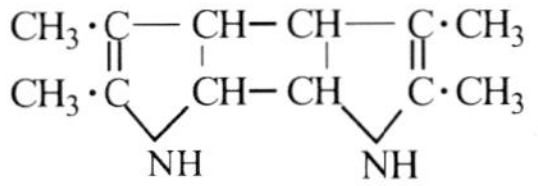

Possibly more significant than revealing a partial structure of bilirubin, the isolation and *identification* of pyrroles as chemical degradation products in *Piloty*'s work clearly illustrates (i) an increasingly rapid ability of chemists to identify newly isolated products from structurally unknown starting materials, *e.g.* natural products, and (ii) the availability of an ever-growing ensemble of fully characterized, synthetically produced small organic compounds, including pyrroles.

α,β-Dimethylpyrrole (*295*) had been discovered in the late 1800s (*316, 317*) by *Dennstedt* in *Dippel's Oil*, also known as Bone Oil, a dark-colored highly viscous, unpleasant-smelling oil or granular substance produced by the destructive distillation of bones. However, *Dennstedt*'s assertion that he had isolated specifically α,β-dimethylpyrrole is problematic. Bone oil was a subject of investigation going back decades before *Dennstedt*. *Runge* isolated pyrrole from it in the early 1830s (*231*). In 1846 *Thomas Anderson* initiated a series of investigations in which he found pyrroles (*234, 235*). In 1880, citing *Anderson*'s five subsequent papers from 1849–1870, *Weidel* and *Ciamician* (*318*) had isolated from *Dippel's Oil* what they believed to be a *C*-dimethylpyrrole (bp 165°C) because it analyzed for C_6H_9N and formed an acetyl derivative upon heating in acetic anhydride-sodium acetate. The last was to be taken as an indication that the pyrrole nitrogen had a hydrogen attached and thus the erstwhile two methyls were on pyrrole carbons (hence the prefix "*C*"). However, there is no clear evidence for the presence of two methyls; the presence of but a single ethyl group would also yield the C_6H_9N formula. And the belief that the presence of the pyrrole NH would lead to N-acetylation (and that C-acylation would not occur) was unsubstantiated.

The *C*-dimethylpyrrole, bp 165°, of *Weidel* and *Ciamician* (*318*), re-isolated by *Dennstedt* (*316*), was shown by *Knorr* (*319, 320*) and *Paal* (*321*) to be identical to their synthetic 2,5-dimethylpyrrole. By the time of *Dennstedt*'s isolations in 1889, pyrrole synthesis methodology (*295*) had been developed by *Knorr* (*319, 230*) and *Paal* (*321*), and thus *Dennstedt* (using *Knorr*'s method) was able to report his synthesis of α,β'-dimethylpyrrole and show that in the presence of conc. HCl it did not dimerize and rearrange to a methylated indole; whereas, the *C*-dimethylpyrrole from *Dippel's Oil* did, as illustrated below (*316*):

Diese Umwandlung des bei 165° siedenden Dimethylpyrrols aus dem Dippl'schen Oel, macht es wahrscheinlich, dass in der betreffenden Fraction, vielleicht neben dem

4.1 Oskar Piloty and the Structure of Bilirubin

$\alpha\alpha'$-Dimethylpyrrol (siehe auch die folgende Mittheilung) das $\alpha\beta$-Dimethylpyrrol enthalten ist. Die Bildung des Tetramethylindols aus dem Tetramethyldipyrrol würde sich in folgender Weise veranschaulichen lassen:

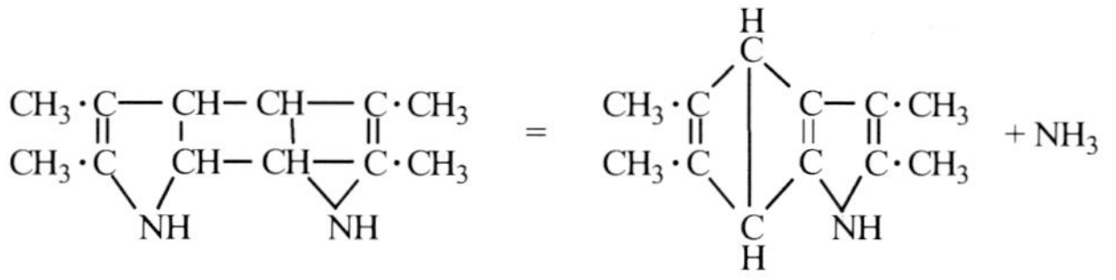

$$\begin{array}{ccc}
CH_3 \cdot C - CH - CH - C \cdot CH_3 & & \\
CH_3 \cdot C \quad CH - CH \quad C \cdot CH_3 & = & \\
\diagdown NH \qquad NH \diagup & &
\end{array} \qquad + NH_3 \qquad [^3]$$

Man würde demnach diese Verbindung zu bezeichnen haben als:
$\alpha\beta$ 3 . 4-Tetramethylindol

oder

Pr. 2 . 3. B . 3 . 4-Tetramethylindol.   16

Structure identification of the pyrroles was greatly facilitated by their earlier rational syntheses (295) due to the pioneering studies of *Knorr*[4] and *Paal*[5], who in 1885 published what later became known as the *Paal-Knorr* synthesis for converting 1,4-diketones into pyrroles, thiophenes and furans (319–323) and of *Knorr*, who showed how to convert β-keto esters into pyrroles (324). Both syntheses remain in use today.

Despite the foregoing assumptions of *Dennstedt* regarding α,β-dimethylpyrrole from *Dippel's Oil*, this pyrrole had apparently not been synthesized until 1912, when *Piloty* and *Wilke* reported its preparation following a *Knorr*-type condensation of 3-aminobutanone with oxalacetate ester (315):

Durch Kombination von Amido-butanon mit Oxalessigester in alkalischer Lösung gelang es, zunächst nach folgender Gleichung:

---

[3] The leftmost structure, as represented here exactly from the drawing in the literature (316), has an unneeded and incorrect bond between the α-carbons of the pyrrole ring on the right.

[4] *Ludwig Knorr* was born on December 2, 1859 in Munich and died on June 4, 1921 in Jena. He received the *Abitur* in 1878 and studied chemistry at the University of Munich under *Jacob Volhard*. After *Volhard* left for Erlangen, *Knorr* was tutored by *Emil Fischer* in Munich, worked for *Bunsen* during a summer semester at the University of Heidelberg in 1880 and later for *von Baeyer*, followed *Fischer* to Erlangen, received the *Doktorat* in 1882, and completed the *Habilitation* in 1885. In 1885 he married *Piloty's* sister, *Elisabeth*. When *E. Fischer* received a call as successor to *Johannes Wislicenus* at the University of Würzburg in winter 1885, *Knorr* followed him there as a. o. Professor of Chemistry, where he investigated pyrrole synthesis, *inter alia*. In 1889, at age 30, he accepted an offer as *Ordinariat für Chemie* to replace the deceased *Geuthers*, where he introduced "modern" organic chemistry.

[5] *Carl Paal* was born on July 1, 1860 in Salzburg and died on January 11, 1935 in Leipzig. He studied organic chemistry in the TH-Berlin under *Liebermann*, then went to the University of Erlangen in 1884 to study organic chemistry under *Emil Fischer*. After the completing *Habilitation* in 1890 in Erlangen, he became a. o. Professor in 1892 and o. Professor and Director of the Laboratory for Applied Chemistry and Pharmacy as successor to *Ernst Beckmann*. In 1912, he accepted a call as o. Professor at the University of Leipzig, where he emeritated in 1929.

$$CH_3 \cdot CO \quad H_2C \cdot COOC_2H_5$$
$$CH_3 \cdot C \quad H \quad OC \cdot COOH \quad NH_2$$

$$\longrightarrow \quad CH_3 \cdot C \text{---} C \cdot COOC_2H_5$$
$$CH_3 \cdot C \quad C \cdot COOH$$
$$NH \qquad \textbf{I.}$$

ein einfaches Derivat des α,β-Dimethyl-pyrrols und aus diesem das Pyrrol-Homologe selbst in solcher Ausbeute zu erhalten, daß das Material zwar nicht billig, aber immerhin zugänglich wurde. Auf ähnliche Weise wie das Dimethyl-pyrrol konnten das β-Mono-methyl-pyrrol und das α-Methyl-β-äthyl-pyrrol dargestellt werden. Die Wichtigkeit dieser Produkte für synthetische Versuche in der Gruppe des Blutfarbstoffs . . . leuchtet ohne weiteres ein. [17]

Thus, heating **I** to 225°C under a blanket of CO_2 gave the corresponding decar-boxylated monoester, then after saponification a second decarboxylation yielded α,β-dimethylpyrrole (*315*):

$$\alpha,\beta\text{-Dimethyl-pyrrol,} \quad CH_3 \cdot C \text{---} CH$$
$$CH_3 \cdot C \quad CH$$
$$NH$$

The synthetic pyrrole proved to be identical with *Dennstedt*'s α,β-dimethylpyr-role from Bone Oil (*317*), and when treated with acid, *e.g.* picric acid, gave the same pyrrole dimer picrate (shown earlier) isolated from treatment of bilirubin with molten KOH (*301*).

Treating bilirubin with HI and PH_4I proved to be more revealing to *Piloty* and *Thannhauser* (*301*). Ultimately more important, it led to the isolation of a cleavage product that they named *Bilinsäure* (bilinic acid). Thus 50 g bilirubin yielded >3.5 g pure, crystalline bilinic acid, which after repeated crystallization from ethyl ace-tate melted sharply at 187°C without decomposition. Although initially colorless, the crystals gradually assumed a dark greenish coloration (in air). Elemental com-bustion analysis gave a good fit to the formula $C_{17}H_{26}N_2O_3$ (*301*):

I.	0,1281 g	gaben	0,3147 CO_2	und	0,0978	H_2O.			
II.	0,1431 g	,,	0,3452 CO_2	,,	0,1049	H_2O.			
III.	0,1266 g	,,	0,3125 CO_2	,,	0,0953	H_2O.			
I.	0,1614 g	,,	13,20 ccm	Stickgas	bei	21°	u.	720 mm	Druck.
II.	0,1310 g	,,	10,70 ccm	,,	,,	19°	,,	723 mm	,, .
III.	0,1215 g	,,	9,90 ccm	,,	,,	17°	,,	713 mm	,, .

	Ber. für	Gef.		
	$C_{17}H_{26}N_2O_3$	I	II	III
C	66,66	66,69	66,63	67,32
H	8,49	8,53	8,30	8,35
N	9,15	8,99	9,02	8,99

The formula corresponds to approximately one-half of the bilirubin structure, assuming a bilirubin formula $C_{33}H_{36}N_4O_6$, based on the *Teeple* and *Orndorff* molec-ular weight studies that had been accepted by *Küster* (*325, 326*).

4.1 Oskar Piloty and the Structure of Bilirubin

This discovery turned out to be one of the most important, if not the most important piece of pigment degradation evidence for learning the structure of bilirubin. To solve the structure of bilinic acid would mean solution of at least half of the bilirubin structure – and perhaps all of it, in the event that the pigment had been cleaved into two identical halves. Bilinic acid gave neither a pine split pyrrole reaction nor a red coloration by reaction with p-dimethylaminobenzaldehyde, *Ehrlich's*[6] reagent[7]. (The latter reaction had been shown to be characteristic of pyrroles absent at least one α-substituent.) From titration of an alcoholic solution, with 0.1% KOH, a neutralization equivalent was found to be 306. From the boiling point rise of a glacial acetic acid solution, the molecular weight was determined to be 299 and 347 (two determinations) for a calculated MW = 306. The collected data thus remained consistent with the formula $C_{17}H_{26}N_2O_3$ for bilinic acid, and *Piloty* was quick to note its similarity to his formula, $C_{17}H_{26}N_2O_2$, for hematopyrrolidinic acid. A similarity that perhaps induced him to write a related structure for bilinic acid (*301*):

Wir halten uns daher für berechtigt, für die *Bilinsäure* folgende Strukturformel aufzustellen:

---

[6] *Paul Ehrlich* was born on March 14, 1854 in Strehlen bei Breslau, Prussia, and died on August 20, 1915 in Bad Hamburg vor der Höhe. He studied at universities in Breslau, Strassburg, Freiburg-im-Breisgau, and Leipzig, and was awarded the Dr. med. degree at the last in 1878 with the dissertation: *Beiträge zur Theorie und Praxis der histologischen Färburg* (Contributions to the Theory and Practice of Histological Staining). In 1878, he was appointed *Assistent* to *von Frerichs* at the Charité in Berlin, and after his clinical education and *Habilitation* in 1887 there, he was appointed *Oberarzt* at the II. Medizinische Klinik, then a. o. Professor in 1891 at the Preußische Institüt für Infektionskrankheit in Steglitz, Berlin. In 1899 the Institut moved to Frankfurt am Main and *Ehrlich* became Director of the Institut für experimentelle Therapie. In 1908, he was awarded the *Nobel* Prize in Medicine with *Ilya Ilyich Mechnikov* in recognition of their work on immunity.

[7] There are two color-forming reactions named after *Paul Ehrlich*: (i) the *Ehrlich* aldehyde reaction, a test for urobilinogen in urine and α-H pyrroles, and (ii) the *Ehrlich* diazo reaction, a test for bilirubin. In (i) mentioned above, also used to assay tryptophan (or α-free pyrroles), a solution of p-dimethylaminobenzaldehyde in 5% aq. HCl or H_2SO_4 imparts a red or red-violet color to the substrate. The aldehyde condenses with α-*H*-pyrroles to form colored benzal derivatives. In (ii), a solution of diazotized sulfanilic acid imparts a red color to bilirubin and has been used in colorimetry to quantitate the pigment – as in the *van den Bergh* test.

The *van den Bergh* reaction, reported in 1913 by *Hijmans van den Bergh* (see Section 10) has been used since the early 1900s to detect and quantitate bilirubin in biological fluids such as serum and urine and is based on *Ehrlich's* diazo reaction (typically with diazotized sulfanilic acid) which converts bilirubin into red or red-violet azo-pigments. The *van den Bergh* test reports a "direct" and an "indirect" reacting bilirubin. The latter following addition of an accelerator, typically alcohol, reports the total bilirubin present (intact pigment + pigment conjugates). The former reports bilirubin conjugates, typically acyl glucuronides (see Section 10).

254   4 A Modern Proof of Bilirubin Structure Emerges

Wir bemerken nur zu dieser Formel, daß möglicherweise in derselben die Äthylgruppe und die mit * bezeichnete Methylengruppe passend zu vertauschen wären.   18

The structure shown is very similar to that which he suggested in 1912, for the hematopyrrolidinic acid obtained following reductive cleavage of hematoporphyrin with HI-PH$_4$I (*301*):

Wir nehmen daher an, daß den Eigenschaften der Hämatopyrrolidinsäure folgende Formel vollständig gerecht wird:

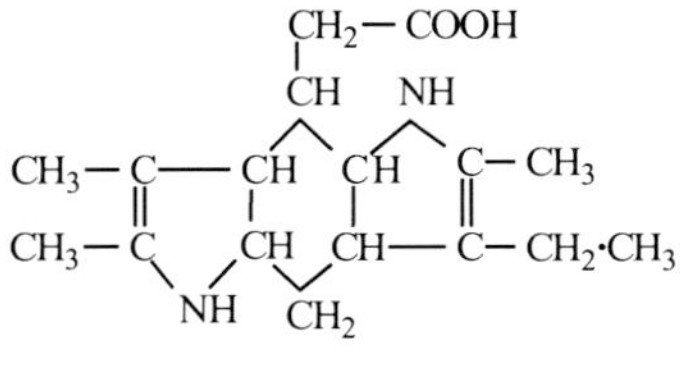

$$CH_3-C\cdots CH \quad CH \quad C-CH_3$$

19

Structural similarities aside, even for the moment allowing the correctness of *Piloty*'s hematopyrrolidinic acid structure, of which there were several different variants written between 1909 and 1912 (*301–304, 306–308*), the structure of bilinic acid was probed further by degradation. From molten KOH, bilinic acid produced a mixture of liquid pyrroles that distilled at 49–80°C/11 mm Hg and contained no detectable *Hämopyrrol*. On the other hand, chromic acid oxidation of bilinic acid produced a green reaction mixture, from which was extracted (into ether), after neutralization, methylethylmaleimide ($C_7H_9NO_2$, mp 68°C) and hematinic acid ($C_8H_9NO_4$, mp 112°C) in approximately equivalent amounts – and in good yield. From reaction of bilinic acid with HNO$_2$ in warm, dilute H$_2$SO$_4$ was obtained, not the expected oxime, but methylethylmaleimide, mp 68°C, and hematinic acid, mp 112°C. Under the right circumstances, it appears conceivable that the proposed bilinic acid structure might be oxidatively cleaved to both imides. This told *Piloty* only that the proposed structure might be viewed as consistent with the structures of its cleavage products, which were also obtained from oxidative cleavage of hematopyrrolidinic acid (*301*):

Wenn man bedenkt, daß bei allen Pyrrolen die α-Kohlenstoffatome die besonders begünstigten Angriffspunkt für oxydative Eingriffe sind, so ist es nicht zu verwundern, daß bei der Oxydation der Hämatopyrrolidinsäure der Sechsring gesprengt und Hämatinsäure bzw. Methyläthylmaleinimid

Hämatinsäure   Methyläthylmaleinimid

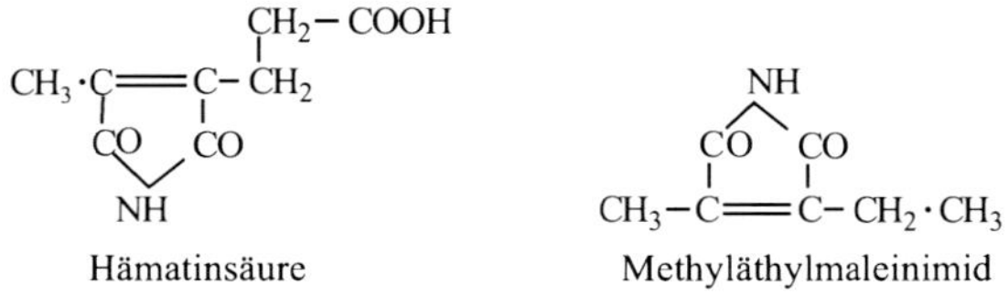

gebildet werden.   20

More information on the structure of bilirubin, and possibly bilinic acid, might be found among the other products from treatment of the former with HI-PH$_4$I.

4.1 Oskar Piloty and the Structure of Bilirubin

After bilinic acid had crystallized out, from the aqueous mother liquor were extracted into ether: (i) a base (meaning a pyrrole), but in quantities then too small to investigate, and (ii) an acid that was isolated in considerable quantity by means of its picrate, mp 146°C. The picrate yielded a pyrrole acid that *Piloty* named *Isophonopyrrolcarbonsäure* (isophonopyrrole carboxylic acid), which he represented as structure **II** (*301*):

I

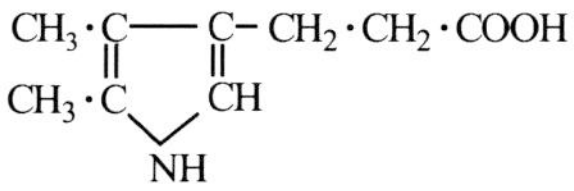

$$CH_3 \cdot C \text{——} C- CH_2 \cdot CH_2 \cdot COOH$$
$$CH_3 \cdot C \diagdown \diagup CH$$
$$NH$$

Phonopyrrolcarbonsäure

II

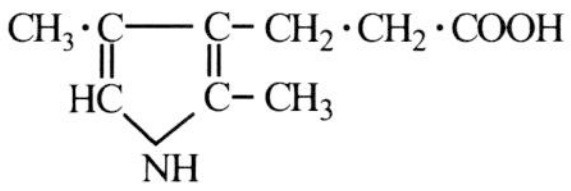

$$CH_3 \cdot C \text{——} C- CH_2 \cdot CH_2 \cdot COOH$$
$$HC \diagdown \diagup C- CH_3$$
$$NH$$

Isophonopyrrolcarbonsäure

III

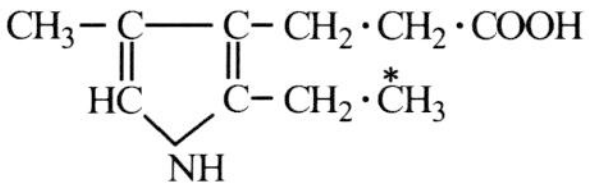

$$CH_3 - C \text{——} C- CH_2 \cdot CH_2 \cdot COOH$$
$$HC \diagdown \diagup C- CH_2 \cdot \overset{*}{C}H_3$$
$$NH$$

Xanthopyrrolcarbonsäure

This pyrrole acid, which crystallized from H_2O in splendid, colorless prisms, mp 126–127°C, with sintering at 105°C, analyzed for $C_9H_{13}NO_2$ (*301*):

> Aus Wasser krystallisiert die Säure in prächtigen, farblosen, prismatischen Nadeln mit zugeschärften Enden. Schmelzp. 126–127°, indem schon bei 105° Sinterung beginnt.

0,1365 g	gaben	0,3247 CO_2 und 0,0984 H_2O.
0,1154 g	,,	8,8 ccm Stickgas bei 16° und 718 mm Druck.

	Ber für $C_9H_{13}NO_2$	Gef.
C	64,58	64,86
H	7,78	8,06
N	8,38	8,50

How did *Piloty* assign the structure to isophonopyrrole carboxylic acid? How did he differentiate it from its isomer, phonopyrrole carboxylic acid (**I**), whose structure he had assigned on the basis of its failure to be converted to skatole and which he had previously called *Hämopyrrolcarbonsäure* (hemopyrrole carboxylic acid)? Of course, they both analyzed for the same formulas, $C_9H_{13}NO_2$, and it was found that their melting points and other properties did not differ in significant ways:

Table 4.1.1 Melting points (mp) of pyrrole carboxylic acids and derivatives

	Phonopyrrolcarbonsäure **I** (*313*)	*Isophonopyrrolcarbonsäure* **II** (*301*)
mp	125°C	126–127°C (sintering at 105°C)
Picrate mp	148°C (sintering at 145°C)	146°C
Mono-oxime mp	246°C (sintering at 221°C)	210°C

Their structures had to be very similar because the mono-oximes from HNO_2 treatment gave the same hematinic acid, whose structure had been proven by logical synthesis (*291*). Compound (**II**) was shown to produce the same mono-oxime as hemin-derived *Xanthopyrrolcarbonsäure* (xanthopyrrole carboxylic acid) (**III**), which (odd to us now) has an α-ethyl group adjacent to the propionic acid – not an α-methyl. Given that treatment of pyrrole acids **I-III** was supposed to produce mono-oximes where the α-alkyl was replaced by NOH, one would expect mono-oximes:

from phonopyrrole carboxylic acid (**I**) and

from isophonopyrrole carboxylic acid (**II**) and xanthopyrrole carboxylic acid (**III**). Although *Piloty* cited that the melting point of **II** was slightly higher than that of **I** (and it apparently did not occur to him to attempt or report mixture melting points here or at the mono-oxime stage), the proof of the structure of **II** depended clearly on that of **III**. So how was the structure of xanthopyrrole carboxylic acid (**III**) determined?

Xanthopyrrole carboxylic acid (**III**) appeared among the cleavage products when, as *Piloty* reported in 1912, hemin was treated with $HI-PH_4I$ in acetic acid. (It may be recalled that 11 years earlier *Nencki* and *Zaleski* found only basic products, including *Hämopyrrol*, by this same reaction (*285*).) *Piloty* and *Dormann* were able to isolate, in addition to *Hämopyrrol*, two monopyrrole acids: (i) the $C_9H_{13}NO_2$ phonopyrrole carboxylic acid (**I**), mp 125°C; picrate mp 157°C.[8] And (ii) a pyrrole acid that analyzed for 10 carbons, had mp 105°C; picrate mp 142.5°C, and which *Piloty* named xanthopyrrole carboxylic acid (*314*). The ratio of the isolated acids, the major acid products from hemin, was 2:1 in favor of phonopyrrole carboxylic acid, which gave a mono-oxime of mp 246°C when reacted with HNO_2 (not the previously reported 242°C [*303*]). Xanthopyrrole carboxylic acid very clearly differed from phonopyrrole carboxylic acid in melting point, picrate and half-oxime (mp 201–202°C). Its mono-oxime, like that from phonopyrrole carboxylic acid, hydrolyzed to the same hematinic acid. Presumably the different mps for the mono-oxime of isophonopyrrole carboxylic acid (mp 210°C) compared to that from xanthopyrrole carboxylic acid was not an issue at that time, but other changes were looming. For *Hans Fischer* had begun to investigate the structure of bilirubin, and

---

[8] The picrate mp differs from that reported in 1909 (*303*) for hemopyrrole carboxylic acid, which was later renamed phonopyrrole carboxylic acid (*313*).

4.1 Oskar Piloty and the Structure of Bilirubin

he and *Richard Willstätter* had begun to publish their findings that *Hämopyrrol* was a mixture that contained at least three different pyrroles (*314*):

> Durch H. Fischer und E. Bartholomäus, . . . und namentlich durch Willstätter und Asahina . . . wurden jüngst aus dem Hämopyrrolgemisch die drei Pyrrole

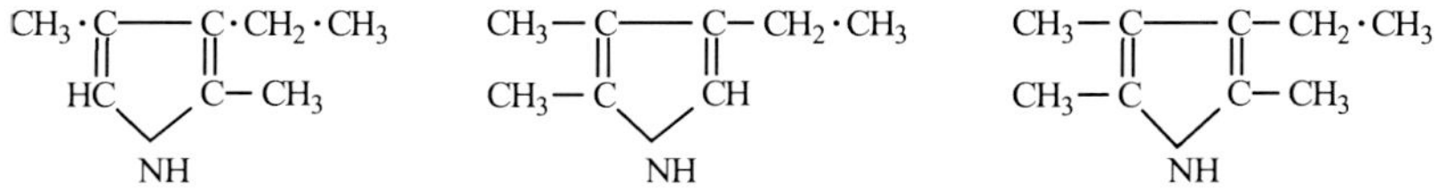

Hämopyrrol — Isohämopyrrol — Phyllopyrrol

> ausgeschieden. 22

As will be shown later, the proof of structure of xanthopyrrole carboxylic acid turned out to be insufficient, and *Piloty*'s isolation of bilinic acid coincided with the entrance of *Hans Fischer* into structural studies of the pigments of bile and blood – and the return of *Küster* to such structure investigations.

While *Piloty*'s group was attempting to clarify the structure of the pigment of blood and the pyrrole compounds, work still continued in 1912 on bilinic acid from HI + PH$_4$I cleavage of bilirubin. Thus, *Piloty* and *Thannhauser* found that bilinic acid was converted by mild 0.1 *N* KMnO$_4$ oxidation at +7°C to an intensely yellow-colored crystalline compound, which they named *Dehydro-bilinsäure* (dehydro-bilinic acid). Dehydrobilinic acid, by elemental combustion analysis, and by molecular weight determination (MW 309) from boiling point elevation in glacial acetic acid fit the formula C$_{17}$H$_{22}$N$_2$O$_3$ (FW 302) (*302*):

> C$_{17}$H$_{22}$N$_2$O$_3$ Ber. C 67.55, H 7.28, N 9.27
> Gef. » 67.20, » 7.73, » 9.33

> Molekulargewichtsbestimmung: 0.3131 Sbst. in 18.11 g Eisessig: Siedepunkts-Erhöhung 0.13%. Ber. Mol.-Gew. 302, Gef. 309. 23

Or nearly the same formula (C$_{17}$H$_{26}$N$_2$O$_3$) as bilinic acid (below) but with four fewer hydrogens (*302*):

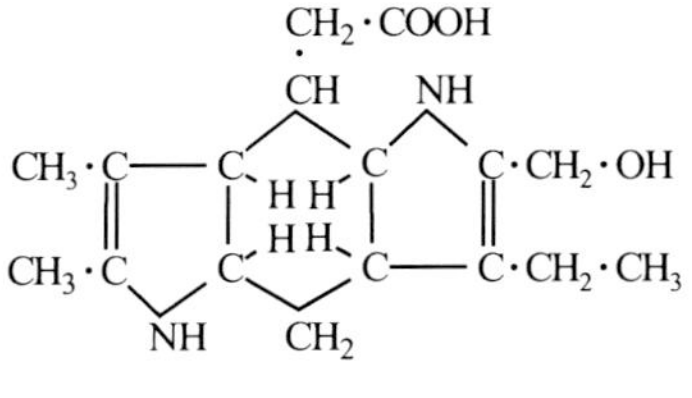

(Clearly, by 1912 "oxidation" also signified something other than the oxygen incorporation "oxygenation" that it had meant in the 19th century, for example, as in the difference between the then proposed formulas for bilirubin and its oxidation product, biliverdin.)

258   4   A Modern Proof of Bilirubin Structure Emerges

Holding to his original structure (*301*), *Piloty* suggested that the structural change in going from colorless bilinic acid to lemon-yellow dehydrobilinic acid was due to the introduction of a system of conjugated double bonds (*302*), as shown by partial structures **I** and **II** (*302*):

> Wir halten es für wahrscheinlich, daß der Farbstoffcharakter dieser Substanz durch ein System konjugierter Doppelbindungen bedingt wird. Solche Systeme kann man sich aus der oben gegebenen Formel der Bilinsäure leicht konstruieren und so einen Versuch der Erklärung des Übergangs der farblosen in die farbige Säure machen, wie folgende Schemata andeuten mögen:

24

Dehydrobilinic acid was further described as appearing in fine-tapered lemon-yellow prisms that decomposed at 260°C without melting; insoluble in H_2O, benzene, petroleum ether; slightly soluble in absolute alcohol and glacial acetic acid. It did not give the typical pine splint pyrrole color reaction, was neutral toward litmus paper (possibly due to its insolubility in H_2O), and produced an intense yellow coloration in alkali and alkaline carbonate. The sodium salt had the color of picric acid and by combustion analyzed for one equivalent of sodium, found as Na_2SO_4 (*302*):

> Die Dehydro-bilinsäure erscheint so als citronengelbe, schlanke, abgeschrägte Prismen, welche Neigung zur Zwillingsbildung haben. Ausbeute an reiner Substanz 0.3 g. Sie zersetzt sich ohne zu schmelzen über 260°; ist in Wasser, Benzol, Petroläther unlöslich, ziemlich leicht löslich in heißem absolutem Alkohol und Eisessig. Sie gibt keine Pyrrolreaktion mit dem Fichtenspane und reagiert nicht mit Dimethylamido-benzaldehyd. Gegen Lackmuspapier ist sie, wohl wegen ihrer Schwerlöslichkeit in Wasser, neutral. Sie löst sich leicht unter intensiver Gelbfärbung in Alkalien und kohlensauren Alkalien. . . .
> Natriumsalz. Sehr charakteristisch ist das Verhalten der Dehydro-bilinsäure gegenüber verdünnter Natronlauge und Sodalösung. In beiden Flüssigkeiten löst sie sich in der Wärme leicht auf. Beim Erkalten aber scheidet sich die Natriumverbindung als pikratgelber, krystallinischer Niederschlag wieder aus. Nach dem Filtrieren und Trocknen kann das Salz aus 70-prozentigem Alkohol umkrystallisiert werden. Es fällt beim Erkalten in konzentrisch gruppierten, schlanken, gelben Nadeln aus.

> 0.1366 g   Subst.:   0.0290 g. Na_2SO_4
>   Ber.   Na 6.88.   Gef.   Na 7.07.   25

Piloty concluded that there is present in bilirubin a complex that resists further oxidation. He acknowledged that *Hans Fischer* and *Heinrich Röse* had also obtained a colorless compound from bilirubin in the same way but had proposed a

4.1 Oskar Piloty and the Structure of Bilirubin

different structure – a structure from which a colored compound could not be formed by the simple loss of hydrogen (*302*):

> Als experimentell wichtiges Ergebnis erscheint das Mitgeteilte auch in sofern, als daraus hervorgeht, daß im Bilirubin ein Komplex enthalten ist, der einer gelinden Oxydation widersteht. . . .
>
> Gleichzeitig mit uns haben H. Fischer und Röse [Berichte (1912) **45**: 1579] auf die gleiche Weise wie wir aus dem Bilirubin die farblose Säure erhalten, dafür aber eine Formel aufgestellt, welche die Entstehung eines farbigen Körpers, wie wir ihn erhielten, durch einfachen Wasserstoffverlust nicht wohl erklären läßt. 26

Piloty's work on bilirubin and his discovery of bilinic acid and dehydrobilinic acid did indeed intersect with *Hans Fischer*'s nearly simultaneous investigations of the bilirubin structure. At the time, with many of the seeds of World War I having already been sown, and the first and second Balkan wars underway, fate it seems had brought *Piloty* and *Fischer* to work on the same problem in natural products chemistry and in the same city, Munich. But from 1899, until his untimely death on a World War I battlefield in 1915, *Piloty* clearly held the more senior academic appointment than *Fischer* and probably had access to greater resources for research. Following 1912 (*301, 302*), he published very briefly only four times more on bilirubin, from 1913 (*327*) to 1914 (*328–331*); rather he appeared to devote his resources to solving what was probably the more attractive research problem: the structure of the blood pigment.

In February 1913, *Piloty* sent a short "clarifying" paper (*327*) to the *Berichte der deutschen chemischen Gesellschaft*, which presented a polite reply and objection to the article by *Fischer* and *Röse* that had appeared earlier in 1913 in the same journal. He reminded the reader, "lest there be confusion in the literature", that he and *Fischer* had isolated the same cleavage product from bilirubin, which he named *Bilinsäure* (bilinic acid), and *Fischer* named *Bilirubinsäure* (bilirubinic acid). Objecting to a potential second obfuscation of nomenclature, *Piloty* also reminded that it was he who first described and characterized exactly a second substance (from reductive cleavage of bilirubin), that he named *Isophonopyrrol-carbonsäure* (isophonopyrrole carboxylic acid), and that it was *Fischer* who found it subsequently and then named it *isomere Phonopyrrolcarbonsäure* (isomeric phonopyrrole carboxylic acid), a name to which he (*Fischer*) held firm.

Remonstrating against *Fischer* and *Röse* further, *Piloty* reaffirmed that it was he who had proposed that the bilirubin structure consisted of three rings, two of which were similar and connected to isophonopyrrole carboxylic acid. However, it was they who claimed for the first time to have provided evidence that bilirubin and *Hemibilirubin* (hemibilirubin) contained a third pyrrole nucleus. *Piloty* further claimed that his dehydrobilinic acid came from bilinic acid by removal of two hydrogen atoms. Yet, the publication cited describes formulas for bilinic acid ($C_{17}H_{26}N_2O_3$) and dehydrobilinic acid ($C_{17}H_{22}N_2O_3$), hence a difference of four hydrogen atoms.

260 4 A Modern Proof of Bilirubin Structure Emerges

Piloty noted that *Fischer* and *Röse* had obtained a yellow pigment from bilirubin that they thought to be identical to dehydrobilinic acid but, according to their statements, failed to identify it as such, due to an insufficiency of material. *Piloty* expressed that the omission could, however, not be recognized as a legitimate reason to give the substance not one but two different names: xanthopyrrole carboxylic acid and *Xanthobilirubinsäure* (xanthobilirubinic acid). Naturally, those names presented two problems to *Piloty*: (i) he had already assigned the former to a different acid in a publication, and (ii) he failed to find an objective basis for assigning the latter name to a compound that had already been assigned and named in the literature. It was a question of avoiding confusion in the literature and who had scientific priority for the discovery of the compound. He asked that the two names, *Xanthopyrrolcarbonsäure* and *Xanthobilirubinsäure*, be dropped and that the name (*Dehydro-bilinsäure*) given by its assignee-discoverer be accepted. He also suggested that it would be appropriate for *Fischer* to become reconciled to the names *Bilinsäure* (rather than *Bilirubinsäure*) and *Isophonopyrrol-carbonsäure*.

Fischer's reaction to all this will be covered shortly, but it remains a fact that *Piloty* did not return to the subject of bilirubin until 1914, and then only very briefly (*328–331*). Instead, in 1912 (*332–336*) and 1913 (*337–342*), he focused his research on the structure of the blood pigment and the structures of its degradation products, especially the mono-pyrroles and hematopyrrolidinic acid (*332–342*), and on dipyrryl-methanes and dipyrryl-methenes (*328–331*). In a paper received in the journal office on January 26, 1914, *Piloty* returned to the structure of dehydrobilinic acid, $C_{17}H_{22}N_2O_3$, for which, two years earlier, he and *Thannhauser* (*301*) had proposed two possible structures with conjugated C=C bonds. In 1914 he proposed a third possible conjugated system (*328*):

> Sie sprachen ihre Meinung dahin aus, daß diese Betrachtung für die Beurteilung der Farbstoffnatur der ganzen Körperklasse von Bedeutung sei. Wir fügen noch hinzu, daß das Schema

III.

gedanklich mit dem Schema I zusammenfällt. 27

This thought presumably had its origin in *Piloty*'s finding that two *Hämopyrrol* molecules, or two phonopyrrole carboxylic acid molecules self condensed through the action of $CHCl_3$ + KOH to form colored products, dipyrrylmethenes (**IV** and **VI**) or dipyrrylmethanes (**V**) that exhibited a great similarity to the bile pigments and to dehydrobilinic acid (*328*):

4.1 Oskar Piloty and the Structure of Bilirubin

IV.

$CH_3 \cdot C$ — $C \cdot CH_2 \cdot CH_3$ $CH_3 \cdot CH_2 \cdot C$ = $C \cdot CH_3$

$CH_3 \cdot C$ C H   Cl — C Cl   H C $C \cdot CH_3$
$\qquad$ Cl   H

NH $\qquad$ NH

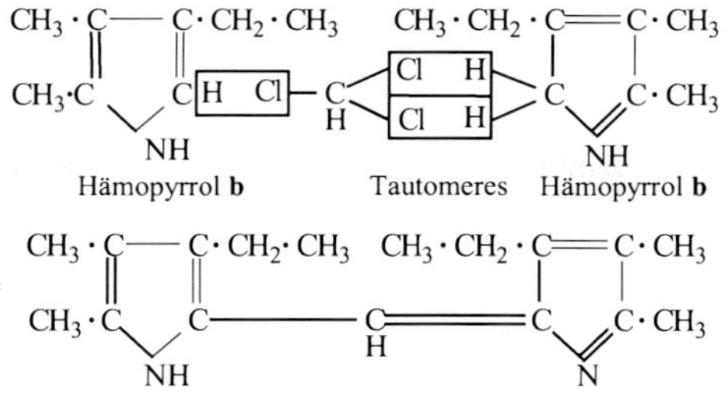

Hämopyrrol **b** $\qquad$ Tautomeres Hämopyrrol **b**

=

$CH_3 \cdot C$ — $C \cdot CH_2 \cdot CH_3$ $CH_3 \cdot CH_2 \cdot C$ = $C \cdot CH_3$

$CH_3 \cdot C$ $\qquad$ C — C = C $C \cdot CH_3$
$\qquad$ H

NH $\qquad$ N

Bi-(äthyl-dimethyl-pyrryl)-methen.

V.

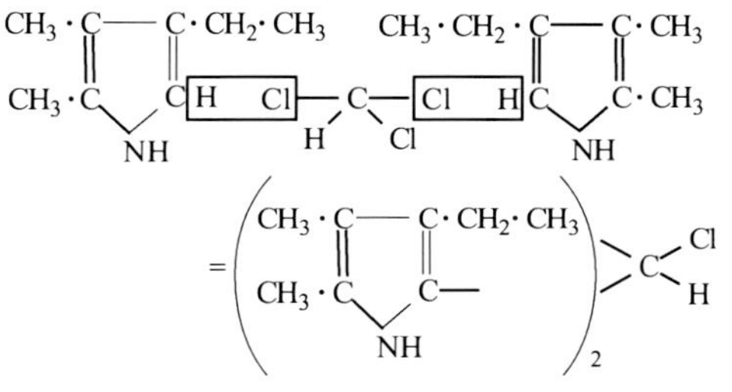

$CH_3 \cdot C$ — $C \cdot CH_2 \cdot CH_3$ $CH_3 \cdot CH_2 \cdot C$ — $C \cdot CH_3$

$CH_3 \cdot C$ H   Cl — C — Cl   H C $C \cdot CH_3$
$\qquad$ H   Cl

NH $\qquad$ NH

=

$\left(\begin{array}{c} CH_3 \cdot C — C \cdot CH_2 \cdot CH_3 \\ CH_3 \cdot C \qquad C — \\ NH \end{array} \right)_2$ C $\begin{array}{c} Cl \\ H \end{array}$

Bi-(äthyl-dimethyl-pyrryl)-chlormethan.

Die Reaktion findet nach **IV** zwischen einem Molekül Chloroform und zwei Molekülen Hämopyrrol **b** unter dreimaligem, nach **V** unter zweimaligem Austritt von Salzsäure statt.

Die Phonopyrrol-carbonsäure **a** reagiert nur, wie es scheint, in einem Sinne, nämlich nach Formel **VI**.

VI.

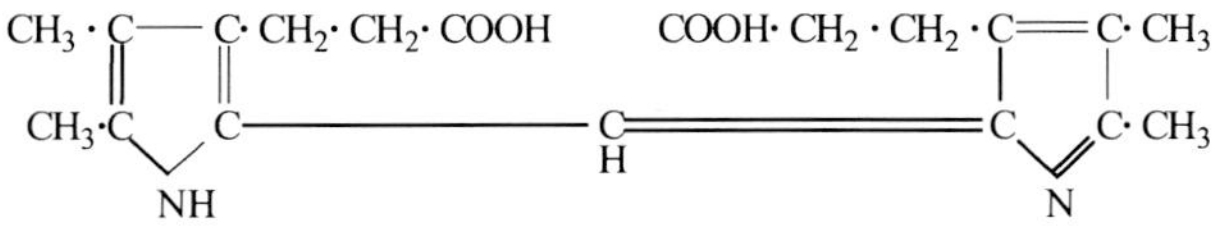

$CH_3 \cdot C$ — $C \cdot CH_2 \cdot CH_2 \cdot COOH$ $COOH \cdot CH_2 \cdot CH_2 \cdot C$ = $C \cdot CH_3$

$CH_3 \cdot C$ $\qquad$ C — C = C $C \cdot CH_3$
$\qquad$ H

NH $\qquad$ N

Bi-(propionyl-dimethyl-pyrryl)-methen. $\qquad$ 28

Piloty reported that dipyrrylmethane **V** had an intense yellow color, similar to that of dehydrobilinic acid and spectrum analysis properties similar to those of bilirubin (*328*):

> Das Bi-(äthyl-dimethyl-pyrryl)chlor-methan (**V**) ist ein intensiv gelb gefärbter Körper, sehr ähnlich der Dehydro-bilinsäure, für welche wir eine der Formel **V** entsprechende Formulierung vorschlagen möchten. . . .
>
> Das spektralanalytische Verhalten der neuen Farbstoffe zeigt eine gewisse Ähnlichkeit mit dem Bilirubin. $\qquad$ 29

Why **V** should have a yellow color is, from today's perspective, unclear – except that possibly some of it had been converted to a little (reddish) dipyrrylmethene **IV**, with an observed resultant yellow coloration from the mixture. From the red color of the dipyrrylmethene Piloty drew a comparison to the core of *Willstätter's*[9] formula (*343, 344*) for the hemin framework (*328*):

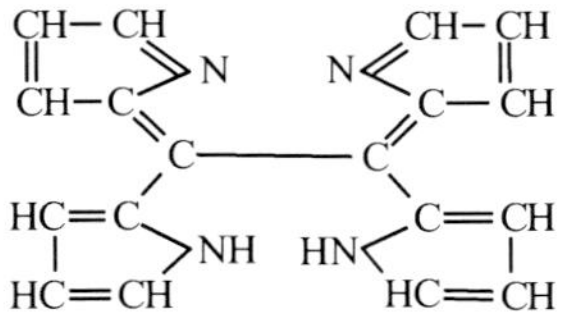

On August 17, 1914, or about a month before his son died on a World War I battlefield on September 25, 1914, *Piloty*'s final paper on the pigments of blood (and bile) was received at the journal (*331*). Only a few weeks earlier, Archduke *Franz Ferdinand* of Austria-Este and heir-presumptive to the Austrian-Hungarian throne had been assassinated in Sarajevo (on July 28, 1914); Austria-Hungary had declared war on Serbia (July 28, 1914). The cascade of events starting World War I had begun: on July 29, 1914, Russia mobilized, on July 30, Germany mobilized, on August 1, France mobilized and Germany declared war on Russia, on August 3, Britain declared war on Germany, *etc. Piloty*'s penultimate publication expressed his thoughts on the relationship of the dipyrrylmethene pigment to bilirubin in the form of a bilirubin structure that he considered to be more likely than that of *Fischer* and *Röse*. Thus, he suggested a partial structure that would resemble the core of bilirubin and explain the various monopyrrole cleavage products as well as its color (*331*):

> Was endlich das Bilirubin betrifft, so halten wir das in der Mitteilung von Piloty und Thannhauser . . . angedeutete Schema seiner Konstitution, ergänzt durch die Folgerung aus unserer Entdeckung der Dehydro-bilinsäure und derjenigen des Zusammenhangs der Dipyrrylmethenefarbstoffe mit dem Bilirubin, wie es folgender Ausdruck veranschaulicht,

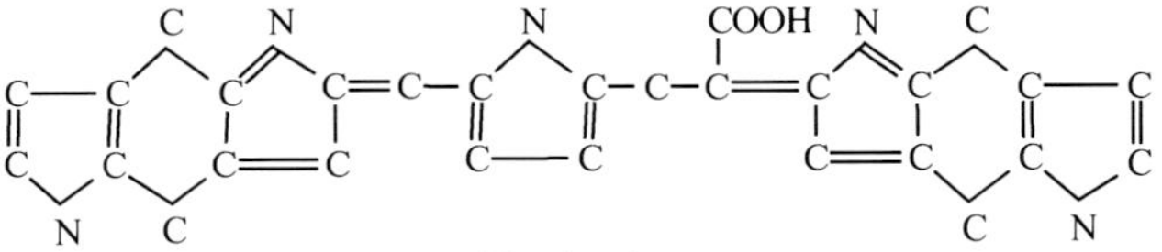

Bilirubinschema

---

[9] *Richard Martin Willstätter* was born on August 13, 1872 in Karlsruhe, Baden, Germany and died on August 2, 1942 in Muralto, Locarno, Switzerland. He began university studies at both the Universität and the Technische Hochschule in Munich in 1890, in the first as a student of *Adolf von Baeyer*, and received the doctorate in 1894. There he commenced his *Habilitation* and in 1896 became *Privatdozent* at the University of Munich, where in 1902 he was named a. o. Professor and head of the Organic Section. In 1905 he left Munich to become o. Professor at the ETH Zürich, and in 1912 he moved to Berlin to become Director of the new Chemical Institute of the Kaiser Wilhelm Gesellschaft zur Förderung der Wissenschaften at Dahlem. Staying only briefly in Berlin, in 1915 he returned to Munich to succeed *von Baeyer* as Professor and Director of the State Chemical Laboratory. In 1915, he was awarded the *Nobel* Prize in Chemistry for his work on plant pigments, including chlorophyll. Then in 1924, in protest of the increasing anti-Semitism in Germany, he resigned his appointment, to the great dismay of his scientific colleagues at home and abroad. In 1938 he fled the Gestapo to Switzerland.

4.2 Hans Fischer and the Early Structures of Bilirubin

für wahrscheinlicher als die Aufstellung von H. Fischer und Röse (a. a. O.). Diese letztere scheint uns allzusehr an die Häminformel angelehnt und die großen Unterschiede im Verhalten der beiden Farbstoffe gar nicht zu berücksichtigen. 30

The structure may be compared with the partial structure of hemin also suggested by him (*215d*):

Häminschema

As it turned out, neither the *Piloty* nor the *Fischer* 1914 structures came close to the actual structures of bilirubin or hemin, though they showed a definite evolution in structure proof methodology over that of 10–20 years earlier. With the passing of *Piloty* in 1915, it was left to *Fischer* and *Küster* to resolve the structures of bilirubin, and hemin. The structure of *Piloty*'s *Hämatopyrrolidinsäure* remains unsolved.

4.2 *Hans Fischer* and the Early Structures of Bilirubin

Hans Fischer[10] and his group, working from 1910 to 1942, over a period of 32 years, solved the then technically difficult problem of the constitutional structure of the yellow pigment of bile, bilirubin, by analysis and especially by total

---

[10] *Hans Fischer* was born on July 27, 1881 in Höchst am Main and died on March 31, 1945 in Munich. Known as the "Father of Pyrrole Chemistry", Fischer solved the structure of bilirubin (*5*). After the *Abitur* in 1899 at the Humanistisches Gymnasium in Wiesbaden, he took up studies of chemistry and medicine at the University of Lausanne for a semester while recovering from poor health, and at the Universität München and the Universität Marburg. Following military service, in 1900 he studied chemistry and medicine at the Universität Marburg. Then, following a two-semester rotation at the Technische Hochshule-München and the Universität München, he concluded his *Promotionsarbeit* at the Universität Marburg under *Th. Zincke* and was awarded the Dr. phil. in chemistry with the thesis title, *Beiträge zur Kenntnis der 4-Oxy-1,2-toluylsäure*. After further studies under *Emil Fischer* at the Kaiser Wilhelm Institute, Dahlem, he returned to Munich with a recommendation from *Zincke* to take up medical studies under the famous clinician-internist, *Friedrich von Müller* (1858–1941), who since 1902 had been Director of the Second Medical Clinic in Munich and was world-famous for his scientific teaching of medicine. *Müller* ran one of the most highly attended clinics and early on recognized the importance of chemistry in clinical diagnosis. In late 1906, *Fischer* passed the government exam and received a license in medicine; in late 1908, he was promoted to Dr. med., with the dissertation: *Zur Kenntnis des carcinomatösen Mageninhalts*. It was *von Müller*'s interest in the blood pigment biological degradation products that ignited *Fischer*'s interest in studies of bilirubin and hemin and thus led him into the realm of pyrroles and pyrrole pigments during his Habilitation studies that followed. In 1909, *Hans Fischer* assumed his first positions at the Second Medical Clinic and in Berlin as

synthesis. Although *Fischer*'s earlier total synthesis of the red pigment of blood, hemin, for which he was the sole recipient of the 1930 *Nobel* Prize in Chemistry, had been accomplished some 13–14 years before the structure elucidation of bilirubin was completed, the hemin work represented a *tour-de-force* of 1920s vintage organic synthesis. Unlike many of his contemporaries in natural products chemistry, *Fischer* maintained a firm belief in structure proof by synthesis, and thus he became a major force in 1920s–1930s synthetic organic chemistry, with the majority of his nearly 500 research papers reporting synthesis chemistry (*12, 345–347*). His monumental series, *Die Chemie des Pyrrols*, published in two volumes and three parts between 1934 and 1940 (*2–4*) and co-authored with *Hans Orth* and *Adolf Stern*, became the long-enduring "bible" of pyrrole chemistry.

4.2.1 Fischer and Piloty Cleave Bilirubin in Half, 1912-1913

Undeterred in 1910 by the major accomplishments in pigment structure from *Nencki* (dec. 1901), *Willstätter*, *Küster*, and *Piloty*, and by the prospect of intense competition from their research groups already up and running, a young and perhaps brash *Hans Fischer* had launched his *Habilitation* work on bilirubin well before the publication of his *Habilitationschrift* in 1912 (*348*) – as his papers

---

Assistent to *Emil Fischer* at the Erstes Chemisches Institut der Universität Berlin, working on peptides, the glycosides of maltose and lactose. Though *Emil Fischer* had asked *Hans Fischer* to remain in Berlin for the *Habilitation*, the mountains of Bavaria were too strong a pull on the younger *Fischer*, who returned in 1910 to the Second Medical Clinic in Munich to work on pyrrole pigments. There he habilitated in internal medicine to the medical faculty at the University of Munich in purely chemical studies with the title of his *Habilitationschrift*: *Über Urobilin und Bilirubin*, and so *Fischer* qualified as a lecturer in 1912.

However, by the time of the publication of his thesis in 1912, *Fischer* had already entered the realm of pyrroles, with 21 research publications in 1911–1912. In 1913, *Fischer* was named *E.F. Weinland*'s successor as lecturer in physiology at the University's Physiological Institute, and in 1915 as a. o. Professor on the medical faculty of the University of Munich. In 1916 he was called to succeed *Windaus* as a. o. Professor of Medicinal Chemistry at the University of Innsbruck and then in 1918 as a. o. Professor of Medicinal Chemistry at the University of Vienna. Yet in the turmoil of World War I and its aftermath there was little opportunity for research and so publications ground to a halt between 1916 and 1921. In 1921, his fortunes improved when he received a call to succeed *Heinrich Wieland* as o. Professor of Organic Chemistry at the TH-München [now Technische Universität (TU-München)]. It was there that his research talents would be focused on three outstanding problems: the structures of the pigments of blood, bile, and green leaves. And there he remained until his self-inflicted death on *Ostersamstag*, March 31, 1945, after having witnessed and anguished over the destruction of his Institute and life's work in the last years of World War II. Considering the impact on university life and research funding of the hyperinflation in Germany of the 1920s, given the rise, intrusions, excesses, and horrors wrought by National Socialism in the 1930s, the loss of students to war, and finally the destruction of German society and infrastructure during World War II, *Fischer*'s research accomplishments in Munich may be viewed as nothing less than astounding.

4.2 Hans Fischer and the Early Structures of Bilirubin

published in 1911 (*349–351*) reveal. In his first paper on bile pigments, submitted June 22, 1911 (*349*), *Fischer* reported on a topic that would not go away (despite *Thudichum*'s best efforts): *Maly*'s hydrobilirubin and its relation to the urobilin of urine and stools. Bilirubin isolated from cattle gallstones and crystallized from $CHCl_3$ or $CHCl_2CHCl_2$ produced a pigment that contained ~1% Cl by combustion analysis – as had been observed by others. Crystallization from dimethylaniline or quinoline also left crystals from which the last traces of solvent were reluctant to leave, as detected by five combustion analyses. However, bilirubin taken up in 0.1 *N* aq. NaOH and precipitated by aq. HCl yielded pigment whose combustion analysis satisfactorily fit the formula $C_{16}H_{18}N_2O_3$ (*349*):

	Berechnet für Bilirubin: $C_{16}H_{18}N_2O_3$	Gefunden: I.	II.	
C	67,09	66,78	66,96	
H	6,34	6,45	6,64	
N	9,79	9,73	9,65	31

Apparently, this confirmed the then accepted bilirubin formula to *Fischer*, although it should have been evident from the earlier molecular weight studies of *Orndorff* and *Teeple* (*211, 213, 214*) that the empirical formula derived from the combustion analysis constituted only about one-half the molecular formula.

Following *Maly*'s procedure exactly for treating bilirubin with Na(Hg), as well as using his own modifications, *Fischer* was led eventually to solid, red hydrobilirubin, which exhibited varying combustion analysis data, none of which matched *Maly*'s (*349*):

Maly erhielt:		C	64,68	H	6,93	N	9,22	
Meine Analysen ergaben:	I.	C	62,94	H	7,45	N	9,06	9,13
	II.	C	65,08	H	7,63	N	9,32	
	III.	C	64,82	H	7,17	N	9,40.	

Aus allen diesen Angaben geht klar hervor, daß das Malysche Urobilin, auch nach der verbesserten Vorschrift, ein kompliziertes Gemisch ist.   32

And from all that, *Fischer* concluded that the Na(Hg) reduction of bilirubin produced mixtures, which, by way of inference, meant that *Maly*'s hydrobilirubin (to which *Maly* ascribed the formula $C_{32}H_{40}N_4O_7$) was a mixture – as was the hydrobilirubin produced by other investigators who followed.

Not one to give up readily, and in keeping with his penchant for careful experimental work, *Fischer* reduced bilirubin in aq. NaOH with 4.6% Na(Hg), during 1–2 hours of agitation and cooling to obtain a 46% yield of colorless prisms (mp 192°C with dec.), after work-up and crystallization in the absence of air. A reasonably consistent set of combustion analysis results of this apparently pure substance fit best to either of two formulas, $C_{16}H_{22}N_2O_3$ or $C_{16}H_{20}N_2O_3$ (*349*):

Berechnet für		Gefunden:			
$C_{16}H_{22}N_2O_3$	$C_{16}H_{20}N_2O_3$:	I.	II.	III.	
C 66,16	66,62	66,7	66,45	66,48	
H 7,64	7,00	7,47	7,86	7,82	
N 9,65	9,72	9,25	9,33	9,27	9,40.

Its molecular weight, found by the melting point depression of naphthalene, was determined to be 460; the FW of $C_{16}H_{20}N_2O_3$ is 288. Despite the apparent disconnect between these data, *Fischer* assumed that the new material represented only one-half of bilirubin and thus he named it *Hemibilirubin* (hemibilirubin) (*349*). Thus, yet another name appeared in bile pigment chemistry lexicon, for a compound obtained by adding two hydrogen atoms to C_{16} bilirubin (*349*):

> . . . da der Körper vielleicht nur die eine Hälfte des Bilirubins darstellt, so schlage ich den Namen Hemibilirubin vor. Nach der Elementaranalyse besitzt das Hemibilirubin die Zusammensetzung $C_{16}H_{20}N_2O_3$.
>
> Es kann sich also um ein einfaches Wasserstoffanlagerungsprodukt von Bilirubin handeln, das dabei 2 Wasserstoffatome aufnehmen würde. . . .   33

Rather remarkably, *Fischer* at first assumed that his hemibilirubin might be the acid anhydride of 4-methylpyrrole-3-propanoic acid ($C_8H_{11}NO_2$), possibly because with HNO_2 it yielded the oxime of *Küster*'s hematinic acid, and with PbO_2 it yielded hematinic acid itself (*349*):

> Was die Zusammensetzung des Körpers anlangt, so wäre die einfachste Erklärung folgende: Das Hemibilirubin ist das Anhydrid einer Säure folgender Konstitution.

$$CH_3C \!\!\!\begin{array}{c} \\ \| \\ HC \end{array}\!\!\! \begin{array}{c} \\ \\ \end{array}\!\!\! \begin{array}{c} C - CH_2 - CH_2 \cdot COOH \\ \| \\ CH \end{array}$$
$$NH$$

> Dementsprechend habe ich das Hemibilirubin mit salpetriger Säure behandelt und in der Tat das Oxim der Küsterschen Hämatinsäure erhalten. Bei der Oxydation mit Bleisuperoxyd erhielt ich Hämatinsäure.

$$CH_3C \!=\! C - CH_2 - CH_2 \cdot COOH$$
$$O \!=\! C \quad C \!=\! O$$
$$NH$$

    34

The anhydride of 4-methylpyrrole-3-propanoic acid would have a formula $C_{16}H_{20}N_2O_3$. The fact that the combustion analysis data (see above) show that the %N found is lower than the 9.72% predicted by the formula of the "anhydride" (and is also too high in %H), *Fischer* explained as due to the sensitivity (instability) of hemibilirubin toward oxidation. And he clearly indicated that the proposed structure could only be settled by degradation (*349*):

4.2 Hans Fischer and the Early Structures of Bilirubin

Stützen für oder gegen diese Anschauung konnten nur durch den Abbau gewonnen werden ..   35

He added further that he could not detect any *Hämopyrrol* coming from reaction of bilirubin with HI-PH$_4$I in glacial acetic acid but found a compound that gave a positive color test with *p*-dimethylaminobenzaldehyde (see Section 4.1, *Ehrlich*). Ether extraction of the aqueous basic reaction work-up solution yielded a liquid pyrrole (as per the aldehyde reaction above) that produced a crystalline compound upon chromic acid oxidation (presumably an imide).

Clearly following an interest in medicine and pathology, *Fischer* and *Meyer-Betz* detected hemibilirubin in pathologic urine and suggested that it may be the parent substance of urobilin (*350*). *Fischer* believed that there were two forms of hemibilirubin: one "acidic", and one "non-acidic" – although both were carboxylic acids. The real value of *Fischer*'s work here was in finding a bilirubin derivative (hemibilirubin) with such favorable solubility in organic solvents that he was able to determine the non-acidic form's molecular weight as 585 and 565 (two independent measurements) in absolute alcohol by the boiling point elevation method – and the acid form as 593. Assuming that they were dibasic acids, titration in alcohol to a phenolphthalein endpoint confirmed the MWs: the neutralization equivalents were 301, 296, 298, 301 (four independent titrations) for the acidic form, and 299 for the non-acidic. Together with combustion analyses, two possible formulas were supported: $C_{33}H_{44}N_4O_6$ and $C_{32}H_{44}N_4O_6$. Though combustion data supported either, they also clearly illustrate the difficulty encountered when analyzing high molecular weight substances that differ only by one or two atoms (*351*):

	Berechnet für:		Gefunden:			
	$C_{33}H_{44}N_4O_6$	$C_{32}H_{44}N_4O_6$:	I.	II.	III.	
C	66.85	66.16	66.64	66.5	66.87	66.60%
H	7.49	7.64	7.68	8.08	7.98	7.73%
N	9.46	9.65	9.55	9.67; 9.48	9.37	9.47%

Employing *Küster*'s chromic acid oxidation method, *Fischer* was able to find both hematinic acid and methylethylmaleimide as oxidative cleavage products from hemibilirubin. It may be recalled that hematinic acid, in contrast, was the only imide found by *Küster* following chromic acid oxidation of bilirubin, biliverdin, hematin or hematoporphyrin. Thus, the appearance of methylethylmaleimide from hemibilirubin apparently led *Fischer* to conclude that the last possessed two differently substituted pyrrole rings. One might think that, by extension, bilirubin too must possess two different pyrrole rings, only one of which was capable of being oxidized to hematinic acid; the other ring in some way being incapable of giving methylethylmaleimide. Including a few new discoveries, such as the imide products, *Fischer* was essentially revealing in print his knowledge of the status of the bile pigments.

Casting a wide net into the structure of pyrrole pigments, between 1911 and 1913, *Fischer* pursued the question of the structure of *Nencki*'s and *Piloty*'s *Hämopyrrol* and other pyrroles cleaved reductively from hemin or hematin.

Meanwhile, not losing focus on the structure of bilirubin, in a manuscript received at the *Berichte* on May 20, 1912 with *Heinrich Röse* (*352*), *Fischer* published his first in a series of papers on its reductive cleavage – using a variation of *Nencki's* HI + PH$_4$I method, *i.e.* with a shorter reaction time. As in *Piloty's* similar study, which was received for publication only five days later on May 25, 1912 (*302*), *Fisher* had found no *Hämopyrrol* but managed to isolate and purify a crystalline product following extensive use of extraction, precipitation, and then recrystallization: **I** from CHCl$_3$-petroleum ether, **II** from CH$_3$OH-H$_2$O, mp 187°C. Analysis after drying to constant weight yielded molecular weights: **I**, 359 and **II**, 301 and neutralization equivalents: **I**, 307 and **II**, 311. The formula C$_{17}$H$_{24}$N$_2$O$_3$ (FW 304) seemed to fit well to these data and to the combustion analyses (*352*):

C$_{17}$H$_{24}$N$_2$O$_3$ (304.21)	Ber.	C 67.05	H 7.95	N 9.21
	Gef.	66.64, 66.78	8.38, 8.50	9.23, 9.30

Fischer suggested the name *Bilirubinsäure* (bilirubinic acid) for the substance, a mono-basic acid, that resisted further reductive cleavage but fell apart into methylethylmaleimide and hematinic acid upon oxidation, either by chromic acid, or better, by PbO$_2$ in aq. H$_2$SO$_4$. The two isolated imides were easily identified, and their structures had been proved earlier by synthesis by *Küster*. It remained only to determine how the two pyrrole-derived oxidative cleavage products might have been assembled in bilirubinic acid. (And eventually to resolve differences that arose with *Piloty*, for apparently neither *Piloty* nor *Fischer* knew of each other's essentially simultaneous work on the reductive cleavage of bilirubin – until their work appeared in print.)

Both *Fischer* and *Piloty* came up with the same formula, C$_{17}$H$_{26}$N$_2$O$_3$, for their bilirubinic acid and bilinic acid, respectively, and observed the same sharp melting point, reactivity, and solubility properties. *Piloty's* oxidation of his bilinic acid also yielded methylethylmaleimide and hematinic acid. However, *Piloty* had also found isophonopyrrole carboxylic acid in the *mother liquor* after retrieving bilinic acid from his HI-PH$_4$I reductive cleavage of bilirubin (*302*); whereas, *Fischer* did not report it. In *Piloty's* research group, efforts were of course being made to clarify the structure of isophonopyrrole carboxylic acid, as well as phonopyrrole carboxylic acid and *Hämopyrrol, etc.* And in this quest *Fischer's* group independently worked on the same goals. *Fischer* (*353*) and *Piloty* (*302*) were both quick to propose (different) structures for their C$_{17}$H$_{26}$N$_2$O$_3$ bilirubin cleavage products, from which one might envision oxidative cleavage to yield the cited imides:

[*Fischer's Bilirubinsäure*
May 20, 1912 (*352*)]

[*Piloty's Bilinsäure*
May 25, 1912 (*302*)]

4.2 Hans Fischer and the Early Structures of Bilirubin

In his publication received at the journal on July 20, 1912 (*302*), in which he showed that bilinic acid was oxidized (by mild $KMnO_4$) to *Dehydro-bilinsäure* (dehydrobilinic acid), *Piloty* acknowledged the *Fischer* and *Röse* study on bilirubinic acid that had been received at the journal on May 20, 1912 (*352*). Whether the two-month difference between May 20 and July 20 reflects a very rapid publication of the *Berichte*, or whether *Piloty* learned of *Fischer*'s work by other means (they were after all both in Munich) is not clear. What is clear is that *Piloty*'s July 20, 1912 paper (*302*) contained essentially the last of his *original* work on the structure of bilirubin.

Fischer and *Röse*, in their follow-up article received at the journal some three months later, on October 20, 1912, acknowledged *Piloty*'s first publication (*301*) on bilinic acid and structure – and wrote that, although it would oxidize to methylethylmaleimide and hematinic acid, they considered the structure to be improbable because the primary hydroxyl should not have survived the reaction conditions (*353*):

> Für die Bilirubinsäure haben wir Formel **I** aufgestellt, während Piloty und Thannhauser Formel **II** diskutieren.

I.   II.

> ... Uns erscheint die Pilotysche Formel nicht wahrscheinlich, da es uns ausgeschlossen erscheint, daß eine alkoholische Hydroxylgruppe der Seitenkette gegen Eisessigjodwasserstoff beständig ist (vgl. Ber. d. Deutsch. chem. Ges., Bd. 45, S. 1580) ...   36

They also chided *Piloty* for having missed the significance of their work (*354*) in which they deduced that two different rings are present in the hemibilirubin produced by reduction of bilirubin with Na(Hg). Polemics aside, a very important part of this *Fischer* work is that $(HI + PH_4I)$ reductive cleavage of bilirubin and hemibilirubin gave the same cleavage products (*353*):

> ... bei der Reduktion geben Hemibilirubin, Körper II und Bilirubin die gleichen Spaltprodukte.   37

Since the reductive cleavage products of bilirubin had been shown to be bilirubinic acid and (more recently) an isomeric acid of phonopyrrole carboxylic acid, hemibilirubin, produced from bilirubin by Na(Hg) reduction would yield the same products. And since bilirubinic acid afforded methylethylmaleimide as well as hematinic acid upon chromic acid oxidation, but bilirubin produced only the latter, one might conclude that the pyrrole ring of bilirubin had become altered by Na(Hg) reduction in such a way that it resembled the pyrrole ring of bilirubinic acid (and hemibilirubin). The failure of bilirubin to yield methylethylmaleimide upon chromic acid oxidation, in contrast to the copious quantities obtained *after* bilirubin had been reduced, remained puzzling to *Fischer* into 1913 (*355*):

4 A Modern Proof of Bilirubin Structure Emerges

> Sollte wirklich «Imid» vorliegen, so bleibt dennoch die Tatsache zu erklären, warum man im Gegensatz zur direkten Oxydation des Bilirubins nach erfolgter Reduktion reichliche Mengen von Methyl-äthyl-maleinimid erhält. 38

Fischer thought that bilirubin consisted of two and probably three rings. If one accepts, as *Fischer* did, that bilirubin has 32 carbons (he believed that hemibilirubin has 33!), he should also have believed that a fourth ring was present too.

Fischer submitted two publications (*353, 354*) within a few months after *Piloty* had submitted his article (received at the journal on July 20, 1912) describing the ($KMnO_4$) oxidation of bilinic acid to yellow dehydrobilinic acid (*302*). Although *Fischer* confirmed that his bilirubinic acid was the same compound as *Piloty*'s bilinic acid, he refuted the *Piloty* structure (*353, 354*) in his October 20, and November 8, 1912 papers (*353, 354*). It was not until his article in the *Berichte* (*356*), received at the journal on February 3, 1913, that he acknowledged *Piloty*'s dehydrobilinic acid. *Fischer* wrote that it was not improbable that the compound was identical to the pigment (which he named) xanthopyrrole carboxylic acid, obtained by treating bilirubinic acid with hot $NaOCH_3$, because it could be reduced back to bilirubinic acid by HI in glacial acetic acid (*356*):

> Auch Bilirubinsäure wurde dem gleichen Verfahren unterworfen, hier aber entstand die tetrasubstituierte Säure nicht, dagegen in relativ guter Ausbeute ein neuer Körper, für den wir den Namen Xanthopyrrol-carbonsäure vorschlagen. Die neue Säure steht in sehr naher Beziehung zu ihrer Muttersubstanz, denn durch Reduktion mit Eisessig-Jodwasserstoff wird die Bilirubinsäure in recht guter Ausbeute zurückgewonnen. Ob der von uns isolierte Körper mit der von Piloty und Thannhauser . . . beschriebenen *Dehydrobilinsäure* identisch ist, erscheint uns nicht unwahrscheinlich, indessen war uns die Herstellung eines Vergleichspräparates nach den Angaben dieser Autoren infolge Mangels an Material nicht möglich. 39

The new compound, obtained as small yellow prisms, analyzed (combustion *micro* analysis) for the same formula as *Piloty*'s dehydrobilinic acid and was also obtained from $NaOCH_3$ treatment of both bilirubin and hemibilirubin (*357*):

> Er bildet ein schwer lösliches Natriumsalz und stimmt in allen seinen Eigenschaften mit der von Piloty und Thannhauser . . . durch Oxydation der Bilirubinsäure erhaltenen Substanz überein, nur daß unser Körper einen ziemlich scharfen Schmelzpunkt besitzt. Die Analysen konnten in Anbetracht der geringen Menge Substanz nur mikroanalytisch ausgeführt werden. . . .

$C_{17}H_{22}N_2O_3$.	Ber.	C	67.51,	H	7.34,	N	9.27	
	Gef.	»	67.01, 68.05,	»	6.64, 7.0,	»	9.61	40

Fischer, whether intentionally or not, made a distracting gaff in his paper (*356*) by naming the oxidized bilirubinic acid as xanthopyrrole carboxylic acid at first in the text, but then later in the text and throughout the Experimental section, referring to it as *Xantho-bilirubinsäure* (xanthobilirubinic acid). About that he would hear later from *Piloty*. *Fischer* also suggested a structure for the oxidized product to account for its constitution (*356*):

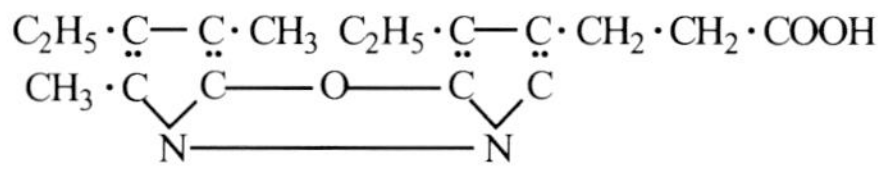

4.2 Hans Fischer and the Early Structures of Bilirubin

$$C_2H_5 \cdot \overset{\cdot}{C} - \overset{\cdot}{C} \cdot CH_3 \quad C_2H_5 \cdot \overset{\cdot}{C} - \overset{\cdot}{C} \cdot CH_2 \cdot CH_2 \cdot COOH$$

[*Fischer*'s *Xantho-bilirubinsäure* (*356*)]

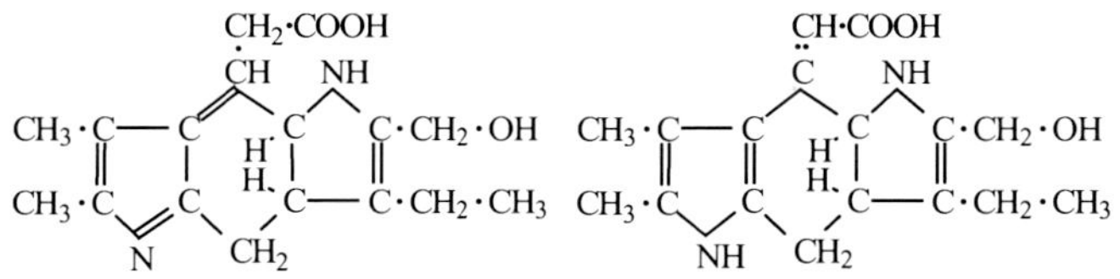

[*Piloty*'s *Dehydro-bilinsäure* (*302*)]

Then the feathers began to fly.

4.2.2 The Piloty-Fischer Polemics

Fischer's assigned names drew an incredibly rapid and pointed response (*327*) from *Piloty*, received on February 27, 1913 in the *Berichte* (*327*):

117. O. Piloty: Bemerkungen zu der Mitteilung der HHrn. Hans Fischer und Heinrich Röse im 3. Heft dieser Berichte[1]).
(Eingegangen am 27. Februar 1913.)

Ich habe in einer gemeinsam mit S. J. Thannhauser veröffentlichten Mitteilung[2]) eine Säure von der Zusammensetzung $C_{17}H_{26}N_2O_3$ beschrieben, welche wir durch Reduktion des Bilirubins mittels Jodwasserstoff erhalten haben und mit dem Namen Bilinsäure belegten. Fast gleichzeitig veröffentlichten H. Fischer und H. Röse[3]) ebenfalls eine Arbeit über diese Säure und nannten sie Bilirubinsäure. Um einer Verwirrung in der Literatur vorzubeugen, möchte ich betonen, daß Bilinsäure und Bilirubinsäure ein und dieselbe Substanz bezeichnen. Ferner habe ich in oben erwähnter Mitteilung (S. 206) zuerst die von mir Isophonopyrrol-carbonsäure genannte Substanz beschrieben und genau charakterisiert. Später fanden H. Fischer und E. Bartholomaeus[4]) diese zweite Substanz ebenfalls und nannten sie »isomere Phonopyrrol-carbonsäure«, an welcher Bezeichnung die Autoren bis jetzt festhalten.

Ferner habe ich in oben zitierter Mitteilung (S. 197) den Nachweis geführt, daß im Bilirubin drei Kerne enthalten sind, von denen zwei einander ähnliche durch die Isophonopyrrol-carbonsäure zusammengehalten werden, da diese Säure durch schmelzendes Alkali sowohl in Form von Dimethylpyrrol als auch als Trimethylpyrrol abgespalten wird. Dies hätte Erwähnung finden müssen, als die HHrn. H. Fischer und H. Röse[5]) schrieben, daß sie »zum erstenmal« den Beweis geliefert haben, daß im Hemibilirubin und Bilirubin ein dritter Pyrrolkern enthalten sein muß.

---

[1]) B. **46**, 439 [1913]. [*356*]
[2]) A. **390**, 191 u. 202 [1912]. [*301*]
[3]) B. **45**, 1579 [1912]. [*301*]
[4]) B. **45**, 1979 [1912]. [*366*]
[5]) B. **46**, 439 [1913]. [*356*]

272 4 A Modern Proof of Bilirubin Structure Emerges

Ferner habe ich[1]) in einer ebenfalls gemeinsam mit Hrn. S. J. Thannhauser veröffentlichten Mitteilung gezeigt, daß aus unserer Bilinsäure durch Wegnahme zweier Wasserstoffatome eine intensiv gelb gefärbte Säure entsteht, die wir Dehydro-bilinsäure nannten. Wir haben auf die entscheidende Rolle hingewiesen, welche dieser Substanz für die Beurteilung der Farbstoffnatur des Bilirubins zukommt, die charakteristischen Unterschiede zwischen der Konstitution dieses Farbstoffs und des Blutfarbstoffs hervorgehoben; kurz die Grundzüge geschaffen für die Beurteilung der Konstitution des Bilirubins.

Nun haben H. Fischer und H. Röse[2]) aus Bilirubin ebenfalls eine gelb gefärbte Säure erhalten, die sie selbst für identisch halten mit unserer Dehydrobilinsäure. Die Autoren haben eine Identifizierung nach ihren Angaben aus Mangel an Material unterlassen müssen. Diese Unterlassung kann aber unmöglich als ein berechtigter Grund anerkannt werden, dieser Substanz nicht nur einen, sondern sogar zwei andere Namen zu geben, nämlich Xanthopyrrol-carbonsäure und Xantho-bilirubinsäure. Ich habe den Namen Xanthopyrrol-carbonsäure bereits einer anderen Säure[3]) beigelegt und halte deshalb eine weitere Verwendung dieses Namens für ausgeschlossen. Aber abgesehen davon, muß ich das publizistische Verfahren der Autoren, einen bereits beschriebenen und benannten Körper mit einem neuen Namen ohne sachlichen Grund zu belegen, nicht für angemessen halten. Denn erstens kann dieses Verfahren nur Verwirrung in der Literatur hervorrufen, und zweitens erweckt es bei dem naturgemäß über das einschlägige Literaturmaterial nur oberflächlich orientierten Leser den falschen Eindruck, als ob der neue Pate auch der Entdecker des bekannten Körpers wäre. Da ich nicht annehmen kann, daß diese beiden Eventualitäten in der Absicht der Autoren gelegen haben, so zweifle ich nicht, daß dieselben die beiden Namen Xanthopyrrolcarbonsäure und Xanthobilirubinsäure einfach wieder fallen lassen und den von dem Entdecker gewählten Namen Dehydrobilinsäure akzeptieren werden. Es wäre zweckmäßig, wenn sich Hr. H. Fischer bei dieser Gelegenheit auch entschlösse, sich mit den Namen Bilinsäure und Isophonopyrrol-carbonsäure zu befreunden.

---

[1]) B. **45**, 2393 [1912]. [*302*]   [2]) B. **46**, 439 [1913]. [*356*]
[3]) A. **388**, 315 [1912]. [*314*]   41

Unsurprisingly, this *Piloty* paper drew a fast and pointed response from *Fischer* (*356*), but before his April 28, 1913 response, he had already indicated a rethinking of the meaning of the ease of interconverting his yellow xanthobilirubinic carboxylic acid and colorless bilirubinic acid by redox reactions. He found an analogy in concurrent work relating mesoporphyrin and porphyrinogen, and in the relationship between the reddish dipyrrylmethene and its reduced, colorless dipyrrylmethane. Then, he slipped in revised structures for bilirubinic acid and xanthobilirubinic acid in a March 9, 1913 paper addressing the construction of the porphyrin (*358*):

$$COOH \cdot CH_2 \cdot CH_2 \cdot C \underset{NH}{\overset{}{\diagdown}} \quad C_2H_5 \cdot C \underset{NH}{\overset{}{\diagdown}}$$

Only a month or so later, upon reading *Piloty*'s comparatively politely chiding article (*327*), *Fischer*, apparently possessing no aversion toward polemics, replied

4.2 Hans Fischer and the Early Structures of Bilirubin

vigorously to *Piloty* in a paper (*359*) received at the *Berichte* on April 28, 1913. *Fischer* claimed priority of his name *Bilirubinsäure* over *Piloty*'s *Bilinsäure* by virtue of his manuscript having arrived first at the publisher on May 20, 1912 *vs.* *Piloty*'s arrival on May 25, 1912. Thus, based on the publication dates, he did not acknowledge a priority for *Piloty*'s *Bilinsäure*, nor did he acknowledge a priority to *Piloty* for the name *Isophonopyrrol-carbonsäure* but claimed priority for his *isomeric Phonopyrrol-carbonsäure* name. He further dismissed the name *Bilinsäure* as having already appeared in a different connection in a late 1800s issue of *Beilstein* (II, p. 2008) (*359*):

> Nun zur »Dehydro-bilinsäure«, die Piloty und Thannhauser[5]) durch Oxydation mit Permanganat aus der Bilirubinsäure erhalten haben.
>
> Über die Zusammensetzung des Körpers scheint Piloty nicht mehr ganz sicher zu sein, denn Ber. **46**, 1001 [1913] sagt er, daß diese Säure durch Wegnahme zweier Wasserstoffatome entsteht; in der Originalpublikation[5]) dagegen verschwinden vier Wasserstoffatome, in Wirklichkeit aber müßten sechs Wasserstoffatome entzogen werden, um das von Piloty[5]) mit Formeln wiedergegebene System konjugierter Doppelbindungen zu erzeugen, das den gebräuchlichen Anschauugen über die Konstitution gefärbter Verdingungen entspricht. Sonderlich wichtig ist dieser Punkt nicht, weil die Pilotysche Bilirubinsäure-Formel ja an sich bereits widerlegt ist[6]). Wichtiger ist, ob die Xanthobilirubinsäure[7]) identisch ist mit der »Dehydro-bilinsäure«. Unsere Xanthobilirubinsäure[8]) entsteht durch Erhitzen mit Natriummethylat auf 220-230°, und nach

---

[5]) B. **45**, 2393 [1912]. [*302*]   [6]) B. **45**, 3274 [1912].] [*354*]   [7]) B. **46**, 439 [1913]. [*356*]
[8]) Der zweite, übrigens nur ein einziges Mal gebrauchte Name »Xanthopyrrolcarbonsäures« beruht natürlich auf einem Druck- bezw. Schreibfehler (vergl. B. **46**, 442 [1913]).

Entstehungsart hielt ich ursprünglich, und jeder andere Chemiker hätte es auch getan, eine Identität mit dem von Piloty erhaltenen Oxydationsprodukt für ausgeschlossen, trotz der außerordentlichen Ähnlichkeit. Daß die Xanthobilirubinsäure wirklich in sehr naher Beziehung zur Bilirubinsäure steht, haben wir bewiesen[1]) durch Rückverwandlung in die Bilirubinsäure. Piloty hat diesen Nachweis für seine Säure nicht erbracht. Es erschien daher keineswegs ausgeschlossen, daß verschiedene Körper vorliegen, zumal ja gerade die Pyrrol-Chemie so außerordentlich reich ist an einander sehr ähnlichen, aber doch nicht identischen Körpern. Übrigens besteht auch, wie wir ausdrücklich hervorgehoben haben[1]), zwischen den beiden Körpern ein Unterschied, den Piloty übergeht. Unser Präparat schmilzt scharf bei 274°, während Piloty und Thannhauser[2]) angeben, daß ihre Substanz über 260° sich zersetzt, ohne zu schmelzen.

Seitdem wir allerdings durch Erhitzen des Porphyrinogens mit Natriummethylat Mesoporphyrin[3]) erhalten haben, womit die oxydierende Wirkung dieses Reagenses zweifellos feststeht, bin ich von der Identität der beiden Körper beinahe überzeugt. Stellt Piloty daher an seinem Präparat nachträglich noch fest, daß dieses in Übereinstimmung mit unserem bei 274° schmilzt, so stehe ich nicht an, den Namen Xanthobilirubinsäure zurückzuziehen und konsequenterweise durch Dehydro-bilinsäure zu ersetzen.

---

[1]) B. **46**, 439 [1913].   [2]) B. **45**, 2393 [1912].   [3]) H. **84**, 262.   42

The challenge to *Piloty* was evidently not accepted, as he absented himself from further active studies of bilirubin and concentrated on the structure of hemin and its

274 4 A Modern Proof of Bilirubin Structure Emerges

cleavage products. All the while, the rumblings which led to World War I, to the death of his son in 1914, and to his own death in 1915 continued unabated. Yet, continuing undeterred, *Fischer* excoriated *Piloty* (*359*) over who had priority over the discovery of *Kryptopyrrol* (kryptopyrrole); over *Piloty*'s failure to acknowledge *Küster*'s discovery of hematinic acid as the first elucidation of a ring component of bilirubin; and over *Fischer*'s discovery of the presence of a second ring in hemibilirubin following the isolation of methylethylmaleimide from oxidative cleavage. Discovery of a third pyrrole ring (as isophonopyrrole carboxylic acid) was due to *Piloty*, but *Fischer* claimed to have provided the first *proof* for a third pyrrole ring in bilirubin and hemibilirubin.

4.2.3 Fischer's Bilirubinsäure, Xanthobilirubinsäure, and Bilirubin Structures ca. 1914

With no new work on bilirubin forthcoming from *Piloty*, or rebuttals, the names given by *Fischer*, *Bilirubinsäure* (bilirubinic acid) and *Xanthobilirubinsäure* (xanthobilirubinic acid), became standard usage and remain so today. In early March 1913, *Fischer* had proposed new and different structures for them (*359*), which were left to him to prove. This he did with *Heinrich Röse* in a long paper received on New Year's eve 1913 at *Hoppe-Seyler's Zeitschrift für physiologische Chemie* and published in 1914 (*357*). Thus in 1914, *Fischer* reminded us of his earlier work that showed kryptopyrrole (**I**)[11] and isophonopyrrole carboxylic acid arose from vigorous reductive cleavage of bilirubinic acid and thus constitute its two component pyrrole rings. Although the structure of isophonopyrrole carboxylic acid had not yet been resolved, it was thought to be one of two isomers, **II** and **III** (*357*):

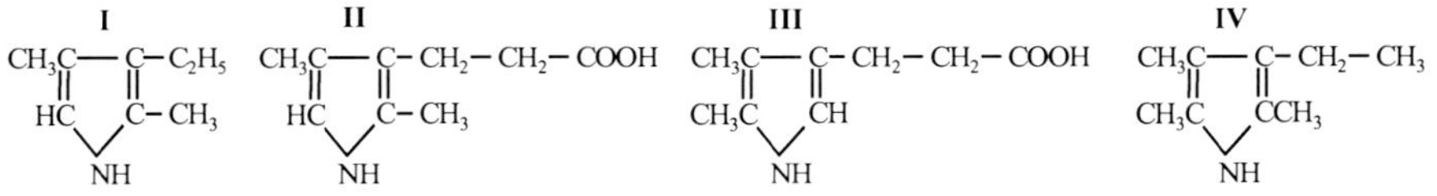

Concluding in favor of **II**, and armed in 1914 with information on the structure and history of *Phyllopyrrol* (phyllopyrrole) (**IV**), *Fischer* was able to replace his earlier structure of bilirubinic acid (**IX**) with structure **VIII**. This then led to his writing the structure of xanthobilirubinic acid as **XI** (*357*):

---

[11] For convenience of back-reference, the Roman numerals assigned in this subsection to the compounds here are the same as those in the *Fischer* and *Röse* publication (*357*), and the original style of structure representation has been preserved as much as possible.

4.2 Hans Fischer and the Early Structures of Bilirubin

VIII

CH$_3$C — C- C$_2$H$_5$ CH$_3$C — C- CH$_2$– CH$_2$– COOH
HOC — C — CH$_2$ — C   C- CH$_3$
      NH   NH

IX

C$_2$H$_5$C — C- CH$_3$ CH$_3$C — C- CH$_2$– CH$_2$– COOH
CH$_3$C   C — O — C   C- CH$_3$
      NH   NH

XI

CH$_3$C = C- C$_2$H$_5$ CH$_3$C — C- CH$_2$– CH$_2$– COOH
O=C   C = CH — C   C- CH$_3$
      NH   NH

It is interesting to note that the *Fischer* structure of 1913–1914 for xanthobilirubinic acid (**XI**) has remained correct in all details to the present day, except for its stereochemistry. In 1914, *Fischer* also knew, or began to suspect, that bilirubin was a tetrapyrrole compound, formed by stitching together two bilirubinic acid or xanthobilirubinic acid units in some fashion, and with the β-positions of the four pyrroles substituted thus by a total of four methyl groups, two vinyl groups and two propionic acids. All this led him to propose two possible tetrapyrrole structures for bilirubin: one doubtless influenced by the porphyrin structure due to *Willstätter* (*343*, *344*), based on a tetrapyrrylethylene; the other, a sensitive-looking macrocycle, inspired perhaps by *Küster*'s 1912 macrocyclic structure of hemin (*360*). Based on the then-accepted molecular formula, which, *mutatis mutandis*, is exactly the molecular formula ($C_{33}H_{36}N_4O_6$) accepted today, both structures were shown graphically as (*357*):

Das Bilirubin selbst würde dann durch folgende Strukturformel dargestellt:

Bilirubin $C_{33}H_{36}N_4O_3$ [N.B. O_3 appears to be an error in print]

CH$_2$= HC- C — C- CH$_3$ CH$_3$C — C- CH= CH$_2$
      I   III
      C   C   C   COH
      O NH   C = C   NH
         NH   NH
      C II C   C IV C- CH$_3$
COOH— CH$_2$– CH$_2$– C — C- CH$_3$ CH$_3$C — C- CH$_2$– CH– COOH

Man könnte für das Bilirubin auch folgende Formel in Betracht ziehen,

CH$_2$=CH-C = C- CH$_3$ CH$_3$C — C- CH$_2$- CH$_2$– COOH   CH$_3$C — C- CH= CH$_2$ CH$_3$C — C- CH$_2$– CH$_2$– COOH
— H$_2$C- C   C = C — C   C — O — C   C — CH$_2$ — C   C- O —
      N   H   NH   NH   NH

die gleichfalls die meisten experimentellen Resultate erklärt, aber für das Hemibilirubin allerdings nur das Entstehen einer Form vorsieht.   43

Although the 1914-era *Fischer* structures fall short of the actual structure of the pigment, they reveal an insight into *Fischer*'s thinking and herald a stubborn clinging to the tetrapyrrylethylene motif for nearly the next decade and a half. Given that xanthobilirubinic acid and bilirubin are both yellow-colored, that the former is a C_{17}

276 4 A Modern Proof of Bilirubin Structure Emerges

and the latter a C_{33} compound, it might have been only a simple leap of faith to conjoin two xanthobilirubinic acid units directly to the same "external" pyrrole α-carbon, thereby yielding a C_{33} product, which when ethyls are replaced with vinyl groups would become an isomer (XIIIα) of natural bilirubin:

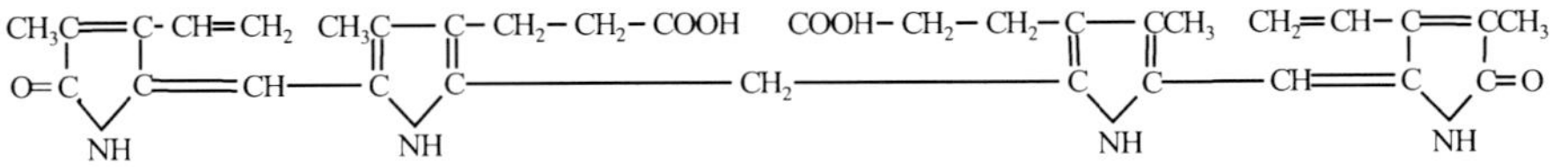

However, *Fischer* was apparently not one to propose structures divined by great leaps of imagination or intuition. Rather, evidence shows that he was very careful and very logical in his scientific reasoning, and he was very much inclined toward proceeding stepwise toward his goals, with a firm footing at each step and a passionate belief in proof by synthesis. Unlike his predecessors who labored in the bile pigment field of the late 1800s and early 1900s, *Fischer* had taken to heart an apparently absolute conviction of the importance of knowledge and understanding of organic chemistry based on chemical structures, proven incontrovertibly by logical design and synthesis. For him, it was the only formula for ultimate success, and in this aspect, he was doubtless the first scientist to bring modern thinking and order to the entire field of pyrroles and pyrrole-based natural products. In order to understand how *Fischer* arrived at the correct constitutional structure of bilirubin, based on its fragmentation products bilirubinic acid and xanthobilirubinic acid, one must step back a few years to the controversies surrounding the structure of hemin and its degradation products.

4.2.4 The Alkylated Monopyrrole Components of Bilirubinsäure and Xanthobilirubinsäure

By 1912, the chemical literature contained a confusing multitude of pyrrole names and uncertain structures coming from fragmentation of bilirubin, hemin, and chlorophyll: *Hämopyrrol, Iso-hämopyrrol, Kryptopyrrol, Phonopyrrol, Phyllopyrrol*; as well as various pyrrole carboxylic acids: *Hämopyrrolcarbonsäure, Phonopyrrolcarbonsäure, Iso-phonopyrrolcarbonsäure, Xanthopyrrolcarbonsäure, Kryptopyrrolcarbonsäure, Phyllopyrrolcarbonsäure*.[12] Taken collectively they constituted the molecular pieces of the puzzle that might allow the structures of the pigments of bile and blood to be defined. In fact, they allowed *Fischer* to assign correctly the structures of bilirubinic acid and xanthobilirubinic acid. And they were instrumental in *Piloty*'s formulation of his (incorrect) structure of hemin in 1910 (*313*) and 1914 (*328*), and hematoporphyrin in 1912 (*314*), in *Willstätter*'s

---

[12] All translate easily into their English-language equivalents. Where the original German names are retained in this chapter, the constitution and homogeneity were as yet unproven.

4.2 Hans Fischer and the Early Structures of Bilirubin

(incorrect) formulation of the hemin structure in 1913 (*343, 344*), in *Fischer*'s (incorrect) structure for hemin in 1914 (*357*), and in *Küster*'s brilliant leap forward in 1912 to formulating a novel structure for hemin that more than 15 years later proved to be correct in nearly every detail (*360*).

The basis for *Fischer*'s understanding the structures of the title compounds goes back a few years, to his 1911 publications (*349–351, 361*), where he reported his efforts to repeat the work of *Nencki* and *Piloty*. At that time he asserted a negative finding: that he could not discover any *Hämopyrrol* following treatment of bilirubin by HI + PH_4I in glacial acetic acid (*349*), the same reaction devised by *Nencki* and *Zaleski* in 1901 (*285*) for cleaving off liquid *Hämopyrrol* from hematin and hematoporphyrin and also used by *Piloty* and *Quitmann* in 1909 for the same purpose (*308*). By applying *Küster*'s chromic acid oxidation (*242–244*) to bilirubin, *Fischer*, like *Küster* before him (*244, 325, 326*), found no methylethylmaleimide; rather he found only hematinic acid imide. In clear contrast, the hemibilirubin (produced from bilirubin by Na(Hg) reduction) gave *Fischer* both imides (*351*), whose structures *Küster* had previously defined by chemical synthesis (*251, 252, 255*). The methylethylmaleimide isolated was identical to the compound obtained by heating (and decarboxylation) of hematinic acid imide (*251, 252*) and was later isolated laboriously from the *Gemisch* following chromic acid oxidation of the *Hämopyrrol* from hematoporphyrin and hematin (*296*).

Both *Küster* (*217, 218*) and now *Fischer* (*361*) began to suspect that the liquid *Hämopyrrol* whose sole source had been natural products, was a mixture. Yet at that time it was clearly not an easy task to separate liquids by distillation, especially when the quantities available were small. For liquid pyrroles, conversion to their (solid) picrates, followed by recrystallization, had become a favorite purification strategy, as it had also become for aromatic compounds. *Piloty* and *Fischer* adapted it to *Hämopyrrol*. In that era, the only sure criterion of sample purity was a sharp melting point, which also served as a physical characteristic of any compound, along with its empirical or molecular composition formula. *Nencki* (*285*) had reported in 1901 that the picrate melted at 108°C; subsequently, *Piloty* (*308*) reported in 1909 that the picrate of refined *Hämopyrrol* (mp 39°C) melted at 108.5°C.

Prior to *Piloty*, in 1907 (*292, 296, 297*), *Küster* suggested that *Hämopyrrol* had one β-ethyl group and two methyl groups, one at a pyrrole α-carbon, the other at a β-carbon. Since both of the two possible structures would still produce methylethylmaleimide upon chromic acid oxidation, nothing more specific about the structure could be said at that time. *Piloty* first assumed *Hämopyrrol* ($C_8H_{13}N$) to be a β-methyl-β'-propylpyrrole (*303*). Then shortly thereafter, he considered *Hämopyrrol* to be either α,β-dimethyl-β'-ethylpyrrole or α,β'-dimethyl-β-ethylpyrrole (*306*) – as *Küster* had suggested a bit less specifically two years earlier (*296*). In 1910, *Piloty* isolated *Phonopyrrol* ($C_8H_{13}N$) and believed that it was isomeric with *Hämopyrrol* but not identical to it. Using a new oxidizing agent, *Piloty* noted that *Hämopyrrol* reacted with HNO_2 to give a mono-oxime (mp 101°C) of methylethylmaleimide (*303, 304, 306–308*), but which of the two possible oximes were formed could not easily be determined. Yet, by 1910, when

278   4   A Modern Proof of Bilirubin Structure Emerges

Fischer had initiated his studies, *Piloty* had begun to denote the *Hämopyrrol* structure as α,β-dimethyl-β′-ethylpyrrole, albeit without proof (*313*). The work apparently caught *Knorr*'s eye (*323*):

> In seiner letzten Arbeit[1]) »Über die Konstitution der gefärbten Komponente des Blutfarbstoffs« gelangt Piloty für das Hämopyrrol und Phonopyrrol zu folgenden Konstitutionsformeln:

I.

$$CH_2 \cdot CH_2 \cdot C\text{---}C \cdot CH_3$$
$$CH_2 \cdot C \diagdown C \cdot H$$
$$NH$$

II.

$$CH_3 \cdot C\text{---}C \cdot CH_2 \cdot CH_3$$
$$CH_3 \cdot C \diagdown C \cdot H$$
$$NH$$

Hämopyrrol ?   Phonopyrrol
α,β′-Dimethyl-*β*-äthylpyrrol   *α,β*-Dimethyl-*β*′-äthylpyrrol
(2,4-Dimethyl-3-äthylpyrrol)   (2,3-Dimethyl-4-äthylpyrrol)

---

[1]) A. **377**, 314 [1910].   44

With *Hess* (*323*), he synthesized the 2,4-dimethyl-3-ethylpyrrole (picrate mp 131–152°C, imide oxime mp 101°C) and concluded that it was not identical to the *Hämopyrrol* (picrate mp 108.5°C, imide-oxime, mp 206-207°C (*303*), 201°C (203d)) of *Piloty* and *Quitmann* (*313*):

> Wir beobachteten unter dem Mikroskop tafelförmige, manchmal konzentrisch gruppierte Krystalle. Das Salz schmilzt unter Dunkelfärbung ziemlich scharf bei 131–132° und erleidet wenig über dem Schmelzpunkt stürmische Zersetzung. . . .
> Wir folgten genau den Angaben von Piloty [1]) und erhielten zunächst das *Monoxim des Methyl-äthyl-maleinsäureimids*, das, aus Wasser umkrystallisiert, den von Piloty und Quitmann angegebenen Schmp. 101° zeigt.

---

[1]) Piloty, A. **366**, 254; Piloty und Quitmann, B. **42**, 4699 [1909].   45

With the structure of *Hämopyrrol* still in doubt, by 1911 *Fischer* had begun his studies (*349*). Like *Piloty*, he too isolated *Hämopyrrol* from hemin by *Nencki*'s reductive cleavage method (*285*), but he found its picrate to melt at 118°C, which was a cause for concern to a firm believer in reliable data. Unsatisfied with the paltry yield of *Hämopyrrol* he modified the *Nencki-Zaleski* HI + PH_4I reductive cleavage method to produce a nearly 21% yield of *Hämopyrrol*, boiling point 96° (12 mm), which gave a picrate with mp 120–122°C (*361*). From hematoporphyrin, using *Piloty*'s Sn + conc. HCl reduction (*303*), in 1911 *Fischer* obtained *Hämopyrrol* that also yielded a picrate of mp 120–122°C and exhibited no mixture melting point depression with that of his earlier isolation (*361*). The fact that *Fischer*'s *Hämopyrrol* picrates still melted 12–14°C higher than those from *Nencki* and from *Piloty*, and given his attention to detail, such a difference between his melting points and those in the literature could not have sat well with him. Added to this concern was the synthesis of 2,4-dimethyl-3-ethylpyrrole reported in 1911 by *Knorr* and *Hess* (*323*), who made its picrate and reported a mp 131–132°. Thus, one might understand *Fischer*'s concern. If the *Knorr* and *Hess* picrate was an accurate

4.2 Hans Fischer and the Early Structures of Bilirubin 279

characteristic of their synthetic pyrrole of known structure, and if the *Nencki-Zaleski* and *Piloty-Quitmann* picrate, which melted some 23°C lower, accurately characterized *Hämopyrrol*, *Fisher*'s *Hämopyrrol* could be neither because its picrate mp rested nearly exactly between the other two.

All this led *Fischer* to repeat the *Knorr-Hess* synthesis of 2,4-dimethyl-3-ethyl-pyrrole from the well characterized 2,4-dimethyl-3-acetylpyrrole, prepared in a synthetically logical, reliable pyrrole synthesis (*323*). *Knorr* and *Hess* had prepared the hydrazone (from reaction with hydrazine) and then reduced it by $NaOCH_2CH_3$ (*Wolff* reduction) at 150–160°C, expecting to distill off 2,4-dimethyl-3-ethypyrrole. In contrast, *Fischer*'s hydrazone synthesis went too far, in fact all the way to the ketazine (1 equiv. NH_2NH_2 + 2 equiv. ketone), which did not reduce to the expected pyrrole (*361*) but produced an oil that at first defied crystallization as a picrate. However, in a subsequent set of experiments (*362*) *Fischer* had sorted out how to make the hydrazone and avoid making the ketazine, and he reduced the hydrazone to the pyrrole that *Knorr* and *Hess* had obtained. Its picrate melted at 131–132°C, as reported by *Knorr* and *Hess*. Thus, *Fischer* had provided proof that his *Hämopyrrol* was not 2,4-dimethyl-3-ethylpyrrole and concluded that it was 2,3-dimethyl-4-ethylpyrrole (known now as hemopyrrole) because it, too, was oxidized to methylethylmaleimide (Scheme 4.2.1):

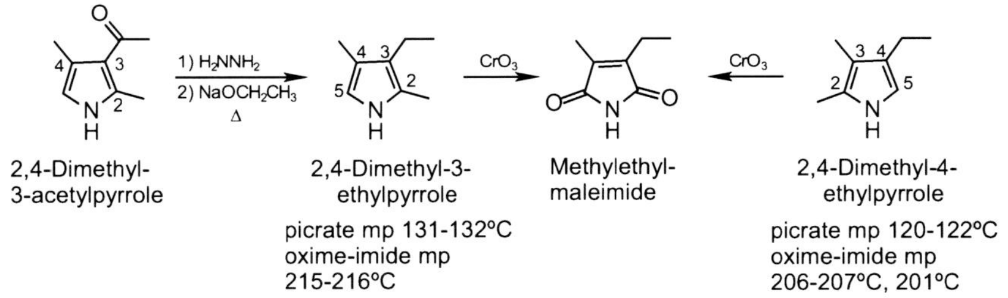

2,4-Dimethyl-3-acetylpyrrole

2,4-Dimethyl-3-ethylpyrrole
picrate mp 131-132°C
oxime-imide mp 215-216°C

Methylethyl-maleimide

2,4-Dimethyl-4-ethylpyrrole
picrate mp 120-122°C
oxime-imide mp 206-207°C, 201°C

Scheme 4.2.1

Fischer's ketazine-derived pyrrole was a mystery that he rapidly solved. It failed to give a solid picrate – an indication to *Fischer* that it might be alkylated at both pyrrole α-carbons. It also failed to react with *p*-dimethylaminobenzaldehyde (*Ehrlich* reaction) and with diazotized *p*-aminobenzenesulfonic acid – indications of a tetra-C-alkylated pyrrole. The mystery of its structure was solved by first repeating the ketazine reduction so as to produce the distilled pyrrole on a sufficiently larger scale. Then, by modifying and improving the picrate formation technique, *Fischer* produced a sharp-melting crystalline picrate, mp 89–90°C.

Thus, *Fischer* felt that he was dealing with a pure substance. Addressing its structure, he deduced that the entire ketazine residue had been split off during the treatment with hot sodium ethoxide and a further substitution (ethylation) at the erstwhile unsubstituted α-carbon of the (dimethyl) pyrrole had occurred. He proposed that the final pyrrole product would probably be 2,4-dimethyl-3,5-diethylpyrrole, achieved by ethylation of the first formed 2,4-dimethyl-3-ethylpyrrole (*362*):

Da jedoch der von uns erhaltene Farbstoff aller Wahrscheinlichkeit nach ein β-Azofarbstoff war, so vermuteten wir, daß ihm ein Pyrrol folgender Konstitution:

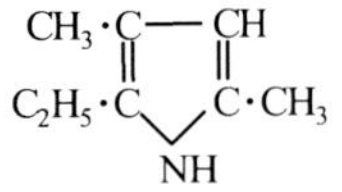

$$CH_3 \cdot \underset{\underset{C_2H_5 \cdot C \diagdown \underset{NH}{} \diagup C \cdot CH_3}{}}{C} — CH$$

zugrunde lag, weil es uns nicht ausgeschlossen schien, daß bei der hohen Temperatur der ganze Ketazinrest abgesprengt würde, nachdem vielleicht zuerst in α-Stellung Äthylierung eingetreten ist.   46

The structure was proved by synthesis, starting from 2,4-dimethyl-5-ethylpyrrole, which was obtained following a *Knorr*-type condensation of 3-oximino-2-pentanone with ethyl acetoacetate using Zn dust in hot glacial acetic acid. The resultant pyrrole 3-carbethoxy ester was saponified and decarboxylated, and the "free" position 3 was ethylated by reaction with $NaOCH_2CH_3$ in CH_3CH_2OH at 220°C to yield the target tetra-*C*-substituted pyrrole (*362*):

Demnach hat das von uns erhaltene Dimethyldiäthylpyrrol folgende Konstitution:

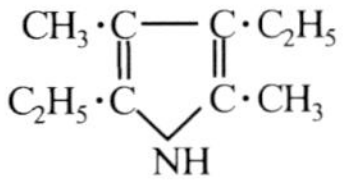

$$CH_3 \cdot C — C \cdot C_2H_5$$

und das von uns erhaltene trialkylierte Pyrrol (aus dem Ketazin):

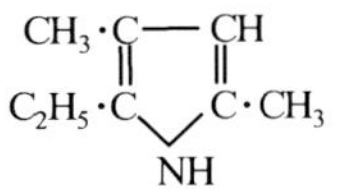

$$CH_3 \cdot C — CH$$
                                                                            47

It gave a crystalline picrate of mp 89–90°C and no mixture melting point depression with the pyrrole picrate following reduction of the ketazine (above) by heating with $NaOCH_2CH_3$ (*Wolff* reduction).

Just how the alkylation took place was a bit of a mystery to *Fischer*. He conceived of a sequence of two steps involving first N-alkylation followed by rearrangement to a pyrrole carbon and noted that alkylation of a vacant α-position preceded any β-alkylation (*362*):

Wie die Reaktion zustande kommt, können wir heute noch nicht sagen. Es ist möglich, dab erst Substitution am Stickstoff stattfindet und dann Wanderung an den Kohlenstoff, ein Verhalten, das ganz der von A. W. Hofmann und Martius entdekten Wanderung der Alkylgruppen in der Benzolreihe analog wäre. . . .

Interessant is auch hier das verschiedene Verhalten der α- und β-Stellung der Pyrrole. Während es leicht gelingt, in α-Stellung Methyl- und Äthylgruppen einzuführen, leistet die β-Stellung erheblichen Widerstand.

Wir haben 2,4-Dimethylpyrrol mit Natriumäthylat auf 220° erhitzt und in nahezu quantitativer Ausbeute 2,4-Dimethyl-5-äthylpyrrol gewonnen.

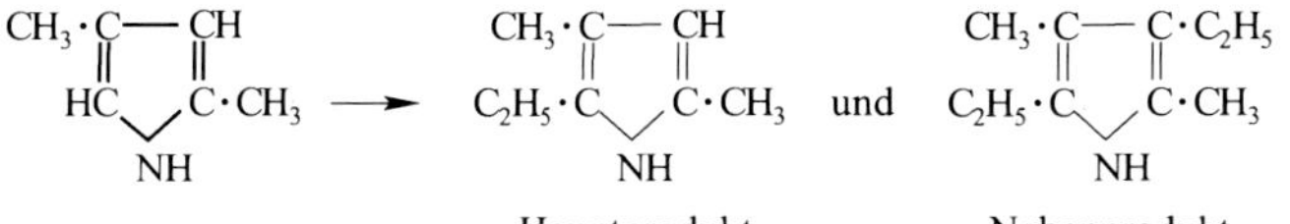

4.2 Hans Fischer and the Early Structures of Bilirubin

$$
\underset{\text{Hauptprodukt}}{
\begin{array}{c}
\text{CH}_3\cdot\text{C}\!\!-\!\!\text{CH} \\
\text{C}_2\text{H}_5\cdot\text{C}\diagdown\text{C}\cdot\text{CH}_3 \\
\text{NH}
\end{array}}
\quad\text{und}\quad
\underset{\text{Nebenprodukt}}{
\begin{array}{c}
\text{CH}_3\cdot\text{C}\!\!-\!\!\text{C}\cdot\text{C}_2\text{H}_5 \\
\text{C}_2\text{H}_5\cdot\text{C}\diagdown\text{C}\cdot\text{CH}_3 \\
\text{NH}
\end{array}}
$$

> Daneben isolierten wir in ganz Menge das uns schon bekannte 2,4-Dimethyl-3,5-diäthylpyrrol. **48**

This unusual pyrrole C-ethylation reaction was explored further by *Fischer* to produce *Phyllopyrrol* (phyllopyrrole, 2,3,5-trimethyl-3-ethylpyrrole) from the action of NaOCH$_3$ on his authentic hemopyrrole (**V**) (*363*):

> Ebenso gelang es uns, aus 2.4.5-Trimethylpyrrol (**II**) mit Äthylat und aus Hämopyrrol (**V**) mit Methylat glatt Phyllopyrrol (**III**) darzustellen. Über seine Synthese aus 2.4.5-Trimethyl-3-acetyl-pyrrol (**IV**) haben wir bereits an anderer Stelle . . . berichtet.

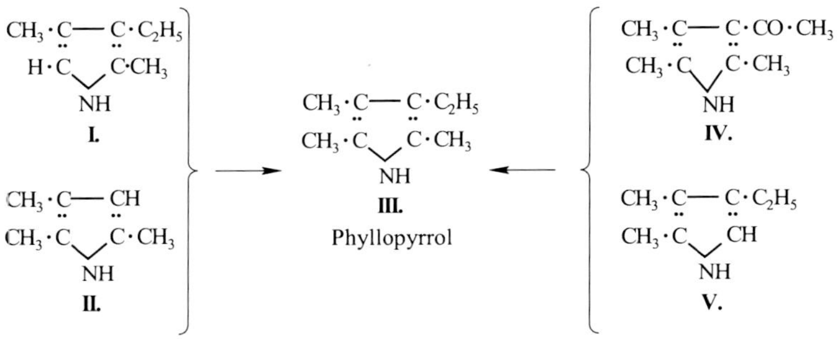

(Scheme showing pyrroles I, II → III (Phyllopyrrol) ← IV, V)

> Wir haben nun in analoger Weise 2.4-Dimethyl-3-äthyl-pyrrol (**I**) mit Natriumäthylat erhitzt und sind so zu dem 2.4-Dimethyl-3.5-diäthyl-pyrrol (**VI**) gelangt. das wir früher bereits sowohl aus 2.4-Dimethyl-3-acetyl-pyrrol und 2.4-Dimethyl-3-äthyl-pyrrol (**I**), wie auch aus 2.4-Dimethylpyrrol erhalten haben. . . . **49**

The experiment thus confirmed the structure of *Fischer*'s hemopyrrole as 2,3-dimethyl-3-ethylpyrrole. Phyllopyrrole, previously isolated by *Fischer* and *Bartholomäus* (*228a*) from the *Hämopyrrol Gemisch* obtained following HI + PH$_4$I treatment of hematin by the *Nencki-Zaleski* method (*285*), is a solid, mp 67–68°C, with a picrate mp 102–103°C. In yet a further confirmation of the phyllopyrrole structure, it was synthesized by *Fischer* from NaOCH$_2$CH$_3$-induced ethylation of 2,4,5-trimethylpyrrole (*363*). Earlier, in 1912, phyllopyrrole had been isolated and named and its correct structure indicated (but without experimental proof) by *Willstätter* and *Asahina* (*364*) from the *Hämopyrrol* produced by treating hemin with HI + PH$_4$I according to *Nencki* and *Zaleski*. Their *Hämopyrrol* (mixture) was separated by fractional crystallization of the picrate(s) into phyllopyrrole (mp 66–67°C, picrate mp 95°C), *Hämopyrrol* (liq., picrate mp 109°C), *Iso-hämopyrrol* (iso-hemopyrrole) (mp 16–17°C, picrate mp 119°C), *etc.* The same pyrroles were said to have resulted from mixed chlorophylls following reaction with Sn + glacial acetic acid, *etc.*

282   4 A Modern Proof of Bilirubin Structure Emerges

At about the same time, in a paper received at the *Berichte* on December 11, 1911 (*365*) *Willstätter* and *Asahina* reiterated their belief that the *Hämopyrrol Gemisch* from hemin and chlorophyll was a mixture – of at least the three pyrroles named above. And they further found that upon reaction with HNO_2, *Isohämopyrrol* (picrate mp 119°C) gave a mono-oxime (mp 218-219°C) of methylethylmaleimide; whereas, their *Hämopyrrol* of picrate mp 109°C gave an oxime of mp 201°C. Both oximes could be hydrolyzed to the same parent imide. Repeating the *Knorr-Hess* synthesis of 2,4-dimethyl-3-ethyl-pyrrole (*323*), they found a picrate of mp 138°C and produced an oxime of methylethylmaleimide with mp 215–216°C. This oxime gave a mixture mp of ~205°C with the oxime (mp 201°C) from their *Hämopyrrol*, and a mixture mp of ~200°C (broad) with the oxime (mp 218–219°C) of *Iso-hämopyrrol*. This led perhaps to more confusion, and to *Willstätter* pursuing other interests.

In his investigations of pyrrole structure, *Fischer* drew upon his certainty of the structures of phyllopyrrole (2,4,5-trimethyl-3-ethylpyrrole), phyllopyrrole carboxylic acid (2,4,5-trimethyl-3-ethylpyrrole), and the 2,4-dimethyl-3-ethylpyrrole synthesized by *Knorr* and *Hess* (*323*). Though he knew in 1911–1912 that the *Hämopyrrol* of *Nencki* and *Zaleski*, and of *Piloty* and *Thannhauser*, formed a picrate of mp 108–109°C, when he repeated its isolation by reductive cleavage (HI + PH_4I) of hemin, he found a *Hämopyrrol* picrate of mp 120–122°C, recrystallized to mp 123°C. This discrepancy was clearly disturbing to *Fischer* because at that time it seemed assured that *Hämopyrrol* was 2,3-dimethyl-4-ethylpyrrole, which was the only possible structure for it because it was not the *Knorr* and *Hess* synthetic isomer (2,4-dimethyl-3-ethylpyrrole) that gave a picrate that melted at 131–132°C. Since both pyrroles gave methylethylmaleimide upon chromic acid oxidation, and both had a molecular formula (C_7H_9N), the only difference between the structures was the location of the pyrrole α-methyl group. To a careful scientist such as *Fischer*, the mixed data for natural *Hämopyrrol* posed a dilemma that was not fully rectified until subsequent analyses of the volatile pyrroles cleaved reductively from bilirubinic acid (*354*), bilirubin (*354*), and hemin (*366*). By fractional crystallization of the picrates of the mixture of alkylated pyrroles, he discovered a fraction that when repeatedly recrystallized to a constant melting point gave a picrate, mp 137–138°C – or an exact match, with no melting point depression upon admixture with the picrate of the *Knorr* and *Hess* synthetic 2,4-dimethyl-3-ethylpyrrole (*323*). *Fischer* named the new pyrrole *Kryptopyrrol* [from χρυπτό ζ (= hidden), kryptopyrrole], because its presence in the basic *Hämopyrrol Gemisch* had been missed. He discovered subsequently that when this picrate was admixed with his mp 123°C synthetic *Hämopyrrol* picrate, the mixture melting point was depressed to 110°C – or to that observed by *Nencki* and *Zaleski*, and *Piloty* and *Thannhauser*, for their *Hämopyrrol* picrate, mp 108–109°C. The discovery of "natural" kryptopyrrole thus solved the earlier mystery of mismatched picrate melting points for the latter and *Fischer*'s *Hämopyrrol*, and it allowed *Fischer* to reconfirm that hemopyrrole is 2,3-dimethyl-4-ethylpyrrole. It also explained the structures of *Willstätter*'s and *Asahina*'s phyllopyrrole (mp 66–67°C, picrate mp 95°C) as 2,3,5-trimethyl-4-ethylpyrrole; their *Hämopyrrol* (picrate mp 131–132°C) as kryptopyrrole; and their *Iso-hämopyrrol* (mp 16–17°C, picrate mp 119°C, oxime mp 218–219°C) as identical to *Fischer*'s hemopyrrole.

4.2.5 The Monopyrrole Propionic Acid Components of Bilirubinsäure and Xanthobilirubinsäure

In 1909, *Piloty* reported the isolation of *Hämopyrrolcarbonsäure*[13] from the "acidic" fraction following reductive cleavage of hematoporphyrin with Sn + hydrochloric acid (*303*), and also by KOH fusion (*306, 307*). He then designated its structure (*307*) as a disubstituted pyrrole, 3-methyl-4-(2-methylpropanoic acid) pyrrole (Scheme 4.2.2). This drew a sharp rebuke from *Küster* (*309*), who pointed out in 1909 that (i) the indicated structure could not give the observed chromic acid oxidation product, *Hämatinsäure* (hematinic acid); and (ii) hemopyrrole carboxylic acid must have an α-CH_3, a β-CH_3, and a β-$CH_2CH_2CO_2H$ (below). However, he could not determine, on the basis of the available evidence, whether the α-CH_3 was adjacent to a β-CH_3 or to a β-$CH_2CH_2CO_2H$ group.

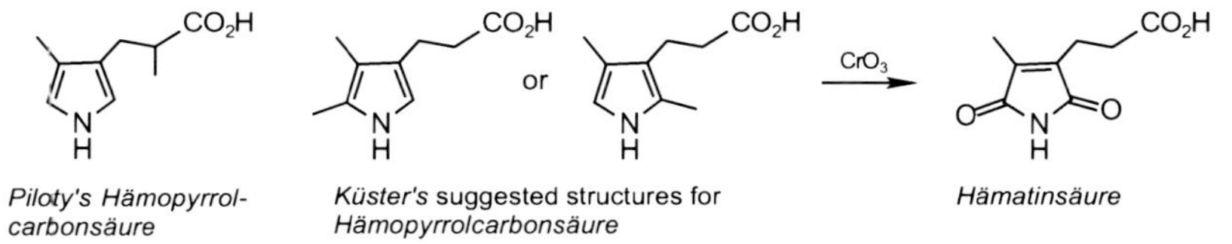

Piloty's Hämopyrrol-carbonsäure

Küster's suggested structures for *Hämopyrrolcarbonsäure*

Hämatinsäure

Scheme 4.2.2

A further complication arose in 1910 when *Piloty* and *Quitmann* (*313*) re-isolated a pyrrole acid from a fresh cleavage (Sn + hydrochloric acid) of hematoporphyrin and named it *Phonopyrrolcarbonsäure*[13] ($\phi\acute{o}\nu\acute{o}\zeta$ = *vergossenes Blut*, shed blood), conceding its structure to be one or the other of the two suggested by *Küster* for *Hämopyrrolcarbonsäure* (Scheme 4.2.2). *Piloty* later isolated *Phonopyrrolcarbonsäure*, in 1912 (*314*) and in 1913 (*337*) by reacting hemin with HI + PH_4I and reported that it had a melting point of 125°C and a picrate melting point of 157°C. Reaction with HNO_2 led to an imide-oxime (two are possible) of hematinic acid that melted at 246°C; reaction with benzenediazonium chloride gave an azo dye ($C_{15}H_{18}N_3O_2Cl$), mp 145–146°C. When heated, *Phonopyrrolcarbonsäure* underwent decarboxylation to give *Phonopyrrol*. Oddly, the latter did not at the time form a crystalline picrate and was not found to produce methylethylmaleimide upon CrO_3 oxidation. Thus, *Piloty* said that *Phonopyrrol* was not *Hämopyrrol*, and wrote its structure as 2,3-dimethyl-4-ethylpyrrole (*313*). Given the two possible structures for *Phonopyrrolcarbonsäure*, upon decarboxylation one would be predicted to yield *Fischer*'s hemopyrrole, and the other kryptopyrrole. Although both were shown by *Fischer* to be present in the *Hämopyrrol Gemisch*, the latter was unknown in 1910, and the former had been isolated only as a mixture.

This peculiar result represented a loose end to *Fischer* in his attempt to understand pyrrole structure, and it prompted him in 1912 to repeat *Piloty*'s *Phonopyrrolcarbonsäure* isolation and decarboxylation. Thus, by early 1912 (*367*,

---

[13] Given the uncertainty of structure and homogeneity, in the early work the original German names are retained until these issues are resolved later in this section.

284   4   A Modern Proof of Bilirubin Structure Emerges

368), *Fischer* and *Bartholomäus* had isolated *Phonopyrrolcarbonsäure* (mp 125–126°C) from the "acidic" fraction following reaction of hemin with HI + PH₄I. As found earlier by *Piloty* (*313*), decarboxylation yielded *Phonopyrrol* (*368*), not the *Hämopyrrol* that *Piloty* had expected – considering that the constitution of *Hämopyrrol* was an issue swirling in controversy at the time of *Piloty*'s work in 1910, and because its homogeneity was being called into question and a structure had not been firmly established. In early 1912, having established the structure of hemopyrrole, *Fischer* investigated the decarboxylation of *Phonopyrrolcarbonsäure*, which led him to believe that *Phonopyrrol* was probably a mixture of dimethylethylpyrrole and possibly trimethylpyrrole (*362, 368*). The peculiarity that *Phonopyrrol* failed to give a picrate, and that its reaction with diazotized *p*-aminobenzenesulfonic acid gave a color-characteristic pyrrole β-azo pigment, not an α-azo pigment (despite the observation that *Phonopyrrolcarbonsäure* gave an α-azo pigment), caused *Fischer* to think of *Phonopyrrol* as 2,3-dimethyl-5-ethyl-pyrrole and/or 2,3,5-trimethylpyrrole – and that its precursor, *Phonopyrrolcarbonsäure*, might in fact be 2,3-dimethyl-pyrrole-5-propionic acid, or even 2,3-dimethyl-pyrrole-5-acetic acid (*368*):

> Durch Destillation der Phonopyrrol-carbonsäure erhielt Piloty . . . unter Kohlensäureabspaltung das »Phonopyrrol«, ein Öl, von dem er kein krystallisiertes Derivat erhielt, und dem er durch eine eigenartige Beweisführung die Formel I zuschrieb.

$$\text{I.} \quad \begin{array}{c} H_3C \cdot \overset{..}{C} - \overset{..}{C} \cdot C_2H_5 \\ H_3C \cdot C \diagdown CH \\ \overline{NH} \end{array}$$

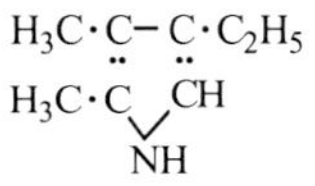

> Wir haben schon früher einen schön krystallisierenden Azofarbstoff . . . aus »Phonopyrrol« erhalten, der nach den Reaktionen der β-Reihe zugehört, wodurch die Formel I für »Phonopyrrol« ausgeschlossen erschien. . . .
> Nach diesem Verfahren konnten wir uns leicht über das Pikrat in den Besitz von reiner Phonopyrrol-carbonsäure setzen. Mit Diazobenzolsulfosäure erhielten wir einen schön krystallisierenden Azofarbstoff, der nach den Reaktionen unzweifelhaft der α-Reihe zugehört.
> Dies Verhalten stand nun im scharfen Gegensatz zu dem Ergebnis der Destillation, und wir synthetisierten daher, um Klarheit zu schaffen, die 2,4-Dimethyl-pyrrol-5-essigsäure und 2.4-Dimethyl-pyrrol-5-propionsäure, womit erstmals homologe, am Kohlenstoff substituierte Pyrrolcarbonsäuren gewonnen sind.
> Wir erweiterten die Knorrsche Pyrrolsynthese dahin, daß wir die Isonitrosoverbindungen von Acetylpropionsäure, CH₃·CO·C(:N·OH)·CH₂·COOH, bezw. Acetylbuttersäure, CH₃·CO·C(:N·OH)·CH₂·CH₂·COOH, mit Acetessigester zu den Pyrrolen II und III kondensierten.

$$\text{II.} \quad \begin{array}{c} H_3C \cdot \overset{..}{C} - - \overset{..}{C} \cdot COOC_2H_5 \\ HOOC \cdot CH_2 \cdot C \diagdown C \cdot CH_3 \\ \overline{NH} \end{array} \qquad \text{III.} \quad \begin{array}{c} CH_3 \cdot \overset{..}{C} - - \overset{..}{C} \cdot COOC_2H_5 \\ HOOC \cdot CH_2 \cdot CH_2 \cdot C \diagdown C \cdot CH_3 \\ \overline{NH} \end{array}$$

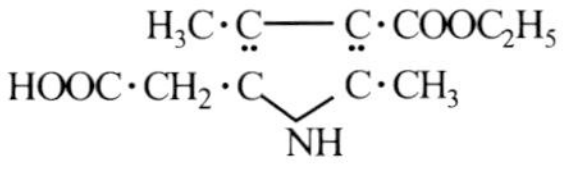

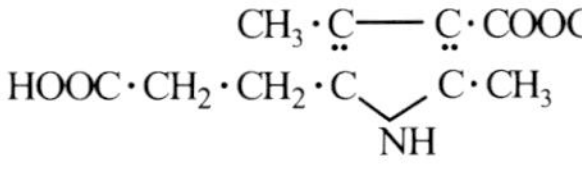

> Die erhaltenen Estersäuren befreiten wir von der am Kern sitzenden Carbäthoxy-Gruppe durch Behandlung mit mäßig konzentrierter Schwefelsäure. Die so entstendenen freien Säuren zeigten leider nur sehr geringe Krystallisationsfähigkeit, die Dimethyl-pyrrol-essigsäure erhielten wir zwar im Vakuum in wohlausgebildeten Krystallen, die aber an der Luft sofort zerflossen, die analoge Propionsäure dagegen krystallisierte nicht. Wir charakterisierten die

4.2 Hans Fischer and the Early Structures of Bilirubin

beiden Säuren daher durch ihre prachtvoll krystallisierenden Azofarbstoffe, die nach allen Reaktionen der β-Reihe zugehören. Die Propionsäure gibt kein Pikrat, die Essigsäure ein überaus leicht lösliches, so daß es zur Identifizierung nicht geeignet ist. Speziell die Propionsäure unterscheidet sich nun in sämtlichen Eigenschaften so scharf von der Phonopyrrol-carbonsäure, daß ihre Nichtidentität mit letzterer zweifelsfrei ist. 50

The two acids were synthesized from **II** and **III**, above, and found *not to be identical* to *Phonopyrrolcarbonsäure*.

The isolation of *Phonopyrrolcarbonsäure*, first accomplished by *Piloty* in 1910 (*313*) from hematoporphyrin was soon followed by the isolation of an isomer, named *Iso-phonopyrrolcarbonsäure*, as described by *Piloty* and *Thannhauser* (*301*) in an article received at the *Berichte* on May 25, 1912. The new acid (mp 126–127°C) was obtained following reaction of bilirubin with HI + PH$_4$I and separated as its picrate, mp 246°C. Subsequently, in August 1912, *Piloty* and *Dormann* (*333*) reported its apparent isolation from the acid fraction following reductive cleavage of hemin with HI + PH$_4$I and fractional crystallization of its picrate mixture to afford a picrate of mp 122°C (imide-oxime mp 227°C (dec.)). Also isolated was *Phonopyrrolcarbonsäure* (mp 129°C), obtained from its mp 157°C picrate. The picrate re-formed from the mp 129°C acid gave a mp of 158–159°C, and the imide-oxime of the acid had a mp of 241°C. Again in 1913, *Piloty* and *Dormann* (*337*) reconfirmed the isolations and melting point data of *Phonopyrrolcarbonsäure* and its isomer, as well as their picrates and their imide-oximes. The melting point of the imide-oxime of *Iso-phonopyrrolcarbonsäure* was corrected, to a mp 210°C. Even a third acid, *Xanthopyrrolcarbonsäure* (mp 105°C, picrate mp 142.5°C, imide-oxime mp 201–202°C) had been isolated by *Piloty* from bilirubin degradation in 1912 (*314*) as well as from hemin (*333*) and again in 1913 (*337*).

By early 1913, however, the structures of the acids were still unproven, at least to *Fischer*, and their melting points were somewhat confusing. *Fischer*, too, had by October 20, 1912 submitted a summary (*353*) of his studies with *Heinrich Röse* on the cleavage of bilirubin and its Na(Hg) reduction product, hemibilirubin, using HI + PH$_4$I. Fractional crystallization of the picrates from the reaction mixture mother liquor following removal of bilirubinic acid yielded an *isomeric Phonopyrrolcarbonsäure* picrate (mp 153°C), which gave a mixture mp 145°C with the picrate of *Phonopyrrolcarbonsäure* (mp 129°C). *Fischer* thought that the picrate of the isomeric acid was probably identical to the picrate of *Piloty*'s *Iso-phonopyrrolcarbonsäure*. Except that the picrate melting points did not match well, and *Piloty*'s picrate might be viewed as more consistent with the mixture of picrates of *Phonopyrrolcarbonsäure* and its isomer.

Later in 1912, in an article received at the *Berichte* on November 8 (*354*), *Fischer* and *Röse* reinvestigated the acidic fraction from cleavage of bilirubinic acid, bilirubin, and hemibilirubin with HI + PH$_4$I. From bilirubinic acid they isolated the picrate (mp 152°C) of the isomeric *Phonopyrrolcarbonsäure* and transformed it into the free acid, which was converted into its oxime (mp 220°C) with HNO$_2$. Following the treatment of the mother liquors with HNO$_2$, the oxime (mp 233°C) of *Phonopyrrolcarbonsäure* was isolated, *not* the oxime of the isomeric acid. A mixture melting point of the latter with the former showed a depression to 210°C. Similar reductive cleavage of bilirubin and hemibilirubin, as reported earlier (*353*), provided an acidic fraction that yielded the picrate of the isomeric *Phonopyrrolcarbonsäure* and (from the mother liquors) its oxime (mp 236 °C). *Fischer* and *Röse* (*354*) further examined the two acids by converting them to their methyl esters by treatment with CH_3OH + HCl phonopyrrole carboxylic acid methyl ester (mp 57–58°C, picrate mp 121–122°C); and the isomeric phonopyrrole carboxylic acid methyl ester (mp 107–108°C, picrate 47–48°C). The investigators took special note that the former picrate had an unusual red-brown color (but regenerated the original ester), while the latter picrate had a bright yellow color, like the other picrates prepared by *Fischer* (*354*):

> Auch dieser Ester gibt ein schön krystallisierendes Pikrat, das auffallenderweise eine dunkelbraunrote Farbe besitzt, während das des isomeren Esters hellgelb ist wie alle Pyrrolpikrate, die wir bis jetzt beobachtet haben. Der Unterschied ist so evident, daß wir anfangs der Ansicht waren, daß durch die Pikrinsäure der Ester Phonopyrrol-carbonsäure irgendwie verändert würde, und wir haben deshalb den Ester aus dem Pikrat wieder regeneriert. Er erwies sich aber als unverändert. 51

Though the data from the two acids and esters, their picrates and oximes were convincing evidence for two different chemical structures, the structure assignments remained uncertain at the end of 1912, at least from *Fischer*'s perspective:

Table 4.2.1 Comparison of melting points of *Fischer*'s and *Piloty*'s pyrrole carboxylic acids

Fischer's *Phonopyrrolcarbonsäure*			*Piloty*'s *Phonopyrrolcarbonsäure*		
mp/°C	Source compound	ref.	mp/°C	Source compound	ref.
Acid 125–126 Picrate 148–149	*Körper* II	(*353*) (*304*)	Acid 125	Hematoporphyrin	(*313, 314*)
Oxime 233	Bilirubin/ Hemibilirubin	(*353*)	Picrate 148	Hematoporphyrin	(*313, 314*)
Oxime 236	Bilirubin/ Hemibilirubin	(*354*)	Picrate 157	Hematoporphyrin	(*314*)
Oxime 238–239	*Körper* II	(*353*)	Oxime 242	Hematoporphyrin	(*313*)
Oxime 233	Bilirubinic acid	(*354*)	Oxime 246	Hematoporphyrin	(*314*)
Methyl ester 57–58	Bilirubinic acid	(*354*)			
Methyl ester – picrate 121–122	Bilirubinic acid	(*354*)			

Fischer's *Iso-phonopyrrolcarbonsäure*			*Piloty*'s *Iso-phonopyrrolcarbonsäure*		
mp/°C	Source compound	ref.	mp/°C	Source compound	ref.
Picrate 153	Bilirubin/ Hemibilirubin/ *Körper* II	(*353*)	Acid 126–127	Bilirubin	(*301*)
Picrate 152	Bilirubinic acid	(*354*)	Acid 118–122	Hemin	(*333*)
Oxime 220	Bilirubinic acid	(*354*)	Picrate 146	Bilirubin	(*301*)
Methyl ester 47–48	Bilirubinic acid	(*354*)	Picrate 146	Hemin	(*333*)
Methyl ester – Picrate 107–108	Bilirubinic acid	(*354*)	Oxime 210	Hemin	(*301*)

4.2 Hans Fischer and the Early Structures of Bilirubin   287

The two acids clearly differed, yet the presence of similar but poorly correlated melting point data and suspect sample homogeneity must have been bothersome to *Fischer*.

In early 1913 (*356*), while exploring a new, cleaner fragmentation reaction (NaOCH$_3$ + CH$_3$OH at 220–230°C, autoclave) *Fischer* and *Röse* discovered *Xanthobilirubinsäure* (xanthobilirubinic acid). And from the reaction mother liquor they isolated an *Ehrlich* reaction-negative (see Section 4.1) monopyrrole carboxylic acid as its picrate, mp 125–126°C. The negative *Ehrlich* reaction indicated the presence of substituents at all four pyrrole carbons. The picrate was identical to that of 2,4,5-trimethyl-pyrrole-3-propionic acid, or phyllopyrrole carboxylic acid (mp 126–127°C), which had been obtained earlier by heating *Phonopyrrolcarbonsäure* in NaOCH$_3$ + CH$_3$OH – and wherein α-methylation had occurred (*369*). Although the transformation provided a structural link between the two acids, the transformation to phyllopyrrole carboxylic acid could also have been viewed as having been conducted by α-methylation of *Iso-phonopyrrolcarbonsäure*. By 1914, the structures of the monopyrrole acids were becoming much clearer.

The earlier discovery (*354*) that the picrate crystals of the methyl esters of *Phonopyrrolcarbonsäure* and its isomer (*Isophonopyrrolcarbonsäure*) differed considerably in color: red-brown or chocolate for the former and bright yellow for the latter, provided *Fischer* with a visual means to effect their separation from the acidic mother liquors following scission of (i) hemin (*367*) and (ii) bilirubin (*352-354, 366*) promoted by HI + PH$_4$I; and of (iii) hemin, promoted by reaction in KOCH$_3$ + CH$_3$OH at 220°C (*370*). The product mixture from hemin (i), following methanol esterification of the base-soluble fractions then vacuum distillation (up to 220°C oil bath temperature), gave a fraction that crystallized in the cold to provide the pure phono-ester, whose picrate melted at 120–121°C. The non-crystalline oily ester fraction residue was converted in ether to a picrate mixture (*a*) that crystallized upon cooling. Evaporation of the mother liquor from *a* afforded a solid picrate, mp 95°C after recrystallization, apparently the picrate of the ester of phyllopyrrole carboxylic acid. Fractional crystallization of picrate mixture (*a*) yielded the known picrate of the phono-ester and a yellow picrate (mp 112–113°C). The latter was converted to the parent acid, mp 140°C. It gave no melting point depression when mixed with authentic *Iso-phonopyrrolcarbonsäure* and in a mixture with *Phonopyrrolcarbonsäure* (mp 131°C) it gave a mixture melting point depression to 110–112°C. Thus, it apparently corresponded to the mp 122°C acid reported in 1913 by *Piloty* (*337*). A sample of *Fischer*'s *Iso-phonopyrrolcarbonsäure* (mp 140°C), was converted to its picrate, mp 155–156°C, and its imide-oxime, mp 219°C; whereas, *Piloty*'s picrate of *Iso-phonopyrrolcarbonsäure* had a mp of 146°C and his oxime had a mp of 210°C.

In contrast, the KOCH$_3$-mediated cleavage (iii) of hemin produced an acid fraction, which after methyl ester formation and vacuum distillation yielded a major liquid fraction that proved to be phyllopyrrole carboxylic acid methyl ester, crystallized as its picrate to mp 95–96°C. In order to determine if the phyllopyrrole carboxylic acid methyl ester (picrate) isolated from the reaction (iii) of hemin with KOCH₃ was formed directly from the pigment, the original oily ester mixture was heated in KOCH₃-CH₃OH at 220–230°C for 4.5 hours, and the product was distilled under vacuum. Extraction of the distillate into ether and picrate formation of the ether solution gave a picrate with a mp of 126–127°C after two recrystallizations, and it proved to be identical to the picrate of synthetic 2,3,4,5-tetramethypyrrole (*371*). The non-ether soluble alkaline distillate was esterified in CH₃OH and distilled under vacuum to yield a sharp boiling point fraction (157–158°C, 10 mm Hg) that produced a mp 97–98°C methyl ester picrate exhibiting no melting point depression in a mixture melting point with an analytical sample of phyllopyrrole carboxylic acid methyl ester picrate. The acid obtained from the ester had a mp of 89°C, with a picrate mp 129–130°C (*vs.* mp 126–127°C found earlier). Subsequently, *Fischer* found that thermal decarboxylation of phyllopyrrole carboxylic acid split off an equivalent of acetic acid to yield 2,3,4,5-tetramethylpyrrole (*372*).

In further contrast, HI + PH₄I cleavage of (ii) bilirubin, after removing bilirubinic acid yielded three fractions after esterification and vacuum distillation of the methyl ester *Gemisch*; one fraction was the ester of isophonopyrrole carboxylic acid while the second and third contained the phono- as well as isophono-ester. Again, fractional crystallization of the ester picrates gave a product of mp 111°C, from which the mp 142°C isophono-acid was freed.

The structure of *Phonopyrrolcarbonsäure* was finally clarified as 2,3-dimethyl-pyrrole-4-propionic acid when, as reported by *Piloty, Stock*, and *Dormann* (*331*) in a paper received on August 17, 1914 at *Liebig's Annalen*, rapid decarboxylation-distillation of the acid at 230°C produced hemopyrrole (not *Phonopyrrol*) that was identified as its picrate, mp 119°C (*331*):

> Was die vier sauren Spaltstücke des Hämins, die Phonopyrrolcarbonsäuren anbetrifft, so haben wir sie sämtlich zuerst neben den Basen aufgefunden (a. a. O.). Wir schließen dieselben aus Bequemlickeitgründen demselben Nomenklaturprinzip an, das wir bei den Basen durchgeführt haben. Wir nennen die Säuren vom Schemlzpunkt 129°, früher Phonopyrrolcarbonsäure, jetzt Phonopyrrolcarbonsäure *a* (Nr.10) [**II**], die Säure mit dem Schmelzpunkt 125°, die ein Pikrat vom Schmelzp. 150-151° liefert, Phonopyrrolcarbonsäure *b* (Nr. 11) [**IV**] (früher Isophonopyrrol-carbonsäure), die Säure vom Schmelzp. 108° Phonopyrrolcarbonsäure *c* (Nr. 12) [**VII**], früher Xanthopyrrolcarbonsäure, und endlich die Säure mit dem Pikrat von Schmelzp. Phonopyrrolcarbonsäure *d* (Nr. 13) [**VIII**]. [See Table 4.2.2 for structures.]   52

This discovery then meant that the isomeric acid, *Iso-phonopyrrolcarbonsäure*, had to be **VI**, 2,4-dimethylpyrrole-3-propionic acid (*331*):

> . . . das Vorhandensein von Hämopyrrol *b* betrachten wir aber als einen Beweis, daß die Phonopyrrolcarbonsäure *a* dieselben Stellen wie Hämopyrrol b besetzt enthält, daß ihr demnach die Formel II zukommt. Daraus folgt weiter, daß die isomere Phonopyrrolcarbonsäure *b*, welche eine den Oxim der Phono-pyrrolcarbonsäure *a* isomeres Oxim liefert, eine durch die Formel VI ausgedrückte Struktur besitzt; namentlich da diese beiden Oxime dieselbe Hämatinsäure (**IX**) liefern.   53

4.2 Hans Fischer and the Early Structures of Bilirubin

Table 4.2.2 *Hämopyrrol* and *Phonopyrrolcarbonsäure* structures assigned by *Piloty* (*331*)

I	III
$CH_3-C\!\!=\!\!C-CH_2\cdot CH_3$ $CH_3-C\quad C-H$ NH	$CH_3-C\!\!=\!\!C-CH_2\cdot CH_3$ $HC\quad CH$ NH
Hämopyrrol b (Isohämopyrrol)	*Hämopyrrol a*

IV	V
$CH_3\cdot C\!\!=\!\!C-CH_2\cdot CH_3$ $HC\quad C-CH_3$ NH	$CH_3-C\!\!=\!\!C-CH_2\cdot CH_3$ $CH_3-C\quad C-CH_3$ NH
Hämopyrrol c (Kryptopyrrol)	*Hämopyrrol d (Phyllopyrrol)*

II	VI
$CH_3-C\!\!=\!\!C-CH_2\cdot CH_2-COOH$ $CH_3-C\quad CH$ NH	$CH_3\cdot C\!\!=\!\!C-CH_2\cdot CH_2\cdot COOH$ $HC\quad C-CH_3$ NH
Phonopyrrolcarbonsäure a (*Hämopyrrolcarbonsäure*) (mp 129°C, picrate mp 162-163°C)	*Phonopyrrolcarbonsäure b* (*Isophonopyrrolcarbonsäure, Kryptopyrrolcarbonsäure*) (mp 125°C, picrate mp 150°C)

VII	VIII
$CH_3-C\!\!=\!\!C-CH_2\cdot CH_2\cdot COOH$ $H\cdot C\quad C-CH_2\cdot CH_3$ NH	$CH_3-C\!\!=\!\!C-CH_2\cdot CH_2\cdot COOH$ $CH_3-C\quad C-CH_3$ NH
Phonopyrrolcarbonsäure c (*Xanthopyrrolcarbonsäure*) (mp 108°C, picrate mp 142°C)	*Phonopyrrolcarbonsäure d* (*Phyllopyrrolcarbonsäure*) (mp 81°C, picrate mp 128°C)

In his long article (*331*), *Piloty* summarized his studies on the separation by (fractional crystallization of picrates) of the "basic" and "acidic" products from cleavage of hemin, hematin, and hematoporphyrin, from which he identified: (i) six different *Hämopyrrole*, a-f, assigned structures to four (a-d or **I, III, IV, V** of Table 4.2.2), left two unassigned, and assigned structures to (ii) four different *Phonopyrrolcarbonsäuren*, a-d or **II, VI, VII,** and **VIII** of Table 4.2.2 (*331*). The yields of these products from cleavage of hemin are presented in the *Tabelle*, below (*331*):

Tabelle über die Ausbeuten an Hämopyrrolen
und Phonopyrrolcarbonsäuren aus Hämin durch
Jodwasserstoff-Eisessig.

Nr.	Name	Gramme aus 100g Hämin	Gramme aus 100 g Basen-bzw. Säuregemisch
1	Hämopyrrol a	0,6	2
2	Hämopyrrol b (Isohämopyrrol Willstätters und Asahinas und Hämopyrrol Fischers)	12,5	40,5
3	Hämopyrrol c (Kryptopyrrol Fischers und Hämopyrrol Knorrs)	2,5	7,5
4	Hämopyrrol d (Phyllopyrrol Willstätters und Asahinas)..........	7,9	25
5	Hämopyrrol e............................	2,6	8,5
6	Hämopyrrol f	unbestimmt	unbestimmt
7	nicht fällbare Hämopyrrole	4,3	14
8	hochsiedendes Hämopyrrol	0.8	2.5
9	Hämopyrrol g	kleine Quant.	kleine Quant.
10	Phonopyrrolcarbonsäure a	16,7	82
11	Phonopyrrolcarbonsäure b	2	10
12	Phonopyrrolcarbonsäure c	0,6	3
13	Phonopyrrolcarbonsäure d	1	5
	Summe sämtlicher Spaltstücke	51,5 g	54

(Geschweifte Klammer für Nr. 1–9: Summe der Basen 31,2 g; für Nr. 10–13: Summe der Säuren 20,3 g)

Given the firmly assigned structure of phonopyrrole carboxylic acid, its earlier conversion to *Phonopyrrol* by a brutal decarboxylation indicated a rearrangement of the sort envisioned by *Fischer* in which the erstwhile β-ethyl formed by decarboxylation had migrated to the neighboring pyrrole α-carbon to produce 2,3-dimethyl-5-ethylpyrrole (*368*):

> Bei der Darstellung des »Phonopyrrols« nach der Pilotyschen Vorschrift erfolgt nun eine Umlagerung derart, daß die bevorzugte α-Stellung besetzt wird. Hierbei muß Abspaltung der Kohlensäure bezw. die Beschaffenheit der Seitenkette in irgendwelcher Weise auf diese Umlagerung begünstigen einwirken, denn es ist uns bis jetzt nicht gelungen, Hämopyrrol durch Erhitzen im zugeschmolzenen Rohr bei entsprechender Temperatur in ein β-freies Pyrrol zu verwandeln. Vielleicht sind auch Reaktionen ganz anderer Art hier im Spiel (auch Ammoniakabspaltung tritt im geringen Maße ein). Durch weitere synthetische Arbeit hoffen wir, hier völlige Klarheit zu schaffen. . . . 55

4.3 Status and Conjectures Regarding Bilirubin and Hemin Structures circa 1916 291

Or as explained in 1916 (*373*):

> Die Natur dieses Phonopyrrols konnten dann Fischer und Bartholomäus[18] [*368*] aufklären, indem es ihnen gelang aus dem Phonopyrrol einen Azofarbstoff zu isolieren, der sich als β-Azofarbstoff erwies. Es war also bei der hohen Temperatur, die bei der Destillation der Phonopyrrolcarbonsäure erforderlich ist, nicht nur Absprengung von Kohlensäure, sondern gleichzeitig Wanderung der Äthylgruppe in α-Stellung eingetreten.

> Es war nach diesem Befund nicht ausgeschlossen, dass die Phonopyrrolcarbonsäure den Propionsäurerest in α-Stellung trüge, bzw. in der Säurefraktion eine dieser Konstitution entsprechende Säure vorhanden sei. Die Synthese der in Frage kommenden Säure[18] [*368*] widerlegte dies Möglichkeit.   56

The same pyrrole had been obtained earlier, in 1912, by *Fischer* and *Bartholomäus* (*362*) by heating 2,3-dimethyl-5-ethyl-4-propionyl-pyrrole in H_2SO_4, a pyrrole synthesized by the same authors from the condensation of 3-oximino-2-butanone with dipropionylmethane in glacial acetic acid using Zn dust (*366*). *Fischer* believed that *Phonopyrrol* was a mixture, *inter alia*, of 2,3-dimethyl-5-ethyl- and 2,3,5-trimethyl-pyrrole – the latter perhaps arising from phonopyrrole carboxylic acid by loss of acetic acid coupled with rearrangement of the newly formed β-methyl to the neighboring α-site. Exact structures of all of the components of *Phonopyrrol* remain unresolved.

4.3 Status and Conjectures Regarding Bilirubin and Hemin Structures circa 1916

Much of the seemingly contradictory properties and uncertain structures assigned to the monopyrroles, *Hämopyrrol*, and *Iso-hämopyrrol*, and the monopyrrole acids, *Hämopyrrolcarbonsäure*, *Phonopyrrolcarbonsäure*, and *Iso-Phonopyrrolcarbonsäure* turned out to be associated with insufficient sample homogeneity. Considering that there were no reliable methods of analysis of sample purity in the first quarter to first third of the 20th century, except sharp melting points and absence of a melting point depression when an authentic and pure sample was admixed, pyrrole identification was problematic – even aside from the notorious sensitivity of alkylpyrroles toward darkening in air (and light). And there was considerable room for error or misjudgment when the existing criterion of purity based on melting points failed, as when a mixture melting point of two unlike substances failed to depress. Of course, certification of sample purity had been a problem of historic proportions and perilous endeavor even in the early days of bilirubin isolation, with progress being thwarted when the yellow pigment presented itself in varying degrees of inhomogeneity. Despite the great advances in organic chemistry since 1800, it may have been even worse when those colorless monopyrrole compounds cleaved from bilirubin, hemibilirubin, hemin, and hematoporphyrin exhibited melting points varying from one sample batch to another and from one laboratory to another. Typically, the liquid pyrroles, *Hämopyrrol* and *Isohämopyrrol* had been isolated by distillation from a volatile *Hämopyrrol Gemisch*, precipitated by picric acid as picrates, which were fractionally and repeatedly recrystallized. The solid acids, obtained from the sodium carbonate soluble fraction of the *Gemisch* or its post-distillation residue were also turned into picrates and recrystallized. Generally the picrate mother liquors were saved, and from them more picrate sample was recovered, often with a relatively sharp melting point, but different from the first crystallization batch. All this *Fischer* learned when he carefully and methodically investigated the monopyrroles and conclusively determined their structures.

Although *Küster*'s publications on bilirubin from 1906 to 1910 provided confirmation of his own and *Nencki*'s earlier investigations, they brought no new breakthroughs on structure. Those were to arrive during the period 1909–1915, which saw enormous advances in the quest to assign the chemical structure of bilirubin (and hemin) by degradation, isolation, and full characterization of the mono- and dipyrrole products thus obtained – and reassembly of the components according to the existing chemical logic and insight. With *Piloty*'s discovery in 1914 (*331*) that rapid distillation-decarboxylation of phonopyrrole carboxylic acid yielded hemopyrrole, its structure became secure, as well as that of its isomer, isophonopyrrole carboxylic acid and as acknowledged by *Fischer* in 1916 (*373*):

> Neuerdings ist es nun Piloty, Stock und Dormann [*331*] gelungen, aus dem Destillat der Phonopyrrolcarbonsäure (bei schneller Destillation) ein krystallisiertes Pikrat vom Schmelzpunkt 119 zu erhalten, das sich als identisch mit dem Salz des Hämopyrrols erwies.
>
> Hiernach würde der Phonopyrrolcarbonsäure folgende Konstitution

$$\begin{array}{c} CH_3C\!\!-\!\!-\!\!CCH_2CH_2COOH \\ \| \qquad \| \\ CH_3C \diagdown \quad CH \\ NH \end{array}$$

> zukommen und es wäre zweckmässig, wiederum den alten Namen Hämopyrrolcarbonsäure einzuführen, wie das Küster schon früher vorgeschlagen hat. Ich glaube aber, dass es sich empfiehlt, dies vorläufig zu unterlassen, weil mir bei Berücksichtigung unserer alten Resultate Umlagerungen nicht ausgeschlossen erscheinen. Selbst eine Umlagerung von α- in α_1-Stellung ist denkbar . . .   57

Given the structure correlation between hemopyrrol and phonopyrrole carboxylic acid, *Fischer* suggested that the latter name be altered to conform more to the former, which was named first (by *Nencki*, although it was then a mixture).

4.3 Status and Conjectures Regarding Bilirubin and Hemin Structures circa 1916

Thus, *Fischer* preferred the name *Hämopyrrolcarbonsäure* (hemopyrrole carboxylic acid), *Piloty*'s original but discarded name (*374*):

> ... zum ersten Male O. Piloty beobachtet [*303, 308*]. Es gelang ihm, aus *Hämatoporphyrin* durch Reduktion mittels Zinn und Salzsäure eine Säure zu gewinnen, die er zuerst als *Hämopyrrolcarbonsäure*, später aber als Phonopyrrol-carbonsäure (von φόνόζ = vergossenes Blut) bezeichnete, weil das bei der Decarboxylierung entstandene Pyrrol („Phonopyrrol") nicht mit dem *Hämopyrrol* identisch war. Die nähere Untersuchung hat dann ergeben, daß bei der raschen Destillation der „Phonopyrrol-carbonsäure" doch Hämopyrrol entsteht [*206b*], während das Phonopyrrol als ein Gemisch von 2,3-Dimethyl-5-äthyl- und 2,3,5-Trimethylpyrrol erkannt wurde. [*368*]. So erhielt die Säure in der Folgezeit wieder ihre ursprüngliche Bezeichnung Hämopyrrolcarbonsäure.   58

Also, *Fischer*'s preference for the name *Kryptopyrrol* (kryptopyrrole) over *Isohämopyrrol* (isohemopyrrole) led him to rename the corresponding acid as *Kryptopyrrolcarbonsäure* (kyrptopyrrole carboxylic acid) (*373*):

> Die Konstitution dieser Isophonopyrrolcarbonsäure wäre, wenn die der Phonopyrrolcarbonsäure im obigen Sinne richtig ist, die dem Kryptopyrrol entsprechende:

$$CH_3C \!-\! CCH_2CH_2COOH$$
$$HC \quad CCH_3$$
$$NH$$

> und ihre käme daher der Name Krytopyrrolcarbonsäure zu, jedoch ist es aus oben genannten Gründen zweckmässig, vorläufig den alten Namen beizubehalten.   59

This left only two acid cleavage products incompletely accounted for: phyllopyrrole carboxylic acid and xanthopyrrole carboxylic acid. The latter, isolated in 1912 by *Piloty* (*333*), could not be obtained subsequently by *Fischer*, despite frequent attempts (*373*). All one knew of it was that its formula was isomeric with that of phyllopyrrole and that it reacted with HNO_2 to give an imide-oxime that was the same as that from isophonopyrrole carboxylic acid. Since the last was undoubtedly inhomogeneous, and the imide-oxime from either source was probably inhomogeneous, one could at the time probably discount xanthopyrrole carboxylic acid as an artifact (*373*):

> Piloty [*333*] beschrieb zuerst neben der Phonopyrrolcarbonsäure eine Xanthopyrrolcarbonsäure, die sich von der Phonopyrrolcarbonsäure durch ein Mehr von einem Kohlenstoffatom unterschied und der folgende Konstitution zukommen soll

$$CH_3C \!-\! CCH_2CH_2COOH$$
$$HC \quad CC_2H_5$$
$$NH$$

> weil sie beim Abbau mit salpetriger Säure das Oxim der Isophonopyrrolcarbonsäure liefert, das bei der Versiefung wiederum Hämatinsäure gibt. [Der] Verfasser hat trotz häufiger Bemühungen niemals Anhaltspunkte für das Vorkommen dieser Säure gewinnen können.   60

Not so with phyllopyrrole carboxylic acid, which had been isolated following HI + PH_4I reductive cleavage of hemin by both *Piloty* (*314, 328, 337*) and *Fischer*

(*369, 372*), and from KOCH$_3$ + CH$_3$OH induced cleavage of hemin by *Fischer* (*375*). The transformation of isophonopyrrole carboxylic acid to phyllopyrrole carboxylic acid by reaction with NaOCH$_3$ in CH$_3$OH confirmed the structure of the latter once that of the former was revealed (*373*):

> Endlich ist von Piloty [*337*] unter den sauren Spaltprodukten des Hämins eine der Xanthopyrrolcarbonsäure isomere Säure isoliert worden, die vorher auf synthetischem Weg von H. Fischer und Bartholomäus [*369*] durch Methylierung der Phonopyrrolcarbonsäure und von Fischer und Röse [*356*] aus Bilirubin erhalten worden war. Danach is die Konstitution dieser Säure die einer Trimethylpyrrolpropionsäure, der also

> der Name Phyllopyrrolcarbonsäure zukommt.   61

Unable to pursue research during the World War I years after 1914, while instructing medical students in 1915–1916 at the University of Munich, following *Piloty*'s death *Fischer* summarized his chemical studies of structures of porphyrins and bile pigments and their cleavage products and expressed his opinions on their interrelationships in a long and important review article (*373*). For the monopyrrole cleavage products, he drew the following structural relationships and (re)assigned names (*373*):

Table 4.3.1 Monopyrroles from cleavage of hemin and their names

Structure	*Piloty*'s Name	*Fischer*'s Name
	Phonopyrrolcarbonsäure (= phonopyrrole carboxylic acid)	*Hämopyrrolcarbonsäure* (= hemopyrrole carboxylic acid)
	Hämopyrrol (= hemopyrrole)	*Hämopyrrol* (= hemopyrrole)
	Isophonopyrrolcarbonsäure (= isophonopyrrole carboxylic acid)	*Kryptopyrrolcarbonsäure* (= kryptopyrrole carboxylic acid)
	Isohämopyrrol (=isohemopyrrole)	*Kryptopyrrol* (= kryptopyrrole)

(continued)

4.3 Status and Conjectures Regarding Bilirubin and Hemin Structures circa 1916 295

Table 4.3.1 (continued)

	Phyllopyrrolcarbonsäure (= phyllopyrrole carboxylic acid)	Phyllopyrrolcarbonsäure (= phyllopyrrole carboxylic acid)
	Phyllopyrrol (= phyllopyrrole)	Phyllopyrrol (= phyllopyrrole)

It was this information that allowed *Fischer* to assign correctly the structures to the dipyrrolic fragments, *Bilirubinsäure* (bilirubinic acid) and *Xanthobilirubinsäure* (xanthobilirubinic acid) obtained from reductive cleavage of bilirubin. His first (flawed) structures for colorless bilirubinic acid and xanthobilirubinic acid, with an ether bridge linking the two pyrroles, were based on their chromic acid oxidation to yield methylethylmaleimide and hematinic acid (Table 4.3.2).

Table 4.3.2 Evolution of *Fischer*'s structures proposed for *Bilirubinsäure* (bilirubinic acid) from 1912

However, the fact that bilirubinic acid could be oxidized to yellow-colored xanthobilirubinic acid rendered problematic *Fischer*'s structures of 1912. The monopyrrole products (kryptopyrrole and kryptopyrrole carboxylic acid) isolated from bilirubinic acid initiated the process of understanding the structure of the dipyrroles. When it was learned that phyllopyrrole carboxylic acid formed the identical imide-oxime as that from phonopyrrole carboxylic acid, *Fischer* began to realize that the kryptopyrrole carboxylic acid unit of bilirubinic acid was probably present as a tetrasubstituted pyrrole, as in phyllopyrrole carboxylic acid, an idea confirmed by the isolation of the latter following $NaOCH_3 + CH_3OH$ reductive cleavage of bilirubinic acid. This led to a new structure proposed for bilirubinic acid in early 1913 (*358*). Thus, he correctly deduced a one-carbon bridge rather than one-oxygen, *e.g.* a dipyrrylmethane for bilirubinic acid, and a dipyrrylmethenone for xanthobilirubinic acid, as illustrated in Table 4.3.3 for the evolution of structures for xanthobilirubinic acid.

Table 4.3.3 Evolution of *Fischer*'s structures of *Xanthobilirubinsäure* (xanthobilirubinic acid) proposed 1913–1916 [4_formula 69]

February 3, 1913 (*356*)	March 9, 1913 (*358*)
$C_2H_5 \cdot C$—$C \cdot CH_3$ $CH_3 \cdot C$—$C \cdot CH_2 \cdot CH_2 \cdot COOH$ $CH_3C \cdot C$ C—O—C $C \cdot CH_3$ N——————N	$COOH \cdot CH_2 \cdot CH_2 \cdot C$—$C \cdot CH_3$ $C_2H_5 \cdot C$=$C \cdot CH_3$ $H_3C \cdot C$ C—CH=C C=O NH NH
December 31, 1913 (*357*)	November 16, 1914 (*376*)
CH_3C—$C \cdot C_2H_5$ CH_3C—C-CH_2-CH_2-$COOH$ HOC C—CH_2—C C-CH_3 NH NH	$COOH \cdot CH_2 \cdot CH_2 \cdot C$—$C \cdot CH_3$ $C_2H_5 \cdot C$=$C \cdot CH_3$ $H_3C \cdot C$ C—CH=C $C{:}O$ NH NH
1916 (*373*)	
CH_3C=C-C_2H_5 CH_3C—C-CH_2-CH_2-$COOH$ O=C C=CH_2—C C-CH_3 NH NH	

The further observation that both bilirubin and its reduction product, mesobilirubinogen, produced phyllopyrrole carboxylic acid and phyllopyrrole by scission upon heating with $KOCH_3$ or $NaOCH_3$ in CH_3OH apparently confirmed the important dipyrrole-methane and methene component structures to *Fischer*.

In 1914, *Fischer* knew that both bilirubin and hemin had four "pyrrole" rings and two propionic acid substituents, and he deduced that each compound had two vinyl groups. This deduction was reached from two pieces of evidence: (i) vinyls were reduced to ethyls when bilirubin ($C_{33}H_{36}N_4O_6$) was hydrogenated catalytically to form mesobilirubin, and (ii) repeated combustion analyses of the latter indicated the formula $C_{33}H_{40}N_4O_6$ (*377*):

4.3 Status and Conjectures Regarding Bilirubin and Hemin Structures circa 1916

Die Mikroanalysen hat Herr Dr. Lieb in Graz ausgefürt . . .

I.	0,1210 g	Sbst	:	0,2989 g	Kohlensäure	und	0,0747 g	Wasser	
II.	4,260 mg	»	:	10,495 mg	»	»	2,62 mg	»	
III.	4,545 mg	»	:	11,30 mg	»	»	2,73 mg	»	
IV.	3,900 mg	»	:	0,333 ccm N bei 733 mm Hg und 22 °,					
V.	0,1436 g	»	:	0,3533 g	Kohlensäure	und	0,0869 g	Wasser	

					I.	**II.**	**III.**	**IV.**
$C_{33}H_{40}N_4O_6$ (588,36)	Ber.	: = C	67,31	Gef. : =	67,37;	67,19;	67,81;	67,10
	»	: = H	6,86	» =	6,91;	6,88;	7,37;	6,77
	»	: = N	9,52	» =	9,53.			

Thus, two equivalents of H_2 had been taken up in the hydrogenation, which could be interpreted as the equivalent of two vinyl groups having been reduced to two ethyls. Further reduction of mesobilirubin with Na(Hg) yielded mesobiliru-binogen ($C_{33}H_{44}N_4O_6$), which was identical in melting point and mixture melting point to the hemibilirubin, $C_{33}H_{44}N_4O_6$ (*350*) from Na(Hg) reduction of bilirubin. Mesobilirubinogen/hemibilirubin was converted back to mesobilirubin, not biliru-bin, by heating with $KOCH_3$ in CH_3OH at 150°C for two hours (*377–379*). More supporting evidence for the structure of mesobilirubinogen (and vinyl groups hav-ing been reduced to ethyls) came from the fact that it yielded methylethylmaleimide (in addition to hematinic acid) upon chromic acid oxidation (*377, 379*); whereas bilirubin gave only hematinic acid (*349*). *Küster*, too, had reached a similar conclu-sion regarding the presence of two vinyl groups in hemin *vs.* two ethyl groups in its porphyrin reduction product, mesoporphyrin (from reduction of hemin): the former gave only hematinic acid upon chromic acid oxidation, the latter gave methylethyl-maleimide in addition (*380*).

Yet a major question remained open: how were the four pyrrole units assembled in bilirubin and also in hemin, which were thought to be built of 31–34 carbons? Based on *Küster*'s assumption that the four pyrrole rings were attached to the two carbons of a carbon-carbon double bond, by 1914 *Fischer* had proposed C_{33} struc-tures for bilirubin, its tetrahydro analog, mesobilirubin (in which two ethyls replaced two vinyls), and octahydro-bilirubin analog, mesobilirubinogen (*373*):

Für das Bilirubin und seine Umwandlungsprodukte habe ich dann ebenfalls unter Annahme der von Willstätter und Max Fischer für Hämin vorgeschlagenen Brücke $\diagdown C=C\diagup$ folgende Formeln aufgestellt,

Bilirubin $C_{33}H_{36}N_4O_6$

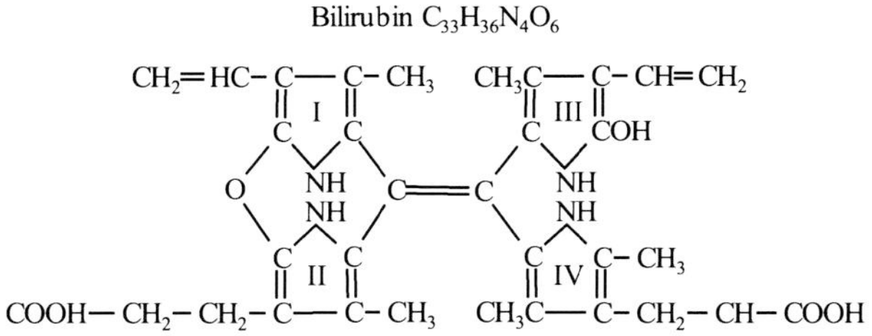

mit dem Bemerken, dass die Formulierung der Vinylgruppen entweder im Hämin oder Bilirubin geändert worden muss wegen des prinzipiell verschiedenen Verhaltens der beiden Farbstoffe gegen Natriumamalgam. Durch dieses Reagens werden nämlich beim Bilirubin Äthylreste erzeugt, während beim Hämin scheinbar keine wesentliche Veränderung in bezug auf die Vinylreste erfolgt, jedenfalls entsteht kein Äthylrest.

63

4.4 Raison d'être Behind *Fischer*'s Failed Tetrapyrrole Structures Prior to 1926

Given that bilirubin degradation studies had revealed to *Fischer* that the pigment consisted of two dipyrryl-methane or methene units, whose structures he had proved, and considering the many possible ways to link them, what led him to believe that the two verified bilirubinic acid or xanthobilirubinic acid units were connected to one another by a carbon-carbon double bond (C=C) or a single bond (CH–CH)? Most likely his thoughts were guided toward such a connection by *Willstätter*'s studies and proposed structures. How could one not pay close attention to *Willstätter*, a giant figure in organic chemistry? He had after all received the *Nobel* Prize in Chemistry in 1911 for his work on chlorophyll and might thus be understood as an expert in pyrrole chemistry. *Willstätter* demonstrated a structural link between chlorophyll and hemin by degrading each to the same decarboxylated large fragment, *Ätioporphyrin* (etioporphyrin), for which he proposed a structure in which the methine bridges of two dipyrrylmethenes were linked by a carbon-carbon single bond (*343*):

... so gelangen wir mit einiger Wahrscheinlichkeit zu folgenden Formeln für Ätioporphyrin und Ätiophyllin:

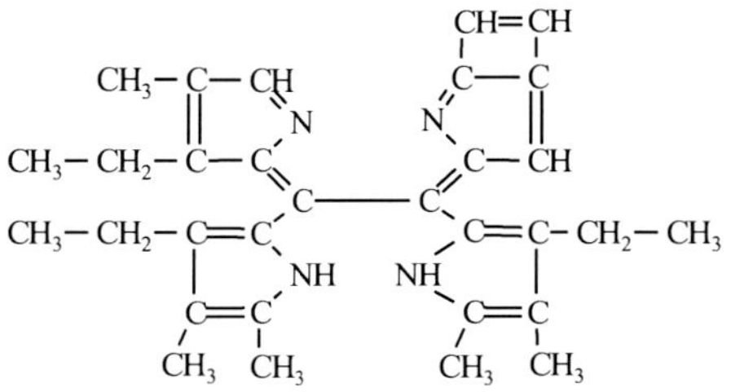

4.4 Raison d'être Behind Fischer's Failed Tetrapyrrole Structures Prior to 1926

Ätioporphyrin $C_{31}H_{36}N_4$

und

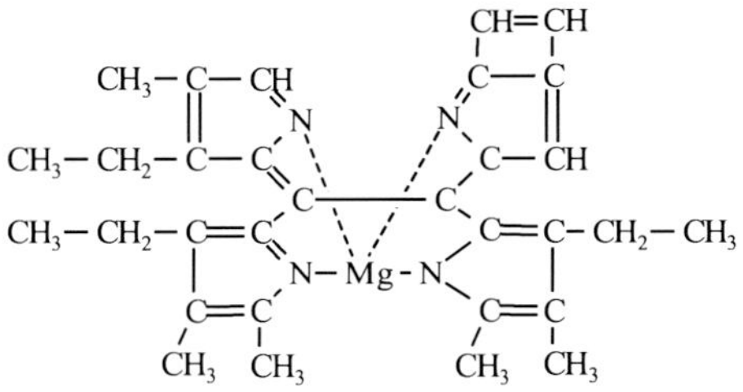

Ätiophyllin $C_{31}H_{34}N_4Mg$

64

Willstätter's then startling finding that he could convert important plant and animal pigments to a common large molecular weight product of 31 carbons was enormously significant, and his structural drawings of 1913 probably bore a *"Nobel"* *imprimatur* to those who learned of them.

Willstätter simultaneously persuaded readers and believers (*343, 344*) that *Küster*'s proposed macrocyclic ring structure of *Hämin* (hemin) (*360*) could not be correct. *Küster*'s structure, proposed in 1912, had four pyrrole rings linked sequentially by methine carbons to form the macrocyclic array familiar to porphyrin chemists, but only since the late 1920s (*380–383*):

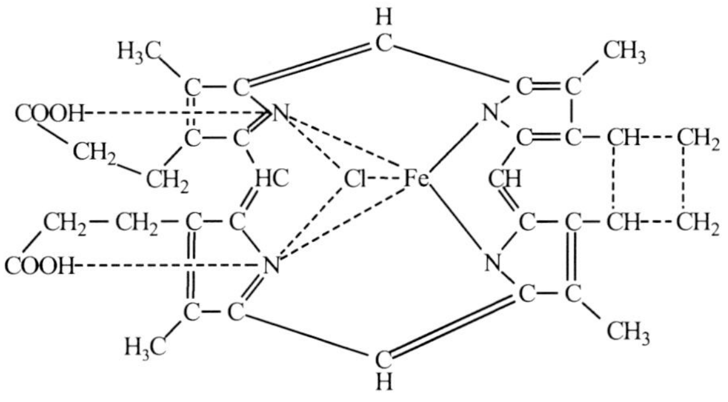

Hämin $C_{34}H_{32}O_4N_4FeCl$.

However, in 1913, in the early days of recognizing and attempting to understand large structural ensembles, *Küster*'s hemin structure was not well received. *Willstätter* objected to it because of the two imino groups bound to Fe: one was present as a pyrrole acid imine, while the other was a basic imino group of a dihydropyrrole. More compelling to *Willstätter* in his rejection of the *Küster* structure was his belief in the improbability of the 16-membered macrocyclic ring containing four nitrogens and 12 carbons, as shown below (*343*):

> Diese Annahme fordert indessen den Einwand heraus, daß von den beiden mit dem Eisen verbundenen Iminogruppen nur die eine als das saure Imin eines Pyrrols dargestellt wird, die andere aber als basische Iminogruppe eines Dihydropyrrols. Das Unwahrscheinliche dieser Formel liegt namentlich in der Annahme eines aus vier Stickstoffatomen und zwölf Kohlenstoffatomen bestehenden 16-gliedrigen Ringgebildes:

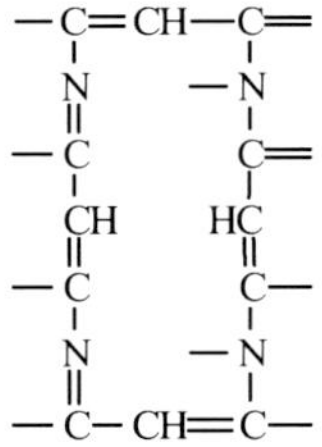

65

While he was offering structural critiques, *Willstätter* also discounted the *Piloty* structures of hemin and bilinic acid as impossible to yield hematinic acid. Rather, he came to the conclusion that a simple connection of four pyrrole rings consisting of two complex-forming and two salt binding sites was the most probable (*343*):

> Als eine einfache Verknüpfung von vier Pyrrolkernen, von nur zwei salzbildenden und zwei komplexbildenden, zum Kern eines Farbstoffs ist uns folgende Formel wahrscheinlich:

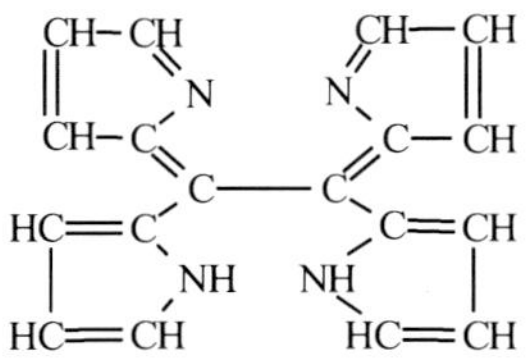

66

It was on this basis that he proposed the etioporphyrin skeletal structure above and followed it with a proposal for the structure of hemin based on the same skeleton (*343*):

> Wenn schon das Ätioporphyrin von der Formel

$$C_{31}H_{36}N_4$$

4.4 Raison d'être Behind Fischer's Failed Tetrapyrrole Structures Prior to 1926

auffallend wasserarm erscheint, so leitet sich Hämin $C_{33}H_{32}O_4N_4FeCl$ von einer noch um 2 Wasserstoffatome ärmeren Grundsubstanz

$$C_{31}H_{34}N_4$$

ab, welche die Annahme von Doppelbindungen und Kohlenstoffringen erzwingt. Auf der Grundlage der angenommenen Verknüpfung von 4 Pyrrolen durch die Gruppe:

$$> C\text{–}C <$$

und der Bindung von 2 Vinylen an Stickstoffatome versuchen wir, eine dem verhalten bei der Oxydation, der Reduktion und der Porphyrinbildung genügende Konstitutionsformel des Hämins zu entwickeln, die noch in mehreren Einzelheiten unsicher und zu verbessern ist, aber, wie wir hoffen, weitere Untersuchungen anzuregen und zu leiten vermag:

Hämin $C_{33}H_{32}O_4N_4FeCl$.

Bei der Bildung des Hämatoporphyrins werden sich die Brücken von den zwei Pyrrolstickstoffen loslösen, worauf sich die mittlere Gruppe

umwandeln kann.   67

In 1914, *Piloty* countered with his own structures for hemin and for bilirubin, although *Willstätter* had not proposed a structure for the latter. They were to become *Piloty*'s final structural contributions before his demise in 1915. These, like his entire array of proposed structures of di- and tetra-pyrroles were undefended and within a decade and a half had been swept into the dustbin of history following his untimely death (*331*):

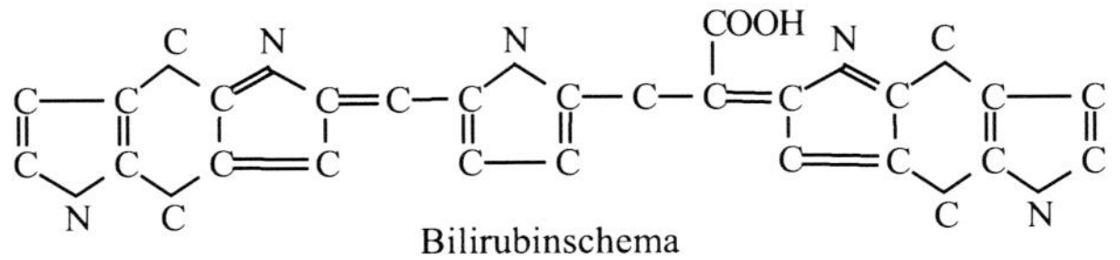

$$ (301) $$

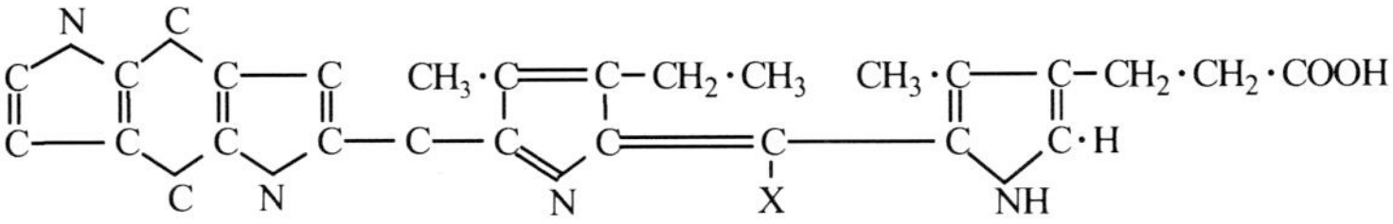

Bilirubinschema

$$ (331) $$

Häminschema

and

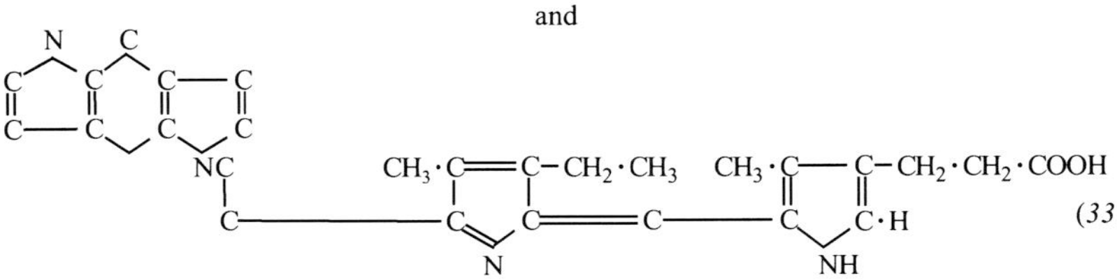

$$ (331) $$

Häminschema

Piloty's structures were denounced by *Küster* in 1915 as untenable (*381*). In 1913 *Willstätter*, too, had published his own objections to the dipyrrole structures (*343*, *344*), but they apparently were unconvincing to *Piloty*, and by 1915, when *Küster* critiqued them (*381*), *Piloty* had passed from the scene.

However, in 1915, *Fischer* had already adopted *Willstätter*'s structure for hemin rather than *Küster*'s, for he too was wary of the unprecedented 16-membered macro-cyclic ring and its unusual Fe binding to nitrogen. He was also confident then (as had been *Küster*) that hemin had 34 carbons, not 33 as represented by *Willstätter*. *Fischer* thus accepted but updated the *Willstätter* hemin formulation, which he modified to incorporate his more symmetric structure with exposed vinyl groups (*373*):

Ich habe dann folgende Formel aufgestellt:

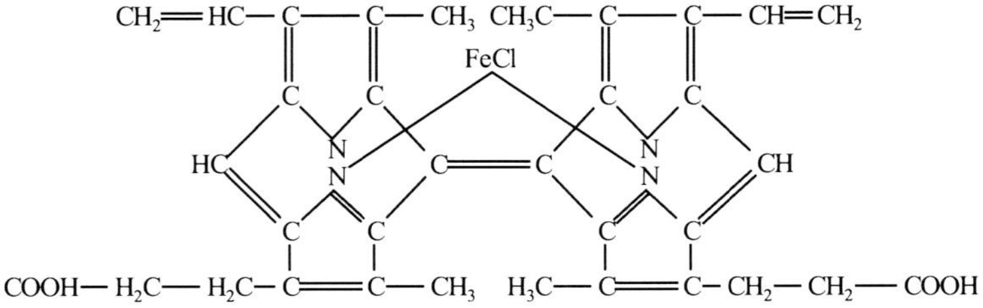

Hämin $C_{34}H_{30}O_4N_4FeCl$.

68

4.4 Raison d'être Behind Fischer's Failed Tetrapyrrole Structures Prior to 1926   303

From such a structure for hemin and *Fischer*'s certainty that bilirubin and biliverdin were its biogenetic transformation products, it was an easy step to assume a hemin-derived structure for the bile pigments, *i.e.* those based on the *Willstätter* tetrapyr-rylethylene skeleton, which *Fischer* had already incorporated into his structures at the end of 1913 (*357*). Thus, *Fischer*'s structure for bilirubin led to his fairly straightforward representations for the vinyl-reduced-to-ethyl structure of meso-bilirubin and the more reduced mesobilirubinogen structures (*373*), *vide supra*. (Biliverdin would then have come from bilirubin by loss of two hydrogens from the NH groups.) With such studies in mind, *Fischer* was prepared to launch his major efforts toward the syntheses of hemin, bilirubin, and biliverdin, and chlorophyll. Efforts that would not, however, commence until 1921–1922.

Fischer's structures, like *Piloty*'s, were found objectionable by *Küster*, albeit for different reasons that doubtless stemmed from his rejection of the *Willstätter* structure of hemin. *Küster* suggested three possible alternative structures for C_{32} bilirubin (*380*):

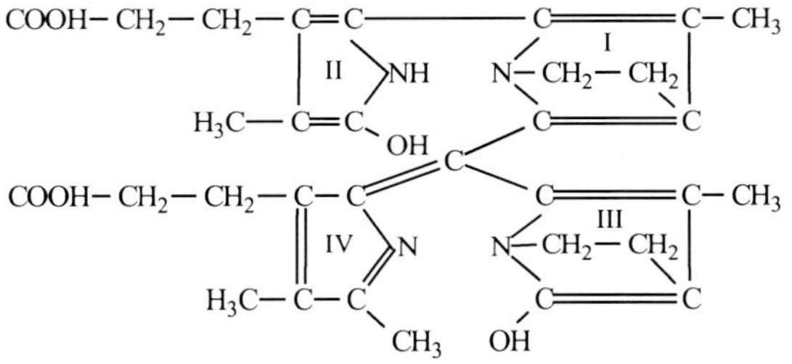

I.

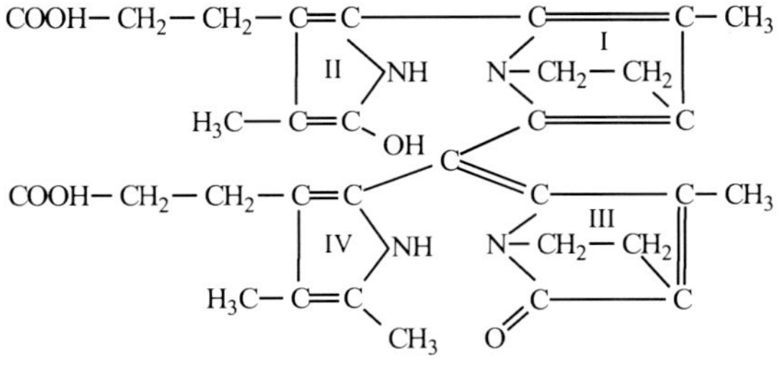

II.

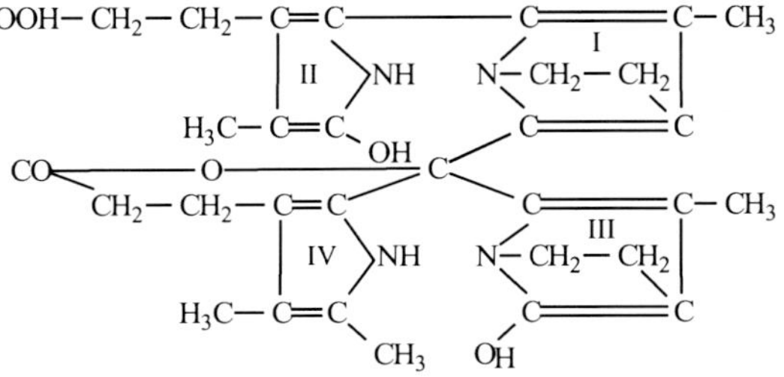

III.

He favored **II**, from which he proposed structures for mesobilirubin and mesobilirubinogen (*380*):

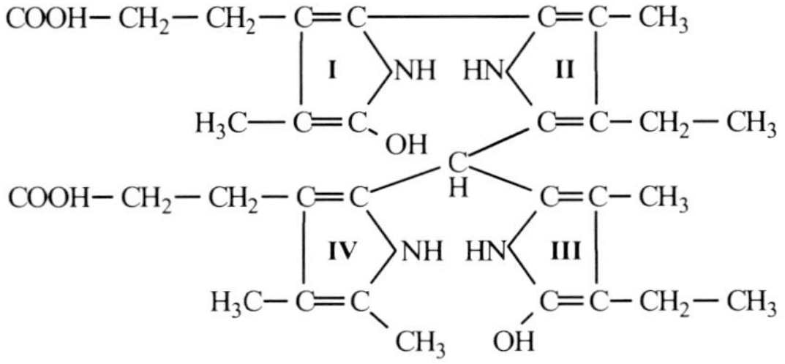

Mesobilirubin

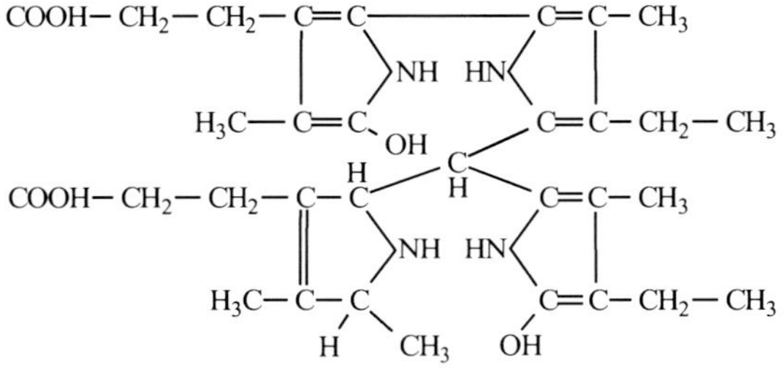

Mesobilirubinogen

Learning from *Fischer*'s work in 1914 (*377*) that bilirubin had 33 and not 32 carbons, *Küster* carried out an intensive study of bilirubin isolation and purification and concluded in favor of bilirubin possessing 33 carbons (*382*):

> Seit meiner letzten Veröffentlichung über das Bilirubin [*383*] ist eine zusammenfassende Arbeit von H. Fischer [*373*] erschienen, in der Formel $C_{33}H_{36}O_6N_4$ für das Bilirubin erneut befürwortet wird auf Grund der analytischen Ergebnisse für das Reduktionsprodukt des Bilirubins, das Bilirubinogen $C_{33}H_{44}O_6N_4$, während die zur bisher üblichen Formel $C_{32}H_{36}O_6N_4$ führenden analytischen Resultate auf einen Gehalt an verunreinigenden, schwefelhaltigen Stoffen zurückgeführt werden, die durch die bisher angewandten Methoden zur Reinigung des Rohbilirubins nicht hatten entfernt werden können, oder auf einen Gehalt an Chlor, der durch die Extraktion der Gallensteine mit Chloroform in das Rohbilirubin hinein gelangt. Ferner wird an die Möglichkeit gedacht, daß das Rohbilirubin ein Gemisch von zwei verschiedenen Stoffen vorstellt, und es wird die Hoffnung ausgesprochen, daß eine Klärung dieser Fragen durch die Untersuchung des von mir entdeckten Bilirubinammoniums zu ermöglichen sein würde. . . .
>
> Und so glaube ich nach allen Erfahrungen annehmen zu können, daß dem Bilirubin tatsächlich die Formel $C_{33}H_{36}O_6N_4$ zuerteilt werden kann . . .   69

This conclusion then induced *Küster* in 1917 to re-evaluate his 1915-vintage bile pigment structures and to analyze how they might arise from C_{34} hematin (*382*). On paper, transforming the *Küster* formula for hemin into that of bilirubin might have been a straightforward proposition if *Küster* had assumed an opening of the porphyrin macrocyclic ring. Instead, the issue was compounded because earlier *Küster* had

4.4 Raison d'être Behind Fischer's Failed Tetrapyrrole Structures Prior to 1926

proposed a structure of C_{32} bilirubin with little resemblance to hemin ($380, 381$), and he was apparently determined simply to modify it to accommodate a C_{33} structure. This he did by proposing a somewhat fanciful pyrrole ring expansion to a pyridine-like ring, and thereby form a rather strained dipyrryl-pyrryl-pyridyl-methene analog of triphenylmethane. This led him to propose five isomeric structures for bilirubin, including lactim and lactam tautomers and C=C isomers (382). In the end, *Küster* settled on the bis-lactam tautomer of bilirubin, but one with $-CH_2-CH_2-$ units linking a pyrrole β-carbon to its nitrogen, *i.e.* without vinyl groups (382):

Bilirubin $C_{33}H_{36}O_6N_4$

For mesobilirubin he proposed a lactam-lactim structure with the erstwhile vinyl groups of bilirubin reduced to ethyls, and for mesobilirubinogen, a bis-lactam (384):

Mesobilirubin $C_{33}H_{40}O_6N_4$

Mesobilirubinogen $C_{33}H_{44}O_6N_4$

306   4   A Modern Proof of Bilirubin Structure Emerges

Küster was not again to revise his bile pigment structures of 1917 for nearly eight years. He did, however, defend his macrocylic structure of hemin, which with its 16-membered ring ligated to Fe, had apparently been unacceptably *avant garde* for its time (*384*):

> Nun hat letzterer meine Formulierung des Hämins mit vier Methinen als verbindende Glieder der Pyrrol- und Pyrrolenkerne, die den Farbstoff-Charakter markieren sollten, als unwahrscheinlich bezeichnet, weil hierdurch ein 16-gliedriger Ring erscheint.   70

Not only did *Willstätter* reject it before a meeting of the German Chemical Society, while advocating his own structure, but he apparently gave an unbalanced account that excluded all but his own structure which was also presented in *Abderhalden*. According to *Küster* (*384*):

> Meine in diesen »Berichten« und der »Zeitschrift für physiologische Chemie«[1]) . . . veröffentlichten Anschauungen über die Konstitution des Hämins und die Bindung des Eisens durch den organischen Teil seines Moleküls haben sich bisher der Anerkennung seitens der Fachgenossen nicht erfreuen können, namentlich hat R. Willstätter[2] in einem Vortrag vor der Deutschen Chemischen Gesellschaft wesentlich andere Ansichten entwickelt. Indem er hierbei mein für das Hämin gegebenes Bild, obwohl er wichtige Teile desselben für das seine verwenden konnte, zusammen mit der von Piloty entwickelten Konstitutionsformel, die sich als nicht zutreffend erwiesen hatte, in einem Satze und mit dem Bemerken verwarf, »dab die Frage hinsichtlich der Art, in der man die Pyrrolkerne verknüpft denken kann, durch diese Vorstellung nicht gelöst worden sei«, um dann seine Ideen vorzutragen, hat er allem Anschein nach den Glauben erweckt, als wäre sein Bild allein geeignet, dem derzeitigen Stand unserer Kenntnisse Rechnung zu tragen. Wenigstens wäre es sonst unverständlich, daß Hjelt in seiner Geschichte der organischen Chemie, Abderhalden in seinem Lehrbuch der physiologischen Chemie Willstätters Konstitutionsformel allein aufführen, während doch der Charakter solcher Bücher verlangt, bei einem in der Entwicklung begriffenen Kapitel unserer Wissenschaft entweder allen Anschauungen Raum zu geben oder keine der zur Diskussion stehenden Meinungen zu vertreten. Zudem hatte ich[3] bereits nachgewiesen, daß die Vorstellungen Willstätters zu Widersprüchen führen . . .

---

[1]) B. **45**, 1935 [1912]. H. **82**, 113 [1912]. [2]) B. **47**, 2821 [1914]. [3]) H. **88**, 377 [1914].   71

Citing considerable contradictory evidence, *Küster* advocated that the *Willstätter* structure be set aside (*384*):

> Alle diese Überlegungen müssen zur Ablehnung des Bildes von Willstätter führen. 72

Yet despite *Küster*'s annoyance and his counter-arguments to justify the macrocyclic porphyrin structure of hemin, the *Willstätter* and *Fischer* formulas were to command center stage. Until 1925, that is, as suspicion was aroused when *Fischer*'s newly synthesized tetrapyrrylmethene did not exhibit the expected spectral properties of etioporphyrin or any natural porphyrin (*385, 386*). Since bilirubin was linked biosynthetically to hemin, as its metabolism product, its structure remained linked to the tetrapyrrylethylene motif until the structure of hemin was completely clarified.

5 Preparing the Way to the Constitutional Structure of Bilirubin

By 1917, the great advances toward the structure of bilirubin of the preceding two decades had ground to a near halt as World War I exacted its toll and *Oskar Piloty* had died, leaving only *William Küster* and *Hans Fischer* as the major investigators. *Küster* continued research on bilirubin throughout the war years and until his death in 1929: between 1911 and 1920 seven of his 36 publications dealt with the pigment, while between 1921 and 1929 six of his 45 publications did. The work, while not insignificant, dealt mainly with the structure of *Hämin* (hemin) and improving the isolation and purification methods for bilirubin (*387–391*) that were subsequently adopted by *Fischer*. Yet, the structure of the latter pigment was apparently not far from his sight, and in the mid-1920s he had realized that his dipyrryl-pyridylmethane structures ought to be abandoned in favor of structures more closely tied to his 1912 structure of hemin. While *Küster* pursued his apparent major interest in the chemistry of hemin through the 1920s, *Fischer* initiated a major synthesis effort toward its structure.

Hans Fischer, still early in his career and without a base for research after 1915, was seriously deterred from his scientific ambitions by the unhappy days of the World War I and its aftermath. His research suffered a hiatus after 1915 when World War I began to take its toll on university and medical school enrollments, and during his call to teach in Innsbruck and Vienna. Unlike *Küster*, *Fischer* was not able to resume research productively until after 1921, the year that he accepted a call to follow *Heinrich Wieland* as o. Professor of Chemistry at the TH-München. The research that recommenced left bilirubin on a "back burner," however, while a major effort was marshaled toward the synthesis of hemin. Thus, among the 145 *Fischer* research papers published between 1921 and 1930, only eight concerned bilirubin directly, while the vast majority dealt with porphyrins. As *Adolph Stern*[1],

---

[1] *Adolph J.C. Stern* was born on February 12, 1900 in Nuremberg and died on April 29, 1992 in New York. After completing studies at the *Gymnasium* in 1917 he served in Germany's South Field Artillery in World War I until 1919. He enrolled in the TH-München to study chemistry and engineering to achieve the *Diplom* in 1923 and the *Dr. ing.* in 1925 in organic chemistry with *Hans Fischer*. After the habilitation he became a lecturer at the TH in 1933 and laboratory director. In 1935 he was promoted to a. o. Prof. and in 1940 published, as the only co-author with *Hans*

a *Fischer* student, noted in his *Fischer* lecture at a porphyrin conference in 1973 sponsored by the New York Academy of Sciences, the early 1920s was a period of intense activity, as *Fischer* brought his laboratory into being (*392*):

> In 1921, Hans Fischer succeeded Heinrich Wieland and took over as head of the Organic-chemical Institute at the Technical University of Munich and immediately an extensive program concerning the systematic exploration of all aspects of the chemistry of pyrrole, bile pigments, porphyrins and related compounds was designed and undertaken. The chemistry of pyrrole, in spite of the large number of naturally occurring derivatives was, at this time, practically untouched. Fischer's first major objective was to elucidate the structures of porphyrins with a view toward synthesizing hemin. In doing so, Fischer's magnetic personality attracted and inspired a great number of graduate students who, from the very beginning of his work in Munich, were his devoted disciples and ready to do anything to cooperate in achieving Fischer's goals. In a short time, the "Fischer School" was to become one of the most remarkable research groups in Germany and well known internationally. He trained numerous excellent chemists, many of whom became leaders in chemical industry and in academic careers at home and abroad. During the period 1921-1925, a very large number of new pyrrole derivatives were synthesized either by substitution methods or ring synthesis.

Even after 1930, when *Fischer* had won the *Nobel* Prize in Chemistry and had begun to solve the structure of bilirubin by total synthesis, he elected to pursue the structure of chlorophyll with greater dedication (*392*):

> A decision had to be made about which direction to continue research – bilirubin or chlorophyll. After discussion with his coworkers, Fischer decided to go ahead vigorously to elucidate the structure of chlorophyll, but not forget the bilirubin problem.

From 1930 until his death in 1945, *Fischer* published another 275 or so papers, the majority aimed at the structure elucidation and total synthesis of chlorophyll, work that remained incomplete despite an enormous effort put forth during the disasterous era of National Socialism and destruction during World War II. Yet between 1930 and 1942, some 25 papers were directed successfully toward the structure elucidation and total synthesis of bilirubin, the latter a more difficult synthesis accomplishment than that of hemin and the second crowning scientific achievement of *Fischer's* life.

In contrast to *Küster*, *Fischer* devoted an enormous amount of energy and every available resource to develop the fundamental organic chemistry of mono-, di-, and tetra-pyrroles during the 1920s. Although the first target was hemin, the synthesis expertise developed in that quest was to become the basis for the synthesis of bilirubin. What was the structure of hemin that *Fischer* set first in his synthesis sight?

Fischer, the second half of volume II (on chlorophyll) of *Die Chemie des Pyrrols* (*4*). In 1938, only days before the *Anschluss*, he fled the Third Reich for London and the U.S., where he worked first as a research associate at the Research Laboratory Children's Fund in Detroit. In 1942, he joined Wagner College as Professor of Chemistry and served with great distinction in many capacities (chairman, 1950; dean, 1952; special assistant to the president) before becoming emeritus on September 1, 1970.

5.1 Evolution of the Bilirubin Structure in the 1920s

Fischer had based his structure of bilirubin in 1914–1916 on the *Willstätter* structure of hemin, which he had adopted. In 1922, when finally returning to investigations of bilirubin, he began to clarify the components representing one-half of the bilirubin tetrapyrrole structure. *Fischer* appeared to be convinced of certain gross features of the structures, determined a decade earlier, of the dipyrrole fragments obtained from cleavage of bilirubin: colorless *Bilirubinsäure* (bilirubinic acid) and its yellow dehydro analog, *Xanthobilirubinsäure* (xanthobilirubinic acid) (*358*). In the former he had found two types of pyrrole units: a hydroxypyrrole and a pyrrole carboxylic acid (*393*):

> Nach der Fischer-Röseschen[1]) [*352*] Auffassung kommt der Bilirubinsäure, dem bimolekularen Spaltprodukt des Bilirubins, folgende Konstitution zu:

> Sie ist also zusammengesetzt aus einem Oxypyrrol und einer Pyrrolcarbonsäure. 1

Bilirubinic acid was believed to comprise one-half of the bilirubin structure; xanthobilirubinic acid, its *yellow* oxidation product, was thought to be an even closer representative, given its color. Closer actually to mesobilirubin, wherein the vinyl groups of bilirubin had been reduced to ethyl groups. By 1923, *Fischer* had abandoned his original *lactam* tautomeric structure of xanthobilirubinic acid (proposed ten years previously) in favor of the *lactim*, which he had believed since 1913 to be present in bilirubinic acid (*394*):

> Von H. Fischer und Röse ist aus Bilirubin-Mesobilirubinogen[1]) [*356*] und von H. Fischer[2]) [*377*] aus Mesobilirubin die Xanthobilirubinsäure als Spaltprodukt der genannten Farbstoffe erhalten worden. Nach den Untersuchungen der genannten Autoren kommt der Xanthobilirubinsäure folgende Konstitution zu:

> Zwar hatten Fischer und Röse der Xanthobilirubinsäure die entsprechende Ketoformel zuerteilt. Nach einer neuen Untersuchung von Fischer und Niemann[3]) [*395*] jedoch gibt die genannte Säure eine Acetylverbindung und demgemäß kommt ihr die obige tautomere Formel zu. 2

Fischer's revision in favor of the lactim tautomer of xanthobilirubinic acid was based on the experimental fact that it formed a beautifully crystalline acetyl compound (acetate ester?), and from that he deduced that it must have a hydroxyl group (*395*):

> Die Existenz einer freien Hydroxylgruppe in der Xantobilirubinsäure hat sich auch experimentell bestätigt, indem der Ester dieser Säure eine schön krystallisierte Acetylverbindung gibt. Die Xanthobilirubinsäure gewinnen wir neuerdings aus Bilirubinsäure durch einfache Luftoxydation. 3

This experiment, repeated some years later and reinforced by the observation that bilirubin converted to (lactim) ethers as well as esters with diazomethane, was probably the reason that he never veered away from writing the α-hydroxypyrrole lactim tautomer in virtually all of his published work. Of course, it was entirely based on the assumption of the presence of an acylatable hydroxyl group and not at all on a consideration of a dynamic lactam-lactim tautomeric equilibrium in which even a trace of lactim might be continually siphoned off by acetylation. At the time, however, there was apparently no basis for assuming that the lactim tautomer might be more (or even less) stable than the lactam. This was (and still is) a tricky issue, and one that would not be settled until many decades later, as described by *Heinz Falk*[2] (*15, 396*). And so *Fischer*'s conclusion was not at all unreasonable for its time, and it also agreed with observations that 2-pyrrolenone tautomers would be acetylated as if they were α-hydroxypyrroles.

Following the decision to adopt the lactim tautomer, again in 1922 as reported in a manuscript received on January 2, 1923 in the journal office (*395*), *Hans Fischer* and *Georg Niemann* described their experiments (*397*) that confirmed their belief that bilirubin consists of two xanthobilirubinic acid units (*373*). Thus, by catalytic hydrogenation of bilirubin in 0.1 N aq. NaOH using Pd-black, uptake of two equivalents of H_2 produced mesobilirubin with an inconsistent melting point, between 300°C and 315°C. The inconsistency was attributed to the presence of tautomers (*395*):

> Die Wasserstoffaufnahme wurde messend verfolgt. Die besten Resultate wurden erzielt, wenn der Katalysator in einzelnen Portionen zugegeben wurde. Im Durchschnitt berechnen sich vom Bilirubin zum Mesobilirubin zwei Moleküle Wasserstoff. . . .
>
> Der Schmelzpunkt des Mesobilirubins ist kein konstanter, obwohl aus Pyridin einheitliche Krystallisation erzielt wird. Der Schmelzpunkt wird zwischen 300 und 315° gefunden. Die Abweichung könnte erklärt werden durch die Annahme tautomerer Formen, die dann auch im Bilirubin vorhanden wären. 4

---

[2] *Heinz Falk* was born on April 29, 1939 in St. Pölten, Austria. After his early schooling in Statzendorf and in Krems an der Donau, he moved to Vienna in 1953 to complete a three-year program of study at the Höhere Bundeslehr-und Versuchsanstalt for Chemical Industry, Rosensteingasse. In 1959, he completed the high school diploma through classes at night school and then began studies in chemistry at the University of Vienna. After passing the *Doktorandum*, he commenced doctoral studies in organic and metallocene stereochemistry under *Karl Schlögl* and received the Dr. phil. in 1966, at which time he became *Assistent* in the university's Institute of Organic Chemistry, and started work leading to the *Habilitation* in 1972, and the qualification to teach organic chemistry at the University. He was a postdoctoral researcher at the ETH-Zürich in 1971 (with *Albert Eschenmoser*) and promoted to a. o. Professor of Physical Organic Chemistry at the University of Vienna (the first such position in Austria). In 1979 he answered a call as o. Professor to found the new Institute of Organic Chemistry at Johannes Kepler University, Linz, where from 1989 to 1991 he served as Dean of the Faculty of Engineering and Natural Sciences. *Falk* is a member of the Austrian Academy of Sciences, and was ranked third among the top 10 scientists in Upper Austria. In 2007, he was promoted to Emeritus Professor. *Falk's* distinguished scientific career, leading to more than 300 research publications on organic structure, synthesis, stereochemistry of pyrrole and other pigments (hypericin, stentorin, *etc.*), as well as their photochemistry, have led him to become one of the world's expert pyrrole and natural organic pigment chemists. His 1989 book on linear tetrapyrroles (*9*) remains a masterpiece of clarity and richness of information. In addition to scientific pursuits, *Falk* is interested in ancient Egyptian language, which he easily interprets.

5.1 Evolution of the Bilirubin Structure in the 1920s

Mesobilirubin formed a dimethyl ester dihydrochloride upon treatment with CH_3OH + HCl (gas) that analyzed for two OCH_3 groups and two HCl equivalents. From that it was concluded that both mesobilirubin and thus bilirubin were di-(propionic) acids (*395*):

> Die Darstellung dieses Esters beweist, in Bestätigung der früheren Untersuchungen, daß in Mesobilirubin und damit im Bilirubin zwei Carboxylgruppen vorhanden sind.

Dimethylester des Mesobilirubins.

> 0,15 g Mesobilirubin wurden in 5 ccm absolutem Methylalkohol suspendiert. Beim Einleiten von trockner Salzsäure tritt Erwärmung und Lösung ein, und nach kurzer Zeit fällt der salzsaure Ester krystallisiert aus.

5,723	mg	Substanz	gaben	12,852	mg CO_2 und 3,742 mg H_2O.
4,217	mg	„	„	2,650	mg AgJ.
4,487	mg	„	„	1,800	mg AgCl.

$C_{35}H_{44}N_4O_6 \cdot 2$ HCl:

Ber.	C	60,96%	H	6,67%	OCH_3	8,99%	Cl	10,3%	
Gef.	„	61,2	„	7,32	„	8,3	„	9,92	5

In addition, the absorption of two equivalents of H_2 was consistent with the presence of two vinyl groups in bilirubin, and the fact that the hydrogenation product (mesobilirubin) yielded two equivalents of methylethylmaleimide upon nitric acid oxidation provided strong support for a tetrapyrrole structure with two easily cleaved pyrrolenone rings (*395*):

> Daß im Mesobilirubin zwei freie Äthylreste vorhanden sind, bzw. sich sehr leicht bilden, haben wir sicher beweisen können. Bei der Oxydation mit Salpetersäure werden zwei Mol Methyläthylmaleinimid erhalten, ein Zeichen dafür, daß die zwei Äthylreste vorgebildet sein müssen, und weiterhin eine willkommene Bestätigung für das Vorkommen von zwei basischen Pyrrolkernen im Bilirubin. . . . 6

These results contrasted with those from *Fischer*'s HNO_3 oxidation of bilirubin, which yielded unexpected results. Although methylvinylmaleimide had been the expected product, *Fischer* was unable to reduce it to methylethylmaleimide. This apparent failure to obtain methylvinylmaleimide as an oxidation product was henceforth to plague *Fischer's* structure proof of bilirubin.

Fischer clearly "got off track" when he compared the oxidation products following reduction of bilirubin and hemin. Keeping in mind that the structures of neither bilirubin nor hemin were known, it was already clear in 1914 (*378*) that hemin possessed two vinyl groups that were reduced to ethyls by catalytic hydrogenation to give mesohemin, from which methylethylmaleinimide could be obtained *via* nitric acid oxidation. Similarly, mesobilirubin afforded methylethylmaleinimide. Since bilirubin was believed to come about in nature from heme, it was logical that bilirubin should have two vinyl groups. Yet, in 1923, *Fischer* and *Niemann* reported that reduction of bilirubin and hemin differed when Na(Hg) was used as reductant. Thus, the unsaturated side chains of bilirubin were reduced and oxidation of the

312 5 Preparing the Way to the Constitutional Structure of Bilirubin

product gave methylethylmaleimide. In contrast, with Na(Hg) hemin lost its color with no reduction of the side chains because no methylethylmaleimide was produced by nitric acid oxidation of the product (*395*):

> Durch die Annahme von Vinylgruppen, die dann zu Äthylgruppen reduziert würden, würde dieses Verhalten erklärlich sein, jedoch müssen die ungesättigten Seitenketten im Bilirubin anders beschaffen sein, als im Hämin, da ein prinzipieller Unterschied der beiden Farbstoffe gegenüber Natriumamalgam besteht. Nur im Bilirubin werden die ungesättigten Seitenketten reduziert, während beim Hämin durch Natriumamalgam wohl Entfärbung eintritt aber keine Reduktion. Denn mit Natriumamalgam reduziertes Hämin gibt bei der Oxydation kein Methyläthylmaleinimid. 7

In 1923, to account for the absence of methylvinylmaleimide, *Fischer* believed first that the vinyl groups were adjacent to the pyrrole α-OH groups, and in such a relationship a furan ring was formed by ring closure between the vinyl and α-OH groups, accompanied by dehydrogenation (*395*):

> Nachdem Fischer und Röse gezeigt haben, daß α-Oxypyrrol im Bilirubin enthalten ist, ist für diese Annahme die experimentelle Grundlage vorhanden . . . . Wir glauben deshalb, daß im Bilirubin die Vinylreste in Beziehung zu den Hydroxylgruppen stehen, dergestalt, daß Furanringe im Bilirubin enthalten sind. Möglicherweise ist dann das wesentliche beim Übergang des Blutgarbstoffes in den Gallenfarbstoff die Oxydation in zwei α-Stellungen. Für die Oxydation der freien CH-Gruppe sind genügen Analogien vorhanden . . . Durch Dehydrierung schließt sich dann der Ring zwischen Vinyl- und Oxygruppen zum Furanring, entsprechend dem Schema:

> Dieses Schema würde natürlich nur bei einer noch unbekannten Hämopyrrolbase im Bilirubin passen, bei Kryptopyrrolkomponente käme man zu zwei stereoisomeren Formen. . . . Die Furanringe erklären dann die außerordentliche Empfindlichkeit des Bilirubins gegenüber Mineralsäuren . . . 8

Accordingly, the presence of furan rings in bilirubin would explain the pigment's failure to yield the expected alkylimide(s) upon oxidation. But their presence would not prevent the conversion of bilirubin to mesobilirubin because, it was thought, the furan rings of the former would still undergo ring opening during Pd-catalyzed hydrogenation or Na(Hg) reduction to produce the two ethyls of the latter (*395*):

> Bei der Reduktion des Bilirubins mit Natriumamalgam oder Palladium-Wasserstoff wird der Furanring aufgespalten und erst dann treten die freien Hydroxylgruppen und Äthylreste auf. 9

Reiterating the published hydrogenation results of 1923 (*395*), in the following year *Fischer* and *Niemann* wrote that the uptake of 2 moles of H_2 in the conversion

5.1 Evolution of the Bilirubin Structure in the 1920s

of bilirubin to mesobilirubin corresponds well with the view of the origin of an ethyl residue from a vinyl group, and a hydrofuran ring from a furan ring (*397*):

> Die Aufnahme von 2 Mol. Wasserstoff vom Bilirubin zum Mesobilirubin stimmt gut überein mit der Auffassung des Entstehens eines Äthylrestes aus einer Vinylgruppe und eines Hydrofuranringes aus einem Furanring. 10

This statement indicates a backing away from the earlier belief (*395*) that the furan ring would open under reduction conditions to give an ethyl group. And by late 1925, in a follow-up publication, *Fischer* and *Postowsky* had revised the view to reflect a belief that the structure of bilirubin was better represented by the presence of both a furan ring *and* a vinyl group that was of course not located adjacent to the lactim OH (*398*):

> Ermutigt durch diese Befunde unterzogen wir den Gallenfarbstoff selbst und seine Reduktionsprodukte Mesobilirubin und Mesobilirubinogen der Untersuchung. Die 3 Körper werden nach der Fischerschen Formulierung durch folgende Konstitutionsformeln wiedergegeben, wobei als Modifikation die verbindenen beiden Kohlenstoffatome in etwas anderer Weise aufgefaßt werden, so daß steroisomere Formen in großer Anzahl denkbar sind, wenn man noch die Tautomeriemöglichkeit von IV in Betracht zieht.

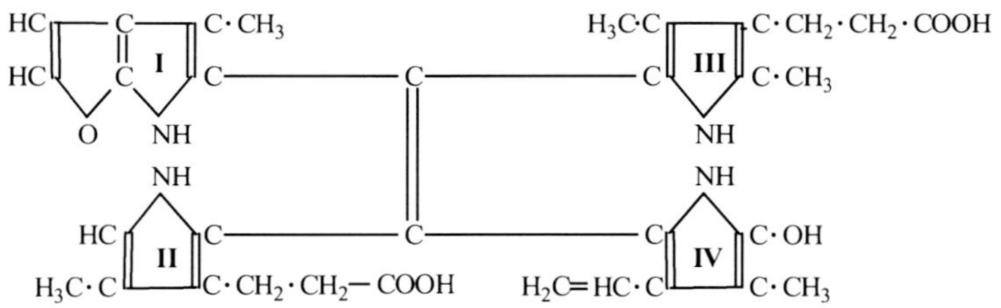

Bilirubin.

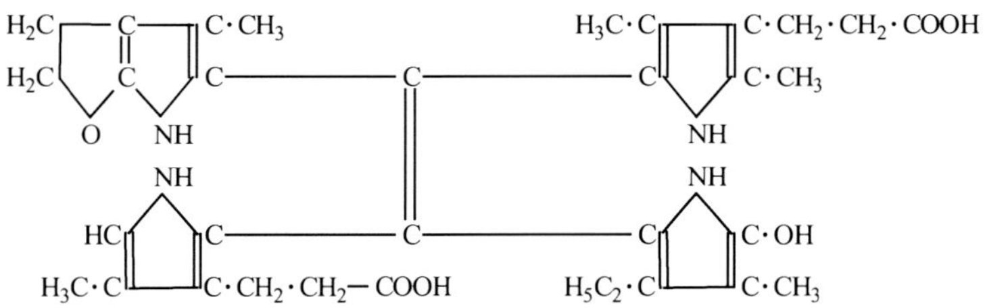

Mesobilirubin.

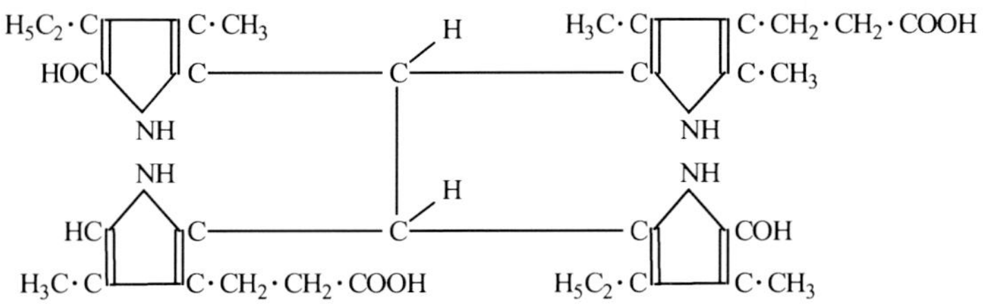

Mesobilirubinogen. 11

314   5 Preparing the Way to the Constitutional Structure of Bilirubin

It may be noted that the *Fischer* formulas above reflect a connection between two dipyrrylmethanes at the "methene" or "methane" carbons, *i.e.* a tetrapyrrylethylene (as in bilirubin and mesobilirubin) and a tetrapyrrylethane (as in mesobilirubinogen). The structures are in accord with *Fischer*'s then decade-old belief that the hemin structure (as suggested by *Willstätter*) is built upon the tetrapyrrylethane motif, and his understanding that bilirubin is derived from hemin biosynthetically (*398*):

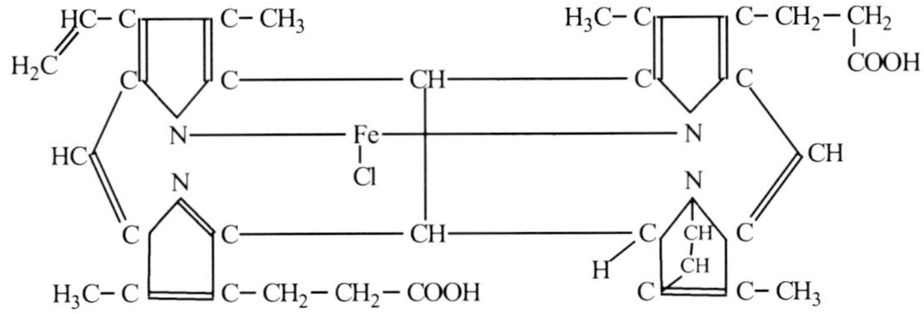

The major subject of the *Fischer-Postowsky* article in 1925 (*398*) was the determination of the number of "active hydrogens" in bilirubin, hemin (and their derivatives), and in pyrroles. Determination of active hydrogens is a classical procedure (*Zerevitinov* determination) to learn the number of NH and OH equivalents from the number of molar equivalents of CH_4 released by reaction with methylmagnesium iodide. The data obtained were consistent with the structures shown above for bilirubin and mesobilirubin (seven active Hs each), mesobilirubinogen (eight active Hs), and hemin (three active Hs). They are also consistent with *Fischer*'s structures of bilirubinic acid (four active Hs), xanthobilirubinic acid (three active Hs), and xanthobilirubinic acid ester (two active Hs). And *Fischer* pointed out that they were not consistent with *Küster*'s tripyrrylmethane motif structure of bilirubin, which has only five active hydrogens (*398*):

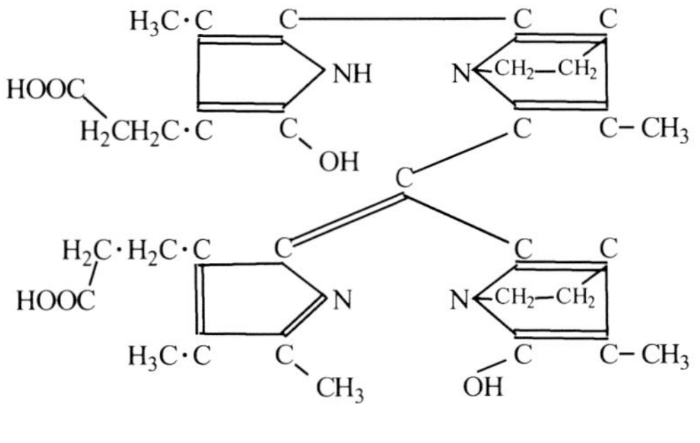

Given *Fischer*'s tetrapyrrylethylene bilirubin structure with one vinyl group and one furan ring, one might expect oxidation to give methylvinylmaleimide from one-half of the molecule and possibly some sort of fused pyrrole-furan amide. An earlier oxidative cleavage of bilirubin by HNO_2 (*372, 377*) gave a beautifully

5.1 Evolution of the Bilirubin Structure in the 1920s

crystalline product whose constitutional clarification could not be achieved because the material was so very unstable that it became transformed into a higher melting compound after but a short time. Upon subsequent reinvestigation, *Fischer* and *Niemann* (*399*) found that reaction of bilirubin in acetic acid with $NaNO_2$ led similarly to a solid, mp 82–86°C. It converted completely to a high molecular weight product (probably a polymer) within 48 hours (but was often stable for eight days when in a dessicator). The oxidation product analyzed by combustion for $C_7H_7NO_2$, and a molecular weight determination by *Pregl* in Graz gave 131 for a formula weight of 137 (*399*):

> Beim Eindunsten zuletzt im Vakuum hinterläbt er die krystalline Kruste des Oxydationsproduktes; die Eigenschaften wurden, wie von H. Fischer beschrieben, gefunden, der Schmelzpunkt zwischen 82 und 86°. Der Übergang in das höher molekulare Produkt vollzieht sich an der Luft rasch binnen 48 Stunden; im Exsiccator ist das ursprüngliche Material oft 8 Tage lang haltbar.
>
> Die Analyse ergab:
>
> | | | | | |
> |---|---|---|---|---|
> | 4,588 | mg | Substanz | gaben | 0,427 ccm N (717 mm, 18°). |
> | 4,515 | mg | „ | „ | 2,130 mg H_2O und 10,180 mg CO_2. |
>
> Methylvinylmaleinimid $C_7H_7NO_2$
>
> | | | | | | | | | |
> |---|---|---|---|---|---|---|---|---|
> | Ber. | C | = | 61,32% | H | = | 5,11% | N | = | 10,29% |
> | Gef. | | | 61,49 | | | 5,27 | | | 10,30 |
>
> Die alten Werte waren folgende: 61,4% C und 5,0% H.[1]) [*234c*].
> Eine Molekulargewichtsbestimmung nach Rast ließ sich nicht durchführen, da sich das Produkt offenbar beim Erhitzen verändert und in Campher nicht mehr löslich ist.
> Die Mikromolekulargewichtsbestimmung nach Pregl ergab folgende Werte. Als Lösungsmittel diente Alkohol.
>
> | | | | | | | | | |
> |---|---|---|---|---|---|---|---|---|
> | I. | 8,223 | mg | ergaben | eine | Depression | Δ | 0,065 °. |
> | | | | | | | M | 104. |
> | II. | 6,750 | | „ | „ | „ | Δ | 0,040 °. |
> | | $C_7H_7[N]O_2$ | mg | | | | Ber. | M137 |
> | | | | | | | Gef. | 131 |

---

[1]) Diese Zs. Bd. 91, S. 192 (1914).

Reduction of the analyzed material using Na(Hg) proved fruitless, but reduction with Al(Hg) yielded a crystalline product of mp 58–60°C that analyzed for $C_7H_9NO_2$, or two hydrogens more than its precursor. Apparently methylethylmaleimide could be reduced under the same conditions but gave a noncrystalline colorless oil (*399*):

> 40 mg des Oxydationsprodukts wurden in 30 ccm Äther gelöst und 2 g Aluminiumamalgam zugesetzt, dann 4 Stunden stehen gelassen. Der Äther hinterließ beim Eindunsten eine krystalline Kruste. Durch Sublimieren wurden 12 mg des Reduktionskörpers rein erhalten. Der Schmelz-punkt liegt bei 58-60°. Die Krystallform is quadratisch. Der

5 Preparing the Way to the Constitutional Structure of Bilirubin

Körper ist farblos und nahezu geruchlos. – Methyl-äthyl-maleinimid, das auf gleiche Weise reduziert wurde, gab ein farbloses Öl, das bisher nicht krystallisierte.

Analysen:

3,880	mg	Substanz	gaben	2,360 mg H_2O,	8,570 mg CO_2.
3,400	mg	„	„	0,303 ccm N (717 mm, 16°).	

$C_7H_9NO_2$	Ber.	C	60,43%	H	6,47%	N	10,07%
	Gef.		60,26		6,81		9,86

Die Molekulargewichtsbestimmung wurde nach Rast ausgeführt. 0,300 mg Substanz gelöst in 2,750 mg Campher gaben eine Depression von 30°. Molekulargewicht ber. 139, gef. 145. [13]

On the basis of these observations, especially the fact that the HNO_2 oxidative cleavage product, $C_7H_7NO_2$, failed to be reduced to the expected methylethylmaleimide, *Fischer* concluded that $C_7H_7NO_2$ was not methylvinylmaleimide. It analyzed (%CHN) for the latter and had an acceptably close experimentally-determined MW; so, *Fischer* deduced that it must have come from the non-vinyl-containing dipyrrylmethane half of bilirubin – the one with a furan ring. *Fischer*'s deduction led to his belief that the $C_7H_7NO_2$ product was a fused ring α-hydroxy-pyrrole-furan (*399*):

> Vor einigen Jahren[1]) wurde von H. Fischer durch Einwirkung von salpetriger Säure auf Bilirubin ein schön krystallisiertes Produkt erhalten, dessen Konstitutionsaufklärung nicht gelungen war. Die Hauptschwierigkeit war die, daß der Körper sehr unbestandig war, indem er sich nach kurzer Zeit in einen höher schmelzenden Körper umwandelte, womit gleichzeitig ein Heruntergehen im Kohlenstoffgehalt verknüpft war. Wir haben den Körper neuerdings dargestellt und können die alten Beobachtungen bestätigen. Wir haben ihn nochmals der Mikronanalyse unterworfen und nach den erhaltenen Zahlen, die gute Übereinstimmung mit den früher gefundenen zeigen, stimmen die Werte sehr gut auf Methylvinylmaleinimid. Auch die Molekulargewichtsbestimmung sprach in demselben Sinne. Trotzdem halten wir es für äußerst unwahrscheinlich, daß Methyl-vinylmaleinimid vorliegt, weil die Hydrierung des Produktes sowohl mit Natriumamalgam als auch besonders gut mit Aluminiumamalgam gelang und zu einem neuen Körper führte, der deutlich wasserstoffreicher war, aber sicher kein Methyl-äthylmaleinimid ist. Wir halten es deshalb für das wahrscheinlichste, daß dem Körper folgende Strukturformel zukommt:

$H_3C-C-C-CH$ / $HO-C,C,CH$ / $NH\ O$

und seinem Hydrierungsprodukt diese:

$H_3C-C-C-CH_2$ / $O=C,C,CH_2$ / $NH\ O$

---

[1]) Diese Zs. Bd. 91, S. 192 (1914) [*372*] und Zs. Biol. Bd. 65, S. 178 (1914) [*377*].

5.2 Tetrapyrrylethylene-Based Hemin and Bilirubin Kaput

Berichtigung.

Diese Zs. Bd. 146, S. 199, Zeile 3-4 [*399*] von unten ist die Formel so richtig:

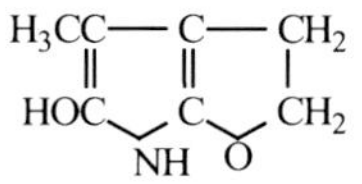

$$H_3CC—C—CH_2$$
$$HOC\diagdown C\diagdown CH_2$$
$$NH \quad O$$

14

Although the structure of the product was not completely proven by the standards that *Fischer* typically set (synthesis), in 1925 he believed that in the event the structure proved to be correct, it would represent a great advance in bilirubin research (*399*):

> Sollten diese Formeln sich als definitiv richtig erweisen, so wäre damit ein großer Fortschritt in bezug auf die Erforschung des Bilirubins gewonnen. 15

Fischer held to the belief until 1941 (*11*) that one end ring of bilirubin had one vinyl group and the other end ring had one furan ring, though the *exo*-vinyl group was never far removed from his concern.

5.2 Tetrapyrrylethylene-Based Hemin and Bilirubin Kaput

While studies of the nature of substituents at the periphery of bilirubin were underway, in the early 1920s *Fischer* initiated his first attempt at a porphyrin synthesis that might pave the way toward the eventual synthesis of bilirubin. For this, he set his sights on a version of *Willstätter*'s *Ätioporphyrin* (etioporphyrin) (*344*):

> Willstätter[4]) [*344*] hat für das Ätioporphyrin die Konstitutionsformel **III**

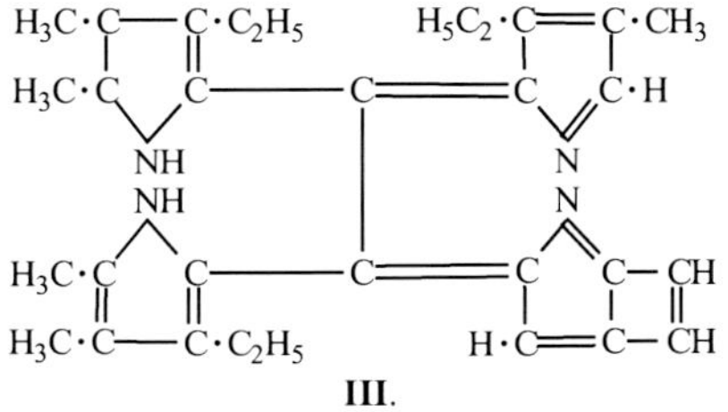

$$H_3C·C—C·C_2H_5 \qquad H_5C_2·C═C·CH_3$$
$$H_3C·C \diagdown C———C═C \diagdown C·H$$
$$NH \qquad N$$
$$NH \qquad N$$
$$H_3C·C \diagup C———C═C \diagup C—CH$$
$$H_3C·C—C·C_2H_5 \qquad H·C═C—CH$$

III.

16

A successful synthesis here might then lead to the synthesis of the structurally simplified analog of bilirubin, *Ätiobilirubin* (etiobilirubin). The synthesis focus was thus a tetrapyrrylethane, which *Fischer* and *Schubert* found to be readily accessible by condensing, *e.g.* 2,4-dimethyl-3-carbethoxy-pyrrole with glyoxal (*400*):

> . . . lag es nahe, Pyrrole mit Glyoxal alkalisch zu kondensieren, um auf diese Weise Tetrapyrryl-äthane zu erhalten. In der Tat gelang diese Reaktion glatt mit 2.4-Dimethyl-3-carbäthoxy-pyrrol, und wir erhielten das schön krystallisierende Tetra-[2.4-dimethyl-3-carbäthoxy-pyrryl-5]-äthan (**I**).

$$\left[\begin{array}{c} H_5C_2OOC \cdot C \underline{\quad} C \cdot CH_2 \quad CH_3 \cdot C \underline{\quad} C \cdot COOC_2H_5 \\ H_3C \cdot C \diagdown C \underline{\quad} CH \underline{\quad} C \diagup C \cdot CH_3 \\ NH \qquad \text{I.} \qquad NH \end{array} \right]_2 \qquad 17$$

The successful reaction here led *Fischer* to recall (*385, 400–403*) that he and *Eismayer* (*386*) had achieved such a condensation with 2,4-dimethyl-3-acetylpyrrole some ten years earlier but had not proven it with certainty (*400*):

> Diese Kondensation wurde schon vor nahezu 10 Jahren von H. Fischer und Eismayer durchgeführt, damals aber nicht veröffentlicht, weil die Analysen nicht gut stimmten und die weitere Untersuchung des analog dargestellten Tetra–[2.4-dimethyl-3-acetyl-pyrryl]-äthans[2]) [*234e, 239b*] zu Resultaten führte, die die Entstehung eines Körpers der genannten Konstitution nicht als gesichert erwiesen.

---

[2]) B. **47**, 2027, 3274 [1914]. 18

Complications were encountered during attempts to convert the tetrapyrrylethane **I** to its more unsaturated analog of *Willstätter*'s etioporphyrin type. Mild oxidation using $FeCl_3$ split the tetrapyrrylethane into two dipyrrylmethenes (*400*):

> . . . daß durch oxydierende Spaltung 2 Mol. Bis-[2.4-dimethyl-3-carbäthoxy-pyrryl-5]-methen (II) entstehen.

$$H_5C_2OOC \cdot C \underline{\quad} C \cdot CH_2 \quad H_3C \cdot C \overline{\underline{\quad}} C \cdot COOC_2H_5 \\ H_3C \cdot C \diagdown C \underline{\quad} CH \overline{\underline{\quad}} C \diagup C \cdot CH_3 \\ NH \qquad \text{II.} \qquad N \qquad 19$$

And in the presence of air, the tetrapyrrylethane turned not the expected red of a porphyrin, but yellow, similar to what happens to mesobilirubinogen (*400*):

> . . . demgemäß sollte bei der Oxydation des Tetra-[2.4-dimethyl-3-carbäthoxy-pyrryl]-äthans das Porphyrin-Spektrum auftreten. Das ist nicht der Fall. Eisenchlorid spaltet, wie bereits erwähnt zum Dipyrryl-methen. An der Luft tritt nach kurzem Liegen Gelbfärbung ein, genau wie beim Mesobilirubinogen. 20

The tetrapyrrylethane behavior thus reminded *Fischer* of a bile pigment, not of a blood pigment, and he concluded that the evidence showed that not the blood pigment (as *Willstätter* assumed) but bile pigments thus have a constitution in which their four pyrrole units are connected to a carbon-carbon bridge, as he had formulated earlier (*400*):

> Alles erinnert an den Gallenfarbstoff, nichts an den Blutfarbstoff, und wir erblicken in den Eigenschaften des synthetisch erhaltenen Tetrapyrryl-äthans eine Bestätigung der Auffassung, daß nicht der Blutfarbstoff, so wie es Willstätter im Ätioporphyrin annimmt, aus vier Pyrrolkernen, verknüpft durch eine C-C-Brücke, aufgebaut ist, sondern, daß diese Konstitution dem Gallenfarbstoff zukommt, entsprechend der Formulierung von H. Fischer. 21

5.2 Tetrapyrrylethylene-Based Hemin and Bilirubin Kaput 319

The study seemed to convince him of the correctness of the concept of his 1914–1915 tetrapyrrylethylene formula so that extrapolating from his structure for hemin (*373*):

Ich habe dann folgende Formel aufgestellt:

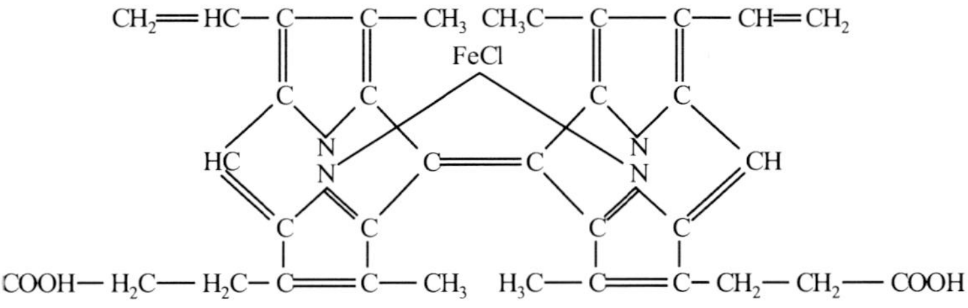

Hämin $C_{34}H_{30}O_4N_4FeCl$.

22

and for bilirubin, with its similarly bridged pyrrole α'-carbons, *Fischer* would come up with the following structure (*357, 373*):

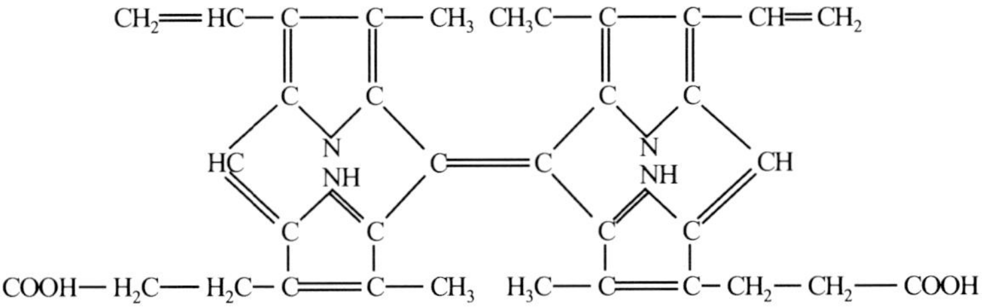

Bilirubin $C_{34}H_{32}O_4N_4FeCl$.

Since bilirubin was formed in the cell from hemin, the latter would also have the *Fischer* structure above (*373*) rather than that originally proposed by *Willstätter*, whose structure of etioporphyrin (see **III**, at the beginning of this section), would have to be revised to accommodate an additional two methine groups (*400*):

> Wir halten es mit Willstätter aber für wahrscheinlich, daß dasselbe System auch im Blutfarbstoff die Grundlage bildet; denn nachdem Fischer und Reindel[5] [*145*] nachgewiesen haben, daß in der Zelle leicht aus dem Blutfarbstoffe Hämatoidin, d. h. Gallenfarbstoff, entsteht, müssen enge, verwandtschaftliche Beziehungen zwischen den beiden Farbstoffen angenommen werden, so, wie es die Häminformel von H. Fischer[6] [*373, etc.*] ausdrückt. Wir nehmen an, daß außer der Äthinbrücke noch 2 Methingruppen in α-Stellung die Pyrrolkerne verknüpfen.

---

[5] H. 127, 299 (1928).
[6] Näheres: Ergebnisse d. Physiologie **15**, 185 bzw. 219, und in Oppenheimer: Handbuch der Biochemie (erscheint demnächst); vergl. auch Küster in Abderhaldens Biolog. Arbeitsmethoden. Abt. J, Teil 2, Heft 8 [1921].

23

5 Preparing the Way to the Constitutional Structure of Bilirubin

In order to prepare a tetrapyrrylethylene structure, *Fischer* and *Beller* discovered that treatment of a solution of the tetrapyrrylethane (**I**, above) in CS_2 with anhydrous $AlCl_3$ effected a smooth dehydrogenation to what they proposed to be **III** below (*385*):

> Wir behandelten das Äthan mit Aluminiumchlorid in Schwefelkohlenstofflösung, wobei eine glatte Dehydrierung erfolgt. Es gelingt so, allerdings mit einiger Mühe, das schön krystallisierte Tetra-(2,3-dimethyl-4-carbäthoxy-pyrryl)-äthylen (**III**) zu erhalten.

III.

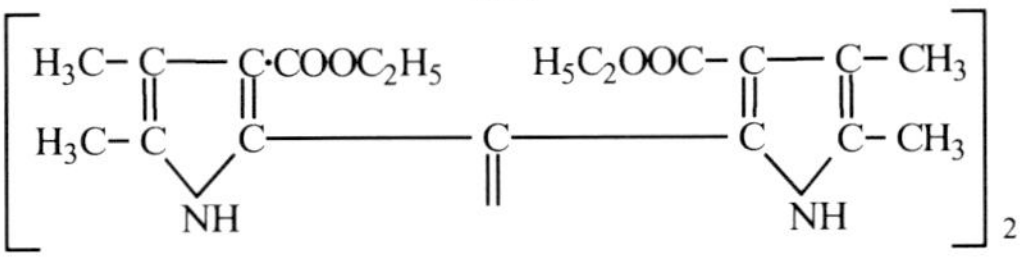

$$\left[\begin{array}{l} H_3C-\overset{|}{C}-\overset{|}{C}\cdot COOC_2H_5 \qquad H_5C_2OOC-\overset{|}{C}-\overset{|}{C}-CH_3 \\ H_3C-\overset{|}{C} \quad \overset{|}{C} \underline{\qquad\qquad} \overset{|}{C} \underline{\qquad\qquad} \overset{|}{C} \quad C-CH_3 \\ \qquad\quad NH \qquad\qquad\qquad C \qquad\qquad\qquad NH \end{array} \right]_2$$

24

The product had an intense yellow color, as in bile pigments, and turned an intense red upon treatment with aq. HCl, consistent with C=C isomerization to a dipyrrylmethene. Since **III** exhibited more of the expected porphyrin spectra (and anyway had the wrong color), *Fischer* deduced that the tetrapyrrylethylene motif, while sufficient for a bile pigment such as bilirubin, was insufficient for porphyrins like etioporphyrin and hemin (*385*):

> Wie zu erwarten, ist das Äthylen intensiv gelb gefärbt; in der Farbe erinnert es an den Gallenfarbstoff. Mit Salpetersäure in Chloroform behandelt, entsteht intensive Rotfärbung, indem offenbar dabei das oben erwähnte Methen entsteht. Soda-alkalisch wird Permanganat sofort entfärbt. Daß hier wirklich ein Körper der genannten Konstitution vorliegt, geht aus den Ergenbnissen der Analyse und der Molekulargewichtsbestimmung, ferner aus der Möglichkeit der Rückführung des Äthylens in das Äthan hervor. Nach den Ergebnissen dieser Untersuchung, insbesondere aus dem Fehlen des Porphyrinspektrums beim Äthylen, ergibt sich klar, daß für das Ätioporphyrin die einfache Verknüpfung der vier Pyrrolkerne durch eine Brücke C=C nicht in Betracht kommt. Vielmehr nehmen wir an, daß außer der Brücke C=C noch zwei weitere Methingruppen die Pyrrolkerne miteinander verknüpfen. Weitere synthetische Arbeiten nach dieser Richtung hin sind im Gange. 25

These required the two additional methine bridges shown in the bilirubin structure above.

Fischer dismissed both a *linear* arrangement of four pyrrole rings and a tripyrrylmethane structure as possible porphyrins (*402*):

> Die lineare Vereinigung von vier Pyrrolkernen, die auch in Betracht gezogen werden konnte, wurde durch eine Arbeit mit Scheyer[1]) [*404*] unwahrscheinlich gemacht, Tripyrrylmethane[2]) [*405, 406*] hatten auch keinerlei porphyrinähnliche Eigenschaften.

---

1) Mit Amman u. Heyse, B. **56**, 2319 (1923) und A. **439**, 246 (1924).
2) A. **439**, 187 (1924). 26

5.2 Tetrapyrrylethylene-Based Hemin and Bilirubin Kaput

This because, together with *Scheyer*, he synthesized the linear structure below and showed that it behaved more like mesobilirubin than a porphyrin (*404*):

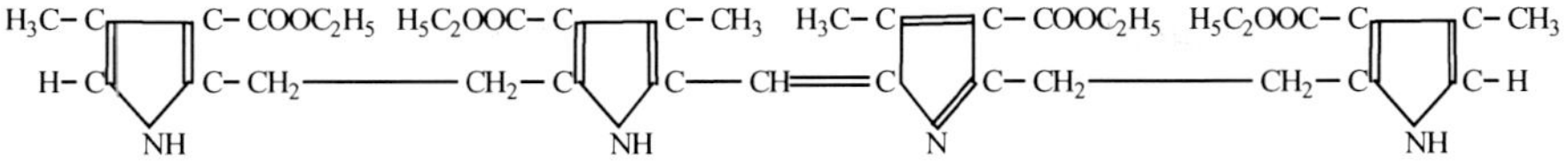

> Es ist dies eine intensiv gelb gefärbte Verbindung, wie das ja auch zu erwarten ist, und die Farbe erinnert außerordentlich an das Mesobilirubin. Die Farbintensität ist sehr groß. Eine charakteristische spektroskopische Absorption ist in Chloroform ebensowenig wie beim Gallenfarbstoff zu beobachten. 27

He also synthesized several tripyrrylmethanes, perhaps stimulated by *Küster*'s related, rather complicated structure for bilirubin (*384*) and found no properties similar to bilirubin or a porphyrin, nor could he achieve the synthesis of a tripyrrylmethane containing a pyrrole lactim structure (*405, 406*). Thus, he apparently became confident that etioporphyrin, which he expected (*401*) to have the structure shown below would be a worthy, simplified synthesis target (*402*):

> In diesem Stadium der Arbeit wurde dann die Annahme vertreten, daß außer der Äthylenbrücke noch zwei Methingruppen dazu kommen müssen, um Ätioporphyrin zu erhalten und dem Ätioporphyrin folgende Konstitutionsformel zugeschrieben . . .:

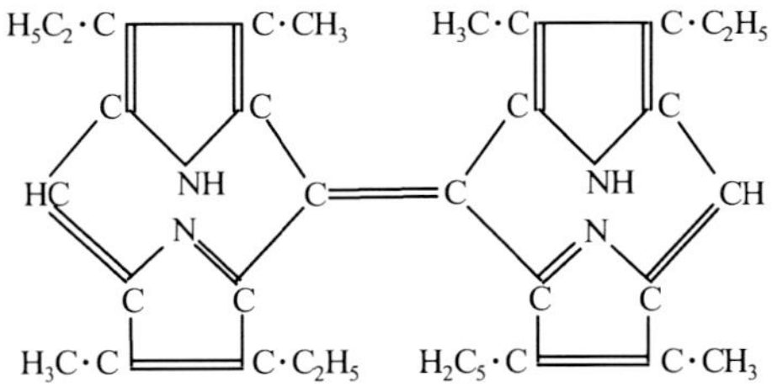

The question of how to assemble it was answered in the following scheme in which *Fischer* envisioned that dibrominated dipyrrylmethene **III** would, under the influence of conc. H_2SO_4 undergo a cyclization to **IV**, which would then self-condense to give etioporphyrin **II** (*402*):

> Wir halten deshalb die Formel III für ziemlich sicher und stellen uns dann den Übergang von hier zum Ätioporphyrin wie folgt vor:

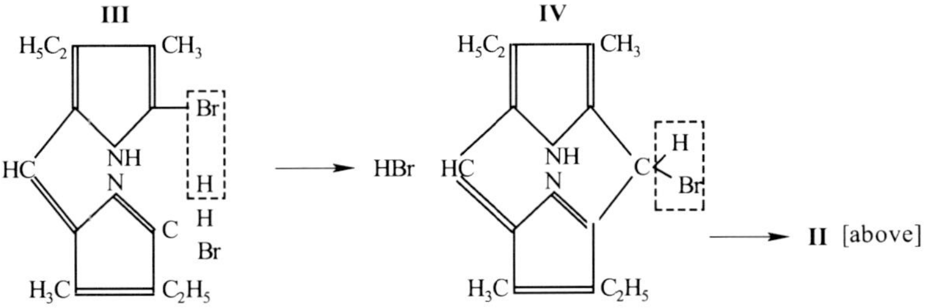

322   5 Preparing the Way to the Constitutional Structure of Bilirubin

The starting material, dibromodipyrrylmethene **III** had in fact been prepared a decade earlier by *Fischer*, who could not then distinguish it from an indigoid structure (*402*):

> Von H. Fischer wurde nun im Jahre 1916 die Entstehung von bromhaltigen, prachtvoll krystallisierten Farbstoffen beim Zusammengießen von Brom mit trisubstituierten Pyrrolen beobachtet[3]). In den letzten Jahren wurde die Konstitution dieser Farbstoffe, für die seiner- zeit Indigoide oder Dipyrryl-methenstruktur angenommen war, eindeutig im Dipyrrylmethensinne entschieden.

---

[3]) Sitz.-Ber. d. Bayr. Akad. d. Wissensch., 1915, 401.   30

Thus, from kryptopyrrole in glacial acetic acid, addition of Br_2 led to **III** above, and after treatment of **III** with conc. H_2SO_4, the isolated crystalline etioporphyrin analyzed (%CHN) for structure **II** and gave the typical porphyrin spectra (*407*) in the visible region. Its crystals were homogeneous; whereas, "analytical" etiopor- phyrin crystals obtained by decarboxylation of natural porphyrins seldom are, thus rendering difficult exact comparisons of crystals of "synthetic" etioporphyrin with the "analytical" (*402*). It is interesting to note that *Fischer*'s etioporphyrin structure (**II**) can exist in two possible stereoisomers, (*Z*) and (*E*), which was apparently not recognized, or not considered to be important – and its preparation should have led to a mixture of isomers. In addition, one might imagine yet other isomers based on N–H prototropy so as to afford "diagonally" oriented NH groups. Yet, only one (crystalline) isomer was apparently found. An oversight (?) that might have other- wise led *Fischer* to suspect the correctness of target structure **II** as etioporphyrin.

In an accompanying publication (*403*), *Fischer* and *Halbig* synthesized iso- etioporphyrin from bis-(3-ethyl-4-methyl-5-carbethoxy-2-pyrryl)-methane by pyr- rolysis of its dicarboxylic acid at 100°C. The yield was low (35 mg from 800 mg diacid) and the crystals, differed from those of etioporphyrin yet gave solutions with similar (visible) spectra. *Fischer* again proposed a reaction sequence that would lead to an ethylene-bridged bis-dipyrrylmethene (*403*):

> Was den Reaktionsmechanismus anlangt, so ist dieser vielleicht folgender:

5.2 Tetrapyrrylethylene-Based Hemin and Bilirubin Kaput

> Wir nehmen an, daß aus 1 Mol. Dicarbonsäure zunächst 1 Mol. Kohlendioxyd abgespalten wird, dann Ringschluß eintritt zu **II**, das sofort Wasser abspaltet und in das Methanketon **III** übergeht. 2 Mol. dieses Methanketons werden dann unter Wasserabspaltung vielleicht zu einem System zusammentreten, wie es in folgender Formulierung wiedergegeben ist, bei der die β-Substituenten der leichteren Übersicht halber weggelassen sind. 31

It seems clear from the available data that *Fischer* had probably synthesized porphyrins (*402, 403*), based on the spectral data and supported by the combustion analyses. However, the available data in no way uniquely supported the expected structures, and while the porphyrin synthesis with *Klarer* (*402*) would today be expected to yield what we now know as a porphyrin structure and not that (**V**, above) proposed by *Fischer*, obtaining such a compound by the reaction of *Halbig* is more problematic. Perhaps even in 1926, *Fischer* was beginning to be suspicious of his porphyrin structural formula. As expected in 1926 (*403*), he began to think in terms of a core structure that he called *Porphin* (**IV**), which by acid-catalyzed isomerization might rearrange to the skeleton of **V**, above (*403*):

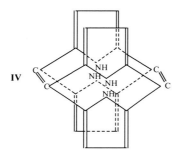

> Wir nehmen dann weiter an, daß unter der Einwirkung von Säuren eine Isomerisation des Ringsystems zu einem Bild eintritt, wie es in **V** skizziert und schon früher formuliert wurde. Zur Diskussion sei dann gestellt, ob vielleicht bei den Porphyrinen in Lösung Gleichgewichte vorliegen können zwischen **IV** und **V**; übrigens sind bei **V** ebenfalls noch Stereoisomere möglich und es ist auch nicht gleichgültig, an welcher der beiden C=C-Bindungen bei **IV** das „Aufklappen" eintritt. 32

But he also was taking into consideration *Küster*'s macrocyclic porphyrin structure as a possibility (*403*):

> Würde man vier Methingruppen zwischen den Pyrrolkernen formulieren, so wie dies Küster getan hat, müßte ein Teil des Moleküls zerfallen, um die nach dieser Reaktion fehlenden zwei Wasserstoffatome zu liefern. Die hier geschilderte Ätioporphyrin-synthese vollzieht sich nun bei 100°, ja schon beim Kochen der Methandicarbonsäure in Salzsäure oder Eisessig tritt das Porphyrinspektrum auf und deshalb halten wir die hier wiedergegebene Auffassung, mit der auch die Brom-Schwefelsäurereaktion im Einklang steht, für wahrscheinlicher. Allerdings haben wir auch mancherlei Beobachtungen gemacht, die im Einklang mit Küsters Formel mit 4 Methingruppen stehen. Die Existenz des Tetrachlor- und des Tetrabrom-Mesoporphyrins sowie die Tatsache, daß das Porphyrinogen 6 H-Atome mehr enthält als Mesoporphyrin, sind mit letzterer Auffassung gut zu erklären, mit der unsrigen allerdings ebenso. 33

When did it dawn on *Fischer* that his tetrapyrrylethylene structure was wrong? Surely suspicion must have been aroused sometime before September 2, 1926, the journal office reception date of two new articles (*408, 409*). One discussed the conversion of his constitutionally well-defined *Bilirubinsäure* (bilirubinic acid), the dipyrrylmethane cleavage product of bilirubin, into mesoporphyrin by heating to 145°C in glacial acetic acid-HBr (*408*) – a rather incredible transformation with a paltry yield. The other described the synthesis of copro- and iso-coproporphyrins from kryptopyrrole carboxylic acid (*409*). In the first (*408*), *Fischer* assumed that under the reaction conditions the lactim OH of bilirubinic acid would convert to an enol bromide (**IV**), which would then follow the scheme (below) to cyclize to **V**. The cyclized dipyrrylmethenes would somehow self-condense involving dehydrogenation to give (with respect to a regioisomeric relationship between the two propionic acid groups) an inside-inside mesoporphyrin (**VI**), or its inside-outside isomer (**VII**) (*408*):

Wahrscheinlicher ist folgender Reaktionsverlauf:

IV.   V.

VI.

VII.

5.2 Tetrapyrrylethylene-Based Hemin and Bilirubin Kaput 325

d.h. es tritt unter Bromwasserstoffabspaltung zunächst ein Dipyrryldimethan **IV** auf, das durch Dehydrierung in **V** übergeht. Dieses „Dichinon"-ähnliche Gebilde kondensiert sich dann sofort zu **VI** bzw. **VII** indem unter Isomerisation zwei „basische" Pyrrolkerne in säure Kerne übergehen. 34

Naturally, other isomers could be possible, including the outside-outside and C=C configurational isomers such as **VIII** (*408*):

Natürlich sind auch noch weitere Kombinationen möglich, z. B. folgende:

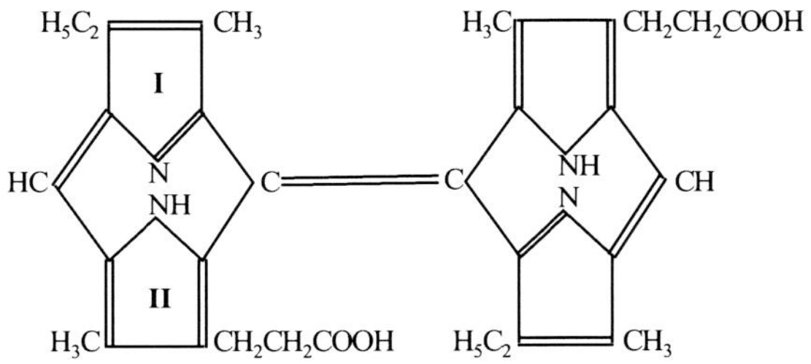

VIII. 35

These improbable stereoselective and regioselective condensation reactions had at least one redeeming feature: the carbon balance. For two C_{17} bilirubinic acid units were said to produce one C_{34} mesoporphyrin, which *Fischer* claimed to have isolated, albeit in very low yield.

The improbably hopeful condensation scheme above may be contrasted with *Fischer*'s less preferred, but nowadays more logical, more direct self-dimerization that would lead directly to the *Küster*-type porphyrin system – and only one isomer (*408*):

Man kann den Reaktionsverlauf auch im Sinne der Küsterschen Blutfarbstofformel deuten derart, daß 4 Methingruppen die 4 Pyrrolkerne miteinander verknüpfen. Folgendes Reaktionsschema ist dann zu verzeichnen:

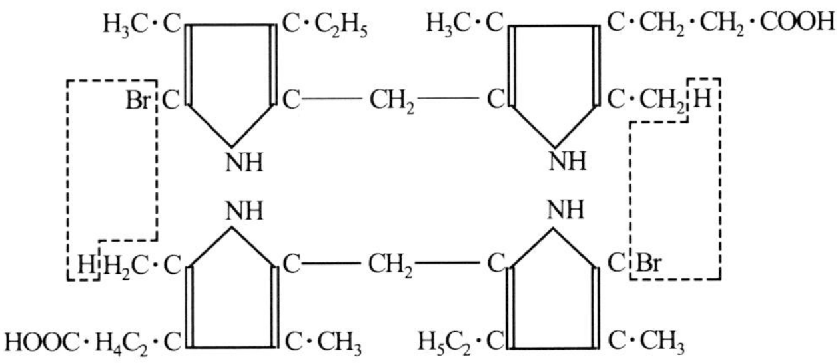

36

The mesoporphyrinogen thus formed directly would then undergo dehydrogenation to the mesoporphyrin, a transformation known to occur under the reaction conditions – and doubtless involved air oxidation.

Fischer also claimed, curiously, that mesobilirubin and bilirubin (less well) could be converted to mesoporphyrin under the same reaction conditions: heating in glacial acetic acid + HBr. Although those starting materials exhibited greater pre-organization with respect to forming the mesoporphyrin macrocycle (in *Küster*'s formulation), they were C_{33} compounds and thus lacked the 34th carbon of the porphyrin. Even better, considering that *Fischer* believed that mesobilirubin and bilirubin shared a tetrapyrrylethylene structural motif, except for the missing 34th carbon of the porphyrin, the high degree of structural pre-organization favored an easier transformation to the very closely related mesoporphyrin.

The uncertainty over which of the distinctly different types of porphyrin structures was correct carried over into work described in the second September 2, 1926 publication (*409*). Here, *Fischer* and *Andersag* converted kryptopyrrole carboxylic acid in glacial acetic acid to the brominated dipyrrylmethene hydrobromide below by reaction with Br_2 (*409*):

> Die Kryptopyrrol-carbonsäure wurde dann in kleinen Mengen in Eisessig unserm Bromierungsverfahren mit 2 Mol. Brom unterworfen und ein prachtvoll krystallisiertes bromwasserstoffsaures Salz eines gebromten Methens erhalten, dem wohl in Analogie mit den bisherigen Versuchen folgende Konstitution zukommt:

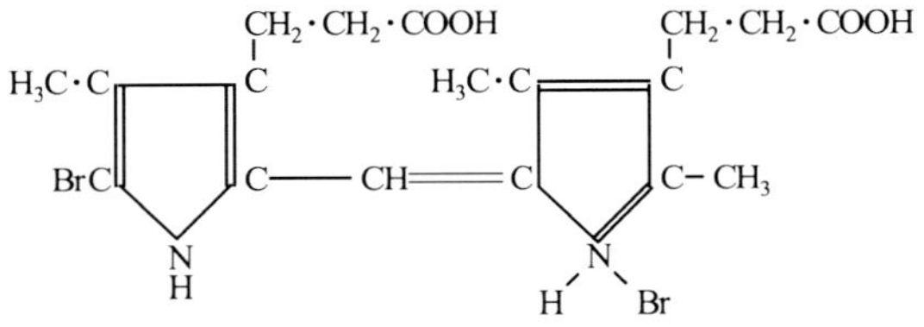

37

Although it yielded no porphyrin upon treatment with H_2SO_4, a small yield of synthetic coproporphyrin was obtained and purified as its tetramethyl ester by crystallization when the dipyrrylmethene solution in glacial acetic acid was heated in an autoclave to 210°C for three hours. The synthetic coproporphyrin gave no mixture melting point depression with "analytical" material from natural sources, and the crystals of the "synthetic" and "analytical" material were identical, as were their spectra (*409*).

> Die nähere Untersuchung steht noch aus. Mit Schwefelsäure gab der Körper kein Porphyrin, wohl aber mit Eisessig unter Druck. Dreistündiges Erhitzen auf 210° erwies sich als sehr günstig. Der Rohrinhalt war nach dieser Zeit neben kohligen Abscheidungen zum Teil krystallisiert. Er wurde durch systematische Behandlung mit Salzsäure, Lauge und Äther in das freie Porphyrin übergeführt, das aus der Ätherlösung häufig direkt krystallisierte. Zur Reinigung führten wir es mit Methylalkohol-Chlorwasserstoff in der üblichen Weise in den Ester über, der aus Chloroform-Methylalkohol krystallisierte und genau so aussah wie Kopro-ester (vgl. Fig. 2, Taf. V). Auch die krystallographische Untersuchung des Herrn Prof. Steinmetz ergab Identität. Der Ester schmilzt bei 243° und gibt mit dem analytischen Kopro-ester keine Depression. Die Elementaranalyse bestätigte die Zusammensetzung. Bei der Projektion der Spektren übereinander war keinerlei Unterschied zwischen dem synthetischen und dem analytischen Präparat zu bemerken. 38

The corresponding free acids, obtained by saponification of their esters, proved to be identical in every way, except for colloidal solubility in water, but *Fischer* could otherwise find no differences and declared them identical (*409*):

> Nur in einer Eigenschaft zeigte das Porphyrin eine Abweichung vom natürlichen insofern, als wir die kolloidale Löslichkeit in Wasser beim synthetischen Präparat vermißten. Indessen hat das natürliche Koproporphyrin auch nicht immer diese Eigenschaft und nach den obigen Feststellungen kann an der Identität der beiden Porphyrine kein Zweifel mehr sein.   39

Though undoubtedly a porphyrin had been synthesized, *Fischer* was uncertain as to whether it had a structure based on his long-held porphyrin core consisting of a tetrapyrrylethylene (**I**) or *Küster*'s more than two-decades old macrocyclic core (**II**) (*409*):

> Nach der Methoxylbestimmung und der Ammoniakzahl . . . sind vier Carboxylgruppen im Koproporphyrin enthalten und demgemäß wurden für dasselbe folgende Konstitutionsformeln aufgestellt, mit denen alle Befunde überreinstimmen . . .:

$$C_{36}H_{36}N_4O_8$$

oder nach Küster (die Anordnung der Seitenketten ist auf Grund der Synthese bereits richtig wiedergegeben):

$$C_{36}H_{38}N_4O_8$$

... Die Synthesen bestätigen restlos die Ergebnisse der analytischen Forschung. Es sind 4 Mol. Pyrrol-propionsäure, aus denen sich das Koproporphyrin aufbaut, es ist eine Tetramethylporphin-tetrapropionsäure. Nicht sicher ist die Struktur des Porphinkerns, wie aus den vorstehenden Mitteilungen des Näheren zu ersehen ist. Der Porphinkern kann indigoid (Formel I) oder nach Küster (Formel II) wiedergegeben werden. In letzterem Falle enthält das Molekül 2 Wasserstoffatome mehr; im Versuchsteil sind deshalb die theoretischen Werte für beide Formeln angegeben. Die Analysen geben keinen Entscheid zwischen beiden Formulierungen.   40

By the end of 1926, *Fischer* appeared to be well on the way toward accepting the *Küster* porphyrin formula. Re-evaluating his porphyrin synthesis from self-condensation of brominated dipyrrylmethenes (**I**), he took into account an alternative (**II**), which he explored further by condensing the dibromodipyrrylmethene with different, *i.e.* β-ethylated, dipyrrylmethene partners (*410*):

Die Porphyrinbildung wird dann durch Zusammentritt von zwei Methenmolekülen bewirkt, derart, daß Methylgruppe und Bromatom miteinander in Reaktion treten unter Bromwasserstoffabspaltung, wobei die Reaktion entweder innerhalb je eines Methenmoleküles sich abspielen kann und dann sekundäre oxydative Kondensation zweier Moleküle zum indigoid formulierten Porphyrin erfolgt [2]) oder es tritt Bromwasserstoffabspaltung zwischen 2 Mol. des Methens I ein und es entsteht ein Sechszehnring mit zwei Methin- und Methylengruppen. Letztere beide müssen dann unter Dehydrierung in Methingruppen übergehen und wir haben dann die Küstersche Porphyrinformel vor uns. Hiernach lag es nahe, zweifach gebromte Methene gemäß **II** mit Dipyrrylmethenen mit zwei α-ständigen Methylgruppen der analogen Reaktion zu unterziehen. Erprobt haben wir die Reaktion zunächst, um neuerdings *Octamethylporphin* zu gewinnen. . .

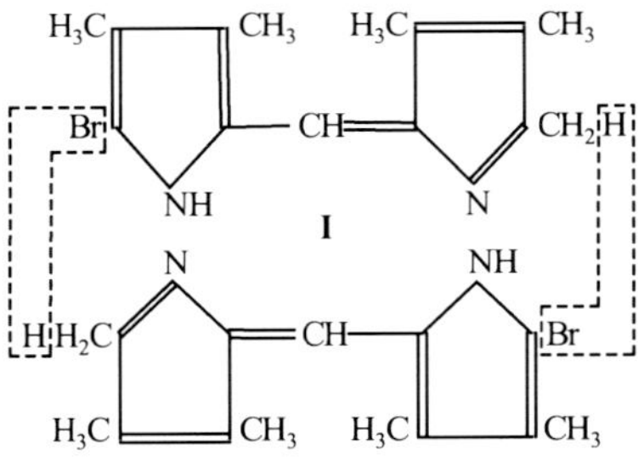

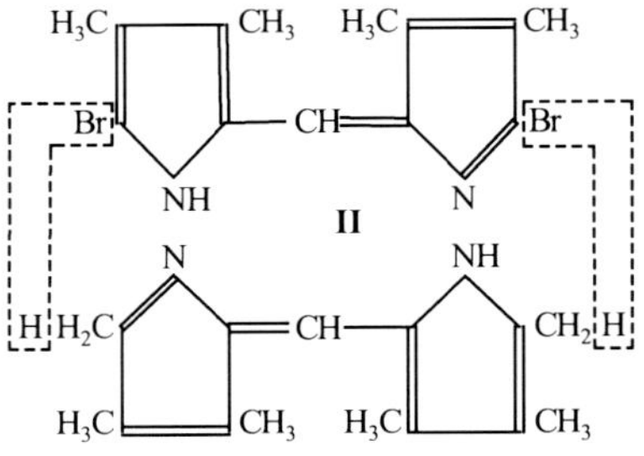

---

[2]) Diese Formulierung ist [in] A. 448, 183 (1926) [*402*] ausführlich wiedergegeben und wir geben deshalb in dieser Arbeit die Umsetzungen nach der Küsterschen Formulierung. Welche der beiden Formeln – eventuell käme auch die mit zwei Neunringen in Betracht – die richtige ist, zu entscheiden, dafür reicht das experimentelle Material noch nicht aus. Auch fehlt noch der exakte Konstitutionsbeweis für die gebromten Methene; hiermit sind wir zurzeit beschäftigt.

Wir haben das oben angegebene Methen **II** mit einer Reihe von methylierten Dipyrrylmethenen in Reaktion gebracht und überall Prophyrinbildung beobachtet.   41

5.3 Küster-Fischer Polemics

And so *Fischer* began to think in terms of a macrocyclic porphyrin structure, and with it the possible substitution patterns at the pyrrole β-carbons (*410*):

> Es ist selbstverständlich, daß alle Kombinationsmöglichkeiten der alkylierten Pyrrole und der Pyrrolcarbonsäuren herangezogen und die Porphyrine hieraus eingehend untersucht werden.[1]) Der übersichtlichen Nomenklatur halber numerieren wir im Porphingerüst die vier Pyrrolkerne mit **I, II, III, IV**.

> Die Substituenten, einerlei, welche Formulierung man auch annimmt, können nur in β-Stellung eintreten und deshalb ist es nur noch notwendig, die β-Stellungen zu bezeichen, wobei man entweder die Substituenten jedes einzelnen Pyrrolkerns für sich anzugeben oder einfacher die β-Substituenten fortlaufend zu numerieren hat mit 1, 2 bis 8, wie das angegebene Schema es zeigt.

> [1]) Anmerkung bei der Korrektur: Diese Synthesen werden über die Konstitution des Hämins wohl die Entscheidung bringen, wie als Erwiderung auf die Ausführungen W. Küsters (H. 163, S. 267) [*412*] bemerkt sei.   42

Clearly *Fischer*'s long held indigoid structure of hemin and with it the tetrapyrrylethylene structure of bilirubin were ready to be abandoned *in perpetuum* in favor of the *Küster* structure for hemin, with its novel macrocyclic porphyrin nucleus. This meant that a re-evaluation of the bilirubin structure would eventually become necessary.

5.3 *Küster-Fischer* Polemics

In his published work describing the synthesis of mesoporphyrin from bilirubinic acid and from mesobilirubinogen, *Fischer* took aim at *Küster*'s Hämin (hemin) structure, with its adjacent unsaturated β-substituents linked by a C–C bond. He stressed that his synthesis of mesoporphyrin from bilirubinic acid proved that the unsaturated side chains could not be bound to one another, as *Küster* formulated (*408*):

> W. Küster[3]) stellte nun vor kurzem für das Hämin folgende Formel auf:

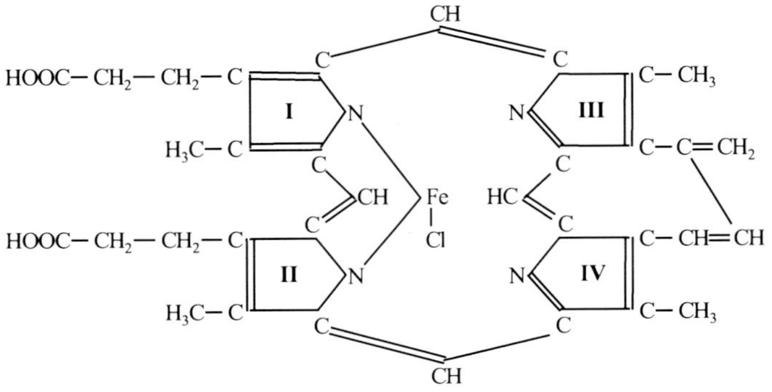

330 5 Preparing the Way to the Constitutional Structure of Bilirubin

Hämin $C_{34}H_{30}N_4O_8FeCl$

in der also die basischen Pyrrolkerne **III** und **IV** in Nachbarstellung stehen. Zu dieser Formulierung zwingt ihn die Annahme der C–C-Bindung zwischen den ungesättigten Seitenketten. Nach der obigen Synthese aus der Bilirubinsäure ist nun dieser Annahme, wenn man nach Küster formuliert, der Boden entzogen. Auf keinen Fall können die Äthylreste einander benachbart stehen und für das Hämin folgt deshalb, daß auch hier die ungesättigten Seitenketten nicht miteinander verbunden sein können, so wie dies Küster formuliert hat.

---

[3]) Chemie der Zelle u. Gewebe. Bd. 12, S. 332 (1926). 43

This apparently did not sit well with *Küster* who had also been re-evaluating the macrocyclic hemin structure he proposed in 1912 (*360*) – and found earlier in the 1907 thesis of one of his students (*411*). The *Küster* macrocyclic structure, having over the succeeding years borne the brunt of criticism from and rejection by *Willstätter* and *Fischer* in favor of a tetrapyrrylethylene motif, was never abandoned by him. Rather *Küster* was fine-tuning it, especially with respect to the adjacency of the unsaturated β-substituents on the periphery of the macrocycle. His modifications came about during studies of the halogenation of hemin ester, which (to *Küster* at least) involved the concomitant breaking apart of the assumed C–C bond between erstwhile neighboring vinyl and acetylene groups. Thus, *Küster*'s analysis of the reaction products convinced him of the validity of his hemin structure (above), with the unsaturated β-substituents linked but capable of being broken apart as indicated below (*412*):

Danach liegt zweifellos der Dimethyläther eines Chlor-Bromhämato-porphyrins vor, welchen Befund ich als vollgültigen Beweis für die Stichhaltigkeit unserer Vorstellung ansehe, wonach im verwendeten Häminester die ungesättigten Seitenketten an benachbarten Pyrrolkernen haften und, in der Form

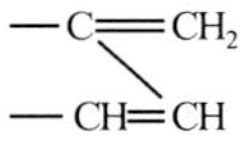

5.3 Küster-Fischer Polemics

reagierend, die zwei Atome Chlor in 1–4-Stellung addieren zu

$$-\overset{\displaystyle C}{\underset{\displaystyle CH}{\big\Vert}}\overset{\displaystyle CH_2Cl}{\underset{\displaystyle CH}{}}$$
$$\underset{\displaystyle Cl}{}$$

Durch Bromwasserstoff-Eisessig bildet sich dann:

$$- CHBr - CH_2Cl$$
$$- CHCl - CH_2Br$$

und durch Einwirkung von Methylalkohol:

$$- CH(OCH_3) - CH_2Cl$$
$$- CH(OCH_3) - CH_2Br.$$

44

Oxidative cleavage of a hemin ester substituted variously with halogens and methoxyl groups led to the isolation of a methylethylmaleimide with those substituents on the ethyl group, *e.g.* (*412*):

$$HN \overset{\displaystyle CO - \underset{\displaystyle \Vert}{C} - CH_3}{\underset{\displaystyle CO - C - CH(OCH_3) - CH_2Br}{}}$$

The same publication, received in the journal office on December 23, 1926 (*412*) also served as a vehicle to chide *Fischer* for his statement (above) that: in no case can the ethyl groups reside next to one another and for hemin it thus follows that here, too, the unsaturated side chains cannot be bonded to one another, as *Küster* has formulated (*409*). To this *Küster* responded forcefully in disagreement by indicating that *Fischer* had not sought some sort of agreement or accommodation on the subject but had tried to discredit his previous results by the very dictatorial wording of his (*408*) publication. As a consequence, *Küster* felt himself forced, regrettably, to clarify *Fischer*'s inconsistencies in a critical consideration of the situation in order to banish the risk of a recurrence of an erroneous tendency in the development of hemin chemistry arising from Munich (*412*):

> Unsere Beobachtungen stehen also im vollkommenen Widerspruch zu der von H. Fischer . . . im Anschluß an die Gewinnung von Mesoporphyrin aus Bilirubinsäure geäußerten Behauptung, daß im Hämin die ungesättigten Seitenketten nicht miteinander in Verbindung stehen und demnach der von mir vertretenen Auffassung, wonach sie an benachbarten Pyrrolkernen stehend im Bilde wiederzugeben sind, der Boden entzogen sei. Da H. Fischer es nicht versucht hat, eine Verständigung mit mir herbeizuführen, sondern durch den diktatorischen Wortlaut einer Veröffentlichung unsere bisherigen Resultate zu diskreditieren trachtet, sehe ich mich leider genötigt, durch eine kritische Betrachtung der Sachlage die Aufklärung des Widerspruchs herbeizuführen und damit die Gefahr des abermaligen Auftretens einer falschen Strömung in der Entwicklung der Häminchemie von München her zu bannen.
>
> 45

Letting it all out, *Küster* argued that from the very beginning a major issue had been the differing views on the hemin structure: his macrocycle structure *vs.* *Willstätter*'s tetrapyrrylethylene, the structural interpretation of various chemical reactions, *e.g.* hematoporphyrin from hemin, *Fischer*'s adoption of the *Willstätter* formulation and his various experiments based upon that representation. According to *Küster*, *Fischer* was, to all appearances, not capable of giving any recognition to the advances made by others in the porphyrin area. Rather, he saw only the unquestionably high merits of his own work while overlooking that such merit and good luck (in the laboratory) are together linked. Citing the example of *Fischer*'s beautiful porphyrin synthesis, *Küster* believed that good luck must have been operating for its success. Specifically, *Küster* referred to the brutal operation of heating bilirubinic acid in glacial acetic acid-HBr to stitch it together in the form of mesoporphyrin. He clearly felt that the sequence of reactions required for that conversion was completely unpredictable and that any successful interconversion must be viewed as absolutely arbitrary (*412*):

> Muß man doch bei seinen schönen Porphyrinsynthesen ebensosehr das experimentelle Geschick bewundern, welches das Herausarbeiten der gesuchten Stoffe ermöglichte, wie das Glück, das ihr vornehmliches Entstehen oder das Entstehen überhaupt begleitete, da manche der angewandten Methoden von vornherein recht wenig geeignet erschienen, zu den gesuchten Stoffen zu gelangen. Denn diese Methoden tragen den Charakter eines brutalen Eingriffs in sich. So ist auch die Gewinnung des Mesoporphyrins aus der Bilirubinsäure an eine Erhitzung desselben im Rohr auf 145° mit Eisessig-Bromwasserstoff geknüpft. Die Vorgänge, die sich hierbei abspielen mögen, sind völlig unübersichtlich, die Deutung der selben muß daher eine durchaus willkürliche sein. 46

Unsurprisingly, the yield of mesoporphyrin was far less from bilirubinic acid than from mesobilirubinogen, although in the latter example, an additional carbon atom was required. Though *Küster* thought that this might be accommodated from partial decomposition of the starting material, he found the conversion from bilirubinic acid to stretch the imagination (*412*):

> Das geht schlagend aus der Tatsache hervor, wonach die Ausbeute an Porphyrin bei der Verwendung von Bilirubinsäure bei weitem geringer ist als die Einsatz von Mesobilirubinogen erreichte, daß aber im letzteren Fall ein fehlendes C-Atom durch partiellen Zerfall des Moleküls zur Verfügung stehend angenommen werden muß. Wie weit muß nun erst der Zerfall der schlechtere Ausbeute liefernden Bilirubinsäure gehen! 47

Küster disputed *Fischer*'s claim to have been the first to produce proof that all of the connecting methines of hemin are located at the pyrrole α-positions. While acknowledging *Fischer*'s experimental proof for it, *Küster* felt that it was equally true that the proof was really not necessary, because his oxidative cleavage of hemin that led to β,β'-substituted imides had already constituted a complete proof that in hemin all of the β-positions are substituted by alkyl or propionic acid residues. From this one could conclude that the connecting methines could only be present at the α-positions. In *Küster*'s view, such a compellingly conclusive deduction required no additional experimental confirmation (*412*):

> Bestreiten muß ich aber die bei wiederholten Gelegenheiten von ihm aufgestellte Behauptung, daß er erst den Nachweis erbracht hätte, wonach im Hämin die verbindenden

5.3 Küster-Fischer Polemics

Methine sämtlich in α-Stellungen eingreifen. Wahr ist, daß er experimentelle Belege hierfür erbracht hat, ebenso wahr ist aber, daß dieselben gar nicht mehr nötig waren, denn es war durch meine Methode der Aboxydation, die zu β,β'-substituierten Imiden führt, schon der volle Beweis erbracht worden, daß in Hämin sämtliche β-Stellungen der Pyrrolkerne durch Alkyle bzw. Propionsäurereste besetzt sind; also konnten die verbindenden Methine nur in der α-Stellung vorhanden sein. Es gibt eben auch Deduktionen, welche zwingende Beweiskraft haben, sie brauchen nicht erst durch das Experiment bestätigt zu werden. 48

Acknowledging that *Fischer's* synthesis of porphyrins from β,β'-substituted pyrroles was consistent with his macrocyclic porphyrin formulation (**I**), *Küster* found even more reason to believe the *Fischer*-modified *Willstätter* structure (**II**) was untenable. For no-one had been able to cleave it into two dipyrrolic halves, as might have been anticipated (*412*):

Der einzige Unterschied nun, der zwischen dieser und meiner Formulierung besteht, nachdem H. Fischer es aufgeben mußte, die ungesättigten Seitenketten ganz oder halb nach Willstätter einzuzeichnen, besteht darin, daß H. Fischer die Pyrrolkerne I und II, sowie III und IV mit je einem Methin verbindet und eine Bindung dieser Methine postuliert, während in meinem Bild die Pyrrolkerne I und IV und II und III durch ein Methin verbunden erscheinen. Den beiden anderen Methinen kommt in beiden Bildern die gleiche Stellung zu, wie dies die Skizze veranschaulichen mag:

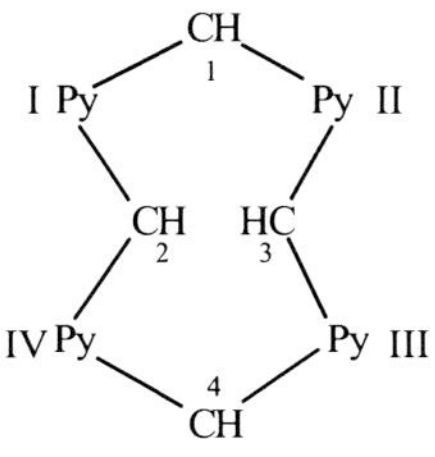

I 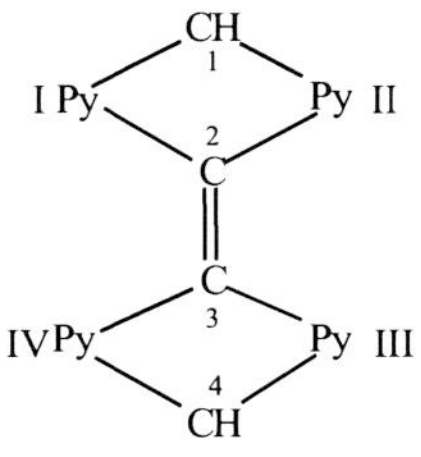   **II**

    Ich halte meine Formulierung für zutreffender, weil es noch nicht gelungen ist, eine Spaltung des Hämins in zwei zweikernige Pyrrolderivate herbeizuführen, welche Möglichkeit sich aus dem Bilde H. Fischers sofort ablesen läßt. 49

However, the suggested scission was still believed possible from bilirubin, which was thought to possess the tetrapyrrylethylene motif (**II**, above), absent methines 1 and 4. *Küster* concluded that *Fischer's* claims were not only risky but unjustified, and he was mystified that *Fischer* could allow himself, on the basis of an unfathomable synthesis, to dispute the argumentative force of his (*Küster's*) results and explanations. He was aggravated at *Fischer's* having cited only for the first time his (*Küster's*) concept of bonding between the hemin side chains and characterized *Fischer* as having an uncritical overestimation of his own results. In conjunction with other remarks, *Küster* found a systematic disparagement by *Fischer* of the results of other workers. For example, in connection with studies by others on porphyrins *Küster* believed that *Fischer* had no right to characterize *Papendieck's* results (see Section 5.4) as meaningless by virtue of the fact that he (*Fischer*) did not recognize an analysis of a porphyrin to be valid simply because

334   5 Preparing the Way to the Constitutional Structure of Bilirubin

the pigment was amorphous. *Küster* saw it as unjust that *Fischer* would at every opportunity demean his (*Küster*'s) results or dismiss an idea of another researcher with the postscript "without experimental proof" (*412*):

> So ist die eingangs erwähnte Behauptung H. Fischers nicht nur gewagt, sie ist ungerechtfertigt, und es ist mir unverständlich, wie H. Fischer, noch dazu auf Grund einer völlig undurchsichtigen Synthese, sich veranlaßt sehen konnte, die Beweiskraft unserer Resultate und Darlegungen, die auf einem fast quantitativ verlaufenden, völlig und leicht erklärbaren Abbau beruhen, in Abrede zu stellen sich erlauben konnte. Verschärfend fällt dabei ins Gewicht, daß H. Fischer meine Vorstellung über die Bindungsart der Seitenketten im Hämin erst in dem Moment zum erstenmal anführt, wo er glaubte, ihr den Boden entzogen zu haben. Solches Beginnen muß ich als kritiklose Überschätzung seiner Resultate kennzeichnen und im Verein mit anderen Aussprüchen empfinde ich es als eine systematische Herabsetzung der Resultate anderer. Hatte H. Fischer schon kein Recht, die Resultate A. Papendiecks als bedeutungslos zu bezeichnen, was m. E. dadurch geschieht, daß die Analyse eines Porphyrins nicht als stichhaltig anerkannt wird, weil es amorph ist, obgleich es aus einem krystallisierten Chlorhydrat dargestellt wurde, so begeht er vollends ein Unrecht, wenn er bei jeder Gelegenheit meine Resultate herabwürdigt oder eine Vorstellung von mir mit den Zusatz „ohne experimentelle Belege" versieht. Denn H. Fischer hat sich, in die Fußstapfen O. Pilotys tretend, eines Arbeitsgebietes bemächtigt, auf dem ich zuvor ausschlaggebende Erfolge erzielt hatte. Diese und die sich daraus ergebenden Folgerungen und Vorstellungen sind denn auch maßgeben gewesen für H. Fischers experimentelle Arbeiten.   50

Extolling *Fischer*'s abilities in organic synthesis, *Küster* indicated that while it was his own idea to combine two dipyrrylmethenes so as to constitute the prosthetic group (of hemoglobin) it was *Fischer* who had brought the idea fully to fruition by synthesis. Though *Fischer*'s preparation of pyrrole aldehydes used in the dipyrrylmethene syntheses had originated from efforts to incorporate a propionic acid group at the β-position of a pyrrole derivative, according to *Küster* the synthesis was made possible only after he (*Küster*) had proven the presence of the β-substituents in *Hämatinsäure* (hematinic acid). He believed that in the essential points *Fischer* had converted his (*Küster*'s) ideas into fact and added, belittlingly, that while he could claim to be the architect, *Fischer*'s only claim was as the builder or construction worker (*412*):

> Ich erinnere nur daran, daß meine Vorstellung von der Kombination zweier Dipyrrylmethene zur prosthetischen Gruppe von H. Fischer ausgebaut worden ist, daß seine mit so schönem Erfolg gekrönte Methode zur Herstellung von Pyrrolaldehyden aus dem Bestreben heraus entstanden ist, den Propionsäurerest an ein Pyrrolderivat in β-Stellung anzugliedern, nachdem ich das Vorhandensein eines solchen in der Hämatinsäure nachgewiesen hatte. So hat H. Fischer in wesentlichen Punkten meine Gedanken zur Tat werden lassen, und ich habe mir als Architekt die wertvolle Hilfe des geschickten Baumeisters gern gefallen lassen. Daß er sich in dieser Rolle nicht gefällt, ist zwar verständlich, aber er muß sich bewußt bleiben, daß diese Stellung sich aus der Natur der Sache heraus ergibt. H. Fischer hätte sich ein neues Arbeitsgebiet wählen sollen, wenn er außer der experimentellen Superiorität auch als selbständiger geistiger Leiter zu gelten sich vorgenommen hatte.   51

According to *Küster*, *Fischer* lacked intellectual leadership in the blood pigment area, having as of 1927 exhibited no new or special (intellectual) contribution – except for his beautiful experimental work in the porphyrin field, where he delved into *Küster*'s concept on the structure of porphyrins. *Küster* fully acknowledged

5.3 Küster-Fischer Polemics

that it would be wrong of him to fail to recognize that *Fischer*'s highly valuable and interesting investigations and his experimental superiority had provided the most practical advances toward expanding the porphyrin field (*412*):

> Auf dem Gebiete der prosthetischen Gruppe des Blutfarbstoffs hat er diese Eigenschaft bisher nicht zu beweisen vermocht, denn weder trägt seine Häminformel den Stempel von etwas Neuem oder Eigenartigem, noch haben seine wunderschönen experimentellen Arbeiten auf dem Porphyringebiet etwas dazu beigetragen, die von mir erhobenen Vorstellungen über den Bau dieser Stoffe zu vertiefen. Dagegen würde es ein Unrecht von mir sein, wollte ich verkennen, daß H. Fischers höchst wertvolle und interessante Untersuchungen das meiste zur Erweiterung des Porphyringebietes geschaffen haben. Wir sind selbstverständlich damit beschäftigt, diese Reaktionen weiter zu verfolgen, und ich möchte jetzt schon hinzufügen, daß ältere und neue Beobachtungen auf eine Verschiedenheit hinweisen, die bereits zwischen dem Eisessighämin und seinem Dimethylester bestehen, mit dem unsere bisherigen Resultate erreicht wurden. Ich glaube schon jetzt behaupten zu dürfen, daß bei der Veresterung schon eine Veränderung stattfindet, und da gerade Ester zur Betainbindung neigen, dürften die Stickstoffatome bei dieser Umwandlung in Mitleidenschaft gezogen sein.

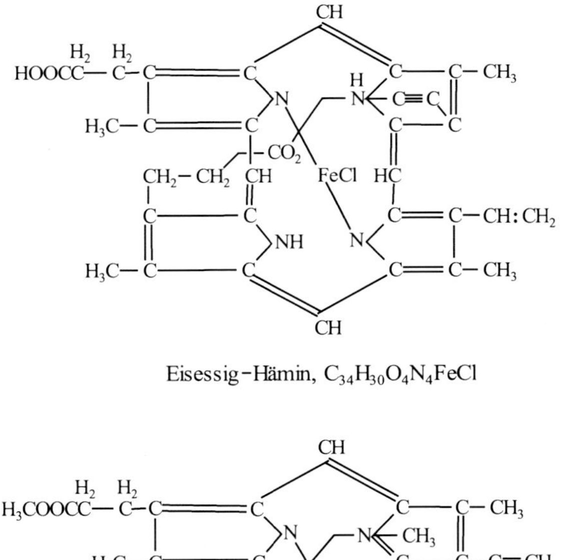

Eisessig–Hämin, $C_{34}H_{30}O_4N_4FeCl$

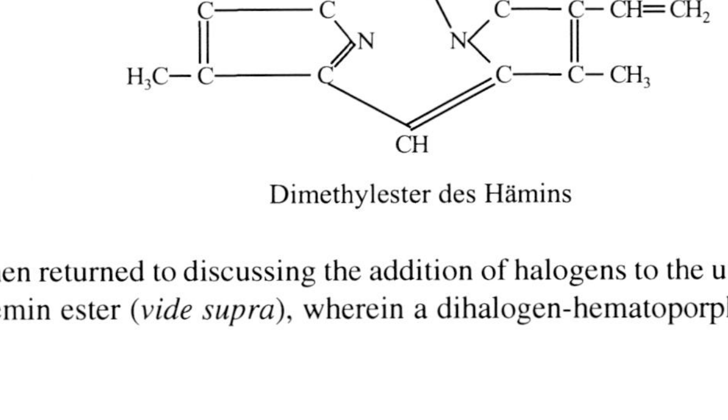

Dimethylester des Hämins

52

Küster then returned to discussing the addition of halogens to the unsaturated side chains of hemin ester (*vide supra*), wherein a dihalogen-hematoporphyrin-dimethyl ether resulted. Oxidation of the latter led to two molecules of an imide each with one halogen (*412*):

> Eine Vertiefung ist dagegen durch die von uns nachgewiesene Möglichkeit der Addition von Halogen an die ungesättigten Seitenketten und die Überführbarkeit in Dihalogen-hämotoporphyrin-dimethyläther erfolgt und dann namentlich durch den Befund der Aboxydationsmöglichkeit zu 2 Molekeln eines Imids, das nur 1 Atom Halogen im Molekül enthält.   53

From the reaction behavior of hemin ester, he concluded in favor of a structure with a vinyl and an acetylene β-substituent, as well as a betaine-like structure at one of the two basic nitrogens (*412*):

> So hat es den Anschein, als ob im Hämin selbst nicht nur das eine Carboxyl, sondern auch noch eine andere Gruppe mit dem einen basischen Stickstoffatom betainartig verknüpft ist, und das kann nur der jetzt als vorhanden nachgewiesene Acetylenrest sein, in welcher Gestalt die eine ungesättigte Seitenkette auftritt. Das Acetylen hat ja saure Eigenschaften. Lassen wir im Hämin selbst diesen Rest und zugleich das eine Carboxyl mit einem Stickstoffatom verknüpft sein, so gewinnen wir eine Formulierung, die sowohl meinen Vorstellungen, wie den von Willstätter-Fischer geäußerten, aber in einer die tatsächlichen Befunde erklärenden Modifikation Rechnung trägt und somit geeignet ist, den Streit über die Konstitution des Hämins zu einem vielleicht alle Beteiligten befriedigenden Ausgleich zu bringen.   54

Why should *Küster* have ever thought that hemin might have an acetylene group, which seems like an unnecessary alteration from his 1912 structure that possessed two vinyl groups? Apparently, he had surmised that the prosthetic group of hemoglobin was covalently bound to the protein at one of the erstwhile vinyl groups *via* an α-hydroxyethylamino linkage. In the conversion to methemoglobin, dissociation from the protein was surmised to leave an acetaldehyde group that would tautomerize to an α-hydroxy vinyl and then dehydrate to an acetylene. The last step should appear to be a high-energy process (*413*):

> Der Übergang in Methämoglobin würde dann darin bestehen, daß diese Bindung gelöst wird[2]) und daß nun eine Wasserabspaltung aus der Oxyvinylgruppe erfolgt, wie es bei der isolierten prosthetischen Gruppe bereits als feststehend bezeichnet werden kann.

$$— CH_2— CHOH— NH\text{-}Glob. \longrightarrow CH_2— CHO\,(NH_2\text{-}Glob.)$$
im Hämoglobin

$$\longrightarrow — CH{=}CHOH \longrightarrow — C{\equiv}CH$$
im Methämoglobin

Übersichtlicher wird die Skizze, wenn alle in Betracht kommenden Gruppen hinzugenommen werden.

$$\underset{\text{im Hämoglobin}}{\overset{-CO_2\cdots\;\cdot}{\diagdown}\underset{Fe}{\diagup}} — CH_2— CHOH— NH\text{-}Glob. \longrightarrow \underset{\text{im Methämoglobin}}{\overset{-COOH}{\diagdown}\underset{\underset{OH}{|}}{Fe}} — C{\equiv}CH\cdot(NH_2\text{-}Glob.)$$

$$\longrightarrow \underset{\underset{OH}{|}}{\overset{-CO———O}{\diagdown\;Fe}} \quad\underset{|}{— CH{=}CH}$$
in der prosth. Gruppe

5.3 Küster-Fischer Polemics

und es ergeben sich damit für das Hämoglobin und für den Übergang in Methämoglobin folgende Bilder:

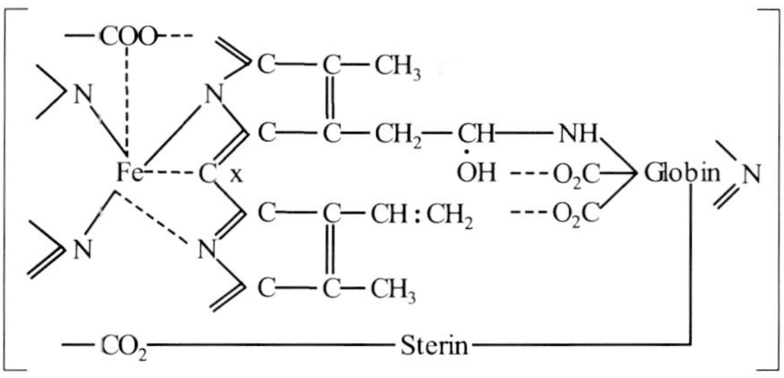

$$
\begin{bmatrix}
\text{—COO---} & \quad \text{C—C—CH}_3 \\
\text{N} & \quad \text{C—C—CH}_2\text{—CH—NH} \\
\text{Fe---C'x} & \quad \text{ÖH ---O}_2\text{C} \;\text{Globin}\; \text{N} \\
\text{N} & \quad \text{C—C—CH:CH}_2 \;\; \text{---O}_2\text{C} \\
& \quad \text{C—C—CH}_3 \\
\text{—CO}_2\text{——————— Sterin———}
\end{bmatrix}
\quad
\begin{array}{l}\text{Hämoglobin, bei}\\ \text{x das Radikal}\end{array}
$$

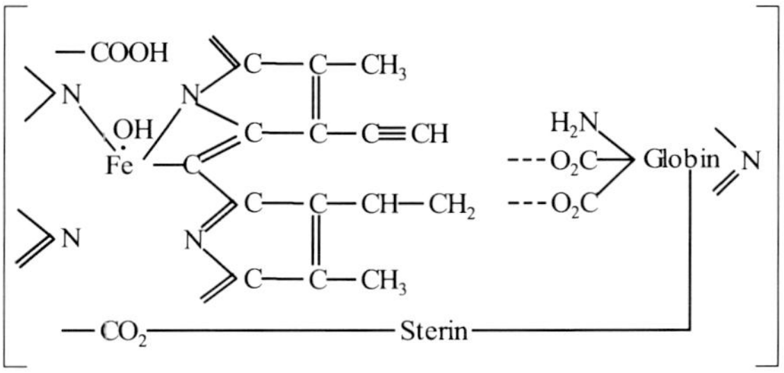

$$
\begin{bmatrix}
\text{—COOH} & \quad \text{C—C—CH}_3 \\
\text{N} & \quad \text{C—C—C}\equiv\text{CH} \qquad \text{H}_2\text{N} \\
\text{OH} \; \text{Fe—C} & \qquad \text{---O}_2\text{C} \;\text{Globin}\; \text{N} \\
\text{N} & \quad \text{C—C—CH—CH}_2 \;\; \text{---O}_2\text{C} \\
& \quad \text{C—C—CH}_3 \\
\text{—CO}_2\text{——————— Sterin———}
\end{bmatrix}
\quad
\begin{array}{l}\text{Methämoglobin,}\\ \text{bei – der}\\ \text{Zusammenschluß zum}\\ \text{Doppelmolekül, von}\\ \text{denen das eine das}\\ \text{Sterin verloren hat}\end{array}
$$

55

---

[2]) Bekannt ist, daß Methämoglobin aus Oxyhämoglobin unter der Einwirkung schwacher Säuren, Hämoglobin aus Methämoglobin durch Reduktion in schwach alkalischer Lösung entsteht, welche Tatsachen durch obige Annahme erklärbar sind. Ich möchte nicht unterlassen, darauf hinzuweisen, daß mit dieser Annahme das von F. Haurowirtz [Diese Zs. Bd. 138, S. 29 (1924)] im organischen Gerüst des Methämoglobins vermutete tautomer reagierende Carbonyl wenigstens in der prosthetischen Gruppe des Hämo- und Oxyhämoglobins vorhanden ist.

According to *Alfred Treibs*[3], a member of the *Fischer* porphyrin synthesis team and later the founder of the field of Organic Geochemistry and a Professor at the

---

[3]*Alfred Treibs* was born on July 21, 1899 in Oberstein, Germany and died in 1983. He was *Assistent* in *Hans Fischer*'s Organisch-Chemischen Institut at the TH-München from 1924 to 1929 and promoted to Dr. ing. In 1925. After the *Habilitation* he was *Konservator* in the TH-München until 1936 when he had to give up his position in view of his posture toward National Socialism. In 1946, right after the reopening, he returned to the TH to his previous position in his old workplace as an *ausserplanmässiger* (adjunct) professor. From 1962 to his emeritation in 1967 he was *o. Prof. für Organische Chemie*. *Treibs'* work with *Fischer* focused on porphyrins and their synthesis. In the 1930s in a seminal paper [Treibs A (1936) Chlorophyll- und Hämin-derivate in organischen Mineralstoffen. Angew Chem **49**:682] he showed the existence of metalloporphyrins in geological materials, and correlated the (mainly) transition metal etioporphyrin complexes, isolated from oil slates mined in southern Bavaria and the Tyrol, with the structures of chlorophylls. This provided definitive evidence that organic matter in fossil fuels has a biological origin. Considered the father of organic geochemistry, *Treibs* opened up the field of geopetroporphyrins in which the *Treibs* Medal is awarded yearly by that society. *Alfred Treibs* wrote an important book summarizing the life and work of *Hans Fischer* (*Das Leben und Wirken von Hans Fischer*), published in 1970 by the *Hans Fischer* Society in Munich.

338   5 Preparing the Way to the Constitutional Structure of Bilirubin

TH-München, *Fischer*'s reaction to *Küster*'s polemics was one of shock or dismay because theretofore he had believed that his relationship with *Küster* was nothing other than cordial (*12*):

> Zu William Küster bestand lange ein gutes Verhältnis, die Sonderabdrucke wurden gegenseitig ausgetauscht, signiert "Mit herzlichen Grüßen", und in [*373*][4] stellte ihm Küster unveröffentlichte Ergebnisse zur Verfügung . . . Zuletzt war allerdings Spannung unverkennbar, die in der verschiedenen Auffassung über die ungesättigten Seitenketten des Hämins ihren Ursprung hatte. Nach den erfolgreichen Porphyrinsynthesen erlaubt sich Küster eine überaus scharfe persönliche Herabsetzung [*412*]: "So hat Hans Fischer in wesentlichen Punkten meine Gedanken zur Tat werden lassen und ich habe mir als Architekt die wertvolle Hilfe des geschickten Baumeisters gerne gefallen lassen. Daß er sich in dieser Rolle nicht gefällt . . . ."
>
> Lange brütete Hans Fischer über einer passenden Antwort – er zitiert in [*410*] in einer Fußnote die Arbeit ganz sachlich – und war schließlich froh, daß ihm diese erspart blieb, denn Küster starb einige Zeit später.   56

Apparently the relationship between *Fischer* and *Küster* in the earlier days, cordial despite *Fischer*'s preference for the *Willstätter* formula for hemin, gave way to *Küster* scolding *Fischer* (*412*) over his comments, *inter alia*, in 1926 (*408*) that the close proximity of the unsaturated β-substituents of hemin, as shown by *Küster*, was untenable. Learning of the polemics of *Küster*'s paper (*412*), received on December 23, 1926 at the journal, *Fischer* believed that he had cited *Küster*'s work objectively and expressed it in a note added in proof to his paper received in the journal office on January 21, 1927 (*410*):

> . . . Anmerkung bei der Korrektur: Diese Synthesen werden über die Konstitution des Hämins wohl die Entscheidung bringen, wie als Erwiderung auf die Ausführungen W. Küsters (H. 163, S. 267) [*412*] bemerkt sei.   57

Though it apparently did little to smooth things over in the relationship between *Fischer* and *Küster*, according to *Treibs* (*12*), *Fischer* brooded long over an appropriate answer and was saved of it only by *Küster*'s death in early 1929. And on the basis of his synthesis work *Fischer* had clearly been rethinking his tenacious hold on the tetrapyrrylethylene formula of hemin.

5.4   *Schumm-Fischer* Polemics

The *Küster-Fischer* polemics of the mid-1920s were about the structure of hemin, not bilirubin. They were very mild compared to the more caustic discourse between *Fischer* and *Schumm*[5], also in the 1920s, but which focused less on structure and

---

[4] The citations in quoted materials have been modified to correspond to the references in this book.
[5] *Otto Schumm* (1874–1952) studied pharmacy at Marburg and received the Dr. rer. nat. at Hamburg in 1920. In 1909 he became chief of the clinical chemistry lab at the Allgemeines Krankenhaus Hamburg-Eppendorf and titular professor in 1919 at the Physiol.-Chem. Institute at the Universität Hamburg. It was at the latter where his polemics with *Fischer* began, and where in 1931 he became a.o. Professor – well after a halt had been called in 1928 to the polemics. A spectroscopic test to determine significant levels of methemalbumin in blood, which could indicate intravascular hemolysis, is known as the *Schumm* Test.

5.4 Schumm-Fischer Polemics

more on spectral analysis of urinary porphyrins excreted in porphyria. And again not on bilirubin.

Küster had referred to the polemics between *Fischer* and *Schumm*, contemporary with his own polemics with *Fischer*, while admonishing *Fischer* for his harshness toward *Schumm*'s co-worker, *Papendieck* (*412*). A dispute had arisen over work related to the identification by spectral analysis of urinary porphyrins, and *Fischer* was unable to repeat a preparation published by *Papendieck* (*414, 415*). *Fischer* chided him for an error of interpretation (*416*), and that apparently did not sit well with *Schumm* who registered his objections in 1924 (*417*), setting off a series of polemical papers back and forth between 1924 and 1926 on the use of spectral measurements for identifying porphyrins. The polemicism advanced more heatedly in 1924, in *Fischer*'s forceful response that the spectro-analytical method is best applied only to pure, crystallized material and that the chemical method of identification of uro- and coproporphyrin from porphyric urine is the only way to ensure future advances on the subject of porphyrinuria (*418*):

> Vor kurzem sind wieder zwei polemische Abhandlungen von Schumm bzw. Papendieck gegen mich erschienen, auf die ich kurz eingehen muß. Herr Schumm (Diese Zs. Bd. 136, S. 243) [*417*][6] meint vor allen Dingen, daß ich neuerdings zu einer viel weitergehenden Anerkennung der spektralanalytischen Methode gekommen sei als früher. Dies ist durchaus nicht der Fall. Im Jahre 1916 habe ich, wie Schumm ganz richtig zitiert, geäußert, daß man die spektroskopische Methode der wissenschaftlichen Arbeiten in der Regel nur auf reines krystallisiertes Material anwenden solle. Diese Angabe bezog sich auf die speziellen Verhältnisse bei der Porphyrinurie. Bei dieser Erkrankung sind meist so viel von Uro- und Koproporphyrin vorhanden, daß ohne große Schwierigkeiten nach meinen Methoden die Reindarstellung aus Harn und Kot gelingt und für ihre Identifizierung ist die chemische Methode heute so einfach, daß Spektroskopie am besten erst auf das reine krystallisierte Material angewendet wird.　58

Fischer disagreed with *Schumm*'s historical explanation of the progress of spectroscopy, stating that it depended on the quality of the instrument and that the differences between *Schumm*'s observations and his resulted from the differing sensitivities of the instruments. He objected to *Schumm*'s reproaching him over a difference of 1–2 nm for he knew of no case in the literature where one author was reproached by another for incorrect measurements based on such small differences – as *Schumm* had done repeatedly (*418*):

> Mit der historischen Auseinandersetzung Schumms über die Fortschritte der Spektroskopie kann ich mich nicht einverstanden erklären. Die Fortschritte der Spektroskopie hängen mit der Güte der Spektroskope zusammen . . . Was die Differenz zwischen den Beobachtungen von Schumm und meinen alten und unseren neuen Beobachtungen anlangt, so sind diese (minimalen)[1] Differenzen teilweise bedingt durch die verschiedene Empfindlichkeit der Apparate, zum größeren Teil aber durch die Art der Ablesung.　59

---

[1] Die Differenzen zwischen den Untersuchungen Schumms und meinen alten Beobachtungen betrugen 1-2 $\mu\mu$. Mir ist in der spektroskopischen Literatur bis jetzt kein Fall bekannt, daß ein Autor dem andern gegenüber bei solch geringen Abweichungen den Vorwurf unrichtiger Messungen erhoben hätte, wie dies von Schumm wiederholt getan worden ist.

---

[6] The citations in quoted materials have been modified to conform to the references in this book.

340   5 Preparing the Way to the Constitutional Structure of Bilirubin

Subsequently, *Fischer* continued to take *Papendieck* to task (*418*):

> Was die Abhandlung des Herrn Papendieck anlangt, so scheint mir Herr Papendieck nicht recht kompetent zu sein in der Frage der Beurteilung, ob im Serum spektroskopisch Koproporphyrin vorkommt oder nicht. Papendieck[1]) [*414*] hat nicht einmal in 170 g Kot Koproporphyrin gefunden und hat ausdrücklich bemerkt[2]) [*415*], daß er zum Trocknen des Lösungsmittels nicht Natriumsulfat angewandt hat (wie ich zu seiner Entschuldigung angenommen hatte); er muß also dabei ebenso wie bei den Seren einen methodischen Fehler begangen haben, der für den Außenstehenden schwer au beurteilen ist. Seine Untersuchungen über Seren beweisen deshalb gar nichts, trotz der vielen schönen und teuren Photographien.

[1]) Diese Zs. Bd. 128, S. 111 (1923).
[2]) Diese Zs. Bd. 133, S. 98 (1924).   60

And then *Fischer* expressed astonishment over *Papendieck*'s "arrogant" criticism regarding *Fischer*'s spectroscopic numbers (*418*):

> Besonders erstaunt war ich über die Ausführungen Papendiecks S. 296, Mitte bis 297 oben. Dort werden unsere spektroskopischen Zahlen einer ausführlichen kritischen Besprechung ohne jegliches experimentelles Material unterzogen und die Kritik gipfelt dann auf S. 297 in der Behauptung: „Für Koprophyrin müßte also eine Lage der Streifen bei etwa 592 bzw. 549,3 festgestellt werden." Es ist doch nun nichts einfacher als diese Behauptung experimentell zu prüfen und es ist charackteristisch für die Polemik des Herrn Papendieck, daß er das nicht tut, sondern statt dessen einfach unsere Zahlen einer höchst anmaßenden Kritik unterwirft.   61

Fischer thus issued a list of objections to *Schumm* and *Papendieck* over methodology, interpretation of data, and their polemics, but he also wrote about establishing a more careful experimental technique. He even offered helpful suggestions regarding whether the porphyrin content of urine of carnivores should not be compared to that from vegetarians, and he tried to warn *Schumm* that using Na_2SO_4 as a drying agent could cause adventitious loss of pigment due to retention on the drying agent (*418*):

> Zum Schlusse möchte ich noch ganz allgemein auf Porphyrin im Serum, Harn und im Kot zu sprechen kommen. Es bestehen Differenzen in den Auffassungen von Schumm und uns. Schumm findet, daß der Porphyringehalt im Harn abhängig ist vom Fleischgenuß, während wir das nicht konstatieren können. Gegen Schumms Resultate ist die fehlerhafte Methodik Schumms einzuwenden, die darin bestand, daß Schumm zum Trocken des Ätherextraktes Natriumsulfat verwendet. Dadurch sind ihm zweifellos oft geringe Mengen von Porphyrin entgangen. Dagegen ist derselbe Einwand natürlich auch gegen seine positiven Befunde zu machen und die Tatsache, daß Schumm dann trotz Trocknen mit Natriumsulfat Porphyrin erhalten hat, spricht für die Schummsche Anschauung. Denn wenn durch die Fleischkost nicht eine Vermehrung des Porphyrins erfolgt wäre, so hätte er es ja nicht finden können, weil ein Teil eo ipso durch Natriumsulfat zurückgehalten wird.   62

Polemics apparently began to run out of control by 1926, leaving *Fischer* frustrated (*419*):

> Eine Diskussion über diese und künftige „Richtigstellungen" erachte ich, insbesondere auch in Anbetracht der im Fettdruck wiedergegebenen „Bemerkung" dieser Zs., Bd. 152, S. 16 (1926) [*420*] für überflüssig und zwecklos.

5.4 Schumm-Fischer Polemics

Als schlagenden Beweis für die Unmöglichkeit einer Polemik mit Herrn Schumm gehe ich auf die während der Drucklegung dieser Notiz neue „Berichtigung" Schumms [diese Zs. Bd. 153, S. 225 (1926)] [*421*] kurz ein. Alle Vorwürfe darin sind Widerholungen alter „Berichtigungen", die ich mehrfach widerlegt habe.

Zu dem also haltlosen Vorwurf der Falschdeutung macht Schumm dann noch eine Anmerkung und wirft mir Befangenheit vor gegenüber der Deutung seiner Serumwerte bei dem Porphyriekranken Petry und behauptet, daß ich seine Zahlen als nicht beweisend ansprechen würde.

Alle Vorwürfe Schumms sind also zum zweitenmal erhoben, obwohl sie von mir bereits restlos widerlegt waren. Neue Momente hat Schumm nicht angeführt. Daß keine unserer einschlägigen Arbeiten zitiert ist, sei nur nebenbei bemerkt. Herr Schumm besitzt kein Recht mehr auf eine Erwiderung. 63

Yet *Schumm* stuck to his guns, accusing *Fischer* of misstatement (*422*):

. . . H. Fischers und H. Hilmers Angabe [*423*], daß ich Pyridinextrakte von Hefe in Porphyrin verwandelt habe, ist durchaus irrtümlich; die ihr folgenden Erörterungen auf S. 169 obiger Abhandlung sind deshalb vollkommen bedeutungslos. Ich erhebe den schärfsten Einspruch dagegen, daß H. Fischer wieder einmal einen Teil meiner Arbeitsergebnisse durch eine unrichtige Angabe schmälert.

In seiner „Bemerkung zu den Richtigstellungen Schumms" (Diese Zs. Bd. 155, S. 96) [*419*] versucht Fischer sich gegen meine Vorwürfe zu verteidigen durch Angaben, welche durchaus der Berichtigung bedürfen.

1. Er wirft mir vor, ich habe über eine frühere falsche Aussage H. Fischers eine neue „Berichtigung" veröffentlicht. In Wahrheit habe ich nicht anders getan als in einer neuen, zusamenfassenden Abhandlung über hiesige Forschungsergebnisse die Lehre vom Hämatoporphyrin des Harns in ihrer geschichtlichen Entwicklung besprochen, wobei selbstverständlich auch H. Fischers Ansicht nebst seiner fehlerhaften Deutung von Garrods Zahlen erwähnt werden mußte.

2. H. Fischer behauptet, mein Vorwurf, er habe MacMunns Zahlen falsch gedeutet, sei unberichtigt und sucht dieses durch Anführung eines späteren Ausspruchs von sich zu beweisen. Hier liegt die Unrichtigkeit darin, daß H. Fischer den Worlaut seines ersten Ausspruchs, gegen den sich meine Kritik gerichtet hat, vorenthält.

3. Meinen Vorwurf, daß H. Fischer meinen Hämatinbefund im Serum Petrys ohne jeden sachlichen Grund verdächtigt hat, halte ich aufrecht. Durch die von Herrn Fischer im ersten Satz seiner „Bemerkung" beliebte Redewendung läßt sich der von mir erhobene Vorwurf nicht im geringsten entkräften.

4. H. Fischers Ausspruch: „Alle Vorwürfe Schumms sind also zum zweiten Mal erhoben, obwohl sie von mir bereits restlos widerlegt waren. Neue Momente hat Schumm nicht angeführt" ist falsch. Mit dem Schlußsatz: „Herr Schumm besitzt kein Recht mehr auf eine Erwiderung" spricht Herr Fischer nicht mir, sondern sich das Urteil.

Auf weitere Auseinandersetzungen mit Herrn Prof. Fischer kann ich mich wegen seiner wiederholten, buchstäblich und dem Sinn nach falschen Angaben über hiesige Arbeitsergebnisse, deren Anerkennung ihn Überwindung zu kosten scheint, vorderhand nicht einlassen. 64

Despite his insistence in 1926 that he could no longer engage in further debates with *Fischer*, in 1928 (*424*), *Schumm*, while working on protoporphyrin, which was also a subject of *Fischer*'s investigations as he strove to complete its total synthesis (*425*), took issue strenuously with *Fischer*'s article in *Berichte* in 1927 (*426*). *Schumm* repeatedly cited disagreements and misinterpretations and concluded (*424*):

Gegen eine derartig weitgehende Außerachtlassung unserer Forschungs Ergebnisse auf dem Gebiete der Porphyrine lege ich Verwahrung ein, nicht minder gegen die jeder

342   5 Preparing the Way to the Constitutional Structure of Bilirubin

sachlichen Berechtigung entbehrende Anfechtung einzelner unserer Arbeitsverfahren und den Versuch, das Verdienst an wichtigen hiesigen Forschungs-Ergebnissen für sich in Anspruch zu nehmen.   65

At this point *Fischer* gave up in a short and repeatable statement expressing futility (*427*):

> Zu den Auslassungen Hrn. O. Schumms in den Berichten[2]) [*424*] zu meinem zusammenfassenden Vortrag „Über Porphyrine und ihre Synthesen" sehe ich mich zu einer Entgegnung nicht veranlaßt. Seit 1926[3]) [*421*] bin ich zu der Überzeugung der Zwecklosigkeit einer Polemik mit Hrn. Schumm gekommen und finde in dem neuerlichen Angriff keinen sachlichen Anhaltspunkt, der mich von meiner Auffassung abbringen könnte.

---

[2]) B. **61**, 784 [1928].
[3]) Ztschr. physiol. Chem. **155**, 96.   66

Polemical rhetoric in chemistry and science in general did not originate in the 20th century with *Küster*, whether toward *Fischer* or *Piloty*. Nor did it originate between *Fischer* and *Schumm*. Rather it has a far longer history stretching back to the burning issues of the 18th and 19th centuries, and doubtless in earlier ages (*22*). We have seen already that *Thudichum* was a scathing polemicist in the last quarter of the 19th century. In the middle of the 19th century, organic chemists argued heatedly at times over atomicity and atomic weights. The revolution in chemistry initiated by *Lavoisier* was resisted in the late 18th century, as were *Berzelius'* ideas in the early 19th century. Inflammatory polemics may even have served better to elicit scientific truths than mere differences of opinion; they certainly attract a broad interest, at least initially. Obviously, polemics were certainly not limited to the field of pyrrole chemistry.

It should not be surprising that such disputes might have fulminated among other chemists of those times. Certainly they did not vanish with the passing of *Piloty*, *Küster*, *Fischer*, and *Schumm* in the first half of the 20th century. In the second half of the 20th century a dispute raged for many years over the structure of the norbornyl carbocation ("classical" *vs.* "nonclassical" carbonium ion) in the physical organic discipline related to mechanism – in the then well-known *Winstein-Brown* debates and polemics. *Saul Winstein* (1912–1969), Professor at University of California, Los Angeles, was one of the founders and giants of physical organic chemistry, a nominee for the *Nobel* Prize in Chemistry, and this author's colleague at UCLA. He left a legacy of understanding the nature of delocalized carbocation intermediates and a host of now common textbook phrases, *e.g.* "neighboring group participation," "solvent participation," "internal return," "anchimeric assistance," "intimate ion pair," "ion-pair-return," "bridged ions", homoaromaticity," "non-classical ions", *etc.* In 1948 and 1952, *Winstein* had proposed a "nonclassical", delocalized norbornyl carbocation to account for the product stereochemistry and kinetics of solvolysis of *exo-* and *endo*-norbornyl brosylate (bicyclo[2.2.1]heptan-2-ol *p*-bromobenzenesulfonate). *Herbert C. Brown* (1912–2004), Professor at Purdue University and co-awardee of the 1979 *Nobel* Prize in Chemistry (with *Georg Wittig*) ("for their development of the use of boron- and phosphorus-containing compounds, respectively into important

5.5 Vindication of the Macrocyclic Porphyrin Structure

reagents in organic synthesis") vigorously and relentlessly advocated the classical norbornyl carbocation while dismissing the nonclassical. Though the existence of the nonclassical ion was subsequently proven directly by nuclear magnetic resonance (*NMR*) measurements in the 1970s (*428*) by *George Olah* (b. 1927), *Nobel* Prize in Chemistry, 1994 ("for his contribution to carbocation chemistry"), the *Winstein-Brown* debates that generated the nonclassical ion controversy drew international attention and enlivened debate throughout the 1960s and 1970s – and even well after *Winstein*'s untimely death on November 23, 1969.

5.5 Vindication of the Macrocyclic Porphyrin Structure

By mid 1927, *Fischer* had accepted *Küster*'s macrocyclic porphyrin structure that he and *Willstätter* had disparaged, and abandoned the indigoid (*429*):

> Für die Konstitution kommen zur Zeit zwei Formeln in Betracht, die Küstersche und die indigoide Formulierung. Die Küstersche Formulierung haben wir, wie aus den letzten Publikationen hervorgeht[2] [..., *410*] für die wahrscheinlichere gehalten. Nach der Synthese aus alkyliertem Dipyrrylmethen und α,α'-bromiertem Dipyrrylmethen[3] [*410*] sollte ein 16-Ring auf jeden Fall primär auftreten.

---

[2] B. **60**, 379 (1927); A. **452**, 268 (1927).
[3] A. **452**, 269 (1927).

From that time forward, the tetrapyrrylethylene framework disappeared from *Fischer*'s publications, to be replaced by something close to the conventional porphyrin structures, as in the various tautomeric forms of his synthetic iso-coproporphyrin. The last he had prepared first in 1926 from self-condensation of bis-(3-propionic acid-4-methyl-5-carboxy-pyrryl)-methane (*409*) and then in 1927 by condensation of an α,α'-dibromodipyrrylmethene (**IV**) with its α,α'-dimethyl counterpart (**V**) (*430*):

> Was nun die Konstitution des Iso-koproporphyrins und der β-Verbindung anlangt, so kommen bei Zugrundelegung der Küsterschen Formulierung und der Annahme, daß bei der Synthese der die zwei Pyrrolkerne verknüpfende Methankohlenstoff erhalten bleibt, folgende drei Formeln in Betracht, die sich nur durch die Verteilung der Pyrrolkerne mit sekundärem und tertiärem Stickstoff unterscheiden.

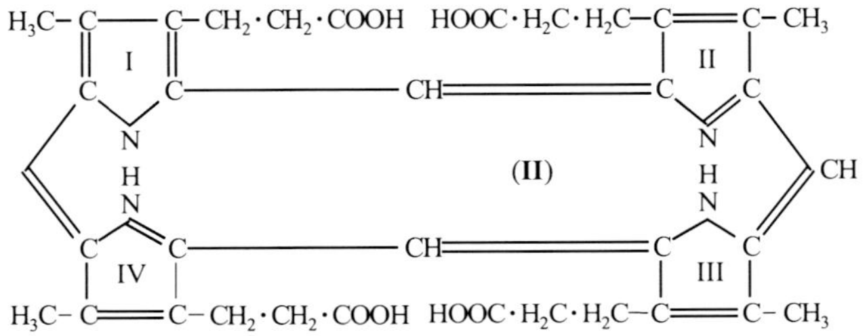

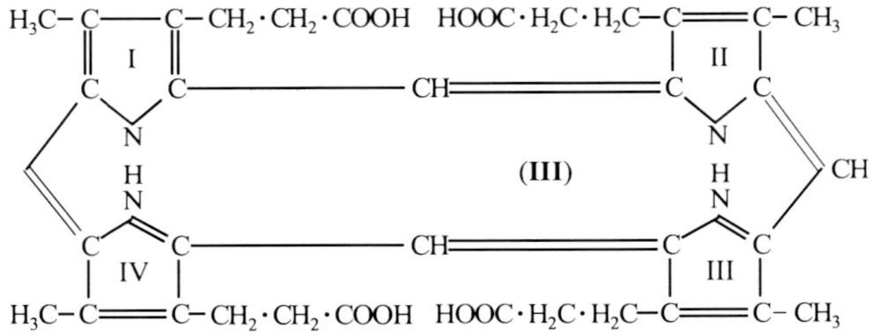

5 Preparing the Way to the Constitutional Structure of Bilirubin

$$H_3C-C\underset{I}{\overset{||}{\underset{C}{\rule{0pt}{0pt}}}}\;C-CH_2\cdot CH_2\cdot COOH \qquad HOOC\cdot H_2C\cdot H_2C-C\underset{II}{=\!=\!=}C-CH_3$$

(II)

$$H_3C-C=\!=\!=C-CH_2\cdot CH_2\cdot COOH \qquad HOOC\cdot H_2C\cdot H_2C-C=\!=\!=C-CH_3$$

(III)

$$H_3C-C=\!=\!=C-CH_2\cdot CH_2\cdot COOH \qquad HOOC\cdot H_2C\cdot H_2C-C=\!=\!=C-CH_3$$

Ein exakter Beweis dafür, daß die Kohlenstoffbrücke erhalten bleibt, ist noch nicht erbracht, aber wir halten es für sehr unwahrscheinlich, daß unter so milden Bedingungen (Erhitzen mit Ameisensäure auf 40°) eine Spaltung erfolgt. Diese Anschauung wird weiter gestützt durch die Tatsache, daß wir das Iso-koproporphyrin noch auf einem anderen Wege . . . synthetisieren konnten, nämlich durch Umsetzung des zweifach gebromten Methens der Kryptopyrrolcarbonsäure **IV** mit dem Methen der Hämopyrrolcarbonsäure **V**, Versuche, die Herr Lamatsch durchgefürt hat.

IV $\quad H_3C\cdot C=\!=\!=C\cdot C_2H_4COOH \quad HOOCH_4C_2\cdot C\!-\!C\cdot CH_3$
$\quad\quad Br\cdot C \quad\quad C=\!=\!=CH\!-\!\!-\!\!-C \quad\quad C\cdot Br$,

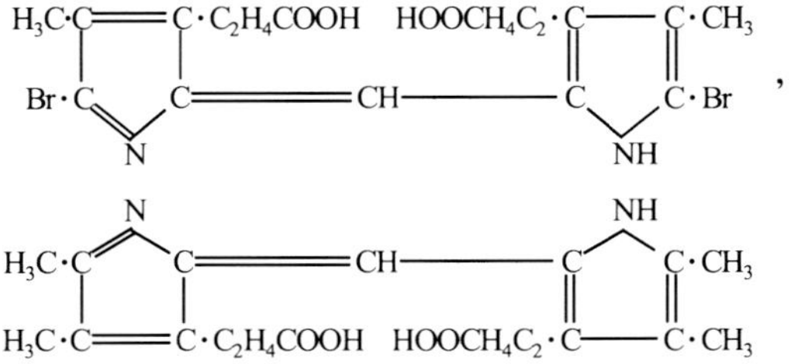

V $\quad H_3C\cdot C \quad\quad C=\!=\!=CH\!-\!\!-\!\!-C \quad\quad C\cdot CH_3$
$\quad\quad H_3C\cdot C=\!=\!=C\cdot C_2H_4COOH \quad HOOCH_4C_2\cdot C\!-\!C\cdot CH_3$

68

Thus, the totally synthetic natural porphyrin, coproporphyrin, would now be shown by *Fischer* as (*430*):

Ein Blick auf die nachfolgend wiedergegebene Koproporphyrinformel zeigt, daß hier die Verhältnisse für das Entstehen von Hämopyrrolcarbonsäure besonders günstig sind (nach dem eben Auseinandergesetzten muß die Spaltung vorwiegend an den vier Strichen erfolgen), eine

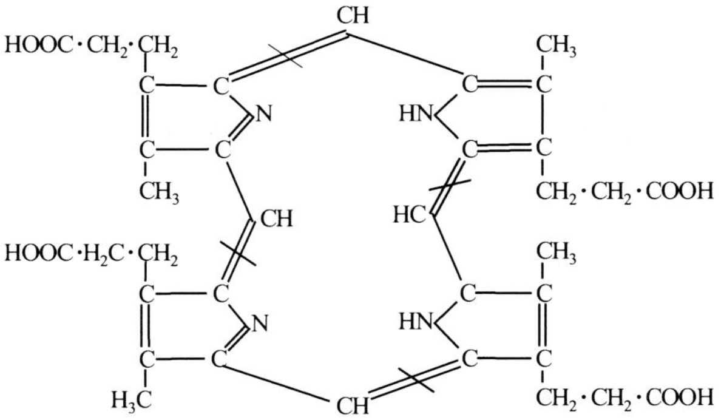

5.5 Vindication of the Macrocyclic Porphyrin Structure

$$[7]$$

$$C_{36}H_{38}N_4O_8$$

Anschauung, die experimentell schon früher[1]) durch das Ergebnis der Reduktion des natürlichen Koproporphyrins sowie jetzt des synthetischen durchaus bestätigt wird. In beiden Fällen trat Hämopyrrolcarbonsäure als Hauptprodukt auf.

---

[1]) H. **98**, 14 (1916).   69

With reference to the *Küster* porphyrin framework, by late 1927, *Fischer* had recognized four possible regio-isomers for etioporphyrin; whereas, the discredited indigoid formula would have seven – with the number of isomers being doubled in each case when NH tautomers were taken into account (*431*):

Es war nun die Möglichkeit vorhanden, daß andere Ätioporphyrine mit den Schmetterlingen[8] identisch seien und wir haben daher zunächst die auf Grund der Küsterschen Formel theoretisch möglichen Ätioporphyrine synthetisiert. Vier sind möglich (nach der indigoiden Formel sieben) von folgender Konstitution:

---

[7] An error in the upper right quadrant of the published coproporphyrin structure of ref. (*430*) has an exocyclic C=C at a pentavalent C. The C=C probably should have been exocyclic to the lower right quadrant ring.

[8] The reference in the German text to *Schmetterlingen* (butterflies) probably refers to the crystal shape.

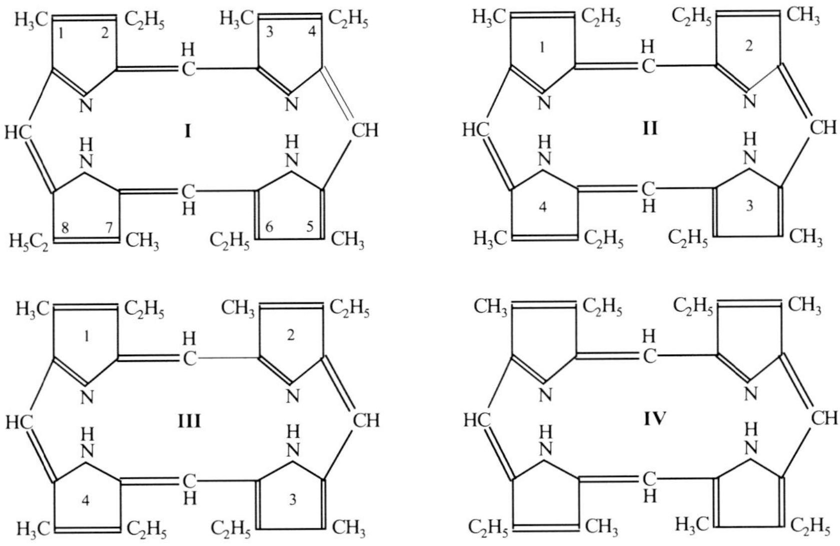

Die Zahl der Isomeren verdoppelt sich, wenn man die Tautomerie des Porphinkerns berücksichtigt.

70

Fischer synthesized all four regio-isomers of etioporphyrin (above). Some of the isomers had been synthesized earlier by *Fischer*: **I** (*402*), **II** (*403, 410*), and **IV** (*410*). Compound **III** was prepared, as were **I**, **II**, and **IV**, by directed synthesis in which structurally well-established brominated dipyrrylmethenes were stitched together – although the condensation of an α,α'-dibromodipyrrylmethene with an α,α'-dimethyl was found to be a more secure route, one less prone to rearrangement (*431*):

> Die bisherigen experimentellen Resultate stimmen alle mit den theoretischen Konstitutionsauffassungen überein, nur eine Feststellung mit Klarer[1]) nicht. Wir versuchten Ätioporphyrin **I** auch aus dem gebromten Hämopyrrolmethen zu synthetisieren, entsprechend folgendem Reaktionsschema und hier entstand nun scheinbar Isoätioporphyrin **II** anstatt Ätioporphyrin **I**.]

---

1) A. **450**, 181 (1926)

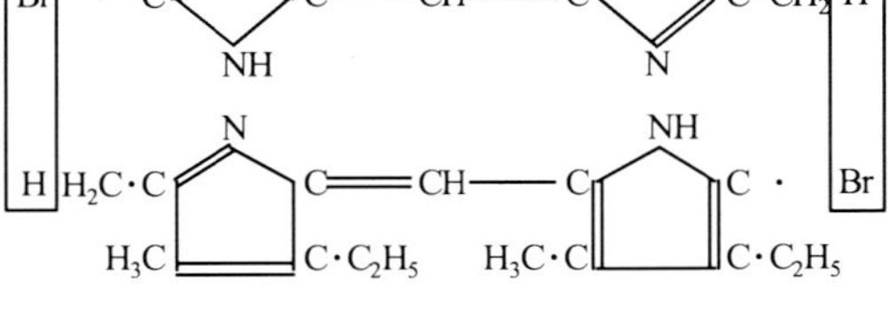

5.5 Vindication of the Macrocyclic Porphyrin Structure

Wenigstens gab das nach der Krystallform gebildete Iso-ätioporphyrin mit dem aus Kryptopyrrol Schmetterlinge. Indessen schmilzt das nach dieser Methode erhaltene Ätioporphyrin bei 400°, genau so wie Ätioporphyrin I. Der letzte Widerspruch, der hier vielleicht liegt, ist jedoch überbrückt durch das Verhalten des Bromierungsproduktes des Hämopyrrolmethens. Wir haben dies bei höherer Temperatur in Eisessig weiter bromiert und hierbei nun ein zweifach gebromtes Methen (vgl. S. 71, Formel **XIII**) erhalten, das mit Ameisensäure Ätioporphyrin I ergab. Mit **II** wurden bei der Mischkrystallisation Schmetterlinge erhalten. Hier ist also die Synthese eindeutige im Sinne der oben angegebenen Formulierung vor sich gegangen.

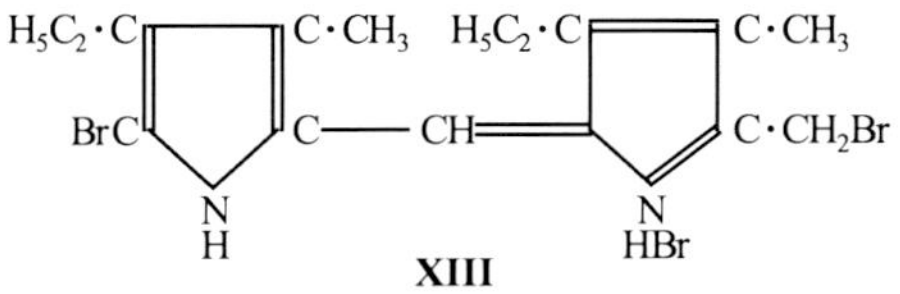

XIII

71

The four regio-isomeric etioporphyrins (see previous page) were distinguishable by their melting points and crystal forms resulting from pairwise co-crystallization (*431*):

Wir haben sie deshalb durch Aufwerfen von wenig Substanz auf den Preglschen Trockenblock . . , der auf eine bestimmte Temperatur erhitzt war, bestimmt, und so erhält man bei den Ätioporphyrinen recht scharfe Schmelzpunkte. Ätioporphyrin **I** zeigte den Schmelzpunkt von 400-405°, Ätioporphyrin **II** 365-370°, Ätioporphyrin **III** (Schmetterlinge) 360-363° und Ätioporphyrin **IV** 355-357°. Die Ausnahmestellung, die **I** einnimmt, zeigt auch der hohe Schmelzpunkt des Präparates an. Mischschmelzpunkt von **I** und **II** ergibt 380°; die durch Zusammenkrystallisieren von **I** und **II** erhaltenen Schmetterlinge haben den gleichen Schmelzpunkt und der Mischschmelzpunkt aus **III** (Schmetterlinge) mit den Schmetterlingen aus **I** und **II** liegt auch bei 380°. Es ist also keine Depression vorhanden. Das oben erwähnte in Schmetterlingen krystallisierende Ätioporphyrin aus einfach bromiertem Hämopyrrolmethen schmilzt auch bei 400°.

Wir haben dann versucht, aus **I, II** und **IV** durch Impfen die Schmetterlingsform zu erzeugen. Diese Versuche verliefen jedoch negativ; es war immerhin möglich, daß das Auftreten der Schmetterlingsform von irgendwelchen zufälligen Krystallisationsbedingungen abhängig sei. Krystallographisch sind nach den Untersuchungen von Prof. Steinmetz sämtliche Schmetterlinge identisch, nach der Schmelzpunktsbestimmung aber nicht. Wir sind nun damit beschäftigt, die Xanthoporphinogene der vier Ätioporphyrine und ihrer Mischformen darzustellen und exakt miteinander zu vergleichen unter Heranziehung ihrer calorimetrischen Werte, sowie der zugehörigen Ätioporphyrine und Koproporphyrine, die in Form ihrer Ester scharfe Schmelzpunkte besitzen, und so wird sich hoffentlich eine restlose Aufklärung ergeben.

72

This important research thus firmly confirmed the *Küster* core porphyrin structure, and, importantly, made possible *Fischer*'s formulating the 15 possible structures for mesoporphyrin, the tetrahydro reduction product (vinyls reduced to ethyls) derived from the porphyrin of hemin. *Fischer* thus diagrammatically represented the four etioporphyrin and 15 mesoporphyrin isomers, as shown below, while indicating that there would be eight regio-isomers of the etioporphyrins with either one or three ethyls replaced by propionic acids (*431*):

Ätioporphyrine.

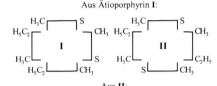

Dicarbonsäuren (Mesoporphyrintyp) (S = Propionsäurest).

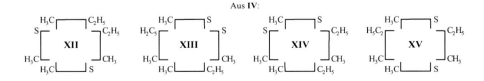

5.5 Vindication of the Macrocyclic Porphyrin Structure

Fischer's work with *Stangler* represented a major conceptual advance in porphyrin structure because mesoporphyrin might be decarboxylated to give one of the four now structurally defined synthetic etioporphyrins. *Fischer* could thus know the structure of the mesoporphyrin derived from the natural product hemin. Unfortunately, decarboxylation of mesoporphyrin gave mixed results; the product of controlled decarboxylation was at the time still being matched to one of the etioporphyrin structures. For *Fischer*, the most direct proof of the structure of mesoporphyrin was clearly synthesis, and in the realm of porphyrin synthesis he had become a master by the late 1920s (*431*):

> Der sicherste Weg war natürlich der der direkten Synthese eines Blutfarbstoff-Porphyrins. Das geeignetste mußte Mesoporphyrin sein, da seine Ester sowie Derivate scharfe Schmelzpunkte haben. 73

But which of the 15 possibilities was to be the target? The two propionic acids situated on the adjacent rings of *Küster*'s 1912 hemin structure (*360*) bore a *syn,syn* relationship to each other, *i.e.* each propionic residue was *syn* to the methine connecting the two propionic acid-bearing rings. In a 1915 structural revision, fostered probably by *Willstätter*'s and *Fischer*'s disputing the 1912 macrocyclic structure, *Küster* bowed to them by connecting two opposing methines to create the subsequently discredited tetrapyrrylethane formula – and he also displaced one of the propionic acids to the *anti,syn* arrangement (*381*):

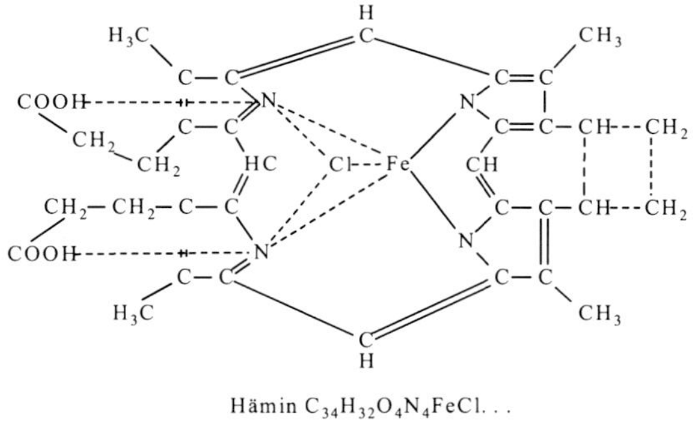

Hämin $C_{34}H_{32}O_4N_4FeCl...$

1912 Structure (*360*)

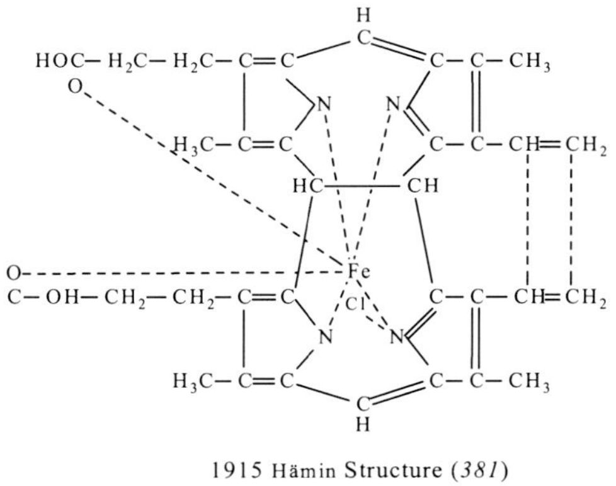

1915 Hämin Structure (*381*)

The *anti,syn* arrangement of the propionic acid residues of the 1915 structure persisted in *Küster*'s 1926 macrocyclic porphyrin structure of the dimethyl ester of hemin (*412*):

1926 *Küster* (*412*)

Or as *Fischer* and *Stangler* represented it in 1927 (*431*):

1927 *Fischer* and
Stangler (*431*)

However, *Fischer* favored the *syn,syn* orientation of the propionic groups of the 1912 *Küster* structure because he had shown that bilirubinic acid could be condensed to mesoporphyrin in glacial acetic acid-HBr (*408*), and so he reasoned that the synthetic dibromodipyrrylmethene (**IX**) should be condensed with a suitable, available α,α'-dimethyldipyrrylmethene, such as **VII**, with its *syn,anti* ethyls, or its analogs with *syn,syn* or *anti,anti* ethyls (*431*):

5.5 Vindication of the Macrocyclic Porphyrin Structure 351

[N.B. In the published structure (**IX**) shown above, the ring on the right shows a horizontal double bond, which should in fact be a single bond.]

The *syn,anti* (**VII**) would lead to porphyrin structure **IX**, the *syn,syn* to **III**, and the *anti,anti* to **XIII**. Condensation of **IX** with **VIII** led to a synthetic mesoporphyrin (now designated mesoporphyrin-**IX**) that had a melting point identical to the "analytical" mesoporphyrin derived from hemin. The synthetic and analytical material gave no mixture mp depression. Likewise, the dimethyl ester of the synthetic porphyrin gave a melting point identical with the dimethyl ester of the analytical material, and no mixture melting point depression "Das gab ihm den Ausschlag [It was decisive]" (*431*):

> *Die Synthese des Mesoporphyrins ist somit in eindeutiger Weise durchgeführt* und es kommt ihm somit folgende Konstitutionsformel zu:

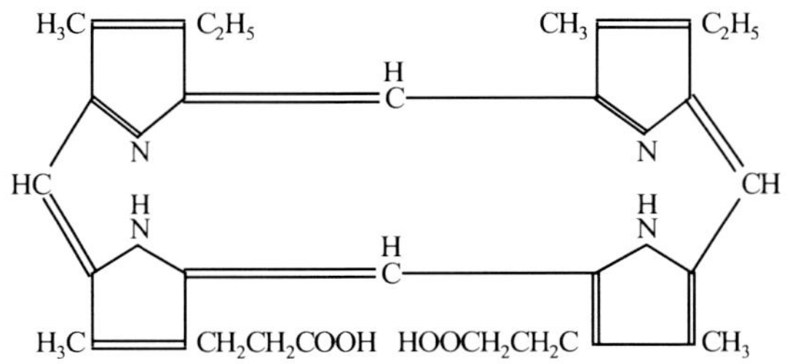

Mesoporphyrin ist also 1,3,5,8-tetramethyl-2,4-diäthyl-6,7-dipropionsäure-porphin. Ätioporphyrin III ist demnach das Ätioporphyrin, das dem Mesoporphyrin zugrunde liegt. 74

In fact *Fischer*'s mesoporphyrin-IX structure was amazingly close to *Küster*'s bridged tetrapyrrylethane representation of mesoporphyrin in 1915 (*381*):

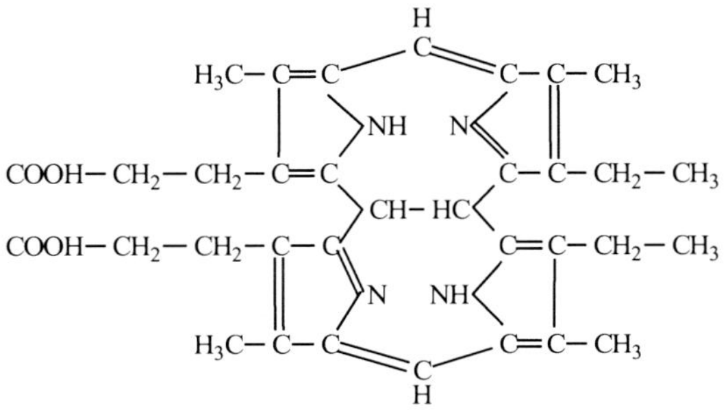

But whether it was an error in print, his porphyrinogen structure with *syn,anti* propionic acids (*381*):

would obviously not be oxidized to his mesoporphyrin with *syn,syn* propionic acids (*381*).

From *Fischer*'s experimental results, one may conclude that the porphyrin of natural hemin (protoporphyrin-IX) has the same formula as *Fischer* and *Stangler*'s 1927 structure (below) of mesoporphyrin (*431*), except with vinyls replacing ethyls. Curiously, perhaps in a concession to *Küster*, *Fischer* drew the protoporphyrin-IX with one vinyl and one acetylene group, while indicating that it was but tentative and synthesis work was underway to clinch the structure (*431*):

> Für das Hämin gilt natürlich die gleiche Anorduung der Seitenketten, denn der Übergang von Hämin in Mesohämin läßt sich sogar durch katalytische Reduktion durchführen und es ist ausgeschlossen, daß unter so milden Bedingungen eine Umlagerung der Seitenketten erfolgt. Wir formulieren das dem Hämin zugrunde liegende Protoporphyrin nach folgender Formel und bemerken, daß die Kennzeichnung der ungesättigten Seitenketten eine vorläufige ist.

> Bewiesen ist der Grad der Absättigung entsprechend der Formulierung C_4H_4. Ob und wieweit die Formulierung der beiden Seitenketten noch zu modifizieren sein wird, muß auch hier die Synthese ergeben, Arbeiten, die von verschiedenen Richtungen her im Gange sind.   75

Thus, *Fischer* had solved, in principle, one of the major historic problems in the realm of natural products, as he summarized at length in late 1927 (*426*). All that

5.5 Vindication of the Macrocyclic Porphyrin Structure

remained to him on this subject was to synthesize hemin itself – which he accomplished with *Karl Zeile* in 1929 (*425*).

For such a *tour de force* of structure proof and total synthesis, *Fischer* was nominated in 1929 for the *Nobel* Prize in Chemistry, which he received in 1930. *Küster*, too, was nominated for the award at the same time, for his novel porphyrin structure was correct in nearly every aspect in 1912, except for the ordering of the vinyl groups. Though *Küster*'s much maligned conception of the porphyrin structure was fully vindicated by *Fischer*'s synthesis, it was *Küster*'s fate that he should die in 1929 while working in his laboratory – and that the *Nobel* Prize is not usually awarded posthumously.

In July 1929, shortly after *Küster*'s death in March 1929, *Paul Pfeiffer*[9] published a memorial tribute and brief summary of his most important work (*432*). *Küster*'s student, *Paul Schlack* delivered a memorial lecture in 1960 at the University of Stuttgart on *Küster*'s life and scientific accomplishments (*411*). *Küster*'s role in determining the structure of porphyrins was summarized by *Lemberg* and *Legge* (*433*):

> Work on the structure of the porphyrins began shortly before the beginning of the twentieth century. Nencki and Zaleski, Piloty, Knorr and Hess, Willstätter and Fischer studied the products of reductive decomposition (pyrrole bases and pyrrole carboxylic acids), while Küster investigated the oxidation products (hematinic acid, methylethylmaleimide). In 1913 [actually 1912] the correct formula for the ring system of the porphyrins was suggested by Küster (*1609*) [*360*]. He assumed four pyrrole rings linked by four single carbon atoms to a sixteen-membered ring system. This formula was based on sound evidence and was stereochemically possible. At that time, however, multimembered ring systems were not yet known, and the conception was so bold that not even Willstätter was prepared to accept it. Willstätter assumed a tetrapyrrylethylene formula, while Fischer for many years defended a stereochemically unlikely formula with two eight-membered rings. By 1921 Küster had dropped his original formula, which was later proved correct by the painstaking research of Fischer and by his brilliant syntheses.
>
> By 1929 Fischer and co-workers had found a complete synthesis of protoporphyrin and hemin and had synthesized a large number of other porphyrins. This work established not only the nature of the side chains, but also their position relative to each other, which is the basis of the isomerism of porphyrins. Hemin was recognized as a complex iron salt and chlorophyll as a complex magnesium salt by Willstätter.

---

[9] *Paul Pfeiffer* was born in Elberfeldt on April 21, 1875 and died in Bonn on March 4, 1951. After studying two semesters at the University of Bonn under *A. Kekulé* and *R. Anschütz*, he transferred to the University of Zürich for his doctoral promotion in 1898 under *A. Werner*, following which he was a postdoctoral fellow in Leipzig with *Ostwald* and in Würtzburg with *Hantzsch*. He re-associated with *Werner* for the *Habilitation* at the University of Zürich and was promoted to a. o. Professor of Theoretical Chemistry there in 1908, *ordinarius* in Rostock in 1916, and at the TH Karlsruhe in 1919. In 1922, he accepted a call to succeed *Richard Anschütz* at the University of Bonn as o. Professor from which he retired in 1947. He is best known for his studies of coordination chemistry and stereoisomerization thereof. The "*Pfeiffer-Effect*" in optical activity is named after him.

5.6 The Structure of Bilirubin Reformulated

In 1929, with the structure of the hemin porphyrin fully clarified as the *Küster* macrocycle and not the earlier favored tetrapyrrylethylene representation of *Willstätter*, the early (since 1914) structures of bilirubin and other bile pigments required revision. For they had been based on the *Willstätter* 1913 structure of etioporphyrin (from hemin) and *Küster's* tripyrrylmethane-like formulations. Now bilirubin, which was known to proceed biosynthetically from hemin, had to show a resemblance to it – absent a macrocyclic ring and lacking one carbon from the C_{34} hemin structure. Like hemin, it should have two propionic acids and four methyls, at least one vinyl group, and (in *Fischer's* eyes) a second vinyl subsumed into a furan ring. The reverse transformation of bile pigments, and especially their dipyrrole fragmentation products, into mesoporphyrin in the laboratory began to make more sense.

In what might seem like a peculiar experiment today, *Fischer* sought to convert the bile pigments mesobilirubinogen and bilirubin, as well as the dipyrrole bilirubinic acid, into a porphyrin (mesoporphyrin). Yet in 1926 this might have seemed less odd because bile pigments and porphyrins were believed to share a common tetrapyrrylethylene or tetrapyrrylethane molecular framework, wherein two pairs of dipyrrylmethanes, *e.g.* two bilirubinic acids, were connected at their central carbons (*408, 409*). The surprising result is that *Fischer* claimed successful conversions, and those results possibly reinforced his belief that the pigment structures were based on a dipyrrylethylene motif – a belief that was then becoming compromised by his directed studies of porphyrins.

Possibly even more surprising were the structures *Küster* proposed in 1925–1926 for bilirubin and mesobilirubin – based more firmly on his long-held belief in a macrocyclic porphyrin. By 1925, he had abandoned his awkward dipyrryl-pyridylmethene motif of the previous decade in favor of a very different structure, one with one oxygen and two methene bridges connecting four pyrroles, that exhibited a dissymmetric ordering of the propionic acids and interconnected adjacent erstwhile vinyl groups, as represented by *Küster's* student *Schlack* (*411*):

Bilirubin nach W. Küster 1925

5.6 The Structure of Bilirubin Reformulated

Illustrating *Küster's* evolution in thought regarding bilirubin, or his uncertainty, in 1926, according to *Schlack, Küster*'s student *Walter Hermann* wrote linear tetrapyrrole structures of a different type in his dissertation (*411*):

Bilirubin nach W. Küster (Ketoform)

Mesobilirubin nach W. Küster

Here, bilirubin was represented as a lactam-hydroxypyrrole structure with erstwhile vinyl groups clamped to the end ring nitrogens, and the mesobilirubin structure was drawn in the lactim-hydroxypyrrole form, with two ethyl groups. The structures, though possessing a symmetric ordering of the pyrrole β-substituents, show an inverted ordering of the methyl and propionic acid groups of the central two rings. Yet *Küster*'s bile pigment structures (above) proved to come closer to the actual structures of the bile pigments than any others proposed in the 1920s.

From the *Fischer* and *Zeile* final structure (below, where $S = CH_2CH_2COOH$) of hemin (*425*), one can imagine four possible connecting methine carbon sites for opening the macrocyclic ring, thereby leading (after loss of a methine bridge) to a bile pigment. But which opening site would lead to the yellow pigment of bile, bilirubin?

6 The Status of Bilirubin in 1930

The year 1930 was both ominous and gratifying to *Hans Fischer*. Gratifying because it was the year in which he was awarded the *Nobel* Prize in Chemistry for his work on hemin and chlorophyll. Ominous because in 1930 the National Socialists won 107 seats in the Reichstag and became the largest political party in Germany. Having achieved major success in solving the structure of hemin by a masterpiece of late 1920s total synthesis and thereby defining the field of porphyrin/pyrrole organic chemistry, *Fischer* was concerned about future research directions. He had successfully solved the structure of the blood pigment and had synthesized numerous other porphyrins associated with metabolism, in addition to many "unnatural" analogs. In 1930, he was apparently confident of being able to synthesize any porphyrin and thus began to contemplate new pigment targets of unproven or incompletely known structure: bilirubin and chlorophyll.

As reported in 1973 by his student and co-author *Adolph Stern*, shortly after *Fischer*'s *Nobel* Prize award he and his students and co-workers came to a decision as to the future direction of research (*392*):

> A decision had to be made about which direction to continue research – bilirubin or chlorophyll. After discussion with his coworkers, Fischer decided to go ahead vigorously to elucidate the structure of chlorophyll, but not forget the bilirubin problem.

By 1930, *Küster* had died, leaving only *Fischer* and *Willstätter* to pursue research in either area. Having consulted *Willstätter* on pursuits related to chlorophyll, *Fischer* was told, in rather complimentary terms to proceed, as he (Willstätter) would no longer be participating in his former field (*434*):

> Bald überzeugte ich mich, daß meine Mitarbeit auf dem alten Gebiet nicht mehr nötig war. Professor Hans Fischer an der Technischen Höchschule München dehnte seine Forschungen über den Blutfarbstoff auch auf den Blattfarbstoff aus, und er behandelte und beherrschte die Konstitutionfragen dieser Pigmente und die synthetischen Aufgaben mit Meisterschaft, mit vollkommener Hingabe an seine Lebensaufgabe und mit größten Mitteln an Arbeitsorganisation und Arbeitskräften. Hans Fischer hat größere, bessere Arbeit geleistet, als ein anderer hätte leisten können. Zugleich war Hans Fischer immer bestrebt, dem Anteil der Vorgänger an der Forschung volle Gerechtigkeit widerfahren zu lassen.

6 The Status of Bilirubin in 1930

Halten wir uns vor Augen, daß selbst chemische Probleme, die uns heute so einfach erscheinen wie die Strukturbestimmung und Synthese der Harnsäure nicht auf einmal gelöst worden sind. Die Forscher aus drei Generationen, Liebig, Baeyer, E. Fischer, haben nacheinander ihre besten Kräfte an dieses Werk gesetzt.

Nach einem der Vorträge von H. Fischer in der Münchner Chemischen Gesellschaft, etwa 1929, über die schwierigen, feineren Einzelheiten der Chlorophyllstruktur wurde der Vortragende von einem großen Kollegen mit den Worten beglückwünscht: „Jetzt erst erleben wir den Anfang der Chlorophyllchemie."

As translated by *Lilli S. Hornig* (*435*):

Soon I convinced myself that my participation in the former field of work was no longer necessary. Professor Hans Fischer at the Munich Institute of Technology expanded his investigations of the blood pigment to include the leaf pigment as well; he was fully conversant with the field, dealing with questions of constitution of these pigments and with synthetic problems in masterful fashion, with complete dedication to his lifework and with superb organization and energy. Hans Fischer's achievements were greater and better than anyone else's could have been. At the same time he always strove to do full justice to his predecessors' share in the work.

Let us remember that even chemical problems which seem as simple today as the structural determination and synthesis of uric acid were not solved all at once. Scientists of three generations – Liebig, Baeyer, and E. Fischer – successively devoted their best efforts to this work.

After one of H. Fischer's lectures to the Munich Chemical Society, around 1929, dealing with the complicated fine details of the chlorophyll structure, a prominent colleague congratulated the lecturer in these words: "Finally we are seeing the beginning of chlorophyll chemistry."

And proceed *Fischer* did, with considerable dedication.

However, the structure of bilirubin was also on *Fischer*'s mind, and in 1930 he summarized his thoughts on the status of porphyrin and bilirubin studies, and, most importantly from a synthesis objective, the probable structure(s) of bilirubin (*436*). He recounted the earlier experimental results that yielded two closely related dipyrroles, bilirubinic acid (**18**) and xanthobilirubinic acid (**19**), representing half of the bilirubin molecule, as well as several monopyrrole components, including kryptopyrrole and kryptopyrrole carboxylic acid, *etc.* (*436*):

Bei der energischen Reduktion entsteht aus Bilirubin neben wenig Kryptopyrrol (Formel **2**, S. 1027) und Kryptopyrrolcarbonsäure (Formel **6**, S. 1027)

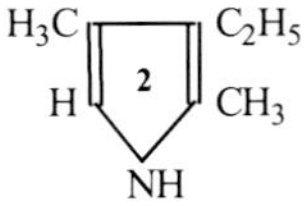

H_3C ⎤ C_2H_5
H ⎣ **2** ⎦ CH_3
NH
Kryptopyrrol

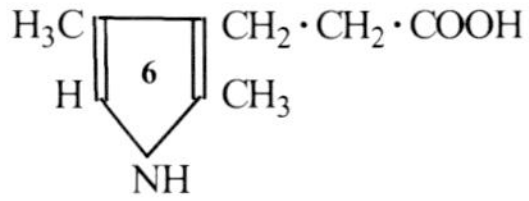

H_3C ⎤ $CH_2 \cdot CH_2 \cdot COOH$
H ⎣ **6** ⎦ CH_3
NH
Kryptopyrrolcarbonsäure

als Hauptprodukt die Bilirubinsäure, deren Konstitution durch das Ergebnis der Oxydation und Reduktion bewiesen ist:

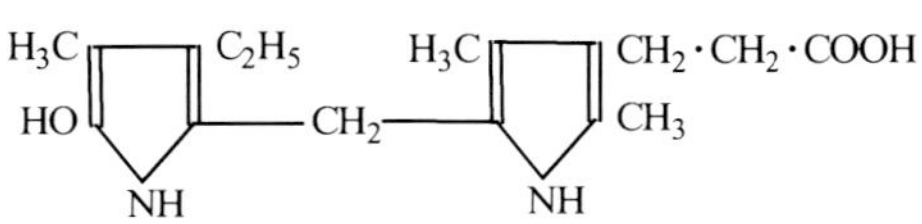

H_3C ⎤ C_2H_5 H_3C ⎤ $CH_2 \cdot CH_2 \cdot COOH$
HO ⎣ ⎦—CH_2—⎣ ⎦ CH_3
NH   NH

18

6 The Status of Bilirubin in 1930

Bilirubinsäure läßt sich zur Xanthobilirubinsäure, einem Farbstoff folgender Konstitution dehydrieren, der auch aus Bilirubin und seinen Derivaten durch Abbau mit Kaliummethylat erhältlich ist:

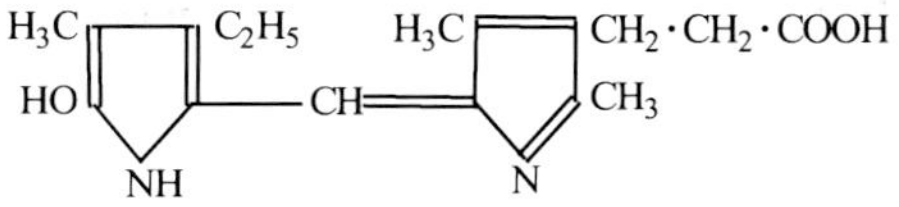

19

Fischer knew that the ethyl groups of the dipyrroles above had come from a vinyl group of bilirubin, and that an unaccounted vinyl was present as a furan (*436*):

Oxydiert man Bilirubin in Eisessig mit salpetriger Säure, so entsteht ein Körper, der nach der Analyse ein Methyl-vinyl-maleinimid sein könnte. Diese Konstitution kann jedoch nicht zutreffen, weil bei der katalystisch Hydrierung unter Aufnahme von 1 Mol Wasserstoff nicht Methyläthylmaleinimid, sondern ein neuer Körper entsteht, dem hiernach nur die Konstitution **21** zukommen kann, während das Ausgangsmaterial die Formel **20** besitzen muß.

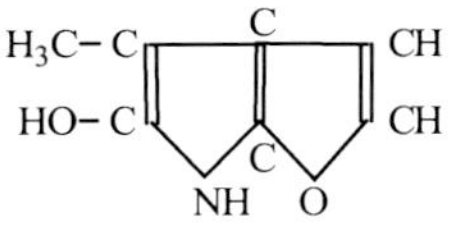

20 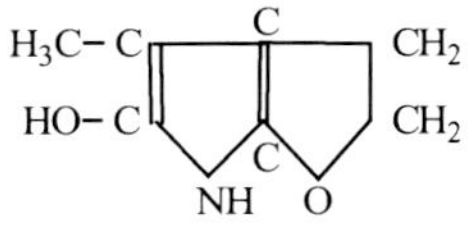   **21**

The latter deduction came from *Fischer*'s apparent inability to reduce the presumed oxidation product, methylvinylmaleimide, to methylethylmaleimide.

He knew that bilirubin contained four pyrrole-type rings, which he believed were each connected to an ethylene, in keeping with the *Willstätter* formula for the biogenetic precursor, hemin, as well as his own disproven structure. Of course, it would have a furan ring. Yet the spectral analysis of synthetic tetrapyrrylethylenes no more resembled that of bilirubin than it did that of a porphyrin. Though the porphyrin structure had been resolved in favor of *Küster*'s macrocyclic picture and not a tetrapyrrylethane-type, it was unclear whether a furancontaining tetrapyrrylethylene might not exhibit the same spectral analysis as bilirubin (*436*):

Durch Molekulargewichtsbestimmung ist die Molekülgröße des Bilirubins zu etwa 600 festgelegt, und aus der Art der Spaltprodukte folgt das Vorhandensein von vier Pyrrolkernen. Die Verknüpfung dieser vier Pyrrolkerne muß andersartig sein wie bei den Porphyrinen, weil das Spektrum fehlt. . . . Bilirubin und seine Derivate besitzen einen hohen Gehalt an aktivem Wasserstoff, und heute ist die wahrscheinlichste Formel die Verknüpfung der vier Pyrrolringe durch eine Äthylenbrücke (s. Formel **22**, S. 1031).

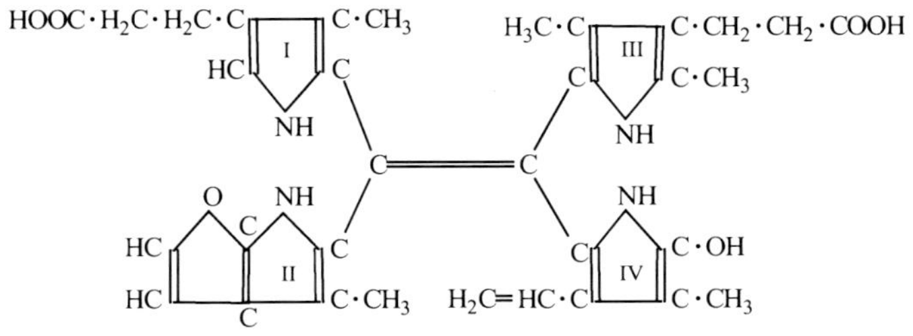

$$360 \qquad\qquad\qquad\qquad\qquad\qquad 6 \quad \text{The Status of Bilirubin in 1930}$$

22

Auf Grund der Konstitution der Spaltprodukte **20** und **21** nehmen wir noch einen Furanring im Bilirubin an, der die Säureempfindlichkeit des Bilirubins gut erklärt und auch alle sonstigen Eigenschaften. Synthetische Tetrapyrryläthylene zeigen keinerlei Porphyrinspektrum, ebensowenig wie Bilirubin. Allerdings geben sie auch keine Gmelinsche Reaktion. Indessen ist die Einführung des Furanrings in Tetrapyrryläthylene bis jetzt noch nicht gelungen. 3

Fischer noted that just as fecal flora reduce bilirubin, $C_{33}H_{36}N_4O_6$, to urobilin, Na(Hg) and catalytic hydrogenation also effect reduction. Addition of two mole equivalents of H_2 yields mesobilirubin, $C_{33}H_{40}N_4O_6$, and addition of two more equivalents of H_2 yields crystallizable mesobilirubinogen, $C_{33}H_{44}N_4O_6$. The latter exhibits a characteristic *Ehrlich* reaction, is also separable from pathologic urine, can be reconverted to mesobilirubin, and does not produce the *Gmelin* reaction because it no longer has a furan ring (*436*):

Oben haben wir die Bildung des Urobilins durch Darmbakterien kennengelernt; mit Hilfe von Natriumamalgam und ebenso durch katalytische Reduktion läßt sich diese Reaktion künstlich nachahmen und durch Aufnahme von zwei Mol Wasserstoff kommt man zu Mesobilirubin $C_{33}H_{40}N_4O_6$ und von da wiederum durch Aufnahmen von 2 Mol Wasserstoff zu Mesobilirubinogen $C_{33}H_{44}N_4O_6$, das durch intensive Ehrlichsche Reaktion ausgezeichnet ist. Dieses krystallisierte Mesobilirubinogen ist auch aus pathologischem Harn abscheidbar, es läßt sich wieder rückwärts in Mesobilirubin überführen. Mesobilirubinogen wird durch folgende Formel wiedergegeben, die das Nichtauftreten der Gmelinschen Reaktion gut erklärt, weil der Furanring in ihr nicht enthalten ist (s. Formel **23**, S. 1032).

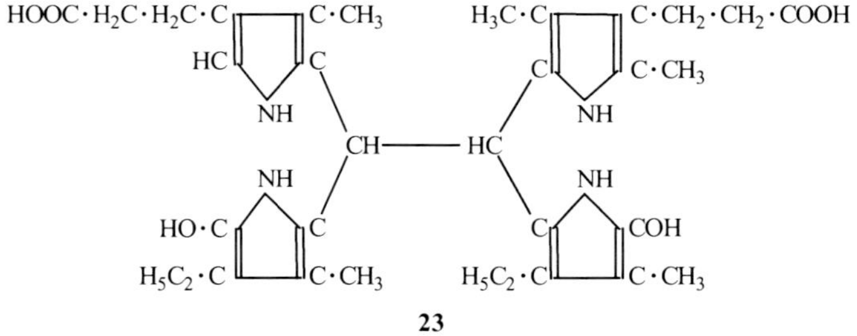

23

4

It seems that *Fischer* was bending over backward to support the "furan structure" of bilirubin and the improbable ring opening to produce mesobilirubinogen.

6 The Status of Bilirubin in 1930

Fischer also considered a linear tetrapyrrole structure (**24**) bilirubin of formula, $C_{33}H_{32}N_4O_6$, which has too many degrees of unsaturation, being short by four hydrogens of the actual formula (*436*):

> Außer der Auffassung des Bilirubins als Tetrapyrryläthylen käme noch eine offene Formel wie folgt in Frage (s. Formel **24**, S. 1032),

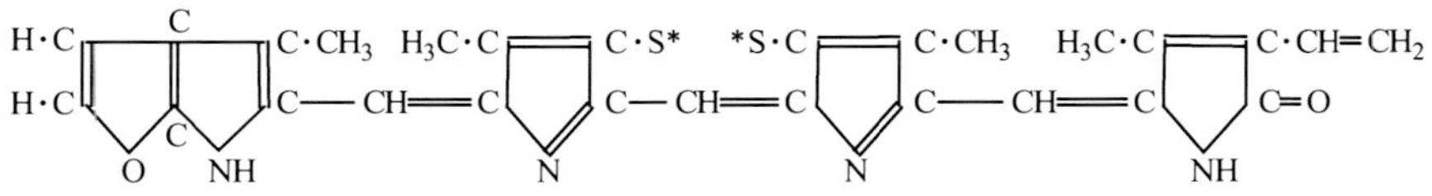

* S = $CH_2 \cdot CH_2 \cdot COOH$.

24

> die den Übergang in Mesobilirubin und Mesobilirubinogen gut erklärt, weniger gut aber die Resultate der Bestimmung des aktiven Wasserstoffes. Die Ehrlichsche Aldehydreaktion müßte man durch Aufspaltung des Furanrings erklären. Dagegen würde sich die Ableitung vom Hämin zwangloser gestalten; auf oxydativem Wege würde nach Enteisenung und Übergang in Protoporphyrin eine Methingruppe aus dem Molekül herausgesprengt und dabei Oxydation an den α-Stellen der durch diese Methingruppe verknüpften Pyrrolkerne erfolgen. 5

According to *Fischer*, for this structure to exhibit an *Ehrlich* reaction, the furan ring would have to be opened. On the other hand, compound **24** could be formed more easily from hemin (or protoporphyrin – see structure **35**, below) – by oxidative cleavage and loss of a methine. In contrast, producing the tetrapyrrylethylene structure (**22**) of bilirubin from the porphyrin (**35**) of hemin involved a far more inventive mechanism. Although *Fischer* did not outline a route using structures, he described in words how such a contorted conversion might arise, with reference to his protoporphyrin below (*436*):

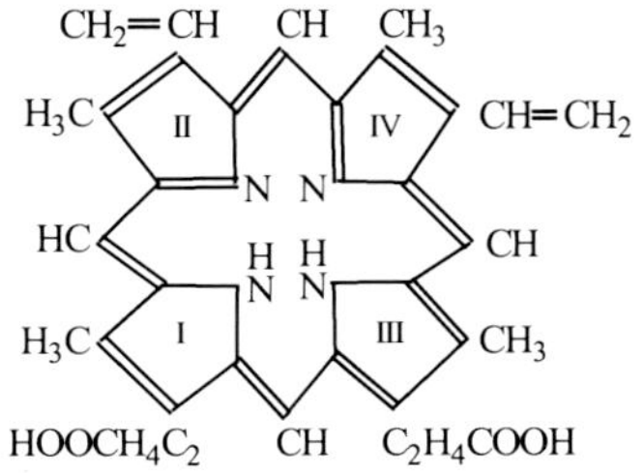

35. Protoporphyrin

First create an imaginary horizontal line (bond) connecting the two methine carbons of protoporphyrin lying between rings I and II, and III and IV. Out of plane rotation by 180° about this line (bond) would place ring II below I, and IV below III. Then dehydrogenation (loss of a hydrogen at each of the newly interconnected methines) and simultaneous conversion of the other two methines to aldehydes by some sort of hydrolytic splitting open of the macrocyclic ring would lead to a

362 6 The Status of Bilirubin in 1930

dialdehyde. A *Cannizzaro* reaction between the aldehydes would convert the newly formed aldehyde groups into a hydroxymethyl and a carboxylic acid group. Reduction of the former would produce a methyl group; decarboxylation of the latter would result in the loss of CO_2 to leave the pyrrole α-H of ring I in the proposed bilirubin structure **22** (*436*):

> Beim Übergang in das Tetrapyrryläthylen (Formel **22**, S. 1031) müßte eine Drehung von Pyrrolkernen erfolgen, die vorstellbar ist, weil ja Pyrrolkerne I und III einfach gebunden sind (vgl. Formel **22**). Primär würde im Protoporphyrin erst eine dehydrierende Vereinigung zweier Methingruppen eintreten und gleichzeitig eine hydrolytische Aufspaltung zum Dialdehyd; zwischen den beiden Aldehydgruppen könnte man dann eine Cannizarosche Reaktion sich vorstellen dergestalt, daß zunächst eine primäre Alkoholgruppe und eine Carboxylgruppe sich bilden würde; die primäre Alkoholgruppe würde dann reduziert werden zur α-Methylgruppe des Pyrrolkerns III, während die Carboxylgruppe (die nicht eingezeichnet ist) in Pyrrolkern I unter Kohlensäureabspaltung die dritte Methingruppe geben würde. So wären drei freie Methingruppen in Pyrrolkern I, IV und II intermediär zustande gekommen, die in Pyrrolkern IV und II in COH-Gruppen übergehen, von da ab ist dann der weitere Übergang zu der Bilirubinformel **22** auf S. 1031 leicht verständlich. 6

Apparently *Fischer* decided that the synthetic results spoke in favor of the (linear) chain formula because he had recently synthesized that type of system and found it to be distinguished by an intense *Gmelin* reaction, saying that it constituted the first time that the *Gmelin* reaction had been observed with synthetic material. Though the complete clarification of the constitutional structure of the bile pigments had not yet been accomplished then, a close kindred relationship to hemin was unequivocal because he had (earlier) shown (*408*) that under rather energetic conditions bilirubin, mesobilirubin, and bilirubinic acid were converted into mesoporphyrin (*436*):

> Für die Kettenformel sprechen synthetische Ergebnisse. Wir haben in neuerer Zeit derartige Gebilde synthetisiert und sie sind ausgezeichnet durch eine intensive Gmelinsche Reaktion. Es ist dies zum erstenmal, daß an synthetischem Material diese Reaktion beobachtet ist. Eine restlose Aufklärung der Konstitution des Gallenfarbstoffes ist noch nicht erfolgt, aber die nahen verwandtschaftlichen Beziehungen zum Hämin stehen eindeutig fest und sie sind weiter erhärtet durch die Uberführung des Bilirubins, Mesobilirubins und der Bilirubinsäure in Mesoporphyrin, ein Reaktion, die sich allerdings erst unter ziemlich energischen Bedingungen vollzieht. 7

Küster scoffed at such a transformation (*412*). Nonetheless, from the standpoint of a target stucture for the synthesis of bilirubin, all this left *Fischer* in doubt about the tetrapyrrylethylene platform and in favor of a linear tetrapyrrole formula. He had after all duped himself into targeting a tetrapyrrylethane-type structure for hemin and was probably hesitant to pursue that structural motif further. He may also have wondered why no dipyrrole analogs (with methyl and ethyl groups switched on the lactim ring) of bilirubinic acid and xanthobilirubinic acid had been found following cleavage of the mesobilirubin or mesobilirubinogen structure related to **24**. The decade of the 1930s would prove to be pivotal for the structure of these pigments and bilirubin.

6.1 The *Fischer* Laboratory in the 1930s and the Structure of Bilirubin

By 1930, *Hans Fischer* had established his research empire as the *Organisch-Chemische Abteilung* at the TH-München. Given that and the confidence that comes with having been awarded a *Nobel* Prize, he initiated studies toward the total synthesis of bilirubin – while the main research focus remained the total synthesis of chlorophyll. It is difficult to know exactly what it was like to work with *Fischer* during the 1930s. There is certainly no-one alive today to reminisce. Yet two of his co-workers, *Alfred Treibs* and *Cecil J. Watson*, colleagues in *Fischer's Institut*, have provided personal insights.

In 1966, *Watson* published his reminiscences of his two years in the *Fischer* laboratory, and these provide an important insight into chemistry in Munich in the early 1930s and especially that of *Fischer (437)*:

> In the fall of 1930 Munich had in some measure regained its equanimity and charm and its usual fervor for the arts and sciences, after the bleak years of World War I and the difficult period of readjustment. Yet it was scarcely affluent, and it was soon to come under a more baneful influence, whose prophet was still regarded, at least in academic circles, as a harmless demagogue or "spellbinder."
>
> Three of the most distinguished organic chemists of that period were highly respected citizens of Munich: Willstätter, already in semi-retirement though still devoting himself to the budding chemistry of enzymes; Wieland, the University Professor, pursuing his studies of the bile acids and of biological oxidation-reduction mechanisms; Hans Fischer, the youngest of the three, the Professor of Organic Chemistry and Director of the Organic Chemical Division of the *Technische Höchschule*. . .
>
> In the fall of 1930 Fischer received the Nobel Prize in Chemistry for his synthesis of hemin. . . Shortly after his return from Stockholm a *Feierabend* was planned by his many graduate students and assistants. With appreciation of their generally straitened finances, Fischer pre-empted the costs of the dinner, which was a happy affair, replete with various items of the Bavarian cuisine, including the *Münchener* beer and *Weisswurst*. The dinner was attended by about ninety of Fischer's graduate students, or *Mitarbeiter*, nearly all of whom were working under his direction in the *Organisch-Chemische Abteilung*. . .
>
> I remember wondering again, when I saw so many of his graduate students gathered for the evening, how Fischer could possibly advise such a large group in respect to their individual research programs. I had asked myself this same question the first time I walked through the enormous laboratories under his direction. There were three of these cavernous rooms, each with bench space for an average of twenty to thirty students. Faintly smoky and heavily redolent of many volatile compounds, the composite picture suggested the old tale of the industrious gnomes in their underground workshops. Several smaller laboratories accommodated the remainder of Fischer's students and staff. Much of the more routine instruction, especially in methodology, depended on the student's willingness to seek information from the more advanced students, or Dozenten. His initiative was highly essential to his progress, and it was stimulated by the knowledge that Fischer required an account of results at reasonably short intervals. Fischer's visits were unannounced and highly informal. . .
>
> His conversations at the laboratory bench were simple and direct, at times even brusque and to the point, never ponderous, loquacious or dogmatic. His manner was unassuming, generally serious. On occasion he displayed an earthy sense of humor, even in the laboratory. With a group of an evening, over a glass of beer, he was a delightful companion and raconteur. . .

364   6   The Status of Bilirubin in 1930

. . . Fischer was the *Herr Geheimrat*, and I never heard him addressed by any other title, nor did his students and staff refer to him in any other way in his absence. I never sensed that the great respect which he enjoyed was based on fear, but rather on his ideas and accomplishments and his manner of treating those who worked with him, not as if they worked under him. While it was clear that he expected his ideas and suggestions to be given careful attention, he knew how to listen and was willing to change his mind on the basis of adequate evidence.

On his visit to a student's laboratory bench, Fischer usually picked up a thread of the problem with an accurate question about a finding or a suggestion discussed at the previous visit. New pyrrolic compounds, both synthetic and analytical, were constantly coming off the massive production line which such a large number of students provided. I am sure it was a matter of wonder to them, as it was to me, that he was able to keep their individual objectives and principal results so well in mind.

For obvious reasons, crystallization was highly important to him, and over the years, as I shall mention again in another context, he had developed a fabulous reputation for his ability to bring substances to crystallization. While he seldom offered specific advice on ordinary techniques, letting the student seek this in other ways and thus develop more initiative and self-reliance, he very often demonstrated the simple methods of crystallization which he had used for many and various compounds. . .

He was always interested in the first melting point determinations on newly crystallized substances and often liked to carry these out himself. He was strongly intrigued by color and might go with the student to the spectroscopic laboratory to observe and measure absorption spectra of colored solutions in the only spectrometer then locally available – a Zeiss, Löwe-Schumm grating type permitting direct comparison, by superposition, of the spectra of two solutions. His obvious love of color may well have been responsible in some measure for his early interest in the bile pigments and porphyrins, and hence for his life-long devotion to this general group.

At the same time it cannot be doubted that his initial work with the pyrrol [sic] pigments was related in part to his *Habilitation* in medicine and the period which he spent in *Friedrich v. Müller's* famous medical clinic of the University of Munich. It seems quite likely that the keen interest of Müller and his associates in bile pigments and porphyrins readily fired Fischer's imagination during the period in which he was earning his M.D. degree in that clinic; I have never known to what extent Müller directly engaged Fischer's interest in the bile pigments. . .

Fischer's vast research program was generously supported by the Rockefeller Foundation. On one occasion during the second year of my stay he called me in to translate a letter from the foundation relating in a general way to future support. I came to a sentence which said, in effect, that if Fischer decided to accept the call to Berlin, there was little doubt that the support of the foundation would be transferred to his program there. At this point Fischer interrupted me: *Halt bitte darüber gar nicht reden; das gäbe eine Schweinerei.* [Watson probably meant: Halt! Bitte darüber gar nicht reden; das gäbe es eine Schweinerei = Stop! Please don't discuss it at all; it would be a dirty joke to be there.] Here the earthy Bavarian spoke; and, indeed, I suspect strongly that his love of Bavaria, its speech, customs, mountains and countryside, not to mention the charms of Munich itself, were important factors in his decision to remain where he was. I heard no talk about the possibility that he might leave; if others knew it, they did not discuss it and the effect on morale which he feared was not discernible.

They also reveal aspects of *Fischer's* persona and history, in addition to *Watson's* relationship with *Fischer* and the turmoil associated with National Socialism.

In 1971 *Treibs* published his recollections in a comprehensive book on *Fischer's* life and work and included many details of *Fischer's* life, his *modus operandi* at the TH-München and the functioning of his co-workers there. Perhaps the most telling of these writings is the expression of gratitude in a *Festschrift* by his students on the occasion of the *Fischer Feierabend*, the *Nobel* Prize celebration (mentioned by *Watson*) arranged by his students and other co-workers shortly before Christmas 1930 (*12*):

FAKSIMILI aus der
NOBELPREIS-FESTSCHRIFT,
WEIHNACHTEN 1930

.... / unnb bas ist bie warhaffte un eigentliche Experienzia, welche ich nach höchstem Fleiß / mit aigener Hand zu Werck gesetzt habe / wie schwer es auch theils ber Vernunfft vorkompt / so will boch bie rechte unb wahre Praxis hierinnen ben Vorzug haben / unb sich burchauß von ber Speculation nit maistern lassen.

Mit Freude und Stolz hat uns Schüler die Nachricht von der großen Ehrung, die unserem verehrten Lehrer zuteil wurde, erfüllt. Hat nun doch auch die internationale Welt der Wissenschaft damit den Ausdruck der Anerkennung und des Dankes für seine wahrhaft große Leistung und sein unermüdliches Schaffen gefunden.

Es drängt uns in dieser Feierstunde unserem Lehrer herzlichst Glück zu wünschen. Mehr aber noch möge uns gestattet sein, die Gelengenheit wahrzunehmen und die Gefühle der Dankbarkeit zu bekunden, die uns bewegen. Die Dankbarkeit für das, was uns der Lehrer als Wissenschaftler übermittelte, für die fast unbegrenzte Geduld und Nachsicht, mit der er unseren Ungeschicklichkeiten und Schwierigkeiten zu begegnen half. Aber auch die stille Dankbarkeit für die stets gütige Teilnahme an dem rein persönlichen Geschick eines jeden von uns. Wie mancher von uns hat durch seinen Rat und seine Hilfe neuen Mut und neue Ziele in dunklen trostlosen Tagen gefunden. Gerade diese Einheit von wissenschaftlicher Größe, pädagogischem Können und hervorragendem Menschentum ist es, die uns durch unserem Lehrer zum beglückenden Erlebnis wurde, das unsere Entwicklung und Bildung entscheidend beeinflußt und gefördet hat. So sind wir alle unserem hochgeschätzten Lehrer für alles, was wir durch ihn empfangen und gelernt haben, in tiefer Dankbarkeit verbunden.

Diese kleine Festschrift aber soll einige Episoden und Ereignisse festhalten, zur Erinnerung an diesen bedeutsamen Tag und an unsere schöne und unvergeßliche Studienzeit im Laboratorium Hans Fischers. 8

The dedication gives an excellent account of the feelings of the *Fischer* students. The motto comes from the "Tin-mastery School" (*Büchsenmeisterey-Schule*) of *Joseph Furttenbach*, Augspurg 1643. *Furttenbach* (1591–1667) was a natural scientist/mathematician and engineer of high rank.

6.2 Serendipitous Discovery of New Dipyrrole Fragments of Mesobilirubin: *Neobilirubinsäure* and *Neoxanthobilirubinsäure*

If the 1920s were a time of focus on the structure of hemin, for *Fischer* it was no less a time to establish his credentials as an organic synthesis chemist *extraordinaire* and to demonstrate his belief that structures were not fully known until the

366   6 The Status of Bilirubin in 1930

compound had been synthesized. Thus, in the 1920s, in a flurry of synthesis endeavors, *Fischer* completed the syntheses of the bilirubin and hemin degradation acids that, *mutatis mutandis*, would become important building blocks in larger molecule synthesis: *Kryptopyrrolcarbonsäure* (kryptopyrrole carboxylic acid) (*409*), *Phyllopyrrolcarbonsäure* (phyllopyrrole carboxylic acid) (*410*), *Xanthopyrrolcarbonsäure* (xanthopyrrole carboxylic acid) (*429*), *Hämopyrrolcarbonsäure* (hemopyrrole carboxylic acid) (*431*) and *Opsopyrrolcarbonsäure* (opsopyrrole carboxylic acid) (*426*) – the last from hemin. The synthetic methodology and the syntheses would later become important in the total synthesis of bilirubin and hemin. The monopyrroles from *Fischer*'s syntheses are shown in Table 6.2.1.

Table 6.2.1 *Fischer*'s monopyrroles, their synthesis precursors, and methods of preparation

Kryptopyrrolcarbonsäure (409)
(mp 153°C)
from ... *Knoevenagel* by condensation with $HOOCCH_2CN$

Phyllopyrrolcarbonsäure (410)
(mp 88°C)
from ... *Knoevenagel* by condensation with $CH_2(CO_2H)_2$

Xanthopyrrolcarbonsäure (429)
(mp 109.5°C)
from ... *Knoevenagel* by condensation with $CH_2(CO_2H)_2$

Hämopyrrolcarbonsäure
(*431*) (mp 130°C)
from ... or from ... by reaction with ...

Opsopyrrolcarbonsäure (426)
(mp 119°C)
from ... *via* the dipyrrylmethane

6.2 Serendipitous Discovery of New Dipyrrole Fragments...

Fischer's synthesis target for bilirubin had been decided in favor of the linear tetrapyrrole, from which *Bilirubinsäure* (bilirubinic acid) and *Xanthobilirubinsäure* (xanthobilirubinic acid) had come following treatment with sodium and potassium methylate. The two dipyrroles differed only by the presence or absence of a carbon-carbon double bond and had 17 of the 33 carbons present in bilirubin. The remaining 16-carbon fraction was unknown then but was thought to contain the furan ring, or some derivative of it.

It is not entirely clear what *Fischer* had in mind when in 1930 he treated bilirubin with hot resorcinol (*438*). *Schumm* had treated hemin in molten resorcinol, a reaction that led to the loss of the two vinyl groups and the production of deuterohemin (*439, 440*). Conceivably, *Fischer* was experimenting on ways to remove any resident vinyl groups from bilirubin, possibly even the furan ring. Perhaps *Fischer* thought that a simplified analog of bilirubin, shorn of vinyl group(s) and furan ring might serve as a potential synthetic intermediate on the way to bilirubin, much as deuteroporphyrin had been when it was converted to hemin. Possibly, he had even considered that cleaving such a stripped-down bilirubin with sodium or potassium methylate would lead to new dipyrrole analogs of bilirubinic acid and xanthobilirubinic acid that would provide important new information on the structure of bilirubin. Whatever his motive, he found, at least initially, that bilirubin in boiling resorcinol (according to *Schumm*) failed to produce satisfactory results. And so, for reasons having little to do with the vinyl group removal, *Fischer* repeated the resorcinol reaction, now for a much shorter, 0.5 minute contact period, with the mesobilirubin prepared by catalytic hydrogenation of bilirubin. The results were quite possibly unexpected, though it is not entirely clear what *Fischer* had expected: he isolated a C_{16} dipyrrole analog of xanthobilirubinic acid. It was very similar to the latter in composition and properties, was reduced by Na(Hg) to a C_{16} analog of bilirubinic acid that gave an intense *Ehrlich* reaction and was oxidized back to the starting material by $KOCH_3$ at 190°C. *Fischer* named it *Neo-Xanthobilirubinsäure* (neoxanthobilirubinic acid), which crystallized from CH_3OH in long, yellow needles, mp 229°C, and analyzed nicely for $C_{16}H_{20}N_2O_3$ (*438*):

1.	5,597 mg	Substanz	gaben	13,685 mg	CO_2,	3,350 mg	H_2O.	
2.	5,625 mg	„	„	13,810 mg	CO_2,	3,465 mg	H_2O.	
3.	4,320 mg	„	„	10,520 mg	CO_2,	2,590 mg	H_2O.	
4.	4,645 mg	„	„	11,340 mg	CO_2,	2,845 mg	H_2O.	

1.	2,943 mg	Substanz	gaben	0,264 ccm	N	(16°, 722 mm).
2.	3,330 mg	„	„	0,299 ccm	N	(21°, 712 mm).
3.	3,430 mg	„	„	0,289 ccm	N	(16°, 629 mm).
4.	2,920 mg	„	„	0,272 ccm	N	(21°, 722 mm).

Für	$C_{16}H_{20}N_2O_8$	Ber.		C	66,62%	H	7,00%	N	9,73	%
		Gef.	1.	„	66,68	„	6,70	„	10,10	
			2.	„	66,96	„	6,89	„	9,77	
			3.	„	66,42	„	6,71	„	9,55	
			4.	„	66,58	„	6,85	„	10,20	

Consistent with the analytical data, the new pigment exhibited a molecular weight (by the boiling point elevation method) in pyridine and acetic acid close to the formula weight (FW = 288). The pigment was soluble in alkali and bicarbonate and formed a methyl ester, mp 190°C, that had a proper analysis for $C_{17}H_{22}N_2O_3$. The ester formed a crystalline azo pigment with benzenediazonium chloride, $C_{23}H_{26}N_4O_3$, suggestive of a pyrrole unit with an α-H. Nitric acid oxidation yielded methylethylmaleimide, mp 67°C. But the confirming proof of structure came from reduction in glacial acetic acid by HI to yield kryptopyrrole and hemopyrrole carboxylic acid, both isolated from the *Gemisch* as picrates, and confirmed by mixture melting points with authentic samples. Catalytic hydrogenation or reduction with Na(Hg) produced the colorless dipyrrole leuco-analog *Neo-Bilirubinsäure* (neobilirubinic acid), mp 183°C (and mp 184°C, with no depression in a mixture melting point with the neobilirubinic acid formed by treating bilirubin with NaOCH₃). Reduction of neoxanthobilirubinic acid in glacial acetic acid with HI-PH₄I gave the same neobilirubinic acid, mp 179°C and no mixture melting point depression. It followed then that from the information above, especially from the monopyrroles obtained by reductive cleavage, *Fischer* was able to assign unique structures (*438*):

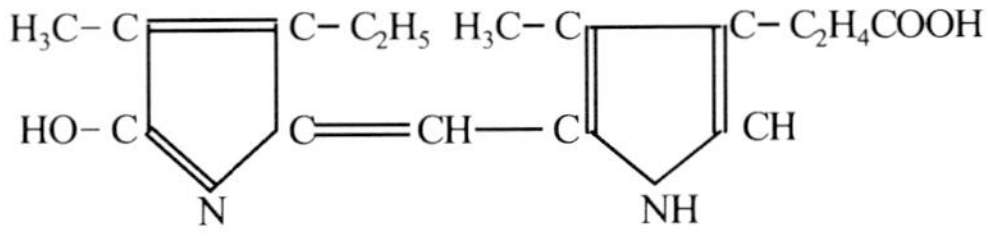

Neo-Xanthobilirubinsäure

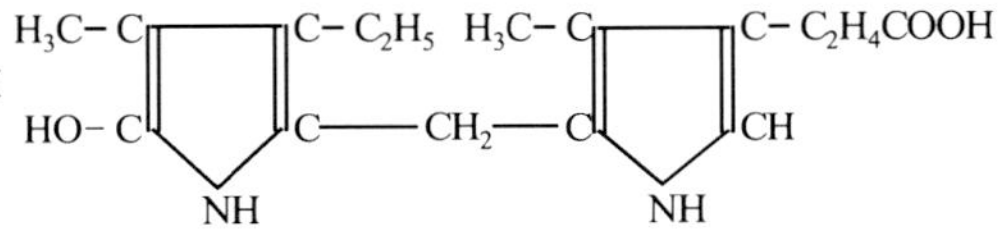

Neo-Bilirubinsäure

Reassembling the newly-obtained dipyrrole pigment into its precursor mesobilirubin must have seemed like a straightforward task because *Fischer* had learned from earlier work that aldehydes would condense with an "α-free" pyrrole, at the α-position. Thus, in a double condensation, neoxanthobilirubinic acid was reacted with formalin, catalyzed by hydrochloric acid, to afford the condensation product, which *Fischer* called *K-Mesobilirubin* (K-mesobilirubin), mp 310°C (*438*). Significantly, its mixture melting point with "natural" mesobilirubin gave no depression. The two pigments behaved entirely similarly in the *Gmelin* reaction and exhibited the same spectral analysis. Combustion analysis was considered to correspond to the expected $C_{33}H_{40}N_4O_6$ formula (*438*):

4,370 mg	Substanz	gaben	10,622	mg	CO_2,	2,695	mg	H_2O.
3,235 mg	„	„	0,287	ccm	N	(19°, 710 mm).		

Für	$C_{33}H_{40}N_4O_6$	Ber.	C	67,31	%	H	6,85	%	N	9,52	%
		Gef.	„	66,29		„	6,90		„	9,69	

6.2 Serendipitous Discovery of New Dipyrrole Fragments...

XII (K-Mesobilirubin)

XIII (K-Mesobilirubinogen)

^{1}) (S = CH$_2$– CH$_2$– COOH)

Esterification in methanolic HCl yielded K-mesobilirubin dimethyl ester, which analyzed correctly as the di-hydrochloride and had mp 190°C (*438*):

4,162 mg	Substanz	Gaben	9,290	mg	CO$_2$,	2,620 mg		H$_2$O.
4,190 mg	„	„	0,303	ccm	N	(20°, 721 mm).		
5,430 mg	„	„	2,101	mg	AgCl.			

| Für C$_{35}$H$_{44}$N$_4$O$_6$ · 2 HCl | Ber. | C | 60,93 % | H | 6,73% | N | 8,13 % | Cl | 10,30 % |
| | Gef. | „ | 61,41 | „ | 7,11 | „ | 7,99 | „ | 9,58 |

> Das Esterchlorhydrat des Mesobilirubins nach H. Fischer-Niemann zeigt starken grünen Oberflächenglanz [*438*] auch in der Farbe mit dem Esterchlorhydrat des K-Mesobilirubins vollständig übereinstimmt. Der Mischschmelzpunkt der beiden Esterchlorhydrate gab keine Depression. 9

A mixture melting point with the dihydrochloride of mesobilirubin dimethyl ester prepared by catalytic hydrogenation of bilirubin followed by esterification gave no depression.

As if even more experiments might turn up a difference between K-mesobilirubin and the natural mesobilirubin obtained by reduction of bilirubin, the former was reduced by Na(Hg) to K-mesobilirubinogen that behaved identically to "natural" mesobilirubinogen in all properties, even showing a very similar crystal form (*438*):

> Wir legten dann nochmals zum Vergleich Herrn Professor Steinmetz Krystallisationen von Mesobilirubinogen und K-Mesobilirubinogen aus Essigester vor, so wie sie bei der direkten Krystallisation entstehen, wobei, wie schon öfter hervorgehoben, verschiedenartige Krystallformen erscheinen, um einen Vergleich der Präparate in toto anstellen zu können. Herr Professor H. Steinmetz teilte uns über seinen Befund Folgendes mit: „Bei beiden Krystallisationen erscheinen in den oberen, d. h. zeitlich zuerst ausgeschiedenen Randpartien mehr dendritisch oder nadelig ausgebildete Krystallanhäufungen, die nach unten von den schärfer ausgebildeten, im Querschnitt schief sechsseitig begrenzten Krystallen abgelöst werden. Es ist anzunehmen, daß aber trotzdem keine verschieden krystallisierenden Substanzen vorliegen, sondern der Habitusunterschied durch die anfänglich mit ausgeschiedenen Schmieren bedingt ist. Die beiden Krystallisationen unterscheiden sich nicht. In beiden sind dieselben Umrißformen und die daraus abzuleitenden Symmetrieverhältnisse vorhanden." 10

370   6   The Status of Bilirubin in 1930

It gave no depression in a mixture mp determination and was cleaved back to the identical neoxanthobilirubinic acid by treating with molten resorcinol. Reductive cleavage of K-mesobilirubin with HI-PH$_4$I or HI alone in glacial acetic acid produced bilirubinic acid. In the first case the product melted at 178°C, in the latter at 188°C, and both samples gave no melting point depression when admixed with the authentic bilirubinic acid produced by similar reduction of bilirubin. From the latter reduction, neobilirubinic acid, mp 180°C, was also separated. It exhibited no depression in a mixture melting point with authentic material. In parallel experiments, natural mesobilirubin was treated similarly and led to (i) isolation of neobilirubinic acid, mp 180°C, that exhibited no depression in a mixture melting point with authentic material, and (ii) isolation of bilirubinic acid, mp 187°C, which again gave no depression in a mixture melting point determination with authentic material. It seemed thus like an open and shut case that K-mesobilirubin was identical to "natural" mesobilirubin, and so *Fischer* was given to comment that (*438*):

> Was die Konstitution des Bilirubins selbst anlangt, so ist wohl auch für dieses die Äthylenformel ausgeschlossen und wird Bilirubin am ehesten durch Formel **XII** wiedergegeben, wobei statt der beiden Äthylreste Vinylgruppen einzusetzen wären. Der Wasserstoffverbrauch bei Amalgam . . . und katalytischer Reduktion würde mit dieser Vorstellung genau übereinstimmen. Aber auch einige Bedenken erheben sich gegen diese Auffassung. In bezug auf die ungesättigen Seitenketten wären dann Bilirubin und Hämin identisch. Beim Hämin und Protoporphyrin aber sind durch Amalgamreduktion die dort experimentell sichergestellten Vinylgruppen nicht reduzierbar, während beim Bilirubin diese Reaktion spielend verläuft. . . .
>
> Weiter (vgl. S. 195) entsteht bei Nitritoxydation des Bilirubins der Körper **III**, dessen Entstehen aus dem Bilirubin Formel **XII** entsprechend kaum denkbar ist, weil die relative Stellung der Vinyl- zu den Oxygruppen einen Furanringschluß, wenigstens in der Tafelebene unmöglich machen. Über all diese Fragen muß weitere analytische und synthetische Arbeit . . . entscheiden und wir wollen nur noch auf einen Punkt hinweisen, nämlich wie sich die Überführung des Blutfarbstoffes in Gallenfarbstoff auf Grund der neuen Erkenntnis erklären kann.

11

Although the new information cast serious doubt on the validity of his long-cherished tetrapyrrylethylene formula for bilirubin, perhaps sending it into oblivion, *Fischer* was also clearly confused by the fact that his logically synthesized K-mesobilirubin had two *endo*-ethyl groups yet was still "identical" to natural mesobilirubin by every measurement then available. Recall that *Fischer*'s understanding of the structure of the latter pigment was that it had a dihydrofuran ring (as in **IV**), in addition to one *endo*-ethyl group. Even more troublesome was that the two-carbon chain, whether of the tetrahydrofuran or an ethyl group was not located *exo*, where it should have been based on his firm knowledge of (i) the locations of the vinyl groups in hemin, and (ii) a knowledge that bilirubin was derived

6.2 Serendipitous Discovery of New Dipyrrole Fragments...

biosynthetically from hemin. Sometimes new information can appear to be more a curse than a blessing – until one understands how easily data can be misinterpreted, especially mixture melting points that exhibit no depression.

6.2.1 Synthetic Dipyrroles, Tetrapyrroles, Inhomogeneity, and Nefarious Mixture Melting Points

It may have seemed odd to *Fischer* that from $NaOCH_3$ or $KOCH_3$ treatments, natural bilirubin, mesobilirubin, and mesobilirubinogen produced only dipyrroles possessing an *endo*-ethyl group (as in xanthobilirubinic acid and bilirubinic acid), with no evidence of analogs with an *exo*-ethyl or with a furan ring. The convenient designations *endo* and *exo*, due apparently to *Bonnett* and *McDonagh* in 1970 (*441*), refer to the location of the specified group, usually vinyl or ethyl, on the ring β-carbons of the lactim or lactam: the *endo* (or inner) substituent is located closer to the dipyrrylmethane core; whereas, the *exo* or outer substituent is located farther away. (This nomenclature has little to do with the stereochemical designations applied to *exo*- and *endo*-borneol, except in the broadest sense, wherein the hydroxyl group of the *exo* is oriented toward the one-carbon bridge and the *endo* is pointed away.) Apparently it also seemed odd that the mesobilirubin reconstituted from these *endo*-ethyl dipyrroles was identical to natural mesobilirubin, and so *Fischer* resolved to investigate the structures further by chemical synthesis. Lacking the analytical methods that render structure proof easier today, in the 1920s *Fischer* had become a firm believer in structure proof by synthesis. He believed that the structures of natural products and their derivatives were not proven absolutely until they had been synthesized *de novo* and were shown to exhibit characteristics identical to the natural material in all measurable ways, including mixture melting points. The compounds prepared by totally synthetic methods would henceforth be labeled "*synthetisch*" ("synthetic") by *Fischer* in order to distinguish them from their counterparts, labeled "*analytisch*" ("natural-analytical") that were derived from the natural product itself, *e.g.* bilirubin.

In work conducted in 1930–1931, *Fischer* thus considered four possible xanthobilirubinic acid isomers (**II**, **V**, **VI**, and **VII**, the numbers assigned by *Fischer*) and their corresponding bilirubinic acid isomers (**-L**), all shown in Table 6.2.1, as targets for synthesis (*442*). By the end of 1932, the synthesis effort had been extended to include the remaining entries (*3, 443*) of Table 6.2.2. The synthetic dipyrroles revealed information to *Fischer* that was both interesting and disconcerting. Interesting because xanthobilirubinic acids **II**, **V**, and **VI** all exhibited the same melting point; yet the melting points of their methyl esters differed, as did the melting points of their bilirubinic acid analogs. Disconcerting because although the melting points of the *synthetischen* (synthetic) material did not match up with the corresponding analytischen (natural-analytical) dipyrroles; yet there was no melting point depression of the *analytischen* upon admixture with *synthetischem*.

Table 6.2.2 Melting points and mixture melting points of X*anthobilirubinsäure* (xanthobilirubinic acid), b*ilirubinsäure* (bilirubinic acid), and analogs

Pigment	Reported as mp/°C	Methyl ester mp/°C	Leuco(**L**)-Pigment	Reported as mp/°C
II H_3C C_2H_5 H_3C $CH_2\cdot COOH$ CH_2 HO N H $C=$ CH N CH_3 $C_{17}H_{22}O_3N_2$			**II-L** H_3C C_2H_5 H_3C $CH_2\cdot COOH$ CH_2 HO N H C H_2 N CH_3 $C_{17}H_{24}O_3N_2$	
(*Xanthobilirubinsäure*)	290–291	212–213	(*Bilirubinsäure*)	187–188
synthetisch	274	212	*synthetisch*	187–188
analytisch	273*		*analytisch*	
mixture mp				
V H_5C_2 CH_3 H_3C $CH_2\cdot COOH$ CH_2 HO N H $C=$ CH N CH_3 $C_{17}H_{22}O_3N_2$			**V-L** H_5C_2 CH_3 H_3C $CH_2\cdot COOH$ CH_2 HO N H C H_2 N H CH_3 $C_{17}H_{24}O_3N_2$	
(*Iso-xanthobilirubinsäure*)	289–290	197–198	(*Iso-bilirubinsäure*)	207
synthetisch	274		*synthetisch*	
analytisch			*analytisch*	

6.2 Serendipitous Discovery of New Dipyrrole Fragments... 373

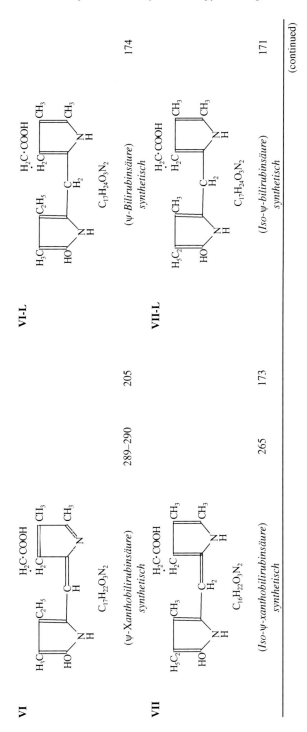

(continued)

Table 6.2.2 (continued)

Pigment	Reported as mp/°C	Methyl ester mp/°C	Leuco(L)-Pigment	Reported as mp/°C
 $C_{16}H_{20}O_3N_2$ (*Neo-xanthobilirubinsäure*) synthetisch analytisch mixture mp	 246 226–227 222–227*	 174 190–191 158*	**L** $C_{16}H_{20}O_3N_2$ (*Neo-bilirubinsäure*) synthetisch analytisch	 179 190-191
 $C_{16}H_{20}O_3N_2$ (*Iso-neo-xanthobilirubinsäure*) synthetisch analytisch mixture mp	 242 227 225–226*	 204 191 200*	**L** $C_{16}H_{22}O_3N_2$ (*Iso-neo-bilirubinsäure*) synthetisch analytisch	 194 190–191

* Mixture melting point of *synthetischen* with *analytischem*.

6.2 Serendipitous Discovery of New Dipyrrole Fragments...

All of the compounds listed in Table 6.2.1 gave satisfactory combustion analyses corresponding to the molecular formulas given (*3, 442*).

Fischer's original plan was to confirm the structures of the natural-analytical dipyrrole products obtained by degradation of natural bilirubin and mesobilirubin, with identification to be made by comparing melting points and mixture melting points with target synthetic material obtained by *de novo* synthesis. In the 1930s and earlier, there was no more certain way to confirm the identity of two organic solids of the same melting point than an undepressed mixture melting point. Spectral analysis (UV-visible spectroscopy) lacked sufficient detail, and even in comparing crystal morphology one had to be concerned with polymorphism.

Thus, synthetic xanthobilirubinic acid (**II**) proved to be identical in every measurable way with the natural-analytical compound, except for melting points; yet, oddly, no depression of the melting point of the analytical compound was observed for a mixture of the natural-analytical and synthetic compounds. The synthetic and natural-analytical methyl esters also showed identical melting points, with no mixture melting point depression. Further identity came from the leuco products: natural-analytical bilirubinic acid showed no melting point depression with synthetic material. Rather curiously, however, the synthetic iso-xanthobilirubinic acid (**V**), which differs from the xanthobilirubinic acid (**II**) only by having an *exo*-ethyl rather than *endo*, also had the same melting point (289–290°C), a melting point shared by the different synthetic isomer **VI**. To make matters even more interesting, which may have rendered *Fischer* (at least momentarily) non-plussed, the mixture melting points of **V** with **II**, and of **VI** with **II** showed a depression of only 1–2°C – not something that would be easily noticed for compounds with such high melting points. On the basis of their having the same melting points, with little or no depression in mixture melting points (and absent a knowledge of how **II**, **V**, and **VI** came about), one might therefore easily believe that they were all the same compound and thus had identical structures – which they clearly do not. Only the melting point of synthetic iso-ψ-xanthobilirubinic acid **VII** stood out as being very different from the other three isomers.

In contrast, the methyl esters of the synthetic acids **II**, **V**, **VI**, and **VII** all had different melting points and exhibited mixture melting point depressions, *i.e.* methyl esters of **V** and **II** (sharp, down to 170–175°C); **VI** and **II** (sharp, down to 180°C). The melting point of the methyl esters of **VII** and **II** was also depressed. Similarly, the corresponding leuco pigments, natural-analytical bilinic acid (**II-L**) and its synthetic analogs, exhibited different melting points and showed melting point depressions upon admixture, *i.e.* leuco pigments **V-L** and **II-L** (sharp, to 181°C); **VI-L** and **I-L** (steep, to 160–165°C); and **VII-L** and **I-L** (sharp). More interesting was the mixture melting point of synthetic bilirubinic acid **II-L** with the natural-analytical material which, like that of its precursor, xanthobilirubinic acid, exhibited no depression.

At the same time, still in 1931, *Fischer* had run into a second peculiarity related to structure. Earlier (*438*), K-mesobilirubin had been prepared by condensing two mole equivalents of natural-analytical neo-xanthobilirubinic acid with one of CH_2O. In order to confirm the structure of K-mesobilirubin, *Fischer* conducted independent synthesis of the pigment from synthetic xanthobilirubinic acid. After a failed attempt to remove the pyrrole α-CH_3 (by tri-chlorination with SO_2Cl_2, hydrolysis, and decarboxylation), he carried out its bromination in glacial acetic acid (*444*). An exothermic reaction ensued, with liberation of HBr and, surprisingly perhaps, production of a synthetic mesobilirubin of mp 312–315°C (dec.) arising by condensation of two mole equivalents of the dipyrrinone and an apparent loss of a carbon. Did this new pigment, however, have a one-carbon or a two-carbon piece linking the two dipyrrinones? That was decided by a combustion analysis. That is, in this then-unprecedented reaction, *Fischer* believed the product to be a C_{33} pigment, not a C_{34}, which would have arisen, perhaps more logically, by doubling the number of carbons (C_{17}) of xanthobilirubinic acid. His belief was supported when its combustion analysis revealed a better fit to C_{33} than to C_{34} (*444*):

4,907 mg	Substanz	gaben	12,110 mg	CO_2,	3,115 mg	H_2O		
3,300 mg	„	„	0,286 ccm	N (18°, 721 mm).				

$C_{34}H_{40}N_4O_6$	Ber.	C	67,96	%	H	6,72	%	N	9,33	%
$C_{33}H_{40}N_4O_6$	Ber.	„	67,31		„	6,85		„	9,52
	Gef.	„	67,31		„	7,10		„	9,65

Because synthetic xanthobilirubinic acid has an *endo*-ethyl group, it followed that this synthetic mesobilirubin could have no other than two *endo*-ethyl groups.

Bromination of xanthobilirubinic acid methyl ester in CH_3OH was (re-)investigated more recently (*445*) and in 1995 was reported to give the dimethyl ester of the identical mesobilirubin (above), *inter alia*, in a study involving three of the author's former postdoctoral fellows (*J.-S. Ma*, now Professor Emeritus at the Institute of Photographic Chemistry, *Q.-Q. Chen*, and *L.-J. Cheng* from the Institute of Photographic Chemistry in Beijing) in collaboration with *Heinz Falk* at Johannes Kepler Universität-Linz. Moreover, in the same year, following reaction of xanthobilirubinic acid methyl ester with Br_2 in CH_2Cl_2, the same authors found the C_{34} pigment akin to what Fischer had dismissed, a *b*-homoverdin (*446*).

Fischer then compared the synthetic mesobilirubin to the natural-analytical mesobilirubin prepared either from natural-analytical xanthobilirubinic acid, or prepared earlier (*438*) by condensing two mole equivalents of natural-analytical neo-xanthobilirubinic acid with one mole equivalent of CH_2O, the so-named K-mesobilirubin. He found two important discrepancies between the totally synthetic mesobilirubin and the natural-analytical K-mesobilirubin that had been identified earlier as natural mesobilirubin. The totally synthetic mesobilirubin (mp 312–315°C, dec.) melted only somewhat higher than natural-analytical mesobilirubin (304–305°C, lying between 307–310°C), and the latter exhibited no mixture melting point depression when synthetic mesobilirubin was admixed (*444*):

> Beim Erkalten der zum Sieden erhitzten Pyridinlösung scheiden sich goldgelbe prismatische Nadeln ab. Schmelzp. 312–315° unter Zersetzung. Gmelinsche Reaktion positiv. Die spektralen Erscheinungen bei der Gmelinschen Reaktion stimmen mit denen beim Mesobilirubin vollkommen überein, wie durch Übereinanderprojizieren der Spektren

6.2 Serendipitous Discovery of New Dipyrrole Fragments... 377

überzeugend festgestellt wurde. Der Mischschmelzpunkt mit einem Analysenpräparat von Mesobilirubin, das einen Schmelzpunkt von 305° zeigte, lag bei 307–310°, gab also keine Depression. 12

In contrast, the dimethyl ester hydrochloride (mp 216°C) of the totally synthetic mesobilirubin melted 26° higher than the dimethyl ester (mp 190°C) of the natural-analytical mesobilirubin. Yet, here too, a mixture melting point of these two meso-bilirubin esters showed no more of a melting point depression than did their parent diacids (*444*):

Das Esterchlorhydrat gibt bei 190° unter Grünfärbung den Chlorwasserstoff ab und schmilzt bei 216°. Mischschmelzpunkt mit einem bei 190° schmelzenden Präparat von analytischem Mesobilirubinester-Chlorhydrat gab keine Depression. 13

Clearly, and despite the observed differences in individual compound melting points, the mixture melting points indicated that the synthetic and natural-analytical acids were identical, as were their dimethyl esters. Disturbingly, synthetic (*endo*-ethyl, *endo*-ethyl)- and (*exo*-ethyl, *exo*-ethyl)-mesobilirubins (known now as mesobilirubins-XIIIα and -IIIα, respectively) exhibited no mixture melting point depression, nor did their dimethyl esters. To complicate matters even further, the natural-analytical mesobilirubinogen (mp 197–200°C) formed by Na(Hg) reduction of natural-analytical mesobilirubin showed no mixture melting point depression with synthetic mesobilirubinogen (mp 194°C) from synthetic mesobilirubin.

Fischer continued to find few differences between synthetic and natural-analytical mesobilirubin even in their cleavage products from reaction with HI in glacial acetic acid. The former gave a separable mixture of neo-bilirubinic acid (mp 179°C) and bilirubinic acid (mp 183°C) neither of which exhibited melting point depressions when admixed with natural-analytical neo-bilirubinic acid (mp 190–191°C) and natural-analytical bilirubinic acid (mp 187–188°C), respectively. However, the products from molten resorcinol treatment of synthetic and natural-analytical mesobilirubin behaved differently.

With the near certainty that the synthetic mesobilirubin possessed the (*endo*-ethyl, *endo*-ethyl) structure of K-mesobilirubin mentioned in the previous section, for comparison purposes *Fischer* synthesized its isomer with both ethyl groups *exo* rather than *endo* by treating the synthetic isomeric xanthobilirubinic acid (iso-xanthobilirubinic acid **V** of Table 6.2.2) with Br_2. The new isomeric synthetic mesobilirubin, dec. 327°C, analyzed correctly for $C_{33}H_{40}N_4O_6$ (*444*):

| 5,833 mg | Substanz | gaben | 14,390 | mg | CO_2, | 3,520 mg H_2O |
| 3,232 mg | „ | „ | 0,284 | ccm | N (17°, 713 mm). | |

| $C_{33}H_{40}N_4O_6$ | Ber. | C | 67,31 | % | H | 6,85 | % | N | 9,52 | % |
| | Gef. | „ | 67,28 | | „ | 6,75 | | „ | 9,70 | |

Yet, upon admixture with synthetic (*endo*-ethyl, *endo*-ethyl)-mesobilirubin, the latter (*exo*-ethyl, *exo*-ethyl)-mesobilirubin exhibited only an insignificant melting point depression of 1–2°, as also happened in the mixture melting point of natural-analytical mesobilirubin when the synthetic (*exo*-ethyl, *exo*-ethyl)-mesobilirubin was admixed (*444*):

> Der neue Körper bildet prächtige goldgelbe, gebogene Nadeln vom Zersetzungspunkt 327°. Mit dem Bromierungsprodukt der Xanthobilirubinsäure, sowie mit Mesobilirubin tritt nur eine unbedeutende Schmelzpunktsdepression von 1–2° ein.   14

Preparation of the dimethyl ester hydrochloride of the isomeric synthetic (*exo*-ethyl, *exo*-ethyl)-mesobilirubin was achieved; it had mp 222°C. Yet, its mixture melting point with the natural-analytical dimethyl ester (mp 193–195°C) gave no depression (mp 193–195°C). On the other hand, when the synthetic isomeric (*exo*-ethyl, *exo*-ethyl)-mesobilirubin dimethyl ester was admixed with the dimethyl ester hydrochloride (mp 216°C) of the synthetic (*endo*-ethyl, *endo*-ethyl)-mesobilirubin (from bromination of synthetic xanthobilirubinic acid) depression of the melting point of the latter to 196°C was observed (*444*):

> Der Schmelzpunkt liegt bei 222°. Der Mischschmelzpunkt mit analytischem Mesobilirubinester liegt bei 193–195°, gibt also keine Depression. Mit dem Ester des synthetischen Mesobilirubins trat eine Depression auf 196° ein.   15

In contrast to the neo-xanthobilirubinic acid (mp 227°C) from natural-analytical mesobilirubin, the neo-xanthobilirubinic acid (mp 237–239°C) obtained in 70% yield from synthetic mesobilirubin showed a lower melting point and was very insoluble in CH_3OH. Raising further doubt as to the identity or homogeneity of the former, the mixture melting point with synthetic neo-xanthobilirubinic acid was 230–232°C. Yet the latter pigment, when reduced nearly quantitatively by Na(Hg) to synthetic neobilirubinic acid (mp 180–181°C), exhibited no mixture melting point depression with natural-analytical neo-bilirubinic acid (mp 190–191°C).

Continuing his efforts to resolve the apparently conflicting data, *Fischer* then reconstructed mesobilirubin from synthetic neo-xanthobilirubinic acid, found it to melt at 308°C, and to give no melting point depression upon admixture with the (synthetic) mesobilirubin derived from bromination of synthetic xanthobilirubinic acid. Treatment of the reconstructed mesobilirubin with molten resorcinol yielded synthetic neoxanthobilirubinic acid, mp 239°C, which showed no melting point depression when admixed with the natural-analytical neo-xanthobilirubinic prepared by resorcinol melt of "natural" mesobilirubin.

As a result of all these studies, at the end of 1931, *Fischer* had concluded that the apparently anomalous melting point behavior was due to isomerism as a result of shifting double bonds (*447*):

> After more than a five-year pause, the costly study of the problems of bile pigment has been carried further by H. Fischer and Richard Hess (27) [*438*]. By mild reduction of bilirubin, mesobilirubin and mesobilirubinogen have been obtained (28) [*397*], and by energetic reduction, bilirubinic acid, which on dehydration becomes xanthobilirubinic acid. Bilirubin has 33 carbon atoms, bilirubinic acid 17. The remaining fraction with 16 carbon atoms was still unknown. In a resorcin "melt" or by glacial-acetic-hydriodic-acid reduction, mesobilirubin gives "neoxantho-bilirubinic acid," which on reduction goes over to "neobilirubinic acid," the latter containing a free methin group in the "acid" pyrrol nucleus. By condensation of "neoxanthobilirubinic acid" with formaldehyde there is formed "K-mesobilirubin," giving the same color play in the Gmelin reaction as mesobilirubin itself.

6.2 Serendipitous Discovery of New Dipyrrole Fragments…

Because of the mode of formation, this substance is ascribed Formula 4 and in its characteristics the substance agrees completely with mesobilirubin. The continuation of synthetic attempts by H. Fischer and Fröwis (29) [448] led to carboxylated xanthobilirubinic acid, very similar to xanthobilirubinic acid itself.

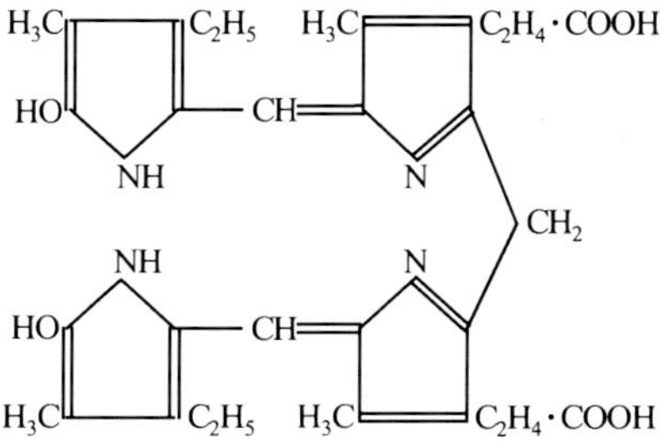

Formula 4

H. Fischer and Kürzinger (9) [Fischer H, Kürzinger A (1931) Über bilirubinoide Farbstoffe und über Koproporphyrin IV. Hoppe-Seyler's Z physiol Chem **196**:213] prepared pigments by the linear coupling of 4-pyrrole nuclei, which give the Gmelin reaction so characteristic for bilirubin and mesobilirubin, and were therefore designated as "bilirubinoid." The Gmelin reaction is not dependent upon the OH-group of bilirubin; much more important is the stabilization of the system, as was shown by Fischer and Adler (30) [442] in the course of further experiments, who, likewise, by continued attempts, carried out the synthesis of bilirubinic acid and xanthobilirubinic acid and, soon thereafter (31) [444], that of neobilirubinic acid and mesobilirubinogen. It is worthy of note that the basic mesobilirubin differs only in the melting-point of its ester from that of the natural, so that it is necessary to accept the existence of isomerism as a result of shifting of the double bonds . . .

6.2.2 A Major Breakthrough in Understanding the Structure of Bilirubin. Resolution of Discordant Melting Points and Synthesis of Mesobilirubin

By mid-July 1931, *Fischer* believed that he had established the structure of natural mesobilirubin definitively (*444*):

Die Konstitution des Mesobilirubins ist daher definitiv im Sinne der früher aufgestellten Formel bewiesen

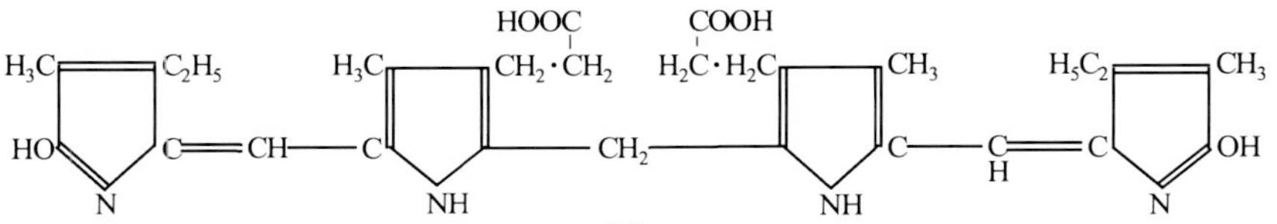

VI

380 6 The Status of Bilirubin in 1930

Previously named *K-Mesobilirubin* (*438*), it was synthesized independently from each of the analytical dipyrroles, xanthobilirubinic acid and neo-xanthobilirubinic acid, exhibited a melting point of 312°C, and gave no mixture melting point depression with natural mesobilirubin (*444*):

> Dieses gab bei der Behandlung mit Pyridin goldgelbe prismatische Nadeln, die intensive Gmelinsche Reaktion zeigten und in allen Eigenschaften mit Mesobilirubin übereinstimmten. Der Schmelzpunkt war 312°; im Mischschmelzpunkt war keine Depression vorhanden. Die Analyse lieferte Zahlen, die gut mit denen des Mesobilirubins übereinstimmten und auf den Austritt eines Kohlenstoffatoms bei der Reaktion hinwiesen.   17

Moreover, K-mesobilirubin underwent the $KOCH_3$ and molten resorcinol cleavage reactions to produce the dipyrroles above. Yet, the melting point of the dimethyl ester of this mesobilirubin was 26° higher than that from natural mesobilirubin itself (*444*):

> Charakteristisch für Mesobilirubin ist sein gut krystallisierter Ester, der auch hier nach der gleichen Methode in schön krystallisiertem Zustand erhalten wurde. Seine Analyse stimmt sehr gut auf Mesobilirubinester, ebenso seine Eigenschaften; jedoch liegt der Schmelzpunkt um 26° höher.   18

It was perhaps *Fischer*'s thoroughness and the unexpected and seemingly random melting point differences that apparently averted his being led astray in his efforts to establish the structure of bilirubin. In early 1931, such a difference might have been ascribed simply to isomerism, but by 1932, the "melting point problem" had begun to reappear repeatedly as *Fischer* strove to reconfirm and more fully establish the structures by synthesis.

Soon after the perplexing discoveries of mismatches between synthetic and natural-analytical mesobilirubin and their dipyrrole cleavage products, *Fischer* understood the root causes of the melting point inconsistencies and mixture melting point peculiarities found between the synthetic and natural-analytical dipyrroles and mesobilirubins of the previous section. By late 1932, he and *Siedel*[1] (*443*) had

---

[1] *Walter Siedel* was born on March 26, 1906 in Sonneberg/Thüringen and died on December 10, 1968. He received the *Diplom. ing.* degree in 1930 at the TH-München, was *Assistent* in *Hans Fischer*'s Organic Chemistry Institute at the TH-München and in January 10, 1933 was awarded the *Dr. ing.* degree, with the thesis title "Neue Synthese des Gamma-Phylloporphyrins, Synthesen des Delta-Phylloporphyrins und Beta-Phylloporphyrins über die Konstitution des Bilirubins, Synthesen der Neoxanthobilirubinsäure und Iso-Neoxanthobilirubinsäure." The *Dr. ing. habil.* was conferred in 1936 (*Habilitationsschrift* "Synthese des Mesobilirubins") and in February 1938 he became *Universitätsdozent* (Lecturer) in organic chemistry and *ausserplanmässiger* (apl.) Professor in 1942. In 1937, he was appointed *Beamter* (civil servant), and from 1937 to 1946 he was appointed *Konservator*. As such he was responsible for various collections of *Fischer*'s Institute, and after World War II he was assigned the task of summarizing research conducted in *Fischer*'s Institute during 1939–1946 as found in ref. (*9*).

Siedel published 22 research papers on pyrroles between 1933 and 1947, including his long summary titled "Gallenfarbstoffe", published in *Zechmeister* in 1939 (*1*), and another 70 papers while in industry. In the mid-1940s, he apparently became the *Fischer* Group spokesman for bile pigments, and especially regarding the total synthesis of bilirubin, publishing as sole author an update of the 1939 *Zechmeister* paper in 1940 in *Angewandte Chemie* (*6*); again in 1943 (*7*), including chlorophyll; and in a summary lecture on bile pigments and porphyrins in October 1943 at a special session of the German Chemical Society meeting in Vienna (*8*). *Siedel* entered the German chemical industry at Farbwerke Hoechst after World War II, where he remained until his death.

6.2 Serendipitous Discovery of New Dipyrrole Fragments...

proved the constitutional structure of natural-analytical mesobilirubin by synthesis. How had this rapid clarification been achieved?

Aside from nagging questions and beliefs related to the probability that bilirubin might possess a furan ring fused to a lactam by cyclization of its *exo*-vinyl group, by the very early 1930s *Fischer* had abandoned his previously favored tetrapyrrylethylene motif for bilirubin and favored a linear tetrapyrrole. Some 20 years earlier he had isolated and proved the structure of two important products from scission of bilirubin representing one-half of the natural pigment: xanthobilirubinic acid and its leuco analog, bilirubinic acid. However, certain problems began to appear in 1931 when a third scission product, neo-xanthobilirubinic acid, was isolated following cleavage in molten resorcinol of the mesobilirubin obtained by catalytic hydrogenation of bilirubin. The structure of neo-xanthobilirubinic acid was simple enough: it was the analog of xanthobilirubinic acid with the pyrrole α-CH$_3$ replaced by an α-H. Where the other half of natural bilirubin or mesobilirubin with the supposed furan ring had disappeared to was open to conjecture. Yet, early on there was little mention in print of its absence or fate.

For example, *Fischer* had come to believe that since bilirubin was derived biosynthetically from hemin, it should possess two vinyl groups, one *endo* and one *exo*. He understood that in bilirubin the *exo*-vinyl had succumbed to the formation of a furan ring. The fact that in such a structure the pigment would have two fewer hydrogens than the most current formula ($C_{33}H_{36}N_4O_6$) was apparently not a major concern. Nor was the mismatch in structure between it and the mesobilirubin with one *endo*-ethyl and one *exo*-ethyl derived from hemin (or mesohemin) and drawn by *Fischer* as (*443*):

$*PS = CH_2 \cdot CH_2 \cdot COOH.$

Though he became apl. Professor at the Universität Frankfurt in 1954, his probable earlier amibitions to achieve o. Professor status were never realized. Whether that was by choice or necessity after World War II is unclear, as his research credentials from the *Fischer* Group were as good as any *Fischer* student and better than most. Yet his *Lebenslauf* indicates he joined the NSDAP as a university student at the TH-München in 1929, clearly one of the *alten Kämpfer* who joined before the Reichstag elections of September 1930 – *aus Überzeugung* (out of conviction), youthful exuberance, or a perception that it was necessary to advance an academic career? Though he may have kept an eye on *Hans Fischer* and the research group, as a *Parteigenosse* he probably undermined the possibility of an academic career in Germany. Apparently his answers to the 131-paragraph denazification questionnaire, compulsory for all Germans of military age shortly after World War II ended, were sufficient to avoid prison because he was charged by the Allied military government of Germany with writing on the research activities in the *Fischer* lab from 1936 to 1945 (*9*).

Given the structural certainty of his synthetic *endo*-ethyl xanthobilirubinic acid, and its *exo*-ethyl isomer called *Iso-xanthobilirubinsäure* (iso-xanthobilirubinic acid), the synthetic mesobilirubins formed by bromination of each of these two dipyrroles could be none other than (in *Fischer*'s new nomenclature): mesobilirubin-XIIIα with two *endo*-ethyls, which came from xanthobilirubinic acid, and mesobilirubin-IIIα with two *exo*-ethyls, which arose from iso-xanthobilirubinic acid (*443*):

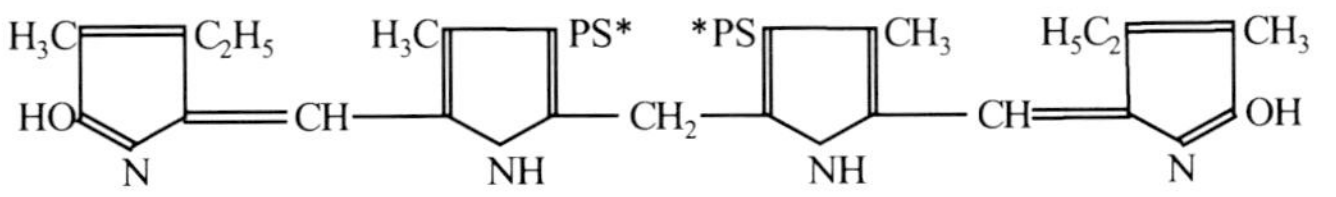

= Mesobilirubin-XIIIα

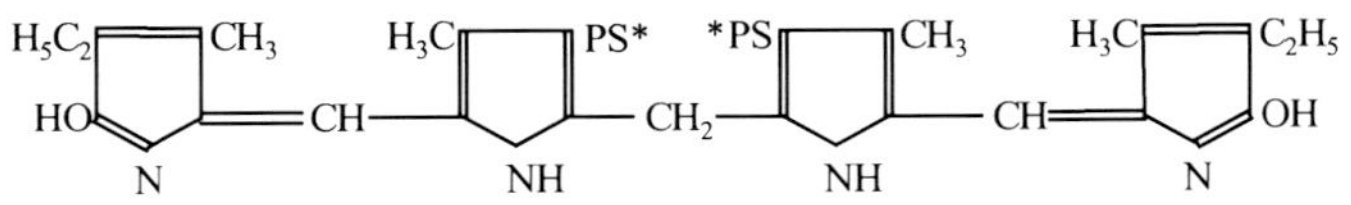

= Mesobilirubin-IIIα

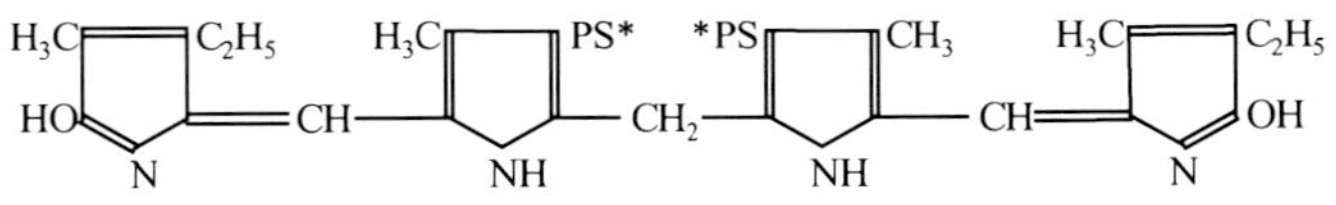

= Mesobilirubin-IXα

* PS = $CH_2 \cdot CH_2 \cdot COOH$.

The mesobilirubin not synthesized (yet) with an *exo*-ethyl and and *endo*-ethyl was named *Mesobilirubin-IXα* (mesobilirubin-IXα). With its "correct" ordering of the two ethyl groups, it would be expected to be the one derived from hemin or mesohemin. The Roman numerals assigned to the termini of mesobilirubin names as shown refer to their hypothetical porphyrin precursors, hemins or mesohemins XIII, III, and IX, and the specific ordering of their pyrrole β-substituents. (See Section 5.5.) The Greek "α" indicates that the porphyrin ring was opened at and with elision of the porphyrin α-methine carbon.

Mesobilirubins-XIIIα and IIIα were again synthesized by an independent route: acid-catalyzed coupling of two mole equivalents of either synthetic neo-xanthobilirubinic acid with one mole equivalent of CH_2O (to give synthetic mesobilirubin-XIIIα), or two mole equivalents of synthetic iso-neo-xanthobilirubinic acid in the same way (to give synthetic mesobilirubin-IIIα). The dipyrroles used came from cleavage of synthetic mesobilirubins-XIIIα and IIIα, respectively, in molten resorcinol.

The proof that mesobilirubin-IXα and natural-analytical mesobilirubin have the same structure was indirect. Given the likelihood that it had one *exo*-ethyl and one *endo*-ethyl, *Fischer* could understand that the (natural-) analytical xanthobilirubinic acid and neo-xanthobilirubinic acid derived from natural mesobilirubin were each in fact mixtures of *endo*-ethyl and *exo*-ethyl isomers: *viz*, xanthobilirubinic acid **II** and iso-xanthobilirubinic acid **V** of Table 6.2.2, and neo-xanthobilirubinic acid and iso-neo-xanthobilirubinic acid of Table 6.2.2. Confirmation of this assumption was provided by their melting points, which were the same as those of analytical xanthobilirubinic acid and neo-xanthobilirubinic acid, and which were not depressed upon admixture with either of their respective component synthetic dipyrroles. The synthetic and analytical methyl esters behaved in parallel fashion, as did the leuco analogs. That is, the mixture of *endo* and *exo* isomers, as well as their esters, melted sharply and did not exhibit a melting point depression when admixed with a pure component of the mixture.

The details of how *Fischer* reached these conclusions reveal refined analytical thought. The mesobilirubin formed by bromination of synthetic *endo*-ethyl xanthobilirubinic acid (**II**) could have no structure other than that of symmetric mesobilirubin-XIIIα with both ethyls *endo* (*443*):

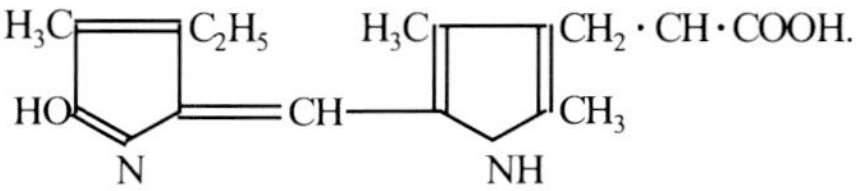

Yet the neo-xanthobilirubinic acid (**III**) produced from the molten resorcinol treatment of this mesobilirubin (called *Mesobilirubin-XIIIα* by Fischer to correlate it with a possible hemin precursor, mesohemin-XIII, whose macrocyclic ring had been opened with elision of the α-methine), had a melting point some 13°C higher than the natural-analytical neo-xanthobilirubinic acid from natural mesobilirubin and also thought to be (*443*):

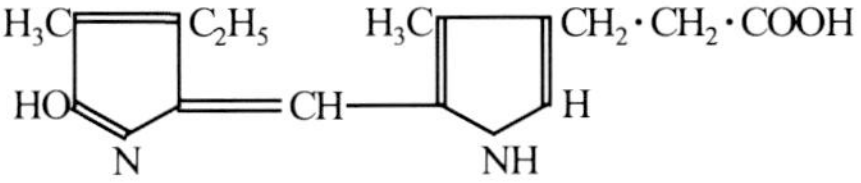

The discrepancy in melting points, previously attributed to keto-enol desmotropy, was put to rest by a chemical synthesis of neo-xanthobilirubinic acid according to the following (*443*):

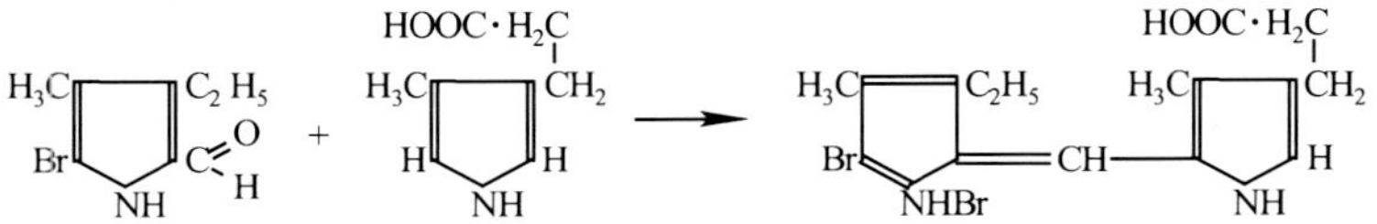

Thus, 3-ethyl-4-methyl-5-bromo-2-formylpyrrole (**V**), obtained by bromination of 3-ethyl-4-methyl-5-carboxyl-2-formylpyrrole, was condensed with *Opsopyrrol* (opsopyrrole) (**VI**) in acetic acid-HBr to give red, crystalline dipyrrylmethene **VII** in almost quantitative yield. In order to replace the α-Br by an α-OH, **VII** was first converted to the crystalline methyl ether (**VIII**), in 50% yield by reaction with $NaOCH_3$ in CH_3OH; then, treatment of the ether with 10% $NaOCH_3$ and H_2O (= aq. NaOH ?) in an autoclave at 150°C liberated the synthetic neo-xanthobilirubinic acid (**III**), mp 246°C, with an *endo*-ethyl (*448*). When added to a natural-analytic sample (mp 227°C), this effected no mixture melting point depression (at mp 226–227°C).

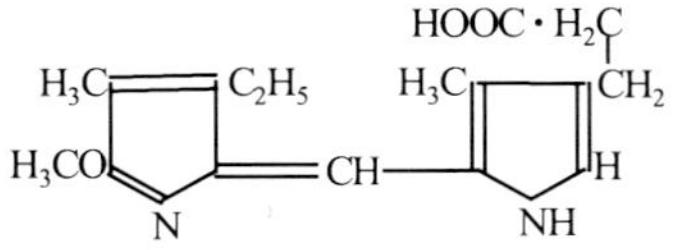

VIII

Reduction of synthetic neo-xanthobilirubinic acid to its leuco analog, neo-bilirubinic acid (**IX**) (mp 179°C) was achieved by heating in glacial acetic acid-HI at reflux (*443*):

> Nach einiger Zeit scheiden sich farblose Krystalle ab. Sie werden abgenutscht, mit Äther gewaschen und aus Essigester umkristallisiert. Schmelzpunkt 179°; Mischschmelzpunkt mit analytischer Neobilirubinsäure (Schmelz. 177°) liegt bei 177,5°, zeigt also keine Depression.

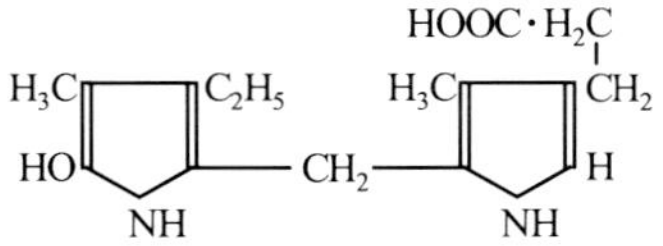

IX

19

It gave no mixture melting point depression (177.5°C) with a natural-analytical sample of mp 177°C. However, the melting point of the benzylidene product from condensation of **IX** with benzaldehyde at the unsubstituted pyrrole α-carbon melted at mp 276°C; whereas, the benzylidene derived from the analytical neo-bilirubinic acid had a different melting point (248°C). Despite the 28° difference in their melting points, the natural-analytical benzylidene depressed the melting point of the "synthetic" benzylidene to 260°C in a mixture melting point determination. Though *Fischer* oddly claimed no melting point depression (*443*):

> Trotz dieser großen Differenz trat in der Schmelze der Mischprobe (Schmelzp. 260°) keine Depression auf.

20

and concluded that while double bond geometric isomerization in the neo-xanthobilirubinic acids might explain the difference between synthetic and natural-analytic samples, isomerization could not explain differences in the benzylidene derivatives of the neo-bilirubinic acid, unless "*cis-trans*" isomerization were involved.

6.2 Serendipitous Discovery of New Dipyrrole Fragments...

Condensation of synthetic neo-xanthobilirubinic acid with formaldehyde-HCl produced mesobilirubin-XIIIα, mp 310°C, or the same mesobilirubin that had been synthesized (surprisingly) by bromination of synthetic xanthobilirubinic acid and reported earlier in 1931 (*444*), thereby illustrating the validity of the unusual bromination-self condensation. Thus, one might have good reason to believe that the simultaneously reported (*444*) bromination of the isomeric *exo*-ethyl xanthobilirubinic acid (**V** of Table 6.2.2) would produce the symmetric mesobilirubin-IIIα with two *exo*-ethyl groups (*443*):

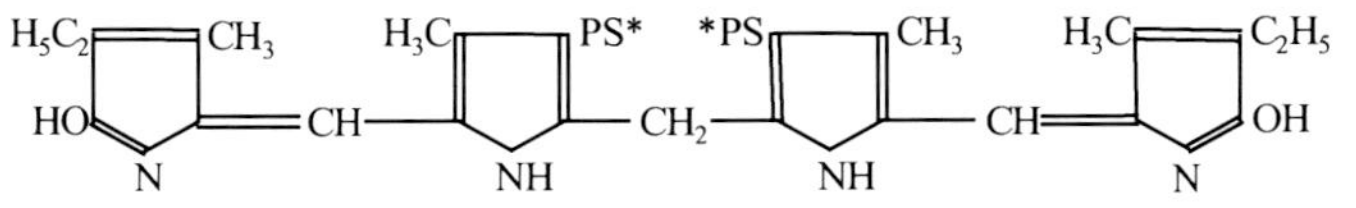

XIII

* PS = CH · CH$_2$ · COOH.

which by treatment with molten resorcinol would yield the *exo*-ethyl analog of neo-xanthobilirubinic acid, or iso-neo-xanthobilirubinic acid (**XII**) (*443*):

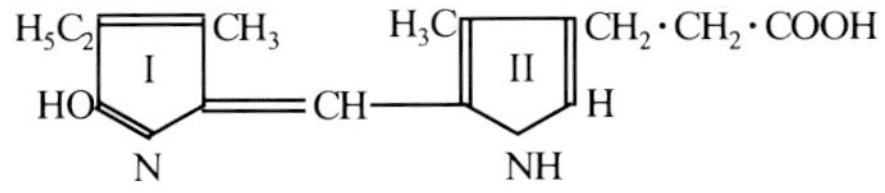

It melted at 242°C when pure, and when added to natural-analytical iso-neo-xanthobilirubinic acid, a mixture melting point commenced at 227°C.

From this, *Fischer* again concluded that there was no melting point depression. However, from all these data, *Fischer* began to suspect that while natural-analytical mesobilirubin might be pure, mixtures of dipyrroles were derived from its cleavage, notably *endo*- and *exo*-ethyl regio-isomers. That is, natural-analytical xanthobilirubinic acid was probably a mixture of synthetic (*endo*-ethyl) xanthobilirubinic acid and its synthetic (*exo*-ethyl) isomer; ditto their methyl esters. And that analytical neo-xanthobilirubinic acid was in fact a mixture of synthetic *endo*-ethyl neo-xanthobilirubinic acid and its synthetic *exo*-ethyl isomer. Likewise their esters. To put this idea to a test, he co-crystallized synthetic neo-xanthobilirubinic acid (*endo*-ethyl) with synthetic iso-xanthobilirubinic acid (*exo*-ethyl) to obtain crystals of mp 226°C, or exactly the melting point of the natural-analytical neo-xanthobilirubinic acid obtained from natural-analytical mesobilirubin by the resorcinol melt cleavage method. An artificial mixture of the dipyrrole pigments was much more soluble in crystallization solvents than either of the the *exo*-ethyl or *endo*-ethyl components and so *Fischer* was unable to separate the mixture into its two components.

Continuing, *Fischer* mixed synthetic mesobilirubin-IIIα and XIIIα and treated the mixture with molten resorcinol, which led to a mixture of cleavage isomers that corresponded to the natural-analytical neo-xanthobilirubinic acid, had the same melting point, 226°C, and was more soluble in CH$_3$OH than either component.

6 The Status of Bilirubin in 1930

Fischer and *Siedel* left no stone unturned in their probings of melting points and mixture melting points. For a better, comprehensive grasp of the situation again, all melting points and mixture melting points are cited below, from *Fischer* (*443*):

Zur besseren Übersicht seien nochmals alle Schmelzpunkte und Mischschmelzpunkte angeführt:

Neo- und Iso-neoxanthobilirubinsäure.

Neoxanthobilirubinsäure...............	Schmelzp.	245°
Iso-neoxanthobilirubinsäure	„	242°
Mischkrystallisation.............	„	226°
Analyt. „Neoxanthobilirubinsäure"	„	227°
Analyt. „N." + synthet. N..............	„	227°
Analyt. „N." + synthet. Iso-n.	„	227°
Analyt. „N." + synthet. Mischkryst.	„	226°

Neo- und Iso-neoxanthobilirubinsäure-methylester.

Neoxanthobilirubinsäure-methylester.............	Schmelzp.	174°
Iso-neoxanthobilirubinsäure-methylester	„	205°
Mischung.............	„	160–161°
Analyt. „ Neoxanthobilirubinsäure-methylester"	„	190–191°
Mischung (umkryst. aus Methylalkohol).............	„	190°
Analyt. „N.-ester" + synthet. N.-ester.............	„	158°
Analyt. „N.-ester" + synthet. Iso-n.-ester	„	200°

Neo- und Iso-neobilirubinsäure-benzylidenverbindung.

Neobilirubinsäure-benzylid.............	Schmelzp.	276°
Iso-neobilirubinsäure-benzylid.	„	273°
Mischkrystallisation.............	„	247–248°
Analyt. „Neobilirubinsäure-benzylid"	„	248°
Analyt. „Neob.-benzylid." + synthet. Neob.-benz.	„	260°
Analyt. „Neob.-benzylid." + synthet. Iso-neob.-benz.............	„	260°
Analyt. „Neob.-benzylid." + Mischkrystallisation.............	„	247–248°
Synthet. Neob.-benzylid. + Mischkrystallisation.............	„	262°
Synthet. Iso-neob.-benzylid. + Mischkrystallisation	„	261°

Xantho- und Iso-xanthobilirubinsäuren

Xanthobilirubinsäure.............	Schmelzp.	287°
Iso-xanthobilirubinsäure	„	289°
Mischkrystallisation.............	„	274–275°
Analyt. „Xanthobilirubinsäure"	„	273°
Analyt. „X." + synthet. X.............	„	273°
Analyt. „X." + synthet. Iso-x.	„	273°
Analyt. „X." + Mischkrystallisation	„	273°
Synthet. „X." + Mischkrystallisation	„	274°
Synthet. Iso-x. + Mischkrystallisation	„	274–275°

6.2 Serendipitous Discovery of New Dipyrrole Fragments…

Xantho- und Iso-xanthobilirubinsäure-methylester.

Xanthobilirubinsäure-methylester	Schmelzp.	212°
Iso-xanthobilirubinsäure-methylester	„	198°
Mischung	„	176°
Mischung (aus Chloroform-Petroläther umkryst.)	„	192–194°
Analyt. „Xanthobilirubinsäure-methylester"	„	212°

21

Choosing the pairs that gave large mixture melting point depressions from the summary table, *Fischer* and *Siedel* keyed in on the methyl esters and successfully separated the methyl ester (mp 190–191 °C) of natural-analytical neo-xanthobilirubinic acid into its two component esters: the less-soluble *exo*-ethyl iso-neo-antho-bilirubinic acid (mp 205 °C) and the more soluble *endo*-ethyl xanthobilirubinic acid (mp 174 °C). From which they concluded that all condensations involving natural-analytical neo-xanthobilirubinic acid to form tetrapyrroles must lead to three products: two symmetric and one unsymmetric, as in the formation of K-mesobilirubin, which in fact is a mixture of mesobilirubins-IIIα, IXα, and XIIIα (*443*):

Der Ester der analytischen Neoxanthobilirubinsäure schmilzt nach etwa zweimaligem Umkrystallisieren aus Methylalkohol bei 190–191°. Der Ester der synthetischen Neoxanthobilirubinsäure, umkrystallisiert aus Methylalcohol, schmilzt bei 174° und der Ester der Iso-neoxanthobilirubinsäure, aus einer Mischung von Methyl- und Äthylalkohol (2:1) umkrystallisiert, schmilzt bei 205°.

Da als Misch-Schmelzpunkt zwischen den beiden synthetischen Estern 160–161° und als Misch-Schmelzpunkt zwischen analytischem Ester und Iso-neoxanthobilirubinsäure-ester 200° festgestellt werden konnte, ging deutlich hervor, daß in dem vermeintlichen analytischen Ester (Schmelzp. 190–191°) bereits das Iso-Isomere der Neoxanthobilirubinsäuren angereichert sein mußte. Diese Feststellung wurde auch durch das mikroskopische Bild bestätigt.

Der Neoxanthobilirubinsäure-methylester (Schmelzp. 174°) ergibt, aus Methylalkohol umkrystallisiert, nur sehr schlecht ausgebildete Krystallindividuen, meist von Spindelform ohne exakt meßbare Winkel.

Der Iso-neoxanthobilirubinsäure-methylester (Schmelzp. 205°), der in Methyl alkohol außerordentlich schwer löslich ist und sich nur aus einer Mischung von Methyl- und Äthylalkohol umkrystallisieren läßt, bildet dagegen sehr schöne, lange Prismen mit scharfen Kanten und dem Winkel von 42–45°.

Das mikroskopische Präparat des analytischen Esters zeigte nun vorwiegend die langen Prismen des Iso-neoxanthobilirubinsäure-methylesters in Übereinstimmung mit dem Schmelzpunkt.

Mit dieser Beobachtung war nunmehr der Weg zur Trennung der analytischen Neoxanthobilirubinsäuren-Mischung vorgezeichnet. Aus den oben angegebenen unterschiedlichen Löslickeitsverhältnissen der synthetischen isomeren Ester mußte eine mehrfache Behandlung des analytischen Gemisches mit Methylalkohol in Kombination mit Äthylalkohol zu einer Lösung des Neoxanthobilirubinsäure-esters und zu einer Anreicherrung des Iso-neoxanthobilirubinsäure-esters in auskrystallisierenden Produkt führen.

Das geschah auch tatsächlich; es konnte schließlich ein Ester vom Schmelzp. 203° erhalten werden, der in der Mischprobe mit Iso-neoxanthobilirubinsäure-ester (Schmelzp. 205°) bei 205° schmolz, also keine Depression ergab. Rückwärts wurde dann aus diesem analytischen Ester (Schmelzp. 203°), durch Verseifung mit NaOH Iso-neoxanthobilirubinsäure

6 The Status of Bilirubin in 1930

vom Schmelzp. 240–241° dargestellt. Mit synthetischer Neoxanthobilirubinsäure gab diese analytische Iso-neoxanthobilirubinsäure eine Schmelzpunktserniedrigung auf 227°.

Mit der Isolierung der Iso-neoxanthobilirubinsäure ist eindeutig bewiesen, daß die Spaltung des Mesobilirubins zu einem Gemisch zweier isomerer Neoxanthobilirubinsäuren führt. Daraus ergibt sich nunmehr, daß alle mit analytischer „Neoxanthobilirubinsäure" vorgenommenen Kondensation mit Aldehyden oder Ketonen zu drei isomeren Produkten führen mußten, zwei symmetrischen und einem unsymmetrischen Reaktionsprodukt.

Das K-Mesobilirubin[1]) ist deshalb als aus drei isomeren Mesobilirubin bestehend anzusehen.

Bewiesen werden konnte diese Annahme an den Reaktionsprodukten der Neoxanthobilirubinsäure mit p-Dimethyl-amino-benzaldehyd.[2]) Das Kondensationsprodukt mit Neoxanthobilirubnsäure schmolz bei 244–245°, dasjenige mit Iso-neoxanthobilirubinsäure bei 246°, die Mischung beider bei 239–240°. Das Kondensationsprodukt mit analytischer „Neoxanthobilirubinsäure" schmolz bei 239°. Dieser Schmelzpunkt beweist das Vorliegen eines Gemisches.

Die Übertragung dieser Anschauung und ihrer Beweisführung auf die analytisch „Xanthobilirubinsäure" und ihrem Veresterungsprodukt deckte auch hier das Vorkommen zweier Isomeren, der Xanthobilirubinsäure (XIV) (Schmelzp. 287°) und Isoxanthobilirubinsäure (XV) (Schmelzp. 289°), auf und beseitigte all Unstimmigkeiten, die sich bisher bei den Schmelzpunkten ergeben haben.

Die Trennung dieser Isomeren gelingt ebenfalls leicht über die Methylester durch Krystallisation aus Chloroform-Petroläther. Diesmal zeichnet sich aber der Isoxanthobilirubinsäure-ester durch leichtere Löslichkeit aus.

Für beide Xanthobilirubinsäuren wurde im Verlauf dieser Arbeit eine neue Darstellungsmethode mittels Natriummethylat eingeführt, die Ausbeuten von durchschnittlich 50-60% liefert.

Mit den obigen Feststellungen ist bewiesen, daß dem Mesobilirubin und damit dem Bilirubin eine unsymmetrische Struktur zukommt. Bilirubin stimmt also in der Anordnung seiner Seitenketten prinzipiell mit Hämin (und Chlorophyll) überein. Die physiologische Entstehung des Bilirubins aus Hämin erfolgt durch Herausoxydation der α-Methinbrücke unter Bildung von 2 Oxygruppen an den Pyrrolkernen I und IV.

---

[1]) *Diese Z.* **194**, 223 (1930)
[2]) *Diese Z.* **194**, 221 (1930)

Die unsymmetrische Struktur des Mesobilirubins erklärt alle Beobachtungen beim Bilirubin und insbesondere auch das Auftreten von Hämopyrrol bei der Reduktion sowie des „Nitritkörpers"[1]), ... der bei Oxydation mit salpetriger Säure aus Bilirubin entsteht und die Konstitution Nr. X, S. 149 besitzen muß. Mit der Tatsache, daß dieser Körper nur aus dem Pyrrolkern IV des Bilirubins entstehen kann, steht die schlecte Ausbeute, in der er erhalten wird, im Einklang.

---

[1]) H. Fischer u. Hess, *Diese Z.* **194**, 195, 207 (1930)

6.2 Serendipitous Discovery of New Dipyrrole Fragments...

With the structure of natural-analytical mesobilirubin having been solved (and shown to be mesobilirubin-IXα), at the beginning of 1934, *Fischer* could summarize (in English) the status of the bilirubin structure for the *Annual Review of Biochemistry (449)*:

> Judging from a series of inconsistencies and a large number of long-known differences in the melting-points of the xanthobilirubin acids, as well as the isolation of haemopyrrol from bilirubin, it was an obvious deduction that bilirubin has a similar side-chain structure as haemin. This view was in the main supported by the isolation of a "nitrite body" of Formula V [Fischer & Hess (14)] [*437*], which could not be explained by the existing bilirubin formula. On the basis of the assumption hitherto prevailing, bilirubin is thought to be derived from haemin xiii,[1] whereas by derivation from the natural haemin ix, mesobilirubin or bilirubin should have Formula **VI** or **VII**, respectively.

VI. Mesobilirubin [(C_2H_5) in rings I and IV]

VII. Bilirubin [(CH = CH_2) in place of (C_2H_5) in rings I and IV]

On the basis of the formula, above, mesobilirubin should give a mixture of two isomeric neoxanthobilirubin acids (formulae VIII and IX) in the resorcin melt.

VIII. Neoxanthobilirubin acid

IX. Iso-neoxanthobilirubin acid

---

[1] For the nomenclature of the haemins or the porphyrins from which they are derived, compare Fischer & Stangler, *Ann.*, 459, 62 (1927). [*431*]

That this really does occur was definitely shown by Siedel & Fischer (15) [*443*] both synthetically and analytically. By a thorough comparison of the melting-points and the mixed melting-points of the synthetic as well as the analytical products of, on the one hand, the free neoxanthobilirubin and *iso*-neoxanthobilirubin acids, and on the other, the condensation products with formaldehyde, it could be shown that mesobilirubin, and therefore also bilirubin, have an unsymmetrical structure. The physiological formation of bilirubin from haemin is thus explained by a mechanism in which the α-methene bridge is oxidised away with the formation of two hydroxy groups on the pyrrol rings I and IV. The assumption of an unsymmetrical structure for mesobilirubin explains all the known facts about bilirubin, and in particular the formation of haemopyrrol by reduction as well as that of the "nitrite body" on oxidation with nitrous acid.

For a discussion of the "nitrite body", see Section 6.3.1.

In this review article (*449*), written with *Hans Orth* prior to 1934, the authors proposed both the lactim-lactam (**X**) and bis-lactam (**XI**) structures for mesobiliverdin-IXα (glaucobilin) – suggesting that tautomeric structures might account for some of the different colors seen in the *Gmelin* reaction (*449*):

Alternative formulae proposed for glaucobilin

Eschewing lactam tautomers for biliverdin-IXα (dehydro-bilirubin-IXα), the authors corrected *Lemberg*'s structure of uteroverdin. In 1932, *Lemberg* isolated uteroverdin from dog placenta (*450*) and in the same year represented its structure (**I**) as biliverdin-XIIIα, while calling it identical to dehydro-bilirubin (*451*):

Uteroverdin ist also Dehydrobilirubin. In der ersten Mitteilung teilte ich diesem Stoffe die Formel **I** zu:

6.2 Serendipitous Discovery of New Dipyrrole Fragments...

– presumably because in 1931, the structure of natural-analytical mesobiirubin was thought to have two *endo*-ethyl groups. In their review (*449*) *Fischer* and *Orth* corrected what they perceived to be an error in *Lemberg*'s structure, making it conform to the IXα skeleton (**XII**) but also indicating that in (*443*) *Fischer* and *Siedel* preferred the tautomeric structure **XIII**, or perhaps its *exo*-vinyl cyclized furan structure (*449*):

> Lemberg (17) [*450*] was able to isolate uteroverdin from the placenta of the dog and ascribed to it (18) [*451*] the constitution of a dehydro-bilirubin (XII). Siedel & Fischer (15) [*443*] prefer Formula XIII . . .

XII. Uteroverdin [Lemberg (18)]

XIII. Uteroverdin (Fischer & Siedel)

Curiously, the lactams of **X** and **XI** were not carried forth into **XIII**, possibly because the presence of vinyl groups in **XIII** necessitated a structure with more C=Cs than found in **X** and **XI**? It is also interesting to note that the tautomeric structures were probably drawn in order to account for differences in color, *e.g.* green, blue, greenish-blue. The actual structure of biliverdin was proved conclusively by *Plieninger*'s and *Fischer*'s total synthesis in 1942 (*10*).

Given their focus on the relationship between colors, tautomers, and carbon-carbon double bond isomers, *Siedel* and *Fischer* proposed structures to explain the display of colors exhibited in the *Gmelin* color test for mesobilirubin (or bilirubin) with nitrous or nitric acid in terms of a series of dehydrogenation steps and double bond migrations (*443*):

> . . . über die oxydativen Vorgänge bei dem Verlauf der Gmelinschen Reaktion. Die Autoren nehmen in der letzten Phase der Reaktion, in der Rotphase, eine Oxydation des Bilirubinoids bis zum holochinoiden Typ in folgender Art an:

* PS = CH₂·CH₂·COOH.

24

Fischer had but two remaining issues related to the structure of bilirubin: should the synthetic target be the bilirubin with both *exo* and *endo* vinyls, or should it have a furan ring in place of the *exo*-vinyl. But first, the absolute proof of the bilirubin-derived natural-analytical mesobilirubin remained to be accomplished by a chemical synthesis.

6.2.3 The Structure of Natural-Analytical Mesobilirubin Confirmed. Synthesis of Mesobilirubin-IXα

In the midst of the Great Depression, in the year 1934, *G.N. Lewis'* student, *Harold Urey*, was awarded the *Nobel* Prize for his discovery of deuterium, *Hitler* was proclaimed *Führer* and *Reichskanzler* of Germany, *Mao Tse-Tung* began the 6,000

6.2 Serendipitous Discovery of New Dipyrrole Fragments...

mile "Long March" a continent away, and *Hans Fischer*, a *Nobel* Prize awardee of four years previous, had begun to lay the groundwork for the total synthesis of bilirubin in his own long march toward its structure determination. But what structure of bilirubin to synthesize? To *Fischer*, the question of whether bilirubin had an *exo*-vinyl group or (he felt more likely) a furan ring would continue to be an important one to answer before attempting a synthesis. *Fisher*'s lab continued to explore and improve the syntheses of xanthobilirubinic acid from kryptopyrrolenone, and iso-neo-xanthobilirubinic acid and iso-xanthobilirubinic acid from hemopyrrolenone (*452*) – a methodology later improved (in 1975) by *Joan Grunewald* at the University of Kansas Medical Center (*453*) and used extensively in the author's research labs at the University of Nevada, Reno. Other routes to neo- and iso-neo-xanthobilirubinic acid were established and explored (*454*) and would subsequently become important in the total synthesis of natural-analytical mesobilirubin.

An improved understanding of the structure of bilirubin had been attained in 1934. By early 1935, *Fischer* and *Haberland* concluded that the structure of bilirubin was best represented by **I**, which explained all findings, except for what was named the *Nitritkörper* (**II**), the suspected and long-sought product of nitrous acid oxidation of bilirubin that came from cleavage of ring **IV** wherein a cyclization of the *exo*-vinyl to a furan had taken place (*455*):

*) Hier und bei den späteren Formeln S=CH$_2$CH$_2$COOH.

II

In order to explore more delicately whether bilirubin has an *exo*-vinyl group or a furan ring, they carefully repeated the catalytic (colloidal Pd) hydrogenation that produced natural-analytical mesobilirubin from bilirubin in 0.1 *N* aq. NaOH. *Fischer* and *Haberland* (*455*) noticed that the uptake of H$_2$ proceeded slowly at first and then suddenly increased very rapidly – usually upon addition of fresh catalyst. Would the two vinyl groups react with differential rates, or might it make more sense to think that a vinyl group and a furan group might behave even more differently? Could they be differentiated, with one reacting rapidly and the other only slowly – and how might one determine it experimentally? An attempt was made to isolate the intermediate dihydrobilirubin on the way to mesobilirubin by breaking off the hydrogenation after one mole equivalent of H$_2$ had been taken up. Fractional crystallization of the product yielded unchanged bilirubin, some mesobilirubin, and dihydrobilirubin in the form of brick-red crystals that gave a positive *Gmelin* reaction and had a melting point of 351°C (or the same as mesobilirubin). Two combustion analyses, both leaving a bit of residual ash, gave acceptable results for $C_{33}H_{38}N_4O_6$ (455):

| $C_{33}H_{38}N_4O_6$ (586) | Ber. | C | 67,59 | H | 6,48 | N | 9,56 |
| | Gef. | „ | 67,89, 68,07 | „ | 6,31, 6,42 | „ | 9,50, 9,35. |

The calculated %CHN values of mesobilirubin: 67.35, 6.80, 9.52; of bilirubin: 67.81, 6.16, 9.59, respectively, are within the range of those found above, especially the latter set. Consistent with a dihydrogenated product, however, oxidative cleavage using conc. HNO_3 gave a 34.2% yield of methylethylmaleimide from dihydromesobilirubin *vs.* 81.5% from mesobilirubin, and so (logically) *Fischer* concluded that dihydrobilirubin was either of two structures (455):

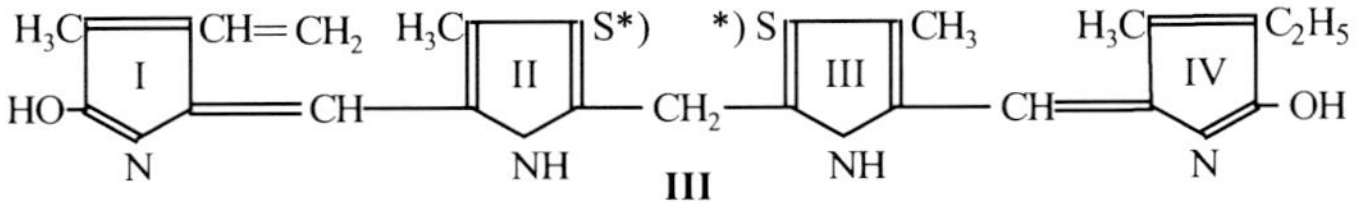

In order to assign the exact structure, dihydromesobilirubin was cleaved in molten resorcinol. Recall that natural-analytical mesobilirubin in molten resorcinol produced a mixture of neo-xanthobilirubinic acid and iso-neo-xanthobilirubinic acid. Dihydromesobilirubin, it was believed, should give only one or the other. In a yield only half of that obtained from natural-analytical mesobilirubin, *Fischer* and *Haberland* isolated the *exo*-ethyl dipyrrole, iso-neo-xanthobilirubinic acid, as its methyl ester, mp 197°C. (The *endo*-ethyl isomer, neo-xanthobilirubinic acid methyl ester melts at 174°C.) Thus, it was the *exo*-vinyl of bilirubin that had undergone the more rapid hydrogenation to give dihydrobilirubin as structure **III** above. The results from catalytic hydrogenation of bilirubin and the belief that it degraded to *Nitritkörper* **II** convinced *Fischer* that the unsaturated side chains in bilirubin could not be formulated in the same way. Accordingly, *Fischer* believed that bilirubin must have structure **IX** below (455):

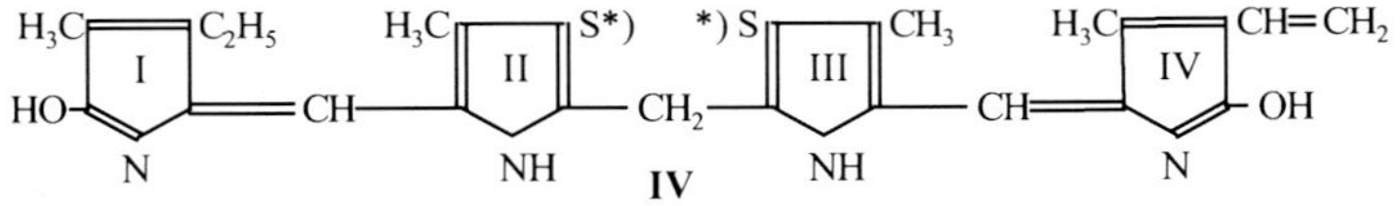

Die ungesättigten Seitenketten des Bilirubins können also nicht in gleicher Weise formuliert werden. Bei Berücksichtigung des Wasserstoffverbrauches bei der katalytischen Hydrierung und der Tatsache der Entstehung des Nitritkörpers (**II**) muß Bilirubin folgender Formel entsprechen,

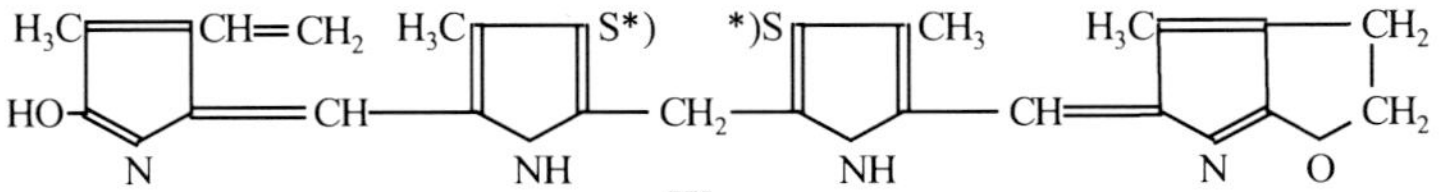

6.2 Serendipitous Discovery of New Dipyrrole Fragments...

> in der am Pyrrolkern IV die Vinylgruppe und die 8′-ständige Hydroxylgruppe durch Isomerisation einen Dihyrofuranring gebildet haben. In dieser Formel muß der ungesättigte Rest am Pyrrolring I als offene Vinylgruppe angenommen werden, denn ein Ringschluß zur Oxoygruppe in 1′-Stellung ist aus sterischen Gründen äußerst unwahrscheinlich. Der Dihydrofuranring, an Pyrrolkern IV angegliedert, muß dann katalytisch sich leichter offnen, als die Hydrierung der Vinylgruppe erfolgt. 25

He assumed its dihydrofuran ring would open more rapidly (hydrogenolysis ?) under the conditions of catalytic hydrogenation than the *endo*-vinyl would undergo hydrogenation. *Fischer* also took the opportunity in this 1935 publication to restate and re-propose structures (*443*) for the series of colorful intermediates of the *Gmelin* color reaction by focussing only on the linear tetracycle, not on vinyl groups or other β-constituents (*455*):

Reaktionsmechansismus der Gmelinschen Reaktion.

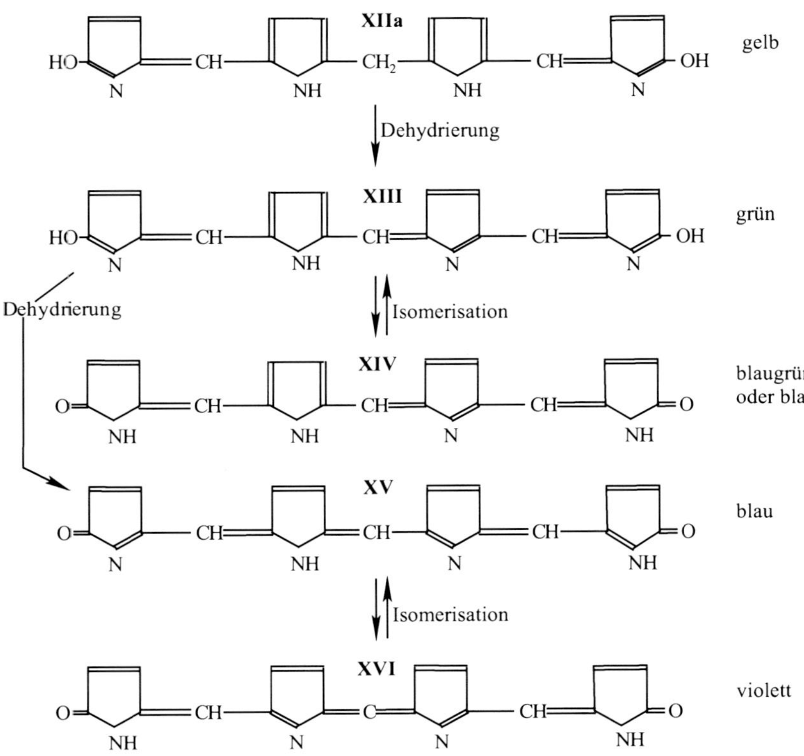

XIIa — gelb

Dehydrierung

XIII — grün

Dehydrierung | Isomerisation

XIV — blaugrün oder blau

XV — blau

Isomerisation

XVI — violett

26

By the end of 1936, *Walter Siedel* in *Fischer*'s group had achieved the total synthesis of natural-analytical mesobilirubin (mesobilirubin-IXα), thereby confirming its non-symmetric structure (*456*). Previously, the synthesis of the symmetric isomers, mesobilirubin-IIIα and XIIIα had been achieved (*444*) by condensing two equivalents of either synthetic neo-xanthobilirubinic acid or synthetic iso-neo-xanthobilirubinic acid, respectively, with CH_2O. However, synthesis of the non-symmetric isomer (IXα) required a different approach, one that might be used for the synthesis of the non-symmetric bilirubin-IXα. The synthesis was designed simply around the coupling of two dipyrrinones, one with a pyrrole α-H to serve as the "receptor" component for a second dipyrrinone possessing an activatable pyrrole α-hydroxymethyl group. In the event that the two dipyrrinone units were differently substituted, *e.g.* one with an *endo*-ethyl, the other with an *exo*-ethyl, a non-symmetric mesobilirubin would result.

Siedel chose as starting materials the two well-known (by then) synthetic dipyrrinones of confirmed structure: neo-xanthobilirubinic acid and iso-neo-xanthobilirubinic acid (*456*):

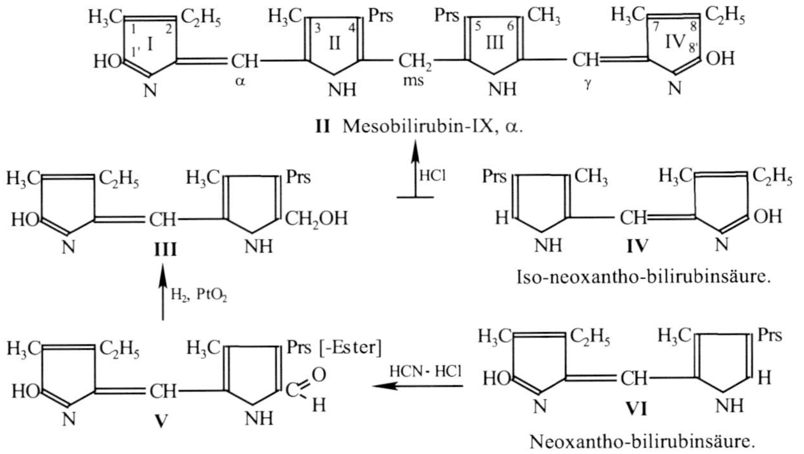

II Mesobilirubin-IX, α.

Iso-neoxantho-bilirubinsäure.

Neoxantho-bilirubinsäure.

To the methyl ester of the former, obtained quantitatively by reaction with diazomethane (*457*), a formyl group was introduced at the pyrrole α-carbon using the then standard formylation procedure (HCN + HCl in $CHCl_3$) to give **V**, mp 205.5°C (*457*):

3,858 mg Subst. (bei 60° i. V. getr.): 9,217 mg CO_2, 2,202 mg H_2O. – 4,012 mg Subst.: 0,316 ccm N (22°, 709 mm). – 5,228 mg Subst.: 3,725 mg AgJ.

$C_{18}H_{22}O_4N_2$ (330,2)	Ber.	C	65,42	H	6,72	N	8,48	OCH_3	9,39
	Gef.	„	65,16	„	6,39	„	8,50	„	9,41.

The aldehyde group of the ester (**V**) in CH_3OH was reduced using H_2/PtO_2 to a hydroxymethyl, as in **III** (*456*):

. . . aus Methanol umkyrstallisiert: gelbe, kleine, teils gerade, teils schief abgeschnittene prismatische Nädelchen mit dem Schmelzpunkt 187° (unkorr.) = 192° (korr.). Sinterung bei 180°. . . .

6.2 Serendipitous Discovery of New Dipyrrole Fragments…   397

4,116, 4,309 mg Subst. (aus Methanol umkrystallisiert bei 30° i. V. getr.): 9,860, 10,265 mg CO_2, 2,580, 2,600 mg H_2O. – 3,473 mg Subst.: 0,270 ccm N (19°, 704 mm).

$C_{18}H_{24}O_4N_2$ (332,2)	Ber.	C	65,02	H	7,28	N	8,43	
	Gef.	„	65,33, 64,97	„	7,01, 6,75	„	8,42.	27

and the hydroxymethyl methyl ester was coupled smoothly in $CHCl_3$-HCl with iso-neo-xanthobilirubinic acid methyl ester to give synthetic mesobilirubin-IXα dimethyl ester dihydrochloride that corresponded to natural-analytical mesobilirubin-IXα dimethyl ester dihydrochloride (from bilirubin) in all properties. The synthetic product melted at 194°C (uncorr.), 199°C (corr.) and upon admixture with the natural-analytical material (mp 192°C, uncorr.) gave a mixture mp of 193°C, which *Siedel* concluded to be "no depression" (*456*):

Schon nach kurzem Stehen tritt prachtvolle Krystallisation des synthetischen Mesobilirubin IX, α-dimethyl ester-dihydrochlorids in den orangefarbenen Prismen mit der Auslöschungsschiefe von 1,5° ein. Diese Prismen stimmen in allen übrigen Eigenschaften mit den S. 268 beschriebenen Krystallen überein. Schmelzp. 194° (unkorr.) = 199° (korr.). Mischschmelzpunkt mit analytischem Material (Schmelzp. 192° unkorr.) = 193°, also keine Depression.   28

The synthetic dimethyl ester dihydrochloride was converted to the simple dimethyl ester, mp 234°C (uncorr.), 240.5°C (corr.). When admixed with the natural-analytical dimethyl ester (mp 233–234°C, uncorr.; 239.5–240.5°C, corr.), it gave a mixture melting point of 233°C, which *Siedel* called "no depression" (*457*). The former gave a mixture melting point at 232°C with the dimethyl ester (mp 248°C) of symmetric mesobilirubin-IIIα, and at 243°C with the dimethyl ester of symmetric mesobilirubin-XIIIα, mp 268°C (uncorr.) (*456*):

In der Auslöschung und im Dichroismus besteht ebenfalls völlige Übereinstimmung. Schmelzp. 234° (unkorr.) = 240,5° (korr.). Die Mischung mit analytischem Mesobilirubin-IX, α-dimethylester (Schmelzp. 233°) schmilzt bei 233°, zeigt also keine Depression. Die Mischung mit dem isomeren Ester-III, α (Schmelzp. 248°C) schmilzt bei 232°, die Mischung mit dem Ester-XIII, a (Schmelzp. 268°) bei 243° (sämtlich unkorr.).

3,521 mg Subst. (mit Methanol extrahiert, bei 50° i. V. getr): 8,839 mg CO_2, 2,202 mg H_2O. – 2,797 mg Subst.: 0,235 ccm N (21°, 717 mm)

$C_{35}H_{44}O_6N_4$ (616,4)	Ber.	C	68,14	H	7,20	N	9,08	
	Gef.	„	68,46	„	7,00	„	9,20	29

Synthetic mesobilirubin-IXα was isolated following saponification of the dimethyl ester dihydrochloride in 5% methanolic KOH to give material of mp 311°C (uncorr.) and 321°C (corr.) that was identical in all properties with the natural-analytical mesobilirubin (mp 310°C). When the two pigments were admixed, the mixture melting point was 310°C, *i.e.* no depression. When synthetic mesobilirubin-IXα was admixed with synthetic mesobilirubin-IIIα (mp 316°C, uncorr.), the latter gave a mixture melting point of 306°C (uncorr.).

With synthetic mesobilirubin-XIIIα (mp 311°C), the mixture melting point was 306°C, uncorr. (*456*):

> Der Rückstand gibt, aus Pyridin umkrystallisiert, schöne gelbe Nadeln, die in allen Eigenschaften denen des analytischen Materials gleichen. Auch die Löslichkeitsverhältnisse sind dieselben. Schmelzp. 311° (unkorr.) = 321° (korr.). Der Mischschmelzpunkt mit analytischem Material (Schmelzp. 310°) liegt bei 310°, zeigt also keine Depression. Mischschmelzpunkt mit Mesobilirubin-III, α (Schmelzp. 316°) = 306° (unkorr.), mit Mesobilirubin-XIII, α (Schmelzp. 311°) = 306° (unkorr.).
> 3,860 mg Subst. (aus Pyridin umkryst., bei 60° i. V. getr.): 9,550 mg CO_2, 2,420 mg H_2O. − 2,977 mg Subst.: 0,265 ccm N (20°, 702 mm).

$C_{33}H_{40}O_6N_4$ (588,32)	Ber.	C	67,31	H	6,85	N	9,52
	Gef.	„	67,47	„	7,02	„	9,58.

30

Thus, the total synthesis of natural-analytical mesobilirubin-IXα firmly established its structure and, thereby, a route for the synthesis of unsymmetrical linear tetrapyrroles had been established. The more difficult problem lay ahead: the total synthesis of bilirubin.

6.3 *Fischer* Solves the Structure of Bilirubin

With the synthesis having been established for (i) symmetric mesobilirubins-IIIα and -XIIIα by acid-catalyzed self-condensation of iso-neo-xanthobilirubinic acid and of neo-xanthobilirubinic acid, respectively, and (ii) unsymmetric mesobilirubin-IXα (or, simply, mesobilirubin) by acid-catalyzed condensation of an α-hydroxymethyl dipyrrinone with an α-H dipyrrinone (*456*), might these routes also be used when the ethyl groups of the dipyrrinones are replaced by vinyls? And how might one obtain the vinyl analogs of neo-xanthobilirubinic acid (**II**) and iso-neo-xanthobilirubinic acid (**III**) needed for such routes? Recall that **II** and **III** had been obtained by synthesis as well as from treatment of mesobilirubin (**I**) with molten resorcinol (*458*):

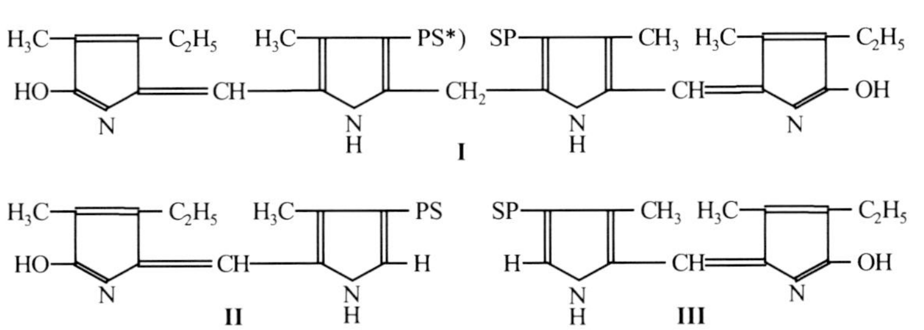

*) PS = CH_2CH_2COOH

In, 1939, *Fischer* and *Reinecke* were prompted to repeat the reaction of bilirubin with molten resorcinol, with the expectation that if the vinyl groups could withstand the reaction conditions of boiling resorcinol, the following products would be expected (*458*):

6.3 Fischer Solves the Structure of Bilirubin

Accordingly, repetition of an earlier "resorcinol melt" reaction afforded $CHCl_3$-extractable yellow needles with a melting point of 234°C, but in only 20% yield. After recrystallization from CH_3OH, the melting point was raised to 245°C, and the new pigment analyzed for $C_{16}H_{18}N_2O_3$, or a formula corresponding roughly to one-half of bilirubin's $C_{33}H_{36}N_4O_6$ formula, less one carbon (458):

> Zur Analyse wurde mit Chloroform extrahiert und die erhaltenen langen feinen Nädelchen bei 70° im Vakuum getrocknet.
>   3,735 mg Subst. – 0,02 mg Asche = 3,715 mg Subst.: 9,110 mg CO_2, 2,060 mg H_2O. – 3,390 mg Subst.: 0,340 ccm N_2 (25°, 716 mm).

$C_{16}H_{18}O_3N_2$ (286,1)	Ber.	C	67,08	H	6,39	N	9,78	
	Gef.	,,	66,88	,,	6,21	,,	9,42.	31

The new pigment formed a mono-methyl ester with diazomethane (458):

> Zur Analyse wurde aus Methylalkohol umkrystallisiert und bei 60° im Vakuum getrocknet.
>   3,417 mg Subst.: 8,515 mg CO_2, 2,040 mg H_2O. – 2,787, 2,470 mg Subst.: 0,258 (25°, 717 mm), 0,215 (23°, 724 mm) ccm N_2. – 4,250 mg Subst.: 3,355 mg AgJ.

$C_{17}H_{20}O_8N_2$	Ber.	C	67,98	H	6,76	N	9,32	OCH$_3$	10,32	
(300,1)	Gef.	,,	67,96	,,	6,88	,,	9,81, 9,56	,,	10,43.	32

Thus, the analyses of both acid and ester indicated preservation of a vinyl group. The ester, admixed with neo-xanthobilirubinic acid methyl ester (mp 174°C) depressed the mixture melting point to 163°C. When the ester was added to iso-neo-xanthobilirubinic acid methyl ester (mp 204°C), the mixture melting point was depressed to 170°C.

In order to prove the structure of this apparently homogeneous acid, which afforded no methylethylmaleimide upon oxidation, it was hydrogenated catalytically to reduce the "vinyl" group as well as the connecting methine to produce only neo-bilirubinic acid. There was no evidence for the formation of iso-neo-bilirubinic acid, and so the "resorcinol melt" product from bilirubin was confirmed as vinyl-neo-xanthobilirubinic acid (IV), with an *endo*-vinyl group. To *Fischer*, this was definitive confirmation of a bilirubin structure with a furan ring rather than an *exo*-vinyl (458):

400   6 The Status of Bilirubin in 1930

> Mithin liegt reine Neo-bilirubinsäure vor und das Ausgangsmaterial ist der entsprechende, reine Vinylkörper **VI**. Wir erblicken in dieser Tatsache eine Bestätigung der Bilirubinformel mit dem Hydrofuranring.   33

It was important to learn whether vinyl groups would survive the known procedures (*438, 442, 444, 447, 448*) for condensing vinyl-dipyrrinones to bilirubins. *Fischer* followed the method used previously (*438*) in transforming neo-xanthobilirubinic acid to mesobilirubin-XIIIα. Accordingly, he carried out a condensation of two mole equivalents of *endo*-vinyl **IV** with one of CH_2O, catalyzed by HCl, which produced bilirubin-XIIIα, mp 312°C, with two *endo*-vinyls. Combustion analysis of this synthetic bilirubin isomer correlated well with the formula $C_{33}H_{36}N_4O_6$ (*458*):

> Es wurde nach der Vorschrift von H. Fischer und R. Hess kondensiert. Man erhält nach dem Umkrystallisieren aus viel Pyridin kleine Rechtecke vom Schmelzp. 312°. Zur Analyse wurde aus Pyridin umkrystallisiert und bei 50° im Vakuum getrocknet.
>
> 4,348 mg Subst.: 10,787 mg CO_2, 2,398 mg H_2O. – 3,658 mg Subst.: 0,334 ccm N_2 (26°, 709 mm).

$C_{33}H_{36}O_6N_4$ (582,3)	Ber.	C	68,01	H	6,23	N	9,27
	Gef.	„	67,66	„	6,17	„	9,72. 34

Fischer, however, would not be satisfied with the *exo*-vinyl or the furan bilirubin structure until he had synthesized the compounds. Nonetheless, despite the successful synthesis of bilirubin-XIIIα, and for reasons unclear, the final investigations took another two to three years. As if to predict slow progress would lie ahead, in August 1939, *Siedel* wrote that the synthesis of bilirubin itself had not yet been accomplished because it was made extremely difficult by the presence of vinyl groups or furan rings (*1*):

> Die *Synthese des Bilirubins* selbst ist noch nicht durchgeführt. Sie ist durch die Existenz der Vinylgruppen, bzw. des Dihydrofuranringes stark erschwert.   35

6.3.1 Der Nitritkörper

The *Nitritkörper* is something of a misnomer, for it is not actually a nitrite compound. Also called the "nitrite body" by *Hans Fischer* in 1934 (*449*), by *Lemberg* in 1949 (*433*), and by *Bonnett* and *McDonagh* in 1969 (*459*), it is a low molecular weight fragment produced by oxidation of bilirubin using nitrous acid in an early attempt at degradation by *Fischer* and *Röse* (*372*). Defying characterization in 1914 (*372*), the cleavage product proceeded to have a long and vexing association with *Fischer*'s attempts to prove the structure of bilirubin. In fact, it proved to be a major distraction, if not obstacle, dating back to the time it was first isolated and when bilirubin was being subjected to various methods of oxidative cleavage. The then-unnamed compound was first reported in mid-1914 by *Fischer* and *Röse* (*372*) and

6.3 *Fischer* Solves the Structure of Bilirubin

Fischer alone (*377*), who employed nitrous acid as an oxidant of bilirubin and isolated a colorless, low molecular weight compound following treatment of 10 g of the pigment in glacial acetic acid with $NaNO_2$. Work-up afforded 0.3 g of hair-like needles, mp 87–88°. Its odor reminded the investigators of methylethylmaleimide and strongly induced sneezing, but it was at first reported to be nitrogen-free. Combustion analysis did not fit any reasonable structure known in 1914. Yet, characteristic of *Fischer*, the unsolved structure of the *Nitritkörper* was never far from his thinking in his dogged pursuit of bilirubin's structure. The repeated failure of this low molecular fragment to be characterized was to become a source of frustration for over 25 years following its first isolation.

The nitrous acid oxidation of bilirubin was apparently of nagging interest to *Fischer* after 1914 and during the following decade led to a reinvestigation in order to improve the yield and prove the structure. By 1925, *Fischer* and *Niemann* (*399*) were able to report oxidation reaction conditions modified from those published (*372*) 11 years earlier. Accordingly, bilirubin (2 g) was first dissolved in 60 cm³ 0.1 N aqueous NaOH and introduced into 160 cm³ glacial acetic acid with vigorous shaking to yield a dispersion of fine particles; then, 0.32 g $NaNO_2$ were added over three hours, during which the mixture underwent a series of color changes. Work-up yielded a crystalline material (melting point between 82°C and 86°C) that rapidly transitioned in air into a higher molecular weight amorphous product during the following 48 hours. (*Fischer* also reported that the original material often lasted eight days in a dessicator.) Analyses were conducted (*399*) as summarized and discussed in Section 5.1. In 1925, the CHN analyses fit methyvinylmaleimide, as did the molecular weight determination; however, *Fischer* was stymied by not being able to reduce the substance to the known methylethylmaleimide that he had expected. As will become evident, this failed conversion delayed the structure proof, and although *Fischer* was biased toward methyvinylmaleimide as the structure of the *Nitritkörper*, when hydrogenation failed to give the expected reduction product, together with *Niemann* he fabricated an entirely different structure (shown below left) – one with the same molecular weight and molecular formula (*399*):

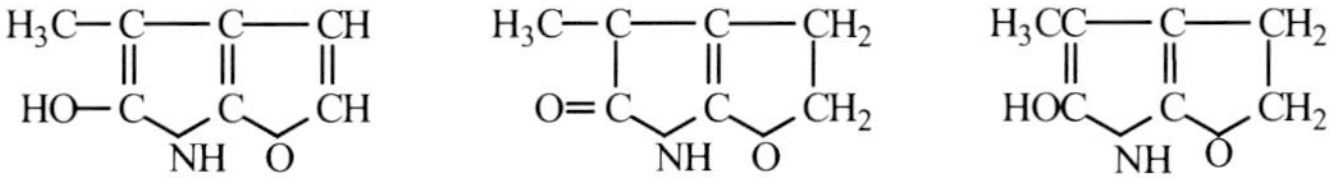

Its projected hydrogenation product, erroneously published first as the middle structure above, was soon thereafter corrected to the rightmost hydroxy-pyrrole structure by *Fischer* and *Postowsky* (*398*). The furan-fused pyrrole structure was then an unknown compound, as was the suggested dihydrofuran hydrogenation product compound. *Fischer* apparently never attempted to synthesize them.

Although the substance was unnamed in 1925 (*399*), it was apparently not forgotten because late in 1930, *Fischer* and *Hess* represented the still unnamed substance again as furano-pyrrole **III** (*438*):

> ... denn bei dem Abbau des Bilirubins mit salpetriger Säure entsteht ein Körper, der nach der Analyse ein Methyl-vinyl-maleinimid sein könnte. Diese Konstitution kann jedoch

nicht zutreffen, weil bei der katalytischen Hydrierung unter Aufnahme von 1 Mol Wasserstoff nicht Methyläthylmaleinimid, sondern ein neuer Körper entsteht, dem hiernach nur die Konstitution (IV) zukommen kann, während das Ausgangsmaterial die Formel (III) besitzen muß.

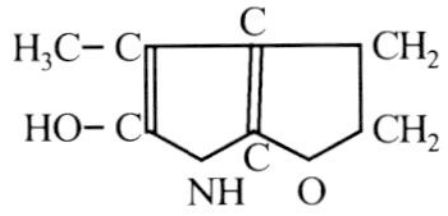

Durch Synthese ist der Beweis noch nicht erbracht. . . .   36

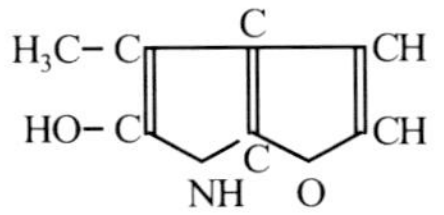

– while explaining that the synthesis of neither **III** nor **IV** had been achieved. Nevertheless, the substance may well have captured the simple, nondescript, originbetraying designation *Nitritkörper* within the *Fischer* group because only two years later, in late 1932, *Siedel* and *Fischer* referred to it in print while also continuing to portray it as a pyrrolo-furan (*443*):

Hierfür sprach auch ganz besonders die Isolierung des Nitritkörpers folgender Konstitution,

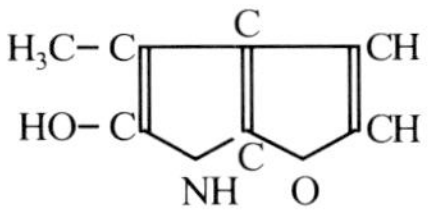

dessen Entstehen die bisherige Bilirubinformel nicht zu erklären imstande ist.[1]

[1] Vgl. die Ausführungen, Diese Z. **194**, 207 (1930). [sic 1931] [*438*]   37

The assigned name and its conjectured structure (would *Thudichum* have objected?) continued to appear in *Fischer*'s publications after 1933, *e.g.* in volume I of his famous book published with *Orth* (2):

„**Nitritkörper**" $C_7H_7O_2N$ entsteht wahrscheinlich bei der Einwirkung von salpetriger Säure auf Bilirubin. Kristallisiert aus heißem Wasser in ziemlich derben Nadeln vom F. 87–88° (unscharf). Erinnert im Geruch an Methyläthyl-maleinimid. Nicht unzersetzt destillierbar. Mit Ätherdämpfen flüchtig. Nimmt bei der Reduktion mit Aluminiumamalgam 2 Atome Wasserstoff auf[3][4]).

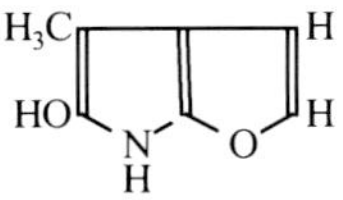

[3] H. Fischer u. H. Röse, H. **91**, 192 (1914) [*372*] vgl. auch Z. Biologie **65**, 178 (1914) [*377*]. –
[4] H. Fischer u. G. Niemann, H. **146**, 199 (1925) [*399*].   38

As seen in Section 6.2.2, it appears in the *Fischer* and *Orth* English language review of 1934 (*449*), where *Nitritkörper* is translated by the authors as "nitrite body"; and it appears again in the article of *Fischer* and *Haberland* of 1935 (*455*). In 1937, it appears yet again in volume II (first half) of "*Fischer and Orth*" (*3*):

6.3 *Fischer* Solves the Structure of Bilirubin   403

Es wurde oben erwähnt, daß das Bilirubin sowohl bei der Oxydation mit salpetriger Säure (in Eisessig) also auch bei der katalytischen Reduktion ein etwas auffälliges Verhalten zeigt. Bei der Nitrit-oxydation insofern, als hierbei in geringer Menge ein Körper auftritt, der seiner empirischen Zusammensetzung nach zunächst als Methylvinylmaleinimid angesehen wurde, in Wirklichkeit aber die nachstehend wiedergegebene Konstitution **III** besitzt, indem bei der nachträglichen katalytischen Hydrierung nicht das erwartet Methyl-äthylmaleinimid, vielmehr ein neuer Körper entstand, dem die Strukturformel **IV** zuerkannt werden mußte. Das eigentliche Nitrit-oxydationsprodukt behielt dann für die folgende Zeit die mehr triviale Bezeichnung "Nitritkorper"[1]).

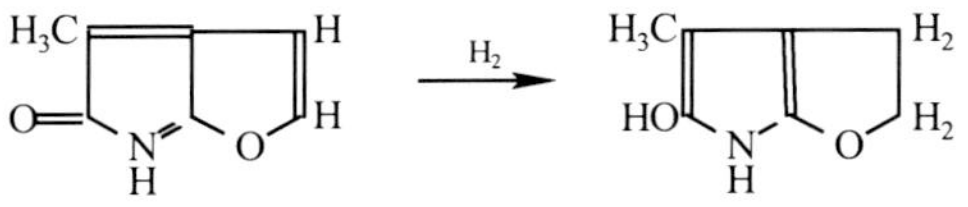

Andererseits nahm die katalytische Reduktion des Bilirubins selbst, welche zum Unterschied von der Natriumamalgam-Reduktion (vgl. dazu wieder das S. 623 aufgeführte Oxydations- und Reduktionsschema) verschiedene Zwischenstufen abzufangen gestattet, insofern einen etwas merkwürdigen Verlauf, als der Übergang vom Bilirubin zum Mesobilirubin, mit anderen Worten die Absättigung der möglicherweise vorhandenen zwei Vinylgruppen im Gallenfarbstoff nicht – wie erwartet – sprunghaft erfolgt, sondern ziemlich träge und auch nur recht ungleichmäßig. In der Tat gelang es in jüngerer Zeit hier gleichfalls noch eine Zwischenstufe, nämlich das S. 637 näher besprochene *Dihydrobilirubin* abzutrennen[2]). Damit war aber die schon lange fraglich gewordene Formulierung des Bilirubins mit zwei Vinylgruppen endgültig gefallen und es trat an deren Stelle die S. 626 wiedergegebene Formel (**V**) mit angegliedertem Furanring.

---

[1]) Vgl. H. Fischer, H. **91**, 192 (1913) [*sic*, 1914] [*372*]; Ztschr. f. Biol. **65** (N. F. **47**), 178 (1914) [*377*].

[2]) Vgl. H. Fischer u. H. W. Haberland, H. **232**, 236 ff. (1935) [*455*].   39

In 1937, *Fischer* (with *Orth*) expressed certainty that natural bilirubin did not have an *exo*-vinyl group but rather possessed a structure where the erstwhile *exo*-vinyl had actually cyclized to form a furan (*3*):

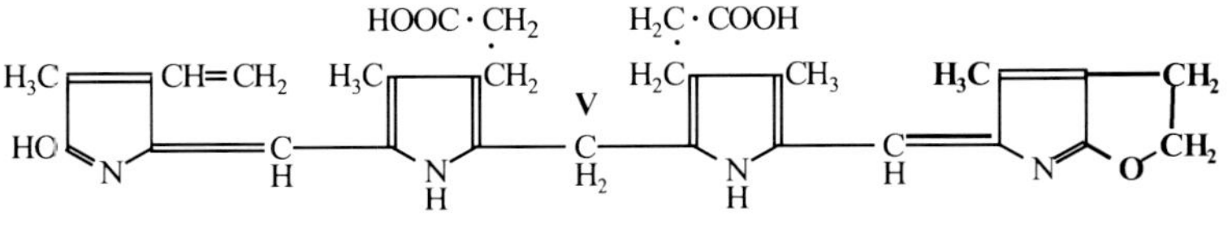

Structure **V** above and the *exo*-vinyl isomer have identical molecular formulas, but how nitrous acid might act upon **V** to produce the *Nitritkörper* with its more unsaturated furan ring would seem to require considerable hydrogen and carbon-carbon double bond migration. Yet 1937 preceeded the blossoming of mechanistic organic chemistry and its thought processes; so, perhaps it never struck *Fischer* as unusual. Perhaps too, it was assumed that any methyvinylmaleimide formed from the left half of **V** would not survive the oxidation conditions, which should also have rendered perplexing the survival of a pyrrolo-furan *Nitritkörper*.

6 The Status of Bilirubin in 1930

Structure **V** above at least seemed to offer an explanation for the observation, reported in 1935 (*455*), that catalytic hydrogenation of bilirubin (using colloidal Pd as catalyst) proceeded by a rapid uptake of one mole equivalent of H_2, followed by a slower uptake of the second mole equivalent (*3*):

> Tatsächlich ist es dann vor wenigen Jahren H. Fischer und H. W. Haberland[1]) geglückt, bei der katalytischen Reduktion des Bilirubins eine Vorstufe des Mesobilirubins, nämlich das *Dihydrobilirubin* zu isolieren, indem die Hydrierung erstmalig abgestoppt wurde, als 1 Mol Wasserstoff verbraucht war. Von den beiden zu erwartenden Beimengungen, unverändertem Bilirubin un dem „Überhydrierungs" produkt Mesobilirubin, ließ sich das Dihydrobilirubin durch fraktionierte Kristallisation aus Pyrdin befreien, um hieraus in Form ziegelroter Kristalle vom Schmelzpunkt 351° zu erscheinen. (Auch das Mesobilirubin schmilzt bei etwa dieser Temperatur, gibt aber dennoch mit dem Dihydrobilirubin im Mischschmelzpunkt eine merkliche Depression.)

---

[1]) Vgl. H. **232**, 236ff. (1935) [*455*].   40

This observation led *Fischer* and *Haberland* to the isolation of a dihydro-bilirubin, following the uptake of one mole equivalent of H_2 by bilirubin, that reacted with molten resorcinol to give only a 50% yield of but one product: (*exo*-ethyl) iso-neo-xanthobilirubinic acid. The simplest rationale would invoke a more rapid catalytic hydrogenation of the *exo*- over the *endo*-vinyl group – if *Fischer* had settled on a bilirubin structure with two vinyl groups, which he had not. Possibly believing that there was no reason for one vinyl group to react faster than the other during catalytic hydrogenation, *Fischer* instead saw the differential rates of hydrogenation as *supporting* a bilirubin structure with an *endo*-vinyl and an erstwhile *exo*-vinyl incorporated into a furan ring (*3*):

> Von den in konstitutioneller Hinsicht aufschlußreichen Um- oder besser Abwandlungen, welchen das Dihydrobilirubin unterworfen wurde, verdienen eigentlich nur zwei besonderer Hervorhebung: Einerseits der oxydative Abbau der Dihydro-verbindung, der im Gegensatz zum Mesobilirubin, das hierbei in einer Ausbeute von etwa 80 Proz. der Theorie *Methyl-äthylmaleinimid* neben Hämatinsäure liefert, kaum die Hälfte dieser Menge an basischem Imid entstehen läßt, und andererseits die Resorcinschmelze, die – wiederum im Gegensatz zum Mesobilirubin – nicht zu einem Gemisch von Neo- und Isoneoxanthobilirubsäure führt (vgl. dazu auch S. 643), sondern lediglich Isoneoxanthobilirubinsäure ergibt, und zwar in etwa 50proz. Ausbeute, bezogen auf die Menge des entsprechenden Isomeren-gemisches bei Mesobilirubin.
> Jedenfals ließ zunächst der Oxydationsbefund einwandfrei erkennen, daß, falls tatsächlich zwei Vinylgruppen im Bilirubinmolekül vorhanden waren, im Dihydrobilirubin eine dieser Gruppen zur Äthylgruppe umgeformt sein mußte, denn das Bilirubin selbst liefert ja bekanntlich bei der Oxydation nur Hämatinsäure. Daß es sich hierbei, entsprechend nachstehender Strukturformel,

6.3 *Fischer* Solves the Structure of Bilirubin 405

um die ungesättigte Gruppe der 8-Stellung des Tetrapyrrensystems handelte (man vgl. dazu das Nomenklaturschema S. 627), darüber unterrichtet dann weiter die Resorcinschmelze, denn das alleinige Auftreten von Iso-neoxanthobilirubinsäure war eben nur mit dieser Auffassung vereinbar. 41

The furan ring would be the source of the "Nitritkörper" mentioned in the following and undergo the hydrogenolysis illustrated below (*3*):

> Zwangsläufig mußte im Bilirubin aber eine der beiden ungesättigten Gruppen in β-Stellung, und nach obigem die der 8-Stellung, irgendwie maskiert und so der Hydrierung im ersten Stadium nicht zugänglich sein. Die Art der Maskierung verriet dann der schon erwähnte „Nitritkörper". Mit anderen Worten, es war dem Bilirubin nicht die einfache, und wie sich später aus den beim Mesobilirubin erhobenen Befunden herausstellte, unsymmetrische Struktur mit zwei Vinylgruppen zuzuerteilen, sondern dem natürlichen Farbstoff kam die S. 626 aufgeführte Formel mit einem angegliederten Dihydrofuranring am Pyrrolkern IV zu, wie sie tatsächlich in früherer Zeit schon von H. Fischer auch interpretiert wurde[1]).
>
> Der Übergang vom Bilirubin zum Dihydrobilirubin beruht mithin nicht auf einer Addition von Wasserstoff an eine in der 8-Stellung des Tetrapyrrensystems befindliche Vinylgruppe, vielmehr besteht er in der reduktiven Öffnung des Dihydrofuranringes, wie dies das nachfolgende schematisierte Formelbild wiedergibt:

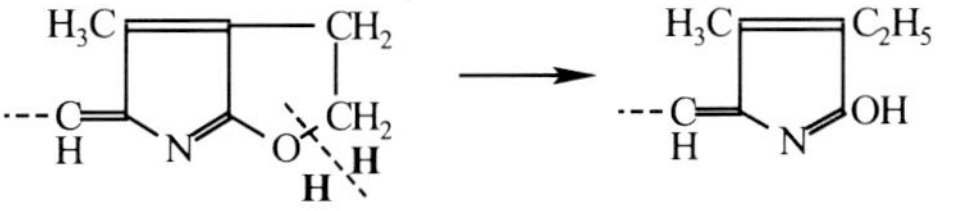

---

[1]) Man vgl. u. a. H. Fischer u. G. Niemann, H. **127**, 310 (1923) [*395*] sowie H. Fischer u. F. Lindner, H. **161**, 9 (1926) [*408*]. 42

If *Fischer* had found (*endo*-ethyl) neo-xanthobilirubinic acid instead of (*exo*-ethyl) iso-neo-xanthobilirubinic acid following resorcinol melt treatment of his dihydro-bilirubin, he might have had a less tenuous basis for concluding that bilirubin has a furan ring, rather than an *exo*-vinyl group. It would be more consistent with the expectation that a vinyl group should hydrogenate more rapidly than a dihydrofuran might undergo ring-opening hydrogenolysis. However, that clearly did not happen. In fact, *Tim Anstine* (Ph. D., 1995) in the *Lightner* group showed in a simple (unpublished) experiment that an *exo*-vinyl group undergoes catalytic hydrogenation more rapidly than an *endo*-vinyl by comparing the catalytic hydrogenation rates of bilirubin-XIIIα (two *endo*-vinyl groups) in methanol (with a trace of NH_3 gas to dissolve the pigment) with 10% Pd(C) to those of mesobilirubin-IIIα (two *exo*-vinyl groups). After four hours, bilirubin-IIIα produced 86% mesobilirubin-IIIα and 14% dihydro-bilirubin-IIIα; whereas, after 24 hours, bilirubin-XIIIα yielded 66% mesobilirubin-XIIIα and 33% dihydro-mesobilirubin-XIIIα. Apparently reduction of an *exo*-vinyl group is faster than that of an *endo*-vinyl.

Fischer had clearly produced (*exo*-ethyl) dihydro-bilirubin-IXα during controlled catalytic hydrogenation of bilirubin. Whether he had given thought to why a dihydrofuran might undergo hydrogenolysis faster than a vinyl group would undergo hydrogenation is unclear. Nowadays, such a distinction might raise more eyebrows than in 1937; yet, it had to be invoked to support the only bilirubin structure as a source of the structure he accepted for *Nitritkörper*. Surprisingly, *Fischer* mentions no examples from the chemical literature for such a hydrogenolysis (because there were none known?) or why it should proceed more rapidly than reduction at the numerous double bond sites. The logic applied in 1937 illustrates how easy it is to fall into a trap in science when attempting to explain evidence, both supporting and contradictory, with the original premise unproven and only conjectured (*Die Conjecturalchemie Thudichums ist niemals gestorben*). *i.e.* the then still unproven structure of the *Nitritkörper* going back to 1914. As will become evident in the following.

Nonetheless, *Fischer*, while proclaiming and favoring the furan ring structure in bilirubin, never fully abandoned a view toward the presence of an *exo*-vinyl group. Perhaps his lack of certainty is attributable to scientific acumen, since the evidence was only negative, *i.e.* he could not hydrogenate the *Nitritkörper* to methylethylmaleimide, nor could he obtain a crystalline hydrofuran. Possibly, too, he was sensitive to *Lemberg*'s admonition. For in 1935, from the Biochemical Laboratory at Cambridge University, *Lemberg* published an article on the "Transformation of Haemins into Bile Pigments" (*460*) in which he pooh-poohed *Fischer*'s logic and suggestion of a furan structure. Perhaps he was annoyed by *Fischer*'s having failed to recognize the published, correct structure differences between biliverdins and biliviolins, for in 1935 *Lemberg* wrote (*460*):

> [1] H. Fischer who had previously confused the blue colour of the neutral biliverdins with the similar colour (although very different spectrum) of the acid biliviolins and assumed that the latter were isomerides of the former, has now [Fischer and Haberland, 1935] recognised his mistake and has arrived at the formula for biliviolins which I proposed in my paper before the Biochemical Society on November 16th, 1934, and which has been published in a short note [1934, 2]. Fischer has overlooked this note.

And specifically on the question of an *exo*-vinyl *vs* a furan structure in bilirubin he wrote (*460*):

> The transformation of haemin into biliverdin is also of interest for the question of the nature of the side-chains of bilirubin. It suggests that bilirubin and biliverdin contain two vinyl groups like haemin. Fischer has often discussed this question and has repeatedly altered his views. Quite recently Fischer and Haberland [1935] [*455*] have come to the conclusion that bilirubin contains one vinyl group, whereas the second unsaturated side-chain is present in the form of a dihydrofuran nucleus (cf. formula. . . [below]). Although it is not quite impossible that such a ring is formed during the reaction leading from haemin to biliverdin by closing a ring between the vinyl group in position b and the hydroxyl group in position a at ring IV, it does not seem probable.

> Moreover, the evidence brought forward by Fischer for the existence of this ring is unconvincing. It is based (1) on the fact that the behaviour of the unsaturated groups in

6.3 *Fischer* Solves the Structure of Bilirubin

bilirubin towards reagents is different from that of these groups in haemin; (2) on the isolation of a substance assumed to be a pyrrofuran compound, obtained in small yield by the action of nitrous acid on bilirubin; and (3) on the isolation of a dihydrobilirubin with one vinyl group and one ethyl group by catalytic hydrogenation of bilirubin.

As regards the first point, these differences concern both unsaturated groups in the same way; since only one can be present in form of a dihydrofuran ring (on account of the position of the unsaturated side-chain in ring I relative to the hydroxyl group), the observed differences of the bilirubin side-chains from those of haemin must be explained in a different way.

The existence of the pyrrofuran compound is very doubtful. In later experiments Fischer [1915] [*377*] tried in vain to reproduce this substance, nor was Dr. Mühlbauer (unpublished observation) able in my laboratory to obtain a substance of the properties described in the first paper by Fischer and Hahn [1914] [*379*]. In a short note Fischer and Hess [1931] [*438*] mentioned that the substance in question reacted with phenylhydrazine and possibly was methylhydroxyvinylmaleinimide (although this substance would yield analytical values quite different from those obtained by Fischer and Hahn). In spite of this uncertainty the substance is now again used as an argument in favour of the existence of a dihydrofuran ring. Previously Fischer had assumed a furan ring, although this was in evident disagreement with the amount of hydrogen necessary for the transformation of bilirubin into mesobilirubin and mesobilinogen.

The third argument of Fischer is the fact that one of the unsaturated groups of bilirubin is more easily reduced to an ethyl group by catalytic hydrogenation. The partially hydrogenated substance has an ethyl group in ring IV and a vinyl group in ring I. Provided that Fischer's hypothesis is correct, it would be necessary to assume that by catalytic hydrogenation the dihydrofuran ring is split open to ethyl and hydroxyl more readily than the vinyl group is saturated. This is very improbable, and Fischer's results are more easily explained by the simple assumption that the vinyl group in the neighbourhood of the α-hydroxyl reacts more rapidly. For these reasons it is suggested that bilirubin contains two vinyl groups like haemin.

– and in his famous text with *Legge*, this was summarized again in 1949 (*433*):

Instead of two vinyl side chains Fischer at first assumed one condensed furane ring in addition to one vinyl group, a structure not in agreement with the fact that only two molecules of hydrogen were taken up in the catalytic hydrogenation of bilirubin to mesobilirubin; later he assumed a dihydrofurane ring condensed to a pyrrole ring:

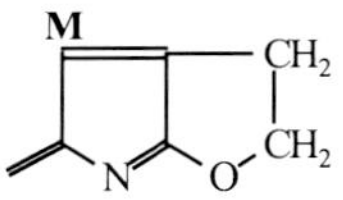

which his book (*861*, p. 625, 637) [*3*] he claimed to have finally established. The evidence was based on the so-called "nitrite body" an ill-defined product of the oxidation of bilirubin by nitrous acid, and on the observation that one of the two molecules of hydrogen added in the catalytic hydrogenation of bilirubin to mesobilirubin was taken up faster than the second. The evidence was discussed by Lemberg (*1681*) [*460*] and rejected. Fischer has now confirmed the formula with two vinyl side chains by synthesis and by the fact that two moles of diazoacetic ester can be added to the vinyl side chains of bilirubin as to those of protoporphyrin (*864*) [*11*]. The "nitrite body" was shown to be methylvinylmaleimide, reducible to methylethylmaleimide, while previously it had been claimed that the reduction yielded a cyclic isomeride of this compound.

Published some four years after *Fischer*'s death, there could be no polemics from him on the *Nitritkörper*; yet, bilirubin continued to be a convenient source of methylvinylmaleimide, after further improvements to the oxidation.

6.3.2 The Exo-vinyl Group of Bilirubin

If progress in research in German universities began to slow due to shortages of chemicals, students, and other research personnel when the seeds of World War II were planted in March 1938 (annexation of Austria) and March 1939 (annexation of the Sudetenland), it must have become even more constrained when World War II broke out in September 1939 with the invasion of Poland, and especially in June 1941 when more than 4.5 million troops invaded the Soviet Union. The war drained German universities of their (largely male) students, and *Hans Fischer* had to devote increasingly limited resources to his primary object: the synthesis of chlorophyll. Nonetheless on February 28, 1941 he was able to report on the constitution of bilirubin, to resolve the long-vexing problem of the structure of the *Nitritkörper* that had led him to propose the presence of a furan ring decades earlier. He was thus able to answer the question as to whether natural bilirubin has an *exo*-vinyl group or whether it has been incorporated into a furan ring (*11*) .

Prior to 1941, *Fischer* believed that bilirubin was represented by either **I** or **II**, but with more support favoring **II**, as summarized in 1941 (*11*):

> Die Konstitution des Bilirubins ist durch unsere Untersuchungen vollkommen aufgeklärt mit Ausnahme der Formulierung der ungesättigten Seitenketten im Kern I und IV. Zur Diskussion standen 2 Vinylgruppen oder eine Vinylgruppe und ein Hydrofuranring, entsprechend den beiden Formulierungen:

> Formel **II** schien vor allem gestützt durch die Isolierung des sog. Nitritkörpers folgender Formulierung:

6.3 *Fischer* Solves the Structure of Bilirubin

Dieser bildet sich bei dem Nitritabbau des Bilirubins leider nur in sehr geringer Ausbeute, so daß sein Entstehen leicht übersehen werden kann [2]).

---

[2]) R. Lemberg, Biochemic. J. 29, 6, 1334 (1935) [*460*]. 43

Although *Fischer* favored **II**, he hedged his bets and had not completely excluded **I** as a possibility; thus, understandably he would prefer to remove all doubt before commencing the total synthesis of bilirubin.

Recall that vinyl-neo-xanthobilirubinic acid, which has an *endo*-vinyl group, was isolated following treatment with bilirubin in boiling molten resorcinol (*458*). Its *exo*-vinyl isomer, vinyl-iso-neo-xanthobilirubinic acid was not found. In a repetition of the resorcinol melt with a larger quantity of bilirubin, the earlier result remained unchanged, *viz.* only *endo*-vinyl-neo-xanthobilirubinic acid was found and no *exo*-vinyl-iso-xanthobilirubinic acid. Repeating the experiment with both bilirubin dimethyl ester and biliverdin dimethyl ester gave the same result: only the methyl ester of vinyl-neo-xanthobilirubinic acid was found. These results, coupled with the fact that natural mesobilirubin in molten resorcinol gave a separable mixture of neo- and iso-neo-xanthobilirubinic acid, both with ethyl groups, and bilirubin-XIIIα (with two *endo*-vinyls) gave a 71% yield of vinyl-neo-xanthobilirubinic acid from molten resorcinol, left *Fischer* attributing the poorer product yield from one-half of natural bilirubin and absence of a dipyrrinone from the other half to either an *exo*-vinyl adjacent to a hydroxyl or a hydrofuran ring (*11*):

> Da die Ausbeuten hier an die beim Mesobilirubindimethylester heranreichen, kann die geringe Ausbeute beim natürlichen Bilirubin nicht auf die Anwesenheit der Vinylgruppen zurück-zuführen sein, vielmehr muß die Ursache in der Konstitution des Bilirubins liegen, wobei ebensogut diese in dem Vorhandensein eines Hydrofuranringes, wie in der Nachbarstellung der Oxy- und der Vinylgruppe im Kern IV erblickt werden kann. Bekannt ist das Cumaronharz, das aus den Teerfraktionen, die reich an Cumaron und Homologen sind, gewonnen wird; ebenso gehen Körper mit benachbarter Oxy- und Vinylgruppe bei der Einwirkung von Resorcin leicht in Hochpolymere über. In der Tat treten bei der Resorcinschmelze des Bilirubins in großer Menge Verharzungsprodukte auf, aus denen trotz vieler Bemühungen kein krystallisiertes Material erhalten werden konnte, auch nicht der „Nitritkörper", der auch aus sehr unreinem Bilirubin entsteht. 44

Fischer then settled the question of a furan ring by repeating his earlier (*399*) oxidative cleavage on 4 g of bilirubin using nitrous acid (*11*) and sublimed the product under high vacuum to give 50 mg of the *Nitritkörper*, mp 86°C. Unlike the earlier failed attempts to reduce the *Nitritkörper* (*372, 377, 399*), catalytic hydrogenation of freshly sublimed material in CH_3OH using a Pd catalyst produced a crystalline substance of mp 65°C (after two sublimations) that was unchanged after 10 days and gave a mixture melting point with authentic methylethylmaleimide (mp 67°C) of 65°C (no depression). An examination of crystals of the analytic and synthetic methylethylmaleimide showed no difference. From these results, in 1941, *Fischer* concluded that the *Nitritkörper* is actually methylvinylmaleimide (**XII**) and not furan **XIII** (*11*):

Nunmehr wurden auch die Arbeiten über den „Nitritkörper"[1,2,3]) wieder aufgenommen. Er konnte ohne weiteres aus Bilirubin leicht, wenn auch in geringer Ausbeute erhalten werden. Durch Sublimation im Hochvakuum erhielten wir prachtvolle, stark lichtbrechende Krystalle. Obwohl die damalige Analyse auf Methylvinylmaleinimid **XII** stimmte, mußte, da es nicht gelang durch Reduktion Methyläthylmaleinimid zu erhalten, eine isomere Formulierung angenommen werden, **XIII**.

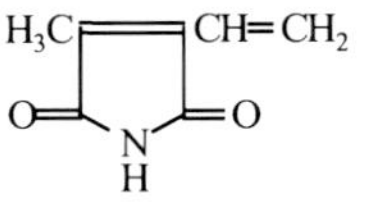

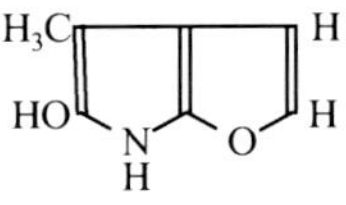

Durch katalytische Reduktion mit Pd-Wasserstoff in Methanol gelang es nun in glatter Reaktion den „Nitritkörper" in ein Produkt überzuführen, das nach 2 maliger Sublimation einen Schmelzpunkt von 65° besaß. Der Mischschmelzpunkt mit Methyläthylmaleinimid (Schmelzp. 67°) ergab keine Depression. Auch in der Krystallform und Auslöschungsschiefe waren die Krystalle identisch mit Methyläthylmaleinimid.

In Gegensatz zum „Nitritkörper" zeigen die Krystalle beim längeren Liegen an der Luft keinerlei Veränderung der Löslichkeitseigenschaften und des Schmelzpunkts. Es unterliegt keinem Zweifel, daß es sich bei der hydrierten Substanz um Methyläthylmaleinimid handelt, während das Ausgangsmaterial selbst alle bereits früher beschriebenen Eigenschaften zeigt. 45

---

[1]) Fischer, H.; Röse, H. *Diese Z.* **1914**, *91*, 191 [*372*]; [2]) Fischer, H.; Niemann, G. *Diese Z.* **1925**, *146*, 197 [*399*]; [3]) Fischer, H., *Z. Biol.* **1914**, *65* (n. F. *47*), 163 [*377*]

which led him then to stress that a bilirubin structure with a hydrofuran ring was unsupported, that only structure **I** (page 408) with one *endo*-vinyl and one *exo*-vinyl should be considered (*11*):

Auf Grund dieser Ergebnisse kann die Bilirubinformel mit angegliedertem Hydrofuranring nicht aufrecht erhalten werden, und für Bilirubin kommt nur noch Formel I in Frage. 46

However, the data did not prove that bilirubin has two vinyl groups. The presence of one *endo*-vinyl group would be sufficient to give the same results. And so *Fischer* proved that bilirubin dimethyl ester has two vinyl groups by reacting it with ethyl diazoacetate to afford, after chromatography on Al_2O_3, a bis-adduct (*11*):

Diazoessigesteranlagerung an Bilirubindimethylester. 2 g amorpher Bilirubinester werden in einem Überschuß Diazoessigmethylester in der Hitze gelöst und auf dem Wasserbad bei 80° etwa 12 Stunden erhitzt. Dann wird in einer mit Äther aufgeschlämmten Aluminiumoxydsäule vom Diazoessigester abgetrennt. . . . Schließlich wird noch eine Chromatographie angeschlossen und der Farbstoff mit Chloroform an Aluminiumoxyd entwickelt. . . . Ausbeute 60 mg.

3,543 mg Subst.: 8,436 mg CO_2, 1,887 mg H_2O. – 4,808 mg Subst.: 0,343 ccm N_2 (25°, 708 mm). – 4,187 mg Subst.: 1,086 ccm n/50 KSCN.

$C_{41}H_{46}O_{10}N_4$ (754,8)	Ber.	C	65,23	H	6,14	N	7,42	OCH_3	16,44	
	Gef.	„	64,94	„	5,96	„	7,61	„	16,27	47

6.3 *Fischer* Solves the Structure of Bilirubin

(The mention of chromatography above leads one to recognize that *Fischer*'s lab practiced column chromatography by early 1941. Though chromatography had been known since the earlier decades of the 20th century, chromatography on a column of Al_2O_3 came into its own in the 1930s through the work of *Richard Kuhn* and colleagues on carotenes and xanthophylls at the Kaiser Wilhelm Institute in Berlin.)

Fischer proposed the following structure for the bis-adduct (*11*):

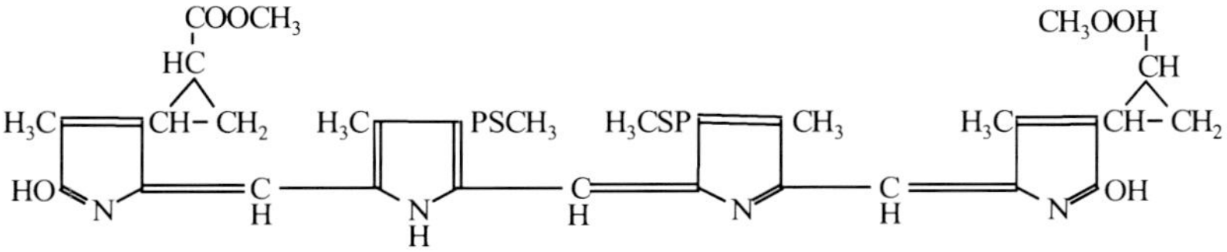

XIV

and he confirmed the generality of the reaction for dipyrrinone β-vinyl substituents by converting vinyl-neo-xanthobilirubinic acid methyl ester and bilirubin-XIIIα dimethyl ester to their mono- and di-adducts, respectively. Thus, unless a hypothetical furan ring of bilirubin had sprung open to reveal a vinyl group under the conditions of the cyclopropanation, there could be little question that the pigment has an *exo*-vinyl group as well as an *endo*-vinyl.

Accordingly, *Fischer* now had bilirubin **I** as the target structure for synthesis. In order to explore whether vinyl-neo-xanthobilirubinic acid (*endo*-vinyl) might undergo α-formylation, as did neo-xanthobilirubinic acid, its methyl ester was reacted with HCN + HCl in $CHCl_3$ to produce a 75% yield of crystalline formyl-vinyl-neo-xanthobilirubinic acid methyl ester, mp 250°C. It could be condensed with vinyl-neo-xanthobilirubinic acid methyl ester (*endo*-vinyl) in CH_3OH-48% HBr to give biliverdin-XIIIα dimethyl ester (**XVIII**), the same pigment as that obtained by condensing 2 mole equivalents of vinyl-neo-xanthobilirubinic acid with HCO_2H-HBr (*11*):

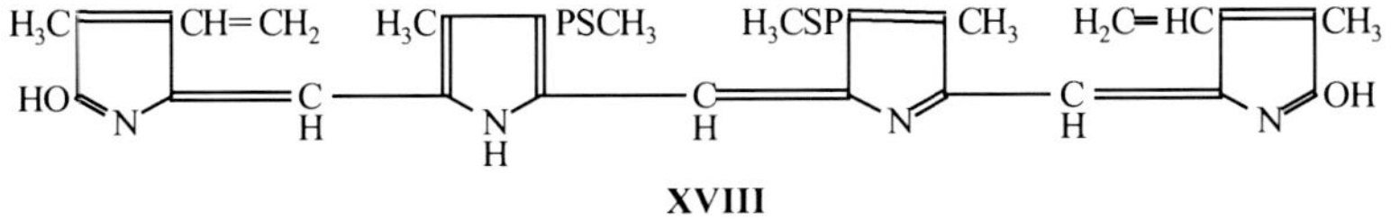

XVIII

Saponification of **XVIII** in methanolic KOH gave the corresponding diacid, which when heated with Zn dust in glacial acetic acid gave bilirubin-XIIIα. Thus, the path to a synthesis of natural bilirubin appeared to have been opened, assuming a source of vinyl-iso-neo-xanthobilirubinic acid (*exo*-vinyl) could be found.

In 1969, *Bonnett* and *McDonagh* reported the first direct structure proof of the *Nitritkörper* as methylvinylmaleimide. They repeated *Fischer*'s nitrous acid oxidation of bilirubin, finding, as did *Fischer*, a low yield (4%) of methylvinylmaleimide and showing that the product was reduced in 54% yield to the more stable

412   6   The Status of Bilirubin in 1930

methylethylmaleimide (459).[2] Also in 1969, the same authors reported a superior oxidation (24% isolated yield after sublimation mp 84–85°C) using CrO_3 as oxidant (459), an oxidizing agent used on bilirubin by *Küster* in 1909 to produce hematinic acid (325) and adapted by *Rüdiger* for the microscale (5). Methylvinylmaleimide, also prepared in the *Lightner* lab at UCLA by *Gary Quistad* (Ph.D., 1972) in 1970–1971 (461), was fully characterized by *Bonnett* and *McDonagh* by IR, UV, and ^{1}H NMR spectroscopy, and by mass spectrometry (459). Absorbed on paper or a silica gel layer, it exhibited a "bright moonlight-white" fluorescence, and "on storage it slowly polymerized to a white horn-like substance" (459).

Polemics aside, by 1941 *Fischer, Plieninger*, and *Weissbarth* (11) were able to isolate methylvinylmaleimide following oxidation of bilirubin and reduce it to methylethylmaleimide, as indicated in the following section.

6.3.3   *Bilirubin Structure Proof and Total Synthesis QED*

With I below having been viewed as the most likely structure of natural bilirubin (273):

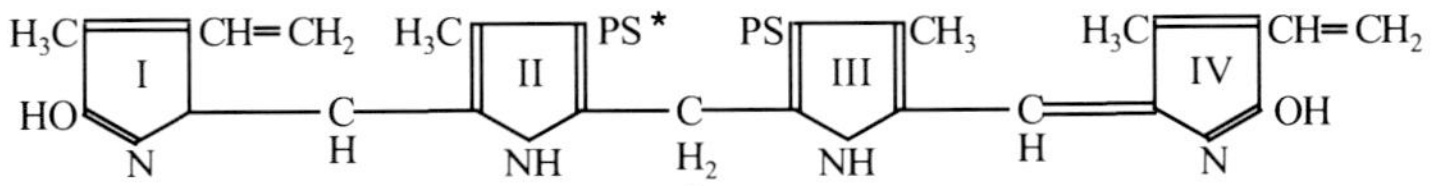

it remained only for *Hans Fischer* and one of his last doctoral students, *Hans Plieninger*,[3] to synthesize it, a feat accomplished by March 5, 1942 (10). That full paper (10) was summarized in the June 19, 1942 issue of *Die Naturwissenschaften*

---

[2] A correction to the *Bonnett-McDonagh* publication (459) by *McDonagh*: "The TLC eluent ratio should read: . . . acetone: 60–80 petroleum ether **1:4** . . . "

[3] *Hans Plieninger* was born on January 17, 1914 in Zürich and died on December 23, 1984 in London. He started his *Diplom* studies at the TH-München in winter semester 1932–1933, which were interrupted by military training from November 1935-December 1937. On December 8, 1938 he was awarded a *Dipl. ing.* degree and was recalled into military service from August 1939 to May 1940. Coming back to the Institute as a research assistant in 1940, he completed his doctoral research, was awarded the *Dr. ing.* degree on July 28, 1941 and worked as a chemical researcher at I.G. Farbenindustrie Werke in Ludwigshafen until January 1, 1945, when he returned to the TH-München as a research assistant until April 30, 1945. From May 1, 1945 to January 1946 he was engaged in external activities (outside of chemistry) and from May 1, 1940 to August 31, 1953 was a research chemist at Knoll AG in Ludwigshafen. He applied for the *Habilitation* in chemistry at the TH-Darmstadt and was granted the *venia docendi* to teach there. In September 1953 he assumed a *Wissenschaftlicher Assistent* (assistant professorship) at the Universität Heidelberg but had to reapply for the *Habilitation (Umhabilitation)* at the Uni-Heidelberg, which was awarded on July 30, 1954, for his *venia docendi* to teach there. He advanced to a. o. Professor on March 26, 1964, a personal *Ordinarius* on May 16, 1967, and o. Professor on July 13, 1970 to March 31, 1979, during which he was excused from administrative duties due to illness.

6.3 Fischer Solves the Structure of Bilirubin

(462), then recounted by *Walter Siedel* in a July 1943 issue of *Angewandte Chemie* (7) and also presented by him in October 1943 at a special session of the German Chemical Society in Vienna (8). Shortly after the end of WWII, German scientists were asked by the U.S. Army to prepare summaries of research conducted in their laboratories between 1939 and 1946. In the *Field Information Agencies Technical* (*FIAT*) report, submitted on 8 January 1947, *Siedel* summarized pyrrole research carried out in the *Fischer* lab, including the total synthesis of bilirubin (9). There were no chemistry *Nobel* Prizes awarded during the World War II years 1940, 1941, and 1942. Had *Fischer* lived past the end of the war, whether his structure proof and synthesis of bilirubin might have merited a second *Nobel* Prize can only be conjectured.

Two strategies for the total synthesis of bilirubin were devised, where special considerations had to be directed toward the eventual introduction of the two vinyl groups. One strategy followed the model used successfully for converting deutero-porphyrin to protoporphyrin, wherein the unsubstituted β-positions of the former were acetylated and the acetyl groups converted to vinyls *via* reduction and alcohol dehydration (425). This was explored by *Plieninger* using a dipyrrinone model having only a β-methyl group on the lactim ring (**IX**) and the biliverdin analog (**X**) prepared from it (463):

Tatsächlich kondensiert das Oxypyrrol (**VII** oder **VIII**) mit dem Opsosäurealdehyd in guter Ausbeute zu einem Oxymethen der Formulierung (**IX**).

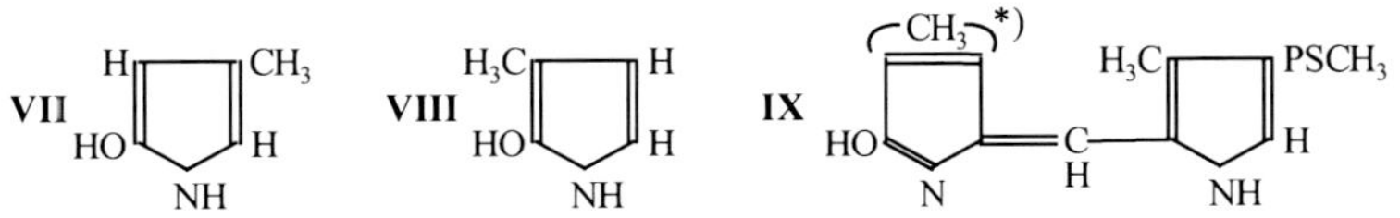

Die Kondensation des Methens zum entsprechenden Bilirubinoid und dessen Dehydrierung mit Eisen-III-chlorid zu dem Glaukobilin der folgenden Formulierung (X) gelant ohne weiteres.

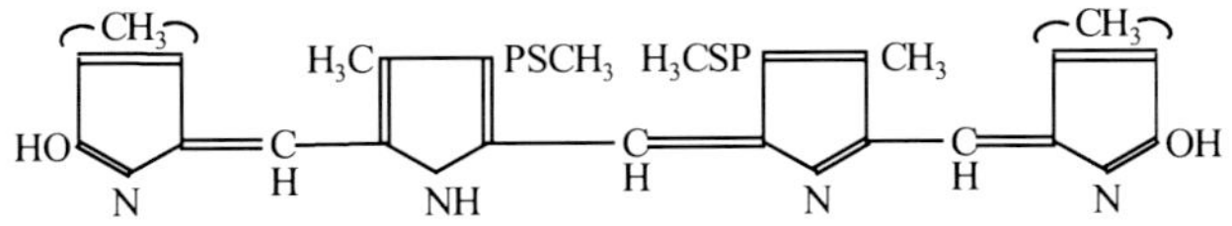

X

---

*) Diese Schreibweise soll angeben, daß die Stellung der Methylgruppe nicht festliegt. 48

The plan was to introduce an acetyl group at the free pyrrole β-position; however, the insertion failed (463):

Eine Einführung von Acetylresten in die freie β-Stellung konnte jedoch in keiner Weise durchgeführt werden. 49

Nonetheless, *Plieninger*'s work provided an important new advance in the synthesis of dipyrrinones: condensation of a pyrrole α-aldehyde with an α'-free α-hydroxypyrrole, such as the hydroxypyrrole mixture prepared by hydrogen peroxide (*464*) oxidation of 3-methyl-4-ethylpyrrole = *Opsopyrrol* = opsopyrrole (*2*):

> Der Name Opsopyrrol ist abgeleitet von ὀ'ψέ = spät und steht in genetischer Beziehung zu der analogen Bezeichnung für die entsprechende Propionsäure, da diese als letzte in den sauren Spaltprodukten des Hämins aufgefunden wurde.[5]

---

[5] Vgl. H. Fischer u. A. Treibs, A. **450**, 141–143 (1926) [*465*]   50

The α-hydroxypyrrole mixture was then condensed with the aldehyde of opsopyrrole carboxylic acid to give the dipyrrinone mixture **III** (*463*):

> Wie schon kürzlich vorläufig mitgeteilt,[2] konnte die Darstellung von Oxypyrromethenen durch Kondensation von α-freien α-Oxypyrrolen mit Pyrrol-α-aldehyden im alkalischen Medium ganz wesentlich vereinfacht werden. Durch Oxydation von 3-Methyl-4-äthylpyrrol (Opsopyrrol) mit Perhydrol konnten H. Fischer und H. Reinecke[3] ein Gemisch der beiden isomeren Oxypyrrole

> erhalten. Durch Kondensation mit dem Aldehyd der Opsosäure in *alkalischem* Medium wurde in hervorragender Ausbeute von 80% ein Gemisch der Neo- und Isoneoxanthosäure erhalten.

> Durch Umkrystallisation des Esters[4] konnte aus dem Gemisch der Ester der Isoneoxanthosäure vom Schmelzp. 201° in reiner Form isoliert werden. Das Oxymethen stimmte in allen Eigenschaften mit dem auf anderem Wege[4] gewonnenen überein.

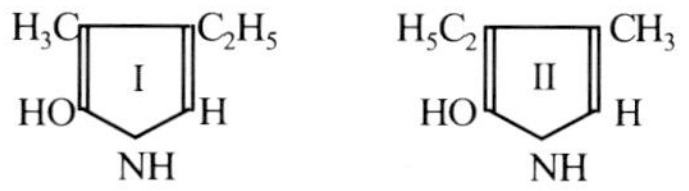

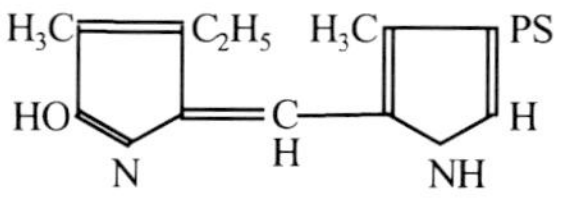

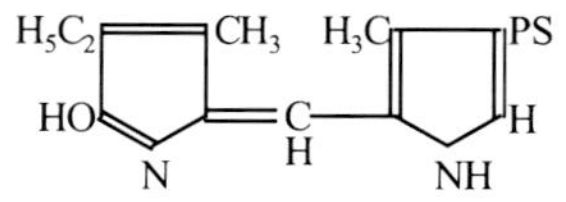

---

[2] Diese Z. **270**, 229 (1941) [*11*]. [3] Diese Z. **259**, 86 (1939) [*464*]. [4] W. Siedel u. H. Fischer, Diese Z. **214**, 164 (1933) [*443*]   51

Consequently, a new and different strategy was developed for the synthesis of bilirubin. This strategy, which ultimately proved to be successful, was based on the conversion of propionic acid groups to vinyl groups, with the early steps illustrated below in the "test-case" conversion of monopyrrole **III** to **VII**, along a path where conversion of **III** to ester **IV** was carried out conventionally (*466*), and the latter was converted to the hydrazide (**V**). Oxidation of the last would nowadays be expected to yield α-keto-azide, represented as a triazene (**VI**) in 1942. *Curtius'* rearrangement of **VI** led to **VII**. The latter, urethane, was projected to be hydrolyzed to the amine, which would undergo a *Hofmann* elimination after being quaternized (*10*):

6.3 *Fischer* Solves the Structure of Bilirubin

However, although the conversion to urethane **VII** succeeded, it could not be hydrolyzed to the amine with either base or acid. The pyrrole ring instead was destroyed (*10*):

> Die Spaltung des Urethan-oxypyrrols (**VII**) zum Amin gelang weder mit Alkalien noch mit Säuren, stets wurde der Pyrrolkern zerstört. 52

Undaunted, *Fischer* and *Plieninger* continued to pursue the strategy of coupling two suitable dipyrrinones. The precursor to **III**, opsopyrrole carboxylic acid, was thus converted to a separable mixture of **II** and **III** by oxidation with H_2O_2 (*10*):

> Durch Oxydation der Opsosäure mit Perhydrol entsteht ein Gemisch der beiden Isomeren Oxyopsosäure (**II**) und „Iso-oxyopsosäure" (**III**). . . .

Opsopyrrolcarbonsäure 53

As above, **III** was converted (below) to urethane **VII**, and **VII** was condensed with opsopyrrole carboxylic acid aldehyde **VIII** (from reaction of opsopyrrole carboxylic acid **III** with HCN-HCl followed by hydrochloric acid to give the imino-aldehyde, then treatment with aq. NaOH (*2*)) by heating briefly (0.5 h) on a hot water bath with aq. methanolic NaOH to afford dipyrrinone **IX** in good yield (*10*):

Dipyrrinone **IX** was converted to its methyl ester by diazomethane and analyzed correctly for $C_{20}H_{27}N_3O_5$ (*10*):

> Das Oxymethen wird wie üblich mit Diazomethan verestert und wegen der besseren Löslichkeitseigenschaften als Methylester zur Analyse gebracht. Schmelzp. 205°.
> 3,780 mg Subst. (bei 70° i. V. getr.): 8,582 mg CO_2, 2,215 mg H_2O. – 3,161 mg Subst.: 0,301 ccm N_2 (22°, 728 mm). – 4,033 mg Subst.: 1,06 ccm n/50-KSCN.

$C_{20}H_{27}N_3O_5$ (389,41)   Ber.   C   61,68   H   6,98   N   10,50 OCH_3 15,92
                               Gef.   „   61,92   „   6,54   „   10,55   „   16,31   54

Continuing "reaction-modeling", when the conditions of based-catalyzed condensation to form dipyrrinone-urethane **IX** were extended by heating at reflux with stronger NaOH for a longer time, the urethane was hydrolyzed without markedly affecting the dipyrrinone core, and the aminoethyl product (**XIV**) was obtained (*10*):

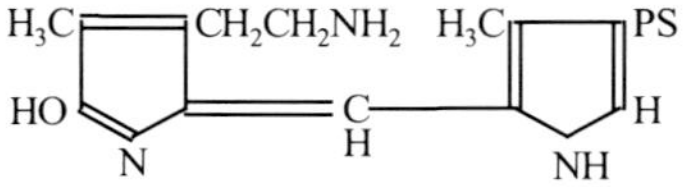

Yet, the real measure of success had to be directed to the tetrapyrrole level, and for that dipyrrinone-urethane **IX** was converted to rubin dimethyl ester **X** by heating in CH_3OH-CH_2O with added methanolic HCl (*10*):

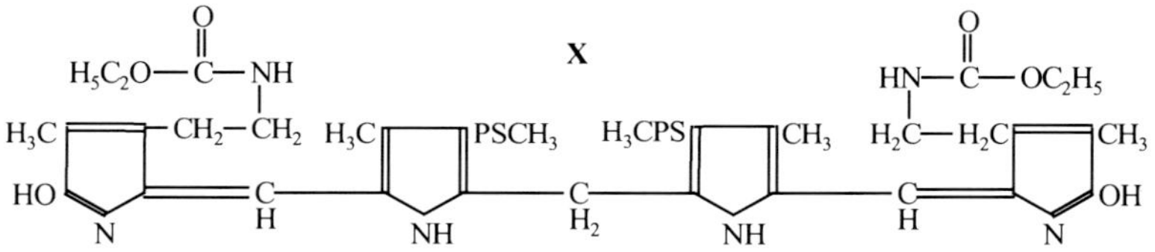

$PSCH_3 = CH_2CH_2COOCH_3$

Apparently yellow, crystalline **X**, mp 250°C, was extremely sensitive to acid and was deemed an inappropriate test vehicle for converting the two urethane groups to vinyl groups. So, it was dehydrogenated in glacial acetic acid to the corresponding verdin dimethyl ester (**XI**) using quinone as the oxidant (*10*):

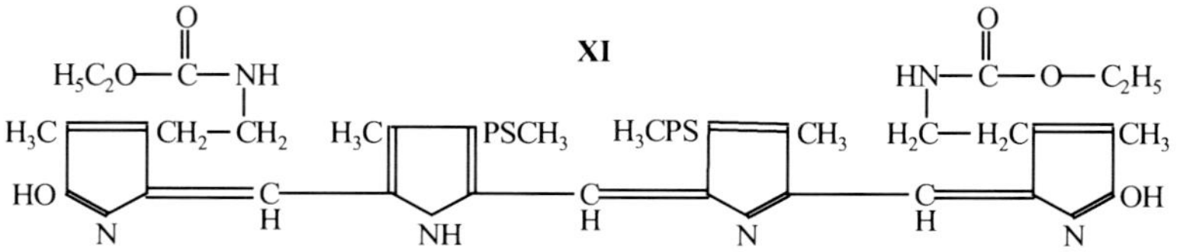

Alternatively diester **XI** was obtained more directly by heating **IX** in acetic anhydride + formic acid, or formic acid + HCl. When **XI** was treated with 18% aq.

6.3 *Fischer* Solves the Structure of Bilirubin 417

HCl in an autoclave for two hours at 135–140°C, its carbamate groups were hydrolyzed to afford the bis-ethylamino hydrochloride salt (**XII**) of the verdin (*10*):

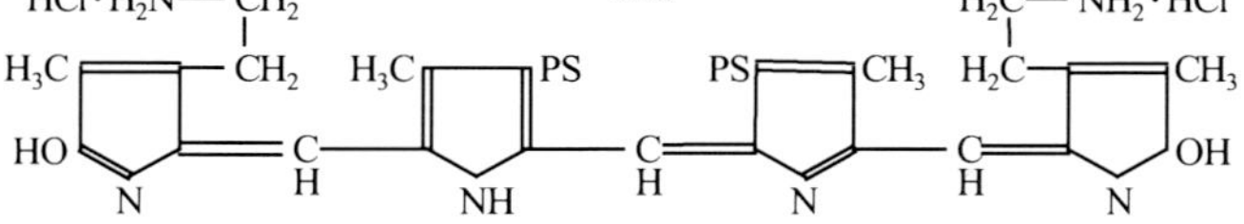

which could also be prepared by condensing dipyrrinone **XIV** with formic acid in acetic anhydride to give verdin amide **XV** (*10*):

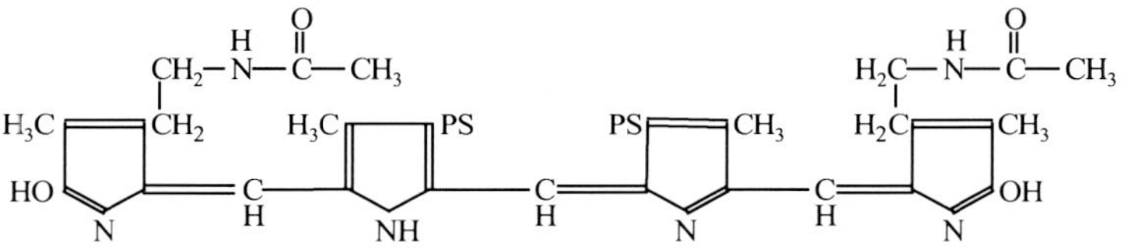

Acid hydrolysis of **XV** followed by permethylation of the primary amine groups of the bis-(aminoethyl)verdin using $(CH_3O)_2SO_2$ or CH_3I led to difficulties and the probable methylation of one or more pyrrole ring nitrogens. In order to prevent the last, verdin **XII** was reacted with dilute base to convert the hydrochloride salts to free amines. Zinc complex **XVI** was isolated following addition of zinc acetate (*10*):

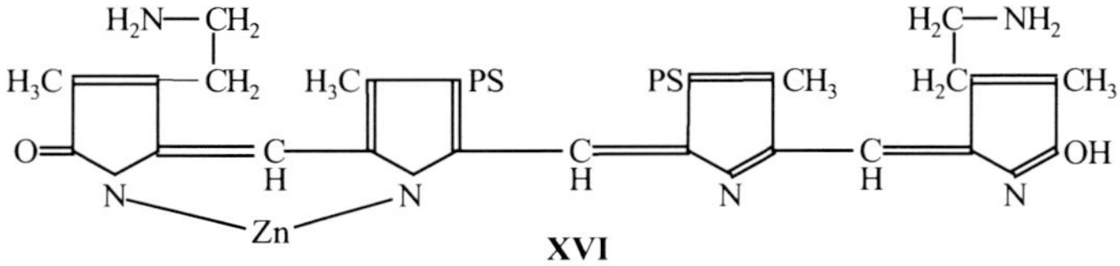

Although the zinc would more likely be placed differently today, **XVI** nonetheless underwent permethylation of the two primary amine groups using $(CH_3O)_2SO_2$ in aq. NaOH, or CH_3I in CH_3OH, and the resultant bis-quaternary ammonium salt then underwent smooth elimination in $CHCl_3 + Na_2CO_3$ to afford a 50% yield of synthetic biliverdin-XIIIα dimethyl ester (**XVII**), mp 240°C (*10*):

> Tatsächlich gelang es nun durch Methylierung dieses Zinksalzes in Natronlange mit Dimethylsulfat oder auch bei längerer Einwirkung von Jodmethyl in methylalkoholischer Kalilauge, zu Körpern zu gelangen, die die gewünschten Eigenschaften besaßen.
> Nach der Methylierung ist die Substanz immer noch spielend in Wasser und Säure löslich und konnte aus diesem Grunde noch nicht isoliert werden. Nach kurzem Erhitzen des Reaktionsgemisches mit starker methylalkoholischer und wäßriger Kalilauge jedoch ändern sich die Löslichkeitsverhältnisse grundlegend. Beim Ansäuern mit verdünnter

418   6   The Status of Bilirubin in 1930

Salzsäure fällt die gesamte Substanz als amorpher Niederschlag aus. Darauf wurde mit Methylalkohol-Chlorwasserstoff verestert und nach Chromatographie an Aluminiumoxyd aus Methylalkohol zur Krystallisation gebracht. Auf diese Weise erhält man aus 600 mg Aminhydrochlorid (XII) 350 mg eines Glaukobilins, das in allen Eigenschaften mit dem schon auf teilsynthetischem Wege gewonnenen Biliverdin XIII a[1]) identisch ist.

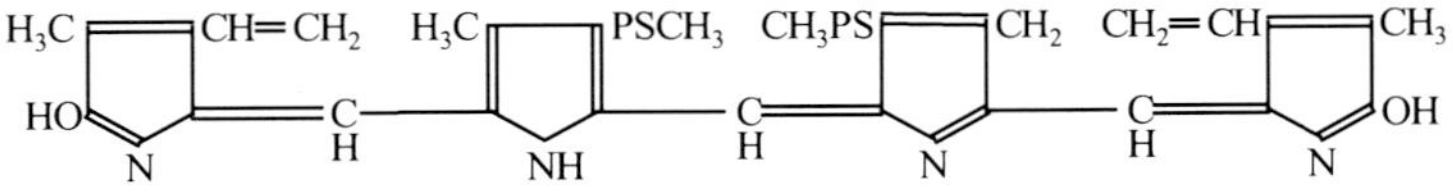

XVII

---

[1]) H. Fischer, H. Plieninger u. O. Weissbarth, Diese Z. **268**, 197 (1941) [*11*]   55

It was identical in every way to the analytical biliverdin-XIIIα dimethyl ester prepared earlier (*11*) by condensing vinyl-neo-xanthobilirubinic acid-methyl ester with formic acid in acetic anhydride, and upon treatment with molten resorcinol it gave vinyl-neo-xanthobilirubinic acid-methyl ester, mp 187°C, which did not depress the mixture melting point (185°C) of the analytical material (mp 185°C).

Thus, for the first time the synthesis of a bile pigment possessing vinyl groups had been achieved, using a methodology that formed the basis for pursuing a path toward the total synthesis of bilirubin itself (*10*):

> *Somit ist zum erstenmal die Totalsynthese eines Gallenfarbstoffs mit Vinylgruppen durchgeführt und der Weg für die Synthese des Bilirubins freigelegt.*   56

In order to explore whether *exo*-vinyl groups might be introduced as easily as *endo* using the same formalism, the other symmetric biliverdin-IIIα with two *exo*-vinyl groups, was synthesized by substituting the isomeric opsopyrrole carboxylic acid (**III**) for **II** (*vide ante*), while following the same reaction scheme via urethane-dipyrrinone (**XXII**) (*10*):

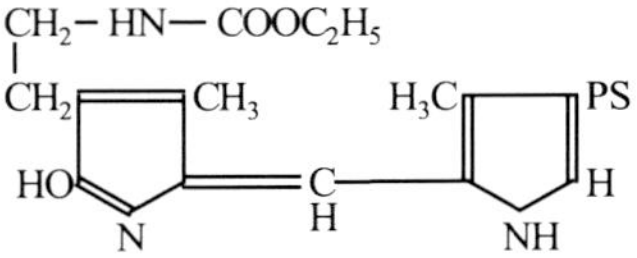

Thus biliverdin-IIIα dimethyl ester (**XXV**), mp 230°C, was synthesized (*10*):

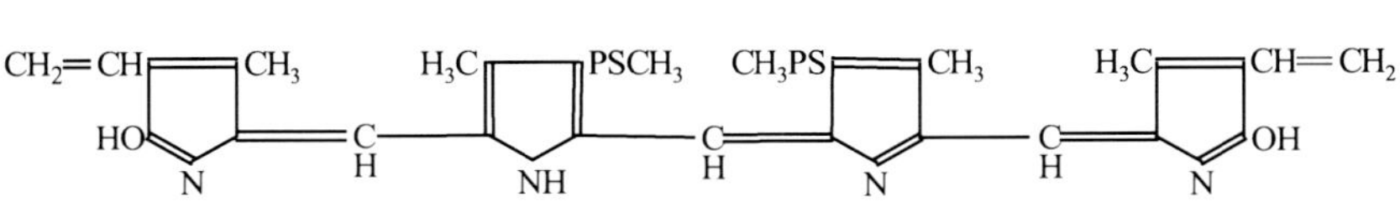

XXV

It was identical to the analytical verdin ester in all properties.

6.3 *Fischer* Solves the Structure of Bilirubin

The crowning achievement of *Fischer*'s bilirubin research, the total synthesis, was accomplished following a modification of the two routes established above to give first biliverdin-IXα dimethyl ester. In one route, the verdin was synthesized by acid-catalyzed condensation of the aldehyde of (*endo*-vinyl) vinyl-neoxanthobilirubinic acid methyl ester (**XXVI**), which had been synthesized earlier (*11*), with the aminoethyl-dipyrrinone (**XXIII**) obtained for the synthesis of biliverdin-IIIα dimethyl ester (*10*):

> Zur Synthese des Biliverdins IXα, dem Dehydrierungsprodukt des Bilirubins, wurde nun ein neuer Weg beschritten unter Verwendung des Aminoäthyl-iso-oxypyrromethens (**XXIII**), das zunächst mit dem Aldehyd der Vinylneoxanthosäure kombiniert wurde[1]) Der Aldehyd wurde, wie früher[2]) beschrieben, dargestellt und der entstandene Formylvinylneoxanthosäureester (**XXVI**) mit dem Aminoäthyl-iso-oxypyrromethen unter Zuhilfenahme von Bromwasserstoff-Methylalkohol kondensiert. Hierbei mub ein einheitliches Aminoglaukobilin der folgenden Formulierung entstehen: (**XXVII**).

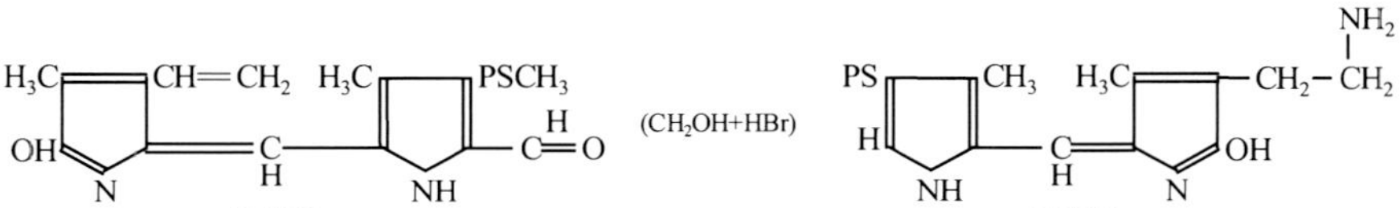

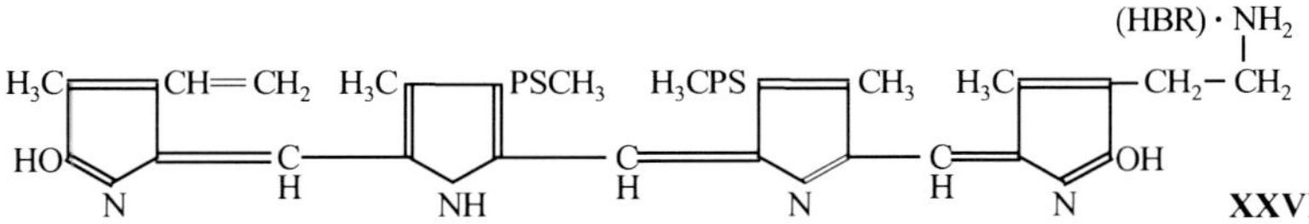

---

[1]) Vgl. hierzu die Glaukobilinsynthese W. Siedels, Diese Z. **237**, 15ff (1935). [*457*]
[2]) a.a.O., S. 10

The resulting hydrobromide (**XXVII**), after purification and conversion to the free amine, was converted to its zinc salt and permethylated using $(CH_3O)_2SO_2$. Heating the latter in methanolic NaOH, followed by esterification and chromatography on Al_2O_3 gave biliverdin-IXα dimethyl ester, mp 206–209°C (*10*):

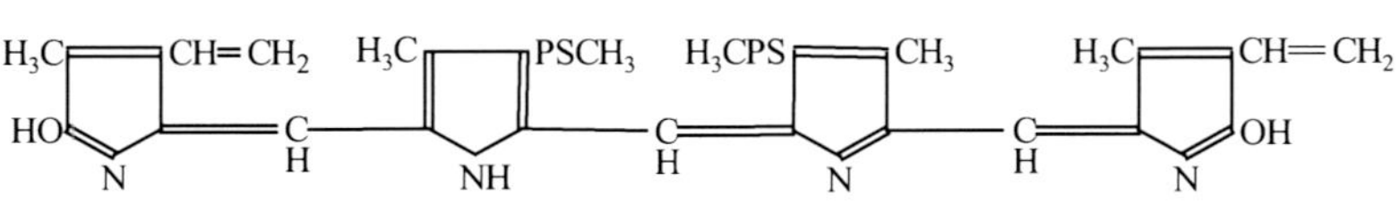

XXVIIa

A mixture melting point with twice-chromatographed natural-analytical biliverdin dimethyl ester (mp 202–204°C) gave no depression (*10*):

> **Biliverdin IXα (XXVIIa)**. 400 mg des „Monovinylaminoglaukobilins" werden in 50 ccm Methylalkohol, 50 ccm Wasser und 5 g Ätznatron auf dem Wasserbad erhitzt, bis alles verseift ist. Nun wird der Methylalkohol im Vakuum abgedampft, mit Zinkacetat und Dimethylsulfat versetzt und wie beim Biliverdin XIIIα aufgearbeitet. Bei der

Chromatographie an Aluminiumoxyd wird ein Lösungsmittelgemisch, Chloroform und Aceton, mit einigen Tropfen Methylalkohol verwendet. Auch hier ist ein roter Vorlauf zu beobachten, der verworfen wird, und ein blauer Nachlauf, von dem sauber abgetrennt werden muß. Der Nachlauf folgt direkt auf die blaugrüne Schicht des Biliverdins und hat ein schwach nach Blau verschobenes Jod-Zink-Spektrum.

Zur Analyse wird mit Chloroform-Methylalkohol extrahiert und noch weitere zweimal aus Methylalkohol umkrystallisiert. So erhält man feine Prismen von Schmelzp. 206-209°. Mit „analytischem", zweimal chromatographiertem Biliverdin (Schmelzp. 202-204°) keine Depression. Auch unter dem Mikroskop (Schmelzp. 224°) wurde gleichzeitiges Schmelzen beobachtet und war keine Mischschmelzpunktsdepression vorhanden. Jod-Zink-Spektrum: 639 mμ.

3,968 mg Subst. (60° i. V. getr.): 9,960 mg CO_2 2,090 mg H_2O. – 5,273 mg Subst.: 0,441 ccm N_2 (23°, 723 mm). – 4,980 mg Subst.: 0,816 ccm n/50-KSCN.

$C_{35}H_{38}O_6N_4$ (610,30) Ber. C 68,81 H 6,28 N 9,17 OCH_3 10,16

Gef. „ 68,48 „ 5,89 „ 9,18 „ 10,17   58

In a second approach, the aldehyde of urethane-dipyrrinone **IX** was prepared and condensed with urethane-dipyrrinone **XXII** to give the verdin bis-urethane **XXIX** possessing the IXα carbon skeleton (*10*):

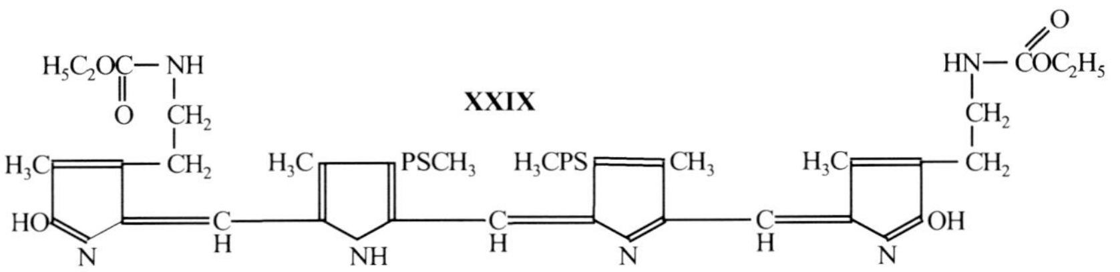

However, this pigment did not deprotect nicely at the urethane sites and was completely destroyed by the previous method (18% aq. HCl in a sealed tube at 135°C). Nonetheless, what might seem like the more drastic conditions (heating three hours in conc. HCl (38%) in a sealed tube at 100°C) seemed to be more effective. The resultant bis-amine hydrochloride, after the usual treatment to effect the *Hofmann Elimination* and subsequent final product isolation, afforded synthetic biliverdin-IXα dimethyl ester, mp 199–200°C, which was identical with the dimethyl ester of the natural material (*10*):

Es besteht somit kein Zweifel, daß auch dieses synthetische Biliverdin mit dem natürlichen Biliverdin IXα identisch ist, womit auch das Ergebnis Elementaranalyse gut übereinstimmt.   59

With the reconversion of a biliverdin to bilirubin having been already described (*11*), reduction of the free acid of biliverdin with Zn in acetic acid in the cold gave a yellow-red product, bilirubin-IXα, which was identical to natural bilirubin in all properties and reactions (*10*):

Die Rückführung des Biliverdins in das Bilirubin haben wir bereits vor kurzer Zeit beschrieben[1]) Durch Reduktion mit Zink-Eisessig in der Kälte konnte hier aus der freien

6.3 *Fischer* Solves the Structure of Bilirubin

Säure des Biliverdins, allerdings in mäßiger Ausbeute, ein gelb-roter Körper isoliert werden, der in allen Eigenschaften und Reaktionen mit Bilirubin identisch war. Das Jod-Zink-Spektrum zeigte einwandfrei, daß die Vinylgruppen noch intakt sind. Auch im Debye-Scherrer-Diagramm, sowie in der Kristallform war völlige Identität vorhanden. Die Reduktion des Biliverdins zu Bilirubin gelang auch durch alkalische kalte Reduktion mit Hydrosulfit. Auch so konnte ein mit Bilirubin IXα identisches Präparat erhalten werden.

---

[1]) H. Fischer, H. Plieninger u. O. Weissbarth, Diese Z. **268**, 197 (1941) [*11*]   60

As per the experimental section (*10*), biliverdin *dimethyl ester* was described as being saponified in methanolic KOH and then reduced to bilirubin by *Natriumhydrosulfit* (= sodium hydrosulfite = sodium dithionite = $Na_2S_2O_4$) (*10*):

> **Rückführung des Biliverdins in Bilirubin mit Natriumhydrosulfit**. 3 g Biliverdinester werden in wäßrig-methylalkoholischer Kalilauge verseift und nach dem Filtrieren der Methylalkohol im Vakuum verdampft. Nun werden in der Kälte etwa 10 g Natriumhydrosulfit zugegeben, worauf beim Schütteln alsbald Farbumschlag nach Gelb auftritt. Es wird mit Essigsäure angesäuert und der ausgefallene Niederschlag abfiltriert. Nach dem Trocknen wird mit Chloroform erschöpfend extrahiert und dies auf etwa 20 ccm eingeengt. Nach Zusatz von wenigen Tropfen Methylalkohol krystallisieren im Laufe einer Nacht etwa 100 mg Bilirubin als schönes braunrotes Pulver aus. Kein Schmelzpunkt. Die Substanz ist halogenfrei und erweist sich in allen Reaktionen identisch mit Bilirubin.   61

and, of course it followed that *Fischer* had no doubt that by his synthesis of biliverdin-IXα, the total synthesis of natural bilirubin had been accomplished (*10*):

> *Somit ist durch diese Synthesen des Biliverdins IXα = Uteroverdin auch die Synthese des natürlichen Bilirubins durchgeführt.*   62

It is a curious fact that the final step of the bilirubin-IXα total synthesis, the saponification and reduction of biliverdin-IXα dimethyl ester **XXXVIIa** is described only cursorily in the *Fischer-Plieninger* paper (*10*). Indeed, very few details of the important final step (reduction) of the total synthesis are given elaboration in the experimental section, *e.g.* was the $Na_2S_2O_4$ added at once; what reactions did *Fischer* mean in the ending phrase ". . . turns out to be identical to bilirubin in all reactions"? Unusual for *Fischer*, none are specified, not even a *Gmelin* reaction (*10, 462*). Oddly, no further information on the totally synthetic bilirubin or its "reactions" was provided. One might assume the *Gmelin* reaction was at least one of them. Also unusual for *Fischer*, no combustion analyses were reported. There were no follow-up data in subsequent publications to explain the low yield of isolated bilirubin (only a few percent) from reduction by sodium hydrosulfite, and there were no experimental details provided in the description of the Zn + glacial acetic acid reduction of biliverdin-IXα to bilirubin-IXα – though they may be found for the analogous reductions of the biliverdin-XIIIα and -IIIα cited in 1941 (*11*). Considering that the synthesis of bilirubin-IXα represented the second crowning achievement of *Fischer*'s research, one might have expected more detail to be provided for its synthesis and reactions.

422   6   The Status of Bilirubin in 1930

What led to the uncharacteristic (for *Fischer*) lack of experimental detail in final publications on the total synthesis of bilirubin is unclear. Not only was a description of the ultimate step sketchy but in ref. *(10)* the diagram for the conversion of dipyrrinone **XXVI** to key tetrapyrrole intermediate **XXVII** lacked clarity, and the experimental section description of verdin **XXVII** lacked the expected combustion analysis.

No further clarification was provided in the *Naturwissenschaften* publication *(462)* in mid-1942. *Fischer* and *Plieninger* mentioned only two reductive methods for converting biliverdin to bilirubin *(462)*:

> *Auf zweifache Art ist somit die Totalsynthese des natürlichen Biliverdins erbracht.*
> Die Rückführung des Biliverdins in Bilirubin haben wir bereits vor kurzem[8]) beschrieben. Durch Reduktion mit Zinkstaub-Eisessig in der Kälte oder alkalische Reduktion bei Zimmertemperatur mit Hydrosulfit ist Überführung möglich. Aus Chloroform tritt bald Kristallisation eines gelb-roten Körpers ein, der in allen Eigenschaften und Reaktionen, auch der Kupplungsreaktion, mit Bilirubin identisch war. Im Debye-Scherrer-Diagramm bestand Übereinstimmung.
> *Somit ist durch obige Synthesen auch die Synthese des natürlichen Bilirubins durchgeführt.*

---

[8]) H. Fischer, H. Plieninger u. O. Weissbarth, Hoppe-Seylers Z., **268**, 197 (1941) [*11*]   63

They cited the difficulties of identification associated with the final product: absence of a melting point, unfavorable solubility properties, inability to detect isomeric impurities at a level of 2–3% *(462)*:

> Diese Synthesen lassen sich mannigfach variieren, und es ist so eine außerordentliche Bereicherung der Synthesenmöglichkeit bilirubinoider Farbstoffe gegeben. Auch wird sich die Frage nach der Einheitlichkeit des Bilirubins am ehesten auf synthetischem Wege beantworten lassen. Bilirubin besitzt ja keinen Schmelzpunkt und außerordentlich wechselnde Löslichkeitsverhältnisse. Infolgedessen ist eine restlose Identifikation schwierig, bzw. es ist schwer, ein Gehalt eines allenfallsigen Isomeren auszuschließen. Ein Gehalt eines fremden Bilirubinoids von 2–3% z. B. wäre schwer oder gar nicht erkennbar.   64

and it is clear that *Fischer*'s thoughts had moved on to other bilirubin isomers, that the syntheses of the **IX** β, γ, and δ isomers of biliverdin (and bilirubin) were under consideration. Such isomers would come from macrocyclic ring opening of hemin at the other three meso sites and might also be found in nature, as in pterobilin (= γ-bilirubin) *(462)*:

> Nun ist die Möglichkeit der Existenz isomerer Bilirubine bzw. Biliverdine durchaus nicht von der Hand zu weisen, denn Hämin obiger Formulierung (i) kann ja an jeder der 4 Methingruppen aufgespalten werden, wie das ja auch bei der chemischen Aufspaltung mindestens teilweise der Fall ist[13,14,15]) Denkbar sind auber Biliverdin IXα der oben angegebenen Formulierung noch die drei folgenden Isomeren (β, γ, δ), deren Synthese in Angriff genommen ist und nach Kenntnis ihrer Eigenschaften wird ein etwaiges Auffinden in der Natur bzw. im Bilirubin möglich sein,

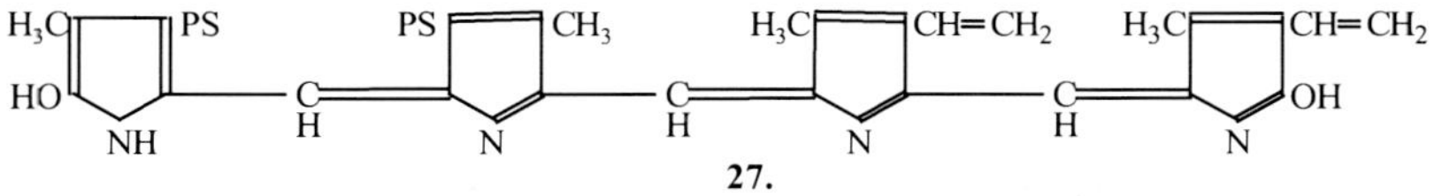

6.3 *Fischer* Solves the Structure of Bilirubin

β:

27.

γ:

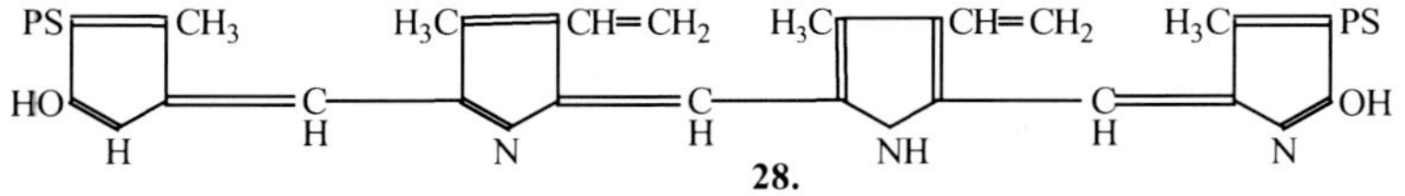

28.

δ:

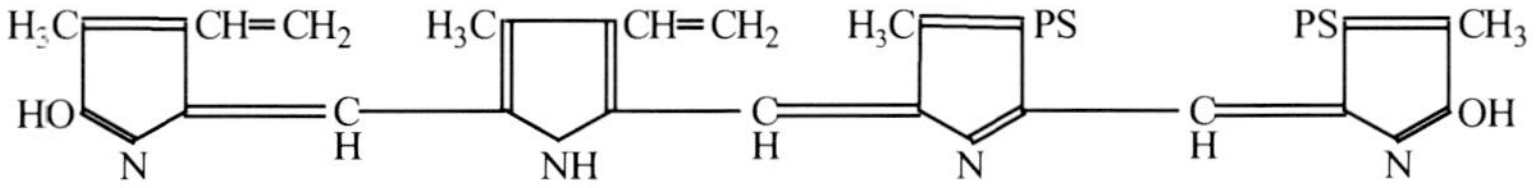

29.

denn ihre Rückverwandlung in das entsprechende Isobilirubin wird kaum Schwierigkeiten bereiten. Möglicherweise ist g-Biliverdin (**28**) bereits in der Natur beobachtet worden, wenigstens schreiben H. Wieland und A. Tartter[1]) dem Pterobilin diese Konstitution zu. Allerdings ist Pterobilin in seiner spektroskopischen Erscheinung gegenüber Zinkacetat-Jod abweichend von Biliverdin IXα, jedoch könnte das eben durch die besondere „γ-Konstitutionsanordnung" bedingt sein, wenn auch auf Grund der bisherigen Untersuchung der Einfluß des Seitenkettensystems in der Gallenfarbstoffreihe wesentlich geringer ist als in der Porphyrinreihe,[16,15]), vgl. aber[17]) Indessen steht eine systematische Untersuchung noch aus, insbesondere in bezug auf die Zinkacetat-Jodreaktion, bei der unter Umständen nicht unbeträchtliche Abweichungen zu konstatieren sind.[16]) [8]) S. 210 ff]

---

[1]) Liebigs Ann. **545**, 197 (1940) . . . [8]) H. Fischer, H. Plieninger u. O. Weissbarth, Hoppe-Seylers Z., **268**, 194 (1941) [*11*]. . . [13]) R. Lemberg, Biochemic. J. **29**, 1322 (1935), [14]) E. Steier, Hoppe-Seylers Z. **272**, 239 (1942), [15]) E. Steier, Hoppe-Seylers Z. **273**, 58 (1942), [16]) H. Fischer u. H. Reinecke, Hoppe-Seylers Z. **259**, 88 (1939), [17]) F. Pruckner u. A. Stern, Z. physik. Chem. (A) **180**, 25 (1937).

65

And so, in mid-1942, *Fischer* exuded complete confidence that any and every question related to bile pigments could then be solved by synthesis, now that he had mastered the introduction of vinyl groups (*462*):

> Zum Schluß sei noch erwähnt, daß bereits seit einigen Jahren die synthetischen Gallenfarbstoffe an Zahl die in der Natur vorhandenen bei weitem übertreffen, und ihre Anzahl kann durch die Einführung der Vinylgruppe auf synthetischem Wege fast beliebig gesteigert werden, so daß auch alle biologischen und chemischen Fragen in der Gallenfarbstoffreihe auf breiter Grundlage bearbeitet werden können. 66

As the devastation imposed by World War II outside the *Deutsches Reich* began to recoil into Germany itself and life in Europe was disintegrating, research at the TH-München was coming to a halt. Then, a slightly different version of the final step in the bilirubin total synthesis was presented by *Walter Siedel* of the *Fischer* research group. In mid-1943, as reported in *Angewandte Chemie*, he recapitulated the total synthesis of bilirubin, as part of a broad summary of progress in the chemistry of pyrroles (*7*), and in his description, the final reduction step that converted biliverdin to bilirubin was achieved by both Zn + glacial acetic acid and in aq. NaOH by $Na_2S_2O_4$ (sodium dithionite, sodium hydrosulfite). The synthesis outlined and diagrammed by *Siedel* thus matched the 1942 publications (*10, 462*).

Later in 1943 (October 25), *Siedel* again presented the total synthesis at a special session of the German Chemical Society in Vienna – a wide-ranging lecture on the chemistry and physiology of the breakdown of the pigment of blood that was published in 1944 (*6*). Here again biliverdin-IXα dimethyl ester was presented as being first saponified, then reduced to bilirubin-IXα by NaOH + $Na_2S_2O_4$ (*8*):

Von. H. Fischer und H. Plieninger[22]) ist im vergangenen Jahre schließlich auch die Totalsynthese des Biliverdins durchgeführt worden (Schema 6). Dabei werden die β-ständigen Vinylgruppen ausgehend von Propionsäurehydrazid- Resten über die Äthylurethan- und Äthylaminogruppen und deren Hofmannschen Abbau gewonnen. Es wurde schließlich (verg. Schema 6) die „Vinylneoxanthobilirubinsäure" dargestellt und diese nach Einführung

Schema 6

(Siedel's graphics reproduced)

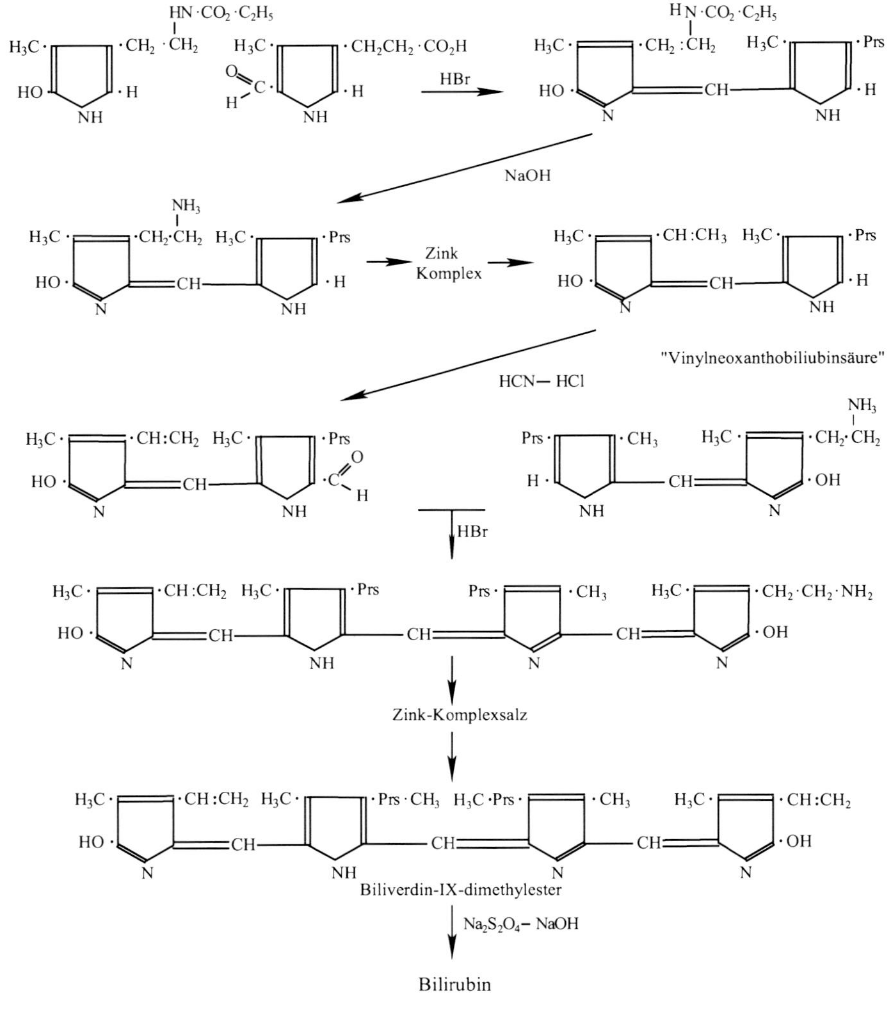

[22]) Ztschr. physiol. Chem. **274**, 231 [1942]

6.3 *Fischer* Solves the Structure of Bilirubin

der Formylgruppe mit der entsprechend synthetisierten Amino-isoneoxanthobilirubinsäure zum vierkernigen Produkt kondensiert. Nach erneutem Hofmannschen Abbau (vorgenommen am Zink-Komplexsalz) und Veresterung mit Methanol-Chlorwasserstoffsäure wurde der Biliverdindimethylester erhalten. Die Reduktion desselben mit alkalischer Natriumdithionit-Lösung führte zum Bilirubin selbst.   67

In his final summary of the bilirubin total synthesis, *Siedel* had again specified $Na_2S_2O_4$ as the reducing agent (*9*):

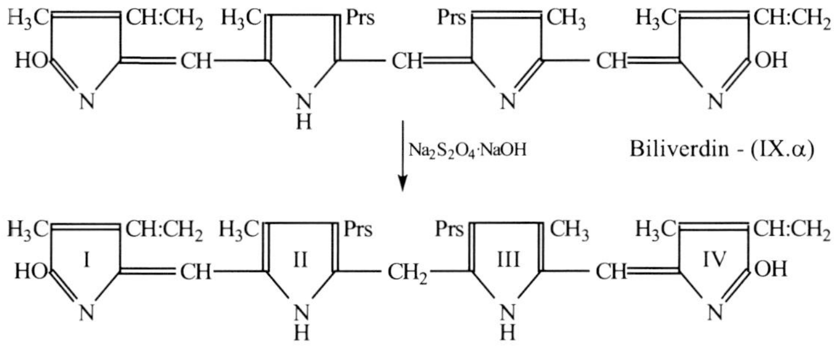

Biliverdin - (IX.α)

Bilirubin

Sodium dithionite (hydrosulfite) was then a well-known reducing agent, capable of converting (blue) indigo to indigo white, and thus it would have been a logical choice for reducing a pigment such as biliverdin. Catalytic hydrogenation was apparently avoided, (although it would have succeeded in reducing biliverdin to bilirubin) because it was known to reduce vinyl groups to ethyl. Sodium hydrosulfite (dithionite) is usually prepared by cathode reduction of bisulfite, or by addition of Na_2SO_3 to ZnS_2O_4, which is prepared by reduction of $NaHSO_3$ by Zn. Since hydrosulfite was shown nearly 25 years prior to the *Fischer-Plieninger* work (*462*) to react with water to produce bisulfite ($2\ Na_2S_2O_4 + H_2O \rightarrow 2\ NaHSO_3 + Na_2S_2O_3$) and thiosulfate (*467*, *468*), and since hydrosulfite reduces oxygen almost instantaneously to yield $NaHSO_3$, it might be assumed that the $Na_2S_2O_4$ used in the *Fischer* lab would form $NaHSO_3$ readily, if it did not already contain some.

In 1982, the *Fischer-Plieninger* total synthesis of bilirubin was again outlined (*14*), with the reducing agent of biliverdin to bilirubin indicated in a footnote as $NaHSO_3$ – a mis-translation of *Natriumhydrosulfit*, or an anticipation of its demise in H_2O. In the same decade, a former postdoctoral fellow of the author's, Prof. *Ma Jinshi*, read the procedure described in 1982 and employed $NaHSO_3$ as an inexpensive means to reduce biliverdin to bilirubin, possibly unaware that the yield of isolated bilirubin reported by *Fischer* was only a few percent. A thorough investigation by *Ma* led to the realization that the major product of the reaction with biliverdin, a yellow pigment, was the bright-yellow bisulfite addition product, with addition occurring at the central C=C, in the dipyrrylmethene core. No bilirubin was formed (*168*). Based on the *Fischer* graphical structure of biliverdin (*10*, *462*), the bisulfite addition product might be shown as:

The $NaHSO_3$ reduction was subsequently repeated with biliverdin by *Ma* in *A.F. McDonagh*'s[4] laboratory at the University of California San Francisco, but this time by using $Na_2S_2O_4$ as reductant. From *Ma*'s research notebook (*169*):

> 10 mg BV in 10 ml MeOH was added to a solution of 10 mg KOH in 1 ml water. Shook, filtered and evap on Buchi to remove MeOH. 30 mg $Na_2S_2O_4$ was added. Solution turned yellow in 2 min. Added HOAc to adjust pH to 4 and centrifuged. Hplc of the supernate showed only $BRSO_3H$ and a minor impurity. [Hplc of the precipitate showed $BRSO_3H$ (87%) and BR (13%)]. The precipitate was taken into $CHCl_3$ (0.2 ml) and the solution was centrifuged. The supernate was blown down to 0.1 ml and 0.1 ml MeOH was added and the solution cooled to –20C and centrifuged. On HPLC the precipitate was mainly BR, with a relatively small amount of $BRSO_3H$; the supernate contained only $BRSO_3H$.

McDonagh commented that (*169*):

> Thus, it is clear that $BRSO_3H$ is overwhelmingly the major product. Most of this would have been discarded by Fischer in the filtrate after adding acetic acid.

> In another experiment I added 50 mg dithionite to a solution of 2.5 mg BV in 2.5 ml 0.1M NaOH. The solution turned orange in about 3-4 min. At that point HPLC showed 87% $BRSO_3H$ and 13% BR (plus some minor peaks). The mole ratio of dithionite to BR in that experiment was (by mistake) 5 times what Fischer used.

> I wrote in a footnote, referring initially to Fischer and Plieninger: ….. "*bilirubin was isolated in 3% yield from biliverdin dimethyl ester. We have found that the principal product from the reaction of sodium hydrosulfite with biliverdin in air is not bilirubin, but bilirubin C-10 sulfonic acid. However, the work-up procedure used in the original synthesis, probably fortuitously, selectively removes the highly polar C10-sulfonic acid and leads to isolation of the relatively minor reaction product, bilirubin.*"

A decade later, bisulfite addition was achieved with a different verdin, and its X-ray crystallographic structure was published (*170*).

---

[4] *Antony F. McDonagh* was born on September 30, 1938 in Manchester, U.K. and died on October 22, 2012 in San Francisco, CA, USA. He graduated in chemistry in 1962 from the Royal College of Advanced Technology, Salford, Lancashire in England and received the Ph.D. in organic stereochemistry and ORD in 1966 from Vanderbilt University, Nashville, Tennessee, USA, working with Prof. *Howard Smith* (a postdoctoral fellow with *Carl Djerassi*). After postdoctoral studies at Queen Mary College, University of London with Prof. *Raymond Bonnett* from 1966 to 1969, *McDonagh* joined the Gastrointestinal (GI) Unit of the University of California, San Francisco Medical School, working with *Rudi Schmid* as a Research Chemist from 1970 to 1976, during which his independent research career in the bilirubin area developed. From 1971 to 1976 he was Assistant Professor in the School of Pharmacy, from 1976 to 1982, Associate Research Chemist in the GI Unit, then from 1982 to 1987 Research Chemist, and from 1987 Adjunct Professor of Medicine. In 2006 he emeritated and remained a consultant and research collaborator as the world's leading authority on bilirubin until his untimely death.

Ma and *McDonagh* were not alone in having used $Na_2S_2O_4$ to effect a reduction of a verdin to a rubin during the post-*Fischer* era. Evidence that sodium dithionite reduced verdins to rubins was published in 1979 by *Werner Kufer* and Prof. Dr. *Hugo Scheer* at the Botanisches Institut der Uni-München. Perhaps following *Fischer*, they achieved $Na_2S_2O_4$ "reduction" of to phycorubins (*471*). And subsequently, in 1982, they reported their studies of the reduction of the verdin-type chromophore of the denatured plant pigment, C-phycocyanin, using both $NaBH_4$ and $Na_2S_2O_4$ to give phycorubins (*472*). That work clearly distinguished between the phycorubins produced by the two reagents. With $NaBH_4$, hydrogen addition at C-10 produced a bilirubin; with $Na_2S_2O_4$, only a trace of the bilirubin was produced, with the major product being the bisulfite addition produced at C-10.

Currently, the best way to achieve reduction of biliverdin (or its dimethyl ester) to bilirubin is probably by the use of $NaBH_4$ in CH_3OH, which dates back to the early 1970s (*13, 169, 469, 470*).

Given the current state of knowledge of bilirubin-IXα, one might wonder whether the synthesis of bilirubin from biliverdin-IXα dimethyl ester, as accomplished by *Plieninger* produced 100% isomerically pure bilirubin-IXα. This could happen only if the opsopyrrole carboxylic acid and iso-opsopyrrole carboxylic acid had been cleanly separated early in the synthesis – and there had been no constitutional isomerization when the final bilirubin was isolated. Although indicated as pure by *Fischer*, a partially scrambled mixture of the IIIα, IXα, and XIIα isomers might have been obtained, for *McDonagh* has shown that bilirubin-IXα is susceptible to constitutional isomerization in base (*473*), and, more effectively, in acid (*474, 475*). In fact, most commercially available samples of natural bilirubin contain at least small amounts of the IIIα and XIIIα isomers (*476*). Though such constitutional isomerization seems unlikely, *Fischer* did not have the means to establish clearly the isomeric purity of his synthetic bilirubin-IXα, especially at levels of a few percent IIIα and XIIIα contamination. All of which further illustrates the unique nature of the pigment. However, *McDonagh* has indicated (*169*) that *Plieninger*'s reaction conditions in the final reduction step, as repeated in his (*McDonagh*'s) laboratory would be unlikely to cause isomerization. And that the bisulfite addition product contained 8.8% of the XIII isomer, but none of the III isomer – consistent with the presence of biliverdin-XIIIα dimethyl ester in the starting material from commercial sources of bilirubin.

6.4 Hans Fischer, Finis Vitae

Doubtless there were many distractions from research facing *Hans Fischer* in 1942 as WWII advanced into western Soviet Russia and ground to a halt with the catastrophic surrender on February 2, 1943, of the German 6th Army at Stalingrad – a battle involving some two million combatants that signaled the beginning of the end of the *Grossdeutschen Reiches*. A drop in civilian morale heightened as increasingly more German cities were subjected to Allied bombing after 1942.

And research, never easy in the best of times, suffered when universities were inexorably drained of *Diplom* students, which led to subsequent losses in research at the doctoral level, *etc.* The *Fischer* labs published only approximately 30 papers from 1942 to 1945, compared to nearly three times as many in the previous three-year period, 1938–1941. Pyrrole research at the TH-München was grinding to a halt as the war was brought to the *Fischer* Institute's doorstep.

Fischer published no papers in 1945. His institute had been devastated by the bombings; research had come to a halt, as (doubtless) had journal printing. *Fischer*'s research empire was reduced to a few demolished rooms, few chemicals and apparatus, and his group members were dispersed or dead. Considering his near-total dedication to research, his despair must have been enormous. And so, as his student *Karl Zeile* wrote in a memorial in 1946 (*345*):

> Am 31. März 1945 hat HANS FISCHER sein Leben und sein Werk beschlossen. Wenn in jenen Tagen, in denen in Deutschland alles der Auflösung zutrieb, die Nachricht vom Tode eines Menschen so besonders tief berührte, wie es den zunächst wenigen, die sie erfuhren, geschah, so war es deshalb weil alle, die HANS FISCHER kannten, in ihm den Begriff der unbeirrbaren Beständigkeit und unverwüstlichen Kraft gesehen hatten. Doch eben jene daraus entspringende und für ihn so bezeichnende Gradlinigkeit des Handelns mag ihn veranlaßt haben, schonungslos die Folgerungen aus einer Situation zu ziehen, die ihm nicht mehr gemäß war. Nach der fast völligen Zerstörung seines Instituts in der Münchener Technischen Hochschule wußte er, daß der Arbeit seines Stils ein für allemal ein Ende gesetzt sei und ohne sie konnte er nichte leben. 68

Likewise, his co-worker *Albert Treibs* wrote in 1946 (*346*):

> Mitten in die Zeitkrise fällt das tragische Ende von Hans Fischer am Ostersamstag, den 31. März 1945. Es war ihm vergönnt, seine umfassenden Arbeiten durch die Synthese des Hämins und Bilirubins zu krönen, die Konstitution des Chlorophylls aufzuklären und seine Synthese fast zu vollenden, jedoch seine ebenso bewundernswerte Schöpfung, sein Institut, die Wirkungsstätte einer großen Schar begeisterter Mitarbeiter, ist dem Krieg zum Opfer gefallen. Nicht allen Kennern seines Werkes wird zum Bewußtsein gekommen sein, wie groß der Anteil der von ihm geschaffenen Organisation an seinen Erfolgen war. Er wußte, daß mit wenigen ihm verbliebenen demolierten Räumen, mit bescheidenen Beständen an Apparaten und Chemikalien ein Weiterarbeiten in seinem Stil so bald nicht möglich war, daß seine gute Tradition verloren zu gehen drohte. Waren ihm sachliche Schwierigkeiten bei seiner Forschung nur ein Anreiz, die jahrelangen Widerwärtigkeiten, die seiner Einstellung gegen das nationalsozialistische System entsprangen und die ihn immer wieder zwangen, sich und sein Werk zu verteidigen, hatten ihn doch ermattet. In seiner Wissenschaft ein unverwüstlicher Optimist, in der Politik hoffnungsloser Pessimist, ein guter Deutscher, der Welt gegenüber aufgeschlossen, hat er die Entwicklung der Dinge lange vorhergesehen. Und das raubte ihm alle Hoffnung, auch für die weitere Zukunft. Eines hat er vielleicht nicht bedacht – wieviel er durch sein internationales Ansehen seiner Hochschule beim Wiederaufbau hätte helfen können. Doch er liebte eine verwaltende Tätigkeit nicht, nie hätte er z. B. das Rektoramt übernommen; seine Organisation war immer ein geniales Improvisieren, ein Schaffen aus dem Vollen im Dienste seiner Wissenschaft. Mit voller Überlegung griff er zur Ampulle, die er für den Fall der Not bereitet hatte. 69

And in the prologue to his affectionate memorial, *Fischer*'s colleague, predecessor at the TH-München, and fellow chemistry Nobelist, *Heinrich Wieland* wrote in 1950 (*347*):

6.4 Hans Fischer, Finis Vitae

Der Krieg hat die Chemie an den Münchener Hochschulen bis tief in Mark hinein getroffen. Nachdem schon im Sommer 1943 die wissenschaftliche Arbeit nahezu zum Erliegen gekommen war, lagen ein halbes Jahr vor dem Zusammenbruch des Dritten Reiches die Institute unserer beiden Hochschulen in Trümmern. Aber am schwersten hat uns das persönliche Schicksal von *Richard Willstätter, Hans Fischer* und *Otto Hönigschmid* getroffen, die uns innerhalb weniger Jahre genommen wurden. Sie sind als Opfer des Dritten Reichs und des Kriegs gestorben; in ruhigen Zeiten würden sie wohl noch unter uns weilen. 70

Finally, from yet another *Fischer* student and co-author of volume II, 2nd half, of the famous *Fischer* and *Orth* trilogy (*Die Chemie des Pyrrols*) (*4*), *Adolph Stern* presented a more recent memorial in 1973 (*392*):

There are few organic chemists who have been so productive (n.b. the three volumes by Fischer and Orth entitled: "Die Chemie des Pyrrols"). Yet, he was careful to prepare or synthesize only those compounds essential for achieving the ultimate goal of his research. Often, he told overenthusiastic doctoral candidates: "remember we do not work here to enrich the Beilstein." The enormous scope of his endeavor was possible because of his exceptional vitality and persistent hard work. Often Fischer was in the laboratory or at the office from early morning till late at night, even on Sundays. He remarked to his admiring students: "what else is there to do on Sunday?" Recreation was somewhat alien to him, and if he took vacation or a rest at all it was of short duration. He enjoyed short ski trips. Besides skiing he had few other diversion other than riding a bicycle, motorcycle or driving. Many who visited his office will remember the bicycle in front of his desk. Art, music, theatre and movies did not attract him. He enjoyed the full devotion of his coworkers, who spared neither time nor effort to support and serve him. There was an aura of greatness around him, which one could sense immediately and which inspired respect and admiration. Fischer's daily visits to the laboratories were always eagerly anticipated, and his advice and interest in the individual research of the students were welcomed. He recalled even the smallest details of individual research, and his knowledge of the chemical literature encompassed its many aspects. He was always friendly and understanding. Modest in dealing with peoples of all walks of life, he never mentioned his many accomplishments and the honors that were bestowed upon him, such as the title "Geheimrat," an honorary doctoral degree from Harvard University (1936) and, of course, the Nobel Prize (1930), to mention a few. He was unassuming in dress, and always enjoyed being mistaken for the janitor by some new visitors to the laboratories.

In 1933, when Hitler came to power, it was expected that the change would be of short duration. Some of Fischer's coworkers joined in the movement; the majority, however, were not interested and kept to themselves. In the years that followed, some of his experienced coworkers and pupils were forced to leave the country, others left voluntarily. This slowed the progress of research. Fischer was against the Nazi party until the end. His great reputation as one of the outstanding scientists spared him too much harrassment. Again, the dark events leading up to World War II and finally, the war itself interrupted his research, as in his early years in Vienna. Several partial syntheses of chlorophyll could still be completed. Fischer kept on working as usual, but the progressive influence of the war, the destruction of buildings and laboratories and personal disillusionment were too much for him. On March 31, 1945, he chose to pass from this world which had abundantly honored his great achievements, but also caused him untold sorrow. One of the great scientists of this century died then. His memory will live forever in the world of science, and in the hearts of all who had the good fortune to know him personally.

Fig. 6.4.1 *Fischer* group photograph from 1934. Names mentioned in this work include *Hans Fischer* (seated, center), *Albert Treibs* (standing front, immediately to the right of *Fischer*), *Adolf Stern* (standing front, second to the left of *Fischer*), *Karl Zeile* (without lab coat, standing immediately to the right of *Treibs*), *Hans Orth* (standing front right, with lab coat open), *W. Neumann* (standing without lab coat, front row, far left), *Walter Siedel* (middle, 4th row, partially hidden third face above *Treibs*), *Hans Haberland* (center, first face to right of *Siedel*). With permission from the *Hans Fischer* Gesellschaft, from ref. (*12*), wherein all other members of this *Fischer* group photo are identified.

National Socialism was devastating no less to *Fischer* and pyrrole chemistry than to Europe and the future of chemistry in Germany, though it led to the enhancement of science elsewhere (*477*):

> The Nazi period stands out, among other atrocities, as a spectacular example of a country's scientific self-destruction. By as early as 1936, 1,617 university professors and researchers had been expelled, not only Jews, of whom 1,160 went to English-speaking countries, 825 of them to the USA. . . . Scientific self-destruction went so far that even sciences essential for the war suffered, as the Nazi leaders themselves finally had to admit. . . . As a consequence of Nazism the country was in ruin and there were virtually no resources left for science. A huge brain drain, again mainly to the USA, followed.

Fischer's passing at the end of World War II also coincided with the final stage of the German language suffering eclipse as the preeminent language of chemistry (*477, 478*). Shortly before World War I, Germany had become the international Mecca of science. In 1927, *Gross* and *Gross* (*479*) tabulated comparisons of citations of chemistry publications by language for the period 1875–1925 and found 52.5% were in German, 35.2% in English, 9.4% in French. They recommended (*479*):

6.4 *Hans Fischer, Finis Vitae*

Certainly it should be insisted that a reading knowledge of German be required of every student majoring in chemistry in college.

And so it was, until the late 1960s when many U.S. colleges and universities dropped their foreign language requirements. Though German exceeded all other languages in natural science publications of the first two decades of the 20th century, the share continually declined after World War I. The decline accelerated greatly during and after World War II, and by the 1990s over 90% of the publications in the natural sciences were in English (*477*):

> A complimentary [sic] blow at German and all other international languages of science except English was dealt by an important change of foreign language requirements at American colleges and universities in the 1960's. Many of them decided to reduce or to abolish language requirements especially for natural science studies after they had "come under mounting criticism as being of doubtful utility" (Wiltsey 1972, Part 1: 7; cf. for details on the discussion Ammon 1998: 13). William Mackey (personal communication) believes that the American Tertiary Institutions' decision to reduce or to abolish foreign language requirements had a major impact on the decline of languages other than English as international languages of science. American scientists who play a dominant role in the international scientific community have thus been encouraged not to read publications in languages other than English, or are no longer able to do so. Scientists from other linguistic backgrounds therefore have to publish in English if they want their findings to be read by their most influential and powerful colleagues.

Thus, the regime and its legacy of devastation that brought destruction to the *Fischer* laboratory, with its enormous productivity and dominance in pyrrole chemistry, led to *Fischer*'s demise at the height of his scientific prominence and also displaced the German language from its position as the language of chemistry – as will become evident in Section 7 and the citations therein. Yet, if pyrrole chemistry serves as an example, it would be foolish to think that one might read and understand the vast written knowledge of pyrrole compounds absent a reading knowledge of German.

7 Evolution of the Constitutional and Stereochemical Structure of Bilirubin

With the death of *Hans Fischer* in 1945, the destruction of his labs, and the dispersal of his research group, investigations of pyrrole natural products diminished significantly. Yet the decades following his death and the demise of his laboratory coincided with major advances in different areas of natural products chemistry, notably steroids. For example, the structure of cholesterol, discovered by *Chevreul* in 1812 and isolated by *Gmelin* in 1828, was elucidated in 1955 (*480*), and its first total synthesis accomplished by *John F. W. Keana* (Ph.D. 1964) and Prof. *William S. Johnson* at Stanford University in 1964 (*481*). The importance of steroids lay in their biological properties, as illustrated by the work of *Carl Djerassi* and colleagues at Syntex Mexico, who in 1951 synthesized the first oral highly active progestin (norethindrone), work that in 1960 led to its approval by the U.S. Food and Drug Administration (FDA) for marketing as a female birth control pill in the U.S. Steroid natural products research flowered in the 1950s and 1960s. Meanwhile, in the decade of the 1950s, bilirubin and bile pigment research was arising from its apparent dormancy. *Cecil J. Watson*, whose association with bile pigments stretched back to the 1930s in *Hans Fischer*'s group, and *Charles Gray* were reporting studies of the more reduced bilirubin metabolites, urobilin and stercobilin. Issues concerning their chemical structures were raised, and especially how the *Fischer* bis-lactim structure of bilirubin translated into the bilirubin metabolites – an issue not fully solved (by others) for another two decades.

Fischer had solved the structure of bilirubin in 1942 by what was then a masterpiece of analysis and total synthesis, though from 1930 until his demise he had dedicated his research to the total synthesis of chlorophyll, an objective not reached. For bilirubin, he provided the following unsymmetrical constitutional structure, drawn as he represented it (*10, 462*):

$$* \text{ PS} = CH_3 - CH_2 - COOH$$

D.A. Lightner, *Bilirubin: Jekyll and Hyde Pigment of Life*, Progress in the Chemistry of Organic Natural Products, Vol. 98, DOI 10.1007/978-3-7091-1637-1_7,
© Springer-Verlag Wien 2013

It seems clear that by 1942, *Fischer* had taken the structure proof of bilirubin as far as was possible for that time. He chose not to speculate on the stereochemistry of the exocyclic carbon-carbon double bonds, presumably because there was at the time no way to prove or disprove that aspect of the pigment's structure. Likewise the question as to whether the pigment adopts the lactim or lactam tautomer could not be answered unequivocally. *Fischer* had based his lactim structure on the best available evidence, observations that *Xanthobilirubinsäure* (xanthobilirubinic acid) could be acetylated and that bilirubin formed lactim methyl ethers upon reaction with diazomethane. Though suggestive of a lactim, the available evidence could not be considered incontrovertible. Perhaps sufficient for 1942, but already viewed as incomplete by other investigators of that age, *e.g. Rudi Lemberg (460)*, the stereochemistry would almost surely have been elaborated further by *Fischer* had he survived another decade, particularly the most obvious missing stereochemical aspect of the 1942 structure, the unspecified "*cis-trans*" (Z,E) geometry at the pigment's ene-amide *exo*-cyclic carbon-carbon double bonds.

Fischer might also have revisited his hydroxy-pyrrole tautomeric structures when infrared spectroscopy began making its entrée into organic chemistry. And he might even have thought about the three-dimensional structure of bilirubin when organic stereochemistry was set upon its course of prominence by the seminal work of *Odd Hassel* (1897–1981) in the early 1940s at the University of Oslo, and Sir *Derek H.R. Barton* (1918–1998) of the Imperial College of Science and Technology (London) that culminated in their 1969 *Nobel* Prize in chemistry. Then, too, hydrogen bonding, an electrostatic property of carboxylic acids and amides, though seemingly unrelated to stereochemistry might have also intrigued *Fischer*, when it captured the attention of chemists of the 1940s and 1950s (*482*). As it turned out, all four of the aforementioned considerations (double bond configuration, lactim-lactam tautomerism, hydrogen bonding, and stereochemistry in three dimensions) became topics of bilirubin investigation following *Fischer*'s death. And all four taken collectively served to define the final aspects of the structure of bilirubin.

These four aspects were resolved only during the advances in natural products structure proof, as well as in all of organic chemistry, that began after World War II with the arrival of new forms of spectroscopy, *i.e.* infrared (IR) and nuclear magnetic resonance (NMR). A spectroscopic advance beyond the "spectrum analysis" (ultraviolet and visible spectroscopy, see Section 2.10) that was used by but predated *Hans Fischer*. Infrared, useful for identifying functional groups, and NMR, useful for learning details of local structure, became increasingly employed, increasingly powerful, and increasingly important in elucidating the details of organic structure and stereochemistry. Yet it was X-ray crystallography, a method older than either IR or NMR, that, with the aid of computers to facilitate diffraction data reduction, settled the unresolved issues with respect to the *Fischer-Plieninger* bilirubin structure: the question of a lactim or lactam structure and the (Z,E) stereochemistry of its exocyclic C=C bonds. In addition, it revealed the conformation of the pigment and the intramolecular hydrogen bonding that served to maintain the pigment's secondary structure.

7.1 Lactim or Lactam Tautomeric Structure?

Unlike bilirubin and bile pigment structure research in the first half of the 20th century that was the focus of but two or three research groups, which by 1945 had ceased activity, research in the second half of the 20th century brought numerous new investigators, with new ideas and new techniques. With the demise of the *Fischer* School and *Hans Fischer's* death in 1945, the missing elements of the bilirubin structure were left to others to provide in the second half of the 20th century. As will be seen in this and the following chapters, no single laboratory brought the bilirubin structure to completion.

In the next section, the lactim-lactam tautomer issue is discussed, then the stereochemistry of the exocyclic double bonds, followed by a discussion of hydrogen bonding.

7.1 Lactim or Lactam Tautomeric Structure?

As illustrated graphically in Section 7, and throughout this volume, *Hans Fischer* preferred to present the end rings of bilirubin in the lactim tautomeric representation. Very early in his studies of the degradation products of bilirubin, when he had correctly assigned the structures of *Bilirubinsäure* (bilirubinic acid) and *Xanthobilirubinsäure* (xanthobilirubinic acid) in 1913, he wrote the former as a hydroxypyrrole and the latter as its lactam tautomer (*358*):

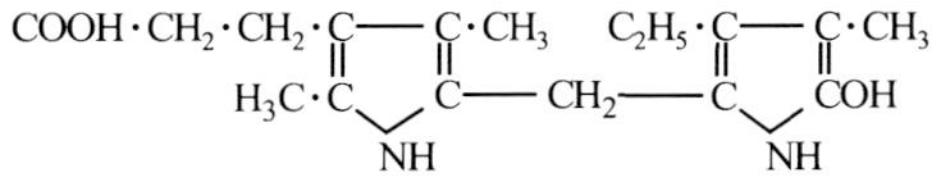

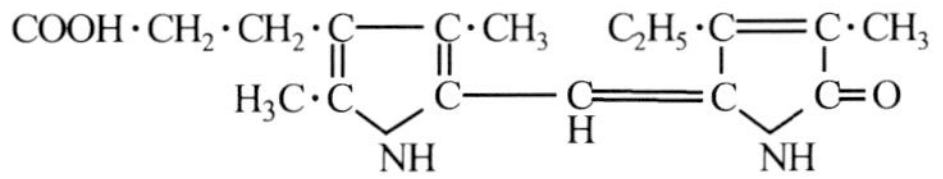

and again in 1914 (*357*), with the stereochemistry of the *exo*-cyclic double bond not specified. While one might understand *Fischer's* representation of bilirubinic acid as a hydroxypyrrole, a similar representation is not possible for xanthobilirubinic acid. *Fischer* continued to represent xanthobilirubinic acid as a lactam from 1914 through 1916 (*373, 376*), then changed to the isopyrrole lactim structure in 1923 (*394*):

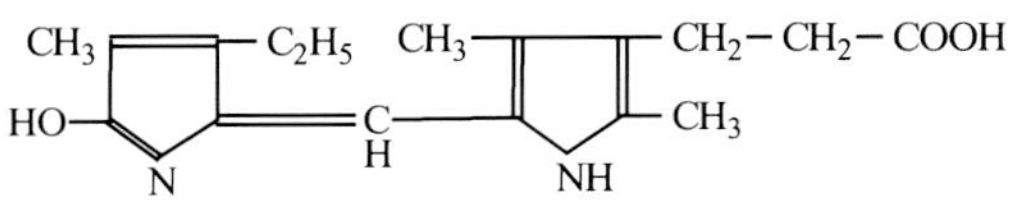

Why? *Fischer* offered the following explanation (*394*):

> Zwar hatten Fischer und Röse der Xanthobilirubinsäure die entsprechende Ketoformel zuerteilt. Nach einer neuen Untersuchung von Fischer und Niemann [Diese Zs. Bd. im Druck] jedoch gibt die genannte Säure eine Acetylverbindung und demgemäß kommt ihr die obige tautomere Formel zu. 1

436　　　　7 Evolution of the Constitutional and Stereochemical Structure of Bilirubin

According to *Fischer*, studies in press at the time by *Fischer* and *Niemann* showed that xanthobilirubinic acid gave an acetyl compound – presumably an acetate ester, as befits the lactim tautomer represented above. Whether real or imagined, one can find no mention of an acetyl compound of xanthobilirubinic acid either in the 1923 publication of *Fischer* and *Niemann* in *Hoppe-Seyler's Zeitschrift* (*395*), or in their subsequent publications of 1924 (*397*) and 1925 (*399*). Possibly the transformation was intended to be in one of the three cited publications with *Niemann*, as there were no others; possibly the information never left *Niemann's* dissertation. In any event, *Fischer* believed that he could form the acetate of xanthobilirubinic acid. One would guess that might have been accomplished from reaction either with acetic anhydride or acetyl chloride. And one would hope that the reaction did not simply produce a mixed anhydride with the pigment's propionic acid – or that it was carried out on an ester of xanthobilirubinic acid. Irrespective, that finding was the key element in persuading *Fischer* to abandon the lactam in favor of the lactim in all subsequent representations of dipyrrinones and linear tetrapyrroles (*3*). The acetate (unstable) was later postulated as an intermediate in the synthesis of xanthobilirubinic acid from monopyrroles (*442*), as was the more stable methyl ether (*443*) – both intermediates coming from replacement of the bromine of the bromopyrrole components of a dipyrrylmethene.

After *Fischer* had recognized as invalid and abandoned his tetrapyrrylethylene formula for bilirubin, and was led thus to a linear tetrapyrrole formula, he continued to express his preference for the lactim tautomer in his graphics. Consistent with his early belief in a fused furan ring in both bilirubin and its biological precursor, biliverdin, as early as 1930, *Fischer* represented the latter as a linear tetrapyrrole (**24**), interestingly with an intact *exo*-vinyl group on a lactam ring. He assumed that a lactim OH would be required for formation of the furan ring, by adding to the erstwhile precursor *exo*-vinyl group of the pigment. However, he also allowed the lactam tautomer in the other end ring, the one with an intact *exo*-vinyl group (*436*):

> Außer der Auffassung des Bilirubins als Tetrapyrryläthylen käme noch eine offene Formel wie folgt in Frage (s. Formel 24, S. 1032).

* $S = CH_2 \cdot CH_2 \cdot COOH$.

24

2

Although **24** looks more like a verdin than a rubin, it is still closer to the actual bilirubin structure than the tetrapyrrylethylene representation (**22**, below) favored by *Fischer* in 1930 (*436*):

7.1 Lactim or Lactam Tautomeric Structure? 437

22

Perhaps *Fischer* was hedging his bets.

The structure of xanthobilirubinic acid continued to be represented as a lactim in 1931, and the structure of mesobilirubin-XIIIα, correct in most details, was represented (**VI**) as the bis-lactim (*438*). But the structure of neo-xanthobilirubinic acid was drawn both as the lactim and as its lactam isomer (**VII**) in 1931 (*444*):

. . . Die Konstitution des Mesobilirubins ist daher definitiv im Sinne der früher aufgestellten Formel

VI   3

. . . Der um 10° höhere Schmelzpunkt der Neoxanthobilirubinsäure aus synthetischem Mesobilirubin muß durch Isomerie bedingt sein, etwa im Sinne folgender Formel:

VII   4

This was done because *Fischer* believed that the mp 227°C analytical neo-xanthobilirubinic acid, formed by treating analytical mesobilirubin (mesobilirubin-IXα) with molten resorcinol, had a lactim structure, as represented in his proposed mesobilirubin-XIIIα (**VI**, above). On the other hand, the mp 237–239°C synthetic neo-xanthobilirubinic acid formed by treating synthetic mesobilirubin (actually mesobilirubin-XIIIα) with molten resorcinol was proposed to have the isomeric lactam structure (**VII**). Though a mixture melting point of these two neo-xanthobilirubinic acids confusingly gave no melting point depression, it was

438   7 Evolution of the Constitutional and Stereochemical Structure of Bilirubin

learned subsequently that the natural-analytical sample was a mixture of neo- and isoneo-xanthobilirubinic acid (Section 5).

In the 1930s and early 1940s, *Fischer* and his students almost never wavered from consistency in drawing the lactim tautomers of bilirubins, biliverdins, mesobilirubins, and the various xanthobilirubinic acids. Other investigators, such as *Lemberg*,[1] did likewise in the 1930s (*460, 483*) and 1940s (*433*), but in 1949 *Lemberg* graphically expressed the possibility of a lactam tautomer in bilirubin (*433*):

> It accounts well for the properties of the rubins, their similarity to α-hydroxypyrromethenes, their very weakly basic character [*383*], and their dehydrogenation to bilatrienes. The weakly basic character is due to the lack of nonhydroxylated pyrrolene nuclei, the α-hydroxylated pyrrolenes tautomerizing to the lactam form:

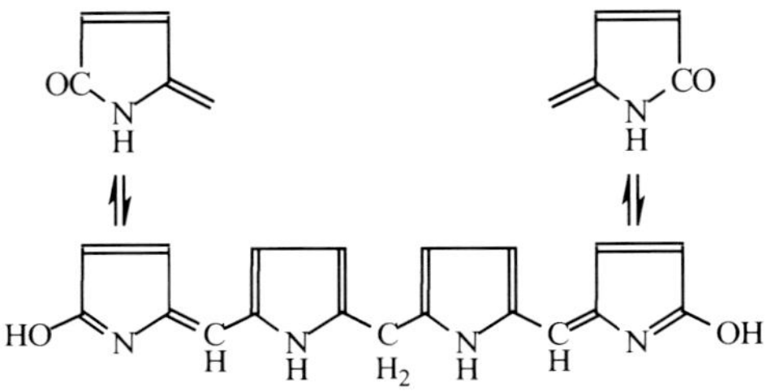

Addressing the question of whether bilirubin adopted the lactim tautomer preferentially, or the lactam, began seriously in the 1950s. The possibility of such

---

[1] *Max Rudolf Lemberg* was born October 9, 1896 in Breslau and died on April 10, 1975 in Sydney, Australia. He entered the Universität Breslau in 1914 to study chemistry, physics, mineralogy, and geology just months before World War I broke out. Rejected for military duty, he spent 1915–1916 in studies at the Universities of Breslau, Munich, and Heidelberg. In 1917 he was accepted for military service, which set him on a firm path of pacifism. He renewed his studies at Breslau and was awarded the Ph.D. at the University in 1921 for work on methyl-substituted uric acid derivatives and then became a private assistant to his Ph.D. advisor, *Heinrich Biltz* from 1921 to 1923. In 1923, he accepted a position at Bayer Pharmaceutical only to return to academic life as *Habilitand* and began research for the *venia legendi* at the Uni-Heidelberg from 1925 to 1930 under *Karl Freudenberg*, where he investigated the chromoproteins of red and blue algae. After the *Habilitation* he went to the University of Cambridge to work with *Gowland Hopkins* at the Sir William Dunn Institute of Biochemistry from 1930 to 1931, where his interest in tetrapyrrole pigments was kindled. He returned to Heidelberg as *Privatdozent* and *Assistent* in the Inorganic Chemistry Department from 1931 to 1933. In his own words: "now the Nazi shadows began rapidly to gather, and in 1933 the *Beamtengesetz* [Nazi-era Civil Service Act] made an end to my academic career in Germany. I had none of the illusions which trapped so many into thinking that they might weather the storm. Thus the second crisis of my life began. I was first dismissed from the assistantship, and a few months later, when already at Cambridge, I was informed of my dismissal as university lecturer by the letter of a German furniture removal firm offering its services." Restitution for the injustice then done was made later by appointment as o. Professor Emeritus at Heidelberg and a corresponding member of the Heidelberg Akademie der Wissenschaften. *Freudenberg* and other Heidelberg colleagues were instrumental in getting *Lemberg* to England, from whence he went to Australia from 1933 to 1975 as director of the biochemical laboratories at the Royal North Shore Hospital in Sydney. The author made *Lemberg*'s acquaintance at the 1968 Gordon Research Conference on Tetrapyrroles at Crystal Mountain, WA.

7.1 Lactim or Lactam Tautomeric Structure?

tautomers was, of course, recognized early on by *Fischer* and by *Küster*, who had reasons to prefer the lactim, based upon chemical reactivity, such as the formation of lactim acetates and methyl ethers. In 1932, lactim-lactam tautomerism in bilirubin was invoked to distinguish the differential reactivity of the pigment in the *van den Bergh* diazo reaction (*484*). (See Sections 4.1 and 10.) So even in 1953, it should not be surprising that in his little book titled *The Bile Pigments* (*485*), *Gray*[2] adopted the *Fischer* lactim and hydroxypyrrole tautomers of bilirubin, other bile pigments, and dipyrroles. Yet, within five years, in 1958, he proposed a nomenclature in which the bis-lactam formula was designated (*486*):

> The 'hydroxypyrroles' in general, and particularly those used in the syntheses of the bile pigments, are in fact pyrrolenones, . . . and it is therefore probable that the bile pigments are also lactam compounds. The present systems of nomenclature . . . make no provision for this fact and do not allow easy derivation of the structures of the bile pigments. In this communication we have named the bile pigments as derivatives of the fully unsaturated structure **I**, to which we give the trivial name 'bilenone'.

(I)

Why did *Gray* undergo a conversion?

Perhaps he was stimulated by *Lemberg* and *Legge*'s earlier assertion in 1949 that the lactam tautomer of bilirubin better fit its weakly basic character (*433*). More likely, he was attentive to *Plieninger* and *Decker*'s publication in 1956 (*487*) and possibly to *Grob*'s in 1949–1955 (*488*, *489*). *Grob* and *Ankli*'s studies in Basel on small molecules cast serious doubt that the α-hydroxypyrrole tautomer would be more stable than the lactam (2-pyrrolenone). And no less an authority on bilirubin than one of the co-authors of its total synthesis, *Hans Plieninger* at the Ruprecht Karl Universität Heidelberg showed, using infrared spectroscopy that the 2-pyrrolenone lactam was more stable than the hydroxypyrrole tautomer. He wrote that the dipyrrinones and bilirubinoids adopt the lactam tautomer preferentially (*487*):

> . . . womit wahrscheinlich wird, daß auch die „Oxydipyrryl-methene" und Bilirubinoide in der Lactamform vorliegen. 5

---

[2] *Charles H. Gray* was born on June 30, 1911 in Erith, Kent, U.K., and died on August 15, 1997 in Leatherhead, Surrey, U.K. He was a distinguished biochemist in the bile pigment and porphyrin field, was trained in chemistry at Imperial College London, moved on to physiology at University College, London, became a demonstrator in biochemistry, then a lecturer in physiology there, worked on bile pigments, and graduated in medicine in 1937. He was invited to a position as Biochemist and Consultant at King's College Hospital District from 1938 to 1976 and Consulting Chemical Pathologist from 1976 to 1981. He held a personal chair as Professor of Chemical Pathology from 1948 to 1976 at London University, which he held at King's College Hospital.

In addition, *Plieninger* showed that attempts to acetylate and trap the hydroxypyrrole tautomer using acetic anhydride resulted in N-acylation rather than O-acetylation – and even the N-acetyl derivative retained the lactam structure because it exhibited no evidence by infrared spectroscopy of an O-H group that would have been present in an *N*-acetyl-2-hydroxypyrrole. Such findings might explain the error of *Fischer*'s claim of some 30–40 years earlier that because xanthobilirubinic acid underwent acetylation, such behavior signified a lactim structure.

In February 1957, *Gray* and *Nicholson* drew the structure of *i*-urobilin as a bis-hydroxypyrrole, with the caveat that (*490*):

> The structure of *i*-urobilin is as shown in (**I**), except that the end rings, in common with all so-called hydroxypyrroles, are probably lactam in nature. . . .

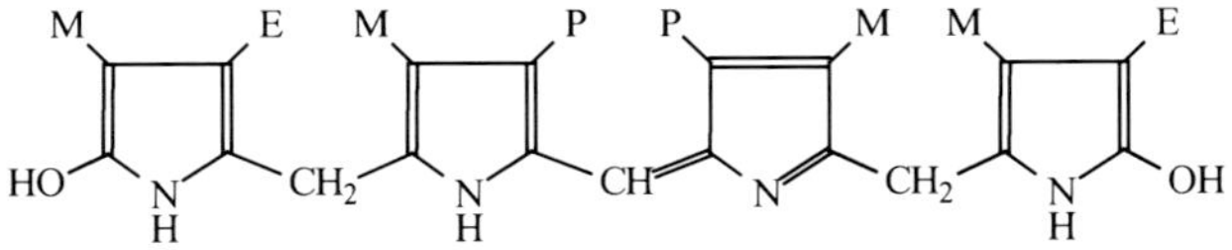

(**I**)

And by August 1957, the structures of the urobilins were bis-lactams according to *Gray* (*491*). Two years later, in 1959, and probably earlier, *Cecil J. Watson* at the University of Minnesota Medical School had also adopted lactam tautomers in his bile pigment structures (*492*). Recall from Section 6 that *Watson* was a postdoctoral fellow in *Hans Fischer*'s lab during 1931–1933 – and thus he was well acquainted with the lactim formula preference in the *Fischer* group.

From the early 1960s, even without direct proof for the lactam tautomer in bile pigments, *Watson* (*492*) as well as *Gray et al.* consistently began to express it (*493, 494*). Others, too, began to profess the bilirubin lactam tautomer; for example, from Oslo in 1963, *Fog* and *Jellum* (*495*); in 1964 *Fog* and *Bugge-Asperheim* (*496*); in 1965, *Fischer*'s student *von Dobeneck* at the TH-München (*497*), *ÓhEocha* in Ireland (*498*); and in 1967, *Brodersen, Flodgaard,* and *Krogh-Hansen* in Copenhagen (*499*) – see Section 7.3. Even from the plant linear tetrapyrrole area, *Colin ÓhEocha*, in his studies of algal pigments, depicted biliverdin as a bis-lactam in 1964 (*498*):

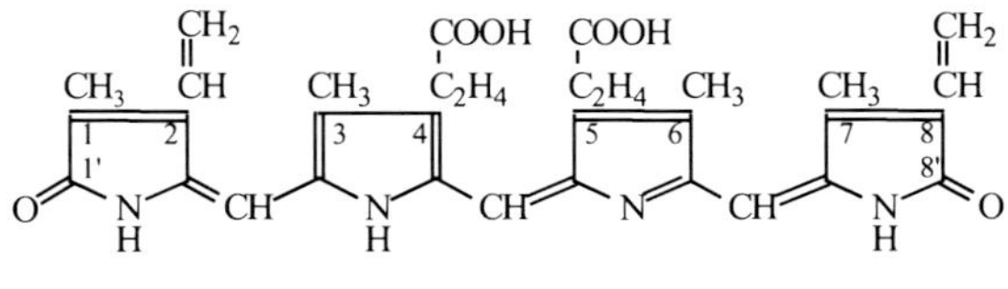

as well as phycoerythrobilin, mesobilirubin, mesobiliverdin, mesobiliviolin, mesobilirhodin, i-urobilin, and stercobilin.

7.1 Lactim or Lactam Tautomeric Structure?

Yet the adoption of the lactam formulation was not acceptable to everyone, as illustrated in the mid-1960s in lectures presented at a conference on bilirubin metabolism held in 1966 at the Royal Free Hospital, London (*500*). In the five lectures where bile pigment structures were drawn, three authors wrote lactims; two, including *Nicholson*, drew lactams. A year or so later, in 1968, in his authoritative book, *Bile Pigments*, covering the literature through 1966, *Torben K. With* presented bis-lactim skeletal structures for bilirubin and biliverdin, explaining that (*501*):

> There is not complete agreement on details concerning the structure of bile pigments. The structures presented . . . are of the classic bis-lactim form employed by Fischer, Siedel, and Lemberg. Modern investigators prefer the bis-lactam notation (*cf. ÓhEocha* [*498*]).

Biladien-(*ac*) skeleton [of Bilirubin]

Bilatriene-(*ac*) skeleton [of Biliverdin]

But apparently not all "modern investigators" (*502, 503*). *Gray* apparently believed that the lactam tautomers are formed during the biosynthesis of bilirubin, and *With* diagrammed how hydroxypyrrole and lactim components can be tautomerized to lactams (*501*):

> It is now believed that a lactim form with OH in rings I and IV is formed during biosynthesis but subsequently tautomerizes to a lactam form with O in both these positions [*494*]. . . .

Lactim-lactam tautomerism of rings I and IV (A and D) of verdins, rubins, and violins.

In late 1968, *A.W. Nichol* and *D.B. Morell* in Australia investigated the lactim-lactam tautomerism in bilirubin and biliverdin (*504*) and determined, on the basis of IR and ^{1}H NMR spectral analysis that bilirubin adopted the lactim tautomer; whereas, biliverdin and its dimethyl ester adopted the bis-lactam tautomer (*504*):

7 Evolution of the Constitutional and Stereochemical Structure of Bilirubin

. . . Infrared and NMR spectrometric studies on bilirubin and biliverdin and their derivatives suggest that whereas bilirubin occurs as the enol-imine tautomer, biliverdin and its dimethyl ester are in the lactam form. . . .

The finding of strong amide carbonyl absorption in the infrared spectrum of bilirubin dimethyl ester together with the presence of four protons in the region expected for N-H groups in the NMR spectrum of this compound confirms the generally accepted lactam structure for this compound. In contrast bilirubin free acid shows very weak amide carbonyl absorption in the infrared spectrum, showing a greater resemblance to α,α'-dimethoxybilirubin. In the NMR spectrum four protons are again found in the downfield region, however, since two of these must be attributed to the two propionic acid carboxyl protons only two N-H protons are present. These data are fully accounted for if bilirubin free acid exists almost entirely in the enol-imine form. The only ambiguity arising from the NMR spectra is the possible assignment of an enol or partial enol form to bilirubin dimethyl ester if one or both of the downfield protons were actually strongly hydrogen-bonded OH groups. Such large downfield shifts are generally only observed when the OH group is hydrogen-bonded to a carbonyl function. This would necessitate a half enol-half lactam structure for the ester. Infra-red evidence makes this unlikely. . . . Molecular models show that hydrogen bonding between the propionic acid side chains tends to hold the molecule in a conformation for hydrogen bonding to occur between the two terminal hydroxyl groups, thus stabilising the enol-imine form (**III**). On formation of the ester, however, the molecule may assume many conformations which are not suitable for hydrogen bonding and thus tautomerism to the lactam form (λ_{max} 405 mμ) occurs in non-polar solvents. On addition of methanol, hydrogen bonding of an intermolecular type may occur between methanol molecules and the terminal hydroxyl groups. With increasing methanol concentration tautomerism occurs to the enol-imine form (λ_{max} 450 mμ) . . . . the absence of isosbestic points in this conversion is probably due to the fact that bilirubin consists of two different conjugated systems isolated by the central methylene group. Expected intermediates in the conversion would thus be the two half lactam-half enol-imine forms.

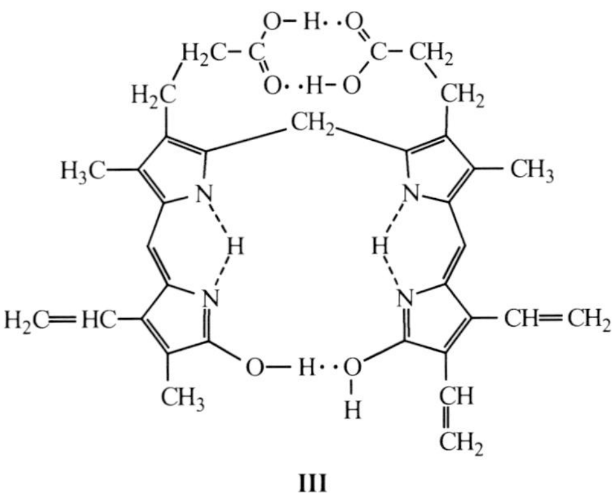

III

By January 1970, *C.C. Kuenzle* at the University of Zürich refuted *Nichol* and *Morell*'s interpretation of bilirubin's ^{1}H NMR data (*504*) and the bis-lactim formula, while concluding in favor of the bis-lactam (*505*):

The n.m.r. spectra presented . . . are consistent with the assigned lactam structures, . . . and give no evidence for tautomeric equilibria occurring in dimethyl sulphoxide. In contrast, Nichol & Morell (1969) suggest that bilirubin occurs as the lactim tautomer. Their major argument for a lactim structure stems from the fact that the n.m.r. signal of the

7.1 Lactim or Lactam Tautomeric Structure? 443

carboxyl protons at 11.89 p.p.m. does not appear in their spectrum. Consequently they misinterpret the signals of the amide protons occurring at 10.45 and 10.49 p.p.m. as arising from the carboxyl protons. The apparent lack of the amide signals thus explains the incorrect conclusion that bilirubin occurs as the lactim. However, a small proportion of bilirubin in solutions of dimethyl sulphoxide and chloroform must be present as the lactim. This is evident from the fact that bilirubin yields some monomethoxy- and dimethoxy-bilirubin dimethyl ester on reacting with diazomethane.

Kuenzle allowed for a small population of the lactim tautomer in $(CH_3)_2SO$ or $CHCl_3$. A year later, *D.W. Hutchinson, B. Johnson*, and *A.J. Knell* in the U.K. wrote only the bis-lactam structure for bilirubin (*506*). By early 1973, it had become clear that most investigators of bilirubin, other bile pigments, and plant linear tetrapyrroles (*5*) had concluded in favor of the lactam tautomer. They did this largely on the basis of IR spectroscopic evidence, as well as on certain solution properties, that the 2-pyrrolenone tautomer is generally more stable than the 2-hydroxypyrrole tautomers. Yet, direct evidence was lacking for the pigments themselves.

In the early 1970s, with the bis-lactam structure of bilirubin only slowly becoming widely adopted, some investigators still favored the bis-lactim formula for bilirubin and biliverdin (*502, 503*). However, by the mid-1970s, as illustrated in lectures presented at a conference on Phototherapy in the Newborn, in February 1973 in Washington, D.C. (*507*), where structures were drawn, the authors (*McDonagh, Lightner, Ostrow*) had represented the bis-lactam structure for bilirubin, the last author having equivocated a bit by writing hydroxypyrroles in some structures. Again, in lectures given at the September 1974 conference on Metabolism and Chemistry of Bilirubin and Related Tetrapyrroles held in Oslo, Norway (*508*), the bis-lactam structure of bilirubin was found in ten contributions, from *A.H. Jackson, R.K. Bramley, J.R. Jackson; P. Manitto, D. Monti; R. Bonnett, J.C.M. Stewart; R.. Bonnett; G.S. Ricca, P. Manitto, D. Monti; A.J. Knell, F. Hancock, D.W. Hutchinson; M.S. Stoll; D.E. Gomes, A.H. Jackson, R.G. Saxton; P. Manitto, D. Monti, R. Forino; H.P. Köst.* As well as in an 11th, from *J.D. Ostrow*, who had only two years earlier advocated the bis-lactim formulation (*502, 503*). If all of those attendees had been asked to vote (as proposed by *Ostrow* to assign the pK_a values of bilirubin at the Bilirubin Workshop in Trieste in 1995!) on whether the lactam or lactim structure was more correct, the answer would almost surely have been in favor of the former. However, the issue was still incompletely resolved even in April 1975, the time of the Fogarty International Symposium on the Chemistry and Physiology of Bile Pigments (*509*) held on the campus of the National Institutes of Health (NIH) in Washington, D.C., where a majority of the presenters used only the lactam formulas (*P. O'Carra; S.B. Brown; D.A. Lightner; B.H. Billing; R. Troxler; C.J. Watson; M.S. Stoll, C.K. Lim, C.H. Gray; M.R. Chedekel, B.V. Crist; M.J. Burke, A. Moscowitz; H.P. Köst, W. Rüdiger*), and a minority (*R. Tenhuenen; R.B. Howe, N.I Berlin, P.D. Berk; E.A. Jones, J.R. Bloomer, P.D. Beck, E.R. Carson, D. Owens, N.I. Berlin*) did not.

Experimental work that should have settled the bilirubin lactim-lactam tautomer question appeared early in the second half of the 1970s and continued until the beginning of the next decade. Thus, incontrovertible spectroscopic evidence for the bis-lactam structure of bilirubin appeared in the 22 July 1976 issue of *Nature,* in which *Bonnett, Davies,* and *Hursthouse* in the U.K. described the structure of crystalline bilirubin (Fig. 7.1.1) from X-ray crystallography *(510).* This seminal communication of far-reaching consequence was followed within two years by *Bonnett*'s[3] comprehensive full paper *(511)* that clearly showed that bilirubin in the crystal had end ring carbon-oxygen bond lengths of 1.25–1.28 Å, which were a better match to the –N-C=O lactam carbon-oxygen double bond length than that of the carbon-oxygen single bond length lactim (–N=C–O), 1.357, 1.347 Å *(512).* This study was one of six to eight related crystallographic studies published between 1976 and 1983 that should have put to rest any question as to whether the *Fischer* lactim structure was an accurate representation of the pigment in the solid state – and by extension in solution. But it apparently did not, as will be shown later.

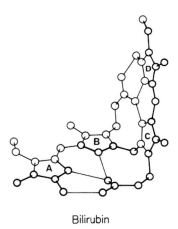

Fig. 7.1.1. *Bonnett*'s diagrammatic representation of bilirubin-IXα in the crystal *(510).* Hydrogen bonds are indicated by *fine lines* (Redrawn from Fig. 3 of ref. *(510)*)

Bilirubin

The differences between the lactam C=O and lactim C–O bond distances were reconfirmed in 1983 by *Kratky, Jorde, Falk,* and *Thirring (513)* in an X-ray crystallographic analysis of a synthetic biliverdinoid with both a lactam C=O distance = 1.216 Å and a lactim C-O distance = 1.339Å. Not surprisingly, *Becker* and *Sheldrick*'s 1978 X-ray crystallographic analysis of mesobilirubin *(514)* in

---

[3] *Raymond Bonnett* was born in London in 1931, received the chemistry B.Sc. degree in 1954 in *R.P. Linstead*'s department at Imperial College, London, and the Ph.D. in 1957 for studies on vitamin B_{12} with Lord *Alexander Todd* and *Alan Johnson* at Cambridge University. After postdoctoral studies at Harvard University from 1959 to 1960 with *R.B. Woodward,* he was Assistant Professor at the University of British Columbia from 1959 to 1961, Lecturer, Reader, and Professor of Organic Chemistry from 1961 to 1994 at Queen Mary College, University of London. In 1994 he became Scotia Research Professor at Queen Mary College and is currently Emeritus Professor.

7.1 Lactim or Lactam Tautomeric Structure?   445

Germany again provided evidence for a short carbon-oxygen bond (1.253 and 1.219 Å) that better fit the lactam C=O bond, rather than the longer lactim C–O bond. The bilirubin lactam was reconfirmed in 1977 by *LeBas et al.* *(315)*, in France, at the 4th European Crystallographic Meeting at Oxford, and again in 1980 by the same group *(516)*. In 1978 *(517)*, it was confirmed even for the bis-isopropylammonium salt (Figs. 7.1.2a and b) of bilirubin (lactam C=O distance = 1.239 Å) by *Mugnoli, Manitto,* and *Monti* in Italy *(517, 518)*. The same finding – the presence of the lactam tautomer and not the lactim in biliverdin, as characterized by the shorter lactam carbonyl carbon-oxygen bond distance (1.234 and 1.219 Å) – resulted from *Sheldrick*'s X-ray crystallographic analysis of biliverdin dimethyl ester *(519)*. Thus, according to the results of X-ray crystallographic investigations, the bis-lactam tautomers represented the stable forms of bilirubin, mesobilirubin, and biliverdin in the solid, as had been assumed since the mid-1950s.

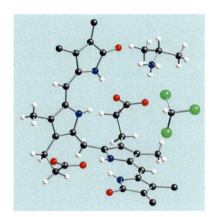

Fig. 7.1.2 (a) Crystal structure of di-isopropylammonium bilirubinate chloroform solvate displayed in its ridge-tile conformation *(517, 518)*. Replotted with permission by Prof. *A.F. McDonagh* using Crystalmaker software from atomic coordinates kindly supplied to him by Prof. *Paolo Manitto*. The =CH$_2$ termini of the vinyl groups showed disorder and were deleted for clarity. One carboxylate O to lactam and pyrrole N distances are 2.78 and 2.95 Å, respectively; the other carboxylate O to isopropyl ammonium N and chloroform C distances are 2.91 and 3.18 Å, respectively *(517)*. The atom color coding is: *black* (C), *red* (O), *blue* (N), *white* (H), *green* (Cl). (Figure courtesy of Prof. *McDonagh*)

Fig. 7.1.2 (b) The five crystallographically independent inter- and intramolecular hydrogen bonds in the di-isopropylammonium bilirubinate chloroform solvate complex (*dashed lines*), as viewed at a 5° tilt from the (vertical) twofold axis through atom [C-10]. Equivalent positions are: (i) 1-*x, y,* ½-*z*; (ii) *x,* 2-*y,* -½+*z*. For clarity, some H atoms not involved in hydrogen bonds have been omitted (Graphics and text of Fig. 2 of ref. *(518)* reproduced with permission. Copyright IUCr journals, http://journals.iucr.org/)

446   7 Evolution of the Constitutional and Stereochemical Structure of Bilirubin

Without question, at least in the crystal, the lactam-lactim tautomer uncertainty of the bilirubin structure had been fully resolved in the late 1970s – and reconfirmed by X-ray crystallographic studies that continued on linear tetrapyrroles into the 1980s. The conclusion in favor of the lactam tautomer was reinforced by X-ray crystallographic analysis of dipyrrinone model compounds for one-half of the bilirubin structure, produced in the *Lightner* (*520*) and *Falk* (*521*) labs. In such dipyrrinones, the ring carbon-oxygen bond distances were consistently in line with that of the lactam C=O: 1.216 Å (*520*), 1.248 and 1.245 Å (*521*), 1.232 Å (*522*) – not the longer lactim C–O bond distance, ~1.34 Å. The shorter lactam C=O bond was evident even in a pyrrolidone ring: 1.223–1.241 Å (*523, 524*), as compared to the lactim C–O distance, 1.326 Å (*524*).

One month prior to *Bonnett*'s communication in *Nature* (*510*) on the structure of bilirubin by X-ray crystallography, in June 1976, spectroscopic evidence from *Heinz Falk* for the prevalence of the lactam tautomer of solid bilirubin arrived at the office of the journal *Monatshefte für Chemie* (*525*). X-ray photoelectron spectroscopy (PES) was used to probe the 1s level electrons at the nitrogens and distinguish between the lactam –NH-C=O and lactim, –N=C-OH. Thus, *Falk, Gergely, Hofer*, and *Grubmayr*, then at the University of Vienna, showed that the two N_{1s} levels of bilirubin are located at 399.2 eV and are scarcely separated, by ~0.4 eV. In contrast, the N_{1s} levels of the bis-lactim methyl ether are separated by ~1.4 eV and appear at 399.2 eV and 387.8 ± 0.1 eV. A similar result was found in the PES spectra of xanthobilirubinic acid methyl ester and its lactim methyl ether (*525*):

> Zur Festlegung der Struktur von **17** [Bilirubin] und darüber hinaus von Pyrromethenonen im kristallinen Zustand bietet sich die Röntgenphotoelektronenespektrometrie an. Wie bereits berichtet, . . . sind die beiden N_{1s}-Niveaus von **17** kaum separiert (399.2 eV), ihr Abstand ist geringer als 0.4 eV. Bei **20** [Dimethoxybilirubin-dimethyl ester] werden dagegen zwei N_{1s}-Niveaus im Abstand von 1.4 eV (399.2 und 387.8 ± 0.1 eV) erhalten, ein Ergebnis, das man auch für die Partialstruktursysteme **11** [Xanthobilirubinsäuremethylester] und **12** [Methoxyxanthobilirubinsäure methyl ester] beobachtet. Dies zeigt mit großer Deutlichkeit, daß auch im festen Zustand ausschließlich die Lactamform vorliegt (auch die Röntgenstrukturanalyse eines Pyrromethenons zeigt dies . . .) und weiterhin, daß die Lactimform eine enge Verwandtschaft zu den Pyrromethenen aufweist (der Unterschied in der Bindungsenergie der beiden N_{1s}-Zuständ beträgt dort ca. 1.7 eV . . .).   6

Again, two N_{1s} levels were found at 399.2 and 387.8 ± 0.2 eV for a simpler dipyrrinone lactim methyl ether (*526*). Previously, in PES studies published in 1974, *Falk, Hofer*, and *Lehner* (*527*) had shown that the N_{1s} level electron ejection energy from a pyrrole NH was in the range 397.9–398.3 eV (± 0.2 eV), and *Falk* and *Hofer* had shown, in CNDO/2 molecular orbital calculations that the predicted difference in N_{1s} binding energies of the lactam and lactim nitrogens is 1.7 eV (*528*).

At the time, it could have been argued that bilirubin in solution might still be a bis- or even a mono-lactim – as unlikely as it might seem. Yet, evidence for the prevalence of the lactam tautomer in solution had been forthcoming at nearly the same time as the definitive X-ray studies on solid bilirubin. Thus, in 1974–1977, *Falk et al.* (*526, 527*) had been determining the pK_a of model N-methylated lactams and O-methylated lactims, then comparing the "standard" data to bilirubin and its lactim methyl ether. The work (*15, 527*) clearly indicated the prevalence of the lactam in bilirubin.

7.1 Lactim or Lactam Tautomeric Structure?

In 1975, shortly after the Oslo meeting in 1974 (*508*), *Manitto* followed up his Oslo presentation involving a ^{13}C NMR analysis of bilirubin (*529*) with a further ^{13}C NMR analysis to conclude that the pigment in CDCl$_3$ and (CD$_3$)$_2$SO solvent was mainly in the lactam tautomeric form, with <15% to as much as 30% lactim (*530*). Subsequently, the ability to distinguish the lactam carbonyl (–NH–C=O) from the lactim "enol" (–N=C–OH) carbon by ^{13}C NMR was shown to be inexact due to the large overlap of the chemical shift regions in which the two types of (^{13}C NMR) signals typically occur (*15*).

In 1984, if there was still any question as to whether the stable tautomer of bilirubin was a lactam or a lactim, it was resolved unequivocally by natural isotopic abundance ^{15}N nuclear magnetic resonance spectroscopy from *Jakobsen et al.* in Aarhus and Zürich (*531*) and *Hansen* and *Jakobsen* in Roshilde and Aarhus (*532*). Thus, two well-separated sets of signals from bilirubin in (CD$_3$)$_2$SO solvent were found in the ^{15}N NMR spectrum: one set near –250 ppm, the other near –130 ppm, both relative to an external standard (CH$_3$NO$_2$). The first corresponds to the pyrrole nitrogens, the second to the lactam nitrogens of the bilirubin-IXα structure (with a nonsymmetric ordering of the end ring methyls and vinyls). The publications containing these results were received in the respective journal offices on January 5, 1984 (*531* and *532*) and were followed only a few months later (June 12, 1984) by a far more comprehensive treatment by *Falk* and *Müller* at Johannes Kepler Universität Linz (*533*). Thus, from a dozen model compounds, in CDCl$_3$ and (CD$_3$)$_2$SO including mono-, di-, and tetrapyrroles it was clear that the ^{15}N NMR resonances fell near –230 ppm for pyrroles, the pyrrolenone lactam appears near –250 ppm, and the lactim near –140 ppm – all relative to an external aqueous solution of 30% enriched K^{15}NO$_3$, corrected to CH$_3$NO$_2$, whose ^{15}N resonance was only –0.05 ppm different from external CH$_3$NO$_2$. The vast differences between the ^{15}N NMR resonances of the nitrogens in lactams and lactims made it highly improbable that bilirubin (and biliverdin) in solution, as in the solid, would adopt anything but the lactam tautomer (*15*):

> Natural isotopic abundance nitrogen-15 NMR spectra of lactam - lactim type fragments are characterized by chemical shifts of about –250 ppm for the lactam type nitrogen and about –140 ppm for the lactim type nitrogen [*533*] .... Therefore, a shift difference of approximately one hundred ppm is provided to differentiate between the two types, as influences from structural variations are of minor importance, i.e., within several ppm only [*533*]. ...
>
> Moreover, due to a direct bond between nitrogen and hydrogen in the lactam form, their nitrogen signals are easily recognized by means of a large coupling constant (J$_{NH}$ = 93.3 Hz) whereas the pyrroleninic type nitrogen of lactim forms may be observed only by using long range polarization transfer experiments.
>
> Of course, other spectroscopic techniques could be used in principle to discriminate between the two tautomeric forms. However, in the case of carbon-13 NMR and IR spectroscopy chemical shift ranges overlap or are spaced only marginally ... and therefore do not allow unequivocal assignments.

As per *Falk* (*15*), ^{13}C NMR chemical shifts of the lactam C=O (~171–179 ppm) and lactim –N=C–O (~168–175 ppm) are less differentiating, as are the corresponding carbon-oxygen infrared stretching vibrations are 1,600–1,650 cm^{-1} and 1,640–1,690 cm^{-1}, respectively.

448   7 Evolution of the Constitutional and Stereochemical Structure of Bilirubin

All of these results, from X-ray crystallography and X-ray PES to pK_a and ^{15}N NMR resonances, with the overwhelming evidence pointing to the lactam structure of bilirubin, both in the solid and in solution, should have put to rest any doubt as to whether the *Fischer* lactim structure represented an accurate pictorial representation of the pigment. But for some, apparently, the evidence fell short or was neglected. Which lends credence to the observation that errors in the literature continue to be propagated, or adopted uncritically, well after firm scientific evidence against them has been marshalled. Nonetheless, it should be no stretch to imagine that the X-ray crystal structure of bilirubin would also prove the (Z)- or (E)-configuration of the pigment's ene-amide exocyclic carbon-carbon double bonds, thereby providing a strong hint toward the configuration of bilirubin in solution – as is revealed in Section 7.2.

7.2   Configuration at the Exocyclic Double Bonds

Fischer and *Plieninger* (*10, 462*) chose not to designate the double bond stereochemistry, although it would have been surprising (given the state of knowledge in 1942 of "*cis-trans*" isomers, as in maleic and fumaric acids, *etc.*) if they had not recognized the possibility of four diastereomers: (Z,Z), (Z,E), (E,Z), and (E,E) in the currently used designations (*534*). The (Z,E) and (E,Z) pairs differ only by the dissymmetric ordering of the methyl and vinyl groups of the end rings, and the entire set of diastereomers is represented below as lactam tautomers. (See Section 7.1 for an analysis of the last.)

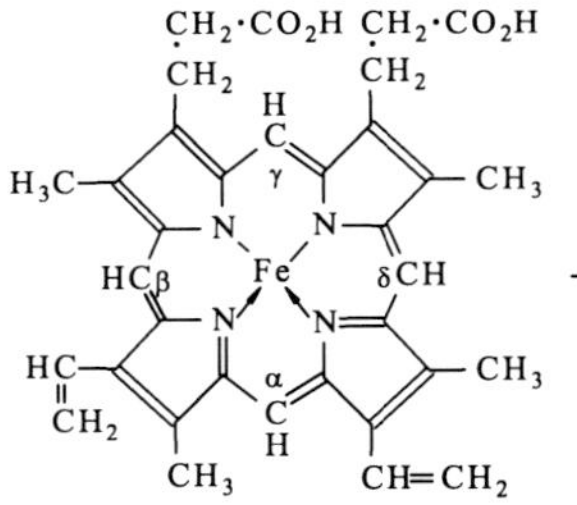

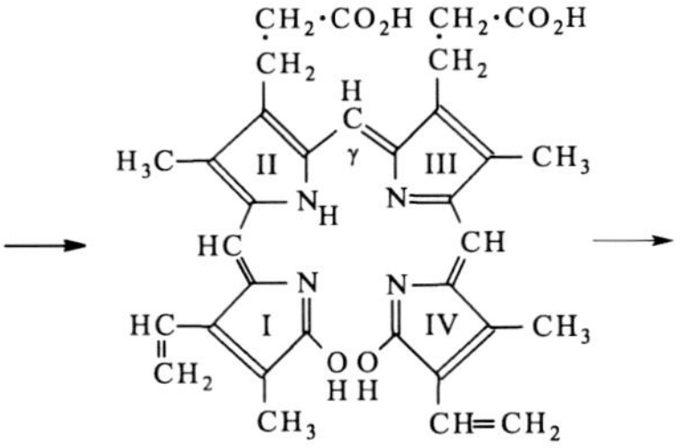

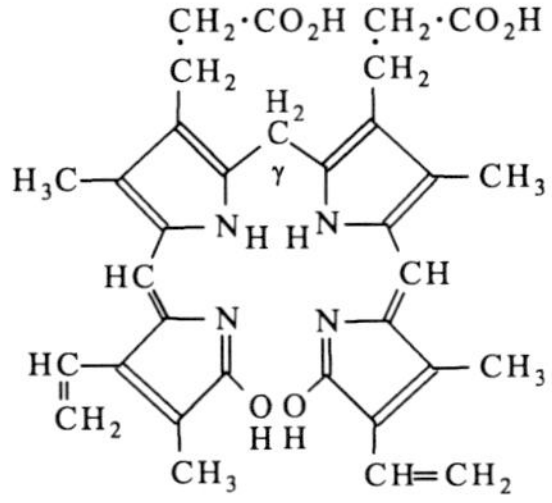

Fig. 7.2.1. The (Z/E) diastereoisomers of bilirubin in linear representation

7.2 Configuration at the Exocyclic Double Bonds

Note that for the (Z)-configuration at the C(4) to C(5) C=C, C(6), which is the higher priority group located at C(5) of the C=C, is *cis* to N, the higher priority group located on C(4) of the C=C. Similarly, at the C(15) to C(16) C=C, C(14) is *cis* to the N on C(16). In the (E)-configuration, C(6) is *trans* to N, and C(14) is *trans* to N.

Even before *Fischer* had completed the total synthesis of bilirubin, its biosynthetic origin from heme had been well established. Which prompted *Lemberg* in 1935, perceptively and logically, to draw a structure of bilirubin (**III**) in which the (Z,Z) double bond configurations were clearly shown, consistent with the bilirubin (**II**) and hemin (**I**) precursors (*460*):

> A glance at the formula now established for bilirubin (**III**) by the investigations of Fischer since 1931 shows that it must arise from the porphyrin nucleus of haemoglobin by oxidative scission. The α CH-group is removed and replaced by two hydroxyl groups, and the porphyrin ring is thus opened. If this reaction leads to bilirubin (**I → III**), the γ CH-group is reduced to CH_2.

Lemberg's investigations with *Freudenberg* led to an association with pyrrole pigments (*535*):

> My instinct told me that they were pyrrole pigments, and the evidence brought forward against this I found unconvincing. I had read all papers by H. Fischer, not an easy task because of Fischer's habit of rapid and sometimes premature publication and his failure to admit an occasional error later. Thus, without ever having had the pleasure of working under him, I became, in a sense, one of his pupils. The same inspiration which I derived from Fischer's chemical work, I derived in the field of biochemistry . . . .

450 7 Evolution of the Constitutional and Stereochemical Structure of Bilirubin

Subsequently, in his important book, *Hematin and the Bile Pigments* (*433*), *Lemberg* stressed that the methine hydrogen of the exocyclic C=C is *"trans"* to the pyrrole nitrogen, without proof, but logically because bilirubin (and biliverdin) arise in nature from heme degradation. Yet he continued to write the linear formula (**II** in the following) for the reasons cited (*433*):

1.4 Stereochemistry

Bile pigments are conventionally written as linear tetrapyrrolic chains . . . Since they are formed, however, by opening the porphyrin ring of hematin, . . . the correct formulas are cyclic rather than linear. . . . In the bile pigments containing methene groups, the double bond linking the carbon to the pyrrolic ring places the methene hydrogen in the *trans* position to the pyrrole nitrogen:

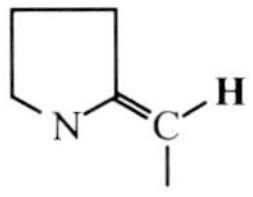

not in the *cis* position:

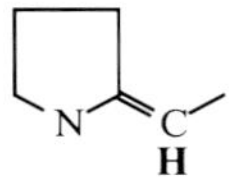

as the linear formulas suggest . . . . The true linear formula for bilirubin would thus be formula **I** rather than formula **II** in Figure 8.

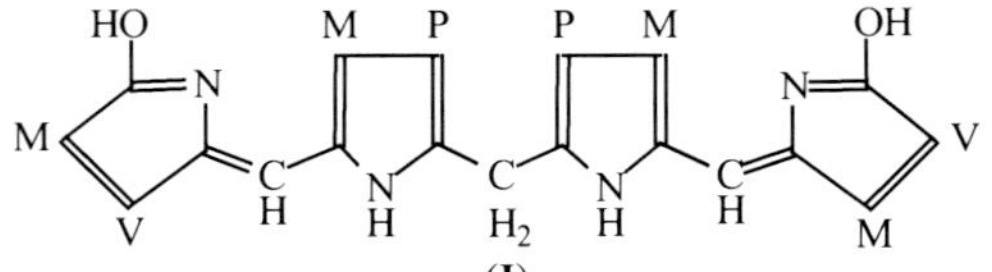

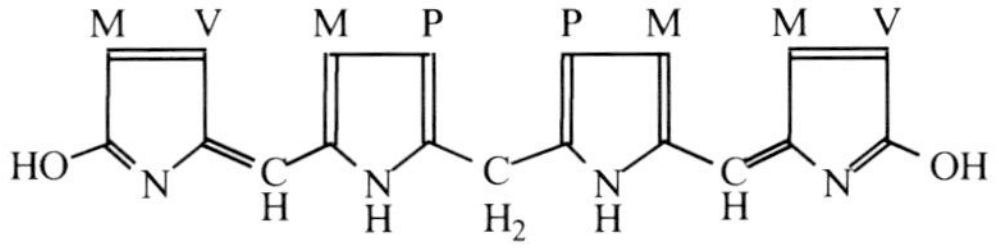

Figure 8. Linear formulas of bilirubin.

We shall nevertheless continue to use the linear formulas as a rule, since they save space and are more readily visualized. For most purposes, the difference is of no significance ...

This conclusion, now considered as correct, regarding the double bond stereochemistry of bilirubin (and biliverdin) was both logical and perceptive, a designated stereochemistry which perhaps *Fischer* could have shown but did not. However, *Lemberg*'s belief that the "difference is of no significance" could not be farther from the truth, as will be seen later. And so *Lemberg*'s linear representation of bilirubin (**II**) with *"trans"*, nowadays (*E*)-configuration double bonds, was propagated

7.2 Configuration at the Exocyclic Double Bonds 451

henceforth, as was his preference for the *Fischer* lactim tautomer. Though he recognized the possibility of a lactam tautomer, he had no basis for choosing it (*433*):

> In bilirubin two α-hydroxypyrromethene groups are linked by the methylene group and, like the α-hydroxypyrromethenes, bilirubins do not form fluorescent zinc salts. This is probably due to the fact that the hydroxylated pyrrolene ring in bilirubin and α-hydroxypyrromethenes occurs in the tautomeric form:

rather than as:

> as given in the formulas. Lactim-lactam isomerism (usually wrongly called keto-enol isomerism) has again and again been called in to explain differences between bile pigment classes or other phenomena, e.g., the "direct" or "indirect" reaction of serum bilirubin ... . Later this explanation was always found to be wrong. As in uric acid and isatin, one would expect an equilibrium between tautomerides rather than the existence of different tautomeric isomerides.

Simultaneously, although *Lemberg* had no experimental evidence for assigning the (4Z,15Z) C=C configuration of bilirubin (and biliverdin), he clearly preferred it and explained in his book with *Legge* in 1949 (*433*) that "linear" bilirubin would be shown with the (4E,15E) stereochemistry because such linear formulas "save space and are more readily visualized". After 1949, the (4E,15E) linear structure of bilirubin became a widely adopted way to represent the pigment, though the *Fischer-Plieninger* noncommittal stereochemical designation of 1942 (*10, 462*) continued to be represented, as in *Gray*'s bilirubin in 1953 (*485*), *Plieninger* and *Decker*'s dipyrrinone of 1956 (*487*), *Watson*'s bile pigments of 1960 (*492*), and *Fog* and *Jellum*'s bilirubin of 1963 (*495*). Apparently following *Lemberg* and *Legge* (*433*), however, in early 1958, *Gray et al.* (*486*) depicted biliverdin in what might be viewed as the (4E,10E,15E) isomer (see Section 7.1) – albeit as a bis-lactam. *Gray* (*493*) and *ÓhEocha* (*498*) continued to adopt the all-(E) configuration structures of bilirubin and biliverdin, as well as plant linear tetrapyrroles, into the 1960s (*536*):

Bilirubin IXα

By 1968, a time gap of nearly 20 years since *Lemberg*'s book (*433*), *Torben With* drew linear bile pigments with the (2′E,7′E) stereochemistry; thus, in the somewhat

452 7 Evolution of the Constitutional and Stereochemical Structure of Bilirubin

controversial style bilirubin and biliverdin had skeletal structures with "*trans*" (N *trans* to C) or (*E*) double bonds (*501*):

Biladien-(*ac*) skeleton [of Bilirubin]

Bilatriene-(*ac*) skeleton [of Biliverdin]

Yet it was still clear that as the non-designated *Fischer* style of representation was being abandoned, the "*trans*"or (*E*)-configuration structures were coming into use (*536*) in the late 1960s – possibly for convenience of representation, as *With* drew for bilirubin including an accidental C=C at C(10) (*501*):

Bilirubin IXα

Some would disagree. In 1970 (*505*) and 1973 (*537*), *Kuenzle* had adopted the (4Z,15Z) configuration for bilirubin; whereas, *Hutchinson et al.* indicated the (4E,15E) configuration in 1971 (*506*) and 1973 (*538*). In 1974, *Manitto et al.* (*529*) showed bilirubin as the (4Z,15Z) diastereomer; whereas, in the same conference (*508*), *Ostrow* kept to the (4E,15E) structure.

In the early 1970s, *Burke*, *Pratt*, and *Moscowitz* (*539*) predicted the (4Z)-configuration of phytochrome (the plant linear tetrapyrrole photosensory pigment) in their analysis of its photochemistry and related spectroscopic properties. But it was not until the mid-1970s that the question of the (*Z*) or (*E*) exocyclic C=C configuration of bile pigments and plant linear tetrapyrroles was on its way to becoming resolved experimentally. In 1974, working with dipyrrinone model compounds for bilirubin and biliverdin, *Heinz Falk* and his research team (*15, 540*), then at the University of Vienna, pointed the way by using nuclear *Overhauser* effect (NOE) nuclear magnetic resonance (NMR) spectroscopy. They found a 30% integral enhancement due to polarization transfer, the NOE effect (*541–543*) of the methine proton NMR signal from an Ar-saturated $CDCl_3$ solution of **I** when the neighboring ring methyl was irradiated (*540*):

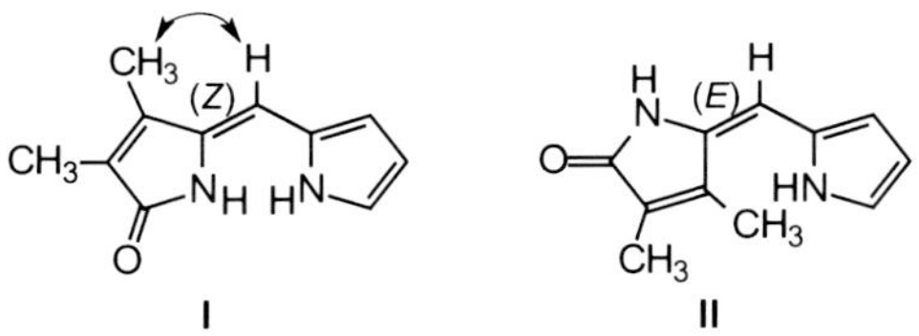

7.2 Configuration at the Exocyclic Double Bonds 453

Fig. 7.2.2. *Falk's* assignment of the (Z) or (E) stereochemistry of dipyrrinones based on their NOEs (*540*). The double-headed arrow represents a NOE

No NOE was observed between the same two proton signals of the (*E*)-diastereomer (**II**) of **I**, prepared by photoirradiation of the latter and separated chromatographically. These early NMR measurements confirmed the NOE method as a highly useful technique for sorting out the stereochemistry at the exocyclic C=C as well as at the attached C–C, and it was used extensively by *Falk (15)* and other investigators of the stereochemistry of synthetic di-, tri-, and tetrapyrroles in the last half of the 20th century, continuing to the present. In early 1981, *Kaplan* and *Navon (544)* at Tel-Aviv University applied the NOE method to natural bilirubin in CDCl$_3$ to show clearly that the pigment adopted the *syn*-(Z) stereochemistry at C(4) and C(15).

Even earlier, in the mid-1970s, the stereochemistry of bilirubin and biliverdin in the solid state had been revealed by X-ray crystallography, which brought clarity and definition to the question of (*Z/E*) stereochemistry. Thus in late 1975, *William S. Sheldrick*, then at the Gesellschaft für Biotechnologische Forschung mbH and currently at the Lehrstuhl für Analytische Chemie at the Ruhr-Universität Bochum, submitted an article to The Chemical Society on the crystal structure of natural biliverdin dimethyl ester (*519*). In it, the stereochemistry of the pigment was clearly revealed as all-(Z)-configuration, (4Z,10Z,15Z), thereby completely confirming *Lemberg's* expectations (*460*) of some 40 years earlier based on the bonding requirements of the porphyrin macrocycle. Within two years' time of the reported X-ray crystallographic structure of biliverdin dimethyl ester, *Falk et al. (15, 545, 546)* confirmed the all-(Z)-configuration for the pigment in CDCl$_3$ solvents by the NMR NOE method (curved double-headed arrows indicate ^{1}H-^{1}H NOEs) and showed that it is characteristic of numerous synthetic analogs of the pigment by the same method:

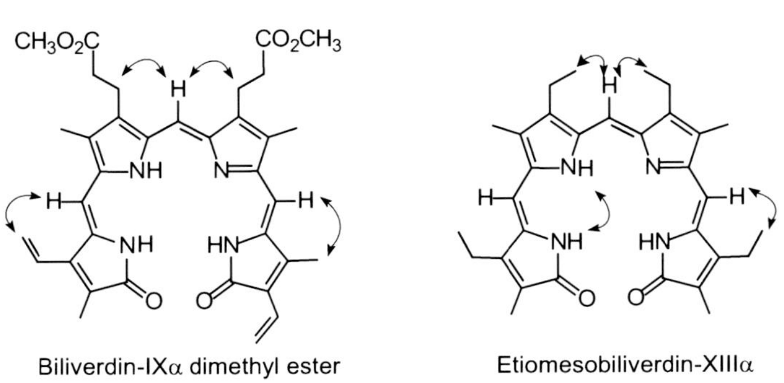

Biliverdin-IXα dimethyl ester   Etiomesobiliverdin-XIIIα

Fig. 7.2.3. *Falk's* assignment of the all-(Z) stereochemistry of verdins. The double-headed arrows represent NOEs

Definition of the exocyclic double bond stereochemistry of solid bilirubin in the solid state followed almost immediately. Thus in May 1976, a communication by *Bonnett et al.* (*510*) was received at the journal office of *Nature* describing an X-ray crystallographic study of crystalline bilirubin that confirmed the (4Z,15Z) configuration of the pigment. The following year, on September 29, 1977 (*511*), the full paper was received at the journal shortly after a presentation by *LeBas, Allegret,* and *De Rango* (the "Paris group") at the 4th European Crystallographic Meeting in Oxford, August 30-September 3, 1977 (*515*). In late 1978, the full X-ray crystallographic study of bilirubin from the Paris group had been completed and it, too, confirmed the (Z)-configuration (*516*). Preceding the latter publication were two related crystal structure determinations, one on the crystal structure of mesobilirubin-IXα (*514*) and another on di-isopropylammonium bilirubinate (*517*). Again, both confirmed the (4Z,15Z)-configuration of the pigment. Thus, early in the penultimate decade of the 20th century any controversy over the "*cistrans*" or "(Z)–(E)" stereochemistry of natural bilirubin (and biliverdin) had been completely resolved in favor of "all-(Z)". At that time the two unresolved issues related to the *Fischer-Plieninger* bilirubin structure of 1942 (*10, 462*) had been settled: the pigment adopted preferentially the bis-lactam tautomer and the *syn,syn*-(4Z,15Z) configuration.

Nonetheless, though *Lemberg* presented a logical rationale for the (4Z,15Z)-configuration of bilirubin as early as 1935 (*460*), in 1949 he chose to draw bilirubin as the (4E,15E) linear representation (*433*). To the present day his qualified adoption of the latter has been the more typical representation – and the more incorrect. It was not until after the X-ray crystal structure of bilirubin had been presented in 1976 by *Bonnett et al.* (*510, 511*) that most researchers adapted their bilirubin representations to indicate (4Z,15Z) stereochemistry. In 1982, *Lightner* wrote of the various (Z,E) and *syn,anti* conformers (*14*). Three years earlier, in 1979, *McDonagh* (*13*) represented structures with (E)-configurations – for convenience of representation and only after explaining that the (Z) was the more correct. Yet the incorrect (E)-stereochemistry has persisted in structural drawings of both natural bilirubin (*547, 548*) and biliverdin (*549*) into the 21st century, as may be found in various sources. In fact, in connection with the preparation of a short article in *Chemistry & Engineering News* in 2003 (*550, 551*), a sampling of biochemistry textbooks and handbooks from the 1990s to 2003, 25 sources in all, found the incorrect (4E,15E) structures drawn for bilirubin-IXα, and even a smattering of bis-lactim tautomers. If the stereochemistry had been put to a vote of textbook and handbook authors, as was suggested for deciding bilirubin's pK_a by one of the organizers of the 1995 Bilirubin Workshop in Trieste to its attendees, the (4E,15E) configuration would have won by a landslide.

7.3 Proposing and Detecting Hydrogen Bonds

By 1963, and perhaps earlier, intramolecular hydrogen bonds were being invoked to explain the relatively greater stability of bilirubin solutions as compared to its esters, whether the glucuronides formed in nature or methyl, ethyl, *etc.*, esters from laboratory conversions. In 1963, *Fog* and *Jellum* (*495*) suggested two types of intramolecularly hydrogen bonding in bilirubin and attempted to justify hydrogen bonding by shifts in the carboxylic acid O–H stretch, and in the amide C=O stretch of the acid relative to its ester (*495*):

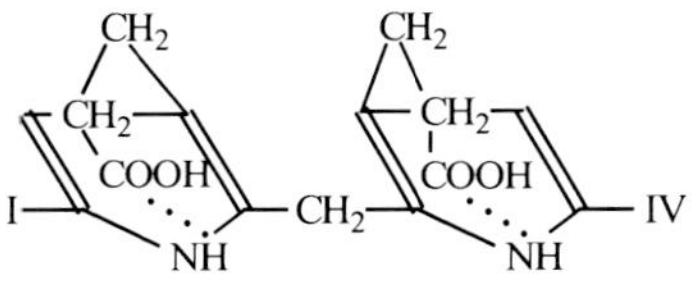

STRUCTURE 1. 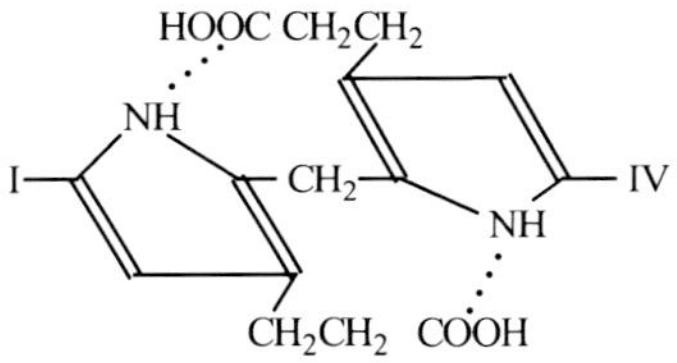   STRUCTURE 2.

In 1964, the earlier partial structure was clarified by *Fog* and *Brugge-Asperheim* (*496*):

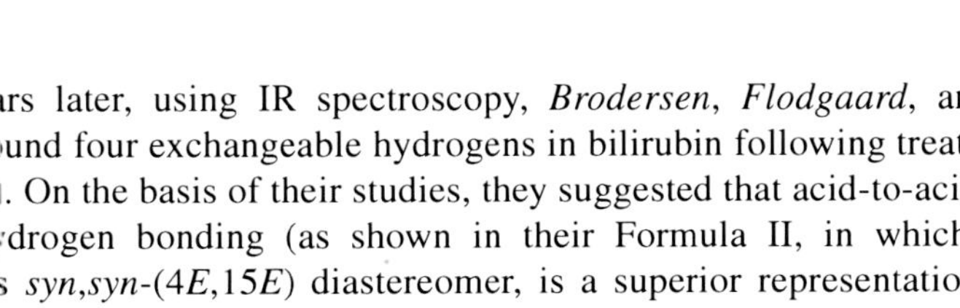

Three years later, using IR spectroscopy, *Brodersen, Flodgaard,* and *Krogh-Hansen* found four exchangeable hydrogens in bilirubin following treatment with D_2O (*499*). On the basis of their studies, they suggested that acid-to-acid intramolecular hydrogen bonding (as shown in their Formula II, in which bilirubin appears as *syn,syn*-(4E,15E) diastereomer, is a superior representation as compared to *Fog* and *Jellum*'s (*495*) or *Fog* and *Brugge-Asperheim*'s (*496*) acid-to-pyrrole hydrogen bonded structure (which *Brodersen* redrew as Formula I as the *anti*-(Z) conformer) (*499*):

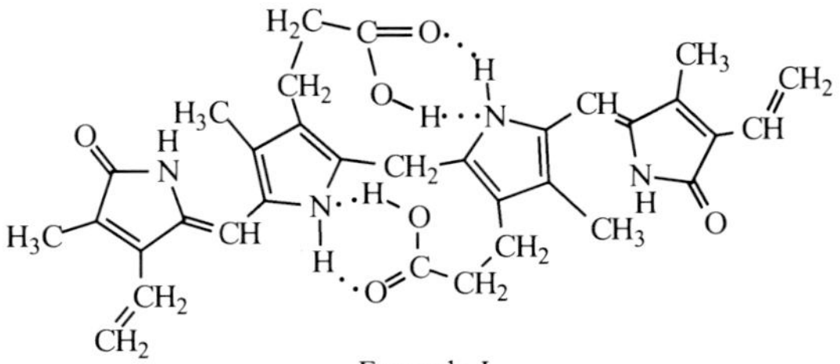

Formula I

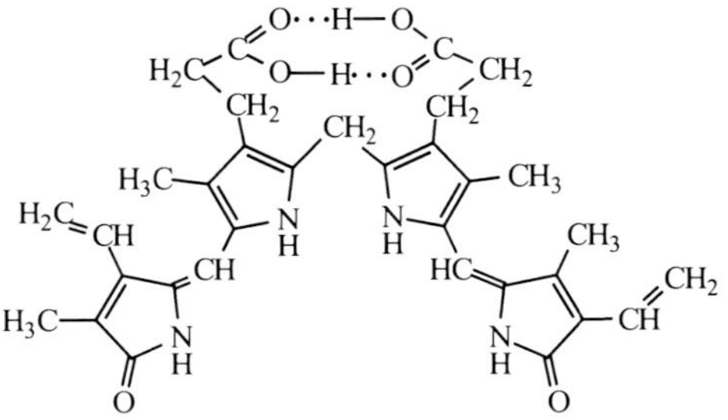

Formula II

By 1969, *Nichol* and *Morell* had considered an intramolecularly hydrogen-bonded bis-lactim formula of the pigment held in a porphyrin-like shape (*504*):

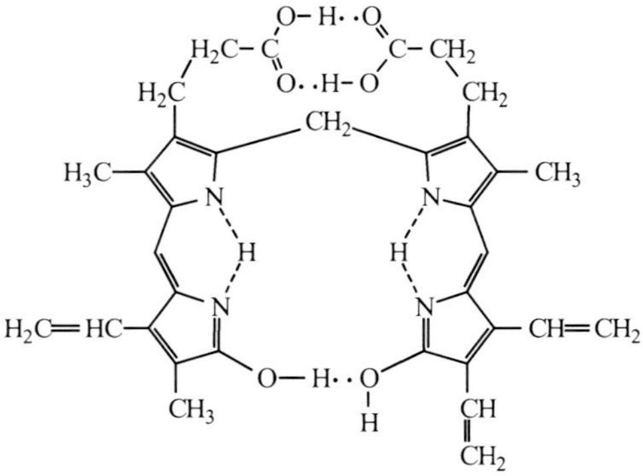

III

Yet other variations of intramolecular hydrogen bonding were expressed by *Hutchinson*, *Johnson*, and *Knell* in 1971 (*506*), who in (**I**), below, combined the (4*E*,15*E*) configuration bilirubin with elements of the *Brodersen*-redrawn Formula II above or *Nichol* and *Morell* hydrogen bonding in **III** above, and in (**II**), below, assumed some of the Formula I hydrogen bonding of *Fog* and *Jellum* (*495*):

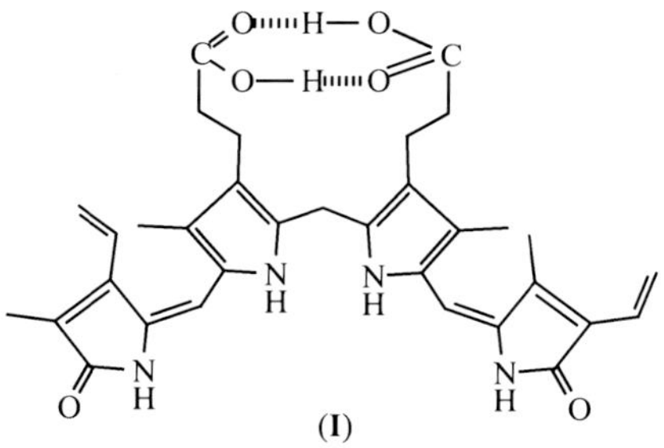

7.3 Proposing and Detecting Hydrogen Bonds

(I)

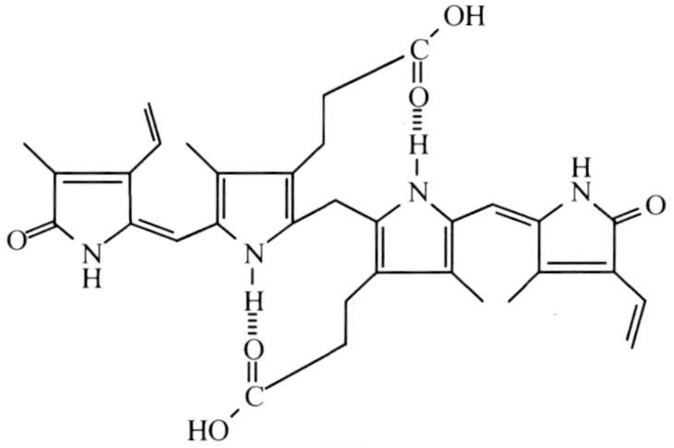

(II)

In mid-1972, a new and different hydrogen bonding pattern for bilirubin was proposed by *Kuenzle, Weibel, Pelloni*, and *Hemmerich* at the University of Zürich and the University of Konstanz (*552*):

> We have accumulated evidence for a novel conformational structure of bilirubin that is maximally stabilized by two pairs of intramolecular hydrogen bonds (C.C. Kuenzle, M. H. Weibel, R. R. Pelloni & P. Hemmerich, unpublished work). Each pair of bonds links one of the two carboxyl groups to the contralateral pyrrolenone end ring, so that the hydrogen atoms are interposed between both oxygen atoms of the carboxyl group on the one hand and the oxygen and nitrogen atoms of the pyrrolenone ring on the other. This arrangement specifies a conformation that is characterized by two molecular planes embracing an angle of approx. 110°. Each plane is occupied by one of the chromophoric moieties of bilirubin, which are separated by a central methylene bridge.

The arguments leading to the formulation above were based on six observations (*553*) and provided a structural picture to understand them: (i) Bilirubin is more stable than its diester. (ii) Reaction of bilirubin with diazomethane gave mono-methoxy- and dimethoxy-bilirubin dimethyl esters, but reaction of bilirubin dimethyl ester did not give its lactim methyl ethers. (iii) From ^{1}H NMR analysis it was argued that bilirubin dimethyl ester in $CDCl_3$ and CCl_4 (*501*) exists as two distinct molecular species. (iv) Contrary to the objections raised earlier by *Kuenzle* (*299a*), the *von Dobeneck* and *Brunner* (*497*) betaine structure deduced from IR and ^{1}H NMR analysis agreed with the structure (**I**) formulated below. (v) Bilirubin is a chiral molecule and can assume an "inherently dissymmetric" conformation (*554–556*) as detected by its circular dichroism (CD) spectrum in solutions with serum

458   7   Evolution of the Constitutional and Stereochemical Structure of Bilirubin

albumin (557–565). (vi) Bilirubin forms chelates with transition metals (552). The novel hydrogen bonding was represented in the following bilirubin structure (**I**) and partial structures (**II–IV**), with the latter corresponding to what were indicated to be mesomeric, or resonance, structures (553):

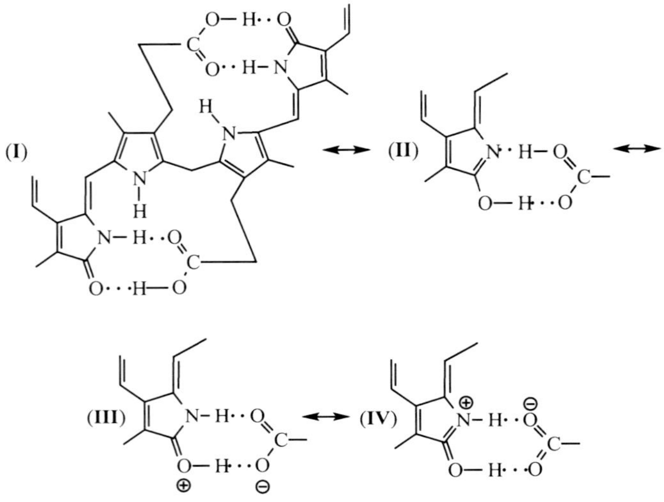

It is not quite clear that what were represented as partial resonance structures **III** and **IV** add much to the picture, but it is clear that on the basis of limited "evidence" in (i)–(vi) above, *Kuenzle et al.* had predicted a nearly correct pattern of hydrogen bonds. Item (i) is consistent with structure (**I**) but it is probably also consistent with earlier patterns of hydrogen bonding. Item (ii), also indirect evidence, might lead one to believe in the bis-lactim tautomer of bilirubin, at least for bilirubin itself – as it did *Hans Fischer*. Presumably, (**II**) would explain those results. Item (iii) provides no direct evidence for structure (**I**). Similarly, item (iv) *might* find IR and ^{1}H NMR data consistent with the picture, but it is difficult to argue that the previously proposed hydrogen bonding was not. Nor does item (vi) serve to define such a structure. It is item (v) that is the most intriguing and most relevant to the proposed structure (**I**). For in 1968 *Gideon Blauer*[4], then on leave to Oregon State University in Corvallis from the Biophysical Chemistry unit at The Hebrew University in Jerusalem, had submitted an extraordinary prediction on the structure of bilirubin (557). Together with *Tsoo E. King*, in a communication describing the ORD (Optical Rotatory Dispersion) spectra of bilirubin in solutions of bovine serum albumin, he explained the high intensity long wavelength ORD *Cotton* effect (566, 567) of albumin bound bilirubin (557) in terms of a helical, dissymmetric conformation of the pigment (558), following the then-existing

---

[4] *Gideon Blauer*, 1918–2008, protein biochemist and Professor of Biophysical Chemistry at The Hebrew University, Jerusalem.

7.3 Proposing and Detecting Hydrogen Bonds

models of hexahelicene (*566, 568–570*) and the urobilins (*571*). He was especially attracted to the latter because its dissymmetric conformation was maintained by intramolecular hydrogen bonds, as explained by *Moscowitz* (*571*). Without invoking hydrogen bonding, *Blauer* designated three possible dissymmetric conformations of bilirubin as responsible for the strong dipole-dipole coupling associated with the intense ORD magnitudes (*557*). All three conformations were illustrated with *Corey-Pauling-Koltun* space-filled molecular models, and one of them was most likely adapted by *Kuenzle et al.* as a three-dimensional stick model in Plate 1 of their 1973 publication (*533*) that showed a conformation bent at the central CH_2 group to accommodate the intramolecular hydrogen bonding indicated in (**I**) above. As *Kuenzle et al.* wrote, prophetically, in the explanation of the Fig. 7.3.1 below (*553*):

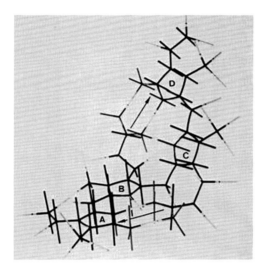

Fig. 7.3.1. Molecular model of bilirubin demonstrating one of the two enantiomeric conformations that ensue from intramolecular hydrogen bonding

Note the two molecular planes separated by the central methylene bridge. Conversion into the enantiomeric conformation requires breakage of all four hydrogen bonds, rotation about the central methylene bridge resulting in a downward orientation of the right-hand side half-molecule and re-formation of the hydrogen bonds. Hydrogen atoms are shown in light grey, carbon atoms in black, oxygen atoms and nitrogen atoms in dark grey. The bars that are not directly involved in bond formation either represent π orbitals or, in the case of oxygen, orbitals that accommodate free electron pairs. The arrows mark the two hydrogen-bond systems and are interposed between the inner and the outer hydrogen bonds in a parallel arrangement; their direction is from the carboxyl groups to the pyrrolenone rings. Rings are labelled A-D [Graphics and text reproduced with permission from Plate 1 opposite page 366 of ref. (*553*). Copyright The Biochemical Society.]

460 7 Evolution of the Constitutional and Stereochemical Structure of Bilirubin

The lactam to propionic acid intramolecularly hydrogen bonded bilirubin structure proposed in 1973 by *Kuenzle et al.* (*553*), and independently in 1973 by *Manitto, Ricca*, and *Monti* (*572*), was shortly thereafter adopted by others, including *Knell, Hancock*, and *Hutchinson* in 1974 (*573*), who in their even more fully hydrogen-bonded structures, proposed that the propionic acid carboxyl groups were tied to the *pyrrole and lactam* groups by hydrogen bonds, as illustrated by the structure of Fig. 7.3.2:

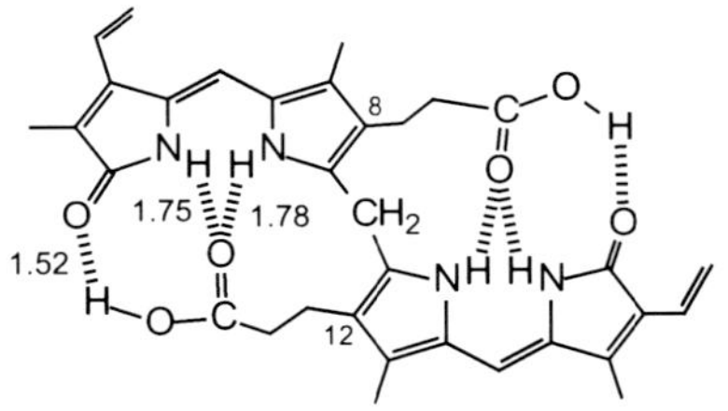

Fig. 7.3.2. Intramolecularly hydrogen-bonded bilirubin, where the six hydrogen bonds are represented by *hatched lines*. The numbers adjacent to the hydrogen bonds are in Ångstroms and correspond to the hydrogen bond distances determined from the X-ray crystal coordinates of bilirubin in ref. (*510, 511*)

With their fully hydrogen-bonded structure, the authors had anticipated by a few years the results of *Bonnett*'s 1976 X-ray crystallographic investigation of the bilirubin structure. It is perhaps even more interesting that the fully intramolecularly hydrogen-bonded bilirubin structure had been presented a few years earlier, at a conference in 1972 (*574*) – one year before *Kuenzle et al.* (*553*) and *Manitto et al.* (*572*) proposed a bilirubin structure with four hydrogen bonds (*553*).

In 1972, *Blauer et al.* (*560*) published on the circular dichroism (CD) spectra of bilirubin in aqueous serum albumin and came to the perceptive conclusion that the CD spectra were best interpreted as from the pigment behaving as a molecular exciton. To explain the spectra, *Blauer* considered three conformations, shown in print in the form of *Corey-Pauling-Koltun* molecular models: a porphyrin-like helical conformation, an extended conformation, and a "screwsense" conformation. Although one was not favored over the others by *Blauer* (*560*), he suggested that the conformations "may be stabilized by internal hydrogen bonding between carboxylate groups and pyrrole nitrogens". Not exactly the picture proposed by *Kuenzle* (*553*), but in the right direction. Yet in a discussion following his lecture in 1974 at the Conference on Bilirubin Metabolism in the Newborn (*561*) held in Jerusalem, *Blauer* claimed to have proposed in 1972 (*560*) a hydrogen-bonded conformation of bilirubin similar to that of *Kuenzle et al.* (*561*):

> Dr. Blauer noted . . . the conformation proposed by Kuenzle and co-workers (*Biochem. J. 133*:364, 1973) was proposed earlier by Dr. Blauer and co-workers for free bilirubin (*Biochim. Biophys. Acta 278*:68, 1972) except for one difference: they considered additional internal hydrogen bonds involving carboxyl groups. This may be true under certain

7.3 Proposing and Detecting Hydrogen Bonds

conditions, but at neutral pH the hydrogen bond between the carboxyl group and the lactam carbonyl may be dissociated into a carboxylate ion. The only hydrogen bond would then be between the propionic acid carbonyl and the pyrrole NH. It occurs twice, as proposed by Dr. Blauer and co-workers, and it would stabilize this extended molecular conformation to some extent.

Though the fully intramolecularly hydrogen-bonded bilirubin structure was also presented at lectures given in Arnhem, The Netherlands in September 1972 at the 7th Annual Meeting of the European Association for the Study of the Liver (573) and in September 1974 (574) at the Conference on Metabolism and Chemistry of Bilirubin (508) in Oslo, Norway, the proof of structure was based only on four observations: six exchangeable hydrogens; an IR band at 1690 cm^{-1} (attributed to the acid C=O); ^{1}H NMR signals in (CD$_3$)$_2$SO at 10.48 (2H, brs), 9.89, and 10.0 (2H doublets); and a pK_a determination. Unquestionably the fully hydrogen-bonded structure was well ahead of its time and seems to have been recognized in publications between 1974 and 1976. The work was not published, or easily available, and was thus not widely adopted or even appreciated. In 1976, when X-ray crystallographic studies began to appear in print: 1976–1978 (510, 511, 514, 515, 517) and 1978–1983 (514, 516–518), those studies clearly indicated that bilirubin in the crystal had its propionic carboxyl groups oriented toward the lactam N-H and C=O and the pyrrole N–H of the opposing dipyrrinones. The X-ray structures thus clearly showed that bilirubin adopted a syn,syn-(4Z,15Z) conformation with its lactam and carboxylic acid groups located within distances appropriate for acid-to-amide hydrogen bonding, and with the acid C=O oxygen located within hydrogen bonding contact distance of the pyrrole N–H. In fact the low temperature crystallographic study by LeBas et al. located the hydrogens (516). (The relevant hydrogen bond distances are shown in Fig. 7.3.1.)

Combined X-ray crystallographic studies had clearly shown that bilirubin and mesobilirubin adopt the intramolecularly hydrogen-bonded type of structure shown in Fig. 7.3.1. But what of bilirubin in solution? To address this question, Navon and Kaplan initiated a series of studies from 1976 to 1984 using NMR spectroscopy.

At the Second International Symposium on Bilirubin Metabolism in the Newborn (II), held April 1–5, 1974 in Jerusalem, Gil Navon (Professor of Chemistry, Tel Aviv University) lectured on the NMR of bilirubin (575) and presented new data explaining the shortcomings in Nichol and Morell's (504) NMR studies that led to their assigning an intramolecularly hydrogen-bonded lactim structure for bilirubin. In 1977, Navon commenced a detailed analysis of the structures of bilirubin and its dimethyl ester (576) in which the ^{13}C resonances were assigned, and spin-lattice carbon relaxation time (T_1) measurements seemed to suggest that (576):

> This result indicates that the propionic side chains in the ester molecule are relatively immobilized due probably to intramolecular hydrogen bonding. Thus, the hydrogen-bonded structure which has been found in crystals of bilirubin [510] seems to exist also for the dimethyl ester and even when it is dissolved in DMSO, a solvent which is known to participate in hydrogen bonding. Our conclusion agrees with the conformation of the ester proposed by Kuenzle [553], but stands in contrast to that of Manitto, Ricca and Monti [572].

462   7 Evolution of the Constitutional and Stereochemical Structure of Bilirubin

The preliminary study (576) was followed a few years later by a detailed T_1 analysis that concluded (577):

> In dimethyl sulphoxide solutions, the segmental motion of the propionic side chains is very limited in bilirubin, bilirubin dimethyl ester and mesobilirubin dimethyl ester, faster in dimethoxybilirubin dimethyl ester and is very fast in the propionic residue of vinylneoxanthobilirubinic acid methyl ester. The relative mobilities of the propionic residues are interpreted on the basis of intramolecular hydrogen bonding in which the solvent participates. The dimeric form of bilirubin dimethyl ester in chloroform solutions is confirmed by the correlation time of tumbling of the backbone. In this solvent the segmental motion of the propionic side chains is rapid, indicating the probable absence of hydrogen-bonding.

In the same year, *Kaplan* and *Navon* presented their ^{1}H NMR analyses (578) to assign the proton resonances of bilirubin, its dimethyl ester, and mesobilirubin in CDCl$_3$. They used ^{1}H{^{1}H} NOE analyses and ^{13}C spin-lattice relaxation times (T_1) to confirm (578):

> The observation of an NOE between the lactam and pyrrol NH protons . . . as well as between methyl and methine CH protons . . . indicates a *syn-Z* structure. . . . The interproton distances as obtained for the molecule of mesobilirubin show unequivocally that its conformation in chloroform solutions involves internal hydrogen-bonding of the COOH protons to the oxygen atoms at C-1 and C-19 of the pyrrolenone rings. The very slow exchange rate of the pyrrole NH protons with methanol indicates that these groups are hydrogen-bonded to the carboxy-groups. In view of the conclusions above, it is very probable that the lactam protons are also hydrogen-bonded to the carboxylic groups. . . . Thus, the present study confirms the suggestion of Kuenzle *et al.* [553] with respect to the structure of bilirubin in solution.

It thus appeared conclusive that the CO$_2$H groups of mesobilirubin in a nonpolar solvent were tethered directly to the pyrrole NH and probably lactam carbonyl and NH by hydrogen bonds – and probably in bilirubin as well.

A reinvestigation of ^{13}C T_1 measurements from bilirubin and its dimethyl ester (579) reconfirmed the earlier data and again indicated less segmental motion in their propionic chains than in the model dipyrrinone for one-half the pigment, vinylneoxanthobilirubinic acid (VBA), an indication that (579):

> Our previous ^{13}C relaxation studies have shown that the motional freedom of the propionic side-chain of bilirubin, bilirubin dimethyl ester and mesobilirubin dimethyl ester in DMSO solutions is very limited (Kaplan & Navon, 1981*b*) [578]. This is in contrast with the independent fast motion of the propionic residue of VBA methyl ester, where the structure does not allow internal hydrogen-bonding. The possibility of direct hydrogen bonding between the propionic residues and the pyrrole and lactam NH groups of bilirubin and its dimethyl ester is invalidated by the large lactam NH–CO$_2$H distance and by the faster exchange of the pyrrole NH protons. We thus conclude that the propionic CO$_2$H or CO$_2$CH$_3$ residues are tied to their nearest pyrrole NH and lactam groups . . . via bound solvent molecules.

In 1983, in their comprehensive summary article, *Kaplan* and *Navon* concluded that the data, taken collectively, support intramolecularly hydrogen-bonded bilirubin and mesobilirubin structures in CDCl$_3$ as compared to (CD$_3$)$_2$SO solvent (580):

> The detailed conformation of bilirubin and *meso* bilirubin in chloroform solutions, as well as interproton distances were found to be very similar to that found by X-ray for the

7.3 Proposing and Detecting Hydrogen Bonds

crystalline state. The conformation of the backbone of bilirubin and bilirubin dimethyl ester molecules in DMSO is similar to that of bilirubin in chloroform. For bilirubin the main difference between the two solvents is the absence of direct hydrogen bonds between the carboxylic protons and the lactam oxygen in DMSO solutions.

Some ten years later and shortly before the beginning of the 3rd millennium, *Daniel Nogales* (Ph.D. 1993) in the *Lightner* research group synthesized 99% ^{13}C-enriched carboxylic acid labeled mesobilirubin-XIIIα and conducted a ^{13}C{^{1}H} heteronuclear NOE analysis between the ^{13}CO$_2$H group and the dipyrrinone NHs in order to explore intramolecular hydrogen bonding in solution to determine whether dicarboxylic acid and its the dicarboxylate ions retain intramolecular hydrogen bonding in CDCl$_3$, (CH$_3$)$_2$SO, and H$_2$O (581). Thus, in 1995, analysis of ^{13}C{^{1}H} heteronuclear NOEs indicated non-bonded pyrrole and lactam N-hydrogen to ^{13}CO$_2$H-carbon distances of 2.72 Å for mesobilirubin-XIIIα in CDCl$_3$, and 2.75 and 2.70 Å, respectively, for bis-tetra-*n*-butylammonium mesobilirubin-XIIIα dicarboxylate ion in CDCl$_3$ (581). Analysis of the same set of NOEs from the latter in pH 7.4 aqueous buffer indicated the same, longer distance: 3.24 Å. These distances may be compared with the distances found in the energy-minimized structures from molecular dynamic calculations using SYBYL: 2.78 and 2.43 Å, respectively, in mesobilirubin-XIIIα and 2.43 Å in its dianion. The 1995 publication (581) also reported pyrrole and lactam N-H to ^{13}CO$_2$H distances of 3.06 and 3.19 Å for mesobilirubin-XIIIα in (CD$_3$)$_2$SO and 2.65 and 3.03 Å for its carboxylate dianion in (CD$_3$)$_2$SO. In 1997, more detailed analyses by Drs. *Thomas Dörner* and *Bernhard Knipp* in the *Lightner* group confirmed the results from CDCl$_3$ that showed that the data from (CD$_3$)$_2$SO required revision (582). (In both studies, mesobilirubin-XIIIα dimethyl ester showed no detectable NOEs in CDCl$_3$ and (CD$_3$)$_2$SO.)

In 1997, a more detailed study reported heteronuclear NOEs for the same ^{13}C-labeled mesobilirubin-XIIIα (582), with relevant non-bonded distances shown in Table 7.3.1:

Table 7.3.1. Carboxyl carbon to dipyrrinone N-hydrogen distances in [$8^3,12^3$-C$_2$]-mesobilirubin-XIIIα as measured by ^{13}C{^{1}H}-NOE NMR spectroscopy

Distance From Carboxyl Carbon to	Diacid			
	Distance to ^{13}CO$_2$H Carbon (Å)			
	CDCl$_3$[a]	10% (CD$_3$)$_2$SO-90% CDCl$_3$	(CD$_3$)$_2$SO[b]	Crystal[c]
Lactam N-H	2.7	2.8	>3.9	2.7
Pyrrole N-H	3.0	3.0	~4.3	2.9

[a] ± 0.1 Å. [b] ± 1 Å. [c] References [511] and [516].

The data in $CDCl_3$ solvent correlate well with the nonbonded distances found in crystalline bilirubin (*511, 516*). The nonbonded distances lengthen when $(CD_3)_2SO$ is present. Though in $(CD_3)_2SO$ the NOEs are only weak, they correlate with the conclusions of *Kaplan* and *Navon* from their NMR studies indicating that when $(CD_3)_2SO$ is present: " . . . from the slowness of the internal motion of the propionic residues of bilirubin and its dimethyl ester . . . it is concluded that these residues are tied to the skeleton via bound solvent molecules" (*579*). Strong $^1H\{^1H\}$-homonuclear NOE studies between the pyrrole and lactam N-Hs confirmed the *syn*-(Z) conformation of the pigment in $CDCl_3$ and $(CD_3)_2SO$ (*580, 582*). To summarize, the various NOE data support: (i) a tightly intramolecularly hydrogen bonded conformation of bilirubin in $CDCl_3$ that is very similar to that found in the crystal (*510, 511*); (ii) a tightly intramolecularly hydrogen bonded structure of the dicarboxylate ion in H_2O that is very similar to that of the bis-isopropylammonium dicarboxylate ion in the crystal (*517, 518*); (iii) the presence of small amounts of DMSO (10%) in $CDCl_3$ have little effect in expanding the matrix of hydrogen bonds; but (iv) increasing amounts of DMSO yield a loosely intramolecularly hydrogen bonded structure, as found by *Kaplan* and *Navon* (*579*).

The X-ray crystallographic studies, in addition to defining the lactam structure of bilirubin and its *syn*-(Z) configuration about the exocyclic carbon-carbon bonds, also defined the three-dimensional shape of the pigment in the crystal. Recent solid state NMR studies, coupled with quantum mechanical calculations are, of course, consistent with the defined structure in the crystal (*583–585*) and with that in solution (*583*). For the pigment in solution, the earlier and the recent NMR studies defined the lactam tautomer and the *syn*-(Z) conformation, and they were instrumental in elucidating the three-dimensional shape of the pigment and its dynamics in solution. The last comes into play, as revealed in Section 8 when considering the chirality of bilirubin, a molecule possessing no stereogenic centers.

8 Analysis of Bilirubin in Three Dimensions

Lemberg's prediction in 1935 (*460*) of the (Z)-configuration at the ene-amide exocyclic C=C double bonds of bilirubin and biliverdin was based insightfully on the then-known structure of heme and its role as the biogenetic precursor to bile pigments: oxidative elision of the heme α-methine forms first biliverdin, which is reduced rapidly in mammals to produce bilirubin (Fig. 8.1a). Molecular models or even simple inspection of the structures reveals that unlike biliverdin, each dipyrrinone of bilirubin can rotate freely about the C-10 methylene and adopt an orientation conducive to intramolecular hydrogen bonding, as shown in the planar projection of Fig. 8.1b. Folding the latter along the vertical dashed line converts it into the three-dimensional shape of Fig. 8.1c. This is the very structure of bilirubin in the crystal.

All of the X-ray crystallographic studies of bilirubin (*510*, *511*, *515–518*) and mesobilirubin (*514*) that were published by four laboratories during the short span of 1976–1983 agreed on but one type of three-dimensional structure of the two pigments in the solid state: one with the two dipyrrinone halves oriented about the central methylene group as if lying on the pages of a half-opened book (Fig. 8.1c) – or in a ridge-tile conformation (Fig. 8.2), as *Bonnett* first coined it (*510*), a description now in common usage.

Fundamental to the stereochemistry of the bilirubin molecule are three aspects that, taken collectively, have a dominating influence on its three-dimensional shape in solution: (i) The (Z)-configuration exocyclic C=C double bonds at C-4 and C-15 are most stable in the *syn* conformation. (ii) An sp^3 hybridized C-10 central methylene constrains the molecule to bend in the middle and allows the two dipyrrinones to rotate independently about the C-9–C-10 and C-10–C-11 single bonds. (iii) The two propionic acid groups at C-8 and C-12 of the dipyrrinones are able to engage the dipyrrinones in hydrogen bonding.

D.A. Lightner, *Bilirubin: Jekyll and Hyde Pigment of Life*, Progress in the Chemistry of Organic Natural Products, Vol. 98, DOI 10.1007/978-3-7091-1637-1_8,
© Springer-Verlag Wien 2013

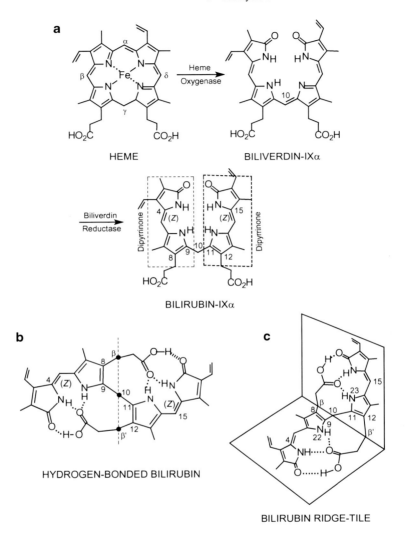

Fig. 8.1 (a) Biogenetic sequence from heme to (4Z,15Z)-bilirubin shown in a porphyrin-like shape. Rotating the bilirubin dipyrrinones about the C-9–C-10 and C-10–C-11 bonds brings them into juxtaposition with the C-8 and C-12 propionic acid groups so as to enable intramolecular hydrogen bonding, as in planar project (**b**). When the latter is folded along the *vertical dashed line*, the 3D structure of bilirubin seen in the crystal (**c**) emerges with a half-opened book or ridge-tile shape with the seam along the line from the propionic acid β and β′ carbons through C-10. The *dotted lines* in (**b**) and (**c**) represent hydrogen bonds

Even in the absence of hydrogen bonds, bilirubin tends to adopt a folded shape that minimizes intramolecular nonbonded steric repulsions and is ideal for intramolecular hydrogen bonding between the dipyrrinones and propionic acids. Engaging hydrogen bonds greatly stabilizes the bilirubin ridge-tile conformation over all others, as will be discussed in Section 8.3, which covers intramolecular motion in the bilirubin molecule.

8.1 Chirality Considerations

Fig. 8.2 (*Left*) typical roofs of different pitches in Dijon, France, with a space-filled molecular model of intramolecularly hydrogen-bonded ridge-tile bilirubin floating above the ridge of similar pitch. (Graphic from Prof. *A.F. McDonagh*.) (*Right*) Molecular model showing intramolecularly hydrogen-bonded bilirubin fitting snugly atop a yellowish ridge tile (Graphic from o. Prof. Dr. *Heinz Falk*)

8.1 Chirality Considerations

Given the constitutional structure of bilirubin and the emergence of organic stereochemistry related to molecular conformation in the second half of the 20th century, it seemed clear to the investigators of bilirubin structure that the pigment was capable in principle of adopting innumerable shapes (conformations) simply by independent rotations about the two carbon-carbon single bonds connecting the two dipyrrinones to the central methylene group. Added to these were conformations involving rotations about the carbon-carbon single bonds within the dipyrrinones themselves, connecting the pyrroles to the methines. Elements of conformational analysis were surely implicit in the early investigations of internal hydrogen bonding of bilirubin. While not widely appreciated, most of the many bilirubin conformations were in fact chiral. That is, except for conformations with coplanar dipyrrinones, every conformation has a non-superimposable mirror image (enantiomeric) conformation, *i.e.* bilirubin generally adopts chiral conformations. And, in principle, separation of bilirubin into enantiomers is possible; yet, the pigment has never been observed to exhibit optical activity in an isotropic medium, nor has the pigment been separated, in the classical sense, into enantiomers. An anisotropic medium presents a different situation, and indeed optical activity from bilirubin was reported in 1966. *Scholtan* and *Gloxhuber*, in their studies of bilirubin binding to albumin to determine a binding constant, observed a molecular rotation at 578 nm, $[\alpha]_{578}$, for pH 7.4 aqueous albumin-bilirubin (*586*). An experiment that suggested selective binding of chiral conformations of the pigment to the protein. Shortly thereafter *Gideon Blauer* began a long series of investigations of protein-bilirubin binding, the influence of such binding on the optical rotatory dispersion (ORD) and circular dichroism (CD) spectra – and thus the conformation of bilirubin.

What stimulated *Blauer* to investigate bilirubin binding and bilirubin conformation?

8.2   Bilirubin as a Chiral Molecule

Late in the seventh decade of the 20th century, *Gideon Blauer* initiated a series of experiments involving ORD and CD spectroscopy that opened the door to our understanding of bilirubin as a chiral molecule during the following decades. Although there is no evidence that *Blauer* had ever worked with those chiroptical spectroscopic tools prior to 1968, his interests as a biophysical chemist from 1951 clearly showed a familiarity with spectroscopy. In 1967–1968, while on a leave of absence to Oregon State University, with *Tsoo E. King*, a prominent protein biochemist, his interests were protein binding. In 1967, *King* had published on the protective influence of bovine serum albumin (BSA) against bilirubin-induced mitochondrial swelling, and in 1966 he had begun to make an entrance into chiroptical spectroscopy by publishing jointly with *John Schellman*[1] at the University of Oregon on the ORD of hemoproteins. During *Blauer*'s leave of absence to *King*'s lab, it probably seemed only logical that bilirubin and BSA should be studied together – and by ORD. Though *Blauer* apparently had little familiarity with ORD or bilirubin then, the investigation resulted in his initial paper in 1968 on the optical activity of (apparently aqueous buffered) solutions of bilirubin-protein complexes (*557*). In this work, he discovered large ORD *Cotton* effects in the 400–500 nm region (where bilirubin absorbs UV-visible light). Three ORD curves at differing pH were published, and the results were interpreted, after discussions with *John Schellman*, as follows (*557*):

> The unusually large Cotton effects of this complex at pH 5 are likely to be due to the formation of twisted dipyrrylmethene chromophores of fixed chirality (not unlike the urobilins, ref. 9 [*571*])[2] with possible dipole-dipole coupling between them. Helical conformations of this kind may be formed by non-covalent and specific interactions between BSA and bilirubin. In addition, various hydrogen bonds may thus be formed within the bilirubin molecule itself. It is then suggested that on change of pH the observed changes of optical rotation and light absorption reflect, at least in part, conformation changes in the bound bilirubin molecule which may result from changes with pH of the state of ionization of groups participating in the binding. These conformations exhibit different degrees of dissymmetry and dipole coupling.

The initial report was followed in greater detail (*558*) and clarity by CD spectroscopic measurements (*559*) with the demonstration that the pigment produced bisignate circular dichroism (CD) *Cotton* effects (Fig. 8.2.1) in the presence of human serum albumin (HSA) near the pigment's 450 nm long wavelength UV-visible absorption (*559*):

---

[1] *John Schellman*, born 1924, received the bachelor's degree in 1948 (Temple University), and the M.S. in 1949, and Ph.D. in 1951 (Princeton University). *Schellman*, a physical biochemist who studied protein conformational changes by chiroptical methods, *inter alia*, was Professor of Chemistry at the University of Oregon, now Emeritus, and Professor at the Institute of Molecular Biology at the University of Oregon.

[2] The original reference numbers (superscripted or otherwise) correspond to the references of this work, placed in square brackets here and elsewhere.

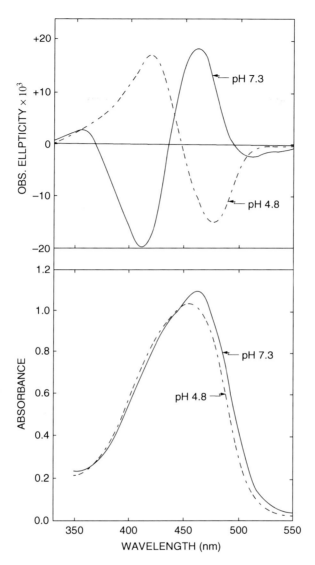

Fig. 8.2.1 The first CD spectra of the bilirubin-HSA complex published by *Gideon Blauer* and represented at two different pH values:
Bilirubin, 2.5×10^{-5} M; HSA, 5.0×10^{-5} M; temp., $28 \pm 1°$; low ionic strength.

Upper part: observed ellipticity in degrees (optical path 1.0 cm), measured after constant values were reached (about 3 hour from preparation of the complex at pH 4.8 ± 0.05, and about one hr at pH 7.3 ± 0.1).

Lower part: absorbance per cm, measured at similar time intervals for ellipticity.

The ellipticity of the protein was practically zero under the experimental conditions used and in the wavelength range given. Calculated on a bilirubin basis, the molar ellipticities of the main band maxima are in the range of $(6-8) \cdot 10^4$ deg. cm^2 per decimole. the light-absorption spectrum of bilirubin in the absence of HSA has been measured in apparently supersaturated aqueous solutions at pH 5 or 7.5. In both cases there is a maximum at about 440 nm with an extinction coefficient of 30–40 mM^{-1} cm^{-1}. . . . Water [was] the reference solvent in all cases. The light absorption of HSA is negligible over the wavelength range given (Fig. 1 of ref. (559), reprinted with permission from Elsevier Science, copyright ©1970)

The data interpretation was again apparently conducted in consultation with *John Schellman*, who knew of the ORD studies of *Albert Moscowitz*[3] *et al.* in 1964 involving urobilin and stercobilin *(571)* and had recently published a review on the rules governing the origin of optical activity, including the dipole-dipole coupling mechanism *(587)*.

Prior to *Blauer*'s work demonstrating induced optical activity from bilirubin, the only previous reports were from two then well-known examples in which the bile pigments *d*-urobilin and *l*-stercobilin exhibited *natural* optical activity, viz., with molecular rotation magnitudes of $[\alpha]_D$ ~5,000° measured at the sodium D-line (589 nm) in $CHCl_3$ *(571, 588)*. It was long known but not at all understood why *l*-stercobilin, the intense optical activity of which was first reported by *Fischer, Halback,* and *Stern* in 1935 *(589)*, or *d*-urobilin, first isolated by *Watson* in 1942, who measured its unusually large $[\alpha]_D$ values *(588)*, each exhibited intense optical activity. Although *Gray et al.* *(591)* had measured the first ORD spectra of the two pigments, the large ORD amplitudes and $[\alpha]_D$ values remained unexplained. That is, until *Cecil J. Watson* at the University of Minnesota Medical School approached *Albert Moscowitz* of the Chemistry Department at Minnesota to solve the problem. Given the data and the pigments' structures (but not the stereochemistry), *Moscowitz* understood the data in terms of the dipyrrylmethene chromophore held in a helical or inherently dissymmetric conformation by intramolecular hydrogen bonds, with the helical sense being determined by the pigments' two stereogenic centers located at the lactam –NH–CH– carbons (see Fig. 8.2.2). *Watson* was advised to add trichloroacetic acid to a $CHCl_3$ solution of the pigments, whereupon the large $[\alpha]_D$ and ORD amplitudes should drop by one or two orders of magnitude. As ridiculous as that experiment might have seemed to *Watson*, he obliged – and giving evidence to the prediction, the ORD magnitudes dropped greatly *(571, 588)*.

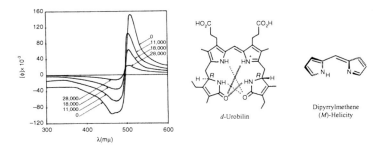

Fig. 8.2.2 ORD curves *(above)* for titration of *d*-urobilin·HCl (see *right*) in $CHCl_3$ with trichloroacetic acid. Numbers near curves give mole ratios of trichloroacetic acid to *d*-urobilin HCl (Fig. 6 of ref. *(571)*)

---

[3] *Albert Joseph Moscowitz* was born in Manchester, New Hampshire on August 20, 1929 and died on September 25, 1997 in Minneapolis, Minnesota. He received the B.A. degree in 1950 from City College of New York and the M.S. (1954) and Ph.D. (1957) in chemical physics from Harvard University. After two years as a postdoctoral fellow at Washington University, St. Louis, he joined the chemistry faculty at the University of Minnesota in 1959, where he remained for his entire career. In his time, *Moscowitz*, an excellent mathematician, was the world's leading theoretician on the subject of optical activity and made many seminal contributions to our understanding of chiroptical spectroscopy and its applications to organic and biological chemistry. The author was a National Science Foundation Postdoctoral Research Fellow with him in 1964–1965, and the two maintained a long, productive collaboration until *Moscowitz*' untimely death.

8.2 Bilirubin as a Chiral Molecule

With knowledge of *Moscowitz*'s work, *Blauer* thought in terms of two "dipyrrylmethene" chromophores in bilirubin "adopting a fixed chirality (not unlike the urobilins . . .)". But he also allowed "possible dipole-dipole coupling between them", a theme that was reiterated in the full paper, submitted in late 1968 but not published until early 1970 (*558*), then from SUNY Albany, New York, where *King* had moved. It was in that expanded treatment in which *Blauer*, who had received "general advice from Professor *J.A. Schellman*" (*558*), explained the large ORD *Cotton* effects from bilirubin on BSA as originating from inherent dissymmetry and dipole-dipole coupling (*558*):

> As in the case of hexahelicene, the extremely large Cotton effect of the bilirubin-BSA complex at pH 5 is very likely associated with a high degree of inherent dissymmetry (see Reference 40) [*571*] and dipole-dipole coupling (41) [*587*] in the bound bilirubin molecule. It can be shown by space-filling molecular models that by rotation around the C—C single bonds of bilirubin connecting the two pairs of conjugated pyrrole rings, numerous conformations of varying degrees of dissymmetry can be produced (for example, Fig. 9 [Fig. 8.2.3]). In some of these helical conformations, dipole-dipole coupling between the dipyrrylmethene chromophores of the bilirubin may be possible by their juxtaposition. A likely and relevant conformation as shown in Fig. 9 [Fig. 8.2.3] (*left*) could be formed at pH 5 by specific and noncovalent interactions of bilirubin attached to a certain site of the BSA molecule. Intramolecular hydrogen bonding involving carboxyl, carbonyl, and nitrogenous groups of pyrrole may further stabilize a given conformation (42) [*495*], especially if the "local" medium around the bound bilirubin is less polar. Moscowitz *et al.* (40) [*571*] have suggested that because conformations are stabilized by such kinds of hydrogen bonds, even free urobilins in chloroform show large Cotton effects (40, 43) [*571*, *587*]. In contrast, free bilirubin either in chloroform or in water at pH 5 is optically inactive, or its rotation is below the limits of detection.

Fig. 8.2.3 Various conformations of the bilirubin molecule constructed by *Blauer* and *King* (*558*) as *Corey-Pauling-Koltun* models. The pyrrole nitrogens are shown nonprotonated; A and B are terminal lactam rings; C and D are the two carboxyl groups. At the left is the proposed model of bilirubin bound to BSA at pH 5. In the middle is a dissymmetric conformation with fewer possible interactions between the dipyrrinone chromophores. Possible hydrogen bonding exists between the carboxyl groups C and D. At the right is the less dissymmetric porphyrin-like conformation (Fig. 9 of ref. (*558*), reproduced with permission from the American Society for Biochemistry and Molecular Biology (ASBMB), copyright © 1970).

One can see the hand of *Schellman* in such an explanation and perhaps *Blauer*'s reading of *Schellman*'s review article, submitted in February 1968 (*587*).

As examples of inherent dissymmetry, *Blauer* appealed to the natural optical activity and large *Cotton* effect of hexahelicene (*566*, *568–570*), the classical example of an inherently dissymmetric chromophore (*554–556*, *566*), and to the large ORD

Cotton effects originating from the urobilins (*556, 571*). In these compounds, however, with each possessing a single chromophore, it is the dissymmetry of the chromophore that gives rise to the large *Cotton* effects, and not dipole-dipole coupling (*554–556, 566*). Nonetheless, moving away from their appeal to a dissymmetry due to "twisted dipyrrylmethene chromophores" (*557*), *Blauer* and *King* described the origin of the optical activity in bilirubin as due to helical conformation of the pigment and coupling between the chromophores in such dissymmetric conformations.

They proposed three possible dissymmetric bilirubin conformations to account for the large *Cotton* effect at pH 5 in aqueous BSA. They favored a stretched dissymmetric conformation at the left in Fig. 8.2.3 in which the propionic acids were too far removed from the dipyrrinones for intramolecular hydrogen bonding. They disfavored a porphyrin-like conformation (Fig. 8.2.3, right) because it was less dissymmetric, and they also disfavored a third dissymmetric conformation (Fig. 8.2.3, middle) because it had "fewer interactions between the dipyrrylmethene chromophores."

Blauer and *King* allowed for the possibility that intramolecular hydrogen bonds "involving carbonyl, and the nitrogenous groups of bilirubin may further stabilize a given conformation"; such an arrangement was not possible in their favored bilirubin conformation (Fig. 8.2.3, left). Nor is there any reason to believe that in isotropic media bilirubin would exhibit a rotation either above or below the detection limit: " . . . free bilirubin in chloroform or in water . . . is optically inactive, or its rotation is below the limits of detection" (*558*).

In 1970, *Blauer* had begun to examine the optical activity of bilirubin on human serum albumin (HSA) using circular dichroism (CD) spectroscopy and found pH-dependent large bisignate *Cotton* effects (*559*). These results he interpreted in terms of exciton splitting from bilirubin in a dissymmetric conformation. The bisignate CD *Cotton* effect sign reversal in going from pH 7.3 to pH 4.8 he viewed as due to a "change in the dissymmetric mode of binding of the bilirubin by HSA (*559*). In the same year, *Woolley* and *Hunter* published their binding studies of bilirubin to HSA and BSA and reported bisignate CD spectra from the pigment but offered no insight into pigment structure (*592*).

By 1976, when it was firmly established that bilirubin existed in enantiomeric ridge-tile conformations in the crystal (*510*), *Blauer* had already published numerous new papers on the optical activity of bilirubin-serum albumin complexes in which bilirubin chiropticity was being used as a probe of protein structure conformation. In the first of them, in 1972 (*560*), the studies were conducted with added clarity by using CD spectroscopy, where the easy-to-observe bisignate nature of the CD curves doubtless influenced *Blauer*'s thinking in terms of exciton splitting. Again thanking *John Schellman*, *Blauer* interpreted the CD spectra (*560*):

> The main Cotton effects observed in the visible region are interpreted as coupling between electric transition dipole moments of bound bilirubin leading to exciton splitting. This explanation is substantiated by computer analysis into Gaussian curves of observed light-absorption and CD spectra. From these data, and on the basis of some assumptions, both distances and relative positions in space of the transition dipoles are estimated for various cases.

8.2 Bilirubin as a Chiral Molecule

And then proceeded to calculate the CD spectra with the result that (*560*):

> The largest possible distance between the chromophores of bilirubin would be achieved in the extended conformation according to the model given in Fig. 4B [Fig. 8.2.4], obtained by rotation of the dipyrromethene chromophores around the center methylene group in either direction. These conformations may be stabilized by internal hydrogen bonding between carboxylate groups and pyrrole nitrogens (see also ref. 32) [*499*]. It is therefore likely that free alkaline bilirubin is present in this type of conformation.

The model space-filled *Corey-Pauling-Koltun* (CPK) structure referred to as Fig. 4B is reproduced below in Fig. 8.2.4 (*560*):

Fig. 8.2.4 CPK model of *Blauer*'s "extended conformation" of bilirubin from ref. (*560*). In this model, the propionic acid carboxyl groups are poised to make intramolecular hydrogen bonds (though not shown) with the dipyrrinones (Fig. 4(B) of ref. (*560*) reproduced with permission from Elsevier Science, copyright © 1972)

Blauer concluded (*560*):

> It appears that the main factor affecting the magnitude of the rotatory strength of bound bilirubin is the angle φ between the transition dipoles as defined above either of the two bilirubin halves or of two molecules. This angle can be varied in a single bilirubin molecule by rotation around the single carbon-carbon bonds at the center methylene group[2] [*558*]. Although the presently proposed interpretation is arbitrary to some degree it is nevertheless able to account for a large part of both splitting and rotation effects observed under various conditions.
> It may be noted that contributions to the optical activity of bound bilirubin, apart from possible interactions with the asymmetric protein environment, could also result from (internal) hydrogen bonding causing the chromophores to assume skewed conformations, as has been suggested for urobilin in chloroform[33] [*571*].

Apparently the exciton model might not have been sufficient, and the explanation of the urobilin optical activity (*571*) weighed in, with internal hydrogen bonding causing the chromophores to assume skewed conformations.

474   8 Analysis of Bilirubin in Three Dimensions

By 1974, in the discussion following *Blauer*'s lecture (*593*) at a Protein-Ligand Interactions Symposium in Konstanz, September 2–6, 1974, Dr. *P. Hemmerich*, then at the Department of Biology, Universität Konstanz, offered (*593*):

> Stimulated by Dr. Blauer's finding of the surprisingly high CD of the bilirubin-BSA complex, we have tried to simulate this behavior in a "chemical" (i.e. low molecular weight) system in order to decide, whether the induced chirality of bilirubin was protein-specific or not. Indeed, we found (Ph.D.-thesis O. Meyer) that the protein effect can be simulated quantitatively in petrol ether solutions of bilirubin containing α-phenethylamine as low-molecular weight asymmetric inducer. C.C. Kuenzle, M.H. Weibel, R.R. Pelloni and P. Hemmerich (Biochem. J. (1973) 133, 364) have verified by NMR that, under aprotic unpolar conditions, bilirubin assumes a skewed conformation, which is stabilized by strong hydrogen bridges between each propionate side chain and the outer cyclic amide group of the opposite pyrromethene half. An asymmetric inducer will select one enantiomeric conformer out of this racemic mixture.

To which *Blauer* replied (*593*):

> According to our calculations and models, such internally hydrogen-bonded structures which had been proposed by us earlier for unbound bilirubin (see above, ref. 4) [*560*], should not exhibit the large optical activity observed in the visible region, since the splitting would be too small in such conformations. However, the bound bilirubin could assume a variety of different conformations on the protein, which would most likely be stabilized mainly by hydrophobic interactions between the protein and the bile pigment. With suitable dihedral angles between the two dipyrromethene chromophores, the observed splitting and large optical activity could be accounted for.

This reply related the absence of (large) optical activity in unbound intramolecularly hydrogen-bonded bilirubin to a small (intramolecular exciton) splitting, and the presence of intense optical activity of the bound pigment to a suitable dihedral angle between the two dipyrrinone chromophores. The first part appears to suggest that bilirubin in an isotropic environment might exhibit intense optical activity if the dihedral angle between the chromophores were suitable large. Actually, the dihedral angle (Fig. 8.3.1) of importance is that between the relevant electric transition dipole moments that each roughly lie along the long axis of the dipyrrinones. And as that dihedral angle approaches $0°$ or $180°$ (wherein the two chromophores lie coplanar and the bilirubin conformation is thus achiral), the CD magnitude approaches zero. The magnitude of the exciton splitting depends on that dihedral angle, the intensity of the electric dipole transition moment, *i.e.* the dipole strength, or intensity of the long wavelength UV-visible absorption (near 430 nm), and the distance between the electric moments. Thus, while one factor is always invariant, *i.e.* the dipole strength of the UV-visible transition moments, the distance between the transition moment vectors will vary in bilirubin according to the dihedral angle (see Section 8.5.2 and Fig. 8.5.8), as will the exciton splitting (energy). However, the exciton splitting energy does not go to zero even if the dipyrrinones of bilirubin were to lie in the same plane; whereas, the rotatory strength will, and thus optical activity or CD would go to zero when the dipyrrinones are coplanar and the conformation is achiral. The organic chemist might find the treatment of exciton chirality and CD (*566*) useful in the book by *Harada* and *Nakanishi* (*594*). A more advanced treatment may be found in the work of *Aage Hansen* (*595*, *596*).

Although *Blauer*'s interest in bilirubin appears to be more related to using the pigment as a probe of protein binding and binding site in the years between 1968 and 1979, as revealed in his summary article in 1983 (*597*), in 1975, together with *Georges Wagnière*, a theoretician at the University of Zürich, he employed molecular orbital calculations to predict that when bilirubin is bound to human serum albumin at neutral or higher pH, the dipyrrinones would be rotated with respect to one another into a right-handed conformation (*598*).

8.3 Bilirubin Conformational Dynamics by NMR Spectroscopy

Subsequent to *Blauer*'s studies of bilirubin chirality, by 1978 it had become clear that the pigment adopted an intramolecularly hydrogen-bonded ridge-tile conformation in the crystal (*510*). From such, it was easy to see that such a conformation had near C_2 symmetry and would therefore exist as either of two non-superimposable mirror image (enantiomeric) structures. Indeed, both enantiomers are present in the crystal, as has been shown repeatedly for bilirubin (*510, 511, 515, 516*) and its analog with ethyl groups replacing the two vinyls, mesobilirubin (*514*), and even for the bis-isopropylammonium salt of bilirubin (*517, 518*). The available evidence suggested that very similar conformations persist in nonpolar solvents, such as CHCl$_3$ (*578–585*), which means that one should assume that bilirubin is present in solution as a (racemic) mixture of interconverting, intramolecularly hydrogen-bonded enantiomers, as represented in Fig. 8.3.1:

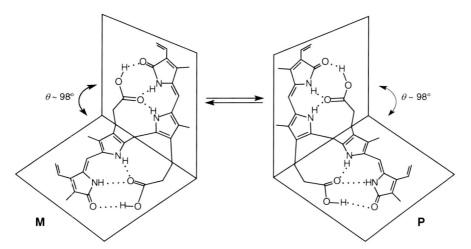

Fig. 8.3.1 Interconverting, intramolecularly hydrogen-bonded bilirubin 3-D conformational structures shaped like ridge-tiles. Their molecular chirality is indicated by **M** and **P** isoenergetic, non-superimposable mirror images (enantiomers); dotted lines represent hydrogen bonds. The dipyrrinones lie in nearly orthogonal intersecting planes, where θ is the dihedral angle of intersection

In order to gain insight into the dynamics of the interconversion, *Manitto* and *Monti* in Milan very briefly described the results from the temperature dependence (*599*) of the proton nuclear magnetic resonance (¹H NMR) spectrum of the thiolacetic acid adduct to the *exo*-vinyl group of bilirubin (*600*) that they had synthesized three years earlier (Fig. 8.3.2).

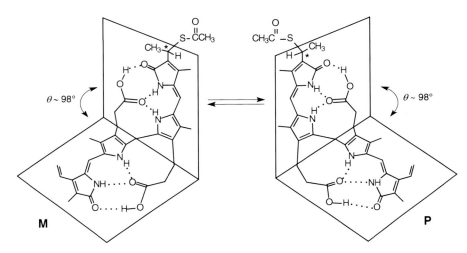

Fig. 8.3.2 Interconverting diastereomeric, intramolecularly hydrogen-bonded bilirubin adducts with HSCOCH₃ at the *exo*-vinyl group, as shown in **M** and **P** ridge-tile conformations. *Dotted lines* represent hydrogen bonds; the asterisk (*) indicates a stereogenic center

Astutely recognizing that the adduct produced a stereogenic center (*) at the former *exo*-vinyl group, and knowing that a second element of chirality (the near C_2 symmetry conformation of the pigment) was also present, *Manitto* and *Monti* then observed two diastereotopic methyl signals in CDCl₃ by ¹H NMR spectroscopy. And by raising the temperature of the solution, coalescence of the diastereotopic methyls was observed. Thus, the room temperature CH₃COS-methyl "doublet" centered at δ 2.18 ppm, Δv 3.1 Hz, merged to a broad singlet at 53 ±3°C, while the room temperature "doublet of doublets" of the –CH–CH₃ methyl centered at 81.50 ppm, Δv 3.4 Hz, J 7.0 Hz, merged to a doublet. Though the authors recognized that implicitly there should have been two different coalescence temperatures, the coalescence nonetheless pointed to two exchanging populations of conformers, and the broadened line shapes observed indicated approximately equal populations of conformers. From this NMR study, the forward rate constant (k), was approximately equal to the reverse (k_{-1}), which was calculated from the coalescence temperature (53 ±3°C) to be 7.2 ±0.4 sec⁻¹ – a fast rate of conformational inversion. The interconversion barrier was determined from the *Eyring* equation to be $\Delta G^\ddagger$ = 74.9 ±2.1 kJ·mol⁻¹ (17.9 ±0.5 kcal·mol⁻¹) based on the value of k and a transmission coefficient of unity. Such an activation barrier is approximately the value of six hydrogen bonds (*482*),

8.4 Conformational Dynamics by Molecular Mechanics Computation 477

but other energetic factors, such as conformational, must also contribute. The synthesis of the adduct and the NMR experiments are easy to reproduce, as was accomplished in the *Lightner* laboratory, with the same result. It is important that the solvent be absolutely anhydrous and free from DCl; otherwise, the coalescence temperature falls. Those studies gave a coalescence temperature closer to 60°C.

The *Manitto-Monti* conformational analysis experiment (*599*) was repeated subsequently by *Kaplan* and *Navon* in work submitted in mid-1983 (*601*). Their work focused on bilirubin itself and on mesobilirubin by examining the $-CH_2-CH_2-$ segment of the propionic acid groups. *Navon et al.* (*578, 580*) showed that the ^1H NMR signals corresponding to the fragment $-CH_2-CH_2-CO_2H$ exhibited an ABCX splitting pattern assigned to $-C_\beta H_A H_X C_\alpha H_B H_C-CO_2H$ for bilirubin in $CDCl_3$ at room temperature rather than the A_2B_2 pattern associated with rapid rotation in the fragment. Analysis of the vicinal coupling constants yielded H–C–C–H torsion angles with good qualitative agreement to those from bilirubin in the crystal at low temperature (*516*). As the solution temperature was raised to 338.7 K and to 379.0 K, the $-CH_2-CH_2-$ fragment signals changed shape and multiplicity reflecting mutual site exchanges of the non-equivalent α-hydrogens and analogous β-protons. The data showed that even at high temperatures fully intramolecularly hydrogen-bonded structures predominate, and that the observed changes seen in the NMR spectrum correlate with the interconversion of enantiomeric conformers shown in Fig. 8.3.1. Line shape analysis yielded the activation parameters $\Delta H^\ddagger = 74.1$ kJ·mol^{-1} and $\Delta S^\ddagger = -8.8 \pm 7.1$ J·deg^{-1}·mol^{-1} for bilirubin, and $\Delta H^\ddagger = 72.4 \pm 1.2$ kJ·mol^{-1} and $\Delta S^\ddagger = -12.5 \pm 3.8$ J·deg^{-1}·mol^{-1} for mesobilirubin. From the data, $\Delta G^\ddagger$ was determined to be ~75 kJ·mol^{-1}, or a value close to that determined by *Manitto* and *Monti* (*599*). The interconversion rate between the two conformational enantiomers has an associated rate constant $k = 3-95$ sec^{-1} for temperatures of 50–95°C (*601*). Mesobilirubin behaves similarly, and in $CDCl_3$ its enantiomers as well as those of bilirubin (Fig. 8.3.1) are therefore in a dynamic conformational equilibrium.

8.4 Conformational Dynamics by Molecular Mechanics Computation

Even before experimental evidence mounted for the intramolecularly hydrogen-bonded ridge-tile conformation of bilirubin in the crystal, *Falk*'s laboratory had initiated computational studies of structure and dynamics of bile pigments (*15*). First came a CNDO/2 molecular orbital calculation analysis of dipyrrylmethene by *Falk* and *Hofer* in 1974 (*602*). Shortly after the structures of bilirubin (*510, 511, 515, 517*) and mesobilirubin (*514*) were revealed by X-ray crystallography, in 1979 *Favini*, *Pitea*, and *Manitto* calculated a conformational energy map for rotation of the pyrroles about the CH_2 of the dipyrrylmethane core of bilirubin using CNDO molecular orbital methods (*603*). But no further molecular modeling studies were conducted until *Falk*'s initial study was followed by a series of conformational analyses from forcefield calculations of biliverdin, published in 1981 (*604–606*), that led in 1981 to forcefield calculations of bilirubin's conformation (*607*). As indicated in 1981 (*607*) and presented in 1983 in a broad summary of forcefield calculations from the *Falk* laboratory (*608*), the computed energy hypersurface (Fig. 8.4.1) for bilirubin pointed to enantiomeric conformations at global energy minima of 35 kJ·mol⁻¹ (8.4 kcal·mol⁻¹), suggestive of the barrier to conformational inversion (*607*):

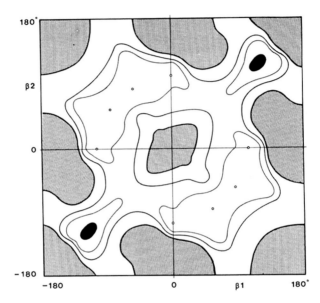

Fig. 8.4.1 *Falk* and *Müller*'s energy hypersurface for rubinoid (Z,Z)-configuration bile pigments. The distance between the contour *lines* is 5 kJ·mol⁻¹, and *shaded* regions lie >15 kJ·mol⁻¹ higher in energy than *dark border lines*. The regions designated in *black* are 35 kJ·mol⁻¹ for bilirubin and 25 kJ·mol⁻¹ for its dimethyl ester, relative to the minimum energies in the margin/border of this region (Reproduced from Fig. 2 of ref. (*607*), with permission of the publisher)

For the dimethyl ester, the minima were 25 kJ·mol⁻¹ (6.0 kcal·mol⁻¹). The global minima of both compounds correspond to conformations with torsion angles N_{22}–C_9–C_{10}–C_{11} (ϕ_1) = C_9–C_{10}–C_{11}–N_{23} (ϕ_2) = 120°. These may be compared with the torsion angles $\phi_1 = \phi_2 = 60°$ of bilirubin (see Fig. 8.1) in the crystal, and thus correspond to a flattened ridge-tile shape.

Falk's analysis of bilirubin's conformation was followed some ten years later by computations at the University of Nevada, Reno that were presented in early January 1991 (*609*) and in September 1991 (*610*), and by *Shelver, Rosenberg*, and *Shelver* on October 10, 1991 (*611*) at North Dakota State University. The former two used a molecular mechanics approach with Tripos Associates' SYBYL

8.4 Conformational Dynamics by Molecular Mechanics Computation 479

forcefield, the last employed a molecular orbital approach using AM1 or PM3 computations. Both analyses reached qualitatively similar conclusions. The approach of the *Lightner* laboratory at the University of Nevada treated bilirubin and its analogs as molecular propellers and determined the total steric energy for a matrix of conformations by molecular mechanics dynamics methods similar to those employed in the 1970s to the early 1980s by *Kurt Mislow et al.*, at Princeton University in their analysis of the diphenylmethane conformation (*612, 613*). Like diphenylmethane and dipyrrylmethane (*15, 603, 608*), bilirubin and its analogs were viewed as molecular propellers with two blades, each consisting of dipyrrinone units connected to, and capable of rotating independently about, the central CH_2 group at C-10. Rotations swept through large spatial volumes to create a potentially infinite array of different conformers possessing different total steric energies. This simplified view, related to the simpler dipyrrylmethane model (*15, 608*) is potentially more complicated by the fact that various conformations also obtain *within* the dipyrrinone blades due to twisting about the C-4–C-5 and C-15–C16 bonds. Such rotation is, however, severely constrained sterically to the *syn-periplanar* (*sp*) or *syn-clinal* (*sc*) conformations, as *Falk* has already shown (*15, 608, 614, 615*). The *sp* conformation (with torsion angles $0°$–$20°$) is found in crystalline bilirubin and mesobilirubin (*510, 511, 514–518*), where intramolecular hydrogen bonding is adopted, and also in intermolecularly hydrogen-bonded dipyrrinone dimers (*15, 608, 614–617*). Such perturbations of structure were seen to be small compared with those conformations generated by rotations about the C-9–C-10 and C-10–C-11 bonds.

In the late 1980s, when studies of the energetics of bilirubin conformational inversion were initiated at the University of Nevada, molecular mechanics methods had become well-accepted in organic chemistry and were becoming widely adopted in molecular modeling work in biochemistry for investigations of peptide and protein conformation. What was the optimum molecular mechanics computer modeling program for bilirubin? It remained to be determined which forcefields would reproduce organic chemical structures faithfully, reveal conformational energy profiles that correlated well with experimental results, and proved to have the right electrostatic potentials for accurate hydrogen bonding. After numerous investigations, the SYBYL program consistently produced the best results for a wide variety of compounds, and specifically for bilirubin. Thus, SYBYL became the standard for all of the molecular modeling studies in the *Lightner* laboratory from 1988 forward. It accurately predicted that the intramolecularity hydrogen-bonded ridgetile structure of bilirubin and mesobilirubin found in the crystal (*510, 511, 514, 515, 517*) would lie at the global energy minimum, and it allowed for an assessment of the barriers to and pathways of the interconversion of the conformational enantiomers. How well the molecular mechanics/dynamics program used at the University of Nevada predicted the structures of bilirubin found by X-ray crystallography is indicated by the comparisons of various molecular parameters shown in Table 8.4.1.

Table 8.4.1 Important structure parameters of bilirubin found by SYBYL molecular dynamics calculations at a global energy minimum conformation compared with those found by X-ray crystallography

torsion angle/° or distance/Å	global minimum from molecular dynamics SYBYL ver. 6.5	X-ray crystallography	
		Bonnett et al. [a]	*Le Bas et al.* [b]
N-22–C-9–C-10–C-11 (ϕ_1)	−59.7	−59.8	−63.3
C-9–C-10–C-11–N-23 (ϕ_2)	−60.2	−63.7	−60.6
θ (interplanar dihedral angle)	88	96,99	97
C-4–C-5–C-6–N-22	−4.7	−17.5	−7.8
N-23–C-14–C-15–C-16	−14.2	2.7	−0.9
N-21–C-4–C-5–C-6	−1.0	10.7	1.8
C-14–C-15–C-16–N-24	−0.9	−5.8	1.3
C-9–C-8–C-8^1–C-8^2	−122.7	−118.4	−113.5
C-11–C-12–C-12^1–C-12^2	−122.7		
C-8–C-8^1–C-8^2–C-8^3	67.8	68.2	67.9
C-12–C-12^1–C-12^2–C-12^3	68.1		
d C-4–C-5	1.34	1.30	1.37
d C-15–C-16	1.34	1.30	1.37
d C-5–C-6	1.48	1.48	1.41
d C-14–C-15	1.48	1.48	1.42
d O-1–O-19	11.28	11.13	11.14
d O-1 ⋯ HOOC	1.54	1.52	1.49
d O-19 ⋯ HOOC	1.54	1.52	1.49
d C-12^3=O ⋯ HN-22	1.56	1.78	1.90
d C-8^3=O ⋯ HN-23	1.56	1.78	1.90
d C-12^3=O ⋯ HN-21	1.55	1.75	1.77
d C-8^3=O ⋯ HN-24	1.55	1.75	1.77

[a] Refs. (*510, 511*); [b] Refs. (*515, 516*)

In the initial analysis of conformation, mesobilirubin-XIIIα with its symmetrically disposed β-substituents was selected at the University of Nevada as a suitable model for studying structure and dynamics. The initial results were presented first in a series of five invited lectures at the Institute of Photographic Chemistry of the Academia Sinica in Beijing in October 1990; a few months later, at a Symposium on Biomolecular Spectroscopy at the January 22–23, 1991, meeting of SPIE (Society of Photo-Optical Instrumentation Engineers) – The International Society for Optical Engineering, held in Los Angeles, California (*609*); and in a lecture at the September 3–9, 1991 International Conference on Circular Dichroism Spectroscopy held in Bochum, FRG (*610*). In these and subsequent comprehensive analyses of bilirubin and its analogs, the pigment's conformational energies were mapped for independent rotations of the two dipyrrinone units about the central methylene, *i.e.* about the C-9–C-10 and C-10–C-11 bonds (*618, 619*).

8.4 Conformational Dynamics by Molecular Mechanics Computation 481

The porphyrin-like conformation (Fig. 8.1) was chosen arbitrarily as a convenient reference point conformation, and it was "driven" into an infinite array of new conformations by progressive, independent $10°$ rotations about ϕ_1 and ϕ_2, which correspond to torsion angles N-22–C-9–C-10–C-11 and C-9–C-10–C-11–N-23, respectively. Thus, after assigning the porphyrin-like structure as $\phi_1 = \phi_2 \sim 0°$, *Richard V. Person* (Ph.D. 1993), then at the University of Nevada in Reno (618, 619) used the SYBYL molecular mechanics program, including "melting" dynamics to locate global and local energy minima and identify the highest energy conformations.

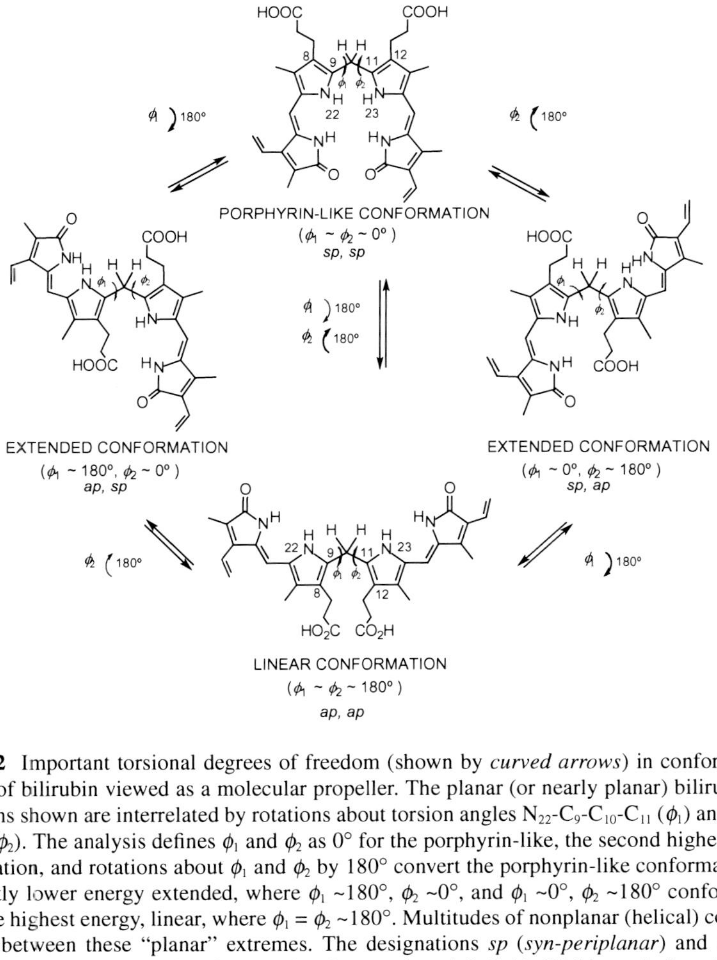

Fig. 8.4.2 Important torsional degrees of freedom (shown by *curved arrows*) in conformational analysis of bilirubin viewed as a molecular propeller. The planar (or nearly planar) bilirubin conformations shown are interrelated by rotations about torsion angles N_{22}-C_9-C_{10}-C_{11} (ϕ_1) and C_9-C_{10}-C_{11}-N_{23} (ϕ_2). The analysis defines ϕ_1 and ϕ_2 as $0°$ for the porphyrin-like, the second highest energy conformation, and rotations about ϕ_1 and ϕ_2 by $180°$ convert the porphyrin-like conformation into the slightly lower energy extended, where $\phi_1 \sim 180°$, $\phi_2 \sim 0°$, and $\phi_1 \sim 0°$, $\phi_2 \sim 180°$ conformations and to the highest energy, linear, where $\phi_1 = \phi_2 \sim 180°$. Multitudes of nonplanar (helical) conformations lie between these "planar" extremes. The designations *sp* (*syn-periplanar*) and *ap* (*anti-periplanar*) refer to the rotational stereochemistry at ϕ_1 and ϕ_2 C-10. Within each dipyrrinone the *syn*-clinal or *syn*-periplanar conformation is retained

The latter are shown in Fig. 8.4.2 as the commonly represented porphyrin-like ($\phi_1 = \phi_2 = 0°$), and extended ($\phi_1 = 0°$, $\phi_2 = 180°$; $\phi_1 = 180°$, $\phi_2 = 0°$) and linear ($\phi_1 = \phi_2 = 180°$) structures that sit ~155, ~138, and 201 kJ·mol^{-1} (~37, ~33, and 48 kcal·mol^{-1}) higher energy, respectively, than the global minimum at $\phi_1 = \phi_2 \sim 60°$.

With such large relative energy differences, even allowing for the fact that the SYBYL molecular dynamics calculations are for molecules in the gas phase, and solvation influence was taken into account only by imposing a solvent dielectric, one can reasonably assume that the high energy forms do not even remotely resemble the actual 3D structure of bilirubin or mesobilirubin in solution. They are essentially inaccessible conformations (*619*):

> [Thus], as seen [even] in CPK space-filled molecular models, the linear conformation is destabilized by a severe buttressing [repulsion] between the propionic acid groups at [C(8)] and [C(12)], the porphyrin-like conformation is destabilized by nonbonded steric interactions between the lactam carbonyl groups; and the extended conformation is destabilized by severe nonbonded steric repulsions between the propionic acid groups and the pyrrole NH groups. . . . [Yet,] between [the] limiting planar [structures] . . . of bilirubin lie a collection of more stable conformers, [which] can [also] be detected even in CPK space-filled models . . . .

Lying among the planar conformations are a very large number of nonpolar conformations, each of which has a nonsuperimposable mirror image (as in $\phi_1 = \phi_2 = -60°$ and $\phi_1 = \phi_2 = 60°$). Though not all are equal energy, all are bent in the middle to varying degrees. Some are capable of intramolecular hydrogen bonding to varying extents; others are incapable of it. When presented as a conformational energy map (energy hypersurface and contour maps), with energy corresponding to the values of ϕ_1 and ϕ_2 (and thus conformations) (Fig. 8.4.3), a series of valleys and canyons separated by peaks and ridges appears (*619*):

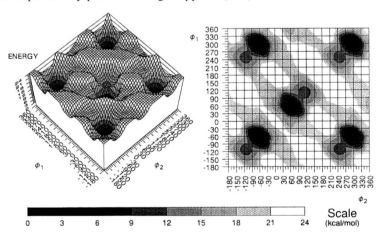

Fig. 8.4.3 Computed energy (kcal·mol^{-1}) hypersurface (*left*) and contour map (*right*) for bilirubin conformations generated by rotating each dipyrrinone independently about the C-9–C-10 and C-10–C-11 carbon-carbon single bonds, torsion angles ϕ_1 and ϕ_2, respectively, in steps of 10°. The computations reveal isoenergetic global minima (set to 0.0 kcal·mol^{-1}). The one corresponding to the (*P*)-chirality enantiomer lies near $\phi_1 = \phi_2 \sim +60°$. Those corresponding to the *M*-chirality enantiomer lie near $\phi_1 = \phi_2 \sim -60°$; $\phi_1 \sim -60°$, $\phi_2 \sim 300°$; $\phi_1 \sim 300°$, $\phi_2 \sim -60°$ and $\phi_1 = \phi_2 \sim 300°$. The molecular dynamics simulations used the SYBYL forcefield, and the energy maps were created using Wingz (Informix) (Fig. 5 of ref. (*619*) reproduced with permission of the American Chemical Society. Copyright 1994.)

The global and local energy minima lie in the valleys. In the complete conformational energy map, where ϕ_1 and ϕ_2 range from $-180°$ to $360°$, isoenergetic global energy minima corresponding to the (*M*)-chirality enantiomer (Fig. 8.3.1) were found in "potholes" at (ϕ_1, ϕ_2) ~(−60°, −60°), and (300°, 300°), as well as (ϕ_1, ϕ_2)

8.4 Conformational Dynamics by Molecular Mechanics Computation

~(−60°, 310°) and (310°, −60°) surrounding the centrally located (ϕ_1, ϕ_2) ~(60°, 60°) corresponding to the (*P*)-chirality enantiomer. Based on their shapes, the first four of these potholes correspond to identical conformations and the last represents their mirror image. These global energy minimum structures lie significantly lower in energy than the planar conformations of Fig. 8.4.2 and correspond almost exactly to the enantiomeric intramolecularly hydrogen-bonded structures found in crystals of bilirubin and mesobilirubin (see Fig. 8.3.1).

The essential special characteristics of molecular structure that are unique to the global energy minima of bilirubin and mesobilirubin are twofold: (i) a ridge-tile shape arising from rotating the dipyrrinones about the C-10 CH_2 and (ii) a full complement of six intramolecular hydrogen bonds. A local minimum located immediately adjacent to each global minimum is detected on the energy contour map of Fig. 8.4.3. The local energy minima sit some 40 kJ·mol^{-1} (9.5 kcal·mol^{-1}) higher than the adjacent global energy minima. For example, there is a local minimum located at $\phi_1 = \phi_2$ ~115° next to the global minimum at $\phi_1 = \phi_2$ ~60°.

When the electrostatic potential for hydrogen bonding is relaxed or turned off, or when the CO_2H groups are absent, the pigment's global energy minimum floor rises considerably such that, relative to the planar conformations, they lie only 42–50, 50–59, and 59-67 kJ·mol^{-1} (10–12, 12–14, and 14–16 kcal·mol^{-1}), respectively, lower than the porphyrin-like, extended, and linear conformations, respectively. The valleys and canyons are thus not as deep, and energy differences between the global and local minima are not large. A good example may be found in the conformational energy maps (*618*) of etiobilirubin-IVγ of Fig. 8.4.4. Four "(*M*)-helical" global minima at (ϕ_1, ϕ_2) ~(−115°, −115°), (245°, −115°) (245°, 245°), and (115°, 245°) surround the centrally located "(*P*)-helical" global minimum at (ϕ_1, ϕ_2) ~(115°, 115°) (*618*):

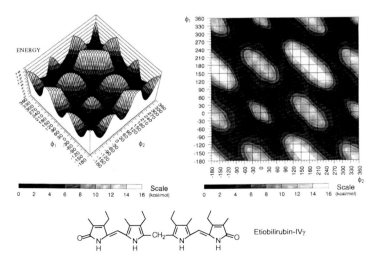

Fig. 8.4.4 The potential energy hypersurface for etiobilirubin-IVγ shows global minima located at (ϕ_1, ϕ_2). = (40°, -80°), (40°, 280°), (80°, 320°), *etc.* and local energy minima some 3.8 kJ·mol^{-1} (0.9 kcal·mol^{-1}) higher in energy at (ϕ_1, ϕ_2) = (40°, 40°), *etc.*, and other local energy minima some 7.5 kJ·mol^{-1} (1.8 kcal·mol^{-1}) higher than the global minimum at (110°, 120°), *etc.* (Reproduced from the Ph.D. dissertation (1993) of *Richard V. Person*, University of Nevada, Reno)

Four local minima sitting some 3.8 kJ·mol⁻¹ (0.9 kcal·mol⁻¹) above the global energy minima at (ϕ_1, ϕ_2) ~(40°, -80°), (80°, 40°), (−40°, 80°), and (−80°, 40°) surround the porphyrin-like conformation $[(\phi_1, \phi_2)$ ~(0°, 0°)]. And two additional local minima at 3.8 kJ·mol⁻¹ (0.9 kcal·mol⁻¹) higher than the global minimum lie at (ϕ_1, ϕ_2) ~ (40°, 40°) and (−40°, −40°). For the reasons stated above, even this pigment, which is incapable of intramolecular hydrogen bonding, is highly unlikely to adopt any of the planar conformations of Fig. 8.4.2, but the designated minimum energy conformations should be accessible. It may be noted that the global energy minimum conformations of this pigment are folded structures with shapes similar to those of the global energy minimum conformations of bilirubin. That is, the dipyrrinones adopt orientations that are predisposed favorably to intramolecularly hydrogen bonding even in the absence of such hydrogen bonding.

The conformational energy maps reveal energetically favored bilirubin structures similar to those predicted by *Falk* (*15, 608*) and *Shelver* (*611*). *Falk*'s molecular mechanics analyses (*15, 608*) predicted somewhat less deep global energy minimum wells for bilirubin, with $\phi_1 = \phi_2$ ~120° – a shape rather like that lying at the local minimum ($\phi_1 = \phi_2$ ~115°) in the bilirubin conformational analyses leading to the map of Fig. 8.4.3, or the global energy minimum where hydrogen bonding is not possible, as in etiobilirubin-IVγ (Fig. 8.4.4). The *Shelver* molecular orbital studies (*354*) also predicted a global energy minimum bilirubin conformation near $\phi_1 = \phi_2$ ~125°, with a heat of formation energy apparently some 33–42 kJ · mol⁻¹ (8–10 kcal · mol⁻¹) below that of the planar conformations – or similar to the energy computed by *Falk* and *Müller* (*607*). The differences may stem from the way hydrogen bonding electrostatics are treated in the computations (*619*).

Even in the absence of conformational stabilization from intramolecular hydrogen bonds, the minimum-energy conformation of bilirubin has the dipyrrinones rotated into a ridge-tile shape in order to minimize nonbonded steric repulsions. Such positions are suitable for hydrogen bonding to the propionic acid carboxyl groups. Rotation of the dipyrrinones into the ridge-tile molecular geometry of bilirubin and mesobilirubin minimizes nonbonded steric repulsions, and "turns on" the intramolecular hydrogen bonding that greatly lowers the global minimum energy by adding a powerful conformation-stabilizing force that is accessible to only a few tetrapyrrole conformations on the conformational energy surface of Fig. 8.4.3. Accordingly, the ability of bilirubin to adopt structures differing from the global energy minimum conformations of Fig. 8.4.3 will depend on the energy available to the system under study. Though their independent existence might be improbable, higher energy conformations on the global energy hypersurface might be accessible to bilirubin by association complexation, *e.g.* with proteins. Naturally, the higher the energy of the bilirubin conformation above the global minimum, the more improbable is its existence. Yet, the fleeting existence or passing of unstable high-energy conformations seems assured because the (*M*)- and (*P*)-chirality enantiomers are able to interconvert at room temperature. Interconversion of the enantiomeric ridge-tile structures of Fig. 8.4.5, shown in Ball and Stick representation (*620*), is accomplished by rotating the dipyrrinones about the C-9–C-10 and C-10–C-11 bonds, torsion angles ϕ_1 and ϕ_2 (*619*):

8.4 Conformational Dynamics by Molecular Mechanics Computation

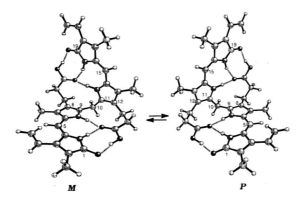

Fig. 8.4.5 Ridge-tile shape (*M*)- and (*P*)-chirality intramolecularly hydrogen-bonded interconverting enantiomers (**M** and **P**) of bilirubin that correspond to the global energy minima identified in Fig. 8.4.3. *Hatched lines* correspond to hydrogen bonds (Fig. 6 of ref. (*619*). Reproduced with permission of the American Chemical Society. Copyright 1984)

In analyzing this interconversion, a further measure of the utility of the molecular mechanics approach that incorporates good electrostatic potentials for treating hydrogen bonding is illustrated in Fig. 8.4.6, which predicts the energy barriers to interconverting the conformational enantiomers (Fig. 8.4.5) of bilirubin. Thus, *Person (618, 619)* located the two lowest energy interconversion pathways denoted as A and B in Fig. 8.4.6. The energy barrier to interconversion in Path A lies 81.6 kJ·mol^{-1} (19.5 kcal·mol^{-1}) above the global energy minima; that of Path B lies 89.5 kJ·mol^{-1} (21.4 kcal·mol^{-1}) above. Path A follows a ridge; Path B crosses a slightly higher ridge in order to access a portion of the same ridge used in the interconversion along Path A. They do not differ significantly in energy (*619*):

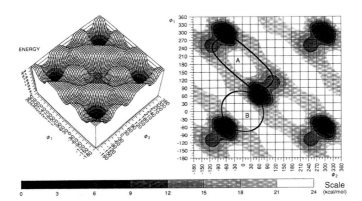

Fig. 8.4.6 Conformational energy hypersurface (*left*) and topographic map (*right*) for mesobilirubin and bilirubin conformers generated by rotating the two dipyrrinone groups independently about C-9–C-10 and C-10–C-11, ϕ_1 and ϕ_2, respectively, in steps of 10°. The two lowest energy routes (*A* and *B*) connecting the **M** and **P** conformational enantiomers (Fig. 8.4.5) are identified on the topographic map (*right*) for interconversion barriers of 81.6 and 89.5 kJ·mol^{-1} (19.5 and 21.4 kcal·mol^{-1}), respectively, between (ϕ_1, ϕ_2) = (60°, 60°) and (ϕ_1, ϕ_2) = (300°, –60°) and between (ϕ_1, ϕ_2) = (60°, 60°) and (ϕ_1, ϕ_2) = (–60°, –60°) for paths A and B, respectively (Fig. 7 of ref. (*619*) Reproduced with permission of the American Chemical Society. Copyright 1994)

The slightly higher energy path (B) is usually seen as the more direct, for interconverting bilirubins. Starting from the (P)-chirality ridge tile conformer at $\phi_1 = \phi_2$ ~60° one might view the conversion into the (M)-chirality enantiomer at $\phi_1 = \phi_2$ ~–60° in steps by climbing up a ridge and then crossing a saddle near ϕ_1 ~–60°, ϕ_2 ~+60°. As one torsion angle rotates from +60° through 0° and on to –60°, only minor changes occur in the second torsion angle. The latter is then conrotated +60° through 0° to –60° and in the middle of Path B, at a point near $\phi_1 = -60°$, $\phi_2 = +60°$ (or $\phi_1 = +60°$, $\phi_2 = -60°$) all six hydrogen bonds are broken and then reformed as the second torsion angle rotates to form in the (M)-chirality enantiomer. The intermediate conformations near the middle of the path are stabilized a bit by the formation of one or two hydrogen bonds between the two propionic acid groups. As noted above, the highest point on Path B sits on a ridge some 90 kJ · mol^{-1} (21.4 kcal · mol^{-1}) above the global energy minimum and corresponds to a conformation with hydrogen bonding between the propionic acid groups. The actual interconversion energy barrier is higher than 90 kJ · mol^{-1} (21.4 kcal · mol^{-1}) because at some point along the path all hydrogen bonds are broken, and that would give a truer energy barrier of about 118 kJ · mol^{-1} (28.1 kcal · mol^{-1}). Selected conformations along Path B are illustrated in the Ball and Stick (620) drawings of Fig. 8.4.7 (619):

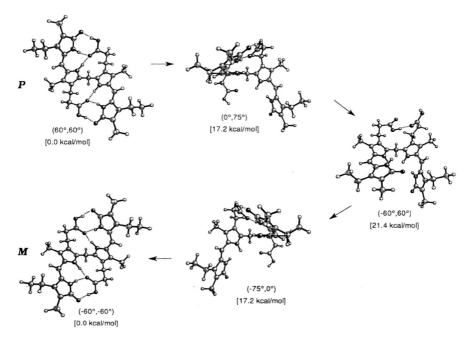

Fig. 8.4.7 Ball and Stick representations of bilirubin and mesobilirubin intermediate conformations (defined by torsion angles ϕ_1, ϕ_2) and energies (kcal · mol^{-1}) lying along interconversion Path B connecting the (P)- and (M)- chirality global energy minimum enantiomeric conformers, **M** and **P**. The transition state for the interconversion at $(\phi_1, \phi_2) = (-60°, +60°)$ lies some 90 kJ · mol^{-1} (21.4 kcal · mol^{-1}) above the global energy minima (Fig. 8 of ref. (619) Reproduced with permission of the American Chemical Society. Copyright 1994)

8.4 Conformational Dynamics by Molecular Mechanics Computation

Interconversion Path A may be less well-known. According to *Person*'s work (*618, 619*), it is the lowest energy path. Starting again from the (*P*)-chirality enantiomer at the global energy minimum $\phi_1 = \phi_2 \sim +60°$, it takes a path (A) toward the (*M*)-chirality enantiomer corresponding to the isoenergetic global energy minimum at $\phi_1 \sim +300°$, $\phi_2 \sim -60°$. In Path A, ϕ_1 and ϕ_2 open synchronously to ascend into a hanging valley and a local energy minimum near $\phi_1 = \phi_2 \sim +115°$. From this point ϕ_1 undergoes a steady positive rotation toward $+300°$, while ϕ_2 continues on a steady negative rotation toward $-60°$. With these rotations about ϕ_1 and ϕ_2, a steep ridge is ascended while the molecule crosses a saddle near $\phi_1 = +180°$, $\phi_2 = +80°$ then slowly descends near points $\phi_1 = +230°$, $\phi_2 = +20°$ and $\phi_1 = +270°$, $\phi_2 = -20°$ before dropping steeply into the enantiomeric global energy minimum at $\phi_1 = +300°$, $\phi_2 = -60°$. Representative changes in conformation along Path A may be seen in the Ball and Stick (*620*) models of Fig. 8.4.8. In Path A all three of the hydrogen bonds of one dipyrrinone are broken, while none of the hydrogen bonds to the other dipyrrinone are broken, *i.e.* all three hydrogen bonds on one half of the molecule may remain intact during the entire enantiomeric interconversion. Along Path A, as ϕ_1 becomes positive, and one set of hydrogen bonds is broken, the retained set of hydrogen bonds apparently helps to reverse the direction of the ϕ_2 rotation and "pull" the rest of the molecule around to the opposite chirality. The energy barrier for Path A is 82 kJ·mol⁻¹ (19.5 kcal·mol⁻¹) (*619*):

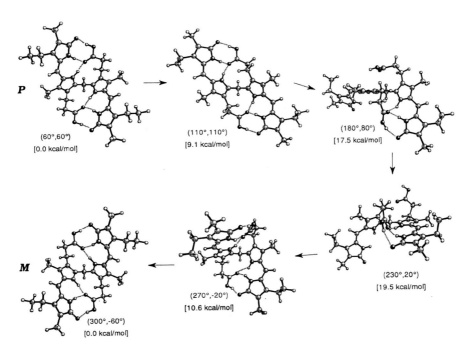

Fig. 8.4.8 Ball and Stick representations of bilirubin and mesobilirubin intermediate conformations (defined by torsion angles (ϕ_1, ϕ_2) and energies (kcal·mol⁻¹) lying along interconversion Path A connecting the (*P*)- and (*M*)- chirality global energy minimum enantiomeric conformers, **M** and **P**. The transition state, at $\phi_1 \sim +230°$, $\phi_2 \sim +20°$, for the interconversion lies some 82 kJ·mol⁻¹ (19.5 kcal·mol⁻¹) above the global energy minima (Fig. 9 of ref. (*619*). Reproduced with permission of the American Chemical Society. Copyright 1994)

488   8 Analysis of Bilirubin in Three Dimensions

The representative conformations (Figs. 8.4.7 and 8.4.8) at points along Paths A and B illustrate the changes in shape that a bilirubin molecule might undergo during conformational enantiomerism. It may be noted that Path A, in which only three hydrogen bonds are broken, is slightly lower energy (8 kJ·mol^{-1}, ~2 kcal·mol^{-1}) than Path B, where all six hydrogen bonds are broken, and that the computed energy barriers for **M** → **P** conformational inversion (~82 kJ·mol^{-1}, 19.5 kcal·mol^{-1}, Path A); (~90 kJ·mol^{-1}, 21.4 kcal·mol^{-1}, Path B) are satisfyingly close to those barriers determined independently by two different ^{1}H NMR methods in CDCl$_3$ solvent: 75–83 kJ·mol^{-1} (18–20 kcal·mol^{-1}) (*599, 601*) for rates of 7.2 ±0.4 sec^{-1} at ~53°C (*348a*) and 3 – 95 sec^{-1} at 50–95°C (*599, 601, 610*). It may be noted, too, that solvation effects are not specifically taken into account, except by imposing a solvent dielectric. In order to take into account solvation effects on the bilirubin conformational inversion, *Alagona et al.* (*621*) reported studies using a polarizable continuum model of the solvent, as implemented in GAUSSIAN 98, using *ab initio* calculations in the HF/6-31G* format. The results from a bilirubin model devoid of methyl groups was that the folded, intramolecularly hydrogen-bonded ridge-tide structures were lowest energy, and the barrier to interconversion was ~83 kJ·mol^{-1} (~20 kcal·mol^{-1}) in the gas phase and in water. Subsequently, *Alagona, Ghio, Agresti*, and *Pratesi* continued their *ab initio* HF/3-21G calculations (*622*) again with the "trimmed-down" bilirubin model to include solvation effects (PCM method) and determine energy differences between the most stable ridge-tile conformation and selected different conformations. The results were qualitatively the same as from the "*in vacuo*" conformation. Molecular mechanics calculations of bilirubin surrounded by solvent spheres were conducted by *D. Timothy Anstine* (Ph.D. 1995), *William Pfeiffer* (Ph.D. 1997), and *Brahmananda Ghosh* (Ph.D. 2003) of the *Lightner* group at the University of Nevada. A solvent sphere of CHCl$_3$ some 20–50 solvent molecules deep had little effect on the relative energies and the barrier to enantiomer interconversion. A solvent sphere of (CH$_3$)$_2$SO reduced the predicted lowest barrier for **M** → **P** interconversion by ~8 kJ·mol^{-1} (~2 kcal·mol^{-1}).

8.5 Bilirubin Chirality Revisited. Chiral Discrimination and Optical Activity

The earliest detection of bilirubin's optical activity came in studies involving albumin as long ago as 1966 (*586*), followed by *Blauer*'s investigations a few years later (*557, 558*) that led to a small explosion of related research articles by *Blauer* in the 1970s (*557-565, 593, 598, 623*) that carried into the 1980s (*565, 597*). Though *Blauer*'s early optical activity studies were directed toward bilirubin-albumin binding (*557, 558*), he was not alone in using circular dichroism (CD) spectroscopy to investigate binding to proteins (*624–626*) and polypeptides (*627*) before 1980. Why should anyone be interested in solutions of bilirubin + albumin or other proteins? For reasons related to its transport, in blood and into the liver for glucuronidation, there had long been an interest in measuring binding tightness of the pigment to the protein, and chiroptical spectroscopy had opened a new door to learning the efficacy

8.5 Bilirubin Chirality Revisited. Chiral Discrimination and Optical Activity

of binding, the binding constant, and detecting and analyzing the influence of pH, ionic strength, *etc.* on the protein structure at the binding site in an unusually sensitive way (*561, 624*). A knowledge of the albumin binding capacity and of the pigment structure in its noncovalent bonded association complex with the protein attracted considerable interest, for serum albumin serves not only as a vehicle for bilirubin transport but also as a biologic buffer against bilirubin encephalopathy or tissue damage in physiologic jaundice of the newborn (*155–157*). The affinity of human serum albumin for bilirubin had been determined independently to be rapid, reversible, and strong, and is known (*13, 628, 630*) to have an association constant of 10^7–10^8 M^{-1} for the first bilirubin and about an order of magnitude less for the second. In the 1970s, it was conjectured that the bilirubin species that binds to albumin is the propionate dianion (*631*), with binding probably through amine (lysine) residues at the binding site on the protein (*631*). The possibility that chiral amines might be involved in the complexation of bilirubin on albumin and responsible for its intense optical activity was the motivation in 1986 to explore the CD behavior of the pigment first with optically active neutral complexation agents, then with optically active amines. In connection with the latter, it may be noted that bilirubin CD induced by the presence of (+)- or (−)-α-phenylamine in organic solvents was cited by *Blauer* in 1975 (*597*) and again in a discussion following a lecture by *Blauer* (*593*), referring to work of *O. Mayer* (Ph.D. thesis, University of Konstanz, 1979). Later observations (*642*) with (−)-(*S*)-α-phenethylamine in $CHCl_3$ solvent gave $\Delta\varepsilon_{max}^{414}$ +6.26 and $\Delta\varepsilon_{max}^{472}$ −9.2 $dm^3 \cdot mol^{-1} \cdot cm^{-1}$ for 4.5 · 10^{-5} M bilirubin, where the amine:bilirubin molar ratio was 15 000:1. In DMSO as the solvent, α-phenethylamine did not induce bilirubin optical activity.

Blauer's early studies of bilirubin-albumin binding in the late 1960s led him to investigate the structure of the pigment on the protein. He interpreted the induced optical activity of the pigment in terms of a dissymmetric, extended bilirubin structure (*558, 559*) and the bisignate CD spectra (*559, 560, 593, 597, 623*) were explained in terms of a dipole-dipole coupling mechanism (*598*). Other chiral complexation agents for bilirubin began to be discovered in the 1970s. In 1971, weak bisignate induced CDs were observed for the pigment in aqueous sodium dexycholate micelles (*632*), results that were confirmed and explained by Dr. *Michael Reisinger* at the University of Nevada some 14 years later (*633*) in terms of displacing the conformational enantiomerism equilibrium (Figs. 8.3.1 and 8.4.5). In 1982, *LeBas et al.* reported (*634*) that aqueous solutions of bilirubin with cyclodextrins produced a similar effect, again with weakly induced bilirubin optical activity. These results were reconfirmed by Prof. *Jacek Gawroński* and Dr. *Krystyna Gawrońska* (both at Adam Mickiewicz University, Poznan) at the University of Nevada in 1985, and again, the induced CD was explained in terms of a shift in the **M** ⇄ **P** conformational equilibrium of interconverting intramolecularly hydrogen-bonded conformational enantiomers (Fig. 8.3.1) (*635*). Subsequent studies by *Kano et al.* in the 1990s (*636, 637*) reconfirmed the earlier observations, and they concluded that inclusion of bilirubin in the cyclodextrin cavity was not essential for inducing the pigment's optical activity (*638*). Even nucleosides were able to induce bilirubin optical activity (*639*).

8.5.1 Bilirubin Circular Dichroism from a First-Order Asymmetric Transformation

With the secondary structure of bilirubin in the crystal well established by the 1980s and the high probability of the same intramolecularly hydrogen bonded enantiomeric structures existing in solution, it was but a short leap to propose that the observed CD of bilirubin in the presence of chiral pertubers such as serum albumin *(640, 641)* was probably due to a displacement from $K_{eq} = 1$ of the **M** → **P** conformational equilibrium of Figs. 8.3.1 and 8.4.5. This was the basis for exploring the CD of bilirubin in the presence of optically active amines *(642–645)* in the 1980s at the University of Nevada. Thus, from complex amines like quinine *(645)* to simple amines such as (+)-(S)-2-aminobutane *(643)*, bisignate CDs were observed (Fig. 8.5.1), and the induced optical activity of bilirubin was later reviewed *(646)*.

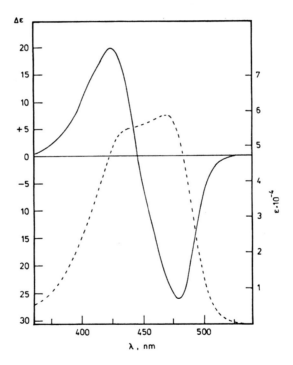

Fig. 8.5.1 Induced circular dichroism (ICD) (——) and UV-visible (- - - -) spectra of $4.6 \cdot 10^{-5}$ M bilirubin in benzene solution containing 0.69 M (+)-(S)-2-aminobutane at 25°C (Fig. 4 of ref. *(642)*. Reproduced with permission of Elsevier Science. Copyright © 1988)

Recall that *Manitto et al.* *(517, 518)* showed that bilirubin adopts a ridge-tile structure even as its bis-isopropylammonium salt. Thus, Fig. 8.5.2 is especially important because it shows that a bilirubin ammonium salt of an optically active primary amine, very close in structure to isopropylamine is capable of giving the bisignate CD spectra of Fig. 8.5.1.

8.5 Bilirubin Chirality Revisited. Chiral Discrimination and Optical Activity

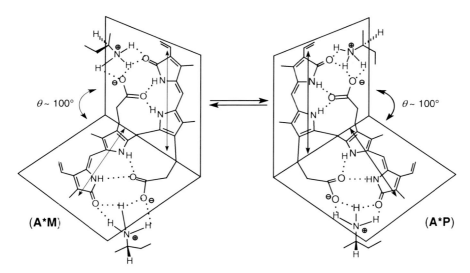

Fig. 8.5.2 Interconverting diastereomeric heteroassociation complexes of bilirubin with optically active α-substituted ethylamines, e.g. (+)-(S)-2-aminobutane

The intensity of the CD *Cotton* effects, while falling rather short of those seen from bilirubin-albumin solutions (*640, 641*), showed a dependency on the relative concentration of the optically active amine (Table 8.5.1), until maximum values were achieved near a ratio of 15,000:1, amine:pigment (*642*).

Table 8.5.1 Circular dichroism (*CD*) and ultraviolet-visible (*UV*) spectral data[a] for $4.6 \cdot 10^{-5}$ M bilirubin-IXα with varying concentrations of (+)-(S)-2-aminobutane in CHCl$_3$[b] at 20°C

Amine:Pigment Molar Ratio	Time/h[c]	CD				ε_{max}	λ (nm)
		$\Delta\varepsilon_{max}$ (λ_1)	λ_2 at $\Delta\varepsilon=0$	$\Delta\varepsilon_{max}$ (λ_3)			
1:1	0.5	+1.1 (420)		«0.1		54,600	453
	24	+1.1 (420)		«0.1			
10:1	0.5	+2.0 (415)	440	−0.6	(470)	52,400	453
	24	+2.0 (415)	440	−0.6	(470)		
100:1	0.5	+3.2 (416)	440	−2.0	(471)	53,100	455
	24	+3.2 (416)	440	−2.0	(471)		
1,000:1	0.5	+6.5 (420)	440	−7.8	(474)	57,600	455
	24	+6.1 (420)	440	−18.0	(472)		
5,000:1	0.5	+11.2 (420)	440	−14.0	(474)	61,500	458
	24	+9.8 (420)	439	−14.0	(472)		
15,000:1	0.5	+14.6 (420)	440	−19.4	(474)	62,400	459
	24	+13.8 (419)	440	−19.4	(472)		

[a]$\Delta\varepsilon_{max}$ and ε_{max} in dm^3·mol^{-1}·cm^{-1}; λ in nm; [b] ethanol-free, stabilized with 1% n-hexane; [c] spectra run 0.5 hours and 24 hours after preparing solutions

And since it was known that bilirubin forms ammonium salts with isopropylamine, it seemed reasonable to assume that the origin of the measured optical activity stems from non-equimolar concentrations of diastereomeric salts $(\mathbf{A}^*\mathbf{M})$ and $(\mathbf{A}^*\mathbf{P})$ formed in the bilirubin (acid) to amine base complexation equilibria of equations (1) and (2):

$$K^1_{eq} \qquad\qquad \mathbf{A}^* + \mathbf{M} \rightleftarrows (\mathbf{A}^*\mathbf{M}) \qquad\qquad (1)$$

$$K^2_{eq} \qquad\qquad \mathbf{A}^* + \mathbf{P} \rightleftarrows (\mathbf{A}^*\mathbf{P}) \qquad\qquad (2)$$

Here, $\mathbf{A}^*$ is the optically active amine base, $\mathbf{M}$ and $\mathbf{P}$ are intramolecularly hydrogen-bonded bilirubin conformational enantiomers of Figs. 8.3.1 and 8.4.5, and $(\mathbf{A}^*\mathbf{M})$ and $(\mathbf{A}^*\mathbf{P})$ are the optically active salts.

The values of K^1_{eq} and K^2_{eq} will be governed by the tendency of $\mathbf{A}^*$ to form the salt complex and its selectivity for either $\mathbf{M}$ or $\mathbf{P}$. Although the concentrations of enantiomers $\mathbf{M}$ and $\mathbf{P}$ are expected to be equal (same ΔG_f°), barring anisotropic solvation effects, the concentrations of diastereomers $(\mathbf{A}^*\mathbf{M})$ and $(\mathbf{A}^*\mathbf{P})$ will not be equal, *i.e.* with different ΔG_f°. Therefore, solutions containing the diastereomeric salts are expected to show a net optical activity of the pigment, and as the equilibria of equations (1) and (2) are driven toward salt (complex) formation by a large excess of amine, the magnitude of the induced CD can be expected to increase. This is reflected in the data of Table 8.5.1. In sum, the reaction of bilirubin $\mathbf{M}$ and $\mathbf{P}$ enantiomers (Figs. 8.2.1 and 8.4.5) with a chiral amine can be viewed as a prime example of a *first-order asymmetric transformation* described long ago by *Turner* and *Harris* (*647*) where one diastereomeric complex is greatly favored over the other in an equilibrium (Fig. 8.5.2) such as:

$$\mathbf{M}\text{-Bilirubin} \cdot (S)\text{-amine} \rightleftarrows \mathbf{P}\text{-Bilirubin} \cdot (S)\text{-amine} \qquad\qquad (3)$$

where the bilirubin component of the complex is subject to $\mathbf{M} \rightarrow \mathbf{P}$ mutarotation.

Optically active amines that bind tightly to and are highly enantioselective in forming the salt complexes are expected to generate the largest CD *Cotton* effects. This is true for any chiral complexing agent and has been noted for tightly-bound (association constant of 10^7–10^8 M^{-1}) (*13, 618–630*) bilirubin heteroassociation complexes with serum albumins, *e.g.* with HSA $\Delta\varepsilon_{max}^{407}$ −26, $\Delta\varepsilon_{max}^{460}$ +49 dm³·mol⁻¹·cm⁻¹ (*640, 641*). With other albumins, $\Delta\varepsilon$ magnitudes as high as 250 dm³·mol⁻¹·cm⁻¹ have been recorded (*623*). The equilibria depicted by equations (1)–(3) can also be influenced by the choice of solvent. Non-polar solvents, *e.g.* benzene, should favor the tightly bound ion-pair salt structures and drive the equilibria of equations (1) and (2) to the right. Polar, hydrogen-bonding solvents should facilitate dissociation of the ion pair. These predictions are borne out by the data of Table 8.5.2, where the $\Delta\varepsilon$ values are largest in solvents like benzene and chloroform, and smallest in dimethyl sulfoxide (*642*).

8.5 Bilirubin Chirality Revisited. Chiral Discrimination and Optical Activity

Table 8.5.2. Circular dichroism (*CD*) and ultraviolet-visible (*UV*) spectral data for $4.6 \cdot 10^{-5}$ *M* solutions of bilirubin-IXα and 0.65 *M* (+)-(*S*)-2-aminobutane[a] at 20°C

Amine	Solvent	Time /h[b]	CD			UV	
			$\Delta\varepsilon\,(\lambda_1)$	λ_2 at $\Delta\varepsilon=0$	$\Delta\varepsilon\,(\lambda_3)$	ε_{max}	λ
(+)-(*S*)-2- Aminobutane	C_6H_6	0.5	+19.5 (423)	445	−26.0 (479)	60,000	469
		24	+19.5 (423)	446	−26.2 (479)	60,700	469
	$CHCl_3$	0.5	+14.6 (420)	440	−19.4 (474)	62,400	459
		24	+13.8 (419)	439	−19.4 (472)	61,700	459
NH_2	Me_2CO	0.5	+15.1 (415)	438	−18.4 (468)	68,400	457
		24	+9.26 (415)	438	−11.1 (470)	65,400	457
	MeCN	0.5	+7.08 (416)	439	−8.17 (466)	68,100	449
		24	+7.30 (414)	439	−8.06 (471)	67,600	449
	MeOH	0.5	+1.1 (420)		«0.1	57,800	451
		24	+1.5 (420)		«0.1	48,200	452
	Me_2SO	0.5	«0.1		«0.1	70,800	460
		24	«0.1		«0.1	70,800	460

[a] 1:15,000 pigment:amine concentration ratio; [b] spectra run 0.5 hours and 24 hours after preparing solutions; [c] $\Delta\varepsilon$ and ε in $dm^3 \cdot mol^{-1} \cdot cm^{-1}$; λ in nm.

In the last, the vanishingly small $\Delta\varepsilon$ values might be attributed to significantly reduced concentrations of the diastereomeric salt, either (**A*M**) or (**A*P**). On the other hand, chiral complexing agents that either do not have a high affinity for bilirubin or exhibit little enantioselectivity for **M** or **P** can be expected to induce only weak optical activity, *e.g.* $\Delta\varepsilon_{max}^{409}$ +2.4, $\Delta\varepsilon_{max}^{460}$ −3.0 $dm^3 \cdot mol^{-1} \cdot cm^{-1}$ for bilirubin in 0.1 *M* Tris buffer at pH 8.0 with α-cyclodextrin, a torus-shaped cycloamylose (*631, 635*).

In order to understand the origin of the CD of bilirubin, two interrelated phenomena were considered in the mid-1980s: molecular structure and electronic structure. The preceding analysis, based on equilibrating diastereomeric ammonium salt of bilirubin with (*S*)-2-aminobutane (*642*), was therefore broadened in a detailed study in 1987 by *Jacek Gawroński*[4] and *Donald Wijekoon* (Ph.D. 1983), both then at the University of Nevada (*645*), who first observed induced CD (Fig. 8.5.3) by an alkaloid (absent primary amine groups).

---

[4]*Jacek Gawroński*, born on April 20, 1943 in Jutrosin, Poland, received the M.Sc. and Ph.D. degrees in organic chemistry at Adam Mickiewiz University, Poznań, Poland in 1966 and 1972, respectively. In 1970, he was a foreign exchange scholar at the University of Kansas with *A.W. Burgstahler*. After the *habilitation* in 1977, he remained at Adam Mickiewiz University, rising through the academic ranks to Professor of Chemistry and Head of the Natural Products Laboratory. In 1978-79, 1982, 1986, and 1988, he was a visiting professor at the University of Nevada, Reno.

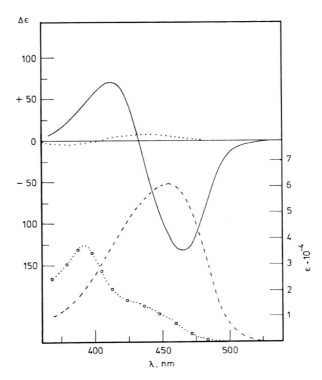

Fig. 8.5.3 Circular dichroism (—) and UV-visible (- - -) spectra of $3.0 \cdot 10^{-5}$ M bilirubin in dichloromethane in the presence of $9.0 \cdot 10^{-3}$ M quinine at 21°C. CD (· · ·) and UV-visible (○) spectra of $3.0 \cdot 10^{-5}$ M mesobilirubin-IVα in dichloromethane in the presence of $9.0 \cdot 10^{-3}$ M quinine at 21°C. The spectra were recorded within 15 min after preparation of the solution and remained essentially invariant for 24 hours at 22°C. A CD spectrum of the same concentration of bilirubin without added quinine falls on the $\Delta\varepsilon = 0$ line (Fig. 5 of ref. (645). Reproduced with permission of the American Chemical Society. Copyright 1987)

Therein it was shown that CH_2Cl_2 solutions of bilirubin with a 300 molar excess of quinine induced very large bisignate CD *Cotton* effects near the pigment's long wavelength (~450 nm) electronic transition(s) (Fig. 8.5.2), but in CH_3OH or $(CH_3)_2SO$ solvents the intensity dropped precipitously (645) due apparently to diminished enantioselective complexation of the pigment. In the absence of chiral perturbers bilirubin solutions in dichloromethane consist largely of equimolar concentrations of interconverting conformation enantiomers (**M** and **P**, Figs. 8.3.1 and 8.4.5) and exhibit no optical activity. However, addition of <1 molar equivalent of a cinchona alkaloid produced a profound change: the solution exhibited CD in the region of the pigment's long-wavelength UV-visible absorption band near 450 nm – a region where the alkaloid does not absorb light. (In contrast, brucine, an alkaloid not of the cinchona family, showed no induced CD at these (2:1) concentration ratios.) Quinine and cinchoinidine induced nearly the same CD behavior; their diastereomers, quinidine and cinchonine, respectively, induced oppositely signed *Cotton* effects of

8.5 Bilirubin Chirality Revisited. Chiral Discrimination and Optical Activity

nearly the same magnitudes. And with a 300:1 quinine:bilirubin ratio, the *Cotton* effects (Fig. 8.5.3) approached the largest values ($\Delta\varepsilon_{max}^{473} -214$, $\Delta\varepsilon_{max}^{423} +130$ dm^3·mol^{-1}·cm^{-1}) seen (*344c*) for $0.83 \cdot 10^{-5}$ *M* bilirubin complexed to human serum albumin (HSA) at pH 4.05, when pigment:(HSA) = 15, and exceeded those recorded (*557–565, 640, 641*) for HSA:bilirubin solutions near physiologic pH [$\Delta\varepsilon_{max}^{460} +49$, $\Delta\varepsilon_{max}^{407} -26$ dm^3·mol^{-1}·cm^{-1}, pH 7.3, (HSA):(pigment) = 2], where it is the bilirubin dianion that is said to bind to HSA (*628–630*). They contrast markedly with the considerably weaker *Cotton* effect magnitudes ($\Delta\varepsilon_{max}^{460} -3.0$, $\Delta\varepsilon_{max}^{409} +2.4$ dm^3·mol^{-1}·cm^{-1}) for pH 8.0 bilirubin solutions with a 1,000-fold molar excess of α-cyclodextrin (*635*). The remarkably intense *Cotton* effects suggest a high degree of enantioselectivity by a chiral complexation agent (*648*) for one of the bilirubin conformational enantiomers and a fairly strong coordination to it. High selectivity and strong binding, which are characteristic of protein-bilirubin complexation (*557–565, 593, 597, 623, 640, 641*), appear to be mimicked here.

The picture was thus enlarged, in view also of the induced CD from cyclodextrins (*635*) and sodium deoxycholate micelles (*633*), to perturbation of the **M ⇄ P** equilibrium of conformational enantiomers simply by a chiral complexation agent. To achieve induced CD, however, a well-defined secondary structure of the pigment appeared to be essential for the large CD *Cotton* effects of Fig. 8.5.3.

As outlined at the beginning of this section, bilirubin owes its well-defined chiral secondary structure (Figs. 8.3.1 and 8.4.5) in the crystal and nonpolar solvents to (i) *syn-periplanar* conformations of the two pyrromethenone chromophores, each possessing (Z)-configuration carbon-carbon double bonds at C-4 and C-15; (ii) two propionic acid groups at C-8 and C-12, each capable of forming *intra*molecular hydrogen bonds with the opposing dipyrrinone lactam carbonyl and NH and pyrrole NH groups; and (iii) an sp^3 carbon at C-10, which keeps the two dipyrrinones 98–104° apart. Even when bilirubin propionic acids are deprotonated, as in salts with amines (*517, 518, 642–644*), or tetra-alkylammonium hydroxides (*649–651*), the pigment retained a marked preference for folded, intramolecularly hydrogen-bonded ridge-tile structures. However, its dimethyl ester exhibited no unusual preference for this conformation (*652–655*), tending instead to hydrogen-bond *inter*molecularly in nonpolar solvents *via* a dipyrrinone-to-dipyrrinone dimeric arrangement similar to that observed in simple dipyrrinone pigments such as xanthobilirubinic acid (*656–662*). Similarly, when the propionic acid groups are relocated away from C-8 and C-12, as in mesobilirubin-IVα, which has erstwhile C-8 and C-12 propionic acids transposed with the C-7 and C-13 methyls, the pigment is poorly capable of expressing the type of intramolecular hydrogen bonding of Figs. 8.2.1 and 8.4.5.

The heteroassociation complexation (equations (1)–(3) of this section) apparently requires acid functional groups on the pigment and is not due uniquely to other forces, *e.g.* micellar, electrostatic, or π-π interactions, because the dimethyl esters of bilirubin and mesobilirubin-IVα give only relatively weak *Cotton* effects with quinine (Fig. 8.5.4), where the consequences of not adopting the intramolecularly hydrogen bonded structure has been seen in the greatly shrunken CD *Cotton* effects. It also requires pigments that can adopt and retain enantiomeric conformations such as **M** and **P** because mesobilirubin-IVα, which is a diacid not capable of adopting

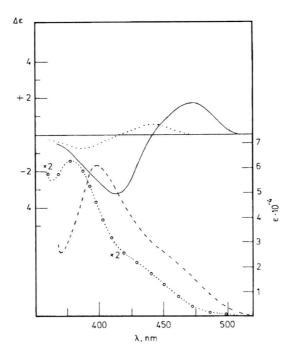

Fig. 8.5.4 Circular dichroism (—) and UV-visible (- - -) spectra of $3.0 \cdot 10^{-5}$ M bilirubin dimethyl ester in dichloromethane in the presence of $9.0 \cdot 10^{-3}$ M quinine at 21°C. CD (⋯) and UV-visible (○) spectra of $3.5 \cdot 10^{-5}$ M mesobilirubin-IVα dimethyl ester in dichloromethane in the presence of $9.0 \cdot 10^{-3}$ M quinine at 21°C. The spectra were recorded within 15 min after preparation of the solution and remained essentially invariant for hours at 22°C. (The UV curve for mesobilirubin-IXα dimethyl ester is scaled ·2 in $\Delta\varepsilon$ values.) A CD spectrum of the concentration of either pigment diester *without added quinine* falls on the $\Delta\varepsilon = 0$ line (Fig. 6 of ref. (*645*). Reproduced with permission of the American Chemical Society. Copyright 1987)

the *intramolecularly hydrogen-bonded conformation* of Figs. 8.2.1 and 8.4.5, shows only relative weak *Cotton* effects (Figs. 8.5.3 and 8.5.4). Although mesobilirubin-IVα is capable of forming amine salts, such diastereomeric complexes can assume a multitude of conformations, apparently with no strong preference for the unique, chiral conformations expressed by bilirubin.

In acting as a chiral template, the strong preferential selectivity of quinine for the **M** enantiomer and not the **P**, or *vice versa*, probably involves a unique dissymmetric arrangement (asymmetric microenvironment) of "multiple-point" binding features, including hydrogen bonding from the secondary hydroxyl group and *van der Waals* attractions between the alkaloid aromatic quinoline nucleus and the pigment's pyrromethenone moiety, in addition to the amine salt ionic bonding. The concept of multiple-point binding was invoked by *Rebek* (*663–665*) in his models for molecular recognition and was advanced by *Pirkle* (*648*) for chiral solvating agents in NMR studies and in resolutions by chromatographic methods.

The reaction of bilirubin (**M** and **P**) with a chiral complexing agent **A*** can be viewed as an example of a first-order asymmetric transformation (*647*), where one

8.5 Bilirubin Chirality Revisited. Chiral Discrimination and Optical Activity 497

complex (**A*****M**) or (**A*****P**) is greatly favored over the other in equilibrium, but where the bilirubin component of the optically active complex rapidly racemizes when free from the chiral complexation agent. Indeed, attempts at the University of Nevada to liberate optically active bilirubin form its complex with quinine at $-20°C$ by washing the 100:1 and 1,000:1 molar ratio (quinine:pigment) solutions in dichloromethane with 2 N HCl gave recovered pigment with no CD detected at 465 nm.

The early 1990s brought forth a variety of chiral discrimination agents capable of inducing bilirubin CD, including (*S*)-ethylmethylsulfoxide as a chiral solvent (*666, 667*) in studies by *Jacek Gawroński* and *Tadeusz Polonski* (Professor, University of Gdansk) while at the University of Nevada, and tripeptides by *Polonski*. These were followed from mid-1990 into the new millennium by observations of CD from bilirubin induced by an eclectic array of chiral complexation agents: from cyclophanes with chiral binding sites (*668, 669*); to polymeric nanoparticles (*670*) and cellulose derivatives (*671*). Even the use of bilirubin as a chiral switch has been proposed in model membranes (*672*). These studies reconfirmed the essential stereochemical picture, as have the electronic and vibrational CD studies of *Goncharova* and *Urbanová* using cyclodextrins (*673*), who also explored binding to peptides, *e.g.* poly(L-lysine) using both electronic and vibrational CD, with similar results (*674–676*). There seems to be no end to the exploration using bilirubin as a sensitive probe of the ability of chiral substance to act as a complexation agent.

8.5.2 Bilirubin as a Molecular Exciton. Absolute Configuration from Exciton Chirality

On the molecular level, it is relatively easy to understand that bilirubin should exhibit optical activity simply by displacing a (racemic) 1:1 equilibrium mixture of equilibrating conformational enantiomers (Figs. 8.3.1 and 8.4.5) toward either the (*M*)- or (*P*)-helical conformer by a chiral complexation agent. This is akin to a classical resolution of a racemic compound by making and separating diastereomers. The results of such a displacement are readily observed by CD spectroscopy (*556, 594, 677*), wherein oppositely signed bisignate CD curves are typically seen in the vicinity of the pigment's long-wavelength electronic transition near 450 nm (*619, 642–644, 646, 649–651*). Simply observing CD might be expected, but why is the CD curve bisignate at the UV-visible absorption band, and why are the magnitudes sometimes huge? From the electronic perspective, both electronic phenomena are clues to understanding the origin of the CD. Long ago, *Moscowitz* proposed an easy-to-understand way to look at the relationship between molecular structure and optical activity (*554–556*), especially as viewed by ORD or CD spectroscopy. He suggested that the molecules be viewed from the perspective of the relevant chromophores contained in them, *viz.*, the UV-visible light absorbing units, and described two limiting situations: (i) inherently dissymmetric and (ii) inherently symmetric but disymmetrically perturbed (*554–556*). As an example of (i), think in terms of hexahelicene (*554–556, 568–570*), where the entire molecule is the chromophore, and is helical, shaped like a lock-washer. Perhaps more relevant to bile pigments are the examples of the linear tetrapyrroles, *d*-urobilin and *l*-stercobilin (*555, 571*) discussed in Section 7.

There, the central two rings comprised of a dipyrrylmethene chromophore, were twisted into a helical shape maintained by intramolecular hydrogen bonds. Strong CD was typically observed because the relevant two components associated with the absorption of light, the induced electric ($\vec{\mu}_e$) and magnetic ($\vec{\mu}_m$) dipole moments are both allowed, and their dot product ($\vec{\mu}_e \cdot \vec{\mu}_m$), which indicates the intensity of the CD *Cotton* effect, is large *(554–556, 566, 567)*. Thus, inherently dissymmetric chromophores (i) typically give large CD *Cotton* effects of an order of magnitude or two greater than those from (ii) inherently symmetric chromophores. Good examples of inherently symmetric chromophores (symmetric in local symmetry) include the ketone carbonyl *(555, 566–570, 677)* of (*R*)-3-methyl-cyclohexanone and the aromatic ring of (*S*)-*sec*-butylbenzene, where the CD *Cotton* effects associated with the long-wavelength electronic transitions are of the order of magnitude of 1.

The large intensity of the bilirubin-albumin CD *Cotton* effects might have suggested an inherently dissymmetric chromophore *(554–556, 566, 567)*, such as a twisted dipyrrinone. However, the bisignate nature of the CD suggests otherwise, namely, that the CD associated with the bilirubin ~450 nm long-wavelength electronic transition comes from dipole-dipole coupling of the electronic transitions associated with the two dipyrrinone chromophores *(598, 645)*. That is, bilirubin behaves like a typical molecular exciton *(594, 678–681)*, characterized in its CD spectrum by a bisignate curve with two oppositely signed *Cotton* effects straddling the (~450 nm) UV-visible electronic excitation due to excited-state interaction in its weakly coupled electronic system *(566, 594, 682, 683, 684)*. This is nicely seen in Figs. 8.5.1 and 8.5.3, and is more generally summarized diagrammatically in Fig. 8.5.5 *(684)*.

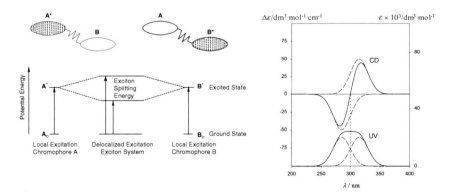

Fig. 8.5.5 (*Left*) Diagrammatic representation of exciton coupling between two chromophores (**A** and **B**) fastened together by covalent bonding or intermolecular forces. Local excitations are shown (*left* and *right*) for the chromophores in their locally excited monomer states (**A*** or **B***). When the two chromophores lie sufficiently close to one another, or the local excitations are sufficiently intense, excitation is delocalized between the two chromophores and the excited state (exciton) is split by resonance interaction of the local excitations in the composite system or molecule (center). Exciton coupling may take place between identical chromophores (**A**=**B**) or nonidentical chromophores (**A**≠**B**) and is less effective when the excitation energies are very different, i.e. when the relevant UV-visible absorption bands do not overlap (Redrawn from Fig. 1 of ref. (*684*).) (*Right*) Exciton coupling detected by simulated UV (*lower*) and CD (*upper*) spectroscopy;

8.5 Bilirubin Chirality Revisited. Chiral Discrimination and Optical Activity

A typical molecular exciton has at least two nonadjacent chromophores, usually π-electron systems, that are not in conjugation. As in all molecular excitons, the requirement here is that the chromophores have sufficiently strongly allowed electronic transitions, *e.g.* with molar absorptivity coefficients $>10,000$ dm^3·mol^{-1}·cm^{-1}, which typically arise from π-π^* excitations. In fact the chromophores of a molecular exciton can lie a long distance apart, as *Matile, Berova*, and *Nakanishi* at Columbia University have shown in bichromophore systems with chromophores separated by as much as 40–50 Å (*685*). Exciton splitting thus gives rise to two long-wavelength UV-visible transitions, one higher in energy and one lower in energy, with the separation dependent on the strength and relative orientation of the (dipyrrinone) chromophore's electric dipole transition moments (*566, 594, 677–684*). In the UV-visible spectrum, the two electronic transitions overlap to give the characteristically broadened (Figs. 8.5.3 and 8.5.5) and ideally seen as split (Fig. 8.5.1) long-wavelength absorption band of bilirubinoids. The CD spectra show the splitting more clearly because the two exciton CD transitions are always oppositely signed, as predicted by theory, and thus produce the bisignate *Cotton* effects observed (Fig. 8.5.5). In contrast the UV-visible exciton absorption curves may show only slight broadening when the exciton splitting energy is small. When two oppositely signed curves overlap in the CD, there is considerable cancellation in the region between the band centers with the net result that the *observed* bisignate *Cotton* effect maxima are displaced from the actual locations of the (separate) CD transitions (*566, 594, 686*) and are typically seen to flank the corresponding UV-visible band(s).

The two dipyrrinone chromophores of bilirubin, with strongly allowed long-wavelength electronic transitions (ε_{max} ~30,000 dm^3·mol^{-1}·cm^{-1}), have only a small interchromophoric electron overlap but interact through their locally excited states by resonance splitting (electrostatic interaction of the local transition moment dipoles). This is amply illustrated with the CD and UV-visible spectra of bilirubin in the presence of quinine (Fig. 8.5.3) and (*S*)-2-aminobutane (Fig. 8.5.1), and may even be recognized in the much weaker bisignate CDs of its dimethyl ester or mesobilirubin-IVα and its dimethyl ester (Fig. 8.5.4). However, it was not observed from the monochromophore molecular analog, xanthobilirubinic acid, which showed only weak monosignate *Cotton* effects with added quinine. Yet, Prof. *Stefan Boiadjiev* and *Tim Anstine* showed that under certain circumstances an intrinsically optically active xanthobilirubinic acid can be induced to partner with its twin and exhibit exciton coupling, may be as shown in Fig. 8.5.5 (*660, 661*). In CH$_3$OH, where the pigment is most likely *monomeric*, only a weak monosignate CD *Cotton* effect ($\Delta\varepsilon^{388}_{max} = +0.4$ dm^3·mol^{-1}·cm^{-1}) is observed in the case of the optically active xanthobilirubinic acid analog with a stereogenic center at the β-carbon of the propionic acid chain (*390*). In marked contrast, a typical bisignate exciton CD ($\Delta\varepsilon^{435}_{max} = -25$, $\Delta\varepsilon^{392}_{max} = +12$ dm^3·mol^{-1}·cm^{-1}) spectrum is observed in CCl$_4$ (Fig. 8.5.6), where dipyrrinones form a π-stacked *dimer* (Fig. 8.5.7) by hydrogen bonding carboxylic acid to dipyrrinone (*660, 661, 687*).

Fig. 8.5.5 (continued) the observed broadened and sometimes split UV curve comes from summing the UV curves for the two exciton transitions. The observed bisignate CD curve arises from summing the two oppositely-signed CD curves from the two exciton transitions (Redrawn from Fig. 2 of ref. (*584*))

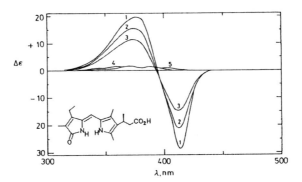

Fig. 8.5.6 Circular dichroism spectrum of (β S)-methylxanthobilirubinic acid in (1) CCl₄, (2) benzene, (3) CHCl₃, (4) CH₃CN, and (5) CH₃OH. Spectra were obtained from $5 \cdot 10^{-5}$ M solution at 23°C (Fig. 8 of ref. (660). Reproduced with permission of the American Chemical Society. Copyright 1995)

In this example the (β S) methyl group lies in a relatively unhindered site in the (M)-helicity π-stacked dimer and in a relatively more crowded situation in the (P)-helicity, as shown in Fig. 8.5.7. Steric crowding is predicted by molecular mechanics computations to render the (M)-helical dimer ~20 kJ·mol⁻¹ more stable than the (P).

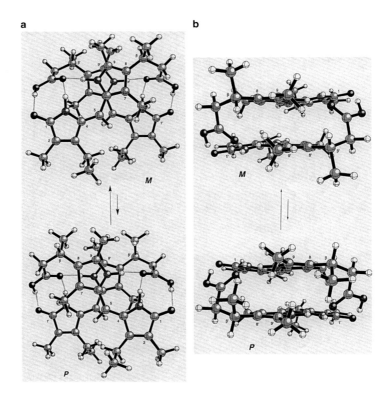

8.5 Bilirubin Chirality Revisited. Chiral Discrimination and Optical Activity 501

To understand the induced exciton chirality CD exhibited by bilirubin, it must be recognized that bilirubin preferentially adopts either of two intramolecularly hydrogen-bonded enantiomers (Figs. 8.3.1 and 8.4.5); that these conformational enantiomers are in dynamic equilibrium at room temperature; that the equilibrium can be displaced toward one or the other enantiomer by a chiral complexing agent (described in general terms more than 60 years ago as a first-order asymmetric transformation (*647*)); and that an excess of one enantiomer leads to a bisignate CD curve. Can that enantiomer be known? As will be seen in the following, the answer is clearly "yes". In treating bilirubin, or any other compound, as a molecular exciton to derive knowledge of its absolute configuration, it is essential to recognize the relative orientation and magnitude of the electric transition dipole moment associated with light-induced electronic excitation of each chromophore. That is, the direction of movement of charge in going from an electronic ground state to an electronic excited state. Fortunately, *Falk et al.* (*614, 615*) showed in 1977 that the intense long-wavelength UV-visible excitation near 400 nm of the dipyrrinone chromophore is oriented along the longitudinal axis of the molecule, studies that confirm the direction determined by *Blauer* and *Wagnière* in 1975 (*598*) using a π-electron molecular orbital calculation (SCF-MO-CI in the PPP approximation). Also, fortunately, *Nobuyuki Harada* (then at Sendai University) and *Koji Nakanishi* (at Columbia University) showed (*594*) how to calculate the signed order and intensity of the bisignate CD *Cotton* effects due to excited-state dipole-dipole coupling using simple principles and a knowledge of the direction and intensity of the chromophore's electric transition dipole moment.

This provided the basis for calculating the exciton UV-visible and CD spectra of bilirubin at the University of Nevada in 1987 (*645*), and from it for assigning the absolute configuration of the bilirubin conformational enantiomer that gives rise to the observed CD. *Kasha* described (*681*) three limiting orientations of the relevant electric transition moments and the spectral consequences of each orientation (Fig. 8.5.8). It can readily be seen that the observed CD from the bilirubin (Figs. 8.2.1, 8.5.1, 8.5.3) is better represented by the oblique orientation of the ridge-tile conformation than either the parallel or in-line.

Fig. 8.5.7 (a) Ball and Stick π-stacked model of the (βS)-methyl xanthobilirubinic acid π-stacked dimer shown in **M** $\leftrightarrows$ **P** equilibrium in edge-view representation and in (b) top-view representation. The relevant electric dipole transition moments (not shown) along the long axis of each chromophore are in an oblique orientation (Figs. 29 and 30 from the Ph.D. dissertation of Dr. *Timothy Anstine*, University of Nevada, Reno, 1995)

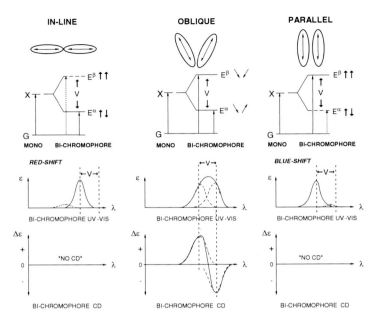

Fig. 8.5.8 Orientation dependence in exciton coupling between two chromophores (*ellipsoids*) and their long-axis transition dipoles (represented by *double-headed arrows*). The *solid single-headed arrows* connecting ground (G) and excited (X) states represent allowed transitions; *dashed arrows* represent forbidden transitions. The consequences of the orientation may be found in wavelength-shifted UV-visible spectra. Two limiting orientations lead to red-shifted (in-line) and blue-shifted (parallel) UV-visible bands: in the oblique orientation, both transitions are allowed and lead to broadened or split UV-visible spectra. Allowed CD transitions are found only when the dipoles have a chiral, oblique orientation forming a chiral array. $V = 2\Delta E_{ij}$, the exciton splitting of Fig. 8.5.10 (Redrawn from Fig. 3 of ref. (*684*))

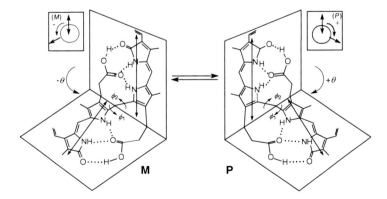

Fig. 8.5.9 Intramolecularly hydrogen-bonded enantiomeric conformations of bilirubin (**M** and **P**) folded into ridge-tile shapes. The **M** → **P** interconversion is accomplished as illustrated in Figs. 8.4.6, 8.4.7, and 8.4.8 by rotating about ϕ_1 and ϕ_2. The dipyrrinone chromophores are planar, and the angle of intersection of the two planes (dihedral angle, θ) is ~100° for $\phi_1 \cong \phi_2 \cong 60°$ in the **M** and **P** enantiomers. The *double-headed arrows* represent the approximate direction and intensity of the dipyrrinone long wavelength electric transition dipole moments. The relative orientations or helicities ((*M*), minus; (*P*), plus) of those vectors are shown (*inset*) for each enantiomer. The (*M*) dipole helicity correlates with (*M*)-molecular chirality of **M** and the (*P*)-helicity with (*P*)-molecular chirality of **P**

8.5 Bilirubin Chirality Revisited. Chiral Discrimination and Optical Activity

This corresponds exactly to the relative orientation of electric transition dipoles shown in Fig. 8.5.9. Exciton coupling theory (594–596, 677–681) provides a way to link the signed order of the CD *Cotton* effects to the (M)- or (P)-helicity of the electric transition dipole moments and thus to the absolute configuration of the conformational enantiomers. To correlate the (M)- and (P)- chirality conformations of bilirubin with the signed order of the exciton CD *Cotton* effects using exciton coupling theory (594–596, 677–681), the pigment is treated as two isolated dipyrrinone chromophores (i and j) connected by single bonds and with no interchomophoric π-orbital overlap (645). Since electronic excitation (from the ground state $\langle 0 |$ to the excited state $|q\rangle$, Fig. 8.5.10) in either dipyrrinone involves a displacement of electronic charge from the chromophore's ground state to its excited state (charge density polarization), an electric transition dipole moment, $\vec{\mu}_{0q} = \langle 0|\Delta|q\rangle$, is induced for each chromophoric excitation. The two transition dipole moments ($\vec{\mu}_{i0q}$ from dipyrrinone i and $\vec{\mu}_{j0q}$ from dipyrrinone j) interact, and the excited states are said to couple or mix. This coupling of the strongly allowed, long-wavelength excited state of dipyrrinone i with the comparable excited state of dipyrrinone j gives a molecular excited state (exciton), which may be viewed as being delocalized between the two chromophores and is split in energy by dipole-dipole interaction (Fig. 8.5.10).

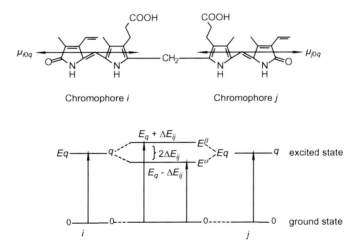

Fig. 8.5.10 State energy levels (0 = ground state, q = excited state) for dipyrrinone chromophores and their exciton interaction to give the bilirubin molecular electronic excited state with a *Davydov* splitting energy equal to $2\Delta E_{ij}$. For simplicity of calculations, the two dipyrrinones (i and j) were taken to be identical

If the dipyrrinone chromophores preserve their individuality in bilirubin, e.g. little or no electron overlap, application of perturbation theory (594, 678–681) will give *molecular* electronic wave functions and energies in terms of the wave functions and energies of the dipyrrinone units. Since the ground-state energy level of each dipyrrinone is taken to be zero and the *van der Waals* interaction between the permanent dipoles is assumed to be very small (678–681), the bilirubin molecular ground state is unsplit and also taken to be zero (Fig. 8.5.10). In contrast, perturbation theory predicts both a nonzero resonance (*Davydov*) splitting and the splitting energy ($2\Delta E_{ij}$) of the bilirubin molecular excited state due to exchange of excitation energy between the two dipyrrinone chromophores (exciton coupling). The center of the splitting will be displaced downward from energy level q (E_q, Fig. 8.5.10) by the dipole-dipole interaction energy between the permanent dipole of one dipyyrinone chromophore in its excited state and the other in its ground state (678–681). This displacement is taken to be negligible here. Consequently, assuming for simplicity that the two dipyrrinone chromophores are identical, the energies of the two different molecular excited states, α and β, arising from exciton splitting will be given by equations (1) and (2):

$$E^\alpha = E_{0q} - \Delta E_{ij} \qquad (1)$$

$$E^\beta = E_{0q} + \Delta E_{ij} \qquad (2)$$

where E_{0q} is the excitation energy $(0 \mid q)$ of the dipyrrinone chromophore, ΔE_{ij} is the dipole-dipole interaction energy between the two transition dipole moments of the dipyrrinones, and $2\Delta E_{ij}$ is the exciton (Davydov) splitting energy. E_{0q} maybe determined experimentally from the UV-visible spectrum or computed by molecular orbital calculations (598, 611), and ΔE_{ij} may be calculated in the point dipole-point dipole approximation from equation (3):

$$\Delta E_{ij} = [\, \vec{\mu}_{i0q} \cdot \vec{\mu}_{j0q} - 3\, R_{ij} (\, \vec{\mu}_{i0q} \cdot \vec{R}_{ij}\,)(\, \vec{\mu}_{j0q} \cdot \vec{R}_{ij}\,)]\, R_{ij}^{-3} \qquad (3)$$

where $\vec{\mu}_{i0q}$ and $\vec{\mu}_{j0q}$ are the electric dipole transition moments of dipyrrinone chromophores i and j, respectively, and R_{ij} is the interchromophoric distance vector connecting the two transition moments. Since the dipyrrinone transition dipole moments $\vec{\mu}_{i0q}$ and $\vec{\mu}_{j0q}$ arise from charge density polarization by the long-wavelength excitation $(0 \mid q)$ and are thus proportional to the intensity of the electronic transition, the exciton splitting $2\Delta E_{ij}$, will be large for this strongly allowed transition, probably π-π^*. The magnitude of the splitting also depends on the relative orientation of the dipyrrinone transition dipole moments and falls off as the inverse cube of the intermolecular distance, R_{ij}. For bilirubins, $2\Delta E_{ij}$ is smallest when the dipyrrinone chromophores are coplanar (180° interplanar angle).

In order to compute ΔE_{ij}, we make the simplifying assumption that the two dipyrrinone chromophores are identical and the same as that of the xanthobilirubinic acid chromophore; therefore, $\mu_{i0q} = \mu_{j0q}$. In the ridge-tile conformations of Fig. 8.5.9 the angle (θ) between the two dipole moments is taken to be 100° since X-ray crystallographic studies (510, 511, 514–518) show that the two dipyrrinones of bilirubin and mesobilirubin are essentially planar and their interplanar (dihedral) angle (θ) is 98–104° (510, 511). With the direction of polarization of μ_{i0q} and μ_{j0q} lying in the plane of the chromophore, approximately along the line connecting C-2 and C-8, R_{ij} ~ 6 Å and $\vec{R}_{ij}$ makes angles of ϕ ~ 50° with $\vec{\mu}_{i0q}$ and ϕ'~ 130° with $\vec{\mu}_{j0q}$. Equation (3) simplifies to (4):

$$\Delta E_{ij} = 5.034 \cdot 10^{15}\, D[\cos \theta - 3 \cos \phi \cos \phi']\, R_{ij}^{-3}\ \mathrm{cm}^{-1} \qquad (4)$$

8.5 Bilirubin Chirality Revisited. Chiral Discrimination and Optical Activity

where 1 erg $= 5.034 \cdot 10^{15}$ cm^{-1} and D is the dipole strength, determined from the UV-visible spectrum of xanthobilirubinic acid to be approximately $44 \cdot 10^{-36}$ cgs. (Because dipyrrinones are known to dimerize in nonpolar solvents, the UV-visible spectrum was run in $(CH_3)_2SO$ to give the relevant measured data for the monomer: $\varepsilon_{410}^{max} = 30{,}000$ dm^3·mol^{-1}·cm^{-1}, $\Delta\sigma = 2{,}250$ cm^{-1}, and D.) Substitution gives $\Delta E_{ij} \sim +1{,}000$ cm^{-1}, and since the value is positive, the α excited state is lower energy (longer wavelength) than the β state (Fig. 8.5.10).

For a chiral molecular exciton, the two excited states (α and β) have been shown to give oppositely signed rotatory strengths, R^α and R^β:

$$R^{\alpha,\beta} = \pm\, (\pi/2)\sigma_0 \cdot \vec{R}_{ij} \cdot (\,\vec{\mu}_{i0q} \cdot \vec{\mu}_{j0q}\,) \tag{5}$$

where the upper and lower signs are associated with the α and β excited states, respectively, and the σ_0 is the excitation energy of the dipyrrinone in cm^{-1}, taken here as 410 nm or 24,390 cm^{-1} for xanthobilirubinic acid. For a bilirubin with C_2 symmetry, the α state corresponds to B symmetry and the β state to A symmetry (*345*). For the **M** enantiomer (Fig. 8.5.9) with the left-handed helicity of ($\vec{\mu}_{i0q} \cdot \vec{\mu}_{j0q}$), the triple product $\vec{R}_{ij} \cdot (\vec{\mu}_{i0q} \cdot \vec{\mu}_{j0q})$ <0 and therefore R^α takes a negative value and R^β takes a positive value. Since the α excited state is lower in energy than the β excited state, a $(-)$ long-wavelength *Cotton* effect will be followed by a $(+)$ short wavelength *Cotton* effect in the exciton couplet E^α, E^β. This corresponds to a negative exciton chirality, where exciton chirality may be defined as $\vec{R}_{ij} \cdot (\vec{\mu}_{i0q} \cdot \vec{\mu}_{j0q})$ ($2\Delta E_{ij}$), as shown in Fig. 8.5.10.

Thus, both here and in the general case (*594*) the handedness of screw sense that the electric dipole transition moments of the coupled dipyrrinone chromophores make with each other (Fig. 8.5.10) correlates with signed order of the bisignate CD *Cotton* effects. A left-handed screw sense (negative chirality) of the transition moments leads to a $(-\!\!-)$ longer wavelength *Cotton* effect followed by a $(+)$ shorter wavelength *Cotton* effect. For a right-handed screw sense (positive chirality) the *Cotton* effect signs are inverted: $(+)$ at the longer wavelength and $(-)$ at the shorter wavelength component of the bisignate *Cotton* effect. Given the direction of the electric dipole transition moment in the dipyrrinone chromophore (*598, 614, 615*) the exciton model can be used to predict the *Cotton* effect signs of the pigment in the structurally well-defined diastereomers (**A*M**) and (**A*P**) (of Section 8.5) or the *Cotton* effect signs of the bilirubin enantiomers, (**M** and **P**) of Fig. 8.5.10. In these intramolecularly hydrogen-bonded conformations, the relative orientations of the two dipyrrinone electric dipole moments constitute a left-handed chirality for (**A*M**) and **M** and a right-handed chirality for (**A*P**) and **P**. Accordingly, theory predicts a predominance of the left-handed diastereomeric complex (**A*M**) for solutions of bilirubin in the presence of quinine since the induced bisignate CD shows a negative *Cotton* effect near 465 nm followed by a positive *Cotton* effect near 415 nm (Fig. 8.5.3). By analogy, an excess of the right-handed diastereomeric complex (**A*P**) is favored in the presence of HSA at pH 7.3 and (**A*M**) at pH 4.8 (Fig. 8.2.1) (*559*).

The exciton calculations (*645*) also provided simulations of the UV-visible and CD spectra that matched those determined experimentally and predicted that CD maxima $\Delta\varepsilon \pm 260$ and ± 190, respectively, for the long- and short-wavelength exciton components, respectively. Furthermore, as shown in Section 8.5.3, the exciton chirality *Cotton* effects can be predicted for every conformation of bilirubin.

8.5.3 Mapping Bilirubin Conformation to Exciton Chirality CD

The last two decades of the 20th century had seen numerous studies of optical activity in bile pigments, some from induced optical activity, others with natural (*15, 687*). Bilirubin stood out as the last to be investigated in full. In a study published in 1994, by *Richard Person* and *Blake Peterson* (currently Regents' Distinguished Professor of Medicinal Chemistry at the University of Kansas), both then in the *Lightner* research group, computed the dependence of bilirubin's exciton chirality CD spectra from the vast array of conformations comprising the conformational energy map of Fig. 8.4.3, while an undergraduate student at the University of Nevada (*619*). To simplify the computations, mesobilirubin-XIIIα, which has a symmetric ordering of its β-substituents, was driven through all possible conformational isomers by rotations about torsion angles ϕ_1 and ϕ_2, as described in Section 8.4, and for each conformation the sign and magnitude of the computed lower energy (greater λ) at the long-wavelength exciton state was plotted vs. ϕ_1 and ϕ_2 to create a three-dimensional map of CD intensities vs. conformation. Thus, using the coupled oscillator formalism (*594*), the CD and UV-visible spectra of mesobilirubin-XIIIα were computed for all of the conformations represented in the conformational energy map of Fig. 8.4.6. The computed CD data are displayed as a three-dimensional graph in Fig. 8.5.11, where $\Delta\varepsilon$ for the lower energy long wavelength exciton state is plotted on the vertical axis and conformational torsion angles ϕ_1 and ϕ_2 are plotted on the horizontal axes. (An essentially identical map would be obtained for bilirubin-IXα. In Fig. 8.5.11, a diagonal line connecting points ($\phi_1 = 360°$, $\phi_2 = 0°$) and ($\phi_1 = 0°$, $\phi_2 = 360°$) separates mirror image enantiomeric conformations.

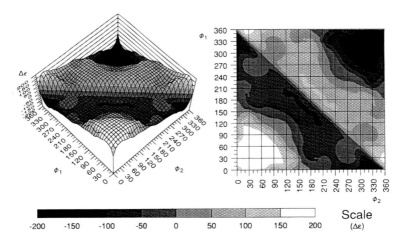

Fig. 8.5.11 Three-dimensional plot (*left*) and "topo" (*right*) map of the variation of computed circular dichroism $\Delta\varepsilon$ sign and unscaled magnitude for the lower energy exciton component (α-state) correlated with mesobilirubin and conformations described by independent rotations about torsion angles ϕ_1 and ϕ_2 (Fig. 8.4.2) from 0° to 360° in steps of 10°. An essentially identical map would be computed for bilirubin-IXα. The diagonal line connects points ($\phi_1 = 0°$, $\phi_2 = 360°$) and ($\phi_1 = 360°$, $\phi_2 = 0°$) separates mirror image conformers (enantiomers), and therefore the sign of $\Delta\varepsilon$ (but not its magnitude) corresponding to a given set of (ϕ_1, ϕ_2) coordinates (pigment conformation) becomes inverted by reflection across that diagonal (Fig. 13 of ref. (*619*). Reproduced with permission of the American Chemical Society. Copyright 1994)

8.5 Bilirubin Chirality Revisited. Chiral Discrimination and Optical Activity

That is, conformations corresponding to points falling below the indicated diagonal line have non-superimposable mirror image conformations lying at points reflected across the diagonal. Reflection across the indicated diagonal thus inverts the sign (but not the magnitude) of the corresponding *Cotton* effect ($\Delta\varepsilon$). In the case of bilirubin, for example, the conformation corresponding to the point ($\phi_1 = +30°$, $\phi_2 = +270°$) has an enantiomer at ($\phi_1 = +330°$, $\phi_2 = +90°$) with equal but oppositely signed bisignate *Cotton* effects. The conformation of bilirubin corresponding to the point ($\phi_1 = +90°$, $\phi_2 = +330°$) is, strictly speaking, not an enantiomer of that at ($\phi_1 = +30°$, $\phi_2 = +270°$), though for CD purposes it behaves like one. Due to its greater molecular symmetry, the ($\phi_1 = +30°$, $\phi_2 = +270°$) conformation of mesobilirubin-XIIIα would have an enantiomer at ($\phi_1 = +330°$, $\phi_2 = +90°$) as well as at ($\phi_1 = +90°$, $\phi_2 = 330°$). In a second example, the point ($\phi_1 = \phi_2 \cong +60°$) corresponds to the (*P*)-helicity enantiomer of Figs. 8.5.4 and 8.5.9 and also to a global minimum structure. A *positive* $\Delta\varepsilon$ is computed for the lower energy longer wavelength half of the bisignate CD curve. Reflection of this point across the indicated diagonal gives the new point ($\phi_1 = \phi_2 \cong +300°$), which corresponds to the mirror image (*M*)-helicity enantiomer having a computed *negative* $\Delta\varepsilon$ for the longer wavelength half of the bisignate CD curve.

Rather importantly, the graph of computed CD *Cotton* effect $\Delta\varepsilon$ maxima also shows that *the sign of $\Delta\varepsilon$ can become inverted without an inversion of molecular chirality*. That is, when ϕ_1 and ϕ_2 increase, *e.g.* from $\phi_1 = \phi_2 \cong +60°$, the magnitude of $\Delta\varepsilon$ is computed to decrease to zero for conformations corresponding to points (ϕ_1, ϕ_2) lying near a contour line running roughly from ($\phi_1 = 180°$, $\phi_2 \cong 0°$) to ($\phi_1 \cong 0°$, $\phi_2 = 180°$) and passing through ($\phi_1 = \phi_2 \cong 120°$). In the region above this contour, the sign of $\Delta\varepsilon$ reverses, and its magnitude gradually increases then rapidly decreases back to zero for conformations at points lying along a diagonal line running between ($\phi_1 = 360°$, $\phi_2 = 0°$) and ($\phi_1 = 0°$, $\phi_2 = 360°$). This line separates enantiomeric conformations. The changes in $\Delta\varepsilon$ along the diagonal path connecting points ($\phi_1 = \phi_2 = 0°$) and ($\phi_1 = \phi_2 = 360°$) are illustrated in Fig. 8.5.12, which displays not only $\Delta\varepsilon$ (black line) but also the conformational potential energy (gray line) values (taken from Figs. 8.5.11 and 8.4.3, respectively). From the solid, dark line of Fig. 8.5.12, which tracks the relationship between $\Delta\varepsilon$ and conformations where $\phi_1 = \phi_2$, it is easy to see that between 0° and 180°, which corresponds to (*P*)-molecular chirality, the helicity of the electric dipole transition moments remains positive (*P*) between $\phi_1 = \phi_2 = 0°$ and 120° but inverts to (*M*) between 120° and 180° for a stretched conformation (see Fig. 8.5.14). Likewise, where $\phi_1 = \phi_2$ lies between 180° and 360°, which corresponds to the (*M*)-molecular chirality, the helicity of the long wavelength electric dipole transition moments is (*M*), except when $\phi_1 = \phi_2$ falls between 180° and 240°, where it inverts to (*P*). Thus, in the regions where $\phi_1 = \phi_2$ lies between 120° and 180°, the expected positive exciton chirality inverts to negative even though the molecular chirality remains unchanged. And where $\phi_1 = \phi_2$ falls between 180° and 240°, the expected negative exciton chirality inverts to positive while the molecular chirality remains (*M*). However, the probability of encountering conformations that would exhibit the "unexpected" exciton chirality CD behavior may be evaluated from the corresponding conformational potential energies, which are also co-plotted in Fig. 8.5.12 (along the same diagonal). They show that the "inverted" type CD may be expected mainly from higher energy ($\phi_1 = \phi_2$) conformations.

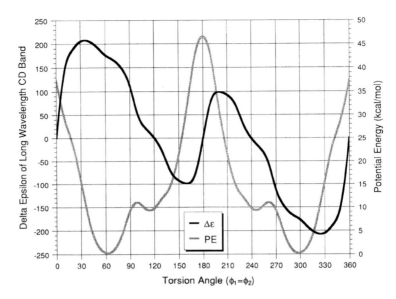

Fig. 8.5.12 (*Black line*) Computed, unscaled $\Delta\varepsilon$ (left vertical axis) for the lower energy, long wavelength exciton state and its variation with torsion angles ϕ_1 and ϕ_2 along the diagonal (of Fig. 8.5.11) stretching from ($\phi_1 = \phi_2 = 0°$) to ($\phi_1 = \phi_2 = 360°$). (*Gray line*) Variation of conformational energy (right vertical axis) – taken from Fig. 8.4.3 or 8.4.6 – along the same diagonal for bilirubin-IXα or mesobilirubin-XIIIα (Fig. 14 of ref. (*619*). Reproduced with permission of the American Chemical Society. Copyright 1994)

Ball and Stick drawings of representative conformations where $\phi_1 = \phi_2$ may be seen in Fig. 8.5.13 for structures taken from points along the gray (potential energy) line of Fig. 8.5.12. In Fig. 8.5.13, as ϕ_1 and ϕ_2 increase in conrotatory fashion from $\phi_1 = \phi_2 \cong 0°$, (i) the conformation relaxes from a very high energy, nearly planar porphyrin-like structure, (ii) passes through lower energy (*P*)-helical ($\phi_1 = \phi_2 = 20°$) and ($\phi_1 = \phi_2 = 40°$) conformations where intramolecular hydrogen bonding can be accessed, and (iii) on to the (*P*)-helical ridge-tile structure ($\phi_1 = \phi_2 = 60°$) lying at a global minimum, **P** of Figs. 8.3.5 and 8.5.9. (iv) Further rotations about ϕ_1 and ϕ_2 increase the potential energy of the molecule allowing the pigment structure to pass through the gabled conformation near ($\phi_1 = \phi_2 = 90°$) and (v) into a series of stretched conformations: one lying near a local minimum near ($\phi_1 = \phi_2 = 110°$) and other more stretched conformations lying near ($\phi_1 = \phi_2 = 120°$), ($\phi_1 = \phi_2 = 140°$), and ($\phi_1 = \phi_2 = 160°$) where weakened residual hydrogen bonding is accommodated by twisting within the dipyrrinone chromophores that yield large increases in potential energy. The linear conformation ($\phi_1 = \phi_2 = 180°$) lies at an energy maximum. Large $+\Delta\varepsilon$ values are predicted for folded helical and ridge-tile conformations, but the magnitude of $\Delta\varepsilon$ is expected to decrease as the ridge-tile conformer begins to unfold and stretch into higher energy conformations, dropping to $\Delta\varepsilon = 0$ at a conformation ($\phi_1 = \phi_2 = 120°$) lying very near the local minimum and lying some 38 kJ·mol^{-1} (9 kcal·mol^{-1}) above the global minimum. Then with increased unfolding and much

8.5 Bilirubin Chirality Revisited. Chiral Discrimination and Optical Activity   509

higher conformational energies, $\Delta\varepsilon$ becomes increasingly negative – before declining rapidly to $\Delta\varepsilon = 0$ in the linear conformation ($\phi_1 = \phi_2 = 180°$). At no point along the diagonal path between ($\phi_1 = \phi_2 = 0°$) and ($\phi_1 = \phi_2 = 180°$), however, does the molecular chirality (P) of the pigment invert.

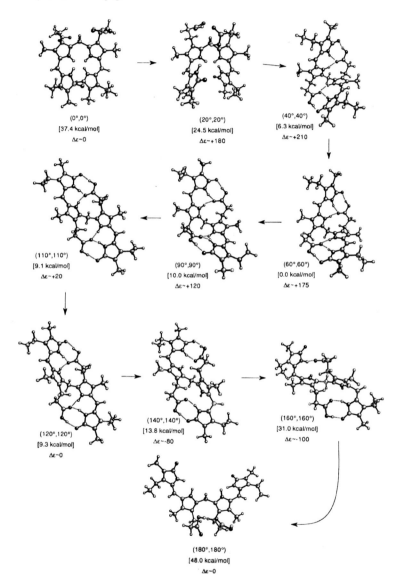

Fig. 8.5.13 Representative conformations of mesobilirubin-XIIIα in Ball and Stick representation for points ($\phi_1 = \phi_2$) lying between 0° and 180° on the conformational potential energy curve of Fig. 8.5.12 (Fig. 15 of ref. (*619*). Reproduced with permission of the American Chemical Society. Copyright 1994)

510   8 Analysis of Bilirubin in Three Dimensions

Inversion of CD *Cotton* effects ($\Delta\varepsilon$) without an accompanying inversion of molecular chirality occurs because the relative orientation of the dipyrrinone electric transition dipoles inverts their relative helicity as the pigment stretches or unfolds (Fig. 8.5.14) into a broad realm of conformers lying in regions close to and on either side of a diagonal line (Fig. 8.5.11) connecting points ($\phi_1 = 360°$, $\phi_2 = 0°$) and ($\phi_1 = 0°$, $\phi_2 = 360°$) and bordered roughly by parallel lines connecting points ($\phi_1 = 180°$, $\phi_2 = 0°$) and ($\phi_1 = 0°$, $\phi_2 = 180°$), and points ($\phi_1 = 360°$, $\phi_2 = 180°$) and ($\phi_1 = 180°$, $\phi_2 = 360°$). Thus, bond rotations about ϕ_1 and ϕ_2, as described by Figs. 8.5.12 and 8.5.13, are predicted to lead to inversion of *Cotton* effect signs accompanied either with or without inversion of molecular chirality.

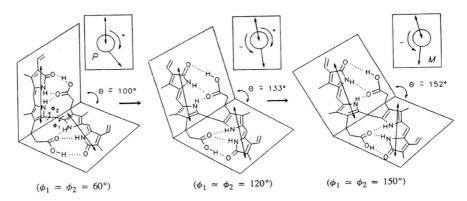

Fig. 8.5.14 (*Left*) (*P*)-Chirality bilirubin in its most stable (global energy minimum) intramolecularly hydrogen-bonded ridge-tile conformation. The intersection of the two dipyrrinone planes makes a dihedral angle, $\theta \cong 100°$ for torsion angles $\phi_1 = \phi_2 \cong 60°$ relative to the planar porphyrin-like conformation where ϕ_1 and ϕ_2 are defined as 0°. This conformation is predicted to show a *positive* $\Delta\varepsilon$ for the low-energy, long wavelength CD exciton state. (*Middle*) Stretched (*P*)-chirality bilirubin arising from rotations about torsion angles ϕ_1 and ϕ_2 to values $\phi_1 = \phi_2 \cong 120°$ (Fig. 8.5.12). At or near this conformation, the pigment is predicted to exhibit "no CD" because $\Delta\varepsilon$ is computed to be zero. This conformation lies some 42 kJ·mol^{-1} (10 kcal·mol^{-1}) above the global minimum. (*Right*) The same conformation and absolute configuration for an even more stretched bilirubin with larger θ and a flatter ridge-tile shape. Here $\phi_1 = \phi_2 \cong 150°$ and in this conformation the pigment is predicted to exhibit a *negative* CD *Cotton* effect for the low-energy, long wavelength state – an inversion of $\Delta\varepsilon$ sign compared with that predicted for the conformation at the far left. This conformation lies some 92 kJ·mol^{-1} (22 kcal·mol^{-1}) above the global minimum. Conformational stretching is accommodated by lengthening (or breaking) hydrogen bonds between the carboxylic acid CO$_2$H and lactam –NHC=O groups and by twisting the individual dipyrrinone chromophores. Consequently, the planes shown in the *middle* and *right* structures are only average planes passing through the dipyrrinones. Significantly, the transition dipole vectors, shown in *inset boxes*, associated with the dipyrrinone long wavelength UV-visible absorption reverse relative orientation from (*P*) to (*M*) without an inversion of *molecular chirality* (Fig. 16 of ref. (*619*). Reproduced with permission of the American Chemical Society. Copyright 1994)

In contrast, conformational changes along a slice of the conformational energy maps of Figs. 8.4.3 and 8.4.6, where ϕ_2 is held at 0° and ϕ_1 is rotated from 0° to 180°, lead to inversion of *Cotton* effect signs *only* with an inversion of molecular chirality.

8.5 Bilirubin Chirality Revisited. Chiral Discrimination and Optical Activity   511

The computed conformational energies along with the computed CD $\Delta\varepsilon$ values (taken from the left edge of Fig. 8.5.11) for the long wavelength component of the bisignate *Cotton* effects are plotted in Fig. 8.5.15. Representative conformations lying at various points ($\phi_2 = 0$, $\phi_1 = 0°$ to $180°$) along the curves of Fig. 8.5.15 are shown in Fig. 8.5.16. The borders represent special cases, however, and in the more general case, inversion of *Cotton* effect signs may occur in absence of inversion of molecular chirality to form enantiomers.

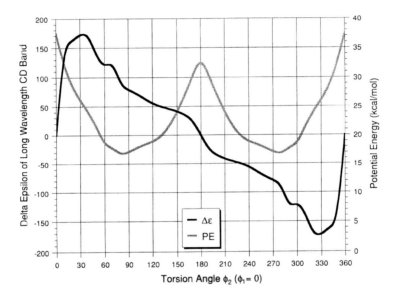

Fig. 8.5.15 (*Black line*) Variation of computed $\Delta\varepsilon$ (left vertical axis) with torsion angle ϕ_2 (from $\phi_2 = 0°$ to $\phi_2 = 360°$), where ϕ_1 is held at $0°$. (*Gray line*) Variation of conformational energy (right vertical axis) – taken from Figs. 8.4.3 or 8.4.6 – along the edge ($\phi_1 = 0°$, $\phi_2 = 0°$ γ $360°$) of the conformational energy map for bilirubin-IXα or mesobilirubin-XIIIα (Fig. 17 of ref. (*619*). Reproduced with permission of the American Chemical Society. Copyright 1994)

Selected computed data (Tables 8.5.2 and 8.5.3) illustrate the concepts and principles discussed above. In Table 8.5.2, which tracks the pathways and conformations illustrated in Figs. 8.5.15 and 8.5.16, the computed UV-visible transitions show the expected (i) blue shift for the porphyrin-like conformations, with parallel transition dipole moments, and (ii) red shift for the extended in-line conformations, with in-line transition dipole moments. The CD *Cotton* effect $\Delta\varepsilon$ values are predicted to reach their largest values for the helical folded conformations and decline to zero in the planar porphyrin-like and the extended in-line conformations. Of course, all of the conformations of Table 8.5.2 are predicted to lie significantly above the global energy minimum ridge-tile conformation, and they are therefore unlikely to be easily accessible.

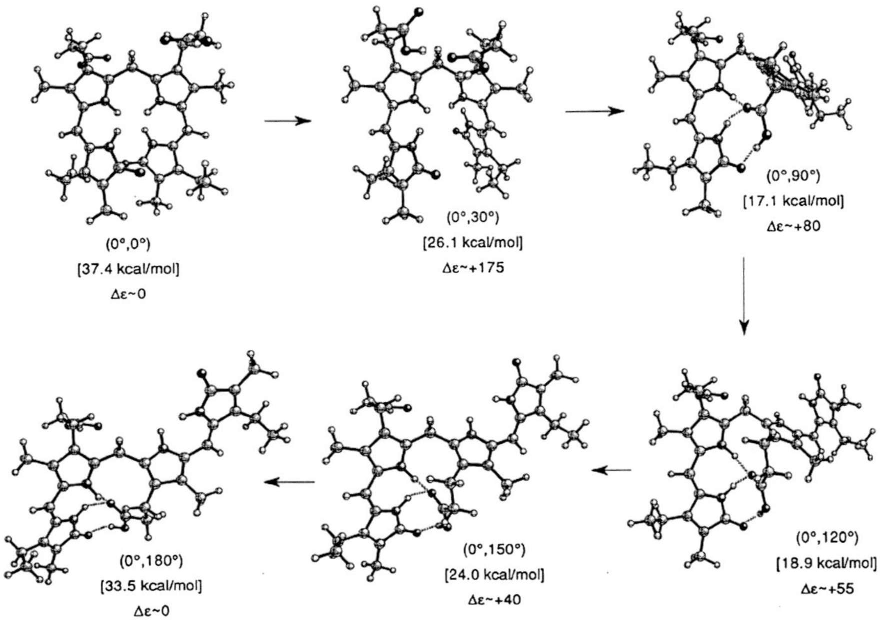

512   8 Analysis of Bilirubin in Three Dimensions

Fig. 8.5.16 Representative mesobilirubin-XIIIα conformations in Ball and Stick representation for points (ϕ_1, ϕ_2) lying between $\phi_1 = 0°$ and $0° < \phi_2 < 180°$ on the conformational potential energy curve of Fig. 8.5.15 (Fig. 18 of ref. (619). Reproduced with permission of the American Chemical Society. Copyright 1994)

Table 8.5.2 Selected computed CD and UV-visible data[a,b] for bilirubin and mesobilirubin-XIIIα conformers: porphyrin-like, helical, folded, ridge-tile, gabled, stretched, and extended in-line (Fig. 8.5.16) for selected points on the curves of Fig. 8.5.15

torsion angle/°		conformation type	potential energy above global min/ kJ·mol⁻¹	theor UV-visible[c]				summed theor UV-visible		computed CD[c]		
ϕ_1	ϕ_2			λ^β	ε_{max}	λ^α	ε_{max}	λ_{max}	ε_{max}	λ_1	λ_2	$\Delta\varepsilon$ at λ_2
~0	~0	porphyrin-like	155	367	74,000	507	1,000	367	74,000	352	479	~0
0	20	helical	134	402	65,000	454	10,000	403	68,000	378	438	+160
0	30	helical folded	109	403	61,000	450	14,000	406	65,000	378	438	+175
0	60	folded[d]	75	414	42,000	438	34,000	425	70,000	381	434	+120
0	90	gabled	71	418	27,000	434	48,000	428	73,000	381	434	+80
0	100	stretched	75	419	23,000	434	53,000	429	73,000	381	434	+75
0	120	stretched	79	420	14,000	433	61,000	430	74,000	382	434	+55
0	140	stretched	84	419	9,000	433	67,000	431	75,000	382	434	+45
0	160	stretched	113	418	4,000	434	71,000	433	75,000	381	434	+35
0	~180	extended in-line	138	418	1,000	434	74,000	434	75,000	381	434	~0

[a] Based on conformations in which the porphyrin-like conformation passes through (P)-molecular chirality (Fig. 8.5.16) on the way to extended in-line; [b] ε and $\Delta\varepsilon$ in dm³·mol⁻¹·cm⁻¹; λ in nm; [c] for the β-state and α-state of the molecular exciton; [d] folded, with intramolecular hydrogen bonding

8.5 Bilirubin Chirality Revisited. Chiral Discrimination and Optical Activity

Table 8.5.3 Selected computed circular dichroism and UV-visible data[a,b] for bilirubin and mesobilirubin-XIIIα conformers: porphyrin-like, helical, folded, ridge-tile, gabled, stretched, and linear (Fig. 8.5.13) for selected points on the curves of Fig. 8.5.12

torsion angle/°		conformation type	potential energy above global min/ kJ·mol^{-1}	theor UV-visible[c]				summed theor UV-visible		computed CD[c]		
ϕ_1	ϕ_2			λ^β	ε_{max}	λ^α	ε_{max}	λ_{max}	ε_{max}	λ_1	λ_2	$\Delta\varepsilon$ at λ_2
~0	~0	porphyrin-like	155	367	74,000	507	1,000	367	74,000	352	479	~0
20	20	helical	100	405	58,000	450	17,000	409	64,000	379	438	+190
40	40	helical folded	25	402	44,000	453	31,000	414	54,000	378	438	+210
60	60	ridge-tile[d] (global min)	0	407	36,000	446	40,000	427	55,000	401	437	+175
90	90	gabled	42	411	25,000	442	50,000	433	67,000	380	435	+120
110	110	local min	38	405	30,000	450	45,000	438	58,000	379	438	+20
120	120	stretched	38	404	34,000	451	41,000	433	54,000	378	438	~0
140	140	stretched	59	399	45,000	456	30,000	406	51,000	377	440	−80
160	160	stretched	130	401	50,000	454	25,000	406	56,000	378	439	−100
~180	~180	linear	201	408	73,000	446	73,000	409	74,000	380	436	~0

[a] Based on conformations in which the porphyrin-like conformation passes through (P)-molecular chirality (Fig. 8.5.13) on the way to the linear; [b] ε and $\Delta\varepsilon$ in dm^3·mol^{-1}·cm^{-1}; λ in nm; [c] for the β-state and α-state of the molecular exciton; [d] with intramolecular hydrogen bonding

Computed CD and UV-visible spectra for the (P)-molecular chirality, ridge-tile conformation are shown in Table 8.5.3, which tracks the pathway and conformations illustrated in Figs. 8.5.12 and 8.5.13. Again, $\Delta\varepsilon$ assumes maximum values for helical and folded conformations, including the ridge-tile global minimum conformation. As the ridge-tile conformation begins to stretch through rotations about ϕ_1 and ϕ_2, the magnitude of $\Delta\varepsilon$ is predicted to decrease substantially as the local minimum conformation is approached. In a stretched conformation ($\phi_1 = \phi_2 = 120°$), $\Delta\varepsilon$ is predicted to be ~0; then, with further stretching by rotation about ϕ_1 and ϕ_2, the sign of $\Delta\varepsilon$ reverses. Reasonably large reversed-sign $\Delta\varepsilon$ values are found in high energy stretched conformations near ($\phi_1 = \phi_2 \cong 140°$) and ($\phi_1 = \phi_2 \cong 160°$). In the planar linear conformation, $\Delta\varepsilon$ is zero.

As indicated in Tables 8.5.2 and 8.5.3, in going from the gabled or the ridge-tile conformation to the planar porphyrin-like conformation (where $\Delta\varepsilon$ is zero), the "observed" UV-visible maximum is computed to shift to shorter wavelengths, toward the blue, as the relevant transition dipoles move toward a parallel alignment (Fig. 8.5.8). On the other hand, as the ridge-tile or the gabled conformation of Table 8.5.3 is stretched, the "observed" UV-visible maximum is predicted to red-shift, until the conformation approaches ($\phi_1 = \phi_2 = 120°$), where the relevant dipyrrinone transition dipoles adopt an in-line orientation where the predicted CD $\Delta\varepsilon$ is zero. With further stretching, the "observed" UV-visible maximum shifts toward the blue due to a realignment of the transition moments. In contrast, as the gabled conformation of Table 8.5.2 stretches into the extended in-line, the "observed" UV-visible maximum is computed to shift toward the red as the relevant transition dipoles move 180° apart – as in the in-line arrangement of Fig. 8.5.8. Thus for mesobilirubin-XIIIα, the porphyrin-like conformation is predicted to have the shortest wavelength UV-visible absorption (~367 nm) with the narrowest band width. The global minimum stretched (Table 8.5.3) and high energy extended in-line (Table 8.5.2) conformations are predicted to have the longest wavelength UV-visible absorption (~435 nm) with the narrowest bandwidth. And the global minimum ridge-tile conformation is predicted to have a broad UV-visible absorption at an intermediate maximum wavelength (427 nm). The last is, in fact,

514   8   Analysis of Bilirubin in Three Dimensions

typically observed in solutions of mesobilirubin in organic solvents (*364, 369*) and in pH 7.4 aqueous buffer (*368*). Qualitatively similar UV-visible and CD spectral characteristics are predicted for bilirubin-IXα – except with all wavelengths red-shifted by ~25 nm due to the presence of vinyl groups located on the lactam rings.

Exciton chirality theory thus allows the *prediction of the absolute configuration* of bilirubin since the relative orientation of the relevant dipyrrinone electric transition dipole moments is known (*618–620, 645*). A typical exciton coupling CD curve is bisignate (Fig. 8.5.3). Enantiomers have mirror image CD curves – either a curve with a long wavelength positive-short wavelength negative or a curve with long wavelength negative-short wavelength positive series of *Cotton* effects. Thus, the two exciton states (UV-visible transitions) predicted in Fig. 8.5.8 for the oblique case (the other two cases predict zero CD) each give rise to two corresponding CD transitions, which are always oppositely signed (*645*). The two possible signed sequences of *Cotton* effects can be correlated with the relative orientation of the transition dipoles. When the dipoles are oriented with a positive (+) torsion angle, corresponding to a positive helicity, the sequence of *Cotton* effect signs in the CD is long wavelength positive-short wavelength negative. When the torsion angle is negative (–), corresponding to a negative helicity, the sequence is long wavelength negative-short wavelength positive – as illustrated in Fig. 8.5.8. The latter orientation clearly corresponds to the (*M*)-molecular chirality global minimum ridge-tile conformer of Fig. 8.5.9; the former corresponds to the (*P*)-molecular chirality of Fig. 8.5.9.

Unfortunately, according to the analysis of the preceding section, a (*P*)-molecular chirality conformation can also exhibit a negative chirality bisignate CD – simply by rotating about ϕ_1 and ϕ_2 so as to access stretched conformations that do not change the molecular chirality but reverse the helicity of the dipyrrinone electric transition dipoles (Fig. 8.5.14). Given this possible ambiguity, it seemed wise to correlate the predictions based on exciton chirality theory of optically active bilirubins of known absolute configuration and well-defined chiral conformations with the CD spectrum of bilirubin. This was accomplished experimentally with synthetic bilirubins wherein (Fig. 8.4.5) **M** or **P** enantioselection was forced by intramolecular allosteric effects, as explained in Section 8.5.4.

8.5.4   *Stereocontrol of Bilirubin Conformational Enantiomerism by Intramolecular Buttressing*

Given the well-established intramolecularly hydrogen-bonded ridge-tile structures of bilirubin, a close examination of the "internal" stereochemistry and nonbonded steric interactions of the conformational enantiomers by Dr. *Gisbert Puzicha*, a Ph.D. student of *Günther Snatzke*[5] at the Ruhr Universität Bochum and a postdoctoral

---

[5]*Günther Snatzke* was born on October 20, 1928 in Hartberg, Austria and died on January 14, 1992 in Bochum, Germany. He studied chemistry at the University of Graz and was promoted to the Ph.D. there in 1953, initiated the *Habilitation* first at the University of Hamburg (*Tschesche*) and completed it at the University of Bonn (*Tschesche*). A leading authority on chiroptical spectroscopy, in 1972 he became one of the first generation of professors at the Ruhr-Uni-Bochum and Director of the *Lehrstuhl für Strukturchemie*.

8.5 Bilirubin Chirality Revisited. Chiral Discrimination and Optical Activity

research fellow at the University of Nevada, proved to be quite revealing. The steric environment of each of the diastereotopic α- and β-methylene hydrogens of the $-C_\beta H_2-C_\alpha H_2-CO_2H$ intramolecularly hydrogen-bonded propionic acids in bilirubin suggested a way to displace the **M** ⇌ **P** equilibrium of Fig. 8.4.5 toward either the **M** or the **P** enantiomer (*688*). In the (*M*)-chirality ridge-tile enantiomers bilirubin or mesobilirubin, the *pro-R* β-hydrogen (but not the *pro-S*) is in close nonbonded proximity to the C-10–CH$_2$, as illustrated in Fig. 8.5.17 for the symmetric bilirubin analog, mesobilirubin-XIIIα. In contrast, in the (*P*)-chirality enantiomer, the *pro-S* β-hydrogen is buttressed against the C-10–CH$_2$. Likewise, in the (*M*)-chirality enantiomer, the *pro-R* α-hydrogen is buttressed against the CH$_3$ group at C-7 or C-13; whereas, in the (*P*)-chirality enantiomer it is the *pro-S* α-hydrogen that is thrust at the C-7 or C-13 CH$_3$. Consequently, when mesobilirubin or bilirubin adopts either of the thermodynamically preferred intramolecularly hydrogen-bonded ridge-tile conformers, one conformational isomer can be destabilized relative to the other by judicious replacement of a hydrogen on each propionic acid chain with a larger group. For example, a methyl group replacing the *pro-S* hydrogen on the β-carbon of the propionic acid would be expected to destabilize the (*P*)-chirality intramolecularly hydrogen-bonded conformational enantiomer by introducing a severe nonbonded β-CH$_3$|C-10 CH$_2$ steric interaction (Fig. 8.5.17). This would result in a shift of the **M** ⇌ **P** conformational equilibrium toward the enantiomer **M**. The same shift would be expected from replacing the *pro-S* α-hydrogen with a CH$_3$ group due to steric buttressing of the α-CH$_3$ against the C-7 or C-13 CH$_3$ of the (*P*)-chirality enantiomer. In contrast, replacing a *pro-R* hydrogen with a methyl group would be expected to destabilize the (*M*)-chirality enantiomer and shift the equilibrium toward the (*P*).

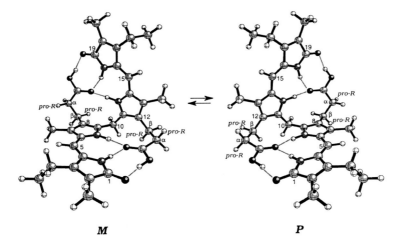

Fig. 8.5.17 Conventional Ball and Stick concave conformational representations for the ridge-tile shape (*M*)- and (*P*)-chirality intramolecularly hydrogen-bonded, interconverting enantiomers of mesobilirubin-XIIIα. In the propionic acid side chains attached to pyrrole ring carbons C-8 and C-12, the hydrogens on the β- and β'–CH$_2$– groups, and the α- and α'–CH$_2$– groups are either *pro-R* or *pro-S* (only one hydrogen of the set is designated). When the (*M*)-chirality conformer inverts into the (*P*)-chirality, steric crowding of the *pro-R* hydrogens is relieved and replaced by similar crowding of the *pro-S* hydrogen (Fig. 3 of ref. (*689*). Reproduced permission of the American Chemical Society. Copyright 1992)

516   8 Analysis of Bilirubin in Three Dimensions

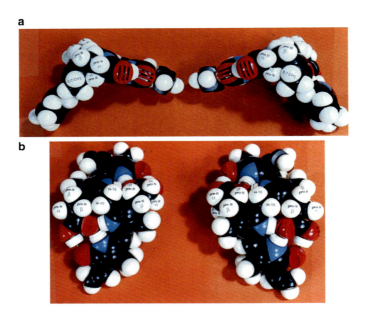

Fig. 8.5.18 CPK space-filled molecular modes of intramolecularly hydrogen bonded bilirubin representing the **M** (*left*) and **P** (*right*) enantiomeric conformations of bilirubin shown in edge (**a**) and backside (**b**) views. The *pro-R* and *pro-S* hydrogens of the propionic chains, as well as the C-10 hydrogens and C-7/C-14 methyl hydrogens are designated. The atoms are color-coded: *red* = oxygen; *blue* = nitrogen; *black* = carbon; *white* = hydrogen

Crucial to the analysis and predictions above was the necessity that the propionic acids remain intramolecularly hydrogen bonded to the opposing dipyrrinones in a ridge-tile pigment conformation. Thus a successful outcome in displacing the **M** → **P** equilibrium would also confirm the hydrogen-bonded conformations. Thus, two such bilirubin derivatives were selected for synthesis at the University of Nevada by Dr. *Gisbert Puzicha* and by *Yuming Pu* (Ph.D., University of Nevada, 1991), and *Stefan Boiadjiev* (postdoctoral research associate and Assistant Professor at the University of Nevada): α,α'-dimethyl- and β,β'-dimethylmesobilirubins **8.5.1** and **8.5.2** (*688, 689*):

8.5.1: α,α'-Dimethylmesobilirubin-XIIIα

8.5.2: β,β'-Dimethylmesobilirubin-XIIIα

8.5 Bilirubin Chirality Revisited. Chiral Discrimination and Optical Activity

A directed synthesis was required; yet, the synthesis of bilirubins and mesobilirubins had seldom been attempted since the *Fischer* syntheses of the 1930s. In fact, from the time of its first reported synthesis by *Plieninger* and *Fischer* in the early 1940s (*10, 462*), bilirubin-IXα had not been resynthesized, except by *Plieninger et al.* in 1972 (*690*). Its total synthesis was required in the earlier decade in order to establish its constitutional structure. Just prior to 1942, the *Fischer* lab had synthesized symmetric bilirubins-XIIα and XIIIα (*11, 458*); in the 1930s, their mesobilirubin analogs had been prepared (*438, 442, 444–448, 452–457*). However, in the nearly three decades that followed *Fischer*'s death and the destruction of his laboratory, bile pigment synthesis languished, and it was not renewed until it was learned that linear tetrapyrroles could be cyclized to porphyrins.

Although a synthetic linear tetrapyrrole was shown to cyclize to porphyrins in 1952 by *Erwin Corwin* (*691*), a type of porphyrin synthesis repeated some nine years later by *Alan Johnson* (*692, 693*) beginning in the late 1960s by *Peter S. Clezy* (*694*) and continuing into the 1990s (*695*) and especially in the enormous number of porphyrins synthesized in the *Kevin M. Smith* laboratories from the 1970s to present (*696–699*), the linear tetrapyrroles used were not bile pigments. Following his return to *academia* in the 1950s, *Hans Plieninger* resumed his studies on the chemistry of pyrrole compounds (*700, 701*), and in 1960s showed a new way to condense dipyrrinones to bile pigments (*702–704*). Yet, bile pigment synthesis did not blossom again until the 1970s, mainly through the work of *Plieninger* (*690, 705, 713*), *Henning von Dobeneck* (*714*) (both former *Fischer* students), *Albert Gossauer* (*713, 715–720*), *Benjamin Frydman* (*721–725*), and in the considerable synthesis work from the *Falk* laboratory (*15*). The typical synthesis path to a linear tetrapyrrole was that pioneered by *Hans Fischer* and involved coupling two dipyrrinones. In 1977, *Falk* introduced a powerful new, simplified synthesis for preparing symmetric biliverdins by an oxidative self-coupling of 9-methyldipyrrinones using DDQ (2,3-dichloro-5,6-dicyanoquinone) in trifluoroacetic acid (*15, 726*). This method was used to great advantage in tetrapyrrole synthesis in the *Falk* laboratory (*15*), and it was this method (with modification) that enabled the *Lightner* laboratory to prepare optically active mesobilirubin-XIIIα derivatives **8.5.1** and **8.5.2** and numerous other linear tetrapyrroles (*618–620, 655, 687–689, 727–730*). The modification exchanged *p*-chloranil for DDQ and formic acid for trifluoroacetic acid, which improved the yield, lessened the number of impurities due to DDQ adducts, and avoided the more difficult task of removing DDQ and its phenol reduction product.

Thus, as a consequence of the advances in linear tetrapyrrole synthesis following the "*Fischer* era", two main approaches toward the synthesis of bile pigments were envisioned (*15, 726*) the "2 + 2" approach (dipyrrinone + dipyrrinone) of *Fischer*, and the "1 + 2 + 1" approach (2 monopyrroles + 1 dipyrrole). Within the realm of the first, two routes have been employed: condensing two 9-*H* dipyrrinones with an aldehyde or ketone (as per *Fischer*) and oxidative coupling of 9-CH$_3$ dipyrrinones (according to *Falk*). In the second, a dipyrrylmethane dialdehyde was condensed with two equivalents of a pyrrolinone. The disproportionation or scrambling, reaction of a bilirubin or mesobilirubin discovered by *A.F. McDonagh* (*13, 731–735*) served as yet a third way to synthesize unsymmetric bilirubinoids (*736, 737*), but it required having the relevant symmetric bilirubinoid reactants on hand.

To recognize the effectiveness of the α- and β-methyl groups in displacing the $\mathbf{M} \rightleftarrows \mathbf{P}$ equilibrium of Fig. 8.5.17, the (R) or (S) stereochemistry at the corresponding α- and β-propionic acid carbons needs to be known. For **8.5.2**, this knowledge came from the X-ray crystal structure of a diastereomeric derivative (brucine salt) of a "resolved" monopyrrole synthetic intermediate (689); for **8.5.1**, it came from the ¹H NMR analysis of its bis-amide derivative with (–)-(S)-α-phenethylamine following resolution of **8.5.1** on a chromatography column packed with a chiral adsorbent or by extraction into $CHCl_3$ from pigment-HSA solutions (688).

The first hint that the concept expressed above was in fact being realized came from the synthesis of totally racemic **8.5.1** and **8.5.2**, meaning syntheses that produced a mixture of *meso* (R,S) and racemic (R,R) + (S,S) diastereomers. Here two types of products of the same constitutional structure were observed and easily separated by chromatographic methods. One ran faster on a silica thin-layer chromatogram, consistent with the diastereomers possessing a full set of intramolecular hydrogen bonds, *viz.* the (R,R) + (S,S). In the syntheses of **8.5.1** and **8.5.2**, the more polar, slower moving diastereomer was extractable from the mixture in $CHCl_3$ into 5% aqueous $NaHCO_3$, consistent with the *meso* (R,S)-diastereomer. The greater polarity of the (R,S)- isomer can be understood in terms of the predicted internal steric conflict in the molecular architecture: the α- or β-(S)-methyl group fits comfortably only into the (M)-helical diastereomer but not the (P). Conversely, the (R)-methyl fits comfortably only into the (P), not the (M). Under the circumstances, one of the propionic acid groups of the *meso*-(R,S) diastereomer will always be only poorly engaged in intramolecular hydrogen bonding, if at all, thereby increasing the pigment's polarity and solubility in bicarbonate. Only in the (S,S) and (R,R) diastereomers do the propionic acids adopt the tight intramolecular hydrogen bonding, as can be seen in CPK molecular models, in (SYBYL) molecular modeling, and reconfirmed most recently in time-dependent density function theory (TDDFT) calucations (738). The stereochemical conclusions were supported well by NOEs observed between the C-10–CH_2 and the β-Hs, but not the β-CH_3 groups of the *racemic* (S,S) + (R,R) diastereomers.

All three diastereomers of β,β'-dimethylmesobilirubin-XIIIα were prepared and examined by CD and $[\alpha]_D$ (393). The (R,S)-diastereomer had $[\alpha]_D = 0°cm^2g^{-1}·10^{-1}$ and exhibited no CD. The *meso* diastereomer does not favor either the (M)- or (P)-chirality conformer. The (R,R) and (S,S) isomers do. If the predicted forced displacement of the pigment's $\mathbf{M} \rightleftarrows \mathbf{P}$ equilibrium was achieved through internal steric interactions introduced by the β,β'-CH_3 groups, then the (βS,β'S) and (βR,β'R) enantiomers would be expected to exhibit intense optical activity from a conformation pre-organized for nearly maximum exciton coupling. If such conformations were not adopted, only modest optical activity would be expected due to some sort of π-π* excitation in the dipyrrinone perturbed simply by dissymmetric vicinal action from the β,β' stereogenic centers. In contrast, if the two dipyrrinones are held in a fixed relative geometry $\mathbf{M}$ or $\mathbf{P}$ through intramolecular hydrogen bonds and if, as predicted, the β,β'-methyl groups force a resolution to give the (M) or the (P) diastereomer, depending on the (R,S) stereochemistry at β and β', then a strong exciton chirality interaction should lead to strong optical activity, as was observed for the (βS,β'S) diastereomer: $[\alpha]_D^{20}$ –5,200° $cm^2g^{-1}·10^{-1}$, which may be compared to $[\alpha]_D$

8.5 Bilirubin Chirality Revisited. Chiral Discrimination and Optical Activity

$+63°$ cm^{2}g$^{-1}\cdot 10^{-1}$ due to dissymmetric *vicinal* action in (βS)-methylxanthobilirubinic acid *(618–620)*. Detection of optical activity from an excess of the (*M*)- or (*P*)-chirality conformer can also be accomplished by CD spectroscopy with the conformational enantiomers predicted *(645)* to exhibit intense bisignate *Cotton* effects, with $|\Delta\varepsilon|$ values computed to approach 270 dm^3·mol^{-1}·cm^{-1} for the pure enantiomers *(465)*. And from the signed order of the CD *Cotton* effects, the absolute configurations can be deduced. Thus, the orientation of the relevant electric transition dipole moments of the (*M*)-helical isomer corresponds to a negative (—) exciton chirality, and the (*P*)-helical corresponds to a positive (+) exciton chirality.

In complete agreement with the predictions of the allosteric model and exciton coupling theory *(645)*, *Stefan Boiadjiev* showed that the ($\beta S,\beta'S$) enantiomer of **8.5.2** exhibited an intense bisignate CD that is characteristic of the (*M*)-helical conformational enantiomer, and the ($\beta R,\beta'R$) showed an equally intense bisignate CD characteristic of the (*P*)-helicity conformer, as seen in Fig. 8.5.19 *(660, 661, 689)*. Nearly identical bisignate CD spectra may be seen from the ($\beta S,\beta'S$)- and ($\beta R,\beta'R$)-dimethylmesobilirubins *(689)*.

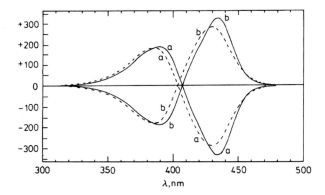

Fig. 8.5.19 Circular dichroism spectra of $1.2 \cdot 10^{-5}$ M solutions of (–)-($\beta S,'S$)-dimethylmesobilirubin-XIIIα (*a*) and (+)-($\beta R,\beta'R$)-dimethylmesobilirubin-XIIIα (*b*) in CHCl$_3$ (*solid line*) and CH$_3$OH (*dashed line*) solvents at 22°C. The $\Delta\varepsilon^{max}$ (λ^{max}) values are –337 (434) nm and +186 (390 nm) in CHCl$_3$ and –285 (431 nm) and +177 (385 nm) in CH$_3$OH for enantiomerically pure (*S,S*) with the corresponding, oppositely signed $\Delta\varepsilon^{max}$ values for (*R,R*) (Fig. 10 of ref. *(689)*. Reproduced with permission of the American Chemical Society. Copyright 1992)

The $\Delta\varepsilon$ values values in nonpolar solvents like CHCl$_3$ ($\Delta\varepsilon^{max}_{434}$ –337, $\Delta\varepsilon^{max}_{390}$ +186 dm^3·mol$^-$·cm^{-1}) are close to the theoretically predicted maximum values *(645)* and not very different from the average $|\Delta\varepsilon|$ values (183 dm^3·mol^{-1}·cm^{-1} for the short-wavelength component and 324 for the long-wavelength) taken from Table 8.5.4 over a wide range of organic solvents (except (CH$_3$)$_2$SO).

The electronic CD data were recently reproduced independently in Prof. *Sergio Abbate*'s laboratory at the University of Brescia *(738)*, where vibrational CD (VCD) measurements were also performed on the (β,β)- and (α,α)-dimethylmesobilirubin-XIIIα. *Ab initio* density functional theory (DFT) computations

520 8 Analysis of Bilirubin in Three Dimensions

reconfirmed intramolecularly hydrogen-bonded ridge-tile conformations as the most stable (738). The intramolecular hydrogen bonding, which is so very important in stabilizing the (M)-chirality conformation of ($\beta S,\beta'S$) and the (P)-chirality conformation of ($\beta R,\beta'R$) and would be expected to be strongest in nonpolar solvents such as chloroform or diethyl ether and weaker in polar aprotic solvents such as acetonitrile or acetone, is surprisingly insensitive to solvent polarity. Thus, the CD intensities in all these solvents were found to be within 10% of the average $|\Delta\varepsilon|$ value. Apparently, even the polar solvents do not significantly disrupt the intramolecular hydrogen-bond network.

Table 8.5.4 Circular dichroism and ultraviolet-visible spectral data from $5 \cdot 10^{-5}$ M ($\beta S,\beta'S$)-dimethylmeso bilirubin-XIIIα and ($\beta R,\beta'R$)-dimethylmesobilirubin-XIIIα at 22°C

Configura-tion	Solvent	Dielectric constant[a]	CD[b]			UV[b]	
			$\Delta\varepsilon^{max}$ (λ_1)	λ_2 at $\Delta\varepsilon = 0$	$\Delta\varepsilon^{max}$ (λ_3)	ε^{max}	λ
($\beta S,\beta'S$)	Carbon tetrachloride	2.2	+179 (392)	406	−393 (434)	59,000	435
($\beta R,\beta'R$)			−176 (392)	406	+385 (434)	60,800	435
($\beta S,\beta'S$)	Dioxane	2.2	+184 (389)	405	−336 (433)	56,600	431
($\beta R,\beta'R$)			−180 (389)	405	+329 (433)	58,300	431
($\beta S,\beta'S$)	Benzene	2.3	+191 (390)	406	−362 (434)	55,400	433
($\beta R,\beta'R$)			−204 (391)	407	+380 (434)	60,000	432
($\beta S,\beta'S$)	Toluene	2.4	+196 (391)	406	−375 (434)	55,800	433
($\beta R,\beta'R$)			−192 (391)	406	+367 (434)	57,500	433
($\beta S,\beta'S$)	Diethyl ether	4.3	+183 (387)	402	−365 (429)	57,500	429
($\beta R,\beta'R$)			−179 (387)	402	+357 (429)	59,200	429
($\beta S,\beta'S$)	Chloroform	4.7	+186 (389)	407	−337 (434)	55,500	431
($\beta R,\beta'R$)			−188 (389)	407	+332 (434)	55,800	431
($\beta S,\beta'S$)	Tetrahydrofuran	7.3	+188 (390)	406	−338 (433)	56,200	431
($\beta R,\beta'R$)			−184 (390)	406	−331 (434)	57,900	431
($\beta S,\beta'S$)	Dichloromethane	8.9	+180 (392)	407	−319 (433)	54,800	430
($\beta R,\beta'R$)			−176 (392)	407	+312 (433)	56,400	430
($\beta S,\beta'S$)	1,2-Dichloroethane	10.4	+193 (389)	407	−332 (433)	55,400	430
($\beta R,\beta'R$)			−189 (389)	407	+325 (433)	57,100	430
($\beta S,\beta'S$)	1-Butanol	17.1	+181 (392)	408	−293 (435)	55,800	427
($\beta R,\beta'R$)			−177 (391)	408	+287 (435)	57,500	427
($\beta S,\beta'S$)	Acetonitrile	36.2	+181 (384)	403	−315 (429)	55,000	423
($\beta R,\beta'R$)			−177 (384)	403	+308 (429)	56,700	423
($\beta S,\beta'S$)	1-Propanol	20.1	+169 (388)	406	−253 (431)	55,900	426
($\beta R,\beta'R$)			−165 (388)	406	+248 (431)	57,600	426
($\beta S,\beta'S$)	Acetone	20.7	+182 (387)	404	−322 (430)	55,400	426
($\beta R,\beta'R$)			−178 (387)	404	+315 (430)	57,100	427
($\beta S,\beta'S$)	Ethanol	24.3	+168 (389)	405	−284 (434)	55,900	426
($\beta R,\beta'R$)			−164 (389)	405	+278 (434)	57,600	426
($\beta S,\beta'S$)	Methanol	32.6	+177 (386)	405	−285 (431)	56,600	425
($\beta R,\beta'R$)			−175 (386)	405	+269 (431)	60,800	425
($\beta S,\beta'S$)	N,N-Dimethylformamide	36.7	+165 (386)	404	−246 (429)	53,100	421
($\beta R,\beta'R$)			−169 (386)	404	+252 (429)	54,000	421
($\beta S,\beta'S$)	Dimethylsulfoxide	46.5	−5.8 (369)	385	+23.0 (425)	55,900	425
($\beta R,\beta'R$)			+4.3 (368)	384	−24.2 (422)	56,700	425
($\beta S,\beta'S$)	Borate buffer, pH 8.5-10.0	(78)[c]	+107 (379)	397	−171 (424)	51,900	418
($\beta R,\beta'R$)			−105 (379)	397	+167 (424)	53,500	418
($\beta S,\beta'S$)	Phosphate buffer, pH 7.4	(78)[c]	+95 (379)	398	−150 (423)	44,500	416
($\beta R,\beta'R$)			−93 (379)	398	+147 (423)	45,800	416
($\beta S,\beta'S$)	N-Methylformamide	182.4	+200 (383)	400	−359 (427)	66,000	426
($\beta R,\beta'R$)			−188 (383)	405	+337 (427)	68,300	427

[a] From Gordon AJ, Ford RA, The Chemist's Companion, Wiley, New York, 1972, pp 4-8; [b] $\Delta\varepsilon$ and ε in $dm^3 \cdot mol^{-1} \cdot cm^{-1}$; λ in nm; [c] pure water

8.5 Bilirubin Chirality Revisited. Chiral Discrimination and Optical Activity

Surprisingly, even in polar, hydroxylic organic solvents such as ethanol or methanol, the $\Delta\varepsilon$ values are not significantly reduced from the $\Delta\varepsilon$ values in nonpolar or other polar solvents. In basified water, the CD magnitudes still remain quite large, although about half of the average value for the organic solvents. Whether this loss of **M** *vs.* **P** conformational selectivity is due to the ionization of the carboxylic acid groups or to the large dielectric constant or to the greater hydrogen-bonding ability of water is unclear. However, previous work has shown that deprotonation of the carboxyl hydrogen does not necessarily lead to substantial disruption of the network of intramolecular hydrogen bonds, since the strength of the remaining hydrogen bonds (to the carboxylate ion) is intensified (*581–585, 663–665, 739*). The very large $\Delta\varepsilon$ values in alkaline pH water support the presence (probably a major presence) of folded intramolecularly hydrogen-bonded conformations in bilirubin carboxylate anions. These observations are of considerable importance to studies of bilirubin in biological systems and in studies of its metabolism.

The CD in $(CH_3)_2SO$ appear to be a major exception. Not only were the CD magnitudes reduced to only 3-7% of the average $|\Delta\varepsilon|$ values but the *Cotton* effect signs were also inverted. $(CH_3)_2SO$ is known to be a potent hydrogen bond acceptor and is thought to insert into the conventional intramolecular hydrogen-bonding matrix of Figs. 8.3.1, 8.5.4, and 8.5.17, with the propionic acid carboxyls linked to the dipyrrinones through $(CH_3)_2SO$ solvent (*579, 580, 601, 649–651*). In this model, where intramolecular hydrogen bonding is swollen through solvent participation, the allosteric effects due to the β,β'-methyls may be substantially decreased and play only a limited role in the selection of the most stable conformational diastereomer. However, the exciton chirality map of Fig. 8.5.11 suggests a more likely explanation. Swelling the matrix of intramolecular hydrogen bonds opens the dihedral angle (θ) and re-orients the relevant ~450 nm electric transition dipoles (Fig. 8.5.9), as in Fig. 8.5.14. Thus, the intercalated $(CH_3)_2SO$ may not fully break open the pigment's ridge-tile structure but only relax it.

An experimental-theoretical study (*738*) from *Sergio Abbate*'s lab in Brescia examined changes in the electronic CD and VCD of ($\alpha R, \alpha' R$) and ($\beta S, \beta' S$) dimethylmesobilirubin-XIIIα in going from 100% $CHCl_3$ to 100% $(CH_3)_2SO$, finding that significant changes do not occur in the electronic CD until 40–50% $(CD_3)_2SO$, with curve flattening at >60% $(CH_3)_2SO$ and weakly inverted CD effects at 100% DMSO. DFT calculations for the pigments in $(CH_3)_2SO$ fit well with a "pseudo-extended conformation" encompassing four intramolecular NH$\cdots$O=C hydrogen bonds. Similar, but more pronounced examples of molecular chirality selectivity are the inversion of the CD spectra of the bilirubin-HSA complex on addition of a few drops of $CHCl_3$ or volatile anesthetic (*740, 741*), or the induction of optical activity in $CHCl_3$ by the addition of isopropyl amine (*742*). Consequently, solutions of bilirubin in $(CH_3)_2SO$ may well be good models for solutions of the pigment in alkaline aqueous solvents or in biological systems.

As predicted, α,α-dimethylmesobilirubin-XIIIα (**8.5.1**) (*688*) behaves similarly to its β,β′ isomer (**8.5.2**) (*689*). Thus, (αS,α′S)-dimethylmesobilirubin-XIIIα gave a greatly enhanced $[\alpha]_D$ −5000°cm²g⁻¹·10⁻¹ (CHCl₃) compared to the monochromophore component, (αS)-methylxanthobilirubinic acid with $[\alpha]_D$ +50° cm²g⁻¹·10⁻¹ (CHCl₃). And the (αR,α′R)-enantiomer showed a (+) exciton chirality in its CD spectrum in CHCl₃ and in CH₃OH, or essentially mirror image spectra to those of (βS,β′S)-dimethylmesobilirubin-XIIIα (Fig. 8.5.19). Thus, for the same range of solvents as shown in Table 8.5.4 for the β,β-dimethylmesobilirubins, the α,α′-dimethyl enantiomers show the same distribution of large *Cotton* effects (Table 8.5.5) resulting from intramolecular allosteric effects with the (*S,S*) and (*R,R*) enantiomer exhibiting nearly mirror image CD spectra (*728*).

Research on intramolecular allosteric effects resulting from substituting a methyl group at each α- or β-position of the propionic acids of mesobilirubin-XIIIα led to many opportunities for exploring steric effects on the **M ⇌ P** conformational equilibrium of bilirubinoid pigments. One question raised in such studies was which methyl substituent, the one at the α-carbon or that at the β-carbon, exerts the larger steric effect; *i.e.* which dominates the conformational equilibrium. For methyl substituents, the question was settled by *Stefan Boiadjiev*, who set (αR)- and (βS)-methyls in opposition by synthesizing (αR,β′S)-dimethylmesobilirubin (*728*). As can be seen from Table 8.5.5, this pigment clearly shows a strong (+) exciton chirality, as predicted for (*R*)-methyls, thereby proving that an α-substitutent exerts a larger steric buttressing effect than a β-substituent and confirming the sense that bilirubins prefer to be intramolecularly hydrogen bonded. Other interesting studies of steric size, as judged by CD spectroscopy of intramolecularly hydrogen-bonded mesobilirubins pit various groups: *t*-butyl, isopropyl, ethyl, phenyl, benzyl (*729*); OCH₃ and SCH₃ (*730*) at the (α′R)-carbon of one propionic acid chain against a methyl group at the (βS)-carbon of the other propionic acid.

Is the structure of bilirubin of only academic interest to stereochemists? One might expect at least a tentative "no" to this question because the pigment is a natural product. And it will become apparent in the following section that chemical structure plays a major role in bilirubin's metabolism and in the widely used therapeutic treatment for jaundice in the newborn – phototherapy.

8.5 Bilirubin Chirality Revisited. Chiral Discrimination and Optical Activity

Table 8.5.5 Circular dichroism and UV-visible spectral data from $(\alpha R,\alpha'R)$, $(\alpha R,\beta'S)$ and $(\beta S,\beta'S)$-dimethylmesobilirubins-XIIIα at 22°C[a]

Pigment Configuration	Solvent	ε[b]	A[c]	CD[d] $\Delta\varepsilon^{max}(\lambda_1)$	$\Delta\varepsilon^{max}(\lambda_2)$	UV[d]		
$(\alpha R,\alpha'R)$	Hexane	1.9	+615	+422 (430)	−193 (390)	$(\alpha R,\alpha'R)$	Hexane	1.9
$(\alpha R,\beta'S)$			+340	+229 (428)	−111 (386)	$(\alpha R,\beta'S)$		
$(\beta S,\beta'S)$			−617	−423 (430)	+194 (388)	$(\beta S,\beta'S)$		
$(\alpha R,\alpha'R)$	CCl$_4$	2.2	+586	+396 (435)	−190 (391)	$(\alpha R,\alpha'R)$	CCl$_4$	2.2
$(\alpha R,\beta'S)$			+319	+216 (433)	−103 (391)	$(\alpha R,\beta'S)$		
$(\beta S,\beta'S)$			−572	−393 (434)	+179 (392)	$(\beta S,\beta'S)$		
$(\alpha R,\alpha'R)$	CHCl$_3$	4.7	+537	+344 (435)	−193 (391)	$(\alpha R,\alpha'R)$	CHCl$_3$	4.7
$(\alpha R,\beta'S)$			+265	+165 (430)	−100 (386)	$(\alpha R,\beta'S)$		
$(\beta S,\beta'S)$			−523	−337 (434)	+186 (389)	$(\beta S,\beta'S)$		
$(\alpha R,\alpha'R)$	Tetrahydrofuran	7.3	+544	+351 (432)	−193 (389)	$(\alpha R,\alpha'R)$	Tetrahydrofuran	7.3
$(\alpha R,\beta'S)$			+115	+70 (433)	−45 (394)	$(\alpha R,\beta'S)$		
$(\beta S,\beta'S)$			−526	−338 (433)	+188 (390)	$(\beta S,\beta'S)$		
$(\alpha R,\alpha'R)$	CH$_2$Cl$_2$	8.9	+526	+334 (432)	−192 (389)	$(\alpha R,\alpha'R)$	CH$_2$Cl$_2$	8.9
$(\alpha R,\beta'S)$			+273	+174 (431)	−99 (388)	$(\alpha R,\beta'S)$		
$(\beta S,\beta'S)$			−499	−319 (433)	+180 (392)	$(\beta S,\beta'S)$		
$(\alpha R,\alpha'R)$	(CH$_3$)$_2$CO	20.7	+515	+328 (430)	−187 (387)	$(\alpha R,\alpha'R)$	(CH$_3$)$_2$CO	20.7
$(\alpha R,\beta'S)$			+172	+107 (426)	−65 (384)	$(\alpha R,\beta'S)$		
$(\beta S,\beta'S)$			−504	−322 (430)	+182 (387)	$(\beta S,\beta'S)$		
$(\alpha R,\alpha'R)$	CH$_3$CH$_2$OH	24.3	+427	+257 (431)	−170 (388)	$(\alpha R,\alpha'R)$	CH$_3$CH$_2$OH	24.3
$(\alpha R,\beta'S)$			+94	+57 (424)	−37 (382)	$(\alpha R,\beta'S)$		
$(\beta S,\beta'S)$			−452	−284 (434)	+168 (389)	$(\beta S,\beta'S)$		
$(\alpha R,\alpha'R)$	CH$_3$OH	32.6	+392	+238 (429)	−154 (384)	$(\alpha R,\alpha'R)$	CH$_3$OH	32.6
$(\alpha R,\beta'S)$			+39	+23 (422)	−16 (380)	$(\alpha R,\beta'S)$		
$(\beta S,\beta'S)$			−462	−285 (431)	+177 (386)	$(\beta S,\beta'S)$		
$(\alpha R,\alpha'R)$	CH$_3$CN	36.2	+490	+310 (428)	−180 (385)	$(\alpha R,\alpha'R)$	CH$_3$CN	36.2
$(\alpha R,\beta'S)$			+157	+99 (428)	−58 (384)	$(\alpha R,\beta'S)$		
$(\beta S,\beta'S)$			−496	−315 (429)	+181 (384)	$(\beta S,\beta'S)$		
$(\alpha R,\alpha'R)$	(CH$_3$)$_2$SO	46.5	−52	−23 (429)	+29 (382)	$(\alpha R,\alpha'R)$	(CH$_3$)$_2$SO	46.5
$(\alpha R,\beta'S)$			−57	−26 (425)	+31 (382)	$(\alpha R,\beta'S)$		
$(\alpha R,\alpha'R)$			+29	+23 (425)	−6 (369)	$(\alpha R,\alpha'R)$		
$(\alpha R,\beta'S)$	CH$_3$NCHO	182.4	+625	+382 (427)	−243 (384)	$(\alpha R,\beta'S)$	CH$_3$NCHO	182.4
$(\alpha R,\alpha'R)$			+131	+80 (427)	−51 (385)	$(\alpha R,\alpha'R)$		
$(\alpha R,\beta'S)$			−559	−359 (427)	+200 (383)	$(\alpha R,\beta'S)$		

[a] The $\Delta\varepsilon$ values cited, which are generally larger than those of Table 8.5.4, arise from pigments with 100% e.e.; conc. ~1.5·10^{-5} M; [b] Dielectric constant taken from Gordon AJ, Ford RA, The Chemist's Companion, Wiley, New York, 1972, pp 4-8; [c] $\mathbf{A} = \Delta\varepsilon^{max}(\lambda_1) - \Delta\varepsilon^{max}(\lambda_2)$; [d] $\Delta\varepsilon$ and ε in dm^3·mol^{-1}·cm^{-1}, λ in nm

9 Understanding and Translating Bilirubin Structure

By the end of the second millennium A.D., nearly 100 years after *Nencki* and *Küster* had initiated the first meaningful degradation of bilirubin (*208, 244*) that set into motion its eventual structure elucidation, the constitutional and stereochemical structure of the pigment had been solved. From the *Fischer-Plieninger* structure of 1942 that left the exocyclic carbon-carbon double bond stereochemistry an open question, until its eventual determination as (4Z,15Z) both in the solid by X-ray crystallography and in solution by NMR methods, there had been no evidence to exclude (*E*)-isomers as the correct structure(s). Surely *Fischer* and *Plieninger* (*10, 462*) understood carbon-carbon double bond stereochemistry, as did virtually every organic chemist in the 1930s. Yet they chose not to indicate the stereochemistry around the exocyclic double bonds of bilirubin and biliverdin. If they had, they would have recognized four possible double-bond stereoisomers (see Fig. 9.1) but were simply disinclined to draw a definitive representation for the natural product. Even with the knowledge of the hemin structure and the required all-(Z) stereochemistry at the *meso* carbons, *Fischer* and *Plieninger* (*10, 462*) were not persuaded to follow *Lemberg*'s logic (*433, 483*) and failed even to predict the all (Z)-stereochemistry in the bile pigments. Which emphasizes again that *Fischer* "hung his hat on experimental evidence" and not the *Conjecturalchemie* (conjectural chemistry) that *Thudichum* disdained (*191*). Yet if space-filled (*CPK*) molecular models had been available and used, they might have raised doubt that (*E*)-isomers are more stable than (Z).

D.A. Lightner, *Bilirubin: Jekyll and Hyde Pigment of Life*, Progress in the Chemistry of Organic Natural Products, Vol. 98, DOI 10.1007/978-3-7091-1637-1_9,
© Springer-Verlag Wien 2013

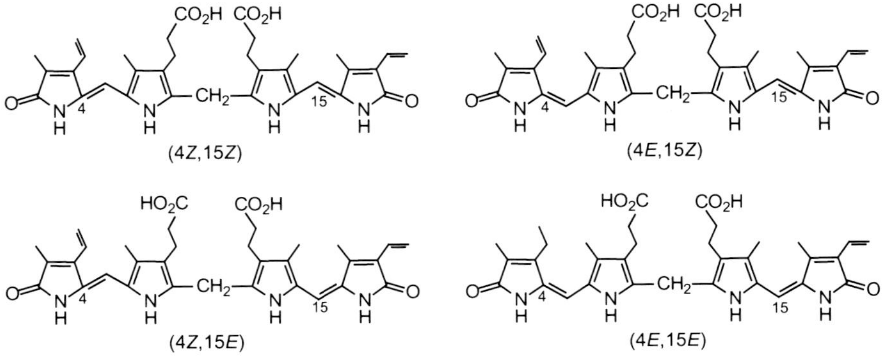

Fig. 9.1. Linear representations of the four (Z,E) diastereoisomers of bilirubin

In fact, four diastereomers of bilirubin: (4Z,15Z), (4E,15E), (4Z,15E), and (4E,15E) can be drawn, as in Fig. 9.1. Only one, the (4Z,15Z) represents the natural isomer, despite the doggedly-represented (E)-isomer structures even today in some biochemistry textbooks, reference sources, and the chemical/biochemical scientific literature of the current millennium. Given the certainty of the (4Z,15Z) configuration of natural bilirubin, it is not clear why the (E)-configuration isomers persist in representations of the natural pigment; nevertheless, it is a fair academic question to ask whether and where bilirubin (E)-isomers might in fact exist, and how they might be detected and studied.

From the current vantage point it seems likely that knowing the final details of the structure of bilirubin, including its (Z,E) stereochemistry might have been viewed as nothing more than an obscure problem of limited academic interest had it not been for two observations from medicine. First, clinical jaundice is the most common medical diagnosis in the neonate, with some 60–70% of healthy full-term babies displaying elevated serum bilirubin during the first few days after birth (743–749). Left untreated, severe unconjugated hyperbilirubinemia (jaundice, icterus) can lead to irreversible neurologic damage when the pigment enters the brain (kernicterus), as manifested in hearing loss, auditory neuropathy, retarded motor development, cerebral palsy, and even death (13, 14, 155–157, 743–749). Even mild hyperbilirubinemia may be unsafe, and optimal management has been controversial. Second, from its first curious report in 1958 (750, 751) that by exposing the jaundiced neonate to sunlight or visible light, the exposed areas of the skin were bleached of the yellow pigment and the levels of circulating bilirubin were reduced, what is now known as neonatal phototherapy spread rapidly because it lessened the need for exchange transfusion to remove the pigments, and because of its simplicity.

In an apparent return to its historic roots couched in mammalian physiology and benefitting from the long, dedicated structural investigations by *Hans Fischer*, interest in bilirubin was thus re-ignited by one of the major advances in neonatal medicine of the last half of the 20th century: a non-invasive, surprisingly simple method to treat a serious condition (insofar as a photon of the right energy penetrating the neonate's skin to the subcutaneous capillaries is medically non-invasive). Understanding how visible light acts upon bilirubin to detoxify the pigment thus constituted one of the great challenges to photobiology and internal medicine from the early 1960s to the mid-1980s.

What ensued from the "discovery" of neonatal phototherapy in 1958 was a flurry of research activity coming from clinicians and basic scientists that led to over 600 research articles and abstracts (*752, 753*) being published in the 20 years following. As revealed in *Chemical Abstracts*, articles seemed to address every possible aspect of phototherapy: assessing the efficacy of the light (wavelength or energy), blue light, green light, thermal effects, dose, skin penetrability and reflectance, Plexiglas incubators, potential retinal damage, circadian rhythms, delayed onset of secondary sexual characteristics, low birth weight, nursing *vs.* non-nursing, Rh negative blood incompatibility of mother and infant light-administered drug compatibility, diarrhea, the premature neonate, eye shields and ocular risks, cellular growth, hemolytic disease, autohemolysis, peripheral blood flow, insensible water loss, "bronze baby", gut transit time, urinary excretion of tryptophan metabolism, serum albumin binding capacity, apnea and brachycardia, platelet injury, changes in serum gonadotropin, nitrogen electrolyte and water balance, luteinizing hormone; elimination of riboflavin and gentamicin; sister chromatid exchange, ionized calcium, *etc.* Among the flurry of activity, from the 1960s through the 1970s, the more chemistry-related articles eventually explained how phototherapy works, as was summarized in the mid-1980s by *McDonagh* and *Lightner* (*155–157*).

9.1 Bilirubin Chemistry Reawakened Thanks to Jaundice Phototherapy

Reinvestigations and further studies of bilirubin were thus driven by a desire to understand how neonatal phototherapy worked, how light caused the levels of circulating bilirubin to decrease, and to answer questions as to how phototherapy might possibly introduce untoward side effects by shunting the pigment off into sensitive tissues or by the formation of photoproducts of a toxic nature. From the early 1960s, the problem attracted investigators ranging from clinicians to basic scientists who learned that understanding how phototherapy works required knowing more about the structure of bilirubin than that presented by *Hans Fischer* and *Hans Plieninger* in 1942 (*10, 462*) – and even more than the state of knowledge of its structure 25 years later. Consequently, a full knowledge of the pigment's structure also became part and parcel of understanding bilirubin metabolism and, not unrelated, its metabolic fate during neonatal phototherapy. History here is both convoluted and confusing, even to this author, who lived and worked through it.

However, early in the sixth decade of the 20th century, as the areas of bilirubin photochemistry, photobiology, and photomedicine emerged synergistically, the structure of the pigment was still incompletely known. Though most thinking organic chemists in 1960 would then have immediately settled on the bis-lactam tautomer, the *Fischer* bis-lactim structure had not been ruled out, the ene-amide exocyclic C=C double bond stereochemistry was as yet unproven, and the notion of intramolecular hydrogen bonding was only obscure. These unknowns all awaited resolution in the seventh decade, and by the eighth had been widely accepted – as revealed in Sections 7 and 8. The search for the molecular basis for jaundice phototherapy, mainly from the early 1960s to mid-1980s, was thus hampered from the start by an insufficient knowledge of the bilirubin structure. The late 1960s to early 1980s was also a time that might be characterized as replete with unnecessary confusion due to misinterpretation, misconceived beliefs, and a mishmash of conflicting and often scientifically unwarranted conclusions. Yet the rebirth of bilirubin research it rekindled provided the molecular basis for how phototherapy works. That rebirth also intersected with the emergence and flowering of the new and exciting field of singlet oxygen ($^1O_2$) in chemistry and biology, and when integrated into bilirubin's domain almost surely delayed the eventual full explanation of the molecular basis for jaundice phototherapy. From the viewpoint of one who worked in the areas of photochemistry and photosensitized oxidation, in retrospect the simplest explanation was detected early but fully grasped only later.

Perhaps the first intriguing observations of a biochemical nature related to phototherapy came in 1960 (*754*) from *Blondheim* in Jerusalem who reported that serum or albumin to which aqueous alkaline bilirubin had been added showed an increase in diazo reactivity upon exposure to visible light as compared to an unexposed control. Exposure to light of serum from severe physiologic jaundice (due to unconjugated hyperbilirubinemia) also increased the direct diazo reactivity and decreased the indirect (see Sections 4.1 and 4.10). A shift in the UV-visible absorption from 450 nm through 420 nm was also seen. However, the results were not viewed as related to the success of jaundice phototherapy but to the pathogenesis of kernicterus, which is what phototherapy was thought to prevent. The investigation was followed by two reports in 1962 (*755, 756*) that a water-soluble diffusable bilirubinoid with UV-visible λ_{max} 420 nm was produced by the light. The focus was again that the pigment might cross into the brain and cause kernicterus (*755*) and that light-induced changes in the jaundiced neonate might destroy excess bilirubin and prevent the onset of kernicterus (*756*).

At about the same time, and seemingly unrelated to phototherapy *Schmid*[1] *et al.* synthesized [^{14}C]-bilirubin (*757*) and used it to follow the "alternate pathways" of

---

[1] *Rudi Schmid* was born on May 2, 1922 in Glarus, Switzerland, the son of two practicing physicians, and died on December 1, 2007 in Kentfield, California. He received the baccalaureate from the Kantonsschule Zürich, and the Dr. med. in 1947, following which he interned at the University of California, San Francisco (UCSF) from 1948-1949, assumed a residency and fellowship leading to a Ph.D. in medical sciences at the University of Minnesota with *Cecil J. Watson* (the "father of hepatology"), under whose tutelage he was introduced to porphyrins and bile pigments. In 1954 he assumed a NIH Fellowship in biochemistry at Columbia University under *David Shemin*,

9.1 Bilirubin Chemistry Reawakened Thanks to Jaundice Phototherapy

its elimination in the *Gunn* rat and *Crigler-Najjar* patient (*758*); *i.e.* pathways not involving glucuronidation. (The *Gunn* rat has a congenital deficiency of the glucuronosyl transferase enzyme and thus cannot conjugate and eliminate its bilirubin by the normal pathway – secretion by the liver into bile, leaving it in a lifelong hyperbilirubinemic state. It serves as an animal model for the jaundiced neonate and for humans with a congenital deficiency of glucuronosyl transferase, the rare *Crigler-Najjar* syndrome.) Synthesis of the labeled bilirubin was to prove important in *Schmid*'s later studies and those of *J. Donald Ostrow*.[2] Thus, in the early 1960s, while investigating the fate of bilirubin in alternate metabolic pathways, *Schmid et al.* found labeled material in the bile of rats following intravenous injection with [^{14}C]-bilirubin (*758*). The bile was found by *Schmid et al.* to contain small amounts of unconjugated bilirubin, but 33–40% of the labeled material appearing in *Gunn* rat bile was in the form of polar, water-soluble bilirubin catabolites, the yellow components of which failed to give a diazo reaction and lacked the UV-visible spectral properties of bilirubin. They were said possibly to be related to the breakdown products on exposure of bilirubin to mild alkali, or to light (*754*), a statement that probably stimulated further investigations, especially those involving (photo) oxidative scission. Whether the experiments were conducted in the dark, or more likely under room lighting, is unclear, and *Schmid et al.* followed the studies seven years later with an analysis of products secreted into bile of photo-irradiated *Gunn* rats. Thus, *in vivo* photo-irradiation studies from the *Schmid* group (*759*) followed late in the 1960s, along with investigations due to two additional research groups from which observations were also made (and published in 1970). Though they were not understood until much later, (i) *Schmid*'s group showed for the first

---

continued on in an intramural fellowship at the NIH, where he solved the puzzle of direct reacting (diazo reaction) bilirubin – showing it to be the glucuronide. After three years in Harvard's Thorndike Laboratory, Boston City Hospital (1959–1962), and at the University of Chicago (1962–1966), he rejoined UCSF in 1966 to help create the Division of Gastroenterology. From 1983-1989 he was Dean of the Medical School before retiring from UCSF in 1995.

[2] *J. Donald Ostrow* was born in New York City in January 1, 1930 and died in Seattle on January 9, 2013, received the B.S. degree in chemistry in 1950 at Yale University, the Dr. med. in 1954 from Harvard Medical School, assumed an internship in 1955 at Johns Hopkins Hospital in medicine and began his residency in 1958 at Peter Brent Brigham Hospital, Harvard in 1958. Following three years as a Research Fellow with *Rudi Schmid* at the Thorndike Memorial Laboratory of Boston City Hospital, in 1962 he moved to Case Western Reserve University where, as assistant professor of medicine and gastroenterologist at University Hospitals, Cleveland, except for a leave of absence in the late 1960s to *Barbara Billing's* lab in the Medical Unit of the Royal Free Hospital, London, where in 1970 he received the M.Sc. degree, and where he remained until 1970 when he assumed a position as Associate Professor of Medicine and Veterans Administration Medical Investigator in the Veterans Administration Hospital in Philadelphia and the Department of Medicine, Gastroenterology Section, of the University of Pennsylvania Medical School. In 1978 he was promoted to Professor of Medicine at the Veterans Administration Lakeside Hospital, Northwestern University Medical School in Chicago, a position he held until achieving emeritus status and moving to the University of Washington Medical School as Affiliated Professor of Medicine in 1998.

530   9   Understanding and Translating Bilirubin Structure

time that phototherapy of jaundiced neonates led to excretion of *diazo-negative*, polar water-soluble photoproducts in bile (*759*). (ii) *Ostrow* and *Branham* claimed to have isolated unconjugated bilirubin from the bile of (jaundiced) *Gunn* rats undergoing photo-irradiation from six 15 W daylight fluorescent lights (*760*). (iii) *Davies* and *Keohane* produced UV-visible absorbance difference (AD) spectra of bilirubin in $CHCl_3$, as well as in aqueous buffered human serum albumin (HSA) solutions, during photo-irradiation with blue light from a filtered 1000 W xenon-mercury arc lamp (*761*). Of course, in (i) and (ii) the bile contained numerous other ^{14}C-labeled photodegradation products. The fact that a major diazo-negative material was found in (i) was explained 12 years later, when a yellow, polar, diazo-negative photoproduct was fully characterized (*762*).

In (ii) above, the purported unconjugated bilirubin isolated was identified as such on the basis of a positive (indirect) diazo reactivity, a positive *Gmelin* reaction, and a UV-visible absorption band centered near 450 nm. As seen in earlier chapters of this book, many bilirubinoid pigments, however, give a positive *Gmelin* reaction, the diazo reaction is not specific for bilirubin, nor is an absorption maximum at maximum ~450 nm uniquely diagnostic of bilirubin – though collectively they are suggestive of it. Further characterization of the biliary pigment as (apparently) natural bilirubin awaited further studies, presented in April 1974 (*763*) in a book containing the lectures from a conference on birth defects. There, *Ostrow et al.* characterized the pigment by its mass spectrum and its conversion to bilirubin dimethyl ester. Discovering natural bilirubin in phototherapy bile must have been viewed as extraordinary then and presented a conundrum because the natural pigment was known to be unable to cross the liver into bile in the absence of conjugation. It raised a vexing issue that *Ostrow* repeatedly attempted to resolve. Despite this intriguing finding, in 1970 it was thought by most investigators that the success of phototherapy was due to photodestruction of bilirubin and was not linked to seemingly obscure observations of unconjugated bilirubin appearing in bile, or the presence in bile of yellow, diazo-negative pigments more polar than bilirubin.

The spectroscopic studies of *Davies* and *Keohane* in (iii) initiated the more chemical investigations of phototherapy when they detected by AD spectroscopy (*761*):

> Irradiation of bilirubin ($6 \cdot 10^{-5}$ *M*) in buffered serum albumin caused rapid production of a chromophore absorbing at 490 nm.

> A similar peak was generated during the destruction of bilirubin in other solvents (e.g. chloroform, carbon tetrachloride) . . .

> A second characteristic of photodecomposition in albumin solutions was the selective destruction of a 420 nm chromophore . . . . This absorbing species was detectable as a shoulder in the original solution . . .

Absorbance difference spectroscopy measures *differences* in absorbance. Thus, the *Davies* and *Keohane* AD study showed that photoirradiation of bilirubin solutions produces at least one photoproduct with (i) greater absorbance in the long wavelength wing of bilirubin's long wavelength absorption band, and (ii) lesser absorbance in the

9.1 Bilirubin Chemistry Reawakened Thanks to Jaundice Phototherapy

short wavelength wing. The authors interpreted the results in terms of the emergence of a photolabile product absorbing at 490 nm during photoirradiation of bilirubin in aqueous buffered HSA and in $CHCl_3$. In the former, they assumed rapid destruction of a chromophore absorbing at 420–430 nm. Unfortunately, the AD spectroscopic results were subsequently misinterpreted by other investigators (764–767), who apparently assumed that photoirradiation of bilirubin produced a photoproduct with a characteristic absorption (λ_{max}) at 490 nm. The AD results led to the belief that photolabile products were produced early in the photo-irradiation. In their final study related to bilirubin photochemistry, in 1973, *Davies* and *Keohane* (768) reconfirmed and extended their studies published in 1970 but produced no new information on what subsequently became known as the "490 product". It was only later revealed by *McDonagh*'s AD studies (769) that the increase in absorbance at 490 nm was due to a broadening of the long wavelength wing of the absorption band of rapidly emerging bilirubin photoisomers whose long wavelength absorption band fell mainly under that of bilirubin itself. On the other hand the "430 nm pigment" might have predicted the ultrafilterable golden yellow pigment absorbing at 430–435 nm that was isolated by *Kapitulnik et al.*, also in 1973 (764, 765). The authors then reported (765) separating two diazo-negative substances, one of which was a pent-dyo-pent negative, $CHCl_3$-soluble yellow pigment absorbing in the UV-visible spectrum at λ_{max} ~430 nm. Though no structural information could be provided at the time, the formation of the "430 nm pigment" was suggestive of photochemical reactions yet to be understood, and the "430 nm pigment" became a topic of some discussion over the next 10 years or so at bilirubin meetings.

Thus, the 1960s ushered in a quest to understand how jaundice phototherapy works by unleashing a small torrent of interest among those with training in the molecular sciences. As indicated above, among the earliest investigators were *Rudi Schmid*, who in 1970 published what was probably the first peer-reviewed work (759) on the effect of phototherapy on pigment excretion in the jaundiced neonate, and *J. Donald Ostrow*, who pursued both photochemical research and photobiological studies on the *Gunn* rat (760). *Ostrow*'s tenacious pioneering studies focused first on the alternate pathways for bilirubin excretion and on bilirubin photo-oxidation products and photocatabolites in the *Gunn* rat to determine whether the compounds from the two pathways were the same (760). This approach stemmed not illogically from the early belief that photodestruction of the pigment accounted for the photobleaching of the neonates' skin during phototherapy.

Earlier, in his initial studies, *Ostrow* reported his investigation to determine whether the yellow derivatives from [^{14}C]-bilirubin photodecomposition were the same as the yellow compounds found in the bile of a *Gunn* rat kept in the dark (770). The family of products, which *Ostrow* named "photo-oxidation products", were found to pass rapidly into bile following intravenous injection but differed from the family of excretion products found normally in the *Gunn* rat. Thus began *Ostrow*'s long association with bilirubin and jaundice phototherapy research, and a career as one of the earliest proponents of bilirubin photo-oxidation. The work was followed shortly thereafter by more studies between 1968 and 1970, and by that of his former mentor, *Rudi Schmid* (*759*), with both groups using [^{14}C]-bilirubin. The work with *Branham* was presented at the June 6, 1969 Symposium on Bilirubin Metabolism in the Newborn (*760*) and also received at the *Gastroenterology* journal office on June 18, 1969 (*771*). The latter article described the decomposition of both oxygenated and deoxygenated aqueous solutions of human serum albumin using six 15 W fluorescent lamps. The studies led to those authors' belief that biliverdin is an early intermediate in bilirubin photodecay (*771*), which turned out to be incorrect (*772*).

In April 1970, *Schmid et al.* submitted to the journal *Pediatrics* the first peer-reviewed work (*759*) on the effect of phototherapy on pigment excretion in the jaundiced neonate. The work, using [^{14}C]-bilirubin, related that phototherapy using daylight fluorescent lamps led to the appearance of the ^{14}C label in aspirated duodenal bile in polar, water-soluble, predominantly diazo-negative bilirubin derivatives. The authors interpreted the data as suggesting (*759*):

> . . . that exposure to light simply accelerates the breakdown of bilirubin by mechanisms similar to those normally operative in patients with the Crigler-Najjar syndrome and in Gunn rats.

A suggestion that was elaborated in greater detail by *Ostrow* and *Branham* with the explanation (*760*):

> This suggests that the photoderivatives excreted by the Gunn rat during exposure to intense illumination are the same as those excreted by this animal under normal lighting conditions. Apparently, phototherapy simply accelerates the alternate pathway(s) of bilirubin catabolism which exist normally in the Gunn rat. . . . Since I have shown previously . . . that the Gunn rat bile pigments differ from the *in vitro* photoderivatives of bilirubin, it is evident that catabolites formed during phototherapy of the jaundiced animal are not the same as the photoderivatives formed from bilirubin in the test tube. The present *in vivo* studies thus constitute the first clear demonstration that the action of phototherapy is to accelerate the catabolism of bilirubin to water-soluble derivatives which are excreted rapidly, principally in the bile.

Why the photochemistry of bilirubin should be different *in vivo* and *in vitro* was not explained and was later shown to be incorrect. A more important finding was also mentioned (*760*):

> During the *control* period, the bile was pale yellow in color and gave a weak, atypical diazo reaction. When the animal was exposed to light, the bile turned a deep golden-brown color. This was accompanied by a seven-fold increase in excretion of diazo-reactive products, and a parallel but even greater increase in biliary excretion of radioactivity.

The presence of bilirubin in phototherapy bile was subsequently ascribed, in 1970, to an alteration of the hepatocyte membrane by phototherapy (*773*):

> It is tentatively suggested that phototherapy alters the state of the pigment and/or the hepatocyte membrane so that the liver becomes freely permeable to passive diffusion of bilirubin from blood to bile. This hypothesis is supported by the finding that, during phototherapy, the concentration of bilirubin in bile rose to a maximum value just below the minimum concentration attained in the plasma. . . .

9.1 Bilirubin Chemistry Reawakened Thanks to Jaundice Phototherapy

Yet *Ostrow* thought not only of biological mechanisms but clearly, without further explanation, he thought in chemical terms in the phrase " . . . phototherapy alters the state of the pigment . . ." As vague as the phrase might seem, it came remarkably close, already in 1970, to suggesting that a photochemically induced change in bilirubin structure might explain how phototherapy works. But by 1972, only the more biological explanation remained (*502*):

> This finding suggests that phototherapy alters the canalicular membrane of the hepatocytes, rendering the liver freely permeable to passive diffusion of bilirubin from blood to bile.

although by 1977, this association was proved to be untenable, and the excretion of unconjugated bilirubin remained a mystery (*766*):

> The second conclusion is that excretion of these polar photoproducts does not mediate the observed outpouring of unconjugated bilirubin that occurs during phototherapy. The mechanism of this remarkable excretion of unconjugated bilirubin remains a mystery.

Another explanation, one involving phototautomerization and an imaginary suppression of carboxylic acid dissociation by hydrogen bonding, had been proposed and discarded in 1974 (*763*):

> Our next postulate to explain the excretion of unconjugated bilirubin was that absorption of light energy might tautomerize the molecule . . . and thus temporarily interrupt the internal hydrogen-bonding which suppresses the ionization of the carboxyethyl side chains. . . . The highly ionized and therefore water-soluble bilirubin phototautomer might then be capable of being excreted into the bile, where it would return to its ground state and once again be indistinguishable from ordinary unconjugated bilirubin. Unfortunately, the evanescence of photoactivated intermediates . . . not only renders this hypothoesis unlikely, but also renders difficult its experimental verification. . . . The mechanism for augmented excretion of unconjugated bilirubin during phototherapy remains unexplained.

And so, by 1976, though the pigment was certified as unconjugated bilirubin, the mystery as to how it got into phototherapy bile was unsolved (*774*). All the while, research on bilirubin photodegradation and its role in phototherapy continued on from the 1960s into the mid 1970s, while other hints accumulated as to why diazo-positive unconjugated bilirubin appeared in bile along with other diazo-negative bilirubinoids.

Of course, *Donald Ostrow* and *Rudi Schmid* were but two of several key figures in the quest to understand phototherapy. Others included *Onishi, McDonagh,* and *Lightner*. By July 1968, *Lightner* had been introduced by *Rudi Schmid* to the unsolved problem of how phototherapy works. This occurred at the Tetrapyrrole Gordon Research Conference at Crystal Mountain, Washington, where *Lightner*, an expert in organic mass spectrometry, lectured on mass spectra of bile pigments, studied in collaboration with *Cecil J. Watson* and *Albert Moscowitz*. *Lightner*'s initial collaborative study with *Schmid* was broadened two years later when *Antony McDonagh* joined the *Schmid* group at the University of California, San Francisco, in September 1970. Some four years later, *Onishi* presented preliminary studies on bilirubin photo-oxidation at the Jerusalem meeting on bilirubin in early April 1974 and published in 1976 (*775*). Yet by 1974–1976, the events surrounding the molecular mechanism of phototherapy were moving beyond photo-oxidation to other, more plausible mechanisms and away from singlet oxygen (*760, 763, 771–778*). As early as 1972, on the basis of his kinetic studies, *Lightner*'s UCLA colleague, *Christopher S. Foote*, one of the fathers of $^1O_2$ chemistry and its reigning guru, who later published an authoritative study of the kinetics of self-sensitized bilirubin photo-oxidation (*776*), expressed doubt to this author that a photo-oxidation mechanism could account for the rapid drop in bilirubin levels in the jaundiced neonate undergoing phototherapy.

The note of skepticism from *Foote* came after the theory was widely believed and seemingly well-established that bilirubin photo-oxidation explained jaundice phototherapy. After all, one could easily see the photobleaching of the jaundiced neonate's skin, and was not photobleaching akin to photodestruction of the pigment? From the mid-1960s to mid-1970s *Ostrow* labored to isolate and identify the bilirubin degradation products coming into *Gunn* rat bile from both photo-irradiation and the alternate pathways of elimination (*502, 503, 760, 763, 766, 767, 771, 773*), as well as those coming from direct photolysis of bilirubin solutions (*502, 503, 760, 766, 771, 780*). In this connection, in late 1971–1972 he depicted a series of chemical structures and outlined how they might arise from alkaline degradation of bilirubin (Fig. 9.1.1) (*502, 503*) and from its photo-oxygenation (Fig. 9.1.2) (*766, 780*) – with the assertion that (*503*): "All products illustrated have been isolated and characterized." However, the "characterization" was unlikely to have satisfied the tenets of organic chemistry structure proof because the structures represented in Fig. 9.1.1 were obviously unstable for a variety of reasons and the "mechanism" of their formation improbable. Analogous reactions (simple hydration of the exocyclic C=C bond) had never been detected in dipyrrinone models. Similarly, Fig. 9.1.2 represents structures that would have been recognized as predictably unstable and their origin was outlined by unknown mechanisms in singlet oxygen chemistry, *i.e.* one involving reduction of dioxetanes to diols under oxidative reaction conditions, and an unlikely phototautomer structure, giving new evidence that *Thudichum*'s *Conjecturalchemie* was still alive. Although depicting organic chemical structures represented an advancement in thought in the *Ostrow* lab, the structures and mechanisms proposed in Figs. 9.1.1 and 9.1.2 were never revised, and new and different structures would not appear from the group until some seven years later.

9.1 Bilirubin Chemistry Reawakened Thanks to Jaundice Phototherapy   535

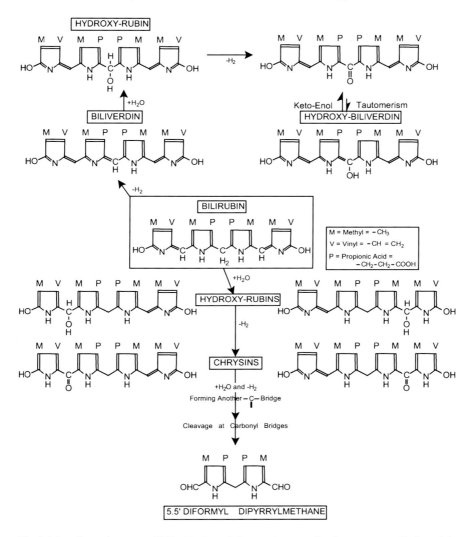

Fig. 9.1.1. *Ostrow*'s proposed bilirubin degradation products coming from aqueous alkali, and the mechanism of their formation (Redrawn from Fig. 1 of ref. (*503*))

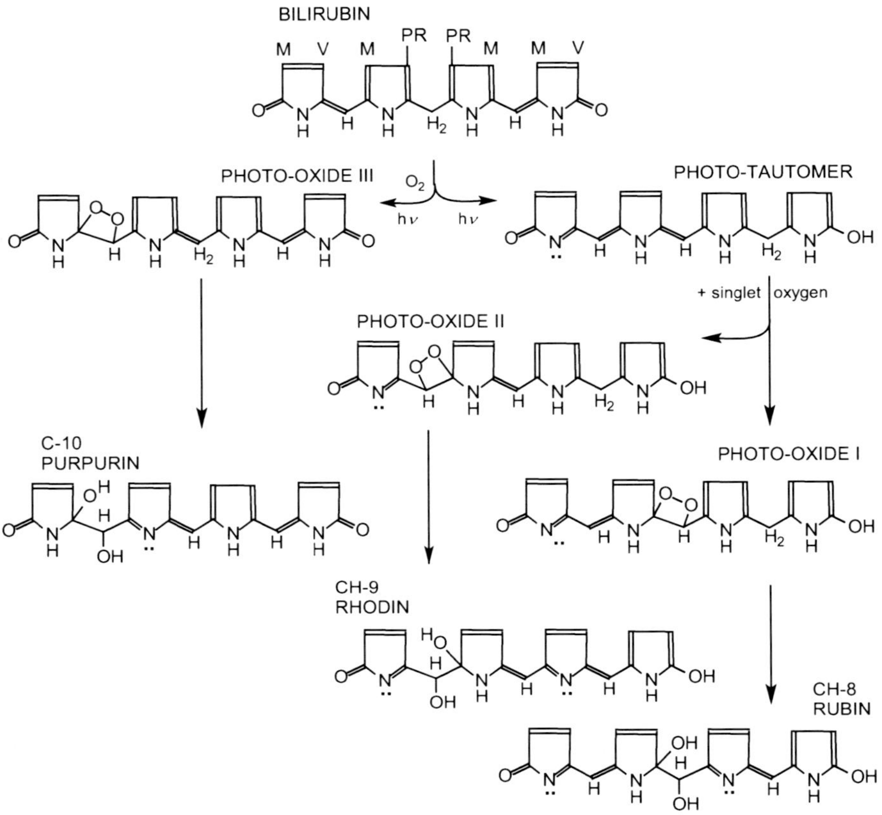

Fig. 9.1.2. *Ostrow*'s proposed dihydroxylated products from photo-oxygenation of bilirubin and his proposed mechanism of their formation (Redrawn from Figs. 5 and 6 of ref. (*780*))

Well before the *Ostrow* studies were published in 1970–1976, other photochemical investigations involving bilirubin had begun in earnest in 1968–1969. *Gary Quistad* (Ph.D. 1972), working in the *Lightner* lab at UCLA, initiated photochemical experiments involving singlet oxygen ($^1O_2$) dye-sensitized photo-oxidation of pyrroles and of self-sensitized and dye-sensitized photo-oxidation of bilirubin under aerobic and anaerobic conditions (*781*). By 1971–1972 the photo-oxygenation products from bilirubin (*461, 782–784*) and biliverdin (*785*) had been separated by chromatography and fully characterized by NMR and mass spectrometry (and color tests where appropriate) as methylvinylmaleimide (**9.1.1**) and hematinic acid (**9.1.2**), and (mainly) the propentdyopents (**9.1.3** to **9.1.5**) shown in Fig. 9.1.3 (*776, 777*), where one may recognize the elusive methylvinylmaleimide [the "nitrite compound" (see Section 6.3.1)] that bedeviled *Hans Fischer* and was not purified and completely characterized until 1969 (*459*).

9.1 Bilirubin Chemistry Reawakened Thanks to Jaundice Phototherapy 537

In 1972, *Bonnett* and *Stewart*, too, published their initial bilirubin photo-oxygenation studies (*786, 787*) in which methylvinylmaleimide (**9.1.1**) and propentdyopents **9.1.3** to **9.1.5**, as well as dialdehyde **9.1.7**, were isolated and characterized in 1974 (*788*). Also in 1974, *Onishi et al.* presented preliminary work on bilirubin photodegradation to propentdyopents and methylvinylmaleimide (*775*).

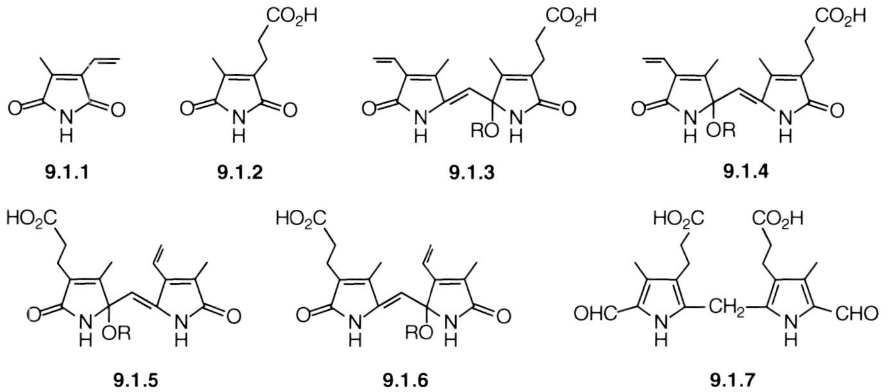

Fig. 9.1.3. The products (9.1.1 to 9.1.6) of self-sensitized and dye-sensitized photo-oxidation of bilirubin-IXα in aqueous base (R=H) and methanol (R=CH₃). Compound 9.1.7, a suggested photo-oxidation product (Fig. 9.1.1), was not found

Although only **9.1.3**, **9.1.4**, and **9.1.5** were found in the initial work, by the 1980s Dr. *Kultar S. Kumar* (working in the *Lightner* lab at the University of Nevada) found and characterized isomer **9.1.6**. Dipyrrole dialdehyde **9.1.7** was not found among the photoproducts. [This compound had been prepared as its dimethyl and diethyl esters by *Clezy* in 1968 (*789*) and turned out not to have the properties assigned the diacid by *Ostrow* (*503*).] The special circumstances of the photo-chemical conversion to biliverdin were also investigated during 1970–1972 (*790*).

In autumn 1970, shortly after taking up his position in the *Schmid* lab at UCSF, *Antony F. McDonagh* initiated his long career in bilirubin chemistry and metabolism with a careful examination of bilirubin photochemistry. By December 1970 he had all the necessary evidence related to the involvement of $^1O_2$ in the photodestruction of bilirubin, and on Tuesday, December 8, 1970, he presented his preliminary work in a seminar at UCLA. When at UCLA he discussed the results and mechanism with *Christopher Foote* and *David Lightner* – the role of $^1O_2$ and also the curious observation of a rapid drop in bilirubin absorbance followed by a slower, more gradual rate of loss that responded to the usual $^1O_2$ quenchers (*772, 778*). Though the graph (Fig. 9.1.4) was not published until 1979 (*769*), it was a component of *McDonagh*'s lectures in the early 1970s. And it showed clearly that there were two rate processes associated with bilirubin photochemistry: an initial very rapid rate, apparently not related to a slower rate associated with photo-oxygenation involving $^1O_2$. As it turned out, in this graph *McDonagh* was the first to detect the fast $(Z) \rightleftarrows (E)$ geometric isomerization of bilirubin, and he also showed that the photo-oxidative degradation of the pigment involved $^1O_2$.

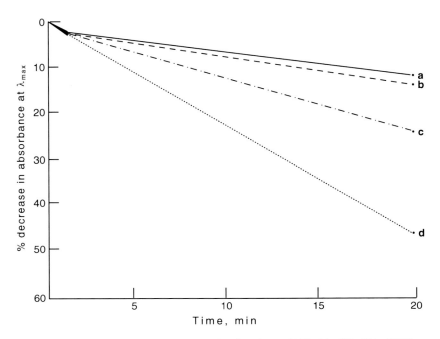

Fig. 9.1.4. Photo-oxidation of bilirubin in CHCl$_3$. Solutions of bilirubin (15 μM in CHCl$_3$ were irradiated with light from a Duro Test R57 400 W high-pressure Hg lamp filtered (Corning no. 3389) to transmit >391 nm light. Loss of pigment was followed spectroscopically by the decrease in absorbance at λ^{max} ($\approx$ 453 nm). Curves: **a**, argon-saturated; **b**, oxygen-saturated and containing 0.15 M 1,4-diazobicyclo[2.2.2]octane; **c**, oxygen-saturated and containing 1.5 mM dimethylbutene; **d**, oxygen-saturated (Redrawn from Fig. 6 of ref. (*769*))

The initial faster rate of bilirubin photochemistry might have been explored more promptly had *McDonagh*'s investigations in early 1971 not turned up the fact that most commercially available bilirubin consisted of more than the IXα isomer (*476*), some as much as 40%. Though a diversion from phototherapy, here and elsewhere it has always been important to know the purity of natural bilirubin used in experiments involving its photochemistry, something all other investigators had failed to do, which meant that they were probably working with an impure commercial starting material. Commercial samples of natural bilirubin were shown to contain the symmetric IIIα and XIIIα isomers that were produced by a constitutional

9.1 Bilirubin Chemistry Reawakened Thanks to Jaundice Phototherapy   539

isomerization of the natural pigment.[3] Detected first in early 1971 by thin layer chromatography (TLC) on silica gel (eluent: CHCl₃ – 1% acetic acid by volume) (476), the IIIα and XIIα isomers were thought to be artifacts of the commercial extraction processes from animal bile and pigment gallstones (a problem unrecognized by *Hans Fischer*). The isomeric scrambling was carefully investigated by *McDonagh*, found to be reversible, and shown to be acid-catalyzed (474). A year later, *Manitto* and *Monti* reported the constitutional isomerization during acid-catalyzed addition of –SH and –OH groups to the *exo*-vinyl group of bilirubin (600). Exploring further, *McDonagh* and *Assisi* found that incubation of bilirubin-IXα in aqueous buffers at pH 7.4–7.6 and at 8.4–8.6 or in 0.1–0.3 *M* in NaOH, pH ~11, also led to constitutional scrambling (473, 475) and required the presence of oxygen or other free radical initiators. Scrambling is greatly slowed when albumin is present (473, 475). It may be induced photochemically in O₂-free aqueous buffers at pH 8.0–8.5 (475).

In the early 1970s, *McDonagh* and *Lightner* were quite aware of the (Z) → (E) photoisomerization. Possibly relevant to bilirubin photochemistry and well known to *Lightner* and *McDonagh*, *Moscowitz* and associates (*Burke* and *Pratt*) at the University of Minnesota had been studying (Z) → (E) photochemistry and analyzing the circular dichroism of the linear tetrapyrrole, phytochrome using molecular orbital theory (791). The work extended *Lightner*'s theoretical studies of the absolute configuration of *d*-urobilin and *l*-stercobilin conducted while he was a postdoctoral fellow in the *Moscowitz* lab during 1964–1965. It lent support to the emerging belief by *McDonagh* and *Lightner* in the early 1970s that isomerization of a bilirubin exocyclic C=C bond might occur more rapidly than any other bilirubin photoreaction. Although in the early 1970s, bilirubin photo-oxidation was a mechanism generally believed to explain the molecular basis of phototherapy and the subject of mechanistic and product structure investigations, it could not explain the apparent appearance of bilirubin in bile. Aware of the (Z) → (E) photoisomerization studies of phytochrome by *Moscowitz et al.* (791), *McDonagh* and *Lightner* thought that a ground state reversible (E) ⇌ (Z) bilirubin isomerization would provide an explanation for the appearance of natural bilirubin in phototherapy bile. Yet, while they believed it to be the only plausible mechanism, for the hypothesis to be believed, the presence of bilirubin in bile had to be fully certified, the photochemical conversion of bilirubin to (E)-isomers and their facile reversion to natural

---

[3]N.B.: Although Dr. *Wooldridge* (working in the *Lightner* lab at the University of Nevada) learned in 1977 (792) how to isolate the bilirubin constitutional isomers by high-performance liquid chromatography (HPLC) on a useful scale, it was not until 1980 that *Joe DiCesare* at Perkin-Elmer Corp., working with the *Lightner* group, showed how to achieve clean separation and isolation of the three isomers on a semi-preparative scale (793) that led to the *Lightner* group being able to cleanly separate 280 mg/day of a constitutionally isomerized mixture of bilirubin-IIIα, -IXα, and -XIIIα, on a preparative scale (20 injections of 14 mg of mixture). Yet, more generally convenient, especially for the isolation of bilirubin-XIIIα, was the method of *Manitto* and *Monti* (794) who showed how to prepare the pure IIIα, IXα, and XIIIα isomers by vinyl addition-extraction methods. The latter was later developed for a preparative scale (795, 796). Subsequently, reverse isomeric scrambling served as an important synthon in the preparation of complex bilirubinoids (727–730, 736, 737, 742, 797).

bilirubin had to be established, and it had to be shown that bilirubin (E)-isomers would cross the liver into bile in contrast to the (Z,Z) isomer. This mechanism was predicted by *McDonagh* in early April 1974 at a bilirubin metabolism conference in Jerusalem (*798*):

> But what of the second effect of phototherapy, the enhancement of unconjugated bilirubin excretion? It has recently been implied[7] [*808*] that this fraction may not really consist of bilirubin but rather of some similar rubinoid, perhaps a photoadduct of bilirubin, which would not be distinguishable from bilirubin with the techniques used in earlier studies[3,4] [*773, 779*]. In confirmation of Ostrow's pioneering work[3] [*773*] we found that there is pronounced excretion of a rubinoid material in Gunn rat bile during phototherapy. This pigment can be isolated simply by freezing and thawing the bile and spinning off the orange amorphous precipitate which forms. The precipitate is insoluble in chloroform, but dissolves in dimethyl sulfoxide to give an absorption spectrum identical with that of bilirubin $IX\alpha$ in the same solvent. By diluting this solution with water and extracting it with chloroform, the pigment can be transferred into chloroform, wherein it displays an absorption spectrum congruent with bilirubin $IX\alpha$ in chloroform. The pigment from the chloroform phase showed only one major spot on two very discriminating, thin-layer chromatographic systems (silica gel, 1% acetic acid in chloroform[26] [*474*]; polyamide, 1% ammonium hydroxide in methanol), and this spot did not separate from bilirubin $IX\alpha$ on admixture. Finally, a sample of the pigment purified by preparative thin-layer chromatography gave a mass spectrum with a molecular ion at m/e 584, similar to the spectrum of a purified sample of authentic bilirubin $IX\alpha$ run under similar conditions on the same spectrometer. Therefore, although we do not know the identity of the original material isolated from the bile (perhaps a calcium salt), we can be reasonably confident that it contained an intact bilirubin $IX\alpha$ chromophore and that it did not contain significant amounts of structures with materials covalently added to the vinyl side chains.

The "freeze-thaw" precipitated pigment almost certainly contained bilirubin photoisomers that reverted to the natural pigment during manipulation. By the end of 1974, *McDonagh* had reported its characterization (*800*):

> The simplest and most straightforward explanation, which is that the pigment is bilirubin itself, was considered to be less plausible. Plausible or not, there is in fact little doubt that the pigment *is* bilirubin and that a major effect of phototherapy is to stimulate the excretion of intact unconjugated bilirubin into the bile. This was first demonstrated by Ostrow in his pioneering study of the effect of phototherapy on rats with unconjugated hyperbilirubinæmia.[4] [*773*] Ostrow showed that these animals, which are considered to be a good model for the jaundiced neonate, excrete considerable amounts of unconjugated bilirubin into the bile on treatment with phototherapy. His identification of the pigment as bilirubin was based largely on its visible absorption spectrum and chromatographic properties, and the chromatographic behaviour of its azo-derivatives; fairly convincing evidence, but not unambiguous. I have confirmed Ostrow's findings and have further identified the pigment as bilirubin IX-α by thin-layer chromatography on two different adsorbents, by absorption spectroscopy, and by field-desorption mass spectrometry.

On December 1, 1975, *McDonagh* conducted an experiment that foretold how phototherapy works. Brief photo-irradiation of a $CHCl_3$ solution of bilirubin, overlaid with 1 M aqueous $NaHCO_3$, and through which a stream of Ar was bubbled, led to yellow coloration of the bicarbonate layer (in which the natural pigment is insoluble). Acidification of the bicarbonate layer drove the yellow

9.1 Bilirubin Chemistry Reawakened Thanks to Jaundice Phototherapy

color into fresh $CHCl_3$. The experiment thus showed that brief photo-irradiation of bilirubin in $CHCl_3$ produced a yellow photoisomer that was far more soluble in bicarbonate than natural bilirubin – and that reverted to natural bilirubin upon acid treatment. The material was called *isobilirubin*; later, in 1977, it was renamed *photobilirubin* (769).

This seminal experiment had been preceded by another of *McDonagh*'s equally instructive but different photochemical experiments. Thus, on April 2, 1973, as reported in a National Academy of Sciences meeting on *Phototherapy in the Newborn: An Overview* (801), he photo-irradiated (typically for six hours with visible light) a solution of bilirubin in aqueous buffered bovine serum albumin (BSA), treated the photolysate solution with acetic acid to precipitate protein, then extracted a yellow pigment into $CHCl_3$. The yellow material was slightly more polar and less lipophilic than natural bilirubin. It was extracted into bicarbonate, which was acidified with acetic acid and then back-extracted (a yellow pigment) into $CHCl_3$. There was no evidence that this "stable" bilirubin photoisomer would revert to natural bilirubin. It exhibited a UV-visible absorption maximum at λ ~430 nm in $CHCl_3$, chromatographed on silica gel TLC with some decomposition and on polyamide TLC with no decomposition. The new bilirubin isomer was then named *compound X*. It was repeatedly formed and isolated in small quantities, using human serum and other albumins, over the next few years. It failed to be detected in the blood of *Gunn* rats undergoing photo-irradiation – for reasons that became clear in 1976–1977, and it was renamed *lumirubin* during the time it was isolated and characterized completely in preparation for publication in 1982 (762).

It was from these two experiments by *McDonagh* that the main molecular mechanism of jaundice phototherapy eventually became understood. Those findings became a focus of study in the *McDonagh* lab during the mid-1970s, while investigations of bilirubin photo-oxygenation began to wane as early as 1974. Thus, at a U.S. National Academy of Sciences-sponsored meeting (*Phototherapy in the Newborn: An Overview*) in 1973, *McDonagh* signaled that a change in perspective might be needed, with the announcement that (801):

> Irradiation of bilirubin IXα at pH 8.5 in aqueous solution with visible light in the presence of serum albumin and the *absence of oxygen* leads to formation of a yellow photoderivative (λ_{max} 442-444 nm in chloroform). This compound is slightly more polar and less lipophilic than bilirubin at slightly basic pH values. Its identity and properties are being investigated.

And again at the U.S. National Institutes of Health (NIH)-sponsored conference on *Phototherapy for Hyperbilirubinemia*, held in Bethesda, Maryland April 24–26, 1974 (802):

Conversion to Unidentified 443-nanometer Pigment

> A yellow pigment with a single visible absorption maximum at 442-444 nm in chloroform or at 433 nm in water at pH 8.5 is formed slowly when solutions of bilirubin IXα in buffer (pH 7.4 or 8.5), aqueous albumin, or rat serum are irradiated with visible light.[11]

542   9   Understanding and Translating Bilirubin Structure

[*798*] Yields of this pigment are generally low, but they are highest when albumin is present and oxygen is excluded from the solution. Under optimal conditions, the yield is about 13 percent (assuming that the compound has a molecular weight of 600 and a molar extinction coefficient of 60,000). The remainder of the product consists of unreacted starting material (20-30 percent) and several other uncharacterized compounds. Like bilirubin, the pigment is soluble in chloroform and can be chromatographed on thin polyamide layers. Unlike the natural pigment, however, it is unstable on silica gel layers, is readily extractable from chloroform solutions into 0.1 M aqueous $NaHCO_3$, and fails to give a positive diazo and pentdyopent reactions. On electron impact or chemical ionization mass spectrometry, the compound failed to give a distinct parent of molecular ion and only weak fragments of low molecular weight could be detected. Following treatment with *p*-toluene sulfonic acid in chloroform, the compound was largely recovered unchanged, indicating the absence of hydroxylated vinyl side chains. At present, the structure of this pigment is unknown. . . .

Formation of Unidentified 430-Nanometer Compound

Irradiation of solutions of bilirubin in aqueous albumin results in formation of unspecified yields of a yellow ultrafilterable pigment with λ_{max} = 430-435 nm in 0.1 M veronal buffer, pH 8.4, as well as several other uncharacterized products.[12] [*764*] This pigment, which is soluble in chloroform and does not give a typical diazo or positive pentdyopent reaction, has not been completely characterized or identified. Its properties are suggestive of a dipyrrylmethene and resemble those of the pigment described in the preceding section, to which it may be identical. Interest in this pigment has been stimulated by the detection of a similar compound in the serum and excreta of an infant with Crigler-Najjar syndrome undergoing phototherapy.[12,13] [*764, 765*] The same material was also detected when the patient was not receiving phototherapy.[13] [*765*] At present, it is an open question whether "430 pigment" is a major or important photoproduct of bilirubin in vivo during phototherapy.

Which meant that doubts were being raised in the mid-1970s, at the *Fogarty International Symposium on the Chemistry and Biology of Bile Pigments* held in Bethesda, Maryland on April 28–30, 1975 as to whether a $^1O_2$ mechanism could fully account for the success of phototherapy (*803*):

Singlet oxygen invariably seems to be formed when bilirubin is irradiated in solution *in vitro* with visible light and it is largely responsible for the photodegradation of the pigment. Undoubtedly it also is formed to some degree in the peripheral tissues of jaundiced neonates when they are irradiated with light, and the degradation of bilirubin that occurs during phototherapy probably is largely dependent on singlet-oxygen formation. However, the contribution of singlet-oxygen processes to the overall total effect of phototherapy is not known and may not, in fact, be large. . . .

It is difficult to see how singlet oxygen could play a role in the second major effect of phototherapy, enhanced bilirubin excretion. A more plausible hypothesis is that bilirubin in subcutaneous lipid (where unsaturated fats would inhibit singlet-oxygen photodegradation) undergoes photochemical conversion to a stable conformational or geometric isomer that can be excreted without conjugation. However, this hypothesis is pure speculation.

Even though bilirubin destruction *in vivo* in the *Gunn* rat can be sensitized by porphyrins administered intravenously (*801*). Again, at the 561st Meeting of the Biochemical Society, held at the University of Leeds on March 24–26, 1976, where,

9.1 Bilirubin Chemistry Reawakened Thanks to Jaundice Phototherapy

even before the X-ray structure of bilirubin had been presented, *McDonagh* provided a rationale on how phototherapy works, based on $(Z) \rightarrow (E)$ photoisomerization (*804*):

> One appealing, yet entirely speculative, theory is that bilirubin can be converted photochemically into a more polar, but less thermodynamically stable, isomer that does not need to be conjugated to be excreted. Cis-trans photo-isomerization of a model dipyrrolylmethene has been demonstrated and the individual isomers isolated (Falk *et al.*, 1965), but analogous reactions of bilirubin have not been reported.

And so, at the end of 1976, in a comprehensive review, *McDonagh* suggested that (*804*):

> On the other hand, the photo-induced excretion of (unconjugated) BR IXα probably does not involve $^1O_2$ and might be rationalized by photochemical conversion of BR IXα to a thermodynamically less stable, but more hydrophilic conformational isomer. This might be accomplished via a $Z \rightleftarrows E$ photoisomerization about the exocyclic ene-amide carbon-carbon double bond(s) . . . .

That is, to *McDonagh* and *Lightner*, the photodegradation rationale of how phototherapy works was being subsumed to a mechanism whereby bilirubin underwent a more rapid, reversible, $(Z) \rightarrow (E)$ photoisomerization to give one or more polar (E)-isomers due to the breaking of intramolecular hydrogen bonds, isomers that crossed the liver into bile without conjugation and reverted to (4Z,15Z)-bilirubin (*805*).

Yet as the faith but not fact based *sola scriptura* emphasis on photo-oxidation began to dissolve and shift away from bilirubin photodegradation to explain how phototherapy works, *Onishi* was publishing some of his important studies on bilirubin photo-oxidation (*430*) following his early work of 1970 (*806*). There was some catching up to do.

As these significant discoveries from San Francisco were being made and becoming understood in the early mid-1970s, and as *Ostrow* reiterated his finding unconjugated bilirubin in *Gunn* rat bile, investigators in Denmark found results similar to *Ostrow's*, now coming from jaundiced neonates undergoing phototherapy. Thus in 1972, *Lund* and *Jacobsen* found unconjugated bilirubin in duodenal bile (*799*) but could offer no better an explanation for its presence than anyone else at the time, nor was there an explanation in the follow-up larger study in 1974 (*807*):

> At present no obvious explanation can be offered to account for the increased concentration of unconjugated bilirubin during phototherapy, which does not seem to bear any clear-cut relationship to the serum bilirubin-reducing effect of this treatment.

Yet to *Lund* and *Jacobsen* (*807*) the formation of *exo*-vinyl adducts by *Garbagnati* and *Manitto* in Milan (*808*) offered a possible explanation.

Clearly the times were changing from the early 1970s, about which *Ostrow et al.* proclaimed that (*809*):

> The effects of phototherapy on bilirubin metabolism were first elucidated in our laboratory in 1971 . . .

544   9 Understanding and Translating Bilirubin Structure

and believed that phototherapy was based on a combination of acceleration of the alternate pathways of elimination and excretion of unconjugated bilirubin (*773*):

> . . . phototherapy both accelerates the alternate pathways of bilirubin catabolism which exist in the Gunn rat under normal conditions (9) [*758*], and dramatically augments biliary excretion of unconjugated bilirubin.

However, their extonious belief that photo-oxidation produced hydroxylated tetrapyrroles introduced yet another potential explanation (*502*):

> . . . fully explains *all* the changes that are observed during decomposition of bilirubin in alkali, both in the dark and under illumination [*771*]. . . . Specifically: (1) the diazoreactivity of the products is traceable to the three hydroxyrubins and to an orange-red pseudodiazo reaction given by the two bilichrysins and the dipyrrole that oxidize in strongly acid solutions to yield bilipurpurins and bilrhodins; (2) the yellow fluorescence and absorption maximum at 315 nm result from the carbonyl groups of the two bilichrysins and the dipyrrole.

Which led to the belief, also in 1972, that (*503*):

> Ostrow et al. have fully characterized the products of bilirubin degradation in alkali in the dark, . . . which are the same as the photoderivatives formed during the illumination of bilirubin at neutral pH in the presence of albumin[31]. . . . [*771*]

That was summarized in 1974 (*766*):

> We have thus identified for the first time several of the bilirubin derivatives formed during phototherapy of the jaundiced Gunn rat. It is fascinating that the Gunn rat also excretes substantial amounts of these same products in the dark. This suggests that these derivatives can be formed by catabolic processes as well as by photochemical reactions, presumably by uncharacterized bilirubin oxygenases that are utilized by the Gunn rat for the alternative pathways of bilirubin metabolism.

However, in 1976, the shifting sands of the mechanism of phototherapy appeared to shift once more when *Ostrow et al.* seemed to discover that phototherapy of patients with alcoholic cirrhosis increased the excretion into bile of *conjugated* bilirubin and they thus wrote (*763*):

> Phototherapy does not augment biliary excretion of bilirubin in normal nonjaundiced rats, but does increase the concentration of conjugated bilirubin in the bile of patients with alcoholic cirrhosis.

As we shall see, in the end *McDonagh*'s then curious discovery of a fast photochemical reaction of bilirubin, coupled with *Ostrow*'s discovery that unconjugated bilirubin appeared in the bile of jaundiced (*Gunn*) rats undergoing photo-irrradiation, would lead to the molecular mechanism for how phototherapy works. For despite the considerable investment in bilirubin photo-oxygenation, realization was dawning in the mid-1970s, that its rate was too slow to account for the effectiveness of neonatal phototherapy, and the much ballyhooed photodestruction mechanism was starting to wither on the vine. It seemed that photooxidation of bilirubin and its relevance to jaundice phototherapy was destined to assume a minor role. Whether photo-oxygenation products appear in bile during

9.2 How Phototherapy Exposed Bilirubin (*E*)-Configuration Isomers

photherapy may never be answered completely because, as one may recall (Section 2), bile contains so very many diverse types of bile pigments and other substances that even the isolation of bilirubin itself was no picnic. Yet, the presence of some photo-oxygenation products would not be surprising, nor would the absence of others, especially pent-dyo-pents (**9.1.4** and **9.1.6** of Fig. 9.1.3) and monopyrroles (**9.1.2** and the diacid of **9.1.1** but not **9.1.7**). These were found by *William Linnane* (M.S. 1982) in the *Lightner* lab, who, with Drs. *Charles Ahlfors* and *Richard Wennberg* (then at the University of California, Davis Medical School), proved that such photoproducts appeared in the *urine* of jaundiced neonates undergoing phototherapy (*811, 810*). Other purported photo-oxidation products have been swept into the dustbin of history.

9.2 How Phototherapy Exposed Bilirubin (*E*)-Configuration Isomers

As noted above, the solution to the "phototherapy problem", at least from a molecular perspective, was detected in *McDonagh*'s earliest work at UCSF, in late 1970. In studying the role of $^1O_2$, when monitoring bilirubin photolysis by UV-visible spectroscopy, he found it odd that the long wavelength absorption dropped more rapidly during the first minute of photo-irradiation than in the next two minutes (*772, 778*). That is, *McDonagh* had, unknowingly at that time, detected two photochemical processes, of which photo-oxidation clearly exhibited the slower rate, as shown clearly in Fig. 9.1.4. But what type of photochemistry might occur faster than photo-oxidation?

Had the data been available to *William Küster*, who in 1912 had shown the enviable ability to extrapolate intuitively to the nearly correct structure of hematin (*360*) (see Section 4.4) on the basis of sketchy evidence at best, the good and bad, but always interesting (if not amusing) work on photo-oxidization might have been unnecessary. For *Küster* might have suggested the simplest reaction in organic photochemistry: the *cis* → *trans i.e.* (*Z*) → (*E*) configurational isomerization (Fig. 9.2.1) at the ene-amide exocyclic C=C bonds, had he realized the double bond stereochemistry of bilirubin. Such a possibility was never far from *McDonagh*'s view of bilirubin photochemistry, for although the exocyclic C=C configuration had not been firmly established before 1976, most organic chemists predicted the (*Z*)-stereochemistry from the structure of the porphyrin precursor, which was shown by *Falk et al.* (*812*) in 1974 to be the more stable stereochemistry in dipyrrinone models for bilirubin, and that photoirradiation inverted the (*Z*) C=C to (*E*).

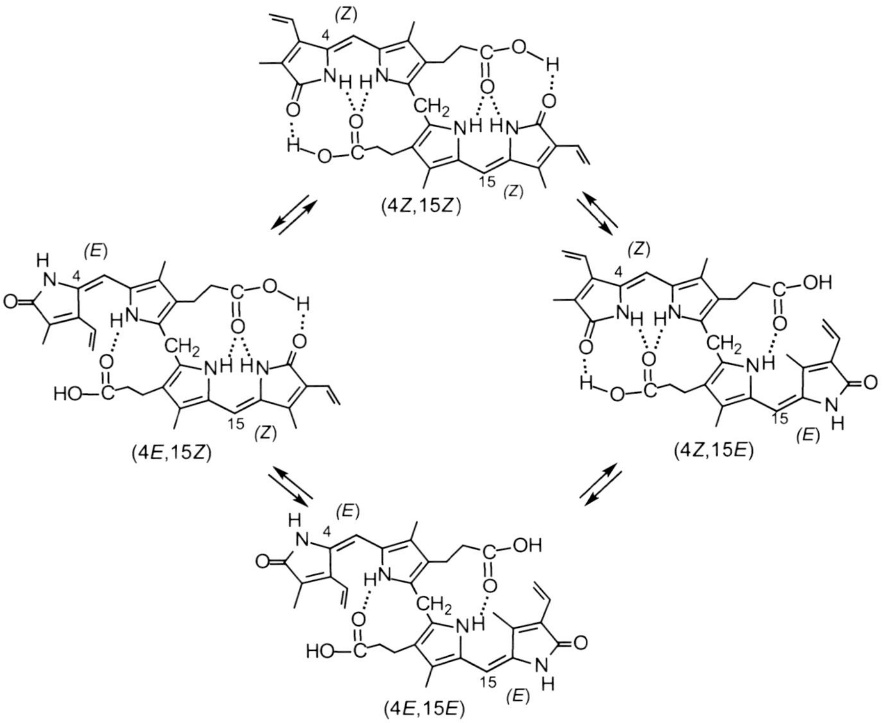

Fig. 9.2.1. Ene-amide C=C inversion of (4Z,15Z)-bilirubin

Even before *Bonnett*'s X-ray crystal structure had defined the (Z)-stereochemistry of bilirubin and its intramolecularly hydrogen bonding in late 1975, *McDonagh* wrote perceptively for his lecture at the February 16–20, 1976 4th Annual Meeting of the American Society for Photobiology in Denver (*813*):

> How phototherapy causes hepatic excretion of bilirubin remains an unanswered and intriguing conundrum. Several theories have been proposed and some rule out experimentally. Perhaps the simplest and most appealing is that bilirubin-IXα can be converted photochemically into a more polar, but less thermodynamically stable, geometrical isomer that can be excreted rapidly. This type of photo-isomerization has been observed with related compounds, but not with bilirubin.

In the sentence above, *McDonagh* was referring to the (Z) → (E) C=C photo-isomerization of dipyrrinones and arylmethylidenepyrrolinones reported in 1975 by *Falk et al.* (*812, 814*) and in 1976 by *Yong-Tae Park* (Ph.D. 1977) in the *Lightner* lab (*815*). Even earlier, in 1973, *McDonagh*'s photolysis experiments of bilirubin in aqueous serum albumin had revealed two photoproducts, one with UV-visible λ^{max} 430–435 nm, the other with λ^{max} at 442–444 nm (*802*). The focus on photo-oxidation as the basis for jaundice phototherapy was waning, and more attention was being directed toward *Ostrow*'s seminal observation reported in 1971 (*773*) that unconjugated bilirubin was found in the bile of photoirradiated *Gunn* rats. But how?

9.2 How Phototherapy Exposed Bilirubin (*E*)-Configuration Isomers — 547

Given experimental verification of the double bond stereochemistry of normal bilirubin (*510*) and information on (*Z*) $\rightleftarrows$ (*E*) C=C isomerization in bilirubin model compounds coming from the *Falk* (*812, 814*) and *Lightner* (*815, 816*) labs, a probable explanation emerged from *McDonagh*'s lab, linking (*Z*) $\rightleftarrows$ (*E*) photoisomerization to his seminal but then puzzling observation in 1970, of a fast photochemical reaction of bilirubin (*769*). During 1974–1978, in order to explore his as yet experimentally unconfirmed predictions of bilirubin's geometric photo-isomerization, *McDonagh* analyzed the bilirubin photolysates by TLC on polyamide. This revealed a new yellow, polar photoproduct that proved to be capable of reverting to the natural pigment by thermal and mild acid catalysis ground state processes. Why the unusual polyamide TLC? Because spotting the photolysate on a typical silica gel TLC plate caused reversion of the fast-formed photoproducts to (4Z,15Z)-bilirubin, as did HPLC on normal phase columns, as shown later.

At the same time, in 1976, Dr. *Timothy Wooldridge* in the *Lightner* lab at the University of Nevada reproduced *McDonagh*'s findings and also found that the photoproduct from bilirubin dimethyl ester could be separated by HPLC, which indicated two, close-running more polar major photoproducts, as well as a third, even more polar photoproduct. The photoproduct eluent was collected in a Pyrex glass vessel and reversion to starting material was observed to occur, whereas, collection in a silanized or Teflon vessel preserved the photoproducts from acid-catalyzed reversion to bilirubin dimethyl ester, thus demonstrating the mild acid sensitivity of the three photoproduct(s). This was all explained in an article received at the *Biochemical and Biophysical Research Communications* journal office on September 6, 1978 (*818*), wherein *McDonagh, Wooldridge,* and *Lightner* described their preliminary results in which bilirubin photolysis in $CHCl_3$-1% CH_3CH_2OH-10% $(C_2H_5)_3N$ was followed by absorbance difference spectroscopy (in Ar-purged $CHCl_3$) by polyamide TLC, and (for the dimethyl ester in toluene-5% ethanol) by HPLC. The last revealed three photoproducts, in addition to (a majority of) starting material, that were more polar than bilirubin. The major photoproducts had similar retention times, longer than bilirubin, but shorter than the least prevalent photoproduct. In contrast, the symmetrical bilirubin-XIIIα and IIIα dimethyl esters each produced only two photoproducts, the more prevalent of which had a retention time similar to the two more prevalent photoproducts from natural bilirubin, and the minor photoproduct with a retention time similar to that of the minor photoproduct from bilirubin. The authors explained these results as follows. The more prevalent photoproducts corresponded to the (4Z,15E) and (4E,15Z) isomers (see Section 9.2.1) and the minor photoproduct corresponded to the (4E,15E) isomer. Taken collectively, they were renamed *photobilirubin* (or PBR), from their earlier in-house name, *iso-bilirubin*, and their importance to phototherapy was explicated (*818*):

> Recently, we reported that brief irradiation of BR-IXα in many solvents yields a photostationary mixture containing mostly BR-IXα and a small proportion of a more polar material which we called photobilirubin (PRB) . . . .
>
> The ease with which PBR is formed *in vitro*, even in the presence of oxygen, . . . suggests that it may well be formed *in vivo* during phototherapy of jaundiced patients or on irradiation of congenitally hyperbilirubinemic rats (Gunn rats) . . . . Indeed, the reaction kinetics make it a more likely reaction than self-sensitized photooxidation of BR which has also been postulated to occur *in vivo* . . . . Furthermore, the solubility properties of PBR,

548   9 Understanding and Translating Bilirubin Structure

which resemble those of biliverdin-IXα, indicate that, like biliverdin . . . , PBR would not require conjugation for its hepatic excretion. We suggest, therefore, that during phototherapy, PBR formed photochemically from BR near the surface of the skin is transported in plasma to the liver where it is taken up and excreted into bile. This would explain the almost instantaneous biliary pigment excretion that occurs when Gunn rats are exposed to light . . . and the presence of much unconjugated BR in extracts of pigmented bile obtained during phototherapy . . . . Due to the facile thermal and photochemical reversion of PBR, solvent extraction of bile under the usual conditions would tend to yield BR rather than its isomers, even if they were present initially. Although the steady-state concentration of PBR in peripheral tissues would remain low during phototherapy, the efficiency of the overall process would be high because the PBR being continuously swept away and excreted would be replenished constantly by newly-formed PBR. Thus, PBR formation provides a particularly plausible rationale for the enhanced biliary excretion of unconjugated BR in jaundiced rats and humans exposed to light. Although excretion of unconjugated BR is not the only process that contributes to the effectiveness of phototherapy, the available data . . . indicate that it is quantitatively the most important. Therefore, if our hypothesis is correct, PBR formation is the most important and fundamental reaction in phototherapy of neonatal jaundice.

In a related publication by the same authors, received at the *Journal of the Chemical Society Chemical Communications* on September 25, 1978, the proposed $(Z) \rightleftarrows (E)$ photo-isomerization was investigated and the inescapable conclusion was stated (*819*):

> The most plausible, of not the only explanation for these observations is geometric isomerisation of BR-IXα and BRDME-IXα at the meso double bonds (Scheme) [Figure 9.2.1]. . . This is consistent with the reversibility of the reaction and the formation of the three products from BRDME-IXα compared to only two from each of the corresponding IIIα and XIIIα isomers. Presumably the main photoproduct detected chromatographically from BR-IXα is an unresolved mixture of *E,Z* and *Z,E* isomers. Under the conditions reported, the photoequilibrium lies well to the left so that [Z,Z] >> [E,Z] = *ca*. [Z,E > [E,E].
>
> This reaction is important for three reasons. (a) It occurs readily in BR-IXα solutions exposed to light but is undetectable by silica chromatography and difficult to detect by simple absorbance measurements. (b) It accompanies other photochemical reactions of BR-IXα and may complicate the kinetics and mechanism(s) of these. (c) It indicates that, despite extensive hydrogen bonding, BR-IXα can be converted into more polar configurational isomers thereby providing an explanation for the sudden biliary excretion of bile pigment observed when jaundiced rats . . . and babies . . . are irradiated with visible light.

Although the exact structures of the photoisomers could not be verified until they had been isolated and fully characterized, the intramolecularly hydrogen-bonded structure of (4Z,15Z)-bilirubin-IXα reported in 1976 by *Bonnett* (*510*) gave further justification to the belief held early by *McDonagh* and *Lightner* that $(Z) \rightarrow (E)$ photoisomerization of bilirubin-IXα would convert it to a more polar photoisomer that could be excreted into bile. As supported by *Bonnett* in autumn 1977 (*511*):

> It may be noted, though, that if either or both of the C5/C15 bridges had the *E*, rather than the *Z*, configuration then the tightly knit system of six intramolecular hydrogen bonds could no longer be formed: the resulting geometrical isomer of bilirubin would be expected to be more soluble in water. This may offer an explanation for the increased levels of bilirubin observed during phototherapy in the bile of infants suffering from neonatal hyperbilirubinemia (Lund & Jacobsen 1972; cf. Ostrow 1971) since the $Z \rightarrow E$ change is known to occur photochemically, and to be readily reversed thermally, in model pyrromethenone systems (Falk *et al.* 1975).

What of the "490 nm pigment" reported by *Davies* and *Keohane* some years earlier? Subsequently, in an article composed in 1977 and received at the *Proceedings of*

9.2 How Phototherapy Exposed Bilirubin (E)-Configuration Isomers

the National Academy of Sciences (U.S.) on November 7, 1978, the authors explained (*769*) an earlier report by *Davies* and *Keohane* (*761, 768*) who in 1970 had reported a photoproduct, detected by absorbance difference (AD) spectroscopy and absorbing at 490 nm, some 40–50 nm shifted bathochromically from bilirubin. The "490 nm pigment" caught the attention of those investigating phototherapy, and it was suggested that it might be a factor in the successful therapy (*764, 765*). In 1974, *Ostrow* suggested that the 490 nm pigment might be a phototautomer (*767*):

> ... we had to postulate initial formation of a phototautomer of bilirubin, which could account for the intermediate absorbing at 490 nm which was detected by Davies and Keohane ... during irradiation of bilirubin in vitro.

In 1979, *McDonagh*, *Wooldridge*, and *Lightner* reported that the 490 nm absorbance found by AD spectroscopy (Fig. 9.2.2) during bilirubin photo-irradiation was not due to a substance with λ_{max} at 490 nm but to the formation of PBR, for which wavelength UV-visible absorption band near 440 nm had a broader, hence more intense, long wavelength wing that that of natural bilirubin (*769*):

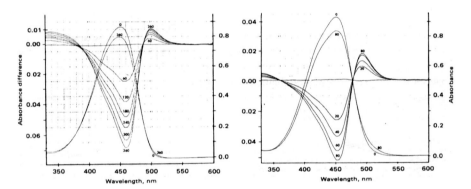

Fig. 9.2.2. Absorbance difference (AD) spectra (sigmoid) obtained from irradiation of 15 μM bilirubin solutions with source A. Cumulative irradiation time(s) is indicated on each scan. Absorption curves of sample solutions before and after irradiation are superimposed on the AD spectra. (*Left*) In chloroform/1% ethanol; (*Right*) in methanol/1% conc. ammonium hydroxide. (Fig. 1 of ref. (*769*))

> The rapid formation of PBR was also noticeable in kinetic studies of BR photooxidation. Fig. 6 [Fig. 9.1.4] shows the rate of self-sensitized photooxygenation of BR in chloroform in the presence of various inhibitory substances .... Although the individual reaction rates were different after the first minute, the loss of BR during the first minute was similar in each case. This early oxygen-independent loss of absorbance is due to PBR formation. We have observed a similar phenomenon during photooxidation of BR in aqueous buffer (pH 7.4), in detergents (Triton X-100, cetyltrimethylammonium bromide, pH 7.4), in rat serum, in aqueous bovine serum albumin (pH 7.4), and in several organic solvents including dimethyl sulfoxide, ammoniacal methanol, benzene, and pyridine. This suggests that rapid formation of PBR occurs in all of these media. Early changes in the spectra of irradiated BR/albumin solutions also have been noted by others...

They also showed the generality of the (*Z*) ⇌ (*E*) photo-isomerization in a wide variety of solvents (*769*):

These findings show that when BR is irradiated in solution it is converted rapidly to an equilibrium mixture containing BR and a novel substance, PBR. The absorption spectra of BR and PBR are rather similar and almost congruent. But there are differences, and in several solvents formation or disappearance of PBR is manifested by complementary changes in the AD spectra near 460 and 500 nm. Formation of PBR does not require oxygen and is faster than the aerobic photooxidation of BR. The reaction occurs in a wide variety of polar and nonpolar media and under neutral or basic conditions, and it is reversible photochemically, thermally, and catalytically. The most plausible explanation for these findings is $Z \rightarrow E$ photoisomerization . . . of BR at the exocyclic enamide carbon-carbon double bonds. In its most stable state, BR has the 5Z,15Z configuration. . . . Photoisomerization should generate a pair of almost identical Z,E isomers and smaller amounts of an E,E isomer . . . (Fig. 7) [Fig. 9.2.1]. We believe that PBR is a mixture of these geometric isomers containing predominantly the 5Z,15E/5E,15Z pair. Further evidence to support this conclusion and to show that PBR formation *in vivo* is crucial for the successful treatment of neonatal jaundice by phototherapy will be reported separately.

These results followed logically from *McDonagh*'s bilirubin photolysis experiments in 1973–1974 and his observation then of two new photoproducts, one with a UV-visible λ_{max} at 442–444 nm (*801, 802*) and the other at 430–435 nm (*769, 765*) and they began to fit nicely into the curious yet seminal observation by *Ostrow* that that unconjugated bilirubin might be excreted into the bile of jaundiced (*Gunn*) rats undergoing light irradiation (*773*) and restated subsequently (*763, 766, 773*).

By mid-1976 not only had *Ostrow et al.* (*763, 766, 767, 773, 809*) reported unconjugated bilirubin in phototherapy bile, but so had *Lund* and *Jacobsen* (*799, 807*). In mid-1977, the *Brodersen* lab in Aarhus weighed in (*820*):

On the other hand, bilirubin is photooxidized during irradiation of its complex with albumin. This process is preceded by a non-oxygen-consuming step, characterized by a small red shift and formation of a pigment which is more polar than bilirubin and is bound less firmly to albumin. These observations are consistent with those of Ostrow . . . who described an anaerobic photodegradation of bilirubin taking place only in the presence of albumin. Spectral evidence for the existence of a similar pigment during photolysis of the bilirubin×albumin complex has also been presented by Davies & Keohane. . . . It should be noted that this reaction is caused by an amount of light thousands of times less than the amount required for photooxidation of bilirubin.

Though that group also reported on *in vitro* photo-oxidation studies in 1977, they too predicted that $(Z) \rightarrow (E)$ photo-isomerization should create a more polar isomer capable of being excreted into bile and urine without conjugation, albeit without proof (*820*):

In the present state of knowledge it seems possible to hypothesize that phototherapy results in formation of a hydrophilic isomer of bilirubin, which is loosely bound to albumin and, at least partially, is excreted without previous conjugation. Photooxidation probably also takes place, although at a lower rate.

From the perspective of the natural products chemist, however, direct proof that bilirubin (*E*)-isomers were excreted into phototherapy bile remained incomplete until 1976 when *McDonagh* developed a powerful new reversed-phase HPLC system for separating and identifying bilirubins. Armed with this methodology, he was able to show unequivocally that the unconjugated bilirubin was secreted into the bile of *Gunn* rats only by the action of light treatment, and in the form of (*E*)-isomers. Thus, the molecular mechanism for phototherapy was starting to become revealed by firm experimental evidence (*821*):

9.2 How Phototherapy Exposed Bilirubin (*E*)-Configuration Isomers   551

> The mechanism for the enhanced excretion of unconjugated bilirubin during phototherapy is unknown. A current hypothesis cites geometric isomerization. . . . Bilirubin IXα, which has two Z-configuration bridge double bonds . . . and requires conjugation for excretion, is thought to be converted by absorption of light to E-Z or E-E isomers that can be excreted without conjugation. Our observations are compatible with this theory and consistent with immediate formation of a compound in the skin that migrates to the blood, is taken up by the liver, and is rapidly excreted in bile.

The "current hypothesis" of 1978 (*821*) was based on the photochemical studies of bilirubin during 1971–1978 by *McDonagh* and *Lightner*, *by Ostrow*'s early observation of unconjugated bilirubin in *Gunn* rat "phototherapy" bile, and by the reports of *Lund* and *Jacobsen* on finding unconjugated bilirubin the bile of jaundiced babies undergoing phototherapy. Thus, bilirubin $(Z) \leftrightarrows (E)$ photo-isomerization was emerging as the most plausible explanation for how unconjugated bilirubin might be found in phototherapy bile, albeit an explanation rejected earlier, in 1976 (*774*):

> Mass spectrometric analysis reveals that the bilirubin excreted during phototherapy is the customary IXα isomer. This precludes the possibility that photo-isomerization of the pigment, to a form incapable of internal hydrogen bonding, accounts for excretion of unconjugated bilirubin in bile during phototherapy.

Nevertheless, the situation of 1976 began to advance rapidly toward an explanation when a yellow, polar bilirubin photoproduct was reported at the May 11–15, 1977 5th Annual Meeting of the American Society for Photobiology in San Juan, Puerto Rico (*822*), and the May 21–26, 1977 Gastroenterology meeting in Toronto (*823*). The photoproduct was found to revert to natural bilirubin by photo-irradiation, in alkaline pH, and in polar media. Significantly, it was excreted, apparently intact, into bile (*823*):

> UCB in chloroform under nitrogen was irradiated with visible light; 7% was converted to a yellow photoproduct more polar than UCB. . . . The pure photoproduct has a λ_{max} in chloroform of 442 nm, is diazo-positive, and forms a verdin on dehydrogenation, suggesting it is an isomer of UCB. The product rapidly reverts to UCB on re-irradiation, at alkaline pH, and in a variety of polar media including bile, but is stabilized by binding to plasma albumin. I.V. injection of this pure photoproduct increased UCB excretion 140-fold. . . . A mechanism for excretion of UCB during phototherapy involves formation of a polar photoproduct which is stabilized in plasma, allowing its rapid and efficient excretion in bile. Being unstable in bile, the product reverts back to UCB after excretion.

The photoproduct was said to consist of three components, with the two most polar (**Ia** and **Ib**) being diazo-negative and unable to revert to bilirubin. In contrast, the less polar (**II**) isomer was said to be diazo-positive and able to revert to bilirubin (*822*):

> In-vitro studies: Bilirubin in CHCl$_3$ under N$_2$ was exposed to intense visible light. Three coupounds [*sic*], isolated by t.l.c., were shown by mass spectrometry to be isomers of bilirubin (m/e = 584). All have a λ_{max} in CHCl$_3$ of 442 nm, and revert to bilirubin on re-irradiation in solution. The two most polar isomers (**Ia+b**) are diazo-negative, yield no verdin on dehydrogenation, and spontaneously form brown products when concentrated. They have not been chemically or thermally reverted to bilirubin. In contrast, isomer **II** is diazo-positive and forms a verdin. In the dark, **II** reverts to bilirubin in a variety of polar media, including bile, but is stabilized by serum albumin. In-vivo studies: These isomers, labelled with ^{14}C, were given i.v. to Gunn rats with a bile fistula, kept in the dark. **Ia+b** engendered no increase in the output of bilirubin in bile, but was rapidly excreted unchanged, with properties like the major photoderivative found in Gunn rat bile during phototherapy. **II** was rapidly cleared, but reverted to ^{14}C-bilirubin after passage into the bile. Conclusions:

552   9   Understanding and Translating Bilirubin Structure

Photoisomerization of bilirubin may account for both the major photoderivative, and the bilirubin, which appear in bile during phototherapy of the Gunn rat.

This was not the only exciting news to appear in 1977, for from the Liver meeting in Chicago an illuminating abstract became available from *W.T. Roos et al. (824)*:

BILIRUBIN ISOMERS AND PHOTOTHERAPY. W.T. Roos, H.H. Wieger, S.C.H. Windler and C.O. Jones, Departments of Pediatric Oncology and the William B. Russell Institute for Molecular Photopharmacology, University of California, San Francisco CA 94143

Irradiation of bilirubin (I) with blue fluorescent lights in dimethylformamide (DMF) in the presence of β-carotene to inhibit photooxidation afforded, after preparative high pressure liquid chromatography on LiChrosorb RP-8, a 20% yield of a yellow diazo-positive pigment (II) (λ_{max} 441 nm in DMF) that was shown to be an isomer of bilirubin by mass spectrometry (M+ 584). II was crystallized from ice-cold propan-2-ol containing a trace of EDTA and its structure was unambiguously established by infrared spectroscopy and Fourier transform nmr, which showed a characteristic A_2B_2 multiplet for the furanoid ring. This was proved by characterization of the corresponding 5-benzoyloxy ethyl anthranilate azopigments. Irradiation of II in veronal buffer (pH 9.7) gave back I along with traces of the E,E isomer of II. In phosphate buffer (pH 7.6) in dark II was converted thermally to I (64%), a more polar hydroxy derivative (III) (17%), and a yellow diazo-negative dipyrrole that polymerized on mass spectrometry giving M+ = 584. Injection of II into Gunn rats and White Leghorn chickens caused rapid excretion of I and III in the bile. These studies indicate that photoisomerization of I to II is important in phototherapy. Attempts to detect II in vivo have been unsuccessful but irradiation of I bound to rat tail collagen in buffer yielded small amounts of a compound similar to II on tlc. Incubation of II (3.9×10^{-3}M) with HTC cells in Swim's medium depressed DNA synthesis by 54% and incubation (2.7×10^{-2}M) with single-stranded calf thymus DNA lowered sedimentation coefficients.

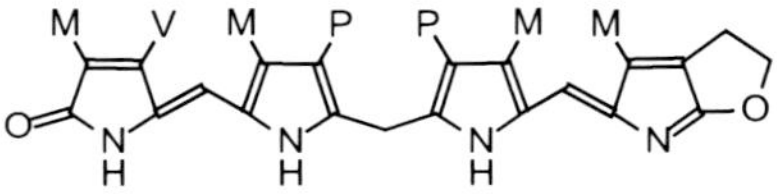

(II)

The news was startling to some investigators of phototherapy, for it proclaimed yet another new and different sort of bilirubin photoproduct, one essentially identical to the long-sought-after *Fischer* furan structure of bilirubin (see Section 6). A structure coming apparently by photochemical cyclization of the *exo*-vinyl with (presumably) a neighboring lactim hydroxyl to give a furan ring. That the article might be a cleverly crafted hoax, written in semi-believable scientific jargon was suggested by the names of its authors: *C.O. Jones*, well known to Spanish speakers, and *S.C.H. Windler*. Anyone unable to detect the German word *Schwindler* (= swindler, liar, con man, *etc.*) in the latter would, perhaps, be unaware of the famous *Wöhler* satirical article (*825, 826*) published in the *Annalen der Chemie und Pharmacie*, of which *Wöhler* and *Liebig* were editors. With a German title and a text in French the article was composed amidst bitter controversy and submitted allegedly from Paris, with an accompanying letter claiming "one of the brilliant facts of organic chemistry". Paris was the center of what *Wöhler* felt were extreme claims regarding "substitution" theory, the successor to *Berzelius'* "dualistic" theory. The former was based on the assumption that an atom of one element could be replaced by an atom of almost any other element and leave no significant alteration in the properties of the substance. Thus, in the *S.C.H. Windler* article manganous acetate was taken

9.2 How Phototherapy Exposed Bilirubin (*E*)-Configuration Isomers 553

through a series of remarkable substitutions to yield a new compound formed of 24 atoms of chlorine and one of water ($Cl_2Cl_2 + Cl_8Cl_6Cl_6$ + aq.).

Despite its obvious shortcoming, the *Roos* abstract, which also illustrated how easy it is to "publish" an abstract on the basis of only a small submission fee, was sufficiently believable to be discussed for its scientific merit in meetings held in the late 1970s and early 1980s. In refereed journal articles (*827, 828*) by at least one author, whose name is an anagram of the first author of the abstract, the claims were cited, discussed, and deemed to be unlikely.

As a consequence of numerous investigations, by the mid-1970s, it had become clear to some investigators, by the end of the decade, that the presence of unconjugated natural bilirubin in phototherapy bile was most likely due to a rapid, reversible photochemically induced, structural change that brought a decreased lipophilicity to the pigment. A change of structure that might be accomplished by (*Z*) → (*E*) inversion at either of the ene-amide exocyclic carbon-carbon double bonds, an inversion demonstrated in dipyrrinones first by *Falk et al.* in 1974 and reported in 1975 (*812, 814*). In the case of bilirubin (*Z*) → (*E*) photo-isomerization would necessarily break a set of intramolecular hydrogen bonds between a propionic acid and dipyrrinone, and that would render the photo-isomer less lipophilic than its parent. In 1978, *Bonnett*, too, suggested basically this same rationale (*511*); he wrote later, in 1981 (*829*), that he had mentioned it even earlier, at the 51st Meeting of the Biochemical Society in Leeds, U.K., in March 1976. Though the suggestion is apparently not found in *Bonnett*'s lecture abstract for that meeting (*830*), it is found in *McDonagh*'s abstract for the same meeting (*831*):

> One appealing, yet entirely speculative, theory is that bilirubin can be converted photochemically into a more polar, but less thermodynamically stable, isomer that does not need to be conjugated to be excreted. Cis-trans photo-isomerization of a model dipyrrolylmethene has been demonstrated and the individual isomers isolated (Falk *et al.* 1975), but analogous reactions of bilirubin have not been reported.

By 1979, there appeared to be common agreement that bilirubin photo-isomers were capable of passing across the liver into bile and that the isomers were also formed and excreted during phototherapy. However, the structures of the photoisomers involved in this attractive rationale were only hypothetical and had yet to be fully characterized. The first record of an attempt to remedy this deficiency came after *Mark Stoll* arrived from the *Charles Gray* lab in the U.K. on leave to the *Ostrow* lab in Philadelphia. The *Ostrow* group examined two pairs of photo-isomers (designated **IA** and **IB**, **IIA** and **IIB**) which were isolated following photo-irradiation of bilirubin in O_2-free $CHCl_3$ (*828*):

> Polar photoisomers of bilirubin were formed by irradiation of bilirubin in chloroform solution in the absence of O_2. Two pairs of compounds were isolated with molecular weights identical with bilirubin. One pair reverted to bilirubin in polar media and gave chemical reactions similar to bilirubin; the other pair were not reconverted into bilirubin by chemical means and gave reactions distinct from those of bilirubin. However, both groups were reconverted into bilirubin by irradiation in chloroform solution in the absence of O_2.

The first pair (**IA** and **IB**), which was separable on silica gel TLC and also faster moving with $CHCl_3:CH_3OH:H_2O$ (40:9:1 by volume eluent) than the second more polar pair (**IIA** and **IIB**) that was separable only on "double development"), reverted to natural bilirubin. (The assignments contradict those presented in 1977 (*822*)). The more polar isomers (**IIA** and **IIB**) proved to be more stable in a variety of solvents and did not revert to natural bilirubin through many attempts. On the basis of infrared and electronic spectroscopy, field-desorption mass spectrometric molecular weights (584 amu), and diazo-reactivity from material of dubious homogeneity, the authors, with some equivocation, assigned chemical (*E*)-isomer structures to **I** and **II**. That is, **IA** + **IB** were assigned to structures shown in Fig. 9.2.3 as (4*E*,15*Z*)- and (4*Z*,15*E*)-bilirubin; the more stable photoisomer (**II**) was assigned to (4*E*,15*E*)-bilirubin, as shown in Fig. 9.2.3 (*828*):

> If the photobilirubins are indeed geometrical isomers of bilirubin, it seems most likely that the less stable photobilirubins **IA** and **IB** are the two isomeric forms of E,Z-bilirubin . . . and that the more stable photobilirubin **II** is the E,E-isomer . . . .

The structures of **IA** and **IB** were assigned, **IA** to (4*Z*,15*E*), **IB** to (4*E*,15*Z*), on the basis of the differences in their UV-visible spectra (*828*):

> Of the two E,Z-isomers, the one with the altered *E*-configuration about the 4,5-double bond would throw the more tightly conjugated endovinyl group out of alignment with the conjugated diene system in the B-ring, giving its absorption maximum a shorter wavelength and lesser intensity than that of the 15*E*,16*E*-isomer. This would suggest that the photobilirubin **IB** is the 4*E*,5*E*-isomer . . . and that photobilirubin **IA**, spectrally more akin to bilirubin, is the 15*E*,16*E*-isomer . . . .

Assuming (*Z*,*E*) isomers, one has a 50:50 chance of guessing the correct structures. Not bad odds for gamblers, although the assignments were reversed in a later publication (*832*).

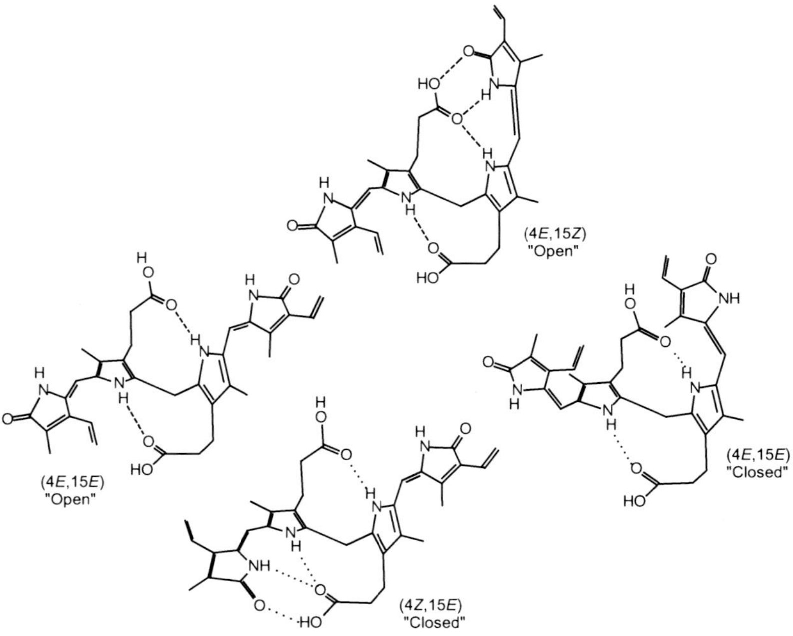

Fig. 9.2.3. The proposed (*828*) photoisomers of bilirubin. (Redrawn from Fig. 2 of ref. (*828*))

9.2 How Phototherapy Exposed Bilirubin (*E*)-Configuration Isomers

However, it should seem odd that an (*E,E*) isomer might be more stable than an (*E,Z*) or *Z,E*), especially when it was claimed that the latter easily reverted to natural bilirubin. As improbable as it might seem, the (*E,E*) structures were assigned two different rotational isomers of (4*E*,15*E*)-bilirubin, with rotation about the C-5–C-6 and C-14–C-15 carbon-carbon single bonds extending from the dipyrrylmethane core. The two rotamers were described as "open" and "closed" rotamers. The "open" has N-21 and N-22 *anti*, and N-23 and N-24 *syn*. The closed has N-21 and N-22 *syn*, and N-23 and N-24 *anti*. Details of "open" and "closed" aside, the assertion of stable rotamers was an open invitation to raise eyebrows of most organic chemists acquainted with stereochemistry. Yet, despite the improbability, the authors rationalized the assignments because (*828*):

> On the high-resolution t.l.c., photobilirubin **II** was separated into two components. However, individual elution and rechromatography of either component always yielded approximately equal quantities of both components. It is therefore postulated that these two components represent two readily interconvertible forms of *E,E*-bilirubin, e.g., the 'closed' and 'open' rotamers . . . . Molecular models indicate that both could exist, and that they would be separated by a steric energy barrier.

And in an unusual outreach to the "photo-oxidation past", the authors linked photobilirubin II to a dihydroxylated bilirubin found a few years earlier (*780*) and assigned to the rhodin (CH-9 of Fig. 9.1.2) said to arise *via* a $^1O_2$ pathway (*828*):

> Thus, the more stable photobilirubin II has spectral and chemical properties that strongly resemble the major photoproduct (called C-9) isolated by Berry *et al.* (1972) and Ostrow *et al.* (1974) from the bile of the Gunn rat during phototherapy.

Subsequent to the report in 1980 by *McDonagh, Palma*, and *Lightner* (*833*) that provided the molecular mechanism for the origin of (4Z,15Z)-bilirubin-IXα in phototherapy bile, in 1981, following his development of a powerful reverse-phase HPLC method of analysis, *McDonagh* summarized succinctly (*834*):

> A current view of the overall mechanism of phototherapy is shown schematically below. In this scheme I represents the normal excretory path for bilirubin (B) which in the neonate is impaired due to deficient formation of conjugated bilirubin (CB). II represents the main photoinduced excretory pathway and III represents a minor pathway that also may play a role. The first step in both pathways is photoexcitation of protein- or lipid-bound bilirubin in extravascular tissue. In the main pathway, this leads, via singlet bilirubin, to formation of 4*E*,15*Z* and 4*Z*,15*E* geometric isomers of bilirubin (PB) which are more hydrophilic than bilirubin and can be excreted via the liver without undergoing conjugation. During their passage from their site of formation to bile, the isomers are stabilized by binding to proteins, but once in the bile, reversion to the normal Z,Z-isomer probably occurs. Evidence for this mechanism is based on, (a) synthesis and properties of authentic Z,E/E,Z isomers of bilirubin and model compounds, (b) clearance studies performed with synthetic and photobiologically-produced isomers, (c) detection of E,Z/Z,E isomers in vivo during phototherapy, and (d) kinetic studies of bile pigment excretion during phototherapy.

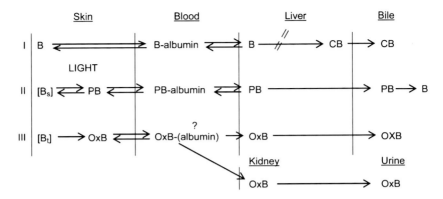

In addition to the photoisomerization pathway, it is likely that some self-sensitized photooxygenation of bilirubin, via triplet bilirubin, also occurs (III). Photooxygenation, rather than isomerization, would account more readily for the putative dipyrrolic photoproducts that are excreted in urine. However, in view of the low triplet yield of bilirubin and the relative quantum yields of the photoisomerization and photooxygenation pathways, it is unlikely that pathway III contributes more than about 10% to the overall process, if indeed it occurs to a significant extent at all.

While *Stoll, Zenone,* and *Ostrow* indicated in one of their final publications on phototherapy, that (*835*):

> By illumination of bilirubin under anaerobic conditions that precluded oxidative reactions, we ... and Onishi et al. ... produced and purified chromatographically a pair of unstable (**I**) and a pair of stable (**II**) photobilirubins. All the reported photobilirubins were more polar than the parent bilirubin IXα, being easily soluble in water and methanol.

Onishi et al. claimed to have found bilirubin photoisomers in a multicomponent HPLC scan of the photoproducts (*836*). *Ostrow*'s suggestion above (*835*) concerning the photoproducts' ease of aqueous solubility was probably overstated and may be an indication that photobilirubins **I** and **II** were impure or structurally misdiagnosed because neither (4E,15Z) nor (4Z,15E)-bilirubin is easily soluble in H₂O. Nonetheless, as *Stoll et al.* asserted earlier (*835*):

> Thus, our experiments suggest strongly that photobilirubins **IA** and **IB** are the precursors of the bilirubin IXα that appears in bile during phototherapy, but it is appreciated that some means of prolonged stabilization of the photobilirubins **I** in pure form is essential to achieve unequivocal documentation of both their in vivo behavior and their structural identification.
>
> Much clearer was our demonstration that, after intravenous administration, the more stable photobilirubin **II** is excreted in the bile and corresponds to the other major photoproduct (C-9) that appears in the bile of the Gunn rat ... and human infant ... during phototherapy.

And so, they were led to summarize (*835*):

> Considered together, all published studies suggest that the photobilirubins are geometric (Z,E) isomers of bilirubin IXα, whose formation may account for the major pathways of bilirubin photocatabolism in the Gunn rat and human neonate. These photoisomers presumably

9.2 How Phototherapy Exposed Bilirubin (*E*)-Configuration Isomers

are formed rapidly when light penetrates to the bilirubin deposits in the skin and subcutaneous tissue, then partitioned to the plasma, taken up by the liver, and excreted into the bile. In bile, the thermally unstable photobilirubins **IA** and **IB**, believed to be the E,Z and Z,E isomers, rapidly revert to the Z,Z configuration, accounting for the augmented excretion of bilirubin during phototherapy of both Gunn rats . . . and human neonates. . . . The more stable photobilirubin **II** remains mostly intact and constitutes the major polar photoproduct, but is in part hydroxylated at the methene bridges, by photo-dependent reactions, to yield other minor derivatives. Together, these photoisomerizations and secondary reactions apparently account for >80% of the accelerated bilirubin catabolism during phototherapy. . . .

Als ob, nunquam reformata quia nunquam deformata. 1

Coming only slightly later to the phototherapy banquet was the *Onishi* group, which like all others first embraced photo-oxygenation, then photoisomerization. As mentioned by *Stoll, Zenone*, and *Ostrow (835)*, *Onishi* had begun to publish his studies on the configurational isomers of bilirubin and their relationship to phototherapy following the studies of *Ostrow, McDonagh*, and *Lightner*. Thus, in late 1979, *Onishi et al.* followed the photo-irradiation of natural bilirubin in CHCl₃ (*836*) by reversed-phase HPLC and showed a complete array of some 24 new peaks representing more polar compounds, as well as peaks for bilirubin-IIIα and XIIIα, in addition to IXα. Yet sampling of the serum of a hyperbilirubinemic neonate during phototherapy showed only one peak representing at least one more polar compound, which was labeled *photobilirubin-IX*a and was believed to constitute a collection of bilirubin (*E*)-isomers reported earlier in 1979 (*769, 818, 819*).

In subsequent studies, reported in February 1980, *Onishi et al.* used reversed-phase HPLC to analyze the photolysate of bilirubin-IXα in pH 7.4 phosphate buffered human serum albumin (HSA) following short periods (3–30 min) of photoirradiation with a bank of four 20 W blue-white fluorescent tubes (*837*). The conditions differed from and were milder than those reported by *Stoll et al.* in 1979 (*828*), who irradiated 100 mg/dm³ bilirubin in N₂-purged CHCl₃ with an unfiltered GE 100 W high-pressure Hg lamp for 2 min/mg bilirubin. In the HPLC scans, *Onishi et al.* marked three new peaks, labeled as 1, 2, and 3, as "unknown", "unknown pigment", and photobilirubin-IXα, respectively (*837*). None were characterized beyond their UV-visible λ_{max} and diazo-reactivity, of which peak 2 was negative and peak 3 was positive. Peak 3, which apparently could be separated further into two peaks by HPLC, was assigned on the basis of its apparent identity with the photobilirubin-IXα found earlier in serum (*836*). Peak 2 was indicated (*837*) to correspond to the "unknown pigment" found in addition to bilirubin and its mono- and diglucuronides, in aspirated duodenal bile of a hyperbilirubinemic neonate undergoing phototherapy, but not in the bile just before phototherapy (*838*). *Onishi* theorized that the "unknown pigment"/peak 2 might be the same substance as the photobilirubin **IIA** and **IIB** of *Stoll et al.* (*828*), albeit without evidence or proof. A year later, in 1981, the "unknown pigment"/peak 2 was separated into two components ("unknown pigment" A and B/ peak 2A and 2B) by HPLC and TLC on silica gel. Their structures (Fig. 9.2.4) were also named (*E,Z*)-unknown pigments A and B and drawn as spiro-lactone constitutional isomers of bilirubin (Fig. 9.2.4) (*839*). Possibly after hearing *Stoll*'s presentation at the April 1981 meeting of the Tetrapyrrole Discussion Group (*840*), *Onishi et al.* then proceeded to assign peak 1, as Peaks 1A and 1B, and provided the spiro-lactone structures for (*E,E*)-unknown pigments A and B (*839*).

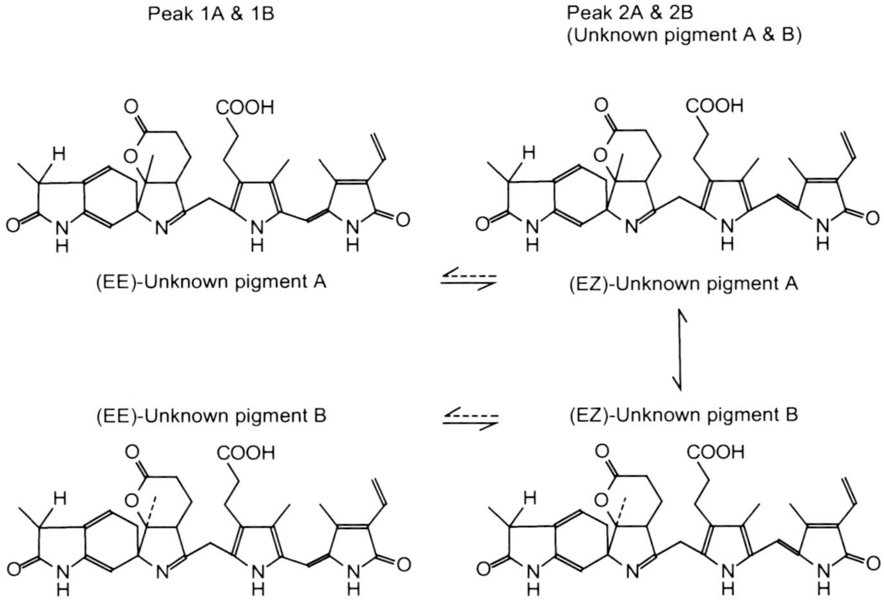

9 Understanding and Translating Bilirubin Structure

Peak 1A & 1B

Peak 2A & 2B
(Unknown pigment A & B)

(EE)-Unknown pigment A   ⇌   (EZ)-Unknown pigment A

(EE)-Unknown pigment B   ⇌   (EZ)-Unknown pigment B

Fig. 9.2.4. HPLC peak structure assignments from *Onishi et al.* (*839*) (Redrawn from ref. (*839*))

Thus, although in late 1980 (*841*) peak 2 was still referred to as an (*E,E*)-photoisomer of photobilirubin, photobilirubin **IIA** and **IIB** of *Stoll et al.* (*828*), in 1981, *Onishi* had adopted the spiro-lactone structure (*840*) and dubbed it *cyclobilirubin* in 1983 (*842*) and in 1985 revised the structures to the vinyl-cyclized rubins (*843*).

In general, the work of *Onishi* appeared to confirm that reported by *Ostrow*, *McDonagh*, and *Lightner*, though the structures of the photobilirubins remained questionable, wrongly assigned or unproven, unproven, that is, until 1982 (*762*, *844*). Between 1976 and 1982, *McDonagh* and *Lightner* set about the tedious and tricky task of isolating sufficiently pure amounts of the supposed bilirubin photoisomers for complete structural characterization. Their instability was well-known, which became understood as other investigators encountered them, as reported by *Ostrow*, *Onishi*, and *Stoll*, in whose hands they proved to be too unstable to isolate and characterize. Thus, it was said of the (4*Z*,15*E*) and (4*E*,15*Z*) photoisomers by *Stoll et al.* in 1979 (*828*):

> Unfortunately their relative instability has thus far precluded satisfactory n.m.r and X-ray diffraction studies.

And again in 1982 (*845*):

> . . . because these substances revert so readily to (4Z,15Z)-bilirubin . . . chemical proof of this is still lacking.

As should have become well-evident from the long history of the bilirubin structure proof, certifiable sample purity is a *sine qua non*. This was no less true for the photo-isomer structure proof. Thus, *McDonagh* was able to achieve the isolation of

9.2 How Phototherapy Exposed Bilirubin (*E*)-Configuration Isomers

(*E*)-isomers of sufficient purity for 360 MHz ^{1}H NMR studies that were investigated in the preliminary stages in the late 1970s, initially at Stanford University (courtesy of Dr. *Lois Durham*) and then completely in 1980–1981 at the U.C. Davis NMR facility (courtesy of Drs. *J.L. Dallas* and *G.B. Watson*). Sample purity was defined by HPLC. A most important adjunct to this endeavor was the ability to assess the composition, distribution, and relative quantities of photoisomers by the sensitive, high resolution reversed-phase HPLC system developed, perfected, and revealed by *McDonagh* by 1981 (*834*), so that by 1982 it could be stated in detail that (*846, 847*):

> We have previously reported that when homozygous Gunn rats labeled with ^{14}C-bilirubin (BR) are exposed to continuous blue light the concentrations of yellow pigment and ^{14}C in bile increase during the first 1.5 h to plateau values which are then maintained for many hours. In the present studies Gunn rats were treated similarly and bile samples, carefully collected without exposure to light, were analysed every 30 min by reverse phase hplc (C-18 column) using 0.1M di-n-octylamine acetate (pH 7.7) in MeOH as eluent. After 1.5-2 h of continuous irradiation the pigment composition of the bile had reached a new steady-state which was maintained until the light was extinguished 2-3 h later. During the photostationary phase the approximate pigment composition of the bile (based on uncorrected peak areas) was 5% E/E-BR, 5% Z-lumirubin, 5% Z-isolumirubin, 49% E/Z-BR, 37% Z/Z-BR. The ratio of 4E/15Z-BR to 4Z/15E BR was = 1:4. Essentially all of the Z/Z-BR present was formed by thermal reversion of E isomers in bile. Therefore the main process responsible for hepatic pigment excretion during phototherapy in the rat is single-photon Z → E isomerization, predominantly at the 15Z bridge. Intramolecular vinyl cyclization to lumirubins and formation of E/E isomers are relatively minor contributors. These observations are consistent with the photochemistry of BR in vitro. [*846*]
>
> We have developed a suite of simple solvent systems for the rapid hplc separation of BR photoisomers and have used these in conjunction with spectroscopic (absorption, mass, nmr) and tlc measurements to study the anaerobic photochemistry of BR and model compounds. Two general reactions occur: Z → E isomerization of exocyclic double bonds and, if there is a vinyl group at C-3 (and/or C-17), intramolecular cyclization. Z → E isomerization is fast and rapidly reversible photochemically or catalytically by H$^+$. Vinyl cyclization is relatively slow and not rapidly reversible by light or H$^+$. Thus, photolysis of Z-xanthobilirubinic acid (a dipyrrylmethenone) yields a photoequilibrium mixture of E and Z isomers. Similarly, irradiation ($\cong$450 nm) of meso-BR XIIIα or BR IIIα gives a photoequilibrium mixture containing only Z/Z, E/Z and E/E isomers (Z/Z > Z/E > E/E). Short-term irradiation of BR gives a pseudo-photostationary mixture containing Z/Z, 4E/15Z, 4Z/15E and E/E isomers in which the 4Z/15E isomer predominates over the 4E/15Z. On continued irradiation, intramolecular vinyl cyclization leads to gradual accumulation of two further products, Z-lumirubin and Z-isolumirubin. These also undergo rapid reversible Z → E isomerization. Consequently, prolonged irradiation of BR (and BR XIIIα) gives mixtures containing Z/Z, E/Z, and E/E bilirubins plus E and Z lumirubins and isolumirubins. The proportion of cyclized products depends on irradiation time, solvent and the presence of binding proteins. [*847*]

9.2.1 *(4Z,15E)- and (4E,15Z)-Bilirubin Diastereomers (Photobilirubin)*

Mindful of the relative ease to which incorrect conclusions may be drawn and published – and then amended, often repeatedly, it was essential to prove the structures

560   9 Understanding and Translating Bilirubin Structure

of the (*E*)-isomers (or at least disprove all other possibilities) and to demonstrate that they are formed *in vivo* and found in serum and in bile – if they are to have any relevance to phototherapy. Yet, as always, it is one thing to speculate and another to know.

Although certification of structure has become a standard fixture in organic and natural products chemistry for over a century, no less a person than *Fischer* himself fell into traps of misunderstanding data, as seen earlier, coming from the inability to isolate methylvinyl-maleimide, the ease in which bilirubin and dipyrrinones form enol ethers, a reliance on mixture melting points, and a belief in structural motifs (heme, bilirubin) proposed by a then better known scientist (*Willstätter*). In fact the entire field of bile pigment and porphyrin chemistry is replete with examples of investigators having led themselves astray with bad data, impure samples, and an inability to interpret data. *Thudichum*, though his invective cut a wide swath of scathing rebuke through many of his fellow investigators for their (perceived) ineptitude and errors, was generally careful, extraordinarily tenacious, and disdainful of *Conjecturalchemie* (see Section 2.12). That is, the propagation of conclusions based on supposition and not fact, and deficient in firm experimental verification.

Firm experimental verification appeared nearly simultaneously from two sources in 1982: *Heinz Falk* at the Johannes Kepler Universität Linz (*848, 849*) and *A.F. McDonagh* at the University of California, San Francisco, together with *D.A. Lightner* and Dr. *Francesc Trull* in the *Lightner* research group at the University of Nevada, Reno (*762, 844*). *Falk*'s group made excellent use of a chemical correlation based on their earlier structural investigations of (*E*)-isomers of biliverdin prepared by photoisomerization of the natural pigment (4Z,10Z,15Z)-biliverdin-IXα and its dimethyl ester. The photoproducts produced thereby were separated and well-characterized spectroscopically as the (4E,10Z,15E) and (4E,10Z,15Z) dimethyl esters, which were stable (*849*). A carefully conducted brief anaerobic photolysis of natural bilirubin-IXα converted it to an equilibrium mixture of (4Z,15E) and (4E,15Z) diastereomers, present together with the (4Z,15Z) starting isomer. The photoisomers were selectively washed into cold CH_3OH, methylated with CH_2N_2, and oxidized to the corresponding verdin esters by chloranil. The last was separated into the components consisting of (4Z,10Z,15E)- and (4E,10Z,15Z)-biliverdin dimethyl esters that were identical with "authentic" material reported earlier (*849*). Hence, a chemical correlation that proved the structures of the bilirubin (*E*)-isomers.

The approach taken by *McDonagh* and *Lightner* rested on a different sort of chemical correlation, one using bilirubin analogs, 360 MHz ^{1}H NMR analyses, and, especially, the availability of an indispensable and powerful reversed-phase HPLC system developed by *McDonagh* in 1981 (*762*) in a long process that included experimentation with numerous ion-pairing salts and methanol-acetic acid-alkylamine systems. With the availability of the last (0.1 *M* di-*n*-octylamine acetate in CH_3OH) the fast anaerobic photochemistry of bilirubin could be followed exquisitely well, and the production, purity, and stability of the separated photoisomers could be monitored, quantitated, and evaluated. Armed with this ability, the approach taken by *McDonagh* and *Lightner* was simple and logical. It eschewed a frontal assault and depended on a small collection of structurally related model compounds, bilirubinoids that had become available by HPLC separation (*793*) and synthesis methods (*795, 796, 850*). Much as dipyrrinone models had served to help investigators understand bilirubin's

9.2 How Phototherapy Exposed Bilirubin (*E*)-Configuration Isomers 561

photochemistry and *Hans Fischer* had used bilirubins-IIIα and -XIIIα and mesobilirubin-XIIIα in his studies, so did these pigments again find use – in ¹H NMR measurements and analyses. Why were the last three pigments important? Because they possessed the essential constitutional and conformational characteristics of the natural pigment (and behaved in metabolism like it); yet, unlike the natural pigment they are structurally symmetric, as shown in the linear representations of Fig. 9.2.5. The structural symmetry of the (4Z,15Z) diastereomers of bilirubin-IIIα and XIIIα, and in mesobilirubin-XIIIα, translates into ¹H NMR spectra simplified by symmetry.

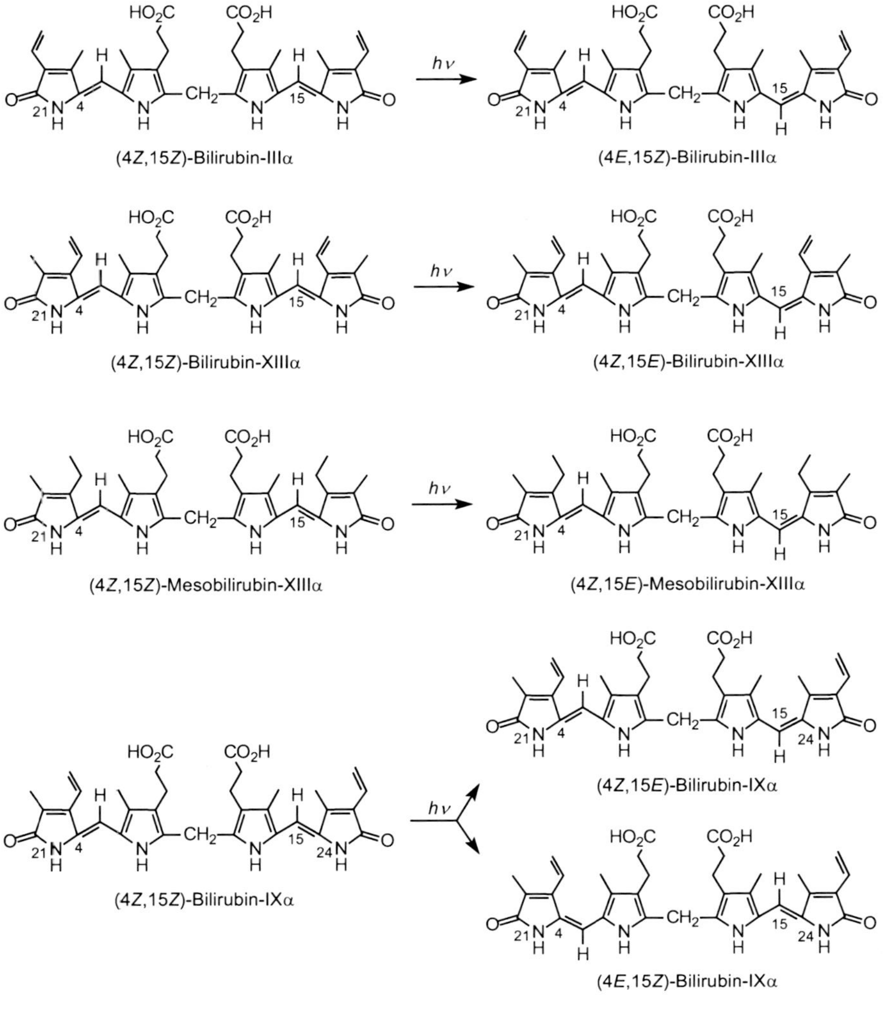

Fig. 9.2.5. Linear representations of bilirubins-IIIα, XIIIα, and IXα, and of mesobilirubin-XIIIα. The first two and last possess symmetric structures that are desymmetrized by inverting either the (4Z) or (15Z) C=C. Bilirubin-IXα has an asymmetric structure and produces two different asymmetric diastereomers by inverting the (4Z) and the (15Z) C=C

That is, the signals from the dipyrrinone on the left half of the tetrapyrrole are identical to those from the dipyrrinone on the right half. In contrast, the ^{1}H NMR spectrum of the asymmetric (4Z,15Z) bilirubin-IXα is more complicated, with two sets of NMR signals appearing – one from the *endo*-vinyl dipyrrinone, the other from the *exo*-vinyl dipyrrinone.

Accordingly, photochemical studies were carried out on (4Z,15Z)-bilirubin-IXα and the symmetrical analogues (4Z,15Z)-bilirubin-IIIα, (4Z,15Z)-bilirubin-XIIIα, and (4Z,15Z-mesobilirubin-XIIIα (Fig. 9.2.6). Solutions ($\sim 10^{-3}$–10^{-4} M) were degassed with argon and photolyzed with blue light (Westinghouse special blue 20 W fluorescent tube, λ_{max} 436, 446 nm) or for wavelength-dependence studies, with 20 nm band-pass light from a variable-wavelength monochromator. Reactions were followed by absorbance difference spectroscopy and HPLC. Products were analyzed by reversed-phase HPLC (Beckman-Altex Ultrasphere-IP or -ODS columns, 5 μm, C_{18}, 25 × 0.46 cm; with a Beckman-Altex ODS precolumn, 4.5 × 0.46 cm) by using 0.1 M di-n-octylamine acetate in CH_3OH as eluent (0.75 cm^3/min, detector 435 or 450 nm). Solutions were considered to have reached a photostationary state if there was no further marked change in composition on doubling the irradiation time. Rapid configurational photoisomerization of (4Z,15Z)-bilirubin-IXα occurs in organic and aqueous albumin solutions. In water, disproportionation to IIIα and XIIIα structural isomers occurs (*851*), and (Z) → (E) isomerization is undetectable.

On photolysis of (4Z,15Z)-bilirubin-IIIα in 50% $CHCl_3$-Et_3N the system rapidly reached a photostationary state with development of a characteristic sigmoidal difference spectrum showing a positive peak (487 nm), a negative peak (456 nm), and *tight* isosbestic points (360 and 476 nm). HPLC at photoequilibrium (Fig. 9.2.6a) revealed, in addition to (4Z,15Z)-bilirubin-IIIα, two rather more polar compounds that reverted to (4Z,15Z)-bilirubin-IIIα slowly on warming and instantly on treatment with a trace of CF_3CO_2H. No new peaks appeared on further prolonged irradiation. (4Z,15Z)-Mesobilirubin-XIIIα behaved almost identically with (4Z,15Z)-bilirubin-IIIα. (4Z,15Z)-Bilirubin-XIIIα behaved somewhat similarly to (4Z,15Z)-bilirubin-IIIα in that brief irradiation led to a pronounced difference spectrum and the appearance of two new acid-labile peaks on HPLC (Fig. 9.2.6c). However, a true photostationary state was not reached, and in contrast to bilirubin-IIIα and bilirubin-XIIIα, continued irradiation led to further changes in the difference spectrum with loss of isosbestic points and the slow emergence of additional peaks on HPLC (marked by asterisks in Fig. 9.2.6c). The behavior of the naturally occurring IXα isomer, (4Z,4Z)-bilirubin-IXα, was intermediate between that of IIIα and XIIIα and HPLC near photoequilibrium showed three new acid-labile products (Fig. 9.2.6b), as expected, each more polar than the parent isomer.

9.2 How Phototherapy Exposed Bilirubin (E)-Configuration Isomers

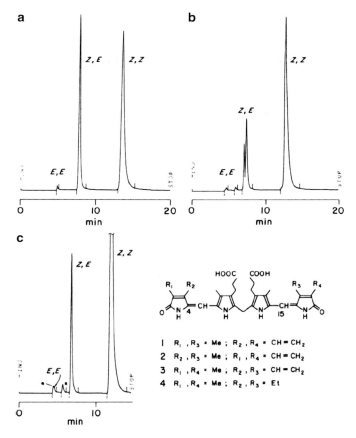

Fig. 9.2.6. HPLC of samples at or near photoequilibrium from the photolyses of (**a**) bilirubin-IIIα [**2**], (**b**) bilirubin-IXα [**1**], and (**c**) bilirubin-XIIIα [**3**] in CHCl$_3$-Et$_3$N. Base-line HPLC separation of the two (Z,E) isomers of bilirubin was obtained with 0.1 M di-n-dodecylamine acetate in CH$_3$OH as eluent (1 cm^3/min) (Fig. 1 of ref. (*844*). Reprinted, with permission of the American Chemical Society. Copyright 1982)

These observations, taken alone or in conjunction with previous work, are consistent with the structural assignments shown in Fig. 9.2.6, and they provide strong evidence that bilirubin-IXα undergoes configurational photoisomerization as follows:

$$(4Z,15Z) \rightleftarrows (4E,15Z) + (4Z,15E) \rightleftarrows (4E,15E)$$

Reversibility also was demonstrated by irradiating purified (Z,E) isomers obtained from (4Z,15Z)-bilirubin-IXα, (4Z,15Z)-bilirubin-IIIα, and (4Z,15Z)-mesobilirubin-XIIIα. In each case the photostationary mixture of isomers obtained was nearly the same as that obtained on irradiation of the corresponding parent (Z,Z) isomer. The possibility of phototautomerism (lactam ⇄ lactim) rather than geometrical configurational photoisomerization may be rejected as follows: (i) the 0.2 ppm deshielding of H-15 to d 6.07 ppm in the (4Z,15E) diastereomer of bilirubin-IIIα falls in the range found for *meso*-H deshieldings of (E) isomers in model dipyrronones and benzalpyrrolinones, not in the region of the more strongly deshielded (d 6.4-6.5 ppm) H-5/H-15 signals of the bis-lactim methyl ether of bilirubin-IXα dimethyl ester. (ii) The N-H signals of the (E) half of (4Z,15E)-mesobilirubin-XIIIα [δ 9.88 (H-21), 10.45 (H-22), 9.68 (H-23), 10.06 (H-24) ppm] are shifted upfield from the parent (4Z,15Z)-mesobilirubin-XIIIα [d 9.83 (H-21= H-24), 10.34 (H-22 = H-23) ppm], whereas the N-H signals [δ 10.95 (H-22 = H-23) ppm] of the bis-lactim methyl ether of bilirubin-IXα dimethyl ester appear farther downfield.

Preparations highly enriched in the polar (E) photoisomers were collected by solvent extraction of crude photolysis products at 0°C. Thus, solutions of (Z,Z) isomers in purified $CHCl_3$-Et_3N (1:1) were irradiated to near photoequilibrium. The odorless residue obtained after evaporation of solvent *in vacuo* was mixed with CH_3OH at 0°C, and the mixture was filtered rapidly into cold $CHCl_3$-Et_3N. Removal of solvent at <25°C gave the enriched isomer preparation (*e.g.* see Fig. 9.2.7). These preparations contained traces of solvent residues that inhibited the autocatalytic thermal (E) → (Z) reversion of (E) isomers in $(CH_3)_2SO$ and facilitated the NMR studies. Nonphotochemical manipulations were done under a safelight. Typical enrichments are shown in Fig. 9.2.7. Comparison of the ¹H NMR spectra of photoisomerized bilirubin-IIIα and its structurally symmetric parent (Z,Z) isomer reveals quite strikingly the anticipated doubling of most of the H resonances due to desymmetrization of the molecule. Thus, there appear two meso-bridge =CH proton signals at C-5 and C-15, two vinyl group multiplets (Fig. 9.2.8), four CH_3 singlets, and four NH proton signals. One set of signals in the photoisomer preparation corresponds closely to the spectrum of the (Z,Z) isomer; the second set, equal in intensity to the first, exhibits chemical shifts characteristic of (E) configuration dipyrrinones (*812*).

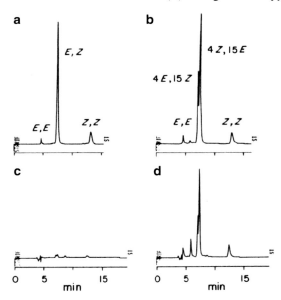

Fig. 9.2.7 HPLC of (**a**) purified (4Z,15E)-bilirubin-IIIα, (**b**) purified (4E,15Z/4Z,15E)-bilirubin-IXα, (**c**) hepatic bile from a jaundiced *Gunn* rat kept in the dark, (**d**) hepatic bile from the same rat after exposing the animal to blue light for two hours (Fig. 2 of ref. (*844*). Reprinted, with permission of the American Chemical Society. Copyright 1982)

9.2 How Phototherapy Exposed Bilirubin (E)-Configuration Isomers

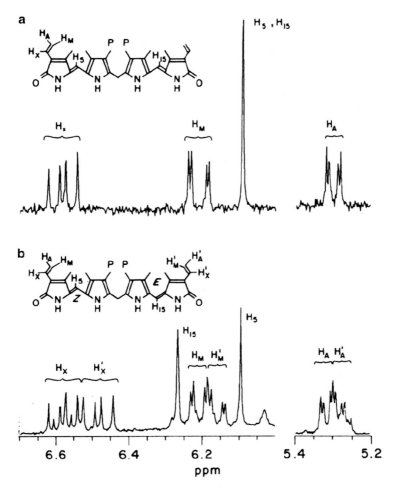

Fig. 9.2.8 Olefinic region of the 360 MHz ¹H NMR (Me₂SO-d₆) of (**a**) bilirubin IIIα, and (**b**) its purified (Z,E) diastereomers (P = CH₂CH₂CO₂H). Proton resonances are in ppm downfield from Me₄Si (Fig. 3 of ref (*844*). Reprinted, with permission of the American Chemical Society. Copyright 1982)

Particular note was taken of the readily observed changes of what in the symmetric (4Z,15Z) rubins is a singlet signal at $\delta \sim 6$ ppm from the C-5 and C-15 methine hydrogens. For example in the (4Z,15Z) bilirubin-IIIα and XIIIα spectra, the methines (=CH) appear at 6.09 ppm. In the (4E,15Z) diastereomer of IIIα, a new signal appears at 6.26 ppm (integration = one proton), corresponding to a change from a *trans* to a *cis* relationship of the H-5 relative to the 21-NH, attending the change in stereochemistry from (4Z) to (4E). While the H-15 signal remains at 6.09 ppm but at half its original intensity, reflecting a change in integration from two protons to one proton. The deshielding of the "(E)-configuration methine" relative to the (Z) is characteristic of dipyrrinones (*812*) and arylpyrromethenones (*814–817*) and *Lugtenburg*'s simplified rubins (*852, 853*). Analogous features (*i.e.* signal doubling and characteristic chemical shifts) were observed in the spectra of purified photoisomers from bilirubin-XIIIα (Fig. 9.2.9d) and mesobilirubin-XIIIα. The photoisomer preparation derived from bilirubin-IXα contains two major components in the ratio of about 1:2 (Fig. 9.2.7b). The NMR spectrum of this material (Fig. 9.2.9b) was essentially a composite of the spectra of the corresponding IIIα and XIIIα photoisomers. Comparison of all of these spectra (Figs. 9.2.8 and 9.2.9) showed unambiguously that the predominant photoisomer in the preparation derived from (4Z,15Z)-bilirubin-IXα is the (4Z,15E) diastereomer, as designated in Fig. 9.2.7b.

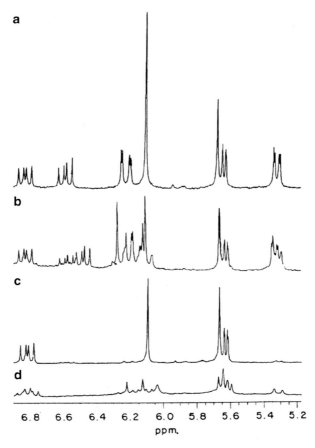

Fig. 9.2.9. Olefinic region of the 360 MHz ^{1}H NMR spectra ((CD$_3$)$_2$SO) of (**a**) (4Z,15Z)-bilirubin-IXα, (**b**) purified (4E,15Z/4Z,15E)-bilirubin-IXα with the (4Z,15E) diastereomer predominating, (**c**) (4Z,15Z)-bilirubin-XIIIα and (**d**) purified (4Z,15E)-bilirubin-XIIIα. The sample used in (**d**) contained 52% of the (Z,E) diastereomer and 34% of the (Z,Z) diastereomer and was contaminated with ~ 10% of the corresponding IXα isomer (Fig. 4 of ref. (*844*). Reprinted, with permission of the American Chemical Society. Copyright 1982)

9.2 How Phototherapy Exposed Bilirubin (*E*)-Configuration Isomers

These data provided conclusive evidence that (*E,Z*) isomers are the main photo-products formed on short-term irradiation of bilirubin *in vitro*, and they confirmed *McDonagh* and *Lightner*'s previous structural assignments for "photobilirubin". Clear evidence that they are the predominant yellow photoproducts excreted by the liver *in vivo* during blue-light irradiation of jaundiced rats (*833, 854, 855*) is shown in Fig. 9.2.7.

The (4*E*,15*E*) diastereomers were assigned to the minor, most polar photoproducts from the various (4*Z*,15*Z*)-bilirubins of Fig. 9.2.5, as follows from their thermal and acid lability, and comparison with analogous (*E,E*) isomers (*818, 852, 853*). And what of the term *photobilirubin*? *McDonagh* and *Lightner* used it (a successor to *isobilirubin*, used by them in the early to mid-1970s) to refer to the mixture of photoisomers obtained on irradiating (4*Z*,15*Z*)-bilirubin-IXα to photoequilibrium [Section 9.2 (*834*)]. Now that the individual isomers had been characterized, they recommended that the *photobilirubin* nomenclature be discontinued for the individual isomers. Further, they noted that several other "photobilirubins" had been reported and designated as configurational isomers of bilirubin. However, many of the structural assignments were inconsistent with the *McDonagh-Lightner* data. For example, *Stoll et al.* (*828*) isolated four substances by preparative TLC on silica and referred to them as photobilirubins **IA, IB, IIA**, and **IIB**. **IA** and **IB** were designated as (4*Z*,15*E*)-bilirubin-IXα and (4*E*,15*Z*)-bilirubin-IXα, respectively. **IIA** and **IIB** were considered to be conformational isomers of (4*E*,15*E*)-bilirubin-IXα. *McDonagh* and *Lightner* were unable to separate or purify authentic (*E,Z*) isomers of bilirubin-IXα by TLC as described. HPLC studies of **IA** and **IB** by *McDonagh* revealed that both were complex mixtures containing only trace amounts of the designated (*E,Z*)/(*Z,E*) isomers and that neither **IIA** nor **IIB** was the (4*E*,15*E*) diastereomer of bilirubin-IXα. The data showed unequivocally that none of the structural assignments in ref. (*828*) was correct. Those incorrect assignments were perpetuated in subsequent papers (*827, 829, 832, 845, 856, 857*).

Onishi et al. detected more than 24 components after irradiating (4*Z*,15*Z*)-bilirubin-IXα anaerobically in $CHCl_3$ (*836*). At least seven of those were stated to be geometric isomers of bilirubin, but structures and adequate supporting data were not presented. Many of the observed products were probably secondary products and artifacts unrelated to the primary photochemistry of bilirubin resulting from overirradiation and radical reactions. *Isobe* and *Onishi* isolated three substances, designated as peaks 1, 2, and 3, from photolysis of (4*Z*,15*Z*)-bilirubin-IXα in aqueous serum albumin (*841*). Peak 1 was not identified, peak 2 was attributed to (4*E*,15*E*), and peak 3 was attributed to a mixture of the two (*E,Z*) isomers. On the basis of data described in this and their following communication, it was clear that peak 2 was not (4*E*,15*E*), and it was probable that peak 3 was almost exclusively (4*Z*,15*E*). With regard to the failure of other workers to characterize the (*E*)-isomers of bilirubin, it was suggested in late 1981 (*832*): ". . . initial attempts to isolate solid photobilirubin **I** and to study its properties at room temperature . . . are now realised to have given misleading results", largely (*845*) ". . . because these substances revert so readily to (4*Z*,15*Z*)-bilirubin . . . chemical proof of this is still lacking".

9.2.2 Lumirubin and Isobilirubin Diastereomers

As indicated in Section 9.1, on April 2, 1973, *McDonagh* found that long term (six hour) irradiation of bilirubin in aqueous albumin produced a "stable" yellow pigment that was soluble in aqueous bicarbonate, could be back extracted into $CHCl_3$ after acidification with acetic acid, had a UV-visible λ_{max} at 430 nm, and a mass spectrometric molecular weight of 584 – the same as bilirubin. It was then called *Compound X* by *McDonagh* and *Lightner*, and it was reported soon thereafter, briefly, in at least two publications stemming from lectures given in early February 1973 (*801*) and late April 1974 (*802*). It proved not to be identical to the unconjugated bilirubin found in the bile of *Gunn* rats undergoing photoirradiation (*773, 798, 799, 800*), and its definitive structure proof was published in 1982 (*762*) next to that of the (*E*)-configurational photoisomer of bilirubin (*844*).

The (*E*) isomers of bilirubin are unstable and revert to the parent isomer photochemically or thermally. At room temperature thermal reversal is highly solvent dependent, being slow in basic organic solvents, serum, or aqueous albumin and rapid in acidic or polar solvents. In 1973–1974 *McDonagh* described the formation of additional yellow products, more stable than the (*E*) isomers, on prolonged photolysis of bilirubin. Small amounts of these products were formed in jaundiced rats exposed to blue light and in humans during phototherapy (*844*). In 1981, the compounds, which were called *lumirubins*, were shown to be structural isomers of bilirubin (*762*). It was also shown that complexation of bilirubin with serum albumin (SA) had a marked species-dependent influence on bilirubin photoisomerization and, in particular, on the regioselectivity of the configurational isomerization.

On photolysis of bilirubin in $CHCl_3$-Et_3N (1:1) (Fig. 9.2.10, a–d) the system reached a photostationary state in ~10 min. HPLC revealed the expected (*E,E*) and (*E,Z*) isomers (R_t 4.8 and 7.4 min, respectively) and a minor peak at R_t 6.1 min. Thus, solutions containing 1.25 mg of pigment/10 cm^3 were photolyzed under Ar with a 20 W blue fluorescent tube. Albumin solutions were made up in 0.1 M phosphate buffer, and the albumin:pigment ratio was 1.1:1. The solvent for reversed-phase HPLC (Beckman-Altex Ultrasphere-IP or Ultrasphere-ODS column, 5 µm, C-18, 25 × 0.46 cm; with a Beckman-Altex ODS precolumn, 4.5 × 0.46 cm) was 0.1 M di-*n*-octylamine acetate in CH_3OH (0.75 cm^3/min, detector 450 nm). Retention volume = 0.75 R_t. For HPLC, albumin solutions were diluted 1:4 or 1:9 with ice-cold mobile phase, and 20 mm^3 of the supernatant was injected. On continued photolysis the 6.1-min peak grew slowly, without affecting the (*E,Z*):(*Z,Z*) ratio, along with a smaller peak (R_t 4.7 min) that ran close to (*E,E*)-bilirubin. All peaks except the R_t 6.1-min peak and a minor unidentified peak at R_t 6.7 min disappeared on treatment of the photolysate with CF_3CO_2H (*481*) (Fig. 9.2.10). Similar rapid configurational isomerization accompanied by slow growth of the 4.7- and 6.1-min peaks was observed when complexes of bilirubin with SA were irradiated at pH 7.4. However, the identity of the albumin had a marked effect on the rates of product formation and on the composition of the photoproducts at a given irradiation time (*cf.* Fig. 9.2.10, f–j).

9.2 How Phototherapy Exposed Bilirubin (*E*)-Configuration Isomers

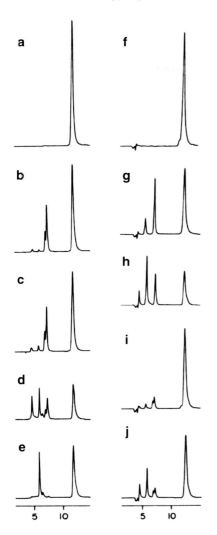

Fig. 9.2.10 Photolysis of bilirubin and bilirubin-serum albumin complexes: photolysis of bilirubin in CHCl$_3$-Et$_3$N (1:1); HPLC of samples at (**a**) 0 min, (**b**) 10 min, (**c**) 30 min, (**d**) 4 h; (**e**) HPLC of sample d after treatment with CF$_3$CO$_2$H; photolysis of bilirubin in aqueous human serum albumin, pH 7.4, HPLC at (**f**) 0 min, (**g**) 30 min, (**h**) 4 h; photolysis of bilirubin in aqueous rat serum albumin HPLC at (**i**) 30 min, (**j**) 4 h. Retention times are in minutes (Fig. 1 of ref. (762). Reprinted, with permission of the American Chemical Society. Copyright 1982)

Lumirubins were isolated by solvent extraction and TLC on silica (solvent, 1% CH$_3$CO$_2$H in 10% CH$_3$OH-CHCl$_3$). The isolated R_t 6.1-min product separated on TLC into two components that exhibited identical absorption spectra (λ^{max} 434 nm, 10% CH$_3$OH-CHCl$_3$) and readily interconverted in base (1% NH$_4$OH-CH$_3$OH), suggesting an isomeric (epimeric) relationship. The components remained homogeneous on two-dimensional TLC and coeluted on the HPLC systems used. The most mobile component was called *lumirubin-IX*; the least mobile *isolumirubin-IX* (Fig. 9.2.11). Methylation (CH$_2$N$_2$) of an unresolved mixture of the two gave a product with a field desorption mass spectrometry parent ion at 612 amu, the

9 Understanding and Translating Bilirubin Structure

same value as for bilirubin dimethyl ester, thereby indicating that all three are isomeric. An analogous pair of compounds (lumirubin-XIII and isolumirubin-XIII, λ^{max} 431 and 432 nm, respectively) was isolated from photolysis of (4Z,15Z)-bilirubin-XIIIα in aqueous human serum albumin or $CHCl_3$-Et_3N. However, similar compounds were not detected during photolysis of bilirubin-IIIα, mesobilirubin-IXα, mesobilirubin-XIIIα, or xanthobilirubinic acid in the same solvents. Since none of these compounds has a vinyl group at C-3, as in bilirubin-IXα or bilirubin-XIIIα, it was concluded that this function is involved in lumirubin formation.

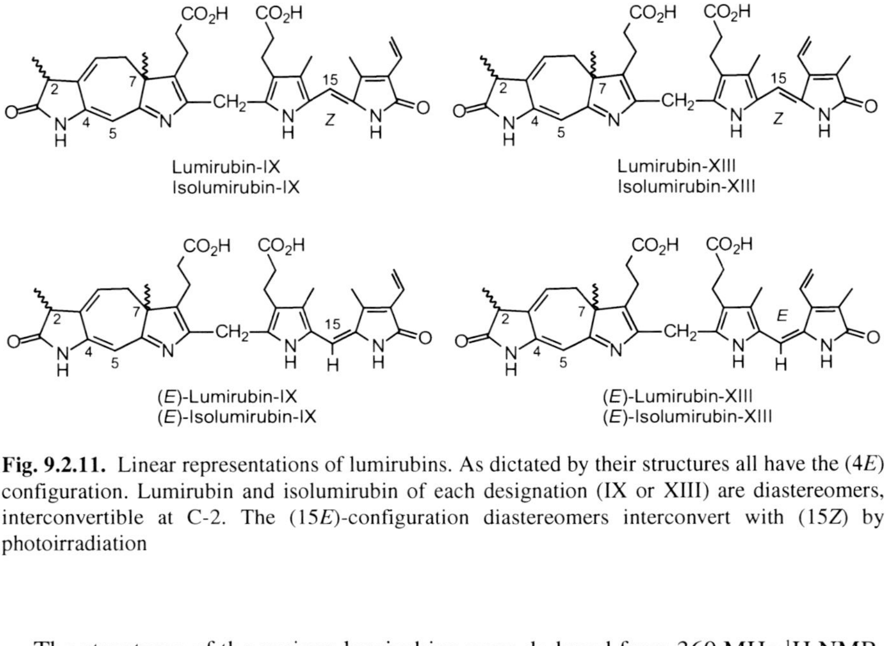

Fig. 9.2.11. Linear representations of lumirubins. As dictated by their structures all have the (4E) configuration. Lumirubin and isolumirubin of each designation (IX or XIII) are diastereomers, interconvertible at C-2. The (15E)-configuration diastereomers interconvert with (15Z) by photoirradiation

The structures of the various lumirubins were deduced from 360 MHz ^{1}H NMR. Fig. 9.2.12 shows the olefinic region of the spectra of lumirubins-IX and -XIII. (In this region, the spectrum of each isolumirubin resembles that of the corresponding lumirubin.) Striking features in the spectrum of lumirubin-XIII, compared to its symmetrical parent (4Z,15Z)-bilirubin-XIIIα are (i) a strong upfield shift from 6.09 to 5.77 ppm of one of the two, formerly equivalent, meso

9.2 How Phototherapy Exposed Bilirubin (E)-Configuration Isomers

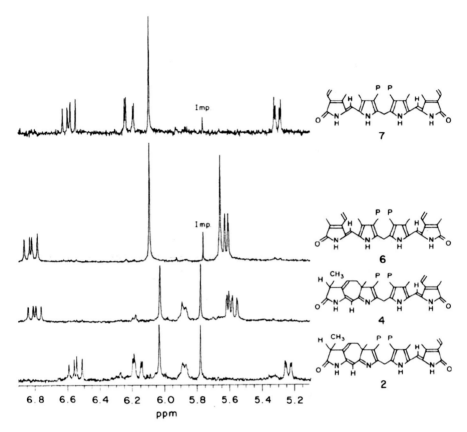

Fig. 9.2.12. Olefinic region of the 360 MHz NMR ((CD$_3$)$_2$SO) of (*top* to *bottom*) bilirubin-IIIα [7], bilirubin-XIIIα [6], lumirubin XIII [4], lumirubin IX [2]. Proton resonances are in ppm downfield from Me$_4$Si (Fig. 2 of ref. (762). Reprinted, with permission of the American Chemical Society. Copyright 1982)

protons at C-5 and C-15, (ii) loss of one of the two identical vinyl groups, and (iii) emergence of a new broad doublet at 5.87 ppm corresponding to one olefinic proton. The olefinic regions of lumirubins-IX and -XIII are similar, except that in the former the remaining unreacted (*exo*) vinyl group has the same chemical shift and appearance as that in bilirubin-IIIα. Therefore, in the conversion of (4Z,15Z)-bilirubin-IXα and (4Z,15Z)-bilirubin-XIIIα to the corresponding lumirubins, half of each molecule remains essentially unchanged. Moreover, the unchanged halves of lumirubins-IX and -XIII differ only in the position of a vinyl substituent. Key evidence for the structure of the unchanged half of lumirubin-XIII is (i) the appearance of two new high-field CH$_3$ signals, one a singlet at 1.07 ppm (1.01 ppm for its iso-compound) and the other a doublet near 1.17 ppm (1.22 ppm in its iso-compound), (ii) loss of signals corresponding to the C-2 and C-7 CH$_3$ groups of the parent compound (4Z,15Z)-bilirubin-XIIIα, and (iii) emergence of a single proton signal near 3.2 ppm coupled to the high-field CH₃ doublet. Similar data were obtained for lumirubin-IX and isolumirubin-IX. These data are consistent with the structures formulated in Fig. 9.2.11 for lumirubins-IX and -XIII. The corresponding isolumirubins are diastereomers related by epimerization at C-2. The structures are among several that were proposed previously for "photobilirubins IIA and IIB", prepared in 2% yield by UV-visible irradiation of bilirubin in (CH₃)₂SO (*828, 845*). It seemed likely that lumirubin-IX and isolumirubin-IX were the same as "photobilirubin II" (*828, 832, 845, 856, 857*) and "unknown pigment", or *Onishi*'s "peak 2" (*837-843*). However, it was not found, as reported for photobilirubin II, that lumirubin-IX and isolumirubin-IX *readily* revert to bilirubin-IXα on photolysis. Photoreversion of lumirubin-IX and isolumirubin-IX to bilirubin-IXα and of lumirubin-XIII and isolumirubin-XIII to bilirubin-XIIIα was observed on relatively *prolonged* irradiation or at high irradiances. Reversion was accompanied by overall loss of pigment, and isosbestic points were not observed in the difference spectra. Lumirubin formation from bilirubin-IXα was not inhibited by O₂. The data indicated that (*E,Z*)-bilirubins were not intermediates in lumirubin formation, suggesting a concerted single-photon process proceeding via singlet excited state bilirubin, though others (*843*) have argued for a two-photon process proceeding via (4*E*,4*Z*)-bilirubin. Therefore the "fast" anaerobic photochemistry of bilirubin could be summarized by the scheme below (*843*):

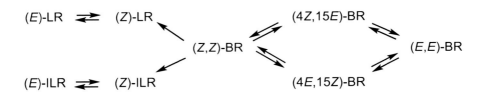

where LR and ILR represent lumirubin- and isolumirubin-IX, respectively, and each step is a one-photon process. On the time scale of (*Z*) ⇌ (*E*) conversions, lumirubin formation is slow and irreversible.

Like other compounds with a dipyrrinone chromophore (*844*), lumirubins and isolumirubins undergo facile reversible configurational isomerization to more polar and thermally unstable (*E*) isomers on exposure to light, reaching (*Z*) ⇌ (*E*) equilibrium without detectable reversion to the parent bilirubins. This reaction is readily detectable by difference spectroscopy (*844*) and HPLC (Fig. 9.2.13), and it accounts for the minor HPLC peak at R_t 4.7 min appearing on prolonged photolysis of bilirubin (Fig. 9.2.10).

9.2 How Phototherapy Exposed Bilirubin (E)-Configuration Isomers

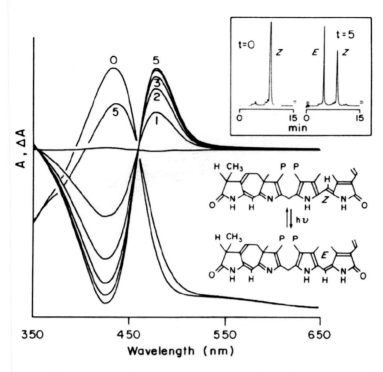

Fig. 9.2.13. Absorbance and absorbance difference spectra on photolysis of isolumirubin IX and (*inset*) HPLC before and after photolysis. The numbers by the curves give the cumulative irradiation time in arbitrary units. The HPLC solvent was 0.1 M di-*n*-dodecylamine acetate in CH_3OH (1 cm^3/min) (P = $CH_2CH_2CO_2H$) (Fig. 3 of ref. (*762*). Reprinted, with permission of the American Chemical Society. Copyright 1982)

Configurational isomerization of bilirubin bound to horse, rat, guinea pig, or bovine serum albumin generates (4*E*,15*Z*)-bilirubin *and* (4*Z*,15*E*)-bilirubin, with a slight preference for the latter (*cf.* Fig. 9.2.10i). In contrast, bilirubin bound to human serum album yields, under the same conditions, only one of the two diastereomeric (*E*,*Z*) isomers, (4*Z*,15*E*)-bilirubin (Fig. 9.2.10g). Interestingly, similarly marked stereoselectivity and regioselectivity are observed in humans during phototherapy, whereas in jaundiced rats both (*E*,*Z*) isomers are formed. Therefore the *in vivo* photochemistry appears to resemble the *in vitro* photochemistry very closely, contrary to earlier claims (*760, 773*).

Also published in 1982 (*845*) was the collaborative study from *Stoll*, *Vicker*, *Gray*, and *Bonnett* revising the structure of the *Stoll-Ostrow* photobilirubin-II (*828*). The material was isolated following photo-irradiation of bilirubin (200 mg) Ar-purged $(CH_3)_2SO$ (200 cm^3) for 60 min with a 100 W medium pressure Hg lamp. It was learned that photobilirubin-II separated into two components, as reported in 1979 (*828*) but that upon isolation and rechromatography the compounds had undergone interconversion to some extent. Irradiation of photobilirubin-II in $CHCl_3$ yielded more polar photobilirubin-III. Seven possible structures (Fig. 9.2.14)

574   9   Understanding and Translating Bilirubin Structure

were considered for photobilirubin-II, all of which were intramolecularly-cyclized structures in which one of the two vinyl groups of natural bilirubin had been incorporated into a ring.

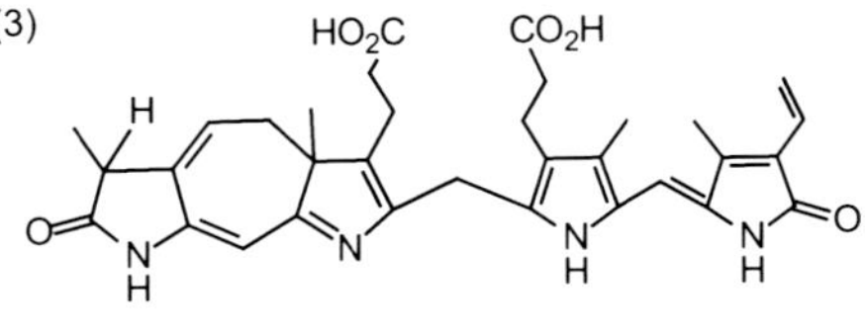

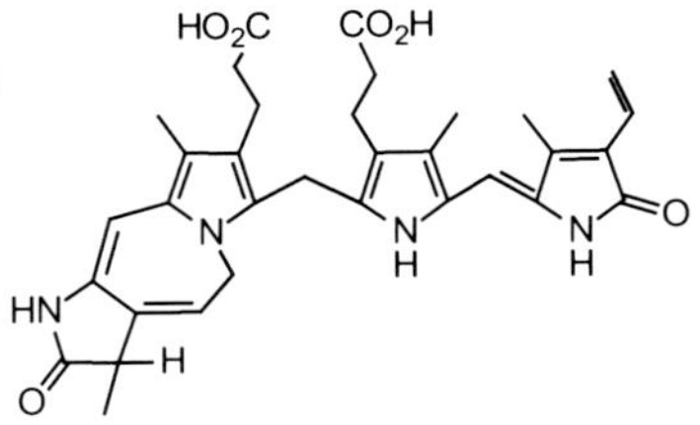

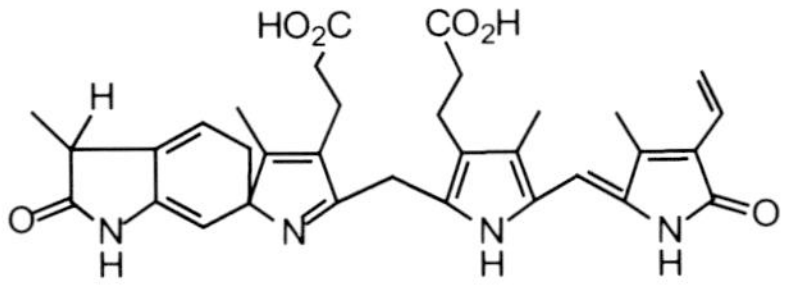

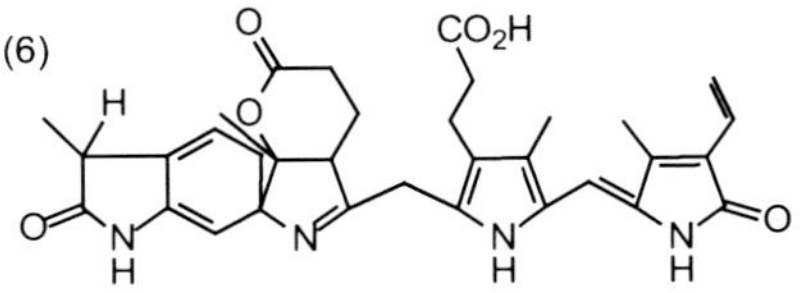

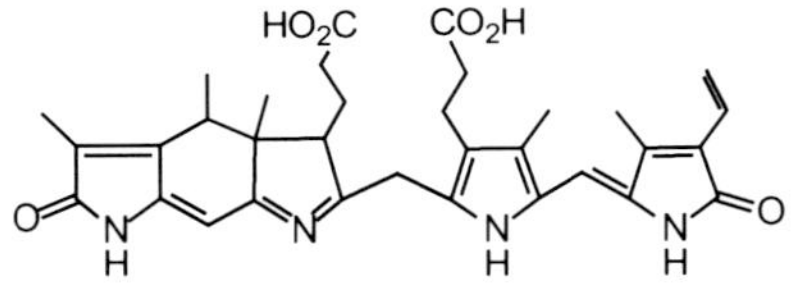

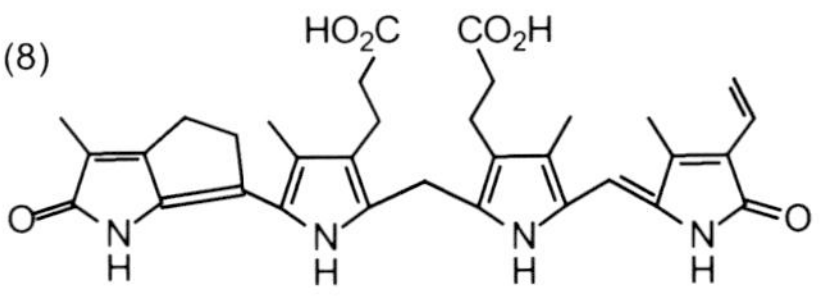

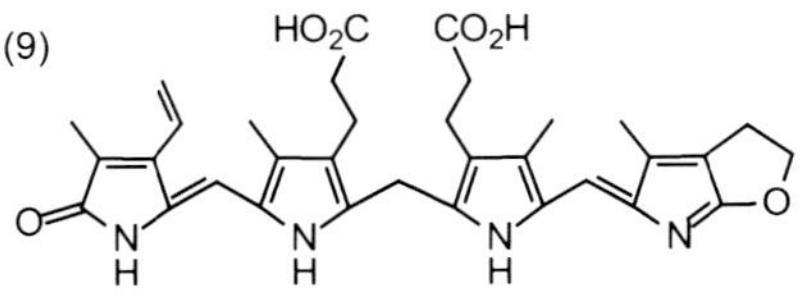

Fig. 9.2.14 Possible structures, with the author's numbering, resulting from cyclization of β-vinyl substituents in bilirubin-IXα, as suggested in ref. (*845*)

Though the list included number (9), the infamous *Roos* structure (*824*), it was discounted because it had been shown that the *endo*-vinyl and not the *exo* had been involved in the cyclization. All other structures, except (3) and (6) were dismissed based on the evidence available. The latter (6) was favored by *Onishi et al.* (*839*), who called it both *spirobilirubin* and *cyclobilirubin* (*842*). However, *Stoll et al.* (*845*) indicated that they preferred "to leave the choice between structures (3) and (6) an open question." And then renamed *photobilirubin-II* as *cyclo-(4E,15Z)-bilirubin-IXα* and *photobilirubin-III* as *cyclo-(4E,15E)-bilirubin-IXα*.

A year later, at the Neonatal Jaundice meeting in Padua in June 1983, *Bonnett* included a brief update to state a preference for structure (3) of Fig. 9.2.14 for photobilirubin-II on the basis of infrared evidence, giving it the systematic name, *3^2,7-cyclo-(15E)-bilirubin*, and that of photobilirubin-III as *3^2,7-cyclo-(15Z)-bilirubin* (*832*), with each indicated as being composed of two pairs of enantiomers. In a subsequent publication, in 1987 (*857*), the stereochemistry of the cyclobilirubins was corrected from (15E) to (15Z) for photobilirubin-II and from (15Z) to (15E) for

9.2 How Phototherapy Exposed Bilirubin (*E*)-Configuration Isomers   575

photobilirubin-III. However, the photoproduct structures had already been clarified in 1982 (*762*), when it was known that (*Z*)-lumirubin formed a dimethyl ester, which fact would discount both structure (6) of Fig. 9.2.14 and *Onishi*'s spirobili-rubin (*839*). And the main events of bilirubin photochemistry and the molecular mechanism of phototherapy were explained by *McDonagh* in his lecture at the same Padua meeting (*854, 855*).

Subsequently, in 1984, *Bonnett et al.* (*856*) reconfirmed their earlier structure of photobilirubin-II as (1) of Fig. 9.2.14, the (15*Z*) isomer (*845*) and not the (15*E*) proposed in 1983 (*832*). The name *photobilirubin-II* was restored, and *cyclobiliru-bin* disappeared. Not so, however, with *Onishi*, who kept to the latter while mar-shalling NMR evidence for structure (1) of Fig. 9.2.14 in late 1983 (*858*) and kept to it in subsequent publications into the late 1980s (*859–863*).

In the meantime, the molecular basis for jaundice phototherapy had already been explained by *McDonagh* in 1982 (*762, 844*) and in 1983 by the first unambiguous evidence that configurational photoisomerization of bilirubin occurs *in vivo* in *Gunn* rats and the jaundiced human neonate (*854, 855, 864*). These were followed soon thereafter by review articles in 1984 (*155*), 1985 (*156, 865*), 1987 (*857*), 1988 (*157, 866*), and 1989 (*867*). And by studies of the chirality control of lumirubin formation (*868*) and especially the photophysics of (*Z*) $\rightleftarrows$ (*E*) isomerization in order to explain the ratio of (4*Z*,15*E*) (4*E*,15*Z*) isomers produced at various wave-lengths of light, the influence of solvent and protein binding (*869–879*).

9.2.3. (*E*)-Configuration Bilirubins from X-Ray Crystallography

Although there have been no crystallographic studies of the lumirubins, nor of the less stable (*E*)-isomers of bilirubin, two studies led to reports in 2008 of X-ray crystal structures of (4*Z*,15*E*)-bilirubin-IXα bound to human serum albumin (*880*), and of an amphiphilic synthetic bilirubin by *Sanjeev K. Dey* (Ph.D. 2008) in the *Lightner* research group (*881*). Curiously, formation of both of these crystalline products was serendipitous. In the first, the (*E*)-isomer is located in an L-shaped binding pocket of the protein sub-domain IB with the (*Z*)-configuration *endo*-vinyl dipyrrinone placed in a predominantly hydrophobic cleft and the (*E*)-configuration *exo*-vinyl dipyrrinone twisted periplanar (*synclinal* conformation) in a more open section of the pocket (Fig. 9.2.15).

The synthetic pigment, with methoxy groups replacing the lactam methyl and vinyl groups of bilirubin, crystallized with three different molecules in the unit cell – all (4*E*,15*Z*) isomers. All three components had an *anti*-(*E*)-configuration exocyclic ene-amide C=C (at C-4); yet, the pigment still maintained a ridge-tile shape, with intramolecular hydrogen bonding between the (15*Z*)-configuration dipyrrinone and the C-8 propionic acid group, as shown in Fig. 9.2.16.

Well before 2008, however, (*E*)-isomers of bilirubin, biliverdin, and model dipyrrinones had become so well established and the proof of pigment stereochem-istry had advanced so far that an (*E*)-isomer X-ray crystal structure lacked the novelty that it would have had 30–35 years earlier.

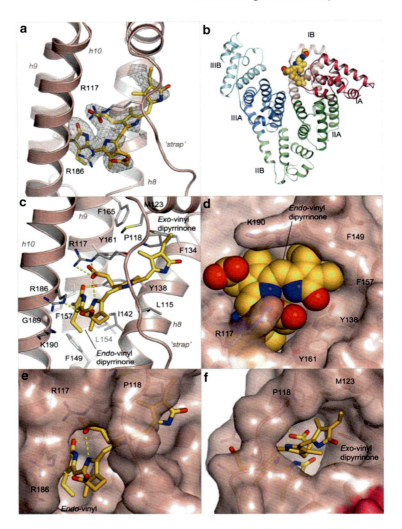

Fig. 9.2.15. Crystal structure of HSA complexed with (4Z,15E)-bilirubin-IXα. (a) Simulated annealing F_o-F_c omit map contoured at 1.75 s showing the pigment bound to sub-domain IB of HSA. (4Z,15E)-bilirubin is shown in a stick representation with atoms colored by atom-type: C, yellow-orange; O, red; N, blue. Sub-domain IB is shown in a ribbon representation (light red). (b) Overall structure of HSA complexed with (4Z,15E)-bilirubin. The pigment is depicted with space-filling *spheres*; the protein secondary structure is shown colored by sub-domain. (c) Detailed view of the interactions of (4Z,15E)-bilirubin with the binding pocket in sub-domain IB. Hydrogen bonds are indicated by broken *yellow lines*. Selected protein sidechains are shown as *sticks* (with carbon atoms colored *gray*). (d) View of the fit of (4Z,15E)-bilirubin (shown as a space-filling *CPK* model) to the pocket in sub-domain IB. (e) View of the fit of (4Z,15E)-bilirubin to the contours of the binding pocket in sub-domain IB (shown as a *pink*, semi-transparent surface). (f) Same as in **d** but rotated ~90° about a vertical axis (Fig. 2 of ref. (*880*). Reprinted with permission of the publisher)

9.2 How Phototherapy Exposed Bilirubin (*E*)-Configuration Isomers

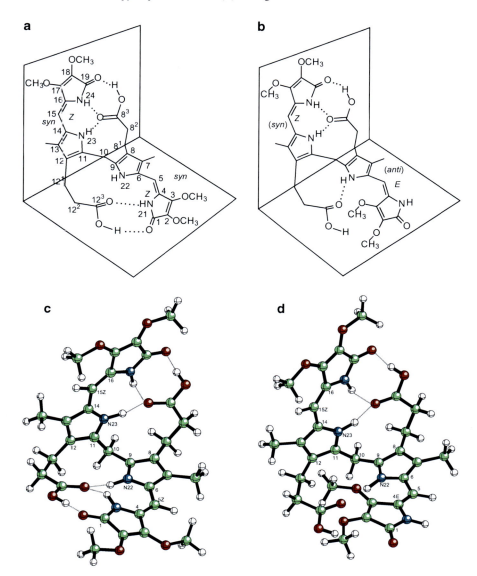

Fig. 9.2.16. (**a**) Line drawings of intramolecularly hydrogen-bonded *syn*-(4Z,15Z)-tetramethoxybilirubin, and (**b**) its *anti*-(4E), *syn*-(15Z)-isomer. (**c**) Ball and Stick representation of the energy-minimum conformations of (**c**) *syn*-(Z,Z) tetramethoxybilirubin and (**d**) *anti*-(*e*), *syn*-Z tetramethoxybilirubin, as determined by molecular dynamics using the SYBYL forcefield; the relevant torsion angles from the X-ray crystal structure, and from molecular dynamics (in parentheses) are as follows, from ref. (*881*): ϕ [N-21 – C-4 = C-5 – C-6] = 170° (170°), ϕ [C-4 = C-5 – C-6 – N-22] = 16° (-26°), ϕ [N-22 – C-9 – C-10 – C-11] = –59° (61°), ϕ [C-9 – C-10 – C-11 – N-23)] = –47° (53°), ϕ [N-23 – C-14] = C-15 – C-16 = –7° (23°), ϕ [C-14 = C-15 – C-16 – N-24] = –3° (1°); interplanar dihedral angle = 96° (91°). (Fig. 2 of ref. (*881*). Reprinted with permission of the American Chemical Society. Copyright 2010)

9.3 The Glue that Binds. Persistence, Recognition, and Role of Hydrogen Bonds

One of the defining features of the structure of bilirubin (Figs 8.1 and 9.2.1) is the web of intramolecular hydrogen bonds, an important design of nature rendering lipophilic a compound which would otherwise be polar. Had the porphyrin macrocycle not been opened at the α-position, but rather at the β, γ, or δ, animals might have had a problem disposing of bilirubin produced in the fetus. The resulting three different biliverdins are very similar in their solution properties to natural biliverdin-IXα, and all four biliverdin isomers would be unlikely to cross the placenta for disposal if they were formed (*882*). On reduction, they would form four different bilirubins, three of which are more polar than bilirubin-IXα and unlikely to cross the placenta readily (*882*) and only one of which, the IXα, can adopt the intramolecularly hydrogen-bonded conformation that confers lipophilicity and facilitates transport across the placenta from the fetal to the maternal circulation. Thus, the intramolecular hydrogen bonds of bilirubin-IXα define not only its structure but also its solution properties and metabolism.

The pattern of intramolecular hydrogen bonds, as unique as it might appear in (4Z,15Z)-bilirubin-IXα, is reproduced in bilirubin analogs and other dipyrrinone-containing synthetic pigments, not just in bilirubins-IIIα and -XIIIα but also in the corresponding mesobilirubins. Numerous synthetic analogs have been designed to probe the extent to which hydrogen bonding might endure certain structural modifications and how such modifications might influence glucuronidation and metabolism (*e.g. 883–897*), and spectroscopic and other probes were used in order to decipher the structural features that render hydrogen bonding effective and durable.

Numerous non-natural product bilirubinoids were built and scrutinized. In some cases the modification of the basic structure of bilirubin was far removed from the hydrogen bonding centers. Thus, whether the β-substituents of lactam rings were replaced by methyl (*898*), ethyl (*881*), *n*-butyl (*886*), phenyl (*890*), fluorophenyl (*899*), or even methoxyl (*881*), the result was invariable the same: the ridge-tile skeleton was scarcely modified and the intramolecular hydrogen bonding was unchanged (Table 9.3.1A). Nuclear *Overhauser* Effect (NOE) measurements in CDCl$_3$ by ^{1}H NMR invariably were found between the acid OH and lactam NH, and the vicinal coupling in the $-CH_2-CH_2-CO_2H$ fragment invariably indicated a fixed staggered geometry at room temperature.

9.3 The Glue that Binds. Persistence, Recognition, and Role of Hydrogen Bonds

Table 9.3.1. Selected molecular parameters of bilirubinoids from X-ray crystallography (*) and from molecular dynamics calculations (SYBYL). The dihedral angles (θ and θ') are de-termined from the intersection of the average planes of the dipyrrinones and from the extended planes of the central pyrroles, respectively

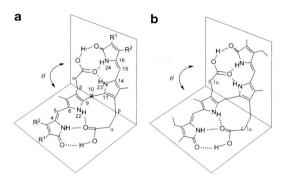

Set A

Bilirubinoid X	R¹	R²	Dihedral Angle/° θ	θ'	Torsion Angle/° φ_1	φ_2	H-Bond Distance/Å LN-H···O	PN-H···O	O-H···O
(4Z,15Z)-Bilirubin-IXα*			95	99	60	64	1.75	1.78	1.52
(4Z,15Z)-Bilirubin-IXα			86	88	60	60	1.55	1.56	1.66
CH₂	CH₃	CH₃	88	87	59	59			
CH₂	CH₃	CH₂CH₃	92	94	64	64	1.56	1.59	1.53
CH₂	CH₂CH₃	CH₂CH₃	92	92	60	61	1.55	1.58	1.54
C=O	CH₃	CH₂CH₃	88	89	60	60	1.61	1.56	1.55
S	CH₂CH₃	CH₂CH₃	86	90	61	62	1.97	2.05	1.79
S	CH₂CH₃	CH₂CH₃*	77	77	61	60	1.54	1.58	1.54
Se	CH₂CH₃	CH₂CH₃	72						
CH₂	OCH₃	OCH₃	88	90	62	62	1.61	1.56	1.51
C(CH₃)₂	CH₃	CH₂CH₃*	99	92	62	63	1.97	2.25	1.89
C(CH₃)₂	CH₃	CH₂CH₃	101	91	60	60	1.65	1.62	1.52
⌬	CH₂CH₃	CH₂CH₃*	94	95	60	64	1.93	2.13	1.77
⌬	CH₂CH₃	CH₂CH₃	94	91	60	60	1.50	1.77	1.53
⌬⌬	CH₂CH₂	CH₂CH₃*	84	84	53	53	1.88	2.12	1.78
⌬⌬	CH₂CH₂	CH₂CH₃	92	92	60	60	1.51	1.70	1.54
CH(iPr)	CH₃	CH₃*	105	104	61	72	1.6	1.7	1.9
CH(iPr)	CH₃	CH₃	90	90	55	65	1.6	1.5	1.6
CH(t-Bu)	CH₃	CH₃	87	86	63	52	1.55	1.60	1.54
CH(Ada)	CH₃	CH₃	90	88	60	60			

Set B	n		θ	θ'	φ_1	φ_2	LN-H···O	PN-H···O	O-H···O
	0	acetic	107	103	77	77	1.65	1.49	1.59
	1	propionic	92	94	64	64	1.56	1.59	1.53
	2	butyric	50	65	46	46	1.74	1.52	1.57
	3	valeric	32	47	34	34	1.55	1.54	1.60
	4	caproic	31	45	29	29	1.52	1.53	1.58
	9	undecanoic	56	92	115	130	1.67	1.53	1.72

Moving the perturbation closer to the hydrogen bonding matrices similarly imposed no gross modifications. One might imagine that replacing the C-10–CH₂ group with a sulfur atom or a selenium atom might induce changes in pigment shape and/or hydrogen bonding, considering that the C–CH₂ bond length is 1.5 Å and the C–CH₂–C bond angle is 106° in bilirubin, and the C–S and C–Se bond lengths are longer (1.77 and 2.2 Å, respectively), C–S–C and C–Se–C bond angles (92° and 88°, respectively) are smaller. Yet these new orange and red bilirubinoids, bloated as they are in the middle, still retain a ridge-tile shape (albeit with a reduced interplanar dihedral angle θ, Table 9.3.1A) and two sets of three intramolecular hydrogen bonds (*887–889*). Again, ^{1}H NMR NOE measurements detect a close proximity of the acid OH and lactam NH hydrogens, and analysis of the vicinal H–H coupling constants of the propionic acids indicates a fixed staggered geometry characteristic of bilirubin itself.

If reducing the size of interplanar angle by a change at C-10 from CH_2 to S and to Se caused no great perturbation of the bilirubin conformation (*895*), neither did an expansion of the angle, imposed by replacing the sp^3-hybridized C-10–CH_2 to the sp^2-hybridization of a (C-10)=O (*885*) and a (C-10)=C(CH$_3$)$_2$ (*900*). A potentially more brutal perturbation on structure by replacing one hydrogen of the C-10–CH_2 group by an isopropyl (*900, 901*), *tert*-butyl (*898*), or adamantyl group (*898*) did not translate into much of a conformation change of the pigment or its intramolecular hydrogen bonding arrangement (Table 9.3.1). Nor did replacing both hydrogens of the C-10 hydrogens with methyl groups (*902–904*) or even by the potentially more effective modification at C-10 by the spirocyclohexyl (*905*) and spirofluorenyl (*905*) groups (Table 9.3.1A).

Whether the collection of intramolecular hydrogen bonds of the synthetic bilirubinoids dictated the near conformational homogeneity or whether a strong conformational preference for a ridge-tile shape enabled hydrogen bonds, despite attempts to modify the shape of the pigment might be answered from an analysis of the synthetic bilirubinoids in which the propionic acids were contracted to acetic acids, or expanded to longer chain alkanoic acids (Table 9.3.1B). Insofar as molecular dynamics studies showed, the pigments endured molecular distortion in order to maintain intramolecular hydrogen bonding, with the butyric acid homolog accommodating the least distortion. Its solubility properties were more like those of the parent, mesobilirubin-XIIIα, while bis-pentanoic and bis-hexanoic homologs were more polar, as was the acetic acid analog, and the undecanoic acid analog, exhibited lipophilic properties.

That the three-carbon acid chain presents the ideal geometry for intramolecular hydrogen bonding can be understood from experiments that impose rigidity along a three-carbon, a four-carbon, and a five-carbon path – as revealed from bilirubin analogs where *o*-, *m*-, and *p*-benzoic acids, respectively, replaced propionic (*906*). The *o*-benzoic acid analog exhibited tightly imposed intramolecular hydrogen bonds and was lipophilic. In marked contrast, the *m*- and *p*-benzoic acid analogs were polar and apparently devoid of intramolecular hydrogen bonds.

9.3.1 Understanding Dipyrrinone-Carboxylic Acid Hydrogen Bonding

Hydrogen bonding in bilirubin requires two partners, a carboxylic acid and an amide embedded in the dipyrrinone framework. Prior to the last decade of the 20th century, there were no reported examples of a carboxylic acid engaging in hydrogen

9.3 The Glue that Binds. Persistence, Recognition, and Role of Hydrogen Bonds

bonding to an amide – except in nature's unique design of (4Z,15Z)-bilirubin-IXα (Fig. 9.3.1a), its IIIα and XIIIα constitutional isomers, and the corresponding mesobilirubins. There was also no reason to believe that amide to carboxylic hydrogen bonds (Fig. 9.3.1c) would be any stronger than the well-studied examples (*482, 907*) of carboxylic acid to carboxylic acid hydrogen bonds (Fig. 9.3.1d), or any weaker than the also well-studied amide-to-amide hydrogen bond (Fig. 9.3.1e) (*482, 907*). Yet, at least one example of a hydrogen bond between an amide and a carboxylic acid may be found in the structures designed and built by *Rebek* (*908*) based on molecular scaffolding used in his molecular studies, which also included acid-acid and amide-amide hydrogen bonding (Fig. 9.3.2). There is a paucity of examples where in solution a carboxylic acid is hydrogen-bonded to the C=O and NH of an amide (not an imide). In bilirubin (Fig. 9.3.1) the amide and carboxylic acid components may be less forced to confront each other than in *Rebek*'s example and more likely to be guided toward each other by conformational pre-organization, where among many possible conformations of the pigments, the one in which nonbonded intramolecular steric repulsions are minimized turns out to be nearly the "right one" for hydrogen bonding.

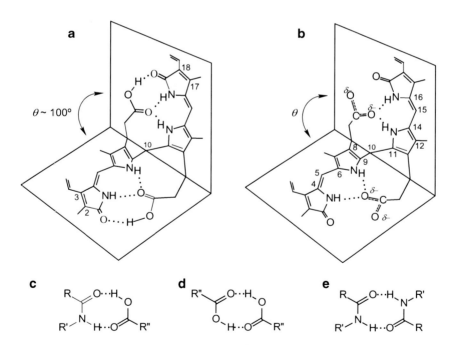

Fig. 9.3.1. Intramolecularly hydrogen bonded (4Z,15Z)-bilirubin-IXα (**a**) and its dicarboxylate anion (**b**). (**c**) Amide-carboxylic acid hydrogen bonding. (**d**) Carboxylic acid-carboxylic acid hydrogen bonding. (**e**) Amide-amide hydrogen bonding

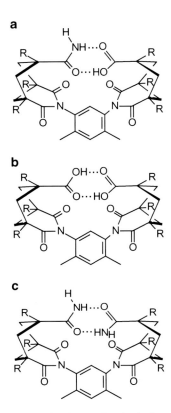

Fig. 9.3.2. *Rebek*'s compounds expressing intermolecular hydrogen bonding between (**a**) amide and carboxylic acid, (**b**) carboxylic acid and carboxylic acid, and (**c**) amide and amide (Redrawn from parts of Fig. 2 in ref. (*908*))

Perhaps the key to understanding the effectiveness of the hydrogen bonding in bilirubin is related to two factors: the avidity with which dipyrrinones engage in it, and, given the predilection of the pigment toward a ridge-tile shape, the cooperativity of two sets of hydrogen bonds that prevents the dipyrrinone and carboxylic acid components from moving far apart. One need only consult the work of *Ducharme* and *Wuest* (*909*) to understand how cooperativity works. In a simple example, the cyclic amide, pyridone, is known to form a hydrogen-bonded homo-dimer (Fig. 9.3.3a), with a moderate association constant in a nonpolar solvent, such as CHCl$_3$, K_{assoc} ~100–300 M^{-1} at 25°C (*909*). Yet when two dipyrrinones are tethered in a particular way, the resulting simple bis-pyridone also forms an intermolecular homo-dimer (Fig. 9.3..3b) but with a much greater association constant in CHCl$_3$, $\Delta G°$ >27 kJ·mol^{-1} (6.5 kcal·mol^{-1}) (*528*), or K_{assoc} >61,000 M^{-1} with a >20 magnification factor indicating that dissociation, breaking all of the hydrogen bonds, to form monomers is not as simple as breaking two equivalents of the pyridone dimer (Fig. 9.3.3a). That is, breaking the one set of hydrogen bonds, though necessary for dissociation to occur, does not alter the fact that the other, intact set of hydrogen bonds still preserves the dimer pre-organized for reforming the broken first set.

9.3 The Glue that Binds. Persistence, Recognition, and Role of Hydrogen Bonds

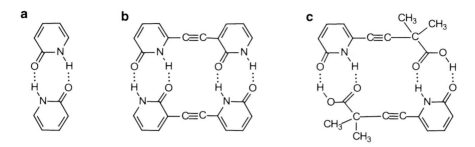

Fig. 9.3.3. Hydrogen-bonded homo-dimers of (**a**) pyridone, (**b**) bis-pyridone, and (**c**) pyridone acid

Using the *Ducharme-Wuest* model, amide to carboxylic acid hydrogen bonding was explored (*910*) in the pyridone acid of Fig. 9.3.3c by *Paul Wash* (Ph.D. 1998) of the *Lightner* group and Dr. *Emily Maverick* of the UCLA Chemistry Department. Like *Wuest*'s bis-dipyridone system, it too exhibited a strong tendency toward dimer formation, being a dimer in the crystal and in CHCl$_3$. Its ^{1}H NMR spectrum showed characteristically strongly deshielded (~15.5 and 13 ppm, respectively) carboxylic acid and amide protons, and vapor phase osmometry (VPO) yielded a molecular weight = 409 in CHCl$_3$ solution, almost exactly twice the molecular weight = 205 of the monomer. Fast atom bombardment mass spectrometry (FAB-MS) gave a prominent *m/z* 411 peak for the dimer. The available measurements indicated a very large K_{assoc}, >10^4 M^{-1} in CHCl$_3$. Again, cooperativity and structural pre-organization appear to play a decisive role. Despite the more favorable entropy associated with intramolecular hydrogen bonding in bilirubin, even intermolecular hydrogen bonding (as in the pyridone-acid association) would seem to indicate that structural pre-organization can play a major role in maintaining hydrogen bonding, and that carboxylic acid to amide hydrogen bonds are not weaker than amide to amide.

Dipyrrinones are also amides, and if one were to estimate a K_{assoc} value for the homo-dimer of a dipyrrinone (Fig. 9.3.4a), and consider hydrogen bonding involving only the lactams, a reasonable first approximation might be the value of pyridone association (K_{assoc} ~ 100–300 M^{-1} in CHCl$_3$ at 25°C). However, that would be well off the mark, for the K_{assoc} of the dipyrrinones xanthobilirubinic acid methyl ester and kryptopyrromethenone were determined by *Daniel Nogales* (Ph.D. 1993) and Prof. *Jin-shi Ma* in the *Lightner* lab to be ~25,000 M^{-1} (*659*). Here again, the K_{assoc} values are a factor of ~20 greater than the simple amide to amide model might predict; yet, in the dipyrrinone dimer, there is no pre-organization involving two sets of hydrogen bonds, as in the bis-pyridone and pyridone-acid dimer of Fig. 9.3.3. In the dipyrrinone dimer of Fig. 9.3.4, amide-amide hydrogen bonding brings the lactam carbonyl within hydrogen bonding distance of the pyrrole NH. Although a pyrrole NH to carbonyl hydrogen bond is normally weaker than a single hydrogen bond of the amide to amide type, the pyrrole NH to lactam carbonyl (Fig. 9.3.4b) apparently adds sufficiently to create a total of three, and not two, intermolecular hydrogen bonds, thereby strengthening the hydrogen-bonded dipyrrinone dimer. Here, too, a different example of structural pre-organization.

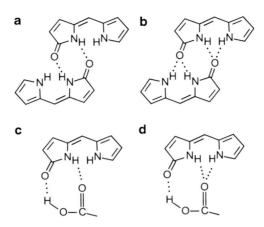

Fig. 9.3.4. Dipyrrinone-dipyrrinone intermolecularly hydrogen-bonded dimers: (**a**) amide to amide; (**b**) amide + pyrrole to amide and pyrrole. Dipyrrinone to carboxylic acid hydrogen bonding: (**c**) amide to carboxylic acid; (**d**) carboxylic acid to amide + pyrrole

Can an isolated dipyrrinone engage a carboxylic acid in hydrogen bonding? Studies of xanthobilirubinic acid itself show that it is possible – even when the carboxylic acid group is covalently attached to the dipyrrinone but incapable of hydrogen bonding to it (*660, 661*). Yet in this example, two molecules of xanthobilirubinic acid do indeed form a dimer, with the propionic acid of one molecule forming intermolecular hydrogen bonds to the second dipyrrinone of the other (Fig. 9.3.5a). In order for this to occur, and to avoid steric buttressing of the C-9–CH$_3$ groups, the two dipyrrinones stack atop one another (Fig. 8.5.7) (*660, 661*), a feat one might think would require a greater reduction in entropy than the face-to-face dipyrrinone-to-dipyrrinone hydrogen bonding arrangement of Fig. 9.3.4, both types being intermolecular associations. Apparently, a dipyrrinone is a better receptor for a carboxylic acid than it is for a second dipyrrinone ($K_{assoc} > 25{,}000\ M^{-1}$ at 25°C).

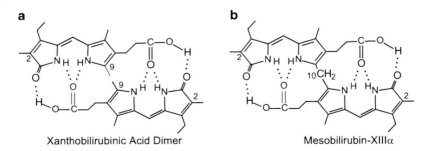

Fig. 9.3.5. Dipyrrinone to carboxylic acid hydrogen bonding in (**a**) xanthobilirubinic acid dimer, where C-9–CH$_3$ to C-9–CH$_3$ buttressing forces the dimer into p-stacking; and (**b**) mesobilirubin-XIIIα, where the two dipyrrinones are conjoined covalently to the C-10 CH$_2$ group

9.3 The Glue that Binds. Persistence, Recognition, and Role of Hydrogen Bonds

In bilirubin, all hydrogen bonds are intramolecular and are guided by a molecular framework inclined toward the pigment's most stable conformation, one where a total of five carbon atoms lie between the C-9 α-carbon of the dipyrrinone host and the COOH guest (Fig. 9.3.6a). Will the two termini, dipyrrinone host and CO$_2$H guest, connect by hydrogen bonds if the pre-organization engendered by a second set of hydrogen bonds is absent? The question was answered in the affirmative by two examples (Fig. 9.3.6b), each with a dipyrrinone covalently connected at C-9 to a six-carbon carboxylic acid that runs through the third ring – a pyrrole ring (*911*), or a benzene ring (*912, 913*). The rings impart a small element of pre-organization by limiting the number of degrees of rotational freedom in the acid chain. The tripyrrole acid adopts a bilirubin-like truncated ridge-tile conformation in the crystal (*911*), with intramolecular hydrogen bonding between acid and dipyrrinone. The same result obtains in CHCl$_3$ solution, as determined (i) by an NOE between the lactam NH and the carboxylic acid OH, and (ii) a solution molecular weight measurement by VPO that shows the pigment to be monomeric. The corresponding compound with a benzene ring, called *hemirubin* (Fig. 9.3.6c), was similarly found to be monomeric and intramolecularly hydrogen bonded in CHCl$_3$.

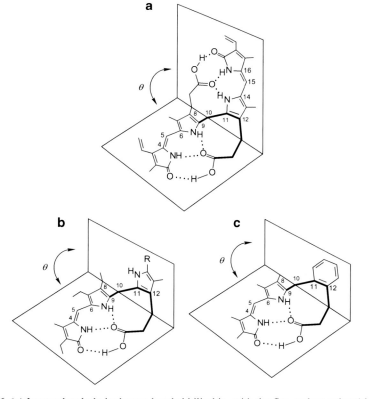

Fig. 9.3.6 (a) Intramolecularly hydrogen-bonded bilirubin, with the five-carbon tether (shown in *bold*) connecting C-9 to the hydrogen bond dipyrrinone host for the CO$_2$H guest. Tripyrrole (**b**) and hemirubin (**c**) synthetic models expressing dipyrrinone to carboxylic acid intramolecular hydrogen bonding, with the same carbon tethering

Generalizing to a dipyrrinone with no particular pre-organization in the hexanoic carboxylic acid chain (*914*) (Fig. 9.3.7a) one might also expect dipyrrinones with (flexible) alkanoic acids of varying lengths (*915, 916*) covalently connected to C-9 (Fig. 9.3.7b) to provide further insights into the ability of the dipyrrinone host to accept the –CO$_2$H guest. As with the tripyrrole and hemirubin above, three possible hydrogen bonding arrangements may be considered: (i) dipyrrinone to dipyrrinone (as in Fig. 9.3.4b), (ii) acid to acid (as in Fig. 9.3.1c), and (iii) intramolecular dipyrrinone to acid (as in Fig. 9.3.4c). Two different hydrogen bonding arrangements might thus be envisioned for polymeric associations involving intermolecular hydrogen bonding: dipyrrinone to acid pairings, or dipyrrinone to dipyrrinone + acid to acid pairings. However, only in representation (iii), the dipyrrinone to carboxylic acid intramolecular hydrogen bonding of Figs. 9.3.4d and 9.3.7 would the species be monomeric in solution. Again, two methods of detection were employed by *Michael Huggins* (Ph.D. 2000): VPO to determine the molecularity in CHCl$_3$ solution, and homonuclear ^{1}H{^{1}H}-NOE NMR to detect an enhancement of the lactam NH and CO$_2$H proton signals and thus their close proximity, as were found in the intramolecularly hydrogen-bonded species of Fig. 9.3.6. In the so-named [n]-semirubins of Fig. 9.3.7b, for alkanoic acid chain lengths from n = 5 (pentanoic) to n = 20 (eicosanoic), the pigment was monomeric in CHCl$_3$ and exhibited an NOE between the lactam NH and carboxylic acid OH hydrogens (*914, 915, 917*). These results conform to a dipyrrinone to CO$_2$H pattern of hydrogen bonding in monomeric species in solution, despite the considerable organization required when n = 20 and possible transannular non-bonded steric repulsions introduced in forming the macrocyclic rings.

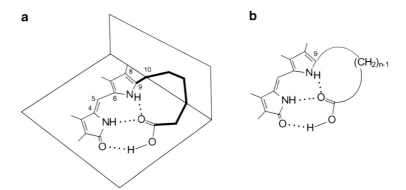

Fig. 9.3.7. (a) [6]-Semirubin and (b) the generalized [n]-semirubin structure, both with intramolecular hydrogen bonding between the dipyrrinone and carboxylic acid termini shown by *dotted lines*

When the carboxylic acid chains are shortened from hexanoic, as in [4]-semirubin, the butyric acid chain is predicted to be too short to engage its appended dipyrrinone in hydrogen bonding (*916*), as found by *Michael Huggins* and *Nicholas Salzameda* (Ph.D. 2006). Nevertheless, an NOE is observed between the dipyrrinone lactam

9.3 The Glue that Binds. Persistence, Recognition, and Role of Hydrogen Bonds 587

and CO_2H groups, while VPO indicates that the species in $CHCl_3$ solution is a *dimer*, presumably a p-stacked dimer akin to that seen for xanthobilirubinic acid in $CHCl_3$ (Fig. 8.5.7). Similarly, the propionic acid chain in [3]-semirubin is predicted to be too short to engage the dipyrrinone in intramolecular hydrogen bonding, either as a monomer or in the π-stacked dimer motif (Fig. 8.5.7) of [4]-semirubin (*916*). Yet, here too, NOE studies indicate a close proximity of the CO_2H and dipyrrinone lactam hydrogens. Unexpectedly, VPO studies indicate that the predominant species in $CHCl_3$ solution is a *tetramer*. Molecular modeling studies by *Michael Huggins* of the *Lightner* group suggest two pairs of parallel p-stacked dipyrrinones, with one dyad oriented approximately orthogonal to the other and the propionic acids of one dyad connected reciprocally in a cyclic manner to the dipyrrinones of the opposing dyad. This rather remarkable, highly organized structure bespeaks of the proclivity of dipyrrinone hosts for carboxylic acid guests.

In contrast, the methyl esters of the [n]-semirubins, of hemirubin and its tripyrrole analog are all dimeric in $CHCl_3$ solution, with dipyrrinone-dipyrrinone intermolecular hydrogen bonding (Fig. 9.3.4b) dominating the self-association. Clearly the presence of the carboxylic acid hydrogen makes an important difference and is a most important component of dipyrrinone to CO_2H hydrogen bonding. A conclusion no less important for bilirubin, because its dimethyl ester is likewise a dimer in $CHCl_3$ (*918–923*), apparently with dipyrrinone-dipyrrinone intermolecular hydrogen bonding (*918–923*). In contrast, bilirubin monomethyl ester is a monomer in $CHCl_3$ (*919*). Like the diacid, even the dianion does not default to the dipyrrinone-dipyrrinone dimer of the diester. Rather it is a monomer in $CHCl_3$ (*919*), like bilirubin, and its dipyrrinones remain hosts even for the carboxylate ions (Fig. 9.3.1b). This is presumably due to strengthened electrostatic attraction between the negative charge delocalized carboxylate anion and the dipyrrinone lactam and pyrrole NHs, as was observed in the X-ray structure of bilirubin bisisopropylammonium salt (*517, 518*) shown in Figs. 7.1.2 and 8.5.2 and in heteronuclear $^{13}C\{^1H\}$-NOE studies of mesobilirubin-XIIIα dianion in $CDCl_3$, $(CD_3)_2SO$, and even H_2O (*581, 582*). The last should have removed any skepticism expressed in an earlier discussion (*924*):

> *Carey*: As a baseline, it is necessary to determine the solution configuration of the dianion. The crystal structure of the isopropyl ammonium salt-chloroform solvate was analyzed by Mugnoli et al. (7) [*517*] and indicates that four of the six internal hydrogen bonds of bilirubin are still intact. By extrapolation, this suggests that this configuration may possibly be the case in aqueous solution at high pH.
>
> *Ostrow*: That may be a wild extrapolation. You are assuming a stabilized side chain, so that the resonance between the two oxygens on the carboxyl groups is no longer possible. I don't think that is valid in solution. I would disagree with that interpretation.

9.3.2 Misconceptions About Hydrogen Bonding

Despite the ability of a dipyrrinone to act like a perfect host toward a carboxylic acid guest, it might seem odd to view the intramolecular hydrogen bonds in natural

588   9 Understanding and Translating Bilirubin Structure

bilirubin as tenacious while also believing that all of them are relatively easily broken (*925*):

> The enantiomers rapidly and reversibly interconvert by transient breakage of all six of the hydrogen bonds, despite their tenacity . . .

At least six nonbonded electrostatic attractions involving O–H and N–H hydrogens to nonbonded oxygens (called hydrogen bonds) are present in bilirubin, which act synergistically within a conformation that is already in nearly its most stable orientation even in the absence of any hydrogen bonds, with the dipyrrinone hosts and propionic acid guests nearly unavoidably juxtaposed. Of course, the hydrogen bonds further stabilize the ridge-tile conformation considerably, but to have believed in 1994 that they impart rigidity to the conformation (*925*):

> The structure of unconjugated bilirubin IXα-Z,Z, diacid (H_2B) . . . consists of two slightly asymmetrical, rigid, planar dipyrrinone chromophores, connected by a central – CH_2– bridge. At the bottom are the two optical enantiomers of H_2B in their rigid, folded, ridge-tile conformations. . . .

> This rigid biplanar structure, with its internal hydrogen bonds, was first demonstrated in the crystalline state by X-ray diffraction . . . but is also the preferred conformation in solutions of UCB in water, alcohols, and chloroform . . .

should have seemed as doubtful then as a decade earlier (*924*):

> Collectively, these two trios of hydrogen bonds stabilize the bilirubin molecule in a rigid conformation that resembles a ridge tile . . .

Similarly, there is an apparent lack of evidence describing its carboxylic monoanion as "semi-rigid" or its dicarboxylate anion as devoid of hydrogen bonds (*926*):

> Dissolution of the semi-rigid BH . . .

> . . . like Na_2B are in the open configuration without hydrogen bonds . . .

"I have said it thrice: What I tell you three times is true" was the claim of the Bellman in *Lewis Carroll*'s nonsense poem "The Hunting of the Snark" (*928*), but in science, repetition is no guarantee of truth.

It was well-established that bilirubin in $CHCl_3$ possesses a network of six intramolecular hydrogen bonds, and it soon became known that the pigment is a very rapidly interconverting mixture of flexible conformational enantiomers in $CDCl_3$ (*599, 601*) (see Section 8). By 1992–1994, it had been shown further that the low energy pathway for bilirubin conformational enantiomerization does not transit through a flat or planar conformation as proposed in 1994 (*925*):

> . . . unbonded transient passes through a flat conformation . . .

> two enantiomers are in rapid equilibrium via the unfolded, planar intermediate . . . in which all six hydrogen bonds are transiently ruptured . . .

> . . . enantiomeric interconversion via a flexible, fully planar intermediate . . .

> . . . enantiomeric interconversion via a fully unbonded planar intermediate . . .

without any supporting evidence and despite evidence to the contrary from simple model building with *Dreiding* or *CPK* models. Indeed, transition *via* a planar conformation is calculated to cost 155–201 kJ·mol⁻¹ (37–48 kcal·mol⁻¹) of

9.3 The Glue that Binds. Persistence, Recognition, and Role of Hydrogen Bonds 589

activation energy *vs.* the lowest energy interconversion pathway, calculated (*610, 618–620, 689*) and found (*599, 601*) to be a barrier of ~75 kJ·mol^{-1} (~18 kcal·mol^{-1}) (Section 8.4.). Along the low energy interconversion path (Fig. 8.4.6), a single propionic acid guest detaches relatively easily from its host dipyrrinone as one conformation enantiomer flexes and transitions over to the other, in a series of folded intermediates (Fig. 8.4.–8.6) (*618–620*), but never a "fully nonbonded planar intermediate" (*925*).

The measured low energy barrier to conformational inversion and the rapid rate (k ~2 sec^{-1} at 25°C) at which the conformational enantiomers interconvert in CDCl$_3$ (*599, 601*) should dispel any notion that the propionic acid OH protons are firmly ensconced, even in a solvent (CHCl$_3$) that clearly supports retention of bilirubin's matrix of hydrogen bonds.

The intramolecular hydrogen bonds in (4Z,15Z)-bilirubin-IXα clearly relate to the pigment's relatively more lipophilic character than either the IXβ, γ, or δ isomers, which cannot engage in intramolecular hydrogen bonding. Yet, despite its unusually nonpolar character, the IXα isomer is not lipophilic in the sense of being soluble in nonpolar organic solvents, such as hydrocarbons, in olive oil, or even in lard (Table 9.3.2) (*928–931*). As the highly respected, careful experimental biophysical chemist *Rolf Brodersen*[4] showed, bilirubin is insoluble in many polar neutral solvents, including hydrogen bonding solvents such as water and many alcohols, as well as polar solvents such as ketones, ethers, and ethyl acetate. It is soluble to the extent of ~1 mg·cm^3 in CHCl$_3$, somewhat less soluble in CH$_2$Cl$_2$, both solvents with modest dielectric constants (ε = 4.7, 8.3, respectively), but it is *insoluble* in CCl$_4$. Apparently the presence of the halogen *per se* in the solvent is unimportant. Its solubility in the much more polar solvent formamide (ε = 109) is comparable to that in CHCl$_3$ but it is less soluble in *N,N*-dimethylformamide (DMF) (ε = 37), the polarity of which greatly exceeds that of CHCl$_3$. According to *Brodersen*'s measurements, it is most soluble in DMSO (ε = 47), whose polarity greatly exceeded that of CHCl$_3$ but is less than that of formamide and not greatly higher than that of DMF. Solubility in DMSO, just as insolubility in most nonpolar solvents, might be seen as unusual for a lipophilic molecule. The solubility in CHCl$_3$ might also be seen as unusual given the pigment's insolubility in alcohols and ketones of comparable dielectric constant. It may be that the polarized C–H and

---

[4] *Rolf Brodersen* was born on March 28, 1921, in Skorshoved, Denmark, and died on May 4, 1998, in Aarhus. He studied biochemistry at the University of Copenhagen, for a semester under the famous *J.N. Brønsted* (of acids-bases), for two semesters in *T. Astrup's* laboratory at the Biological Institute of the Carlsberg Foundation, and received the Magister Conference in 1944. From 1944 to 1947 he was employed in the Department of General Pathology of the University of Copenhagen, was at the University of Chicago Medical School from 1947 to 1948 (with Professor *C.P. Miller*). He was awarded the Danish Doctorate in 1949 at the University of Copenhagen. From 1948 to 1954, he was Department Leader at Leo Pharmaceutical Products (Løvens Kemish Fabrik). From 1954 to 1961, he worked at Ndolage Mission Hospital in Tanzania. Then in 1961 he joined the Institute of Biochemistry at the University of Copenhagen, first as Scientific Assistant, from 1962 as Department Leader, and from 1963 as Associate Professor. In 1969, he was appointed Professor of Medical Biochemistry at the University of Aarhus, from which he emeritated on March 31, 1991.

S–O bonds of CHCl₃ and DMSO coordinate with the pigment to enhance its solubilization. On the basis of these solubility studies, *Brodersen* suggested that bilirubin should not be viewed as lipophilic (*929*). However, it is unquestionably lipophilic in the usual biological sense, in that it readily partitions from aqueous solution to lipophilic solvents such as olive oil (Fig. 9.3.8), or *n*-octanol (*931*).

Table 9.3.2. Approximate solubility of bilirubin at 25°C, according to *Brodersen* (*928*)

Solvent	Solubility/ μM	Solvent	Solubility/ μM
n-Hexane	0	Methanol	0
Cyclohexane	0	Ethanol	0
n-Heptane	0	1-Propanol	0
Liquid paraffin	0	2-Propanol	0
		1-Butanol	0
Benzene	22	1-Pentanol	0
Toluene	50	Cyclohexanol	0
Xylene	45		
Pyrrole	150	Ether	1
Pyridine	580	2-Methoxyethanol	15
Dichloromethane	1,800	Acetone	10
Chloroform	2,500	Butanone	22
Carbon tetrachloride	40	2-Pentanone	25
		4-Methyl-2-pentanone	15
Formamide	2,300		
N,N-Dimethylformamide (DMF)	800	2,6-Dimethyl-4-heptanone[b]	8
Dimethyl sulfoxide (DMSO)	>10,000	Ethyl acetate	9
Water, pH 7.0	~0.001[b]	Olive oil	1.3
Water, alkaline	>10,000	Lard	2[b]

[a] Ten micromoles of purified bilirubin was equilibrated with 1 cm³ of solvent. [b] Extrapolated.

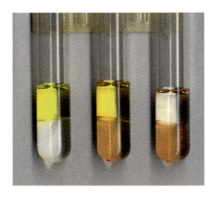

Fig. 9.3.8. (*Left*) Extraction of bilirubin from pH 7.4–7.8 aqueous buffer into the upper layer of olive oil. (*Middle*) Separation of bilirubin from a mixture of it with protoporphyrin-IX dissolved in pH 7.4–7.8 aqueous buffer. (*Right*) Partitioning of protoporphyrin-IX between pH 7.4–7.8 aqueous buffer and olive oil (Photograph courtesy of Prof. A.F. McDonagh. See ref. (*931*))

9.3   The Glue that Binds. Persistence, Recognition, and Role of Hydrogen Bonds   591

The spectral properties of bilirubin have probably been studied most often in CHCl$_3$ (or CDCl$_3$) and DMSO (either (CD$_3$)$_2$SO or (CH$_3$)$_2$SO). While it is abundantly clear that all studies in CHCl$_3$ or CDCl$_3$, chiefly by NMR, conclude that the pigment is intramolecularly hydrogen bonded (see Sections 7 and 8), studies in DMSO solvent have been somewhat less clear. Proton NMR studies from the *Lightner* lab at the University of Nevada showed a fairly consistent pattern of NH chemical shifts for bilirubin and its intramolecularly hydrogen-bonded analogs in CDCl$_3$ (*932*): lactam NH (~10.5–10.8 ppm), pyrrole NH (~9.2–9.3 ppm). These differed from those of intermolecularly hydrogen-bonded dipyrrinones: lactam NH (~1.1–11.2 ppm), pyrrole NH (~10.0–10.7 ppm). However, the bilirubin and dipyrrinone NH chemical shifts in (CD$_3$)$_2$SO were remarkably similar: lactam NH (~9.7–10 ppm), pyrrole NH (~10.2–10.6 ppm). On the basis of the chemical shifts in (CD$_3$)$_2$SO it was thought then that a DMSO molecule was intermolecularly hydrogen bonded, S–O$\cdots$H-N, to a dipyrrinone, whether alone or in a bilirubin, thereby implying that the multidentate hydrogen bonds of bilirubin are disrupted by coordination to DMSO. Yet, that is only one possible explanation. Though not unreasonable, it is unsupported by experimental proof that DMSO disrupts the intramolecular hydrogen bonds, and structures suggesting that it does do not reflect reality and even provide a misleading picture (*932–936*).

Resonance *Raman* studies of bilirubin hydrogen bonding in DMSO led to the conclusion that the internal hydrogen bonds are ruptured and replaced by hydrogen bonds to the (CH$_3$)$_2$SO oxygen – without a change in the overall conformation, in which the dipyrrinones remain at a fixed angle (*935*). But more detailed experimental evidence favors retention of a ridge-tile conformation in DMSO with intramolecular bonds between the CO$_2$H carbonyl oxygens and the dipyrrinone NHs (*580*). Early ^{13}C NMR studies of *Navon* and *Kaplan*, who measured T_1 relaxation times, found in 1977 that the propionic acids of bilirubin were relatively immobilized, consistent with intramolecular hydrogen bonding (*576*). Subsequent NOE and T_1 studies and analyses predicted a bilirubin conformation similar to that in the crystal but with the propionic acids tied to the dipyrrinones by bound DMSO molecules (*578*). In 1982, *Kaplan* and *Navon* concluded that (*579*):

> Our previous ^{13}C relaxation studies have shown that the motional freedom of the propionic side chain of bilirubin . . . in DMSO solutions is very limited. This is in contrast with the independent fast motion of the propionic residue of VBA methyl ester [vinyl-neoxanthobilirubinic acid methyl ester], where the structure does not allow internal hydrogen bonding

and that (*579*):

> . . . from the slowness of the internal motion of the propionic residues of bilirubin . . . it is concluded that these residues are tied to the skeleton via bound solvent molecules.

In their final analysis, *Kaplan* and *Navon* concluded that the conformation of bilirubin in DMSO is similar to that in CHCl$_3$, differing only in the former by the absence of *direct* hydrogen bonds between the CO$_2$H protons and the lactam oxygen (*580*):

> The result is attributed to immobilization of the propionic acid side chains by internal hydrogen bonds in which solvent participates.

Carbon-13 NMR T_1 measurements of ^{13}C-enriched ^{13}CO$_2$H-labeled mesobilirubin-XIIIα and analogs in the mid-1990s (*581–585*) reinforced the conclusions of *Navon* and *Kaplan* by finding $T_1 = 2.4$ sec in (CD$_3$)$_2$SO and 3.6 sec in CDCl$_3$. For comparison, T_1 of xanthobilirubinic acid methyl ester in (CD$_3$)$_2$SO is 6.0 sec, and mesobilirubin-XIIIα dimethyl ester in CDCl$_3$ is 5.9 sec. Yet, it probably should also be understood that the A$_2$B$_2$ coupling pattern in (CD$_3$)$_2$SO of bilirubin's propionic acid –CH$_2$–CH$_2$–CO$_2$H segments signifies more conformational mobility than in CDCl$_3$, where an ABCX pattern is seen at room temperature.

Thus, although the exact structure of bilirubin in DMSO may not be completely understood, and whether coordination of the CO$_2$H hydrogen to DMSO prevents its electrostatic attraction to the lactam oxygens, it seems probable that the most stable pigment conformation is the same in DMSO as in CHCl$_3$. This overall picture of structure and hydrogen bonding also comes from heteronuclear ^{13}C{^{1}H}-N*O*E studies of ^{13}CO$_2$H labeled mesobilirubin-XIIIα in DMSO and CDCl$_3$. These studies support a structure of bilirubin in which the acid carboxyl carbon has moved in DMSO only slightly distant from its position in CDCl$_3$ (*581*). Thus, the available experimental evidence is consistent with a structure of bilirubin in DMSO that is not much different from bilirubin in CDCl$_3$, or in the crystal, except that in the former, the carboxylic acid hydrogen is also hydrogen-bonded to DMSO. In CHCl$_3$, which might be considered to be a polar solvent, the solvent's hydrogen might engage the pigment by hydrogen bonding to the lactam carbonyl, but whether the solvent donates or receives a hydrogen bond from bilirubin, it dissolves the pigment. The effect of DMSO on the pigment may relax the ridge-tile conformation by lengthening the intramolecular hydrogen bonds. A similar situation was found in 10,10-dimethylmesobilirubin-XIIIα, for which the ^{1}H NMR spectrum in CDCl$_3$ revealed persistent intramolecular hydrogen bonding (as in bilirubin). The X-ray crystal structure indicated comparatively longer hydrogen bonds in a ridge-tile conformation (Table 9.3.3). Remarkably, the *gem*-dimethyl pigment is more soluble in CHCl$_3$ than bilirubin and it is surprisingly soluble in CH$_3$OH, wherein bilirubin is insoluble. It is tempting to ascribe the former's enhanced solubility in all of the conventional organic solvents to the longer but still persistent intramolecular hydrogen bonds, which might explain the greater solubility of bilirubin in DMSO than in CHCl$_3$.

Table 9.3.3. Comparison of the torsion angles (ϕ/°), dihedral angles (θ/°), and hydrogen bond lengths/Å in 10,10-dimethylmesobilirubin-XIIIα and bilirubin-IXα from X-ray crystallography[a]

Pigment	φ	θ_1	θ_2	CO$_2$H···O=C(L)	LNH···O=COH	PNH···O=COH
10,10-Dimethylmesobilirubin-XIIIα	62	99	101	1.83, 1.95	1.95, 2.36	2.13, 1.98
Bilirubin-IXα	60	95	99	1.50	1.80	1.80

[a] θ_1 = interplanar angle between the average planes through the two dipyrrinones. θ_2 = interplanar angle between the pyrroles. The two sets of hydrogen bond distances indicate a slight dissymmetry of the ridge-tile conformation, also predicted by molecular dynamics calculations. See ref. (*904*)

Consequently, it may be a misconception or misunderstanding to concede that all of the available hydrogen bonds in bilirubin are broken in DMSO solvent (*935*), contrary to the author's earlier depictions of bilirubin in the presence of optically

9.3 The Glue that Binds. Persistence, Recognition, and Role of Hydrogen Bonds 593

active sulfoxides (*933, 934*), and contrary to the incautious assertions expressed here and there that the hydrogen bonds are broken in DMSO, and, by extension, the unsubstantiated assertion that they are also broken in DMF (*925*):

> When this internal bonding is broken, e.g., by dissolution of UCB in dimethylformamide (55, 56) or dimethyl sulfoxide (57, 58) . . .

Curiously, the internal references (55 = ref. (*937*)) cited for DMF say nothing about the presence or absence of hydrogen bonding in bilirubin, which might be expected because they were published in 1974, before the X-ray crystal structure revealed it (*510*). The one internal reference that might be relevant for bilirubin hydrogen bonding in DMSO (57 = ref. (*925*)) appears to overextend the findings of *Navon* and *Kaplan* (*579*). In the other (58 = ref. (*938*)), hydrogen bonding is not mentioned. Oddly, in the mid-1990s, believers in the bond-breaking effects of DMSO and DMF (*925*):

> When this internal bonding is broken. e.g., by dissolution of UCB in dimethyl sulfoxide or dimethylformamide, titrimetric and ^{13}C-NMR studies yield, as expected, pKa values between 4.5 and 5.0 . . .

did not apparently extend the same effect to H_2O, where (*925*):

> These unexpectedly high pK'a values are probably the result of retarded dissociation of the protons due to the internal hydrogen bonds involving each –COOH group . . .

The belief that DMSO decreases hydrogen-bond formation in bilirubin also found its way into the photochemistry of the pigment when it was asserted that photoisomerization is facilitated in DMSO (*827*):

> Facilitation of photoisomer formation by factors that decrease hydrogen bonding of the – COOH side chains of UCB-IXα (methyl esterification, increased ionization at higher pH values, dimethyl-sulfoxide) . . .

However, it is only slightly facilitated, as determined by *Kanna et al.* (*936*), who found the quantum yields of photoisomerization to be only a factor of two better in DMSO than in $CHCl_3$ and about the same as in tetrahydrofuran. Yet the feat was attributed to (*828*):

> . . . there was a higher yield of photobilirubin II than IA and IB when bilirubin was irradiated in dimethyl sulphoxide, whereas only traces of photobilirubin II were generated on irradiation of bilirubin in chloroform. This is compatible with the greater hydrogen-bondbreaking power of dimethyl sulphoxide, which should thermodynamically favour formation of the *E,E*-isomer (II), which has fewer hydrogen bonds than the *E,Z*-bilirubins.

No evidence was presented to support the assertion that fewer hydrogen bonds might render greater stability to an (*E,E*)-isomer than an (*E,Z*)-isomer, which has more hydrogen bonds, a mystery unexplained by the authors (*828*).

Hydrogen bonding is clearly important to bilirubin's secondary structure and metabolism, but repeated assertions that it is disrupted by DMSO fly in the face of a large body of experimental evidence on bilirubin and related compounds, as well as evidence from closely related systems (*939, 940*). Nonetheless, an intransigent faith (or *Überzeugungstreue*) in the disruptive power of $(CH_3)_2SO$, but not H_2O, has greatly bedeviled the interpretation of measurements of the acid dissociation constants (pK_as) of bilirubin.

594   9 Understanding and Translating Bilirubin Structure

9.3.3   Intramolecular Hydrogen Bonds and pK_a. Un Cordon Sanitaire?

Belief in the transcendent strength of hydrogen bonds may have arisen from misinterpretation, or perhaps even faith, evidence to the contrary notwithstanding, *i.e.* the rapid conformational enantiomerism exhibited by bilirubin in $CDCl_3$ at room temperature (*601*). Yet the supposed difficulty in breaking hydrogen bonds that corral the carboxyl groups, especially the perceived difficulty in ionizing the CO_2H hydrogen, slipped into late 20th century theories and dialectics related to its acidity (Section 9.3.2), *viz.* the pK_as of the pigment's propionic acids. Even before *Bonnett*'s X-ray crystal structure of bilirubin had been published with convincing evidence for intramolecular hydrogen bonding in solid bilirubin (*510*), it had been suggested by *Ostrow et al.* in 1974 (*763*) that internal hydrogen bonding suppresses ionization of the propionic acids. Oddly, the work of *Hutchinson et al.* of 1971 (*506*) was cited in support; however, that work actually reported an extremely rapid exchange of hydroxylic protons of bilirubin by H_2O, and from the observation that all six protons are exchangeable, concluded (*506*) only that "hydrogen bonding does not alter with medium." A year later, *Knell, Johnson,* and *Hutchinson* concluded that (*941*):

> The pKa of the propionic carboxyl groups is probably less than 4 whereas the expected value is 4.8. Bilirubin is probably a *bis*-anion between pH values of about 5 and 7 . . .

In 1974, *Knell, Hancock,* and *Hutchinson* commented again on bilirubin pK_a (*574*):

> . . . it is most exceptional for a carboxylic acid to have a pK_a greater than 6 . . .

> . . . The pK_a of the carboxyl groups of bilirubin should be less than 4.8, by comparison with propionic acid and anticipating the acid strengthening effect usually detected if there is intramolecular hydrogen bonding.

Whether the last referred to earlier efforts to determine pK_a is unclear. Attempts to determine the pK_as of bilirubin usually refer to the 1955 work of *Overbeek et al.* (*942*), who did not determine the pK_a values of bilirubin but assumed them to be ~4.5 and 5.0 for the first and second ionization, respectively, in order to determine the aqueous solubility of the pigment. His experimental and calculated titration curves found very satisfactory agreement and were made to coincide at half-neutralization and pH 7.95 – a number subsequently misunderstood as pK_a.

Attempts to measure the pK_a of bilirubin had begun by 1960 in *Charles Gray*'s lab, where spectrophotometric titrations measured apparently pK ($\sim pK_{a1} + pK_{a2}$) values of about 7.1, 7.2, and 7.3 for bilirubin, dihydrobilirubin, and mesobilirubin, respectively (*943*). A dozen years later, *Lathe* (*944*) attributed an incorrect and misleading bilirubin $pK_a = 8.0$ to *Overbeek et al.* (*942*):

> Overbeek et al.[39] [*942*] titrated bilirubin and found a single pK_a at 8.0 (cf. pK_a of propionic acid 4.9).

A year later, *Krasner* and *Yaffe*'s aqueous titration yielded a *combined* apparent pK ~7.55 (*945*). In 1974, *Lee, Daly,* and *Cowger* (*937*), recognizing that bilirubin's aqueous insolubility complicated with measurements of pK_a, titrated bilirubin and a series of aliphatic dicarboxylic acids spectrophotometrically and by emf in dimethylformamide (DMF) in order to solubilize the pigment, thereby to estimate pK_1 and pK_2 values of 4.3 and 5.3, respectively, for the pigment. A year later, *Kolosov* and *Shapovalenko* were able to confirm the lower pK_a values as 4.5 and 5.9 from their spectrophotometric titrations of bilirubin (*946, 947*). (The widely separated pK_as seem odd, given that the carboxyl groups are widely separated.) One of the most thoughtful and carefully executed pK_a determinations of bilirubin was carried out by 1978 and reported by *Hansen, Thiessen*, and *Brodersen* (*938*), who conducted a titration of bilirubin in DMSO, monitored by emf, spectrophotometry, and ^{13}C NMR to reach a value of 4.4 for both carboxyl groups. Thus, the carboxylic acids of the pigment exhibited the expected ordinary carboxylic acid pK_a, as predicted earlier by *Knell, Johnson*, and *Hutchinson* (*941*) and by *Lee, Daly*, and *Cowger* (*937*).

However, by 1980, a belief that hydrogen bonding suppresses ionization was beginning to take root, despite any scientific supporting evidence. It was apparently stimulated by misinterpretation of the 1955 study (*942*) of *Overbeek, Vink*, and *Deenstra*. *Overbeek et al.* were interested in the aqueous solubility of the pigment and noted that when a crude suspension of it was basified to pH 11, it reacted with NaOH only slowly and on back titration with hydrochloric acid, the bilirubin (BH$_2$) that precipitated below pH 8 consisted of fine, reactive flocculates. Back titration starting from pH 5 with aq. NaOH gave a curve practically coincident with that from the HCl titration. The titration curve was reproduced with very good agreement with the calculated curve when K_{a1} and K_{a2} were assumed to be half and twice the value of pyrrole propionic acid ($K_a = 2 \times 10^{-5}$), respectively, and the calculated curve adjusted (Fig. 9.3.9) to coincide with the experimental at pH = 7.95.

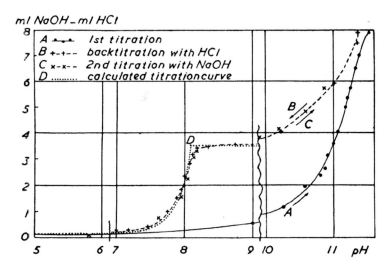

Fig. 9.3.9. Titration of 50 mg bilirubin, suspended in 250 cm^3 H$_2$O with 0.0510 *N* NaOH and 0.0510 *N* HCl (Fig. 1 of ref. (*942*) reproduced with permission of the publisher)

Overbeek et al. worked from the first and second ionization equilibrium expressions for bilirubin, $[BH_2]$: $K_1 = [BH^-][H^+]/[BH_2]$ and $K_2 = [B^=][H^+]/[BH^-]$, which are combined to give $[B^=] = [BH_2] \cdot K_1 \cdot K_2[H^+]^{-2}$, or $[BH_2] = [B^=][H^+]^2/(K_1 \cdot K_2)$. The numerator of the right side $= 2.15 \cdot 10^{-20}$ when 50 mg BH_2 was titrated to neutralization (to $B^=$) of the two carboxyl groups by 3.68 cm^3 0.0510 N NaOH and pH = 7.95. So, equation (1): $[BH_2] = 2.15 \cdot 10^{-20} (K_1 \cdot K_2)$ and the solubility of bilirubin at pH 4 is $[BH_2] \sim 5 \cdot 10^{-11}$ M when K_1 and K_2 are taken to be one half and twice the K_a value $(2 \cdot 10^{-5})$ of the model compound, pyrrole propionic acid, and the solubility at pH 7 $([BH_2] + [BH^-] + [B^=] = 5 \cdot 10^{-11} + 2 \cdot 10^{-8} + 2 \cdot 10^{-6})$ is calculated to be $2 \cdot 10^{-6}$ M. The titration plot was reproduced nearly exactly in 1961 by *Lucassen*, who acknowledged *Overbeek*, titrated bilirubin in 1:1 acetone-water and compared the behavior to *Overbeek*'s in a plot showing the consumption of two equivalents of NaOH over an essentially invariant pH $\sim$7.7 (*948*).

Although it is clear that *Lucassen* and *Overbeek et al.* considered the acid dissociation constants to be within the normal range, their work was misleadingly cited by *Lathe* in 1972 (*944*), and in 1980 by *Cohen* and *Ostrow* (*827*) as support for the claim that the pK_a of bilirubin is 7.95 and abnormally high because of hydrogen bonding (*827*):

> . . . the ionization of the two –COOH groups is suppressed, accounting for the high negative logarithm of their ionization constant (pK_a = 7.95) . . .

The perceived but unsubstantiated "suppression" was expressed even in an article on bilirubin photochemistry (*828*):

> As a result, the ionization of the two carboxyl groups is suppressed, accounting for the high pK_a (7.95) . . .

And in 1981, though it had been pointed out previously (*949*) that there was no evidence to support a pK_a of 7.95, the response was (*950*):

> All evidence in the literature indicates that the pK_a of the propionic acid groups in bilirubin is in the vicinity of 7.9 when the molecule is in the fully internally hydrogen-bonded structure, first described by Bonnett et al.[13] [*511*]. The pK_a values quoted by Lightner and McDonagh were obtained in solvents, such as DMSO, in which the internal hydrogen bonding is broken, allowing full expression of the –COOH ionization in the usual range of pK_a, 4.0 to 5.0[14] [*938*]. Apparently internal hydrogen-bonding does not suppress the ionization of the –COOH groups of bilirubin.

Despite the apparent recanting on "suppression" in the last quoted sentence, what seems to have gone unrecognized is that an overall apparent pK of $\sim$8 for a dicarboxylic acid would be consistent with normal pK_a values of $\sim$4.5 for the individual carboxyl groups, and there was no evidence in the literature by 1981 for a bilirubin pK_a value of 7.9. Yet, in 1984, not only had *Ostrow* proposed in a discussion following his lecture (*924*):

> Ostrow: . . . My data suggest that the pK_a of one carboxyl group is around 6.5, and the other one may be close to 9. I cannot be certain of that.

the discussion also elicited the following (*924*):

> *Mysels*: It might be better to ask a good organic chemist what he thinks the ionization constant of that carboxyl group in that position with that environment is than to try to make complicated measurements which are then impossible to interpret.

Ostrow: I asked Heinz Falk, and he said he did not know.

Attempts to learn the true pK_as of bilirubin continued with reports in 1984 that the pK_{a1} of bilirubin in aqueous taurocholate is 6.6–6.7 and 7.2 in aq. NaCl (*951*). The data come from an "isoextraction" method promoted by *Pasupati Mukerjee* in the School of Pharmacy, University of Wisconsin, possibly based on a method proposed in 1901 by *Farmer*, who cautioned that while (*952*):

> ... advantage has been taken of the fact that a substance distributes itself between two immiscible solvents in a constant ratio, independent of the dilution, that is, if we except the case of substances which exist in one of the two solvents in a state of abnormal molecular aggregation.

And so in 1984 a collaboration was embarked upon (*951*):

> *Ostrow*: I read your paper, and we tried your method. The problem is we do not have a solvent that solubilizes bilirubin at the concentrations we require.
> *Mukerjee*: You mean you cannot choose a good organic solvent?
> *Ostrow*: We have tried many of them; either the bilirubin is too soluble, the bilirubin is too insoluble or the solvent messes up the micelle.
> *Hofmann*: Dr. Moore and I do not agree with Dr. Ostrow. We think he can use that method if he consults with you.
> *Ostrow*: So far I have not been successful, so I would be happy to work with Dr. Mukerjee on that approach.

Unfortunately, the organic solvent chosen was $CHCl_3$, which had been shown in 1977 to be an inappropriate solvent for partition studies of bilirubin between albumin-free aqueous and organic phases because it extracts the pigment too completely over the pH range of interest (*953*, *954*). *n*-Heptane, possibly with a few percent methyl isobutyl ketone (*955*) was shown to be superior (*953*), as indicated in Fig. 9.3.10.

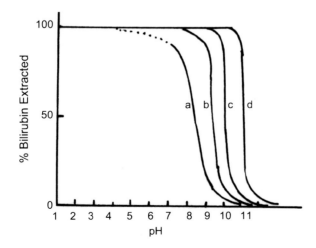

Fig. 9.3.10. Variation in the yield (%) of extracted bilirubin from H_2O *vs.* pH. The solvents are: *n*-heptane + (a) 10%; (b) 20%; and (c) 100% methyl isobutyl ketone; (d) 100% $CHCl_3$ (Redrawn from ref. (*953*))

Thus, it was recommended that for bilirubin extraction from aqueous media (*954*):

> Le solvant choisi doit présenter les caractérisques précédemment décrites mai aussi ne pas extraire de trop grandes quantités de bilirubine afin que la concentration de celle-ci dans la phase aqueuse après équilibre reste suffisamment grande pour être mesurable. Au cours d'essais préliminaires [32 = (*954*)], le chloroforme s'est montré trop bon solvant de la bilirubine car il l'extrait pratiquement en totalité de sa solution aqueuse, même en présence d'albumine. La méthyl-isobutyl-cétone (MIC), déjà utilisée par Girard et coll. [33 = (*955*)], est un moins bon solvant et, additionnée en proportions variable de n-heptane, apolair et miscible à la MIC, elle donne une gamme de mélanges dont la polarité et donc le pouvoir solvant vis-à-vis de la bilirubine décroissent à mesure que le pourcentage en MIC diminue. ⟨2⟩

Attempts to learn the true pK_as of bilirubin continued nonetheless. In 1985, *Moroi et al.* (*956*), using solubility measurements and UV-visible spectrophotometric curve analysis of bilirubin in aqueous phosphate and borate buffers of varying pH, reported pK_a values of 6–6.5 and 7.3–7.7. In the following year, 1986, *Brodersen* wrote authoritatively (*929*):

> A controversy has previously existed over the pK value of bilirubin acid. A chemist would expect a pK of about 4, as for other carboxylic acids, and this has in fact been confirmed by several investigators [*930*]. Others have proposed a pK around 8, based upon a misinterpretation of the classic paper of Overbeek et al. [*942*], in which alkaline bilirubin solutions were titrated with acid. In these experiments, bilirubin dianions are not only titrated as the acid is formed, but the acid itself is precipitated also, whereby the acid-base equilibrium is shifted. A detailed study of bilirubin acidity has shown that the pK values of both carboxyl groups are around 4 [*938*].
>
> Two arguments have sometimes been raised against this plain understanding of bilirubin acidity. It might be expected that the intramolecular hydrogen bonds could stabilize the acid and thus shift the pK to a higher value. There is no experimental evidence for such an effect; the dianion probably also has intramolecular hydrogen bonds; and other internally or externally hydrogen bonded carboxylic acids have pK values in the neighborhood of 4. Another argument says that the two carboxyl groups could have widely differing pK values, owing to electrostatic attraction of the monoanion for hydrogen ions. This is true, but only to a slight extent. Oxalic acid has in fact two very different pK values. On the other hand, adipic acid, in which the distance between the two carboxyl groups is about the same as in bilirubin, has two almost equal acidities, and this is probably so for bilirubin as well, at least in aqueous solutions. In micellar solutions, however, a large proportion of bilirubin theoretically may be present as the monoanion . . . .

The analyses and views of *Brodersen*, a famous solution thermodynamicist and biophysical chemist, a student of *Brønsted*, typically commanded attention. Not every one listened. In the event that *Brodersen*'s message was intended to serve as a caution or increpation to future investigators, it failed. For, in 1986, *Hahm, Ostrow, Mukerjee*, and *Webster* (*951*) suggested that the first pK_a of bilirubin is >6.5. This was followed up in 1988 by a study that determined pK_a values of 5.6–6.8 and >9.2 for the individual carboxyl groups. Those remarkably high values were attributed, without evidence, to hydrogen bonding (*926*):

> . . . the internal hydrogen bonding of the –COOH groups in UCB (42, 45) [*15, 510*] that retards the dissociation of the bonded protons.

9.3 The Glue that Binds. Persistence, Recognition, and Role of Hydrogen Bonds

and the ordinary carboxylic acid pK_a values obtained earlier (*938, 943, 945*) were then rationalized on the same basis (*926*):

> ... pK'a values below 5.0, expected of carboxylic acids, are obtained when UCB is titrated in solvents (e.g., DMSO or dimethylformamide) which interfere with formation of hydrogen bonds (22, 46) [*924, 938*].

Yet by 1988, when the above was written, there was no experimental evidence as to whether hydrogen bonds were broken or retained in DMF, and the best evidence left the bilirubin structure in DMSO not strongly perturbed (*580–585*). Nonetheless, in 1988, the then "latest" reported (*926*) pK_a values were $pK_{a1} = 7.25$, $pK_{a2} = 8.98$, until a year later when *Tiribelli* and *Ostrow* reported that (*957*):

> When a solid UCB [unconjugated bilirubin] phase was present, the two pK_a values were 6 to 7 and >9 respectively ... By contrast, in a new solvent partition and spectroscopic studies, which lacked a solid UCB phase, both carboxyl groups had similar pK_a values (between 7.1 and 7.8) ...

Recalling that bilirubin is insoluble in H_2O (*942*) and forms aggregates at pH <9 (*958*), wherever solids are present or bilirubin solubility is in question, it should be noted that the pK being measured is an *apparent* pK_a value coming from an aggregation or solubility equilibrium coupled to the acid dissociation equilibrium. Thus, in most cases involving bilirubin, the apparent pK that is measured reflects the actual pK_as plus the disaggregation pK. This explains the normal carboxylic acid pK_as of bilirubin found in DMSO and the elevated measured (apparent) carboxylic acid pK_as in water, as in 1992, when *Hahm, Ostrow, Mukerjee*, and *Celic*, arrived at a new set of bilirubin pK_a values (8.12 and 8.44) based upon solvent partitioning (*952*) and which were acknowledged as unexpectedly high and attributed to retarded dissociation of the protons (*959*):

> Our unexpectedly high pK_a values are probably the result of retarded dissociation of the protons due to the trio of internal hydrogen bonds between each carboxyl group and the – CO–NH– and >NH groups of the opposite dipyrrylmethene rings ...

whereas, the more conventional carboxylic acid pK_a values were again attributed to the power of the solvent (DMSO, DMF) to break an alleged mighty fortress of internal hydrogen bonds (*959*):

> When this internal bonding is broken, e.g., by dissolution of UCB in dimethyl sulfoxide or dimethylformamide, titrimetric and ^{13}C-NMR studies yield, as expected, pK_a values between 4.5 and 5.0 ...

Given a belief in the doctrine of a *rigid* ridge-tile structure of bilirubin (*924, 925*), apparent misunderstanding of *Overbeek*'s work, and neglect of *Brodersen*'s writings (*929*), it is perhaps not surprising that the mantra that intramolecular hydrogen bonding suppresses bilirubin ionization became an often repeated exegesis for the high pK_a values found in some studies, and that rupture of hydrogen bonds by solvent, especially DMSO, became a reason for dismissing other studies which had found normal pK_a values (*960*):

> Internal hydrogen bonding of the carboxyl groups presumably suppresses their ionization, accounting for the unexpectedly high pK'a values. ... UCB in DMSO, in which the hydrogen bonds are ruptured, exhibit pK'a values below 5.0, as expected for –COOH groups ....

or in 1994 (*925*):

> These unexpectedly high pK'a values are probably the result of retarded dissociation of the protons due to the internal hydrogen bonds involving each –COOH group . . . . [T]itrimetric and ^{13}C-NMR studies yield, as expected, pK'a values for UCB between 4.5 and 5.3.

Unfortunately, "retarded dissociation of the protons" of bilirubin remains (*auf die Weltanschauung von* pK_a) only an unsubstantiated theory and contradicts: (i) the experimental determination that a propionic acid rapidly disembraces an opposing dipyrrinone during conformational enantiomerism, and (ii) the best available evidence from NMR studies that indicate that intramolecular hydrogen bonding in bilirubin is not completely broken in $(CD_3)_2SO$ (*580*). Furthermore, the proof that DMSO disrupts intramolecular hydrogen bonds to such a degree that it has a marked effect on their dissociation conflicts with a large body of evidence, not just on bilirubin, but on other intramolecularly hydrogen-bonded carboxylated compounds (*944*).

Ei incumbit probatio qui dicit non qui negat. [3]

The DMSO brouhaha continued to resurface as the pK_a values of bilirubin, numerous model compounds, dipyrroles, monopyrroles, dicarboxylic and monocarboxylic acids were determined by ^{13}C NMR in DMSO-laced H_2O. The over 40-year-old observation of a difference of ~5 ppm between the ^{13}C NMR chemical shift of carboxylic acid and carboxylate ion carbons (*961*) offered a useful way to determine pK_a (*962*). The technique became even more sensitive when the carboxylic acid CO_2H carbon was highly enriched in ^{13}C. (The natural isotopic abundance of ^{13}C in carbon is only ~1.1%.) Thus *Hamilton et al.* (*963–966*) were able to determine the pK_a values of H_2O-insoluble fatty acids and learned that the observed high pK_a values were associated with aggregation (*963–967*) and extrapolated to ordinary values in pure H_2O. The promise of these studies led *Darren Holmes* (Ph.D., 1994) in the *Lightner* lab to investigate the pK_as of mesobilirubin-XIIIα 99% ^{13}C-enriched in its two carboxyl groups (*958, 968–971*). Preliminary determinations in the mid-1990s of $^{13}CO_2H$-enriched monocarboxylic acids, including monopyrrole propionic and acetic acids and xanthobilirubinic acid (*972*), and dicarboxylic acids (*958, 968–971*) showed, as *Hamilton* predicted, that aqueous pK_a values could readily be derived from measurements in $(CD_3)_2SO$-H_2O solutions of $(CD_3)_2SO$ mole fractions from 0.004 to 0.3. *Holmes* was able to measure pK_{a1} and pK_{a2} for mesobilirubin-XIIIα (4.2 and 4.9), mesobiliverdin-XIIIα (3.9 and 5.3), 12-ethylmesobilirubin-XIIIα (4.3), and 8-ethylmesobiliverdin-XIIIα (4.5), among others (*970*). Although the pK_a values were contested (*973*) on the basis of mesobilirubin's insolubility in DMSO-H_2O), the concerns were shown to be unfounded (*971*).

Subsequently, in 2004, it was shown that aggregation of bilirubin leads to elevated pK_a (*972*), as in fatty acid aggregation (*963–967*), suggesting a possible explanation for some of the high values previous reported in the literature (*925, 957, 959, 960, 974–976*). Most recently, in 2010, a partially water-soluble bilirubinoid with a full set of six intramolecular hydrogen bonds was shown by a standard method for determining the pK_a of pharmaceuticals, vacuum-assisted

9.3 The Glue that Binds. Persistence, Recognition, and Role of Hydrogen Bonds 601

multiplexed capillary electrophoresis in CH_3OH-H_2O, to yield $pK_aS \sim 4.9$ (562). The intramolecularly hydrogen-bonded model for one-half of bilirubin, a [6]-semi-rubin (see Fig. 9.3.7b) with a 99% ^{13}C-labeled CO_2H group was synthesized by *Suchitra Datta* (Ph.D. 2008) in the *Lightner* group and shown also to have a $pK_a = 4.3$, or the same as the carboxyls of mesobilirubin (978).

These results are consistent with studies on other carboxylic acids that are related to bilirubin by their molecular geometry-enabled hydrogen bonding between carboxylic acid and amide groups. In 1986, *Rebek et al.* (979) showed that pK_{a1} of the dicarboxylic acid of Fig. 9.3.2b was normal (4.8), though pK_{a2} was high (11.1), consistent with the stabilization of carboxylate anion by strong intramolecular hydrogen bonding. Subsequently, in 1996 *Rebek et al.* (908) reported a pK_a of ~4.3 for dissociation of the carboxylic acid hydrogen from a CO_2H group that is intramolecularly hydrogen-bonded to an amide (Fig. 9.3.2a).

Some ten years later, *Ueyama*'s group showed (980) that a carboxylate ion forms a strong hydrogen bond with a proximal amide (H–N–C=O) hydrogen, consistent with the expectation that stabilization of the bilirubin mono- and dicarboxyl-ate anions by intramolecular hydrogen bonding to the dipyrrinone NHs would, if anything, lower the corresponding pK_aS. Accordingly, the schemozzle over whether the hydrogen bonding in the diacid is tight or broken may be irrelevant.

Does DMSO disrupt hydrogen bonds? According to *Kidrič et al.* (939):

> . . . intramolecular H-bonding appears to persist in DMSO solution.

and, according to *Tolstoy et al.* (940):

> In aprotic solvents such as dimethyl sulfoxide (DMSO) . . . the intramolecular hydrogen bond is usually retained. . . .

Drueckhammer and *Schwartz* (981) provided persuasive evidence favoring the retention of intramolecular hydrogen bonding in a butenedioic acid mono-anion in both DMSO and $CHCl_3$ by factors of 160:1 and 1050:1, respectively, for hydrogen-bonded to non-intramolecularly hydrogen-bonded species. In CH_3OH, the ratio is reversed (1:2), yet the data from DMSO studies emphasize that intramolecular hydrogen bonding is unlikely to be disrupted in that solvent. The ratio favoring intramolecular hydrogen bonding inverts somewhat for the mono-anion of the mono-amide presumably due to the greater proton acidity of the residual CO_2H group relative to the residual $CONH_2$ group.

Experimental studies (982) of DMSO-H_2O mixtures indicate that, while small amounts of DMSO solute have little or no effect on water structure even at high mole fractions of DMSO ($\chi_{DMSO} = 0.35, 0.21$), the local tetrahedral network of H_2O-H_2O hydrogen bonds is preserved though the percentage of water molecules hydrogen-bonded H_2O to H_2O is reduced at higher χ_{DMSO}; thus (982):

> Water structure is not found to be strongly affected by the presence of DMSO. However, the percentage of water molecules which hydrogen bond to themselves is substantially reduced compared to pure water, with a large proportion of the hydrogens available for bonding associated with the lone pairs on the DMSO.

That is, DMSO molecules are hydrogen-bonded to H_2O molecules. This is supported by theoretical studies from molecular dynamics simulations (983) that show that the average number of hydrogen bonds per water molecule decreases with increasing DMSO mole fraction (χ_{DMSO}), and that H_2O molecules are hydrogen-bonded to DMSO molecules in a strong 2:1 $DMSO:H_2O$ hydrogen-bonded complex that exists over a wide range of χ_{DMSO}. Just how competitive bilirubin is for hydrogen bonding to DMSO in $DMSO-H_2O$ solutions, where c_{DMSO} is small, is unknown. It remains unlikely that intramolecular hydrogen bonds are disrupted in the aqueous DMSO solutions used in pK_a determinations.

What, then, are the pK_as of bilirubin? Since the pigment has two non-interacting propionic acid groups, the individual pK_a values would be predicted to lie between 4.5 and 5, the normal value for alkanoic acids; most determinations have found pK_a values in that range. Aggregation would be expected to lead to higher values due to field effects, as observed for high molecular weight fatty acids, whereas hydrogen bonding would be expected to have only a weak effect.

In retrospect, it is intriguing to trace the origin and evolution of the theory that the propionic carboxyl groups of bilirubin are anomalously weaker acids compared to other aliphatic acids. An apparent misunderstanding or misinterpretation of the work of *Hutchinson et al.* (941) and *Overbeek et al.* (942) that led to the conclusion (827, 950) that the pK_a of the carboxyl groups of bilirubin is 7.95. (Whether 7.95 was thought to be the pK_a of each of the two individual carboxyl groups in bilirubin, or whether it represented the overall pK_a ($\sim pK_{a1} + pK_{a2}$) of bilirubin, in which case the individual pK_as would be in the normal range, was not made clear.) The conclusion is astonishing because it ignored the fact that both *Hutchinson et al.* and *Overbeek et al.* clearly considered the individual pK_a values for the carboxyl groups to be within the normal range for aliphatic acids, *i.e.* ~4.5. Nevertheless, early speculation based on the supposed rigidity and properties of the hydrogen bonds and solvent effects (that lacked experimental basis and later turned out to be largely incorrect), were advanced to justify the unusually high $pK_a(s)$ of bilirubin. Experimental support for the theory seemed to come in 1988 (926) when pK_{a1} and pK_{a2} values of 6.8 and >9.3, respectively, in buffered saline were found for the individual carboxyl groups of bilirubin based on solubility measurements. Unremarked was that a pK_a >9.3 is something of a record value for a simple aliphatic carboxylic acid group. Subsequently, those pK_a values were withdrawn, only to be replaced by similarly extraordinary pK_{a1} and pK_{a2} values of 8.1 and 8.4, respectively, or nearly the average of pK_as 6.8 and 9.3. The new pK_a values were based on a solvent partitioning method (959) done first with unlabeled bilirubin and later with radio-labeled bilirubin (974) with similar results. Thus, the results of a single experimental method on a single compound with no apparent standard reference compounds, and the discredited (953, 954) use of $CHCl_3$ as the sole extraction solvent, contradict many independent determinations on bilirubin, mesobilirubin, and a large number of model compounds by many investigators using a variety of other methods.

9.4 Extrapolating from Bile Pigment Stereochemistry... 603

One might use equation (1) mentioned earlier in this section to estimate some consequence of the high pK_a values. When the equation based on *Overbeek et al.* (*942*), $[BH_2] = 2.15 \cdot 10^{-20}$ $(K_1 \cdot K_2)$, is used to calculate the aqueous solubility of bilirubin from the pK_a values (pK_{a1} ~8.1, pK_{a2} ~8.4) most recently favored by *Ostrow et al.* (*960, 974–976*), one finds $[BH_2] = 2.15 \times 10^{-20}/(3.16 \times 10^{-17}) = 6.8 \cdot 10^{-4}$ *M*. Thus, at p*H* 7, $[BH^-]$ ~ $5 \cdot 10^{-5}$ and $[B^=]$ ~$2.7 \cdot 10^{-5}$, and the solubility of bilirubin calculates at $7.6 \cdot 10^{-4}$ *M* – or higher than the values ($2 \cdot 10^{-6}$ and ~0.6 n*M*) published by *Overbeek et al.* (*942*) and *Brodersen et al.* (*929*), respectively. The aqueous solubility at pH 7, calculated on the basis of pK_a values of 8.1 and 8.4 (*960, 974-976*) also exceeds the solubilities predicted by *Ostrow et al.*: 66 n*M* (*959*) and 51 n*M* (*974*).

What then are the pK_as of bilirubin? Given contradictory results, it seems that the interested reader must "pay yer money and take yer choice". Unquestionably, the propensity of bilirubin to form intramolecular hydrogen bonds has a marked effect on its chemical and biochemical properties and its metabolism. In general the strength of hydrogen bonds is weak, but in bilirubin it is the combined effect of six hydrogen bonds in the free acid or four hydrogen bonds in the dianion that leads to its anomalous properties compared to, say biliverdin, including the possible formation of a stable bilirubin hydrate (*984*). But whether the weak hydrogen bonding of a single propionic acid hydrogen is sufficient to suppress dissociation to such an unprecedented degree that the pK_a rises to ~8.5 seems highly unlikely. Recent DFT computations of bilirubin flexibility, hydrogen bonding, and pK_as in water (*985*) give pK_{a1} and pK_{a2} that are close (4.80 and 5.17) to the more conventional lower experimental values (~4.5). On the basis of these calculations (*985*):

> Although UCB [= unconjugated bilirubin] present [sic] the –COOH/–COO– groups in a crowded microenvironment and establishing intramolecular hydrogen bonds that lead [one] to think that rotation and solvation are difficult, our findings showed that: (a) rotation of [the] –COO– group is not restricted; (b) new stabilizing interactions appear after an H-bond is cleaved at each ionization step; and (c) solvent accessibility is not inhibited. Due to the above, these factors are no longer credible/acceptable and pK$_a$ values rationalization based on them[13] [*975*] may be misleading.

A principio non ita fuit. 4

9.4 Extrapolating from Bile Pigment Stereochemistry and Hydrogen Bonding to Synthetic Bilirubinoids

With the repeated bombing of Munich in World War II and the resultant destruction of the *Fischer* Institute at the TH-München, the synthesis of pyrrole compounds slowed to a trickle then ceased with *Hans Fischer*'s death in the waning months of the war. Though the total syntheses of hemin, bilirubin, and biliverdin had been achieved, along with hundreds of other pyrrole compounds, and the total synthesis of chlorophyll was well underway in 1945, for the following several decades there followed scant interest in linear tetrapyrrole synthesis. Revived to some extent in the late 1960s by *Fischer*'s student *Hans Plieninger* (*690, 700–712*), at the University of Heidelberg, more dedicated synthesis efforts toward bile pigments endured a nearly 25-year hiatus while awaiting the research investments of *Albert Gossauer* (*713, 715–720, 986*) in Germany and Switzerland and *Heinz Falk* (*15*) in Austria who, together with *Hans Plieninger* (*690, 700–712*) in the 1970s, created a multitude of novel dipyrroles and linear tetrapyrroles on the way to understanding bile pigment stereochemistry, phytochrome, biliverdin, urobilins, *etc.* When the configurational and conformational structure of bilirubin had been established and the presence of intramolecular hydrogen bonds understood, doors opened toward probing the molecular details of bilirubin chirality and the role of hydrogen bonding, while phototherapy for neonatal jaundice promoted investigations into bilirubin binding, hepatic transport, glucuronidation, and excretion through the design and synthesis of bilirubin analogs. In the U.S., the *Lightner* group's association with bile pigments dates back to 1964 with theoretical calculations of the absolute configuration of *d*-urobilin and *l*-stercobilin [determined later by degradation, synthesis, and spectroscopy (*708*)] when the author was a postdoctoral research fellow in the *Moscowitz* lab at the University of Minnesota, and with experimental work in the *Watson* lab at the University of Minnesota Medical School. (Recall that *Watson* was a postdoctoral research fellow from 1930 to 1932 in *Hans Fischer*'s lab). In 1965, the synthesis of pyrroles was initiated in the *Lightner* lab at the University of California, Los Angeles (UCLA). By 1969, research had begun to provide an understanding of the molecular mechanism of phototherapy for neonatal jaundice. Pyrrole research by the *Lightner* group continued in the mid-1970s at the University of Nevada with the synthesis of numerous dipyrrinones and their analogs, model compounds for bilirubin. By the mid-1980s, a serious effort to synthesize tetrapyrrole analogs of bilirubin was well underway, in part to understand the importance of intramolecular hydrogen bonding to pigment metabolism, in part to understand the stereochemistry/chirality of bilirubin and the importance of the relationship between stereochemistry and hydrogen bonding. Toward achieving these objectives, many novel bilirubin analogs, new bilirubinoids, were designed and synthesized between 1985 and 2010, a partial listing of which is found in the following, structures **9.4.1** through **9.4.53**.

What follows is a description of the research, including a sampling of the syntheses of numerous dipyrrinones and their analogs, model compounds for bilirubin, which took place at the University of Nevada in the *Lightner* lab. Research was conducted by graduate students or postdoctoral fellows, unless otherwise noted. All metabolism studies cited here were carried out by Professor *A.F. McDonagh* at the University of California, San Francisco.

In 1987, Dr. *Roger Franklin* and Prof. *Francesc Trull* of the *Lightner* lab produced mesobilirubin-IVα (**9.4.1**), a polar pigment incapable of intramolecular hydrogen bonding, as well as mesobilirubin-VIIIα (**9.4.2**), and the bilirubinoid (**9.4.3**) with but one propionic acid (*732, 850*), synthesized along with mesobilirubin-XIIIα

9.4 Extrapolating from Bile Pigment Stereochemistry…   605

(**9.4.4**). The last had been synthesized some 50 years earlier by *Fischer* (*438, 444*) (see Section 6.2.2). Pigments **9.4.2** and **9.4.3** are capable of intramolecular hydrogen bonding at only one dipyrrinone and are more polar than **9.4.4**, in which both dipyrrinones engage in intramolecular hydrogen bonding.

9.4.1: Mesobilirubin-IVα

9.4.2: Mesobilirubin-VIIIα
$(R^1 = CH_3, R^2 = CH_2CH_2CO_2H)$
9.4.3: $R^1 = CH_2CH_3, R^2 = CH_3$
9.4.4: Mesobilirubin-XIIIα
$(R^1 = CH_2CH_2CO_2H, R^2 = CH_3)$

Pigment **9.4.1** is excreted intact, without glucuronidation by the liver in the *Sprague-Dawley* rat (a rat bred for scientific research that readily glucuronidates and excretes bilirubin hepatically); **9.4.2** is excreted both intact and as a glucuronide; **9.4.3** is excreted only as a glucuronide, as is **9.4.3** (*987*).

The decade of the 1990s brought forth a more dedicated effort in synthesis aimed toward understanding the limits of intramolecular hydrogen bonding and its control of bilirubin's conformation. The work of *Young-Seok Byun* (Ph.D. 1990) (*883*) produced a bilirubin analog with propionic acid shortened to carboxyl (**9.4.5**), which turned out to be very polar (*883*), and the synthesis efforts of *David Shrout* (Ph.D. 1991) and *Richard Person* (Ph.D. 1998) yielded analogs with acetic through heptanoic acid chains (*988*). Although the acetic acid analog (**9.4.6**, n=1) was more polar than mesobilirubin-XIIIα (**9.4.6**, n = 2), the butyric acid analog (**9.4.6**, n = 3) was comparably lipophilic to the latter. Counter-intuitively, as the chains were lengthened, the polarity of the pigment increased, though molecular dynamics modeling predicted intramolecularly hydrogen-bonded minimum energy conformations, which differed in shape from those of bilirubin (*989*). With extreme alkanoic acid chain lengthening, Dr. *John Chiefari*'s undecanoic acid bilirubin analog (**9.4.6**, n = 10) was more lipophilic (*990*), as was Dr. *Thomas Thyrann*'s eicosanoic acid pigment (**9.4.6**, n = 19), which had wax-like properties (*991*).

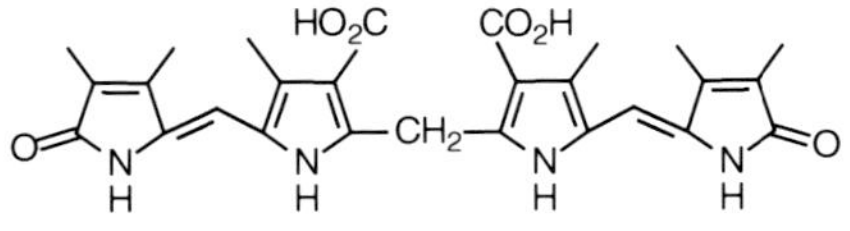

9.4.5

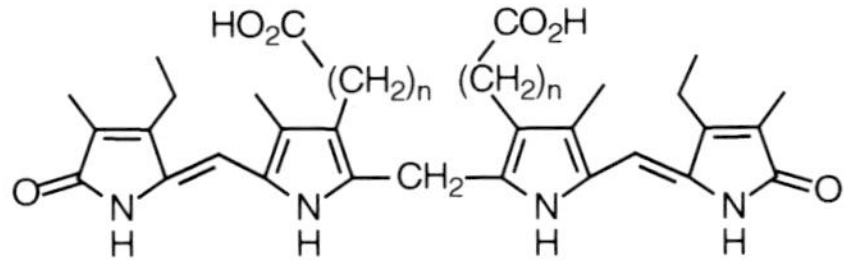

9.4.6: n = 1-6, 10, 19
(n = 2: Mesobilirubin-XIIIα)

Rubins **9.4.6**, n = 1, 4, and 5 were excreted by the liver in the *Sprague-Dawley* rat, partially intact and partially as acylglucuronides (*896*); whereas **9.4.6**, n = 2 and 3 were excreted as acyl glucuronides (*896*) and **9.4.6**, n = 10 and 19 were not excreted (*169, 893*).

Attempts to perturb the ridge-tile conformation of bilirubin by internal buttressing steric interactions were investigated by *Meiqiang Xie* (Ph.D. 1992) and *Darren Holmes* (Ph.D. 1994) in 10,10-dimethylmesobilirubins **9.4.7/9.4.9** (*902, 903*), later by *Brahmananda Ghosh* and Dr. *Bin Tu* in **9.4.8** (*904*), and in 10,10-spiro analogs **9.4.10/9.4.11** by *Ghosh* (*905*). From X-ray crystallographic analyses (*904, 905*) and NMR, it was learned that the pigments retain the intramolecularly hydrogen-bonded ridge-tile conformation, though with longer hydrogen bonds (*904*), and in the case of **9.4.7/9.4.9**, improved solubility in all organic solvents, *e.g.* soluble in CH_3OH, in which bilirubin is insoluble.

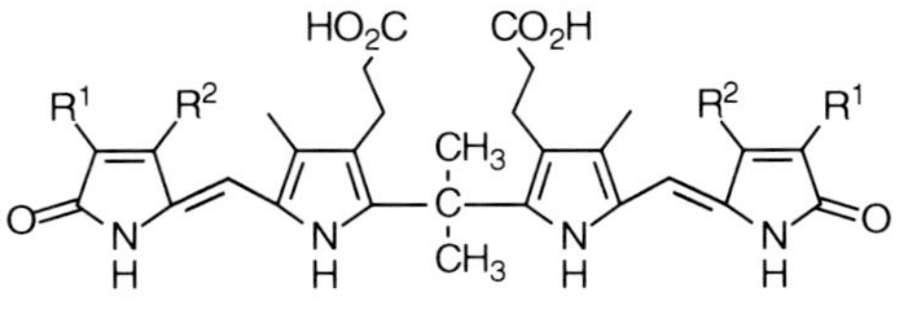

9.4.7: $R^1 = R^2 = CH_3$
9.4.8: $R^1 = CH_3$, $R^2 = CH_2CH_3$
9.4.9: $R^1 = R^2 = CH_2CH_3$

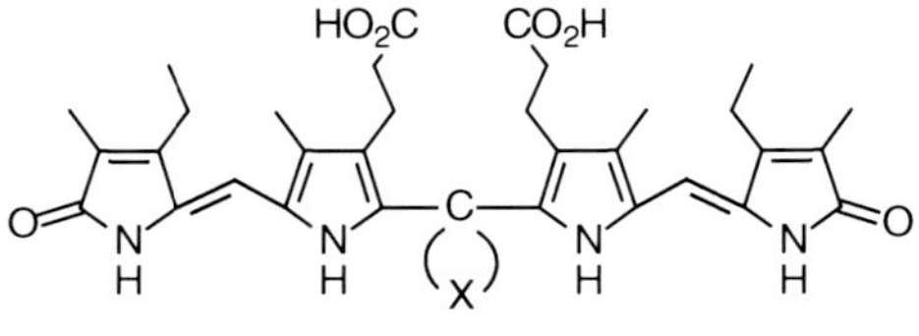

9.4.10: X = $(CH_2)_5$

9.4.11: X =

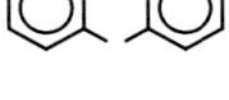

9.4 Extrapolating from Bile Pigment Stereochemistry...

Rubins **9.4.7** and **9.4.8** were excreted as glucuronides in the *Sprague-Dawley* rat (*893, 992*).

Attempts to interfere with or to force a break of intramolecular hydrogen bonding by intermolecular steric buttressing were made by *Ari Kar* (Ph.D. 1998) in his synthesis and analysis of C-10 isopropyl, isopropylidene (*900, 901*), *tert*-butyl and 1-adamantyl (*898*) substituents (**9.4.12** to **9.4.15**). Despite mild, nonsymmetric distortions of the ridge-tile and H-bond lengthening, these pigments displayed the usual two sets of intramolecular hydrogen bonds. A significant change in conformation, however, attended the introduction of a second CH$_2$ group in the middle of pigment (**9.4.16**), prepared by *William Pfeiffer* (Ph.D. 1997), who found that this "10-homorubin" folded into U-shaped and Z-shaped conformations, while maintaining intramolecular hydrogen bonds (*993*) and a lipophilic nature.

9.4.12: X =CH–CH(CH$_3$)$_2$
9.1.13: X =C=C(CH$_3$)$_2$
9.4.14: X =CH-C(CH$_3$)$_3$
9.4.15: X =CH–(1-Ada)

9.4.16

Pigment **9.4.16** was excreted as mono and diglucuronides, the same as bilirubin, in the *Sprague-Dawley* rat (*169*). Pigment **9.4.15** was excreted unchanged, **9.4.14** mostly unchanged, and **9.4.12** and **9.4.13** only slowly as a monoglucuronide (*169*).

Exploring the possibility that a wider separation of the two dipyrrinones might decouple hydrogen bonds, *Daniel Nogales* prepared a series of bilirubinoids with pyrrolyl and xylyl spacers (**9.4.17** to **9.4.19**) and found that all three were more polar than bilirubin and do not engage in intramolecular hydrogen bonding (*994*). *D. Timothy Anstine* (Ph.D. 1995) modified the general motif of **9.4.17** to **9.4.19**, keeping to a fifth pyrrole ring in structures **9.4.20** to **9.4.22** (*995*). Although molecular modeling computations determined that all three bilirubinoids should engage in intramolecular hydrogen bonding, only **9.4.21** and **9.4.22** were sufficiently soluble in CDCl$_3$ to enable verification. Like **9.4.21**, **9.4.20** was insoluble in CDCl$_3$ and only slightly soluble in CH$_3$OH; whereas **9.4.22** was insoluble, and **9.4.21** was slightly soluble. All were soluble in (CH$_3$)$_2$SO. The ^{1}H NMR chemical shifts of both **9.4.21** and **9.4.22** Cl_3 were consistent with the prediction that the acid groups engage the dipyrrinones in intramolecular hydrogen bonding.

Pigments **9.4.17** to **9.4.19** were excreted intact in the *Sprague-Dawley* rat (*891*).

Additional studies aimed at perturbations at C-10 included the preparation of 10-oxo-mesobilirubin-XIIIα (**9.4.23**) by Dr. *Qingqi Chen* (*885*) and the syntheses of thia-rubin **9.4.24** by *Adrianne Oldham-Tipton* (Ph.D. 1999) (*887, 888*) and *Brahamananda Ghosh* (Ph.D. 2003) (*889*), and the selena-rubin **9.4.25** by *Ghosh* (*514*). X-ray crystallographic studies of the thia-rubins (*888*) confirmed their ridge-tile structures with a full set of six intramolecular hydrogen bonds, despite the long C–S bonds in the middle and smaller interplanar angle (*θ*) (see Fig. 9.3.1a). Similarly, selena-rubin **9.4.25**, with its even longer C–Se bonds and more acute interplanar angle (*θ*), exhibited intramolecular hydrogen bonding. Both were non-polar, in contrast to the oxo-rubin (**9.4.23**) which, though more polar than bilirubin, gave every indication in the ^{1}H NMR NH and CO_2H chemical shifts of being intra-molecularly hydrogen bonded (*885*).

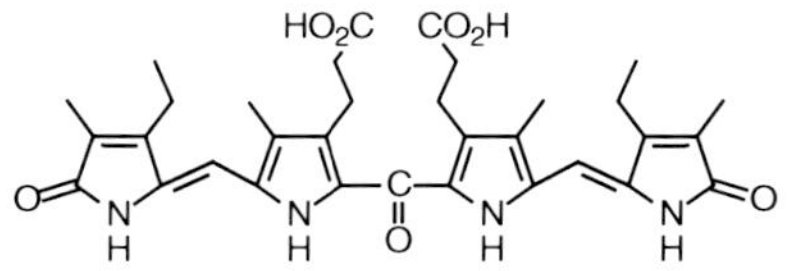

9.4 Extrapolating from Bile Pigment Stereochemistry…

9.4.23

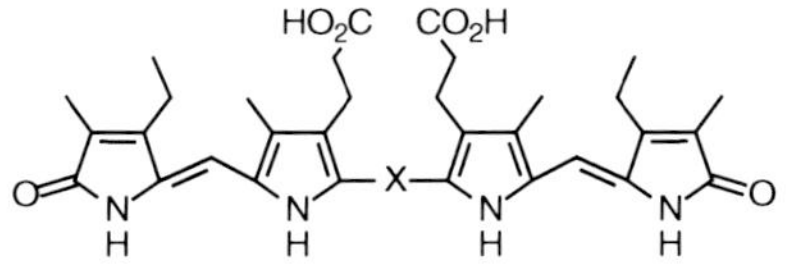

9.4.24 X = S
9.4.25 X = Se

Oxo-rubin **9.4.23** was excreted both intact and as acylglucuronides in *Sprague-Dawley* rats, indicating that small changes in polarity at the right site on the molecule can produce sufficient polarity for excretion intact (*885*). In contrast, **9.4.24** and **9.4.25** were both excreted but only as glucuronides (*887, 895*).

Perturbation of the acidity of the propionic acid groups by substitution at the α-carbon with electronegative groups (**9.4.26** to **9.4.28**) produced interesting results and are of potential importance when considering the pK_as of bilirubin. With the most electronegative atom, fluorine, at the α-carbon, the α,α'-difluoromesobilirubin-XIIIα (**9.4.26**) synthesized by Professor *Stefan Boiadjiev*[5] (Ph.D., D.Sc.) in the *Lightner* group (*996*) was found to be more polar than its parent, *water-soluble* and insoluble in CHCl₃. The last prevented the ¹H NMR studies that might have given experimental verification to its intramolecularly hydrogen-bonded structure; yet, there is probably no reason to assume that the pigment's conformation is altered or that intramolecular hydrogen bonds are absent. One can expect a drop of two pK units from α-F substitution in the propionic acids (*e.g.* CH_3CO_2H $pK_a = 4.76$; FCH_2CO_2H $pK_a = 2.58$), due to the electronegative influence of the F. Apparently the acid dissociation is sufficiently great to enable the pigment's aqueous solubility despite its high molecular weight, and much like the sulfonic acid analog (**9.4.29**) of mesobilirubin-XIIIα that was synthesized by Dr. *Boiadjiev* (*892*).

---

[5]*Stefan Boiadjiev* was born on May 11, 1956 in Pleven, Bulgaria, into a family of physicians. He received the M.Sc. degree in organic chemistry in 1981 from the University of Sofia and the Ph.D. in organic chemistry from the Bulgarian Academy of Sciences in 1988 before joining the *Lightner* group at the University of Nevada in 1989 as a postdoctoral research fellow and advancing to a research professorship, during which time he made major advances in our knowledge of pyrrole and bile pigment synthesis and stereochemistry. He joined the faculty of the Medical University in Pleven in 2008, where he is currently full professor.

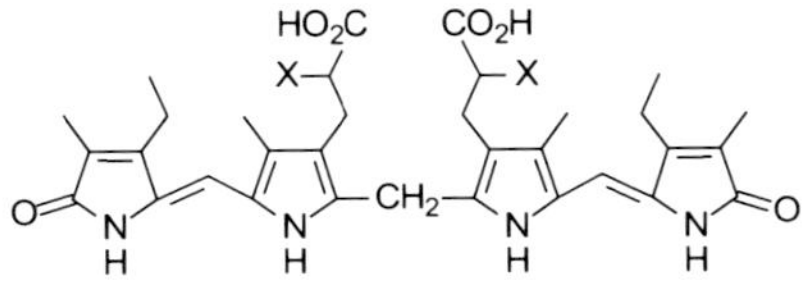

9.4.26: X = F
9.4.27: X = OCH₃
9.4.28: X = SCH₃

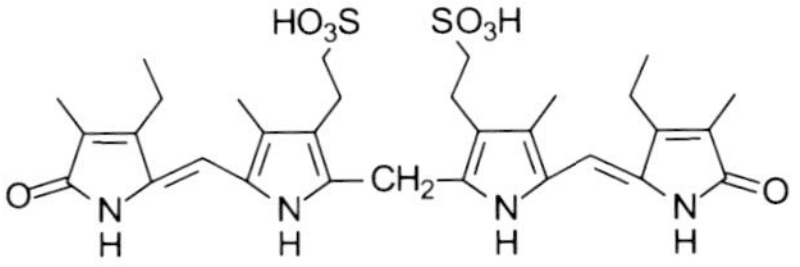

9.4.29

Given that OCH$_3$ groups are less electron-withdrawing inductively and drop the pK_a of acetic acid by only about one pK unit, the expected pK_a drop of α,α'-dimethoxy- (**9.4.27**) and α,α'-bis-methylthio (**9.4.28**) mesobilirubin-XIIIα is also expected to be about one pK unit. These two substituted mesobilirubins, synthesized by Dr. *Boiadjiev* (*997*), actually exhibited different solution properties. The dimethoxy analog **9.4.27** was soluble in aqueous NaHCO₃, in which the parent rubin is insoluble, and in which, surprisingly, the bis-methylthio analog **9.4.28** is also insoluble. Both are insoluble in pH 7 water but they are both soluble in CDCl₃, from which solvent ¹H NMR measurements confirm the intramolecularly hydrogen-bonded ridge-tile structure in the (separable) racemic diastereomers. The water-soluble sulfonic acid **9.4.29** and the α,α'-dimethoxy pigment were shown to cross the liver into bile without conjugation in *Sprague-Dawley* rats (*169*).

Aqueous solubility was the rationale for *Sanjeev Dey*'s (Ph.D. 2008) synthesis of mesobilirubins with CH₃O groups (**9.4.30**) on the lactam β-carbons (*881*) and with methyl-capped mono-, di-, and tri-ethylene glycol (**9.4.31**, to **9.4.33**) (*977*). Pigment **9.4.30** was soluble in CH₃OH, unlike mesobilirubin-XIIIα and more soluble in CHCl₃ than the latter. Pigments **9.4.31** to **9.4.33** were soluble in CHCl₃, CH₃OH, and H₂O, with H₂O solubility increasing with polyethylene glycol chain length. All four pigments adopted preferentially the intramolecularly hydrogen-bonded ridge-tile conformation as determined by ¹H NMR measures in CDCl₃, CH₃OH, and H₂O.

9.4 Extrapolating from Bile Pigment Stereochemistry...

9.4.30: R = OCH_3, $O(CH_2CH_2O)_nCH_3$
9.4.31: R = $OCH_2CH_2OCH_3$
9.4.32: R = $O(CH_2CH_2O)_2CH_3$
9.4.33: R = $O(CH_2CH_2O)_3CH_3$

9.4.34 $R^1 = (CH_2)_3CH_3$, $R^2 = CH_3$ **9.4.35**: $R^1 = CH_3$, $R^2 = (CH_2)_3CH_3$
9.4.36: $R^1 = C_6H_5$, $R^2 = CH_3$ **9.4.37**: $R^1 = CH_3$, $R^2 = C_6H_5$
9.4.38: $R^1 = o\text{-}FC_6H_4$, $R^2 = CH_3$ **9.4.39**: $R^1 = CH_3$, $R^2 = o\text{-}FC_6H_4$

Attempts to improve the lipophilicity of bilirubin by the placement of lipid, H_2O-insoluble groups at the lactam β-carbons were addressed by *Justin Brower*'s (Ph.D. 2001) synthesis that produced **9.4.34/9.4.35** with *n*-butyl groups (*886*); **9.4.36/9.4.37** with phenyl groups (*890*); and **9.4.38/9.4.39**, with four *o*-fluorophenyl groups (*899*). In all cases, the intramolecularly hydrogen-bonded folded conformation was preserved. The *n*-butyl analogs (**9.4.34/9.4.35**) showed an increased solubility in $CHCl_3$, as anticipated, as did the phenyl analogs (**9.4.36/9.4.37**) and the *o*-fluorophenyl analogs (**9.4.38/9.4.39**).

The location of the *n*-butyl groups in **9.4.34/9.4.35** influenced the pigments' metabolism in *Sprague-Dawley* rats considerably. With *endo-n*-butyls, **9.4.35** was excreted by glucuronidation rather like mesobilirubin-XIIIα and bilirubin-IXα. With *exo-n*-butyls, **9.4.34** was glucuronidated and excreted rather poorly (*886*). Despite the presence of bulky phenyl groups, **9.4.36/9.4.37** are metabolized, like bilirubin, to mono- and di-glucuronides and excreted (*509*), with the ratio of mono- to di-glucuronides dependent on the *exo vs. endo* location of the phenyl groups, with the proportion of diglucuronides being higher for the *endo*-phenyl isomer (**9.4.37**).

In order to bring the two dipyrrinones of mesobilirubin-XIIIα closer, connecting them as a 2,2′-bipyrrole by removing the C-10 CH_2 group, *Edward Nikitin* (Ph.D. 2006) synthesized "imploded" 10-*nor*-bilirubin analogs (*998*). These bipyrrole pigments (**9.4.40/9.4.41**) cannot fold into a ridge-tile conformation; yet, the dipyrrinones still engage the two propionic acids (**9.4.40** less well) and butyric acids (**9.4.41** very well). In the latter, the dipyrrinones are rotated into a twisted, *anticlinal*, conformation about the bipyrrole core. Interestingly, with acetic acid side chains, fully intramolecularly hydrogen bonded, the bipyrrole is planar.

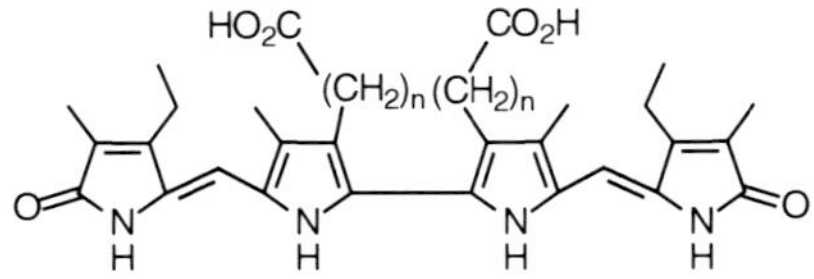

9 Understanding and Translating Bilirubin Structure

9.4.40: n = 2
9.4.41: n = 3

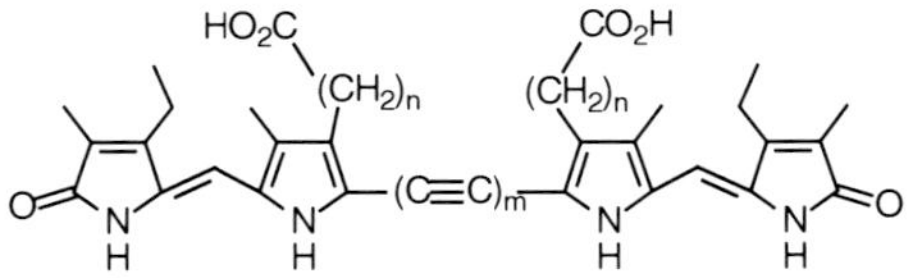

9.4.42: m = 1, n = 2
9.4.43: m = 1, n = 3
9.4.44 m = 2, n = 2
9.4.45: m = 2, n = 5

Moving in the other direction, in order to more widely separate the dipyrrinones of mesobilirubin-XIIIα, while retaining only two degrees of rotational freedom, as is present about C-10 in the parent, the C-10 CH_2 was replaced with a stiff $-C{\equiv}C-$ unit (**9.4.42/9.4.43**) by Dr. *Bin Tu* (*999*) to create "linearized" pigments. Only one of the two propionic acids of **9.4.41** can engage a dipyrrinone with intramolecular hydrogen bonds. The pigment adopts a twisted shape, not a ridge-tile shape, and is insoluble in $CHCl_3$ and CH_3OH. However, with two butyric acids (**9.4.43**) replacing the propionic, both CO_2H groups are fully intramolecularly hydrogen-bonded and adopt a conformation with the extended planes intersecting along the $-C{\equiv}C-$ axis at an angle of 136°. The pigment exhibits a limited solubility in $CHCl_3$. Neither **9.4.42** nor **9.4.43** is soluble in 5% aqueous $NaHCO_3$. Wider separation of the dipyrrinones was achieved in *Tu*'s syntheses of tetrapyrroles with a diacetylene $-C{\equiv}C-$ $C{\equiv}C-$ spacer in **9.4.44/9.4.45** (*1000*). In the diacetylene rubin (**9.4.44**), intramolecular hydrogen bonding is impossible as the acid chain length is too short to allow either propionic acid to engage the opposing dipyrrinones. In contrast, in the hexanoic acid analog (**9.4.45**) the acid chains are sufficiently long for both CO_2Hs to be engaged in intramolecular hydrogen bonding to the dipyrrinones, with the extended planes of the dipyrrinones being rotated so as to intersect the $-C{\equiv}C-C{\equiv}C-$ axis at an angle of 102°. This pigment is soluble in most organic solvents, including $CHCl_3$ and CH_3OH. Curiously, **9.4.45** but not **9.4.44** is extracted from $CHCl_3$ into 5% aqueous $NaHCO_3$. Metabolism studies of **9.4.42** to **9.4.45** in *Sprague-Dawley* rats (*896*) indicate that **9.4.42** is excreted partly intact, partly as acylglucuronides, whereas **9.4.43** is excreted only as acylglucuronides. In contrast, **9.4.44** is excreted intact, and **9.4.45** is not excreted. These and previously cited examples illustrate very vividly the controlling effect of conformation and intramolecular hydrogen bonding on the metabolism and excretion of tetrapyrroles as well as bilirubin itself.

Benzoic acid analogs (**9.4.46** to **9.4.48**) of bilirubin (*906*) were synthesized by *Boiadjiev* in order to examine possible intramolecular hydrogen bonding in the

9.4 Extrapolating from Bile Pigment Stereochemistry...

ortho analog (**9.4.46**), the equivalent of having placed a (*Z*)-configuration C=C in the propionic side chains. In fact, **9.4.46** was fully intramolecularly hydrogen bonded in a ridge-tile pigment conformation, whereas the *meta* (**9.4.47**) and *para* (**9.4.48**) isomers were incapable of it. Yet it still adopted ridge-tile conformations. Thus, as might be expected, **9.4.46** was soluble in CHCl$_3$ but **9.4.47** and **9.4.48** were insoluble and polar. On the other hand, whereas **9.4.46** is insoluble in CH$_3$OH, **9.4.47** and **9.4.48** exhibited partial solubility.

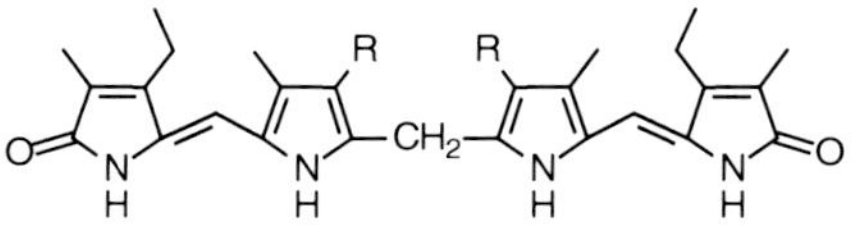

9.4.46: R =

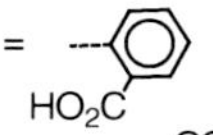

9.4.47: R =

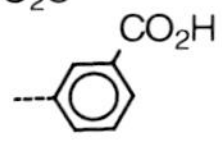

9.4.48: R =

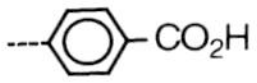

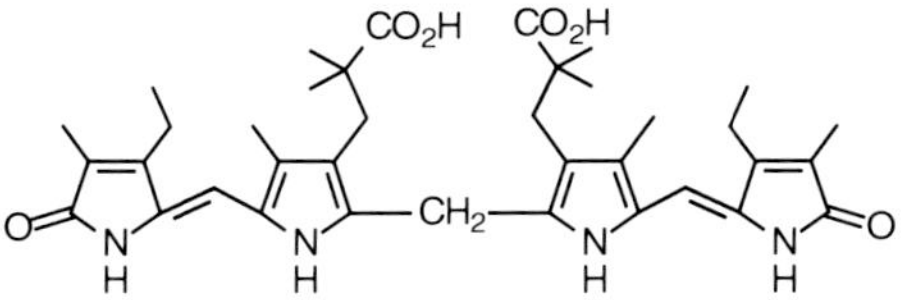

9.4.49

Attempts to force the ridge-tile to open by intramolecular buttressing effects arising from permethylated a-carbons of mesobilirubin-XIIIα were examined following the synthesis of **9.4.49** by Drs. *Boiadjiev* and *Holmes* (*1001*). In this bilirubin analog, both propionic acid groups have been replaced by pivalic acids. These results further indicate the sensitivity of the UGT1A1 enzyme for glucuronidation of an intramolecularly hydrogen-bonded CO$_2$H group and the 3D shape of the pigment. **9.4.46** was not excreted in a *Sprague-Dawley* rat but its (*E,Z*) photoisomer was excreted intact. In contrast, **9.4.47** and **9.4.48** were excreted intact as (*Z,Z*)-isomers (*515*).

In the late 1970s, when it became known through its X-ray crystal structure that bilirubin is a racemic mixture (*510, 511*), and, later, in the 1980s when it was shown that bilirubin in solution is a rapidly interconverting racemic mixture of conformational enantiomers, Dr. *Gisbert Puzicha*, in the *Lightner* group, carefully inspected the local stereochemistry of intramolecularly hydrogen-bonded ridge-tile bilirubin, which revealed a way to displace the conformational equilibrium toward one enantiomer or the other by taking advantage of intramolecular non-bonded steric interactions (*688*). Inspection of intramolecularly hydrogen-bonded *CPK* space-filled molecular models showed that in the (*P*)-helicity enantiomer the *pro-S* α-hydrogens of the propionic acids are forced against the C-7 or C-13 CH$_3$ group, while the *pro-R* α-hydrogens remain relatively uncrowded. In the (*M*)-helicity conformer the situation is pairwise inverted with the *pro-R* hydrogens being sterically compressed and the *pro-S* uncompressed. Likewise, in the (*P*)-helicity enantiomer, the *pro-S* β-hydrogens of the propionic acids are sterically compressed into the C-10 CH_2; whereas, in the (*M*)-helicity enantiomer, the *pro-R* hydrogens are compressed. (See Section 8 and Fig. 8.5.17.)

The stereochemical consequence of replacing the *pro-R* hydrogens of either the α- or β-carbons of the propionic acids by a CH_3 substituent is such that in the (*M*)-helicity conformational enantiomer a steric compression is introduced between that CH_3 group and either the C-7/C-13 CH_3 or the C-10 CH_2. The net effect is to destabilize the (*M*)-helical conformation relative to the (*P*)-helical conformational enantiomer in which the (α*R*)-CH_3 or (β*R*)-CH_3 is less sterically compressed. For *Puzicha*, this suggested the importance of synthesizing mesobilirubin-XIIIα analogs with stereogenic centers produced by replacing a *pro-R* or *pro-S* hydrogen with a CH_3 group at the propionic acid α- or β-carbons. This would lead to three possible isomers: (*R,R*), (*S,S*), and (*R,S*) = (*S,R*), with the last being the *meso* diastereomer and the first two being the *racemic* diastereomers. The first target was the totally racemic mixture, which was separated into diastereomers and enantiomers.

The predictions, based on intramolecularly hydrogen-bonded conformations, were fulfilled by the synthesis talents of *Gisbert Puzicha, Yuming Pu* (Ph.D. 1991), and Dr. *Stefan Boiadjiev*, who prepared, resolved, and determined the absolute configuration of optically active (α*R*,α'*R*) and (α*S*,α'*S*), and (β*S*,β'*S*) and (β*R*,β'*R*)-dimethylmesobilirubins-XIIIα (*687–689, 728, 1002*) shown in **9.4.50** and **9.4.51**. Synthesized also were their more polar, optically inactive *meso* (*R,S*)-diastereomers. The results were fully in accord with the stereochemical predictions of steric buttressing leading to a forced intramolecular displacement of the **M** ⇄ **P** conformational enantiomerism consistent with the predictions and correlations from exciton chirality theory and CD spectroscopy (*619, 687, 1003, 1005*), as discussed in Section 8.

9.4.50: R = CH_3

9.4.51: $R^1 = R^2 = CH_3$
9.4.52: $R^1 = CH_3$, $R^2 = OCH_3$, SCH_3
9.4.53: $R^1 = CH_3$, $R^2 = Et$, *i*-Pr, *t*-Bu, C_6H_5

9.4 Extrapolating from Bile Pigment Stereochemistry... 615

Professor *Stefan Boiadjiev* used his extraordinary synthesis talents in the *Lightner* lab to prepare numerous optically active bilirubinoids (*609, 610, 660, 661, 687, 689, 727–730, 797, 892, 897, 918, 919, 958, 971, 972, 996, 997, 1001–1008, inter alia*). Some served to explore the stereochemical demands of various alkyl and other substituents at the propionic acid α- and β-carbons, and produced a new set of conformational A-values (*742, 797, 1003*). In others, the relative steric demand of α-CH$_3$ *vs.* β-CH$_3$ was explored in (αR,$\beta' S$)-dimethylmesobilirubin-XIIIα (*1006*). In yet others, the importance of a single chiral center toward maintaining the designated helicity was explored in (βS)-methylmesobilirubins with an ordinary propionic acid, its methyl ester, or a simple *n*-propyl group at C-12, *etc.* (*797, 1005*). In still other investigations, it could be shown that the presence of certain aliphatic amines in solutions of (βS,$\beta' S$)-dimethylmesobilirubin-XIIIα (*742*) led to an inversion of the signed order of the exciton CD *Cotton* effects, hence a relaxation of the ridge-tile dihedral angle so as to reverse the angle of intersection of the dipyrrinone electric dipole transition moments. This was also postulated to account for the inversion of the exciton chirality CD of bilirubin on human serum albumin upon addition of a few drops of CHCl$_3$ (*740*), or a volatile anesthetic (*741*). Such observations are consistent with the reduced magnitude exciton chirality CD *Cotton* effects of the optically active bilirubinoids in DMSO (*738, 1008*), and *Navon*'s thesis that the ridge-tile conformation of bilirubin is retained in DMSO, with the propionic acids linked to the dipyrrinones *via* solvent molecules. *Boiadjiev*'s accomplishments, a vast collection of organic syntheses, stereochemistry, and metabolism projects published in over 60 peer-reviewed articles covering nearly 20 years, represent some of the most intriguing and difficult encountered in modern bilirubinoid chemistry.

Yet this work and all bile pigment research of the past 70 years, as in all of natural products and organic chemistry, depended on the even more difficult accomplishments of the previous 100 years – and those in earlier centuries. As *John of Salisbury* (*Johannes Parvus, ca.* 1115-1117, Anglo-Saxon Bishop of Chartres) reminded, when writing in Book 3 of his *Metalogicon* in 1159, when comparing the then 12th century (modern) scholars and the ancient scholars of Greece and Rome (*1009*):

Dicebat Bernardus Carnotensis nos esse quasi nanos, gigantium humeris insidentes, ut possimus plura eis et remotiora videre, non utique proprii acumine, aut eminentia corporis, sed quia in altum subvenimus et extolliumur magnitudine gigantae.

As translated into English by *Daniel Doyle McGarry* (*1010*):

Bernard of Chartres used to compare us to (puny) dwarfs perched on the shoulders of giants. He pointed out that we see more and farther than our predecessors, not because we have keener vision or greater height [literally: sweated profusely], but because we are lifted up and borne aloft by their gigantic stature.

10 Bilirubin. *Quod Erat Faciendum*

In current times, the complete proof of structure of a natural product may take weeks to months, less often days, occasionally a year. Yet, before the advent of modern organic spectroscopy techniques and powerful isolation-purification methods, structure proof tended to consume years, if not decades. The structure proof of bilirubin stands in a unique position by having spanned nearly two centuries, dating back before the second half of the 19th century, when the concept of organic structure began to be recognized and its practical importance understood by chemists, and reaching completion only recently. And, like the notable changes in spellings in the German quotations and the scientific phraseology used in both German and English over the past two centuries, so did the bilirubin structure proof witness the evolutionary changes in natural product structure proof and the birth of organic synthesis.

The history of the yellow pigment of bile stretches back millenia, with recorded attempts to isolate it apparently beginning late in the 18th century. From the early, primitive studies of the colorant(s) of bile and bile stones, the route to pigment structure (as we now understand the concept of organic structure) was rocky and arduous, full of pitfalls in the numerous and varied initial steps to isolate and later purify the pigment(s). Investigations of a "chemical nature" began to be described early in the 19th century, at a time coinciding with the first performance of *Haydn*'s *Kaiserhymne*[1] "*Gott erhalte Franz den Kaiser*" in Vienna's *altem Burgtheater*, the original *k.k. nächst der Burg* on February 12, 1797 on the occasion of the 29th birthday of the last Holy Roman Emperor of the House of Habsburg, *Franz II* (*Franz Josef Karl*).

Mammalian bile is known today to contain the yellow pigment (bilirubin) in the form of its sensitive conjugated metabolites, principally acyl glucuronides, *e.g.* bilirubin diglucuronide (Fig. 10.1); sensitive because the bilirubin glucuronides are easily hydrolyzed and intramolecularly transesterified, and because bilirubin glucuronides are readily oxidized by air to

---

[1] In 1841, *August Heinrich Hoffman von Fallersleben* set *Haydn*'s melody into the then revolutionary lyrics of *Das Lied der Deutschen* that became one of the symbols of the March 1848 revolution in the German states. With its opening line (*Deutschland, Deutschland über alles*) it was adopted as the national anthem of Germany in 1922, then banned in 1945 at the end of World War II. The third stanza became the national anthem of West Germany in 1952 and remains thus today in the reunified Germany.

D.A. Lightner, *Bilirubin: Jekyll and Hyde Pigment of Life*, Progress in the Chemistry of Organic Natural Products, Vol. 98, DOI 10.1007/978-3-7091-1637-1_10, © Springer-Verlag Wien 2013

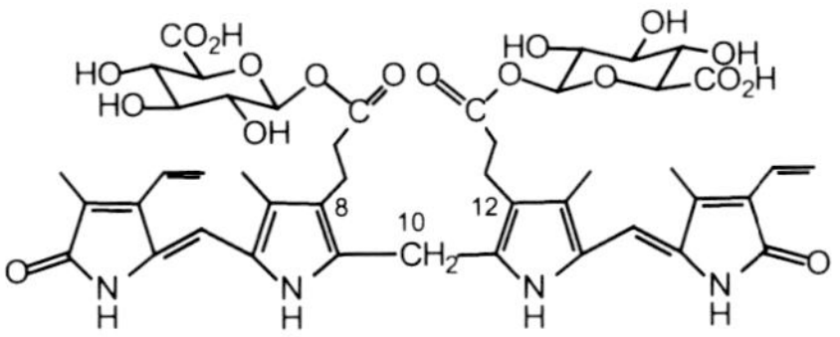

Fig. 10.1 Linear representation of bilirubin diglucuronide

the corresponding green biliverdin conjugates. The difficulties in isolating either pigment from bile or from gallstones (which were learned to be a more tractable source of bilirubin) exemplify not only the sensitivity of the pigments toward alteration during any manipulation, but also the fact that from the 19th century into the 20th there were no isolation methods that guaranteed sample purity, and significantly, no means to ascertain sample purity with reasonable certainty. A major stumbling block in structure elucidation was the widely held assumption that the yellow pigments of bile and jaundiced blood are identical, a view that was undermined by *van den Bergh's*[2]

---

[2] *Abraham Albert Hijmans van der Bergh* was born in Rotterdam on December 1, 1869 and died there in 1943. He became Dr. med. in 1896 at Leiden, studied in 1900 in Breslau with *Adalbert Czerny* and returned to Rotterdam as physician-in-chief at the city hospital at Coolsingle. In 1912, he became Professor of Medicine in Groningen and later, Professor in Utrecht, from which he retired in 1938. In 1918 he became the personal physician to Kaiser *Wilhelm II* until the latter's death in 1942.

In 1913 (*1011*), *van den Bergh* and *Snapper* applied the sensitive *Ehrlich* diazo color reaction [of aryldiazonium ions, typically diazotized sulfanilic acid, used to detect bilirubin by rapidly giving a reddish color in neutral solution and a bluish color in acid solution (see Section 4.1)] to serum, bile, urine, and other biological fluids. He described how to quantitate the bilirubin present by this incredibly sensitive color test. Later, in 1916 (*1012*), *van den Bergh* and *Müller* reported a seminal mistake. In his investigations of bilirubin reacting with aryldiazonium salts, *Ehrlich* carried out reactions on chloroform solutions of the pigment to which alcohol had been added in order to maintain homogeneity. Apparently for no other reason than *Ehrlich* had done it, subsequent investigators, including *van den Bergh*, always added alcohol to the fluid being analyzed for bilirubin. Except when *van den Bergh* accidentally forgot to add the alcohol to a sample of bile being analyzed and watched a very rapid *Ehrlich* reaction ensue, followed by a much slower reaction, which as accelerated upon addition of alcohol. The total bilirubin was the same as when alcohol had been added to a new sample of the same bile.

To examine whether alcohol was necessary for the reaction, *van den Bergh* prepared an aqueous alkaline solution of bilirubin, neutralized it with acid, and added the diazonium salt. He found a negative *Ehrlich* reaction but a positive reaction if alcohol was added first. Addition of alcohol was therefore absolutely necessary. In contrast, ordinary human bile at any dilution always gave the *Ehrlich* reaction without any addition of alcohol. The results suggested two types of bilirubin to *van den Bergh*: what he named "direct", a fast-reacting bilirubin requiring no alcohol accelerator, and "indirect", a bilirubin reacting so slowly that it requires an accelerator (*1012–1015*). The labels "direct" and "indirect" quickly caught on in medicine to differentiate different types of jaundice (*1016*):

The value of van den Bergh's test in the differentiation of obstructive from what may be termed functional and haemolytic jaundice . . .

10 Bilirubin. Quod Erat Faciendum

accidental discovery in 1913 (*1011*) that two kinds of bilirubin occur in mammals (*13, 484, 501, 1011–1024*) and conclusively disproved with the demonstration that the pigments in bile are mainly glucuronic acid esters (*1017–1024*).

However, significant advances in *detection* of bilirubin (and biliverdin) were made early in the 19th century by *Tiedemann* and *Gmelin*, who discovered (or rediscovered) that yellow bile underwent a spectrum of color changes, from green to blue, violet, red, and (finally) yellow in succession upon addition of HNO_3. The *Gmelin* reaction (or test) became the first "qualitative analysis" method of detecting bilirubin in the presence of other materials and pigments, though biliverdin also gave a positive *Gmelin* reaction. Although the concept of molecular structure was unknown at the time, each stage of the progression of colors, starting from the initial green of the biliverdin, indicated a different oxidation state of bilirubin. For quantitative analysis, combustion fit the bill, and through its refinements in the 19th and early 20th centuries it remained an absolutely essential analytical tool in structure analysis, though its usefulness depended essentially on the purity of the sample under investigations, which was then and remains a fundamental factor in the analysis of organic structure. Irrespective of sample purity, extracting empirical

---

The chief clinical value of the test is that by its use jaundice due to obstruction of the main bile ducts by carcinoma, hepatic cirrhosis, obstruction in the portal fissure, or gall stone in the common bile duct, can be clearly differentiated from jaundice of haemolytic origin or due to functional derangement of the liver cells. In this latter category are now included the various forms of haemolytic and acholuric jaundice, and also functional jaundice, such as catarrhal jaundice, toxic jaundice in infective diseases (typhoid fever, pneumonia), icterus neonatorum, etc.

Such is the practical value of the *van den Bergh* reaction that it was used in medicine for nearly 40 years without an understanding of its molecular basis (*501*):

Some assumed that the direct and indirect reactions are caused by two closely related forms of bilirubin; others thought that the phenomenon occurs because of the binding of bilirubin to different substances; and still others believed that the real cause was not associated with the bilirubin itself but with its surroundings, i.e., variation in the amount of accelerators or retarders, or differences in the colloids of the serum. Finally, a few authors assumed that the main cause is the variation in total bilirubin concentration.

And it remains in continued use today for quantitating two types of bilirubin using colorimetric spectrophotometry (*13, 484, 501*). It was not until nearly 40 years had elapsed since *van den Bergh*'s classification that *Cole, Lathe,* and *Billing* (*1017*) separated "direct" and "indirect" reacting bilirubin fractions from serum, bile, and urine in 1953, which were later shown to be bilirubin acyl glucuronides and intact bilirubin, respectively, by *Billing et al.* (*1018, 1019*) a few years later. At the same time, *Schmid* (*1020, 1021*) showed that the "direct" reacting bilirubin is actually its conjugate, mainly β–glucuronides, and it is intact bilirubin that requires the accelerator (as *Ehrlich* showed). *Talafant* came to the same conclusion (*1023, 1024*) at the same time.

Thus, with an accelerator added to a sample of bile, serum, or urine, the "indirect" quantitative *van den Bergh* reaction gives the total bilirubin present (intact and conjugated); whereas, in the absence of the accelerator only conjugated bilirubin is measured. The difference between the two measurements gives the amount of intact bilirubin present. Direct bilirubin refers to the conjugate, indirect bilirubin to the intact pigment.

formulas from combustion analyses required a knowledge of atomic mass weights, which were in dispute for much of the 19th century.

Bile is an incredibly complex conglomerate, of which bilirubin might be only a minor, though its most visible, component. There was no way to know that, however, in the 19th century. And so early attempts to isolate the yellow pigment from wet or dried bile depended almost exclusively on precipitation filtration, and dissolution. The common solvents used to effect the first and last were alcohol, ether, acetic acid, and mineral acids. Attempts were made to separate the pigment using ammonia, KOH, and NaOH, and precipitants such as barium, lead, silver, or calcium salts. Of course, none of these were specific for bilirubin or even the biliverdin that was inevitably formed during such manipulations, and it may have been that they generated an even more complicated mixture (array of pigments) than was originally present. The primitive separatory funnel did not then exist or had not yet come into extensive use; the notion of extraction using immicible solvents came late in the 19th century, and the investigators were thus left to precipitate, separate, wash, redissolve, and reprecipitate as they manipulated bile (and later, gallstones).

Numerous misunderstandings and pitfalls attended the 19th century bile pigment studies, *e.g. Berzelius'* belief that biliverdin and chlorophyll were the same substance; that the *Bilifulvin* (orange pigment) of bile and *Cholepyrrhin* (yellow pigment of gallstones) differed. Which they may well have, as there was no criterion of purity for either. As always, the *Schleim* of bile had to be dealt with, and a multiplicity of names for the pigments was introduced as new investigators began their attempted isolations. The names varied for bilirubin: *Harnfarbstoff* (from jaundiced urine), *Bilifulvin, Cholephäin, Cholophaeine, Gallenroth, Biliphäin* (from bile), *Cholepyrrhin*, and *Gallenbraun* (from gallstones). *Städeler* coined the name *Bilirubin*, and it stuck; while for biliverdin only *Cholechlorin, Gallengrün*, and *Berzelius'* name *Biliverdin* came into use, and the last name stuck. And the long disputed correlation between bilirubin and hematoidin, the yellow-orange pigment found in blood exudates, remained unresolved for decades, though the pigments were identical.

A great advance was achieved in the mid-1800s when $CHCl_3$ was introduced as a solvent for isolating bilirubin by washing or by extraction. Though $CHCl_3$ also extracted other material, it offered a reasonable way to achieve better separation of bilirubin than had been previously available. This advance coincided with the arrival of a new analytical tool, the then-named "spectroscopy analysis." Like almost any new analytical tool today, spectrum analysis began to be applied widely, including to bile pigments. Indeed, since this was a colorimetric spectroscopic analysis, it was ideal for the bile pigments, and so differences seen by the naked eye could be transcribed into a more scientific spectral description. The technique eventually morphed into UV-visible or electronic spectroscopy, and in the late 1800s and early 1900s there was doubtless a need for every productive investigator to have access to "spectrum analysis." Yet, real progress toward the *structure* elucidation of bile pigments had to await the emergence of organic structural chemistry in the mid-1800s that led to the ability to synthesize small molecules and represent structures on paper that were based on the atomic-scale molecular structure. The structure of the yellow

10 Bilirubin. Quod Erat Faciendum

pigment, for which many different names had been proposed during the 19th century and which is known today as *bilirubin* thanks to *Städeler*, began to be pried apart by *William Küster* at the beginning of the 20th century. *Küster* showed that bilirubin and biliverdin could be degraded oxidatively to a fragment identified (by synthesis) as hematinic acid imide. A principle of structure based on degradation to small components, structure assignment of the degradation products and reassembly of the natural product's structure from the separated component parts (by a *Gedanken* experiment) was thus on the way to becoming established.

It was not until the last years of the long reign (67 years and 355 days) of the Austrian Emperor *Franz Josef*, that chemical structure investigations were well underway by three major investigators: *Oskar Piloty, William Küster*, and *Hans Fischer*. These investigations were abruptly slowed when the assassination of *Franz Josef*'s heir apparent and nephew, *Franz Ferdinand*, precipitated World War I. *Piloty* lost his life early in the war, leaving only *Küster* and *Fischer* as the major investigators. *Küster* died in his laboratory in 1929, leaving *Fischer*, who had already demonstrated his prowess in pyrrole natural products synthesis by his proof of the structure of *Hämin* (hemin) using total synthesis – for which he was awarded the *Nobel* Prize in Chemistry in 1930. *Fischer*'s proof of the structure of bilirubin by total synthesis, a masterpiece of 1930s-era synthetic organic chemistry, was completed and published during the upheavals of World War II, the second of two world wars that coincided with the beginning and the end of *Fischer*'s illustrious career. Although a second total synthesis of bilirubin-IXα was published in 1972 by *Plieninger et al.* (*690*), the remainder of the 20th century following *Fisher*'s death on Easter Saturday, March 31, 1945, produced mainly refinements of bilirubin stereochemistry and conformation that were missing from the original *Fischer-Plieninger* constitutional structure. These refinements revealed the importance of intramolecular hydrogen bonding for the secondary structure of bilirubin and led to an improved understanding of the widely used phototherapy for physiological jaundice in newborn babies. The latter was a subject that might have appealed to *Fischer*, given his Dr. med. degree; it clearly led to an improved understanding of the pigment's metabolism and solution properties.

While much is known currently of the bilirubin structure, challenges remain in understanding this pigment fully, including the relationship of its structure and flexibility in animal metabolism, and in view of the recent detection of bilirubin in the Bird of Paradise plant (*Strelitzia reginae*) (*1025, 1026*), possibly in plant metabolism (Fig. 10.2). The typical metabolic route to bilirubin involves a biliverdin precursor, although the mechanism of biliverdin reduction is not understood. *Falk* suggested that reduction might involve prior addition of a sulfhydryl group to C-10 of biliverdin to give a yellow bilirubinoid (*1027*), and *Zhou, Scheer et al.* (*1028*) have reported bilirubinoid conjugates linked apparently *via* the *endo*-vinyl group and C-10 to cysteines of cyanobacteriochromes — the so-named *phycorubins* (*471, 472*). However, as of this writing no free bilirubin has been found in plants, except *Strelitzia reginae*. Still not understood are: the mechanisms of bilirubin toxicity; bilirubin transport across biological membranes; the structure of bilirubin when bound to albumin and proteins *in vivo*; the origin and role of bilirubin in plants; the significance of bilirubin as an antioxidant, a subject of interest dating back at least 25 years (*1029–1031*) that has provoked disputed explanations for the mechanism of biological action concerning a bilirubin-biliverdin cytoprotection redox cycle (*1032–1038*).

"The past is never dead. It's not even past."
— *William Faulkner (1951), Requiem for a Nun*, Act I, Scene III

Fig. 10.2 (*Left*) Bird of Paradise (*Strelitzia reginae*) flower, photographed under ambient light in a San Francisco, CA garden. The orange color of the flower is not bilirubin. (*Right*) Seeds of Bird of Paradise photographed in daylight with an obsolete, pre-euro 10-groschen Austrian coin for size comparison. The color of the orange tufts (arils) is due to bilirubin. Trituration of the arils in warm DMSO yields a yellow supernate that on HPLC with detection at 450 nm shows a single peak corresponding to bilirubin-IXα. (Photographs courtesy of Prof. *A.F. McDonagh*, UCSF)

11 Translations[1]

Numerous passages from the German scientific literature and from Latin and French appear scattered throughout the text and are translated into English in the following. Errors of translation are entirely the author's. Let the reader beware: a translation is typically less than the original; sometimes more.

11.1 Translations to Section 2

1 I can thus conclude that the bile is a true soap, composed of animal fat, and of the alkaline base of sea-salt; that it contains also a salt of the nature of sugar of milk [lactose], and a calcareous earth [lime] which is slightly ferruginous.

2 Most remarkable, as *Herr Seger* has already noticed, is that by the addition of spirit of nitre [dilute nitric acid], salt, and oil of vitriol [sulfuric acid] the bile turns green *almost instantly*.

3 "Nitric acid [nitric spirit, spirit of nitre] compels the bile more efficaciously to precipitate green clots. It transforms yellow or red (bile) into green. Sometimes with strong (acidic) water, straw-like strands appear, and the lot falls to the bottom. In other purer (acid) samples the bile is transformed into a bitter clot similar to a green resin."

4 Nitric acid shows the same effect, but instantaneously and progressing further, doubtless because it alone supplies the oxygen necessary for the color change. In this respect, all types of bile, from red-blooded animals as well as birds, amphibians, and fish undergo coloration with gradual addition of nitric acid, first green, then blue, then violet, then red, and of course all of these take place within a few seconds of addition of acid. Destruction of the red color occurs a few hours afterward, or after a few minutes with a larger excess of acid, whereupon the fluid appears yellow.

5 We have detected the pigment of bile in pathologic blood serum, chylus serum, and urine by means of these conditions, and perhaps it even has medical importance because it is the surest way to detect bile . . .

6 Dog bile, for example, is treated with sufficient nitric acid so that blue coloration occurs and is supersaturated then with potash. Then oil of vitriol [sulfuric acid] is added in a sufficient amount so that the tints of the *rainbow* are displayed; namely above the colorless oil of vitriol [sulfuric acid] is found a rose-red layer, above it a blue, then a green, and uppermost a yellow-green . . .

---

[1]Here and in all subsequent translations, the references cited in the original publication are re-numbered to correspond to the references of this book.

D.A. Lightner, *Bilirubin: Jekyll and Hyde Pigment of Life*, Progress in the Chemistry of Organic Natural Products, Vol. 98, DOI 10.1007/978-3-7091-1637-1_11, © Springer-Verlag Wien 2013

624   11 Translations

7   *Fourcroy* assumed already a colored component in bile, and although it appeared doubtful
in a few subsequent experiments whether a characteristic substance of the sort existed.
Inasmuch as the coloration of bile would be attributed to part of the bile pigment, so *Thenard*
[Traité de chimie, 4th ed, vol 4, p. 580] assumed that a characteristic yellow material exists in
almost all animals. He assumed that this pigment entirely constitutes the gallstones of oxen and
is contained in almost all of human (gallstones). Our experiments lead us to concur fully with
this view. That actually a very characteristic distinctively colored compound is present in the
bile of all animals is proven in the following:

    1)   If *Schleim* [mucus] were the colored principle, so must it remain if it is evaporated to
dryness. If bile is extracted with spirit of wine [ethyl alcohol], all of the color should
remain in the insoluble mucus residue; however, exactly the opposite ensues. If the
mucus is precipitated by acid, this drags down a somewhat larger amount of pigment
with it, though the largest amount of it remains dissolved.

    2)   All of the remaining compounds of bile still have a little color and the latter can
therefore still be regarded as nothing less than the colored principle.

    3)   The pigment shows, for example as it presents itself in bile, highly conspicuously; yet,
there are still no sufficiently known reactions that on the whole distinguish it from all
known substances. The following are the most important.
If one treats the yellow-brown bile of dogs with hydrochloric acid . . .

   The bile of animals possesses a different color depending on its type and individuality. It is
usually yellow-brown and only a little green from dogs; from oxen brownish green; from birds
mostly a vivid emerald or grass-green. Conclusions can probably be reached from this on the
more alkaline or more acidic conditions of bile, and on the more deoxidized or more oxidized
condition of the pigment itself.

8 4) We also investigated oxen bile stones, which can be ground easily to a mostly reddish-
brown powder. Heating in absolute spirit of wine [ethanol] easily colors it very pale, takes up
however only a little solid fat, which cannot be obtained crystalline. When the residue is acted
upon by ammonia, it assumes a somewhat more intense coloration, which was yellow, turned
grass-green in air, however, and became pale red with nitric acid and colorless with chlorine.

   The greatest part of the powder remained undissolved but was dissolved completely by
continuous digestion with potash, with a few flakes of potassium phosphate remaining. The
solution was initially yellow-brown and also became greenish brown overnight. With nitric
acid it produced the color changes noted above. With hydrochloric acid it gave a copious
precipitate of dark green flakes, and after this had completely settled, the initially still green
supernatant liquid became pale yellow. After drying, the thusly precipitated green flakes gave a
pale red solution with nitric acid that soon became yellow. They dissolved completely with an
emerald green color in conc. hydrochloric acid. The hydrochloric acid solution did not become
cloudy with added water, turned yellow with ammonia, red with nitric acid, and colorless with
chlorine. The green flakes that precipitated from the posash solution by hydrochloric acid also
dissolved in ammonia with a grass-green color. The pigment of bile appears to be made soluble
in hydrochloric acid and in ammonia after oxidation, which it undergoes in potash solution.

   According to these experiments, we might regard the gallstones that we studied as nearly
pure pigment of bile, which was co-mixed with only a small amount of fat and calcium salts,
possibly also some *Schleim*, and because of its nitrogen content, we might first suppose this
pigment is similar to indigo.

9 My analysis of the bile in question is not nearly ready. On June 22 I was afflicted by an oncom-
ing attack of gout, which however was relieved easily and without consequences. I had, however, to go into the country and drink the waters of Marienbad. Owing to my official functions
I had to be in the city twice a week, and these days would like a little time to pursue experiments. However in the warm season, the damned flies always drowned themselves in my solutions, regardless of all attempts to deter them. As one with years of bad memories I was
prompted to leave everything behind, until October when I am settled in my apartment in the
city again and can thus deal with it again on a daily basis. *Demarçay's* transformation of bile

11.1 Translations to Section 2

into taurine and *Gallenharz* [bile resin] is entirely correct. I have also obtained cholic acid according to my method. It works better with carbonated potash [potassium carbonate] than with caustic potash [potassium hydroxide]. His choleic acid is an artifact that you can never get twice. It contains at least four different organic substances, since the inorganic acid, with which it is precipitated, has been separated. Choleic acid contains two resinous acids that are also contained in choleic acid and a neutral resin. The main component of bile is my old bile pigment, which I will name *Bilin*. (*Thénard's* picromel, *Gmelin's Gallenzucker* [sugar of bile]), which is not acidic and which metamorphoses with the greatest ease so that it is extremely difficult to obtain pure. It is soluble in water and alcohol in all proportions. However, bile also contains acid components that have cost me a lot of trouble to separate, and although I have obtained at least four different components, I cannot say at this moment which are metamorphosis products and which are not. From old *Bilis bubula spissasta* [ox bile], the acid components wherein the bilin is largely metamorphosed can easily be separated, and I became acquainted with them. However, they are difficult to draw out from fresh bile because they are always masked as bound to bilin. None resemble the crystalline substance that you were so good to send me. It is a matter of how it behaves toward bases that I will put to the test.

10   *Demarçay's* experiments on bile led me to a revision of the analysis of bile. You will recall that in my chemistry the idea was expressed that *Gmelin's* constituents of bile are all metamorphosed, which now *Demarçay* has proven. But *Demarçay's* new acid, *choleïque* acid, is also a metamorphosed product. It was clear from his discussion because the new acid is not separable from bile by acetic acid, but separates well from its components with alkalis. I was fortunate that the pigment of bile precipitated out and that it was obtained pure. The bases in bile separated out, and an intensively bitter acid was obtained that was soluble in water and alcohol in all proportions and which was changed passably well into *Demarçay's choleïque* acid when dissolved in water by sulfuric acid or hydrochloric acid. It has the characteristic of dissolving in fat to a higher degree than in soap, was soluble in ether, was never separable from fat. The compounds are soluble in all bases presently attempted, even silver hydroxide. As soon as I am finished with my experiments, I will communicate my work in your journal [the *Annalen*]. I call the new acid *acidum bilicum* [bile acid]. In inspissated bile from an apothecary, I found a new crystalline acid as the main component. However, I don't know yet whether it is contained in all old bile extracts. It appears to be found in the fresh bile acid extracts.

11   A fifth (*Demarçay*) has done a lot of work on bile. His results are still too uncertain to be communicated, even though he has been working on it for six months, alone it appears; so, all assumptions up to now have been overturned.

12   I'm now going to describe the characteristic acid of bile, which I call *Choleinsäure* (from χολη, gall) [choleic acid], and its three decomposition products: the solid, nitrogen-free substance which I call *Choloidinsäure* (from χολειδηζ, gall-like) [cholidinic acid]; *Taurin* [taurine]; and the crystallizable, ether-soluble acid for which I have retained the name *Cholsäure* [cholic acid], for it is, I believe the same compound that was described under this name.

13   The question now arises: Are all these substances found in bile, or have they been produced under the influence of the reagents on one or several components of bile, whose composition is easily altered? More than 20 years ago when I investigated the chemical make-up of a few animal materials, I found that they undergo changes with certain reagents and new products arose, and I hold especially applicable heating with water, ether, or alcohol as a few since the latter two produced a fat with a characteristic repugnant odor from albumin, fibrin, gelatin among others (see the *Jahresbericht* (1826) p. 277). These ideas have been disputed by *Chevreul*, and *Leopold Gmelin* assumed *Chevreul's* view to be the more correct.

As everyone knows, bile freed by acids, preferably by sulfuric acid, from gallbladder *Schleim* [mucus], is decomposed in a manner such that at a certain concentration the acid precipitates as a resinous substance that is somewhat soluble in water and completely soluble in alcohol. Only flesh extract and salts are thereby left behind in the acidic liquid. From analysis, which I employed in this way more than 20 years ago for bile, I found that bile has an entirely simple composition. It is, namely, the albuminous components of blood that would be changed into characteristic substances, which in every way, would have the property of being

626   11   Translations

precipitated by mineral acids but not glacial acetic acid. The fluid in which this substance was dissolved would be nearly of the same nature wherein albumin and fibrin are dissolved in blood. This substance can be obtained back from the compound from sulfuric acid by digestion with carbonated baryta [barium carbonate], whereby it would, as with its former properties, be soluble again in water. I named it *Gallenstoff*. The experiments of *Gmelin* and *Tiedemann* were repeated. They found that sulfuric acid precipitated *Gallenstoff*; however, the solution from digestion with carbonated baryta [barium carbonate] contained barium, and that *Gallenharz* remained behind undissolved in the excess carbonated baryta [barium carbonate] used. They concluded that sulfuric acid had precipitated glacial acetic acid along with the resin (a conclusion certainly entirely unfounded), that barium acetate had dissolved, and that the substance that I named *Gallenstoff* is a composite of *Gallenharz*, pigment, *Gallenzucker*, *Asparagin*, *Gallenfett*, margaric acid, oleic acid, *etc.*, and barium acetate. This conclusion cannot be correct because, as was found in my work, the composition of bile is not so simple that it can be stated with certainty that seven different organic substances are not combined with each other in order to produce a single substance of such definite character that forms a resin with sulfuric acid and other mineral acids and is not precipitated by glacial acetic acid; and as oleic acid and margaric acid should follow (immediately) in such a compound because their compound with baryta [barium oxide] is insoluble. This observation is correct as a far as the barium compound is concerned, not only for baryta [barium oxide] but also calcareous earth [calcium oxide], and lead oxide[1] which removes the sulfuric acid, binds to the substance, leaves everything behind, and dissolves in water. When digestion with an excess of base is not continued too long, whereby an insoluble compound forms, it is not the acid but the animal material that dissolves in the base. In these cases it is similar to many other organic substances, especially glycyrrhizic acid, which with sulfuric acid and acids in general forms resinous combinations, and by its decomposition with a carbonate base, *e.g.* carbonated baryta [barium carbonate], baryta [barium oxide] dissolves and is soluble in water. If the similarity in taste between *Gallenstoff* (bile substance) and glycyrrhizic acid still exists then the correspondence is still more striking.

If *Asparagin* is present, dissolved in bile, the substance would remain behind undissolved with *Schleim* [mucus]. If dried bile is dissolved in alcohol, that does not happen, however. *Gmelin* and *Tiedemann* noted there is not a single case when bile treated with glacial acetic acid and evaporated to dryness can be dissolved in alcohol; therefore, the affinity of the acid must have dissolved the linkage that one might think unites these substances in combination. It is fairly certain that *Asparagin* is not found in bile prior to the influence of certain reagents; however, at the same time if *Asparagin* arises from any one component of bile, other substances must also be formed from it and in conclusion cannot be shown to be contained in bile.

However, I note herewith that if the composition of bile were simpler than it would appear from the previous experiments, it therefore cannot be disputed that the most interesting aspect of our knowledge of bile is the connection with the principal transformations that it undergoes by reagents outside of the body. By means of which we are able to anticipate some of the changes that it undergoes in the living body by the digestion process.

---

[1] *Lychnell* made a few attempts to ascertain the difference in analytical results arising from the dissimilar treatment of bile. In one of these experiments sulfuric acid bile material [*Gallenstoff*] was dissolved in alcohol and digested with carbonated baryta [barium carbonate] until the liquid was neutral. After evaporation the solution left behind a substance that is completely soluble in water and similar to the bile that left carbonated baryta [barium carbonate] behind upon combustion. The same happened with lead carbonate; however the solution was not neutral. Dilution with water precipitated sulfuric acid *Gallenstoff*, and after filtration and evaporation the same substance remained, as previously, but contained no lead oxide. When potassium carbonate was added to a solution of the acid compound in alcohol, potassium sulfate and regenerated bile resulted. I hope in the future to be able to communicate the results of *Lychnell's* experiments.

11.1 Translations to Section 2

14 *VII. Pigments.* As everyone knows, the bile from all internal compartments of the gallbladder is colored yellow, liver disease gets it yellow color, *etc.*, from absorbed bile. This is due to a characteristic pigment in enervated bile, for which no extraction method has yet been discovered, but whose existence can nevertheless be proven. *Thénard* believed he had found that this pigment constitutes the bulk of ordinary gallstones from oxen. When it occurs in them, it forms a yellow-brown, easily pulverized mass. Boiling water removes a little non-crystalline fat that is colored pale yellow. Caustic ammonia removes more from it, and the fluid is yellow, changes to grass-green in air, becomes light red with nitric acid, and loses color with chlorine. It dissolves best in aq. potash. The extract is yellow-brown and gradually becomes greenish. With hydrochloric acid, an emerald green precipitate forms that with nitric acid dissolves with a rose-red color that gradually goes over to yellow. The precipitate from hydrochloric acid is easily dissolved by caustic ammonia.[2] These results show up with bile. If dog bile is mixed with hydrochloric acid in an inverted glass tube over quicksilver [mercury], then the color does not change. However if oxygen is introduced, a portion of it is absorbed, and the liquid turns green. In the same way, all bile mixed with acid turns green during evaporation in air. Every type of bile, mixed in small portions with nitric acid turns green at first, then blue, violet, and afterwards, red, and certainly after a few seconds. After a longer time or with more acid it finally becomes yellow. The presence of bile in serum and in urine during sickness can be detected by this reaction. If dog bile, turned green by nitric acid, is saturated with potash, it then becomes brownish-yellow, into greenish. If it is blue or violet, then when sulfuric acid is added to the yellow-green alkaline liquid, the first color is produced again. If blue-colored bile produced by added nitric acid is saturated with lime [calcium oxide] and conc. sulfuric acid is added without stirring, then from above the acid, which has sunk to the bottom, different colors are produced. Namely, the one lying first above the acid is red, then blue, then green, and finally yellow-green.

15 *9) Pigment.* The greenish color of oxen bile belongs, in all probability, to a characteristic substance that is dissolved in bile along with the remaining material, and which certainly still cannot be separated in an analytical way with certainty. However, it is occasionally deposited in bile in such great amounts during illness that it forms a type of gallstone from which, even in the isolated form, one can come to know with respect to its characteristic reactions. It is the same material which in jaundice colors a great part of the body yellow, such as the skin and the whites of the eyes, *inter alia*, and is the cause of the yellow color that has been found in the gallbladder and in the surrounding parts after death. *Thénard* first called attention to that. He found it in human bile precipitated out in the form of a yellow powder, which he named *matière jaune de la bile* [yellow material of bile] and which he showed to be the same substance that is found in the gallstones of oxen and also has been found in an accumulated mass of 1.5 pounds weight in the bile duct of an elephant that died in *le Jardin du Roi* [in Paris].

To explain the nature of this substance, I was led to *Gmelin's* investigation of ox bile stones, of which it is the main component. It can be ground into a bright red-brown powder. Boiling alcohol extracted only a little fat and was colored yellow. Caustic ammonia dissolves a small amount of it. The best solvent for it was, however, aq. caustic potash [potassium hydroxide]. The solution obtained by digestion was bright yellow and became greenish-brown by absorption of oxygen from air. Strongly saturated with nitric acid, the solution shows a reaction that is characteristic for the pigment of bile. If too much acid is not added at once and is not well mixed, the liquid is at first green, then blue, violet, and finally red. This color change takes place within a few seconds. After a while the red color also disappears, the liquid is yellow, and the characteristics of the pigment have now changed completely. It requires only a very small amount of material in order to make this reaction distinctly noticeable, and it takes place not only with bile but also with blood serum, chyle serum, urine, and other fluids that had assumed a yellow color due to jaundice. It is therefore the surest means of discovering the presence of

---

[2] The same results have been noted by *Lassaigne* and *Leuret* from the yellow pigment in the skin and fluids of children who were born with icterus (jaundice). Journ. de Ch. Med. II p 264.

bile or its pigments. A solution of the pigment in aq. potash precipitates into thick dark green flakes, and afterward the liquid shows only a weak green tinge. After washing and drying, the precipitated pigment dissolves in nitric acid with a red color, without blue or violet in between, and the red color quickly goes over to yellow. A dark green precipitate from treatment with hydrochloric acid dissolves very easily, with a grass-green color, in ammonia as well as in aq. potash. The cause of the color changes in bile that often proceed from yellow to brown and green, appears to depend on oxidation of the pigment, whereby it goes over from yellow to green and becomes more easily soluble in alkali. Bile, treated with acid and left exposed to air becomes completely green after a few days. Gmelin mixed dog bile, which is yellow-brown, with hydrochloric acid in a glass tube sealed at one end with the other end over Hg. In this way, before air was introduced, the color of the mixture remained unchanged; however, after oxygen introduced it turned green, at first at the point of contact with the gas and later through and through, during which the bile had absorbed half its volume with the gas. Chlorine brings about the same display of colors as nitric acid but less vividly; the blue is scarcely noticeable but the colors proceed likewise from green to red, and an excess of chlorine destroys the color of bile completely and bleaches the same with formation of a white turbidity.

16   This view became the prevailing one, and all analyses undertaken later emanate from the idea that bile is composed principally of picromel and *Gallenharz* (bile resin).

17   As was mentioned, analysis of bile can proceed in two different ways, namely with sulfuric acid or with lead salts, but metamorphosis must be avoided as much as possible, with alterations as the precautions applied so far are employed.

18   1. Analysis of bile by sulfuric acid – Oxen bile was evaporated over sulfuric acid in a water bath or in an open chamber in which the temperature of the chamber was kept at 100–110°C so that the mass was thereby dried, and could be pulverized. Then it was covered with ether. If the ether was wet, the pulverized dried bile took up water and "melted" together [merged]. The ether extracts all the fat that is not bound up with alkali as a soap. After digesting two or three times with ether, the powder was dissolved in anhydrous alcohol, which left behind *Schleim* [mucus], sodium chloride, and other things as alcohol-insoluble salts and animal material. In comparison, a combination [compound] of the bitter components of bile, oleic acid, and magaric [heptadecanoic] acid, the pigment of bile in a similar association, *etc.*, all dissolved with alkali. The solution obtained was filtered, and the undissolved material was treated first with anhydrous alcohol, which was then added to the filtered solution, then with 85% alcohol, which dissolves certain substances, and which was itself taken up. The solution in anhydrous alcohol was then mixed, in small portions and with shaking, with a solution of barium chloride in water until a dark green precipitate was formed, which was filtered off and washed with alcohol which however should was not be anhydrous. Then baryta water [aq. barium hydroxide] was added dropwise into the filtered solution. The precipitate that formed was initially dark grey, but it changed to green after a few moments. The baryta water was added until the solution became turbid. Soon the precipitate was no longer green but at first brownish-yellow and finally only yellowish, whereupon the solution had for the most part lost its color and was still inclined only toward yellow. The precipitate was filtered and washed with 84% alcohol.

19   *Gmelin* also found these components, namely *Gallenharz* [bile resin], *Gallenzucker* [bile sugar], in addition to taurine, cholic acid, cholesterol, oleic acid, margaric [heptadecanoic] acid, pigment, meat extract, a substance similar to that extracted from urine, a material analogous to gliadin [vegetable glue – glycoprotein], cheesy material, ptyalin [saliva material/amylase], albumin, *Schleim* [mucus], sodium carbonate, sodium acetate, sodium and potassium oleate, margarate, cholate, sulfate and phosphate, sodium chloride, and calcium phosphate.

20   *Demarçay* denies entirely the existence of a *Gallenzucker* and assumes that *Gmelin's Gallenzucker* and *Thénard's picromel* are identical to choleic acid.

        Accordingly, in this way we have gone in a circle during a period of more than 30 years with respect to the main concept of the nature of bile, admittedly without significantly increasing our knowledge. We stand at the same point and, regardless of all practical knowledge which we have obtained by cited work, it is still not possible without new investigations to

11.1 Translations to Section 2 629

give an only somewhat correct concept of the composition of bile. I now intend to discuss it according to the investigations which I recently undertook with oxen bile.

21 The first precipitate with barium chloride contains material that gives bile its green color and is bound up with barium oxide. I am naming it *Biliverdin* (from *Bilis*, bile, and *verdire*, becoming green). The other, or the precipitate with aq. baryta water [barium hydroxide], contains beside biliverdin a reddish-yellow pigment which I am naming *Bilifulvin* (from *Bilis*, bile and *fulvus*, reddish-yellow), an extractable substance and a characteristic, nitrogen containing animal material, to which I will return later.

22 6. *Biliverdin*, bile-green. The still damp precipitate was overlaid with dilute hydrochloric acid, which extracted barium oxide and left biliverdin behind. It is mixed with only a little fat, which is extracted into ether, in which, however, also a little component of biliverdin is dissolved at the same time. That which was left behind was treated with cold anhydrous alcohol, which thereupon became a green-brown color and left behind a green residue that was insoluble in cold alcohol. The anhydrous alcoholic extract, produced by spontaneous digestion, left behind biliverdin in the form of a nearly black-brown earth-like compound. By evaporation with warming, it formed a shiny translucent dark green film.

23 These properties of biliverdin coincide so totally with those of chlorophyll that I regard them to be identical. I have obtained biliverdin from various biles in all three modifications of chlorophyll. What is now said is valid naturally only for biliverdin from ox bile, possibly even for the bile of all herbivores. However, in the bile of carnivores it possesses quite different properties, or it is bound up therein with yet a different pigment, which up to now has not been separated. Since I have not yet had the opportunity to do a few experiments with it, I must thus give an account according to the results of others.

24 7 *Bilifulvin* is the name I have given to a still problematic, crystalline, reddish-yellow substance from *Bilis bubula spissata* [condensed bovine bile], a substance that I have not yet had the opportunity to study appropriately. After an alcoholic solution of bile was precipitated with barium chloride, a few added drops of baryta water [aq. barium hydroxide] gives a new precipitate which at first is brown. However, its color changes and it becomes green, whereupon precipitates brown, and in the end brownish-yellow. If it is taken onto a filter and washed, at first with alcohol and then with water, a large part is dissolved, and barium-biliverdin remains behind on the filter.

The filtrate, treated with sugar of lead [lead(II) acetate], gives a dark green precipitate and becomes reddish-yellow. Now it is precipitated with vinegar of lead [aq. solution of basic lead acetate]; however, it cannot be precipitated such that its color is lost entirely. When the precipitate has sunk to the bottom, it appears to be a mixture of two precipitates, one of which is reddish-yellow and heavy. It lies lowermost. Above it sits an only slightly yellower and lighter precipitate that is still not mechanically separable even with care. When it is filtered off, washed, and decomposed with sulfuric acid, a yellow solution results that with evaporation leaves behind a reddish-brown extract. If that is dissolved in alcohol and the solution is allowed to evaporate freely, then at first small reddish-yellow crystals are formed, about which a reddish-brown extract then forms during continued evaporation. I have named the crystals *Bilifulvin*.

25 The green material precipitated from hydrochloric acid by water behaves toward alcohol, ether, hydrochloric acid, sulfuric acid, glacial acetic acid, and alkali completely like *Blattgrün* of the first modification. However, when precipitated by carbonated lime [calcium carbonate], it behaves entirely like the *Blattgrün* of the second modification. And if the precipitate from treatment of the alkali solution with hydrochloric acid is heated in alcohol, a dark green powder remains behind. The last is still green in alcohol, in which it is difficultly soluble. It is yellow in hydrochloric acid, and therefore also like *Blattgrün* contains the third modification. However, the green obtained from gallstones showed a difference in that with nitric acid it formed a red liquid that was not present with the biliverdin separated from ox bile. All three modifications obtained from gallstones gave a red liquid with nitric acid. I had kept a small leftover portion of the *Blattgrün* of the second modification. I overlayed this with pure nitric acid, and it formed a deep-red liquor, for only a moment; however, it became yellow as nitrogen oxide gas was liberated. The red remains much longer with the green from gallstones.

630   11   Translations

26   I have no interest in investigating anything that hydrochloric acid does not precipitate since I intend only to compare the green material to the biliverdin that in the presence of air is taken up into alkali, from the flaming yellow disease-producing product from the bile of various animals.

27   These experiments show that the yellow compound which occurs in this type of gallstone is transformed by contact with air and under the influence of alkalis as well as by acids. They show that *Blattgrün* is formed by this transformation, formed so to say by an artificial process. This is a new example that we add to several already known, that such material which living nature brings forth also brings forth other material artificially by metamorphosis. I hereby regard it as demonstrated that biliverdin and *Blattgrün* are essentially identical compounds, and they are a product of metamorphosis of the characteristic pigment of bile, which has been transformed during analysis. The pigment deserves a characteristic name: it can be called *Cholepyrrhin* (from $\chi o\lambda\eta$, gall, bile, and $\pi\upsilon\rho\rho o\zeta$, *brandgelb*, flaming yellow)

28   The nature of cholepyrrhin in the unaltered state still remains to be determined, and the products, in addition to biliverdin, that are formed during its metamorphosis also remain to be investigated. Although in the experiments above with ox bile I have not been able to extract the same yellow compound contained in gallstones, it is clear that it is still present in it. This is because fresh bile filtered from *Schleim* [mucus] appears barely noticeably green but is yellow or brownish-yellow. In the early stages of evaporation it is dark and green, during which biliverdin is thus formed and cholepyrrhin has metamorphosed. Biliverdin is therefore a product of the metamorphosis of bile; likewise taurine, cholic acid, *etc.*

29   ... It consists of a green, easily pulverized, resinous mass which is insoluble in water but readily soluble in alcohol. It is barely soluble in ether, which is even more difficult if the ether contains a little alcohol. It is odorless and has a somewhat bitter taste. It is insoluble in hydrochloric acid and sulfuric acid but easily soluble in potash and ammonia, whereby the green color is transformed into yellow. The green color also fades with heating and goes over into yellow. With warm aq. potash, it evolves ammonia. If this does not come from a different compound, then the bile pigment cannot be identical to *Blattgrün*, as *Berzelius* thinks.

30   Finally, I will make a few more remarks on the pigment of bile. Bile, of course, becomes green little by little with acids, if air is allowed to enter in at the same time. The color change of the bile is, however, instantaneous with nitric acid. At first it turns green, the blue, violet, and finally yellow. Later on, nitric acid destroys the pigment. In my experience however, barely adding air little by little causes the same color changes. If an alcoholic solution of bile is set out into air for a longer time, so that it is green at first, it goes over by and by into a red color, however, without doubt these color changes thus occur by a progressive oxidation of the pigment. The pigment prepared by *Berzelius* from bile with the assistance of barium contains no nitrogen. However, that of mine prepared in the way described above is, however, nitrogen-containing. This subject might also be recommended for further examination.

31   Fresh urine was filtered to remove *Schleim* [mucus] and, if necessary, the precipitated uric acid. It is thereupon treated with barium chloride. The bright green precipitate obtained was then washed with water, filtered and, afterward, the bile pigment was separated from the same by two different methods.

32   ... Cholepyrrhin exhibits all blends of colors, from saffron yellow through dark brown to blackish green, and although it can appear in nearly all tissues, it is found mainly in the constituent principles of the bile duct. Every obstruction of bile in its excretory path depends initially on an infiltration of liver cells situated around the bile duct, a partial icterus, so that in all cases where general jaundice depends on biliary obstruction, icterus of the liver precedes icterus of the body. Infiltration of liver cells with cholepyrrhin is at first evenly diffuse. Pigment very soon collects as a small insoluble brownish or greenish substance that lies very abundantly in clusters near the nucleus.

33   It is known that nitric acid is a reagent much used to detect the presence of bile in any fluid. It is stated that such fluids become colored thereby, at first green, then blue, violet, red, and finally yellow. This is entirely correct in most cases; however, nitric acid does not in *all* cases produce every color change in the presence of bile constituents. To begin with, it must be kept in mind that

11.1 Translations to Section 2   631

that reaction is not induced by the characteristic intrinsic compound of bile but by the *Gallenbraun* which *Simon* named *Biliphäin* and which *Berzelius*, in contrast, named *Cholepyrrhin*. When no color change can be produced by nitric acid, that is due only to the absence of *Gallenbraun*, not, however, to the other intrinsic component of bile.

However, that color change which nitric acid brings about in fluids that contain *Gallenbraun* remains therefore at least a true indication of the presence of this material when it actually occurs in each case.

34

Bile Pigment.
Chemical Behavior.

Properties. This material belongs, like so many pigments, to the chemicals that still have been chemically little investigated. This is due in part to the ability to procure the material in only very small quantities, and in part to its great changeability. This causes it to be found in various modifications, not only in animal organisms but also already changed during the simplest chemical treatment. The most characteristic modification, which also appears to be the original substance of the bile pigments from higher animals, is the so-called *Gallenbraun*, cholepyrrhin (*Berzelius*), *Biliphäin* (*Fz. Simon*). It forms a red-brown amorphous powder that is tasteless and odorless, is not soluble in water and very slightly soluble in ether, but more soluble in alcohol. In the last, it becomes yellow. It is more easily soluble in caustic potash [potassium hydroxide] than in caustic ammonia, with the bright yellow alkaline solutions gradually becoming greenish-brown in air. It is this modification of the bile pigment on which the well-known spectrum of color changes in animal fluids depends. With gradual addition of nitric acid (especially if it contains some nitrous acid, as per *Heintz*), the yellow solution of these pigments turns green at first, then blue (which however is scarcely noticeable due to its rapid change to violet), and then red. After a longer time, the red color goes over into yellow again; however, at that point the pigment is completely transformed. It [the original pigment] is precipitated green from aqueous potash by hydrochloric acid. This precipitate dissolves in nitric acid with a red color and appears thereby to convert completely into the green modification. The pigment contained in fresh bile becomes colored green by added acids. *Gmelin* found that this coloration does not occur without the addition of oxygen; therefore, it is highly likely that most of each color change rests in a gradual oxidation. Chlorine gas has the same effect on the pigment as nitric acid, only somewhat faster. Larger amounts of chlorine bleach the pigment completely and precipitate white flakes.

The brown pigment is very much inclined to bind to bases, and of course not only with alkalies but also with metal oxides and alkaline earths. With the last it also forms insoluble compounds, which is why the material is often taken to be insoluble.

Gallengrün, Biliverdin (from *Berzelius*) is a dark green amorphous substance, tasteless and odorless, insoluble in water, only a little soluble in alcohol, but soluble in ether with a red color. It is soluble in fat, hydrochloric acid, and sulfuric acid with a green color; soluble in glacial acetic acid and alkalies with a yellowish-red color. The compound does not melt with heating and decomposes while evolving no noticeable ammonia but leaves behind a little carbon. *Berzelius* took this compound to be completely identical to the chlorophyll from leaves and also believed that he had found all three modifications of chlorophyll in different biles. This green pigment no longer has the ability to undergo color changes with nitric acid; however, now and then greenish bile pigments are found that still possess that property. Mostly, after treatment with alkalies or acids, the pigment of bile shows characteristics different form the original compound, in part because it enters into different compounds with this material alone, partly however because it is also so easily modified.

On this basis, the reports on the properties of this material differ; see perhaps *Berzelius*[1] [71], *Scherer*[2] [92], *Hein*[3] [105], *Platner*[4] [87] and *Andre*.

Berzelius also found alcohol-soluble material precipitating from bile in small reddish-yellow crystals (which he named *Bilifulvin*). I have the same only in solution but have not been able to obtain it in a solid form. I often found it conspicuously precipitated in bile by neutral and basic acetate of lead oxide so that it therefore is not precipitated by these metal salts, or appears rather to be redissolved in an excess of the basic salt.

632 11 Translations

Composition. Given our lack of familiarity with pure unaltered bile pigments, it is not surprising that its elemental composition is still not known. *Scherer* and *Hein* have investigated bile pigments, but from their analyses it appears that they had very different substances at hand, and *Scherer* in particular showed that the bile pigment lost much carbon and water by treatment with air, alkalies, and acids. About 7–9% nitrogen was found in the bile pigments.

Preparation. As recommended earlier, usually for the preparation of bile pigment, preferably bile concretions [gallstones] available as such are extracted with water and ether. The residue as a rule does not have the property cited above as being able to dissolve in alcohol since it is insoluble when bound up with lime [calcium oxide] (as *Bramson*[1] [*Bramson*, Zeitschr. f. rat. Med. Bd. 4, S. 193–208] has stated correctly, and any unbiased observer can easily convince himself) that the largest component in such concretions consists of cholesterol.

Bramson's method of investigation, which I have often repeated, appears to me to leave absolutely no doubt as to the correctness of his views. Moreover, it also agrees with the gallstone analyses of *Schmid* [*Schmid*, Arch. der Pharm. Bd. 42, S. 291–293] and *Wackenroder* [*Wackenroder*, ebendas, S. 294–296].

Berzelius prepared biliverdin from cattle bile, an alcoholic extract of which he precipitated with $BaCl_2$. The precipitate was washed first with alcohol, then with water and decomposed by hydrochloric acid, which extracted the barium. The residue is freed of fat using ether and then dissolved in alcohol.

Platner precipitated the bile pigment from bile by digesting the bile with stannous oxide hydrate [= $Sn(OH)_2$] to give a bright green precipitate which, after washing well with water, is shaken with sulfuric acid-containing spirit of wine [ethyl alcohol]. The pigment is precipitated in green flakes from the filtered green solution by adding water.

Scherer separated the pigment from bile pigment-containing urine using barium chloride, producing it in two ways: either (1) by decomposing the barium compound with sodium carbonate and precipitating the pigment from the sodium solution by hydrochloric acid, where it was then purified by dissolving it in ether-containing alcohol and washed with water, *etc.*; or (2) the barium compound was extracted with hydrochloric acid-containing alcohol, the solution was evaporated, extracted with water, and then treated as above.

Assay. If the presence of bile pigment in a liquid is not too scant, then nitric acid (especially if it contains some nitrous acid) gives the very characteristic display of colors mentioned above. With small amounts of pigment, however, nitric acid often correctly gives no characteristic reaction such as when the pigment is already partly modified. *Schwertfeger* [*Schwertfeger*, Jahrb. f. prakt. Pharm. Bd 9, S. 375] recommends in such cases precipitating the liquid with basic acetylated lead oxide, and extracting the precipitate with sulfuric acid-containing alcohol, which becomes green due to the presence of pigments. *Heller* [*Heller*, Arch. f. phys. u. pathol. Ch. Bd. 2, S. 95] advises adding soluble albumin to the liquid to be investigated, assuming that it does not already contain it and then precipitating with excess nitric acid. The coagulated protein is thus colored bluish or greenish-blue by the pigment. According to *Heller* a red cake forms above the surface of the liquid by careful addition of ammonia to urine that already contains unaltered pigment.

Physiological Behavior.

Occurrence. Bile pigments are usually present as dissolved in fresh bile, though often only suspended; they nearly always form the nucleus of gallstones. At times one also finds gnarled, knobby concretions in the gallbladder and in the bile ducts, which almost always originate from bile pigments. This bile pigment has been found, not coincidentally, in the bile of humans and cattle but also in other meat- and plant-eating animals, however in different modifications, as the different colors of bile show not only different genera but also different individuals of the same species. Thus, dog bile is yellow-brown, cattle bile is brownish-green, the bile of birds, fish, and amphibians is mostly emerald green.

11.1 Translations to Section 2   633

35 Indeed, humors that have been dried and roasted too much condense into the shape of stones, to a point where they cannot be crushed/dissolved any more.

36 According to *Gmelin's* experiments, this transformation happens at the expense of oxygen in air. In order to obtain *Biliphäin* in pure condition from it, each dissolution and precipitation must happen in an oxygen-free space.

37   After this apparatus was assembled, the hydrogen gas generation was begun with enough gas generated so that it could be assumed that oxygen [was replaced] and could no longer be present in the flask containing the sodium carbonate solution. The flask was opened and the red *Gallenbraun* contained in it was washed rapidly by shaking with hydrochloric acid and water. After all this the stopper was immediately placed on the flask, and hydrogen gas was allowed to flow through the apparatus for several more hours until no more oxygen could be found, even in the bell jar. Then the sodium carbonate solution was heated for a long time under a continuous stream of gas. After the dissolution of the *Gallenbraun* had been completely achieved, the tube that conducted the gas from the flask containing the solution was sunk so deeply in it that the gas stream must have driven the *Gallenbraun* solution over into the bell jar. It was collected by the filter located underneath, and the clear dark-brownish-black liquid driven over from it was filtered directly into dilute hydrochloric acid. Decomposition occurred with the release of carbon dioxide. *Gallenbraun* precipitated in dark flakes. After the entire amount of solution was clearly drained in this way into dilute hydrochloric acid, the flask was removed, shaken rapidly, glass-stoppered, and let stand a little while until the solution clarified. The precipitate thus obtained now no longer absorbed oxygen from the air as easily as the initial solution. It was washed with hot water.

Pure *Gallenbraun* in the dried state, which I will name *Biliphäin*, has a dark-brown color tending toward olive green.

38   The [combustion analysis] numbers that I found for the composition of *Biliphäin* and biliverdin differ significantly from those that had been reported earlier by *Scherer* and *Hein*. The earlier study cannot give correct results because from the pigment prepared from gallstones neither ash (which of course can contain carbonated or caustic calcereous earth [calcium carbonate] or calcium oxide) nor the epithelium had been removed. Also, according to the method used for its preparation from icteric urine, it cannot contain pure bile pigment, as was assumed *a priori*. *Hein's* experiments with the compound that he called *Gallenbraun* and which he obtained as an insoluble residue by boiling the life out of crude *Gallenbraun* in ammonia, argue against the first mentioned experiments of *Scherer*. However, the green material that *Hein* investigated must have had an accidental impurity since it melted at 140–145°C. It probably still contained some fat or cholesterol. The deviation of the results of his analysis (he had carried out only a C,H determination) is thereby explained. It is hoped that other chemists with appropriate material at their disposal, which I now lack, would repeat my experiments in order to confirm or modify my conclusions.

39   Mr. *Valentiner* discovered a new solvent, chloroform, for animal pigments. First he succeeded in obtaining a yellow solution by chloroform digestion of pulverized gallstones that had been exhaustively extracted with alcohol and ether. By evaporation of the chloroform (avoiding considerable introduction of air), red and reddish-brown crystals separated, the majority with the characteristics of hematoidin. They were lancet-like and rhomboidal little plates and prismatic crystals in glandular grouping. In order to obtain the larger crystals pure, it was beneficial before evaporation to add some animal fat to the chloroform solution, and this was washed out promptly from the residue with ether. A crystalline pigment was also often obtained by ether extraction (*Frerichs, Atlas zur Klinik der Leberkrankheiten* [Atlas for the Clinic of Liver Diseases], Plate I, Fig. 7), as well as in many cases crystals directly from the chloroform solution, crystals in color and shape that seem to be different from hematoidin.

40   Fatty, icteric livers, the best of which are the icteric fatty livers of the highest grade, were rendered on a hot water bath, during which the fat separated in a layer; numerous hematoidin crystals are formed in the still fat-impregnated pieces of parenchyma. They were wrested free by repeated purification and recrystallization into nearly rectangular platelets of considerable thickness with very flat pyramids on them which appeared to the eye only by diagonally intersecting

634   11 Translations

lines. After purification, they are elongated rhomboidal plates, sometimes with rounded angles, sometimes put together dumbbell-like, or one sees the well-known unsymmetric prisms with rhombic end surfaces or, during rapid evaporation, fine rhombic needles with short, almost rectangular surfaces. The pure substance is insoluble in water, alcohol, and ether. In the last, the crystals decay (after being exposed a long time in diffuse daylight) to a loose amorphous green powder. Essential and fatty oils are ineffectual. Pure concentrated sulfuric acid dissolves [them] with a rapid color change and decomposition to a granular flocculent mass with a predominant brownish color. If the decomposition is interrupted by addition of water while the solution is a uniform green color, then, after dissolution in ammonia and evaporation, an amorphous green color is obtained. At this point the solution separates into compact green granules and delicate shapeless films. Impure nitric acid quickly decomposes the solution with initially green, then blue-green, green, and finally reddish-yellow and pale yellow colors, until loss of all color. After addition of hydrochloric acid to give a blue-green color, addition of nitric-sulfuric acid, even after long standing of the hydrochloric acid solution, develops the usual progression of color changes observed with this reagent, *i.e.* green, bluegreen, blue, red, yellow, pale yellow.

41   In December of last year, Dr. *Valentiner* reported in *Günzburg's Zeitschrift* that a crystalline substance is obtained by means of chloroform from gallstones and from bile, in addition to icteric livers and often from other tissues as well. The substance is different from the previously known bile pigments and corresponds in all its properties to hematoidin. With nitric acid, the chloroform solution gave the known color changes of the *Gmelin* bile test in an especially beautiful way. In comparison "after removing the chloroform-soluble pigment, the still dark-green pigmented bile no longer contains the substrate for the bile pigment reaction". Dr. *Valentiner* therefore proposes to demonstrate whether small amounts of bile pigment in a fluid can be detected by shaking the fluid again [and again] with chloroform and testing the latter with nitric acid after each successive extraction.

42   Apart from a few of Dr. *Valentiner's* reported experiments, which I undertook with crystals, I directed my attention next to [learn] whether in fact bile that is exhaustively extracted with chloroform no longer shows the color changes with nitric acid. On a water bath, I evaporated to dryness a portion of the bile decanted from chloroform, pulverized it, extracted it [the powder] with chloroform, filtered the same, and emptied the filtration residue into a flask. Fresh chloroform was poured over it, then as much water as necessary so was added that the dried bile dissolved in it. After all that, I extracted it further by shaking while replacing the chloroform again with fresh chloroform from time to time. The chloroform always took up less pigment. The color changes shown by added nitric acid were ever weaker and finally unobservable. From the now decanted bile a small quantity was diluted with much water and it was subjected to the *Gmelin* test. *It showed a very beautiful color change.* I have repeated this experiment several times and employed the original form of the test reported by *Gmelin*, in part with the modification that I brought to it ten years ago and which consists of adding only dilute nitric acid, followed by conc. sulfuric acid, which sinks to the bottom and from beneath initiates the decomposition process so that one can observe all of the colors simultaneously in superimposed layers. I always obtained the same results.

43   This fact contradicts Dr. *Valentiner's* report, and the question is raised as to how I should clarify the contradiction. Bile that had been exhausted by chloroform formed a green solution which did not turn yellow with the addition of caustic potash [potassium hydroxide] solution, but turned only a little more yellowish-green – and with hydrochloric acid a more blue-green. I presume therefore that of the two bile pigments known as *Biliphäin* and biliverdin, which are objects of the *Gmelin* test, one of them, *Biliphäin*, is soluble in chloroform, and the other is not. This therefore suggests investigating whether the crystals obtained from chloroform are crystalline *Biliphäin* or a even crystalline compound containing *Biliphäin*. This would by no means exclude its identity with hematoidin, as maintained by Dr. *Valentiner*. Eleven years ago, *Virchow* called attention to the correlation with *Biliphäin* (cholepyrrhin), which his hematoidin presented by the action of certain reagents.

44   Mr. *Brücke* at first repeated part of the preceding experiment of Mr. *Valentiner* in order to see whether bile exhausted by chloroform no longer reacts. He still found that this bile, too,

11.1 Translations to Section 2

exhibited the color change of the *Gmelin* test, and the question raised now is whether the crystals obtained are not *Biliphäin* or a compound of it. In fact he obtained yellow-brown flakes from hydrochloric acid added to an ammoniacal solution of the crystals, flakes that presented all of the properties of *Biliphäin* (*Heintz*), and from which crystals can again be obtained after distilling off the chloroform used to dissolve the flakes and form a yellow solution. *Brücke* concluded thus that the new method is an excellent way to separate *Biliphäin* and biliverdin. The latter can be obtained pure from the red crystals by dissolving them in aq. sodium carbonate and allowing the solution to absorb oxygen from air, then precipitating the solution with hydrochloric acid. The filtrate was washed out and any possible residue of *Biliphäin* was extracted by chloroform.

45   With respect to blood vessels, I will indicate the few cases where one might estimate with certainty the presence of crystalline hematoidin.

46   I consider the first observation of the same from *Everard Home*. In his latest work (A Short Tract on the Formation of Tumours. Lond. 1830) in its first figure may be seen very beautiful illustrations of clots from aneurismatic sacs which leave no doubt that he has really seen crystals from altered hematin. Unfortunately he gives no further descriptions of it but refers only to crystallization of blood salts (p. 22). In the explanation of the figure it says: *The figure shows the different shades of colours of the layers, according to the length of time they had been deposited, and the crystallized salts as they appear in different parts of the coagulum.*

47   Let us therefore turn back again to a comparison of our pigments with the bile pigment. We cannot suppress commenting that each observer of the individual cases where the pigment was obtained in the crystalline state is certainly better convinced that a derivative of the same pre-formed bile pigment cannot be established, as we might do here by lengthy deductions. The difference that we have noted between both types of pigments serves as a distinction, according to the present state of chemistry. Yet if one looks more closely, it will be easy to see that they are not absolute, but, strictly speaking, trace back more to differences of cohesion so that even an extraordinarily great similarity between the two pigments cannot be denied. We come now to another problem that is of the greatest significance to the physiology of healthy and sick bodies, namely whether the bile pigment might be assumed to be a product of the breakdown of red cells, from separated and altered hematin. This view, which has been proposed by various parties for a long time, always completely hypothetically however, appears to be disputed by *Berzelius*, based on the similarity of such bile pigments (which he designated as *Biliverdin*) to the green plant pigment, which emphasizes chlorophyll . . .

The characteristic change of colors, which comes from extravasated blood formed in the skin after a contusion, has already served as an argument for the transformation of hematin to a substance similar to the bile pigment. But it must be conceded that one such type of proof, if it was not accompanied by an actual investigation of the extravates, is worth nothing at all. The question will, however, be settled completely at the moment in which we are able to conduct the proof exactly by a chemical experiment, that neither a yellowish or greenish substance arises from hematin but one identical to the bile pigment. I flatter myself that the previously reported investigations have paved the way to a final determination of the question. I would have gladly sought this very determination if my numerous occupations had not forced me to pursue too many matters at the same time. I wish therefore to hand over the facts above to other investigators to use them further. Of special interest, it appears to me, is the investigation of *Bilifulvin* crystals. If one can obtain crystals from bile that are identical with those occurring in aged blood, then nothing more would remain to be desired. Pathological anatomy appears not to be able to make such a proof.

48   The hematoidin that crystallizes in prisms as well as needles is rather hard, brittle, and refracts light strongly under a microscope. Internally the crystals are bright orange-red (corn poppy red) colored, at the edges and corners of a dark carmine-red color. In incident light, crystals freed from all impurities have a color similar to mercuric iodide or alizarin. They possess a strong coloration, are somewhat heavier than water, and form voluminous masses. The angles of the prisms are 118° and 62°.

By heating in air it developed at first an animal-like odor, like nitrogenous compounds and burning horns, then caught fire and burned with a bright/brilliant flame while leaving behind a puffed up, voluminous charcoal that in the end completely disappeared. It was therefore difficult to analyze the compound in a combustion apparatus. By heating the substance in the absence of air a foul-smelling gas was evolved, an animal-like substance distilled and also left behind voluminous charcoal.

The crystals are insoluble in water, alcohol, ether, glycerol, essential oils, and glacial acetic acid. Its solution in concentrated ammonia is amaranth red [purplish-red] and quickly assumes a saffron yellow and brownish color. In contact with caustic potash [potassium hydroxide] and soda [sodium carbonate] the crystals of hematoidin break down and gradually dissolve, however in smaller quantities in ammonia the solution is reddish. The same are dissolved moderately quickly with a dark red color in nitric acid, with the evolution of gas bubbles if the nitric acid is concentrated. They are also dissolved in hydrochloric acid, however in smaller quantities. The solution is golden yellow or reddish-yellow. In incident light the remaining undissolved crystals have a brown ochre [earthy brown color], a reddish yellow color under the microscope. They are not dissolved in sulfuric acid, which makes the crystals merely darker and if the crystals still contain traces of alkali or iron-containing compounds, they assume a green color.

49   I used crystals purified by water, alcohol, and ether, after I was convinced beforehand, using a microscope, that all impurities could be removed in this way. Thus, I obtained the following results:

	I	II	III
Carbon	65.0460	65.8510	–
Hydrogen	6.3700	6.4650	–
Nitrogen	–	–	10.5050
Acid	18.0888	17.1788	–
Ash	0.0002	0.00002	–

On the basis of five analyses, hematoidin is

$$C_{44}H_{22}N_3O_6Fe$$

or from 100 parts

Carbon	65.84
Hydrogen	5.37
Nitrogen	10.40
Acid	11.75
Iron	6.64

50   It is thus easy to understand that hematoidin is none other than the pigment of blood, or hematin, in which one equivalent of iron is replaced by one equivalent of water.

51   More recently, the science of pigments has found a more careful treatment, and one has come to accept more and more that the hematin of blood constitutes the basis of all pigments. It was inevitable to observers that *Senac's* idea, icteric coloration of skin, which as in pyemia[3], in putrid infection and related processes not involving the liver, results from a direct transformation of hematin to a yellow pigment similar to or identical with the bile pigment.

---

[3]A diseased state in which pyogenic (pus-forming) bacteria circulate in blood, leading to abscesses in various organs.

11.1 Translations to Section 2

52   The origin of the bile pigment is with complete certainty traced back to the content of pigmented blood cells, as is the origin of all pigments formed in animal organisms under normal or pathological conditions. Apart from that we need to regard the supposed pigment of the contents of blood cells, the so-named hematin, as preformed therein. We have already explained above that the crystalline *Biliphäin* produced from bile according to *Valentiner* and *Brücke* is identical to hematoidin, that other crystalline transformation product of the blood pigment, while *Virchow* had indicated earlier both pigments as indeed still different yet similar to each other and genetically related to each other. This identity or close relationship arises just as evidently from an earlier observation made simultaneously by *Zenker* and myself. *Virchow* had found a reddish-yellow pigment from stagnant bile under pathological conditions that crystallized in broken needles and groups of needles (*Funke*, Atlas, 2nd ed. Table/Diagram IX, Fig. 3), which he named *Bilifulvin* because he took it to be identical with one of *Berzelius'* so-called (third) bile pigments. *Virchow* has already called attention to the similarity of *Bilifulvin* to hematoidin in its behavior to reagents; *Zenker* and I showed that *Bilifulvin* is transformed either by itself or by treatment with ether changes into beautiful large crystals that have all the properties of hematoidin, and are therefore identical to it. We add to *Brücke's* cited observations above, it is probably not to be doubted that this *Bilifulvin* of *Virchow* is none other than crystalline *Biliphäin*. In short, the normal brown pigment of bile that turns green by oxidation is identical to one formed from a transformation of the pigmented content of red cells in stagnant blood. From his investigations of icterus, *Kuehne* provided a further excellent proof for the origin of bile pigments from the blood pigment. *Frerichs* had observed that after injection of pigment-free bile or pure bile acid salts into blood, bile pigment appears in urine and concluded from that that bile acids are transformed into bile pigment in the blood, a transformation that he also achieved synthetically by digestion of bile acids with sulfuric acid. *Kuehne* provided a different, better-supported explanation for the cited facts. Bile acid salts already have the above-mentioned characteristic ability to dissolve red blood cells completely; thus, also after injection of the same into blood, mainly bloody-colored urine is secreted. This blood pigment, freed from dissolved blood cells, not the bile acid itself, which appears in urine in addition to the pigment, is converted into bile pigment and thus passes into urine.

53   It can be assumed to be established that in icteric urine, if it is rich in pigment, no bile acids or even a trace thereof is present. We ourselves could discover no bile acids in it from experiments repeated earlier, which is the same result that *Griffith*, *Pickford*, *Gorup-Resanez*, and *Scherer* arrived at. In contrast, *Lehman* observed that bile acids are often present in large amounts in weakly pigmented urine from diagnosed icterus.

We decided that this observation, whose accuracy probably cannot be doubted, seemed to indicate that a close relationship is present between the acids and the pigments of bile, and that with hindered efflux of bile, the acids arrive in urine either undecomposed or undergo a transformation into pigment or before the blood turns orange.

54   If pure sodium glycocholate [sodium salt of cholic acid amide with glycine] is covered with conc. sulfuric acid, it sticks together into a colorless, resinous mass which dissolves in the cold to give a saffron yellow color, and by heating to give a strong fire-red to brownish-red color. Colorless, greenish, or brownish flakes are precipitated by water depending on the temperature at which the solution takes place.

The glycocholic acid that was altered by sulfuric acid has the property of rapidly taking up oxygen from air and by so doing is transformed into a beautifully colored compound. If the colorless, amorphous mass resulting from sulfuric acid, after it has been freed as much as possible from adhering sulfuric acid, is placed on a piece of filter paper, it melts and leaves a rubin-red spot that soon shows a blue periphery and after a short time is pure indigo-blue. After a few days this color also disappears and the spot is light brown. The paper material seems to have no influence in this reaction because we observed a very similar color change in the deliquescence of the amorphous mass either on glass or on porcelain, where in these cases it occurs a little less quickly.

The solution of glycocholic acid in conc. sulfuric acid contains the same dissolved chromogen; however, excess acid delays the oxidation and the related coloration. If the solution is precipitated with water and warmed gently in the water bath to separate the flakes from the weakly acidic liquid, after a few seconds they turn violet and blue. One can also see a very beautiful color orange if a piece of filter paper is moistened then coated with the acid solution and dried over a lamp. If the sulfuric acid acts on the separated acid for a longer time at the temperature of the water bath then the spot becomes green in the same way as produced on the paper.

55   Syrupy bile was mixed with 3–4 volumes of conc. sulfuric acid, whereby it assumes a brownish-red color with spontaneous warming. After a half hour of heating on the water bath, the mass was a deep reddish-brown and reflected light with a strong grass-green color. Added water precipitated brown flakes that became indigo-blue by heating in air. The blue mass was insoluble in water. Heating to boiling yielded a brown solution from which a decomposition product separated during evaporation as a dark brown film. Evaporation of a grass-green alcoholic solution of the blue pigment leaves behind a greenish-blue residue, which was yellow-brown in caustic lye, without dissolving in significant amounts. Acids, even dilute acetic acid, reproduced the original color.

Essentially the same result was obtained after heating the mixture of bile and sulfuric acid for six hours. Here, too, the blue mass became brown-colored upon addition of caustic potash [potassium hydroxide], hardly dissolved in an excess, and became greenish-blue upon addition of glacial acetic acid. A yellow-brown solution arose with hot acetic acid and upon addition of nitric acid immediately turned deep blue-green, then violet, and finally dirty (*schmutzig* = schmutzy) yellow. Acetylated lead(II) oxide produced a small quantity of colored precipitate in the brown acetic acid solution which by the addition of nitric acid likewise showed the change of colors.

After the mixture of bile and sulfuric acid had been heated eight days on a moderately hot water bath, a dark green mass consisting of small, microscopic globules separated out. It was insoluble in acidified water, but soluble in pure water with a deep green color. It dissolved completely in dilute caustic lye with a pure bile-brown color and, upon the addition of nitric acid became first green, then reddish, and finally yellow-colored.

The behavior of these decomposition products from nitric acid is reminiscent of the natural bile pigments, however, in which the change of colors is less vivid than is seen from the mixing of highly pigmented icteric urine with nitric acid. However, we obtained more favorable results when we treated the amorphous precipitate, arising preferably from sodium taurocholate with sulfuric acid, that we had precipitated with ether from an alcohol solution of decolorized ox bile.

56   For the present we limit ourselves to drawing attention to the similarity of the natural bile pigments to our decomposition products from bile acids. However we are now able to express with certainty that the chromogen from which the blue pigment arises by oxidation is present now and then in the liver and apparently also in the pancreas. We had already drawn attention to this pigment in the past. At that time we did not know that it had such a simple relationship to bile acids. Also, the blue pigment that separates from human urine upon addition of acids, and is changed according to *Sicherer's* experiments into a compound that is completely similar to indigo, is perhaps a decomposition product of bile acids. We expressed the view earlier . . . that this pigment can arise as a by-product in the formation of glycocholic acid, in which tyrosine is decomposed in the liver into glycine and saligenin [salicyl alcohol].

57   Recent experience has of course confirmed this. We injected a dog with about a drachma [1 dram = 1/16 ounce = 1.77 grams] of the pure, colorless oxen bile, which was dissolved in distilled water. Six hours later the animal expelled 3 ounces of dark brown urine of specific gravity 1.015, and a very weak alkaline reaction. Upon standing, it precipitated an appreciably thick layer of green flakes, which under the microscope appeared as brownish-green granules. Upon the addition of nitric acid, it showed a most beautiful the color change reaction characteristic of a bile pigment. The *Pettenkofer* test gave a negative result.

58   *W. Kühne* has expressed the claim, supported by a series of experiments, that bile acids, which enter the bloodstream, undergo no alteration and are removed from the body in urine.

59   Here however, only the most successful color reaction was achieved since the greatest influence is whether the mixing with H_2SO_4, with its associated temperature fluctuations, is faster or

11.1 Translations to Section 2

639

slower. A quantitative colorimetric determination of bile acids therefore cannot be attained with the help of the *Pettenkofer* reaction.

60   The restrictions of the reaction were broadened significantly when each process was modified somewhat. I observed that a single drop of a 1/20% solution of cholic acid or glycocholic acid still gave a beautiful purple-violet color if it was diluted in a porcelain dish and mixed with a *trace* of sugar solution and a few drops of dilute sulfuric acid (4 parts OH = [H_2O] + 1 part $HOSO_3$ = [H_2SO_4]), followed by careful evaporation with gentle warming and rotation over a small alcohol lamp. After standing for some time, the sample took on a considerably intense color. Since 1 cm^3 amounts to nearly 8 drops, even only 6/100 mg bile acid can be proven with complete accuracy. A greater concentration of the solution is naturally not inconvenient; with a greater dilution one has to evaporate the test solution beforehand to one or two drops. In the manner indicated, 1 cm^3 of a 1/100% solution of both acids still gave the splendid purple-violet color while, with the same dilution and application of 3 cm^3 solution, *Pettenkofer's* procedure gave no result [was negative].

61   Confirmation of the presence of bile acids was accomplished only by the last [revised] method, thus for brevity in the following will be indicated "proof in the porcelain dish".

62   These experiments show that even with the use of a not insignificant amount of bile acids, the *Hoppe* method gives only ambiguous results and that it is completely useless for detecting small quantities.

63   According to this method, 1/1,000% glycocholic acid could be detected in urine, while with *Hoppe's* procedure experiments detection at 1/50% was scarcely possible. That method is useful therefore for the detection of small quantities of bile acids. I must add to that that *Hoppe's* method is uncertain and therefore appears to be unsuitable.

64   Since now in *Kühne's* first-mentioned article a transformation of bile acids in blood was completely denied, claiming that the same acids were eliminated from the body in the urine, it thus appears to be of interest to physiology as well as to pathology, to subject *Kühne's* reports to further examination partly by investigating icteric urine, partly by injection experiments on animals.

In the following I report the results of the proposed investigations.

Even with this urine, at most a negative or highly questionable result was obtained from the usual *Pettenkofer* test while according to our modified procedure the least traces of bile acids were detected unambiguously.

65   In this case no trace of bile acid could be detected in urine using a careful application of the usual method.

66   ... In no cases, however, was a bitter taste noticed in the sodium compound finally obtained; in no cases was bile acid proven with certainty using the customary *Pettenkofer* procedure; and only in two cases was a characteristic staining noticed using the "porcelain dish test".

These facts show that bile acid which enters the blood can go over into urine only in trace amounts, which thereby satisfactorily refutes *Kühne's* statement: "The sodium compound of glycocholic, cholic, and choloidic acid leaves the bodies of animals via the kidneys following injection into their veins." *Kühne* has repeatedly been content to try the *Pettenkofer* test directly with urine which, only if necessary, was freed from protein. Evidently in such cases a misunderstanding occurred as a result of the existing pigmented and extractive material present, because the addition of sulfuric acid alone to human and dog urine seldom induced red and even violet colors.

67   Sometimes the urine of dogs, which had been injected in blood with sodium glycocholate, contained first larger, then smaller amounts of bile pigment. *Frerichs* conducted 29 experiments, of which 19 gave positive results. Usually urine contains some undissolved protein and dissolved red blood at the same time. Of the seven injection experiments that I conducted, in one the pigment occurred in such quantity that it partly precipitated in flakes. In two different cases only dissolved pigment was present, leading to a negative result. In the experiments communicated by *Kühne*, bile pigment was present instead of the supposed bile acids.

68   From these observations, coming from entirely different views on pigment formation, one may conclude from introduction of bile acids into blood that bile acids artificially introduced

as well as those normally in the bloodstream change into chromogenes and finally into pigments. However, the observed exceptions are too numerous to deny that a transformation of bile acids into bile pigment can take place only under especially favorable circumstances. It would appear to me that a certain degree of irritation is necessary for this transformation because bile pigment appeared in urine in three of my experiments. The first time it occurred by chance, the other times by intentionally jerky injections. These experiments failed to be reproduced in dogs.

Kühne rejected entirely the transformation of bile acids into bile pigments, although he reported a large number of experiments in which regularly, after injection of bile, pigment appeared in bile. He upholds the view that all bile pigments are derived from the blood pigment and of course the hematin going into solution from spontaneous destruction of red cells should undergo transformation into bile pigment. This view, however, received no experimental support, for when *Kühne* injected dissolved hematin into veins, no bile pigment arose in urine; whereas if he simultaneously injected hematin and bile acid, formation of pigment was observed. *Kühne* therefore sees himself also compelled to attribute to the bile acid particular, a yet mysterious, influence on the dissolved hematin.

69 I am far from accepting that destruction of hematin in the body cannot give rise to the formation of bile pigment, although this is not yet proven by experiment. On the other hand, however, it has not been disproved by *Kühne's* experiment that the bile acids coming into the bloodstream can go over into bile pigment under certain circumstances. *Frerichs* has already expressed that here a void is still to be filled before one may regard the transformation as firmly established. Frequent repetition of the experiments and unbiased interpretation of the acquired results will lead us gradually to the truth.

70 This formula is assumed for hematoidin; therefore, it would be derived from tyrosine:

$$\underbrace{2\ C_{18}H_{11}NO_6}_{\text{tyrosine}} \quad + \quad 2\ O \quad = \quad \underbrace{C_2O_4}_{\text{carbonic acid}} \quad + \quad \underbrace{C_4H_4O_4}_{\text{acetic acid}} \quad + \quad \underbrace{C_{30}H_{18}N_2O_6}_{\text{hematoidin}}$$

The basic pigment of blood, hematoidin, could therefore arise by an oxidation process from tyrosine, which itself comes from destruction of proteins in organisms.

71 Moreover it should in no way be maintained that hematoidin and *Erythrosin* are identical; I assume it to be possible, of course, and I will follow up this question further as soon as time permits.

72 An analysis of *Gallenroth* has not been made, and if one compares the formula from *Robin's* analysis of hematoidin ($C_{30}H_{18}N_2O_8$) with the formula of biliverdin ($C_{16}H_9NO_5$ or $C_{32}H_{18}N_2O_{10}$), it appears that the latter contains more carbon in relation to nitrogen than hematoidin, that therefore if *Robin's* analyses are correct, biliverdin cannot arise by oxidation of hematoidin.

73 Differences of opinion prevail about whether bile acids are transformed directly into pigments in the bloodstream, or whether pigment formation must be attributed to the decomposing influence of these acids on hematin. From the simple injection experiments, as accomplished so far, the question apparently cannot be answered satisfactorily, while the synthetic and natural bile pigments await a definitive explanation from a comparative chemical investigation.

In order to be able to undertake this comparison, I occupied myself at first with an investigation of the natural pigment. As I report the results obtained, at the same time I need to express my thanks herewith to all friends and colleagues who have aided this research by sending material.

74 1) *Bilirubin*. – In order to purify this pigment, which occurs in large amounts in human gallstones, it was dissolved a few times in chloroform, the filtered solution evaporated, and the residue washed with ether and alcohol. The drained-off alcohol was always colored more or less green to greenish-brown, while bilirubin remained behind as a bright red to orange-red granular-crystalline powder.

Analysis of the thusly purified pigment gave numbers that do not correspond sufficiently to an acceptable formula. From this an impurity must be inferred. I succeeded in removing such by letting the chloroform solution evaporate only until bilirubin begins to separate, and

11.1 Translations to Section 2

then precipitated it by the addition of alcohol. In this way the bilirubin was obtained as an amorphous orange-colored powder; in doing so, a fairly significant loss was unavoidable.

75   The pigment obtained burned on platinum foil without leaving a residue. After drying several days over sulfuric acid at 100°C it lost nearly 1% of its weight. With further heating from 120°C to 150°C the weight remained constant. Bilirubin melted when heated in a glass tube; it swelled up and released a vile-smelling vapor that turned lead paper black [a test for the presence of hydrogen sulfide]. In contrast, following combustion of 0.176 g of substance with lime and saltpeter, no turbidity was observed when the red-hot mass was dissolved in dilute hydrochloric acid and barium chloride was added. The trace of sulfur shown by the lead paper test was also detected in all the pigments above.

The bilirubin used in the following analyses had been obtained from two preparations.

I.   0.3765 g, dried at 120°C, gave 0.927 g CO_2 and 0.2125 g H_2O.

        0.2563 g, dried at the same temperature gave a quantitative sal ammoniac [NH_4Cl] from burning with soda-lime [mixture of $Ca(OH)_2$, ~75%; NaOH, ~3%; KOH, ~1%; H_2O, ~20%] from which 0.252 g AgCl was precipitated with $AgNO_3$.

II.   0.3105 g dried at 130°C gave 0.764 g CO_2 and 0.171 g H_2O

From these data, the bilirubin formula is calculated to be $C_{32}H_{18}N_2O_6$.

			calculated		I.	II.
32	aq.	C	192	67.13	67.15	67.11
18	"	H	18	6.29	6.27	6.12
2	"	N	28	9.79	9.59	–
6	"	O	48	16.79	16.99	–
			286	100.00	100.00	

76 A solution diluted 30,000 to 40,000 times still colors the skin distinctly yellow. With such extraordinary coloring ability, the yellow coloration of the eyes and the skin is easily explainable due to the occasional rapid appearance of icterus. One may infer approximately a 20,000 to 25,000 times dilution of the pigment from the coloration of the eyes during intensive icterus.

77   The communicated analysis of color intensity was made with an ammonia solution of bilirubin. Such solutions bleach, rather rapidly but not completely, in direct sunlight, while in diffuse light they decompose only slowly. They gradually become light-brownish yellow and gradually lose the capacity to be precipitated by hydrochloric acid, while from the undecomposed solution, even in very dilute solutions, additional hydrochloric acid precipitates bilirubin in orange flakes.

78 0.2549 g ash left from combustion and moistened with ammonium hydroxide and dried at 130°C, yielded 0.0414 g calcium carbonate, corresponds to $C_{32}H_{17}CaN_2O_6$. The calculation requires 9.18% calcium oxide; found was 9.10%.

79   No significant action can be seen if dilute nitric acid, which contains 20% hydrate, is poured over bilirubin. In contrast, with warming it changes into dark violet resinous flakes that become light brown with further action and dissolves with a yellow color by heating. Acid of 30% hydrate forms resinous, red-colored flakes in the cold. The flakes disappear by warming, and the solution becomes yellow. If pure nitric acid hydrate is used, bilirubin dissolves quickly in the cold with a deep red color and, after a little while, or by heating, the solution lightens, but after standing several days it becomes a vivid cherry-red color.

80   If solutions of bilirubin are mixed with commercial conc. nitric acid, to which is added appropriately some red, fuming acid, then the well-known bile pigment reaction is outstandingly obtained. Best, alkaline solutions are employed and mixed before the addition of acid with approximately an equal volume of alcohol. Upon the addition of alcohol, a beautiful reaction also occurs when the acid used contains a little nitrogen peroxide [NO_2], and the test is not turned turbid by separated flakes of pigment. The yellow color first goes over into green,

642   11 Translations

is then blue, violet, ruby-red, and finally dirty [schmutzy] yellow. If the solution is not shaken, then all these colors appear simultaneously as layers one above each other. ¼ mg bilirubin in 4 cm³ solution still brings forth a splendid display of colors. The limit of the reaction occurs at 70,000 to 80,000 times dilution.

81   The blue pigment arising from the specified reaction can be isolated without difficulty. If a not-too-dilute ammonia solution of bilirubin is mixed dropwise to the acid mixture specified above, and too great an excess of nitric acid is removed from time to time by bringing it closer to neutral with ammonia, then a green flocculent precipitate is obtained at first that gradually becomes blue. After washing with water, it can be separated from the co-mixed green pigment by alcohol, and that then leaves behind a deep black-blue powder. It is assumed that this blue is pigment is related to the indigo content of urine. Unfortunately I did not have enough material to be able to undertake experiments in this direction.

A beautiful blue can also be obtained using chloroform. If a yellow chloroform solution of bilirubin is mixed with one or two drops of nitric acid and shaken, the liquid becomes very dark, passing quickly through violet and then becoming ruby-red. If a lot of alcohol is added quickly, as soon as the violet color appears, when mixing is complete the solution is deep blue and changes its color only slowly. In the same way a beautiful green or red can also be produced. The color depends on an earlier or later addition of alcohol.

82   Reducing agents act very energetically on bilirubin. If a deep, red-brown alkaline solution of the pigment is mixed with sodium amalgam, the color rapidly wanes and the solution becomes pale yellow. This hue does not disappear, even with heating. I have not been able to investigate the resulting compound, which probably has the same relationship to bilirubin as indigo-white to indigo-blue. If the indicated relationship is correct, then the yellow compound would have the formula $C_{32}H_{20}N_2O_6$.

83   2)   *Biliverdin.* – If a solution of bilirubin in excess caustic soda solution is layered on a flat plate and exposed to air, or is shaken with air, it rapidly takes up oxygen and the solution becomes green. When this color has reached its maximum intensity, a bright green precipitate appears upon addition of hydrochloric acid. It is insoluble in ether and in chloroform, but it dissolves very easily in alcohol with a beautiful green color. Some co-mixed unchanged bilirubin is left behind in orange-colored flakes. Nitric acid colors the green solution first blue, then violet, red, and finally dirty [schmutzy] yellow.

This green pigment is without any doubt the same as the biliverdin analyzed by *Heintz*, for which he proposed the formula $C_{16}H_9NO_5$ or $C_{32}H_{18}N_2O_{10}$.

If this formula is assumed to be correct, then the formation of biliverdin from bilirubin is based on the simple oxidation:

$$\underbrace{C_{32}H_{16}N_2O_6}_{\text{bilirubin}} \quad + \quad 4\,O \quad = \quad \underbrace{C_{32}H_{18}N_2O_{10}}_{\text{biliverdin}}$$

However, I have made a few observations that cast doubt on the correctness of this formula.

84   The carbon and nitrogen content found corresponds better to the formula $C_{32}H_{20}N_2O_{10}$, while the hydrogen found lies in the middle between the two formulas:

	$C_{32}H_{20}N_2O_{10}$	found	$C_{32}H_{18}N_2O_{10}$
Carbon	60.00	60.04	60.38
Hydrogen	6.25	5.84	5.66
Nitrogen	8.75	8.53	8.80
Oxygen	25.00	25.59	25.16
	100.00	100.00	100.00

Probably the biliverdin analyzed by *Heintz* was not completely pure since it was obtained from a pigment mixture, from the so-named *Biliphäin*, by dissolving it in aq. sodium carbonate

11.1 Translations to Section 2

and allowing spontaneous oxidation. I regret ever so much not at present being in possession of a sufficient quantity of pure bilirubin in order to be able to subject biliverdin to a new analysis.

85   I note still that I have not run across completely formed biliverdin in gallstones. It does not occur at all in them and so can be present only in traces. Probably it is transformed to *Biliprasin* in alkaline bile by the uptake of water.

86   Bilifuscin forms a nearly black, shiny, brittle mass that upon pulverizing gives a dark brown powder exhibiting a somewhat olive color. It is free of ash constituents, behaves in heating just like bilirubin and gives just as beautiful a bile pigment reaction with nitric acid.

    0.2655 g of the substance dried at 120°C gave by combustion 0.614 g CO_2 and 0.1575 g water, corresponding to the formula $C_{32}H_{20}N_2O_8$.

				calculated		found
32	aq.	carbon	192	63.16		63.07
20	,,	hydrogen	20	6.58		6.59
2	,,	nitrogen	28	9.21		–
8	,,	oxygen	64	21.05		–
			304	100.00		

Accordingly, the analysis shows that bilifuscin has a very simple relationship to bilirubin. It differs from it in composition only by the elements of two equiv. water, of which it contains more

$$C_{32}H_{18}N_2O_6 \qquad\qquad C_{32}H_{20}N_2O_6$$

bilirubin   bilifuscin.

87   0.301 g of the pigment dried at 100°C gave by combustion 0.627 [g] carbon dioxide and 0.1765 [g] water

    Nitrogen was determined in the same way as with bilirubin. 0.096 g gave 0.073 [g] silver chloride.

    These results lead to the formula $C_{32}H_{22}N_2O_{12}$.

				Calculated		Found
32	Aq.	C	192	56.81		56.81
22	"	H	22	6.51		6.52
2	"	N	28	8.28		7.42
12	"	O	96	28.40		29.25
			338	100.00		100.00

    The deviation in nitrogen content is not surprising considering that only a very small amount of pigment was available for the experiment.

88   I have not conducted an elemental analysis since I could not persuade myself that the compound is pure, and because the material at hand was insufficient for further purification. I note only that purified bilihumin is not completely soluble in ammonia or is very slowly soluble. In contrast it dissolves easily upon heating in dilute soda lye, and when the resulting dark brown solution is mixed with alcohol and then NO_4 [NO_2]-containing nitric acid, it shows an entirely pretty play of colors. The red, namely, is very pure and intense while the previously occurring colors in the dark brown solution are not distinctly recognizable.

89   Without doubt the formula of bilihumin stands in a similar relationship to that of *Biliprasin*, as the formulas of the analyzed compounds stand with each other. I also take it to be very probable that dark, insoluble pigment substances present in living organisms, the so-called *Melanin*, join with bilihumin and perhaps have the same origin.

90 *Human Bile*

No chemical argumentation is required to justify that the same pigment is present in human bile as in the concretions that form in it. The experiment that I have carried out with human bile has thus a different purpose. As mentioned earlier, the crystalline form of bilirubin is much inferior to the purer solutions from which it is in contact, while impure chloroform solutions yield quite ordinary crystalline bilirubin. The crystalline precipitate appears to require or to be very much promoted by the presence of certain foreign material, possibly just as the presence of any kind of metal chloride is in the precipitation of *Teichmann's* crystalline hemin from glacial acetic acid solution. I therefore selected bile in order to render the bilirubin in a more measurable form. If the red pigment present in it really is identical to hematoidin, as *Valentiner* assumes, then it must be able to be obtained even in the more normally occurring form of hematoidin by a properly selected treatment.

91

If bile is shaken with chloroform, during slow evaporation of the chloroform one can observe, as *Valentiner* found, the formation of orange-colored elliptical leaflets, or very small nearly rectangular plates whose angle proportions are quite substantially different from those of hematoidin. The result was almost always the same in repeated experiments. That rhomboid configuration with small differences of the sides and angles was always observed, from which the diagonals of the rhomboids were marked by divergent coloring. Nearly without exception, occasionally a form was an isolated that was close to the usual hematoidin form.

92

The chloroform solution had an intensive green color, and unassisted evaporation left behind a more violet sticky residue. Cholesterol and fat were removed by treatment with ether; alcohol took up the green pigment, in addition to other substances. According to its properties with respect to alkali, the green pigment appeared to be biliverdin, and bilirubin was obtained as a residue, however, not in good crystallinity but in orange-colored crystalline granules and flakes that were mixed with the rhomboid form described.

93

The carbon disulfide solution had a pure golden-yellow color. Evaporation in air left behind a reddish, crystalline mass from which ether and alcohol extracted cholesterol, fat, and perhaps also some bile acid, while bilirubin remained behind in dark red microscopic crystals. The crystals appeared as clino-rhombic prisms with basal surface where the anterior angle was very sharp and the prism faces were curved convex so that the view of the basal surface showed ellipses. The convex surface of the inspected crystals had a rhomboidal appearance with considerably greater differences of sides and angles than with the crystals from chloroform. One frequently finds prismatic crystals tied up in the center, which appears to be explained by twinning. The diagonals were marked in the same way as the crystals from chloroform. The angle relationships of these crystals shows a similarity to those of hematoidin, but exact measurements and comparisons of the convexity of the surfaces were, however, unfortunately not possible due to the smallness of the hematoidin crystals at my disposal.

94

To begin with, convex faces on hematoidin have still never been observed while the same on bilirubin are so prominent that, like uric acid, one can easily observe them from a cursory inspection. The primary weight must however be placed on the results of analysis, and since such a large variation in composition is shown in the following summary it is impossible to attribute the difference to small amounts of impurities[*] as an unavoidable analysis error.

	Bilirubin		*Hämatoïdin*	
C	67.15	67.11	65.85	65.05
H	6.27	6.12	6.47	6.37
N	9.59		10.51	
O	16.99		17.17	
	100.00		100.00	

---

[*]From sufficiently purified bilirubin I found the following % composition: C, 66.52; H, 6; N, 8.7; O, 18.78

11.1 Translations to Section 2 645

95   From that analysis, *Robin* calculated the formula $C_{14}H_9NO_3$ for hematoidin; however I already made note several years ago that this formula does not correspond to *Robin's* analysis, and that a more correct calculation leads to formula $C_{30}H_{18}N_2O_6$. Only the hydrogen is found to be about 1/10 to 2/10% smaller in this case than that corresponding to the formula. That bilirubin and hematoidin are closely related compounds is shown by the great similarity of the formulas. If hematoidin contained two fewer equivalents of hydrogen, it would therefore have the formula $C_{30}H_{16}N_2O_6$. Thus it would belong to a homologous series with bilirubin, $C_{32}H_{18}N_2O_6$, and given that the multiple similarities in properties would be sufficiently explained. However, this can only be determined by new analyses.

96 This pigment was probably also only a decomposition product arising from the influence of hydrochloric acid on the original pigment. In any case it was not pure, as revealed by the higher carbon and hydrogen content in addition to the smaller nitrogen content.

97   Since a relationship between the synthetic pigment and the natural bile pigment seems to be indicated by this pigment reaction, and moreover since we, as was already stated above, still observed that after injection of bile and salts into a vein, nearly normal bile pigment appears in urine; thus, it was certainly not premature when we concluded that even bile acids in the bloodstream can undergo transformation into pigment. This transformation has moreover never been advanced as a completely proven and incontrovertible fact because some, if only a few, cases appeared where after bile injection no pigment could be detected in urine. I have now succeeded in transforming nitrogen-free cholic acid into pigment in the same way as glycocholic acid and taurocholic acid. Therefore, neither are the nitrogen-containing bile pigments derived from nitrogen-free compounds, nor are bile acids transformed into bile pigments.

98 The question still remains unanswered as to what role that bile entering into blood plays in the production of bile pigment, for the assumption that bile acids *only* dissolve red cells and that the dissolved hemin then goes over into bile pigment appears to me not yet justified. Of course it is not the case that bile pigments must always appear in urine after one injection of bile acid, and moreover that injection of water must bring about the same result as injection of bile acids. This too is not the case. *Röhrig* has shown that if 100 cm³ water is injected by catheter into the jugular vein of a small dog whose blood content amounted to 130 cm³, it is observed that the urine expelled from the dog is rich in the blood pigment but contains no bile pigment.

99   According to these observations I assume it likely that in this enormous circulatory disturbance, which must naturally also be associated with a large perturbation in chemical transformation, we have to seek the reason for pigment formation after introduction of bile acids into blood. It would also explain that pigment formation does not occur constantly because animals of various ages and sizes, of weaker and stronger constitutions cannot be affected in the same way by the same amount. Accordingly, bile pigment formation after bile injection would be a secondary consequence when a bile acid is brought into the blood. If this is the case, then it remains to be seen if different substances, which similarly bring about destruction of cardiac activity, must likewise give rise to formation of bile pigment. We possess such a substance in digitalis, with which I have undertaken a few experiments.

100   These two experiments are at variance with each other. Therefore the problem posed is still not settled. It can be answered only by a larger series of experiments, and I regret that other work prevents me from devoting the additional attention to this subject that it appears to deserve.

101   Before the discovery of the use of chloroform as a solvent for this pigment, only brown modifications of the same had been obtained and were called *Biliphäin* and *Cholephäin*. However, after the red pigment had been obtained using chloroform, it was generally assumed that the brown color of the earlier preparation was a sign of its impurity. The red pigment, under the names *Bilirubin* or *Cholerythrin*, was assumed to be the only form of the pure bile pigment.

102 . . . In numerous operations that I undertook in order to isolate pure bile pigment, I always obtained two modifications which of course were chemically the same, of which, however, one was red-brown, the other pure red, like mercuric oxide. A microscopic investigation showed that the dark brownish-red pigment was composed of numerous little crystalline particles with many complete crystals. In contrast, bilirubin consisted almost entirely of small amorphous granules. Only when it had been precipitated with alcohol did it contain small yellow rhombic prisms. If I let a mixture consisting of a saturated chloroform solution of *Cholephäin* and absolute alcohol stand, from which the first precipitate of bilirubin had

646 11 Translations

been removed by filtration, and I gradually added more alcohol, then I usually obtained a second precipitate as half amorphous and red and half crystalline and brown. The crystals were often clustered in aggregates. They could be separated from the longer-suspended remaining bilirubin by precipitation with alcohol in which they settled rapidly.

103   The smallest bilirubin crystals showed the same form and yellow color. By careful recrystallization the red modification could always be partially changed into the brown. It is thus clear that the crystallized or crystalline purple-brown *Cholephäin* or *Biliphäin* is only a different state of aggregation of amorphous red bilirubin or *Cholerythrin*. I will therefore regard *Cholephäin* and bilirubin as chemically identical in the following. I add, however, that if one or the other name is used in the description of a process, the modification that has been used in the process is thereby designated.

104   The material is completely insoluble in water and slightly soluble in boiling absolute alcohol, with yellow coloration. If this solution is filtered through [filter] paper, the pigment adheres to the paper fibers and the alcohol portion flows through nearly colorless. It is poorly soluble in ether and somewhat more soluble in carbon disulfide and in benzene. The best solvent is chloroform, whereupon 1,000 parts dissolve 17 parts; thus, 586 parts dissolve one part bilirubin. The solution is colored a beautiful dark red. Sunlight changes the color of the solution to brown and black, probably from formation of hydrochloric acid. Addition of aq. hydrochloric acid produces a precipitate in the solution. If dry hydrogen chloride gas is bled into the solution to saturation and the chloroform and acid are distilled off completely, a mixture of two brilliant green residual compounds is left. The mixture cannot be separated by alcohol, in which both are soluble, however, but by using ether, in which only one is soluble. Bilirubin is completely transferred into these new compounds.

105   *Comparison of empirical and theoretical elemental composition of Cholephäin.*

	I.	II.	III.	IV.	V.	VI.	average
C	66.02	66.41	65.61	–	–	–	66.01
H	5.97	6.13	5.95	–	–	–	6.01
N	–	–	–	9.05	9.49	8.56	9.03
O	–	–	–	–	–	–	18.95
							100.00

These numbers lead to the formula $C_9H_6NO_2$, which according to theory compares with the above facts as follows:

Atom	Atomic Weight	Theory in %	Average of the Analysis
C_9	108	66.26	66.01
H_9	9	5.52	6.01
N	14	8.59	9.03
O_2	32	19.63	18.95
	163	100.00	100.00

106   That the formula above is correct, and that the true atomic weight [molecular weight] of *Cholephäin* or bilirubin is 163, I will prove in more detail in the following by a long series of remarkable compounds as well as several interesting transformations of this substance by the influence of various acids and alkalis.

107 The average of this determination is 37.39% Ag.

     If one now takes into consideration that the analyses of *Cholephäin* or bilirubin leads to the empirical formula $\mathsf{C}_9H_9NO_2$, then there can be no doubt that the amount of silver in the silver salt described above corresponds exactly to that afforded by a neutral hydrated compound of formula $C_9H_{10}AgNO_3$. As anomalous as a silver salt with one atom of water might always be, it is now certain that the elemental composition and the molecule of *Cholephäin* is expressed by the formula $\mathsf{C}_9H_9NO_2$.

11.1 Translations to Section 2

Comparison of theoretical and empirical [values] of vacuum-dried silver Cholephäinate.

Symbol	Atomic Weight	Theory in %	Found			Average
			a.	b.	c.	
C_9	108	37.30	–	–	–	–
H_{10}	10	3.47	–	–	–	–
Ag	108	37.50	37.63	37.52	37.03	37.39
N	14	–	–	–	–	–
O_3	48	–	–	–	–	–
	288					

108   A small amount of *Cholephäin*, which had been repurified by dissolving in chloroform and in alcoholic potash solution, was dissolved in ammonia and mixed with silver nitrate. Since no precipitate appeared, more silver nitrate solution was added, and the entirety taken to neutrality by adding nitric acid. The precipitate now appearing left a colorless supernatant. It was washed with water and dried under vacuum.

109   In this way it is found that silver *Cholephäinate* is freely soluble in ammonia and that if this solution in the presence of excess silver nitrate is toned down to a certain level of neutrality bordering on alkalinity, the basic salt precipitates. His theory, derived from facts known of the free *Cholephäin* and *Cholephäin* bound simply to silver, is confirmed by analyses; its formula is $C_9H_7Ag_2NO_2$. In this compound two hydrogen atoms have been replaced by silver atoms. I will describe later an analogous lead compound in which two atoms of hydrogen are replaced by one divalent lead atom. Its formula is $C_9H_7PbNO_2$, and it is an important theoretical support for the assumption that the basic silver salt described above is really a definite compound and not only an accidental mixture.

Comparison of theoretical and empirical values of the basic silver Cholephäinate

Symbol	Atomic Weight	Theory in %	Found			Average
			a,a.	b.	c.	
C_9	108	28.69	–	–	–	–
H_7	7	1.85	–	–	–	–
Ag_2	216	57.29	56.81	56.41	55.86	56.27
N	14	–	–	–	–	–
O_2	32	–	–	–	–	–
	377					

110   These facts satisfy the requirements of theory for a barium salt exactly analogous to the neutral silver salt but where divalent barium ion replaces two hydrogen atoms of *Cholephäin*. In addition, two molecules of water are present in the compound.

111   In this analysis, the tube broke at the end of the operation when potash in the flask backed up in the safety apparatus so that the residual carbon dioxide and water could not be aspirated out. As for the rest, this analysis shows quite clearly that in this *Cholephäinate* one atom of barium is bound to three molecules of *Cholephäin*.

If we add one molecule of *Cholephäin* to the neutral barium *Cholephäinate* dihydrate described above, as here:

1	Barium-*Cholephäinat,*	$C_{18}H_{29}BaN_2O_6$	=	497	Atomic weight
1	Cholephäin	$C_9H_9NO_2$	=	163	Atomic weight
	thus we obtain	$C_{27}H_{29}BaN_3O_8$	=	660	Atomic weight

Analyses of the compounds described above now correspond rather completely to theory.

Symbol	Atomic weight	Theory in %	Found				Average
			a.	b.	c.	d.	
C_{27}	324	49.09	–	–	51.50	49.76	50.63
H_{29}	29	4.39	–	–	4.65	4.09	4.37
Ba	137	20.75	20.66	20.60	–	–	20.66
N_3	42	–	–	–	–	–	–
O_8	128	–	–	–	–	–	–
	660						

112   The formula derived from these data leads to a neutral calcium *Cholephäinate* $C_{18}H_{20}CaN_2O_6$, which is analogous in every connection to the barium compound described above. One must also assume the existence of two molecules of water, which are not released at a temperature of 100°C.

The following comparison of theory for this compound with the analytical data easily makes this conclusion clear.

Symbol	Theory		Found			
	Atomic Weight	Theory in %	a.	b.	c.	d.
C_{18}	216	54.00	–	–	–	52.35
H_{20}	20	5	–	–	–	5.04
Ca	40	10	9.63	9.88	9.92	–
N_2	28	–	–	–	–	–
O_6	96	–	–	–	–	–
	400					

	Found				Average
	e.	f.	g.	h.	
C	–	–	54.96	54.26	53.86
H	–	–	5.03	4.65	4.9
Ca	11.02	10.44	–	–	10.17

. . . The compound is therefore analogous to the already described half-acid barium *Cholephäinate* and has the formula $C_{27}H_{29}CaN_3O_8$. From this perspective, the results of the analysis remain as follows:

Symbol	Theory		Experimental		
	Atom	%	a.	b.	c.
C_{27}	324	57.54	–	–	60.37
H_{20}	29	5.15	–	–	5.74
Ca	40	7.1	7.03	6.79	–
N_3	42	–	–	–	–
O_8	128	–	–	–	–
	563				

11.1 Translations to Section 2

From the following compilation, differences in composition of the neutral calcium *Cholephäinat* on one side and the half-acid on the other side are very evident.

	Neutral Calcium-*Cholephäinat*,			Half-acid Calcium-*Cholephäinat*,	
	$€_{18}H_{20}€aN_2Θ_6$			$€_{27}H_{29}€aN_3Θ_8$	
	Atomic weight = 400			Atomic weight = 563	
	Theory	Found		Theory	Found
$€$	54	53.86	$€$	57.54	60.37
H	5	4.9	H	5.15	5.74
$€a$	10	10.17	$€a$	7.1	6.91

113   In his investigations on the pigment of human gallstones, *Städeler* prepared a calcium compound of bilirubin which yielded 9.1% calcium oxide by analysis. Proceeding from the assumption that this compound is a normal neutral salt, he determined the molecular weight of bilirubin according to the latter. He thus rejected his earlier analysis of crystalline *Cholephäin* that was communicated in *Frerich's Klinik der Leberkrankheiten* [liver diseases], as well as the empirical formula $C_{18}H_9NO_4$. He substituted $C_{32}H_{18}N_2O_6$ for the formula of free bilirubin and $C_{32}H_{17}CaN_2O_6$ for that of calcium bilirubinate [the preceding three formulas are given in the old notation]. Now these formulas are supported by only one unsatisfactory atomic weight determination based on the weight of calcium oxide. On the other hand, however, all of *Städeler's* analyses of bilirubin and *Cholephäin* can be brought into complete harmony with my results. So I cannot hesitate to explain as erroneous the formulas that this researcher has given to bilirubin and calcium bilirubinate.

The bilirubinate analyzed by *Städeler* was evidently the half salt.

Theory for $€_{27}H_{29}€aN_2Θ_6$	*Städeler*
requires	found
$€aO$ 9.94 %	9.1 %
$€a$ 7.1 "	6.5 "

With *Städeler's* formula for bilirubin fall the formulas of all other derivatives of bile pigments described by him, namely biliverdin, *Biliprasin*, bilifuscin, and bilihumin.

114   According to these data it is clear that the zinc salt is composed analogously to the half salts of barium and calcium, as follows:

1	Neutral zinc-*Cholephäinat*	$€_{18}H_{16}Zn\,N_2\,Θ_4$
2	Water	$H_4\quad\quad Θ_2$
1	*Cholephäin*	$€_9\,H_9\quad N\,Θ_2$
1	Molecule of half-acid zinc- *Cholephäinat*	$€_{27}H_{29}Zn\,N_3\,Θ_8$

The results of the analysis harmonize with this interpretation as follows:

		Theory		Found	
		Atom	%	a.	b.
$€_{27}$		324	–	–	–
H_{29}		29	–	–	–
Zn		65	11.05	12.03	11.30
N_3		42	–	–	–
$Θ_8$		128	–	–	–
		588			

A neutral *Cholephäinate* of formula $C_{18}H_{16}ZnN_2O_4$, molec. wt. = 389 would have yielded 16.70% zinc.

11 Translations

115   This compound can be interpreted as a basic *Cholephäinate* or as *Cholephäin* in which two atoms of hydrogen are replaced by a divalent lead atom.

	Theory		Found	
	Atom	%	a.	b.
C_9	108	29.4	–	–
H_7	7	1.9	–	–
Pb	207	56.5	58.38	57.91
N	4	–	–	–
O_2	32	–	–	–
	68			

This compound corresponds to the basic silver *Cholephäinate* or twice silver *Cholephäin* $C_9H_7Ag_2NO_2$, which has been described above.

116   *Composition and average of the analyses.*

	a.	b.	c.	d.	e.	Average
C	–	63.08	62.09	–	62.14	62.43
H	–	6.25	6.12	–	6.00	6.13
N	9.32	–	–	9.36	–	9.34
O	–	–	–	–	–	22.10
						100.00

These findings are compared with the facts concerning bilirubin:

	Bilirubin		Biliverdin	
	Theory in %	Average of the Analyses		Theory
C	66.26	66.01	62.43	63.57
H	5.52	6.01	6.13	5.96
N	8.59	9.03	9.34	9.27
O	19.63	18.95	22.10	21.20

117   ... thus it is found that in order to go over into biliverdin, bilirubin lost four carbons, gained a little hydrogen, its nitrogen content was increased somewhat, and the oxygen content was increased almost as much as carbon is reduced. According to *Heintz's* theory, biliverdin is an oxide of *Cholephäin*; according to *Städeler's* hypothesis it is a hydrated oxide. If biliverdin were a simple oxide of bilirubin ($C_9H_9NO_2 + O =)C_9H_9NO_3$, then it would require 60.33% C, 5.02% H, and 7.82% N. Comparisons of these numbers with those that analysis of biliverdin gave contradicts the view completely that biliverdin is an oxide of bilirubin. *Städeler's* hypothesis is still much less applicable since the formula $C_9H_9NO_2 + O + H_2O$ requires 54% C and a correspondingly lesser amount of hydrogen and nitrogen. If biliverdin consists of two molecules of *Cholephäin* bound with one atom of water, or $2(C_9H_9NO_2) + H_2O = C_{18}H_{20}N_2O_5$, then the following percentages would be required: C, 62.81%; H, 5.81%, and N, 8.13%. Even if the detected %C allowed this assumption, the percentage of H and N would be completely rejected. Biliverdin is neither an oxide, nor a hydrated oxide, nor a hydrate of bilirubin.

118   The average of the elemental analyses leads to a calculation of the formula of the following proportions.

$$N_1 : H_{9.2} : C_{7.8}.$$

11.1 Translations to Section 2

This gives the formula $C_8H_9NO_2$ for biliverdin. The analyses come out right for this theory as follows.

Atom	Atomic weight	Theory in %	Averaged/empirical
C_8	96	63.57	62.43
H_9	9	5.96	6.13
N	14	9.27	9.34
O_2	32	21.20	22.10
	151	100.00	100.00

119   Compilation from analysis and comparison with theory for 2 × calcium + 9 × biliverdin

	gr-Atom	%	a.	b.	c.	d.	Average
					Analyses		
C_{72}	864	60.20	–	–	61.33	63.06	62.19
H_{77}	77	5.36	–	–	5.68	5.8	5.74
Ca_2	80	5.56	5.52	5.77	–	–	5.64
N_9	126	–	–	–	–	–	–
O_{18}	288	–	–	–	–	–	–
	1435						

(Theory spans the gr-Atom and % columns.)

120   From the amount of calcium found, an atomic [molecular] weight 709 is calculated which, however, must be doubled so that the atomic [molecular] weight of biliverdin comes out even with a simple quotient in the residual. $\dfrac{1418-80+4}{9}=149$, which is as good as not being different from the atomic [molecular] weight, 151, of biliverdin directly found. The compound is composed of nine atoms of biliverdin and two atoms of calcium. If the grounds were present to assume the loss of one atom [molecule] of water from the compound, then one would have absolute agreement between theory and analysis. That this compound is actually a unique compound and not a mixture of a calcium compound with free biliverdin follows, among other things, from the fact that it is insoluble in alcohol. If it contained free biliverdin, alcohol would easily have extracted out the latter.

121   The average (22.41%) of the bariumfound leads to an atomic weight = 611, which by the operation $\dfrac{611-137+2}{151} = 3$ and a residual of 23, which means that one atom [molecule] of water can probably be added. A little support for this assumption can be obtained from the composition of the barium salt of *Cholephäin*, which contains the same two molecules, however with one atom of barium and one [molecule] of water. According to this assumption, barium biliverdinate is a half-acid mono-hydrate consisting of

1 molec.	Neutral biliverdinate	....	$C_{16}H_{16}BaN_2O_4$
1 molec.	Biliverdin	....	C_8H_9 NO_2
1 molec.	water	....	H_2 O
1 molec.	Half-acid barium biliverdinate	....	$C_{24}H_{27}BaN_3O_7$

	Atom	%	a.	b.	c.	d.
	Theory		Experiments			
C_{24}	288	47.52	–	–	49.39	48.20
H_{27}	27	4.45	–	–	4.43	4.34
Ba	137	22.60	22.15	22.67	–	–
N_3	42	6.93	–	–	–	–
O_7	112	18.50	–	–	–	–
	606	100.00				

652 11 Translations

122   If one water is left out of the calculation, then the theoretical %C fits better to the experimental, but the [%] barium less well.

			Average
C_{24}	288	48.97	48.79
H_{25}	25	4.25	4.38
Ba	137	23.29	22.41
N_3	42	7.14	–
O_6	96	–	–
	588		

123   . . . This behaves like an amide toward alkali, *i.e.* ammonia is released from it, while the residue combines with base to form yellow or green salt-like compounds.

All the *Cholepyrrhin* used for the experiments was twice recrystallized; for the time being, with exception of the following analyses, I divided it up:

Alcoholic or aqueous caustic potash [potassium hydroxide] caused *Cholepyrrhin* to evolve ammonia at the usual temperature. This was colored red for a short time and then became greenish-yellow.

Aqueous soda lye [sodium hydroxide] did likewise.

124   Biliverdin is an acid, *Cholepyrrhin* is its amide (biliverdin diamide). The former belongs to a water type, the latter to an ammonia type; or biliverdin and *Cholepyrrhin* behave like carbonic acid and urea.

125   The contents of such tubes was poured into water; a dark green chloroform layer collected below, while the water took up the acetic acid. The former layer was washed with water sufficiently to remove acidity. Then the aq. solutions were combined and taken to dryness on a water bath. The residue contained ammonium acetate in concentric white rings. It was thus a part of the nitrogen in *Cholepyrrhin* that split off in the form of ammonium acetate from the influence of acetic acid. The chloroform layer that had been washed with water and freed of acetic acid gave, after the solvent was evaporated, a dark, almost blackish-green residue of pure biliverdin.

126   This and the previous reactions show unmistakably that *Cholephäin* is an amide (an ammonium salt would not have produced such a long-lasting effect). As is characteristic of amides, hydrolysis occurs by alkali as well as by acids, leading to the corresponding acid (here, biliverdin) and ammonia, which in the first case escapes (as a gas) and in the second is present as a simple ammonium salt.

Biliverdin is an acid; its amide (biliverdin amide) is *Cholepyrrhin*. The former belongs to a water-type, the latter to an ammonia-type, or they behave like carbonic acid and urea.

127   Ammonia and caustic alkali dissolve *Cholepyrrhin* with a brownish-red color. From application of the latter, I observed earlier the evolution of ammonia. This error was, however, caused by a not-entirely pure sample of the substance obtained from human bile. I thus retract the conclusion derived from it at that time. At present, after many more extensive observations I have, moreover, reached the conclusion, based on evidence established later, that no ammonia is split off during the conversion of *Cholepyrrhin* into biliverdin and that the same atomistic amount of nitrogen is contained in the latter compound, as well as in the former.

128   Analysis

I.   0.2770 g human gallstones gave 0.681 g CO_2 and 0.545 g H_2O by combustion

II.   0.2734 g *Cholepyrrhin* from oxen gallstones gave 0.1532 g water. These gave the following percentages:

	I.	II.
%C	67.16	–
%H	6.18	6.22

11.1 Translations to Section 2

653

These values show agreement for with $C_{16}H_{18}N_2O_3$, and with *Städeler's* analytical average values:

	Calculated for $C_{16}H_{18}N_2O_3$	Average of *Städeler* (*i.e.*)
%C	67.13	67.13
%H	6.29	6.19

of so great a correspondence that I affirm completely the composition of these compounds, and no additional material needs to be sacrificed.

129   Accordingly, I consider the question of oxygen uptake to be settled. The uniqueness of oxygen from the air being absorbed and binding chemically, as we know, is not entirely odd. Indigo-white, gallic acid, and pyrogallic acid in alkaline solution act just like *Cholepyrrhin*.

130   The *preparation* of biliverdin can, according to the foregoing, take several paths. 1) Either the chloroform solution containing *Cholepyrrhin* is heated together with glacial acetic acid in a sealed tube, and the acetic acid is washed away with water; or 2) an alkaline solution is left standing in air for a few days, precipitated with hydrochloric acid and washed with water. The biliverdin was always purified further by dissolving in a little strong or absolute cold alcohol, filtered from some remaining brown flakes, and completely precipitated with water. The flocculent black-green precipitate thus obtained was washed with water and finally with ether.

    3) The influence of lead superoxide [lead dioxide] mentioned above, as well as that of Br_2, is also useful for the preparation of biliverdin. If lead superoxide is slowly mixed into an alkaline solution of *Cholepyrrhin* until it tests acidic, a pure green precipitate results. Then the entirety is slightly supersaturated with dilute acetic acid, whereby biliverdin-lead precipitates with complete decolorization of the solution, which is filtered. It is then washed until the washings are lead-free, decomposed with alcohol containing sulfuric acid, filtered, and precipitated with water.

131   Pure biliverdin is a glossy black compound and when powdered becomes dark green. It is tasteless and odorless and is wetted difficultly with water. Dried at 100°C, it exhibits a somewhat hygroscopic moistness. Its weight remains unchanged at this temperature; yet, it is very hygroscopic when dried this way.

    The purest dried biliverdin dissolves in alcohol not with a bright green color but with more of a sap-green color. However, when only a trace of acid (hydrochloric, sulfuric, glacial acetic) is added to this solution it becomes a splendid pure green.

    After addition of a little ammonia, the alcoholic solution of biliverdin turns a dark green; added calcium chloride forms a water-insoluble precipitate; with added silver nitrate a flocculent dark brown precipitate forms leaving complete discoloration of the liquid [supernatant]. The silver-biliverdin formed is insoluble in water but readily soluble in ammonia leaving a dark chestnut-brown color. Lead-biliverdin, prepared in a similar way using sugar of lead [Pb(II) acetate] appears brownish-green and flocculent.

    Trituration with conc. sulfuric acid dissolves biliverdin to yield a green color [solution], which upon the addition of water precipitates unchanged in alcohol-soluble green flakes.

    In alkali carbonates and hydroxides it dissolves with a sap-green or brownish-green color. It is taken up into ether in only insignificant amounts but dissolves very easily in chloroform as soon as a few drops of alcohol are added. It also dissolves in glacial acetic acid to give an especially beautiful coloration to the solvent.

    Biliverdin is insoluble in benzene and carbon disulfide, very poorly soluble in amyl alcohol and iodoethane, but completely soluble in either of the last two if a little ethyl alcohol is added to it.

    Methyl alcohol dissolves biliverdin as easily as common alcohol.

132   **Analysis**

I.   0.2400 g biliverdin gave 0.561 g CO_2 and 0.129 g H_2O.

II.   0.2905 g substance from a different preparation gave 0.1585 g H_2O.

11 Translations

III.   0.3356 g substance from a third preparation gave with red-hot soda-lime, etc., 0.204 g Pt.

IV.   0.3465 g of a fourth preparation gave a 0.210g Pt, leaving behind a quantity of ammonium hexachloroplatinate.

These results, after subtracting ~2% ash from III and IV (the substance in I and II was ash-free), correspond to the following percentages:

	I.	II.	III.	IV.
C	63.74	–	–	–
H	5.97	6.05		–
N	–	–	8.77	8.74

If *Cholepyrrhin* had taken up an atom of oxygen (at wt. 16) when it converted to biliverdin:

$$C_{16}H_{18}N_2O_3 + O = C_{16}H_{18}N_2O_4,$$

then the formula of biliverdin would be $C_{16}H_{18}N_2O_4$, and this would correspond to the calculation:

	%
C	63.58
H	5.96
N	9.26
O	21.19

which is only a little different in %N from that found.

If *Cholepyrrhin*, as *Städeler* reports, has also taken up a molecule of water:

$$C_{16}H_{18}N_2O_5 + O + H_2O = C_{16}H_{20}N_2O_5,$$

then the %C of biliverdin drops to 60.00%. I therefore believe that the former formula must be the more correct. Complete exhaustion of my material, from which the conclusion of this first article is occasioned, prevents me temporarily from a final yet necessary control analysis of biliverdin.

$C_{16}H_{18}N_2O_4 =$ Bilirubin + O requires	My earlier analysis gave		*Thudichum* found loc. ait.			*Städeler's* Formula $C_{16}H_{20}N_2O_5$ = Bilirubin + H_2O + O needs	
	I	II	I	II	III		
C 63.58	63.74		63.08	62.09	62.14	C	60.00
H 5.96	5.97	6.05	6.25	6.12	6.00	H	6.25
N 9.26	N {8.77 & 8.74}		N {9.32 & 9.36}			N	8.75
O 21.19						O	25.00.

My earlier analysis showed a difference only in the %N, which had been determined from NH_3. Now, however, a number of sources, such as *Ritthausen* and *Kreusler* – and especially from *Nowak* – reveal that certain compounds give up their entire nitrogen content while being heated red-hot with copper oxide. So this time the %N was determined by the *Dumas* method.

1.   0.2785 g biliverdin, dried at 100°C, gave 0.6516 g CO_2 and 0.1452 g H_2O.
2.   0.369 g of a different preparation gave 31.5 cm³ wet nitrogen at 15°C and 27.35%

	Found		Calculated for $C_{16}H_{18}N_2O_4$
%C	63.82		63.58
%H	5.80		5.96
%N		9.35	9.26.

11.1 Translations to Section 2

The modified %N-determination thus caused the small deficit in %N to disappear for biliverdin. Thus, therefore correlation with respect to the different preparations from the basis of *Thudichum's* and my analyses is entirely complete, and the composition of this compound may be regarded as definitively established.

135 Bilirubin was dissolved in a very dilute soda [sodium carbonate] solution and allowed to stand for a few days with occasional introduction of oxygen. The biliverdin [so formed] was precipitated with hydrochloric acid, dried, and collected on a weighed filter and washed until the washings gave a negative chloride test. It was then dried at 110°C and weighed. The green-yellow filtrate was evaporated and the content of organic substance was determined from the residue that had been dried at 125°C. The wash water from the first filtration also showed a trace of yellow-green coloration in a thicker layer and was estimated colorimetrically. Thus, one obtains:

Bilirubin used (dried at 110°C)	0.4558	g
Biliverdin filtered off (dried at 110°C)	0.4458	"
Organic substance in the filtrate	0.0223	"
Total biliverdin	0.4681	g

136 In any case the analysis and in   in weight are   eement and both lead to a biliverdin formula $C_{16}H_{18}N_2O_4$, which uiiieis from bilirubm oy an excess content of O.

137 . . . Without going into it in more detail, I mention only that it is in fact very much richer in oxygen than *Cholepyrrhin* or biliverdin, while the carbon and hydrogen content is lower. The following numbers show this:

	%Oxygen	%Carbon
Cholepyrrhin content }	16.79	67.13
Biliverdin " }	21.19	63.58
New compound "	30.39	55.23

This new compound might not have been obtained completely pure, as its careful analysis shows that oxidation has proceeded beyond the formation of biliverdin.

138 Consequently, there can be no doubt that other oxides of *Cholepyrrhin* are present in the bile pigment samples and lie between biliverdin and the 30% oxygen compound of the wine-red solution formed by a series of multiple oxygen incorporations. In any case a blue and a red compound and the light brown end product are still present, while the violet is probably a mixture of the red and blue.

Accordingly I will seek to expand the direction of my research by using bromine as a means to its preparation and purification.

139 Using spongy platinum reduces the time of biliverdin production from a few days to a few hours. From a reddish-brown solution in a shallow dish, one can see the color changes emanating from inside the spongy platinum.

140 Samples were taken for analysis of the substance stemming from three different preparations. They were dissolved in alkali and rapidly precipitated by acid, and the process was repeated first once, then twice, and three times.

1. 0,2193 g substance gave 0.523 CO_2 and 0.1404 H_2O.
2. 0,2652 g substance gave 0.1646 H_2O.
3. 0,2262 g substance gave 0.1474 Platin.
4. 0,2483 g substance gave 0.5886 CO_2 and 0.1559 H_2O.
5. 0,2174 g substance gave 0.5142 CO_2 and 0.1347 H_2O.

Shown in %:

	1.	2.	3.	4.	5.	Average
C	64.89	–	–	64.65	64.50	64.68
H	7.09	6.80	–	6.98	6.87	6.93
N	–	–	9.22	–	–	9.2.

These numbers correlate so well with each other that the reaction must be characterized as very smooth. The substance is poorer in carbon and richer in hydrogen than bilirubin, corresponding to its formation. It can have arisen therefore only by bonding to hydrogen. If one assumes that water has also been added in, of course one water with two bilirubins in addition to hydrogen, then the compound $C_{32}H_{40}N_4O_7$ would result, according to the equation:

$$2\ C_{16}H_{18}N_2O_3 + H_2 + H_2O = C_{32}H_{40}N_4O_7$$

and this requires:

		Found (Average)
C_{32}	64.86	64.68
H_{40}	6.75	6.93
N_4	9.45	9.22
O_7	–	–

which is in good agreement with the results obtained. The new compound thus arose from the uptake of hydrogen and water, with a doubling of the bilirubin molecule (if not, as is perhaps more likely, the bilirubin formula is twice as large as usually written) and should henceforth be designated as *hydrobilirubin*.

141 By repeating these passages from *Jaffe's* article, I have that at best shown the properties of my hydrobilirubin and *Jaffe's* urobilin are identical, and thus the substances themselves are identical. The name *hydrobilirubin* that I introduced for my substance (more correctly, for both) is based on its synthetic and natural formation, and secondly at least some of its constitution is expressed by the name.

142 ... what is not far removed from the composition of hydrobilirubin, and would suggest that *Scherer's* preparation at least contains no large amounts of contaminant. Other analyses of course gave more prominent results*).

---

*) The alleged iron content of the most important binary pigment is at least disproved. Also Dr. *Schlemmer* has recently again sought in vain in my laboratory for iron in large quantities of urine.

143 As we have seen the close relationship that the orange pigment of bile and the (major) urinary pigment, at least in humans, exhibit toward each other, the circulation of this pigment emerges, and many unrelated facts are nicely set straight. The bilirubin poured forth into the intestine from bile undergoes changes during its travel down the colon, and there the compound takes up hydrogen and water under the influence of hydrogen-releasing processes. Biliverdin behaves quite similarly: I have treated an alcoholic solution of biliverdin with sodium amalgam and soon obtained a brown solution identical with that of bilirubin.

Hydrobilirubin is absorbed from the gut and ultimately goes into urine in order to complete its cycle there. Because hydrobilirubin plays no evident role in the gut, and its absorption is only a means to bring the compound out of the organism, there is no reason to think that bile pigments should have any other role except mainly that. At present one has to view the bile pigment as none other than a useless by-product of liver chemistry.

11.1 Translations to Section 2 657

144   I have mentioned that in addition to acids and bases there is yet a third series of compounds that produces biliverdin from *Cholepyrrhin*; these are the halogens, bromine and iodine. The transformation by means of bromine is especially astonishingly beautiful. If *Cholepyrrhin* is placed under a bell jar that contains bromine vapor mixed with damp air, then its color soon darkens, and it is no longer soluble in chloroform but is soluble in alcohol with a pure green color. However, the since action of bromine easily goes a bit too far, one can adjust the experiment more advantageously in the following manner. Mix a yellow chloroform solution containing *Cholepyrrhin* with a very dilute alcoholic solution of bromine. The first drops turn the liquid a dark sap-green, and with further careful addition of bromine the solution reaches a point where it exhibits a splendid fiery green color.*) At this moment all of the *Cholepyrrhin* has gone over into biliverdin, and the liquid can stand for weeks without change.

---

*) Biliverdin remains dissolved in this mixture of chloroform and a little alcohol.

145   I am probably correct to assert from such agreement that the previous conventional formula of bilirubin, which was expressed as $C_{16}H_{18}N_2O_3$, cannot be maintained but that it must be doubled. It then becomes $C_{32}H_{36}N_4O_6$ and one has

Bilirubin $C_{32}H_{36}N_4O_6$
Tribromobilirubin $C_{32}H_{33}Br_3N_4O_6$
Hydrobilirubin $C_{32}H_{40}N_4O_7$

146   Absorption Spectra of Bile Pigments.
A chloroform solution of *Cholepyrrhin* placed before the slits of a spectral apparatus extinguished the blue and violet entirely up to line 70 on the *Bunsen* Scale. Very dilute, barely yellow solutions still extinguish the violet.
Solutions of *Cholepyrrhin* in aqueous ammonia behave similarly. If they have the color of a concentrated solution of potassium dichromate, the field of vision from the violet end to nearly the sodium line (50) is completely darkened [extinguished], and delimited more or less sharply. If the solution is diluted, it gradually begins to appear yellow and green by degrees and is somewhat indistinct. Even solutions that are so dilute that they appear nearly colorless in lamplight, as with the coloring power of barely detectable traces of *Cholepyrrhin* in ammonia solutions, a good part of the violet is still extinguished.

147   Biliverdin in alcohol shows absorption at both ends of the spectrum. Only green light passes through strongly colored layers. In somewhat more dilute solutions there first appear yellow, orange, and some of the red, later blue and violet. The most extreme red is still removed from very dilute solutions.

148   If hydrobilirubin is dissolved in dilute alcohol, or if it is added to such a dilute alkaline solution (in ammonia or sodium phosphate, *etc.*) so that a little added hydrochloric or glacial acetic acid no longer causes precipitation when taken to acidity, *i.e.* so far that the solution loses its yellow color and becomes reddish-yellow or rose-colored; then, thinner layers (0.5–2 cm) placed before the spectral slit show a very strong and marked absorption of the spectrum between green and blue, and of course in my larger apparatus (when Li is at 102.5, Na at 120, and K-β at 219.5) within the graduation marks 146 to 160, or more generally expressed exactly between *Fraunhofer* lines b and F. Just as it is stronger when the solution is acidified, in contrast, ammonia makes the band disappear, leaving only a weak, diffuse absorption between green and blue. However, with addition of acid, the black band returns, together with reddish color.

149   In contrast, an ammonia solution of the pigment, when it contains a little dissolved zinc salt (also cadmium), gives especially beautiful [spectral] bands. A couple of drops of added zinc chloride or zinc sulfate to a conc. ammonia solution of hydrobilirubin (whereby the emerging precipitate is easily dissolved again) is sufficient for bringing it before the apparatus. Or one can dissolve the precipitated zinc-hydrobilirubin in ammonia and dilute it. Both

658   11   Translations

solutions are rose-red and give a [spectral] band distinguished by sharpness and darkness, which in relation to the acid solutions appear shifted toward the left, from 142 of my scale, thus a little sharply delimited before b, disappearing broadly toward the right, according to the concentration of the solution, that however always appears darkest from 142 to 155, that is from b to the middle of the spectral region b to F. The entire appearance is at least just as sensitive as that of acidic solutions of the pigment.

150   If a concentrated solution is brought before the slits of the spectral apparatus, the spectrum thus appears completely dark from the violet end to approximately line b. Upon dilution the extinguished part lightens and finally an absorption line (γ) with somewhat indistinct edges remains at the often cited place between *Fraunhofer* lines b and F . . .

The dilution at which fluorescence appears in a urobilin solution is enormous. Solutions which are nearly colorless in transmitted light show a striking yet distinct green shimmer, especially if they are exposed to direct sunlight.

151   This absorption band lies, as already described, between lines b and F, but closer to line b than the line in acidic solution (γ). Furthermore, it is darker, more sharply delineated than the latter, and still remains visible at the greatest degree of dilution.

152   Therefore this compound in feces, which, by the way, has never been isolated, is entirely different from that of urine, and all reports on its identity, *etc.* are erroneous.

153   Mr. *Jolles'* article begins with the entirely unconditional statement that (an alcoholic solution of) bilirubin is converted into biliverdin by iodine. This assertion is, however, only an analogy deduced from the alleged and disproved reactions of bromine, and it thus falls with their model. Also no experiment has been completed to substantiate the alleged reaction. No products have been isolated or analyzed. I could be content therefore with the results that since Mr. *Jolles'* premises do not exist at all, his theses necessarily have this same fate.

154   Spectroscopic Results. On page 3 of his article Mr. *Jolles* said he had proven spectroscopically the identity of his green product (obtained from bile or bilirubin by iodine) with biliverdin. However, the biliverdin prepared from bilirubin under the influence of aq. soda [sodium carbonate] as solvent and air as oxidizing agent has no specific absorption shadow [extinction] in its spectrum. From that alone, as absorption shows, it follows that *Jolles'* green product obtained from bile of bilirubin by iodine, is not identical with bilirubin.

155   Unsubstantiated Formulas. Thus, the various formulas repeated in the article should disappear, whereby one molecule of so-named bilirubin takes up four atoms of iodine and two molecules of water for oxidation to biliverdin. Since the reaction does not exist at all, the formulas designated for the pigments are not valid.

156   Mr. *Jolles* repeats the erroneous reports that *Maly* and *Städeler* had "determined" the formula of bilirubin as $C_{32}H_{36}N_4O_6$. Not only had *Maly* determined no formula at all for bilirubin, he not once analyzed it completely and especially had neither measured nor weighed the nitrogen. He was therefore not quite in a position to calculate a formula. Only *Städeler* had attributed the formula $C_{16}H_{18}N_2O_3$ to bilirubin; it was doubled around [the year] 1870 in order to be able to represent the material as a hexabasic acid. This entirely unwarranted attempt is completely faulty.

157   After the publication of my investigations, *Städeler* also abandoned this second formula of bilirubin, and with it all other formulas of his biliverdin, *Biliprasin*, and bilifuscin. He doubled his stipulated formula for bilirubin a second time, to $C_{32}H_{36}N_4O_6$, and declared it to be a hexabasic acid. The then-recent discovery of mellitic acid had perhaps misled him. He is now constrained by my results into this hypothesis without carrying out a single preparation or analysis. No single one of *Städeler's* formulas and no single element of any one of his formulas can be derived from my preparations and analyses. His calculated amount of metal in the compounds amounts to all of 1–6% less than the amount I determined. I can characterize this method of *Städeler's* as merely the result of desperation in all his work. According to my conviction his first preparation was pure and his analysis correct, although he has abandoned it without explanation, erroneously being led astray by a misleading calcium determination.

158   Before I go in more detail into the substantive content of Mr. *Thudichum's* objections, I must first express my astonishment that *Thudichum* has apparently read through my article

11.1 Translations to Section 2 659

cnly quite hostilely, to which he quite arbitrarily and unjustifiably gave an account of my work: that it is based on the alleged discovery of Prof. *Maly* in 1868, whereby the transformation of bilirubin into "biliverdin" by means of bromine should be based on an oxidation process. I must, however, categorically reject this imputation for I have not brought up the results of *Maly's* investigation as evidence for the correctness of my results. Rather – as usual – because it appears in the literature, and appears as well to be tangential to my work, I felt obligated to cite *Maly's* published results of 1868. Mr. *Thudichum* errs greatly if he believes that *Maly's* second investigation of 1872, in which he revoked the first investigation, was not known to me. However, *Maly* had followed quite different experimental conditions in his second investigation, which were chosen more advantageously for the formation of the brominated substitution product, and it can be taken for granted that the change in experimental conditions just from the reaction of halogens can lead directly to entirely different results, and on the other hand . . . .

159 . . . that in my investigations I have no other purpose than to make known a method which allows the bilirubin present in animal bile to be determined quantitatively, it appeared to me to be superfluous to point out *Maly's* second investigation in the references. The quintessence of my investigation was of course only to show that by observing defined experimental conditions, bilirubin is transformed quantitatively into a green pigment, whereby four atoms of iodine were expended to one mole of bilirubin so that this process presents a means to determine quantitatively the bilirubin content in animal bile and urine. This method satisfies my purpose to prove that my numerous analyses provide the experimental proof which previously no-one had refuted and which has actually been recognized in handbooks as the method *Huppert* recommended for the quantitative determination of bilirubin in the 10th edition of his "Introduction to Qualitative and Quantitative Analysis of Urine" (p. 865).

160 *Thudichum* regarded the green pigment as an iodinated substitution product; however, I regarded it as an oxidation product, and certainly as biliverdin. I admit that *Thudichum* is correct to the extent that I have not provided analytical proof for the identity of the green pigment with biliverdin. However, I note that insufficient material was present for publication. On the basis of that I am entitled to hold to the view that the action of a dilute alcoholic solution of iodine on bilirubin is an oxidation process and the initially formed green pigment is biliverdin.

161 What Mr. *Thudichum* now primarily disputes is the fact that the influence of iodine solution cn bilirubin is an oxidation process. According to him it proceeds as simply and solely as a substitution process. Mr. *Thudichum* puts into his rejoinder the specific point of view that all his assertions relative to bile pigments should be regarded formally as irrefutable dogma.

162 Moreover, both researchers remained indebted to us for the incontestable proof of whether the action of bromine on bilirubin under the specified conditions proceeds in parallel to the substitution product and also to an oxidation product. Furthermore, if oxidation does not take place at first and occurs from further action following substitution, the compound obtained by the authors would, on one hand, be a mixture of oxidation and substitution products and, on the other hand, could be regarded as brominated derivatives of the oxidation products of bilirubin. One must wonder whether, on the basis of such as yet rather incomplete work, the influence of halogens on bilirubin is to be viewed as a closed field . . .

163 The chief support for his contention that an iodine substitution product emerges from the influence of iodine on bilirubin is a conclusion formed by the analogy that because bromine effects substitution on bilirubin, this must also doubtless be the case with iodine. Such a conclusion is generally not allowed for iodine and bromine. It is still to be taken into account that iodine dissolved in solution especially when extraordinarily dilute, does not generally effect substitution on dissolved organic substances, which of course *Kekulé* first pointed out in detail.

164 I. 0.1846 g substance dried at 100°C gives 0.3928 g CO_2 and 0.0963 g H_2O.
II. 0.1708 g substance dried at 100°C gives 14.4 cm³ N_2 at 728 mm and 20°C.

Calculated for bilirubin, $C_{16}H_{18}N_2O_4$:		Found:
%C	63.58	62.76
%H	5.96	6.27
%N	9.26	8.44

660   11   Translations

165   From the results indicated above, the fact emerges with certainty thereby that, under the specified experimental conditions, the product arising from the influence of an alcoholic iodine solution on bilirubin is neither an iodine substitution product nor an iodine addition product, but only an oxidation product. And that, based on the results of the elemental analysis as well as the characteristic properties of the compound, is clearly to be regarded as biliverdin.

166   Thus, a 57-page article from Dr. *Adolf Jolles* in Vienna was completely refuted. Nevertheless, he has sought in this journal to revive a few of his earlier conjectures to this purpose by changing his position so as to compel the reader who does not follow the subject to be oriented toward them.

167   The foundation on which he based his submitted work from existing analogies, namely the oxidation of bilirubin by bromine, though long proven as non-existent, has now slipped out of his hands. All of the false information that he indicated in the formulas, results, and processes allegedly from *Rödeler* and *Maly* are also omitted from the new text. Therefore, *e.g.* instead of $C_{32}H_{36}N_4O_6$, which is found by no-one for the formula of bilirubin but only presented on paper and is false in every connection, he now gives the likewise entirely false formula $C_{16}H_{18}N_2O_3$, with no analysis or supported connection. He has not, for example, analyzed bilirubin or defined compounds or investigated the biliverdin prepared by *Zuntz's* method (however not by iodine), but argues for the preparation of biliverdin from a method, supposedly *Maly's*, that does not exist at all.

168   I have shown that the spectra assigned by Dr. *Jolles* to bilirubin and biliverdin, as shown in a table with similar compounds, are not properties of these compounds. He just now admits that his spectra were obtained from a "substance", which he purchased pure from a manufacturer while it is shown in fact to be impure. The absorption spectra were sketched by him and are therefore products of impurities in his preparations, and not at all the substance under investigation. On page 3 of his article in *Pflüger's Archiv* he claims to have shown spectroscopically the identity of his green product "obtained from iodine reacting with bile or the putative bilirubin" with biliverdin. All that he had proven, without knowing it, was the impurity of all his preparations. It is thus entirely beyond dispute that from his mistakes he would create only confusion if anyone believes them. In any case, however, his alleged quantitative estimations, designated as falsely determined, possess no value.

169   In his article on this subject in the *Vienna Monatsheften*, is essentially a repetition of the paper in this journal. Mr. *Jolles* said I had not analyzed my bromine substitution product. This statement is not about an investigation by Mr. *Jolles* himself but it is plagiarized/copied from a paper by Prof. *Maly*. It is completely unfounded. The theory of formation of dibromobilirubin has not only been proven by the increase [of weight] of bilirubin by bromine, and the evolution of hydrogen bromide, but also by complete elemental analyses of two preparations, one of which weighed over 20 g. Long ago I refuted *Maly's* false data, and herewith once again I protest the carelessness with which Mr. *Jolles* treats the literature.

170   I have shown many examples of this from the chemistry of urine, of bile, of the brain, and other organs and parts of the body mainly in articles in this journal, and in more than 30 articles in English medical periodicals. In order not to trouble you with repetitions, at this point I call attention to three of my most recent communications, which have appeared in *Virchow's Archive for Pathological Anatomy and Physiology and for Clinical Medicine* **150** (1897), **586** *ibid*, **153** (1898) 154, and **156** (1899) 284. Two of them concern mainly the errors I refuted 25 years ago: that the so-named urobilin, the one isolated from urine, an as yet unanalyzed substance, is identical with the hydrobilirubin, the same one obtained as a mixture of pigments from sodium amalgam reacting with bilirubin. This false statement begat by *Maly* was later schlepped along by physiological chemists until it had received another, and it is hoped, final rejection from investigations of Mssrs. *F.G. Hopkins* and *A.E. Garrod* in London. More recently it has been shown by elemental analysis that the urobilin is nothing more than *Omicholin* discovered by me in 1864 and described exactly from eight preparations. Their urobilin contained 4.11% N; my *Omicholin* 4.18% N. In contrast hydrobilirubin has 9.75% N. Accordingly, the differences above need no further explanation.

171   Just because *Heller* now mentions the name *uroxanthin* for a sufficiently characterized educt, and the name has come into use in the literature, I consider its application to a different product as forbidden according to the laws of literature ethics.

11.2 Translations to Section 3

172   As further proof for the careless way with which many authors of physiological articles deal with the best discoveries of their predecessors, I mention the fate of *Heller's Urohodin*, a product first produced by him. It was interpreted by conjectural chemists as indigo-red or *Indirubin* because it was obtained in addition to indigo-blue and had a red color. Only the elemental analysis put an end to this assumption. My *Experiment über das Urohodin* in *Pflüger's Archiv* **15** (1877) 346 showed that it is obtained from uncolored *Urohodingen* by strong hydrochloric acid and can in no way be an indigo-blue isomer because it contains no nitrogen and probably 80% C. Its ether solution shows a specific absorption spectrum in which all of the green is extinguished by a dark band when red and blue are transmitted.

173   He has then written an entire chapter describing the preparation of alleged crystalline bilirubin, without having produced even a single crystal. I therefore reject the appearance of the products of Mr. *Küster* and his conclusion derived based on their purity, and assert that all products of bilirubin from this process are impure. He has led himself to a proof from his elemental analysis of samples whose %C is 0.7–1.3% too high, but whose %N, from 1.53% to 2.89%, is particularly too high when compared to the compound prepared according to my method and to macroscopic crystalline bilirubin.

174   Finally, I must caution the reader of a few failures made by Prof. *Maly* in his Annual Report, as evidence of the continuation of his attack against the truth of my research on the brain, which he had proven previously by plagiarism, so I am compelled straightaway to convey publicly. The "reports" of Mr. *Maly* are incorrect from beginning to end and thus are not believed by the informed reader.

11.2   Translations to Section 3

1   In any case the results obtained from bilirubin speak in favor of *Städeler's* simple formula, $C_{16}H_{18}N_2O_3$. As can be seen, the non-integral values from our research are caused by the sparing solubility and easy decomposition of both pigments. They are sufficient, however, to determine whether the simplest formula deduced from the analysis is correct.

2   As everyone knows, *Maly* found that bilirubin is transformed into urobilin by sodium amalgam. A very similar pigment is obtained from hematoporphyrin following reaction with tin and hydrochloric acid, or iron and acetic acid. However, as later investigators showed, the urobilin from hematoporphyrin is not identical to the urobilin from bilirubin.

3   Recently *Nencki* and *Rotschky* sought to answer the interesting, controversial question whether the composition of bilirubin corresponds to the simple formula, $C_{16}H_{18}N_2O_3$, or the doubled, $C_{32}H_{36}N_4O_6$, by *Raoult's* method. Their plan can hardly be considered successful because bilirubin is poorly soluble in the collection of solvents under consideration, or is changed thereby, as *e.g.* by acetic acid. Phenol proved to be the best solvent for bilirubin. A saturated solution of bilirubin in phenol contains approx. 0.4% pigment. So from only a content of 0.3–04%, corresponding to one molecule of bilirubin dissolved in 1,000 molecules of phenol, the [molecular weight] determination gave a value corresponding to the formula $C_{16}H_{18}N_2O_3$ . . . The experiments of *Nencki* and *Rotschky*, despite the insolubility of this pigment, do not show [even] what [mp] depressions [might have been obtained] corresponding to the measured molecular weight that had not one molecule but, for example, 2, 3 up to 10 [been] dissolved in phenol.

4   The determination of bilirubin's molecular weight by *Nencki* and *Rotschky*, as has already been emphasized at the outset, cannot be regarded as decisive for one or the other formula because of the pigment's sensitivity to decomposition and (sparing) solubility. As everyone knows, [of course] *Maly* obtained hydrobilirubin by reduction of bilirubin with sodium amalgam, the former a pigment identical with the urobilin obtained by *Jaffe* from urine. *Maly's* analysis is in good agreement with the formula $C_{32}H_{40}N_2O_7$. Precisely the composition of hydrobilirubin was an occasion for *Maly* to double *Städeler's* bilirubin formula because the formation of the reduction product of bilirubin itself in the simplest way yields:

$$C_{32}H_{36}N_4O_6 + H_2 + H_2O = C_{32}H_{40}N_2O_7.$$

Since hydrobilirubin has the desirable property that it is soluble in phenol as well as in glacial acetic acid, and it was of great interest to see whether this pigment, the formula of which, based on elemental analysis is not divisible, would follow *Raoult's Law*. At the request of Prof. *Nencki, Herr* Prof. *Maly* kindly presented us with hydrobilirubin, sending somewhat over 1.5 g to me. The preparation was eventually dissolved in ammonia and precipitated with HCl. Before using it, I dried it over H_2SO_4 under vacuum to constant weight.

5 Upon the invitation of *Hrn*. Prof. *Hüfner*, I thus took up the study of hemin and want to report in the following the results obtained over the summer semester 1891.

6 Hemoglobin is cleaved rapidly into hematin and globulin by caustic lye of fixed alkalinity and likewise by acids, *e.g.* glacial acetic acid and tartaric acid, and more slowly by concentrated ammonia. In the presence of HCl compounds, by treatment in a large excess of glacial acetic acid at the usual (room ?) temperature and more quickly by heating, hemoglobin is cleaved into globulin and hemin, which crystallizes out and contains 4% chlorine. Hemin in the hydrochloride of hematin. By dissolving the former in caustic alkali and partially by gaseous ammonia possessing sufficient moisture, it is split into metallic chloride and a compound of alkali + hematin. Pure hematin is obtained from hemin crystals by evacuating its solution in (aqueous) ammonia, to complete dryness, extracting the residue with water and drying. Hematin possesses the same dark gray-blue color as hemin.

7 If the formula above for hematin is compared with that which *Städeler**) indicated recently for bilirubin in his nice work on bile pigments, the simple relationship results:

$$2 (\Theta_{48}H_{51}N_6Fe_3\Theta_9) + 3\ H_2\Theta = 6\ (\Theta_{16}H_{18}N_2\Theta_3) + 3\ Fe_2\Theta$$
$$\text{Hämatin} \qquad\qquad\qquad \text{Bilirubin}$$

or three molecules of bilirubin arise by substitution of hydrogen for iron in one molecule of hematin. On the basis of physiology, which supports the assumption of the formation of bile pigments from hematin, I have already indicated amply in earlier work on bile substances the physiologic basis that supports assumptions on the formation of bile pigments and will not return to that again here.

---

*) Communications from the analytical laboratory in Zürich 1863. I. On the Pigments of Bile.

8 . . . the preparation obtained from bilirubin by treatment with sodium amalgam or tin and hydrochloric acid by *Hrn*. Dr. *Kistiakowsky* in my laboratory led to the conviction that *Jaffe's* urobilin on *Maly's* hydrobilirubin is identical with my reduction product of hematin. Since the same pigment is easily obtained by treatment of intact hemoglobin with tin and hydrochloric acid in alcoholic solution it follows that the pigment of normal fecal material and of urine may be understood as a blood pigment cleavage product changed by reduction, that bilirubin and biliverdin represent intermediates of this transformation, or at least remain in a close relationship to the blood pigment . . .

9

			C	H	Cl	Fe	N	
I.	Cattle blood	1)	62.73	5.69	5.29	8.95	8.99	%
		2)	62.75	5.71	5.28	8.72	–	"
II.	Horse blood	3)	62.81	5.86	5.38	8.61	9.13	"
		4)	62.90	5.98	5.30	8.96	–	"
III.	Pig blood	5)	62.72	5.72	–	–	–	"
IV.	Human blood	6)	–	–	5.22	8.96	–	"

11.2 Translations to Section 3 663

With the hindsight that the crystals contain a molecule of amyl alcohol [1-pentanol], the percent composition corresponds to the formula $(C_{32}H_{30}N_4FeO_3HCl)_4C_5H_{12}O$; which requires 63.09% C, 5.69% H, 5.59% Cl, 8.86% Fe, and 8.86% N.

10 The elemental analyses of the hematin prepared from hemin crystals later yielded the following values:

			C	H	Fe	N	
I.	Cattle blood	1)	64.98	5.61	9.35	9.49	%
II.	Horse blood	2)	64.99	5.62	9.29	9.34	"
		3)	64.68	5.37	–	–	"
III.	Pig blood	4)	65.04	5.53	9.29	–	"
		5)	65.13	5.55	9.31	–	"

The values obtained for the hematin prepared from hemin crystals correspond to the formula: $C_{32}H_{32}N_4FeO_4$, which requires: 64.68% C, 5.40% H, N and Fe [both] 9.47%.

11 By dissolving hemin crystals in alkali, not only are HCl amyl alcohol split off, but H_2O is also taken up in the molecule, corresponding to the equation:

$$(C_{32}H_{30}N_4FeO_3HCl)_4C_5H_{12}O + (NaOH)_4 = (C_{32}H_{32}N_4FeO_4)_4 + C_5H_{12}O + (NaCl)_4.$$

12 We will therefore designate the compound $C_{32}H_{30}N_4FeO_3$ with the name *Hämin. Teichmann's* crystals are the hydrochloride of the same compound. The latter is transformed into hematin by dissolving hemin in alkali, and its composition $C_{32}H_{32}N_4FeO_4$.

13 ... analyses of hematoporphyrin yield a composition corresponding to the formula $C_{32}H_{32}N_4O_5$, and its formation happens according to the following equation:

$$C_{32}H_{32}N_4O_4Fe + SO_4H_2 + O_2 = C_{32}H_{32}N_4O_5 + SO_4Fe + H_2O.$$

The yield of hematoporphyrin from hematin is not great.

14 Formation of hexahydrohematoporphyrin from hemin crystals occurs with the simultaneous uptake of water and hydrogen in the molecule, according to the equation:

$$C_{32}H_{30}N_4O_3FeHCl + 2\ H_2O + HCl + H_2 = C_{32}H_{38}N_4O_5 + FeCl_2.$$

15 By heating in alcoholic potassium hydroxide, hexahydrohematoporphyrin is converted to a product which is readily soluble in aqueous base and which has a great similarity to urobilin.

16 The theoretical requirement is also in accord with this; the transformation of hemin into urobilin depends chiefly on a hydration:

$$C_{32}H_{30}N_4FeO_3 + 4\ H_2O + 2\ HCl = C_{32}H_{40}N_4O_7 + FeCl_2.$$

17 ... in such a way, one arrives at the formula of pyrrole as C_4H_5N, so that indole would be assembled from benzene and pyrrole exactly like naphthalene from two benzene rings.

18 III.

The Relationship of the Pigment of Blood to Bile Pigments.

Bilirubin is constituted according to the formula $C_{32}H_{36}N_4O_6$. It follows from *Maly's* investigations that this formula is to be preferred over the older one from *Städeler*: $C_{16}H_{18}N_2O_3$. Bilirubin is transformed into urobilin (hydrobilirubin = $C_{32}H_{40}N_4O_7$) by uptake of H_2O and H_2. This transformation is understandable only by doubling *Städeler's* formula. Furthermore, bilirubin gives tribromobilirubin (= $C_{32}H_{33}N_4O_6$) with bromine, which is converted into biliverdin by alkali.

$$C_{32}H_{33}Br_3N_4O_6 + 3\ KHO = C_{32}H_{33}(HO)_3N_4O_6 + 3\ KBr$$

Tribromobilirubin.   Biliverdin.

11 Translations

When the blood pigment is converted into the bile pigment, it loses iron and the molecule takes up water.

$$C_{32}H_{32}N_4O_4Fe + 2\ H_2O - Fe = C_{32}H_{36}N_4O_6$$
Hämatin.   Bilirubin.

19   With this simple equation chemistry fulfills an old requirement of pathology, that a genetic relationship must exist between the blood and bile pigments. As often as blood extravasates from the wall of a living vessel into the surrounding tissue, bilirubin (hematoidin) is either deposited in the tissues or increased excretion of urobilin occurs. The formation of bile pigments from the blood pigment under physiological conditions is now understandable. The converse of the situation is also possibly the case, namely that in its formation in the liver cell bilirubin is incompletely [formed] hemin.

20   We have calculated the formula that we advanced for hematoporphyrin based on the elemental analysis. From two differently produced preparations we found:

C	69.57	and	69.44	%
H	6.20	"	6.13	"
N	9.67,	9.83	and	10.17 %.

The formula suggested by us requires: C, 69.55%; H, 5.80%; N, 10.14%; O, 14.51%.

The values obtained are in even better accord with the formula: $C_{32}H_{34}N_4O_5$, which requires: C, 69.31%; H, 6.13%, N, 10.10%.

It differs from that above only by the addition of H_2. The origin of the hematoporphyrin from hemin would therefore be very simple:

$$C_{32}H_{32}FeN_4O_4 + H_2O - Fe = C_{32}H_{34}N_4O_5.$$

21   The formula of the hydrochloride salt [$= C_{16}H_{18}N_2O_3 \cdot HCl$] obtained from the experimental numbers requires:

			Experimental values.		
Theoretical values			I.	II.	III.
C_{16}........	59.53%	C.........	59.80%	59.79%	59.57%
H_{19}........	5.89	H.........	6.16	5.89	6.29
N_2.........	8.68	N.........	8.5	–	8.41
Cl	11.00	Cl........	10.77	10.73	10.9

The hematoporphyrin formula [$= C_{16}H_{18}N_2O_3$] requires:

				Experimental values	
Theoretical values		I.		II.	III.
C	67.13%	C.........	66.84% and	67.16% 66.98%	66.85%
H	6.29	H.........	6.32 "	6.56 6.21	6.53
N	9.79	N.........	9.77	– –	9.51

22   The elemental analysis of the hydrochloride salt as well as the free hematoporphyrin prepared from it yielded the surprising fact that the latter is composed according to the formula: $C_{16}H_{18}N_2O_3$, that is isomeric with the bile pigment – with bilirubin.

23                      $(C_{16}H_{18}N_2O_3)2 = C_{32}H_{34}N_4O_5 + H_2O.$

11.2 Translations to Section 3 665

The hematoporphyrin obtained by means of conc. sulfuric acid would therefore be an anhydride of the newly obtained hematoporphyrin, and this assumption seems the most probably to us. The hemin crystals are composed according to the formula $C_{32}H_{31}ClN_4O_3Fe$. Hydrogen would have to be released during the formation of hematoporphyrin corresponding to the equation:

$$C_{32}H_{31}ClN_4O_3Fe + (BrH)_2 + 3\ H_2O = (C_{16}H_{18}N_2O_3)_2 + FeBr_2 + HCl + H_2$$

24 As a result of our first investigation on hematin we formulated the proposition that if the blood pigment is transformed into bile pigments, it must occur with the removal of iron and the uptake of two molecules of water.

$$C_{32}H_{32}N_4O_4Fe + 2\ H_2O - Fe = C_{32}H_{36}N_4O_6$$

Hämatin.   Bilirubin.

The equation is realized in the preparation of hematoporphyrin by means of hydrogen bromide. Simultaneously, *Städeler's* original simple bilirubin formula thus appears to be the more correct. From the composition of hematoporphyrin hydrochloride as well as its metal complex, it follows with certainty that the formula $C_{16}H_{18}N_2O_3$ belongs to hematoporphyrin. *Maly* doubled *Städeler's* bilirubin formula. The composition of tribromobilirubin, biliverdin, and urobilin occasioned this conclusion.

It is certain that formation of the cited products is much more simply explained from the $C_{32}H_{36}N_4O_6$ bilirubin formula.

25 For us, *Thudichum's* assertions are not worth a serious consideration and any competent chemist who has taken pains to read through his publications is certain to hold our view.

26 My preparation, for example, yielded the value from analysis that would best fit formula $(C_{32}H_{31}ClN_4FeO_3)_2C_5H_{12}O$.

It requires:

	C		H		Cl		N		Fe		
$C_{32}H_{31}ClN_4FeO_3)_{16}C_5H_{12}O$:	C	62.96,	H	5.155,	Cl	5.74	N	and	Fe	9.09	%
$(C_{32}H_{31}ClN_4FeO_3)_8C_5H_{12}O$:	"	63.006,	"	5.23,	"	5.69,	"	"	"	9.01	"
$(C_{32}H_{31}ClN_4FeO_3)_4C_5H_{12}O$:	"	63.08	"	5.376[1],	"	5.59,	"	"	"	8.85	"
$(C_{32}H_{31}ClN_4FeO_3)_2C_5H_{12}O$:	"	63.27	"	5.66,	"	5.45,	"	"	"	8.55	"
Nencki and *Sieber[2]* obtained											
on average [2])	C	62.78,	H	5.79	Cl	5.294	N	9.06	Fe	8.84	%
My analysis yields:											
Preparation I	"	62.95,	"	6.04,	"	5.04,	"	8.91,	"	8.28	»
Preparation II	"	62.94,	"	6.07,	"	5.08,			"	8.57	»
									"	8.48	»

---

[1]) 5.376% H, not 5.69 as per *Nencki*.
[2]) Calculated from the totality of this analytical result.

27 I therefore view the investigation concerning the preparation and empirical constitution of "hemins" as not yet settled. Likewise, the process that takes place during the transition of hemin into hematin still requires further clarification. Under each condition it does not appear to proceed as smoothly as *Nencki* and *Sieber* assume. My numerous analyses always gave for carbon, sometimes also for hydrogen, too low a value in relation to the formula $C_{32}H_{32}N_4FeO_4$ advanced by *Nencki*, so that I do not assume it to be impossible [for] an oxidation of hematin in alkaline solution to take place by oxygen in air. This idea also leads me to assume further oxidation experiments of hematin, whether *Hoppe-Seyler* as well as *Nencki* had obtained no noteworthy results, from which the constitution of hematin could be derived.

I used an oxidizing agent, chromic acid, which acts more gently on many compounds, such that an aqueous solution of sodium dichromate at water-bath temperature[1]) acted on hematin dissolved in glacial acetic acid. By this procedure I obtained an ether-soluble mixture of acids from which presently two chemically distinct compounds could be isolated that I have named *dibasic Hämatinsäure* and the *tribasic Hämatinsäure anhydride*. The former has a mp $112°–113°$ and has the formula $C_8H_{10}O_5$

Analysis: Calculated for $C_8H_{10}O_5$

	Percent:	C	51.61,	H	5.37.
Found		"	"	51.68,	"	5.29.

This acid is dibasic because it gives a silver salt, $C_8H_8Ag_2O_5$ that calculates for:

	Percent	Ag	53.84.
Found		"	"	53.76.

Also the formula corresponds to the molecular weight.

Calculated for $C_8H_{10}O_5$:	186.
Found:	194, 200.

The tribasic hematinic acid arises from the first in any case by further oxidation; it has the composition $C_8H_{10}O_6$ and easily forms an anhydride of mp $94.5°$ and is gifted with great crystallizablity, apparently however, with the inclusion of some water.

Analysis:	Calculated for $C_8H_{10}O_6$		$C_8H_8O_6$ + 1/8 H_2O	$C_8H_8O_5$
Percent: C	47.52,		51.54,	52.17.
H	4.95,		4.43,	4.35.
Found "	C	51.24, 51.35, H	4.44, 4.50.	

Here too the assumed formula corresponds to the molecular weight because it calculates for $C_8H_8O_5$.

Molecular weight:	184.
Found:	199, 192, 202.

The salts are derived from the tribasic acid.

---

[1]) Chromic acid works even at room temperature; I am about to investigate the product originating thereby.

28   In addition to the acids, an iron-containing compound is yielded as a further oxidation product that is soluble in sodium carbonate. In this solution the iron is not detectable by the usual analytical means. I am also occupied in a further investigation of this compound, which perhaps is related to pyrrole.

29   It was assumed to be nearly certain that the bile pigment arises from hemoglobin and especially from the iron-contäining part of the same, hematin, because among other things the first is found only in red-blooded animals.

A stronger proof for this assumption is lacking insofar as the chemical relationship of the two pigments is not yet clarified. It could be possible however in this particular case to clarify the chemical relationship experimentally.

11.2 Translations to Section 3

667

30 As everyone knows, various facts exist from which it can be concluded that the blood and bile pigments are chemically different compounds. Namely, *Nencki* has shown that hematoporphyrin, which is transformed into hematin by the action of hydrogen bromide and glacial acetic acid, is isomeric with bilirubin: both compounds have the empirical composition $C_{16}H_{18}N_2O_3$. Furthermore, certain color changes occur through the action of nitric acid that remind one of the *Gmelin* color change reaction, and reduction affords a compound resembling a reduction product of bilirubin. [1]These pigments are doubtless chemically related compounds and must be composed of similar atom complexes.

---

[1] In the Arch. f. exp. Pathol. Pharmkol. 24, 441, *Nencki* said: "this much is certain, a product identical to the urobilin pigment does not arise in this process."

31 . . . and thus obtains the bile pigment oxidation product in the form of slightly yellowish colored needles of mp 100°–101°. Analysis leads to the following formula: $C_8H_9NO_4$.

Analysis: Calculated for: Percent: C 52.45, H 4.92, N 7.65.
Found   "   "   52.12, " 5.52, "   7.81.

On the basis of titration with 1/5 *N* ammonium hydroxide, the substance, whose aqueous solution reacts after all like a strong acid, behaves in the cold like a monobasic acid. To neutralize 0.083 g requires 1.72 cm^3, calculated 1.8 cm^3.

The silver salt of the acid, prepared by precipitation of its ammonia neutralized solution with an ethanolic solution of silver nitrate, contains two metal atoms, however.

Analysis: Calculated for $C_8H_7Ag_2O_4N$.
           Percent: Ag 54.33.
Found: "   "   53.96.

32 In this case a direct relationship to hematinic acid would also be possible, as already shown from the empirical formulas $C_8H_8O_5$ and $C_8H_9NO_4$. The latter ultimately exhibits a very simple relationship to biliverdin and if the isolated acid were the sole product of oxidation, its origin would be presented by the equation:

$$C_{16}H_{18}N_2O_4 + 4\ O = 2\ C_8H_9NO_4.$$

33   This result merely confirmed the earlier. Different portions of hematin and hematoporphyrin, in contrast, afforded preponderantly only nitrogen-containing acids, and this gave a readily soluble calcium salt that does not precipitate by heating, [and] which can be purified by recrystallization. The acid regenerated now has mp 111°–113°.[2]

Those are however the properties of dibasic hematinic acid.
Analysis now leads to the following formula: $C_8H_8NO_4$.

$C_8H_9NO_4$. Calc'd C   52.45,   H 4.92,   N 7.65.
            Horse hematin Cattle hematoporphyrin   Hematoporphyrin. . .
      Found C   52.49, 52.52, 52.12,   " 5.16, 5.30, 506, " 7.59.

---

[2] Different preparations give: 110–113°; 111.5–112.5°; softens from 103°.

11 Translations

34 I was led therefore to the formulation of the false formula $C_8H_{10}O_5$ by the previous failure of the qualitative test for nitrogen and the accidentally good agreement of the carbon [combustion] value of $C_8H_{10}O_5$, while all previously found values of hydrogen fit the correct formula. A later proof of the still available amounts of the previously designated and still present dibasic hematinic acid confirms the nitrogen content.

35 I do not exclude the possibility that the $C_8H_9NO_4$ compounds (dibasic hematinic acid – *Biliverdinsäure*) do not represent chemically distinct compounds but mixtures of isomeric compounds, which however are transformed into one and the same compound, $C_8H_8O_5$.

36 From these corrections of my earlier work, the relationship of the previously obtained cleavage products of hematin with the latter has become clearer.

As *Nencki* has shown, hematin (**I**) decomposes into hematoporphyrin (**II**) by release of iron and uptake of water. Oxidation of the latter, as well as isomeric bilirubin, gives the dibasic hematinic acid = *Biliverdinsäure* (**III**), and from this the anhydride of tribasic hematinic acid (**IV**) arises finally by loss of ammonia.

I		**II**		**III**		**IV**
$C_{32}H_{32}N_4FeO_4$	$\rightarrow$	$C_{16}H_{18}N_2O_3$	$\rightarrow$	$C_8H_9NO_4$	$\rightarrow$	$C_8H_8O_5$

The unsaturated compound $C_8H_8O_5$ is transformed finally into a potassium permanganate-resistant substance, $C_8H_{12}O_6$, by reduction with hydrogen iodide, which indeed gave accurate values by analysis and proved to be a tribasic acid. However, from different preparations it was not the same and showed no sharp melting point – and therefore should probably be regarded as a mixture of isomeric compounds.

37 At a little higher temperature, from about 120° upward, carbon dioxide release starts; at 130° it gives $C_8H_9NO_4$ as well as $C_8H_8O_5$ and a new compound, $C_7H_9NO_2$, which has a mp 72–73° and displays all of the properties (odor of iodoform, volatility, sublimability) as for an imide from the maleic acid series.

$C_7H_9NO_2$. Calculated C 60.43, H 6.48, N 10.07, Molecular weight 139.
Found " 60.20, " 6.52, " 10.36, " 136.

38 The compound $C_7H_8O_3$ is therefore to be viewed as an anhydride of a dibasic acid.

The second barium salt, obtained from saponification of the imide, belongs to an unsaturated acid, $C_7H_{10}O_4$, mp 175° (with partial decomposition), that is sparingly soluble in cold water and crystallizes in wide needles from absolute alcohol.

$C_7H_{10}O_4$. Calculated for C 53.17, H 6.32.
Found " 53.18, " 6.29.

39 The analysis gave the formula $C_7H_8O_3$
0.1755 g gave 0.3847 CO_2 and 0.0969 H_2O.

	Calculated for $C_7H_8O_3$	Found
C	60.00	59.64
H	5.71	6.11.

Accordingly, the compound is not the acid of the barium salt but its anhydride.

11.2 Translations to Section 3   669

40   0.1738 g of substance dried in vacuum gave 0.8044 g CO_2 and 0.1080 g H_2O

Calculated for	$CH_3-COH-COOH$	:	Found:
	$C_2H_5-CH-COOH$		
C	47.78		47.89%
H	6.81		6.92 "

41   0.8068 g. oil gave 0.0750 g CO_2 and 0.1603 g H_2O.

Calculated for	CH_3-C-CO C_2H_5-C-CO >O	Found:
C	60.00	59.97%
H	5.71	5.80 "

42   This anhydride is identical with the compound that *Bischoff* prepared from methylethylsuccinic acid, and that *Fittig* and *Parker* prepared from pyruvic acid and sodio-methylsuccinic acid (sodium pyruvate ?).

43   ... that hematin can be regarded as arising from two symmetrical parts that are linked by iron:

$$\boxed{R}-\boxed{R'}-Fe-\boxed{R'}-\boxed{R} \, ,$$

therefore all observations point to the complex designated as R as the source of hematinic acid.

If now, as it has indeed already shown to be the most likely, hematinic acid is a derivative of maleic acid, the complex R must contain the following group:

$$\begin{array}{c} \backslash \\ C-C \\ | \quad \quad >NH \, , \\ C-C \\ / \end{array}$$

therefore that means a pyrrole ring. That however is an assumption that on the basis of numerous observations of reactions, from which hematin of course must undergo a most substantial decomposition, has already been repeatedly made. Thus, *Hoppe-Seyler* reported: dry distillation of hematin produced pyrrole abundantly.

44   *Nencki* and *Sieber* reduced hemin with tin and alcohol and obtained among other things *a volatile, water soluble compound that imparted an intense red color to a moistened pine splint*. The same researchers heated 20 g of hematin with a five-fold weight of molten potassium hydroxide to [achieve] complete decomposition, whereby a fairly large amount of pyrrole evolved. It speaks thus to hematin and its hematoporphyrin as sources of the pyrrole.

45   I believe therefore that the assumption of the complex R in hematin – corresponding to my scheme – has the greatest probability of containing a substituted pyrrole group that is transformed by oxidation into the imide of tribasic hematinic acid. It finally provides a view into the pigment nature of hematin and therefore shows that pigments can be obtained by a combination of pyrrole with ketones.

46   Through the beautiful investigations of *W. Küster* and his coworkers we know that hematoporphyrin can be oxidized in a yield of about 50% to an acid of formula: $C_8H_9NO_4$, which is converted to the acid $C_8H_8O_5$ by the uptake of H_2O and release of NH_3 by heating with alkali. By reduction with hydrogen iodide, the latter ($C_8H_8O_5$) is transformed into *M. Kölle's* tribasic hemotricarboxylic acid = $C_8H_{12}O_6$, which according to a comparative summary of these authors, is alike in all elements to the ethyltricarballyl acid prepared synthetically by *Auwers, Köbner*, and *Meyenburg*.

670   11 Translations

47   A greater number of experiments was carried out with fumaric acid ester. Condensation of this ester with sodio-malonic ester, or the sodium compound of the alkyl malonic ester, leads easily to tricarballyl acid and to the as yet not prepared mono-alkyl tricarballyl acid:

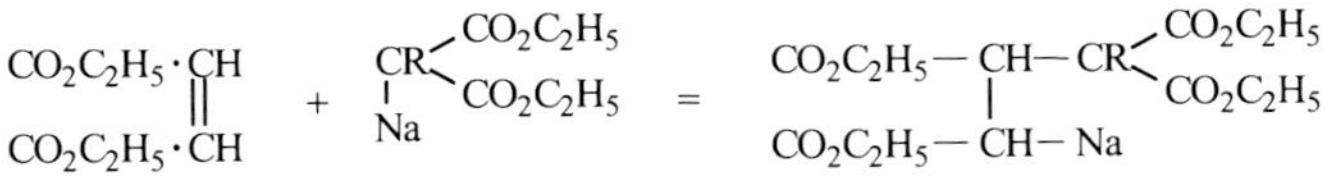

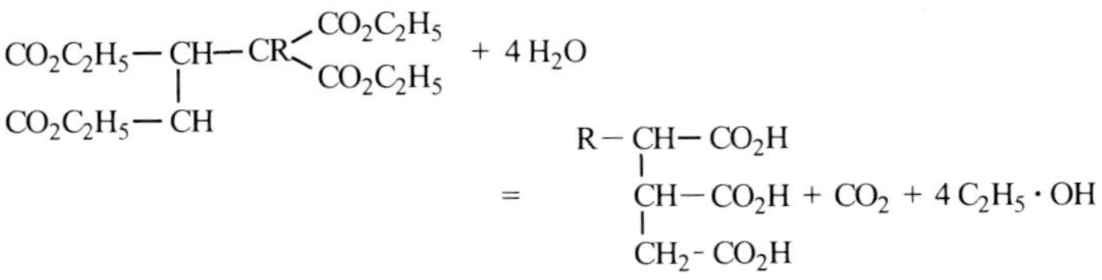

$$CO_2C_2H_5 \cdot CH \\ \| \\ CO_2C_2H_5 \cdot CH \quad + \quad CR \overset{CO_2C_2H_5}{\underset{\underset{Na}{CO_2C_2H_5}}{\diagdown}} \quad = \quad CO_2C_2H_5 - CH - CR \overset{CO_2C_2H_5}{\diagdown} \\ \underset{CO_2C_2H_5 - CH - Na}{\big|}$$

$$CO_2C_2H_5 - CH - CR \overset{CO_2C_2H_5}{\diagdown}_{CO_2C_2H_5} \quad + \quad 4\,H_2O \\ \underset{CO_2C_2H_5 - CH}{\big|}$$

$$= \quad \underset{CH_2 - CO_2H}{\overset{R - CH - CO_2H}{\big|}} \\ CH - CO_2H + CO_2 + 4\,C_2H_5 \cdot OH$$

48   All of our experiments speak in favor of the assumption that *Hämopyrrol* is either a butyl or a methylpropyl pyrrole.

49   After we had discovered *Hämopyrrol* it was clear to the likes of us that it must bear a close relationship to *Küster's* hematinic acid. Proceeding from entirely different viewpoints, we arrived at the same opinion as *Küster*, that hematinic acid arises from oxidation of a pyrrole nucleus. Helping to conclude which carbon atom, outside of the two pyrrole carbons, is oxidized to carboxyl is the observation of *Kölle*, whereby the anhydride of *Küster's* tribasic acid, $C_8H_8O_5$, is reduced by HI to the acid, $C_8H_{12}O_6$. The last has the same melting point and other properties as the ethyltricarballyl acid prepared synthetically by *Auwers*, *Köbner*, and *Meyenburg*. Accordingly, *Hämopyrrol* must have the following structure

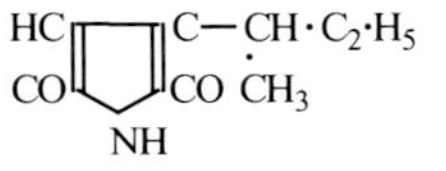

$$\underset{NH}{\overset{HC \quad \quad C - CH \cdot C_2 \cdot H_5}{CO \quad \quad CO \; \dot{C}H_3}}$$

and by oxidation to *Küster's* acid of formula $C_8H_9O_4N$, the methyl of the side chain is oxidized to carboxyl

$$(1) \quad \underset{NH}{\overset{CH \quad \quad C - CH \cdot C_2 \cdot H_5}{CO \quad \quad CO \; \dot{C}O_2H}}$$

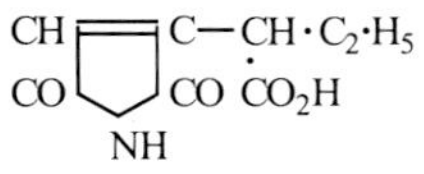

= partial imide of ethyl aconic acid = *Küster's* acid, $C_8H_9O_4N$.

$$(2) \quad \underset{O}{\overset{CH \quad \quad C - CH \cdot C_2 \cdot H_5}{CO \quad \quad CO \; \dot{C}O_2H}}$$

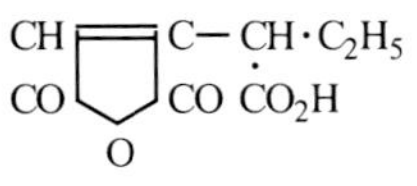

= partial anhydride of ethyl aconic acid = *Küster's* acid $C_8H_8O_5$; from the latter by reduction

$$(3) \quad \underset{\dot{C}O_2H \quad \dot{C}O_2H \; \dot{C}O_2H}{H_2C - CH - CH - C_2H_5} \quad ,$$

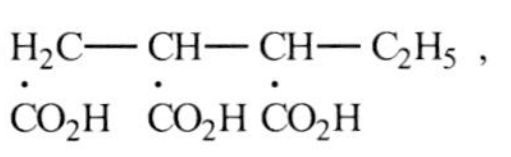

ethyltricarballyl acid of *Auwers*.

50   The imide $C_8H_9O_4N$ is transformed into the imide of propylmaleic acid by splitting off CO_2

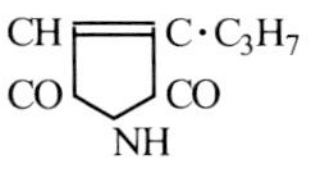

$$\underset{NH}{\overset{CH \quad \quad C \cdot C_3H_7}{CO \quad \quad CO}}$$

11.2 Translations to Section 3   671

According to *Küster*, it is however more likely that this imide is identical with the imide of *Fittig's* methylethylmaleic acid. If *Küster's* view is correct and also no basis is available for molecular rearrangement, then *Hämopyrrol* would be a methylpropylpyrrole,

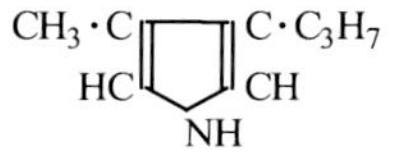

$$CH_3 \cdot C \underset{HC \quad CH}{\overset{C \cdot C_3H_7}{\bigsqcup}}$$
NH

which of the two formulas befits *Hämopyrrol* is no doubt to be distinguished in the immediate future. The molecular construction of bilirubin and the three porphyrins known presently will depend on this representation.

51   If *Hämopyrrol* is an isobutylpyrrole, then the configuration of the porphyrin is presentable very simply by linking two *Hämopyrrol* molecules, for example according to the following scheme:

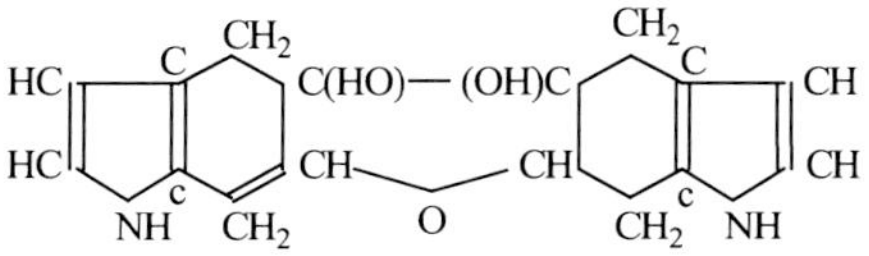

Hematoporphyrin, $C_{16}H_{18}O_3N_2$

52   According to whether we regard *Hämopyrrol* as butyl- or methylpropylpyrrole, hemin could have the following structure:

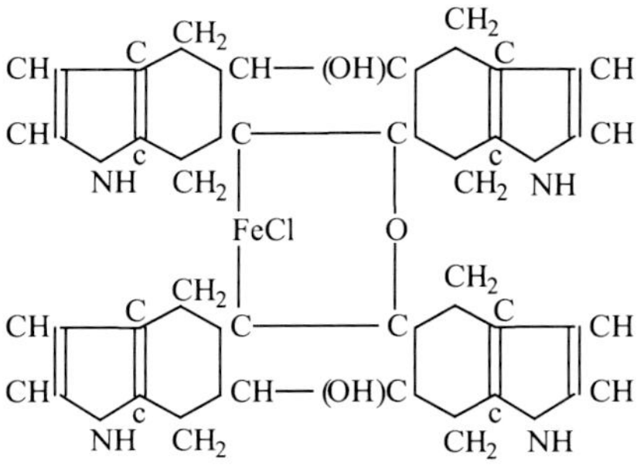

or

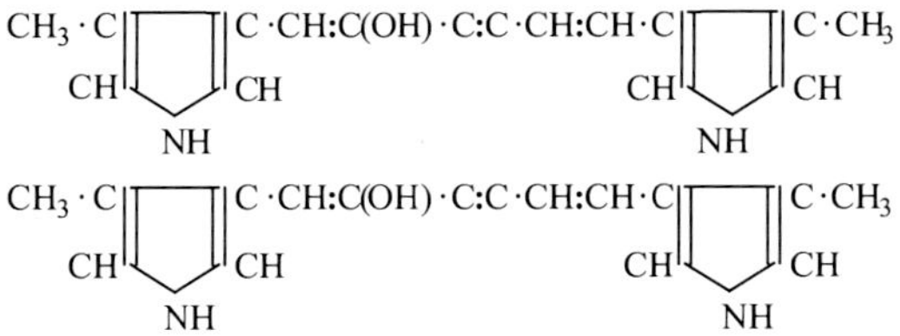

53   The results of all observations induce us therefore to double all of the formulas cited above and to designate the chemical composition of these compounds as follows:

$C_{34}H_{38}O_4N_4 \cdot 2\ HCl$	Mesoporphyrin hydrochloride
$C_{34}H_{38}O_6N_4 \cdot 2\ HCl$	Hematoporphyrin hydrochloride
$C_{34}H_{30}O_4N_4(C_2H_5)_2$	Ethyl ether of mesoporphyrin
$C_{34}H_{36}O_4N_4Zn$	Zinc or copper salt of mesoporphyrin
$C_{34}H_{34}O_4N_4(C_2H_5)_2Cu$	Copper salt of mesoporphyrin ethyl ether
$C_{34}H_{38}O_4N_4$	Free mesoporphyrin

We now assume the formula for hemin hydrochloride to be $C_{34}H_{33}O_4N_4ClFe$ . . .

11 Translations

54 Then we can formulate the formation of hematoporphyrin from hemin by means of hydrogen bromide according to the following equation:

$$C_{34}H_{33}O_4N_4ClFe + 2\ HBr + 2\ H_2O = C_{34}H_{38}O_6N_4 + FeBr_2 + HCl.$$

This equation finds proof in that no hydrogen can be detected in the thrice-conducted investigations of gases that are liberated in this reaction.

55 From these data, the objections that *Thudichum* raised against my method of preparing pure and crystalline bilirubin might be regarded as settled.

56 According to this, the partial anhydride of tribasic hematinic acid, $C_8H_8O_5$, may be represented as a carboxylated methylethylmaleic acid; so, only the following three formulas can be considered:

I.
$$\begin{array}{c} COOH\cdot CH_2\cdot C\cdot CO \\ \parallel \qquad\qquad\quad >O \\ H_3C\cdot CH_2\cdot C\cdot CO \end{array}$$

II.
$$\begin{array}{c} H_3C\cdot C\cdot CO \\ \parallel \qquad\quad >O \\ H_3C\cdot CH\cdot C\cdot CO \\ \cdot \\ COOH \end{array}$$

III.
$$\begin{array}{c} H_3C\cdot C\cdot CO \\ \parallel \qquad\qquad\quad >O \\ COOH\cdot CH_2\cdot CH_2\cdot C\cdot CO \end{array}$$

The following experiments favor formula **III**.

57 ... which is perhaps that of pyruvic acid, which must, by oxidation, result in a compound constituted according to formula **III**. Taking the origin of succinic acid as a basis of formulas **I** and **II**, however, would be interpreted as highly constrained.

Reduction by hydrogen iodide in a sealed tube at 150°, as mentioned earlier, leads to a non-sharply melting mixture of optically active acids. In fact reduction of an acid for formula **III** gives two inactive, racemic acids because of course by the addition of two hydrogen atoms two carbon atoms become asymmetric.

The apothecary [pharmacist], Mr. *O. Metzger*, of the National Institute, succeeded in separating the reduction mixture into two inactive acids, $C_8H_{12}O_6$, by a very difficult fractional crystallization from water, for which I would like to retain the collective name (selected by *Kölle*) *Hämotricarbonsäure*, for the sake of conciseness and to show its origin. According to their constitution they are β,γ,ε-tricarboxylic acids ...

58 Accordingly, there is no doubt that oxidation of *Hämopyrrol* has led to a substituted maleic acid via an imide. Only larger quantities are required to prove that the acid obtained, or its anhydride, $C_8H_{10}O_3$, is identical with the expected methyl-*n*-propylmaleic anhydride.

59 It melts at 56°–57° and sublimes without decomposition; it is easily soluble in ether, alcohol, chloroform, benzene, and ethyl acetate, is sparingly soluble in cold water but easier in hot water and in hot ammonia-containing water.

The odor is reminiscent of iodoform.

0.1117 g of substance.: 0.2567 g CO_2, 0.0755 g H_2O. – 0.1222 g of substance.: 10.1 cm^3 N (16°, 729 mm).

$C_8H_{11}O_2N$.	Calculated	C	62.7,	H	7.2,	N	9.2.
	Found	"	62.7,	"	7.5,	"	9.2.

60 At present 2 g of a still-colored syrup still remained with us that must, according to its odor, contain the sought-for imide. In fact, after months crystallization began, and now further purification could be realized after taking up hydrochloric acid and shaking with ether, whereby brown, clustered needle-like crystals were isolated. These were finally dissolved in hot water and the solution decolorized with animal charcoal; extraction with ether now afforded a small quantity of colorless needles of a characteristic odor. The melting point of the same was 63°–64°; thus about seven degrees higher than that of the synthetically produced methyl-propyl-maleic acid imide. Unfortunately the quantity of pure product obtained was insufficient for analysis.

11.2 Translations to Section 3

61 Consequently, our investigation has not yet produced the positive result striven for. According to all appearances the imide of methyl- propylmaleic acid is not identical to the imide that can be obtained from *Hämopyrrol*. Now since the latter, owing to its properties belongs to the class of disubstituted maleic acid imides, and, on the basis of only one analysis of course, one from its barium salt coming about from saponification, contains eight carbon atoms in the molecule, the possibility might be imagined that it would be identical with methyl-isopropyl-maleic acid imide or with xeronimide [diethylmaleimide].

We prepared and analyzed it.

$C_8H_{11}O_2N$.   Calculated   C   62.7,   H   7.02,   N   9.2.
       Found   "   62.65,   "   7.3,   "   9.24.

It produced a fibrous crystalline mass, whose properties and also odor were extraordinarily similar to methyl-propyl-maleic acid imide. The melting point of the preparation was 44°–45°; therefore it might not be identical to the imide from *Hämopyrrol*.

62 We are occupied at this time with the preparation of xeronic acid imide. We also plan to investigate the oxidation of *Hämopyrrol* again, if the difficulty in isolating the pure imide from the oxidation products of hemopyrrole is not trifling. Also, we don't exclude the possibility that *Hämopyrrol*, which previously could not be isolated in a pure state, may prove to consist of a mixture of isomers.

63 This brings forth the proof that a γ-penten-α,γ,δ-tricarboxylic acid of the following constitution

$$H_3C-\overset{\delta}{C}-COOH$$
$$HOOC\cdot H_2C-H_2C-\overset{||}{C}-COOH$$
$$\quad\quad\;\;\alpha\quad\;\beta\;\;\gamma$$

must be present in the tribasic hematinic acid, whose anhydride produces the acid $C_8H_8O_5$ and whose imide produced the imide $C_8H_9O_4N$. Certainly the adjacent carboxyls are those from which water leaves and from which imide formation is accomplished.

64 ... whereby the sought-after product should come about by condensation of methyl-succinic acid with pyruvic acid in the presence of acetic anhydride and anhydrous sodium acetate:

$$H_3C-\underset{\underset{CH_2-COOH}{|}}{CH}-COOH - 2\,H_2O - CO_2 \;=\; H_3C-CH_2 ...$$

$$+\,H_3C-CO-COOH$$

$$H_3C\cdot CO-\underset{\underset{COOCH_3}{|}}{CH}-C_2H_5 \longrightarrow H_3C\cdot\underset{\underset{CN}{|}}{C}OH-\underset{\underset{COOOCH_3}{|}}{CH}-C_2H_5 \longrightarrow$$

$$H_3C\cdot C\cdot OH-\underset{\underset{COOH}{|}}{CH}-C_2H_5 \longrightarrow H_3C\cdot C\!=\!\!=\!\!C-C_2H_5$$
$$\underset{COOH}{|} \qquad\qquad\qquad \underset{CO\cdot O\cdot CO}{|}$$

65 Given the difficulty associated with purification of the crude imide, we deduced that a complicated mixture was present and also therefore that *Hämopyrrol* may not be a single compound.

66 In the purification that was finally attained, in which the ether solution of the imide was extracted with conc. hydrochloric acid, this quantity is considerably reduced. Indeed a preparation in the form of nearly colorless needles was now obtained, however all the properties of an imide of disubstituted maleic acid and showing a 67°–68° melting point. Accordingly only methyl-ethylmaleic acid imide can be present. A mixture melting point with a synthetically obtained sample began to melt at 64°, and its analysis also afforded only poor values.

0.1334 g of substance (dried under vacuum): 0.2985 g CO_2, 0.083 g H_2O. – 0.1205 g of substance: 11.1 cm^3 N (20°, 746 mm).

$C_7H_9O_2N$.	Calculated	C	60.40,	H	6.50,	N	10.10
	Found	"	61.00,	"	6.91,	"	10.34.

When the purification was repeated with a small, available quantity of 0.5 g, it gave a still less good result.

0.1447 g of substance: 0.3260 g CO_2, 0.1065 g H_2O = 62.44 % C, 8.29 % H.

67 The total yield of crude imide was 1.62 g from six experiments. The imide crystallized beautifully and purification with conc. hydrochloric acid yielded 0.65 g of a preparation that melted at 69°–70°.

0.1315 g of substance (dried at 75° for ½ hour): 0.2907 g CO_2, 0.080 g H_2O.

$C_7H_9O_2N$.	Calculated	C	60.40,	H	6.50
	Found	"	60.29,	"	6.76

Accordingly there can be no further doubt that methyl-ethyl-maleic acid imide is the main product from oxidation of the "acidic *Hämopyrrol.*"

68 . . . Repetition of the purification process yielded 0.7 g of faintly yellow colored crystals arranged in clusters. It had a mp 64°–66°. A mixture melting point of the imide from "acidic" and "basic" *Hämopyrrol* melted at 63°.

0.1333 g of substance (vacuum dried): 0.294 g CO_2, 0.083 g H_2O. – 0.1231 g of substance: 11.5 cm^3 N (24°, 746 mm). – 0.0936 g of substance: 0.2068 g CO_2, 0.0638 g H_2O. – 0.1178 g of substance: 10.9 cm^3 N (18°, 728.6 mm).

$C_7H_9O_2N$.	Calculated	C	60.40,		H	6.50,		N	10.10.	
	Found	"	60.15,	60.26,	"	6.92,	7.46,	"	10.27,	10.34.

Despite the two-degree lower melting point found, there is no doubt here too of its identity as methyl-ethyl-maleimide.

69 From the cited observation it appears that *Hämopyrrol* is a mixture in which two pyrrole derivatives are present, with both having a basic character. In addition, the "acidic *Hämopyrrol*" has weakly acidic characteristics. The latter is smoothly oxidized, and I assume that it contains β,β-methyl-ethylpyrrole. In the second, which exhibits only basic properties, we have in contrast either a β,β'-ethyl-pyrroline or an α,β'-dimethyl-β'-ethyl-pyrrole or -pyrroline, that loses its α-methyl by oxidation.

11.3   Translations to Section 4

1   The sample for analysis was dried to constant weight at 60°C. Preparations of different origin were used for all three of the following analyses.

I.	0.1491 g	gave	0.2717 CO_2 and 0,0679 H_2O.
	0.1143 g	"	15.2 ccm N_2 at 17° and 712 mm pressure.
II.	0.1520 g	"	0.2764 CO_2 and 0.0704 H_2O.
	0.1384 g	"	18.4 ccm N_2 at 16° and 108 mm pressure.
III.	0.1394 g	"	0.2537 CO_2 and 0.0657 H_2O.

11.3 Translations to Section 4

$$\text{Calculated for}$$

	$(C_{17}H_{28}N_2O_2)_2(C_6H_3N_3O_7)_3$	$(C_{14}H_{22}N_2O_2)C_6H_3N_3O_7$
C	49.1	50.1
H	5.11	5.22
N	14.32	14.61

Found

	I	II	III
C	49.70	49.59	49.63
H	5.09	5.18	5.27
N	14.78	14.42	–

As can be seen, these analyses are not able to distinguish between hematopyrrolidinic acid of constitutions $C_{17}H_{28}N_2O_3$ and $C_{17}H_{27}N_2O_2$.

2   [10]) The nature and position of the side chains on the pyrrole ring are still not firmly established. I present with caution this formula as well as the remaining formulas used in the following, only for the purpose of assisting the reader with a tentative orientation. Also, this investigation of these simple cleavage products of hemin is from my perspective still inconclusive and, in reference to this communication, only provisional. I have taken up a more careful investigation of this compound jointly with candidate in chemistry Mr. *Quitmann* and hope very soon to be able to communicate on it.

3   Therefore it follows from the investigations reported above that the hemopyrrole carboxylic acid component of hematoporphyrin is the parent substance of hematinic acid. So it follows from the appearance of hematinic acid from this cleavage that hemopyrrole carboxylic acid [below], is a part of hematopyrrolidinic acid, where it is condensed with an oxygen-free basic residue.

4   Formula **I**, assumed by *Piloty* on the basis of its empirical formula, and the constitution of hematinic acid might however be better represented by formula **II**:

I.

II.

I believe that the latter at least easily explains the origin of hematinic acid **III**, as indeed according to *Plancher* [293, 294] the α-substituents of pyrroles are lost during oxidation in order to be transformed into imides of maleic acid.

III.

5   Of course a *Hämopyrrol*, such as *Piloty* has represented, *i.e.* a β-methyl-β′-*n*-propylpyrrole, can also arise from an isoindole ring (**I**), which would then give hematinic acid.

I.

    However, this formula, which *Piloty* considers probable, especially from my work, I plainly showed to be untenable by the oxidation of my *Hämopyrrol Gemisch* to the imide of β-methyl-β′-ethylmaleic acid!

6   A distinction between these two formulas can still not yet be made.

7   One of us has been able to produce *Hämopyrrol* in such larger quantities and purity by reduction of hematoporphyrin using tin and hydrochloric acid that careful investigation of its constitution became possible. In addition to *Hämopyrrol*, a carboxylic acid was discovered from the same procedure, which by virtue of its similar properties and its same origin as *Hämopyrrol*, was named *hemopyrrole carboxylic acid*. The name evokes the idea that this acid is a carboxylated *Hämopyrrol*. Further studies have shown, however, that the acid is not based on *Hämopyrrol* but is a different pyrrole derivative. Thus we see ourselves forced to change the name of the acid and propose the name *phonopyrrole carboxylic acid*. (The name is formed from φόνος = overlooked blood).)

8   *Phonopyrrole carboxylic acid $C_9H_{13}NO_2$*

In an earlier communication [*308*] we raised the question relative to the constitution of phonopyrrole carboxylic acid, whether one of the two following formulas can be proven with certainty

and it remained undecided only whether the α-CH_3 was in the α or α′ position. Indeed we can resolve this because the acid from treatment with nitric acid loses a CH_3 group and is transformed into the oxime of hematinic acid, $C_8H_9NO_4$, an oxime that with subsequent hydrolysis goes over into *Küster's* hematinic acid. And we are able to prove that an α-CH_3 is always split off in this procedure.

    It follows therefore that a carboxyl-free pyrrole based on the carboxylic acid precursor must have one of two formulas:

    From the communication cited above it follows also that *Hämopyrrol* must correspond to one of these two formulas.

11.3 Translations to Section 4   677

9 The definitive proof of the constitution of phonopyrrole carboxylic acid and phonopyrrole thus simultaneously includes *Hämopyrrol*, in the event that it and phonopyrrole are not identical. Thus, if phonopyrrole has one structure, then *Hämopyrrol* must accordingly have a different constitution.

Phonopyrrole is obtained by loss of carbon dioxide from phonopyrrole carboxylic acid; it is not identical with *Hämopyrrol*.

10   Analysis of *Hämopyrrol*:

$$0.1253 \text{ g gives } 0.3614 \text{ } CO_2 \text{ and } 0.1246 \text{ } H_2O$$

	Calculated for $C_8H_{13}N$	Found
C	78.05	78.35
H	10.47	11.01

Mixing an ether solution of *Hämopyrrol* with a moist ether solution of picric acid led to the formation of the picrate, obtained in obliquely cut rhombic prisms that after recrystallization from alcohol melted at 108.5°C.

$$0.1116 \text{ g gives } 0.1960 \text{ } CO_2 \text{ and } 0.0484 \text{ } H_2O.$$

	Calculated for $C_{14}H_{16}N_4O_7$	Found
C	47.72	47.89
H	4.55	4.81

One gram of *Hämopyrrol* was dissolved in dilute sulfuric acid and the solution was treated with sodium nitrite as described previously. The oxime of methylethylmaleimide precipitated with cooling and gave the correct mp 201°C after recrystallization form water.

11   Phonopyrrole therefore has the same composition as *Hämopyrrol* yet is not identical to it. While *Hämopyrrol* is characterized by its easily and beautifully crystallized picrate, we could not obtain such a derivative in crystalline form from phonopyrrole either from ethereal or aqueous solution. The former becomes colored upon addition of most ethereal picric acid solution. Just darkens without precipitation. An oily picrate deposits yet does not solidify even in the cold. Very characteristic of *Hämopyrrol* is the exceedingly easy formation of the oxime of methylethylmaleimide by treatment of its sulfuric acid solution with sodium nitrite. Phonopyrrole affords a syrupy maleic acid imide derivative in very small amounts whose nature has not yet been identified. Even exposure to chromic acid in sulfuric acid solution, a procedure followed according to *Plancher* and *Cattadori* [*293, 294*] did not lead to a crystalline derivative.

Phonopyrrole in air becomes brown colored at first, then later red.

12   One would be entitled to expect that if phonopyrrole carboxylic acid had the α-CH_3 group on the same side of the pyrrole ring as the propionic acid residue, as in the following structure,

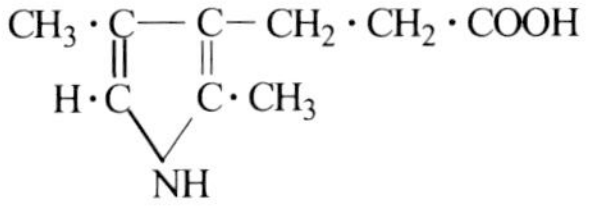

$$CH_3 \cdot \underset{\underset{H \cdot C}{\|}}{C} - \underset{\underset{C \cdot CH_3}{\|}}{C} - CH_2 \cdot CH_2 \cdot COOH$$
$$NH$$

then it would have the tendency to form an indole derivative (skatole) by dehydration.

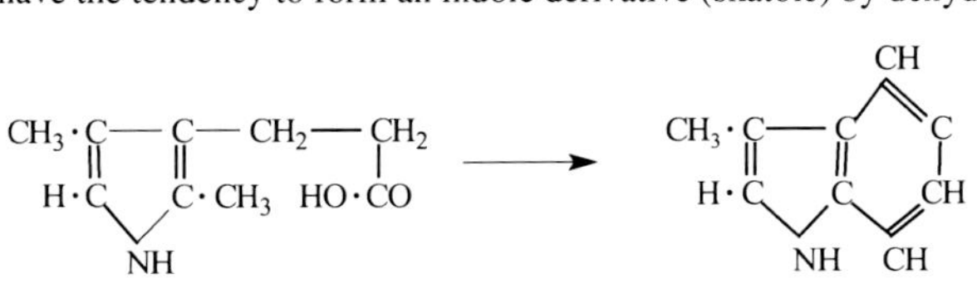

$$CH_3 \cdot C - C - CH_2 - CH_2 \longrightarrow CH_3 \cdot C - C \overset{CH}{\diagdown} C$$

678   11   Translations

Skatole formation could, however, not be brought about. If it is therefore unlikely that the α-CH_3 group of phonopyrrole carboxylic acid occupies the position adjacent to the propionic acid residue, we would still not base an opinion on this negative result alone. It was supported however by a different analysis.

13   Phonopyrrole carboxylic acid and phonopyrrole thus have the following structures:

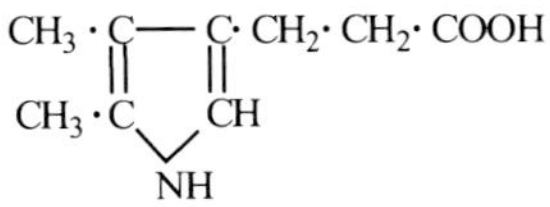

$$CH_3 \cdot \underset{CH_3 \cdot C}{\overset{CH_3 \cdot C}{\underset{NH}{\vee}}} \cdot CH_2 \cdot CH_2 \cdot COOH$$

Phonopyrrole carboxylic acid
α,β-Dimethyl-β'-propionylpyrrol

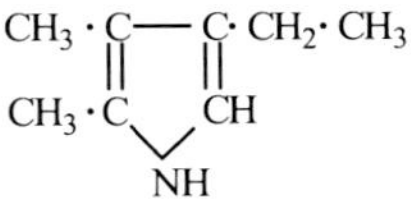

$$CH_3 \cdot \underset{CH_3 \cdot C}{\overset{C}{\underset{NH}{\vee}}} \cdot CH_2 \cdot CH_3$$

Phonopyrrol
α,β-Dimethyl-β'-ethylpyrrol

It follows directly from the supposition above that *Hämopyrrol* possesses the constitution expressed by the following formula;

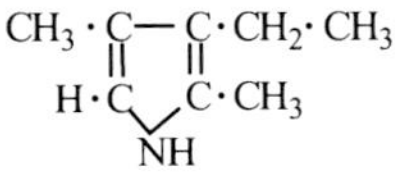

$$CH_3 \cdot \underset{H \cdot C}{\overset{C - C}{\underset{NH}{\vee}}} \cdot CH_2 \cdot CH_3$$

Hämopyrrol
α,β'-Dimethyl-β-ethylpyrrol

14   α,β,β'-Trimethylpyrrole, $$CH_3 \cdot \underset{H \cdot C}{\overset{C - C \cdot CH_3}{\underset{NH}{\vee}}} \cdot CH_3$$

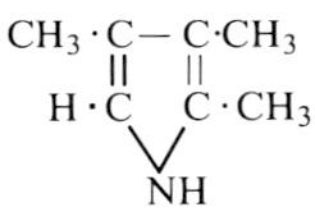

The picrate from Part II of mp 104°C was not transformed into the free base due to a lack of material, but analyzed only as the salt, and compared with the picrate from α,β,β'-trimethylpyrrole obtained by synthesis, whose description will also follow later. The mixture melting point of the preparations was 104°C.

     0.1196 g   gave   0.2030 CO_2 and 0.0461 H_2O.

     0.1171 g   "   17.8 ccm N_2 at 20° and 705 mm pressure.

	Analyzed for $C_{13}H_{14}N_4O_7$	Found
C	46.16	46.28
H	4.15	4.31
N	16.57	16.66

The picrate dissolves readily in hot alcohol and can be crystallized therefrom.

15   After recrystallization from petroleum ether the substance melts at 84–85°C. It is easily soluble in ether, alcohol, benzene, glacial acetic acid, hot petroleum ether, and water.

     0.1254 g   gives   0.3476 CO_2 and 0.1079 H_2O.

     0.1134 g   "   15.20 ccm N_2 at 20° and 721 mm pressure.

     0.2382 g   "   in 14.565 benzene 0.3495 mp-depression.

	Analyzed for $C_{12}H_{18}N_2$	Found
C	75.79	75.61
H	9.47	9.62
N	14.74	14.81
M	190	234

11.3   Translations to Section 4

α,β-Dimethylpyrrole,

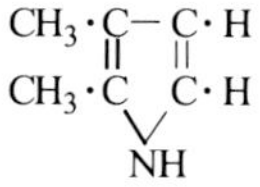

$$\begin{array}{c} CH_3 \cdot \underset{\parallel}{C} - \underset{\parallel}{C} \cdot H \\ CH_3 \cdot C \quad C \cdot H \\ \diagdown NH \diagup \end{array}$$

The picrate of mp 148°C is not a normal pyrrole salt but contains only 1 mole of acid with 2 moles of base. The basic component of the mixture is a solid compound and, as the molecular weight determination and analysis yield, a polymer of α,β-dimethylpyrrole, a bis-α,β-dimethylpyrrole. Picric acid thus has a polymerizing effect toward simple pyrrole derivatives. The α,β-dimethyl pyrrole prepared by synthesis behaves exactly similarly. Its preparation and properties will be reported in a later communication. Also, the mixture melting point of the two preparations confirms the identity.

16   This transformation of the bp 165°C dimethylpyrrole from *Dippel's* oil renders it probable that in the fraction in question perhaps α,β-dimethylpyrrole is present, in addition to α,α'-dimethylpyrrole (see also the following communication). The formation of tetramethylindole from tetramethyldipyrrole would be illustrated in the following way:

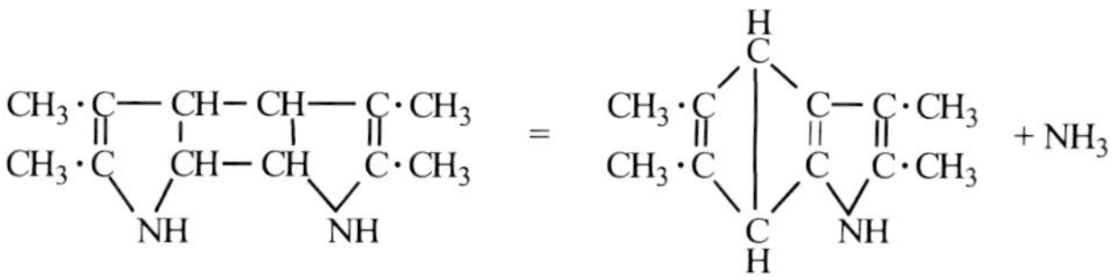

$$\begin{array}{ccc} CH_3 \cdot \underset{\parallel}{C} - CH - CH - \underset{\parallel}{C} \cdot CH_3 & & \\ CH_3 \cdot C \quad CH - CH \quad C \cdot CH_3 & = & \\ \diagdown NH \diagup \quad \diagdown NH \diagup & & \end{array} \quad \begin{array}{c} H \\ C \end{array} \quad + NH_3$$

One would therefore have to refer to this compound as

α,β-3,4-tetramethylindole

or

Pr. 2,3.B.3,4-Tetramethylindole

17   To begin with, according to the following equation, a simple derivative of α,β-dimethylpyrrole could be formed by combining aminobutanone and oxalacetic ester:

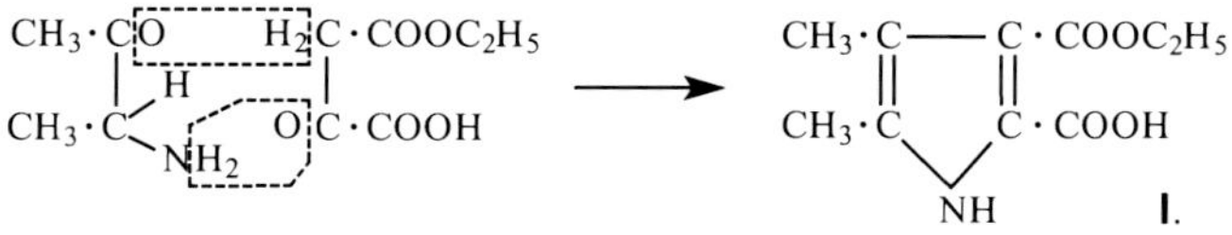

$$\begin{array}{ccc} CH_3 \cdot CO \quad H_2C \cdot COOC_2H_5 & & CH_3 \cdot C - C \cdot COOC_2H_5 \\ | \; H \; | & \longrightarrow & \parallel \quad \parallel \\ CH_3 \cdot C \quad O \; C \cdot COOH & & CH_3 \cdot C \quad C \cdot COOH \\ NH_2 & & \diagdown NH \diagup \quad I. \end{array}$$

and from it the pyrrole homolog itself was obtained in such a yield that the material was of course not cheap but was still accessible. β-Methylpyrrole and α-methyl-β-ethylpyrrole could be prepared in a similar way as dimethylpyrrole. The importance of this product for synthesis research in the blood pigment group [301, 334] is clear without further ado.

18   Accordingly, we feel justified in formulating the following structure formula for bilinic acid.

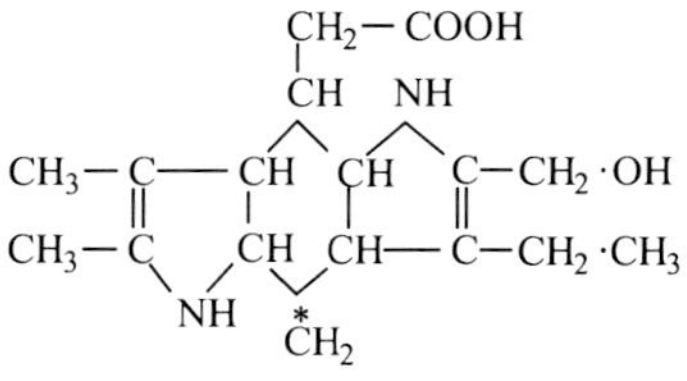

$$\begin{array}{c} CH_2 - COOH \\ | \\ CH \quad NH \\ CH_3 - \underset{\parallel}{C} - CH \quad CH \quad C - CH_2 \cdot OH \\ CH_3 - C \quad CH \; CH - C - CH_2 \cdot CH_3 \\ \diagdown NH \diagup \quad \overset{*}{CH_2} \end{array}$$

To this formula we note the possibility that the ethyl group might be interchanged with an appropriate methylene group designated by *.

19 We assume therefore that the following formula does complete justice to the characteristics of hematopyrrolidinic acid:

$$CH_3-C \Bigg\|{}\atop\Bigg\| \, CH_3-C$$

CH₂—COOH
|
CH   NH

CH₃—C——CH CH C—CH₃
CH₃—C   CH CH——C—CH₂·CH₃
       NH   CH₂

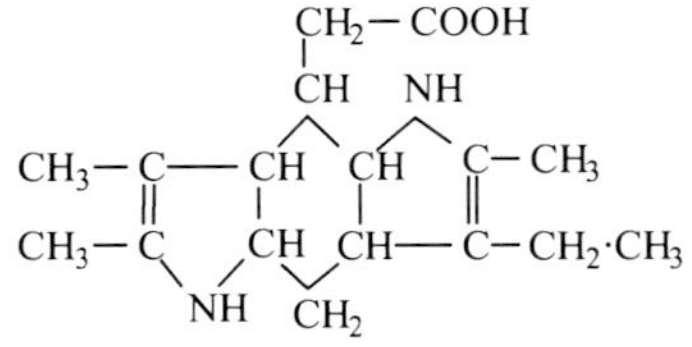

20 If one considers that with all pyrroles the α-carbons are especially favored points of attack by oxidative action, it is not surprising that the six-membered ring of hematopyrrolidinic acid is ruptured and hematinic acid and methylethylmaleimide are formed.

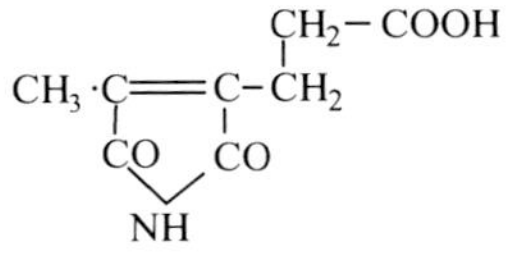

CH₂—COOH
|
CH₃·C═C-CH₂
   CO CO
    NH

Hematinic acid

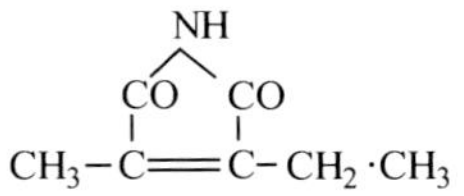

NH
CO CO
CH₃—C═C-CH₂·CH₃

Methylethylmaleimide

21 The acid crystallizes from H_2O in beautiful, colorless, prismatic needles with pointed ends. Mp 126–127°C, during which sintering begins at 105°C.

0.1365 g   gives   0.3247 CO_2 and 0.0984 H_2O.
0.1154 g   "   8.8 ccm N_2 at16° and 718 mm pressure.

	Analyzed for $C_9H_{13}NO_2$	Found.
C	64.58	64.86
H	7.78	8.06
N	8.38	8.50

22   The three pyrroles were separated recently from the *Hämopyrrol* mixture by *H. Fischer* and *E. Bartholomäus* [*361*] and especially by *Willstätter* and *Asahina* [*364*].

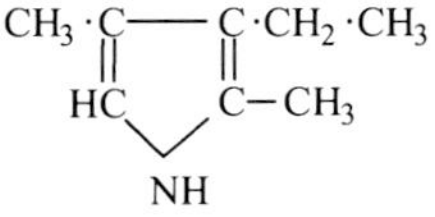

CH₃·C——C·CH₂·CH₃
HC   C—CH₃
    NH

Hämopyrrol

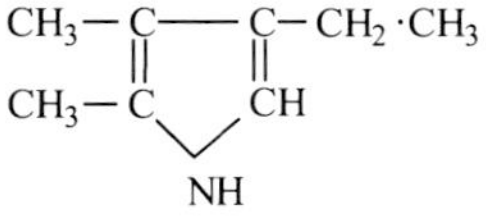

CH₃—C——C-CH₂·CH₃
CH₃—C   CH
    NH

Isohämopyrrol

CH₃—C——C-CH₂·CH₃
CH₃—C   C—CH₃
    NH

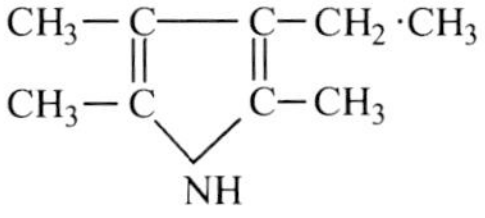

Phyllopyrrole

23

$C_{17}H_{22}N_2O_3$ Analyzed   C   67.55,   H   7.28,   N   9.27
              Found   "   67.20,   "   7.73,   "   9.33

Molecular weight determination: 0.3131 g of substance in 18.11 g of glacial acetic acid: boiling point elevation 0.130°C – calculated molecular weight 302, found 309.

24 We assume it likely that the pigment character of this substance is due to a system of conjugated double bonds. Such a system can easily be constructed from the formula given above for bilinic acid and thus an attempt can be made to explain the transformation of a colorless acid into colored acid, as the following scheme might suggest:

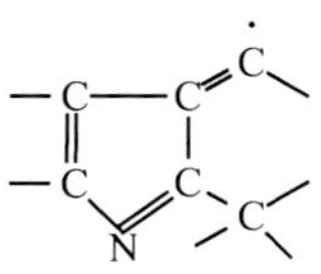

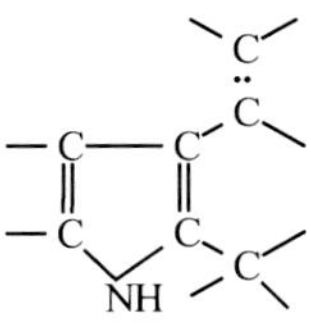

11.3 Translations to Section 4

25 The dehydrobilinic acid thus appeared as lemon yellow, fine tapered prisms, which have a tendency toward twinning. Yield of pure substance 0.3 g. It decomposes without melting at 260°C; is insoluble in water, benzene, petroleum ether; slightly soluble in hot absolute alcohol and in glacial acetic acid. It does not give a pyrrole reaction with a pine splinter and does not react with dimethylaminobenzaldehyde. It is neutral toward litmus paper, probably owing to its insolubility in water. It is readily soluble, with intense yellow coloration in alkalis and alkaline carbonates. . .

Sodium salt. The behavior of dehydrobilinic acid is very characteristic between dilute sodium hydroxide and sodium carbonate solution. In both liquids it dissolves easily upon warming. Upon cooling however, the sodium compound falls out again as a picrate-yellow precipitate. After filtration and drying the salt can be recrystallized from 70% alcohol. It precipitates with cooling in concentrically grouped fine yellow needles.

$$0.1366 \text{ g} \quad \text{Subst.:} \quad 0.0290 \text{ g.} \quad Na_2SO_4$$
$$\text{Analyzed for} \quad Na\ 6.88. \quad \text{Found} \quad Na\ 7.07.$$

26 The communication shows an experimentally important result that also in bilirubin a complex is present that resists mild oxidation. . .

Simultaneous with us, *H. Fischer* and *Röse* [*352*] obtained a colorless compound from bilirubin in the same way that we did. But they instead showed a formula, from which the appearance of a colored compound, as we maintain it, cannot be completely explained by simple loss of hydrogen.

27 They expressed their opinion there that this analysis is of significance for determining the pigment nature of the entire compound class. We add also that the scheme

III.

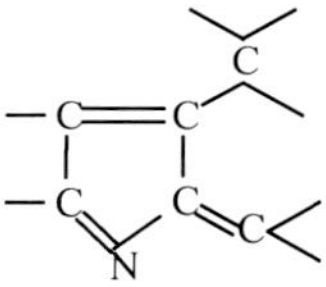

falls in conceptually together with Scheme I.

28

IV.

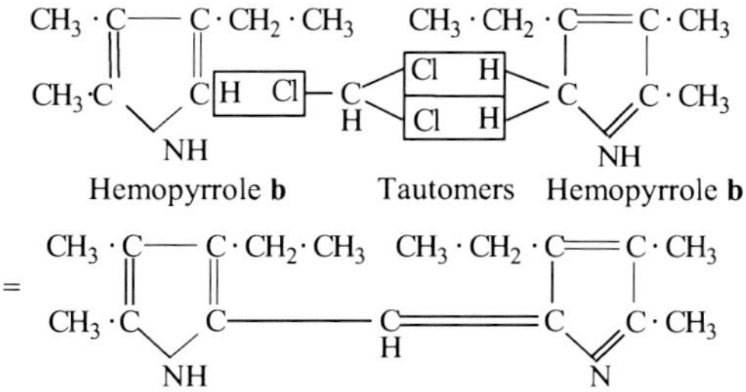

Hemopyrrole **b** Tautomers Hemopyrrole **b**

Bi-(ethyl-dimethyl-pyrryl)-methene.

V.

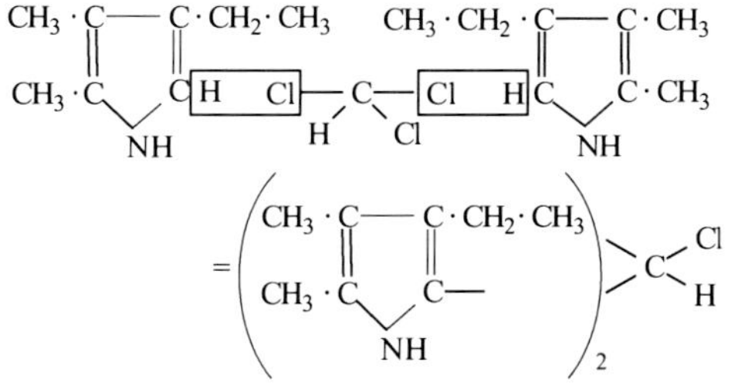

Bi-(ethyl-dimethyl-pyrryl)-chlormethane.

682　　　　　　　　　　　　　　　　　　　　　　　　　　　11　Translations

Reaction to **IV** takes place between one molecule of chloroform and two molecules of hemopyrrole **b** by loss of three hydrogen chloride molecules, and to **V** by loss of two.

Phonopyrrole carboxylic acid **a** reacts only, as it seems, in a sense, namely according to formula **VI**.

VI.

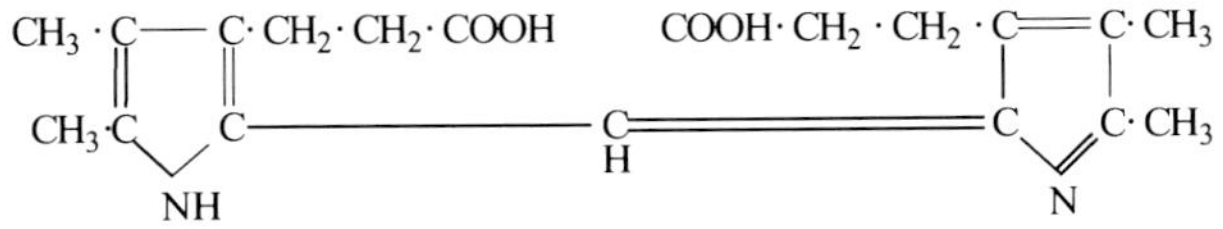

Bi-(propionyl-dimethyl-pyrryl)-methene.

29　　Bis-(ethyl-dimethylpyrryl)chloromethane (**V**) is an intensely yellow-colored compound, very similar to dehydrobilinic acid, for which we might suggest a formula corresponding to structure **V**. . . .

Spectral analytical characteristics of the new pigment show a certain similarity to bilirubin.

30　　Finally, concerning bilirubin, we take the structure shown by *Piloty* and *Thannhauser* in their communication of 1912 [*301*] to be more likely than that drawn up by *Fischer* and *Röse* in 1914 [*357*]. Our proposed structure comes from our discovery of dehydrobilinic acid and the relationship of the dipyrrylmethene pigment to bilirubin. The *Fischer* and *Röse* formulation does not appear to us to follow from the formula of hemin and does not at all take into account the great differences in behavior of the two pigments.

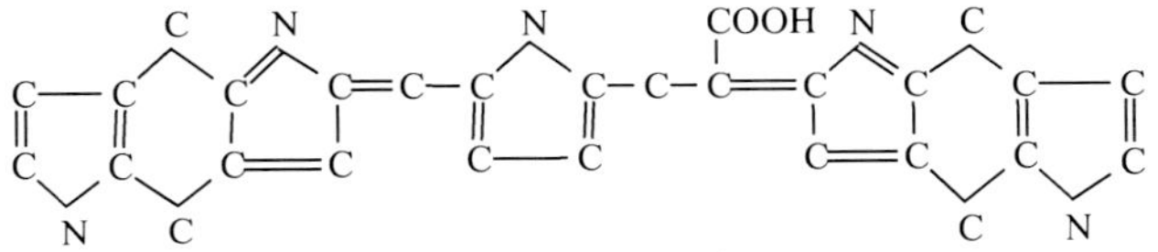

Bilirubin Scheme

31

	Analyzed for Bilirubin:	Found:	
	$C_{16}H_{18}N_2O_3$	I.	II.
C	67.09	66.78	66.96
H	6.34	6.45	6.64
N	9.79	9.73	9.65

32

Maly obtained:		C 64.68	H 6.93	N 9.22	
My analysis gave:	I.	C 62.94	H 7.45	N 9.06	9.13
	II.	C 65.08	H 7.63	N 9.32	
	III.	C 64.82	H 7.17	N 9.40.	

From all of these results it follows clearly that *Maly's* urobilin, even after improved instruction, is a complicated mixture.

33　　. . . because the compound represents perhaps only one half of bilirubin, I propose the name *hemibilirubin*. From elemental analysis, hemirubin has the composition $C_{16}H_{20}N_2O_3$.

It can therefore be a simple hydrogen addition product of bilirubin that has incorporated two hydrogen atoms . . .

34　　Concerning the composition of the compound, the simplest explanation would be the following: Hemibilirubin is the anhydride of an acid of the following constitution.

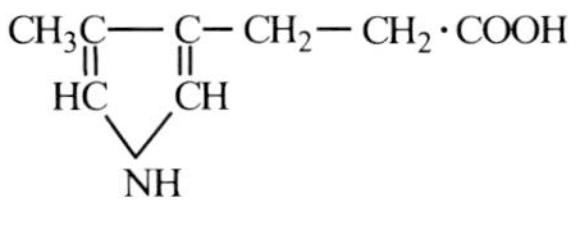

11.3 Translations to Section 4

Accordingly, I treated hemibilirubin with nitrous acid and in fact obtained the oxime of *Küster's* hematinic acid. I obtained hematinic acid from oxidation with lead superoxide.

$$CH_3C=C-CH_2-CH_2 \cdot COOH$$
$$O=C \quad C=O$$
$$NH$$

35 Support for or against this view can be obtained only by degradation. . . .

36 We have proposed formula **I** for bilirubinic acid, while *Piloty* and *Thannhauser* discussed formula **II**.

$$H_3C \cdot C - C \cdot C_2H_5 \quad HOOC \cdot CH_2 \cdot CH_2 \cdot C - C \cdot CH_3$$

I. **II.**

. . . It appears to us that *Piloty's* formula is improbable because it seems out of the question that an alcohol hydroxyl group would be resistant to hydrogen iodide-acetic acid (see [352]) . . .

37 . . . Hemibilirubin, *Körper* II, and bilirubin give the same reductive cleavage products.

38 Should the "imide" actually be present, then the fact remains to be clarified why, in contrast, copious amounts of methylethylmaleimide are obtained by direct oxidation of bilirubin after it was reduced.

39 Bilirubinic acid was also subjected to the same procedure; here however the tetrasubstituted acid did not arise, but a new compound in relatively good yield did, for which we propose the name x*anthopyrrole carboxylic acid*. The new acid has a very close relationship to its parent substance, because bilirubinic acid is recovered in very good yield by reduction with HI-acetic acid. Whether the compound isolated by us is identical with the dehydrobilinic acid described by *Piloty* and *Thannhauser* [302] appears not improbable to us; however, a comparison with a compound prepared according to the directions of those authors was not possible due to lack of material.

40 It forms a difficultly soluble sodium salt and corresponds in all its properties to the substance obtained by *Piloty* and *Thannhauser* [302] from oxidation of bilirubinic acid, except that our compound possesses a moderately sharp melting point. Its analysis could be conducted only microanalytically considering the small amount of substance available.

$C_{17}H_{22}N_2O_3$. Analyzed for C 67.51, H 7.34, N 9.27
 Found " 67.01, 68.05, " 6.64, 7.0, " 9.61

41
117. *O. Piloty's* Reply to the Communication of *Mssrs. Hans Fischer* and *Heinrich Röse*
In Issue No. 3 of this Journal [*Chem. Ber.* **1913**, *46*, 439][4]
(Received on February 27, 1913)

[From the Chemical Laboratory of the Royal Academy of Sciences in Munich]

In a communication [301] published jointly with *S. J. Thannhauser*, I described an acid of composition $C_{17}H_{26}N_2O_3$, which was obtained by HI reduction of bilirubin and given the name bilinic acid. At nearly the same time *H. Fischer* and *H. Röse* [356] also published[4] a study on this acid, and they named it *bilirubinic acid*. In order to guard against confusion in the literature,

---

[4][*Piloty's* article was received at *Liebig's Annalen* on May 25, 1912; *Fischer's* at the *Berichte* on May 20, 1912.]

684 11 Translations

might I emphasize that bilinic acid and bilirubinic acid denote one and the same substance. Further, in my above-mentioned article (p. 206), I first described and exactly characterized a substance I named as *isophonopyrrole carboxylic acid*. Later, *H. Fischer* and *E. Bartholomäus* [*366*] also found this second substance and named it *isomeric phonopyrrole carboxylic acid*, a designation to which the authors adhere up to the present.

Furthermore, in my article cited above (p. 197), I proved that bilirubin contains three rings of which two are similar to one another and connected to isophonopyrrole carboxylic acid, since this acid is split up into dimethylpyrrole and trimethylpyrrole upon heating in molten alkali. This should been mentioned when *Mssrs. Fischer* and *Röse* [*356*] wrote that they "for the first time" have provided proof that a third pyrrole nucleus must be contained in hemibilirubin and bilirubin.

In addition, I have shown, also in work published jointly with *S. J. Thannhauser* [*302*], that an intensively yellow colored acid, which we named dehydrobilinic acid, results from removal of two hydrogens from our bilinic acid. We have shown the definitive role which this substance presents toward the evaluation of the pigment nature of bilirubin; emphasized the characteristic difference between the constitution of this pigment and the blood pigment; and in a short time established the basic concepts evaluating the constitution of bilirubin.

H. Fischer and *H. Röse* [*356*] have now also obtained a yellow colored acid from bilirubin that they considered to be identical to our dehydrobilinic acid. According to their statements, the authors failed in making an identification due to a lack of material. This omission can, however unlikely, be accepted as a legitimate reason, for this substance is given not only one but two different names, namely *xanthopyrrole carboxylic acid* and *xanthobilirubinic acid*. I already attributed the name *xanthopyrrole carboxylic acid* to a different acid [*314*] and therefore ask that further appropriation of this name be avoided. However, apart from that, I consider it unsuitable that, without any objective basis, the authors' published procedure assigned a new name to a compound already described and named. In the first place this only introduces complexity into the literature; and in the second place it gives the false impression to the unsophisticated reader of related material that the new sponsor [*Fischer*] is also the discoverer of a new compound. While I cannot assume that these two eventualities are the purpose of the authors, I do not doubt that the same two names, *xanthopyrrole carboxylic acid* and *xanthobilirubinic acid*, should simply be dropped, and the name assigned (*dehydrobilinic acid*) by the discoverer should be accepted. It would be appropriate if at this opportunity Herr *H. Fischer* also decided to become reconciled to the names *bilinic acid* and *isophonopyrrole carboxylic acid*.

42   Now to dehydrobilinic acid, which *Piloty* and *Thannhauser* (*302*) obtained by oxidation of bilirubinic acid with permanganate.

Concerning the composition of the compound, *Piloty* showed that it appears to be no longer entirely certain. Then [*327*] he said that this acid arises by elimination of two hydrogen atoms. In contrast, in the original publication [*302*] four hydrogen atoms had disappeared. Actually, however, six hydrogen atoms must be removed to produce the system of conjugated double bonds rendered in *Piloty's* formula [*327*] that corresponds to the usual view on the constitution of colored compounds. However, this point is not particularly important because *Piloty's* formula for bilirubinic acid has already been refuted [*354*]. More important is whether xanthobilirubinic acid [*356*] is identical to "dehydrobilinic acid". Our xanthobilirubinic acid [The name *Xanthopyrrolcarbonsäure* was, by the way, misused only one time and is of course based on a printing or clerical error (*356*)] arises from heating bilirubin with sodium methylate at 220–230°C, and according to this manner of formation, as any other chemist would also have done, I took its identity with the oxidation product obtained by *Piloty* as out of the question, despite the extraordinary similarity. By reconverting it to bilirubinic acid we have shown [*356*] that xanthobilirubinic acid bears a very close relationship to bilirubinic acid. *Piloty* has not produced this sort of proof for his dehydrobilinic acid. The possibility that these are different compounds cannot be excluded, especially because pyrrole chemistry is extraordinarily rich enough to have compounds similar to one another but still not be identical. By the way, as we have also expressed previously [*356*], one difference between the two compounds does exist, which *Piloty* disregards. Our preparation melts sharply at 274°C, while *Piloty* and *Thannhauser* indicate that their substance decomposes without melting above 260°C [*302*].

11.3 Translations to Section 4

Of course since we have obtained mesoporphyrin by heating porphyrinogen with sodium methylate [*358*], which is doubtless caused by the oxidizing effect of this reagent, I am almost convinced of the identity of the two compounds. If *Piloty* later establishes for his preparation that it melts in agreement with ours at 274°C, then I will not hesitate to retract the name *Xanthobilirubinsäure* and replace it consequently with *Dehydro-bilirubinsäure*.

43 Bilirubin itself would then be represented by the following structure:

Bilirubin $C_{33}H_{36}N_4O_3$ [N.B. O_3 appears to be an error in print]

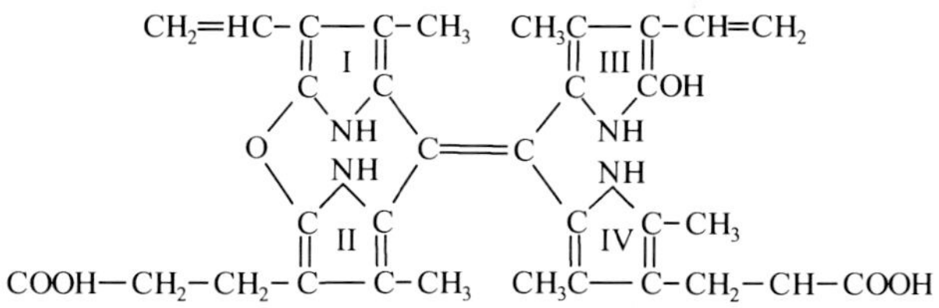

The following formula for bilirubin can also be taken into consideration, which likewise explains most of the experimental results and provides for the formation of only one form of hembilirubin.

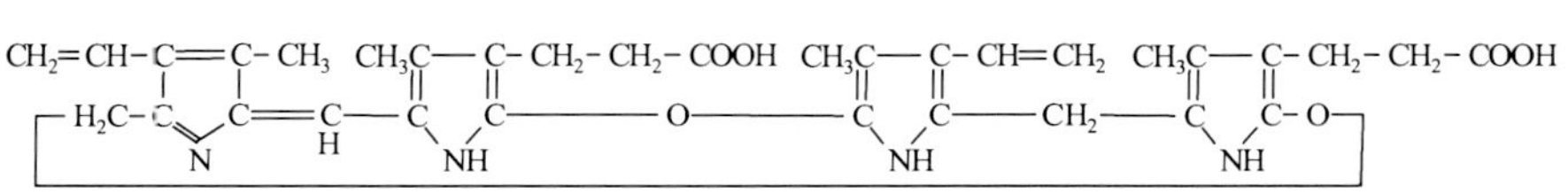

44 In his last work [*313*], "On the constitution of the colored compounds of the blood pigment", *Piloty* reached the following constitutional formulas for *Hämopyrrol* and *Phonopyrrol*:

I.

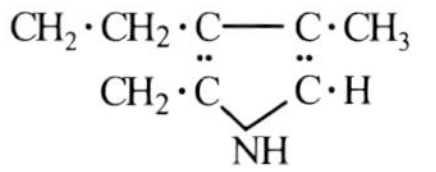

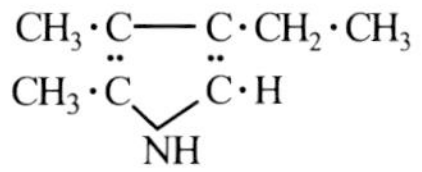

Hämopyrrol?
α,β'-Dimethyl-β-ethylpyrrole
(2,4-Dimethyl-3-ethylpyrrol)

II.

Phonopyrrol
α,β-Dimethyl-β'-ethylpyrrole
(2,3-Dimethyl-4-ethylpyrrole)

45 We observed lamellar, sometimes concentrically grouped crystals under the microscope. The salt melts with darkening quite sharply at 131–132°C and suffers a little violent decomposition above the melting point. . . .

We followed *Piloty's* instructions [*303*, *308*] exactly and obtained first of all the *monooxime of methylethylmaleimide*, which, after recrystallization from water, gave the mp 101°C reported by *Piloty* and *Quitmann*.

46 Since the pigment obtained by us was in all probability a β-azopigment, we conjecture that it is essentially a pyrrole of the following constitution:

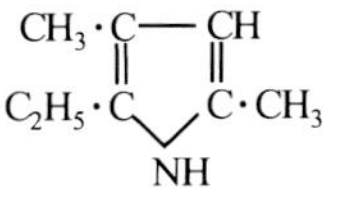

because it does not seem out of the question that at the higher temperature the entire ketazine was broken off, after which perhaps ethylation is introduced first in the α-position.

47 Consequently the dimethylethylpyrrole that we obtained has the following constitution:

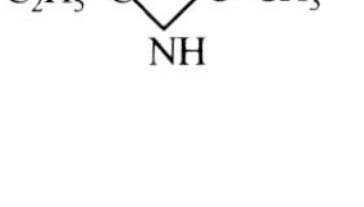

and the trialkylated pyrrole obtained by us from the ketazine:

$$\begin{array}{c} CH_3 \cdot C \!\!-\!\! CH \\ \| \quad \| \\ C_2H_5 \cdot C \diagdown \diagup C \cdot CH_3 \\ NH \end{array}$$

48 How the reaction comes about, we still cannot say today. It is possible that substitution first takes place on nitrogen and then rearrangement to carbon, a behavior that would be entirely analogous to the rearrangement of alkyl groups in the benzene series discovered by *A.W. Hofmann* and *Martius.* . . .

The differing behavior here of the α and β positions of pyrrole is also interesting. While it is easy to introduce methyl and ethyl groups in the α-position, the β-position gives considerable resistance.

We have heated 2,4-dimethylpyrrole with sodium ethylate at 220°C and obtained 2,4-dimethyl-5-ethylpyrrole in nearly quantitative yield.

$$\begin{array}{c} CH_3 \cdot C \!\!-\!\! CH \\ \| \quad \| \\ HC \diagdown \diagup C \cdot CH_3 \\ NH \end{array} \longrightarrow \begin{array}{c} CH_3 \cdot C \!\!-\!\! CH \\ \| \quad \| \\ C_2H_5 \cdot C \diagdown \diagup C \cdot CH_3 \\ NH \end{array} \quad \text{and} \quad \begin{array}{c} CH_3 \cdot C \!\!-\!\! C \cdot C_2H_5 \\ \| \quad \| \\ C_2H_5 \cdot C \diagdown \diagup C \cdot CH_3 \\ NH \end{array}$$

Main Product   By-product

In addition, we isolated in quite small quantities the 2,4-dimethyl-3,5-diethylpyrrole, already known to us.

49 In the same way, we succeeded in smoothly preparing phyllopyrrole from 2,4,5-trimethyl-pyrrole (**II**), with ethylate, and from hemopyrrole (**V**) with methylate. We have already reported its synthesis from 2,4,5-trimethyl-3-acetylpyrrole (**IV**) in a different place [*362*].

$$\begin{array}{c} CH_3 \cdot C \!\!-\!\! C \cdot C_2H_5 \\ H \cdot C \diagdown \diagup C \cdot CH_3 \\ NH \\ \textbf{I.} \end{array}$$

$$\begin{array}{c} CH_3 \cdot C \!\!-\!\! CH \\ CH_3 \cdot C \diagdown \diagup C \cdot CH_3 \\ NH \\ \textbf{II.} \end{array}$$

$$\begin{array}{c} CH_3 \cdot C \!\!-\!\! C \cdot C_2H_5 \\ CH_3 \cdot C \diagdown \diagup C \cdot CH_3 \\ NH \\ \textbf{III.} \\ \text{Phyllopyrrole} \end{array}$$

$$\begin{array}{c} CH_3 \cdot C \!\!-\!\! C \cdot CO \cdot CH_3 \\ CH_3 \cdot C \diagdown \diagup C \cdot CH_3 \\ NH \\ \textbf{IV.} \end{array}$$

$$\begin{array}{c} CH_3 \cdot C \!\!-\!\! C \cdot C_2H_5 \\ CH_3 \cdot C \diagdown \diagup CH \\ NH \\ \textbf{V.} \end{array}$$

In an analogous way we heated 2,4-dimethyl-3-ethylpyrrole (**I**) with sodium ethylate and thus obtained 2,4-dimethyl-3,5-diethylpyrrole (**VI**) which we had as well obtained earlier from 2,4-dimethyl-3-acetylpyrrole and 2,4-dimethyl-3-ethylpyrrole (**I**), and also 2,4-dimethylpyrrole [*362*].

50 *Piloty* obtained *Phonopyrrol* by distillation-decarboxylation of phonopyrrole carboxylic acid. *Phonopyrrol* is an oil, from which he obtained no crystalline derivative and to which he assigned formula **I** by a peculiar reasoning.

$$\textbf{I.} \quad \begin{array}{c} H_3C \cdot C \!\!-\!\! C \cdot C_2H_5 \\ H_3C \cdot C \diagdown \diagup CH \\ NH \end{array}$$

11.3 Translations to Section 4

We had already obtained a beautifully crystallizing azopigment earlier from *Phonopyrrol*, which according to its reactions belonged to the β-series, whereby formula **I** appears to be out of the question . . .

From the picrate, we were left with possession of pure phonopyrrole carboxylic acid. We obtained a beautifully crystalline azopigment [from it] by reaction with diazobenzenesulfonic acid, which according to the reaction doubtless belonged to the α-series.

This behavior stands in sharp contrast to the results of distillation, and in order to provide clarification we therefore synthesized 2,4-dimethylpyrrole-5-acetic acid and 2,4-dimethyl-pyrrole-5-propionic acid, whereby for the first time homologous pyrrole carboxylic acids substituted at the carbon are obtained.

We extended the *Knorr* pyrrole synthesis by condensing the isonitroso-compound of acetylpropionic acid, $CH_3COC(=NOH)CH_2CO_2H$, or acetylbutyric acid, $CH_3COC(=NOH)CH_2CH_2CO_2H$ with acetoacetic ester to form pyrroles **II** and **III**.

$$\textbf{II.} \quad \underset{NH}{\underset{\diagdown\diagup}{HOOC\cdot CH_2\cdot C}}\overset{\displaystyle H_3C\cdot \underset{\cdot\cdot}{C}\text{———}\underset{\cdot\cdot}{C}\cdot COOC_2H_5}{}C\cdot CH_3$$

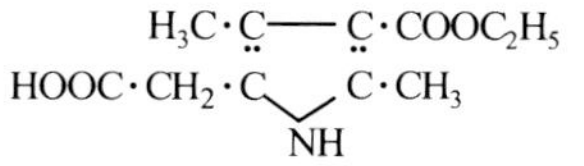

$$\textbf{III.} \quad \underset{NH}{\underset{\diagdown\diagup}{HOOC\cdot CH_2\cdot CH_2\cdot C}}\overset{\displaystyle CH_3\cdot \underset{\cdot\cdot}{C}\text{———}\underset{\cdot\cdot}{C}\cdot COOC_2H_5}{}C\cdot CH_3$$

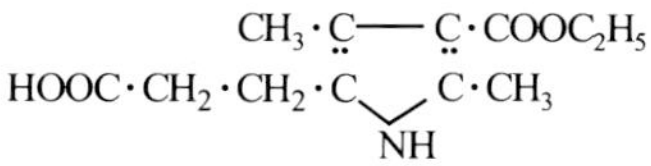

We removed the ring carbethoxy group of the ester-acid by treatment with fairly concentrated sulfuric acid. The resulting free acids unfortunately showed only very little crystallizability. Of course, under vacuum, we obtained the dimethylpyrrole acetic acid in well-formed crystals. We therefore characterized the two acids by their splendidly crystallizing azopigments, which consistent with all of their reactions belong to the β-series. The propionic acid does not give a picrate; the acetic acid gives an exceedingly easily soluble picrate so that it is not suited to identification. The propionic acid especially is differentiated so sharply from phonopyrrole carboxylic acid in all properties collectively that its non-identity with the latter is free from doubt.

51 This ester also gives a beautifully crystalline picrate, which strikingly possesses a dark red-brown color while that of the isomeric ester is bright yellow, like all pyrrole picrates that we have observed up to now. The difference is so evident that at first we were of the opinion that picric acid had somehow altered the ester of phonopyrrole carboxylic acid, and we therefore again regenerated the ester from the picrate. However, it proved to be unchanged.

52 Concerning the four acid cleavage products of hemin, the phonopyrrole carboxylic acids, we have first discovered all of them along with the bases (*ibid.*). For convenience, we relate them using the same principle of nomenclature that we had employed for the bases. [See Table 4.2.2.] We previously named the acid of mp 192°C, *Phonopyrrolcarbonsäure*, now we call it *phonopyrrole carboxylic acid a* (Nr. 10); the acid of mp 125°C, which gives a picrate of mp 150–152°C, is now *phonopyrrole carboxylic acid b* (Nr. 11) (previously *Isophonopyrrolcarbonsäure*): the acid of mp 108°C, now *phonopyrrole carboxylic acid c* (Nr. 12), was previously *Xanthopyrrolcarbonsäure*; and finally the acid of picrate mp 128° we call *phonopyrrole carboxylic acid d* (Nr. 13).

53 . . . however, we viewed the presence of hemopyrrole *b* as proof that phonopyrrole carboxylic acid *a* is substituted at the same positions as hemopyrrole *b*, as befitting formula **II**. It follows further that the isomeric phonopyrrole carboxylic acid *b*, which gives an oxime that is isomeric to the oxime of phonopyrrole carboxylic acid *a*, possesses the structure expressed by formula **VI**, especially since these two oximes afford the same hematinic acid (**IX**).

[Formulas **II**, **VI**, and **IX** are shown below for ease of reference.]

$$\begin{array}{ccc}
\textbf{II} & \textbf{VI} & \textbf{IX} \\[4pt]
\underset{NH}{\underset{\diagdown\diagup}{\overset{\textstyle CH_3- C\text{——}C- CH_2\cdot CH_2- COOH}{CH_3- C\quad CH}}} & \underset{NH}{\underset{\diagdown\diagup}{\overset{\textstyle CH_3\cdot C\text{——}C- CH_2\cdot CH_2\cdot COOH}{HC\quad C- CH_3}}} & \underset{NH}{\underset{\diagdown\diagup}{\overset{\textstyle CH_3- C\text{==}C- CH_2\cdot CH_2\cdot COOH}{OC\quad CO}}}
\end{array}$$

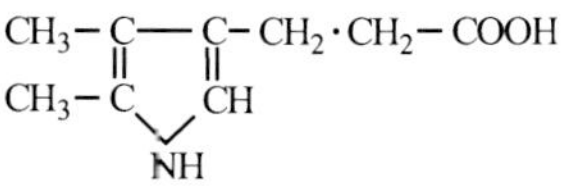

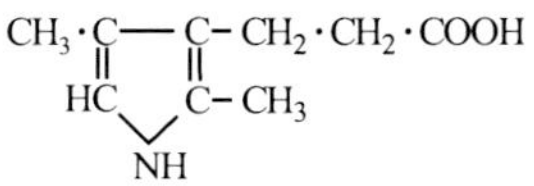

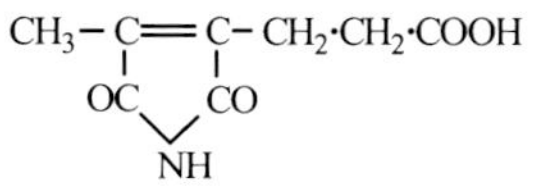

54	Table of the yields of hemopyrroles and phonopyrrole carboxylic acids from hemin by HI-glacial acetic acid

Nr.	Name	Grams from 100 g hemin		Grams from 100 g of base or acid mixture
1	Hemopyrrole a	0,6		2
2	Hemopyrrole b (*Willstätter's* and *Asahina's* isohemopyrrole and *Fischer's* hemopyrrole)	12,5		40,5
3	Hemopyrrole c (*Fischer's* kryptopyrrole and *Knorr's* hemopyrrole)	2,5		7,5
4	Hemopyrrole d (*Willstätter's* and *Asahina's* phyllopyrrole)	7,9	Sum of the bases 31.2 g	25
5	Hemopyrrole e	2,6		8,5
6	Hemopyrrole f	undetermined		undetermined
7	Non-precipitable hemopyrrole..........	4,3		14
8	High boiling hemopyrrole	0.8		2.5
9	Hemopyrrole g	small quantity		small quantity
10	Phonopyrrole carboxylic acid a	16,7		82
11	Phonopyrrole carboxylic acid b........	2	Sum of the acids 20.3 g	10
12	Phonopyrrole carboxylic acid c	0,6		3
13	Phonopyrrole carboxylic acid d........	1		5

55   In the preparation of "phonopyrrole" according to *Piloty's* directions [313], a rearrangement follows in such a way that the α-position indicated earlier becomes substituted. Loss of carbon dioxide must thereby act in some way to promote the disposition of the side chain to rearrange, for we have not yet succeeded in rearranging hemopyrrole to a β-free pyrrole by heating in a sealed tube at the corresponding temperature. Perhaps a reaction of an entirely different type also comes into play (even loss of ammonia occurs in small quantities). We hope to bring complete clarity here by further synthetic work.

56   *Fischer* and *Bartholomäus* [368] clarified the nature of this *Phonopyrrol*, from which they were able to isolate an azopigment that proved to be a β-azopigment. Not only did loss of carbon dioxide occur but also simultaneous migration of the resultant ethyl group to the α-position at the high temperature that is reached during the distillation of phonopyrrole carboxylic acid.

$$CH_3C\!\!-\!\!CCH_2CH_2COOH$$
$$CH_3C\quad CH \quad -CO_2 \longrightarrow$$
$$NH$$

$$CH_3C\!\!-\!\!CH$$
$$CH_3C\quad CC_2H_5$$
$$NH$$

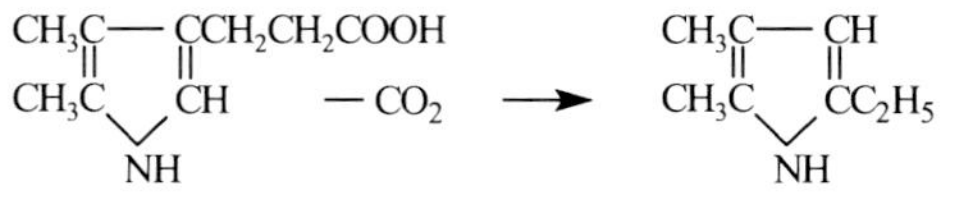

According to this finding, it could not be ruled out that phonopyrrole carboxylic acid bears the propionic acid in the α-position, or that in the acid fraction an acid of that constitution is present. The synthesis of the acid in question [368] refutes this possibility.

11.3 Translations to Section 4 689

57   Recently *Piloty, Stock,* and *Dormann* [*331*] succeeded in obtaining a crystalline picrate of 119°C from the distillate of phonopyrrole carboxylic acid (by rapid distillation). It proved to be identical to the picrate salt of hemopyrrole.

Accordingly, phonopyrrole carboxylic acid would have the following constitution:

$$CH_3C{=}CCH_2CH_2COOH$$
$$CH_3C\quad CH$$
$$NH$$

and it would be appropriate to introduce the old name *hemopyrrole carboxylic acid* again, as *Küster* has already proposed earlier. I believe however that it is advisable to refrain from this temporarily because from a consideration of our old results, it appears to me that rearrangements do not appear to be excluded. A rearrangement from the α to the α'-position is conceivable . . .

58   . . . *Piloty* first observed [*303, 308*]. He succeeded in obtaining an acid by reduction of hematoporphyrin with tin and hydrochloric acid that he at first designated hemopyrrole carboxylic acid, later however as phonopyrrole carboxylic acid (from $\varphi\acute{o}\nu\acute{o}\zeta$ = shed blood) because the pyrrole ("phonopyrrole") from decarboxylation was not identical to hemopyrrole. The latest investigation has shown that hemopyrrole indeed does result from rapid distillation of "phonopyrrole carboxylic acid" [*313*], while phonopyrrole would be understood as a mixture of 2,3-dimethyl-5-ethyl- and 2,3,5-trimethylpyrrole [*368*]. Thus the acid was subsequently returned to its original designation, hemopyrrole carboxylic acid.

59   If phonopyrrole carboxylic acid is correct in the sense above, the constitution of isophonopyrrole carboxylic acid would be that corresponding to kryptopyrrole:

$$CH_3C{=}CCH_2CH_2COOH$$
$$HC\quad CCH_3$$
$$NH$$

and it would be assigned the name *kryptopyrrole carboxylic acid*. However, it is appropriate for the reasons mentioned above to retain the old name provisionally.

60   *Piloty* [*333*] first described a xanthopyrrole carboxylic acid, in addition to phonopyrrole carboxylic acid, that differs by one more carbon atom and should have the following constitution

$$CH_3C{=}CCH_2CH_2COOH$$
$$HC\quad CC_2H_5$$
$$NH$$

because with nitrous acid it affords the oxime of isophonopyrrole carboxylic acid, which gives hematinic acid by hydrolysis. Despite much effort the authors have never been able to obtain evidence for the occurrence of this acid.

61   Finally, *Piloty* [*337*] was able to isolate an acid isomeric to xanthopyrrole carboxylic acid from the acidic cleavage products of hemin. This acid had been obtained by *H. Fischer* and *Bartholomäus* [*369*] by methylation of phonopyrrole carboxylic acid and by *Fischer* and *Röse* [*356*] from bilirubin. Accordingly, the constitution of this acid is that of a trimethylpyrrole propionic acid that

$$CH_3C{=}CCH_2CH_2COOH$$
$$CH_3C\quad CH \quad + CH_3ONa$$
$$NH$$

$$\longrightarrow$$

$$CH_3C{=}CCH_2CH_2COOH$$
$$CH_3C\quad CCH_3$$
$$NH$$

therefore is assigned the name *phyllopyrrole carboxylic acid*.

62 Dr. *Lieb* in Graz carried out the microanalyses ...

I.	0.1210 g	Substance	:	0.2989 g	CO_2	and	0.0747 g	H_2O
II.	4.260 mg	"	:	10.495 mg	"	"	2.62 mg	"
III.	4.545 mg	"	:	11.30 mg	"	"	2.73 mg	"
IV.	3.900 mg	"	:	0.333 ccm N at 733 mm Hg and 22°,				
V.	0.1436 g	"	:	0.3533 g	CO_2	and	0.0869 g	H_2O

			I.	**II.**	**III.**	**IV.**
$C_{33}H_{40}N_4O_6$ (588.36)	Calc. : = C	67.31	Found : = 67.37;	67.19;	67.81;	67.10
	" : = H	6.86	" = 6.91;	6.88;	7.37;	6.77
	" : = N	9.52	" = 9.53.			

63 I have also proposed the following formulas for bilirubin and its transformation products, with $\diagup C=C \diagdown$ the assumption of the bridge proposed by *Willstätter* and *Max Fischer* for hemin ..., with the proviso that

Bilirubin $C_{33}H_{36}N_4O_6$

the formulation of vinyl groups either in hemin or bilirubin must be modified to account for the fundamentally different behavior of the two pigments toward Na(Hg). Namely, ethyl residues are produced in bilirubin by this reagent, while in hemin apparently no real change takes place in its vinyl groups, and in any case no ethyl residues are formed.

Mesobilirubin $C_{40}H_{40}N_4O_6$

Mesobilirubinogen $C_{33}H_{44}N_4O_6$

11.3 Translations to Section 4

64   ... we thus arrive at a few probable formulas for etioporphyrin and etiophyllin:

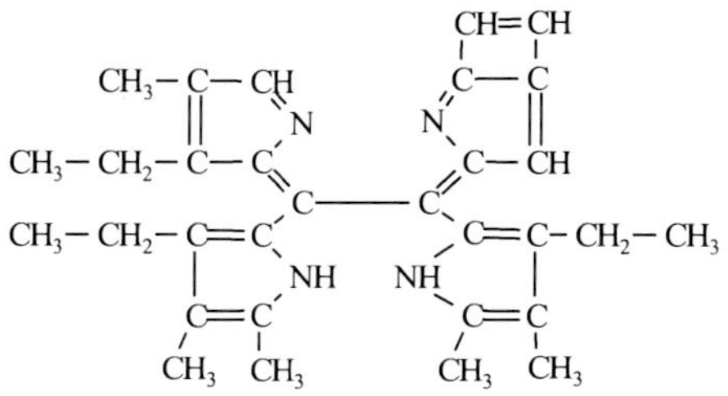

$$
\begin{array}{c}
CH_3-C-CH \\
\parallel \quad \searrow N \\
CH_3-CH_2-C-C \\
\qquad C \quad \text{---} \quad C \\
CH_3-CH_2-C=C \qquad C=C-CH_2-CH_3 \\
\mid \quad NH \quad NH \\
C=C \qquad C=C \\
CH_3 \; CH_3 \qquad CH_3 \; CH_3 \\
\end{array}
$$

Etioporphyrin $C_{31}H_{36}N_4$

and

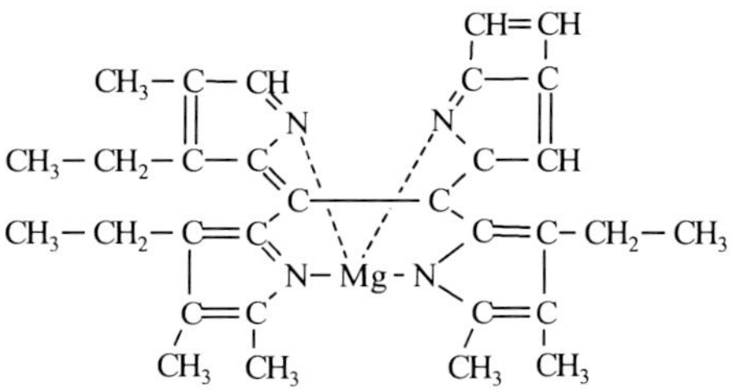

$$
\begin{array}{c}
CH_3-C-CH \qquad CH=CH \\
CH_3-CH_2-C-C \qquad C-CH \\
CH_3-CH_2-C=C \qquad C=C-CH_2-CH_3 \\
N-Mg-N \\
C=C \qquad C=C \\
CH_3 \; CH_3 \qquad CH_3 \; CH_3 \\
\end{array}
$$

Etiophyllin $C_{31}H_{34}N_4Mg$

65   This assumption prompts the objection, however, that of the two imino groups bound to iron only one is represented as the acidic imine of a pyrrole, but the other as a dihydropyrrole basic imino group. The improbability of this formula is due particularly to the assumption of a 16-membered ring containing four nitrogen atoms and 12 carbon atoms.

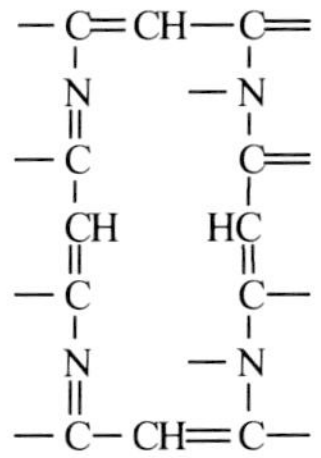

$$
\begin{array}{c}
-C=CH-C= \\
N \qquad -N \\
-C \qquad C= \\
CH \qquad HC \\
-C \qquad C- \\
N \qquad -N \\
-C-CH=C- \\
\end{array}
$$

66   The following formula for the nucleus of the pigment seems likely: a simple linkage of the four pyrrole rings, two of which are salt-forming and two of which are complex-forming.

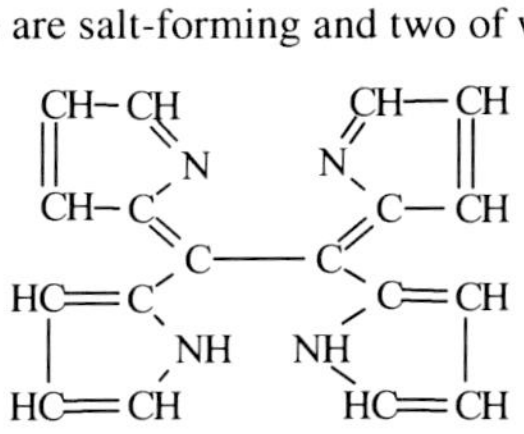

$$
\begin{array}{c}
CH-CH \qquad CH-CH \\
CH-C \qquad C-CH \\
\qquad C \text{---} C \\
HC=C \qquad C=CH \\
NH \quad NH \\
HC=CH \qquad HC=CH \\
\end{array}
$$

67   If the etioporphyrin of formula $C_{31}H_{36}N_4$ seems conspicuously low in water, then hemin ($C_{33}H_{22}O_4N_4FeCl$) is derived from a basic substance that is still about two hydrogen atoms poorer ($C_{31}H_{34}N_4$), which requires the assumption of double bonds and carbon rings. On the basis of the assumed connection of four pyrroles by the >C—C< group, and the formation of

two vinyls, we attempt to develop a constitutional formula of hemin sufficient for oxidation, reduction, and porphyrin formation, one that is still improving and uncertain in several details, however, as we hope to stimulate further investigation, and be able to lead to:

Hemin $C_{33}H_{32}O_4N_4FeCl$.

In the formation of hematoporphyrin, the bonds are severed from the two pyrrole nitrogens whereupon the central $>C=C<$ group can be transformed into §

68   I have then proposed the following formula:

Hemin $C_{34}H_{30}O_4N_4FeCl$.

69   Since my last publication on bilirubin [383] a comprehensive paper by *H. Fischer* [373] appeared, in which the formula $C_{33}H_{36}O_6N_4$ for bilirubin was recommended anew on the basis of the analytical results from the reduction product of bilirubin, bilirubino-gen ($C_{33}H_{44}O_6N_4$). While the previously accepted formula ($C_{32}H_{36}O_6N_4$) coming from analyti-cal results can be ascribed to a level of sulfur-containing contaminants that had not been able to be removed by the methods used earlier for purification of crude bilirubin, or from a content of chlorine, which occurred in the crude bilirubin during extraction of gallstones with chloro-form. Furthermore, it is also possible that crude bilirubin represents a mixture of two different substances, and the hope is expressed that a clarification of these questions would become possible by investigating the bilirubin-ammonium [salt] that I discovered.

And so I believe I should be able to assume, according to all available knowledge, that the formula $C_{33}H_{36}O_6N_4$ really can be assigned to bilirubin.

70   Now my latter formulation of hemin with four methines linking the pyrrole and pyrrolene rings, which should give it pigment character, has been termed unlikely because this appears as a 16-membered ring.

71   My views, published in this *Berichte* and in the *Zeitschrift für physiologische Chemie*[1]) on the constitution of hemin and the binding of iron to the organic part of its molecule have not thus far been able to enjoy the recognition of my professional colleague, namely *R. Willstätter*[2]), who developed a substantially different view in an address before the German Chemical Society in which he rejected my structure of hemin. Although he could use signifi-cant parts of it for his own [structure], together with the constitutional formula developed by *Piloty* (which has proven to be incorrect), in rejecting the proposition and with the remarks "that the question regarding the manner in which the pyrrole rings are linked has not been

11.4 Translations to Section 5   693

resolved by this concept", he has, to all appearances, brought forth the belief that his structure alone is best suited to account for the present status of our knowledge. At least it would be otherwise incomprehensible that *Hjelt* in his "History of Organic Chemistry", *Abderhalden* in his "Textbook of Physiological Chemistry" included *Willstätter's* constitutional formula alone, while for an evolving science, the nature of such books requires either giving room to include all ideas or to represent none for the opinions discussed. Moreover, I[3]) had already proven that *Willstätter's* representation leads to contradictions.

---

[1]) Küster W (1912) Über die Konstitution des Hämins. 45: 1935–1946. Küster W (1912) Über die Methylierung des Hämins. Hoppe-Seyler's Zeitschrift für physiologische Chemie 88: 113–159.
[2]) Willstätter R (1914) Überflanzenfarbstoffe. Hoppe-Seyler's Zeitschrift für pysiologische Chemie 88: 2831–2874
[3]) Küster W (1914) Über die Konstitution des Hämins. Hoppe-Seyler's Zeitschrift für pysiologische Chemie 88: 377–388

72   All of these considerations must lead to rejection of *Willstätter's* picture.

11.4   Translations to Section 5

1   According to the *Fischer-Röse* interpretation [*352*] bilirubinic acid, the bimolecular cleavage product of bilirubin has the following constitution:

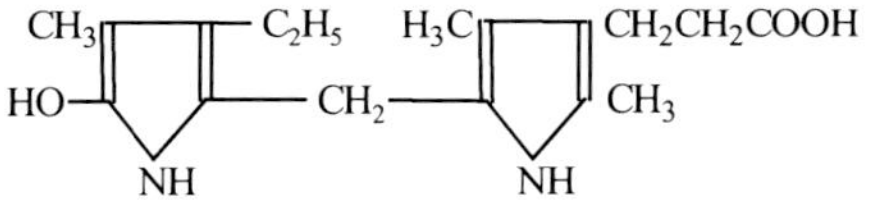

It is therefore composed of a hydroxypyrrole and a pyrrole carboxylic acid.

2   Xanthobilirubinic acid has been obtained as a cleavage product from the cited pigments: the bilirubin-mesobilirubinogen of *Fischer* and *Röse* [*356*] and the mesobilirubin of *Fischer* [*377*]. According to the research of the cited authors, xanthobilirubinic acid has the following constitution:

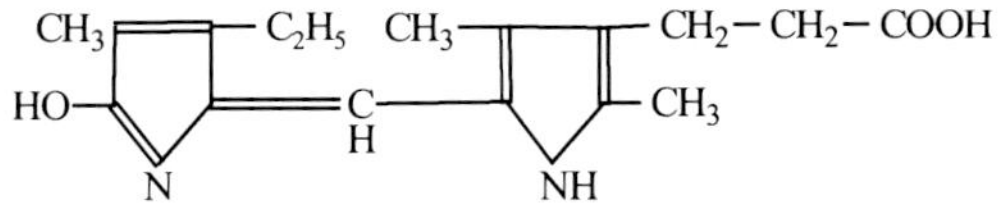

Of course, *Fischer* and *Röse* assigned the corresponding keto-formula to xanthobilirubinic acid. After a recent investigation, *Fischer* and *Niemann* [*395*], however, prepared an acetyl compound from xanthobilirubinic acid and accordingly arrived at the tautomeric formula above.

3   The existence of a free hydroxyl group in xanthobilirubinic acid was also confirmed experimentally, in which the ester of this acid forms a beautifully crystalline acetyl compound. We have recently obtained xanthobilirubinic acid from bilirubinic acid simply by air oxidation.

4   Hydrogen uptake was measured. The best results were obtained when the catalyst was added in a single portion. On average, two molecules of hydrogen were taken up in going from bilirubin to mesobilirubin.

The melting point of mesobilirubin is not constant although uniform crystallization was achieved from pyridine. The melting point is found between 300°C and 315°C. The variation can be explained by the assumption of tautomeric forms that would then also be present in bilirubin.

5   The preparation of this ester proves, in confirmation of earlier work, that two carboxyl groups are present in mesobilirubin and therefore in bilirubin

Dimethyl Ester of Mesobilirubin

0.15 g mesobilirubin were suspended in 5 cm^3 methyl alcohol. Warming and dissolution occurs by introduction of dry hydrogen chloride, and after a short while the hydrochloride of the ester crystallizes.

5.723	mg	Substance	gave	12.852	mg CO_2 and 3.742 mg H_2O.
4.217	mg	"	"	2.650	mg AgI.
4.487	mg	"	"	1.800	mg AgCl.

$C_{35}H_{44}N_4O_6 \cdot 2$ HCl:

		C		H		OCH_3		Cl	
Calculated for		60.96%	H	6.67%	OCH_3	8.99%	Cl	10.3%	
Found	"	61.2	"	7.32	"	8.3	"	9.92	

6   We have been able to prove with certainty that two free ethyl residues [groups] are present in mesobilirubin, *i.e.* they form very easily. Two mole equivalents of methylethylmaleimide were obtained by oxidation with nitric acid, an indication that two ethyl groups must have been preformed, and is, moreover a welcome confirmation for the presence of two basic pyrrole rings in bilirubin.

7   This behavior would be explained by the assumption of vinyl groups, which are then reduced to ethyl groups. However, the unsaturated side chains in bilirubin are constituted differently from hemin since a major distinction between the two pigments occurs in their reaction with sodium amalgam. Only in bilirubin are the unsaturated side chains reduced, while with hemin complete discoloration occurs with sodium amalgam – but no reduction. For the sodium amalgam-reduced hemin gives no methylethylmaleimide upon oxidation.

8   The experimental basis for this assumption was present since *Fischer* and *Röse* have shown that an α-hydroxypyrrole is present in bilirubin. We believe therefore that the vinyl groups in bilirubin lie in close proximity to the hydroxyl groups in such a way that furan rings are components of bilirubin. Oxidation at the two α-positions during the transformation of the pigment of blood to the pigment of bile is then perhaps essential. Dehydrogenation then closes up the vinyl and hydroxyl groups to a furan ring, according to the Scheme.

This scheme would naturally fit only with the presence of an as yet unknown *Hämopyrrol* base in bilirubin; one would arrive at two stereoisomeric forms with the kryptopyrrole component. The [presence of] furan rings then explains the extraordinary sensitivity of bilirubin to mineral acids.

9   The furan ring is opened by reduction of bilirubin with sodium amalgam or hydrogen/palladium and only then yields the free hydroxyl groups and ethyl residues.

10   The uptake of two mole equivalents of H_2 in going from bilirubin to mesobilirubin corresponds well with the interpretation of the origin of an ethyl group from a vinyl group and a hydrofuran ring from a furan ring.

11   Encouraged by these findings, we subjected the bile pigment itself and its reduction products, mesobilirubin and mesobilirubinogen, to investigation. According to the *Fischer* formulation, the three compounds are represented by the following constitutional formulas, in a modification wherein the two connected carbon atoms are interpreted in a somewhat different way so that stereoisomers in greater numbers can be imagined if one still takes tautomer possibilities from ring **IV** into consideration.

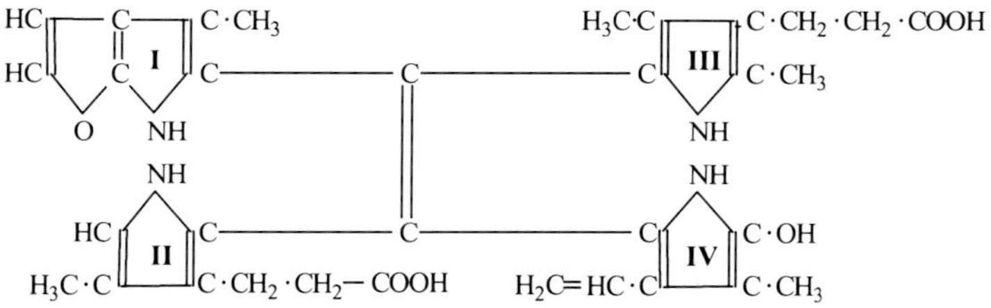

11.4 Translations to Section 5

Bilirubin.

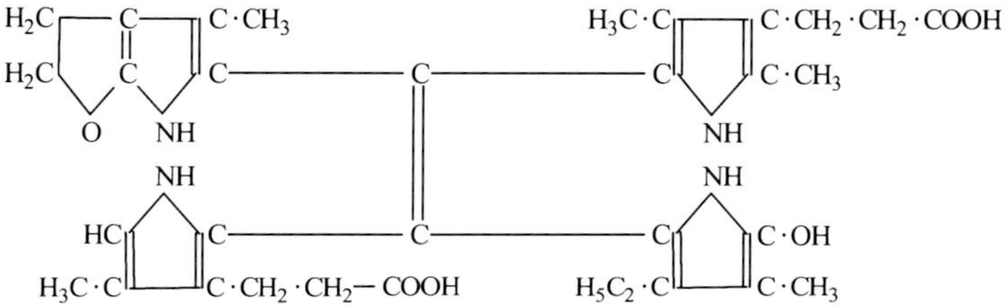

Mesobilirubin.

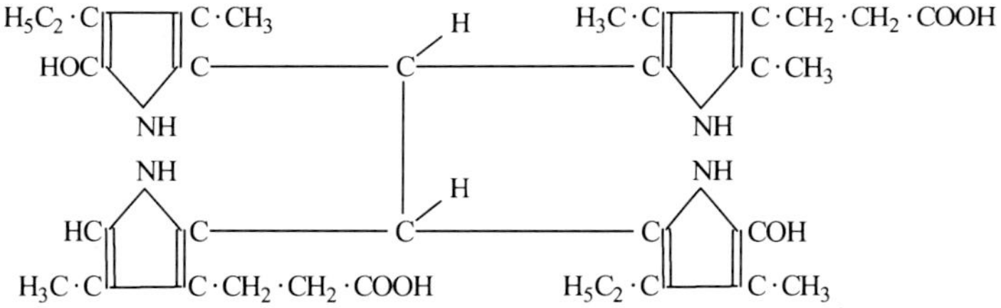

Mesobilirubinogen.

12 Finally, concentration in vacuum left a crystalline encrustation of the oxidation product, the properties of which were found, as *H. Fischer* described, with a melting point between 82°C and 86°C. Transition into a higher molecular weight product took place rapidly in air within 48 hours. The original material often keeps in a desiccator for as long as eight days.

The analysis gave:

$$4.588 \quad \text{mg} \quad \text{Substance} \quad \text{gave} \quad 0.427 \text{ ccm N (717 mm, 18°).}$$
$$4.515 \quad \text{mg} \quad \text{''} \quad \text{''} \quad 2.130 \text{ mg } H_2O \text{ and } 10.180 \text{ mg } CO_2.$$

Methylvinlymaleimide $C_7H_7NO_2$

	C	H	N
Calculated for	= 61.32%	= 5.11%	= 10.29%
Found	61.49	5.27	10.30

The old values were 61.4% C and 5.0% H [*Diese Z.* 1914, 91, p 192] [*234c*].

A molecular weight determination according to *Rast* was not achievable since the product clearly changed upon heating and is no longer soluble in camphor.

A micromolecular weight determination according to *Pregl* gave the following values. Alcohol served as solvent.

I.	8.223	mg	gave	a	Depression	Δ		0.065°.
						M		104.
II.	6.750		''	''	''	Δ		0.040°.
	$C_7H_7[N]O_2$	mg				Calculated for	M	137
						Found		131

13   40 mg of oxidation product were dissolved in 30 cm³ ether and 2 g aluminum amalgam were added, then let stand four hours. Evaporation of the ether leaves behind a crystalline crust. Pure reaction compound (12 mg) was obtained pure by sublimation. The melting point lies between 58°C and 60°C. The crystal form is tetragonal. The compound is colorless and almost odorless. Methylethymaleimide, which was reduced in a similar way, gave a colorless oil, that is as yet not crystalline.

Analysis:

3.880	mg	Substance	gave	2.360 mg H_2O,	8.570 mg CO_2.
3.400	mg	"	"	0.303 ccm N (717 mm, 16°).	

$C_7H_9NO_2$ Calculated for C 60.43% H 6.47% N 10.07%

                Found   60.26   6.81   9.86

The molecular weight determination was carried out according to *Rast*. 0.300 mg substance dissolved in 2.750 mg camphor gave a melting point depression of 30°C. Calculated MW 139; found 145.

14   A few years ago [*372, 377*] a beautifully crystalline product was obtained by *H. Fischer* from treatment of bilirubin with nitrous acid; clarification of its constitution was not successful. The main difficulty was that the compound was very unstable and was transformed after a short time into a higher melting compound; an accompanying decrease in carbon content was linked to this. We have prepared the compound anew and can confirm the old observations. We have subjected it to micro-analysis again, and according to the values obtained, which showed good agreement with those found earlier, the analysis remains in good agreement with methylethylmaleimide. A molecular weight determination also gave the same results. However, we consider it extremely improbable that methylvinylmaleimide is present because although hydrogenation of the product was successful with sodium amalgam, and even went especially well with aluminum amalgam, they led to a newly formed product that was clearly rich in hydrogen but is certainly not methylethylmaleimide. We therefore consider it most probable the compound has the following structural formula:

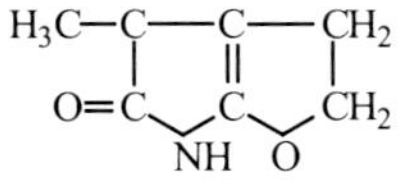

$$H_3C-\underset{\underset{O=C}{|}}{C}\overset{}{\underset{\underset{NH}{}}{-}}\underset{\underset{C}{\|}}{C}\underset{\underset{O}{}}{-}\underset{\underset{CH_2}{|}}{CH_2}$$

_____

Correction

and its hydrogenation product is:

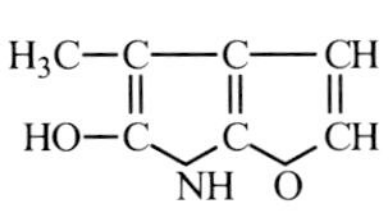

$$H_3C-\underset{\underset{HO-C}{\|}}{C}\overset{}{\underset{\underset{NH}{}}{-}}\underset{\underset{C}{\|}}{C}\underset{\underset{O}{}}{-}\underset{\underset{CH}{\|}}{CH}$$

The second structure above was corrected to that below on p. 312 of ref [*399*].

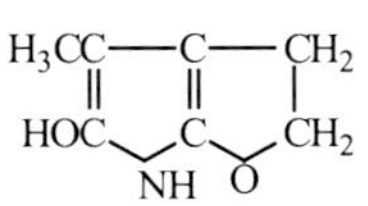

$$H_3CC-\underset{\underset{HOC}{\|}}{C}\overset{}{\underset{\underset{NH}{}}{-}}\underset{\underset{C}{\|}}{C}\underset{\underset{O}{}}{-}\underset{\underset{CH_2}{|}}{CH_2}$$

15 Should these formulas prove to be definitively correct, then a great advance would therefore be made in connection with bilirubin research.

11.4 Translations to Section 5

16   *Willstätter* [218b] has constitutional formula **III** for etioporphyrin

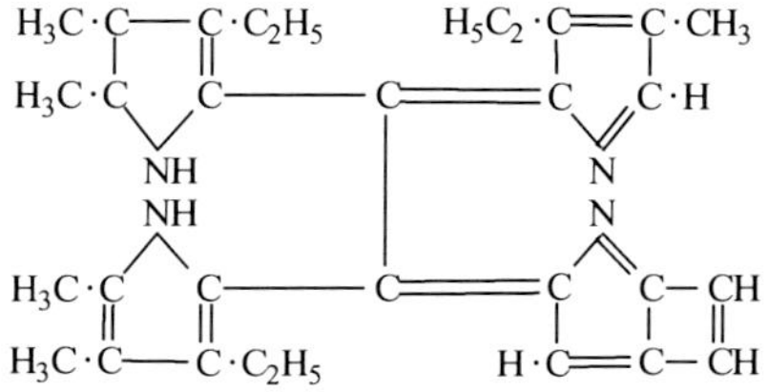

III.

17   ... it is obvious that tetrapyrrylethanes can be obtained by alkaline condensation of pyrroles with glyoxal. In fact this reaction proceeds smoothly with 2,4-dimethyl-3-carbethoxypyrrole, and we thereby obtained the beautifully crystalline tetra(2,4-dimethyl-3-carbethoxy-5-pyr-ryl), ethane (**I**).

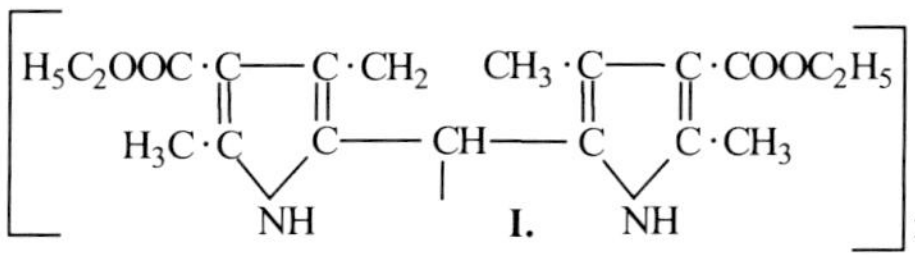

18 *H. Fischer* and *Eismayer* carried out this condensation nearly 10 years ago, but did not however publish it then, because the analysis was not good and further investigation of the tetra(2,4-dimethyl-3-acetyl-5-pyrryl)ethane analog prepared led to results that the formation of a compound of the indicated constitution is not proven as certain.

19   ... that two moles of bis(2,4-dimethyl-3-carbethoxy-5-pyrryl)methene (**II**) arise by oxidative cleavage.

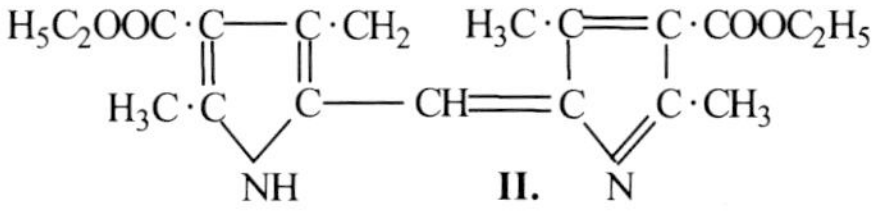

II.

20   ... accordingly, a porphyrin spectrum should follow from oxidation of tetra(2,4-dimethyl-3-carbethoxypyrryl)ethane. But that is not the case. Iron chloride [$FeCl_3$] cleaves it to a dipyrrylmethene, as already mentioned. A yellow coloration occurs after brief exposure to air.

21   Everything is reminiscent of a bile pigment, nothing of a blood pigment, and we see in the properties of the synthetically obtained tetrapyrrylethane a confirmation of the view that the blood pigment, as *Willstätter* assumes in etioporphyrin, is not constructed from four pyrrole rings connected by a C–C bridge, but that this constitution belongs to the bile pigment, corresponding to *H. Fischer's* formulation.

22   I had proposed the following formula then:

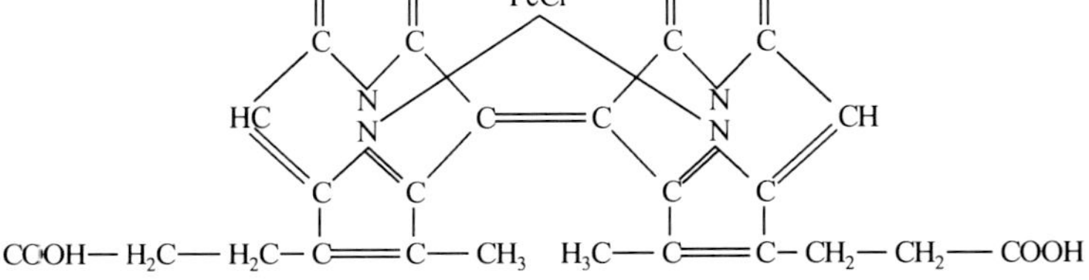

Hemin $C_{34}H_{30}O_4N_4FeCl$.

698   11   Translations

23   We side with *Willstätter*, however, that it is probable that the same system also forms the foundation of the blood pigment. For according to *Fischer* and *Reindel* [*145*], who have demonstrated that hematoidin, *i.e.* bile pigment, easily arises in the cell from the blood pigment, and thus a familiar relationship between the two pigments must be assumed as expressed in *H. Fischer's* hemin formula [*373, etc.*]. We assume that in addition to the ethine bridges two methine groups link the pyrrole rings at the α-position.

24   We treated the ethane with aluminum chloride in carbon disulfide, whereby a smooth dehydrogenation occurred. Thus, we succeeded in obtaining, with a little trouble of course, the beautifully crystalline tetra-(2,3-dimethyl-4-carbethoxy-pyrryl)-ethylene (**III**).

III.

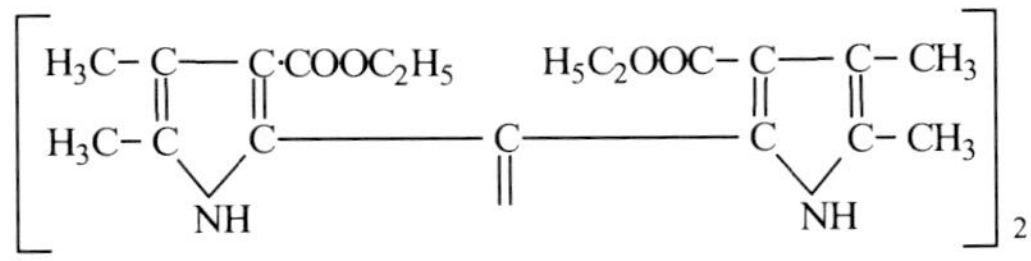

25   As to be expected, the ethylene has an intense yellow color, which reminds one of the bile pigment. When subjected to nitric acid in chloroform, a red coloration occurs, which evidently originates from the above-desired methene. Permanganate is quickly decolorized by a solution [of the pigment] in sodium carbonate. That a compound of the indicated constitution is actually present here comes from the results of analysis and molecular weight determination, and in addition, from the possibility of converting the ethylene back to the ethane. After the results of this investigation, especially from the lack of a porphyrin spectrum from the ethylene compound, it became clear that in etioporphyrin a simple linking of four pyrrole rings to a C=C bridge cannot be considered. Moreover, we assume that in addition to the C=C bridge, two additional methine groups link the pyrrole rings with each other. Further synthesis work in this direction is in progress.

26   The linear union (assembly) of four pyrrole rings, which can also be taken into consideration, was made improbable by work with *Scheyer* [*404*]. Tripyrrylmethanes also have no porphyrin-like characteristics [*405, 406*].

27

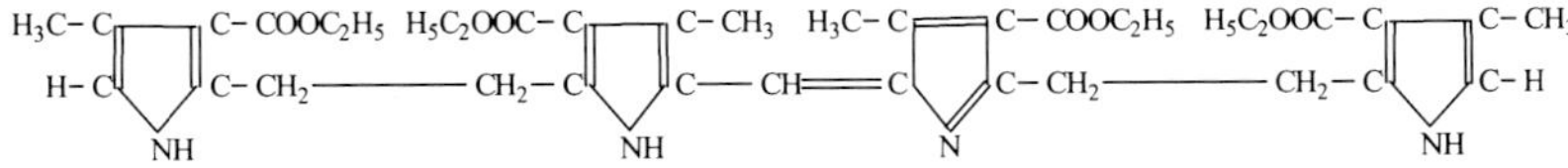

It is an intensely yellow-colored compound, as was of course already expected, and the color is unusually reminiscent of mesobilirubin. The intensity of the color is very great. A characteristic spectroscopic absorption in chloroform is observed just as with bile pigments.

28   In this stage of the work the assumption was then represented that two methine groups must be present in addition to the ethylene bridge in order to obtain etioporphyrin, and the constitutional formula of etioporphyrin can be ascribed to the following:

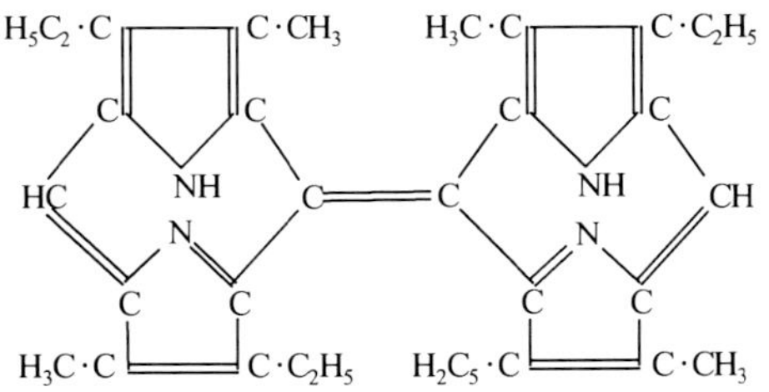

29   We therefore take formula **III** as rather more certain and then represent the transformation from here to etioporphyrin as follows:

11.4 Translations to Section 5

30   In 1916, *H. Fischer* observed the formation of bromine-containing, splendidly crystalline pigments obtained from mixing together bromine with tri-substituted pyrroles. Recently, the constitution of this pigment, assumed at that time to have an indigoid or dipyrrylmethene structure, was definitely decided in the sense of a dipyrrylmethene.

31   As regards the reaction mechanism, see perhaps the following:

We assume that from one mole of dicarboxylic acid first one mole of carbon dioxide is eliminated, then ring closure leads to **II**, which rapidly expels water and goes over into methano-ketone **III**. Two moles of the last are then perhaps combined by loss of hydrogen to a system as is represented again in the following formulation, from which the β-substituents are omitted for easier viewing.

32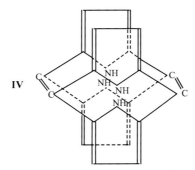

700   11   Translations

We assume then that, under the influence of acids, an isomerization of the ring system takes place to a structure as is sketched in **V** and was formulated earlier. It can then be raised for discussion as to whether perhaps an equilibrium between **IV** and **V** can be present for the porphyrins in solution. After all, stereoisomers of **V** are also possible, and it is also not immaterial on which of the two C=C bonds of **IV** the opening occurs.

33   If one formulates four methine groups between the pyrrole rings, as *Küster* did, a part of the molecule must dissociate in order to produce the two hydrogen atoms lost according to this reaction. The etioporphyrin synthesis here is accomplished at 100°C. Of course a porphyrin spectrum arises during heating the methanedicarboxylic acid in hydrochloric or glacial acetic acid. Therefore, we propose as more probable the concept represented here, with which the bromine-sulfuric acid reaction stands in accord. To be sure, we have also made many observations that are in agreement with *Küster's* formula with methine groups. The existence of the tetrachloro- and tetrabromo-mesoporphyrin as well as the fact that the porphyrinogen has six more hydrogens than mesoporphyrin, is explained well by the latter perception, just as with that of ours of course.

34   The following reaction process more probable:

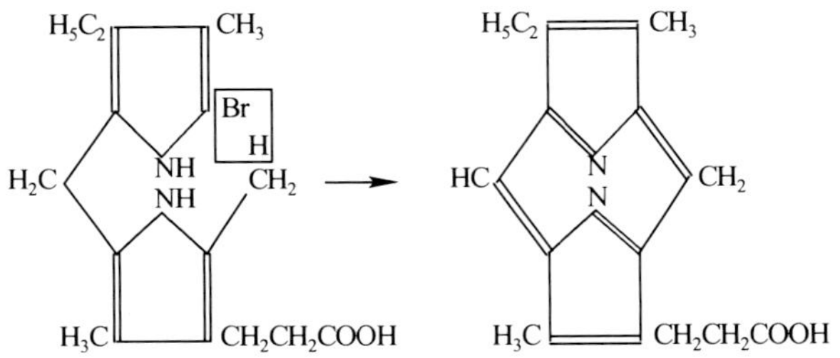

H_5C_2 … CH_3 … Br/H/CH_2 … NH NH … H_2C … H_3C … CH_2CH_2COOH
IV.

H_5C_2 … CH_3 … HC … N N … CH_2 … H_3C … CH_2CH_2COOH
V.

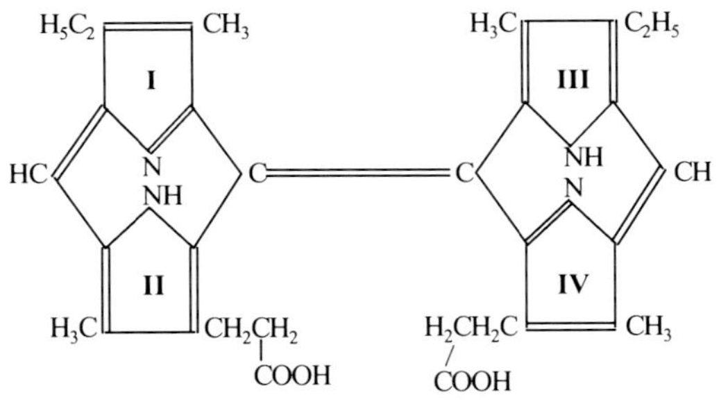

H_5C_2 … CH_3 — I — HC … N NH — II — H_3C … CH_2CH_2 | COOH — C══════C — H_3C … C_2H_5 — III — NH N — CH — IV — H_2CH_2C … CH_3 / COOH
VI.

or

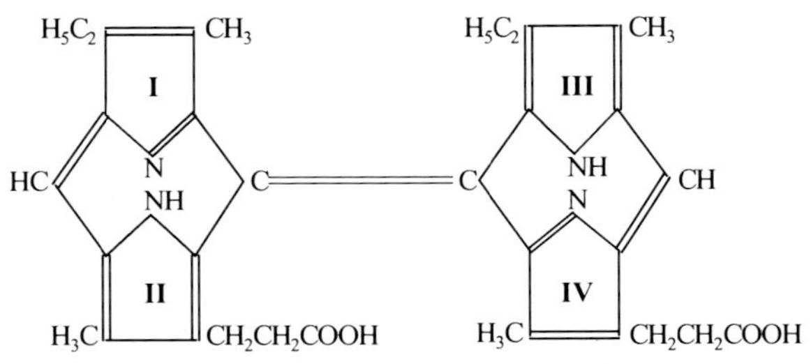

H_5C_2 … CH_3 — I — HC … N NH — II — H_3C … CH_2CH_2COOH — C══════C — H_5C_2 … CH_3 — III — NH N — CH — IV — H_3C … CH_2CH_2COOH
VII.

11.4 Translations to Section 5 701

i.e. dipyrrylmethane **IV** arises first by loss of hydrogen bromide and goes over into **V** by dehydrogenation. This "diquinone"-like structure self-condenses at once to **VI** or **VII**, while during isomerization two "basic" pyrrole rings are transformed into acid rings.

35 Naturally, two additional combinations are also still possible, i.e. the following:

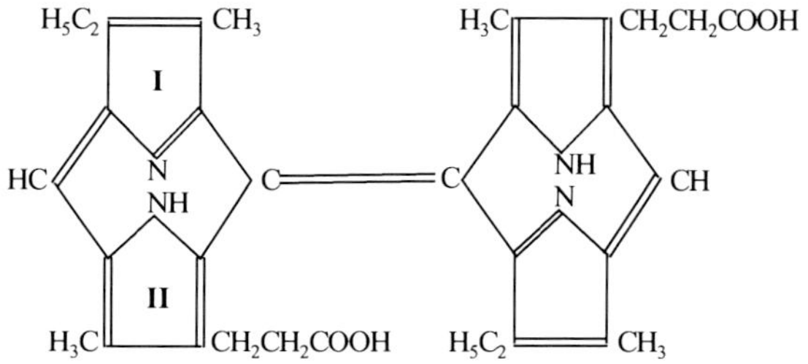

VIII.

36 The course of the reaction can also be interpreted in the sense of *Küster's* blood pigment formula in such a way that four methine groups link the four pyrrole rings to each other. The following reaction scheme is then to be noted:

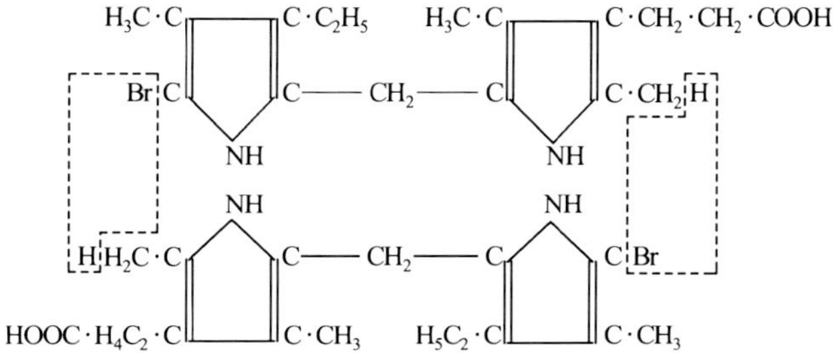

37 Small amounts of kryptopyrrole carboxylic acid in glacial acetic acid were subjected to our bromination procedure with two moles of bromine, and a beautiful crystalline hydrobromide salt of a brominated methene was obtained, which leads, in complete analogy to the previous experiments, to the following constitution:

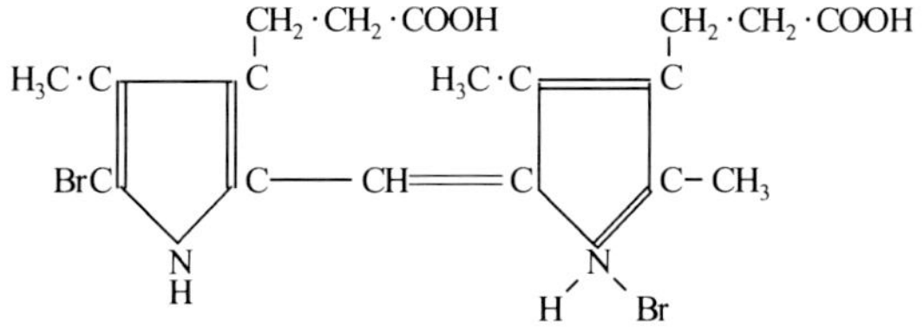

38 A more detailed investigation is still lacking. The compound did not give a porphyrin with sulfuric acid but, however, with glacial acetic acid under pressure. Heating at 210°C for three hours proved to be quite optimal. After this time, the contents of the sealed tube were partly crystallized alongside a carbonaceous sediment. It was converted into the free porphyrin by systematic treatment with hydrochloric acid, alkali, and ether. The free porphyrin often crystallized directly from the ether solution. To purify it we converted it into its methyl ester in the usual way with methanolic hydrogen chloride, and crystallized the product from chloroform-methanol, thus exactly as stated for the copro-ester (see Fig. 2, plate V). Crystallographic investigations by Prof. *Steinmetz* also showed identity. The ester melts at 243°C and with analytical copro-ester gives no depression. The elemental analysis confirms the composition. Projection of their spectra on top of each other showed no difference between the synthetic and analytical preparations.

702 11 Translations

39 Only in a few properties did the porphyrin show a divergence in this regard from the natural, as when we were unable to observe the colloidal solubility in water of the synthetic preparation. However, natural coproporphyrin also does not always show this property and according to the observations above there can no longer be any doubt that the two porphyrins are identical.

40 According to the methoxyl determination and the ammonia content (%N), four carboxyl groups are present in coproporphyrin; accordingly, the following constitutional formula is proposed for the pigment, which takes into account all findings:

$$C_{36}H_{36}N_4O_8$$

Or, according to *Küster* (on the basis of synthesis the ordering of the side chains is already correctly displayed):

$$C_{36}H_{38}N_4O_8$$

... The synthesis completely confirms the results of the analytical research. Coproporphyrin is composed of four mole equivalents of pyrrole propionic acid; it is a tetramethylporphin-tetrapropionic acid. The structure of the porphyrin nucleus is not certain, as may be seen from the details of the preceding communication. The porphyrin nucleus can be rendered by the indigoid (formula **I**) or that according to *Küster* (formula **II**). In the latter case, the molecule has two more hydrogen atoms; the theoretical values for both formulas are therefore given in the experimental section. The analysis shows no difference between both formulations.

41 Construction of the porphyrin was then brought about by condensing two methene molecules in such a way that methyl groups and bromine atoms come together in a reaction that liberates hydrogen brominde in which the reaction clearly takes place either within a methene molecule and yields the indigoid-formulated porphyrin[2] [*402*] by a secondary oxidative condensation of two molecules, or it occurs between two molecules of methene **I** by loss of hydrogen bromide

11.4 Translations to Section 5

and results in a 16-ring with two methine and two methylene groups. The latter two must then be transformed by dehydrogenation into methine groups, and we then have *Küster's* porphyrin formula. In keeping with this, it is obvious that a debrominated methene would undergo an analogous reaction with a dipyrrylmethene possessing two α-methyl groups, as in **II**. We first tested the reaction in order to obtain octamethylporphyrin.

---

[2] This formulation is provided in detail in *Annalen* 1926, 448: 183 [*402*], and for that reason we show the transformation according to *Küster's* formulation in this work. There is insufficient experimental material in order to decide which of the two formulas is correct (including possibly the one with two 9-rings). Even the exact constitution of the dibrominated methene is lacking, with which we are currently occupied.

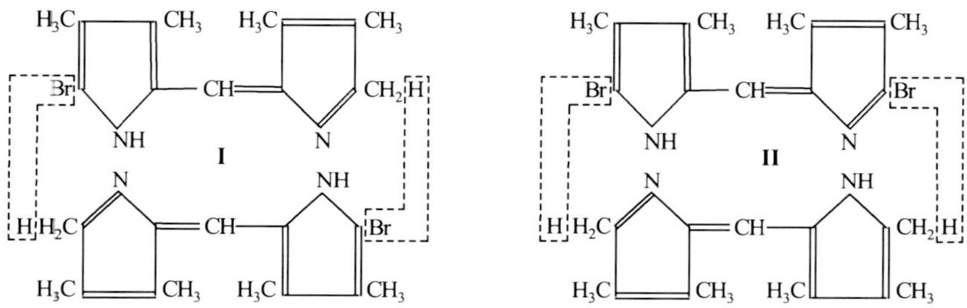

We have reacted methene **II** above with a series of methylated dipyrrylmethenes and observed porphyrin formation universally.

42 It is obvious that every possible combination of alkylated pyrroles and pyrrole carboxylic acid was considered and the porphyrins therefrom were investigated.[1] With a view toward a general nomenclature, we number the four pyrrole rings of the porphyrin framework as **I, II, III, IV**.

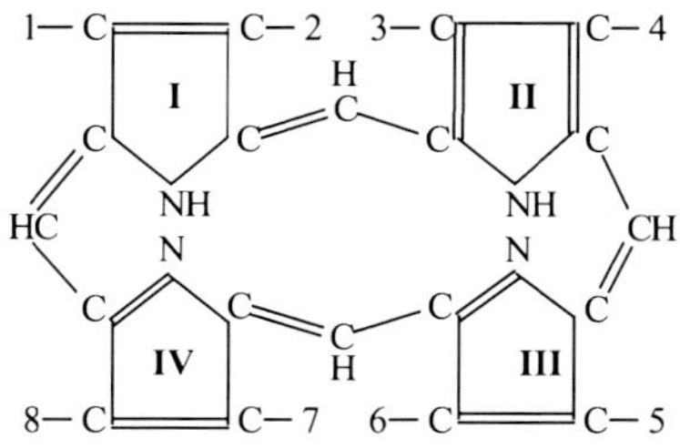

Regardless of which formulation is assumed, the substituents can appear only at the β-positions indicated. At what point either the substituents of each individual pyrrole ring are to be specified or, more simply, the β-substituents have to be numbered continuously as 1, 2 to 8, as shown in the scheme above.

---

[1] Note of correction: this synthesis would presumably bring a decision to the construction of hemin as was commented on in the reply to *W. Küster's* remarks (H. 163, p. 267) [*394*].

43 W. Küster recently proposed the following formula for hemin:

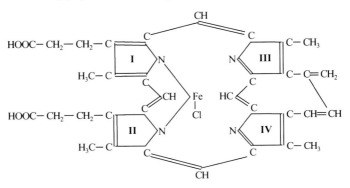

Hemin C$_{34}$H$_{30}$N$_4$O$_8$FeCl

in which the basic pyrrole rings (**III** and **IV**) were adjacent. To accommodate this formulation he was forced into the assumption of a C–C bond between the unsaturated side chains. According to the synthesis above from bilirubinic acid, this assumption, as formulated according to *Küster*, has been stripped away. In no event can the ethyl residues be adjacent to each other, and for hemin it therefore follows that here too the unsaturated side chains cannot be bonded to each other, as *Küster* has formulated.

44 Accordingly, the dimethyl ether of a chloro-bromo-hematoporphyrin is present, whose finding I view as a fully valid proof for the correctness of our presentation, whereby in the hemin ester used, the unsaturated side chains on the adjoining pyrrole rings are responsible, and in the form

$$-C{=\!=}CH_2$$
$$-CH{=\!=}CH$$

reacting to add two chlorine atoms in the 1–4 position

$$-C{-}CH_2Cl$$
$$-CH{-}CH$$
$$Cl$$

with hydrogen bromide-glacial acetic acid then forms:

$$-CHBr-CH_2Cl$$
$$-CHCl-CH_2Br$$

and under the influence of methanol:

$$-CH(OCH_3)-CH_2Cl$$
$$-CH(OCH_3)-CH_2Br.$$

45 Our observations therefore stand in complete contradiction to the claim expressed by *H. Fischer* in connection with obtaining mesoporphyrin from bilirubinic acid, that in hemin the unsaturated side chains are not bonded to each other and therefore the basis for the view expressed by me would be stripped away for the representation in which they were shown in the structure as being on adjacent pyrrole rings. Since *H. Fischer* has not attempted to the latter communicate with me but tried to discredit our previous results by the dictatorial working of a published communication, it is unfortunately necessary for me to clarify the situation by a critical reflection on the circumstances, and by that to avert the danger of further occurrence from Munich of false trends in the development of the hemin chemistry.

11.4 Translations to Section 5

705

46   Given his porphyrin synthesis, surely one must admire just as much his experimental skill that made possible the preparation of the sought-after material, as good luck accompanies its formation or generation, since some of the methods used appear from the outset to be very poorly suited to arrive at the target material. For these methods have the characteristics of brutal force. Such as obtaining mesoporphyrin from bilirubinic acid by heating the latter in glacial acetic acid + hydrogen bromide in a sealed tube at 145°. The reactions which might take place are completely obscure; explanation of the same must therefore be completely arbitrary.

47 That results impressively from the fact whereby the yield of porphyrin from transformation of bilirubinic acid is far smaller than is reached when starting from mesobilirubin. However, in the last case the missing carbon atom must be assumed to be available by decomposition of the molecule. How far now must the poor yield-delivering bilirubinic acid go until decomposition!

48 I must dispute however his assertion on repeated occasions that he had first produced the proof whereby in hemin the methines collectively are bonded in the α-positions. It is true that he produced experimental proof for this, just as it is true however that the same was by no means necessary any longer because the complete proof that in hemin the β-positions of the pyrrole rings are collectively substituted by alkyl or propionic acid groups had been produced by my method of oxidative cleavage that led to β,β'-substituted imides. There are also compellingly conclusive deductions, that need not be confirmed only by experiment.

49 Now the only difference that exists between this and my formulation must according to *H. Fischer* be drawn with entirely unsaturated side chains, or half according to *Willstätter*. *H. Fischer* postulated that pyrrole rings I and II as well as III and IV are each connected by a methine and a bond between these methines, while in my structure pyrrole rings I and IV and II and III appear to be bonded to one methine. The two different methines in both structures are in the same position, as may be illustrated in this sketch.

I   **II**

    I hold my formulation to be more correct because cleavage of hemin into two bicyclic pyrrole derivatives which possibility, as may be picked out immediately from *H. Fischer's* structure, has still not yet been successful.

50   Thus, *H. Fischer's* claim mentioned at the outset is not only risqué, it is unwarranted, and it is not understandable to me how *H. Fischer* can still feel compelled toward it on the basis of an unfathomable synthesis and can allow himself to dispute the argumentative force of our results and explanations, which rest on a nearly quantitatively complete and easily explainable analysis. Adding fuel to the fire *H. Fischer* cited my idea of the type of bonding of the side chains in hemin for the first time only when he believed that he had ruled out the grounds for it. I must characterize such undertaking as an uncritical overestimation of his results. And in conjunction with the other statements, I feel it to be a systematic disparagement of the results of others. *H. Fischer* had no right to characterize *A. Papendieck's* results as meaningless, which in my opinion is done when the analysis of a porphyrin is not recognized as valid because it is amorphous, although it was derived from a crystallized hydrochloride. He thus commits quite an injustice when he demeans my results at every opportunity or wrongs an idea of mine with the postscript "without experimental proof". Following in the footsteps of *O. Piloty*, *H. Fischer* has assumed possession of an area of research in which I had previously achieved decisive success. These and the conclusions and ideas produced therefrom were also decisive for *H. Fischer's* experimental work.

51 I recall therefrom that my idea to combine two dipyrrylmethenes to [form] the prosthetic group have been completed by *H. Fischer* with his method, crowned by such beautiful success in the synthesis of pyrrole aldehydes, that arose from efforts to incorporate a propionic acid residue in the β-position of a pyrrole derivative – after I had proven the presence of such in hematinic acid. Thus, in the essential points, *H. Fischer* makes facts out of my ideas, and I have gladly let the mantle fall to me as architect of the product of the skilled builder. It is certainly understandable that he does not enjoy playing this role; however, he must remain aware that this position arises from the nature of the situation. *H. Fischer* would have to choose a new research field if, aside from his experimental superiority, he had also made up his mind to be counted as the independent intellectual leader.

52 In the [research] area of the prosthetic group of hemoglobin, he had not been able to prove this characteristic because his hemin formula bears neither the stamp of something new or original, nor does it contribute, by his beautiful experimental work in the porphyrin field, to the ideas that I raised on the construction of this material. On the other hand, it would be wrong of me if I fail to recognize that *H. Fischer's* highly valuable and interesting investigations provided the most toward expanding the porphyrin field. We are of course obviously occupied with pursing these reactions further, and I might add to this now by pointing out old and new observations that there already exists a difference between glacial acetic acid hemin and its dimethyl ester, as was obtained in our previous results. I might assert now that a change already takes place during esterification and since the ester is drawn directly into betain bonding, the nitrogen atoms are drawn into involvement in this transformation.

Glacial Acetic Acid-Hemin, $C_{34}H_{30}O_4N_4FeCl$

Dimethyl Ester of Hemin

53 One involvement, however, concerns the possibility of addition of halogen to the unsaturated side chains and the convertability into dihalogenated hematoporphyrin dimethyl ether, and especially by the finding of oxidation to two molecules of an imide that contains only one halogen per molecule.

11.4 Translations to Section 5

54 Thus it appears that hemin itself has not only one carboxyl but also yet another group with a betain-type of nitrogen, and that can only be the acetylene residue as presently shown, in which one unsaturated side chain is represented. The acetylene of course has acid properties. If we let this residue bond at the same time to a nitrogen atom, then a formulation is obtained that is manifest in my representation as well as that of *Willstätter-Fischer*. However, in one of the actual findings clarifying modification is taken into account and is thereby well-suited to bring about satisfactory agreement, perhaps to all involved, to the dispute over the constitution of hemin.

55 The transformation into methemoglobin would then consist in dissolving this linkage,[2]) and now loss of water from the hydroxyvinyl group succeeds, as can be show to be already established by the isolated prosthetic group.

$$-CH_2-CHOH-NH\text{-Glob.} \longrightarrow CH_2-CHO\,(NH_2\text{-Glob.})$$
$$\text{in hemoglobin}$$

$$\longrightarrow -CH{=}CHOH \longrightarrow -C{\equiv}CH$$
$$\text{in methemoglobin}$$

The sketch is clearer when all groups are taken into consideration

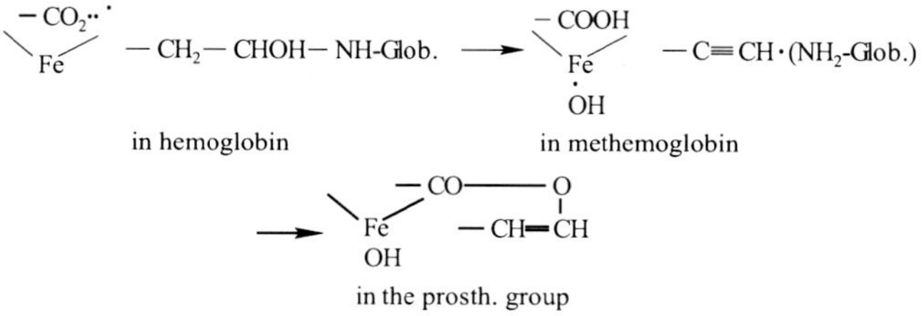

$$\text{in hemoglobin} \qquad\qquad \text{in methemoglobin}$$

$$\text{in the prosth. group}$$

and the following picture thus appears for hemoglobin and its transformation into methemoglobin:

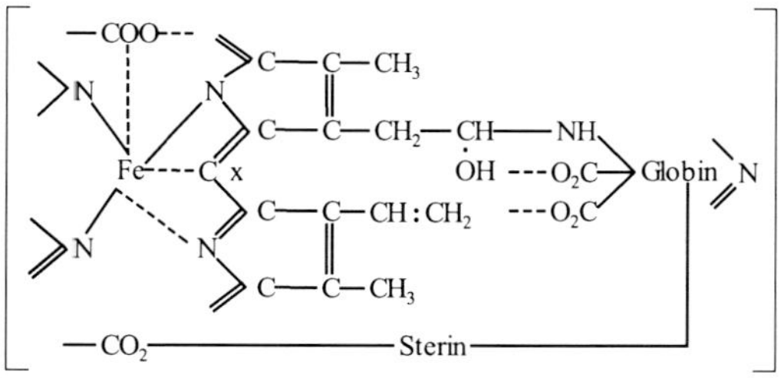

Hemoglobin with radical at x

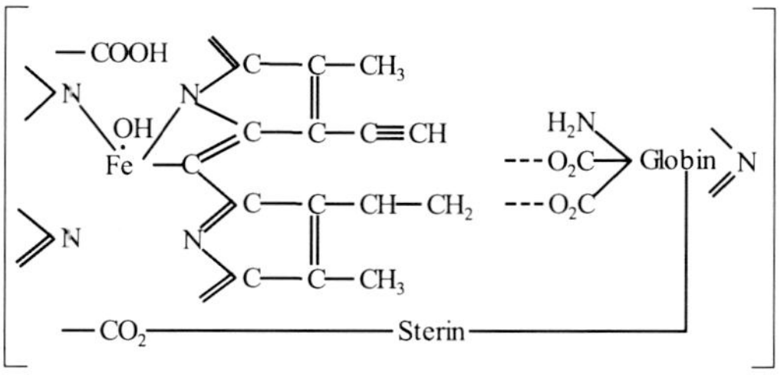

Methemoglobin from the combination at the double molecule from which it has lost one of the sterols

---

[2]) It is known that methemoglobin arises from oxyhemoglobin under the influence of weak acids, also hemoglobin from methemoglobin by reduction in weakly alkaline solution, which facts are also explained by the assumption above. I cannot fail to indicate that with this hypothesis in the organic framework of methemoglobin the tautomeric reacting carbonyl conjectured by *F. Haurowirtz* [This *Zeitschrift*, vol. 138, p. 29 (1924)] is at least present in the prosthetic groups of hemo- and oxymethemoglobin.

56   A good relationship long existed with *William Küster*; reprints were mutually exchanged, signed *Mit herzlichen Grüssen*, and in ref. 75 [*373*] *Küster* made available unpublished results to *Fischer*. It is true that tension, which had its origin in their differing views of the unsaturated side chains of hemin, eventually became unmistakable. After *Fischer's* successful porphyrin syntheses, *Küster* allowed himself to issue an extremely sharp personal belittling [*412*]: "Thus *Hans Fischer* has put the essential points of my idea as architect into fact, and as architect I have gladly allowed myself the valuable help of the master builder. That he did not enjoy this role . . ."

    Hans Fischer brooded long over an appropriate answer — he had cited the work entirely objectively in a footnote in ref. 172 [*410*] – and was finally happy to be spared from answering him when *Küster* died a little while later. [N.B. *Küster* died in his lab in 1929, shortly before the *Nobel* Prize nominations were due.]

57   . . . Note added in proof: These syntheses are to bring the constitution of hemin to complete determination rather than to be perceived as an answer to *W. Küster's* comments [*412*].

58   A short while ago two polemic communications again appeared against me by *Schumm* as well as *Papendieck*, to which I must give attention. *Herr Schumm* [*417*] first and foremost thinks that I may have come recently rather than earlier to a complete acceptance of the spectro-analytical method. This is absolutely not the case. In 1916, I gave my view, as *Schumm* quite correctly cited, that the spectroscopic method of scientific work should be applied as a rule only to pure, crystallized material. This statement refers to the special circumstances of porphyria. So much of uro- and coproporphyrin are mostly present with this illness that purification from urine and feces was successful without great difficulty according to my method. The chemical method for their identification is so simple today that spectroscopy is best used only for pure crystallized material.

59   I cannot declare agreement with *Schumm's* historical explanation on the progress of spectroscopy. Spectroscopic progress depends on the quality of the spectroscope . . . The difference between *Schumm's* observations and my earlier and our recent observations, are minimal differences[1]) that depend on the differing sensitivity of the apparatus, and to a greater extent however by the type of readout.

---

[1]) The difference between *Schumm's* investigations and my old observations amounts to 1–2 $\mu\mu$. At present I know of no case in the spectroscopic literature in which one author is reproached by another for incorrect measurements for such small variations, as has been done repeatedly by *Schumm*.

60   Concerning the communication of *Herr Papendieck*, it appears to me that *Herr Papendieck* is not correct in the question of evaluating whether or not spectroscopic coproporphyrin occurs in serum. *Papendieck* [*414*] did not [even] one time find coproporphyrin in 170 g of feces and has explicitly remarked [*415*] that he had not used sodium sulfate to dry the solution (*which I had accepted as his excuse*). Just as with serum, he must therefore have made an error as regards methods, which is difficult to evaluate at a distance. Despite the many beautiful and expensive photographs, his investigations on serum therefore simply prove nothing.

61   I was especially astonished by *Papendieck's* comments from p. 296 middle to 297 top. Our spectroscopic numbers there were subjected to a detailed critical review, without [his providing] any experimental evidence, and the criticism then culminated on p. 297 with the assertion: "Therefore a band positioned from about 592 to 549.3 must be detected for fecal porphyrins." This claim is however not as simple to prove experimentally, and it is characteristic of the polemic of *Herrn Papendieck* that he does not do that but instead simply subjects our number to a highly presumptuous criticism.

62   In conclusion, might I speak still quite generally about porphyrins in serum, blood, and feces. There are differences of interpretation between *Schumm* and us. *Schumm* finds that the porphyrin content of urine is dependent on consumption of meat, but we cannot verify that. In opposition to *Schumm's* results is my objection to his faulty method, in which he used sodium

11.4 Translations to Section 5

sulfate to dry the ether extract. Small amounts of porphyrin often were undoubtedly thereby missed. On the other hand, naturally the same objection is also to be made against his positive findings, and the fact that he had obtained porphyrins in spite of drying with sodium sulfate speaks to *Schumm's* view. Because if an increase in porphyrins had not resulted from a meat diet, then he would certainly not have been able to find it because a fraction had been *eo ipso* (by the very fact) retained by sodium sulfate.

63   I deem a discussion on these and future "corrections" wasteful and useless, especially in consideration of the remarks rendered in bold face in this journal [*420*].

As striking proof of the impossibility of a polemic with *Herr Schumm*, I recently took notice of *Schumm's* latest "correction" [*421*] during publication of this note. All accusations therein are repetitions of old "corrections" which I have repeatedly refuted.

To the baseless accusation of misinterpretation, *Schumm* still makes a remark in which he reproaches me in relation to the interpretation of his serum value from *Petry*, who was sick from porphyria, and asserts that I regarded his number as unproven.

All of *Schumm's* reproaches were thus raised for a second time, irrespective of whether they had already been refuted by me. *Schumm* has not cited new facts. Let it be noted only incidentally that none of our relevant work is cited. *Herr Schumm* no longer has a right to a reply.

64   *H. Fischer's* and *H. Hilmer's* statement [*423*] that I converted a pyridine extract of yeast into porphyrin, is entirely erroneous; the discussions related to it on p. 169 of the communication above are thus completely meaningless. Once again I raise the most pointed protest against *H. Fischer* declaiming with an untrue statement against a part of the results of my work.

In his "Remarks on *Schumm's* Corrections" [*419*] *Fischer* tries to defend himself against my accusations by statements that require correction throughout.

1. He reproaches me for having published a new "correction" to an earlier false statement. In truth I have done nothing other than discuss in a new comprehensive article on research results from this place a study of urinary hematoporphyrin and its historical development, in which connection obviously *H. Fischer's* opinion must be mentioned together with his erroneous interpretation of *Garrod's* number.

2. *H. Fischer* asserts that my reproaching him for falsely interpreting *MacMunn's* numbers is unjustified, and he tries to prove that by citing a later statement of his. Herein lies the error that *H. Fischer* withholds from the wording of his first statement, against which my criticism was directed.

3. I adhere to my accusation that *H. Fischer* has without any objective basis cast suspicion on my finding of hematin in *Herrn Petry's* serum. The argument I raised is not refuted in the slightest by *Herrn Fischer's* favorite expression in the first sentence of his "Remarks".

4. *H. Fischer's* remark is false that: "All of *Schumm's* accusations are thus raised a second time, irrespective of whether they were already totally disproven by me. *Schumm* has not produced new facts". With the concluding sentence: "*Herr Schumm* possesses no right any longer to reply." *Herr Fischer* spoke not mine but his judgment.

I cannot at present become involved in further debates with *Herrn Prof. Fischer* regarding his repeated, literal and in a sense false statements on results of our work, acknowledgment of which appears to cost him quite an effort.

65   Against such extensive disregard of our research results in the porphyrin field, I protest, no less against each of the relevant contestations dispensed by *Fischer* impugning each of our methods and experiments and claimed for himself the important research accomplishments resulting from this institution.

66   I do not feel compelled to make a rebuttal to the utterances of *Hrn O. Schumm* [*424*] in the *Berichte* [*424*] on my summary lecture "On Porphyrins and Their Syntheses." Since 1926 [*419*] I have come to the conviction of the futility of a polemic with *Hrn. Schumm* and find in the most recent attack no factual evidence that would dissuade me from my view.

67   At present two formulas come into consideration for the constitution, that of *Küster* and the indigoid formulation. As comes forth from the last publication [Fischer H, Treibs A (1927)

Synthesen der Hämopyrrolcarbonsäure. Berichte der deutschen chemischen Gesellschaft 60: 377] [*410*], we have held the *Küster* formulation to be the more probable. According to the syntheses from an alkylated dipyrrylmethene and an α,α'-dibrominated dipyrrylmethene [*410*], a 16-ring should occur primarily in every case.

As to the constitution of iso-coproporphyrin and the β-compound, given the basic *Küster* formulation and the assumption that in the synthesis the two pyrrole rings connected to a methane carbon are preserved, the following three formulas come into consideration. They are distinguished only by a distribution of secondary and tertiary nitrogens in the pyrrole rings.

$$\text{(I)}$$

$$\text{(II)}$$

$$\text{(III)}$$

An exact proof that the carbon bridges remain preserved is not yet produced; however, we accept it as very improbable that under such mild conditions (heating in formic acid at 40°C) would cause cleaving. This view is further supported by the fact that we are able to synthesize iso-coproporphyrin in another way, namely by reaction of the twice brominated methene of kryptopyrrole carboxylic acid (**IV**) with the methene of hemopyrrole carboxylic acid (**V**), which Mr. *Lamatsch* has carried out.

11.4 Translations to Section 5

69   A glance at the coproporphyrin formula reproduced in the following shows an especially favorable relationship to the origin of hemopyrrole carboxylic acid (in keeping with the explanation just provided, the predominant cleavage must occur at the four strike lines).

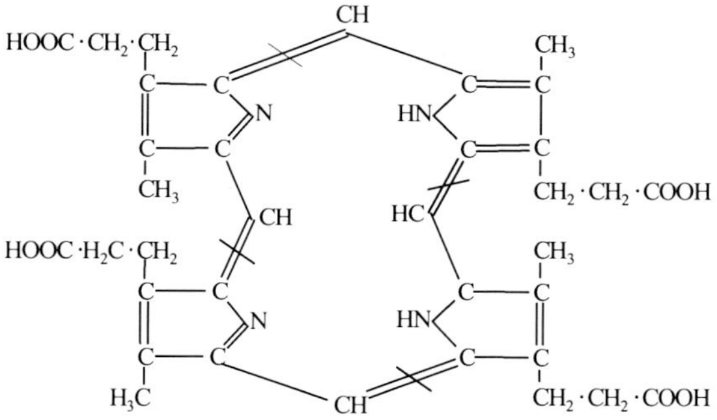

$$C_{36}H_{38}N_4O_8$$

A view that was thoroughly confirmed earlier[1] by the yield from reduction of natural coproporphyrin and now by synthesis. In both cases hemopyrrole carboxylic acid appears as the major product.

---

[1] Fischer H (1916) Über die Konstitution des Kotporphyrins. Hoppe-Seyler's Z physiol Chem 98: 14

70   The possibility was now present that different etioporphyrins are identical with butterflies,[a] and to begin with we have therefore synthesized theoretically possible etioporphyrins on the basis of *Küster's* formula. Four are possible (seven according to the indigoid formula) with the following constitutions:

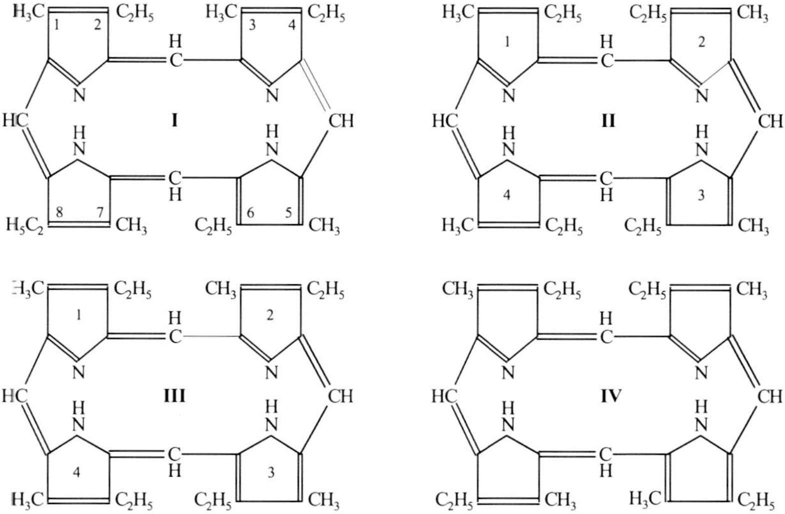

The number of isomers is doubled if one takes tautomeric porphyrin rings into consideration.

---

[a] The German word (*Schmetterlinge*) for butterflies most likely refers to the shape of the crystals.

11 Translations

71   The previous experimental results all correlate with the theoretically conceived constitution, not only a determination by *Klarer*.[1]) We also tried to synthesize etioporphyrin **I** from the brominated hemopyrrole-methene corresponding to the following reaction scheme and here only apparently iso-etioporphyrin **II** was formed instead of etioporphyrin **I**.

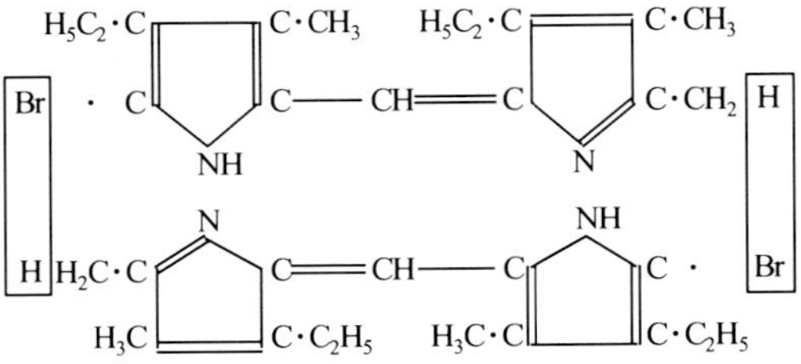

    At least the iso-etioporphyrin formed from kryptopyrrole gave the butterfly crystal form. However, the etioporphyrin obtained according to this method melts at 400°C, exactly as etioporphyrin **I**. The latter inconsistency that perhaps lies here is however reconciled by the behavior of the bromination product of hemopyrryl-methene. We have brominated it again at a higher temperature in glacial acetic acid and obtained thereby a second brominated methene (see p. 71, formula **XIII**) which gave etioporphyrin **I** in formic acid.

---

[1]) Fischer H, Klarer J (1926) Synthese des Ätioporphyrins, Ätiohämins und Ätiophyllins. Annalen der Chemie 448(1): 178–193 [*402*].

Butterflies[a] were obtained from a mixture crystallization. The synthesis proceeds therefrom unequivocally in the sense of the formulation given above.

---

[a] Referring most likely to the shape of the crystals.

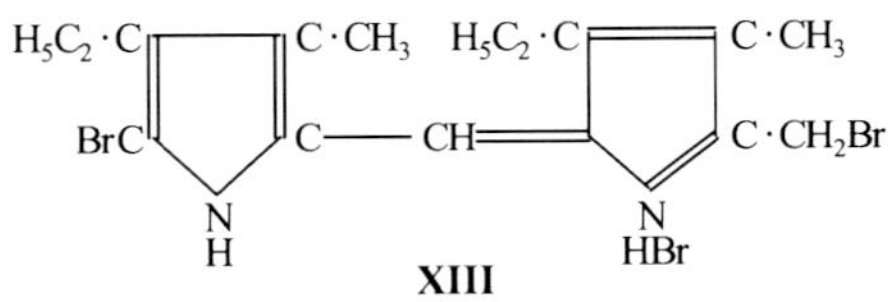

72   We have thus determined [the melting point] by tossing a little substance onto the *Pregl* drying block that was heated to a fixed temperature, and thereby a correct, sharp melting point is obtained. Etioporphyrin **I** shows a mp = 400–405°C, etioporphyrin **II** 365–370°C, etioporphyrin **III** (butterflies) 360–363°C, and etioporphyrin **IV** 355–357°C. The special position that **I** occupies also indicates the high melting point of the preparation. The mixture melting point of **I** and **II** is 380°C; the butterflies from co-crystallization of **I** and **II** also have the same melting point, and the mixture melting point from **II** (butterflies) with butterflies from **I** and **II** is also at 380°C. No depression is therefore present. The etioporphyrin mentioned above that crystallizes in butterflies and came

11.4 Translations to Section 5

We then attempted to generate **II** and **IV** from **I** by seeding the butterfly form. This experiment ended up negative, however. It was at least possible that the appearance of the butterfly form was dependent on some kind of accidental crystallization conditions. According to the investigations of Prof. *Steinmetz* the butterflies are crystallographically identical. We are now therefore busy preparing the xanthoporphyrinogens of the four etioporphyrins and their mixed forms, to compare their calorimetric values with each other exactly, as well as the related etioporphyrins and coproporphyrins that as esters possess sharp melting points and thus hopefully will give a complete clarification.[a]

Ätioporphyrine.

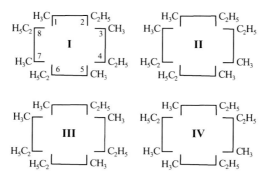

---

[a] This structure did not appear at this point in the original German text, but is included here for clarification.

73   The most certain way was naturally that of the direct synthesis of a blood pigment porphyrin. The most suitable must be mesoporphyrin, since its ester as well as derivatives have sharp melting points.

74   *The synthesis of mesoporphyrin was therefore conducted in an unequivocal manner* and it led accordingly to the following constitutional formula:

Mesoporphyrin is therefore 1,3,5,8-tetramethyl-2,4-diethyl-6,7-dipropionic acid-porphin. Etioporphyrin **III** is therefore the etioporphyrin that serves as the basis for mesoporphyrin.

11 Translations

75 Hemin of course has the same ordering of the side chains because the transformation of hemin into mesoporphyrin is carried out even by catalytic reduction, and it can be excluded that such mild conditions would lead to a rearrangement of the side chains. We formulate that the protoporphyrin formation lies at the basis for hemin according to the following formula and note that the characterization of the unsaturated side chains is tentative.

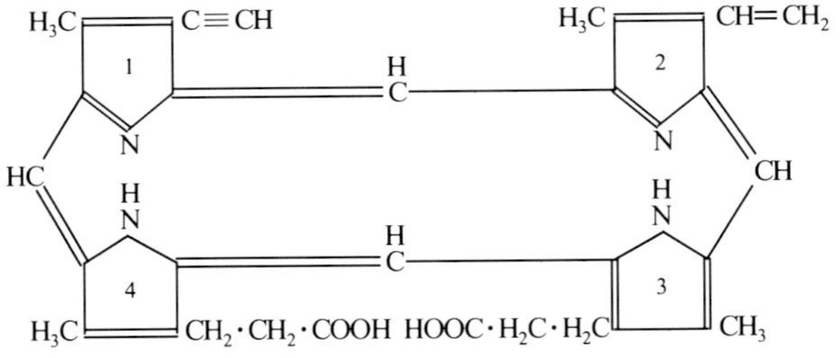

The degree of unsaturation corresponding to the formulation C_4H_4 is proven. Whether and how for the formulation of the two side chains is still to be modified must yield here to synthesis, work that is in progress in different directions.

11.5 Translations to Section 6

1 In addition to a little kryptopyrrole (**2**) and kryptopyrrole carboxylic acid (**6**),

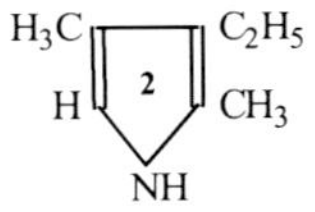

Kryptopyrrole 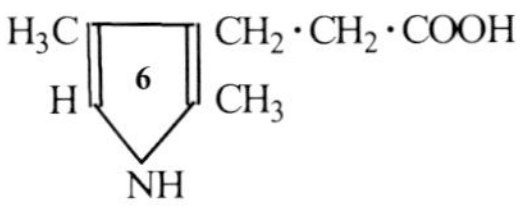   Kryptopyrrole carboxylic acid

energetic reduction of bilirubin gives bilirubinic acid as the major product, whose constitution (**18**) is proven from the results of oxidation and reduction.

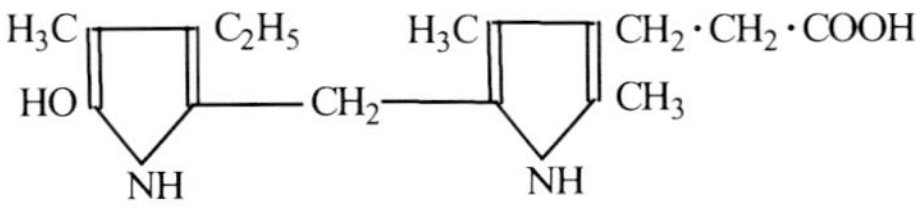

18

Bilirubinic acid (**18**) is dehydrogenated to xanthobilirubinic acid (**19**), a pigment that is also obtained from bilirubin and its derivatives by degradation with $KOCH_3$ and has the following constitution.

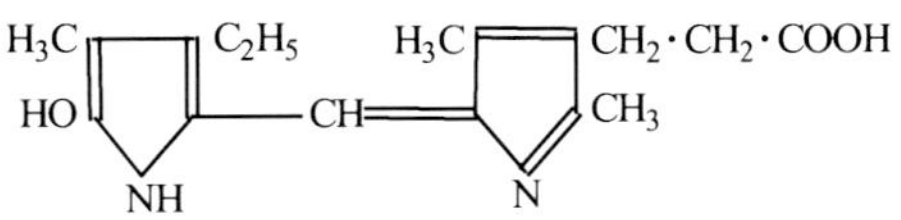

19

2 Oxidation of bilirubin in glacial acetic acid with HNO_2 leads to a compound that, according to its combustion analysis could be methylvinylmaleimide. This constitution cannot be correct, however, because methylethylmaleimide does not arise by catalytic hydrogenation with the

11.5 Translations to Section 6

715

uptake of one equivalent of H_2. However, a new compound is formed, which can only be due to constitution **21**, while the starting material must possess formula **20**.

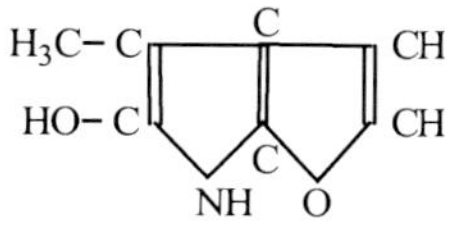

20 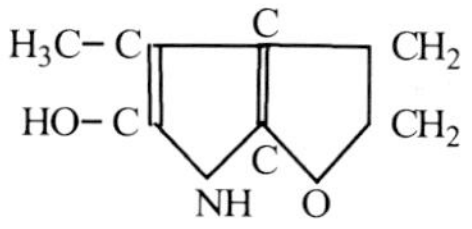   **21**

3   The magnitude of the molecular weight of bilirubin was fixed at about 600 by a molecular weight determination, and the presence of four pyrrole rings follows from the type of cleavage products. The linkage of these four pyrrole rings must differ from that of porphyrins because the spectrum is wrong. Bilirubin and its derivatives possess a greater active hydrogen content, and currently the most likely formula has the four pyrrole rings. linked by an ethylene bridge (see formula **22**, below).

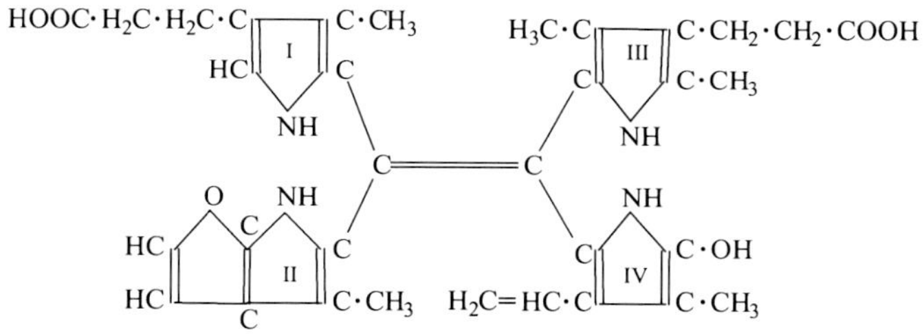

22

On the basis of the constitution of cleavage products **20** and **21**, we still assume a furan ring in bilirubin, which explains well the acid sensitivity of bilirubin and also other properties. Just as with bilirubin itself, synthetic tetrapyrrylethylenes do not show any sort of porphyrin spectrum. Of course, they also give no *Gmelin* reaction. However, the introduction of a furan ring in tetrapyrrylethylenes has at present not succeeded.

4   We became acquainted above with the formation of urobilins by intestinal bacteria; with the assistance of sodium amalgam and likewise by catalytic hydrogenation, this synthetically imitative reaction leads to the formation of mesobilirubin ($C_{33}H_{40}N_4O_6$) by the uptake of two mole equivalents of hydrogen, and to mesobilirubinogen ($C_{33}H_{44}N_4O_6$) by the uptake again of two mole equivalents of hydrogen. The last is distinguished by an intense *Ehrlich* reaction. This crystalline mesobilirubinogen is separable from pathologic urine and is convertible back into mesobilirubin. Mesobilirubinogen is reproduced by the following formula, which nicely explains the non-occurrence of the *Gmelin* reaction because the furan ring is not present in it (see formula **23**, below).

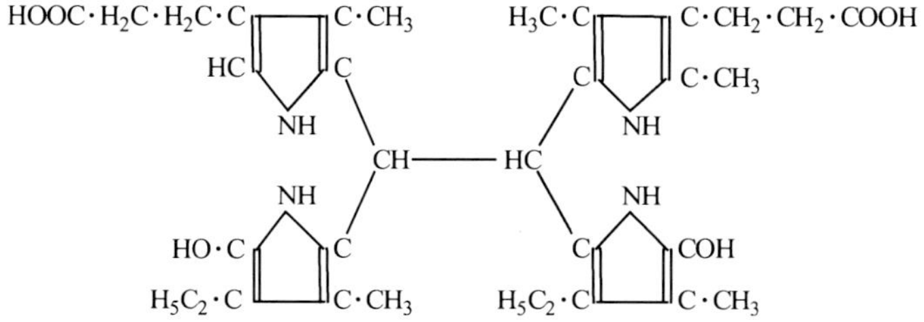

23

716 11 Translations

5   In addition to the concept of bilirubin as a tetrapyrrylethylene, an open formula (as in **24** below) would come into consideration as it nicely explains the transformation into mesobilirubin and mesobilirubinogen. However, it explains the results of active hydrogen determination less well. The *Ehrlich* aldehyde reaction is explained by cleavage of the furan ring. On the other hand, its derivation from hemin would be formed more easily. After loss of iron and transformation into protoporphyrin, a methine group is oxidatively released from the molecule, and oxidation occurs thereby at the α-positions of the pyrrole rings that are linked to this methine group.

* $S = CH_2 \cdot CH_2 \cdot COOH$.

24

6   In order to be transformed into the tetrapyrrylethylene (formula **22**), a rearrangement of pyrrole rings must take place, which is conceivable because of course pyrrole rings I and III are singly bonded (see formula **22**). In the protoporphyrin, primarily a dehydrogenative union of two methine groups must occur at first, and at the same time a hydrolytic cleavage to a dialdehyde. A *Cannizzaro* reaction between the two aldehyde groups can be imagined such that a primary alcohol group and a carboxyl group would be formed next; the primary alcohol group would then be reduced to an α-methyl group of pyrrole ring III, while the carboxyl group (which is not shown) in pyrrole ring I would provide the third methine group following decarboxylation. Thus, the three free methine groups in intermediary pyrrole rings I, IV, and II would come about and go over into CHO groups in pyrrole rings IV and II. From there on the further transformation to the bilirubin formula **22** is easily understandable.

7   The synthesis results favor the chain formula. We have recently synthesized the same type of structure, and it is distinguished by an intense *Gmelin* reaction. It is the first time that this reaction has been observed on a synthetic compound. A complete explanation of the constitution of bile pigments has not yet succeeded; however, its close relationship to hemin remains firmly unambiguous and is further confirmed by the transformation of bilirubin, mesobilirubin, and bilirubic acid into mesoporphyrin, a reaction which of course is accomplished only under considerably energetic conditions.

8

Facsimile from the *Nobel* Prize *Festschrift*, Christmas 1930

. . . / and that is the true and actual experience which I set into action with my own hands according to the highest industriousness (diligence) how difficult it both as a matter of reason / in this way, indeed, should the correct and true practical work (conduction of experiments) be preferred here ("have the day") / and now allow itself to be mastered by speculation in the least . . .

With joy and pride we students are filled with the news of great honor, which our admired leader was granted. The international world of science has now indeed also expressed recognition and thanks for his truly great achievement and his untiring creativity.

We cannot refrain from wishing our teacher heartiest congratulations in this ceremony. Moreover, may we be permitted to observe the opportunity and to express the feeling of gratitude which has been on our minds. Gratitude for that which the instructor conveyed to us as scientists, for the nearly limitless patience and forbearance with which he helped to obviate our ineptitudes and difficulties. However, also quiet gratitude for his continued kind interest in the clearly personal destiny of each one of us. How many of us have found, with his advice and his help new courage and new goals in dark, hopeless days. Exactly this union of scientific greatness, pedagogic skill and outstanding humanity is what led to a happy experience from our

11.5 Translations to Section 6

teacher that had definitely supported and influenced our development and education. Thus, we are all obligated to our highly esteemed teacher in profound appreciation for all that we have learned and received from him.

This small festival should however record a few episodes and events for the memory of this significant day and unforgettable time of study in *Hans Fischer's* laboratory.

9 According to *H. Fischer-Niemann* [*395*], the ester hydrochloride of mesobilirubin shows a strong green surface luster and also corresponds completely in color to the ester hydrochloride of K-mesobilirubin. The mixture melting point of the two ester hydrochlorides gave no depression.

10 In order to be able to make a comparison of preparations *in toto*, we again submit for comparison Prof. *Steinmetz'* crystallization of mesobilirubinogen and K-mesobilirubinogen from ethyl acetate in such a way as they are produced by direct crystallization, whereby, as already emphasized repeatedly, dissimilar crystal forms appear. Herr Prof. *H. Steinmetz* communicated his findings to us as follows: "From both crystallizations in the above there appears, *i.e.* chronologically at first more dendritic or needle-like crystal aggregates separate out at margins. These separated out below as six-sided crystals defined by an oblique cross section. It is likely however that in spite of no differently crystallizing substances being present, the difference in habit in the initial states is due to oils. The two crystallizations do not differ. Both have the same outline of form and the consequent symmetric relationship is presented by that."

11 Concerning the constitution of bilirubin itself, the [tetrapyrryl] ethylene motif can thus probably be excluded, and bilirubin is most likely represented by formula **XII** [see above], in which the two ethyl groups would be replaced by vinyl groups. Hydrogen consumption during Na(Hg) and catalytic reduction would correspond exactly with this concept. However, a few objections also advise against this interpretation. With respect to the unsaturated side chains, bilirubin and hemin would then be identical. However, the experimentally established vinyl groups of hemin and protoporphyrin are not reducible by Na(Hg), while this reaction proceeded easily with bilirubin.

Furthermore, compound **III** arose from oxidation of bilirubin by nitrite, yet it is hardly conceivable for it to originate from the bilirubin of formula **XII** because the relative position of the vinyl groups to the hydroxy group renders a furan ring closure impossible, at least in the plane of projection. Further analytical and synthesis work is needed to settle all these questions and on this point we would only still call attention namely to how the transformation of the blood pigment into bile pigment can be explained on the basis of this new understanding.

12 Golden yellow prismatic needles (mp 312–315°C, dec.) separate out upon cooling the pyridine solution, which had been taken to boiling. Positive *Gmelin Reaction*. The spectral appearance in the *Gmelin* reaction corresponds completely to that from mesobilirubin, as the spectra were completely convincing by projection on top of each other [superposition]. The mixture melting points lay between 307°C and 310°C, with a mesobilirubin prepared for analysis that showed a melting point of 305°, thus giving no depression.

13 The ester hydrochloride gave off HCl with greening at 190°C and melting at 216°C. A mixture melting point with a 190°C-melting preparation of analytical mesobilirubin ester hydrochloride gave no depression.

14 The new compound forms splendid golden yellow needles with a 327°C decomposition point. With the bromination product of xanthobilirubinic acid, as well as with mesobilirubin, and insignificant melting point depression of only 1°–2° occurred.

15 The melting point lies at 222°C. The mixture melting point with analytical mesobilirubin diester lies at 193–195°C and therefore gives no depression. With the ester of synthetic mesobilirubin a depression to 196° occurred.

16 The constitution of mesobilirubin is thus proven definitively in the sense of the formula advanced earlier

$$\text{VI}$$

17 This gave golden yellow prismatic needles from treatment with pyridine, which showed an intense *Gmelin* Reaction and corresponded in all properties with mesobilirubin. The melting point was 312°C; no depression was present in the mixture melting point. Analysis afforded a composition that corresponded well with that of mesobilirubin and points to the loss of a carbon atom.

18 Characteristic of mesobilirubin is a nicely crystallizable ester that was also obtained here in a beautifully crystalline condition by the same method. Its analysis agrees very well with meso-bilirubin ester, likewise its properties; however, the melting point is around 26°C higher.

19 After a little while, colorless crystals separated out. They were suctioned off, washed with ether, and recrystallized from ethyl acetate. Mp 179°C; mixture melting point with analytical neobilirubinic acid (mp 177°C) lies at 177.5°C; thus shows no depression.

20 Despite this great difference no depression occurs in the melting of the mixture sample (mp 260°C).

21 For a better overview, all melting and mixture melting points are introduced:

Neo- and iso-neo-xanthobilirubinic acids

Neo-xanthobilirubinic acid	Melting point	245°
Iso-neo-xanthobilirubinic acid..	"	242°
Mixed crystallization	"	226°
Analyt. "Neo-xanthobilirubinic Acid"....................................	"	227°
Analyt. "N." + synthet. n..	"	227°
Analyt. "N." + synthet. iso-n..	"	227°
Analyt. "N." + synthet. mixed cryst......................................	"	226°

Neo- and iso-neo-xanthobilirubinic acid methyl esters

Neo-xanthobilirubinic acid methyl ester	Melting point	174°
Iso-neo-xanthobilirubinic acid methyl ester	"	205°
Mixture	"	160–161°
Analyt. "Neo-xanthobilirubinic acid methylester"	"	190–191°
Mixed (recrystallized from methyl alcohol)	"	190°
Analyt. "N.-ester" + synthet. n. ester	"	158°
Analyt. "N.-ester" + synthet. iso-n. ester	"	200°

Neo- and iso-neo-bilirubinic acids benzylidene compound.

Neo-bilirubinic acid benzylid.	Melting point	276°
Iso-neo-bilirubinic acid benzylid.	"	273°
Mixed crystallization	"	247–248°
Analyt. "Neobilirubinic acid benzylid"	"	248°
Analyt. "Neo-b.-benzylid." + synthet. neo-b. benz.	"	260°
Analyt. "Neo-b.-benzylid." + synthet. iso-neo-b. benz.	"	260°
Analyt. "Neo-b.-benzylid." + mixed crystallization	"	247–248°
Synthet. Neo-b.-benzylid. + mixed crystallization	"	262°
Synthet. Iso-neo-b.-benzylid. + mixed crystallization	"	261°

11.5 Translations to Section 6

Xantho- and iso-xanthobilirubinic acids

Xanthobilirubinic acid	Melting point	287°
Iso-xanthobilirubinic acid	"	289°
Mixed crystallization	"	274–275°
Analyt. "Xanthobilirubinic acid"	"	273°
Analyt. "X." + synthet. x.	"	273°
Analyt. "X." + synthet. iso-x.	"	273°
Analyt. "X." + mixed crystallization	"	273°
Synthet. "X." + mixed crystallization	"	274°
Synthet. iso-x. + mixed crystallization	"	274–275°

Xantho- and iso-xanthobilirubinic acid methyl esters

Xanthobilirubinic acid methylester	melting point	212°
Iso-xanthobilirubinic acid methylester	"	198°
Mixture	"	176°
Mixture (recryst. from chloroform-petroleum ether)	"	192–194°
Analyt. "Xanthobilirubinic acid methyl ester"	"	212°

22   The ester of the analytical neo-xanthobilirubinic acid melted at 190–191°C after about two recrystallizations from methyl alcohol. The ester of synthetic neo-xanthobilirubinic acid recrystallized from methanol, melted at 174°C, and the ester of iso-neo-xanthobilirubinic acid, recrystallized from a mixture of methyl and ethyl alcohol (2:1), melted at 205°C

Since the mixture melting point of the two synthetic esters was 160–161°C, and the mixture melting point of the analytical ester and iso-neo-xanthobilirubinic ester could be established as 200°C, it clearly followed that the supposed analytical ester (mp 190–191°C) must already be enriched in the iso-isomer of neo-xanthobilirubinic acid. This conclusion was also confirmed by examination under the microscope.

Neo-xanthobilirubinic acid methyl ester (mp 174°C), recrystallized from methyl alcohol, yielded only very poorly-formed individual crystals, mostly of a spindle form without an exact measurable angle of inclination

Iso-neo-xanthobilirubinic acid (mp 205°C), which is unusually difficultly soluble in methanol and could be recrystallized only from a mixture of methyl and ethyl alcohol. In contrast, it formed very beautiful long prisms with sharp edges and an inclination angle of 42–45°C

The microscopic preparation of the analytical ester showed a preponderance of the long prisms of iso-neo-xanthobilirubinic acid methyl ester, in agreement with its melting point.

With this observation, a route to the separation of the analytical neo-xanthobilirubinic acid was thus apparent. From the differing solubility properties indicated above for the synthetic isomeric esters, repeated treatment of the analytical mixture with methyl alcohol in combination with ethyl alcohol should lead to a solution of neo-xanthobilirubinic acid ester and an enrichment of the iso-neo-xanthobilirubinic acid ester in the product that crystallizes out.

That actually also happened. An ester of mp 203°C could finally be obtained which melted at 205°C in a mixture melting point with iso-neo-xanthobilirubinic acid ester (mp 205°C), and thus gave no depression. Retrospectively then, iso-neo-xanthobilirubinic acid of mp 240–241°C was produced from the analytical ester (mp 203°) by saponification with NaOH. With synthetic neo-xanthobilirubinic acid, this analytical iso-neo-xanthobilirubinic acid gave a melting point depression to 227°C.

With the separation of iso-neo-xanthobilirubinic acid from the mixture, it was clearly proven unambiguously that the cleavage of natural mesobilirubin led to a mixture of two isomeric neo-xanthobilirubinic acids. From that it followed that all condensations of analytical neo-xanthobilirubinic acid undertaken with aldehydes or ketones must lead to three isomeric products, two symmetric and one unsymmetric. Condensation with formaldehyde would thus provide a mixture of mesobilirubins-IIIα, XIIIα, and IXα; therefore K-mesobilirubin [ref. *438*, p. 223] must be regarded as a mixture of three isomeric mesobilirubins.

720   11   Translations

This hypothesis could be proven from the reaction products of neo-xanthobilirubinic acid with *p*-dimethylaminobenzaldehyde [ref. *438*, p. 221]. The condensation product with neo-xanthobilirubic acid melted at 244–245°C; that with iso-neo-xanthobilirubinic acid at 246°; the mixture of the two at 239–240°C. The condensation product with analytical "Neo-xanthobilirubinic acid" melted at 239°C. This melting point proved the presence of a mixture.

*Transferring this view and its reasoning to analytical xanthobilirubinic acid and its esterification product also revealed the presence of two isomers, xanthobilirubinic acid (**XIV**) (mp 287°C) and iso-xanthobilirubinic acid (**XV**) (mp 289°C), and removed all uncertainty which up to then resulted from the melting points.*

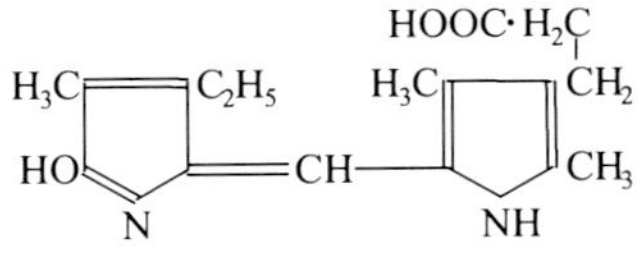

XIV 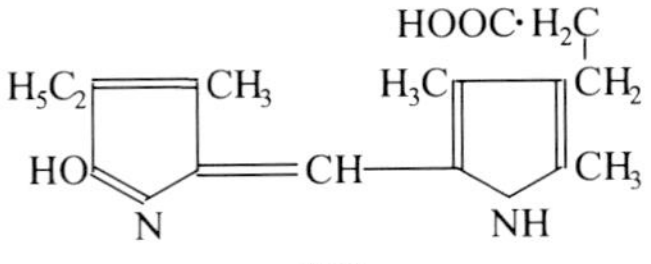   **XV**

Separation of these isomers likewise easily succeeded with the methyl esters by crystallization from chloroform-petroleum ether. This time however iso-xanthobilirubinic acid ester was distinguished by its easier solubility.

A new method of preparation by means of sodium methylate was introduced for both xanthobilirubinic acid in the course of this work, giving yields of 50–60% on the average.

With the observations above, it was proven that mesobilirubin and thus bilirubin belong to an unsymmetric structure. Bilirubin therefore correlated in principle with hemin (and chlorophyll) in the arrangement of its side chains. The physiological origin of bilirubin from hemin takes place by oxidative elision of the α-methine bridge with formation of two hydroxyl groups at pyrrole rings I and IV.

The unsymmetrical structure of mesobilirubin explains all observations concerning bilirubin, especially the formation of hemopyrrole from reduction, as well as the *Nitritkorper* [ref. *438*, pp. 195, 207] which arose from bilirubin by oxidation with nitrous acid and must possess constitution **X**. The poor yield in which it was obtained remained in accord with the fact that this compound could arise only from pyrrole ring IV of bilirubin.[a]

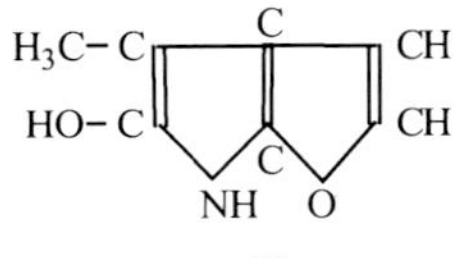

20

---

[a] This structure did not appear at this point in the original German text, but is included here for clarification.

23   Uteroverdin is therefore dehydrobilirubin. In the first communication I assigned the formula of this material to:

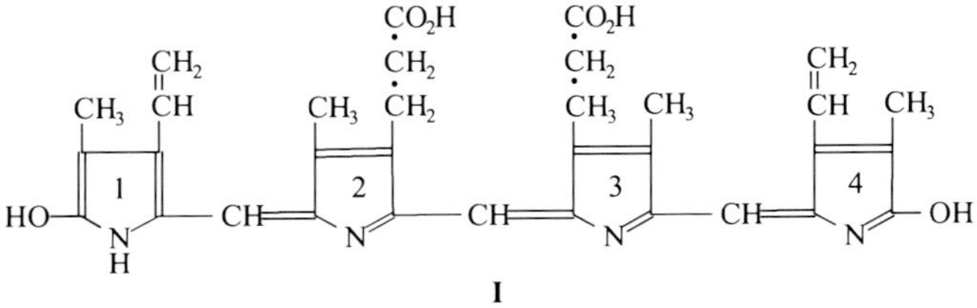

I

11.5 Translations to Section 6

24 ... about the oxidative processes during the course of the *Gmelin* reaction. The authors assume that, in the last phase of the reaction, the red phase, an oxidation of the bilirubinoids to a holoquinoid type in the following way:

* PS = $CH_2 \cdot CH_2 \cdot COOH$.

25   The unsaturated side chains of bilirubin could not therefore be formulated in the same way. From consideration of the hydrogen uptake during catalytic hydrogenation and the fact of the formation of the *Nitritkörper* (**II**) [see **X** of Note 22], bilirubin must correspond to the following formula,

in which the vinyl group and the 8'-hydroxyl group have formed a dihydrofuran ring by isomerization. In this formula the unsaturated part must clearly be assumed to be a vinyl group on pyrrole ring I for a ring closure to the hydroxyl group at 1' is extremely unlikely on steric grounds. The dihydrofuran ring attached to pyrrole ring IV must then open more easily catalytically than hydrogenation of the vinyl group.

26   **Reaction Mechanism of the *Gmelin* Reaction**

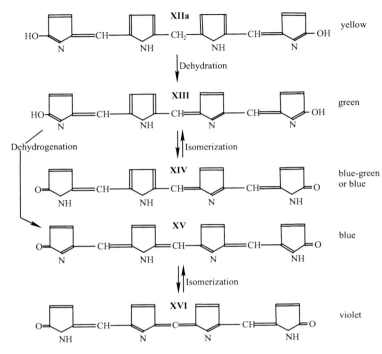

27 ... recrystallized from methanol: yellow, small, partly evenly, partly crookedly cut prismatic little needles with mp 187°C (uncorr.) = 192°C (corr.). Sintering at 180°C.
4.116, 4.309 mg subst. (recrystallized from methanol, dried at 30° *in vacuo*): 9.860, 10.265 mg CO_2, 2.580, 2.600 mg H_2O. – 3.473 mg subst.: 0.270 ccm N (19°, 704 mm).

$C_{18}H_{24}O_4N_2$ (332.2)	Calculated for	C	65.02	H	7.28	N	8.43
	Found	"	65.33, 64.97	"	7.01, 6.75	"	8.42.

28 Soon after standing briefly, a splendid crystallization of synthetic mesobilirubin-IX, α-dimethyl ester-dihydrochloride occurs in orange colored prisms with an extinction angle of 1.5°. These prisms correspond in all remaining properties with the crystals described on p. 268. Mp 194°C (uncorr.) = 199°C (corr.). Mixture melting point with analytical material (mp 192°C, uncorr.) = 193°C, thus no depression.

29 There is complete agreement also in the extinction and in the dichroism. Mp 234°C (uncorr.) = 240.5°C (corr.). A mixture with analytical mesobilirubin-IX, α-dimethyl ester (mp 233°C) melts at 233°, thus showing no depression. A mixture with the IIIα isomeric ester (mp 248°C) melts at 232°C, a mixture with the XIIIα ester (mp 268°C) at 243°C (all without correction).
3.860 mg subst. (recrystallized from pyridine, dried at 50° *in vacuo*): 9.550 mg CO_2, 2.420 mg H_2O – 2.977 mg subst.: 0.265 ccm N (20°, 702 mm).

$C_{33}H_{40}O_6N_4$ (588.32)	Calculated for	C	67.31	H	6.85	N	9.52
	Found	"	67.47	"	7.02	"	9.58.

30 The residue gives, after recrystallization from pyridine, beautiful yellow needles that resemble the analytical material in all properties. Even the solubility proportions are the same. Mp 311°C (uncorr.) = 321°C (corr.). A mixture melting point with analytical material (mp 310°C) is at 310°C, showing thus no depression. A mixture melting point with mesobilirubin-IIIα (mp 316°C) = 306°C (uncorr.), with mesobilirubin-XIIIα (mp 311°C) = 306°C (uncorr.).

11.5 Translations to Section 6

3.860 mg subst. (recrystallized from pyridine, dried at 60° *in vacuo*): 9.550 mg CO_2, 2.420 mg H_2O. – 2.977 mg subst.: 0.265 ccm N (20°, 702 mm).

		C	H	N
$C_{33}H_{40}O_6N_4$ (588,32)	Calculated for	67.31	6.85	9.52
	Found "	67.47 "	7.02 "	9.58.

31 For analysis, it was extracted into $CHCl_3$, and the long, fine needles obtained were dried at 70°C under vacuum.

3.735 mg subst. – 0.02 mg ash = 3.715 mg subst.: 9.110 mg CO_2, 2.060 mg H_2O. – 3.390 mg subst.: 0.340 ccm N_2 (25°, 716 mm).

		C	H	N
$C_{16}H_{18}O_3N_2$ (286.1)	Calculated for	67.08	6,39	9.78
	Found "	66.88 "	6,21 "	9.42.

32 For analysis, it was recrystallized from CH_3OH and dried at 60° under vacuum.

3.417 mg subst.: 8.515 mg CO_2, 2.040 mg H_2O. – 2.787, 2.470 mg subst.: 0.258 (25°, 717 mm), 0.215 (23°, 724 mm) ccm N_2. – 4.250 mg subst.: 3.355 mg AgJ.

		C	H	N	OCH₃
$C_{17}H_{20}O_8N_2$	Calculated for	67.98	6.76	9.32	OCH_3 10.32
(300.1)	Found "	67.96 "	6.88 "	9.81, 9.56 "	10.43.

33 Consequently, pure neobilirubinic acid was available, and the starting material is the corresponding pure vinyl compound **VI**. We recognize, in fact, a confirmation of the bilirubin formula with the hydrofuran ring.

34 It was condensed according to the procedure of *H. Fischer* and *R. Hess* [*438*]. Small rectangles of mp 312°C were obtained after recrystallization from considerable amounts of pyridine. They were recrystallized from pyridine and dried at 50°C under vacuum.

4.348 mg subst.: 10.787 mg CO_2, 2.398 mg H_2O. – 3.658 mg subst.: 0.334 ccm N_2 (26°, 709 mm).

		C	H	N
$C_{33}H_{36}O_6N_4$ (582.3)	Calculated for	68.01	6.23	9.27
	Found "	67.66 "	6.17 "	9.72.

35 The *synthesis of bilirubin* itself has not yet been carried out. It is complicated by the existence of the vinyl groups as well as the dihydrofuran rings.

36 ... thus from the decomposition of bilirubin by nitrous acid a new compound arises that could be methyvinylmaleimide, according to its analysis. This constitution, however, cannot be correct because catalytic hydrogenation with the absorption of one mole [equivalent] of hydrogen gives a new compound, not the expected methylethylmaleimide. Therefore, the new compound can have only constitution **IV**, while the starting material must possess formula **III**.

Proof by synthesis has still not been produced.

37 Relevant to the question of two vinyl side chains of bilirubin is *Nitritkörper* of the following constitution,

whose formation is not able to explain the previous bilirubin formula.[1])

---

[1]) *Cf.* the explanation in *Diese Z.* **194**, 207 (1930). [*sic* 1931] [*438*]

38 "**Nitritkörper**" $C_7H_7O_2N$, probably formed by the action of nitrous acid on bilirubin. Crystallizes from hot water in rather coarse needles of mp 87°–88° (not sharp). Reminds [one] of the odor of methylethylmaleimide. Not distillable without decomposition. Volatile like ethereal vapor. Two atoms of hydrogen are taken up during reduction with aluminum amalgam.[3,4]

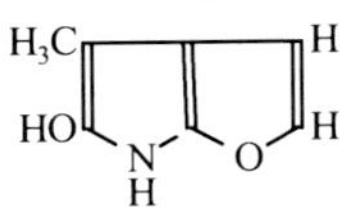

---

[3]) H. Fischer u. H. Röse, H. **91**, 192 (1914) [*372*] cf. also Z. Biologie **65**, 178 (1914) [*377*].
[4]) H. Fischer u. G. Niemann, H. **146**, 199 (1925) [*399*].

39 It was shown previously that bilirubin exhibited a somewhat striking behavior by nitrous acid oxidation (in acetic acid) as well as by catalytic reduction. With respect to the nitrite oxidation, from which a small amount of substance arises therefrom, its empirical composition was regarded at first as methylvinylmaleimide, but in reality, however, possesses the structure shown as **III** below. In subsequent catalytic hydrogenation, it must yield not the expected methylethylmaleimide, but rather a new substance, which should be assigned structure formula **IV**. The actual nitrite oxidation product then retains for the time being the more trivial designation "Nitritkörper"[1] [nitrite body, nitrite compound or nitrite substance].

---

[a] HN= double bond of **III** appears to be an error in ref. (*3*).

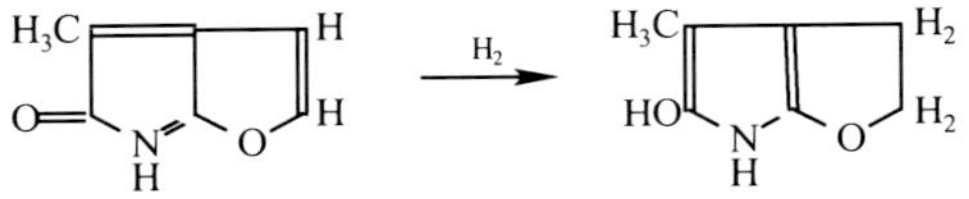

Catalytic reduction of bilirubin itself takes place differently, distinct from sodium amalgam reduction (for this see again the oxidation and reduction schemes of p. 643 [of ref. *3*]) in which it was possible to isolate various intermediates, insofar as a somewhat remarkable progression in the transition from bilirubin to mesobilirubin. In other words, saturation of the two vinyl groups possibly present in the bile pigment occurred not suddenly – as expected – but more or less sluggishly and also rather unevenly. Indeed, earlier there was also success in separating out an intermediate product, namely the dihydrobilirubin[2] discussed earlier, see also on p. 637 [of ref. (*3*)]. With that the long questioned [doubtful] formulation of bilirubin with two vinyl groups had definitely fallen out of favor, and Formula **V** with an attached furan ring stepped into its place (see below).

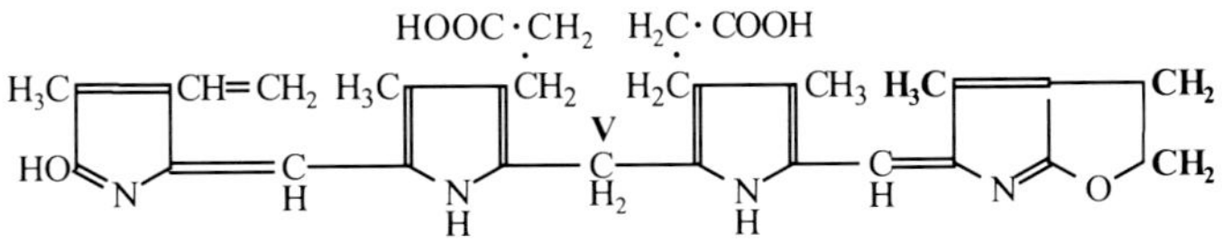

---

[1]) See H. Fischer, H. **91** 192 (1913) [sic, 1914] [*372*]; *Ztschr. f. Biol.* **65** (N. F. **47**) 178 (1914) [*377*].
[2]) See H. Fischer u. H.W. Haberland, H. **232**, 236 ff (1935) [*455*].

11.5 Translations to Section 6

40   Actually, a few years ago *H. Fischer* and *H.W. Haberland* [455] succeeded, in the catalytic reduction of bilirubin, in isolating a precursor of mesobilirubin, namely dihydrobilirubin – because the hydrogenation was first stopped when one mole of hydrogen had been consumed. Dihydrobilirubin can be separated from the two expected impurities, intact bilirubin and the over-hydrogenation product, mesobilirubin, by fractional crystallization from pyridine. The isolated dihydrobilirubin appears as brick red crystals with mp 315°C. (Mesobilirubin also melts at approximately this temperature. It gives, however, an appreciable melting point depression on admixture with dihydrobilirubin.)

41   From the constitutional perspective, the informative transformation or (better) modification, to which dihydrobilirubin was subjected, actually deserves only two special emphases: on the one hand, whereas oxidative degradation of mesobilirubin affords *methylethylmaleimide* in a yield of about 80% of theory, in addition to hematinic acid; in contrast, oxidative degradation of the dihydro compound with respect to the basic imide, produces scarcely half of this amount. On the other hand, molten resorcinol treatment of the dihydro compound, again in contrast to mesobilirubin, leads *not* to a mixture of neo- and iso-neo-xanthobilirubinic acid (see in addition also p. 643 [of ref (*3*)]) but yields only iso-neo-xanthobilirubinic acid, in about 50% yield – with reference to the quantity of the corresponding isomer mixture from mesobilirubin.

    In any case, the oxidation result particularly allows one to decide objectively whether two vinyl groups are actually present in the bilirubin molecule. In dihydrobilirubin (see structure below) one of these groups must be transformed into an ethyl group, because bilirubin itself, as is well known, yields only hematinic acid by oxidation. According to the following structural formula,

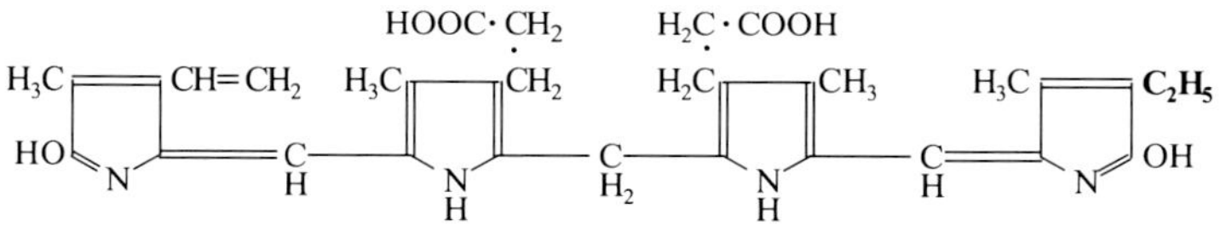

the unsaturated group is located at the 8-position of the tetrapyrrene system (see the nomenclature of p. 627 [of ref. (*3*), below]) because the occurrence of iso-neo-xanthobilirubinic acid from the resorcinol melt was exactly consistent only with this concept.

42   Of necessity, one of the two unsaturated groups in the β-position in bilirubin must, according to the foregoing, be at the 8-position and somehow masked so that hydrogenation is not accessible at the first stage. The type of masking then betrayed the already-mentioned "Nitritkörper". In other words, the simple unsymmetric structure with two vinyl groups was not assigned to bilirubin, and as turned out later from the findings raised from mesobilirubin, but the natural pigment arrived at the formula of p. 626 [of ref. (*3*), shown below] with an attached dihydrofuran ring on pyrrole ring IV, as was already interpreted earlier by *H. Fischer*[1]).

    The transformation of bilirubin to dihydrobilirubin is not therefore an addition of hydrogen to a vinyl group located in the 8-position of the tetrapyrrole system, rather it consists more so in the reductive opening of the dihydrofuran ring, as is represented in the following schematically shown structural formula:

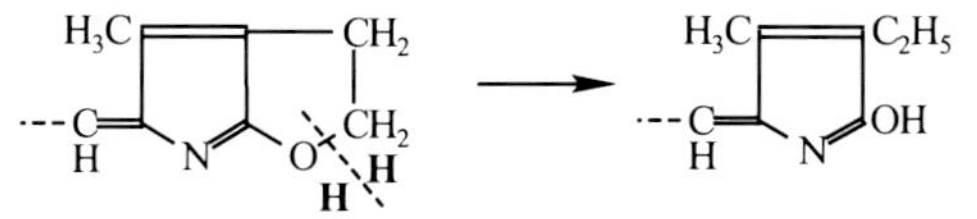

---

[1]) See among others H. Fischer u. G. Niemann, H. **127**, 319 (1923) [*395*] as well as H. Fischer u. F. Lindner, H. **161**, 9 (1926) [*408*].

726   11   Translations

43   The constitution of bilirubin is completely clarified by our investigations, with the exception of the formulation of the unsaturated side chains in rings I and IV. This discussion is concerned with the presence of two vinyl groups, or one vinyl group and one hydrofuran ring, corresponding to the formulations:

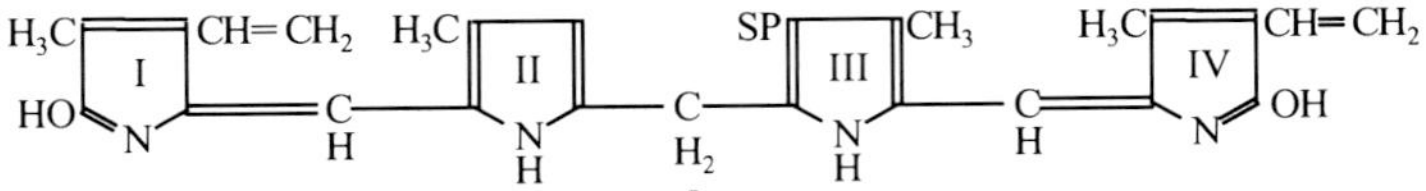

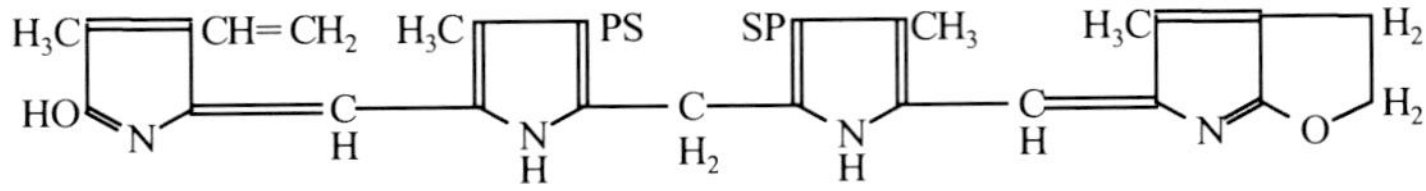

* PS = CH_2-CH_2-COOH

Formula **II** seems to be supported especially by the isolation of the so-named *Nitritkörper* (nitrite compound) of the following formulation:

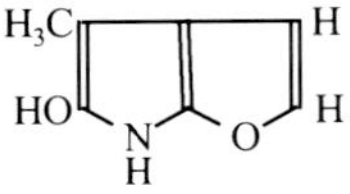

Unfortunately, this is formed only in a very small yield by nitrite degradation of bilirubin; so its formation can easily be overlooked[2]).

---

[2]) Lemberg, R. *Biochemic. J.* **1935** *29*, 6, 1334] [*460*].

44   Since the yield increases with mesobilirubin-XIIIα dimethyl ester, the small yield from natural bilirubin cannot be attributed to the presence of the vinyl groups. More likely, the cause lies in the constitution of bilirubin, which can be recognized by the presence of a hydrofuran ring just as well by the adjacent positions of the hydroxy and vinyl groups on ring IV. It is known that coumarone resin, which is obtained from the tar fraction rich in coumarone [benzofuran] and homologs; likewise, the compound with neighboring hydroxyl and vinyl group goes over easily into high polymers by the action of resorcinol. In fact the resinification product occurs in greater quantity from the resorcinol melt of bilirubin, from which no crystalline material could be obtained in spite of much effort, not even the "Nitritkörper" which also arises from very impure bilirubin.

45   Work on the "Nitritkörper" [*372, 377, 399*] was now resumed. It could, without further ado, be easily obtained from bilirubin even if in small yield. By sublimation in high vacuum we obtained magnificent, strongly light-refracting crystals. Although the analysis at that time agreed with methylvinylmaleimide (**XII**), an isomeric formulation (**XIII**) had to be assumed because methylethylmaleimide could not be obtained by reduction.

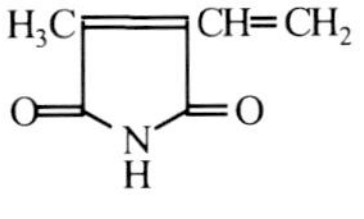

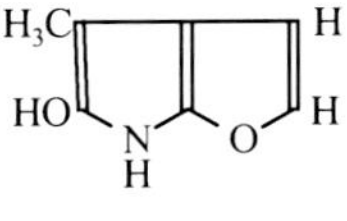

From catalytic reduction with H_2/Pd in methanol, in a smooth reaction, the "Nitritkörper" was successfully converted to a product that after two sublimations possessed a melting point

11.5 Translations to Section 6

of 65°. The mixture melting point with methylethylmaleimide (mp 67°) gave no depression. The crystals were also identical with methylethylmaleimide (Fig. 1) in crystal form and extinction angle.

In contrast to the "Nitritkörper", the crystals do not show any change in solubility properties and melting point after remaining in air for a long time. There is no doubt that methylethylmaleimide is the hydrogenated substance, while the starting material itself shows all the properties already described previously.

46   *On the basis of these results the bilirubin formula with an attached hydrofuran ring cannot be supported, and for bilirubin now only formula **I** is taken into consideration.* [See Note 43, above, for I.]

47   **Diazoacetic Ester Addition to Bilirubin Dimethyl Ester.** Two grams of amorphous bilirubin ester were dissolved in an excess of diazoacetic acid methyl ester with heating and heated on a water bath at 80°C for about 12 hours. Then the product was separated from diazoacetic acid ester on a column packed with aluminum oxide in ether . . . Finally, chromatography was carried out, and the pigment liberated from aluminum oxide by chloroform . . . Yield 60 mg.

3.543 mg subst.: 8.436 mg CO_2, 1.887 mg H_2O. – 4.808 mg subst.: 0.343 ccm N_2 (25°, 708 mm). – 4.287 mg subst.: 1.086 ccm n/50 KSCN.

$C_{41}H_{46}O_{10}N_4$ (754.8) Calculated for C 65.23 H 6.14 N 7.42 OCH_3 16.44
                              Found   " 64.94 " 5.96 " 7.61 "   16.27

48   In fact hydroxypyrrole **VII** or **VIII** was condensed with opsopyrrole aldehyde in good yield to a hydroxymethene of formula **IX**.

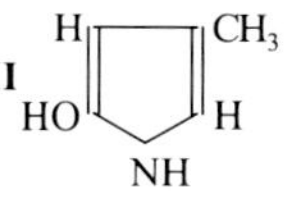

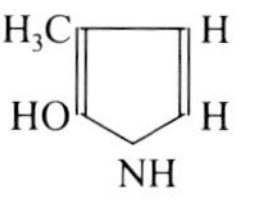

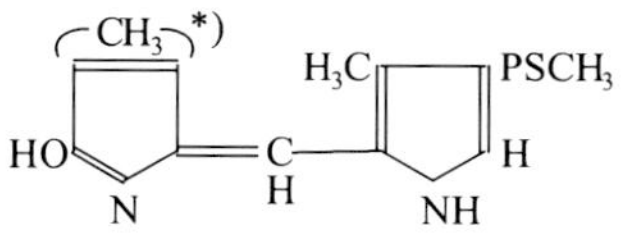

Condensation of the methene to the corresponding bilirubinoid and its dehydrogenation with $FeCl_3$ to the glaucobilin of the following formula (**X**) succeeded without further ado.

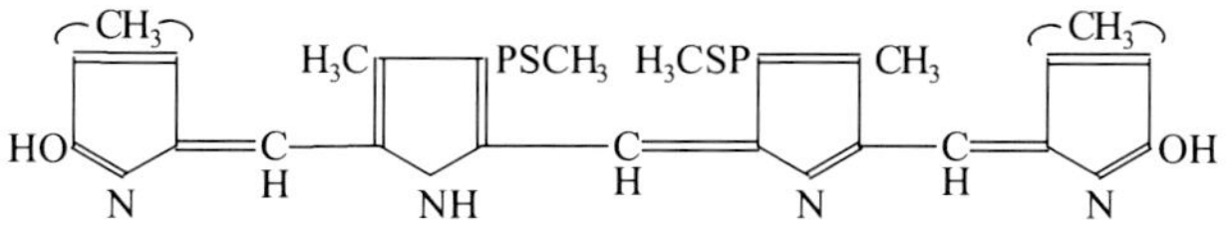

---

* This way of writing indicates that the position of the methyl group is not defined.

49   Introduction of the acetyl residue at the free β-position could not however in any way be accomplished.

50   The name "opsopyrrole" follows from oʹψέ = late and is genetically related to the analogous designation of the corresponding propionic acid, since it was discovered last in the acidic cleavage products of hemin.

51   As already communicated recently,[2]) the preparation of hydroxypyrromethenes can be simplified quite importantly by condensation of α-free α-hydroxypyrroles with pyrrole α-aldehydes in alkaline medium. By oxidation of 3-methyl-4-ethypyrrole (opsopyrrole) with H_2O_2, *H. Fischer* and *H. Reinecke*[3]) obtained a mixture of the two isomeric opsopyrroles.

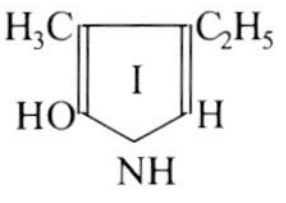

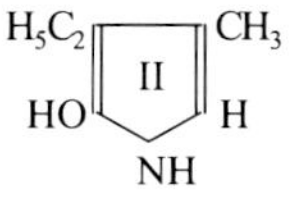

By condensation with the aldehyde of opsopyrrole in *alkaline* medium, a mixture of neo- and iso-neo-xantho acids was obtained in a yield above 80%.

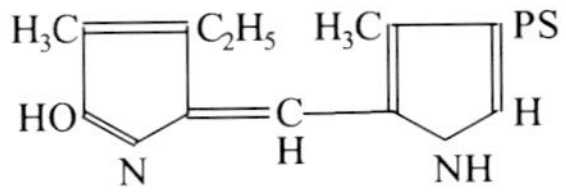

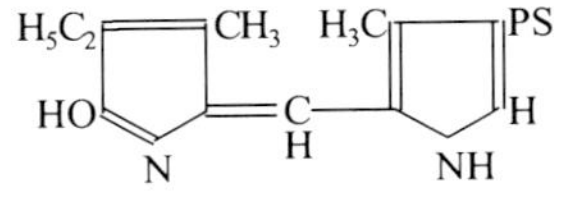

By recrystallization of the esters,[4] the ester of the iso-neo-xantho acid of mp 201°C could be isolated in pure form from the mixture. The hydroxymethene corresponds in all characteristics with that obtained in a different way.

---

[2]) Diese Z. **270**, 229 (1941) [*11*]. [3]) Diese Z. **259**, 86 (1939) [*464*]. [4]) W. Seidel u. H. Fischer, Diese Z. **214**, 164 (1933) [*443*]

52 Cleavage of the hydroxypyrrole urethane (**VII**) to the amine succeeded neither with alkalis nor with acids, rather the pyrrole ring was destroyed.

53 A mixture of the two isomers hydroxyopsopyrrole (**II**) and hydroxyiso-pyrrole (**III**), was formed by oxidation of the opso-acid with hydrogen peroxide. . .

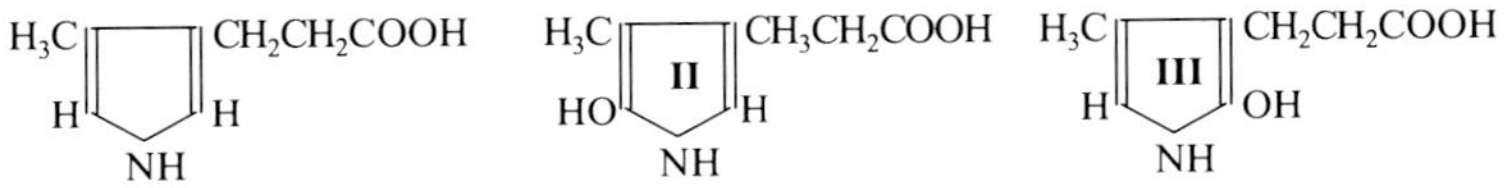

Opsopyrrole carboxylic acid

54 The hydroxymethene was esterified as usual with diazomethane and brought to analysis as the methyl ester on account of its better solubility properties. Mp 205 °C.

3.780 mg subst. (dried at 70° *in vacuo*): 8.582 mg CO_2, 2.215 mg H_2O. – 3.161 mg subst.: 0.301 ccm N_2 (22°, 728 mm). – 4.033 mg subst.: 1.06 ccm n/50-KSCN.

$C_{20}H_{27}N_3O_5$ (389.41) Calculated for C 61.68 H 6.98 N 10.50 OCH_3 15.92

Found ″ 61.92 ″ 6.54 ″ 10.55 ″ 16.31

55 In fact the compound which possesses the desired properties was obtained by methylation of the zinc salt in sodium hydroxide with dimethyl sulfate or even by longer treatment with methyl iodide in methanolic potassium hydroxide.

After methylation, the substance is still easily soluble in water and is acid-soluble, and as a result still cannot be isolated. After heating the reaction mixture for a short time in methanolic and aqueous potassium hydroxide, however, the solubility properties change fundamentally. After acidification with dilute hydrochloric acid the entire substance falls out as an amorphous precipitate. It was thereupon esterified with methyl alcohol-hydrochloric acid and crystallized from methyl alcohol after chromatography on aluminum oxide. In this way 350 mg of a glaucobilin is obtained from 600 mg of amine hydrochloride (**XII**) and it is identical in all properties with the biliverdin-XIIIα [*11*] already obtained by a partial synthesis route.

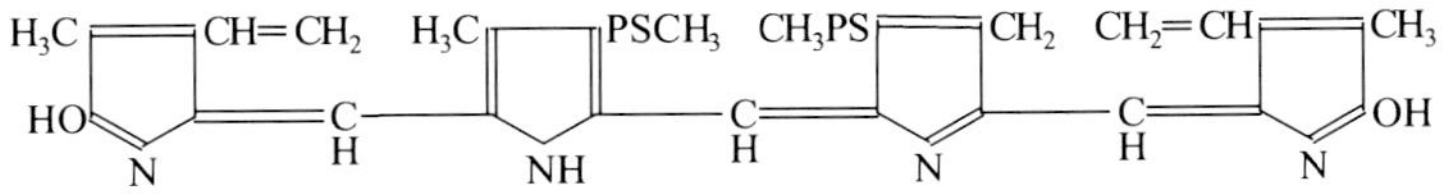

XVII

11.5 Translations to Section 6

56   *Thus for the first time the total synthesis of a bile pigment with vinyl groups is accomplished, and the pathway for the synthesis of bilirubin laid open.*

57   For the synthesis of biliverdin IXα, the dehydrogenation product of bilirubin, a new route was now described for transforming the aminoethyl-iso-hydroxypyrromethene (**XXIII**), which was next combined with the aldehyde of the vinyl-neo-xanthobilirubinic acid [See hereto the glaucobilin synthesis (*457*)]. The aldehyde was, as described previously [*loc. cit.*, p. 10], prepared and the corresponding formyl-vinylneoxantho-acid ester (**XXVI**) condensed with the aminoethyl-iso-hydroxypyrromethene under the recourse of hydrogen bromide-methyl alcohol. A single aminoglaucobilin must herewith correspond to the following formulation (**XXVII**):

58   **Biliverdin IXα (XXVIIa).** Four hundred milligrams of "monovinyl-amino-glaucobilin" were heated on a water bath in 50 cm³ methanol, 50 cm³ water, and 5 g caustic soda [Na_2CO_3] until complete saponification. Then the methanol was removed under vacuum, and [the remaining solution] was treated with zinc acetate and dimethyl sulfate then worked up as with biliverdin-XIIIα. A mixed eluent of chloroform and acetone with a few drops of methanol was used for chromatography on aluminum oxide. Here too a red first eluent was seen which was discarded, and a blue, later eluent must be cleanly separated. The next elution follows directly from the blue-green band of biliverdin and has an iodine-zinc spectrum slightly shifted toward the blue.

For analysis, the material was extracted with chloroform-methanol and recrystallized again a further two times. Thus fine prisms of mp 206–209°C were obtained. With analytical, twice-chromatographed biliverdin (mp 202–204°C) no melting point depression was observed. Also, under the microscope the 224°C melting point was simultaneously observed and no mixture melting point depression was present. Iodine-zinc-spectrum: 639 nm.

3.968 mg subst. (dried at 60° *in vacuo*): 9.960 mg CO_2 2.090 mg H_2O. – 5.273 mg subst.: 0.441 ccm N_2 (23°, 723 mm). – 4.980 mg subst.: 0.816 ccm n/50-KSCN.

$C_{35}H_{38}O_6N_4$ (610.30)	Calculated for	C	68.81	H	6.28	N	9.17	OCH_3	10.16
	Found	"	68.48	"	5.89	"	9.18	"	10.17

59   There is no doubt that this synthetic biliverdin is of course identical to biliverdin-IXα, that combustion analysis results also coincide well.

60   We have already described the reconversion of biliverdin to bilirubin a short while ago [*11*]. From reduction of the free acid of biliverdin with zinc-glacial acetic acid in the cold, a yellow-red compound could be isolated, of course in moderate yield. It was identical to bilirubin in all properties and reactions. The iodine-zinc-spectrum showed beyond doubt that the vinyl groups are still intact. Complete identification was also at hand in the *Debye-Scherer-Diagram*, as well as in the crystal form. Reduction of biliverdin was successful even in cold, alkaline hydrosulfite. In this way, material identical to bilirubin-IXα could be obtained.

61   **Conversion of Biliverdin into Bilirubin by Sodium Hydrosulfite.** Three grams biliverdin ester were saponified in aqueous methanolic potassium hydroxide and after filtering, the methanol was evaporated under vacuum. Then, in the cold, approximately 10 g of sodium

730 11 Translations

hydrosulfite was added whereupon a color change to yellow appeared immediately upon shaking. It was acidified with glacial acetic acid, and the resulting precipitate collected by filtration. After drying it was exhaustively extracted (washed) with chloroform, and the washings were concentrated to approximately 20 cm³. After addition of a few drops of methanol, 100 mg of bilirubin crystallized out overnight as a beautiful red-brown powder. It exhibited no melting point. The substance is halogen-free and turns out to be identical to bilirubin in all reactions.

62   *Therefore, with this synthesis of biliverdin-IX α, the synthesis of natural bilirubin was also accomplished.*

63   *The total synthesis of natural biliverdin was achieved consequently in a two-fold way.*

We have already described the conversion of biliverdin into bilirubin a short while ago [*11*]. The transformation is possible by reaction with zinc dust in glacial acetic acid in the cold or by alkaline reduction at room temperature with sodium hydrosulfite ($Na_2S_2O_4$). A yellow-red compound soon arises in crystallization from chloroform that is identical with bilirubin in all its properties and reactions, even coupling reactions. There is agreement in the *Debye-Scherrer* Diagrams.

Therefore the synthesis of natural bilirubin is also accomplished in the synthesis above.

64   These syntheses allow manifold varying, and an extraordinary enrichment of synthesis possibilities of bilirubinoid pigments is given. Also the problem of the homogeneity of bilirubin can be answered most easily by synthesis. Bilirubin of course possesses no melting point and extraordinary variable solubility circumstances. Consequently, a total identification is difficult, *i.e.* it is difficult at most to rule out the presence of isomers. A content of a foreign substance of 2–3%, for example, would be difficult, if not impossible, to discern.

65   Now the possible existence of isomeric bilirubins or biliverdins was not entirely rejected out of hand because the above-mentioned hemin formulation (**I**) can of course be cleaved at any of the four methine groups, even if the case is at least partly a chemical cleavage. The following three isomers (α, β, δ) are conceivable in addition to biliverdin-IXα of the formulation above, whose syntheses are underway.

β: **27.**

γ: **28.**

δ: **29.**

A knowledge of their properties, after a possible discovery in nature, or as the corresponding bilirubins becomes possible, as transformation into the corresponding isobilirubins is scarcely difficult to accomplish. γ-Biliverdin (**28**) has possibly already been observed in nature. At least *H. Wieland* and *A. Tartter*[1]) have ascribed this constitution to pterobilin. However, pterobilin differs from biliverdin-IXα in its spectroscopic appearance, in comparison with zinc acetate-iodine; but that could be due to the special "γ-constitution arrangement," because on the basis of the previous investigations the influence of the side-chain system in the bile pigment series is essentially smaller than in the porphyrin series[16]), [15]); see, however,[17]). Yet a systematic investigation is still pending, especially in relation to the zinc acetate-iodine reaction, which under circumstances not insignificant differences are ascertained.[16]) [8]) p. 210 ff]

---

[1]) Liebigs Ann. **545**, 197 (1940) . . . [8]) H. Fischer, H. Plieninger u. O. Weissbarth, Hoppe-Seylers Z., **268**, 194 (1941) [*11*] . . . [13]) R. Lemberg, Biochemic. J. **29**, 1322 (1935) [*460*],

11.5 Translations to Section 6

[14]) E. Steier, Hoppe-Seylers Z. **272**, 239 (1942), [15]) E. Steier, Hoppe-Seylers Z. **273**, 58 (1942), [16]) H. Fischer u. H. Reinecke, Hoppe-Seylers Z. **259**, 88 (1939) [*464*], [17]) F. Pruckner u. A. Stern, Z. physik. Chem. (A) **180**, 25 (1937).

66 It should be mentioned in closing that a number of synthetic bile pigments were found a few years ago to exist already exist in nature, and the number can be increased almost at will by the introduction of vinyl groups using syntheses. Thus all biological and chemical matters in the bile pigment series can be dealt with on a broad basis.

67 The total synthesis of biliverdin was finally achieved by *H. Fischer* and *H. Plieninger* [*10*] during the past year (Scheme 6).

Scheme 6
(Siedel's graphics reproduced)

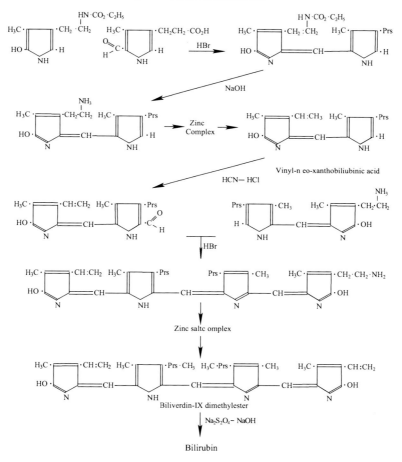

The β-vinyl groups were thus obtained from propionic hydrazide groups via the ethylurethane- and ethyl-amino groups, followed by a *Hofmann* elimination. Finally, vinyl-neo-xanthobiliru- binic acid was prepared [see Scheme 6] and, after introduction of a formyl group, was condensed with correspondingly synthesized amino-iso-neo-xanthobilirubinic acid to a tetra- pyrrole. Again after a *Hofmann* elimination (undertaken on the zinc complex) and esterifica- tion with methanol-hydrochloric acid, biliverdin dimethyl ester was obtained. Reduction of the verdin with alkaline sodium dithionite solution led to bilirubin itself.

732   11 Translations

68   On 31 March 1945 [the day after Soviet forces entered Austria] *Hans Fischer* brought his life and his work to a close. When each day everything in Germany is disintegrating, the news of the death of so special a person touches deeply, as happened to the first few who learned it; it was thus because everyone who knew *Hans Fischer* had seen in him the embodiment of unwavering steadiness and irrepressible energy. Yet precisely arising from that, and for him such characteristic straightforward behavior might have allowed him to bluntly draw the conclusion that for him the situation was no longer tenable. After the nearly complete destruction of his Institute at Technische Hochschule München, he knew that his style of research had once and for all time come to an end, and without it he could not live.

69   In the midst of days of crisis, *Hans Fischer* succumbed to a tragic end on Easter Saturday, the 31st of March 1945. He was granted the privilege to crown his extensive work through the synthesis of hemin and bilirubin, to clarify the constitution of chlorophyll and to nearly complete its synthesis; however, his admirable creation, his Institute, the workplace of a large flock of co-workers, was also the sacrificial victim of the War. Not all experts on his work have been aware of how great a contribution his self-created organization was to his success. He knew that with the few demolished rooms remaining to him, with modest apparatus and few chemicals still remaining, further work in his style was not possible in the future, that his good tradition was threatened to be lost. Practical difficulties from his work were only an incentive, the repugnancy, which for years, sprang from his attitude against the National Socialist system and forced him again and again to defend himself and his work, had exhausted him. In his science, an indestructible optimist, in politics a hopeless pessimist, a good German, the world broke down before him; he had long foreseen the unfolding of it. And that robbed him of all hope even for the more distant future. He had perhaps not once considered how much he would be able to have had help in the reconstruction of his *Hochschule* given his international reputation. Yet he was not fond of a supervisory occupation; never had he, for example, taken over the position as rector; his organization was always a genial improvisation, a creative activity complete in the service of his science. With full consideration he seized the ampule that he had prepared in case of necessity.

70   The war struck deep into the core of the chemistry at Munich's higher education institutions. After summer 1943, scientific work had already been brought nearly to a halt. Half a year before the collapse of the Third Reich, the institutes of our two colleges lay in ruin. However, what hit us most severely was the personal fate of *Richard Willstätter, Hans Fischer*, and *Otto Hönigschmid* who were taken from us within [the course of] a few years. They died victims of the Third Reich and the war; given a time of peace they would have been with us still.

11.6   Translations to Section 7

1   It is true that *Fischer* and *Röse* assigned the corresponding keto formula. According to a recent investigation of *Fischer* and *Niemann* (cited in this journal as *Hoppe-Seyler's Zeitschrift für physiologische Chemie*, volume in press) however, the indicated acid forms an acetyl derivative, and according to that it has the tautomeric structure above.

2   Apart from the view of bilirubin as a tetrapyrrylethylene, the following formula (**24**) still remains an open question.

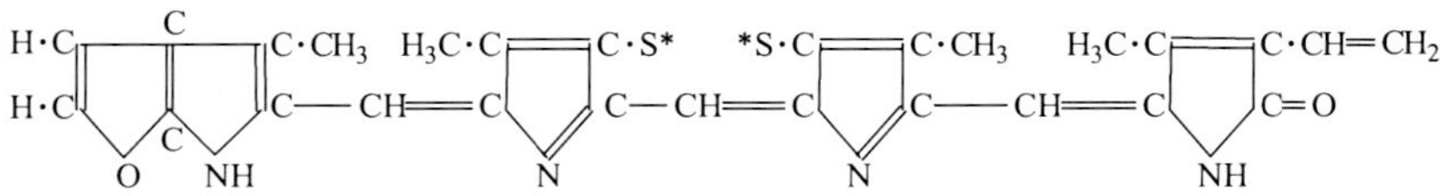

* $S = CH_2 \cdot CH_2 \cdot COOH.$

24

11.7 Translations to Section 9

3

H$_3$C—C$_2$H$_5$, HO, N ═CH— H$_3$C—C$_2$H$_4$COOH, H, NH

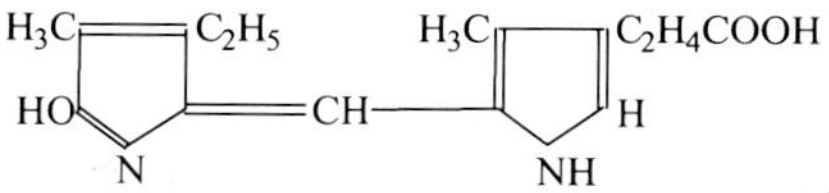

... The constitution of mesobilirubin is therefore definitive in the sense of the formula proposed earlier.

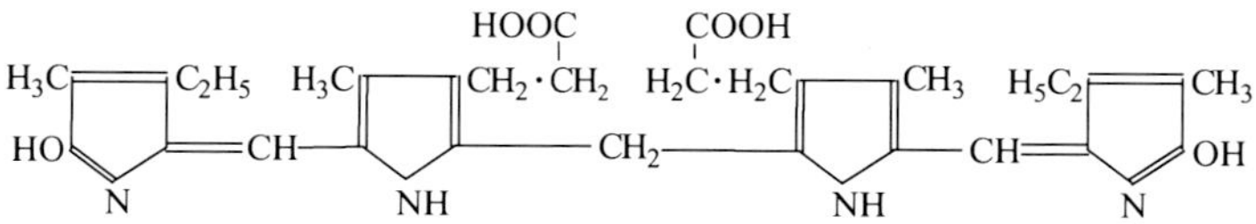

H$_3$C—C$_2$H$_5$, HO, N ═CH— H$_3$C—CH$_2$·CH$_2$ (HOOC, COOH), NH —CH$_2$— H$_2$C·H$_2$C—CH$_3$, NH —CH═ H$_5$C$_2$—CH$_3$, N, OH

VI

4 ... The approximately 10° higher melting point of neo-xanthobilirubinic acid derived from synthetic mesobilirubin must be caused by isomerization, for example in the sense of the following formula:

VII H$_3$C—C$_2$H$_5$, O═, NH ═CH— H$_3$C—C$_2$H$_4$COOH, H, NH

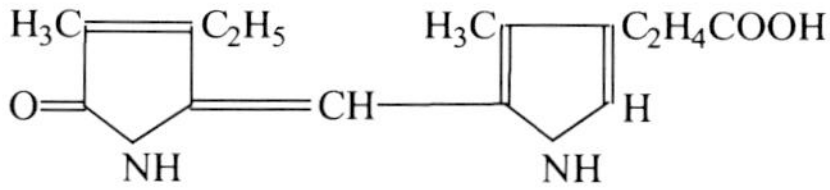

5 ... which is probably that the "hydroxydipyrrylmethene" and bilirubinoids are present in the lactam formula 5.
6 X-ray PES offers a way to define the structure of **17** as well as pyrromethenones. As already reported, the two N$_{1s}$ levels of **17** are barely separated (399.2 eV); their difference is smaller than 0.4 eV. In contrast, with **20**, two N$_{1s}$ levels are found separated by 1.4 eV (399.2 and 387.8 ± 0.1 eV), a result that can also be seen for partial structure systems **11** and **12**. This shows clearly that the lactam form exclusively is present in the solid state (an X-ray structure analysis of a pyrromethenone also shows this) and furthermore, that the lactim form exhibits a close relationship to pyrromethenes (the difference in the binding energy of the two N$_{1s}$ states amounts to about 1.7eV).

11.7 Translations to Section 9

1 As if, never reformed because never in need of reform.
2 The chosen solvent should present the precedingly described characteristics but also not extract too large a quantity of bilirubin so that the concentration of the latter in the aqueous phase remains sufficiently large in order to be measurable after equilibration. In the course of preliminary assaying [32 = (954)], chloroform proved to be too good a solvent for bilirubin because it extracts it practically totally from an aqueous solution, the same in the presence of albumin. Methyl isobutyl ketone (MIC), already used by *Girard et al.* [33 = (955)] is a less good solvent and, when added in varying portions to *n*-heptane, which is apolar and miscible with MIC, it provides a range of mixtures whose polarity and therefore solvation ability *vis-á-vis* the decrease of bilirubin according to the percentage decrease in MIC.
3 The proof lies upon him who affirms, not upon him who denies.
4 From the beginning it was not so.

References

1. Siedel W (1939) Gallenfarbstoffe. Fortschr Chem Org Naturst 3:81
2. Fischer H, Orth H (1934) Die Chemie des Pyrrols, Bd. I. Akademische Verlagsgesellschaft, Leipzig
3. Fischer H, Orth H (1937) Die Chemie des Pyrrols, Bd. II, Pyrrolfarbstoffe. 1. Hälfte. Akademische Verlagsgesellschaft, Leipzig
4. Fischer H, Stern A (1940) Die Chemie des Pyrrols, Bd. II, Pyrrolfarbstoffe. 2. Hälfte. Akademische Verlagsgesellschaft, Leipzig
5. Rüdiger W (1971) Gallenfarbstoffe und Biliproteide. Fortschr Chem Org Naturst 29:60
6. Siedel W (1940) Gallenfarbstoffe. Angew Chem 53:397
7. Siedel W (1943) Fortschritte der Chemie der Pyrrole. Angew Chem 56:185
8. Siedel W (1944) Chemie und Physiologie des Blutstoff-Abbaues. Ber Dtsch Chem Ges 77:A21
9. Fischer H, Siedel W (1947) Naturfarbstoffe II. Pyrrolsynthesen und Gallenfarbstoffe. In: Kuhn R (Sr author) FIAT Reviews of German Science 1939-1946, Biochemistry, Part I, Office of Military Government for Germany, Field Information Agencies Technical, Dieterichsche Verlagsbuchhandlung, Wiesbaden, p 109
10. Fischer H, Plieninger H (1942) Synthese des Biliverdins (Uteroverdins) und Bilirubins, der Biliverdine IIIα und IIIα sowie der Vinylneoxanthosäure. XXXV. Mittlg. Zur Kenntnis der Gallenfarbstoffe. Hoppe-Seyler's Z Physiol Chem 274:231
11. Fischer H, Plieninger H, Weissbarth O (1941) Über die Konstitution des Bilirubins und Über Bilirubinoide Farbstoffe. XXX. Mittlg. Zur Kenntnis der Gallenfarbstoffe. Hoppe-Seyler's Z Physiol Chem 268:197
12. Treibs A (1971) Das Leben und Wirken von *Hans Fischer*. Hans-Fischer-Gesellschaft, München
13. McDonagh AF (1979) Bile Pigments: Bilatrienes and 5,15 Biladienes. In: Dolphin D (ed) The Porphyrins, vol VI. Academic Press Inc, New York, p 293
14. Lightner DA (1982) Structure, Photochemistry, and Organic Chemistry of Bilirubin. In: Heirwegh KPM, Brown SB (eds) Bilirubin, vol. I. CRC Press Inc, Boca Raton, Florida, p 1
15. Falk H (1989) The Chemistry of Linear Oligopyrroles and Bile Pigments. Springer-Verlag, Wien
16. Антин ЕВ, Румянцев ЕВ (2009) Химия билирубина его аналогов. Красанд, Москва, Руссия [Antina EV, Rumajntzev EV (2009) Chemistry of Bilirubin and Its Analogs. Krasand Publishers, Moscow]
17. Ebers G (1875) Papyros *Ebers*. Das Hermetische Buch Über die Artzneimittel der Alten Ägypter in Hieratischer Schrift Herausgegeben mit Inhaltsangabe und Einleitung Versehen von *Georg Ebers*. Mit Hieroglyphisch-lateinischem Glossar von *Ludwig Stern*. 2. Bände, Leipzig
18. Joachim H (1890) Papyros *Ebers*. Georg Reimer, Berlin

D.A. Lightner, *Bilirubin: Jekyll and Hyde Pigment of Life*, Progress in the Chemistry of Organic Natural Products, Vol. 98, DOI 10.1007/978-3-7091-1637-1,
© Springer-Verlag Wien 2013

736 References

19. Westerdorf W (1999) Handbuch der Altägyptischen Medizin. Brill, Leiden, Boston, Köln
20. Breasted JH (1930) The *Edwin Smith* Surgical Papyrus: Hieroglyphic Transliteration, Translation and Commentary, vol 1. The University of Chicago Press, Chicago, Illinois
21. Renouard PV (1856) History of Medicine from Its Origin to the Nineteenth Century. (Comegys CG trans from French). Moore, Wilstach, Keys & Co, Cincinnati, Ohio
22. Ihde AJ (1964) The Development of Modern Chemistry. Harper & Row, New York
23. Moran BT (2005) Distilling Knowledge. Alchemy, Chemistry, and the Scientific Revolution. Harvard University Press, Cambridge, Massachusetts
24. Lavoisier AL (1789) Traité Élémentaire de Chemie, vol II. Cruchet, Paris, chap VII, p 493
25. Willstätter R (1949) Aus Meinem Leben. Verlag Chemie GmBH, Weinheim
26. Gossauer A (2003) Monopyrrolic Natural Compounds, Including Tetramic Acid Derivatives. Prog Chem Org Nat Prod 86
27. Partington JF (1962) A History of Chemistry, vol III. Macmillan & Co, London, p 6, 468
28. Kekulé A (1861) Lehrbuch der Organischen Chemie, Bd. 1. Ferdinand Enke, Erlangen
29. Coe T (1757) A Treatise on Biliary Concretions. Or Stones in the Gall-Bladder and Ducts. D Wilson, T Dorham, London
30. Thomson T (1817) A System of Chemistry, in Four Volumes, vol IV, 5th edn Baldwin, Craeddock and Joy, London
31. Turner E (1835) Elements of Chemistry. Desilver, Thomas & Co, Philadelphia
32. Cadet LC (1767) Expériences sur la Bile de l'Homme et des Animaux. Mémoires de l'Académie Royale des Sciences, Paris, p 471
33. Thomson W (1842) A Practical Treatise on the Diseases of the Liver and Biliary Passages. Barrington ED, Haswell GD, Philadelphia, p 27
34. Vallisneri A (1733) Opere Fisico-mediche, in three volumes, vol 3. Coletti, Venice, p 594
35. von Haller A (1764) Elementa Physiologiae Corporus Humani, vol VI. Bernae Sumptibus Societatia Typographicae, p 554
36. Macquer PJ (1778) Dictionnaire de Chymie, in four volumes, vol 2. Dichot, Paris, p 195
37. Bills CE (1935) Physiology of the Sterols, Including Vitamin D. Physiol Rev 15:1
38. Conradi BGF (1775) Dissertatio Sistens Experimenta Nonnulla cum Calculis Vesiculae Felleae Humanae Instituta. Jena
39. Delius HF (1782) De Cholelithis. Observationes et Experimenta. Walther, Erlangen
40. Dietrich GS (1788) Dissertatio Inauguralis Medica Continens Duas Observationes Rariores Circa Calculus in Corpore Humano Inventos, Prooemium xvi. Halle, p 66
41. Gren FAC (1789) Zerlegung Eines Gallensteins. Crells Beytr zu den Chem Ann 4:19
42. de Fourcroy AF (1789) Examen Chimique de la Substance Feuilletée et Crystalline Contenue dans les Calculs Biliaries, et de la Natur des Concretions Cystiques Cristallisées. Ann Chim Phys, Series 1, 3:242
43. Johnson WB (1803) History of the Progress and Present State of Animal Chemistry, vol II. J Johnson, London, p 351
44. Chevreul ME (1823) De la Cholestérine. Recherches Chimiques sur les Corps Gras d'Origine Animale. Levault FG, Paris, chap IX, p 155
45. Chevreul ME (1815) Récherches Chimiques sur Plusiers Corps Gras. Annales de Chimie; ou, Recueil de Mémoires Concernant la Chimie et les Arts qui en Dépendent et Spécialement la Pharmacie, xcv, 95:5
46. Craven R (1908) Cholesterin: Some Account of Its Chemical and Biological Relations. In: Brockbank EM (ed) Dreschfield Memorial Volume. Manchester University Press, Sherbatt & Hughes Publishers to Victoria University, Manchester, p 125
47. Heuermann G (1753) Der Arzney-Gelahrheit Doktors Physiologie, dritter Theil. Friedrich Christian Pelt, Copenhagen and Leipzig, p 786
48. Tiedemann F, Gmelin L (1826) Über den Farbstoff der Galle. In: Die Verdauung nach Versuchen. K. Groos, Heidelberg, chap 10, p 79
49. Rosenfeld L (1999) Four Centuries of Clinical Chemistry. Gordon and Breach Science Publishers, Amsterdam

References 737

50. de Fourcroy AF (1801) Systéme des Connaissances Chimique, in ten volumes, vol 7. Badouin, Paris, p 44
51. Holmes FL, Levere TH, eds (2000) Instruments and Experimentation in the History of Chemistry. MIT Press, Cambridge, Massachusetts
52. Wöhler F (1828) Über Künstliche Bildung des Harnstoffs. Ann Phys Chem 88:253
53. Thenard LJ (1807) Mémoire sur la Bile; lu á l'Institut le 2 Floréal, An 13. In: Mémoires de Physique et de Chemie de la Société d'Arcueil, Part 1. Paris, p 23
54. Thenard LJ (1807) Mémoire sur la Bile; lu á l'Institut le 2 Floréal, An 13. In: Mémoires de Physique et de Chemie de la Société d'Arcueil, Part 1. Paris, p 40
55. Thenard LJ (1807) Mémoire sur la Bile; lu á l'Institut le 2 Floréal, An 13. In: Mémoires de Physique et de Chemie de la Société d'Arcueil, Part 1. Paris, p 59
56. Thenard LJ (1807) Mémoire sur la Bile; Mémoires de Physique et de Chimie de la Société d'Arcueil 64:103
57. Berzelius JJ (1806, 1808) Föreläsningar i Djukemien, 2 vols. Stockholm
58. Berzelius JJ (1812) General Views of the Composition of Animal Fluids. Medico-Chimirgical Trans 3:198
59. Berzelius JJ (1813) A View of the Progress and Present State of Animal Chemistry, (Brunnmark G, translation of Berzelius' Early Book on Animal Chemistry (57) Into English). J. Shirren Publ, London, p 64
60. Berzelius JJ (1815) Übersicht der Fortschritte und des Gegenwärtigen Zustandes der Thierischen Chemie, (Sigwart GCL, trans of (59) into German). JL Schrag, Nürnberg
61. Thenard LJ (1827) Traité de Chemie Elementaire Theoretique et Pratique, 5th edn. Crochard Libraire – Editeurs des Ann Chim Phys (Paris), p 602
62. Watson CJ (1977) Historical Review of Bilirubin Chemistry. In: Berk PD, Berlin NI (eds) Chemistry and Physiology of Bile Pigments. Fogarty International Center Proceedings No. 35, DHEW Publication No. (NIH) 77-1100. US Dept of Health, Education and Welfare, Public Health Service, US National Institutes of Health, Bethesda, Maryland, p 3
63. Demarçay H (1838) Über die Natur der Galle. Liebig's Ann Pharm 27:270
64. Loir Dr (1835) Chemische Beobachtungen Über die Harn- und Gallensteine. Ann Chem Pharm 13:213
65. Loir Dr (1834) Observations Chimiques sur les Calculs Urinaires et Bilinaires. J Chim Médicale Pharm Toxol 10:515
66. Liebig J, Berzelius JJ (1893) *Berzelius* und *Liebig*. Ihre Briefe von 1831-1845 mit Erläuternden Einschaltungen aus Gleichzeitigen Briefen von *Liebig* und *Wöhler* sowie Wissenschaftlichen Nachweisen, vol I, Carrière J (ed). JF Lehmann, München u Leipzig
67. Curran JO (1846) Medical Periscope. The Physiology of Digestion as Illustrated by Modern Experiments and Observations. In: The Dublin Quarterly Journal of Medical Science; Consisting of Original Communcations, Reviews, Retrospectives and Reports, vol II. Hodges and Smith, Dublin
68. Berzelius JJ (1828) Bestandtheile der Galle. In: Jahres-Bericht Über die Fortschritte der Physischen Wissenschaften, 2nd edn, 7th volume. Heinrich Laupp, Tübingen, p 302
69. Berzelius JJ (1831) Lehrbuch der Chemie, 2nd edn, vol 4, part 1, Lehrbuch der Thier-Chemie. Arnoldischen Buchhandlung, Dresden
70. Berzelius JJ (1833) Traité de Chimie, 2e Partie-Chimie Organique, vol 7. Firmin Didot Frères, Paris
71. Berzelius JJ (1840) Lehrbuch der Chemie, 3rd edn, vol 9 (Aus der Schwedischen Handschriff des Verfassers Übersetzt von *F. Wöhler*), Dritte Umgearbeitete und Vermehrte Original-Auflage. Arnoldischen Buchhandlung, Dresden u Leipzig
72. Berzelius JJ (1842) Galle. In: Wagner F (ed) Handwörterbuch der Physiologie, Bd. I. F Vieweg und Sohn, Braunschweig
73. Berzelius JJ (1840) Ueber die Zusammensetzung der Galle. Ann Chem Pharm 33:139
74. Berzelius JJ (1842) Ueber die Analyse der Ochsengalle und die Characterisirenden Eigenschaften Ihrer Bestandtheile. Ann Chem Pharm 43:1

75. Berzelius JJ (1840) Ueber die Galle. J prakt Chem 20:73
76. Berzelius JJ (1842) Ueber die Galle. J prakt Chem 27:153
77. Partington JF (1962) A History of Chemistry, vol IV. Macmillan & Co, London, p 142
78. Dulk FP (1834) Alkalische Absonderungen aus d. Blute. In: Handbuch der Chemie, zum Gebrauch bei Seinen Vorlesungen und zum Selbstunterricht, Part 2 Chemie der Organischer Körper. A Rücker, Berlin, p 461
79. Berzelius JJ (1838) Untersuchungen des Blattgrüns (Chlorophylls). Ann Chem Pharm 27:296
80. Farrar WV (1977) *Edward Schunck*, FRS, A Pioneer of Natural Product Chemistry. Notes and Records of the Royal Society of London, vol 31, no 2, p 273
81. Pelletier PJ, Caventou JB (1817) Notice sur la Matière Verte des Feuilles. L. Colas, Paris
82. Liebig J (1843) Die Galle. Ann Chem Pharm 47:1
83. Thomson W (1842) A Practical Treatise on the Diseases of the Liver and Biliary Passages. ED Barrington & Geo D Haswell, Philadelphia, p 27
84. Turner E (1835) Bile; Biliary Concretions. In: Elements of Chemistry, Including the Recent Discoveries and Doctrines of the Science. Desilver, Thomas & Co, Philadelphia, p 620
85. Strecker A (1848) Untersuchungen der Ochsengalle. Ann Chem Pharm 65:1
86. Strecker A (1848) Untersuchungen der Ochsengalle. Ann Chem Pharm 67:1
87. Platner EA (1844) Mittheilung Über die Galle. Ann Chem Pharm 51:105
88. Platner EA (1844) Noch Etwas Über die Galle. Archive Anatomie, Physiol Wissenschaftliche Med 2:522
89. Simon JF (1845) II. The Bile. In: Day GE (transl and ed) Animal Chemistry, with Reference to the Physiology and Pathology of Man, in two volumes, vol I. Sydenham Society, C and J Adland, London
90. Simon JF (1840) Analysen Animalischer Substanzen in Flüssigkeiten. Archiv Pharm 74:35
91. Simon JF (1846) II. The Bile. In: Day GE (transl and ed) Animal Chemistry, with Reference to the Physiology and Pathology of Man, in two volumes, vol II. Sydenham Society, C and J Adland, London, 43
92. Scherer JJ (1845) Ueber die Zusammensetzung und Eigenschaften der Gallenfarbstoffes. Ann Chem Pharm 53:377
93. Scherer JJ (1846) Ueber die Extractivstoffe des Harns. Ann Chem Pharm 57:180
94. Virchow R (1847) Die Pathologischen Pigmente. In: Virchow R, Reinhardt B (eds) Archiv für pathologische Anatomie und Physiologie und für Klinische Medicin (Virchow's Archiv), 1:379, 1:407
95. Heintz WH (1846) Notiz Über die Salpetersäure als Reagens auf Gallenbraun. In: Johannes Müller's Archiv für Anatomie, Physiologie und Wissenschaftliche Medicin, Veit et Comp, Berlin
96. Heintz WH (1847) Notiz Über die Salpetersäure als Reagens auf Gallenbraun. Ann Physik Chem (Poggendorffs Annalen) 146:136
97. Heintz WH (1851) Ueber den in den Gallenstein Enthaltenen Farbstoff. Ann Physik Chem (Poggendorffs Annalen) 169:106
98. Lehmann CG (1850) Gallenfarbstoff. In: Lehrbuch der Physiologischen Chemie, erster Band, Zweite Auflage. Wilhelm Engelmann, Leipzig, p 320
99. Legg JW (1880) Chemistry of the Bile; The Bile Pigments. In: On the Bile, Jaundice and Bilious Diseases. D Appleton & Co, New York, chap 1 and 2, p 1, 26
100. Gmelin L (1858) Lebersecrete, Galle. In: Lehman CG, Rochleder DR (eds) Handbuch der Chemie. Universitäts-Buchhandlung von Karl Winter, Heidelberg, p 38
101. Thudichum JLW (1881) The Specific Biliary Acids of the Ox, of Man, of the Hog, and the Goose. Annals of Chemical Medicine, vol II. Longmans, Green, and Co, London, chap. XXI, p 251
102. Thudicum JLW (1863) On the Composition of Gallstones. J Chem Soc:39
103. Thudichum JLW (1863) Digest of Historical Literature. A Treatise on Gall-stones. John Churchill and Sons, London
104. Scherer JJ (1843). Gallensteine aus Gallenfarbstoff. In: Chemische und Microskopische Untersuchungen zur Pathologie. HL Brönner, Frankfurt, p 105

References

105. Heir JA (1847) Chemische Versuche Über Gallensteine und Gallenfarbstoff. J prakt Chem 40:47
106. Bley LF (1847) Chemische Prüfung von Menschlichen Gallensteinen. In: Wachenroder H, Bley LF (eds) Archiv u Zeitung des Apotheker-Vereins in Norddeutschland, vol 1. Hann'schen Hofbuchhandlung, Hannover, p 271
107. Valentiner W (1858) Zur Kenntniss der Animalischen Pigmente. Günzburg's Zeitschr klin Med Neue Folge I, 46
108. Virchow R (1859) Auszüge und Besprechungen. 1. *Valentiner*, Zur Kenntniss der Animalischen Pigmente (Günsburg's Zeitschrift für klinische Medicin. Neue Folge. I. 5. 46). In: Virchow's Archiv Abt A Pathologische Anatomie. Pathology 17:200
109. Brücke EW (1859) Ueber Gallenfarbstoffe und ihre Auffindung. J prakt Chem 77:22
110. Schindler H (1957) Notes on the History of the Separatory Funnel. J Chem Educ 34:528
111. Virchow R (1859) Auszüge und Besprechungen. E. Bücke, Ueber Gallenfarbstoffe und ihre Auffindung (Sitzungberichte der math. naturwiss. Classe der Wiener Akademie. Bd. XXV. S. 13). In: Virchow's Archiv Abt A pathologische Anatomie. Pathologie 17:202
112. Saunders W (1803) A Treatise on the Structure, Economy and Diseases of the Liver, 3rd edn. William Phillips, London, p 147
113. Home E (1830) A Short Tract on the Formation of Tumours. Longmans, Rees, Orme, Brown and Green, London, Fig 3, p 28
114. Wedl C (1855) 14, Colouring Matters. In: Rudiments of Pathological Histology (Busk G, trans). Printed by Adeland JE for the Sydenham Soc, London
115. Robin C-P (1856) Zusammensetzung des Hämatoidins. J prakt Chem 67:161
116. Robin C-P (1855) Zusammensetzung des Hämatoidins. Compte Rendues t. XLI (no. 14) p 506
117. Mulder GJ (1844) Ueber Eisenfreies Hämatin. J prakt Chem 32:186
118. Frerichs FT (1858) Klinik der Leberkrankheiten, 1st edn, vol I of 2 volumes. Friedrich Vieweg u Sohn, Braunschweig
119. Frerichs FT (1861) Klinik der Leberkrankheiten, 2nd edn of vol I (zweite verbesserte Auflage). Friedrich Vieweg u Sohn, Braunschweig
120. Frerichs FT (1861) A Clinical Treatment on Diseases of the Liver, vol II of 2 volumes (Murchison C, trans from German into English). The New Sydenham Society, JW Roche, London
121. Frerichs FT (1860) A Clinical Treatment of Diseases of the Liver, vol I of 2 volumes (Murchison C, trans of improved 2nd edn of vol I of *Klinik der Leberkrankheiten* from German into English). The New Sydenham Society, Roche, London,
122. Pettenkofer M (1844) Notiz Über Eine Neue Reaction auf Galle und Zucker. Ann Chem Pharm 52:90
123. Kühne FW (1858) Beitrage zur Lehre vom Ikterus. In: Virchow R (ed) Virchow's Archiv für pathologische Anatomie 15:310
124. Kühne FW (1866) Die Galle. In: *Kühne's* Lehrbuch der Physiologischen Chemie. Wilhelm Engelmann, Leipzig, p 72
125. Kühne FW (1868) Die Galle. In: *Kühne's* Lehrbuch der Physiologischen Chemie, Wilhelm Engelmann, Leipzig, p 88
126. Funke O (1860) Lehrbuch der Physiologie für Akademische Vorlesungen und zum Selbstudium, volume 1, 3rd revised edn. Leopold Voss, Leipzig, 245
127. Frerichs FT, Städeler G (1855) Ueber das Vorkommen von Leucin und Tyrosin in der Menschlichen Leber. Mittheilungen der Naturforschenden Gesellschaft in Zürich, vol III: 454
128. Frerichs FT, Städeler G (1856) I. Ueber das Vorkommen von Leucin und Tyrosin im Thierischen Organismus. Mittheilungen der Naturforschenden Gesellschaft in Zürich, vol IV:80
129. Frerichs FT, Städeler G (1856) Ueber die Umwandlung der Gallensäuren in Farbstoffe. Mittheilungen der Naturforschenden Gesellschaft in Zürich, vol IV:100

130. Neukomm J (1860) Mittheilung aus dem Laboratorium der Prof. *Städeler* in Zürich (März 1860) I. Ueber die Nachweisung der Gallensäuren und die Umwandlung Derselben in der Blutbahn. Ann Chem Pharm 116:30

131. Taylor ER (1871) Prize Essay: On the Chemical Constitution of the Bile. Am Med Assoc Trans. Collins Printer, Philadelphia, p 1

132. Jaffe M (1862) Ueber die Indentität des Hämatoidins und Bilifulvins. In: Virchow R (ed), Virchow's Archiv für pathologische Anatomie 23:192

133. Jaffe M (1864) On the Identity of Haematoidin and Bilifulvin. In: Ranking WH, Radcliffe CB (eds), The Half-Yearly Abstract of the Medical Sciences 40 (July-December):271

134. Hoppe F (1862) Über die Anwesenheit von Gallensäuren im Icterischen Harne und die Bildung des Gallenfarbstoffes. Virchows Archiv für pathologische Anatomie 24:1

135. Städeler G (1864) Ueber die Farbstoff der Galle. Ann Chem Pharm 132:328

136. Holm F, Städeler G (1867) Untersuchung Über das Hämatoidin. J prakt Chem 100:142

137. Salkowski EL (1868) Hoppe-Seyler's Medicinisch-Chemische Untersuchungen, herausgeg. von Hoppe-Seyler. Berlin Heft 3:436

138. Preyer WT (1871) Die Blutkrystalle. Mauke's Verlag, Jena, p 185

139. Wagner E (1876) A Manual of General Pathology (van Duyn J, Seguin EC, trans, from 6th German edn into English). William Wood & Co, New York

140. Kingzett CT (1878) Bile Pigments. In: Animal Chemistry or the Relations of Chemistry to Physiology and Pathology. Longmans, Green and Co, London, p 100

141. Hermann L (1883) Die Bildung der Specifischen Gallenbestandtheile. In: Vogel FWC, Handbuch der Physiologie, Fünfter Band, 1. Theil. Leipzig, p 231

142. Ewald CA (1891) Lectures on the Diseases of the Digestive Organs, vol I (Saundby R, trans, into English from 3rd German edn, 1890). The New Sydenham Society, London, p 133

143. Landois L (1892) Constituents of the Bile. In: A Text-book of Human Physiology, 4th edn. (Stirling W, trans from 7th German edn, with additions). P Blakiston, Son & Co, Philadelphia, 333

144. Cohen JB (1919) The Proteins. In: Organic Chemistry for Advanced Students, part III, Synthesis, 2nd edn, 2nd impression. Longmans, Green & Co, New York

145. Fischer H, Reindel F (1923) Über Hämatoidin. Hoppe-Seyler's Z physiol Chem 127:299

146. Rich AR (1925) The Formation of Bile Pigment. Physiol Rev 5:182

147. Rocke AJ (2010) Image and Reality. *Kekulé, Kopp* and the Scientific Imagination. University of Chicago Press, Chicago and London

148. Everts S (2010) When Science Went International. Chem Eng News, 6 September 2010, 88:60

149. Kauffman GB, Adloff J-P (2010) The 150th Anniversary of the First International Congress of Chemists, Karlsruhe, Germany, September 3-5, 1860. Chem Educator 15:309

150. Maly RL (1864) Vorläufige Mittheilungen Über die Chemische Natur der Gallenfarbstoffe. Anzeiger der kaiserlichen Akademie der Wissenschaften I:81

151. Städeler G (1860) Ueber das Tyrosin. Ann Chem Pharm 116:57

152. Thudichum JLW (1868) Chemische Untersuchungen Über die Gallenfarbstoffe. J prakt Chem 104:193

153. Heynsius A, Campbell JFF (1871) Die Oxydationsproducte der Gallenfarbstoffe und ihre Absorptionstreifen. In: Pflüger EWF, Archiv für die Gesammte Physiologie der Menschen und Thiere, vierter Jahrgang, erstes Heft. Max Cohen & Sohn, Bonn

154. Wood ES (1873) Report on Medical Chemistry. In: Warren JC, Dwight T, Draper FW (eds) The Boston Medical and Surgical Journal, vol 88, David Clapp & Son, Boston, p 14

155. Lightner DA, McDonagh AF (1984) Molecular Mechanisms of Phototherapy for Neonatal Jaundice. Acc Chem Res 17:417

156. Lightner DA, McDonagh AF (1985) Like a Shrivelled Blood Orange. Bilirubin, Jaundice and Phototherapy. Pediatrics 75:443

157. Lightner DA, McDonagh AF (1988) Phototherapy and the Photobiology of Bilirubin. Sem Liv Dis 8:272

References

158. Thudichum JLW (1875) Further Researches on Bilirubin and its Compounds. In: Papers Read Before the Chemical Society, J Chem Soc 28:389
159. Thudichum JLW (1876) On Some Reactions of Biliverdin. The Chemical News 33:198 (12 May)
160. Thudichum JLW (1876) On Some Reactions of Biliverdin. In: Papers Read Before the Chemical Society, J Chem Soc 30:27
161. Thudichum JLW (1876) Offenes Schreiben an die kaiserliche Akademie der Wissenschaft zu Wien Enthaltend eine Beluchtung der Untersuchen Über die Gallenfarbstoffe von *Richard Maly* in Graz. In: Pflüger's Archiv Anat Physiol (Pflüger's Archiv Eur J Physiol) 13:213
162. Thudichum JLW (1876) State of Animal Chemistry in Austria. (Open Letter to the Imperial Academy of Sciences at Vienna, Containing an Examination of the Researches on the Colouring Matter of the Bile, by *Richard Maly* of Graz.) The Chemical News 32/33:154 (April)
163. Thudichum JLW (1884) A Treatise on the Chemical Consitution of the Brain. Balliere, Tindal, and Cox, London
164. Maly RL (1864) Vorläufige Mittheilungen Über die chemische Natur der Gallenfarbstoffe. Ann Chem Pharm 132:127
165. Maly RL (1864) Über die Chemische Natur der Gallenfarbstoffe. In: Sitzungberichte der Mathematisch-Naturwissenschaftlichen Classe der kaiserlichen Akademie der Wissenschaften, Herr Prof *R Maly*, Assistent der Physiologie und der Grazer Universität Übersendet Einige "Vorläufige Mittheilungen Über die Chemische Natur der Gallenfarbstoffe", XIII. Sitzung vom 12 Mai 1864, vol 49, KK Hof- und Staatsdruckerei, Wien, p 463
166. Maly RL (1868) Untersuchungen Über die Gallenfarbstoffe. J prakt Chem 104:28
167. Maly RL (1868) Intersuchungen Über die Gallenfarbstoffen. In: Sitzungberichte der Mathematisch-Naturwissenschaftlichen Classe der kaiserlichen Akademie der Wissenschaften, Herr Prof *R Maly* Übermittlet die Erste Abtheilung Seiner "Untersuchungen Über die Gallenfarbstoffe", IV. Sitzung vom 6 Feb 1868, vol 57, Heft IV, KK Hof- und Staatsdruckerei, Wien, p 227
168. Ma J, Yan F, Wang C, Chen J (1990) Addition of Sodium Bisulfite to Biliverdin. Chin Chem Lett 8:23
169. McDonagh AF (2010) Personal Communication
170. Senge MO, Ma J, McDonagh AF (2001) Sodium Etiobilirubin-IVγ-C10-sulfonate: A Highly Solvated Bile Pigment Structure Containing Two Different Non-ridge-tile Conformers in the Unit Cell. Bioorg Med Chem Lett 11:875
171. Maly R (1874) Untersuchungen Über die Gallenfarbstoffe. In: Sitzungberichte der Mathematisch-Naturwissenschaftlichen Classe der kaiserliche Akademie der Wissenschaften, Herr Prof *R Maly*, Assistent der Physiologie und der Grazer Universität Übersendet Einige "Vorläufige Mittheilungen Über die Chemische Natur der Gallenfarbstoffe", XVII. Sitzung vom 9 July 1874, vol 70, KK Hof- und Staatsdruckerei, Wien, p 72
172. Maly RL (1875) Untersuchungen Über die Gallenfarbstoffe. Ann Chem Pharm 175: 76
173. Falk H (1989) The Chemistry of Linear Oligopyrroles and Bile Pigments. Springer-Verlag, Wien, Fig 4-6, p 47
174. Maly RL (1869) Untersuchungen der Gallenfarbstoffen. In: Sitzungberichte der Mathematisch-Naturwissenschaftlichen Classe der kaiserliche Akademie der Wissenschaften, Herr Prof *R Maly*, Assistent der Physiologie und der Grazer Universität Übersendet Einige "Vorläufige Mittheilungen Über die Chemische Natur der Gallenfarbstoffe", IX. Sitzung vom 1 April 1869, I. Untersuchungen der Gallenfarbstoffen von Prof *R Maly* in Olmütz, vol 59, Heft 4, KK Hof- und Staatsdrukerei, Wien, p 509
175. Maly RL (1872) Künstliche Umwandlung von Bilirubin in Harnfarbstoff. Ann Chem Pharm 161:368

176. Maly RL (1872) Untersuchungen Über die Gallenfarbstoffe. III. Umwandlung von Bilirubin in Harnfarbstoff. Ann Chem Pharm 163:77
177. Maly RL (1875) Untersuchungen Über die Gallenfarbstoffe. In: Sitzungberichte der Mathematisch-Naturwissenschaftlichen Classe der kaiserlichen Akademie der Wissenschaften, Herr Prof *R Maly*, Assistent der Physiologie und der Grazer Universität Übersendet Einige "Vorläufige Mittheilungen Über die Chemische Natur der Gallenfarbstoffe", III. Sitzung vom October 1875, vol 72, KK Hof- und Staatsdruckerei, Wien
178. Maly RL (1876) Untersuchungen Über die Gallenfarbstoffe. Ann Chem Pharm 18:106
179. Watts H (1868) A Dictionary of Chemistry, vol V. Longman's Green and Co, London, p 377
180. Hoppe-Seyler F (1862) Ueber das Verhalten des Blutstoffes im Spectrum des Sonnenlichtes. Virchow's Archiv pathol Anat klin Med 23:446
181. Kohlrausch WG (1874) An Introduction to Physical Measurements. (Walter TH, Proctor HR, trans from the 7th German edn into English), D Appleton and Company, New York, p 97, 231
182. Beer A (1852) III. Bestimmung der Absorbtion des Rothen Lichts in Farbigen Flüssigkeiten. Poggendorff's Ann Phys Chem 86:78
183. Bunsen R, Roscoe H (1857) IV. Photochemische Untersuchungen. Poggendorff's Ann Phys Chem 13:235
184. Rinsler MG (1981) Spectroscopy, Colorimetry, and Biological Chemistry in the Nineteenth Century. J Clin Pathol 34:287
185. Jaffe M (1869) Zur Lehre von den Eigenschaften und der Abstammung der Harnpigmente. Virchow's Archiv pathol Anatomie klin Medicin 47:405
186. Thudichum JLW (1872) A Manual of Chemical Physiology. William Wood & Company, New York, p 71
187. Thudichum JLW (1896) Ueber die Reactionen des Bilirubins mit Jod und Chloroform. J prakt Chem 53:314
188. Thudichum JLW (1897) Das Sogenannte Urobilin und Damit in Verbindung Gesetzen Physiologischen und Pathologischen Hypothesen. Virchow's Archiv pathol Anat klin Med 150:586
189. Thudichum JLW (1898) Ueber das Urobilin. Virchow's Archiv pathol Anat klin Med 153:154
190. Thudichum JLW (1899) Ueber den Chemischen Process der Gallensteinkrankheit beim Menschen und in Thieren. Virchow's Archiv pathol Anat klin Med 156:384
191. Thudichum JLW (1900) Einige Wissenschaftliche und Ethische Fragen der Biologischen Chemie. J prakt Chem 61:568
192. Grimm F (1893) Ueber Urobilin im Harne. Virchow's Archiv pathol Anat klin Med 132:246
193. Thudichum JLW (1877) A Treatise on the Pathology of the Urine, 2nd edn. J&A Churchill, London
194. Hopkins FG, Garrod AE (1898) On Urobilin. Part II. The Per-centage Composition of Urobilin. J Physiol 22:451
195. Garrod AE, Hopkins FG (1896) On Urobilin. Part I. The Unity of Urobilin. J Physiol 20:112
196. Jolles A (1894) Beitrage zur Kenntnis der Gallen und Über Eine Quantitative Methode zur Bestimmung des Bilirubins in der Menschlichen und Thierischen Galle. Pflüger's Archiv Physiol (Eur J Physiol) 57:1
197. Jolles A (1899) Ueber die Reaction des Bilirubins mit Jod und Chloroform. J prakt Chem 59:308
198. Küster W (1899) Beiträge zur Kenntniss der Gallenfarbstoffe. I. Hoppe-Seyler's Z physiol Chem 26:314
199. Küster W (1899) Ueber den Blut- und den Gallen-farbstoff. Ber Dtsch Chem Ges 32:677

References 743

200. Breathnach CS (2001) *Johann Ludwig Wilhelm Thudichum* 1829-1901, Bane of the Protagonizers. History of Psychiatry vol XII:283
201. Gamgee A (1893) A Text-book of the Physiological Chemistry of the Animal Body, including an Account of the Chemical Changes Occurring in Disease. II. MacMillan & Co, London and New York, p 315
202. Kekulé A (1859) Lehrbuch der Organischen Chemie, vol 1. Ferdinand Enke, Erlangen
203. Rambert PJ (2010) Imagination in Chemistry. Science 329:280 (a review of ref *145*)
204. Nencki M, Sieber N (1888) Über das Hämatoporphyrin. Monatsh Chem 9:115
205. Nencki M, Sieber N (1888) Ueber des Hämatoporphyrin. (Naunyn and Schmiedeberg's) Archiv exp Pathol Pharmacol 24:430
206. Raoult FM (1899) The General Law of Freezing Solvents. In: Ames JS (ed) Harper's Scientific Memoirs IV. The Modern Theory of Solutions (Jones HC, transl and ed, from the original French: Ann Chim Phys 1884 2:66). Harper & Brothers, New York and London
207. Beckmann E (1905) Modifikation des Thermometers für die Bestimmung von Molekulargewichten und Kleinen Temperatur Differenzen. Z Phys Chem 51:329
208. Nencki M, Rotschky A (1889) Zur Kenntniss des Hämatoporphyrins und des Bilirubins. Monatsh Chem 10:568
209. Rast K (1922) Mikro-Molekulargewichts-Bestimmung im Schmelzpunktsapparat. Ber Dtsch Chem Ges 55:1051
210. Abel JJ (1890) Bestimmung des Moleculargewichtes der Cholalsäure, des Cholesterins und des Hydrobilirubins nach der *Raoult*'schen Methode. Monatsh Chem 11:61
211. Teeple JE (1903) On Bilirubin, the Red Coloring-Matter of the Bile. A Dissertation submitted to the University Faculty for the Degree of Doctor of Philosophy at Cornell University, Ithaca, New York
212. von Zumbusch LR (1901) Ueber das Bilifuscin. Hoppe-Seyler's Z Physiol Chem 31:446
213. Orndorff WR, Teeple JE (1901) On Bilirubin, the Red Coloring-matter of the Bile. Amer Chem J 26:86
214. Orndorff WR, Teeple JE (1905) On Bilirubin, the Red Coloring-matter of the Bile. Amer Chem J 32:215
215. Küster W (1902) Beiträge zur Kenntniss der Gallenfarbstoffe. III. Ber Dtsch Chem Ges 35:1268
216. Küster W (1894) Ueber Chlorwasserstoffsaures und Bromwasserstoffsaures Hämatin. Ber Dtsch Chem Ges 27:572
217. Küster W. Haas K (1904) Beiträge zur Kenntniss des Hämatins. Ber Dtsch Chem Ges 37:2470
218. Küster W (1904) Haematin. J Chem Soc Abst 86:647
219. Hünefeld FL (1840) Der Chemismus in der Thierischen Organisation. Brockhaus, Leipzig
220. Hoppe-Seyler F (1864) Ueber die Chemischen und Optischen Eigenschaften des Blutfarbstoffs. Virchow's Archiv: An Internat J Pathol 29:233
221. Reichert ET, Brown AP (1909) The Differentiation and Specificity of Corresponding Proteins and Other Vital Substances in Relation to Biological Classification and Organic Evolution. The Crystallography of Hemoglobins. Carnegie Institution of Washington, DC
222. Reichert KE (1849) Beobachtungen Über eine Eiweissartige Substanz in Krystallform. Müller's Archiv Anat Physiol Wiss Med:198
223. Funke O (1851) Ueber das Milzvenenblut. Z rationelle Med NF 1:172
224. Teichmann L (1853) Ueber die Krystallisation der Organischen Bestandtheile des Bluts. Z rationelle Med für Henle und Pfeuffer 3:375; Ueber das Hämatin ibid 8:141 (1857)
225. Hoppe-Seyler F (1864) Ueber die Chemischen und Optischen Eigenschaften des Blutfarbstoffs. Virchow's Archiv: An Internat J Pathol 29:597
226. Mulder GJ (1844) Ueber Eisenfreies Hämatin. J prakt Chem 32:186
227. Hoppe-Seyler F (1871) Beiträge zur Kenntniss des Blutes des Menschen und der Wirbelthiere (Schluss). Medicin-chemische Untersuchungen, herausgeben von Dr. Felix Hoppe-Seyler, erstes Heft, August Hirschwald, Berlin, p 523

744 References

228. Hoppe-Seyler F (1874) Einfache Darstellung von Harnfarbstoff aus Blutfarbstoffs. Ber Dtsch Chem Ges 7:1065
229. Nencki M, Sieber N (1884) Untersuchungen Über den Blutfarbstoff. Ber Dtsch Chem Ges 17:2267
230. Nencki M, Sieber N (1884) Untersuchungen Über den Blutfarbstoff. (Naunyn and Schmiedeberg's) Archiv exper Pathol Pharm 18:401
231. Runge FF (1834) Ueber Einige Produkte der Steinkohlendestillation. Poggendorff's Ann Phys 31:65; ibid 32:308
232. Anft B (1955) *Friedlieb Ferdinand Runge*: A Forgotten Chemist of the Nineteenth Century. J Chem Educ 32:566
233. Anderson HJ (1995) From Dippel to Du Pont. J Chem Educ 72:875
234. Anderson Th (1858) Ueber die Producte der Trockenen Distillation Thierische Matieren. Ann Chem Pharm 105:335
235. Anderson Th (1846) Ueber Picolin; eine Neue Basis aus dem Steinkohlen-Theeröl. Ann Chem Pharm 60:86
236. Schwanert H (1860) Ueber Einige Zersetzungsproducte der Schleimsäure. Ann Chem Pharm 116:257
237. Baeyer A (1870) Ueber die Wasserentziehung und ihre Bedeutung für das Pflanzenleben und die Gährung. Ber Dtsch Chem Ges 3:63
238. Fittig R (1871) Ueber die Constitution der Sogenannten Kohlenhydrate. LF Fues, Tübingen
239. Baeyer A (1870) Reduction des Isatins zu Indigblau. Ber Dtsch Chem Ges 3:514
240. Cason J, Rapoport H (1950) Laboratory Text in Organic Chemistry. Prentice-Hall, Inc, Englewood Cliffs, New Jersey, p 186
241. Nencki M, Sieber N (1886) Ueber das Hämin. (Naunyn and Schmiedeberg's) Archiv exp Pathol Pharmacol 20:325
242. Küster W (1896) Beiträge zur Kenntniss des Hämatins. Ber Dtsch Chem Ges 29:821
243. Küster W (1897) Ueber Oxydationsproducte des Hämatoporphyrins und die Zusammensetzung des nach Verschiedenen Methoden Dargestellen Hämins. Ber Dtsch Chem Ges 30:105
244. Küster W (1897) Ueber ein Spaltungsproduct des Gallenfarbstoffs, die Biliverdinsäure. Ber Dtsch Chem Ges 30:1831
245. Lassaigne JL (1843) Mémoire sur un Procédé Simple pour Constater las Présence d'Azote dans des Quantités Minimes de Matière Organique. Comptes Rendues Hebdomadaires des Sé Acad Sci 16:387
246. Lassaigne JL (1843) Nachweisung Ausserst Geringer Mengen Stickstoff in Organischen Matieren. Ann Chem Pharm 48:367
247. Gower RP, Rhodes IP (1969) A Review of Techniques in the *Lassaigne* Sodium Fusion. J Chem Educ 46:606
248. Küster W (1899) Spaltungsproducte des Hämatins. Hoppe-Seyler's Z Physiol Chem 28:1
249. Küster W, Kölle M (1899) Ueber Darstellung und Spaltungsproducte des Hämatoporphyrins. Hoppe-Seyler's Z Physiol Chem 28:34
250. Küster W (1900) Spaltungsprodukte des Hämatins (II. Mittheilung). Hoppe-Seyler's Z Physiol Chem 29:185
251. Küster W (1900) Ueber die Constitution der Hämatinsäure. Ber Dtsch Chem Ges 33:3021
252. Küster W (1901) Ueber die Constitution des Hämatinsäuren. Liebig's Ann Chem 315: 174
253. Fittig R, Glaser F (1899) Umlagerung Zweibasische Ungesättiger Säuren. III. Aethylitaconsäure und Isomere Säuren. Liebig's Ann Chem 304:178
254. Bischoff CA (1890) Weitere Beiträge zur Kenntniss der Homologen der Maleïnsäuregruppe. Ber Dtsch Chem Ges 23:3413
255. Bischoff CA (1891) Weitere Beiträge zur Kenntniss der Fumarsäurereihe. Ber Dtsch Chem Ges 24:2001
256. Fittig R, Landolt A (1877) Untersuchungen Über die Ungesättigen Säuren. 4) Beiträge zur Kenntnis der Additionsproducte aus Ita-, Citra- und Mesaconsäure. Liebig's Ann Chem 188:71

References

745

257. Fittig R, Jayne HW (1882) Ueber das Phenyl-Butyrolacton und die Phenyl-Paraconsäure. Liebig's Ann Chem 216:97
258. Fittig R (1889) Ueber Lactonsäure, Lactone und Ungesättigte Säuren. Liebig's Ann Chem 255:1
259. Aronheim B (1874) Synthese des Phenylbutylens. Liebig's Ann Chem 171:219
260. Fittig R, Penfield SL (1883) Ueber die Phenylhomoparaconsäure. Liebig's Ann Chem 216:119
261. Fittig R, Delisle A (1889) Propionaldehyd und Bersteinsäure. Liebig's Ann Chem 255:56
262. Fittig R, Parker G (1892) Ueber die Condensation der Brenztraubensäure mit Zweibasischen Säuren. Liebig's Ann Chem 267:204
263. Michael A, Tissot G (1891) Zur Kenntniss der Homologen der Aepfelsäure. Ber Dtsch Chem Ges 24:2544
264. Michael A, Tissot G (1892) Beiträge zur Kenntniss Einiger Homologen der Aepfelsäure. J prakt Chem 46:285
265. Bischoff CA, Voit E (1890) Ueber die Beziehungen der Beiden Symmetrischen Dimethylbernsteinsäuren zur Pyrocinchonsäure. Ber Dtsch Chem Ges 23:644
266. Bischoff CA (1891) Weitere Beiträge zur Kenntniss der Substituirten Bernsteinsäuren. Ber Dtsch Chem Ges 24:1064
267. Hilditch TP (1911) A Concise History of Chemistry. D Van Nostrand Co, New York
268. von Richter V (1916) Organic Chemistry, or Chemistry of the Carbon Compounds, Vol I, Chemistry of the Aliphatic Series (Spielman PE, trans and rev from German edn), P. Blakiston's Son & Co, Philadelphia
269. Liebig J, Poggendorff JC, Wöhler Fr (1859) Handwörterbuch der Reinen und Angewandten Chemie, zweite Auflage, zweiter Band, zweite Abteilung (neu bearbeitet in Verbindung mit mehreren Gelehrten und redigirt von Fehling H), Friedrich Vieweg u Sohn, Braunschweig
270. Frankland E, Duppa BF (1866) Synthetische Untersuchungen Über Aether. Ann Chem Pharm 138: 204, 328
271. Frankland E, Duppa BF (1868) Synthetische Untersuchungen Über Aether. Ann Chem Pharm 145:78
272. Geuther A (1883) Ueber die Constitution des Acetessigesters (Aethyldiacetsäure) und Über Diejenige des Benzols. Liebig's Ann Chem 219:119
273. Geuther A (1885) Zur Geschichte des Acetylacetessigäthers. Liebig's Ann Chem 227:383
274. Baeyer A (1885) Ueber die Synthese des Acetessigäthers und des Phloroglucins. Ber Dtsch Chem Ges 18:3454
275. Nef JU (1891) Zur Kenntniss des Acetessigäthers. Liebig's Ann Chem 266:52
276. Claisen L, Lowman O (1887) Ueber eine Neue Bildungsweise des Benzoylessigäthers. Ber Dtsch Chem Ges 20:651
277. Claisen L (1892) Ueber die Constitution des Acetessigäthers und der Sogenannten Formylverbindungen der Säuräther und Ketone. Ber Dtsch Chem Ges 25:1776
278. Claisen L (1905) Ueber den Verlauf der Natracetessigester – Synthese. Ber Dtsch Chem Ges 38:709
279. Dessaigne M (1858) On a New Acid Obtained by Oxidation of Malic Acid. The Chem Gazz 382:341
280. Conrad M (1880) Synthesen Mittelst Malonsäureester; 1. Darstellung des Malonsäureesters von *M. Conrad* und *C.A. Bischoff*. Liebig's Ann Chem 204:121
281. Zelinsky N (1887) Ueber eine Bequeme Darstellungsweise von α-Brompropionsäureester. Ber Dtsch Chem Ges 20:2026
282. Nencki M, Zaleski J (1900) Untersuchungen Über den Blutfarbstoff. Hoppe-Seyler's Z Physiol Chem 30: 384
283. Auwers K (1891) Ueber Synthesen Alkylirter Tricarballylsäuren und Anderer Mehrbasischer Fettsäuren. Ber Dtsch Chem Ges 24:307
284. Auwers K, Köbner E, von Meyenburg F (1891) Synthesen Mehrbasischer Fettsäuren. Ber Dtsch Chem Ges 24:2887

746 References

285. Nencki M, Zaleski J (1901) Ueber die Reductionsproducte des Hämins durch Jodwasserstoff und Phosphoniumjodid und Über die Constitution des Hämins und Seiner Derivate. Ber Dtsch Chem Ges 34:997
286. Nencki M, Marchelewski L (1901) Zur Chemie des Chlorophylls. Abbau des Phyllocyanins zum Hämopyrrol 34:1687
287. Zaleski J (1902) Untersuchungen Über das Mesoporphyrin. Hoppe-Seyler's Z Physiol Chem 37:54
288. Küster W (1902) Beiträge zur Kenntniss des Hämatins. Ber Dtsch Chem Ges 35:2948
289. Küster W (1904) Über die nach Verschiedenen Methoden Hergestellten Hämine, das Dehydrochlorhämin und das Hämatin. Hoppe-Seyler's Z Physiol Chem 40:391
290. Küster W (1905) Beiträge zur Kenntniss des Hämatins. Hoppe-Seyler's Z Physiol Chem 44:391
291. Küster W (1906) Ueber die Constitution der Hämatinsäuren. Liebig's Ann Chem 345:1
292. Küster W, Haas K (1906) Ueber die Constitution des Hämopyrrols. Liebig's Ann Chem 346:1
293. Plancher G, Cattadori F (1903) Sull' ossidazione del Dimetilpirrolo Asimmetrico. Atti della Accademia nationale dei Lincei, Classe di Scienze Fisiche, Matematiche e Naturali, Rendiconti 12:10
294. Plancher G, Cattadori F (1903) Sull' Ossidazione del Dimetilpirrolo Asimmetrico. Gazz Chim Ital 33:402
295. Schmidt J (1904) Die Chemie des Pyrrols und Seiner Derivate. Verlag von Ferdinand Enke, Stuttgart
296. Küster W (1907) Ueber das Hämopyrrol. Ber Dtsch Chem Ges 40:2017
297. Küster W, Haas K (1906) Constitution of Haemopyrrole. J Chem Soc Abst 90:693
298. Harries C (1920) *Oskar Piloty*. Ber Dtsch Chem Ges 53:152
299. Remarque EM (1929) Im Westen Nichts Neues. Propyläen-Verlag, Berlin, p 91, 260
300. Remarque EM (1958) All Quiet on the Western Front. (Wheen AW, trans into English for Little, Brown & Co, 1929, 1930). Ballentine Books, Random House, New York, p 87, 263
301. Piloty O, Thannhauser SJ (1912) Über die Konstitution des Blutfarbstoffs. Liebig's Ann Chem 390:191
302. Piloty O, Thannhauser SJ (1912) Dehydro-bilinsäure, ein Gefärbtes Oxydations-Produkt der Bilinsäure. Ber Dtsch Chem Ges 45:2393
303. Piloty O (1909) Ueber den Farbstoff des Blutes. Liebig's Ann Chem 366:237
304. Piloty O (1911) Sur le Coulerant du Sang. Le Moniteur Scientifique du Docteur Quesnville. J Sciences Pures Appliquées, Compte Rendus Académies Sociétés Savantes T4:711
305. Hahn O (1970) My Life. (Kaiser E, Wilkins E, trans into English of *Mein Leben*). Herder and Herder, New York, p 95
306. Piloty O, Merzbacher S (1909) Ueber die Sogenannte Hämatopyrrolidinsäure. Ber Dtsch Chem Ges 42:3253
307. Piloty O, Merzbacher S (1909) Über eine Neue Aufspaltung des Hämatoporphyrin. Ber Dtsch Chem Ges 42:3258
308. Piloty O, Quitmann E (1909) Über die Konstitution des Hämopyrrols und des Hämopyrrolcarbonsäure. Ber Dtsch Chem Ges 42:4693
309. Küster W (1909) Beiträge zur Kenntniss des Hämatins. (Bemerkunges zu *O. Pilotys* Arbeit Über den Farbstoff des Blutes und Über die Oxydation des Hämatoporphyrins). Hoppe-Seyler's Z Physiol Chem 61:164
310. Küster W (1908) Beiträge zur Kenntniss des Hämatins. Hoppe-Seyler's Z Physiol Chem 55:505
311. Goldmann H, Hetper J, Marchelewski L (1905) Studien Über den Blutstoff. V. Vorläufige Mitteilung. Hoppe-Seyler's Z Physiol Chem 45:176
312. Piloty O (1910) Synthese von Pyrrolderivaten: Pyrrole aus Succinylobernstein-säureester, Pyrrole aus Azinen. Ber Dtsch Chem Ges 43:489
313. Piloty O, Quitmann E (1910) Über die Konstitution der Gefärbten Komponente des Blutfarbstoffes. Liebig's Ann Chem 377:314

References

747

314. Piloty O, Dormann E (1912) Über die Konstitution des Blutfarbstoffs. Liebig's Ann Chem 388:313

315. Piloty O, Wilke K (1912) Über das α,β-Dimethylpyrrol. Ber Dtsch Chem Ges 45:2586

316. Dennstedt M (1889) Ueber die *c*-Dimethylpyrrole. Ber Dtsch Chem Ges 22:1920

317. Dennstedt M (1889) Ueber die im *Dippel*'schen Oel Enthaltenen *c*-Dimethylpyrrole. Ber Dtsch Chem Ges 22:1924

318. Weidel H, Ciamician GL (1880) Studien Über Verbindungen aus dem Animalischen Theer. Ber Dtsch Chem Ges 13:65

319. Knorr L (1884) Synthese von Pyrrolderivaten. Ber Dtsch Chem Ges 17:1635

320. Knorr L (1886) Synthetische Versuche mit dem Acetessigester. Überführung des Diacetbernsteinsäure-esters und des Acetessigesters in Pyrrol-Derivate; Einwirkung von Ammoniak und Primären Aminen auf den Diacetbernsteinsäure-ester. Überführung des Acetessigesters in Pyrrole-Derivate. Liebig's Ann Chem 236:290

321. Paal C (1885). Synthese von Thiophen- und Pyrrolderivaten. Ber Dtsch Chem Ges 18:367

322. Knorr L, Lange H (1902) Ueber die Bildung von Pyrrolderivaten aus Isonitrosoketonen. Ber Dtsch Chem Ges 35:2998

323. Knorr L, Hess K (1911) Synthese des 2,4-Dimethyl-3-äthyl-pyrrols, ein Beitrage zur Lösung der Konstitution des Hämopyrrols. Ber Dtsch Chem Ges 44:2758

324. Knorr L (1885) Einwirkung des Diacetbernsteinsäureesters auf Ammoniak und Primäre Aminbasen. Ber Dtsch Chem Ges 18:299

325. Küster W (1909) Beiträge zur Kenntniss der Gallenfarbstoffe. Ueber Bilirubin, Biliverdin und ihre Spaltungsprodukte. Hoppe-Seyler's Z Physiol Chem 59:63

326. Küster W (1906) Beiträge zur Kenntniss der Gallenfarbstoffe. Hoppe-Seyler's Z Physiol Chem 47:294

327. Piloty O (1913) Bemerkungen zu der Mitteilung der HHrn. *Hans Fischer* und *Heinrich Röse* im 3 Heft dieser Berichte. Ber Dtsch Chem Ges 46:1000

328. Piloty O, Stock J, Dormann E (1914) Zur Konstitution des Blutfarbstoffs; Dipyrrylmethan-Derivate mit Farbstoff-Charakter. Ber Dtsch Chem Ges 47:400

329. Piloty O, Stock J. Dormann E (1914) Zur Konstitution des Blutfarbstoffs; Dipyrrylmethan-Derivate mit Farbstoff-Character. II. Ber Dtsch Chem Ges 47:1124

330. Piloty O, Krannich W, Will H (1914) Zur Konstitution des Blutfarbstoffs: Dipyrrylmethan-Derivate mit Farbstoff-Charakter. III. Ber Dtsch Chem Ges 47:2531

331. Piloty O, Stock J, Dormann E (1914) Über die Konstitution des Blutfarbstoffs. Über das Hämopyrrol and die Phonopyrrolcarbonsäuren. Liebig's Ann Chem 406:342

332. Piloty O, Fink H (1912) Über die Molgröße des Hämins und Hämoglobins. Ber Dtsch Chem Ges 45:2495

333. Piloty O, Dormann E (1912) Über die Phonopyrrol-carbonsäure und Ihre Begleiter. Ber Dtsch Chem Ges 45:2592

334. Piloty O, Hirsch P (1912) Über die Hämatopyrrolidinsäure. Ber Dtsch Chem Ges 45:2595

335. Piloty O, Blömer A (1912) Über Eine Synthese des Hämopyrrol b. Ber Dtsch Chem Ges 45:3749

336. Piloty O, Stock J (1912) Über die Konstitution des Blutfarbstoffs. IV. Über Hämopyrrol. Liebig's Ann Chem 392:215

337. Piloty O, Dormann (1913) Über die Sauren Spaltstücke des Hämins. Preliminary Communication. Ber Dtsch Chem Ges 46:1002

338. Piloty O, Stock J (1913) Über das Hämopyrrol. Ber Dtsch Chem Ges 46:1008

339. Piloty O, Wilke K (1913) Tetramethyl-pyrrindochinon und Andere: Derivate des Dimethyl-2.3-pyrrols. Ber Dtsch Chem Ges 46:1597

340. Piloty O, Fink H (1913) Über das Phonoporphyrin, ein Neues Spaltstück des Hämins. Ber Dtsch Chem Ges 46:2020

341. Piloty O, Will H (1913) Über Kondensation von Oxalester mit Acetyl-pyrrolen. Ber Dtsch Chem Ges 46:2607

342. Piloty O, Hirsch P (1913) Pyrrol-Synthesen aus Aminoketonen mit Ketonen und Ketonsäureestern. Liebig's Ann Chem 395:63

748 References

343. Willstätter R, Fischer M (1913) Untersuchungen Über den Blutfarbstoff. I. Mitteilung. Über den Abbau des Hämins zu Porphyrinen. Hoppe-Seyler's Z Physiol Chem 87:423
344. Willstätter R, Stoll A (1913) Untersuchungen Über Chlorophyll. Julius Springer, Berlin
345. Zeile K (1946) Das Lebenswerk *Hans Fischers*. Naturwissenschaften 33:289
346. Treibs A (1946) In Memoriam. *Hans Fischer*. Z Naturforsch 1:476
347. Wieland H (1950) *Hans Fischer* und *Otto Hönigschmid* zum Gedächtnis. Angew Chem 62:1
348. Fischer H (1912) Über Urobilin und Bilirubin. Habilitationsschrift zu Erlangung der *Venia Legendi*. München, Kgl Hof. und Univ. Buchdruckerei Dr. C Wolf & Sohn
349. Fischer H (1911) Zur Kenntniss des Gallenfarbstoffs. I. Mittlg. Hoppe-Seyler's Z Physiol Chem 73:204
350. Fischer H, Meyer-Betz F (1911) Zur Kenntniss des Gallenfarbstoffs. II. Mittlg. Über das Urobilinogen des Urins und das Wesen der *Ehrlich*'schen Aldehyd-Reaktion Hoppe-Seyler's Z Physiol Chem 75:232
351. Fischer H, Meyer P (1911) Zur Kenntniss des Gallenfarbstoffs. III. Mittlg. Über Hemibilirubin und die bei der Oxydation des Hemibilirubins Entstehenden Spaltprodukte. Hoppe-Seyler's Z Physiol Chem 75:339
352. Fischer H, Röse H (1912) Über Bilirubinsäure, ein Neues Bilirubin-Abbauprodukt Ber Dtsch Chem Ges 45:1579
353. Fischer H, Röse H (1912) Zur Kenntnis des Gallenfarbstoffs. IV. Mittlg. Hoppe-Seyler's Z Physiol Chem 82:391
354. Fischer H, Röse H (1912) Über den Abbau des Bilirubins und der Bilirubinsäure. Ber Dtsch Chem Ges 45:3274
355. Fischer H (1913) Bemerkung zu der Publikation *W. Küsters* "Beiträge zur Kenntniss des Bilirubins und Hämins". Hoppe-Seyler's Z Physiol Chem 83:170
356. Fischer H, Röse H (1913) Einwirkung von Natriummethylat auf Bilirubinsäure, Bilirubin, und Hemibilirubin. Ber Dtsch Chem Ges 46:439
357. Fischer H, Röse H (1914) Zur Kenntnis der Gallenfarbstoffe. V. Mittlg. Über die Konstitution der Bilirubinsäure u. des Bilirubins. Hoppe-Seyler's Z Physiol Chem 89:255
358. Fischer H, Bartholomäus E, Röse H (1913) Zur Kenntniss des Porphyrin Bildung. II. Mittlg. Über Porphyrinogen und seine Beziehungen zum Blutstoff und Dessem Derivativen. Hoppe-Seyler's Z Physiol Chem 84:262
359. Fischer H (1913) Erwiderung auf die Bemerkungen von *O. Piloty* im 5. Heft dieser Ber Dtsch Chem Ges 46:1574
360. Küster W (1912) Beiträge zur Kenntniss des Bilirubins und Hämins. Hoppe-Seyler's Z Physiol Chem 82:463
361. Fischer H, Bartholomäus E (1911) Zur Hämopyrrol-frage. Ber Dtsch Chem Ges 44:3313
362. Fischer H, Bartholomäus E (1912) Einwirkung von Natriumalkoholat auf Pyrrolderivate. I. Mittlg. Hoppe-Seyler's Z Physiol Chem 77:185
363. Fischer H, Bartholomäus E (1912) Synthesen des Phyllopyrrols. Ein Beitrag zur Hämopyrrolfrage. Ber Dtsch Chem Ges 45:466
364. Willstätter R, Asahina Y (1911) Untersuchen Über Chlorophyll: XVII. Reduction von Chlorophyll I. Liebig's Ann Chem 385:188
365. Willstätter R, Asahina (1912) Zur Hämopyrrol-Frage. Ber Dtsch Chem Ges 44:3707
366. Fischer H, Bartholomäus E (1912) Die Lösung der Hämopyrrolfrage. Ber Dtsch Chem Ges 45:1979
367. Fischer H, Bartholomäus E (1912) Gewinnung von Phonopyrrolcarbonsäure aus Hämin. Ber Dtsch Chem Ges 45:1315
368. Fischer H, Bartholomäus E (1912) Synthesen von 2,4-Dimethylpyrrol-5-essigsäure und 2,4-Dimethylpyrrol-5-propionsäure. Ber Dtsch Chem Ges 45:1919
369. Fischer H, Bartholomäus E (1913) Experimentelle Studien Über die Konstitution des Blut- und Gallenfarbstoffs. Hoppe-Seyler's Z Physiol Chem 83:50
370. Fischer H, Röse H (1914) Gewinnung der Isophonopyrrolcarbonsäure aus Hämin und eine Neue Isolierungsmethode der Sauren Spaltprodukte des Hämins und Bilirubins. Ber Dtsch Chem Ges 47:791

References

371. Fischer H, Bartholomäus E (1913) Experimentelle Studien Über die Konstitution des Blut- und Gallenfarbstoffs. II. Mittlg. Hoppe-Seyler's Z Physiol Chem 87:255
372. Fischer H, Röse H (1914) Über die Destillation Einiger Pyrrolcarbonsäuren. Hoppe-Seyler's Z Physiol Chem 91:184
373. Fischer H (1916) Über Blut- und Gallenfarbstoff. Ergebnisse der Physiologie, biologischen Chemie und Experimentellen Pharmakologie. Rev Phys Biochem Pharm 15:185
374. Fischer H, Orth H (1934) Die Chemie des Pyrrols. Bd. I, Akademische Verlags-gesellschaft M.B.H., Leipzig, p 278
375. Fischer H, Röse H (1913) Einwirkung von Alkoholaten auf Hämin und seine Derivate. I. Mittlg. Hoppe-Seyler's Z Physiol Chem 87:38
376. Fischer H (1914) Bemerkung zu der Abhandlung von *O. Piloty, W. Krannich* und *H. Will*, Zur Konstitution des Blutfarbstoffs: Dipyrryl-methene. Derivate mit Farbstoff-Struktur. Ber Dtsch Chem Ges 47:3266
377. Fischer H (1914) Zur Kenntnis der Gallenfarbstoffe. VI. Mittlg. Über Mesobilirubin und Mesobilirubinogen. Z Biol 65:163
378. Fischer H (1914) Über Mesobilirubin. Ber Dtsch Chem Ges 47:2330
379. Fischer H, Hahn A (1914) Über Brommesoporphyrin und die Reduktion von Blut- und Gallenfarbstoff bei Gegenwart von Kolloidalem Palladium. Hoppe-Seyler's Z Physiol Chem 91:174
380. Küster W (1915) Ueber den Chemismus der Bildung des Gallenfarbstoffs aus der Eisenhaltigen Componente des Blutfarbstoffs. Archiv Pharm 253:457
381. Küster W (1915) Über die Konstitution des Hämins und des Bilirubins. Hoppe-Seyler's Z Physiol Chem 95:152
382. Küster W (1917) Über das Bilirubinammonium und Über Modifikationen des Bilirubins. Hoppe-Seyler's Z Physiol Chem 99:86
383. Küster W (1915) Beiträge zur Kentniss der Gallenfarbstoff. VIII. Mitteilung. Über das Bilirubin. Hoppe-Seyler's Z Physiol Chem 94:136
384. Küster W (1920) Über die Bindung des Eisens in der Prosthetischen Gruppe des Blutfarbstoffs und die Konstitution des Hämins. Ber Dtsch Chem Ges 53:623
385. Fischer H, Beller H (1925) Synthese Eines Tetrapyrryl-äthylens und Einiger Derivate des 2,3-Dimethyl-pyrrols. Liebig's Ann Chem 444:238
386. Fischer H, Eismayer K (1914) Experimentelle Studien Über die Konstitution des Blut- und Gallen-Farbstoffs. III. Ber Dtsch Chem Ges 47:2019
387. Küster W (1917) Über das Bilirubinammonium und Über Modifikation des Bilirubins. X. Mitteilung zur Kenntnis der Gallenfarbstoffe. Hoppe-Seyler's Z Physiol Chem 99:86
388. Küster W (1921) Einige Neue Beobachtungen beim Studium des Bilirubins. Angew Chem 34:246
389. Küster W (1922) Beiträge zur Kenntnis der Gallenfarbstoffe. XII. Mitteilung. Über die Einwirkung von Diazomethan auf Bilirubin und Biliverdin, die Oxydation des Bilirubins in Alkalischer Lösung und die Einwirkung von Bromwasserstoff – Eissig auf Bilirubin. Hoppe-Seyler's Z Physiol Chem 121:94
390. Küster W, Haas R (1924) Über die Aufarbeitung von Rindergallenstein. 14. Mittlg. Über Gallenfarbstoff. Hoppe-Seyler's Z Physiol Chem 141:279
391. Küster W, Heeß W (1925) Beiträge zur Kenntnis der Prosthetischen Gruppe. Ber Dtsch Chem Ges 58:1022
392. Stern AJ (1973) *Hans Fischer* (1881-1945) Annals NY Acad Sci 206:752
393. Fischer H, Herrmann M (1922) Einige Beobachtungen Über Pyrrole und Pyrrolaldehyde. Hoppe-Seyler's Z Physiol Chem 122:1
394. Fischer H, Loy E (1923) Synthetische Versuche Über die Konstitution des Gallenfarbstoffs. I. Mittlg. Hoppe-Seyler's Z Physiol Chem 128:59
395. Fischer H, Niemann G (1923) Zur Kenntnis des Gallenfarbstoffs. 7. Mittlg. Hoppe-Seyler's Z Physiol Chem 127:317
396. Falk H, Müller N (1985) On the Chemistry of Pyrrole Pigments. Magn Reson Chem 23:353

750 References

397. Fischer H, Niemann G (1924) Zur Kenntnis des Gallenfarbstoffs. 8. Mittlg. Mesobiliviolin, Mesobiliviolinogen, Mesobilirubinogen mit Aldehyden unter Bildung von Neuen Spaltprodukten Diazofarbstoff des Mesobilirubins. Hoppe-Seyler's Z Physiol Chem 137:293
398. Fischer H, Postowsky J (1926) Bestimmung des "Aktiven Wasserstoffs" im Hämin und Bilirubin, Einigen ihrer Derivate und in Pyrrolen. Hoppe-Seyler's Z Physiol Chem 152:300
399. Fischer H, Niemann G (1925) Zur Kenntnis des Gallenfarbstoffs (IX). Kleiner Mitteilungen. Hoppe-Seyler's Z Physiol Chem 146:196
400. Fischer H, Schubert M (1923) Über Tetrapyrrylethane (I.). Ber Dtsch Chem Ges 56:2379
401. Fischer H (1925) Über Blutfarbstoff und Einige Porphyrine. Angew Chem 38:981
402. Fischer H, Klarer J (1926) Synthese des Ätioporphyrins, Ätiohemins, und Ätiophyllins. Liebig's Ann Chem 448:178
403. Fischer H, Halbg P (1926) Synthese des Iso-Ätioporphyrins, Seines „Hämins" und „Phyllins". Liebig's Ann Chem 448:193
404. Fischer H, Scheyer H (1924) Einwirkung von Halogen auf Substituierte Pyrrole; ein Dipyrryläthan und Synthese Eines Farbstoffes aus Vier Pyrrolkernen. Liebig's Ann Chem 439:185
405. Fischer H, Heyse M (1924) Über Tripyrrylmethane. Liebig's Ann Chem 439:246
406. Fischer H, Amman H (1923) Einige Umsetzungen des 2,4-Dimethyl-3-acetyl-pyrrols und Über Tripyrryl-methane (I). Ber Dtsch Chem Ges 56:2319
407. Fischer H, Pützer B (1926) Kenntnis der natürlichen Porphyrine. XIX. Mitteilung. Überführung von Hämin in Protoporphyrin und Eine Neue Darstellung der Mesoporphyrins. Hoppe-Seyler's Z Physiol Chem 154:39
408. Fischer H, Lindner F (1926) Überführung von Gallenfarbstoff und Bilirubinsäure in Mesoporphyrin. Gewinnung von Hämopyrrol aus Mesobilirubinogen. Hoppe-Seyler's Z Physiol Chem 161:1
409. Fischer H, Andersag H (1926) Synthese des Kopro- und Iso-Koproporphyrins. Liebig's Ann Chem 450:201
410. Fischer H, Halbig P, Walach B (1927) Über eine neue Porphyrinsynthese, Oxydation von Porphyrinen und Einige Kleinere Beobachtungen. Liebig's Ann Chem 452:268
411. Schlack P (1960) *William Küster*. Lecture at the Uni-Stuttgart
412. Küster W (1927) Über den Chemismus der Porphyrinbildung und die Konstitution des Hämins. (Eine Entgegnung zu Bemerkungen von *H. Fischer*.) Hoppe-Seyler's Z Physiol Chem 163:267
413. Küster H, Kimmich K (1927) Über den Blutfarbstoff. Hoppe-Seyler's Z Physiol Chem 172:199
414. Papendieck A (1923) Über das Porphyrin der Menschlichen Fäzes. Hoppe-Seyler's Z Physiol Chem 128:109
415. Papendieck A (1924) Über das Porphyrin der Menschlichen Fäzes. (II Mitteilung) Hoppe-Seyler's Z Physiol Chem 133:97
416. Fischer H, Schneller K (1924) Zur Kenntnis der Natürlichen Porphyrine. VI. Mitteilung. Über die Verbreitung der Porphyrine in Organen. Nachweis Eines Porphyrins in des Hefe. Hoppe-Seyler's Z Physiol Chem 135:253
417. Schumm O (1924) Spektroskopisch-chemische Reaktionen Einiger Porphyrine und Ihrer Methylester. Hoppe-Seyler's Z Physiol Chem 136:243
418. Fischer H (1924) Bemerkungen zu den Abhandlungen von *Schumm* und *Papendieck* in Dieser Zeitung Bd. 136, Heft 5 und 6. Hoppe-Seyler's Z Physiol Chem 138:307
419. Fischer H (1926) Bemerkung zu den Richtigstellungen *Schumms*. Hoppe-Seyler's Z Physiol Chem 155:96
420. Schumm O (1926) Spektrochemische Untersuchungen an Porphyrinen und Hämatinen. (Fortsetzung.) I. Über den Nachweis von Koproporphyrin und die Bedeutung Spektrochemischer Methoden für die Porphyrin- und Porphyratinforschung. II. Über das im Blutserum Kranker Auftretende Hämatin. Hoppe-Seyler's Z Physiol Chem 152:1

References

751

421. Schumm O (1926) Weitere Beiträge zur Kenntnis der Natürlichen Porphyrine und der Porphyrinatine. Hoppe-Seyler's Z Physiol Chem 153:225

422. Schumm O (1926) Richtigstellung zur Abhandlung von *Hans Fischer* und *Hans Hilmer*: "Koproporphyrin-Synthese durch Hefe und Ihre Beeinflussung" und der "Bemerkung" von *Hans Fischer*. Hoppe-Seyler's Z Physiol Chem 156:159

423. Fischer H, Hilmer H (1926) Über Koproporphyrin-Synthese durch Hefe und Ihre Beeinflussung. IV. Mitteilung. Hoppe-Seyler's Z Physiol Chem 153:167

424. Schumm O (1928) Bemerkung zu dem Vortrag von *Hans Fischer*: "Über Porphyrine und Ihre Synthese". Ber Dtsch Chem Ges 161:784

425. Fischer H, Zeile K (1929) Synthese des Hämatoporphyrins, Protoporphyrins, und Hämins. Liebig's Ann Chem 468:98

426. Fischer H (1927) Über Porphyrine und Ihre Synthesen. Ber Dtsch Chem Ges 60:2611

427. Fischer H (1928) Bemerkung zu Meinen vor der Deutschen Chemischen Gesellschaft Gehalten Vortrag. Ber Dtsch Chem Ges 61:1596

428. Olah GA, Prakash GKS, Saunders M (1983) Conclusion of the Classical-nonclassical Ion Controversy Based on the Structural Study of the 2-Norbornyl Cation. Acc Chem Res 16:440

429. Fischer H, Treibs A (1927) Über Ätioxantho- und Mesoxantho-porphyrinogen. Liebig's Ann Chem 457:209

430. Fischer H, Andersag H (1927) Synthese des β-Iso-Koproporphyrins und der Opsopyrrolcarbonsäure. Liebig's Ann Chem 458:117

431. Fischer H, Stangler G (1927) Synthese des Meosporphyrins, Mesohämins und Über die Konstitution des Hämins. Liebig's Ann Chem 459:53

432. Pfeiffer P (1929) *William Küster*, 1863-1929. Angew Chem 42:785

433. Lemberg R, Legge JW (1949) Hematin Compounds and Bile Pigments. Interscience Publishers, New York and London

434. Willstätter R (1949) Aus Meinen Leben. Verlag Chemie, GmbH, Weinheim, p 184

435. Willstätter R (1965) From My Life. Hornig LS (transl). WA Benjamin, Inc, New York, p 194

436. Fischer H (1930) Hämin, Bilirubin und Porphyrine. Naturwissenschaften 18:1026

437. Watson CJ (1965) Reminiscences of *Hans Fischer* and His Laboratory. Perspect Biol Med 8:419

438. Fischer H, Hess R (1931) Über Neo-, Xantho-neobilirubinsäure und Partialsynthese des Mesobilirubins und Mesobilirubinogens (Urobilinogens). Hoppe-Seyler's Z Physiol Chem 194:193

439. Schumm O (1928) Über Einige Neue Verfahren zur Gewinnung von Häminderivaten. Hoppe-Seyler's Z Physiol Chem 176:122

440. Schumm O (1928) Über die Gewinnung von Häminderivaten durch Brenzreaktionen. 2. Mitteilung. I. Darstellung des Pyratins aus Hämins durch die Resorcinschmelze. II. Eisenung und Eigenschaften des Pyroporphyrins. III. Eisenung von Porphyrinen in der „Eisen-Phenolschmelze" und Umwandlung von Eisenporphyratinen durch Phenole und Phenolschwefelsäure. Hoppe-Seyler's Z Physiol Chem 178:1

441. Bonnett R, McDonagh AF (1970) Oxidative Cleavage of the Haem System: The Four Isomeric Biliverdins of the IX Series. J Chem Soc Chem Commun:238

442. Fischer H, Adler E (1931) Synthese des Bilirubin- und Xanthobilirubinsäure und Ihre Isomeren, sowie Synthese von Tripyrranen und Bilirubinoiden Farbstoffen. Hoppe-Seyler's Z Physiol Chem 197:237

443. Siedel W, Fischer H (1933) Über die Konstitution des Bilirubins, Synthesen der Neo- und der Iso-neoxanthobilirubinsäure. Hoppe-Seyler's Z Physiol Chem 214:145

444. Fischer H, Adler (1931) Synthese des Mesobilirubinogens und der Neobilirubinsäure, eines Mesobilirubins und einer Neoxanthobilirubinsäure sowie von (1,8)-Dioxy-tripyrro-di-enen. Hoppe-Seyler's Z Physiol Chem 200:209

445. Ma, J-S, Chen Q-Q, Wang C-Q, Liu Y-Y, Yan F, Cheng L-J, Jin S, Falk H (1995) A Novel 16,24-Dehydrobiladiene-ab System; The Reaction of Xanthobilirubic Acid Methyl Ester with Bromine. Monatsh Chem 126:201

752 References

446. Chen Q-Q, Ma, J-S, Wang C-Q, Liu Y-Y, Yan F, Cheng L-J, Jin S, Falk H (1995) 2-*bis*-Dipyrrinone-9-yl-ethene – a Novel *b-homo*-Verdin Chromophore: the Reaction of 9-Methyl-10*H*-dipyrrinones with Bromine. Monatsh Chem 126:983

447. Fischer H, Neumann FW (1932) The Chemistry of Animal Pigments. Ann Rev Biochem 1:527

448. Fischer H, Fröwis W (1931) Synthese von Oxypyrromethenen und Über Einige Derivate des Koproporphyrins I. Hoppe-Seyler's Z Physiol Chem 195:49

449. Fischer H, Orth H (1934) The Structural Chemistry of the Animal Pigments. Ann Rev Biochem 3:410

450. Lemberg R, Barcroft J (1932) Uteroverdin, the Green Pigment of the Dog Placenta. Proc Roy Soc London 110B:362

451. Lemberg R (1932) Über Dehyro-bilirubin. Liebig's Ann Chem 499:25

452. Fischer H, Hartmann P (1934) Synthese des Oxyhämopyrrols, der Iso-neo-, der Isoxanthobilirubinsäure und Über den Kryptopyrroläther. Hoppe-Seyler's Z Physiol Chem 226:116

453. Grunewald JO, Cullen R, Bredfeldt J, Strope ER (1975) An Efficient Route to Xanthobilirubic Acid, an Oxodipyrrylmethene. Org Prep Proc Int 7:103

454. Siedel W (1935) Neue Synthesen der Neo- und Iso-neoxanthobilirubinsäure. Vorarbeiten zur Synthese Natürlicher Bilirubinoide. Hoppe-Seyler's Z Physiol Chem 231:167

455. Fischer H, Haberland H (1935) Über die Konstitution des Bilirubins Sowie die Seiner Azofarbstoffe und die Gmelinsche Reaktion. Hoppe-Seyler's Z Physiol Chem 232:236

456. Siedel W (1937) Synthese des Mesobilirubins (= Mesobilirubin-IX, α). Hoppe-Seyler's Z Physiol Chem 245:257

457. Siedel W (1935) Synthese des Glaukobilins, Sowie Über Urobilin und Mesobiliviolin. Hoppe-Seyler's Z Physiol Chem 237:8

458. Fischer H, Reinecke H (1939) Synthese Zweier Vinyl-substituierter Bilirubinoide. Hoppe-Seyler's Z Physiol Chem 258:9

459. Bonnett R, McDonagh AF (1969) Methylvinylmaleimide (the 'Nitrite Body') from Chromic Acid Oxidation of Tetrapyrrole Pigments. Chem & Ind (London):107

460. Lemberg R (1935) CLXI. Transformation of Haemins into Bile Pigments. Biochem J 29:1322

461. Lightner DA, Quistad GB (1972) Methylvinylmaleimide from Bilirubin Photooxidation. Science 175:324

462. Fischer H, Plieninger H (1942) Synthese des Biliverdins (Uteroverdins) und Bilirubins. Naturwissenschaften 30:382

463. Plieninger H, Lichtenwald H (1942) Über Eine Neue Darstellung von Oxypyrromethenes Durch Alkalische Kondensation von Oxypyrrolen mit Pyrrol-α-aldehyden, Sowie Über Weitere Versuche zur Synthese von acetylsubstituierten Gallenfarbstoffen und Über Tripyrrene. Hoppe-Seyler's Z Physiol Chem 273:206

464. Fischer H, Reinecke H (1939) Über Tripyrrene. Hoppe-Seyler's Z Physiol Chem 259:86

465. Fischer H, Treibs A (1926) Über Abbau des Blutfarbstoffs und Wiederaufbau von Porphyrinen aus Opsopyrrol und Opsopyrrolcarbonsäure. Liebig's Ann Chem 450:132

466. Fischer H, Reinecke H, Lichtenwald H (1939) Über Oxy-amino-pyrromethene, ein Beitrag zur Kenntnis der Pentdyopent-Reaktion. Hoppe-Seyler's Z Physiol Chem 257:190

467. Jellinek K, Jellinek E (1919) Chemische Zersetzung und Electrolytische Bildung von Natriumhydrosulfit. Z Naturforsch 93:325

468. Lisler MW, Garvie RC (1959) Sodium Dithionite, Decomposition in Aqueous Solution and in the Solid State. Can J Chem 37:1567

469. Stoll MS, Gray CH (1977) The Preparation and Characterization of Bile Pigments. Biochem J 163:59

470. Blanckaert N, Heirwegh KPM, Compernolle F (1976) Synthesis and Separation by Thin Layer Chromatography of Bilirubin-IX Isomers. Biochem J 155:405

471. Kufer W, Scheer H (1979) Chemical Modification of Biliprotein Chromophores. Z Naturforsch C: J Biosci 34:776

References

753

472. Kufer W, Scheer H (1982) Rubins and Rubinoid Addition Products from Phycocyanin. Z Naturforsch C: J Biosci 37C:179
473. McDonagh AF (1975) Thermal and Photochemical Reactions of Bilirubin. Ann NY Acad Sci 244:553
474. McDonagh AF, Assisi F (1972) The Ready Isomerization of Bilirubin-IX-α in Aqueous Solution. Biochem J 129:797
475. McDonagh AF, Assisi F (1972) Direct Evidence for the Acid-catalyzed Isomeric Scrambling of Bilirubin IX-α. J Chem Soc Chem Commun 117
476. McDonagh AF, Assisi F (1971) Commercial Bilirubin: A Trinity of Isomers. FEBS Lett 18: 315
477. Ammon U (2004) German as an International Language of the Sciences – Recent Past and Present. In: Gardt A, Hüppauf B (eds) Globalization and the Future of German. Mouton de Gruyter, Berlin
478. Sanders RH (2010) German. Biography of a Language. Oxford University Press, New York
479. Gross PLK, Gross EM (1927) College Libraries and Chemical Education. Science 66:385
480. Fieser LF, Fieser M (1959) Steroids. Reinhold Publishing Corp, Waverly Press, Inc, Baltimore
481. Keana JFW, Johnson WS (1964) Racemic Cholesterol. Steroids 4:457
482. Pimentel GC, McClellan AL (1960) The Hydrogen Bond. WH Freeman and Co, San Francisco and London
483. Lemberg R (1932) Über Dehydro-bilirubin. Liebig's Ann Chem 499:25
484. Fowweather FS (1932) Bilirubin and the *van den Bergh* Reaction. Biochem J 26:165
485. Gray CH (1953) The Bile Pigments. Methuen, London
486. Gray CH, Nicholson DC, Nicolaus RA (1958) The IX-α Structure of the Common Bile Pigments. Nature 187:183
487. Plieninger H, Decker M (1956) Eine neue Synthese für Pyrrolone, insbesondere für "Isooxyopsopyrrol" und "Isooxypyrrolecarbonsäure". Liebig's Ann Chem 598:198
488. Grob CA, Ankli P (1954) Die Tautomerie der α-Amino and α-Oxypyrrole. Untersuchungen in der Pyrrolreihe. 6. Mitteilung. Helv Chim Acta 37:1256
489. Grob CA, Ankli P (1949) Über α-Pyrrolone. 2. Mitteilung. Helv Chim Acta 32:2010
490. Gray CH, Nicholson DC (1957) Structure of Stercobilin. Nature 179:264
491. Gray CH, Nicholson D (1957) Structure of *d*-Urobilin. Nature 180:336
492. Watson CJ, Weimer M, Hawkinson V (1960) Differences in the Formation of Mesobiliviolin and Glaucobilin from *d*- and *i*-Urobilins. J Biol Chem 235:787
493. Gray CH, Kulczycka A, Nicholson DC (1961) Prototropy of Bilirubin to a Verdinoid Pigment. J Chem Soc:2268
494. Gray CH (1961) Bile Pigments in Health and Disease. Thomas, Springfield, Illinois
495. Fog J, Jellum E (1963) Structure of Bilirubin. Nature 198:88
496. Fog J, Bugge-Asperheim B (1964) Stability of Bilirubin. Nature 203:756
497. von Dobeneck H, Brunner E (1965) Über Eine Ordnung der Dipyrromethene und Über die Betainstruktur des Bilirubins. Hoppe-Seyler's Z Phys Chem 341:157
498. ÓhEocha C (1965) Phycobilins. In: Goodwin TW (ed) Chemistry and Biochemistry of Plant Pigments. Academic Press, New York
499. Brodersen R, Flodgaard H, Krogh-Hansen J (1967) Intramolecular Hydrogen Bonding in Bilirubin. Acta Chem Scand 21:2284
500. Bouchier IAD, Billing BH (eds) (1967) Bilirubin Metabolism. Blackwell Scientific Publications, Oxford and Edinburgh
501. With TK (1968) Bile Pigments. Chemical, Biological and Clinical Aspects. Kennedy JP (trans). Academic Press, New York
502. Ostrow JD (1972) Mechanisms of Bilirubin Photodegradation. Semin Hematol 9:113
503. Ostrow JD (1972) Photochemical and Biochemical Basis of the Treatment of Neonatal Jaundice. In: Popper H, Schaffner F (eds) Progress in Liver Disease, vol IV. Grune & Stratton, Inc, New York, p 447

504. Nichol AW, Morell DB (1969) Tautomerism and Hydrogen Bonding in Bilirubin and Biliverdin. Biochim Biophys Acta 177:599
505. Kuenzle CC (1970) Bilirubin Conjugates of Human Bile. Nuclear-magnetic-resonance, Infrared and Optical Spectra of Model Components. Biochem J 119:395
506. Hutchinson DW, Johnson B, Knell AJ (1971) Tautomerism and Hydrogen Bonding in Bilirubin. Biochem J 123:483
507. Odell GB, Schaffer R, Simopoulos AP (eds) (1974) Phototherapy in the Newborn: An Overview. US National Academy of Sciences, Washington, DC
508. Bakken AF, Fog J (eds) (1975) Metabolism and Chemistry of Bilirubin and Related Tetrapyrroles. Pediatric Research Institute, Rikshospitalet, Oslo, Norway. Proceedings of the Bilirubin Meeting in Hemsedal, Norway, September 1974
509. Berk PD, Berlin NI (eds) (1977) Chemistry and Physiology of Bile Pigments. DHEW Publ No (NIH) 77-1001, US Dept of Health, Education and Welfare, National Institutes of Health, Bethesda, Maryland
510. Bonnett R, Davies JE, Hursthouse MB (1976) Structure of Bilirubin. Nature 262:326
511. Bonnett R, Davies JE, Hursthouse MB, Sheldrick GM (1978) The Structure of Bilirubin. Proc Royal Soc London B202:249
512. Sheldrick WS, Becker W (1979) Crystal and Molecular Structure of Diethoxybilirubin Diethyl Ester. Z Naturforsch 34b:1542
513. Kratky C, Jorde C, Falk H, Thirring K (1983) Crystal Structure of the Mono-lactim Ether of a Bilatriene-*abc* Derivative at 101 K. Tetrahedron 39:1859
514. Becker W. Sheldrick WS (1978) Bile Pigment Structures. II. The Crystal Structure of Mesobilirubin-IXα-bis(chloroform). Acta Cryst B34:1298
515. LeBas G, Allegret A, DeRango C (1977) Conformation and Relations with Optical Rotation Properties of Bilirubin Molecule. Proceedings of the Fourth European Crystallographic Meeting, Aug 30-Sept 3, Abst PI107:310
516. LeBas G, Allegret A, Mauguen Y, DeRango C, Bailly M (1980) The Structure of Triclinic Bilirubin Chloroform-methanol Solvate. Acta Cryst B36:3007
517. Mugnoli A, Manitto P, Monti D (1978) Structure of Di-isopropylammonium Bilirubinate. Nature 273:568
518. Mugnoli A, Manitto P, Monti D (1983) Structure of Bilirubin-IXα (Isopropylammonium Salt) Chloroform Solvate, $C_{33}H_{34}N_4O_6^{2-} \cdot 2C_3H_{10}N^+ \cdot 2CHCl_3$. Acta Cryst C39:1287
519. Sheldrick WS (1976) Crystal and Molecular Structure of Biliverdin Dimethyl Ester. J Am Chem Soc Perkin Trans II:1457
520. Cullen DL, Black PS, Meyer Jr EF, Lightner DA, Quistad GB (1977) The Crystal and Molecular Structure of an Oxodipyrromethene Related to Bilirubin. Tetrahedron 33:477
521. Cullen DL, Pèpe G, Meyer Jr EF, Falk H, Grubmayr K (1979) *syn* and *anti*-Conformation in Oxo-dipyrromethenes: Crystal and Molecular Structure of 3,4-Dimethyl-2,2'-pyrromethen-5(1*H*)-one and its *N*-Methyl Derivative. J Chem Soc Perkin Trans II:999
522. Hori A, Mangani S, Pèpe G, Meyer Jr EF (1981) Crystal and Molecular Structure of a Photoisomer of an Oxo-dipyrromethene. The *E*-isomer of 3,4-Dimethyl-2,2'-pyrrometh-5(1*H*)-one. J Chem Soc Perkin Trans II:1525
523. Gossauer A, Blacha M, Sheldrick WS (1976) Synthesis and X-ray Crystal Structure of Two Stereoisomeric Derivatives of 3,4-Dihydropyrromethen-5(1*H*)-one. J Chem Soc Chem Commun:764
524. Sheldrick WS, Borkenstein A, Blacha-Puller M, Gossauer A (1977) Stereoisomerism in Partial Bile Pigment Structures. The Crystal Structures of Z and E Isomers of 5'-Ethoxycarbonyl-3,4-dihydro-3',4'-dimethyl-5(1*H*)-2,2'-pyrromethenone and Their Reaction Products with $Et_3O^+BF_4^-$. Acta Cryst B33: 3625
525. Falk H, Gergely S, Grubmayr K, Hofer O (1977) Die Lactam-Lactim Tautomerie von Gallenpigmenten. Liebig's Ann Chem 1977:565
526. Falk H, Grubmayr K, Thirring K, Gurker H (1978) Beiträge zur Chemie der Pyrrolpigmente, 21. Mitt. Röntgenphotoelektronenspektrometrische Untersuchungen des N_{1s}-Niveaus von Gallenpigmenten. Monatsh Chem 109: 1183

References

527. Falk H, Hofer O, Lehner H (1974) Beiträge zur Chemie der Pyrrolpigmente, 2. Mitt. Röntgenphotoelektronenspektren Einiger Pyrrolpigmente. Monatsh Chem 105:366

528. Falk H, Hofer O (1974) Beiträge zur Chemie der Pyrrolpigmente, 4. Mitt. Die N-H Tautomerie von Substituierten Pyrromethenen: Zur Dynamik der Protonentransfers aus der Sicht Eines Quantenchemischen Verfahrens (*CNDO/2*). Monatsh Chem 105:995

529. Severini-Ricca GS, Manitto P, Monti D (1974) The Carbon-13 Nuclear Magnetic Resonance Spectra of Bilirubin and its Derivatives. A Confirmation of the bis-Lactam Form for these Biladiens (1974). In: Bakken AF, Fog J (eds) Proceedings of a Conference on the Metabolism and Chemistry of Bilirubin and Related Tetrapyrroles. Pediatric Research Institute Rikshospitalet, Oslo, Norway

530. Severini-Ricca G, Manitto P, Monti D, Randall EW (1975) The Carbon-13 Nuclear Magnetic Resonance Spectra of Bilirubin and its Derivatives. Gazz Chim Ital 105:1273

531. Jakobsen JH, Bildsøe H, Dønstrup S, Sorensen OW (1984) Simple One-dimensional NMR Experiments for Heteronuclear Chemical-shift Correlation. J Magn Reson 57:324

532. Hansen PE, Jakobsen JH (1984) A Natural Isotopic Abundance Nitrogen-15 Investigation of Bilirubin IX-α. Org Magn Reson 22:688

533. Falk H, Müller N (1985) On the Chemistry of Pyrrole Pigments 60 – Natural Isotopic Abundance ^{15}N NMR Spectra of Verdinoid Bile Pigments and Their Partial Structures. Magn Reson Chem 23:353

534. Blackwood JE, Gladys CL, Loening KL, Petrarca AE, Rush JE (1968) Unambiguous Specification of Stereoisomerism About a Double Bond. J Am Chem Soc 90:509

535. Rimington C, Gray CH (1976) *Max Rudolf Lemberg*, 1896-1976. Biographical Memoirs of Fellows of the Royal Society 22:255

536. ÓhEocha C (1968) The Formation of Bile Pigments. In: Goodwin TW (ed) Porphyrins and Related Compounds. Academic Press, New York, p 91

537. Kuenzle CC, Weibel MH, Pelloni RR (1973) The Reaction of Bilirubin with Diazomethane. Biochem J 133:357

538. Hutchinson, DW, Johnson B, Knell AJ (1973) The Synthesis of Esters of Bilirubin. Biochem J 133:493

539. Burke MJ, Pratt DC, Moscowitz A (1972) Low-temperature Absorption and Circular Dichroism Studies of Phytochrome. Biochemistry 11:4025

540. Falk H, Grubmayr K, Herzig U, Hofer O (1974) The Configurations of the Isomeric 3.4-Dimethyl-5-(1*H*)-2,2'-pyrromethenones. Tetrahedron Lett:559

541. Overhauser A (1913) Polarization of Nuclei in Metals. Phys Rev 92:411

542. Neuhaus D, Williamson MP (2000) The Nuclear *Overhauser* Effect in Structural and Conformational Analysis, 2nd ed. John Wiley & Sons, Inc, New York

543. Huggins MT, Billimona F (2007) Nuclear *Overhauser* Effect Spectroscopy. J Chem Educ 84:471

544. Kaplan D, Navon G (1981) Nuclear Magnetic Resonance Studies of the Conformation of Bilirubin and Its Derivatives in Solution. J Chem Soc Perkin Trans II:1374

545. Falk H, Grubmayr K, Haslinger E. Schlederer T, Thirring K (1978) Die Stereoisomeren (Geometrisch Isomeren) Biliverdindimethylester – Struktur, Konfiguration und Konformation. Monatsh Chem 109:1451

546. Falk H, Grubmayr K, Magauer K, Müller N, Zrunek U (1983) Phytochrome Model Studies: The Tautomerism at N_{22}-N_{23} of Unsymmetrically Substituted Bilatrienes-*abc* and 2,3-Dihydrobilatrienes-*abc*. Isr J Chem 23:187

547. Barañano DE, Rao M, Ferris CD, Snyder SH (2002) Biliverdin Reductase: A Major Physiologic Cytoprotectant. Proc Natl Acad Sci USA 99:16093

548. Joule JA, Mills K (2010) Heterocyclic Chemistry, 5th ed. J Wiley & Sons, Inc, New York, p 635

549. Frankenberg N, Lagarius JG (2003) Phycocyanobilin: Ferredoxin Oxidoreductase of *Anabaena* sp. PCC7120. J Biol Chem 278:9219

550. McDonagh AF, Lightner DA (2003) Attention to Stereochemistry. Chem Eng News 81:2

551. Ritter SK (2003) Mind Your *E*'s and *Z*'s. Chem Eng News 81:29

552. Kuenzle CC, Pelloni RR, Weibel MH (1972) A Proposed Novel Structure for the Metal Chelates of Bilirubin. Biochem J 130:1147

553. Kuenzle CC, Weibel MH, Pelloni RR, Hemmerich P (1973) Structure and Conformation of Bilirubin. Opposing Views that Invoke Tautomeric Equilibria, Hydrogen Bonding and a Betaine May Be Reconciled by a Single Resonance Hybrid. Biochem J 133:364

554. Moffitt W, Moscowitz A (1959) Optical Activity in Absorbing Media. J Chem Phys 30:648

555. Moscowitz A (1967) Some Remarks on the Interpretation of Natural and Magnetically Induced Optical Activity Data. Proc Royal Soc London, Series A, Math Phys Sci 297:16

556. Deutsche CW, Lightner DA, Woody RW, Moscowitz A (1969) Optical Activity. Ann Rev Phys Chem 20:407

557. Blauer G, King TS (1968) Optical Rotatory Dispersion of Bilirubin Bound to Serum Albumin. Biochem Biophys Res Commun 31:678

558. Blauer G, King TE (1970) Interactions of Bilirubin with Bovine Serum Albumin in Aqueous Solution. J Biol Chem 245:372

559. Blauer G, Harmatz D, Naparstek A (1970) Circular Dichroism of Bilirubin Human Serum Albumin Complexes in Aqueous Solution. FEBS Lett 9:53

560. Blauer G, Harmatz D, Snir J (1972) Optical Properties of Bilirubin-Serum Albumin Complexes in Aqueous Solution. I. Dependence on pH. Biochim Biophys Acta 278:68

561. Blauer G (1976) Physicochemical Methods in the Study of Bilirubin Binding: Optical Rotation and Circular Dichroism. In: Bergsma D, Blondheim SH (eds), Bilirubin Metabolism in the Newborn (II), Birth Defects. Original Article Series, vol XII, no. 2:134

562. Blauer G, Harmatz D (1972) Optical Properties of Bilirubin-Serum Albumin Complexes in Aqueous Solution. II. Effects of Electrolytes and of Concentration. Biochim Biophys Acta 278:89

563. Blauer G, Lavie E, Silfen J (1977) Relative Affinities of Bilirubin for Serum Albumins from Different Species. Biochim Biophys Acta 492:64

564. Lavie E, Blauer G (1979) Circular Dichroism and Hydrodynamic Measurements of Ternary Protein – Two Ligand Systems: Serum Albumin-Bilirubin-Drug or Alcohol. Arch Biochem Biophys 193:191

565. Blauer G (1986) Complexes of Bilirubin with Proteins. Biochem Biophys Acta 884:602

566. Lightner DA, Gurst JE (2000) Organic Conformation Analysis and Stereochemistry from Circular Dichroism Spectroscopy. Wiley-VCH, New York

567. Djerassi C (1960) Optical Rotatory Dispersion. McGraw-Hill, New York

568. Newman MS, Darlak RS, Tsai L (1967) Optical Properties of Hexahelicene. J Am Chem Soc 89:6191

569. Lightner DA, Hefelfinger DT, Frank GW, Powers TW, Trueblood KN (1971) The Absolute Configuration of Hexahelicene. Nature Phys Sci 232:124

570. Lightner DA, Hefelfinger DT, Frank GW, Powers TW, Trueblood KN (1972) Hexahelicene. The Absolute Configuration. J Am Chem Soc 94:3492

571. Moscowitz A, Krueger WC, Kay IT, Skewes G, Bruckenstein S (1964) On the Origin of the Optical Activity in the Urobilins. Proc Natl Acad Sci USA 52:1190

572. Manitto P, Severini-Ricca GS, Monti D (1974) The Conformation of Bilirubin and Its Esters. Gazz Chim Ital 104:633

573. Knell AJ, Hutchinson DW, Johnson B (1972) Intramolecular Hydrogen Bonds in Bilirubin. Abstracts of the 7th Annual Meeting of the European Association for the Study of the Liver (Arnhem, The Netherlands, 7-9 September 1972), Digestion 6:288

574. Knell AJ, Hancock F, Hutchinson DW (1975) Bilirubin Revisted. In: Bakken AF, Fog J (eds) Proceedings of a Conference on the Metabolism and Chemistry of Bilirubin and related Tetrapyrroles. Pediatric Research Institute Rikshospitalet, Oslo, Norway, p 234

575. Navon G (1976) The Nuclear Magnetic Resonance Spectrum of Bilirubin. In: Bergsma D, Blondheim SH (eds) Bilirubin Metabolism in the Newborn (II) – Birth Defects: Original Article Series, vol XII, no 2, Elsevier, New York

References

576. Kaplan D, Panigel R, Navon G (1977) Carbon-13 NMR Spectrum of Bilirubin. Spectros Lett 10:881
577. Kaplan D, Navon G (1981) Carbon-13 Nuclear Magnetic Resonance Study of the Motional Behaviour of Bilirubin and of Some of Its Derivatives. Org Magn Reson 17:79
578. Kaplan D, Navon G (1981) Nuclear Magnetic Resonance Studies of the Conformation of Bilirubin and Its Derivatives in Solution. J Chem Soc Perkin Trans II:1374
579. Kaplan D, Navon G (1982) Studies of the Conformation of Bilirubin and Its Dimethyl Ester in Dimethyl Sulphoxide Solutions by Nuclear Magnetic Resonance. Biochem J 201:605
580. Kaplan D, Navon G (1983) NMR Spectroscopy of Bilirubin and Its Derivatives. Isr J Chem 23:177
581. Nogales D, Lightner DA (1995) On the Structure of Bilirubin in Solution. J Biol Chem 270:73
582. Dörner T, Knipp B, Lightner DA (1997) Heteronuclear NOE Analysis of Bilirubin Solution Conformation and Intramolecular Hydrogen Bonding. Tetrahedron 53:2697
583. Rohmer T, Matysik J, Mark F (2011) Solvation and Crystal Effects Studied by NMR Spectroscopy and Density Functional Theory. J Phys Chem A 115:11696
584. Metzroth T, Lenhart M, Gauss J (2008) A Quantum Mechanical Investigation of the Geometry and NMR Chemical Shifts of Bilirubin. Appl Magn Reson 33:457
585. Brown SP, Zhu XX, Saalwächter K, Spiess HW (2001) An Investigation of the Hydrogen-bonding Structure in Bilirubin by ^{1}H Double-quantum Magic-angle Spinning Solid-state NMR Spectroscopy. J Amer Chem Soc 123:4275
586. Scholtan N, Gloxhuber C (1966) Die Eiweißbindung von Bilirubin und Bromsulfalein. Arzneim-Forsch (Drug Research) 16:520
587. Schellman JA (1968) Symmetry Rules for Optical Rotation. Acc Chem Res 1:144
588. Lightner DA (1979) Derivatives of Bile Pigments. In: Dolphin D (ed) The Porphyrins, vol VI. Academic Press, Inc, Chap 8
589. Fischer H, Halbach H, Stern A (1935) Über Stercobilin und Seine Optische Aktivität. Liebig's Ann Chem 519:254
590. Schwartz S, Watson CJ (1942) Isolation of a Dextrorotatory Urobilin from Human Fistula Bile. Proc Soc Exper Biol Med 49:641
591. Gray CH, Jones PM, Klyne W, Nicholson DC (1959) Optical Activity of Stercobilin and d-Urobilin. Nature 184:41
592. Woolley PV III, Hunter MJ (1970) Binding and Circular Dichroism Data on Bilirubin-Albumin 0in the Presence of Oleate and Salicylate. Arch Biochem Biophys 148:197
593. Blauer G, Lavie E (1975) The Interaction of Low-molecular Weight Compounds with Serum Albumin as Measured by Optical Methods. Circular Dichroism of Bilirubin-human Serum Albumin Complexes in the Presence of Alcohols in Aqueous Solution. Protein-Ligand Interactions. Walter de Gruyter & Co, Berlin
594. Harada N, Nakanishi K (1983) Circular Dichroic Spectroscopy-Exciton Coupling in Organic Stereochemistry. University Science Books, Mill Valley, California
595. Hansen AaE (2005) Molecular Exciton Approach to Anisotropic Absorption and Circular Dichroism. I. General Formulation. Monatsh Chem 136:253
596. Hansen AaE (2005) Molecular Exciton Approach to Anisotropic Absorption and Circular Dichroism. II. The Partial Optic Axis and Its Application in Molecular Exciton Theory. Monatsh Chem 136:275
597. Blauer G (1983) Optical Activity of Bile Pigments. Isr J Chem 23:201
598. Blauer G, Wagnière G (1975) Conformation of Bilirubin and Biliverdin in Their Complexes with Serum Albumin. J Am Chem Soc 97:1949
599. Manitto P, Monti D (1976) Free-energy Barrier of Conformational Inversion in Bilirubin. J Chem Soc Chem Commun:122
600. Manitto P, Monti D (1973) Acid-catalyzed Addition of Alcohols and Thiols to Bilirubin. Experientia 29:137
601. Navon G, Frank S, Kaplan D (1984) A Nuclear Magnetic Resonance Study of the Conformation and Interconversion Between the Enantiomeric Conformers of Bilirubin and Mesobilirubin in Solution. J Chem Soc Perkin Trans II:1145

602. Falk H, Hofer O (1974) Beiträge zur Chemie der Pyrrolpigmente, 4. Mitt. Die N-H-Tautomerie von Substitutieren Pyrromethenen: Zur Dynamik des Protontransfers aus der Sicht Eines Quantenchemischen Verfahrens (*CNDO*/2). Monatsh Chem 105:995
603. Favini G, Pitea D, Manitto P (1979) Conformational Studies on Bilirubin. I. Intramolecular Energy Map for 3,3'-Dimethyl-dipyrrol-2-ylmethane. Nouveau J Chim 3:299
604. Falk H, Höllbacher G, Hofer O (1979) Konformationsanalyse von Gallenfarbstoffen mit Hilfe von Kraftfeldrechnungen. Monatsh Chem 110:1025
605. Falk H, Höllbacher G, Hofer O (1981) Beiträge zur Chemie der Pyrrolpigmente, 39. Mitt. Ein Kraftfeldmodell zur Konformationsanalytischen Untersuchung von Gallenfarbstoffen. Monatsh Chem 112:391
606. Falk H, Müller N (1981) Beiträge zur Chemie der Pyrrolpigmente, 41. Mitt. Kraftfeldrechnungen an Gallenfarbstoffen: Die Energiehyperfläche Verdinoider Pigmente. Monatsh Chem 112:791
607. Falk H, Müller N (1981) Beiträge zur Chemie der Pyrrolpigmente, 42. Mitt. Kraftfeldrechnung an Gallenfarbstoffen: Die Energiehyperfläche Rubinoide Pigmente. Monatsh Chem 112:1325
608. Falk H, Müller N (1983) Forcefield Calculations on Linear Polypyrrole Systems. Tetrahedron 39:1875
609. Lightner DA, Person RV, Peterson BR, Puzicha G, Pu Y-M, Boiadjiev SE (1991) Conformational Analysis and Circular Dichroism of Bilirubin, the Yellow Pigment of Jaundice. Biomolec Spectrosc II, SPIE 1432:2
610. Person RV, Boiadjiev SE, Peterson BR, Puzicha G, Lightner DA (1991) Circular Dichroism from Exciton Coupling. Conformation Analysis of Bilirubin, the Neurotoxic Yellow Pigment of Jaundice. 4th International Conference on Circular Dichroism Spectroscopy, Bochum, FRG, 9-13 September, p 55
611. Shelver WL, Rosenberg H, Shelver WH (1992) Molecular Conformation of Bilirubin from Semiempirical Molecular Orbital Calculations. Int J Quantum Chem 441:141
612. Barnes JC, Paton JD, Damewood JR Jr, Mislow K (1981) Crystal and Molecular Structure of Diphenylmethane. J Org Chem 46:4975
613. Gust D, Mislow K (1973) Analysis of Isomerization in Compounds Displaying Restricted Rotation of Aryl Groups. J Am Chem Soc 95:1535
614. Falk H, Grubmayr K. Höllbacher G, Hofer O, Leodolter A, Neufingerl E, Ribó JM (1977) Beiträge zur Chemie der Pyrrolpigmente, 18. Mitt. Pyrromethenone – Partialstrukturen von Gallenpigmenten: Struktur und Eigenschaften in Lösung. Monatsh Chem 108:1113
615. Falk H, Vormayr G, Margulies L, Metz S, Mazur Y (1986) A Linear Dichroism Study of Pyrromethene, Pyrrometheneone- and Bilatriene-abc Derivatives. Monatsh Chem 117:849
616. Trull FR, Ma JS, Landen GL, Lightner DA (1983) Hydrogen Bonding of Bilirubin and Pyrromethenones in Solution. Isr J Chem 23:211
617. Nogales DF, Ma J-S, Lightner DA (1993) Self-association of Dipyrrinones Observed by 2D-NOE NMR and Dimerization Constants Calculated from ^{1}H-NMR Chemical Shifts. Tetrahedron 49:2361
618. Person RV (1993) Conformational Analysis of Bilirubin and Its Analogues. Ph.D. Dissertation, University of Nevada, Reno
619. Person RV, Peterson BR, Lightner DA (1994) Bilirubin Conformational Analysis and Circular Dichroism. J Am Chem Soc 116:42
620. Müller N, Falk A (1992) "Ball and Stick" Computer Program for the Macintosh. Cherwell Scientific, Oxford Science Park, Oxford OX4 4GA, UK
621. Alagona G, Ghio C, Monti S (2000) Continuum Solvent Effects on Various Isomers of Bilirubin. Phys Chem Chem Phys 2:4884
622. Alagona G, Ghio C, Agresti A, Pratesi R (1998) *Ab Initio* Relative Stability of a Few Conformers of Bilirubin *in vacuo* and in Aqueous Solution (PCM). Int J Quantum Chem 70:395
623. Harmatz D, Blauer G (1975) Optical Properties of Bilirubin-Serum Albumin Complexes in Aqueous Solution. A Comparison Among Albumins from Different Species. Archiv Biochem Biophys 170:375

References

759

624. Beaven GH, d'Albis A, Gratzer WB (1973) The Interaction of Bilirubin with Human Serum Albumin. Eur J Biochem 33:500
625. Woolley PV III, Margaret J, Arias IM (1976) Bilirubin and Biliverdin Binding to Rat Y Protein (Ligandin). Biochim Biophys Acta, Protein Struct 446:115
626. van der Eijke JM, Nolte RJM, Richters VEM, Drenth W (1980) Polymeric Model System for Protein-Bilirubin Interaction. Biopolymers 19:445
627. Marr-Leisy D, Lahiri K, Balaram P (1985) Bilirubin Binding to Polypeptides and Chiral Amines. Int J Peptide Protein Res 25:290
628. Brodersen R (1962) Physical Chemistry of Bilirubin: Binding to Macromolecules and Membranes. In: Heirwegh KPM, Brown SB (eds) Bilirubin, vol I. CRC Press Inc, Boca Raton, Florida, p 75
629. Brodersen R (1986) Aqueous Solubility, Albumin Binding, and Tissue Distribution of Bilirubin. In: Ostrow JD (ed) Bile Pigments and Jaundice. Marcel Dekker, New York, p 157
630. Wennberg RP (1975) The Pathogenesis of Kernicterus: Factors Influencing the Distribution of Bilirubin with Albumin and Tissue. In: Bakken AF, Fog J (eds) Proceedings of a Conference on the Metabolism and Chemistry of Bilirubin and Related Tetrapyrroles. Pediatric Research Institute, Rikshospitalet, Oslo, Norway
631. Jacobsen C (1978) Lysine Residue 240 of Human Serum Albumin Is Involved in High-affinity Binding of Bilirubin. Biochem J 171:453
632. Perrin H, Wilsey M (1971) The Induced Optical Activity of Bilirubin in the Presence of Sodium Deoxycholate. J Chem Soc Chem Commun:769
633. Reisinger M, Lightner DA (1985) Bilirubin Conformational Enantiomer Selection in Chiral Sodium Deoxycholate Chiral Micelles. J Inclusion Phenomena 3:479
634. LeBas G, DeRango C, Tsoucaris G (1982) Chiral Conformation of Bilirubin, Biliverdin and Benzil in Association with Cyclodextrin. In: Szejtli, J (ed) Proceedings of the 1st International Symposium on Cyclodextrins. D Reidel, Dordrecht, The Netherlands, p 245
635. Lightner DA, Gawroński JK, Gawrońska K (1985) Bilirubin Enantiomer Selection by Cyclodextrins. J Am Chem Soc 107:2456
636. Kano K, Tsujino N, Kim M (1992) Mechanisms for Steroid-induced Conformational Enantiomerism of Bilirubin in Protic Solvents. J Chem Soc Perkin Trans II:1747
637. Kano K, Arimoto S, Ishimura T (1995) Conformational Enantiomerism of Bilirubin and Pamoic Acid Induced by Protonated Aminocyclodextrins. J Chem Soc Perkin Trans II:1661
638. Kano K, Yoshiyasu K, Yasuoka H, Hata S, Hashimoto S (1992) Chiral Recognition by Cyclic Oligosaccharides. Enantioselective Complexation of Bilirubin with β-Cyclodextrin through Hydrogen Bonding in Water. J Chem Soc Perkin Trans II:1265
639. Kano K, Yoshiyasu K, Hashimoto K (1990) Molecular Recognition by Saccharides. Asymmetric Complexation Between Bilirubin and Nucleosides. Chem Lett:21
640. Lightner DA, Reisinger M, Landen GL (1986) On the Structure of Albumin Bound Bilirubin. J Biol Chem 261:6034
641. Lightner DA, Wijekoon WMD, Zhang MH (1988) Understanding Bilirubin Conformation and Binding. Circular Dichroism of Human Serum Albumin with Bilirubin and Its Esters. J Biol Chem 263:16669
642. Lightner DA, An JY, Pu Y-M (1987) Circular Dichroism of Bilirubin-Amine Heteroassociation Complexes. Tetrahedron 43:4287
643. Lightner DA, An JY, Pu YM (1988) Circular Dichroism of Bilirubin-Amine Association Complexes. Insights into Bilirubin-albumin Binding. Arch Biochem Biophys 262:543
644. Pu YM, Lightner DA (1989) Intramolecular Exciton Coupling and Induced Circular Dichroism from Bilirubin-Ephedrine Heteroassociation Complexes. Stereochemical Models for Protein Binding. Croatica Chem Acta 62:301
645. Lightner DA, Gawroński JK, Wijekoon WMD (1987) Complementarity and Chiral Recognition: Enantioselective Complexation of Bilirubin. J Am Chem Soc 109:6354
646. Bouvier M, Brown GR (1988) The Induced Optical Activity of Bound Bilirubin. A Review. Can J Spectrosc 33:83
647. Turner EE, Harns MM (1948) Asymmetric Transformation and Asymmetric Induction. Quarterly Rev Chem Soc 1:299

648. Pirkle WH, Hoover DJ (1982) NMR Chiral Solvating Agents. In: Allinger NH, Eliel EL, Wilen SH (eds) Topics in Stereochemistry, vol 13. Wiley Interscience, New York, p 263

649. Lightner DA, Ma J-S (1986) Intramolecular Hydrogen Bonding in Bilirubin bis-*tetra-n-*Butylammonium Salts. Spectrosc Lett 19:311

650. Lightner DA, An J-Y (1987) Circular Dichroism and Conformation of Bilirubin bis-*N,N,N-*Trimethyl-α-phenethylammonium Salts. Spectrosc Lett 20:491

651. Pu YM, Lightner DA (1991) On the Conformation of Bilirubin Dianion. Tetrahedron 47:6163

652. Lightner DA, Trull FR (1983) Intermolecular Hydrogen Bonding and Dimeric Structures in Bilirubin Esters from ¹H-NMR Spectroscopy. Spectrosc Lett 16:785

653. Zhang M-H, Lightner DA (1987) Conformation of Bilirubin Esters from Circular Dichroism Spectroscopy. Tetrahedron Lett 28:4033

654. Brower JO, Huggins MT, Boiadjiev SE, Lightner DA (2000) On the Molecularity of Bilirubins and Their Esters and Anions in Chloroform Solution. Monatsh Chem 131:1047

655. Ghosh B, Catalano VJ, Lightner DA (2004) Crystal Structure and Conformation of a Bilirubin Ester. Monatsh Chem 135:1503

656. Trull FR, Ma JS, Landen GL, Lightner DA (1983) Hydrogen Bonding of Bilirubin and Pyrromethenones in Solution. Isr J Chem 23:211

657. Lightner DA, Reisinger M, Wijekoon WMD (1987) Optically Active Pyrromethenone Amides. Exciton Coupling in Hydrogen-bonded Dimers. J Org Chem 52:5391

658. Byun YS, Lightner DA (1991) Exciton Coupling from Dipyrrinone Chromophores. J Org Chem 56:6027

659. Nogales DF, Ma J-S, Lightner DA (1993) Self-association of Dipyrrinones Observed by 2D-NOE NMR and Dimerization Constants Calculated from ¹H-NMR Chemical Shifts. Tetrahedron 49:2361

660. Boiadjiev SE, Anstine DT, Lightner DA (1995) Intermolecular Hydrogen Bonding in π Facial Dipyrrinone Dimers as Molecular Capsules. J Am Chem Soc 117:8727

661. Boiadjiev SE, Anstine DT, Maverick E, Lightner DA (1995) Hydrogen Bonding and π-Stacking in Dipyrrinone Acid Dimers of Xanthobilirubic Acids and Chiral Analogs. Tetrahedron: Asymm 6:2253

662. Huggins MT, Lightner DA (2001) Hydrogen-bonded Dimers in Dipyrrinones and Acyl Dipyrrinones. Monatsh Chem 132:203

663. Rebek J Jr (1987) Model Studies in Molecular Recognition. Science 235:1478

664. Rebek J Jr (2009) Molecular Behavior in Small Spaces. Acc Chem Res 42:1660

665. Rebek J Jr (1999) Asymmetric Phenomena in Studies of Encapsulation and Assembly. In: Palyi G, Zucchi C, Caglioti L (eds) Advances in Biochirality. Elsevier Science, Amsterdam, p 315

666. Gawroński JK, Polonski T, Lightner DA (1990) Sulfoxides as Chiral Complexation Agents. Conformational Enantiomer Resolution and Induced Circular Dichroism of Bilirubins. Tetrahedron 46:8053

667. Trull FR, Shrout DP, Lightner DA (1992) Chiral Recognition by Sulfoxides. Induced Circular Dichroism from Symmetric Mesobilirubin Analogs. Tetrahedron 48:8189

668. Murakami Y, Hayashida O, Nagai Y (1994) Enantioselective Discrimination by Cage-type Cyclophanes Bearing Chiral Binding Sites in Aqueous Media. J Am Chem Soc 116:2611

669. Hayashida O, Matsuura S, Murakami Y (1994) Specific Molecular Recognition by Cage-type Cyclophanes Having a Helically Twisted and Cylindrical Internal Cavity. Tetrahedron 50:13701

670. Chung TW, Cho KY, Nah JW, Akaike T, Cho CS (2002) Chiral Recognition of Bilirubin by Polymeric Nanoparticles. Langmuir 18:6462

671. Yashima E, Fukaya H, Sahavattanapong P, Okamoto Y (1996) Chiral Discrimination of Bilirubin by Cellulose Trisphenylcarbamate Derivatives in Chloroform and in Film. Enantiomer 1:193

672. Bombelli C, Bernardini C, Elemento G, Mancini G, Sorrenti A, Villani C (2008) Concentration as the Switch for Chiral Recognition in Biomembrane Models. J Am Chem Soc 130:2732

References

673. Goncharova I, Urbanova M (2007) Bile Pigment Complexes with Cyclodextrins: Electronic and Vibrational Circular Dichroism Study. Tetrahedron: Asymm 18:2061
674. Julinek O, Goncharova I, Urbanova M (2008) Chiral Memory and Self-replication of Porphyrin and Bilirubin Aggregates Formed on Polypeptide Matrices. Supramol Chem 20:643
675. Goncharova I, Urbanova M (2008) Stereoselective Bile Pigment Binding to Polypeptides and Albumins: A Circular Dichroism Study. Anal Bioanal Chem 392:1355
676. Goncharova I, Urbanova M (2009) Vibrational and Electronic Circular Dichroism Study of Bile Pigments: Complexes of Bilirubin and Biliverdin with Metals. Anal Biochem 392:28
677. Berova N, Povalarapu P, Nakanishi K, Woody RW (eds) (2011) Comprehensive Chiroptical Spectroscopy. John Wiley & Sons Inc, New York
678. Kasha M, Rawls HR, El-Bayoumi MA (1965) The Exciton Model in Molecular Spectroscopy. Pure Appl Chem 11:371
679. McRae EG, Kasha M (1964) The Molecular Exciton Model. In: Augenstein L, Mason R, Rosenberg B (eds) Physical Processes in Radiation Biology. Academic Press, New York, p 23
680. Kasha M (1963) Energy Transfer Mechanisms and Molecular Exciton Model for Molecular Aggregates. Radiation Res 20:55
681. Kasha M, El-Bayoumi MA, Rhodes W (1961) Excited States of Nitrogen Base Pairs and Polynucleotides. J Chim Phys Physicochim Biol 58:916
682. Berova N, Nakanishi K (2000) Exciton Chirality Method: Principles and Applications. In: Berova N, Nakanishi K, Woody RW (eds) Circular Dichroism. Principles and Applications, 2nd ed. John Wiley & Sons Inc, New York, p 337
683. Lightner DA (ed) (2005) Exciton Chirality: Fundamentals and Frontiers. Monatsh Chem 136, issue 3
684. Boiadjiev SE, Lightner DA (2005) Exciton Chirality. (A) Origins of and (B) Applications from Strongly Fluorescent Dipyrrinone Chromophores. Monatsh Chem 136:489
685. Matile S, Berova N, Nakanishi K, Fleischhauer J, Woody RW (1996) Structural Studies by Exciton Coupled Circular Dichroism Over a Large Distance: Porphyrin Derivatives of Steroids, Dimeric Steroids, and Brevitoxin. J Am Chem Soc 118:5198
686. Wellman KM, Laur PHA, Briggs WS, Moscowitz A, Djerassi C (1965) Optical Rotatory Dispersion Studies. XCIX. Superposed Multiple *Cotton* Effects of Saturated Ketones and Their Significance in the Circular Dichroism Measurement of (–)-Menthone. J Am Chem Soc 87:66
687. Boiadjiev SE, Lightner DA (1999) Optical Activity and Stereochemistry of Linear Oligopyrroles and Bile Pigments. Tetrahedron: Asymm 19:607
688. Puzicha G, Pu Y, Lightner DA (1991) Allosteric Regulation of Conformational Enantiomerism. Bilirubin. J Am Chem Soc 113:3585
689. Boiadjiev SE, Person RV, Puzicha G, Knobler C, Maverick E. Trueblood KN, Lightner DA (1992) Absolute Configuration of Bilirubin Conformational Enantiomers. J Am Chem Soc 114:10123
690. Plieninger H, El-Barkawi F, Ehl K, Kohler R, McDonagh AF (1972) Neue Synthese und ^{14}C-Markierung von Bilirubin-IXα. Liebig's Ann Chem 758:195
691. Corwin AH, Coolidge EC (1952) A Stepwise Porphyrin Synthesis. J Am Chem Soc 74:5196
692. Johnson AW, Kay IT (1961) The Formation of Porphyrins by the Cyclization of Bilenes. J Chem Soc:2418
693. Harris RLN, Johnson AW, Kay IT (1966) A Stepwise Synthesis of Unsymmetrical Porphyrins. J Chem Soc C:22
694. Clezy PS, Liepa AJ (1969) Porphyrins from Electronegatively-substituted Bilenes. J Chem Soc D:767
695. Clezy PS, Salek A (1996) The Chemistry of Pyrrolic Compounds. LXXI. Synthesis of the Ring C and the Ring D Demethyl Analogs of Deoxophylloerythroetioporphyrin (dpep). Aust J Chem 49:265

696. Jackson AH, Smith KM (1973) The Total Synthesis of Pyrrole Pigments. In: ApSimon J (ed) The Total Synthesis of Natural Products. Wiley-Interscience, New York, p 143
697. Hudson MF, Smith KM (1975) Bile Pigments. Chem Soc Rev 4:363
698. Kadish K, Smith KM, Guilard R (eds) (1999) The Porphyrin Handbook, vol I. Academic Press, New York
699. Kadish K, Smith KM, Guilard R (eds) (2010) Handbook of Porphyrin Science. World Scientific Publishing Co, Hackensack, New Jersey
700. Plieninger H, Decker M (1956) Eine Neue Synthese für Pyrrolone, Insbesondere für „Isooxyopsopyrrol" und „Isoopsopyrrol-carbonsäure". Liebig's Ann Chem 598:198
701. Plieninger H, Bühler W (1959) Eine Synthese für α,α'-Unsubstituierten Pyrrole. Angew Chem 71:163
702. Plieninger H, Lerch U (1966) Totalsynthese Zweier Racemischer Stercobiline-IXα. Liebig's Ann Chem 698:196
703. Plieninger H (1968) Synthesen in der Pyrrolon- und Pyrrolreihe. Angew Chem 80:242
704. Plieninger H, Steinsträsser R (1969) Synthese Zweier Vinyl-substituierter Urobiline IXα. Liebig's Ann Chem 723:149
705. Plieninger H, Ruppert J (1970) Synthese des (–)-Stercobilins IXα („nat." Stercobilin) und Anderer Optisch Aktiver Stercobiline. Liebig's Ann Chem 735: 43
706. Plieninger H, Ehl K, Tapia A (1970) Gallenfarbstoffsynthesen IV. Die Synthese Optisch Aktiver Urobiline. Liebig's Ann Chem 736:62
707. Plieninger H, Ehl K, Klinga K (1971) Gallenfarbstoffsynthesen V. Über Mesobilivioline. Liebig's Ann Chem 743:112
708. Brockmann H Jr, Knobloch G, Plieninger H, Ehl K, Ruppert J, Moscowitz A, Watson CJ (1971) Absolute Configuration of Natural (–)-Stercobilin and other Urobilinoid Compounds. Proc Natl Acad Sci USA 68:2141
709. Preuss I, Plieninger H (1982) Ein Neuer Zugang zu 3,4-Dihydropyrromethenonen als Bausteine für Phytochrom-Modell Verbindungen. Tetrahedron Lett 23:43
710. Plieninger H, Preuss I (1983) Ein Neuer Synthetischer Weg zu Phytochromobilin-ähnlichen Gallenfarbstoffen. Liebig's Ann Chem:585
711. Böhm H, Gottschall K, Plieninger H (1984) Darstellung Konfigurationsisomerer 3,4-Dihydrobilinsäurederivate als Bausteine für Gallenfarbstoffsynthesen. Liebig's Ann Chem:1441
712. Gottschall K, Plieninger H (1984) Darstellung des rac-Stercobilins IXα mit der Konfiguration des Naturproduktes Sowie Einer Reihe von Urobilinen, Halbstercobilinen und Stercobilinen Eindeutiger Konstitution und Konfiguration. Liebig's Ann Chem:1454
713. Gossauer A, Plieninger H (1979) Synthesis, Purification and Characterization of Bile Pigments and Related Compounds. In: Dolphin D (ed) The Porphyrins, vol VI, Biochemistry, Part A. Academic Press, Inc, New York, p 585
714. von Dobeneck H, Brunner E, Sommer U (1977) Das Verhalten von 9-Brom-11H-dipyrromethen-1(10H)-onen in Essigsäure und Über 5-(1-Oxo-1,10-dihydro-11H-dipyrromethen-9-yl)-1H-dipyrromethen-1,9(11H)-dione. Liebig's Ann Chem:1435
715. Gossauer A, Miehe D (1974) Totalsynthese des Mesobilirhodin-dimethylesters. Liebig's Ann Chem:352
716. Gossauer A, Kühne G (1977) Stereospezifische Totalsynthesen Diastereomerer Mesobilirhodine und Isomesobilirhodine. Liebig's Ann Chem:664
717. Gossauer A, Hinze R-P (1978) An Improved Chemical Synthesis of Racemic Phycocyanobilin Dimethyl Ester. J Org Chem 43:283
718. Gossauer A, Weller J-P (1978) Totalsynthese des (+)(4R,16R)- und (–)(4R,16S)-[18-Vinyl] mesourobilin IXα-dimethyl esters. Ber Dtsch Chem Ges 111:486
719. Gossauer A, Klahr E (1979) Totalsynthese des racem. Phycoerythrobilin-dimethylesters. Ber Dtsch Chem Ges 112:2243
720. Gossauer A (1983) Synthetic Methods in Bile Pigment Chemistry. Isr J Chem 23:167
721. Valasinas A, Diaz L, Frydman B, Friedmann HC (1985) Total Synthesis of Urobiliverdin Isomers. Identification of Bactobilin as Urobiliverdin I. J Org Chem 50:2398

References

722. Awruch J, Frydman B (1986) The Total Synthesis of Biliverdins of Biological Interest. Tetrahedron 42:4137
723. Iturraspe JC, Bari S, Frydman B (1989) Total Synthesis of "Extended" Biliverdins. The Relation Between Their Conformation and Their Spectroscopic Properties. J Am Chem Soc 111:1525
724. Iturraspe J, Bari S, Frydman B (1995) Synthesis of Biliverdins with Stable Extended Conformations. Part I. Tetrahedron 51:2243
725. Bari S, Iturraspe J, Frydman B (1995) Synthesis of Biliverdins with Stable Extended Conformations. Part II. Tetrahedron 51:2255
726. Falk H, Grubmayr K (1977) Eine Neue Synthese von C_{2V}-Symmetrisch Substituierten a,b,c,-Bilatrienen. Synthesis:614
727. Boiadjiev SE, Lightner DA (1994) Synthetic Strategies for Understanding Bilirubin Stereochemistry. Synlett:777
728. Boiadjiev SE, Lightner DA (2001) An Enantiomerically Pure Bilirubin. Absolute Configuration of $(\alpha R,\alpha'R)$-Dimethylmesobilirubin-XIIIα. Tetrahedron: Asymm 12:2551
729. Boiadjiev SE, Lightner DA (2000) Steric Size in Conformational Analysis. Steric Compression Analyzed by Circular Dichroism Spectroscopy. J Am Chem Soc 122:11328
730. Boiadjiev SE, Lightner DA (2000) Relative Steric Size of SCH_3, OCH_3 and CH_3 Groups from Circular Dichroism Measurements. Chirality 12:204
731. McDonagh AF, Assisi F (1972) The Ready Isomerization of Bilirubin IX-α in Aqueous Solution. Biochem J 129:797
732. McDonagh AF, Assisi F (1972) Direct Evidence for Acid-catalyzed Isomeric Scrambling of Bilirubin-IXα. J Chem Soc Chem Commun:117
733. McDonagh AF (1995) Thermal and Photochemical Reactions of Bilirubin. Ann NY Acad Sci 244:553
734. Bonnett R, McDonagh AF (1970) The Isomeric Heterogeneity of Biliverdin Dimethyl Ester Derived from Bilirubin. Ann NY Acad Sci 244:238
735. Bonnett R, Buckley DG, Hamzetash D, McDonagh AF (1983) Pyrrole Exchange Reactions in the Bilirubin Series. Isr J Chem 23:173
736. Trull FR, Rodríguez M, Lightner DA (1993) A General Method for Synthesizing Unsymmetric from Symmetric Bile Pigments. Synth Commun 23:2771
737. Trull FR, Lightner DA (1988) Synthesis of Unsymmetric Bile Pigments: Mesobilirubin-VIIIα, 17-Desvinyl-17-ethyl-bilirubin-VIIIα and 12-Despropionic Acid-12-ethyl-mesobilirubin-XIIIα. J Heterocycl Chem 25:1227
738. Abbate S, Lebon F, Longhi G, Boiadjiev SE, Lightner DA (2012) Vibrational and Electronic Circular Dichroism of Dimethyl Mesobilirubins-XIIIα. J Phys Chem B 116:5628
739. Pu Y-M, Lightner DA (1991) On the Conformation of Bilirubin Dianion. Tetrahedron 47:6163
740. Pu Y-M, McDonagh AF, Lightner DA (1993) Inversion of Circular Dichroism with a Drop of Chloroform. Unusual Chiroptical Properties of Aqueous Bilirubin-Albumin Solutions. J Am Chem Soc 115:377
741. Pu Y-M, McDonagh AF, Lightner DA (1992) Effect of Volatile Anesthetics on the Circular Dichroism of Bilirubin Bound to Human Serum Albumin. Experientia 48:246
742. Boiadjiev SE, Lightner DA (2000) Chirality Inversion in a Molecular Exciton. J Am Chem Soc 122:378
743. Maisels MJ, McDonagh AF (2008) Phototherapy for Neonatal Jaundice. New Engl J Med 358:920
744. Maisels MJ, Watchko JF (eds) (2001) Neonatal Jaundice. Harwood Academic Publications, Amsterdam
745. Dennery PA, Seidman DS, Stevenson DK (2001) Neonatal Hyperbilirubinemia. New Engl J Med 344:581
746. Chowdury JR, Wolkoff AW, Chowdury NR, Arias IM (2001) Hereditary Jaundice and Disorders or Bilirubin Metabolism. In: Scriver CR, Beaudet AL, Sly WS, Valle D (eds) The

Metabolic and Molecular Bases of Inherited Disease, vol II. McGraw-Hill, Inc, New York, p 3063

747. Ennever J (1988) Phototherapy for Neonatal Jaundice. Photochem Photobiol 47:871

748. Levine R, Maisels MJ (1983) Hyperbilirubinemia in the Newborn. Report of the 85th Ross Conference on Pediatric Research, Ross Laboratory Publishers

749. Brown AK, McDonagh AF (1980) Phototherapy for Neonatal Hyperbilirubinemia: Efficacy, Mechanism and Toxicity. In: Barnes LA (ed) Advances in Pediatrics, vol 27, Year Book Medical Publishers, Inc, New York

750. Cremer RJ, Perryman PW, Richards DH (1958) Influence of Light on the Hyperbilirubinaemia of Infants. Lancet 271:1094

751. Dobbs RH, Cremer RJ (1975) Phototheraphy. Arch Dis Childhood 50:833

752. Bergsma D, Blondheim SH (eds) (1976) Bilirubin Metabolism in the Newborn (II). National Foundation – March of Dimes, New York

753. Neonatal Phototherapy (1982) Bibliography. Olympic Medical Corp, Seattle, Washington

754. Blondheim SH, Lathrop D, Zabriskie J (1960) Diffusible Bilirubin(oid) Produced by Exposure of Jaundiced Serum to Light. Gastroenterology 38:798

755. Blondheim SH, Lathrop D, Zabriskie J (1962) Effect of Light on the Absorption Spectrum of Jaundiced Serum. J Clin Lab Med 60:31

756. Blondheim SH, Kaufmann NA (1962) The Effect of Light on the Direct Diazo Reaction of Conjugated Bilirubin. Pediatrics 65:659

757. Ostrow JD, Hammaker L, Schmid R (1961) The Preparation of Crystalline Bilirubin-C^{14}. J Clin Invest 40:1442

758. Schmid R, Hammaker L (1963) Metabolism and Disposition of C^{14}-Bilirubin in Congenital Nonhemolytic Jaundice. J Clin Invest 42:1720

759. Callahan EW Jr, Thaler MM, Karon M, Bauer K. Schmid R (1970) Phototherapy of Severe Unconjugated Hyperbilirubinemia: Formation and Removal of Labeled Bilirubin Derivatives. Pediatrics 46:841

760. Ostrow JD, Branham RV (1970) Photodecay of Bilirubin *in vitro* and in the Jaundiced (Gunn) Rat. In: Bergsma D, Hsia DY-Y, Jackson C (eds). Bilirubin Metabolism in the Newborn. Birth Defects: Original Article Series vol VI, no 2. Williams and Wilkins Co, Baltimore, p 93

761. Davies RE, Keohane SJ (1970) Some Aspects of the Photochemistry of Bilirubin. Boll Chim Farm 109:589

762. McDonagh AF, Palma LA, Lightner DA (1982) Phototherapy for Neonatal Jaundice. Stereospecific and Regioselective Photoisomerization of Bilirubin Bound to Human Serum Albumin and NMR Characterization of Intramolecularly Cyclized Photoproducts. J Am Chem Soc 104:6867

763. Ostrow JD, Berry CS, Knodell RG, Zarembo JE (1976) Effect of Phototherapy on Bilirubin Excretion in Man and the Rat. In: Bergsma D, Blondheim SH (eds) Bilirubin Metabolism in the Newborn (II). Birth Defects: Original Article Series, vol XII, no 2. American Elsevier, New York, p 81

764. Kapitulnik J, Blondheim SH, Grunfeld A, Kaufmann NA (1973) Photocomposition of Bilirubin: Ultrafilterable Derivatives. Clin Chim Acta 47:159

765. Kapitulnik J, Kaufmann NA, Blondheim SH (1976) Characteristics of a Photoproduct of Bilirubin Vound *in vivo* and *in vitro*, and Its Effect on Bilirubin Binding Affinity of Serum. In: Berk PD, Berlin NI (eds) Chemistry and Physiology of Bile Pigments (Fogarty International Proceedings, no 35). U.S. Government Printing Office, Washington, DC, p 53

766. Ostrow JD, Berry CS, Zarembo JE (1974) Studies on the Mechanism of Phototherapy in the Congenitally Jaundiced Rat. In: Odell GB, Schaffer R, Simopoulos AP (eds) Phototherapy in the Newborn: An Overview. Nat Acad Sci USA, Washington, DC, p 74

767. Ostrow JD (1975) Some Aspects of Bilirubin Photometabolism *in vivo*. In: Bakken AE, Fog J (eds) Metabolism and Chemistry of Bilirubin and Related Tetrapyrroles. Pediatric Research Institute, Rikshopitalet, Oslo, p 199

References

768. Davies RE, Keohane SJ (1973) Early Changes in Light-irradiated Solutions of Bilirubin: A Spectrophotometric Analysis. Photochem Photobiol 17:303
769. Lightner DA, Wooldridge TA, McDonagh AF (1979) Photobilirubin: An Early Bilirubin Photoproduct Detected by Absorbance Difference Spectroscopy. Proc Natl Acad Sci USA 76:29
770. Ostrow JD (1967) Photo-oxidative Derivatives of [^{14}C] Bilirubin and Their Excretion by the Gunn Rat. In: Bouchier IAD, Billing BH (eds) Bilirubin Metabolism. Blackwell Scientific Publications, Oxford, UK
771. Ostrow JD, Branham RV (1970) Photodecomposition of Bilirubin and Biliverdin. Gastroenterology 58:15
772. McDonagh AF (1972) Evidence for Single-oxygen Quenching by Biliverdin-IXα Dimethyl Ester and Its Relevance to Photo-oxidation. Biochem Biophys Res Commun 48:408
773. Ostrow JD (1971) Photocatabolism of Labeled Bilirubin in the Congenitally Jaundiced Gunn Rat. J Clin Invest 50:707
774. Ostrow JD (1976) Effects of Phototherapy on Organic Ion Excretion. 4th Annual Meeting, American Society for Photobiology, February 16-20, 1976, Denver, Abstract WAM-B3, p 60
775. Onishi S, Fujikake M, Ogawa Y, Ogawa J (1976) Photodegradation Products of Bilirubin Studied by High Pressure Liquid Chromatography, Gel Permeation Chromatography, Nuclear Magnetic Resonance and Mass Spectrometry. In: Bergsma D, Blondheim SH (eds) Bilirubin Metabolism in the Newborn (II). American Elsevier, New York, p 41
776. Foote CS, Ching T-Y (1975) Chemistry of Singlet Oxygen. XXI Kinetics of Bilirubin Photooxygenation. J Am Chem Soc 97:6209
777. Lightner DA (1977) Products of Bilirubin Photodegradation. In: Berk PD, Berlin NI (eds) Chemistry and Physiology of Bile Pigments. Proceedings of the Fogarty International Symposium, no 35, p 93
778. McDonagh AF (1971) The Role of Singlet Oxygen in Bilirubin Photo-oxidation. Biochem Biophys Res Commun 44:1306
779. Lightner DA (1974) In vitro Photooxidation Products of Bilirubin. In: Odell GB, Schaffer R, Simopoulos AP (eds) Phototherapy in the Newborn: An Overview. Nat Acad Sci USA, Washington DC, p 34
780. Berry CS, Zarembo JE, Ostrow JD (1972) Evidence for Conversion of Bilirubin to Dihydroxyl Derivatives in the Gunn Rat. Biochem Biophys Res Commun 49:1366
781. Quistad GB (1972) The Photooxidation of Alkylpyrroles and Bilirubin. Ph.D. Dissertation, University of California, Los Angeles
782. Lightner DA, Quistad GB (1972) Methylvinylmaleimide from Bilirubin Photooxidation. Science 175:324
783. Lightner DA, Quistad GB (1972) Hematinic Acid and Propentdyopents from Bilirubin Photooxidation in vitro. Fed Europ Biochem Soc (FEBS) Lett 25:94
784. Lightner DA, Quistad GB (1972) Imide Products from Photo-oxidation of Bilirubin and Mesobilirubin. Nature, New Biol 236:203
785. Lightner DA, Crandall DC (1973) The Photooxygenation of Biliverdin. Tetrahedron Lett:953
786. Bonnett R, Stewart JCM (1972) Singlet Oxygen in the Photo-oxidation of Bilirubin in Hydroxylic Solvents. Biochem J 130:895
787. Bonnett R, Stewart JCM (1972) Photo-oxidation of Bilirubin in Hydroxylic Solvents. Propentdyopent Adducts as Major Products. Chem Commun:596
788. Bonnett R, Stewart JCM (1975) Photooxidation of Bilirubin in Hydroxylic Solvents. J Chem Soc Perkin Trans I:224
789. Chong R, Clezy PS, Liepa AJ, Nichol AW (1969) The Chemistry of Pyrrolic Compounds VII. Synthesis of 5,5'-Diformyldipyrrylmethanes. Aust J Chem 22:229
790. Lightner DA, Crandall DC, Gertler S, Quistad GB (1973) On the Formation of Biliverdin During Photooxygenation of Bilirubin in vitro. FEBS Lett 30:309

791. Burke MJ, Pratt DC, Moscowitz A (1972) Low-temperature Absorption and Circular Dichroism Studies of Phytochrome. Biochemistry 11:4025

792. Wooldridge TA, Lightner DA (1978) Separation of the III-α, IX-α and XIII-α Isomers of Bilirubin and Bilirubin Dimethyl Ester by High Performance Liquid Chromatography. J Liq Chromat 1:653

793. DiCesare JL, Vandemark FL (1981) High Resolution Preparative LC Separation of Bilirubin Isomers. Chromat Newsletter 9:7

794. Manitto P, Monti D (1981) A Simple Procedure for Preparing Bilirubin-XIIIα. Syn Commun 11:811

795. Ma J-S, Lightner DA (1984) Facile Preparation of Symmetric Bilirubins-IIIα and XIIIα from IXα. J Heterocycl Chem 21:1005

796. Reisinger M, Lightner DA (1985) Large Scale Synthesis of Bilirubin-XIIIα from IXα. J Heterocycl Chem 22:1221

797. Boiadjiev SE, Lightner DA (1997) Stereogenic Competition in Bilirubin Conformational Enantiomerism. Tetrahedron: Asymm 8:2115

798. McDonagh AF (1976) Photochemistry and Photometabolism of Bilirubin-IXα. In: Bergsma D, Blondheim SH (eds) Bilirubin Metabolism in the Newborn (II). Birth Defects: Original Article Series, vol. XII, no 2. American Elsevier, New York, p 30

799. Lund HT, Jacobsen J (1972) Influence of Phototherapy on Unconjugated Bilirubin in Duodenal Bile of Newborn Infants with Hyperbilirubinemia. Acta Paediatrica Scand 61: 693

800. McDonagh AF (1975) Phototherapy and Hyperbilirubinaemia. The Lancet, February 8, 339

801. McDonagh AF (1974) Photochemistry and Photometabolism of Bilirubin. In: Odell GB, Schaffer R, Simopoulos AP (eds) Phototherapy in the Newborn: An Overiew. Nat Acad Sci USA, Washington, DC, p 56

802. McDonagh AF (1974) Photochemistry and Photometabolism of Bilirubin. In: Brown AK, Showacre J (eds) Phototherapy for Neonatal Hyperbilirubinemia, DHEW Publication no (NIH) 76-1075, p 171

803. McDonagh AF, Palma LA (1977) Mechanism of Bilirubin Photodegradation: Role of Singlet Oxygen. In: Berk PD, Berlin NI (eds) Chemistry and Physiology of Bile Pigments. (Fogarty International Proceedings, No 35) U.S. Government Printing Office, Washington, DC, p 81

804. McDonagh AF (1976) Phototherapy of Neonatal Jaundice. Biochem Soc Trans 4:219

805. Lightner DA (1977) On the Photochemistry of Bilirubin and Related Pyrroles. Photochem Photobiol 26:427

806. Onishi S, Yamakawa T (1970) Influences to Neonatal Bilirubin Metabolism by Phototherapy. Acta Paediatrica Jpn 12:36 (English abstracts)

807. Lund HT, Jacobsen J (1974) Influence of Phototherapy on the Biliary Excretion Pattern in Newborn Infants with Hyperbilirubinemia. J Pediatr 85:262

808. Garbagnati E, Manitto P (1973) A New Class of Bilirubin Photo-derivatives Obtained *in vitro* and Their Possible Formation in Jaundiced Infants. J Pediatr 83:109

809. Ostrow JD, Knodell RG, Cheney HG, Berry CS (1974) Effects of Phototherapy on Hepatic Excretory Function in Man and the Rat. In: Brown AK, Showacre J (eds) Phototheraphy for Neonatal Hyperbilirubinemia. DHEW Publication no (NIH) 76-1075, p 151

810. Lightner DA, Linnane WP III, Ahlfors CE (1984) Bilirubin Photooxidation Products in the Urine of Jaundiced Neonates Receiving Phototherapy. Pediatr Res 18: 696

811. Lightner DA, Linnane WP III, Ahlfors CE (1984) Photooxygenation Products of Bilirubin in the Urine of Jaundiced Phototherapy Neonates. In: Rubaltelli FF, Jori G (eds) Neonatal Jaundice. New Trends in Phototherapy. Plenum Press, New York, p 161

812. Falk H, Grubmayr K, Herzig U, Hofer O (1975) The Configurations of the Isomeric 3,4-Dimethyl-5-(1H)-2,2'-pyrromethenones. Tetrahedron Lett:559

813. McDonagh AF (1976) Jaundice Phototherapy: Photochemical Aspects. 4th Annual Meeting, American Society for Photobiology, February 16-20, 1976, Denver. Abstract WAM-B1, p 51

References

814. Falk AH, Grubmayr K, Hofer O, Neufingerl F (1975) Beiträge zur Chemie der Pyrrolpigmente, 9. Mitt.: Bildung, Struktur und Konfiguration von Z- und E-Arylmethyliden-3,4-dimethyl-3-pyrrolin-2-onen. Monatsh Chem 106:301

815. Lightner DA, Park Y-T (1976) Dye-sensitized Photooxygenation of Oxopyrromethenes Related to Bilirubin. Tetrahedron Lett: 2209

816. Lightner DA, Park Y-T (1977) Synthesis, Photooxidation and $Z \rightleftarrows E$ Photoisomerization of Benzalpyrrolinones. J Heterocycl Chem 14:415

817. Park Y-T, Lightner DA (1978) Sensitized Photoisomerization of Benzalpyrrolinones. J Kor Chem Soc 22:417

818. Lightner DA, Wooldridge TA, McDonagh AF (1979) Configurational Isomerization of Bilirubin and the Mechanism of Jaundice Phototherapy. Biochem Biophys Res Commun 86:235

819. Lightner DA, Wooldridge TA, McDonagh AF (1979) Geometric Isomerization of Bilirubin-IXα and Its Dimethyl Ester. J Chem Soc Chem Commun:110

820. Pederson AO, Schønheyder F, Brodersen R (1977) Photooxidation of Human Serum Albumin and Its Complex with Bilirubin. Eur J Biochem 72:213

821. McDonagh AF, Ramonas L (1978) Jaundice Phototherapy: Micro Flow-cell Photometry Reveals Rapid Biliary Response of *Gunn* Rats to Light. Science 201:829

822. Stoll MS, Zenone EA, Ostrow JD, Zarembo JE (1977) Photoisomers of Bilirubin and Their Excretion by the *Gunn* Rat. 5th Annual Meeting of the American Society for Photobiology, May 11-15, San Juan, Puerto Rico, Abstract SAM-D7, p 97

823. Zenone EA, Stoll MS, Ostrow JD (1977) Mechanism of Excretion of Unconjugated Bilirubin (UCB) During Phototherapy. Gastroenterology 72:A-157, p 1180

824. Roos WT, Weiger HH, Windler SCH, Jones CO (1977) Bilirubin Isomers and Phototherapy. Gastroenterology 73:A-45, p 1243

825. Windler SCH (1840) Ueber das Substitutionsgesetz und die Theorie der Typen. Ann Chem Pharm 33:308

826. Friedman HB (1930) The Theory of Types – A Satirical Sketch. J Chem Educ 7:633

827. Cohen AN, Ostrow JD (1980) New Concepts in Phototherapy: Photoisomerization of Bilirubin IXα and Potential Toxic Effects of Light. Pediatrics 65:740

828. Stoll MS, Zenone EA, Ostrow JD, Zarembo JE (1979) Preparation and Properties of Bilirubin Photoisomers. Biochem J 183:139

829. Bonnett R (1981) Oxygen Activation and Tetrapyrroles. Essays in Biochem 17:1

830. Bonnett R (1976) Mechanisms of Photodegradation of Bilirubin. Biochem Soc Trans 4:222

831. McDonagh AF (1976) Phototherapy of Neonatal Jaundice. Biochem Soc Trans 4:219

832. Bonnett R (1984) Recent Advances in the Chemistry of Bile Pigments. In: Rubaltelli FF, Jori G (eds) Neonatal Jaundice. New Trends in Phototherapy. Plenum Press, New York, p 111

833. Palma LA, McDonagh AF, Lightner DA (1980) Blue Light and Bilirubin Excretion. Science 208:145

834. McDonagh AF (1981) Phototherapy of Hyperbilirubinemia: Molecular Mechanisms. 9th Annual Meeting of the American Society for Photobiology, June 14-18, 1981, Williamsburg, Virginia. Abstract TPM 3, p 121

835. Stoll MS, Zenone EA, Ostrow JD (1981) Excretion of Administered and Indigenous Photobilirubins in the Bile of the Jaundiced *Gunn* Rat. J Clin Invest 68:134

836. Onishi S, Itoh S, Kawade N, Isobe K, Sugiyama S (1979) The Separation of Configurational Isomers of Bilirubin by High Pressure Liquid Chromatography and the Mechanism of Jaundice Phototherapy. Biochem Biophys Res Commun 90:890

837. Onishi S, Kawade N, Itoh S, Isobe K, Sugiyama S (1980) High Pressure Liquid Chromatographic Analysis of Anaerobic Photoproducts of Bilirubin-IXα *in vitro* and Its Comparison with Photoproducts *in vivo*. Biochem J 190:527

838. Onishi S, Isobe K, Itoh S, Kawade N, Sugiyama S (1980) Demonstration of a Geometric Isomer of Bilirubin-IXα in the Serum of a Hyperbilirubinaemic Newborn Infant and the Mechanism of Jaundice Phototherapy. Biochem J 190:533

768 References

839. Onishi S, Itoh S, Isobe K, Sugiyama S (1981) Photoisomers of Bilirubin in Jaundice Phototherapy. Photomed Photobiol 3:59
840. Stoll MS (1981) Transformations of Bilirubin. Biochemical Society. Tetrapyrrole Discussion Group Meeting on Linear Tetrapyrroles, 10th April, Queen Mary College, London, p 6
841. Isobe K, Onishi S (1981) Kinetics of the Photochemical Interconversion Among Geometric Isomers of Bilirubin. Biochem J 193:1029
842. Isobe K, Itoh S, Onishi S, Yamakawa T, Ogino T, Yokoyama T (1983) Kinetic Study of Photochemical and Thermal Conversion of Bilirubin-IXα and Its Photoproducts. Biochem J 209:695
843. Itoh S, Onishi S (1985) Kinetic Study of the Photochemical Changes of (Z,Z)- Bilirubin-IXα Bound to Human Serum Albumin. Biochem J 226:251
844. McDonagh AF, Palma LA, Trull FR, Lightner DA (1982) Phototherapy for Neonatal Jaundice. Configurational Isomers of Bilirubin. J Am Chem Soc 104:6865
845. Stoll MS, Vicker N, Gray CH, Bonnett R (1982) Concerning the Structure of Photobilirubin II. Biochem J 201:179
846. McDonagh AF, Palma LA (1982) Photometabolism of Bilirubin. Continuous Irradiation of *Gunn* Rats Leads to a Photostationary Equilibrium Mixture of Pigments in Bile. 10th Annual Meeting of the American Society for Photobiology, June 27-July 1, 1982, University of British Columbia, Vancouver, BC, Canada. Abstract TAM-D5, p 110
847. McDonagh, Lightner DA (1982) Anaerobic Photochemistry of Bilirubin (BR) and Related Compounds. Separation and Characterization of Primary and Secondary Products. 10th Annual Meeting of the American Society for Photobiology, June 27-July 1, 1982, University of British Columbia, Vancouver, BC, Canada. Abstract TAM-D, p 111
848. Falk H, Müller N, Ratzenhofer M, Winsauer K (1982) The Structure of "Photobilirubin". Monatsh Chem 113:1421
849. Falk H, Grubmayr K, Haslinger E, Schlederer T, Thirring K (1978) Beiträge zur Chemie der Pyrrolpigmente, 24. Mitt. Die Diasteomeren (Geometrisch Isomeren) Biliverdin Dimethylester – Struktur, Konfiguration und Konformation. Monatsh Chem 109:1451
850. Trull FR, Franklin RW. Lightner DA (1987) Total Synthesis of Symmetric Bile Pigments: Mesobilirubin-IVα, Mesobilirubin-XIIIα and Etiobilirubin-IVγ. J Heterocycl Chem 24:1573
851. McDonagh AF (1975) Thermal and Photochemical Reactions of Bilirubin-IXα. Ann New York Acad Sci 244:553
852. de Groot JA, van der Steen R, Fokkens R, Lugtenberg J (1982) Synthesis and Photoisomerisation of 2,3,17,18,22-Pentamethyl-10,23-dihydro-1,19-[21H,24H]-bilidione, an Unsymmetrical Bilirubin Model Compound. Recl Trav Chim Pays-Bas 101:219
853. de Groot JA, van der Steen R, Fokkens R, Lugtenberg J (1982) Synthesis and Photochemical Reactivity of Bilirubin Model Compounds. Recl Trav Chim Pays-Bas 101:35
854. McDonagh AF, Palma LA, Lightner DA, Ennever JF, Thaler MM (1983) Why Phototherapy Works. Identification of Bilirubin (BR) Photoproducts *in vivo*. Pediatric Res 17:326A
855. McDonagh AF (1984) Molecular Mechanisms of Phototherapy of Neonatal Jaundice. In: Rubaltelli FF, Jori G (eds) Neonatal Jaundice. New Trends in Phototherapy. Plenum Press, New York, p 173
856. Bonnett R, Buckley DG, Hamzetash D, Hawkes GE, Ioannou S, Stoll MS (1984) Photobilirubin II. Biochem J 219:1053
857. Bonnett R, Ioannou S (1987) Phototherapy and the Chemistry of Bilirubin. Molec Aspects Med 9: 457
858. Onishi S, Miura I, Isobe K, Itoh S, Ogino T, Yokoyama T, Yamakawa T (1984) Structure and Thermal Interconversion of Cyclobilirubin-IXα. Biochem J 218:667
859. Itoh S, Onishi S (1985) Kinetic Study of the Photochemical Changes of (Z,Z)-Bilirubin-IXα Bound to Human Serum Albumin. Biochem J 226:251
860. Onishi S, Itoh S, Yamakawa T, Isobe K, Manabe M, Toyota S, Imai T (1985) Comparison of the Kinetic Study of the Photochemical Changes of (Z,Z)-Bilirubin-IXα Bound to Human Serum Albumin with That Bound to Rat Serum Albumin. Biochem J 239:561

References 769

861. Onishi S, Itoh S, Isobe K (1986) Wavelength-dependence of the Relative Rate Constants for the Main Geometric and Structural Photoisomerization of Bilirubin-IXα Bound to Human Serum Albumin. Biochem J 236:23

862. Onishi S, Isobe K, Itoh S, Manabe M, Saski K, Fukuzaki R, Yamakawa T (1986) Metabolism of Bilirubin and Its Photoisomers in Newborn Infants During Phototherapy. J Biochem 100:789

863. Onishi S, Itoh S, Isobe K, Ochi M, Kunikata T, Imai T (1989) Effect of the Binding of Bilirubin to Either the First Class or the Second Class of Binding Sites of the Human Serum Albumin Molecule on Its Photochemical Action. Biochem J 257:711

864. McDonagh AF (1983) Mechanism of Action of Phototherapy. Hyperbilirubinemia in the Newborn. Report of the Eighty-fifth Ross Conference on Pediatrics Research, September 1983. Ross Laboratories (publisher), Columbus, OH, p 47

865. McDonagh AF (1985) Light Effects on Transport and Excretion of Bilirubin in Newborns. Ann New York Acad Sci 453:65

866. Ennever JA (1988) Phototherapy for Neonatal Jaundice. Photochem Photobiol 47:871

867. Pratesi R, Agati G, Fusi F (1989) Phototherapy for Neonatal Hyperbilirubinemia. Photodermatology 6:244

868. McDonagh AF, Lightner DA, Reisinger M, Palma LA (1986) Human Serum Albumin as a Chiral Template. Stereoselective Photocyclization of Bilirubin. J Chem Soc Chem Commun:249

869. McDonagh AF, Lightner DA (1985) Mechanisms of Phototherapy of Neonatal Jaundice. Regiospecific Photoisomerization of Bilirubins. In: Sund H, Blauer G (eds) Optical Properties and Structure of Tetrapyrroles. Walter de Gruyter, Berlin, p 297

870. McDonagh AF, Lightner DA (1985) Intramolecular Energy Transfer in Bilirubins. In Bensasson RV, Jori G, Land EJ, Truscott TG (eds). Primary Photoprocesses in Biology and Medicine. Plenum Press, New York, p 321

871. McDonagh AF, Lightner DA, Agati G (1998) Induction of Wavelength-dependent Photochemistry in Bilirubins by Serum Albumin. Monatsh Chem 129:649

872. Lamola AA (1988) Effects of Environment on Photophysical Processes in Bilirubin. In: Sund H, Blauer G (eds) Optical Properties and Structure of Tetrapyrroles. Walter de Gruyter, Berlin, p 310

873. McDonagh AF, Agati G, Fusi F, Pratesi R (1989) Quantum Yields for Laser Photocyclization of Bilirubin in the Presence of Human Serum Albumin. Dependence of Quantum Yield on Excitation Wavelength. Photochem Photobiol 50:305

874. Agati G, Fusi F, Pratesi R, McDonagh AF (1992) Wavelength-dependent Quantum Yield for $Z \rightarrow E$ Isomerization of Bilirubin Complexed with Human Serum Albumin. Photochem Photobiol 55:185

875. Troup GJ, Agati G, Fusi F, Pratesi R (1996) Photophysics of the Variable Quantum Yield of Asymmetric Bilirubin. Aust J Phys 49:673

876. McDonagh AF, Lightner DA, Agati G (1998) Induction of Wavelength-dependent Photochemistry in Bilirubins by Serum Albumin. Monatsh Chem 129:649

877. Mazzoni M, Agati G, Troup GJ, Pratesi R (2003) Analysis of Wavelength-dependent Photoisomerization Quantum Yields in Bilirubin by Fitting Two Exciton Absorption Bands. J Optics A: Pure Appl 5:S374

878. Zietz B, Gillbro T (2007) Initial Photochemistry of Bilirubin Probed by Femtosecond Spectroscopy. J Phys Chem B 111:11997

879. Plaveskeĭ VY, Mostovnikov VA, Tret'yahova AI, Mostovnikov GR (2007) Photophysical Processes that Determine the Photoisomerization Selectivity of Z,Z-Bilirubin-IXα in Complexes with Albumins. J Opt Technol 74:446

880. Zunszain PA, Ghuman J, McDonagh AF, Curry S (2008) Crystallographic Analysis of Human Serum Albumin Complexed with 4(Z),15(E)-Bilirubin-IXα. J Molec Biol 381:394

881. Dey SK, Lightner DA (2008) Toward an Amphiphilic Bilirubin. The Crystal Structure of a Bilirubin E-Isomer. J Org Chem 73:2704

882. McDonagh AF, Palma LA, Schmid R (1981) Reduction of Biliverdin and Placental Transfer of Bilirubin and Biliverdin in the Pregnant Guinea Pig. Biochem J 194:273
883. Byun YS, Lightner DA (1991) Synthesis and Properties of a Bilirubin Analog with Propionic Acid Groups Replaced by Carboxyl. J Heterocycl Chem 28:1683
884. McDonagh AF, Lightner DA (1994) Hepatic Uptake, Transport and Metabolism of Alkylated Bilirubin in Gunn Rats and Sprague-Dawley Rats. Cell Molec Biol 40:965
885. Chen Q, Huggins MT, Lightner DA, Norona W, McDonagh AF (1999) Synthesis of a 10-Oxo-bilirubin: Effect of the Oxo Group on Conformation, Transhepatic Transport and Glucuronidation. J Am Chem Soc 121:9253
886. Brower JO, Lightner DA, McDonagh AF (2000) Synthesis of a New Lipophilic Bilirubin. Conformation, Transhepatic Transport and Glucuronidation. Tetrahedron 56:7869
887. Tipton AK, Lightner DA, McDonagh AF (2001) Synthesis and Metabolism of the First Thia-Bilirubin. J Org Chem 66:1832
888. Tipton A, Lightner DA (2002) Crystal Structure and Conformation of a 10-Thia-Bilirubin. Monatsh Chem 133:707
889. Ghosh B, Lightner DA (2003) 10-Thia-Mesobilirubin-XIIIα. J Heterocycl Chem 40:1113
890. Brower JO, Lightner DA, McDonagh AF (2001) Aromatic Congeners of Bilirubin. Synthesis, Stereochemistry, Glucuronidation and Hepatic Transport. Tetrahedron 57:7813
891. McDonagh AF, Lightner DA, Nogales DF, Norona W (2001) Biliary Excretion of a Stretched Bilirubin in UGT1A1-Deficient (*Gunn*) and Mrp2-Deficient (TR⁻) Rats. FEBS Lett 506:211
892. Boiadjiev SE, Lightner DA (2001) A Water-soluble Synthetic Bilirubin with Carboxyl Groups Replaced by Sulfonyl. Monatsh Chem 132:1201
893. Lightner DA, McDonagh AF (2001) Structure and Metabolism of Natural and Synthetic Bilirubins. J Perinatol 21:S13
894. Lightner DA, McDonagh AF, Kar AK, Norona WS (2002) Hepatobiliary Excretion of Biliverdin Isomers and C10-Substituted Biliverdins in Mrp2-Deficient (TR⁻) Rats. Biochem Biophys Res Commun 293:1077
895. Ghosh B, Lightner DA, McDonagh AF (2004) Synthesis, Conformation and Metabolism of a Selenium Bilirubin. Monatsh Chem 135:1189
896. McDonagh AF, Lightner DA (2007) Influence of Conformation and Intramolecular Hydrogen Bonding on the Acyl Glucuronidation and Biliary Excretion of Acetylenic bis-Dipyrrinones Related to Bilirubin. J Med Chem 50:480
897. McDonagh AF, Boiadjiev SE, Lightner DA (2008) Slipping Past UGT1A1 and Multidrug Resistance-Associated Protein 2 in the Liver. Effects of Steric Compression and Hydrogen Bonding on the Hepatobiliary Elimination of Synthetic Bilirubins. Drug Metab Dispos 36, 930
898. Kar A, Lightner DA (1998) Synthesis and Properties of Bilirubin Analogs with Conformation-perturbing C(10) *tert*-Butyl and Adamantyl Groups. Tetrahedron 54:12671
899. Brower JO, Lightner, DA (2001) Fluorophenyl Bilirubins. Synthesis and Stereochemistry. Monatsh Chem 132:1527
900. Kar A, Lightner DA (1998) Synthesis and Properties of C(10)-Isopropyl and Isopropylidene Analogs of Bilirubin. Tetrahedron 54:5151
901. Kar A, Tipton AK, Lightner DA (1999) Crystal Structure and Conformation of a 10-Isopropyl Bilirubin. Monatsh Chem 130:833
902. Xie M, Lightner DA (1993) Synthesis and Unusual Properties of C(10)-*gem*-Dimethyl Bilirubin Analogs. Tetrahedron 49:2185
903. Xie M, Holmes DL, Lightner DA (1993) Bilirubin Conformation and Intramolecular Steric Buttressing. C(10)-*gem*-Dimethyl Effect. Tetrahedron 49:9235
904. Tu B, Ghosh B, Lightner DA (2004) The *gem*-Dimethyl Effect: Amphiphilic Bilirubins. Tetrahedron 60:9017
905. Ghosh B, Catalano VJ, Lightner DA (2004) Syntheses, Crystal Structure and Conformations of 10,10-Spiro-Bilirubins. Monatsh Chem 135:1305

References

906. Boiadjiev SE, Lightner DA (2003) Novel Benzoic Acid Congeners of Bilirubin. J Org Chem 68:7591
907. Joesten MD, Schaad LJ (1974) Hydrogen Bonding. Marcel Dekker, Inc, New York
908. Kato Y, Toledo LM, Rebek J Jr (1996) Energetics of a Low Barrier Hydrogen Bond in Nonpolar Solvents. J Am Chem Soc 118:8575
909. Ducharme Y, Wuest JD (1988) Use of Hydrogen Bonds to Control Aggregation. Extensive, Self-complementary Arrays of Donors and Acceptors. J Org Chem 53:5787
910. Wash PL, Maverick E, Chiefari J, Lightner DA (1997) Acid-Amide Intermolecular Hydrogen-bonding. J Am Chem Soc 119:3802
911. Tipton AK, Lightner DA (1999) An Intramolecularly Hydrogen-bonded Dihydrotripyrrinone. Monatsh Chem 130:425
912. Chen Q, Lightner DA (1998) Hemirubin. An Intramolecular Hydrogen Bonded Analog for One-half Bilirubin. J Org Chem 63:2665
913. Huggins MT, Lightner DA (2000) Stereochemistry and Conformational Analysis of Hemirubin. Tetrahedron 56:1797
914. Huggins MT, Lightner DA (2000) Semirubin. A Novel Dipyrrinone Strapped by Hydrogen Bonds. J Org Chem 65:6001
915. Huggins MT, Lightner DA (2001) Intramolecular Hydrogen Bonding Between Remote Termini. Tetrahedron 57:2279
916. Huggins MT, Salzameda NT, Lightner DA (2011) Amide to Acid Hydrogen Bonding. The Dipyrrinone Receptor. Supramolec Chem 23:226
917. Dey SK, Lightner DA (2009) Amphiphilic Dipyrrinones. Methoxylated [6]-Semirubins. Tetrahedron 65:2399
918. Boiadjiev SE, Lightner DA (2000) On the Aggregation of Bilirubin Using Vapor Pressure Osmometry. J Heterocycl Chem 37:863
919. Brower JO, Huggins MT, Boiadjiev SE, Lightner DA (2000) On the Molecularity of Bilirubins and Their Esters and Anions in Chloroform Solution. Monatsh Chem 131:1047
920. Falk H, Schlederer T, Wolschann P (1981) Beiträge zur Chemie der Pyrrolpigmente, 38. Mitt. Zur Assoziation von Gallenpigmenten. Monatsh Chem 112: 199
921. Falk H, Müller N (1982) Beiträge zur Chemie der Pyrrolpigmente, 43. Mitt. Die Temperaturabhängigkeit der Licht-Absorption von Bilirubin und Einigen Seiner Derivate. Monatsh Chem 113:111
922. Holzwarth AR, Langer E, Lehner H, Schaffner K (1980) Absorption Luminescence, Solvent-induced Circular Dichroism and ^{1}H NMR Study of Bilirubin Dimethyl Ester: Observation of Different Forms in Solution. Photochem Photobiol 32:17
923. Crusats J, Delgado A, Farrera JA, Rubires R, Ribó JM (1998) Solution Structure of Mesobilirubin-XIIIα Bridged Between the Propionic Acid Substituents. Monatsh Chem 129:741
924. Ostrow JD, Celic L (1984) Bilirubin Chemistry, Ionization and Solubilization by Bile Salts. Hepatology 4:38S
925. Ostrow JD, Mukerjee P, Tiribelli C (1994) Structure and Binding of Unconjugated Bilirubin: Relevance for Physiological Function. J Lipid Res 35:1715
926. Ostrow JD, Celic L, Mukerjee P (1988) Molecular and Micellar Associations in the pH-Dependent Stable and Metastable Dissolution of Unconjugated Bilirubin by Bile Salts. J Lipid Res 29:33S
927. Carroll L (1898) The Hunting of the Snark. An Agony in Eight Fits. Chatto & Windus, London; reprinted 1941, The Macmillan Company of Canada, Ltd, Toronto
928. Brodersen R (1979) Bilirubin. Solubility and Interaction with Albumin and Phospholipid. J Biol Chem 254:2364
929. Brodersen R (1986) Aqueous Solubility, Albumin Binding, and Tissue Distribution of Bilirubin. In: Ostrow JD (ed) Bile Pigments and Jaundice. Marcel Dekker, New York, p 157
930. Brodersen R (1982) Physical Chemistry of Bilirubin: Binding to Macromolecules and Membranes. In: Heirwegh KPM, Brown SB (eds) Bilirubin, vol I. CRC Press, Inc, Boca Raton, Florida, p 75

931. McDonagh AF (2002) Lyophilic Properties of Protoporphyrin and Bilirubin. Hepatology 36:1028
932. Trull FR, Ma JS, Landen GL, Lightner DA (1983) Hydrogen Bonding of Bilirubin and Pyrromethenones in Solution. Isr J Chem 23:211
933. Gawroński JK, Polonski T, Lightner DA (1990) Sulfoxides as Chiral Complexation Agents. Conformational Enantiomer Resolution and Induced Circular Dichroism of Bilirubins. Tetrahedron 46:8053
934. Trull FR, Shrout DP, Lightner DA (1992) Chiral Recognition by Sulfoxides. Induced Circular Dichroism from Symmetric Mesobilirubin Analogs. Tetrahedron 48:8189
935. Hsieh Y-Z, Morris MD (1988) Resonance Raman Spectroscopic Study of Bilirubin Hydrogen Bonding in Solutions and in the Albumin Complex. J Am Chem Soc 110:62
936. Kanna Y, Arai T, Tokumara K (1993) Photoisomerization of Bilirubins and the Role of Intramolecular Hydrogen Bonds. Bull Chem Soc Jpn 66:1482
937. Lee JJ, Daly LH, Cowger ML (1974) Bilirubin Ionic Equilibria; Their Effects on Spectra and on Conformation. Res Commun Pathol Pharmacol 9:763
938. Hansen PE, Thiessen H, Brodersen R (1979) Bilirubin Acidity. Titrimetric and ^{13}C NMR Studies. Acta Chem Scand B33:218
939. Kidrič J, Mavri J, Podobruk M, Hadži D (1990) Intramolecular Hydrogen Bonding in Acid Malonates. Infrared, NMR and *ab initio* MO Investigations. J Molec Spectros 237:265
940. Guo J, Tolstoy PM, Koeppe B, Denisov GS, Limbach H-H (2011) NMR Study of Conformational Exchange and Double-well Proton Potential in Intramolecular Hydrogen Bonds in Monoanions of Succinic Acid and Derivatives. J Phys Chem A 115:9828
941. Knell AJ, Johnson B, Hutchinson DW (1972) Intramolecular Hydrogen Bonds in Bilirubin. Digestion 6:288
942. Overbeek JTG, Vink CLJ, Deenstra H (1955) The Solubility of Bilirubin. Recl Trav Chim Pays-Bas 74:81
943. Gray CH, Kulczycka A, Nicholson DC (1961) The Chemistry of the Bile Pigments. Part IV. Spectrophotometric Titration of the Bile Pigments. J Chem Soc:2276
944. Lathe GH (1972) Formation and Excretion of Bilirubin. Essays in Biochem 8:107
945. Krasner J, Yaffe SJ (1973) The Automatic Titration of Bilirubin. Biochem Med 7:128
946. Kolosov IV, Shapovalenko (1977) Acid-Base Equilibria in Solutions of Bilirubin. Zh Obshch Khim 47:2149 (original article submitted November 14, 1975)
947. Kolosov IV, Shapovalenko EP (1977) Study of Acid-Base Equilibria in Aqueous Solutions of Bilirubin by the Solubility Method. Zh Obshch Khim 47:2155
948. Lucassen J (1961) The Diazo Reaction of Bilirubin and Bilirubin Glucuronide. Doctoral dissertation, Rijksuniversiteit Utrecht, p 46
949. McDonagh AF, Lightner DA (1981) New Concepts in Phototherapy. Pediatrics 67:929
950. Cohen AN, Ostrow JD (1981) New Concepts in Phototherapy. Pediatrics 67:930
951. Hahm JS, Ostrow JD, Mukerjee P, Webster CC (1986) Solubility and pK_a of Unconjugated Bilirubin in Aqueous Buffer and Bile Salt Solutions, Determined by Isoextraction from Chloroform. Hepatology 6:1185
952. Farmer RC (1901) A New Method for the Determination of Hydrolytic Dissociation. J Chem Soc 79:868
953. Yonger J, Dang VB (1977) Étude de la Liaison Bilirubine-Albumine par Extraction à l'Acide d'un Solvant non Miscible à l'Eau. Pharmacol Perinatale, INSERM 73:199
954. Irollo R, Casteran M, Dang VB, Yonger J (1979) Étude de la Liaison Bilirubine-Albumine I – Étude *in vitro* par une Méthode de Partage de la Bilirubine entre une Solution Aqueuse et un Solvant Organique Non Miscible à l'Eau. Ann Biol Clin 37:331
955. Girard M-L, Paologgi F, Frappier F (1965) Détermination de la Bilirubine Extractible par les Solvants: Intérêt de la Méthyl-isobutylcétone. Ann Biol Clin 23:279
956. Moroi Y, Matuura R, Hisadome T (1985) Bilirubin in Aqueous Solution. Absorption Spectrum, Aqueous Solubility, and Dissociation Constants. Bull Chem Soc Jpn 58:1426
957. Tiribelli C, Ostrow JD (1990) New Concepts in Bilirubin Chemistry. Transport and Metabolism: Report of the International Bilirubin Workshop, April 6-8, 1989, Trieste, Italy. Hepatology 11:303

References

773

958. Trull FR, Boiadjiev SE, Lightner DA, McDonagh AF (1997) Aqueous Dissociation Constants of Bile Pigments and Sparingly Soluble Carboxylic Acids by ^{13}C-NMR in Aqueous Dimethyl Sulfoxide: Effects of Hydrogen Bonding. J Lipid Res 38:1178

959. Hahm J-S, Ostrow JD, Mukerjee P, Celic L (1992) Ionization and Self-association of Unconjugated Bilirubin, Determined by Rapid Partition from Chloroform, with Further Studies of Bilirubin Solubility. J Lipid Res 33:1123

960. Tiribelli C, Ostrow JD (1993) New Concepts in Bilirubin Chemistry, Transport and Metabolism: Report of the Second International Bilirubin Workshop, April 9-11, 1992, Trieste, Italy. Hepatology 17:715

961. Hagen R, Roberts JD (1969) Nuclear Magnetic Resonance Spectroscopy. ^{13}C Spectra of Aliphatic Carboxylic Acids and Carboxylate Anions. J Am Chem Soc 91:4504

962. Choi PJ, Petterson KA, Roberts JD (2002) Ionization Equilibria of Dicarboxylic Acids in Dimethyl Sulfoxide as Studied by NMR. J Phys Org Chem 15:278

963. Cistola DP, Small DM, Hamilton JA (1982) Ionization Behavior of Aqueous Short-chain Carboxylic Acids: A Carbon-13 NMR Study. J Lipid Res 23:795

964. Small DM, Cabral DJ, Cistola DP, Parks JS, Hamilton JA (1984) The Ionization Behavior of Fatty Acids and Bile Acids in Micelles and Membranes. Hepatology 4:775

965. Hamilton JA (1994) ^{13}C NMR Studies of the Interactions for Fatty Acids with Phospholipid Bilayers, Plasma Lipoproteins, and Proteins. In: Beckmann N (ed) Carbon-13 NMR Spectroscopy of Biological Systems. Academic Press, Inc, New York, p 117

966. Parks JS, Cistola DP, Small DM, Hamilton JA (1983) Interactions of the Carboxyl Group of Oleic Acid with Bovine Serum Albumin: A ^{13}C-NMR Study. J Biol Chem 258:9262

967. Cistola DP (1985) Physicochemical Studies of Fatty Acids in Model Biological Systems. PhD Dissertation, Boston University. Chem Abstr 105:148490

968. Holmes DL, Lightner DA (1995) Synthesis and Acidity Constants of $^{13}CO_2$H-labelled Mono and Dipyrrole Carboxylic Acids. pK_a from ^{13}C-NMR. Tetrahedron 51:1607

969. Holmes DL, Lightner DA (1996) Synthesis and Acidity Constants of $^{13}CO_2$H-labelled Dicarboxylic Acid. pK_as from ^{13}C-NMR. Tetrahedron 52:5319

970. Holmes DL, McDonagh AF, Lightner DA (1996) On the Acid Dissociation Constants of Bilirubin and Biliverdin. pK_a Values from ^{13}C NMR Spectroscopy. J Biol Chem 271:2397

971. McDonagh AF, Phimister A, Boiadjiev SE, Lightner DA (1999) Dissociation Constants of Carboxylic Acids by ^{13}C-NMR in DMSO/Water. Tetrahedron Lett 40:8515

972. Boiadjiev SE, Watters K, Lai B, Wolf S, Welch W, McDonagh AF, Lightner DA (2004) pK_a and Aggregation of Bilirubin. Biochemistry 43:15617

973. Mukerjee P, Ostrow JD (1998) Effects of Added Dimethylsulfoxide on pK_a Values of Uncharged Organic Acids and pH Values of Aqueous Buffers. Tetrahedron Lett 39:423

974. Mukerjee P, Ostrow JD, Tiribelli C (2002) Low Solubility of Unconjugated Bilirubin in Dimethylsulfoxide-Water Systems; Implications for pK_a Determinations. BMC Biochem 3:17

975. Ostrow JD, Mukerjee P (2007) Revalidation and Rationale for High pK_a Values of Unconjugated Bilirubin. BMC Biochem 8:7; http://www.biomedcentral.com/1471-2091/8/7

976. Mukerjee P, Ostrow JD (2010) Review: Bilirubin pK_a Studies; New Models and Theories Indicate High pK_a Values in Water, Dimethylformamide and DMSO. BMC Biochem 11:15; http://www.biomedcentral.com/1471-2091/11:15

977. Dey SK, Lightner DA (2010) Lipid- and Water-soluble Bilirubins. Monatsh Chem 141:101

978. Datta S, Lightner DA (2009) $^{13}CO_2$H-Labeled Semirubin. Hydrogen Bonding and pK_a. Monatsh Chem 140:1229

979. Rebek J Jr, Duff RJ, Gordon WE, Parris K (1986) Convergent Functional Groups Provide a Measure of Stereoelectronic Effects at Carbonyl Oxygen. J Am Chem Soc 108:6068

980. Onoda A, Yamada Y, Takeda J, Nakayama Y, Okamura T, Doi M, Yamamoto H, Ueyama N (2004) Stabilization of Carboxylate Anion with a NH···O Hydrogen Bond: Facilitation of the Deprotonation of Carboxylic Acid by the Neighboring Amide NH Groups. Bull Chem Soc Jpn 77:321

981. Schwarz B, Drueckhammer D (1995) A Simple Method for Determining the Relative Strengths of Normal and Low-barrier Hydrogen Bonds in Solution: Implications to Enzyme Catalysis. J Am Chem Soc 117:11902

982. Soper AK, Luzar A (1992) A Neutron Diffraction Study of Dimethyl Sulphoxide-Water Mixtures. J Chem Phys 97:1320

983. Vaisman II, Berkowitz ML (1992) Local Structural Order and Molecular Associations in Water-DMSO Mixtures. Molecular Dynamics Study. J Am Chem Soc 114:7889

984. McDonagh AF, Palma L (1981) Bilirubin Hydrate. Hepatol Rapid Lit Rev 11:1670

985. Vega-Hissi EG, Estrada MR, LaVecchia MJ, Pis Diez R (2013) Computational Chemical Analysis of Unconjugated Bilirubin Anions and Insights into pK_a Values Clairification. J Chem Phys 138:035101

986. Gossauer A (2003) Synthesis of Bilanes. In: Kadish KM, Smith KM, Guilard R (eds). The Porphyrin Handbook, vol 13, Academic Press, New York, p 237

987. McDonagh AF, Lightner DA (1991) The Importance of Molecular Structure in Bilirubin Metabolism and Excretion. In: Bock KW, Gerok W, Matern S (eds). Hepatic Metabolism and Disposition of Endo and Xenobiotics (Falk Symposium No. 57), Kluwer, Dordrecht, The Netherlands, chap 5, p 47

988. Shrout DP, Puzicha G, Lightner DA (1992) An Efficient Synthesis of Symmetric Bilindiones. Mesobilirubin-XIIIα and Analogs with Varying Alkanoic Acid Chain Lengths. Synthesis:328

989. Trull FR, Person RV, Lightner DA (1997) Conformational Analysis of Symmetric Bilirubin Analogs with Varying Length Alkanoic Acids. Enantioselectivity by Human Serum Albumin. J Chem Soc Perkin Trans II:1241

990. Chiefari J, Person RV, Lightner DA (1992) Synthesis and Conformation of a Bilirubin Analog with Propionic Acid Side Chains Extended to Undecanoic Acid. Tetrahedron 48:5969

991. Thyrann T, Lightner DA (1996) Synthesis and Properties of a Lipid Bilirubin Analog. Tetrahedron 52:447

992. McDonagh AF, Lightner DA (1994) Hepatic Uptake, Transport and Metabolism of Alkylated Bilirubin in *Gunn* Rats and *Sprague-Dawley* Rats. Cell Molec Biol 40:965

993. Pfeiffer WP, Lightner DA (1994) Homorubin. A Centrally Homologated Bilirubin. Tetrahedron Lett 35:9673

994. Nogales D, Anstine DT, Lightner DA (1994) Synthesis and Unusual Properties of Expanded Bilirubin Analogs. Tetrahedron 50:8579

995. Anstine DT (1995) Pentapyrrole Analogs of Bilirubin and Hydrogen Bonding. PhD Dissertation, University of Nevada, Reno

996. Boiadjiev SE, Lightner DA (1997) Synthesis of the First Fluorinated Bilirubin. J Org Chem 62:399

997. Boiadjiev SE, Lightner DA (1998) Altering the Acidity and Solution Properties of Bilirubin. Methoxy and Methylthio Substituents. J Org Chem 63:6220

998. Nikitin EN, Lightner DA (2009) Imploded Bilirubins. Synthesis and Properties of C(10)-nor-Mesobilirubin-XIIIα and Analogs. Monatsh Chem 140:97

999. Tu B, Ghosh B, Lightner DA (2003) A New Class of Linear Tetrapyrroles: Acetylenic 10,10a-Didehydro-10a-homobilirubins. J Org Chem 68:8950

1000. Tu B, Ghosh B, Lightner DA (2004) Novel Linear Tetrapyrroles: Hydrogen Bonding in Diacetylenic Bilirubins. Monatsh Chem 135:519

1001. Boiadjiev SE, Holmes DL, Anstine DT, Lightner DA (1995) Synthesis and Unusual Properties of an 8,12-bis-Pivalic Acid Analog of Bilirubin. Tetrahedron 51:10663

1002. Bauman D, Killet C, Boiadjiev SE, Lightner DA, Schönhofer A, Kuball H-G (1996) Study of β,β′-Dimethylmesobilirubin-XIIIα Oriented in a Nematic Liquid Crystal by Linear and Circular Dichroism Spectroscopy. J Phys Chem 100:11546

1003. Boiadjiev SE, Lightner DA (1996) Stereocontrol of Bilirubin Conformation. Tetrahedron: Asymm 7:1309

1004. Boiadjiev SE, Pfeiffer WP, Lightner DA (1997) Synthesis and Stereochemistry of Bilirubin Analogs Lacking Carboxylic Acids. Tetrahedron 53:14547

References 775

1005. Boiadjiev SE, Lightner DA (1999) Intramolecular Hydrogen Bonding and Its Influence on Conformation. Circular Dichroism of Chiral Bilirubin Analogs. Tetrahedron: Asymm 10:2535

1006. Boiadjiev SE, Lightner DA (1997) Exciton Chirality of Bilirubin Homologs. Chirality 9:604

1007. Boiadjiev SE, Lightner DA (2001) Chirality Inversion in the Bilirubin Molecular Exciton. Chirality 13:251

1008. Abbate S, Longhi G, Lebon F, Longhi G, Boiadjiev SE, Lightner DA (2011) ECD and VCD Studies of Dimethylmesobilirubins as Models for Understanding Structural Properties of Bilirubin in Solution. 12th International Congress on Circular Dichroism in Spectroscopy, Oxford, UK

1009. Saresberiensis I (1991) Metalogicon. In: Hall JB, Keats-Rohan KSB (eds) Corpus Christianorum Continuatio Mediaevalis, xcvii, Turnhout: Brepols, Book III, Chap IV. See also: http://abaelard.de/abaelard/060013metalogicon.htm

1010. Salisbury J (1995) The Metalogicon of John Salisbury [Johannes Saresberiensis]: A Twelfth Century Defense of the Verbal and Logical Arts of the Trivium. McGarry DD (English trans). University of California Press, Berkeley

1011. Hijmans van den Bergh A, Snapper I (1913) Die Farbstoffe des Serums. Dtsch Arch klin Med: 540

1012. van den Bergh AAH, Muller P (1916) Über eine Direkte und Eine Indirekte Diazoreaction auf Bilirubin. Biochem Z 77:90

1013. Hijmans van den Bergh A (1918) Die Gallenfarbstoffe im Blute. 1st ed, SC van Doesburgh, Leiden and JA Barth Leipzig; 2nd ed (1928) van Doesburgh, Leiden and Barth, Leipzig

1014. van den Bergh AAH (1954) The Bile Pigment in the Blood. In: Kalmar S (Engl trans), Elton NW (ed) Issue 40 of Medical Laboratories Special Report, Chemical Corps, Army (US) Medical Center, Maryland. Publication Control No 5030-40

1015. Hijmans van den Bergh A (1921) La Recherche de la Bilirubine dans le Plasma Sanguin par la Méthode de la Reaction Diazoique. Presse Med 14:441

1016. McNee JW (1922) The Use of the *van den Bergh* Test in the Differentiation of Obstructive Jaundice from Other Types of Jaundice. Brit Med J May 6:716

1017. Cole PG, Lathe GH, Billing BH (1954) Separation of Bile Pigments of Serum, Bile and Urine. Biochem J 57:514

1018. Billing BH, Lathe GH (1956) The Excretion of Bilirubin as an Ester Glucuronide, Giving the Direct *van den Bergh* Reaction. Biochem J 63:6P

1019. Billing BH, Cole PG, Lathe GH (1957) The Excretion of Bilirubin as a Diglucuronide Giving the Direct *van den Bergh* Reaction. Biochem J 65:774

1020. Schmid R (1956) Glukuronsäure-konjugiertes Biulirubin, das «direkt reagierende» Bilirubin in Serum, Harn und Galle. Schweizerischen Medizin Wachenschr 86:775

1021. Schmid R (1956) Direct-Reacting Bilirubin, Bilirubin Glucuronide, in Serum, Bile, and Urine. Science 124:76

1022. Schmid R (1957) Identification of Direct-Reacting Bilirubin as Bilirubin Glucuronide. J Biol Chem 229:881

1023. Talafant E (1956) Properties and Composition of the Bile Pigment Giving a Direct Diazo Reaction. Nature 178:312

1024. Talafant E, Appelf G (1967) The Nature and Properties of the 'Ether-Extractable' Bile Pigment of Serum. In: Bouchner IAD, Billing B (eds) Bilirubin Metabolism. Blackwell, Oxford, UK, p 103

1025. Pirone C, Quirke JME, Priestap HA, Lee DW (2009) Animal Pigment Bilirubin Discovered in Plants. J Am Chem Soc 131:2830

1026. Pirone C, Johnson JV, Quirke JME, Priestap HA, Lee D (2010) The Animal Pigment Bilirubin Identified in *Strelizia reginae*, the Bird of Paradise Flower. HortScience 45:1411

1027. Falk H, Marko H (1991) Reduction of Bilindione-10-thiol Adduct as a Model of the Reduction Step of the Biliverdin Reductase System. Monatsh Chem 122:319

1028. Ma Q, Hua H-H, Chen Y, Liu B-B, Krämer AL, Scheer H, Zao K-H, Zhou M (2012) Rising Tide of Blue-absorbing Biliprotein Photoreceptors: Characterization of Seven Such Bilinbinding GAF Domains in *Nostoc sp.* PCC7120. Fed Eur Biochem Soc (FEBS) doi: 1111/febs. 12003

1029. Stocker R, Yamamoto Y, McDonagh AF, Glazer AN, Ames BN (1987) Bilirubin Is an Antioxidant of Possible Physiological Importance. Science 235:1043

1030. Stocker R, Glazer AN, Ames BN (1987) Antioxidant Activity of Albumin-Bound Bilirubin. Proc Natl Acad Sci USA 84:5918

1031. Stocker R, McDonagh AF, Glazer AN, Ames BN (1990) Antioxidant Activities of Bile Pigments: Biliverdin and Bilirubin. Methods Enzymol 186:301

1032. Baranano DE, Rao M, Ferris CD, Snyder SH (2002) Biliverdin Reductase: A Major Physiologic Cytoprotectant. Proc Natl Acad Sci USA 99:16093

1033. Sedlak TW, Saleh M, Higginson DS, Paul BD, Juluri KR, Snyder SH (2009) Bilirubin and Glutathione Have Complementary Antioxidant and Cytoprotective Roles. Proc Natl Acad Sci USA 106:5171

1034. Sedlak TW, Snyder SH (2009) Circling the Wagons for Biliverdin Reductase. J Biol Chem 284:le 11

1035. Stocker R (2004) Antioxidant Activities of Bile Pigments. Antioxid and Redox Signal 6:841

1036. Stocker R (2005) Antioxidant Defense Mechanisms by Heme Oxygenase and Bile Pigments. In: Otterbein LE, Zuckerbraun BS (eds) Heme Oxygenase. Nova Science Publishers, Inc, New York, p 313

1037. Maghzal GJ, Leck M-C, Collinson E, Li C, Stocker R (2009) Limited Role for the Bilirubin-Biliverdin Redox Amplification Cycle in the Cellular Antioxidant Protection by Biliverdin Reductase. J Biol Chem 284:29251

1038. McDonagh AF (2010) The Biliverdin-Bilirubin Antioxidant Cycle of Cellular Protection: Missing a Wheel? Free Rad Biol Med 49:814

Author Index

A

Abbate S, 519, 521, 763, 775
Abel JJ, 182, 185, 186, 743
Adler E, 751
Adloff JP, 112, 740
Agati G, 769
Agresti A, 488, 758
Agricola, 216
Ahlfors CE, 545, 766
Akaike T, 760
Alagona G, 488, 758
Allegret A, 454, 754
Allinger NH, 760
Ames BN, 776
Amman H, 750
An JY, 759, 760
Andersag H, 326, 750, 751
Anderson HJ, 744
Anderson T, 198, 250, 744
Andrenacus JG, 50
Anft B, 744
Ankli P, 439, 753
Anschütz R, 11, 353
Anstire DT, 488, 499, 607, 760, 774
Antina EV, 735
Appelf G, 775
Arai T, 772
Arias IM, 759, 763
Arimoto S, 759
Aronheim B, 213, 745
Arouet FM, 10
Asahina Y, 281, 282, 748
Assisi F, 753, 763
Astrup T, 589
Augenstein L, 761
Auwers K, 220, 669, 670, 745

Avogadro A, 5, 112
Awruch J, 763

B

Bader A, 10
Baeyer JFWA von, 11, 112, 199, 216, 238,
   239, 251, 262, 358, 744, 745
Bailly M, 754
Bakken AF, 754, 759
Balaram P, 759
Barañano DE, 755, 776
Bari S, 763
Barnes JC, 758
Bartholomäus E, 281, 284, 291, 680, 684, 688,
   689, 748, 749
Barton DHR, 434
Bauer K, 764
Bauman D, 774
Beaudet AL, 763
Beaven GH, 759
Beck PD, 443
Becker W, 754
Beckmann EO, 184, 251, 743
Beer A, 152, 742
Beller H, 320, 749
Benedict FG, 188
Bensasson RV, 769
Bergsma D, 756, 764, 765
Berk PD, 443, 754, 764–766
Berkowitz ML, 774
Berlin NI, 764–766
Bernardini C, 760
Berova N, 499, 761
Berry CS, 764–766
Berthelot M, 210

Berthollet CL, 18
Berzelius JJ, 18, 20, 22, 24–27, 29–50, 52, 58, 60, 63, 73, 74, 89, 630, 631, 632, 635, 637, 737, 738
Bildsøe H, 755
Billimona F, 755
Billing BH, 443, 529, 619, 753, 775
Bills CE, 736
Biltz H, 438
Bischoff CA, 212, 214, 669, 744, 745
Blacha M, 754
Blacha-Puller M, 754
Black PS, 754
Blackwood JE, 755
Blanckaert N, 752
Blauer G, 458–460, 467–475, 488, 489, 501, 756–758, 769
Bley LF, 52, 54, 739
Blömer A, 747
Blondheim SH, 528, 756, 764, 765
Bloomer JR, 443
Bock KW, 774
Böhm H, 762
Boiadjiev SE, 499, 516, 519, 522, 609, 612–615, 758, 760, 761, 763, 766, 770–775
Bombelli C, 760
Bonnett R, 371, 400, 411, 412, 426, 443, 444, 446, 454, 460, 465, 480, 537, 546, 548, 553, 573–575, 594, 596, 751, 752, 754, 763, 765, 767, 768
Borkenstein A, 754
Bouchier IAD, 753, 765
Bouvier M, 759
Bramley RK, 443
Branham RV, 530, 532, 764, 765
Breasted JH, 736
Breathnach CS, 743
Bredfeldt J, 752
Briggs WS, 761
Brockbank EM, 736
Brockmann H Jr, 762
Brodersen R, 440, 455, 456, 550, 589, 590, 595, 598, 599, 603, 753, 759, 767, 771, 772
Brønsted JN, 589, 598
Brower JO, 611, 760, 770, 771
Brown AK, 764, 766
Brown AP, 195, 743
Brown GR, 759
Brown SB, 443, 735, 759, 771
Brown SP, 757
Brücke EW Ritter von, 59–62, 66, 71, 73, 74, 90, 91, 94, 95, 100, 111, 133, 175, 634, 635, 637, 739

Bruckenstein S, 756
Brunner E, 457, 753, 762
Buckley DG, 763, 768
Bugge-Asperheim B, 440, 455, 753
Bühler W, 762
Bunsen RWE, 149, 742
Burgstahler AW, 493
Burke MJ, 443, 452, 755, 766
Busk G, 64
Bysshe Shelley P, 20
Byun YS, 760, 770
Cabral DJ, 773

C

Cadet de Grassicourt LC, 12
Cadet LC, 736
Caglioti L, 760
Callahan EW Jr, 764
Campbell JFF, 158, 740
Cannizzaro S, 112
Capranica S, 166
Carey MC, 587
Carroll L, 588, 771
Carson ER, 443
Cason J, 744
Casteran M, 772
Catalano VJ, 760, 770
Cattadori F, 230, 677, 746
Cavanaugh GW, 188
Caventou JB, 38, 738
Celic L, 599, 771, 773
Celsus, Aulus Cornelius, 3
Chedekel MR, 443
Chen J, 741
Chen QQ, 376, 608, 751, 752, 770, 771
Chen Y, 776
Cheney HG, 766
Cheng LJ, 376, 751, 752
Chevreul ME, 14, 29–31, 35, 41, 65, 433, 625, 736
Chiefari J, 605, 771, 774
Ching TY, 765
Cho CS, 760
Cho KY, 760
Choi PJ, 773
Chong R, 765
Chowdury JR, 763
Chung TW, 760
Ciamician GL, 250, 747
Cistola DP, 773
Claisen L, 216, 745
Clezy PS, 517, 537, 761, 765
Coe T, 12, 51, 736

Author Index

779

Cohen AN, 596, 767, 772
Cohen JB, 740
Cole PG, 619, 775
Collinson E, 776
Compernolle F, 752
Conrad M, 217, 745
Conradi BGF, 14, 736
Coolidge EC, 761
Corwin AH, 761
Corwin E, 517
Cowger ML, 594, 772
Crandall DC, 765
Cranmer T 50
Craven R, 736
Cremer RJ, 764
Crist BV, 443
Crum Brown A, 11
Crusats J, 771
Cullen DL, 754
Cullen R, 752
Curran JO, 29, 35, 41, 737
Curry S, 769
Czerny A, 618

D
d'Albis A, 759
Dallas JL, 559
Dalton J, 5, 112
Daly, 594
Daly LH, 772
Damewood JR Jr, 758
Dang VB, 772
Darlak RS, 756
Darwin C 57
Datta S, 601, 773
Davies JE, 444, 754
Davies RE, 530, 531, 548–550, 764, 765
Day GE, 43
de Groot JA, 768
De Rango, 454
Decker M, 452, 753, 762
Deenstra H, 595, 772
Deisenhofer J, 5
Delgado A, 771
Delisle A, 213, 745
Delius HF, 14, 736
Demarçay MH, 25–28, 35, 37, 41, 624, 625, 628, 737
Democritus, 4
Denisov GS, 772
Dennery PA, 763
Dennstedt M, 250, 252, 747
DeRango C, 754, 759

Desaga P, 149
Dessaigne M, 217, 745
Deutsche CW, 756
Dey SK, 575, 610, 769, 771, 773
Diaz L, 762
DiCesare JL, 766
Dietrich GS, 14, 736
Djerassi C, 426, 433, 756, 761
Dobbs RH, 764
Dobeneck H von, 440, 457, 517, 753, 762
Doi M, 773
Dolphin D, 735
Dønstrup S, 755
Dormann E, 256, 285, 689, 747
Dörner T, 463, 757
Draper FW, 740
Drenth W, 759
Drueckhammer D, 601, 774
Ducharme Y, 582, 583, 771
Duff RJ, 773
Dulk FP, 35, 738
Dulong PL, 112, 208
Dumas JPA, 4, 89, 654
Duppa BF, 745
Durham L, 559
Dwight T, 740

E
Ebers G, 735
Edward VI, 50
Ehl K, 761, 762
Ehrlich P, 192, 253
Einstein A, 5
Eismayer K, 697, 749, 697
El-Barkawi F, 761
El-Bayoumi MA, 761
Elemento G, 760
Eliel EL, 760
Elton NW, 775
Ennever JA, 769
Ennever JF, 768
Erlenmeyer E, 10
Eschenmoser A, 310
Estrada MR, 774
Everts S, 740
Ewald CA, 86, 740

F
Falk A, 758, 767
Falk H, 3, 310, 376, 444, 446, 447, 452, 453, 467, 477–479, 484, 501, 517, 543, 545–548, 553, 560, 597, 604, 621, 735,

749, 751, 752, 754, 755, 758, 763, 766, 768, 771, 775
Farmer RC, 597, 772
Farrar WV, 738
Farrera JA, 771
Faulkner W, 622
Favini G, 477, 758
Fehling H von, 10
Ferris CD, 755, 776
Fieser LF, 753
Fieser M, 753
Fink H, 747
Fischer E, 11, 19, 238, 251, 263, 264
Fischer H, 1, 5, 23, 24, 87, 182, 194, 233, 236–239, 248, 251, 256, 271–441, 444, 448–452, 454, 458, 433, 470, 517, 525–528, 536, 539, 552, 560, 561, 603–605, 621, 680–684, 688–699, 704–712, 717, 723–732, 735, 740, 748–752
Fischer M, 748
Fischer N, 757
Fittig WR, 199, 212–214, 669, 744, 745
Fleischhauer J, 761
Flodgaard H, 440, 455, 753
Fog J, 440, 451, 455, 456, 753, 754, 759
Fokkens R, 768
Foote CS, 534, 765
Ford RA, 520, 523
Forino R, 443
Fourcroy AF le Comte de, 10, 13, 17, 20, 21, 24, 35, 51, 114, 624, 736, 737
Fowweather FS, 753
Frank GW, 756
Frank S, 757
Frankenberg N, 755
Frankland E, 745
Franklin RW, 604, 768
Franz Ferdinand, Archduke, 262, 621
Franz II (Franz Josef Karl), 617, 621
Franz Joseph I, 50
Frappier F, 772
Fraunhofer J von, 150, 154
Frerichs FT von, 57, 67, 70–78, 82–84, 88, 90–95, 105, 109, 125, 175, 253, 633, 637, 639, 640, 649, 739
Fresenius, 184
Freudenberg K, 438, 449
Friedman HB, 767
Friedmann HC, 762
Frommherz C, 35
Fröwis W, 752
Frydman B, 517, 762, 763
Fujikake M, 765

Fukaya H, 760
Fukuzaki R, 769
Funke O 73, 74, 195, 739, 743
Furttenbach J, 365
Fusi F, 769

G
Galen, 3
Galler H, 228
Galletti G, 51
Gamgee A, 179, 743
Garbagnati E, 766
Garrod AE Sir, 162–164, 172, 660, 709, 742
Garvie RC, 752
Gauss J, 757
Gawrońska K, 489, 759
Gawroński JK, 489, 493, 497, 759, 760, 772
Gay-Lussac, JL, 4, 10, 18
Gergely S, 446, 754
Gerok W, 774
Gertler S, 765
Geuther A, 216, 251, 745
Ghio C, 488, 758
Ghosh B, 488, 606, 608, 760, 770, 774
Ghuman J, 769
Gillbro T, 769
Girard ML, 598, 733, 772
Gladys CL, 755
Glaser F, 212, 213, 744
Glazer AN, 776
Gmelin L, 15, 16, 17, 20, 24, 25, 27, 28, 30–34, 37, 40–42, 46–49, 51, 53, 55, 59–61, 63, 65–67, 70, 77, 79, 80, 89, 98, 112, 121, 136, 149, 175, 181, 433, 619, 625–628, 631, 633, 635, 667, 715–718, 721, 736, 738
Goldmann H, 746
Gomes DE, 443
Goncharova I, 497, 761
Gordon AJ, 520, 523
Gordon WE, 773
Gossauer A, 517, 604, 736, 754, 762, 774
Gottlieb J, 217
Gottschall K, 762
Gower RP, 744
Gratzer WB, 759
Gray CH, 433, 439–441, 443, 451, 470, 553, 573, 594, 752, 753, 755, 757, 768, 772
Gren FAC, 14, 736
Grimm F, 162, 742
Grob CA, 439, 753
Gross EM, 753

Author Index

Gross PLK, 430, 753
Grubmayr K, 446, 754, 755, 758, 763, 766–768
Grunewald JO, 393, 752
Grunfeld A, 764
Gugert H, 35
Guilard R, 762, 774
Guo J, 772
Gurker H, 754
Gurst JE, 756
Gust D, 758

H

Haas R, 749
Haas K, 228, 229
Haberland HW, 393, 402–404, 407, 430, 724, 725, 752
Hadži D, 772
Hagen R, 773
Hahm JS, 598, 599, 772, 773
Hahn A, 749
Hahn O, 238, 239, 746
Halbach H, 757
Halbig P, 322, 323, 750
Hall JB, 775
Haller A von, 13, 15, 51, 114, 736
Hamilton JA, 600, 773
Hammaker L, 764
Hamzetash D, 763, 768
Hancock F, 443, 460, 594, 756
Hansen AaE, 447, 474, 595, 757
Hansen PE, 755, 772
Harada N, 474, 501, 757
Harmatz D, 756, 758
Harns MM, 759
Harries C, 746
Harris RLN, 492, 761
Hartmann P, 752
Hartmut M, 5
Hashimoto K, 759
Hashimoto S, 759
Haslinger E, 755, 768
Hassel O, 434
Hata S, 759
Haurowirtz F, 337, 707
Hawkes GE, 768
Hawkinson V, 753
Hayashida O, 760
Haydn, 617
Heeß W, 749
Hefelfinger DT, 756
Hein JA, 48, 52–56, 111, 633, 739
Heinz H, 51

Heintz WH, 41, 46–50, 52, 54–57, 60, 62, 94, 100, 101, 111, 114, 115, 129, 130, 140, 175, 177, 631, 635, 642, 650, 738
Heirwegh KPM, 735, 752, 759, 771
Heller JF, 49, 173, 632
Hemmerich P, 457, 474, 756
Henry VIII, 50
Hermann L, 86, 740
Hermann W, 355
Herrmann M, 749
Herzig U, 755, 766
Hess K, 747
Hess R, 278, 279, 282, 353, 378, 388, 389, 400, 401, 407, 723, 751
Hetper J, 746
Heuermann G, 15, 736
Heynsius A, 158, 740
Heyse M, 750
Higginson DS, 776
Hijmans van den Bergh A, 775
Hilditch TP, 745
Hilmer H, 709, 751
Hinze RP, 762
Hippocrates of Cos, 3
Hirsch P, 747
Hofer O, 446, 477, 754, 755, 758, 766, 767
Hoffman von Fallersleben AH, 617
Hofmann A von, 10
Hofmann AW, 214, 686
Höllbacher G, 758
Holm F, 85, 86, 740
Holmes DL, 600, 606, 613, 770, 773, 774
Holmes FL, 18, 737
Holzwarth AR, 771
Home E, 62, 63, 635, 739
Hönigschmid O, 429, 732
Hoover DJ, 760
Hopkins FG Sir, 162–164, 172, 660, 742
Hoppe F, 740
Hoppe-Seyler EFI, 79, 80, 85, 86, 88, 150, 154, 162, 176, 177, 195, 196, 198, 204, 219, 665, 669, 742–744
Hori A, 754
Hornig LS, 358
Howe RB, 443
Hsia DYY, 764
Hsieh YZ, 772
Hua HH, 776
Huber R, 5
Hudson MF, 762
Hüfner G von, 19, 193, 662
Huggins MT, 586, 587, 755, 760, 770, 771
Hünefeld FL, 194, 743
Hunter MJ, 472, 757

782   Author Index

Hursthouse MB, 444, 754
Hutchinson DW, 443, 452, 456, 460, 594, 595,
    602, 754–756, 772

I

Ihde AJ, 17, 736
Imai T, 768, 769
Ioannou S, 768
Irollo R, 772
Ishimura T, 759
Isobe K, 567, 767–769
Itoh S, 767–769
Iturraspe JC, 763

J

Jackson AH, 443, 762
Jackson C, 764
Jackson JR, 443
Jacobsen C, 759
Jacobsen J, 543, 548, 550, 551, 766
Jaffe M, 84, 86, 88, 145, 154–162, 186, 196,
    656, 661, 662, 740, 742
Jakobsen JH, 447, 755
Jayne HW, 745
Jellinek E, 752
Jellinek K, 752
Jellum E, 440, 451, 455, 456, 753
Jin S, 751, 752
Joachim H, 735
Joesten MD, 771
John of Salisbury (Johannes Parvus), 615
Johnson AW, 517, 761
Johnson B, 443, 456, 594, 754–756, 772
Johnson JV, 775
Johnson WB, 736
Johnson WS, 433, 753
Jolles A, 164–171, 660, 742
Jones CO, 552, 767
Jones EA, 443
Jones PM, 757
Jorde C, 444, 754
Jori G, 766, 767, 769
Joule JA, 755
Julinek O, 761
Juluri KR, 776

K

Kadish KM, 762, 774
Kalmar S, 775
Kanna, 593
Kanna Y, 772

Kano K, 489, 759
Kapitulnik J, 531, 764
Kaplan D, 453, 461, 462, 464, 477, 591–593,
    755, 757
Kar A, 607, 770
Karon M, 764
Kasha M, 501, 761
Kato Y, 771
Kauffman GB, 112, 740
Kaufmann NA, 764
Kawade N, 767
Kay IT, 756, 761
Keana JFW, 433, 753
Keats J, 20
Keats-Rohan KSB, 775
Kekulé von Stradonitz FA, 10, 11, 112, 121,
    659, 736, 743
Keohane SJ, 530, 531, 548, 549, 764, 765
Kidrič J, 601, 772
Killet C, 774
Kim M, 759
Kimmich K, 750
King TE, 458, 472, 756
King TS, 756
Kingzett CT, 86, 740
Kirchhoff GR, 149
Kistiakowsky GB, 196, 662
Kjeldahl JGCT, 4, 19, 190
Klahr E, 762
Klarer J, 323, 712, 750
Klinga K, 762
Klyne W, 757
Knell AJ, 443, 456, 460, 594, 754–756, 772
Knipp B, 463, 757
Knobler C, 761
Knobloch G, 762
Knodell RG, 764, 766
Knorr L, 238, 250, 251, 278, 747
Köbner E, 220, 669, 670
Koeppe B, 772
Kohler R, 761
Kohlrausch WG, 742
Kolbe H, 184
Kölle M, 220, 669, 670, 672
Kolosov IV, 595, 772
Kopp HFM, 10
Köst HP, 443
Krämer AL, 776
Krannich W, 747
Krasner J, 594, 772
Kratky C, 444, 754
Krogh-Hansen J, 440, 455, 753
Krueger WC, 756
Kuball HG, 774

Author Index

Kuenzle CC, 442, 443, 452, 457, 458, 460–462, 474, 754–756
Kufer W, 427, 752, 753
Kuhn R, 735
Kühne FW, 69, 71–74, 78–85, 88, 638–640, 739
Kühne G, 762
Kulczycka A, 753, 772
Kumar KS, 537
Kunikata T, 769
Kürzinger A, 379
Küster H, 750
Küster W, 173, 174, 179, 182, 190–194,, 200–244, 246, 252, 257, 263–268, 274, 275, 277, 283, 292, 297–308, 314, 319–355, 357, 359, 362, 412, 439, 525, 545, 621, 661, 669–671, 676, 689, 693, 700–704, 708–711, 742–750

L

Lagarius JG, 755
Lagrange JL Comte de, 10
Lagrangia GL, 10
Lahiri K, 759
Lai B, 773
Lamatsch W, 344, 710
Lamola AA, 769
Land EJ, 769
Landen GL, 758–760, 772
Landois L, 740
Landolt A, 744
Lange H, 747
Langer E, 771
Lassaigne JL, 33, 208, 627, 744
Lathe GH, 594, 619, 772, 775
Lathrop D, 764
Laur PHA, 761
Lauroguais, 216
LaVecchia MJ, 774
Lavie E, 756, 757
Lavoisier AL de, 4, 9, 10, 11, 18, 342, 736
LeBas G, 445, 454, 461, 489, 754, 759
Lebon F, 763, 775
Leck MC, 776
Lee DW, 775
Lee JJ, 594, 772
Legg JW, 49, 52, 407, 439, 451, 738, 751
Lehmann CG, 47, 77, 638, 738
Lehner H, 446, 755, 771
Lemberg R, 353, 390, 391, 400, 406, 407, 409, 423, 434, 438, 441, 449–451, 453, 454, 465, 751–753
Lenhart M, 757
Leodolter A, 758

Lerch U, 762
Leucippus, 4
Leuret F, 33, 627
Levere TH, 18, 737
Levine R, 764
Lewis GN, 392
Li C, 776
Lichtenwald H, 752
Liebermann M, 251
Liebig J von, 4, 10, 18, 19, 26–29, 41, 42, 44, 71, 89, 113, 173, 181, 199, 358, 552, 737, 738, 745
Liepa AJ, 761, 765
Lightner DA, 405, 412, 443, 446, 454, 463, 477, 479, 488, 506, 517, 527, 533–539, 543–551, 555, 557, 558, 560, 567, 568, 575, 583, 587, 591, 596, 600, 601, 604, 609, 613, 615, 735, 740, 752, 754–761, 763–775
Lim CK, 443
Limbach HH, 772
Lindner F, 750
Linnane WP III, 545, 766
Linstead RP, 444
Lisler MW, 752
Liu BB, 776
Liu YY, 751, 752
Loening K, 755
Loir Dr, 25, 51, 737
Longhi G, 763, 775
Lord Byron, GG, 20
Loschmidt JJ, 11
Lowman O, 745
Loy E, 749
Lucassen J, 772
Ludwig KFW, 71
Lugtenberg J, 566, 768
Lund HT, 543, 548, 550, 551, 766
Luzar A, 774
Lychnell LP, 626

M

Ma JS, 136, 376, 426, 427, 583, 741, 751, 752, 758, 760, 766, 772
Ma Q, 776
Mackey W, 431
Macquer PJ, 14, 736
Magauer K, 755
Maghzal GJ, 776
Maisels MJ, 763, 764
Maly RL, 50, 90, 93, 132, 133–179, 181, 184–186, 190–196, 200–206, 265, 658, 661–663, 665, 682, 740–742

Manabe M, 768, 769
Mancini G, 760
Mangani S, 754
Manitto P, 443, 445, 447, 452, 460, 476, 477, 490, 754–758, 766
Marchelewski L, 243, 244, 746
Margaret J, 759
Margulies L, 758
Mark F, 757
Marko H, 775
Marr-Leisy D, 759
Marx K, 29
Mason R, 761
Matern S, 774
Matile S, 499, 761
Matsuura S, 760
Matuura R, 772
Matysik J, 757
Mauguen Y, 754
Maverick E, 583, 760, 761, 771
Mavri J, 772
Mayer O, 489
Mazur Y, 758
Mazzoni M, 769
McClellan AL, 753
McDonagh AF, 136, 371, 400, 411, 412, 426, 427, 443, 445, 454, 467, 517, 527, 531, 533, 537–558, 560, 567, 568, 575, 590, 596, 604, 622, 735, 741, 751–753, 755, 761, 763–770, 772–774, 776
McGarry DD, 615
McNee JW, 775
McRae EG, 761
Mendeleev DI, 214
Merklein F, 108
Merzbacher S, 746
Metz S, 758
Metzger O, 672
Metzroth T, 757
Meyenburg F von, 220, 669, 670, 745
Meyer EF Jr, 754
Meyer P, 748
Meyer ESC von, 184
Meyer-Betz F, 748
Mezger O, 228
Michael A, 214–216, 226, 745
Miehe D, 762
Miller CP, 589
Mills K, 755
Mislow K, 479, 758
Miura I, 768
Moffitt W, 756
Monti D, 443, 445, 460, 461, 476, 477, 539, 754–757, 766

Monti S, 758
Moran BT, 736
Morell DB, 441, 442, 456, 461, 754
Moroi Y, 598, 772
Morris MD, 772
Morse HN, 188
Moscowitz AJ, 443, 452, 459, 470, 471, 497, 533, 539, 604, 755, 756, 761, 762, 766
Mostovnikov GR, 769
Mostovnikov VA, 769
Mugnoli A, 445, 587, 754
Mukerjee P, 597–599, 771–773
Mulder GJ, 66, 196, 739, 743
Müller F von, 24, 263, 364
Müller N, 447, 478, 484, 749, 755, 758, 768, 771
Muller P, 775
Murakami Y, 760
Murchison C, 68, 71, 73, 78, 92

N
Nagai Y, 760
Nah JW, 760
Nakanishi K, 474, 499, 501, 757, 761
Nakayama Y, 773
Naparstek A, 756
Napoléon Bonaparte, 19
Navon G, 461, 462, 464, 477, 591–593, 615, 755–757
Nef JU, 216, 745
Nencki WM, 179, 182–186, 189, 193–206, 209, 217–233, 236–244, 256, 264, 267, 268, 277–282, 292, 353, 525, 661, 662, 665, 667–669, 743–746
Neufingerl E, 758, 767
Neuhaus D, 755
Neukomm J, 69, 72, 78–83, 94, 105, 110, 740
Neumann FW, 752
Newman MS, 756
Nichol AW, 441, 442, 456, 461, 754, 765
Nicholson DC, 440, 441, 753, 757, 772
Nicolaus RA, 753
Niemann G, 310, 312, 401, 405, 693, 717, 724, 725, 732, 749, 750
Nikitin EN, 611, 774
Nogales DF, 463, 583, 607, 757, 758, 760, 770, 774
Nolte RJM, 759
Norona W, 770

O
O'Carra, P, 443
Ochi M, 769

Author Index

Odell GB, 754, 765, 766
Ogawa J, 765
Ogawa Y, 765
Ogino T, 768
ÓhEocha C, 440, 441, 451, 753, 755
Okamoto Y, 760
Okamura T, 773
Olah GA, 343, 751
Oldham-Tipton A, 608
Onishi S, 533, 537, 543, 556–558, 567, 572, 574, 575, 765–769
Onoda A, 773
Orndorff WR, 182, 252, 743
Orth H, 1, 264, 390, 402, 403, 429, 430, 735, 749, 752
Ostrow JD 443, 452, 529–537, 540–558, 573, 587, 594–599, 603, 753, 764–767, 771–773
Ostwald W, 5, 19, 184, 212
Overbeek JTG, 594–599, 602, 603, 772
Overhauser A, 755
Owens D, 443

P

Paal C, 250, 747
Palma LA, 555, 764, 766–770, 774
Palyi G, 760
Panigel R, 757
Paologgi F, 772
Papendieck A, 333, 334, 339, 340, 705, 708, 750
Paracelsus, 3, 9, 11
Park YT, 546, 767
Parker G, 215, 669, 745
Parks JS, 773
Parris K, 773
Partington JF, 736, 738
Paton JD, 758
Paul BD, 776
Pavlov IP, 183
Pederson AO, 767
Pelletier PJ, 38, 738
Pelloni RR, 474, 457, 755, 756
Penfield SL, 745
Pèpe G, 754
Perrin H, 759
Perrin JB, 5
Perryman PW, 764
Person RV, 481, 483–487, 506, 605, 758, 761, 774
Peterson BR, 506, 758
Petit AT, 112
Petrarca AE, 755

Pettenkofer MJ von, 71, 78–80, 638, 639, 739
Petterson KA, 773
Pfeiffer P, 353, 751
Pfeiffer WP, 488, 607, 774, 775
Pflüger EWF, 740
Philippus Aureolus Theophrastus Bombastus von Hohenheim (Paracelsus), 3, 9, 11
Phimister A, 773
Piloty O, 182, 233, 236–294, 300–303, 306, 307, 334, 342, 353, 621, 675, 676, 682–689, 692, 705, 746, 747
Pimentel GC, 753
Pirkle WH, 496, 760
Pirone C, 775
Pis Diez R, 774
Pitea D, 477, 758
Plancher G, 230, 677, 746
Platner EA, 41, 42, 48, 49, 631, 632, 738
Plaveskeiĭ VY, 769
Plieninger H, 391, 412, 730, 731, 735, 752, 753, 761, 762
Podobruk M, 772
Poggendorff JC, 745
Polonski T, 497, 760, 772
Postowsky JJ, 313, 314, 401, 750
Poulletier de la Salle, FPL 13, 14, 21
Povalarapu P, 761
Powers TW, 756
Prakash GKS, 751
Pratesi R, 488, 758, 769
Pratt DC, 452, 755, 766
Pregl F, 4, 19, 315
Preuss I, 762
Preyer WT, 85, 740
Priestap HA, 775
Priestley J, 9
Pu YM, 614, 758–761, 763
Puikynĕ JE, 57
Pützer B, 750
Puzicha G, 514, 613, 614, 758, 761, 774

Q

Quirke JME, 775
Quistad GB, 412, 536, 752, 754, 765
Quitmann E, 241, 244, 245, 277–279, 283, 675, 685, 746

R

Radcliffe CB, 740
Ramberg PJ, 182
Rambert PJ, 743
Ramonas L, 767

Randall EW, 755
Ranking WH, 740
Rao M, 755, 776
Raoult FM, 182, 183, 185, 186, 239, 661, 662, 743
Rapoport H, 744
Rast K, 185, 695, 743
Ratzenhofer M, 768
Rawls HR, 761
Rebek, 496, 601
Rebek J Jr, 760, 771, 773
Reichert ET, 195, 743
Reichert KE, 195, 743
Reindel F, 87
Reinecke H, 398, 727, 752
Reinhardt B, 738
Reisinger M, 759, 760, 766, 769
Remarque EM, 236, 746
Renouard PV, 736
Rhodes IP, 744
Rhodes W, 761
Ribó JM, 758, 771
Ricca GS, 443, 460
Rich AR, 87, 89, 740
Richards DH, 764
Richter V von, 745
Richters VEM, 759
Rimington C, 755
Rinsler MG, 742
Ritter SK, 756
Roberts JD, 773
Robin, CP, 62, 65, 66, 91, 640, 645, 739
Rochleder DR, 738
Rocke AJ, 182, 740
Rödeler, 170, 660
Rodríguez M, 763
Rohmer T, 757
Rokitansky, C, 62
Roos, WT, 552, 574, 767
Roscoe H, 112, 742
Röse H, 46, 258–260, 262, 263, 268, 269, 271, 272, 283, 285–287, 294, 309, 400, 402, 410, 435, 681–684, 689, 693, 694, 724, 732, 748, 749
Rosenbach OEF, 16
Rosenberg B, 478, 761
Rosenberg H, 758
Rosenfeld L, 736
Rotschy A, 182, 184–186, 189, 661, 743
Rubaltelli FF, 766, 767
Rubires R, 771
Rüdiger W, 1, 412, 443, 735
Rumajntzev EV, 735
Runge FF, 198, 250, 744

Ruppert J, 762
Rush JE, 755

S

Saalwächter K, 757
Sahavattanapong P, 760
Saleh M, 776
Salek A, 761
Salisbury J, 775
Salkowski EL, 85, 86, 740
Salzameda NT, 586, 771
Sanders RH, 753
Saresberiensis I, 775
Saski K, 769
Saunders M, 751
Saunders W, 62, 739
Saxton RG, 443
Schaad LJ, 771
Schaffer R, 754, 765, 766
Schaffner K, 771
Scheele CW, 9, 216
Scheer H, 427, 621, 752, 753, 776
Schellman JA, 467, 471, 757
Scherer JJ, 41, 44, 45, 48, 49, 52–54, 56, 62, 63, 75, 94, 108, 114, 145, 175, 631–633, 637, 656, 738
Scheyer H, 321, 750
Schindler H, 739
Schlack P, 353–355, 750
Schlederer T, 755, 768, 771
Schmid R, 426, 528, 529–532, 619, 764, 770, 775
Schmidt J, 746
Schneller K, 750
Schlögl K, 310
Scholtan N, 467, 757
Schønheyder F, 767
Schönhofer A, 774
Schönlein JL, 67
Schubert M, 317, 750
Schumm O, 338, 339–342, 708, 709, 750, 751
Schwanert H, 199, 744
Schwartz S, 601, 757
Schwarz B, 774
Scriver CR, 763
Sedlak TW, 776
Seidel W, 728
Seidman DS, 763
Senac JB, 67, 636
Senge MO, 741
Severini-Ricca GS, 755, 756
Shapovalenko EP, 595, 772
Sheldrick GM, 754

Author Index

Sheldrick WS, 444, 453, 754
Shelver WH, 478, 484, 758
Shelver WL, 758
Shemin D, 528
Showacre J, 766
Shrout DP, 605, 760, 772, 774
Sicherer HV, 196, 638
Sieber N, 183, 196–204, 219, 238, 665, 744
Siedel W, 1, 380, 386, 387, 395, 731, 735, 751, 752
Silfen J, 756
Simon JF, 41, 43, 44, 46, 47, 49, 50, 52, 63, 110, 135, 631, 738
Simopoulos AP, 754, 765, 766
Simpson, 216
Skewes G, 756
Sly WS, 763
Small DM, 773
Smith H, 426
Smith KM, 517, 762, 774
Snapper I, 775
Snatzke, G, 514
Snir J, 756
Snyder SH, 755, 776
Sommer U, 762
Soper AK, 774
Sorensen OW, 755
Sorrenti A, 760
Spielman PE, 745
Spiess HW, 757
Städeler GKA, 46, 49, 67, 68, 72–78, 91–120, 125, 126, 130, 136, 140, 143, 149, 160, 165–167, 177–179, 181, 184–186, 189, 190, 191, 195, 200, 203, 620, 621, 649, 650, 654, 658, 661–663, 665, 740
Stangler G, 389, 751
Steinheil CA von, 149
Steinmetz H, 326, 347, 369, 713, 717
Steinsträsser R, 762
Stern AJC, 264, 307, 357, 423, 429, 430, 470, 731, 735, 749, 757
Stevenson DK, 763
Stewart JCM, 443, 537, 765
Stock J, 288, 689, 747
Stocker R, 776
Stokvis BJ, 158
Stoll A, 748
Stoll MS, 443, 553, 556–558, 567, 573, 574, 752, 767, 768
Strecker A, 10, 42, 738
Strope ER, 752
Sugiyama S, 767, 768
Sund H, 769

T

Takeda J, 773
Talafant E, 619, 775
Tapia A, 762
Taylor ER, 84, 740
Teeple JE, 179, 182, 187, 252, 743
Teichmann LK, 195, 197, 644, 663, 743
Tenhuenen R, 443
Thaler MM, 764, 768
Thannhauser SJ, 252, 257, 260, 273, 282, 682–684, 746
Thenard LJ, 4, 18, 20–27, 29, 30, 32–37, 39, 41, 44, 51, 114, 175, 624–628, 737
Thiessen H, 595, 772
Thirring K, 444, 754, 755, 768
Thomson T, 21, 41, 736
Thomson W, 736, 738
Thudichum JLW, 49, 85, 86, 90, 91, 93, 110, 113–133, 140–149, 155–179, 181–185, 190–196, 203, 211, 225, 265, 342, 402, 406, 525, 534, 560, 655, 658, 659, 665, 672, 738, 740–742
Thyrann T, 605, 774
Tiedemann F, 15–17, 24, 30–32, 41, 47, 136, 175, 619, 626, 736
Tipton AK, 770, 771
Tiribelli C, 599, 771–773
Tissot G, 214–216, 226, 745
Todd A, 444
Tokumara K, 772
Toledo LM, 771
Tolstoy PM, 772
Toyota S, 768
Trallianus A (Alexander of Tralles), 50
Treibs A, 337, 338, 363, 364, 414, 428, 430, 709, 735, 748, 751, 752
Tret'yahova AI, 769
Troup GJ, 769
Troxler R, 443
Trueblood KN, 756, 761
Trull FR, 560, 604, 758, 760, 763, 768, 772–774
Truscott TG, 769
Tsai L, 756
Tsoucaris G, 759
Tsujino N, 759
Tu B, 606, 612, 770, 774
Turner E, 492, 736, 738, 759

U

Ueyama N, 601, 773
Urbanová M, 497, 761
Urey H, 392

V

Vaisman II, 774
Valasinas A, 762
Valentiner W (Valentin GG), 57–62, 66, 69, 72–74, 85, 86, 91, 94, 105, 106, 133, 175, 633, 634, 637, 644, 739
Valle D, 763
Vallisneri A, 13, 51, 736
van den Bergh H, 253, 439, 618, 619, 775
van der Eijke JM, 759
van der Steen R, 768
van Goudoever, 196
van't Hoff JH, 11
Vandemark FL, 766
Varentrapp F, 4
Vauquelin LN, 14, 20
Vega-Hissi EG, 774
Vicker N, 573, 768
Vicq D'Azur F, 14
Villani C, 760
Vink CLJ, 595, 772
Virchow RLK, 41, 45, 46, 58, 60, 61, 63, 64, 71, 73, 74, 84, 86–88, 150, 172, 175, 176, 634, 637, 738–740
Vogel FWC, 740
Voit E, 745
Voit C von, 10
Volhard J, 251
Voltaire (François-Marie Arouet), 10
Vormayr G, 758

W

Wackenroder H, 48, 54, 739
Wagner E, 85, 740
Wagner R, 39, 41, 67
Wagnière G, 475, 501, 757
Walach B, 750
Walden P, 212
Wallace R, 57
Wang CQ, 741, 751, 752
Warren JC, 740
Wash PL, 771
Watchko JF, 763
Watson CJ, 23, 24, 100, 144, 363–365, 433, 440, 443, 451, 470, 528, 533, 559, 604, 737, 751, 753, 757, 762
Watson GB, 559
Watters K, 773
Watts H, 742
Webster CC, 598, 772
Wedl C, 62, 64, 739

Weibel MH, 457, 474, 755, 756
Weidel H, 250, 747
Weiger HH, 767
Weimer M, 753
Weinland EF, 264
Weissbarth O, 412, 418, 421–423, 730, 735
Welch W, 773
Weller JP, 762
Wellman KM, 761
Weltzien K, 112
Wennberg RP, 545, 759
Werner A, 353
Westerdorf W, 736
Wheen AW, 237
Wieger HH, 552
Wieland H, 264, 307, 428, 730, 748
Wijekoon WMD, 493, 759, 760
Wilen SH, 760
Wilke K, 747
Will H, 4, 747
Williamson AW, 10, 112, 121, 181
Williamson MP, 755
Willstätter RM, 5, 6, 194, 235, 257, 262, 264, 275, 276, 281, 282, 298–300, 330–333, 336, 338, 349, 353, 354, 357, 359, 363, 429, 560, 680, 688, 690–693, 697, 698, 705, 732, 736, 748, 751
Wilsey M, 759
Windler SCH, 552, 767
Winkler L, 4
Winsauer K, 768
Winstein S, 174, 214, 342
Wislicenus J, 46, 193, 251
With TK, 441, 451, 753
Wittig G, 343
Wöhler F, 19, 29, 67, 71, 72, 75, 149, 173, 212, 552, 737, 745
Wolf S, 773
Wolkoff AW, 763
Wollaston WH, 112
Wolschann P, 771
Wood ES, 740
Woodward RB, 5, 444
Woody RW, 756, 761
Wooldridge TA, 547, 549, 765–767
Woolley PV III, 472, 757, 759
Wuest JD, 582, 583, 771
Wurtz, 213

X

Xie M, 606, 770

Author Index

Y

Yaffe SJ, 594, 772
Yamada Y, 773
Yamakawa T, 766, 768, 769
Yamamoto H, 773
Yamamoto Y, 776
Yan F, 741, 751, 752
Yashima E, 760
Yasuoka H, 759
Yokoyama T, 768
Yonger J, 772
Yoshiyasu K, 759

Z

Zabriskie J, 764
Zaleski J, 219, 221–224, 228, 230, 231, 233, 236, 239, 244, 256, 277, 281, 282, 353, 745, 746
Zao KH, 776
Zarembo JE, 764, 765, 767
Zechmeister L, 1, 380
Zeile K, 317, 353, 355, 428, 430, 748, 751
Zelinsky N, 217, 745
Zenker R, 73
Zenone EA, 556, 557, 767
Zhang MH, 759, 760
Zhou M, 621, 776
Zhu XX, 757
Zietz B, 769
Zincke T, 239
Zrunek U, 755
Zucchi C, 760
Zumbusch LR von, 190, 743
Zunszain PA, 769
Zuntz N, 171
Zwicky F, 62

Subject Index

A

Acetoacetic ester, 216
N-Acetyl-2-hydroxypyrrole, 440
Acidum cholonicium, 36
Acidum fellicum, 36
Acyl glucuronides, 253, 617
Adipocire, 14, 21
Albumin, 23, 164, 173, 458, 488, 568, 621,
  625, 628, 632, 733
  Heller's ring test, 173
AM1 computations, 479
Amide-amide hydrogen bonding, 581
Amide-carboxylic acid hydrogen bonding, 581
(*S*)-2-Aminobutane, 490, 493
Aminoethyl-dipyrrinone, 419
Ammonia, 4, 32, 97, 99–105, 116, 154, 227,
  620, 624, 627, 657, 662, 688
Ammonium cholate, 79, 80
Ammonium cyanate, 20
Ammonium hexachloroplatinate, 654
Ammonium hydroxide, 25, 51–53, 540, 549,
  641, 667
Antioxidants, 622
Asymmetric transformation, first-order, 490,
  492, 501
Atomic weights, 5, 28, 89, 111, 121, 198, 342

B

Bacteriochlorophyll, 5
Baryta water, 36, 37, 626–629
Beckmann thermometer, 184
Beer-Lambert law, 153, 154
Benzenediazonium chloride, 87
Bile, 7
Bile acids, 28, 35, 42, 82, 625, 637–640,
  644, 645

Bile pigments, 43, 45, 47, 50, 86, 94, 108,
  173, 235, 354, 465, 506, 525, 604,
  620, 624, 645, 655–667, 694, 698,
  729–731
Bile pigments, absorption spectra, 657
  chemical behavior, 631
  cleavage, 294
  computational studies, 477
  separation, 115
  spectroscopy, 149, 155, 506, 533
  synthesis, 517
Bile, bilirubin-biliverdin separation, 28
Biliflavin, 128
Bilifulvin, 37, 39, 48, 49, 58, 62, 73, 84, 629,
  631, 635, 637, 643
Bilifuscin, 95, 100–105, 110, 111, 125, 155,
  643, 649, 658
Bilihumin, 95, 102–105, 110, 111, 125, 649
Bilin, 44
Bilinic acid, 252–254, 257–259, 268, 375
Biliphäin, 43, 44, 46, 47, 49, 52–56, 58,
  60–63, 67, 101, 133
Biliprasin, 95, 101–105, 108, 110, 111, 125,
  136, 649, 658
Bilipurpurin, 128
Bilirubin, albumin bound, 458
  chelates with transition metals, 458
  chemical structure, 182
  Cholepyrrhin, 40
  cleavage, 264
  composition/formula (Heintz), 177
  conformational dynamics by NMR, 475
  degradation products, 535
  diglucuronide, 617, 618
  dimethyl ester, 727
  double bond stereochemistry, 450
  (*E*)-configuration isomers, 545

792 Subject Index

Bilirubin, albumin bound (*cont.*)
  elemental combustion analyses, 187
  exciton chirality CD spectra, 506
  exo-vinyl group, 408
  Gallenroth, 93–95
  kaput, 317
  molecular exciton, 497
  molecular fragmentation, 193
  NMR, 442
  optical activity, 488
  Piloty, 238
  solubility, 590
  structure, 5, 309, 354, 398, 412
  SYBYL molecular dynamics
    calculations, 480
  synthesis, 363, 398, 412
  (*Z,E*) diastereoisomers, 526
  IIIα, 405, 400, 539, 557, 561–565, 570, 571
  IXα, 451, 578, 729
    synthesis, 420, 421
    (*Z/E*) diastereoisomers, 448
  XIIIα, 405
    dimethyl ester, 411
Bilirubin-albumin CD Cotton effects, 498
Bilirubin-HSA complex, 470
Bilirubinic acid, 259, 268, 274, 276, 295, 309,
    324, 358, 367, 435, 693
  isomers, 371
Bilirubinuria, 72
Bilis, 36
  bubula spissata, 27
Biliverdate salt, 131
Biliverdin, 5, 37, 40, 49, 88, 100, 406,
    629–668,
  amide, 134
  conversion to bilirubin, 729
  dimethyl ester, 445
  synthesis, 391, 730, 731
  IXα, 421, 729
    dimethyl ester, 419, 421, 424, 453
    XIIIα dimethyl ester, 417–419, 427
Biliverdin-chlorophyll correlation
    (Berzelius), 39
Biliverdinamid (biliverdin amide), 134
Biliverdinsäure (hematinic acid imide), 207,
    209, 225, 668
Biliviolins, 406
Bilixanthin, 172
Bird of Paradise plant (*Strelitzia reginae*), 621
Bis(2,4-dimethyl-3-carbethoxy-5-pyrryl)
    methene, 697
Bis(3-ethyl-4-methyl-5-carbethoxy-2-pyrryl)
    methane, 322
Bis(3-propionic acid-4-methyl-5-carboxy-
    pyrryl)-methane, 343

Bis(aminoethyl)verdin, 417
Bisdipyrrylmethene, 322
Bishydroxypyrrole, 440
Bis-isopropylammonium salt, 445
Bis-lactam, 444
Bis-lactim, 443
  methyl ether, 446
Blattgrün (chlorophyll), 40
Body fluids, 3
Bone oil, 252
Bovine serum albumin (BSA), 458, 468, 541,
    549, 573
Brownian motion, 5
Bunsen-Kirchhoff scale, 153, 154
Bunsen-Kirchhoff-von Steinheil
    spectroscope, 154
(2-Butenyl)benzene, 213

C
Cannizzaro reaction, 362, 716
Carbon–carbon double bonds, exocyclic, 434
Carboxyl carbon–dipyrrinone, N-hydrogen
    distances, 463
Carboxylic acid–carboxylic acid, hydrogen
    bonding, 581
Caustic ammonia (NH_4OH), 32, 33
Caustic potash (KOH), 33
Chirality, 467, 475, 488, 575, 604
Chlorophyll, 5, 6, 38–40, 42, 43, 48, 63, 182,
    221, 235, 262, 276, 281, 282, 298,
    303, 308, 337, 353, 357, 363, 380,
    388, 408, 428, 429, 433, 604, 620,
    629, 635, 720, 732
  synthesis, 363
Cholaric acid, 44
Choleic acid (Choleinsäure), 28, 625
Cholephäin, 50, 645, 646
Cholephäinate (Sesquicholiphäinat), 123–126,
    646–649
Cholepyrrhin, 40, 41, 43, 46, 47, 49, 53–56,
    58, 60–63, 84, 91, 110, 133, 630,
    631, 634, 652–657
Cholerythrine, 115, 118, 119, 645
Cholesterol, 11, 13, 26, 28, 51, 58, 62, 94,
    106, 117, 135, 186, 433, 628,
    633, 644
Choletelin, 166
Cholethalline, 156
Choleverdin, 158
Cholic acid, 28, 30, 37, 41, 76, 79, 80, 109,
    625, 628, 639
Cholinsäure, 36
Cholochloine, 114, 116, 127
Cholochrome, 114, 116

Subject Index

Cholochromic acid, 116
Cholidinic acid (Choloidinsäure), 28, 625
Choloidic acid, 639
Cholophæine (= Cholepyrrhin = Biliphäin), 114–122, 126, 130
Cholsäure, (cholic acid), 28, 30
Chromogen, 70
Circular dichroism (CD), 457, 467, 472, 488, 490, 492
Citraconimide, 230, 244
Citric acid, 213
Claisen condensation, 216
CNDO/2 molecular orbital calculation, 477
Color diagnostics, 15
Colorimetry, 149
Combustion analyses, 11, 17, 30, 49, 50, 56, 65, 89, 187, 633, 714, 729
quantitative, 17
Combustion train (*Kaliapparat*), 18
Conformational dynamics, 477
Conformational energy maps, 484
Conjecturalchemie (conjectural chemistry), 162, 170, 525, 560
Contour map, 482
Coproporphyrin, 326, 339, 344, 702, 708–713
Corey-Pauling-Koltun space-filled molecular models, 459, 473
Cotton effect, 458, 469, 472, 491, 505, 510, 514, 519, 522, 615
Curtius' rearrangement, 414
Cyanobacteriochromes, 621
Cyclobilirubin, 575
Cyclodextrins, 489, 493, 495, 497
Cytochrome, 5

D

Dehydrobilinic acid, 257–261, 269
Dehydrobilirubin, 390
Dehydrochloridhämin, 232
Diazo reactivity, visible light, 529
Diazonium salts, 618
Diazotized sulfanilic acid, 253
α,α′-Dibromodipyrrylmethene, 322, 328, 343, 346, 350
Diethyl ethylmalonate, 217
Diethylmaleimide, 227, 230, 242, 673
Diethylmalonate alkylations, 217
Dihalogen-hematoporphyrin-dimethyl ether, 335
Dihydrobilirubin, 393, 724, 725
IIIα, 405
Dihydrofuran, 405, 721, 723, 725
Dihydromesobilirubin, 394
XIIIα, 405

Dihydropyrrole, 691
Di-isopropylammonium bilirubinate, 454
chloroform solvate complex, 445
2,4-Dimethyl-3-acetylpyrrole, 279, 318
2,4-Dimethyl-3-carbethoxy-pyrrole, 317, 697
2,4-Dimethyl-3,5-diethylpyrrole, 279
2,3-Dimethyl-5-ethyl-4-propionyl-pyrrole, 291
2,3-Dimethyl-4-ethylpyrrole, 282, 283, 685
2,4-Dimethyl-3-ethylpyrrole, 279, 282, 674, 678, 685
2,3-Dimethyl-5-ethylpyrrole, 290
2,4-Dimethyl-5-ethylpyrrole, 280, 284
p-Dimethylaminobenzaldehyde, 253, 267, 681, 720
Dimethylmaleic anhydride, 214, 226
Dimethylmesobilirubin-XIIIα, 516
Dimethylpyrrole, 230, 244, 249–252, 271, 280, 679, 684, 686
2,3-Dimethylpyrrole-5-acetic acid, 284, 687
α,β-Dimethyl-β-ethylpyrrole, 232, 243, 277, 278
2,3-Dimethylpyrrole-4-propionic acid, 288
2,3-Dimethylpyrrole-5-propionic acid, 284, 687
α,β-Dimethyl-β-pyrryl-propionic acid, 244,
Dinitrotyrosine, 91
Dippel's oil, 250
Di(propionic) acids, 311
Dipyrrinone-carboxylic acid hydrogen bonding, 580
Dipyrrinone-dipyrrinone dimers, hydrogen-bonded, 584
Dipyrrinone-urethane, 416
Dipyrrinones, 376, 414–416
Dipyrroles, 296, 358, 367, 371, 380–385, 439, 600
synthetic, 371
Dipyrrolic halves, 333
Dipyrrylmethanes, 260, 314, 701
Dipyrrylmethenes, 260, 262, 298, 318, 324, 334, 384
brominated, 326, 328, 346
CNDO/2, 477
dibrominated, 321
Dipyrryl-pyridyl-methane, 307
Dipyrryl-pyrryl-pyridyl-methene, 305
Double bonds, exocyclic, 448
Dyslysin, 36

E

Ehrlich reaction, 360
Energy hypersurface, 482
Erythrosin, 91
Ethyl acetoacetate, 215, 216

Ethyl α-bromopropionate, 217
(endo-Ethyl, endo-ethyl)-mesobilirubin, 378
(exo-Ethyl, exo-ethyl)-mesobilirubin, 378
(exo-Ethyl)iso-neoxanthobilirubinic acid, 404
Ethylitaconic acid = propylidene-succinic
    acid, 213
Ethylmesaconic acid, 212
8- Ethylmesobiliverdin-XIIIα, 600
3-Ethyl-4-methyl-5-bromo-2-formylpyrrole,
    384
3-Ethyl-4-methyl-5-carboxyl-2-formylpyrrole,
    384
(S)-Ethylmethylsulfoxide, 497
(endo-Ethyl)neo-xanthobilirubinic acid, 405
Ethylparaconic acid, 213
endo-Ethyl xanthobilirubinic acid, 387
exo-Ethyl xanthobilirubinic acid, 385
Etiobilirubin, 317, 483
Etiobilirubin-IVg, potential energy
    hypersurface, 483, 484
Etiomesobiliverdin-XIIIα, 453
Etioporphyrin, 298, 300, 319, 345, 346, 349,
    354, 691, 700
Etioporphyrin, isomers, 347
Etiophyllin, 691
Exciton chirality, 497
Exocyclic double bonds, 448

F

Farbstoff, 30, 32
Fellinic acid (Fellinsäure from Fel fellis,
    bile), 36, 44
Fleischextract (meat extract), 30
Furans, 251, 312–317, 354, 403, 552
Furano-pyrrole, 401

G

Galactaric acid, 199
Gallenbraun, 46, 47, 49, 52–56, 63, 77, 103,
    118, 119, 175, 620, 631, 633
Gallenfarbstoff (bile pigment), 43, 45, 47, 52
Gallenfett (Cholesterin cholesterol), 28, 626
Gallengrün, 38, 43, 47, 49, 56, 110, 631
Gallenharz (bile resin), 26, 30, 625, 628
Gallenroth, 93–95, 110, 620, 640
Gallensäure (Acidum cholicum, Cholsäure) or
    cholic acid, 30
Gallenspäragin, 30
Gallenstoff, 23, 26, 626
Gallenzucker, 26, 30, 31, 36, 37, 625, 628
Gallstones, 10-14, 19–90, 93–156, 175, 187, 191,
    206, 224, 235, 265, 539, 618, 620

bile pigments, 50, 235, 539, 620
classification, 116
fat, 14
Glaucobilin, 390
Gliadin, 30
Global energy, 483
Glucuronidation, 530
Glucuronides, 50, 253, 455, 529, 557,
    605–609, 617, 619
Glucuronosyl transferase, 530
Glycine, 77
Glycocholic acid, 76, 77, 80, 637
Gmelin reaction, 25, 42, 84, 98, 102, 104, 108,
    110, 142, 147, 158, 175, 182, 360,
    362, 368, 379, 390, 393, 421, 530,
    715–718, 721, 722

H

Hämatin (hematin), 46, 58, 65, 182, 195
Hämatoidin, see, hematoidin
Hämatolin, 195, 198
Hämatopyrrolidinsäure, 263
Hämatosin (= Hämatin), 65
Hämin, see, hemin
Hämopyrrol, 217, 221–223
Hämopyrrolcarbonsäure, 283
Hämotricarbonsäure, 226
Harnfarbstoff (urea), 45
Hematin, 46, 58, 65, 83, 195, 201, 204,
    635–637, 640, 662–669
Hematinic acid, 204, 207, 228, 254, 266, 283,
    295, 300, 334, 666–676, 680, 683,
    689, 706, 725
  dibasic, 217
  imide, 228, 233, 235, 242, 243
Hematoidin, 58, 60, 62, 64, 67, 85, 91, 182,
    633–637, 640, 644, 645, 664, 698
Hematoporphyrin, 87,183, 194, 196, 201, 222,
    267, 278, 283, 332
Hematopyrrolidinic acid, 239, 241, 244, 254, 260
Heme, 5, 71, 78, 449, 465, 466, 560
Heme oxygenase, 71
Hemibilirubin, 259, 266, 270, 297, 682–685
Hemin, 5, 182, 195, 200, 235, 307, 354, 357,
    367, 371, 381, 449, 525, 603, 621,
    697, 704
  dimethyl ester, 706
  ester, oxidative cleavage, 331
  structure, 299–302
  tetrapyrrylethylene-based, 317
  unsaturated β-substituents, 338
Hemoglobin, 63, 71, 78, 88, 94, 183, 193–196,
    206, 334, 337, 707

Subject Index

Hemopyrrole, 281, 288, 294, 673, 688, 689, 720
  carboxylic acid, 239, 240, 244, 255, 283, 293, 294, 366, 687, 710, 711
Hexachloroplatinate method, 45
Hexahelicene, 459, 471
Hexahydrohematoporphyrin, 198, 663
Hofmann elimination, 414, 420
Human serum albumin (HSA), 468, 472, 489, 495, 530, 557, 570, 576, 615
Humin, 100
Hydrazide, 414
Hydrobilirubin, 144, 154, 159, 161, 162, 172, 185, 196
Hydrofuran, 406, 410
Hydrogen bonding, 434, 442–445, 455, 482–522, 528, 533, 543–553, 575–614, 621
  misconceptions, 587
  intramolecular, 594
Hydroxydipyrrylmethene, 733
Hydroxypyrrole, 309, 440, 443
  tautomer, 439
  lactim, 310
α-Hydroxypyrromethenes, 438, 451
Hyperbilirubinemia, 526

I

Icteric urine, 56, 67, 73, 74, 80, 90, 103, 108, 118, 633, 637–639
Indican, 173
Indigoblau (indigo blue), 173, 624, 637, 642
Indigogen, 173
Indirubin, 173, 661
Indole, 199, 250, 663, 677
Infrared (IR), 434, 439–442, 552, 554, 574
Intramolecularly hydrogen-bonded conformation, 496
Isohemopyrrole, 281, 293, 294
Isoindole, 242, 243
Isolumirubins, 568
Isomeric phonopyrrole carboxylic acid, 259, 273, 285, 292
Iso-neo-bilirubinic acid, 399, 718
Iso-neo-xanthobilirubinic acid, 383, 385, 398, 438, 718–720, 725
  methyl ester, 399
Isophonopyrrole carboxylic acid, 255, 256, 259, 268, 273, 285, 293, 684, 687, 689
Isopropylamine, 492
Isopyrrole lactim, 435
Iso-xanthobilirubinic acid, 375, 383
Iso-ψ-xanthobilirubinic acid, 375

J

Jaundice (icterus), 2, 23, 33, 44, 64, 67, 71, 97, 176, 489, 522, 526, 574, 618, 627–630
  neonatal, phototherapy, 97, 526, 527, 544, 548, 550, 604

K

K-mesobilirubin, 368, 370, 375, 380
  dimethyl ester, 369
K-mesobilirubinogen, 369
Käsestoff (cheesy material), 30
Kernicterus, 526, 529
Kryptopyrrole, 274, 282, 283, 293, 294, 296, 322, 358, 688, 689, 694, 712, 714
  carboxylic acid, 293, 294, 296, 324, 358, 366, 701, 710, 714
Kryptopyrrolenone, 393
Küster's maleimides, 217
Küster-Fischer polemics, 329

L

Lactam, 305, 309, 435, 563, 733
Lactam-hydroxypyrrole, 355
Lactic acid, 216
Lactim, 305, 309, 435, 461, 528, 563, 733
Lactim-lactam tautomerism, 435, 563
Lactose, 199
Lead oxide, 20
Leucine, 76
Lumirubins, 568
Lutein (xanthophyll), 85

M

Maleic acid, 212, 216, 668, 672, 675
  imide, 673, 674, 677
Maleic anhydride, 215
Malonic acid, 217
Margaric acid (Margarinsäure), 30, 626, 628
Matière jaune de la bile, 34, 39
Meconium (cholesterol, picromel, etc.), 43
Melanin, 105, 643
Melting points, discordant, resolution, 379
Mesobilirhodin, 440
Mesobilirubin, 24, 296, 303–305, 310, 311, 369–440, 444–523, 561–614
  IIIα, 382, 385, 396–398, 405, 722
  IVα, 495
  IXα, 382, 389, 392, 395–398, 437, 454, 570, 722
  dimethyl ester, 496

Mesobilirubin (*cont.*)
  synthesis, 392
  XIIIα, 382, 385, 396, 400, 405, 437, 463,
      480, 506–513, 517, 522, 561–566,
      570, 580, 584, 587, 592, 600,
      604–614, 722, 726
  dimethyl ester, 496, 694
  dipyrrole fragments, 365
  synthesis, 379
Mesobilirubinogen, 24, 297, 303–305, 360,
      369, 695
Mesobiliverdin, 440
  IXα (glaucobilin), 390
  XIIIα, 600
Mesobiliviolin, 24, 440
Mesohemin, 311, 381–383
  XIII, 383
Mesoporphyrin, 87, 223, 224, 230, 272, 297,
      323–326, 329–332, 347–354, 362,
      671, 685, 700, 704, 705, 713–716
  isomers, 347
  synthesis, 713
  IX, 351
Mesoporphyrinogen, 325
Methemoglobin, 707
β,β′-Methyl-äthyl-pyrrol, 232
β,β′-Methyl-äthyl-pyrrolin, 232
Methyl ethyl maleic acid, 212
  anhydride, 212
3-Methyl-4-ethylpyrrole = Opsopyrrol =
      opsopyrrole, 414
Methylethylmaleic anhydride, 214, 217,
      228, 229
Methylethylmaleimide, 216, 228, 243, 254,
      267, 269, 277, 295, 359, 412, 696
Methylethylsuccinic anhydride, 214, 215
2-Methylhexandisäure-3-methylsäure, 226
Methyl isobutyl ketone (MIC), 733
Methylisopropylmaleimide, 227
3-Methyl-2-(2-methylpropanoic acid)
      pyrrole, 283
Methyl-n-propylmaleic acid, 226
  anhydride, 226
Methylpropylmaleimide, 227, 231, 241, 243
Methylpropylpyrrole, 222, 231, 241, 243, 277
N-Methylpyrrole, 200, 679
4-Methylpyrrole-3-propanoic acid, 266
Methylsuccinic acid, 213, 214, 216, 669, 673
Methylvinylmaleimide, 359, 408, 411,
      537, 696
(βS)-Methylxanthobilirubinic acid, 500
Microanalysis, 19
Molecular exciton, 497
Molecular mechanics computations, 477

Molecular weight, 121, 183, 187
Mono-oximes, 256, 685
Monopyrrole propionic acid, 283
Monopyrroles, cleavage of hemin, 294
  Fischer, 366
Mucic acid, 199, 200
Mucus (Schleim), 12, 23, 30, 37, 44, 52, 62,
      71, 624–630

N
Nencki's Hämopyrrol, 217, 221, 226, 238
Neo-bilirubinic acid, 366, 377
Neo-xanthobilirubinic acid, 366, 378, 383,
      384, 398, 400, 437, 718, 719, 733
  ester, 399, 719
Nitritkörper (nitrite body/nitrite compound),
      400, 724, 726
Nuclear magnetic resonance (NMR), 434,
      461, 475
  ^{13}C, 447
  ^{1}H, 476
Nuclear Overhauser effect (NOE) nuclear
      magnetic resonance (NMR)
      spectroscopy, 452

O
Octahydrobilirubin, 297
Oleic acid (Oelsäure), 28, 30, 626, 628
Omnicholin, 163, 172, 660
Opsopyrrole, 384, 414, 727
  carboxylic acid, 366, 414, 415, 418,
      427, 728
ORD (optical rotatory dispersion) spectra,
      458, 467
Osmazom, 30
3-Oximino-2-pentanone, 280

P
Pettenkofer test, 82
Phenylparaconic acid, 213
Phlogiston, 9
Phonopyrrole, 246, 247, 278, 283, 290, 677,
      678, 688
  carboxylic acid, 245, 256, 260, 283, 285,
      290, 292, 293, 296, 676, 682–689
Photobilirubins, 541, 547, 554–559, 567,
      572–575, 593
Photodegradation, 537
Photoisomerization pathway, 556
Photoisomers, 554
Photolysis, 534, 545–547, 562, 567–573

Subject Index

Photo-oxygenation, 536
Phototherapy, 97, 443, 522, 526–577, 604, 621
Phycoerythrobilin, 440
Phycorubins, 427, 621
Phyllopyrrole, 274, 281, 282, 293, 295,
  680, 686
  carboxylic acid (2,3,5-trimethylpyrrole-3-
    propionic acid), 282, 295, 296,
    366, 689
  methyl ester, 287
Picric acid, 248
Picromel, 21, 30, 35–37, 43, 51, 625, 628
Pigment separation, 20
Pigments of life, 5
PM3 computations, 479
Polemics, 342
Porphin, 323
Porphyrinogen, 272
Porphyrins, 323–364, 389, 449, 453, 517, 545,
  560, 578, 671, 692, 730
  spectral analysis, 339, 697, 698
  synthesis, 337, 344
Porphyrinuria, 339
Potential energy hypersurface, 483
Propendyopents, 536
Propionic acids, 217, 347, 354, 460, 472, 677,
  687–689, 705, 727
Propylfumaric acid, 212
Propylmaleic anhydride, 213
Propylsuccinic acid, 212
Protoporphyrin, 341, 352, 361, 370, 407, 413,
  714–717
  IX, 352, 590
Pterobilin (= γ-bilirubin), 422
Pyridone, hydrogen-bonded homo-dimers, 583
Pyrrofuran, 407
Pyrrol, 198, 219
Pyrrole, 7, 198, 219, 413, 663, 676, 679
  β-carbons, 329
Pyrrole acid imine, 300
Pyrrole 3-carbethoxy ester, 280
Pyrrole carboxylic acids, 286, 309, 687
Pyrrole picrate, 280
2-Pyrrolenone tautomers, 439, 443,
3-Pyrrolenone tautomers, 310,
Pyrrolidone ring, 446
Pyrrolin, 232
Pyrrolo-furan, 402
Pyruvic acid, 214–216, 669, 672, 673

R

Resorcinol melt, 399
Rubin dimethyl ester, 416

S

Saligenin, 77
Schumm-Fischer polemics, 338
Self-dimerization, 325
[6]-Semirubin, 586
Singlet oxygen, 528
Skatole, 247, 255, 677
Sodium amalgam, 642, 656, 661, 662, 694,
  696, 715, 724
Sodium deoxycholate micelles, 495
Sodium dithionite, 427, 731
Sodium glycocholate, 76, 81, 637, 639
Sodium hydrosulfite = sodium dithionite, 421,
  423, 425, 729, 730
Sodium methylsuccinate, 216
Sodium taurocholate, 70, 76, 637, 638
Soxhlet extractor, 140
Spectrum analysis, 149
Speichelstoff (ptyalin), 30
Spermaciti, 14
Spin-lattice carbon relaxation time, 461
Spirobilirubin, 574
Stercobilin, 440, 470, 497
Strelitzia reginae, 621
Succinic acid, 212, 213, 216, 225, 228,
  229, 672
Sulfuric acid, 35
SYBYL forcefield, 478

T

Tartaric acid, 216, 662
Taurine, 26, 28, 37, 41, 44, 70, 77, 625,
  628, 630
Taurocholic acid, 638, 645
Tetra-(2,3-dimethyl-4-carbethoxy-pyrryl)-
  ethylene, 697, 698
Tetrahydrophthalic anhydride, 242
Tetramethylindole, 251
Tetrapyrroles, 2, 355, 678, 725, 731
  synthetic, 371
Tetrapyrrylethanes, 317, 318, 320, 697
Tetrapyrrylethylene, 275, 306, 314,
  319, 320, 327, 329, 361, 370,
  715–717, 732
Tetrapyrrylmethene, 306
Torsional degrees of freedom, 481
2,3,5-Trimethyl-3-ethylpyrrole, 281
2,3,5-Trimethyl-4-ethylpyrrole, 282
Trimethylpyrrole, 248, 271, 284, 684, 689
2,3,5-Trimethylpyrrole, 284, 293, 689
2,4,5-Trimethylpyrrole, 281, 686
2,3,5-Trimethylpyrrole-3-propionic acid, 282
2,4,5-Trimethylpyrrole-3-propionic acid, 287

U

Ultraviolet-visible (UV) spectra, 491, 493
Urea, 19, 20, 43, 44, 652
Urerythrin, 159
Urethane, 414, 415, 420, 728
Urethane-dipyrrinone, 418, 420
Urobilinogen, 23, 253
Urobilins, 145, 154, 159, 163, 166, 265, 267, 360, 433, 440, 459, 470–473, 497, 539, 604, 656, 658, 660–667, 682, 715
Urochrome, 145, 159, 172
Urohodin, 173, 661
Uromelanin, 173
Uropittin, 163
Uroporphyrin, 339
Urorhodin, 163
Uroxanthin, 172, 660
Uteroverdin, 390, 720

V

Verdins, 417, 427, 441, 453, 731
    reduction to rubins, sodium
        dithionite, 427
Verdin amide, 417
Verdin bis-urethane, 420
Verdin dimethyl ester, 416
Vinyl-dipyrrinones, 400
Vinyl-iso-neo-xanthobilirubinic acid, 409
exo-Vinyl-iso-xanthobilirubinic acid, 409
Vinyl-neo-xanthobilirubinic acid, 399, 409, 731
endo-Vinyl-neo-xanthobilirubinic acid, 409

Tripyrrylmethanes, 321
Tyrosine, 74, 76, 78, 91–93, 638, 640

Vinyl-neo-xanthobilirubinic acid methyl ester, 411, 418
(endo-Vinyl) vinyl-neoxanthobilirubinic acid methyl ester, 419
Vitalism, 20

W

Weighed quantities, 9
Wurtz reaction, 213

X

Xanthobilirubinic acid, 260, 270, 274, 275, 276, 287, 295, 296, 298, 309, 358, 367, 371, 383, 434, 435, 499, 684, 693, 717–720
    dimer, 584
    ester, 314
    methyl ester, bromination, 376
        PES spectra, 446
    synthetic, 375
Xanthopyrrole carboxylic acid, 256, 270, 285, 293, 366
Xeronimide (diethylmaleimide), 227, 230, 242, 673
X-ray crystallography, 446, 453, 465, 479, 575, 579
X-ray photoelectron spectroscopy (PES), 446

Y

Yellow "magma", 23

Z

Zerevitinov determination, 314